Optical Fiber Telecommunications V A

About the Editors

Ivan P. Kaminow retired from Bell Labs in 1996 after a 42-year career. He conducted seminal studies on electrooptic modulators and materials, Raman scattering in ferroelectrics, integrated optics, semiconductor lasers (DBR, ridge-waveguide InGaAsP, and multi-frequency), birefringent optical fibers, and WDM networks. Later, he led research on WDM components (EDFAs, AWGs, and fiber Fabry-Perot Filters), and on WDM local and wide area networks. He is a member of the National Academy of Engineering and a recipient of the IEEE/OSA John Tyndall, OSA Charles Townes, and IEEE/LEOS Quantum Electronics Awards. Since 2004, he has been Adjunct Professor of Electrical Engineering at the University of California, Berkeley.

Tingye Li retired from AT&T in 1998 after a 41-year career at Bell Labs and AT&T Labs. His seminal work on laser resonator modes is considered a classic. Since the late 1960s, he and his groups have conducted pioneering studies on lightwave technologies and systems. He led the work on amplified WDM transmission systems and championed their deployment for upgrading network capacity. He is a member of the National Academy of Engineering and a foreign member of the Chinese Academy of Engineering. He is also a recipient of the IEEE David Sarnoff Award, IEEE/OSA John Tyndall Award, OSA Ives Medal/Quinn Endowment, AT&T Science and Technology Medal, and IEEE Photonics Award.

Alan E. Willner has worked at AT&T Bell Labs and Bellcore, and he is Professor of Electrical Engineering at the University of Southern California. He received the NSF Presidential Faculty Fellows Award from the White House, Packard Foundation Fellowship, NSF National Young Investigator Award, Fulbright Foundation Senior Scholar, IEEE LEOS Distinguished Lecturer, and USC University-Wide Award for Excellence in Teaching. He is a Fellow of IEEE and OSA, and he has been President of the IEEE LEOS, Editor-in-Chief of the IEEE/OSA J. of Lightwave Technology, Editor-in-Chief of Optics Letters, Co-Chair of the OSA Science & Engineering Council, and General Co-Chair of the Conference on Lasers and Electro-Optics.

Optical Fiber Telecommunications V A

Components and Subsystems

Edited by

Ivan P. Kaminow
Tingye Li
Alan E. Willner

ELSEVIER

AMSTERDAM • BOSTON • HEIDELBERG • LONDON
NEW YORK • OXFORD • PARIS • SAN DIEGO
SAN FRANCISCO • SINGAPORE • SYDNEY • TOKYO

Academic Press is an imprint of Elsevier

Academic Press is an imprint of Elsevier
30 Corporate Drive, Suite 400, Burlington, MA 01803, USA
525 B Street, Suite 1900, San Diego, California 92101-4495, USA
84 Theobald's Road, London WC1X 8RR, UK

This book is printed on acid-free paper. ∞

Copyright © 2008, Elsevier Inc. All rights reserved.

No part of this publication may be reproduced or transmitted in any form or by any means, electronic or mechanical, including photocopy, recording, or any information storage and retrieval system, without permission in writing from the publisher.

Permissions may be sought directly from Elsevier's Science & Technology Rights Department in Oxford, UK: phone: (+44) 1865 843830, fax: (+44) 1865 853333, E-mail: permissions@elsevier.com. You may also complete your request on-line via the Elsevier homepage (http://elsevier.com), by selecting "Support & Contact" then "Copyright and Permission" and then "Obtaining Permissions."

Library of Congress Cataloging-in-Publication Data
Application submitted

British Library Cataloguing-in-Publication Data
A catalogue record for this book is available from the British Library.

ISBN: 978-0-12-374171-4

For information on all Academic Press publications
visit our Web site at www.books.elsevier.com

Printed in the United States of America
08 09 10 11 12   8 7 6 5 4 3 2 1

**Working together to grow
libraries in developing countries**

www.elsevier.com | www.bookaid.org | www.sabre.org

ELSEVIER   BOOK AID International   Sabre Foundation

For Florence, Paula, Leonard, and Ellen with Love—IPK
For Edith, Debbie, and Kathy with Love—TL
*For Michelle, our Children (Moshe, Asher, Ari, Jacob),
and my Parents with Love—AEW*

Contents

Contributors   ix

Chapter 1   Overview of OFT V Volumes A & B   1
Ivan P. Kaminow, Tingye Li, and Alan E. Willner

Chapter 2   Semiconductor Quantum Dots: Genesis—The Excitonic Zoo—Novel Devices for Future Applications   23
Dieter Bimberg

Chapter 3   High-Speed Low-Chirp Semiconductor Lasers   53
Shun Lien Chuang, Guobin Liu, and Piotr Konrad Kondratko

Chapter 4   Recent Advances in Surface-Emitting Lasers   81
Fumio Koyama

Chapter 5   Pump Diode Lasers   107
Christoph Harder

Chapter 6   Ultrahigh-Speed Laser Modulation by Injection Locking   145
Connie J. Chang-Hasnain and Xiaoxue Zhao

Chapter 7   Recent Developments in High-Speed Optical Modulators   183
Lars Thylén, Urban Westergren, Petter Holmström, Richard Schatz, and Peter Jänes

Chapter 8   Advances in Photodetectors   221
Joe Charles Campbell

Chapter 9   Planar Lightwave Circuits in Fiber-Optic Communications   269
Christopher R. Doerr and Katsunari Okamoto

Chapter 10	III–V Photonic Integrated Circuits and Their Impact on Optical Network Architectures *Dave Welch, Chuck Joyner, Damien Lambert, Peter W. Evans, and Maura Raburn*	343
Chapter 11	Silicon Photonics *Cary Gunn and Thomas L. Koch*	381
Chapter 12	Photonic Crystal Theory: Temporal Coupled-Mode Formalism *Shanhui Fan*	431
Chapter 13	Photonic Crystal Technologies: Experiment *Susumu Noda*	455
Chapter 14	Photonic Crystal Fibers: Basics and Applications *Philip St John Russell*	485
Chapter 15	Specialty Fibers for Optical Communication Systems *Ming-Jun Li, Xin Chen, Daniel A. Nolan, Ji Wang, James A. West, and Karl W. Koch*	523
Chapter 16	Plastic Optical Fibers: Technologies and Communication Links *Yasuhiro Koike and Satoshi Takahashi*	593
Chapter 17	Polarization Mode Dispersion *Misha Brodsky, Nicholas J. Frigo, and Moshe Tur*	605
Chapter 18	Electronic Signal Processing for Dispersion Compensation and Error Mitigation in Optical Transmission Networks *Abhijit Shanbhag, Qian Yu, and John Choma*	671
Chapter 19	Microelectromechanical Systems for Lightwave Communication *Ming C. Wu, Olav Solgaard, and Joseph E. Ford*	713
Chapter 20	Nonlinear Optics in Communications: From Crippling Impairment to Ultrafast Tools *Stojan Radic, David J. Moss, and Benjamin J. Eggleton*	759
Chapter 21	Fiber-Optic Quantum Information Technologies *Prem Kumar, Jun Chen, Paul L. Voss, Xiaoying Li, Kim Fook Lee, and Jay E. Sharping*	829
Index to Volumes VA and VB		**881**

Contributors

Dieter Bimberg, Institut fuer Festkoerperphysik and Center of Nanophotonics, Berlin, Germany, bimberg@physik.tu-berlin.de

Misha Brodsky, AT&T Labs – Research, Middletown, NJ, USA, Brodsky@research.att.com

Joe Charles Campbell, School of Engineering and Applied Science, Department of Electrical and Computer Engineering, University of Virginia, Charlottesville, VA, USA, jcc7s@virginia.edu

Connie J. Chang-Hasnain, Department of Electrical Engineering and Computer Sciences, University of California, Berkeley, CA, USA, cch@eecs.berkeley.edu

Jun Chen, Center for Photonic Communication and Computing, EECS Department, Northwestern University, Evanston, IL, USA

Xin Chen, Corning Inc., Corning, NY, USA, chenx2@corning.com

John Choma, Scintera Inc., Sunnyvale, CA, USA, jchoma@scintera.com

Shun Lien Chuang, Department of ECE, University of Illinois, Urbana, IL, USA, s-chuang@uiuc.edu

Christopher R. Doerr, Alcatel-Lucent, Holmdel, NJ, USA, crdoerr@alcatel-lucent.com

Benjamin J. Eggleton, ARC Centre of Excellence for Ultrahigh-bandwidth Devices for Optical Systems (CUDOS), School of Physics, University of Sydney, Australia, egg@physics.usyd.edu.au

Peter W. Evans, Infinera Inc., Sunnyvale, CA, USA, pevans@infinera.com

Shanhui Fan, Ginzton Laboratory, Department of Electrical Engineering, Stanford, CA, USA, shanhui@stanford.edu

Joseph E. Ford, Department of Electrical and Computer Engineering, University of California, San Diego, CA, USA, jeford@ucsd.edu

Nicholas J. Frigo, Department of Physics, U.S. Naval Academy, Annapolis, MD, USA, frigo@usna.edu

Cary Gunn, Chief Technology Officer, Luxtera, Inc., Carlsbad, CA, USA, cary@luxtera.com

Christoph Harder, HPP, Etzelstrasse 58, Schindellegi, Switzerland, harder@charder.ch

Petter Holmström, Department of Microelectronics and Applied Physics, Royal Institute of Technology (KTH), Kista, Sweden, petterh@kth.se

Peter Jänes, Proximion Fiber Systems AB, Kista, Sweden, peter.janes@proximion.com

Chuck Joyner, Infinera Inc., Sunnyvale, CA, USA, cjoyner@infinera.com

Ivan P. Kaminow, 254M Cory Hall #1770, University of California, Berkeley, CA, USA, kaminow@eecs.berkeley.edu

Karl W. Koch, Corning Inc., Corning, NY, USA, kochkw@corning.com

Thomas L. Koch, Center for Optical Technologies, Sinclair Laboratory, Lehigh University, Bethlehem, PA, USA, tlkoch@lehigh.edu

Yasuhiro Koike, Keio University ERATO Koike Photonics Polymer Project, Yokohama, Japan, koike@appi.keio.ac.jp

Piotr Konrad Kondratko, Department of ECE, University of Illinois, Urbana, IL, USA, kondratko@gmail.com

Fumio Koyama, Microsystem Research Center, P&I Lab, Tokyo Institute of Technology, Nagatsuta, Midori-ku, Yokohama, Japan, koyama@pi.titech.ac.jp

Prem Kumar, Technological Institute, Northwestern University, Evanston, IL, USA, kumarp@northwestern.edu

Damien Lambert, Infinera Inc., Sunnyvale, CA, USA, dlambert@infinera.com

Kim Fook Lee, Center for Photonic Communication and Computing, EECS Department, Northwestern University, Evanston, IL, USA

Ming-Jun Li, Corning Inc., Corning, NY, USA, lim@corning.com

Tingye Li, Locust Place, Boulder, CO, USA, tingyeli@aol.com

Contributors

Xiaoying Li, Center for Photonic Communication and Computing, EECS Department, Northwestern University, Evanston, IL, USA

Guobin Liu, Department of ECE, University of Illinois, Urbana, IL, USA, g-liu5@students.uiuc.edu

David J. Moss, ARC Centre of Excellence for Ultrahigh-bandwidth Devices for Optical Systems (CUDOS), School of Physics, University of Sydney, Australia, dmoss@physics.usyd.edu.au

Susumu Noda, Department of Electronic Science and Engineering, Kyoto University, Kyoto, Japan, snoda@kuee.kyoto-u.ac.jp

Daniel A. Nolan, Corning Inc., Corning, NY, USA, nolanda@corning.com

Katsunari Okamoto, Okamoto Laboratory, 2-1-33 Higashihara, Mito-shi, Ibaraki-ken, 310-0035, Japan, katsu@okamoto-lab.com

Maura Raburn, Infinera Inc., Sunnyvale, CA, USA, mraburn@infinera.com

Stojan Radic, Department of Electrical and Computer Engineering, University of California, San Diego, La Jolla, CA, USA, radic@ece.ucsd.edu

Philip St. John Russell, Max-Planck Research Group (IOIP), University of Erlangen-Nuremberg, Erlangen, Germany, russell@optik.uni-erlangen.de

Richard Schatz, Department of Microelectronics and Applied Physics, Royal Institute of Technology (KTH), Kista, Sweden, rschatz@imit.kth.se

Abhijit Shanbhag, Scintera Inc., Sunnyvale, CA, USA, ags@scintera.com

Jay E. Sharping, University of California, Merced, CA, jsharping@ucmerced.edu

Olav Solgaard, Department of Electrical Engineering, Edward L. Ginzton Laboratory, Stanford University, Stanford, CA, USA, solgaard@standford.edu

Satoshi Takahashi, The Application Group, Shin-Kawasaki Town Campus, Keio University, Kawasaki, Japan, takahasi@koikeppp.jst.go.jp

Lars Thylén, Department of Microelectronics and Applied Physics, Royal Institute of Technology (KTH), Kista, Sweden, lthylen@imit.kth.se

Moshe Tur, School of Electrical Engineering, Tel Aviv University, Ramat Aviv, Israel, tur@eng.tau.ac.il

Paul L. Voss, Center for Photonic Communication and Computing, EECS Department, Northwestern University, Evanston, IL, USA

Ji Wang, Corning Inc., Corning, NY, USA, wangji@corning.com

Dave Welch, Infinera Inc., Sunnyvale, CA, USA, dwelch@infinera.com

James A. West, Corning Inc., Corning, NY, USA, westja@corning.com

Urban Westergren, Department of Microelectronics and Applied Physics, Royal Institute of Technology (KTH), Kista, Sweden, urban@imit.kth.se

Alan E. Willner, Ming Hsieh Department of Electrical Engineering, Viterbi School of Engineering, University of Southern California, Los Angeles, CA, USA, willner@usc.edu

Ming C. Wu, Berkeley Sensor and Actuator Center (BSAC) and Electrical Engineering & Computer Science Department, University of California, Berkeley, CA, USA, wu@eecs.berkeley.edu

Qian Yu, Scintera Inc., Sunnyvale, CA, USA, qyu@scintera.com

Xiaoxue Zhao, Department of Electrical Engineering and Computer Sciences, University of California, Berkeley, CA, USA, xxzhao@eecs.berkeley.edu

1

Overview of OFT V volumes A & B

Ivan P. Kaminow[*], Tingye Li[†], and Alan E. Willner[‡]

[*]University of California, Berkeley, CA, USA
[†]Boulder, CO, USA
[‡]University of Southern California, Los Angeles, CA, USA

Optical Fiber Telecommunications V (OFT V) is the fifth installment of the *OFT* series. Now 29 years old, the series is a compilation by the research and development community of progress in optical fiber communications. Each edition reflects the current state-of-the-art at the time. As Editors, we started with a clean slate of chapters and authors. Our goal was to update topics from *OFT IV* that are still relevant as well as to elucidate topics that have emerged since the last edition.

1.1 FIVE EDITIONS

Installments of the series have been published roughly every 6–8 years and chronicle the natural evolution of the field:

- In the late 1970s, the original *OFT* (Chenoweth and Miller, 1979) was concerned with enabling a simple optical link, in which reliable fibers, connectors, lasers, and detectors played the major roles.
- In the late 1980s, *OFT II* (Miller and Kaminow, 1988) was published after the first field trials and deployments of simple optical links. By this time, the advantages of multiuser optical networking had captured the imagination of the community and were highlighted in the book.
- *OFT III* (Kaminow and Koch, 1997) explored the explosion in transmission capacity in the early-to-mid 1990s, made possible by the erbium-doped fiber amplifier (EDFA), wavelength division multiplexing (WDM), and dispersion management.

Optical Fiber Telecommunications V A: Components and Subsystems
Copyright © 2008, Elsevier Inc. All rights reserved.
ISBN: 978-0-12-374171-4

- By 2002, *OFT IV* (Kaminow and Li, 2002) dealt with extending the distance and capacity envelope of transmission systems. Subtle nonlinear and dispersive effects, requiring mitigation or compensation in the optical and electrical domains, were explored.
- The present edition of *OFT, V,* (Kaminow, Li, and Willner, 2008) moves the series into the realm of network management and services, as well as employing optical communications for ever-shorter distances. Using the high bandwidth capacity in a cost-effective manner for customer applications takes center stage. In addition, many of the topics from earlier volumes are brought up to date; and new areas of research which show promise of impact are featured.

Although each edition has added new topics, it is also true that new challenges emerge as they relate to older topics. Typically, certain devices may have adequately solved transmission problems for the systems of that era. However, as systems become more complex, critical device technologies that might have been considered a "solved problem" previously have new requirements placed upon them and need a fresh technical treatment. For this reason, each edition has grown in sheer size, i.e., adding the new and, if necessary, reexamining the old.

An example of this circular feedback mechanism relates to the fiber itself. At first, systems simply required low-loss fiber. However, long-distance transmission enabled by EDFAs drove research on low-dispersion fiber. Further, advances in WDM and the problems of nonlinear effects necessitated development of nonzero dispersion fiber. Cost considerations and ultra-high-performance systems, respectively, are driving research in plastic fibers and ultra-low-polarization-dependent fibers. We believe that these cycles will continue. Each volume includes a CD-ROM with all the figures from that volume. Select figures are in color. The volume B CD-ROM also has some supplementary Powerpoint slides to accompany Chapter 19 of that volume.

1.2 PERSPECTIVE OF THE PAST 6 YEARS

Our field has experienced an unprecedented upheaval since 2002. The *irrational* exuberance and despair of the technology "bubble-and-bust" had poured untold sums of money into development and supply of optical technologies, which was followed by a depression-like period of over supply. We are happy to say that, by nearly all accounts, the field is gaining strength again and appears to be entering a stage of *rational* growth.

What caused this upheaval? A basis seems to be related to a fundamental discontinuity in economic drivers. Around 2001, worldwide telecom traffic ceased being dominated by the slow-growing voice traffic and was overtaken by the rapidly growing Internet traffic. The business community over-estimated the growth rate, which generated enthusiasm and demand, leading to unsustainable expectations. Could such a discontinuity happen again? Perhaps, but chastened investors now seem to be following a more gradual and sensible path. Throughout the "bubble-and-bust" until the present, the actual demand for bandwidth has grown at a very healthy ~80% per year globally; thus, real traffic demand experienced no bubble at all. The growth and capacity needs are real, and should continue in the future.

As a final comment, we note that optical fiber communications is firmly entrenched as part of the global information infrastructure. The only question is how deeply will it penetrate and complement other forms of communications, e.g., wireless, access, and on-premises networks, interconnects, satellites, etc. This prospect is in stark contrast to the voice-based future seen by *OFT*, published in 1979, before the first commercial intercontinental or transatlantic cable systems were deployed in the 1980s. We now have Tb/s systems for metro and long-haul networks. It is interesting to contemplate what topics and concerns might appear in *OFT VI*.

1.3 OFT V VOLUME A: COMPONENTS AND SUBSYSTEMS

1.3.1 Chapter 1. Overview of OFT V volumes A & B (Ivan P. Kaminow, Tingye Li, and Alan E. Willner)

This chapter briefly reviews herewith all the chapters contained in both volumes of OFT V.

1.3.2 Chapter 2. Semiconductor quantum dots: Genesis—The Excitonic Zoo—novel devices for future applications (Dieter Bimberg)

The ultimate class of semiconductor nanostructures, i.e., "quantum dots" (QDs), is based on "dots" smaller than the de Broglie wavelength in all three dimensions. They constitute nanometer-sized clusters that are embedded in the dielectric matrix of another semiconductor. They are often self-similar and can be formed by self-organized growth on surfaces. Single or few quantum dots enable novel devices for quantum information processing, and billions of them enable active centers in optoelectronic devices like QD lasers or QD optical amplifiers. This chapter covers the area of quantum dots from growth via various band structures to optoelectronic device applications. In addition, high-speed laser and amplifier operations are described.

1.3.3 Chapter 3. High-speed low-chirp semiconductor lasers (*Shun Lien Chuang, Guobin Liu, and Piotr Konrad Kondratko*)

One advantage of using quantum wells and quantum dots for the active region of lasers is the lower induced chirp when such lasers are directly modulated, permitting direct laser modulation that can save on the cost of separate external modulators. This chapter provides a comparison of InAlGaAs with InGaAsP long-wavelength quantum-well lasers in terms of high-speed performance, and extracts the important parameters such as gain, differential gain, photon lifetime, temperature dependence, and chirp. Both DC characteristics and high-speed direct modulation of quantum-well lasers are presented, and a comparison with theoretical models is made. The chapter also provides insights into novel quantum-dot lasers for high-speed operation, including the ideas of p-type doping vs tunneling injection for broadband operation.

1.3.4 Chapter 4. Recent advances in surface-emitting lasers (*Fumio Koyama*)

Vertical cavity surface-emitting lasers (VCSELs) have a number of special properties (compared with the more familiar edge-emitting lasers) that permit some novel applications. This chapter begins with an introduction which briefly surveys recent advances in VCSELs, several of that are then treated in detail. These include techniques for realizing long-wavelength operation (as earlier VCSELs were limited to operation near 850 nm), the performance of dense VCSEL arrays that emit a range of discrete wavelengths (as large as 110 in number), and MEMS-based athermal VCSELs. Also, plasmonic VCSELs that produce subwavelength spots for high-density data storage and detection are examined. Finally, work on all-optical signal processing and slow light is presented.

1.3.5 Chapter 5. Pump diode lasers (*Christoph Harder*)

Erbium-doped fiber amplifiers (EDFAs) pumped by bulky argon lasers were known for several years before telecom system designers took them seriously; the key development was a compact, high-power semiconductor pump laser. Considerable effort and investment have gone into today's practical pump lasers, driven by the importance of EDFAs in realizing dense wavelength division multiplexed (DWDM) systems. The emphasis has been on high power, efficiency, and reliability. The two main wavelength ranges are in the neighborhood of 980 nm for low noise and 1400 nm for remote pumping of EDFAs. The 1400-nm band is also suitable for Raman amplifiers, for which very high power is needed.

This chapter details the many lessons learned in the design for manufacture of commercial pump lasers in the two bands. Based on the performance developed for telecom, numerous other commercial applications for high-power lasers have emerged in manufacturing and printing; these applications are also discussed.

1.3.6 Chapter 6. Ultrahigh-speed laser modulation by injection locking (*Connie J. Chang-Hasnain and Xiaoxue Zhao*)

It has been known for decades that one oscillator (the slave) can be locked in frequency and phase to an external oscillator (the master) coupled to it. Current studies of injection-locked lasers show that the dynamic characteristics of the directly modulated slave are much improved over the same laser when freely running. Substantial improvements are found in modulation bandwidth for analog and digital modulation, in linearity, in chirp reduction and in noise performance.

In this chapter, theoretical and experimental aspects of injection locking in all lasers are reviewed with emphasis on the authors' research on VCSELs (vertical cavity surface-emitting lasers). A recent promising application in passive optical networks for fiber to the home (FTTH) is also discussed.

1.3.7 Chapter 7. Recent developments in high-speed optical modulators (*Lars Thylén, Urban Westergren, Petter Holmström, Richard Schatz, and Peter Jänes*)

Current high-speed lightwave systems make use of electro-optic modulators based on lithium niobate or electroabsorption modulators based on semiconductor materials. In commercial systems, the very high-speed lithium niobate devices often require a traveling wave structure, while the semiconductor devices are usually lumped.

This chapter reviews the theory of high-speed modulators (at rates of 100 Gb/s) and then considers practical design approaches, including comparison of lumped and traveling-wave designs. The main emphasis is on electroabsorption devices based on Franz–Keldysh effect, quantum-confined Stark effect and intersubband absorption. A number of novel designs are described and experimental results given.

1.3.8 Chapter 8. Advances in photodetectors (*Joe Charles Campbell*)

As a key element in optical fiber communications systems, photodetectors belong to a well developed sector of photonics technology. Silicon p–i–n and avalanche photodiodes deployed in first-generation lightwave transmission systems operating at 0.82-µm wavelength performed very close to theory. In the 1980s, InP photodiodes were developed and commercialized for systems that operated at 1.3- and 1.5-μm wavelengths, albeit the avalanche photodiodes (APDs) were expensive and nonideal. Introduction of erbium-doped fiber amplifiers and WDM technology in the 1990s relegated APDs to the background, as p–i–n photoreceivers performed well in amplified systems, whereas APDs were plagued by the amplified spontaneous emission noise. Future advanced systems and special applications will require sophisticated devices involving deep understanding of device physics and technology. This chapter focuses on three primary topics: high-speed waveguide photodiodes for systems that operate at 100 Gb/s and beyond, photodiodes with high saturation current for high-power applications, and recent advances of APDs for applications in telecommunications.

1.3.9 Chapter 9. Planar lightwave circuits in fiber-optic communications (*Christopher R. Doerr and Katsunari Okamoto*)

The realization of one or more optical waveguide components on a planar substrate has been under study for over 35 years. Today, individual components such as splitters and arrayed waveguide grating routers (AWGRs) are in widespread commercial use. Sophisticated functions, such as reconfigurable add–drop multiplexers (ROADMs) and high-performance filters, have been demonstrated by integrating elaborate combinations of such components on a single chip. For the most part, these photonic integrated circuits (PICs), or planar lightwave circuits (PLCs), are based on passive waveguides in lower index materials, such as silica.

This chapter deals with the theory and design of such PICs. The following two chapters (Chapters 11 and 12) also deal with PICs; however, they are designed to be integrated with silicon electronic ICs, either in hybrid fashion by short wire bonds to an InP PIC or directly to a silicon PIC.

1.3.10 Chapter 10. III–V photonic integrated circuits and their impact on optical network architectures (*Dave Welch, Chuck Joyner, Damien Lambert, Peter W. Evans, and Maura Raburn*)

InP-based semiconductors are unique in their capability to support all the photonic components required for wavelength division multiplexed (WDM) transmitters and receivers in the telecom band at 1550 nm. Present subsystems have connected these individual components by fibers or lenses to form hybrid transmitters and receivers for each channel.

Recently, integrated InP WDM transmitter and receiver chips that provide 10 channels, each operating at 10 Gb/s, have been shown to be technically and economically viable for deployment in commercial WDM systems. The photonic integrated circuits are wire-bonded to adjacent silicon ICs. Thus a single board provides optoelectronic regeneration for 10 channels, dramatically reducing interconnection complexity and equipment space. In addition, as in legacy single-channel systems, the "digital" approach for transmission (as compared to "all-optical") offers ease of network monitoring and management. This chapter covers the technology of InP photonic integrated circuits (PICs) and their commercial application. The impact on optical network architecture and operation is discussed and technology advances for future systems are presented.

1.3.11 Chapter 11. Silicon photonics (*Cary Gunn and Thomas L. Koch*)

Huge amounts of money have been invested in silicon processing technology, thanks to a steady stream of applications that justified the next stage of processing development. In addition to investment, innovative design, process discipline and large-volume runs made for economic success. The InP PICs described in the previous chapter owe their success to lessons learned in silicon IC processing.

Many people have been attracted by the prospects of fabricating PICs using silicon alone to capitalize on the investment and success of silicon ICs. To succeed one requires a large-volume application and a design that can be made in an operating silicon IC foundry facility. A further potential advantage is the opportunity to incorporate on the same photonic chip electronic signal processing. The application to interconnects for high-performance computers is a foremost motivation for this work.

While silicon has proven to be the ideal material for electronic ICs, it is far from ideal for PICs. The main shortcoming is the inability so far to make a good light source or photodetector in silicon. This chapter discusses the successes and challenges encountered in realizing silicon PICs to date.

1.3.12 Chapter 12. Photonic crystal theory: Temporal coupled-mode formalism (*Shanhui Fan*)

Photonic crystal structures have an artificially created optical bandgap that is introduced by a periodic array of perturbations, and different types of waveguides and cavities can be fabricated that uniquely use the band gap-based confinement. These artificially created materials have been of great interest for potential optical information processing applications, in part because they provide a common platform to miniaturize a large number of optical components on-chip down to single wavelength scale. For this purpose, many devices can be designed using such a material with a photonic bandgap and, subsequently, introducing line and point defect states into the gap. Various functional devices, such as filters, switches, modulators and delay lines, can be created by controlling the coupling between these defect states. This chapter reviews the temporal coupled-mode theory formalism that provides the theoretical foundation of many of these devices.

1.3.13 Chapter 13. Photonic crystal technologies: Experiment (*Susumu Noda*)

Photonic crystals belong to a class of optical nanostructures characterized by the formation of band structures with respect to photon energy. In 3D photonic crystals, a complete photonic band gap is formed; the presence of light with frequencies lying in the band gap is not allowed. This chapter describes the application of various types of materials engineering to photonic crystals, with particular focus on the band gap/defect, the band edge, and the transmission band within each band structure. The manipulation of photons in a variety of ways becomes possible. Moreover, this chapter discusses the recent introduction of "photonic heterostructures" as well as recent developments concerning two- and three-dimensional photonic crystals.

1.3.14 Chapter 14. Photonic crystal fibers: Basics and applications (*Philip St John Russell*)

Photonic crystal fibers (PCFs)—fibers with a periodic transverse microstructure—first emerged as practical low-loss waveguides in early 1996. The initial demonstration took 4 years of technological development, and since then the fabrication techniques have become more and more sophisticated. It is now possible to manufacture the microstructure in air–glass PCFs to accuracies of 10 nm on the scale of 1 µm, which allows remarkable control of key optical properties such as dispersion, birefringence, nonlinearity and the position and width of the photonic band gaps (PBGs) in the periodic "photonic crystal" cladding. PCF has in this way extended the range of possibilities in optical fibers, both by improving well-established properties and introducing new features such as low-loss guidance in a hollow core.

In this chapter, the properties of the various types of PCFs are introduced, followed by a detailed discussion of their established or emerging applications. The chapter describes in detail the fabrication, theory, numerical modeling, optical properties, and guiding mechanisms of PCFs. Applications of photonic crystal fibers include lasers, amplifiers, dispersion compensators, and nonlinear processing.

1.3.15 Chapter 15. Specialty fibers for optical communication systems (*Ming-Jun Li, Xin Chen, Daniel A. Nolan, Ji Wang, James A. West, and Karl W. Koch*)

Specialty fibers are designed by changing fiber glass composition, refractive index profile, or coating to achieve certain unique properties and functionalities. Some of the common specialty fibers include active fibers, polarization control fibers, dispersion compensation fibers, highly nonlinear fibers, coupling or bridge fibers, high-numerical-aperture fibers, fiber Bragg gratings, and special single mode fibers. In this chapter, the design and performance of various specialty fibers are discussed. Special attention is paid to dispersion compensation fibers, polarization-maintaining and single-polarization fibers, highly nonlinear fibers, double clad fiber for high-power lasers and amplifiers, and photonic crystal fibers. Moreover, there is a brief discussion of the applications of these specialty fibers.

1.3.16 Chapter 16. Plastic optical fibers: Technologies and communication links (*Yasuhiro Koike and Satoshi Takahashi*)

Plastic optical fiber (POF) consists of a plastic core that is surrounded by a plastic cladding of a refractive index lower than that of the core. POFs have very large core diameters compared to glass optical fibers, and yet they are quite flexible. These features enable easy installation and safe handling. Moreover, the large-core fibers can be connected without high-precision accuracy and with low cost. POFs have been used extensively in short-distance datacom applications, such as in digital audio interfaces. POFs are also used for data transmission within equipment and for control signal transmission in machine tools. During the late 1990s, POFs were used as the transmission medium in the data bus within automobiles. As we move into the future, high-speed communication will be required in the home, and POFs are a promising candidate for home network wiring. This chapter describes the POF design and fabrication, the specific fiber properties of attenuation, bandwidth and thermal stability, and various communications applications, concluding with a discussion of recent developments in graded-index POFs.

1.3.17 Chapter 17. Polarization mode dispersion (*Misha Brodsky, Nicholas J. Frigo, and Moshe Tur*)

Polarization-mode dispersion (PMD) has been well recognized for sometime as an impairment factor that limits the transmission speed and distance in high-speed lightwave systems. The complex properties of PMD have enjoyed scrutiny by theorists, experimentalists, network designers, field engineers and, during the "bubble" years, entrepreneurial technologists. A comprehensive treatment of the subject up to year 2002 is given in a chapter bearing the same title in *Optical Fiber Telecommunications IVB, System and Impairments*. The present chapter is an overview of PMD with special emphasis on the knowledge accumulated in the past 5 years. It begins with a review of PMD concepts, and proceeds to consider the "hinge" model used to describe field test results, which are presented and analyzed. The important subject of system penalties and outages due to first-order PMD is then examined, followed by deliberations of higher-order PMD, and interaction between fiber nonlinear effects and PMD.

1.3.18 Chapter 18. Electronic signal processing for dispersion compensation and error mitigation in optical transmission networks (*Abhijit Shanbhag, Qian Yu, and John Choma*)

Dispersion equalization has its origin in the early days of analog transmission of voice over copper wires where loading coils (filters) were distributed in the network to equalize the frequency response of the transmission line. Digital transmission over twisted pairs was enabled by the invention of the transversal equalizer which extended greatly the bandwidth and reach. Sophisticated signal processing and modulation techniques have now made mobile telephones ubiquitous. However, it was not until the mid 1990s that wide deployment of Gigabit Ethernet rendered silicon CMOS ICs economical for application in high-speed lightwave transmission. Most, if not all lightwave transmission systems deployed today, use electronic forward error correction and dispersion compensation to alleviate signal degradation due to noise and fiber dispersive effects.

This chapter presents an overview of various electronic equalization and adaptation techniques, and discusses their high-speed implementation, specifically addressing 10-Gb/s applications for local-area, metro, and long-haul networks. It comprises a comprehensive survey of the role, scope, limitations, trends, and challenges of this very important and compelling technology.

1.3.19 Chapter 19. Microelectromechanical systems for lightwave communication (*Ming C. Wu, Olav Solgaard, and Joseph E. Ford*)

The earliest commercial applications of microelectromechanical systems (MEMS) were in digital displays employing arrays of tiny mirrors and in accelerometers for airbag sensors. This technology has now found a host of applications in lightwave communications. These applications usually require movable components, such as mirrors, with response times in the neighborhood of 10^{-6} s, although fixed

1. Overview of OFT V Volumes A & B   11

elements may be called for in some applications. Either a free-space or integrated layout may be used.

This chapter describes the recent lightwave system applications of MEMS. In telecommunications, MEMS switches can provide cross-connects with large numbers of ports. A variety of wavelength selective devices, such as reconfigurable optical add–drop multiplexers (ROADM) employ MEMS. More recent devices include tunable lasers and microdisk resonators.

1.3.20 Chapter 20. Nonlinear optics in communications: from crippling impairment to ultrafast tools (*Stojan Radic, David J. Moss, and Benjamin J. Eggleton*)

It is perhaps somewhat paradoxical that optical nonlinearities, whilst having posed significant limitations for long-haul WDM systems, also offer the promise of addressing the bandwidth bottleneck for signal processing for future optical networks as they evolve beyond 40 Gb/s. In particular, all-optical devices based on the 3rd order $\chi^{(3)}$ optical nonlinearity offer a significant promise in this regard, not only because the intrinsic nonresonant $\chi^{(3)}$ is nearly instantaneous, but also because $\chi^{(3)}$ is responsible for a wide range of phenomena, including 3rd harmonic generation, stimulated Raman gain, four-wave mixing, optical phase conjugation, two-photon absorption, and the nonlinear refractive index. This plethora of physical processes has been the basis for a wide range of activity on all-optical signal processing devices.

This chapter focuses on breakthroughs in the past few years on approaches based on highly nonlinear silica fiber as well as chalcogenide-glass-based fiber and waveguide devices. The chapter contrasts two qualitatively different approaches to all-optical signal processing based on nonphase-matched and phase-matched processes. All-optical applications of 2R and 3R regeneration, wavelength conversion, parametric amplification, phase conjugation, delay, performance monitoring, and switching are reviewed.

1.3.21 Chapter 21. Fiber-optic quantum information technologies (*Prem Kumar, Jun Chen, Paul L. Voss, Xiaoying Li, Kim Fook Lee, and Jay E. Sharping*)

Quantum-mechanical (QM) rules are surprisingly simple: linear algebra and first-order partial differential equations. Yet, QM predictions are unimaginably precise and accurate when compared with experimental data. A "mysterious" feature of QM is the superposition principle and the ensuing quantum entanglement. The fundamental difference between quantum entanglement and classical correlation lies in the fact that particles are quantum-mechanical objects which can exist not only in states $|0>$ and $|1>$ but also in states described by $\alpha|0> + \beta|1>$, while classical objects can only exist in one of two deterministic states (i.e., "heads" or "tails"), and not something in between. In other words, the individual particle in quantum entanglement does not have a well-defined pure state before measurement.

Since the beginning of the 1990s, the field of quantum information and communication has expanded rapidly, with quantum entanglement being a critical aspect. Entanglement is still an unresolved "mystery," but a new world of "quantum ideas" has been ignited and is actively being pursued. The focus of this chapter is the generation of correlated and entangled photons in the telecom band using the Kerr nonlinearity in dispersion-shifted fiber. Of particular interest are microstructure fibers, in which tailorable dispersion properties have allowed phase-matching and entanglement to be obtained over a wide range of wavelengths.

1.4 OFT V VOLUME B: SYSTEMS AND NETWORKS

1.4.1 Chapter 1. Overview of OFT V volumes A & B (Ivan P. Kaminow, Tingye Li, and Alan E. Willner)

This chapter briefly reviews herewith all the chapters contained in both volumes of OFT V.

1.4.2 Chapter 2. Advanced optical modulation formats (Peter J. Winzer and René-Jean Essiambre)

Today, digital radio-frequency (rf) communication equipment employs sophisticated signal processing and communication theory technology to realize amazing performance; wireless telephones are a prime example. These implementations are made possible by the capabilities and low cost of silicon integrated circuits in high-volume consumer applications. Some of these techniques, such as forward error correction (FEC) and electronic dispersion compensation (EDC) are currently in use in lightwave communications to enhance signal-to-noise ratio and mitigate signal degradation. (See the chapter on "Electronic Signal Processing for Dispersion Compensation and Error Mitigation in Optical Transmission Networks" by Abhijit Shanbhag, Qian Yu, and John Choma.) Advanced modulation formats that are robust to transmission impairments or able to improve spectral efficiency are being considered for next-generation lightwave systems.

This chapter provides a taxonomy of optical modulation formats, along with experimental techniques for realizing them. The discussion makes clear the substantial distinctions between design conditions for optical and rf applications. Demodulation concepts for coherent and delay demodulation are also covered analytically.

1. Overview of OFT V Volumes A & B

1.4.3 Chapter 3. Coherent optical communication systems (*Kazuro Kikuchi*)

The first generation of single-channel fiber optic networks used on-off keying and direct detection. Later, coherent systems, employing homodyne and heterodyne detection, were intensely researched with the aim of taking advantage of their improved sensitivity and WDM frequency selectivity. However, the quick success of EDFAs in the 1990s cut short the prospects for coherent systems.

Now, interest in coherent is being renewed as the need for greater spectral efficiency in achieving greater bandwidth per fiber has become apparent. This chapter reviews the theory of multilevel modulation formats that permit multiple bits/s of data per Hz of bandwidth. (See the chapter on "Advanced Modulation Formats" by Winzer and Essiambre.) The growing capabilities of silicon data signal processing (DSP) can be combined with digital coherent detection to provide dramatic improvements in spectral efficiency. Experimental results for such receivers are presented.

1.4.4 Chapter 4. Self-coherent optical transport systems (*Xiang Liu, Sethumadhavan Chandrasekhar, and Andreas Leven*)

As stated above, coherent detection transmission systems were investigated in the 1980s for their improved receiver sensitivity and selectivity, and for the promise of possible postdetection dispersion compensation. However, the emergence of EDFAs and amplified WDM systems relegated the technically difficult coherent technology to the background. Now, as high-speed signal processing technology becomes technically and economically feasible, there is renewed interest in studying coherent and self-coherent systems, especially for their capability to increase spectral efficiency through the use of advanced multilevel modulation techniques and, more important, for the possibility of implementing postdetection equalization functionalities.

Self-coherent systems utilize differential direct detection that does not require a local oscillator. With high-speed analog-to-digital conversion and digital signal processing, both phase and amplitude of the received optical field can be reconstructed, thus offering unprecedented capability for implementing adaptive equalization of transmission impairments. This chapter is a comprehensive and in-depth treatment of self-coherent transmission systems, including theoretical considerations, receiver technologies, modulation formats, adaptive equalization techniques, and applications for capacity upgrades and cost reduction in future optical networks.

1.4.5 Chapter 5. High-bit-rate ETDM transmission systems (*Karsten Schuh and Eugen Lach*)

Historically, it has been observed that the first cost of a (single-channel) transmission system tended to increase as the square root of its bandwidth or bit rate. This observation has prompted the telecom industry to develop higher-speed systems for upgrading transport capacity. Indeed, there is a relentless drive to explore higher speed for multichannel amplified WDM transmission where, for a given speed of operation, the total system cost is roughly proportional to the number of channels plus a fixed cost. It is important to note that the cost of equalizing for signal impairment at higher speeds must be taken into account.

This chapter is an up-to-date review of high-speed transmission using electronic time division multiplexing (ETDM), a time-honored approach for upgrading system capacity. The emphasis is on 100-Gb/s bit rate and beyond, as 40-Gb/s systems are already being deployed and 100-Gb/s Ethernet (100 GE) is expected to be the next dominant transport technology. The chapter includes a basic treatment of ETDM technology, followed by a description of the concepts of high-speed ETDM systems. Requirements of optical and electronic components and the state-of-the-art technologies are then examined in detail, and an up-to-date overview of ultra-high-speed systems experiments is presented. Finally, prospects of the various approaches for rendering cost-effective 100 GE are contemplated.

1.4.6 Chapter 6. Ultra-high-speed OTDM transmission technology (*Hans-Georg Weber and Reinhold Ludwig*)

The expected increase of transmission capacity in optical fiber networks will involve an optimized combination of WDM and TDM. TDM may be realized by electrical multiplexing (ETDM) or by optical multiplexing (OTDM). Dispersion impairment notwithstanding, OTDM offers a means to increase the single-channel bit rate beyond the capability of ETDM. Thus OTDM transmission technology is often considered to be a research means with which to investigate the feasibility of ultra-high-speed transmission. Historically, the highest speed commercial systems have been ETDM systems. Latest examples are 40 G systems being deployed at present and (serial) 100 G systems expected to be commercially available in a few years. In the past 10 years, OTDM transmission technology has made considerable progress towards much higher bit rates and much longer transmission links.

This chapter discusses ultra-high-speed data transmission in optical fibers based on OTDM technology. The chapter gives a general description of an OTDM system, the OTDM transmitter, the OTDM receiver, and the fiber transmission line. WDM/OTDM transmission experiments are also described.

1.4.7 Chapter 7. Optical performance monitoring (*Alan E. Willner, Zhongqi Pan, and Changyuan Yu*)

Today's optical networks function in a fairly static fashion and are built to operate within well-defined specifications. This scenario is quite challenging for higher-capacity systems, since network paths are not static and channel-degrading effects can change with temperature, component drift, aging and fiber plant maintenance. In order to enable robust and cost-effective automated operation, the network should be able to: (i) intelligently monitor the physical state of the network as well as the quality of propagating data signals, (ii) automatically diagnose and repair the network, and (iii) redirect traffic. To achieve this, optical performance monitoring should isolate the specific cause of the problem. Furthermore, it can be quite advantageous to determine when a data signal is beginning to degrade, so that the network can take action to correct the problem or to route the traffic around the degraded area.

This chapter explores optical performance monitoring and its potential for enabling higher stability, reconfigurability, and flexibility in an optical network. Moreover, this chapter describes the specific parameters that a network might want to monitor, such as chromatic dispersion, polarization-mode dispersion, and optical SNR. Promising monitoring techniques are reviewed.

1.4.8 Chapter 8. ROADMs and their system applications (*Mark D. Feuer, Daniel C. Kilper, and Sheryl L. Woodward*)

As service providers begin to offer IPTV services in addition to data and voice, the need for fast and flexible provisioning of mixed services and for meeting unpredictable traffic demand becomes compelling. Reconfigurable optical add/drop multiplexers (ROADMs) have emerged as the network element that can satisfy this need. Indeed, subsystem and system vendors are rapidly developing and producing ROADMs, and carriers are installing and deploying them in their networks.

This chapter is a comprehensive treatment of ROADMs and their application in WDM transmission systems and networks, comprising a review of various ROADM technologies and architectures; analyses of their routing functionalities and economic advantages; considerations of design features and other requirements; and discussions of the design of ROADM transmission systems and the interplay between the ROADM and transmission performance. The chapter ends with some thoughts on the remaining challenges to enable ROADMs to achieve their potential.

1.4.9 Chapter 9. Optical Ethernet: Protocols, management, and 1–100 G technologies (*Cedric F. Lam and Winston I. Way*)

As the Internet becomes the de facto platform for the delivery of voice, data, and video services, Ethernet has become the technology of choice for access and metro networks, and for next-generation long-haul networks. As stated concisely by the authors, "The success of Ethernet is attributed to its simplicity, low cost, standard implementation, and interoperability guarantee," attributes that helped the "networking community it serves to prosper, hence producing the economy of scale."

This chapter is an in-depth review of the evolution and development of Ethernet technology for application in optical fiber telecommunications networks. Topics covered include: point-to-point Ethernet development, Layer-2 functions, Carrier Ethernet, Ethernet in access PONs, Ethernet OAM (Operation, Administration, Maintenance), development of 10 GE for PON and 100 GE for core applications, and examples of high-speed Ethernet.

1.4.10 Chapter 10. Fiber-based broadband access technology and deployment (*Richard E. Wagner*)

One of the earliest long-haul commercial optical fiber telecom systems was the AT&T Northeast Corridor link from Boston to New York to Washington in 1983. In this application, the large capital investment could be amortized among many users. The prospect of economically bringing fiber all the way to a large number of end users, where cost sharing is not available, has continuously appealed to and challenged the telecom industry. Presently, technology advances and volume manufacture are reducing costs/user, fabulous broadband applications are luring subscribers, and government legislation and subsidies are encouraging growth worldwide. This chapter tracks the history of broadband access, compares the competing access technologies, and projects the roadmap to future deployment in the US, Asia, Europe and the rest of the world. The economic driver for widespread deployment is the explosive growth of Internet traffic, which doubles annually in developed countries and grows even faster in developing countries, such as China. In developed countries, growth is due to new broadband applications; in developing countries, both new users and new applications drive traffic growth.

This chapter focuses on the fiber-based approaches to broadband access worldwide, including some of the drivers for deployment, the architectural options, the capital and operational costs, the technological advances, and the future potential of these systems. Three variants of fiber-based broadband access, collectively called FTTx in this chapter, have emerged as particularly important. They are: hybrid-fiber-coax (HFC) systems, fiber-to-the-cabinet (FTTC) systems, and fiber-to-the-home (FTTH) systems.

1.4.11 Chapter 11. Global landscape in broadband: Politics, economics, and applications (*Richard Mack*)

The technology of choice that predominates in a specific telecom arena depends ultimately upon the competitive economics: the normalized capital and operational costs in dollars per unit bandwidth (per unit distance). For metro, regional, and long-haul arenas, lightwave technology is indisputably the king. However, in the access arena, the competitive unit cost of lightwave technology has not favored rational deployment. Indeed, the history of FTTH has followed a tortuous path; the early trials in the 1980s and 1990s did not lead to massive deployment. Globally, Japanese and Korean telecom companies have been leading the installation of FTTH (with as yet unknown economic consequences). In the meantime, the cost of FTTH equipment has been decreasing steadily. Recently, relief from "unbundling" (exemption from requirement for incumbent carriers to share facilities with competitive carriers, as ruled by FCC) and competition from cable TV companies have prompted incumbent carriers in the US to install FTTH with competitive (triple-play) service offerings. As the demand for broadband services grows and revenue improves, the return from the vast investment in FTTH may be realized in the not-to-distant future.

This chapter is a fascinating, data-laden account of the history of deployment of optical fiber telecommunications, with emphasis on economics, growth landscape, and broadband services in the access arena. The discussion includes historical highlights, demographics, costs and revenues, fiber installations, services scenarios, competition and growth, regulatory policies, applications and bandwidth requirements, technology and network architecture choices, market scenarios, etc. The interplay of these issues is discussed and summarized in the concluding section.

1.4.12 Chapter 12. Metro networks: Services and technologies (*Loukas Paraschis, Ori Gerstel, and Michael Y. Frankel*)

Metropolitan networks operate in the environs of a major city, transporting and managing traffic to meet the diverse service needs of enterprise and residential applications. Typically, metro networks have a reach below a few hundred kilometers with node traffic capacities and traffic granularity that require amplified dense WDM technology with optical add/drop, although the more economical coarse WDM technology has also been deployed. At present, convergence of IP services and traditional time-division multiplexed (TDM) traffic with low operational cost is an important issue.

This chapter reviews the architecture and optical transport of metro networks, which have evolved to meet the demand of various applications and services, including discussions of the evolution of network architecture, physical building blocks of the WDM network layer, requirements of network automation, and convergence of packetized IP with traditional TDM traffic, and ending with a brief perspective on the future outlook.

1.4.13 Chapter 13. Commercial optical networks, overlay networks, and services (*Robert Doverspike and Peter Magill*)

As service providers are the ultimate users of novel technologies and systems in their networks, it is important that the innovators have a sound understanding of the structure and workings of the carriers' networks: the architecture and layers, traffic and capacity demands, management of reliability and services, etc. Commercial networks are continuously upgraded to provide more capacity, new services and reduced capital and operational cost; seamless network evolution is essential for obvious economic reasons. Even when "disruptive" technologies and platforms are introduced, smooth integration within the existing infrastructure is imperative.

This chapter reviews the important aspects of current commercial optical networks in all three segments or layers: access, metro, and core. Topics include 1. relationship of services to layers, covering service requirements, layer technology, quality of service, Service Level Agreements, network availability, and network restoration; and 2. network and services evolution, covering demand and capacity, and applications and technologies in all three segments of the network. In the summary section, the authors point out that while much of the "industry focuses on advanced optical technologies for the long-distance network, most of the investment and opportunity for growth resides in the metro/access [sector]."

1.4.14 Chapter 14. Technologies for global telecommunications using undersea cables (*Sébastien Bigo*)

The introduction of WDM has enabled a tremendous capacity growth in undersea systems, both by the increase in the number of carrier wavelengths and by the increase in the channel bit rate. Starting from 2.5 Gbit/s in the mid-1990s, the bit rate was upgraded in commercial products to 10 Gbit/s at the end of the last century. The next generation of undersea systems will likely be based on 40-Gbit/s bit-rate channels. However, transmission at 40 Gbit/s is significantly more challenging than at 10 Gbit/s.

This chapter gives an overview of the specificities of submarine links with respect to terrestrial links and provides a few examples of recently deployed undersea systems. Moreover, it describes the key technologies involved in undersea systems, explaining particular implementations of early optical systems, today's 10-Gbit/s systems, and future 40-Gbit/s links. The key technologies are related to fiber selection and arrangement, to amplifier design, to modulation formats, to detection techniques, and to advanced impairment-mitigation solutions.

1.4.15 Chapter 15. Future optical networks (*Michael O'Mahony*)

In the past few years Internet traffic, doubling annually, has dominated the network capacity demand, which has been met by the advances in lightwave communications. The transformation from a circuit-switched, voice-centric to a packet-switched, IP-centric network is well underway; amplified WDM transmission systems with terabits-per-second capacity are being deployed; rapid reconfigurable networking and automatic service provisioning are being implemented. The drive to reduce cost and increase revenue has being inexorable. In the meantime, carriers are installing FTTH and offering IPTV services, which will undoubtedly change network traffic characteristics and boost the traffic growth rate. What will the future networks look like?

This chapter reviews the growth of the data traffic and evolution of the optical network, including user communities, global regional activities, and service requirements. Discussions cover the diversity of architectures, the evolution of switching, cross-connecting and routing technologies, and the transformation to carrier-grade (100 G) Ethernet. The author notes that device integration "will enable the realization of many of the key functionalities for optical networking."

1.4.16 Chapter 16. Optical burst and packet switching (*S. J. Ben Yoo*)

Optical switching has the potential of providing more-efficient and higher-throughput networking than its electronic counterpart. This chapter discusses optical burst and packet switching technologies and examines their roles in future optical networks. It covers the roles of optical circuit, burst, and packet switching systems in optical networks, as well as their respective benefits and trade-offs. A description is given of the networking architecture/protocols, systems, and technologies pertaining to optical burst and packet switching. Furthermore, this chapter introduces optical-label switching technology, which provides a unified platform for interoperating optical circuit, burst, and packet switching techniques. By exploiting contention resolution in wavelength, time, and space domains, the optical-label switching routers can achieve high-throughput without resorting to a store-and-forward method associated with large buffer requirements. Testbed demonstrations in support of multimedia and data communications applications are reviewed.

1.4.17 Chapter 17. Optical and electronic technologies for packet switching (*Rodney S. Tucker*)

The Internet provides most of the traffic and growth in lightwave systems today. Unlike the circuit-switched telephone network, the Internet is based on packet switching, which provides statistical multiplexing of the many data streams that pass through a given switch node. A key element of any packet switch is the ability to buffer (or store) packets temporarily to avoid collisions. Since it is easy to store bits as electronic charge in silicon, commercial packet switches are electronic.

This chapter introduces the basics of practical packet routers, indicating the requirements of a switch based on either electronics or photonics. The author compares the physical limits on routers based on storing and switching electrons vs photons. While many in the optics field would like to see photonic switches dominate, it appears that the physical inability to provide a large optical random access buffer means that packet switches will continue to be optoelectronic rather than all-optical for some time.

1.4.18 Chapter 18. Microwave-over-fiber systems (*Alwyn J. Seeds*)

The low-loss, wide-bandwidth capability of optical transmission systems makes them attractive for the transmission and processing of microwave signals, while the development of high-capacity optical communication systems has required the import of microwave techniques in optical transmitters and receivers. These two strands have led to the development of the research area of microwave photonics. Following a summary of the historic development of the field and the development of microwave photonic devices, systems applications in telecommunications, and likely areas for future development are discussed. Among the applications reviewed are wireless-over-fiber access systems, broadband signal distribution and access systems, and communications antenna remoting systems.

1.4.19 Chapter 19. Optical interconnection networks in advanced computing systems (*Keren Bergman*)

High-performance computers today use multicore architectures involving multiple parallel processors interconnected to enhance the ultimate computational power. One of the current supercomputers contains over one hundred thousand individual PowerPC nodes. Thus the challenges of computer design have shifted from "computation-bound" to "communication-bound." Photonic technology offers the promise of drastically extending the communications bound by using the concepts and technologies of lightwave communications for interconnecting and networking the individual multi-processor chips and memory elements. Silicon photonics (see the chapter on "Silicon Photonics" by C. Gunn and T. L. Koch) looms as a possible subsystem technology having the cost-effectiveness of silicon CMOS manufacturing process.

1. Overview of OFT V Volumes A & B 21

This chapter is a review of the subject of interconnection networks for high-performance computers. Performance issues including latency, bandwidth and power consumption are first presented, followed by discussions of design considerations including technology, topology, packet switching nodes, message structure and formation, performance analysis, and evaluation. Network design implementation, architectures, and system demonstrations are then covered and thoughts are offered on future directions, including optical interconnection networks on a chip.

1.4.20 Chapter 20. Simulation tools for devices, systems, and networks (*Robert Scarmozzino*)

The ability of optical fiber telecommunications to satisfy the enormous demand for network capacity comes from thorough understanding of the physics and other disciplines underlying the technology, as well as the abilities to recognize sources for limitation, develop ideas for solution, and predict, test, and demonstrate those ideas. The field of numerical modeling has already been an important facilitator in this process, and its influence is expected to increase further as photonics matures.

This chapter discusses the broad scope of numerical modeling and specifically describes three overarching topics: (i) active and passive device/component-level modeling with emphasis on physical behavior, (ii) transmission-system-level modeling to evaluate data integrity, and (iii) network-level modeling for evaluating capacity planning and network protocols. The chapter offers an overview of selected numerical algorithms available to simulate photonic devices, communication systems, and networks. For each method, the mathematical formulation is presented along with application examples.

ACKNOWLEDGMENTS

We wish sincerely to thank Tim Pitts and Melanie Benson of Elsevier for their gracious and invaluable support throughout the publishing process. We are also deeply grateful to all the authors for their laudable efforts in submitting their scholarly works of distinction. Finally, we owe a debt of appreciation to the many people whose insightful suggestions were of great assistance.

We hope our readers learn and enjoy from all the exciting chapters.

2

Semiconductor quantum dots: Genesis—The Excitonic Zoo—novel devices for future applications

Dieter Bimberg

Institut fuer Festkoerperphysik and Center of Nanophotonics, Berlin, Germany

2.1 PREFACE

Decisive for the development of civilization of mankind across the last ten thousand years was the discovery, development, and use of novel materials and technologies. Replacing stone by bronze and bronze by iron enabled new tools like the plough, stimulating quantum leaps in agriculture, and handicraft. The information age started with the discovery and use of silicon—a rather new material when measuring time in bits of a thousand years.

Our knowledge of properties of elements occurring in nature is almost complete. The laws of physics governing the interactions of atoms to form liquids or solids with composition-dependent properties are established to a large extent. Still, new materials based on chemical architecture will continue in the future to present the basics of quantum leaps for many new technologies.

By entering now the nanoage, developing nanotechnologies, we realize that size and shape is more than just another subject of researchers' curiosity in ultrasmall and beautiful objects. Nanotechnologies enable us to **modify** the properties of semiconductors to a large extent **without** changing the composition.

The ultimate class of semiconductor nanostructures, "quantum dots" (QDs), have lateral dimensions in all three directions that are smaller than the de-Broglie-wavelength. They constitute nanometer-size coherent clusters that are embedded in the dielectric matrix of another semiconductor. They are often self-similar and can be formed by self-organized growth on surfaces. **Single or few** QDs enable novel devices for quantum cryptography, quantum information processing, and novel DRAM (Dynamic Random Access Memory)/flash memory cells. **Billions** of them present the active centers in optoelectronic devices like QD lasers or QD optical amplifiers, revolutionizing communication, consumer electronics, measurement techniques, and more.

2.2 THE PREHISTORIC ERA—OR WHY DID A PROMISING APPROACH ALMOST DIE

Besides fundamental interest in novel effects, semiconductor research is strongly driven by the prospect of potential applications. The dawn of nanotechnology in semiconductor physics as well as in optoelectronics is closely related to the work of two physicists working at Bell Labs in Murray Hill, USA. Ray Dingle and Charles Henry [1] applied for a patent on quantum well (QW) lasers in 1976. Here, they listed the benefits of reducing the dimensionality of the active area of a semiconductor laser when changing from a three-dimensional double heterostructure (three-dimensional structure) to a QW (two-dimensional structure) and finally to a quantum wire structure (1D structure). Such a reduction in dimensionality heavily affects the electronic properties of the respective semiconductor (e.g., GaAs in AlGaAs) as demonstrated in Figure 2.1. Both the energy eigenvalues and the density of states become a function of the lateral dimension in x-, y-, and z-direction.

The density of states of a single QD is given by a δ-function which, even when occupied by carriers and at high temperatures, does not show thermal broadening. One-, two-, and three-dimensional structures do show such broadening due to the continuous energy dispersion. The threshold current density of a laser was predicted by Dingle and Henry to be drastically lowered when the dimensionality of the active region is reduced. Indeed, today 95% of the semiconductor lasers produced utilize a QW as an active region. Without such already sophisticated QW structures, the penetration of semiconductor lasers to such diverse fields like optical drives, barcode scanners, printers, or telecommunication would not have been possible.

In 1982, Arakawa and Sakaki [2] finally considered the advantages of zero-dimensional structures triggered by a proposal of Nick Holonyak's group and by their own experiments on double heterostructures in high-magnetic fields. They predicted the threshold current density to be almost temperature independent for semiconductor lasers containing zero-dimensional structures. Additional benefits were proposed by Asada and co-workers [3]. They calculated an enhancement in material gain and a reduction of threshold current density by a factor of 20 (!) for GaInAs/InP and GaAs/GaAlAs QDs as compared to the three-dimensional case. Those and further predictions in the 1980s gave rise to an enormous amount of investigations of zero-dimensional structures for a decade [4]. All that work focused almost exclusively on the small group of heterostructures consisting of materials with close to identical lattice constants. It was the prevailing opinion that lattice match is a prerequisite to obtain defect-free QD and wire heterostructures. Huge intellectual and economic efforts were made—resulting in no superior devices. Hirayama et al. [5] reported the "best" GaInAs/InP QD laser based on this approach in 1994. The threshold current density was 7.5 kA/cm^2 for pulsed excitation at 77 K. Today, we know that the highly complex

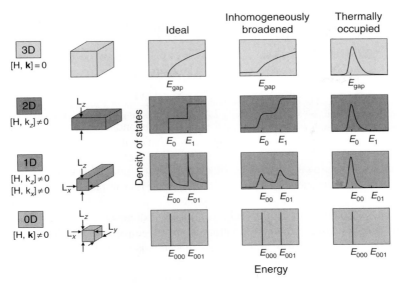

Figure 2.1 The impact of changes of dimensionality on the electronic density of states in a semiconductor (schematical). In a three-dimensional (volume) semiconductor, the wave vector **k** is a good quantum number and the ideal density of states is proportional to the square root of the energy. Inhomogeneous broadening leads to smearing out of the ideal density of states. A continuous density of state results in a broad temperature-dependent distribution of a given number of charge carriers. For quantum wells (2D) and quantum wires (1D), the components of the wave vector in the direction of quantization are no good quantum numbers any more. Yet, the density of states is continuous and carriers obey a thermal occupation. Quantum dots show a complete quantization resulting in discrete energy levels and no broadening upon occupation by charge carriers at finite temperatures (this figure may be seen in color on the included CD-ROM).

process used for device fabrication, including epitaxial growth, two-dimensional lithography, dry etching (all on a nanometer scale!), overgrowth, and additional processing steps, led to a high density of "lethal" defects. Consequently, the quantum efficiency was very low and the modal gain was insufficient. After many such discouraging reports, some theorists started to warn proceeding to work on low-dimensional structures for photonic devices. The low-quantum efficiency was claimed to be inherent to such structures, resulting from orthogonal electron and hole wave functions, and slow carrier capture and relaxation times of more than nanoseconds [6]. The latter effect was referred to as "phonon bottleneck" and attracted attention of many scientists for a decade. Thus, the nanoscientists started to fundamentally question the new field of nanosemiconductor physics and its technologies.

2.3 A NEW DAWN AND COLLECTIVE BLINDNESS

Surface physicists classify the growth modes for the coherent deposition of material 1 on material 2 (Figure 2.2) into three groups. For close to identical lattice constants, a

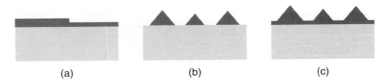

Figure 2.2 The three coherent surface growth modes for semiconductor 1 (dark gray) grown on top of semiconductor 2 (light grey). Layer-by-layer-growth is also called "Frank–van der Merwe" growth. The "Volmer–Weber" growth mode (b) is a three-dimensional one of nonconnected coherent clusters. For "Stranski–Krastanow" growth, coherent three-dimensional structures develop on a thin wetting layer.

two-dimensional growth (monolayer by monolayer) occurs that is called "Frank–van der Merwe" growth mode. This mode is observed, for example, for growth of GaAs on AlGaAs. In case of different lattice constants, the layers of the different materials are strained. When reaching a critical layer thickness of the newly deposited layer, the strain energy is reduced by the formation of defects, in particular of dislocations. An alternative way to minimize strain energy, which was originally disregarded, is the self-organized growth of **coherent** three-dimensional clusters. Depending on a complex interplay of volume energies and orientation-dependent energy contributions by surfaces and edges, the clusters form directly on the substrate ("Volmer–Weber" growth mode) or on a wetting layer with a typical thickness of one to two monolayers ("Stranski–Krastanow" growth mode) [4, 7]. Stranski and Krastanow introduced this universal growth mode on a meeting of the Vienna Academy of Sciences in 1937.

Already in 1985 Goldstein et al. [8] reported on electron microscopy investigations of InAs/GaAs heterostructures that revealed vertically correlated InAs nanoclusters. However, they were lacking information on the electronic properties of these clusters and presented no information on whether these clusters were free of defects. A proof of defect-free growth of such structures was given 5 years later by Madhukar et al. [9] at the University of Southern California and Sasaki et al. [10] at the University of Kyoto. Still, there was no experimental evidence for a delta-like density of states in QDs (see Figure 2.1) and no novel or superior device utilizing QDs was demonstrated. Therefore, the reports [8–10] received little attention for a long time.

2.4 DECISIVE BREAK-THROUGHS

In 1994/1995, we reported four break-throughs that prepared the ground for the subsequent explosion of research on semiconductor nanostructures.

- Cathodoluminescence investigations by Marius Grundmann et al. [11] revealed the delta-function-like emission characteristics of excitons in InAs/GaAs QDs. The samples were grown by molecular beam expitaxy (MBE) at the Ioffe Institute in St Petersburg by Victor Ustinov and Nicolai Ledentsov. Similar observations were reported using photoluminescence by Moison et al. [12] and Petroff et al. [13].

2. Semiconductor Quantum Dots

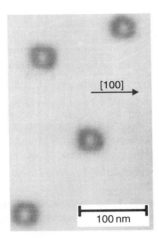

Figure 2.3 High-resolution top-view transmission electron microscopy of a quantum dot layer (four quantum pyramids). The basis of the squares (the pyramids) are oriented parallel to [100].

- Vitali Shchukin et al. [14] developed a theoretical model for the self-similarity of QD sizes and shapes and for the self-organized growth based on thermodynamical arguments [4, 7]. The model also explained the observations reported in Refs [12, 13]. Brilliant high-resolution transmission electron micrographs of InAs/GaAs QDs (Figure 2.3) that were grown by Heinrichsdorff et al. [15] using metalorganic chemical vapor deposition (MOCVD) revealed indeed a close to perfect self-similarity of QDs and confirmed Shchukins thermodynamic approach. Further important theoretical work included kinetic aspects [16] and contributed to a more detailed understanding of QD growth.
- Efficient carrier capture into QDs on a picosecond time scale was demonstrated by Heitz et al. [17] by time-resolved and resonant photoluminescence spectroscopy.
- Nils Kirstaedter et al. [18] succeeded in producing the first injection laser based on coherently grown QDs. Two theoretically predicted properties of QD lasers of fundamental importance were confirmed by this work: reduced threshold current density and improved temperature stability of the threshold current.

2.5 PARADIGM CHANGES IN SEMICONDUCTOR PHYSICS AND TECHNOLOGY

The abrupt change of fundamental, technological, and physical paradigms that were not questioned for decades led to an out-bursting development of the research on zero-dimensional structures for the years after 1994:

- Lattice-mismatched semiconductors have to be used for the epitaxial growth of defect-free QD structures to initiate strain-driven self-organization.

Such QD formation is observed for almost all IV/IV-, III/V-, and II/VI-heterostructures that meet the conditions. One of the underlying processes of formation is the "Stranski–Krastanow" growth mode; spinoidal decomposition or submonolayer (SML) deposition are other ones [7].

- Charge carriers and excitons are strongly localized in real space for QD volumes smaller than 10^3 nm^3.
- There is no conservation of the wave vector **k** in zero-dimensional structures. Consequently, there is no polarization bottleneck like for structures of higher dimensionality.
- The energy levels of charge carriers or excitons in single dots are discrete like in atoms, showing homogeneous (Lorentzian) broadening depending on temperature only.
- Recombination/absorption and gain are purely (bi)-excitonic.
- Carrier capture, relaxation, and recombination at low temperatures or for strong confinement have to be described by master equations of microstates [19]. A global Fermi level is nonexistent.

Figure 2.1 displays the density of states for three-, two-, one-, and zero-dimensional systems. When inhomogeneous broadening and thermal occupation are included, the unique properties of zero-dimensional systems as compared to systems of higher dimensionality are visualized best. All higher dimensional systems resemble each other qualitatively. Zero-dimensional systems are distinct: QDs resemble giant atoms in a dielectric cage more than solids.

2.6 ANYTHING SPECIAL ABOUT THE ELECTRONIC AND OPTICAL PROPERTIES?

Shape and composition of QDs can be revealed by combining transmission electron microscopy and scanning tunneling microscopy (plane view and cross-section). For determining the precise distribution of atoms in a strained QD, one needs sophisticated algorithms to treat the micrographs based, for example, on molecular dynamical calculations [20].

Once the material distribution inside and outside a QD is known, one can calculate the strain field by, for example, continuum theory or the valence-force-field method [21]. Detailed knowledge of the strain field allows for the determination of piezoelectric potentials and finally for the calculation of the electronic structure using a numerical 8-band k.p model [21]. Alternative methods include empirical pseudopotential calculations as well as tight-binding approaches. Figure 2.4 gives the energy levels and wave functions in a pyramidal InAs/GaAs QD of 13.3 nm base length as obtained by our 8-band k.p model.

It is of decisive importance for many applications that in a QD of zinc-blend structure, the energy levels are spin degenerate only. Consider a QD whose structure in the base plane has C_{4V} symmetry. Due to the strain-induced

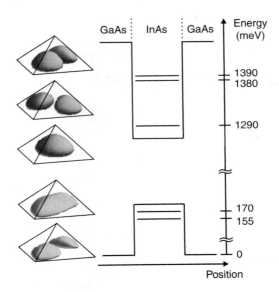

Figure 2.4 Wave functions and energy levels for the three lowest electron and two lowest hole eigenstates in a pyramidal InAs/GaAs quantum dot. The calculation was done using 8-band k.p theory (this figure may be seen in color on the included CD-ROM).

piezoelectric potentials, the symmetry of the confining potential is lowered to C_{2v} resulting, for example, in a splitting of the first excited p-state into two spin-doublets [21]. The centers of mass are not identical for electron and hole wave functions but depend on the actual structure and chemical composition of the QD. For the example given in Figure 2.4, the electron wave function is located on top of the hole wave function. An inhomogeneous In concentration in a ternary InGaAs QD [22] can, however, result in the reverse ordering of wave functions [23]. Our 8-band k.p approach also allows calculations of wave functions in sheets of vertically coupled QDs—in analogy to super-lattices. The device-relevant ratio of TE to TM polarization can be tuned by a variation of the spacer thickness [23].

A beautiful demonstration of the tuneability of the emission of (InGa)As/GaAs QDs by varying the structure is given in Figure 2.5.

The experimentally measured peak energies for MOCVD-grown material range from about 1050 nm to more than 1400 nm. Emission beyond 1600 nm was demonstrated for growth on metamorphic buffers by MBE [24]. Deposition of the QDs within a large-bandgap AlGaAs matrix results in a shift of the emission wavelength toward the visible range. Thus, GaAs-based QD technology allows for an emission and absorption wavelength tuning between the red and infrared. Hence, the "spectral holes" between 1100 nm and 1250 nm where high-performance lasing was hard to obtain before for standard GaAs and InP technologies are filled.

Radiative low-temperature recombination in "classical" semiconductors is dominated by impurity-related-emission processes such as bound excitons and donator–acceptor pair recombination. In QDs, excitonic recombination processes

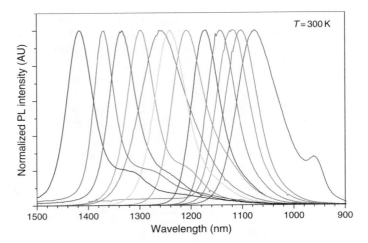

Figure 2.5 The wavelength of emission from the excitonic ground state of a quantum dot can be varied by varying its size and composition. The figure shows the variation of the normalized PL spectra of various InGaAs ensembles in a GaAs matrix (this figure may be seen in color on the included CD-ROM).

are exclusively observed, as electron and hole wave functions are confined within the same small volume. Besides the simplest excitonic complex—the exciton itself—also decays of higher complexes like trions, neutral and charged biexcitons, and multi-excitons are found and shown in Figure 2.6. A whole excitonic zoo reveals itself to the observer when performing laterally high-resolved cathodoluminescence or microphotoluminescence spectroscopy on single QDs. Figure 2.6 gives such an example. The spectrum consists of a few sharp emission lines where the observed half-width is given by the spectral resolution of the setup. For the actual

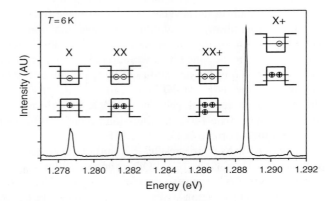

Figure 2.6 Cathodoluminescence spectra of a single-InAs quantum dot in a GaAs matrix. The emission lines are from different excitonic complexes: X, exciton; XX, biexciton, XX +, positively charged biexciton; X +, positively charged exciton. In all cases, an electron recombines with a hole, and the Coulomb interaction with additional charge carriers leads to a shift of the emission relative to the neutral exciton.

2. Semiconductor Quantum Dots

line width, one needs (sub-) microelectronvolt spectral resolution or has to investigate the dephasing in detail [25]. Excitons in (InGa)As QDs exhibit long dephasing times close to 1 ns at low temperatures [25]. Consequently, the homogenous line width is on the order of 2 µeV. Such long dephasing times (coherence times) are of greatest importance for quantum information processing.

A close look on the excitonic and biexcitonic recombination lines shows that they consist of linearly polarized doublets. In QDs Coulomb interaction, anisotropic exchange and correlation have a large impact on the excitonic states due to the strong localization of the charge carriers. The actual contribution of each term depends strongly on the shape, size, and material composition of the dot. The electron–hole exchange interaction results in a fine-structure splitting of the exciton ground state and leads to a doublet structure in the exciton and biexciton recombination [26]. Figure 2.7 demonstrates how the fine-structure splitting as well as the biexciton-binding energy (energy difference between biexcitonic and single-excitonic recombination) depend on the size of the QD. For both values, we observe a monotonous dependence on QD size. A transition from positive to negative fine-structure splitting and binding energy is shown. So both values can be zero for a QD of the right size and shape. That observation is beyond our experience from classical semiconductor and atomic physics—as, for example, there is no "antibinding" H_2 molecule.

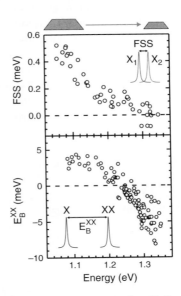

Figure 2.7 Excitonic fine-structure splitting (top) and biexciton-binding energy (bottom) as a function of the emission energy of the neutral exciton and of quantum dot size. The fine-structure splitting is derived from the energetic difference of the two linearly polarized components of the exciton doublet (see inset on the top). The biexciton binding energy is derived from the energetic distance between excitonic and biexcitonic emission (inset on the lower left). Small (large) excitonic emission energies are found for large (small) quantum dots (see scheme on top). All experiments were done at $T = 6\,\text{K}$ (this figure may be seen in color on the included CD-ROM).

2.7 ARE SINGLE QDs GOOD FOR ANYTHING?

Our daily life depends more and more on secure data communication. Increasingly complex cryptographic techniques are developed using elegant mathematical methods based on powerful computers. At the same time, the advancement in computational power threatens even the most complex codes and allows them to be broken [27]. An important shortcoming of conventional data communication is the possibility of undetected eavesdropping. A way out of this dilemma is based on combining basic principles of quantum mechanics with most modern information theory. In 1984, Bennett and Brassard first introduced a quantum communication scheme (BB84 protocol) proposing a secure way to transmit crypto-keys. In July 2004, the "Quantum Information Science and Technology Road Map" was published. This roadmap was developed on behalf of the "Advanced Research and Development Activity" (ARDA) agency of the US government. The roadmap asks for the development of inexpensive practical sources of single photons of well-defined polarization, called q-bits, for the realization of the BB84 and more advanced protocols and/or for sources of entangled photon pairs.

The essential part of a single-photon emitter (SPE) is a quantized system having discrete energy levels. Many different systems have been tried in the past [27]. A single semiconductor QD with its discrete ground state, allowing for nonresonant excitation, has decisive advantages compared to any other system, for example, isolated atoms. Being embedded in a suitable p–i–n structure, an electrically driven high-frequency-pulsed source of single photons can be realized (q-bit on demand). If the operating temperature is not too low, a module, based on such devices, would not cost much more than a "normal" semiconductor laser module for data communication.

A first step toward such a structure was the controlled realization of epitaxial layers with extremely low densities of QDs of $10^8 \, \text{cm}^{-2}$. Such layers are embedded in p-i-n diode structures. For electrical excitation of only one single QD, an oxide aperture ($\varnothing < 1 \, \mu\text{m}$) is fabricated on top of the dots such that the current path is restricted to the region of one QD only (Figure 2.8).

The spectrum in Figure 2.8 shows the emission of the device for a spectral region of more than 400 nm. Only one emission line from the neutral exciton ground state at about 950 nm is observed. Not even the wetting layer, the GaAs matrix, or the AlGaAs barriers exhibit emission!

This device can operate without any spectral filter that would reduce the actual data rate. True single-photon emission was verified by the observation of antibunching in correlation measurements [28]. Further improvements will be based on resonant microcavities to enhance the emission rate through the Purcell effect to have directed emission and a larger output yield. In addition, novel types of high-sensitivity detectors have to be developed. SPEs and emitters of entangled photons will be in demand at a variety of emission wavelengths. For free-space intersatellite communication systems, the wavelength will be different from that of a fiber-based system.

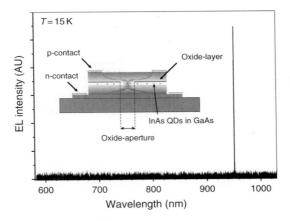

Figure 2.8 Electroluminescence spectrum of a single photon source based on a single-quantum dot and schematic picture of the p–i–n diode. It is remarkable that across a range of 400 nm, only the excitonic emission of a single quantum dot is observed (this figure may be seen in color on the included CD-ROM).

In the beginning of 2006, the actual edition of the "International Technology Road Map for Semiconductors" [29] was published. This biennially appearing strategic plan anticipates the upcoming demands and challenges of the next decade for silicon-based integrated circuits such as flash or DRAM memories. In the second half of the next decade, the number of charge carriers or spins in a single cell has to be dramatically reduced to further increase maximum storage capacity and minimize power consumption. Additional functionality will be gained by ultrafast photonic intrachip communication.

Two different types of memory modules dominate presently:

(1) The DRAM offers read and write times smaller than 20 ns. The memory retention time, however, is on the order of milliseconds, and continuous refreshing cycles are necessary.
(2) The "Flash Memory" is nonvolatile and retains stored information for more than a decade. Write cycles, however, are extremely slow (~ 1 ms) and the endurance is limited to about 10^6 write cycles.

A QD-based memory might combine the best of DRAM and flash: fast access times, long memory-retention times, more than 10^{15} write cycles, and a high-storage density of up to terabit/inch^2. Table 2.1 compares the properties of classical DRAM, flash, and the future nanoflash (QD-Flash).

Two of the major obstacles that have to be surmounted on the way to a nanoflash are as follows:

(1) Is it possible to obtain a sufficiently long retention time of a single-charge carrier in a QD at room temperature?
(2) Can we address a few QDs electrically?

Table 2.1
Comparison of present properties of DRAM and flash memory cells with a possible future QD-Flash.

Memory	Write	Read	Storage	Electrons	Endurance
DRAM	~20 ns	~20 ns	~ms	> 10,000	> 10^{15}
Flash	~1 ms	~20 ns	> 10 years	~1000	~10^6
QD-Flash	<1 ns	<10 ns	> 10 years	<1000	> 10^{15}

Recently, we managed to develop prototypes of QD-based memory cells with retention times of several seconds at room temperature, as was demonstrated by time-resolved capacitance spectroscopy [30]. The increase of retention time by a factor of 10^{10} as compared to previous work results from the insertion of an AlGaAs potential barrier close to the QDs and from choosing holes for storage instead of electrons. Further increase to years is expected for using type-II QD structures like InGaSb/AlGaAs localizing only holes. The task of reading out the single QDs' charge state might also be solved soon by using structures containing a 2D electron gas shown schematically in Figure 2.9.

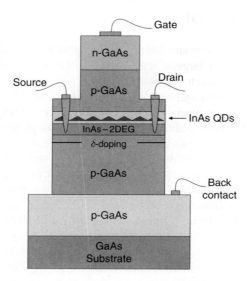

Figure 2.9 Future nanoflash memory cell based on (QDs) (this figure may be seen in color on the included CD-ROM).

2. Semiconductor Quantum Dots

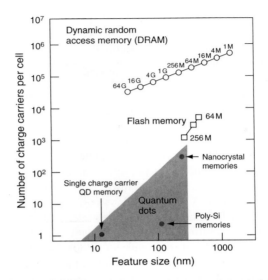

Figure 2.10 Development of feature size of DRAMs and flash memories and of the number of charge carriers needed for one cell (from top right to bottom left) (this figure may be seen in color on the included CD-ROM).

The prospective future of memory devices with respect to feature sizes and the number of charge carriers per memory cell is given by Figure 2.10.

2.8 UTILIZATION OF MANY QDs

Besides our interest in the very small, the development of nanosemiconductor physics and technology was to a large extent driven by the prediction of fundamental improvements of semiconductor lasers. During the decade since the first presentation of a QD laser based on self-organized growth by Kirstaedter et al. [18], indeed lasers exhibiting properties much superior to classical lasers and at novel wavelengths, not covered yet by the respective heterostructure, were demonstrated. The most important advantages as compared to classical double heterostructure or QW lasers include [31]

- Lasing wavelengths in the 1.3-µm spectral range, both for edge and surface emitters using GaAs substrates [32, 33]. A 1.5-µm emission wavelength of GaAs-based edge-emitting lasers using metamorphic buffers [34].
- Very low-transparency current density (<6 A/cm^2 per QD sheet) and internal losses (∼1.5 cm^{-1}), and high-internal quantum efficiency of 98% for a triple sheet QD laser at 1.15 µm. A 12-W output power, equivalent to a power density of 18.2 MW/cm^2, for six-fold MOCVD-grown stack. In lifetime tests at 1.0 W, 1.5 W, and 50°C heat sink temperature, no aging of these lasers within 3000 hours could be observed [35, 36].

- Stability enhancement by 23 dB for external optical feedback at 1.3 µm [37, 38].
- Large tuning range of >200 nm [39].
- Improved radiation hardness and suppressed facet overheating, increasing the castastrophic optic mirror damage (COMD) level [40, 41].
- Complete suppression of filamentation for the transverse ground mode up to stripe widths of 9 µm at 1.3 µm leading to strongly increased coupling efficiency into fibers [42].
- Deep mesa lasers with superb index-guided performance down to very narrow stripe widths (1 µm) and completely spherical far field [43] are opening new opportunities for cost-efficient photonic crystal or distributed feedback (DBF) applications and low-cost coupling to fibers.
- A 12-GHz modulation bandwidth at room temperature [44].
- A 10-Gb/s error-free data modulation (error rate 10^{-12}) obtained at −2-dBm receiver power, 1.3-µm emission wavelength [43–46].
- Passive mode-locking in the range of 5–80 GHz at wavelengths around 1.3 µm [43, 47] with pulse widths below 1 ps. Hybrid mode-locking [47] at frequencies up to 40 GHz yields a significant improvement of the pulse timing jitter and enables external synchronization.
- A 40-Gb/s error-free amplification (error rate 10^{-11}) with a fiber net gain of 8 dB, 1.3-µm emission wavelength.
- A 1.2-mW/2-mW output power and a slope efficiency of 64% at 300 K for vertical-cavity surface-emitting laser (VCSELs) with fully oxidized/semiconductor mirrors at 1.3 µm [48].

Several epitaxial improvements were proposed and partially realized to achieve the abovementioned results, that is, growth of InGaAs/GaAs QDs on template layers [49], overgrowth of QDs with QW layers [50], stacking of QDs [51], close stacking of QDs leading to vertical coupling of the QD layers [52], defect-reduction techniques [53], introduction of strain relaxation layers [54], p-doping of the GaAs barrier layers [55], and tunnel injection of carriers into the QDs through a thin barrier layer [56].

In the next section, we present in detail the high-speed properties of QD-based directly modulated and mode-locked lasers and amplifiers before we turn to VCSELs.

2.9 HIGH-SPEED NANOPHOTONICS

2.9.1 Directly Modulated QD Lasers

Directly modulated laser diodes are the key component of fiber-based datacom, converting digital electrical signals into digital optical signals at a rate of 10 Gb/s or more. Main issues of research are the improvement of modulation speed, reduction of power consumption, reduction of temperature sensitivity, and the

2. Semiconductor Quantum Dots

simplification of processing and mounting of the laser diodes. As presented in the previous section, QD lasers comprise several advantages making them ideal devices for fiber-based datacom [57].

Device Structure

The $Al_{0.35}Ga_{0.65}As$/GaAs laser structures incorporating a 10-fold stack of InGaAs QDs emitting at 1.3-μm wavelength were grown by MBE [58]. The wafers were then processed into ridge waveguide (RW) structures with stripe widths from 1 to 4 μm by dry etching through the active layer to suppress current spreading and provide strong index guiding of the optical mode [59].

A 1000-μm long, 1-μm wide RW diode with 95% HR coating on the rear facet and backside n-contact was mounted in a fiber-optic module comprising a temperature-controlled heat sink, a microwave port with a network, and a single-mode fiber (SMF) pigtail. At room temperature, the QD laser module had a threshold current density of 270 A/cm^2, emission wavelength of 1280 nm, and a small-signal modulation bandwidth of about 7 GHz. Temperature-dependent measurements were done with a device of 500 μm length, 4 μm ridge width, and HR-coated rear facet.

Eye Pattern and Bit-Error Rate Measurements on a QD Laser Module

Eye pattern measurements were carried out back-to-back with the QD module biased at 5–7 times the threshold current density and a nonreturn-to-zero (NRZ) pseudorandom binary sequence (PRBS, word length of $2^{15}-1$) with 2.5 V_{p-p} amplitude (12 dBm). The average output power into the fiber was 1–3 mW. The inset of Figure 2.11(a) shows clearly open 10 Gb/s eye patterns, with a signal-to-noise (S/N) ratio of 6.8, an extinction ratio of 4.9 dB, and a peak-to-peak timing jitter of

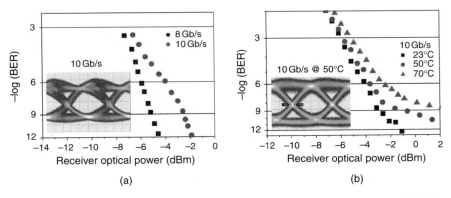

Figure 2.11 (a) BER measurement of QD laser module at 8 Gb/s and 10 Gb/s, and (b) at 10 Gb/s for different temperatures, inset shows the corresponding eye patterns (this figure may be seen in color on the included CD-ROM).

30 ps. Open eye patterns were observed up to 12 Gb/s. Due to the strong damping of relaxation oscillations in QD lasers, all eye patterns show very little overshoot.

Bit error rate (BER) measurements were carried out at data rates of 8, 10, 11, and 12 Gb/s, keeping the eye pattern measurement settings. We inserted a semiconductor optical amplifier (SOA) between laser and BER tester to compensate for optical losses due to a low laser-to-fiber-coupling efficiency of 10%. Figure 2.11(a) shows the BER measurements for the QD laser module. Both for 8 and 10 Gb/s, we achieve error-free operation (BER $< 10^{-11}$) at −4.5 and −2 dBm receiver power, respectively. No error floor could be detected. There is a considerable power penalty of 2.5 dB when moving from 8 to 10 Gb/s data rate in agreement with the moderate bandwidth of 7 GHz.

The BER curve at 8 Gb/s follows a straight line, whereas the data for 10 Gb/s show a curvature that is unexpected. A possible reason for this effect might be a saturation of the radio frequency (RF) amplifier used to amplify the electrical signal at the BER tester.

Temperature-Dependent BER Measurements

One of the main advantages of QD carrier confinement is the decreased temperature sensitivity of the threshold current and quantum efficiency. Cost-effective packaging of QD lasers is feasible only if this temperature insensitivity can also be demonstrated for the dynamic properties of QD lasers. Although temperatures below room temperature generally improve the dynamic behavior, high temperatures might lead to rollover of modulation speed. Preliminary eye pattern measurements at elevated temperatures have shown qualitatively that 10 Gb/s digital modulation seems possible up to 70° C without current adjustment [60]. However, low-error data modulation (BER $< 10^{-9}$) cannot be judged by eye pattern measurements. Recently, we carried out the first BER measurements at 10 Gb/s and elevated temperatures (50° C and 70° C) without adjustment of any of the driving parameters (bias current, RF power, bit rate). For room temperature, we achieved a BER below 10^{-12} only limited by measurement time, comparable to the results depicted in Figure 2.11(a). Figure 2.11(b) indicates a temperature-dependent error floor for the BER, thus limiting the BER at 70° C to values of about 10^{-9}. The error floor is due to the decrease of modulation bandwidth at higher temperatures and could be completely eliminated by an increase of the modulation bandwidth of the QD lasers by a few gigahertz. The small power penalty of less than 2 dB associated with the increase in temperature indicates the decreased temperature sensitivity of the threshold current and output power.

2.9.2 Mode-Locked QD Lasers

Mode-locking of monolithic semiconductor lasers is an efficient method to generate a regular pattern of short optical pulses at high-repetition rate and narrow spectral width with a small footprint device. The application of mode-locked lasers in optical datacom often requires time or wavelength multiplexing of the optical pulses. Therefore, the pulses should be short, have a narrow spectral width (ideally close to the Fourier limit), high-peak power, and a low-timing jitter.

QD lasers offer a broad gain spectrum (>50 nm) leading to ultrashort pulses with sub-ps width, feasibility of long cavities (∼1 cm) for low-repetition-rate applications [43, 61], and a low α-factor [62] for low chirp, Fourier-limited pulses. At the same time, the mode-locked QD lasers comprise the advantages mentioned in the previous sections.

Device Structure

QD devices for mode-locking were grown and processed similar to the laser structures described in the previous section, with 15 layers of QDs for maximum gain and shortest cavities. The samples for mode-locking were processed into two-section devices by defining a metallization and contact layer gap of 20 µm between the sections yielding an insulation resistance of more than 10 kΩ. The length ratio between reverse bias section and gain section was set to 1:9. The devices were mounted p-side up on a copper heat sink and were electrically connected with a double probe head.

The samples with 4-µm ridge width had lengths between 2000 and 500 µm corresponding to round trip frequencies of 20–80 GHz. All samples with length less than 1 mm were HR coated (95% reflectivity) on the absorber section facet to ensure ground state lasing and enhance pulse auto-collision effects in the absorber section. All measurements were carried out at room temperature (297 K) and continuous wave.

Passive Mode-Locking

The 500-µm long laser was passively mode-locked at currents between 10 and 60 mA and reverse bias voltages between 0 and −10 V. The mode-locking frequency was 80 GHz, which is to our knowledge the highest ML frequency achieved for QD lasers. Figure 2.12(a) shows the corresponding autocorrelation trace, along with the full temporal range including the neighboring cross correlation peaks. Figure 2.12(b) depicts the reverse bias vs. current scan of the auto-correlation traces and the corresponding full width at half maximum (FWHM) pulse widths. With increasing reverse bias, the onset of lasing shifts to larger currents due to the increasing absorption within the waveguide. The onset of lasing occurred abruptly as mode-locking, we observed no transition region. With increasing current at constant reverse bias, the pulses became broader until we observed a cw offset, that is, incomplete mode-locking. At even higher currents, we observed a transition region with all kinds of complex pulse patterns until all intensity fluctuations flattened out to cw lasing.

The minimum pulse width we achieved with this device was 1.5 ps. The corresponding spectrum yields a time-bandwidth product of 1.7, which is well above the Fourier transform limit of 0.32 for sech^2-shaped pulses, indicating the

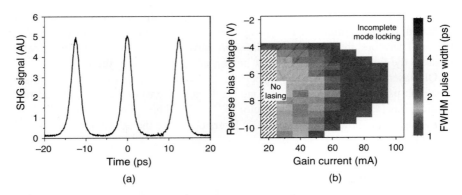

Figure 2.12 (a) Autocorrelation trace of a passively mode-locked quantum dot laser at 1.3 μm and 80 GHz repetition rate. The side peaks correspond to the cross-correlation of two successive pulses, whereas the middle peak presents the autocorrelation of a pulse. (b) Field scan of autocorrelation traces with color-coded FWHM pulse widths of a 80 GHz passively mode-locked QD laser. Three regimes of operation can be distinguished (this figure may be seen in color on the included CD-ROM).

large potential to further reduce the pulse width, for example, by optimization of the passive section.

Similar pulse and Fourier product characteristics were found for other devices at frequencies of 20 and 40 GHz. The smallest pulse width to period ratios we achieved at frequencies between 20 and 80 GHz were in the range between 2% and 10%.

A first hint on the intrinsic limitations of mode-locking in QD lasers is given by the increasing displacement of the center wavelength of the saturable absorber and the center wavelength of the lasing emission that we observed for high currents and large reverse bias voltages. The red shift of the absorber wavelength is possibly due to the quantum-confined Stark effect, whereas the blue shift of the gain spectrum is due to filling of QDs emitting at shorter wavelength at higher pump currents. Both effects lead to an intrinsic limitation of the spectral overlap of both sections and might be responsible for the increase of pulse widths for larger reverse bias voltages (Figure 2.12(b)) and the large time–bandwidth product.

Besides a small pulse width, a low-pulse jitter is desirable for the application of QD mode-locked lasers as optical clocks. We investigated the timing jitter of passively mode-locked QD lasers with repetition rates up to 40 GHz by means of optical cross-correlation measurements and sideband noise measurements using a 50-GHz optical detector and a 50-GHz spectrum analyzer. Comparison of autocorrelation and cross-correlation (e.g., Figure 2.12(a)) allowed us to estimate the uncorrelated jitter to be less than 1 ps. However, the main jitter contribution is expected to be correlated jitter, that is, jitter depending on the temporal position of hundreds or thousands of preceding pulses. Sideband noise measurements of a 40-GHz passively mode-locked QD laser yielding a root mean square (RMS) timing jitter of 5–15 ps showed the dominating influence of correlated jitter. The timing jitter decreases with the gain current.

Hybrid Mode-Locking

Most applications of mode-locked lasers require synchronization with an external electrical signal. This can be done by applying a RF signal to one of the laser sections, typically with a frequency in the vicinity of the cavity round-trip frequency. We found that the application of RF power to the absorber section was much more efficient than modulation of the gain section. The large series resistance of the reverse-biased absorber section caused a doubling of the RF voltage amplitude and a large modulation of the reverse bias voltage, whereas the small series resistance of the forward-biased gain section led to a current modulation which was strongly damped by the RC bandwidth ($\sim$7 GHz).

Hybrid mode-locking of QD lasers yielded a small improvement of the pulse width. The shortest deconvoluted pulse width obtained for the 20 GHz device, best fitted by a sech^2-shaped pulse, was 710 fs (Figure 2.13), as compared to 900 fs for the passively mode-locked device. The improvement of hybrid mode-locking for the 40-GHz device was only marginal (a few percent). The locking range, that is, the frequency range where the pulse repetition frequency locks to the RF signal, was found to be 90 and 7 MHz for 20 and 40 GHz repetition frequency, respectively.

The timing jitter, however, decreased from over 10 ps for passively mode-locked operation to values between 0.5 and 2 ps for the 40 GHz device when applying 14 dBm of RF power to the absorber section. The RMS jitter was calculated from the sideband noise measurement between 1 kHz and 1 GHz. We observe a trade-off behavior, that is, shortest pulse width corresponds to largest jitter and vice versa. The smallest jitter observed was 400 fs, which lies already in the range of state-of-the-art monolithic QW devices [63].

We checked the significance of the RMS jitter values measured by sideband noise integration by comparing to time domain oscilloscope measurements. A 70-GHz bandwidth oscilloscope with a 40-GHz precision time base and a 50 GHz optical detector was used to analyze the 40-GHz pulse trace from the

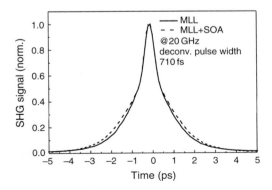

Figure 2.13 Autocorrelation measurement of 20 GHz, 710 fs pulse amplification in QD SOA; we compare both the input and output signal.

mode-locked QD laser. Oscilloscope jitter values differed by less than 10% from the values found by sideband noise integration.

2.9.3 QD Amplifier Modules

For optical networks, optical amplifiers are of largest importance serving as boosters, in-line or preamplifiers. They also perform regenerative (2R), wavelength-conversion or switching tasks for optical signal processing.

QD amplifiers offer large advantages as compared to classical ones: broad bandwidth due to QD-size distribution, ultrafast gain recovery ($\sim$100 fs [64]) for high-speed amplification, reduced chirp due to a low α-factor [62], enabling fast switching, and intrinsic, strongly damped relaxation oscillations [65] yielding low-patterning effects.

Device Structure

QD amplifier structures were grown and processed similar to the laser structures, with either 10 or 15 layers of QDs. The cavity facets were tilted by 7° and antireflection coated with a residual reflectivity of 10^{-3} to prevent the formation of longitudinal modes. The tenfold stacked samples with 4 μm width and 2 mm length were mounted in a fiber-coupled module comprising two SMF ports, a temperature-controlled heat sink and a DC supply. Due to large fiber-coupling losses of about 12 dB, the maximum fiber net gain of the module was 10 dB, corresponding to 22 dB chip gain at a current density of 600 A/cm^2. The −3 dB saturation output power of the amplifier module was +2.5 dBm for the same drive current. The maximum gain was centered at a wavelength of 1290 nm.

Eye Pattern and BER Measurements

Bit-patterned optical input for the amplifier was generated using a tunable external cavity laser (ECL) combined with an electro-optical modulator and a bit pattern generator. The eye pattern measurements were carried out with the QD amplifier module biased at 400–600 A/cm^2, with an optical NRZ pseudo-random bit sequence (NRZ PRBS) with a word length of $2^{31} - 1$ and −5.8 dBm average optical power. The average optical output power from the QD amplifier module was +2.5 dBm. The amplifier output was measured with a fast photodetector using a SOA as optical preamplifier.

Clearly, open eye patterns were observed at a bit rate of 40 Gb/s (inset of Figure 2.14), yielding an extinction ratio of 6 dB, a S/N ratio of 7.7, and a RMS timing jitter of 1.4 ps. BER measurements were performed by varying the optical input power at the preamplifier and for two different drive currents. Error-free modulation (BER $< 10^{-10}$) was achieved for 120 mA drive current, only limited by measurement time. For lower drive current, a power penalty of 2 dB and an error floor at BER $< 10^{-9}$ were found.

2. Semiconductor Quantum Dots

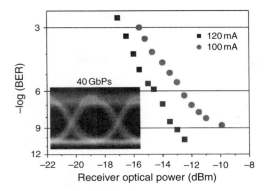

Figure 2.14 BER measurement of QD amplifier module at 40 Gb/s for different drive currents; the inset shows the eye pattern for 120 mA drive current (this figure may be seen in color on the included CD-ROM).

Amplification of ultra-short pulses with subpicosecond width was checked using the QD ML laser at 20 GHz presented in the previous section. The output from the ML laser was fiber-coupled to a SOA of the same QD gain material. The 710 fs pulses in a 20-GHz pulse train were amplified with a fiber net gain of 8 dB by the 4-mm long QD SOA. The maximum chip gain of this device was 26 dB. No degradation of the pulse width by the SOA performance could be observed as demonstrated in Figure 2.13, comparing the autocorrelation traces of the input and the output.

2.9.4 Outlook for High-Speed Edge Emitters

The dynamical properties of InAs–GaAs-based QD devices do not yet reflect the full potential of three-dimensional carrier confinement, mainly due to the limitation of the gain/differential gain by the QD-size dispersion and due to moderate carrier relaxation rates into and within the QDs [66, 67]. QD lasers using tunnel injection of carriers into the QDs show an improved modulation bandwidth [68] due to faster carrier capture into the QDs and the reduction of thermal escape from the QDs. Still, QD tunnel injection lasers have to prove their benefit for optical signal transmission in the telecom wavelength range through eye pattern and BER measurements. Tunnel injection resonances may cause signal patterning effects and impose limits on the data transfer performance. They have to be avoided by careful design of the tunnel injection layers.

For the use as high-performance transmitters in optical networks, directly modulated QD lasers will have to be wavelength stabilized and single mode. First investigations of DFB QD lasers indicate a possible limitation of the modulation bandwidth by gain compression [69]. However, the broad gain spectrum of QD lasers enables a large tuning range of single-mode emission, for example, for coarse wavelength division multiplexing applications. SML QDs offer an alternative approach of three-dimensional carrier confinement besides "Stranski–Krastanow" grown QDs. SML

QDs provide high-modal gain while keeping the excitonic gain mechanism, which makes them suitable for fast single-mode devices like VCSELs [70] and DFB lasers.

Realization of polarization insensitivity of optical amplification in SOAs along with a large spectral gain bandwidth is one of the most promising features of vertically coupled QDs. The implementation of vertical coupling of stacked QD layers will probably also be advantageous for the modulation bandwidth of directly modulated QD lasers.

2.9.5 VCSELs

Next generation short-distance optical networks like chip to chip or backplane applications require temperature-robust ultra-high bit rate sources with low-integration costs. InGaAs-based VCSELs are believed to be the best candidates [71, 72]. As an alternative to InGaAs QWs and "Stranski–Krastanow" grown QDs (SK-QDs) for the wavelength range around 1 µm, we have investigated SML-grown QDs (SML-QDs) [73, 74] as active medium in VCSELs. SML-QDs provide a much higher modal gain as compared to conventional "Stranski–Krastanow" QDs; however, the excitonic gain mechanism and the reduced lateral transport of nonequilibrium carriers characteristic to conventional QDs is probably kept. Lasers based on SML-QDs can exhibit improved temperature stability as compared to QW lasers. This motivates their use in VCSELs, where avoiding gain saturation is crucial to reach high bandwidth/temperature operation.

The structures were grown by MBE on n + GaAs (100) substrates. In SML-QD growth, the initial InAs distribution on the surface is nonuniform, and InAs-rich regions are formed [73, 74]. After overgrowth with the matrix material, the deposition of the next InAs SML is controlled by the nonuniform strain distribution caused by underlying InAs islands. SML deposition leads simultaneously to a significant lateral compositional modulation and high-QD density, leading to a high-material and modal gain. SML-QDs were formed here using ∼0.5-ML InAs deposition cycles separated by 2.3-ML GaAs layers repeated 10 times and providing PL at ∼980 nm. The QDs were stacked three times using 13-nm-thick GaAs spacers and inserted into a GaAs region confined by AlGaAs barriers. The VCSEL structure was realized in an anti-waveguiding design [75] with a high-Al-content cavity and doped bottom and top-distributed Bragg reflectors with 32 and 19 pairs. A single AlAs-rich aperture layer, being partially oxidized, was placed in a field-intensity node on top of the $3\lambda/2$ cavity.

High-speed and high-efficiency devices with a co-planar layout were processed. Figure 2.15 shows static device characteristics for a 6-µm-aperture multimode laser.

The output power exceeds 10 mW at 20° C; the differential efficiency and threshold current are hardly dependent on temperature over a very broad range. From 25° C to 85° C, the differential efficiency reduces only from 0.71 to 0.61 W/A, the threshold current even reduces from 0.29 to 0.16 mA as a result of a reduced cavity-gain detuning.

Figure 2.16 shows the modulation bandwidth of a 6 µm aperture laser for different bias currents at 25° C and 85° C. The maximum bandwidths are 15 and 13 GHz, respectively. Due to a smaller cavity-gain detuning at 85° C for small

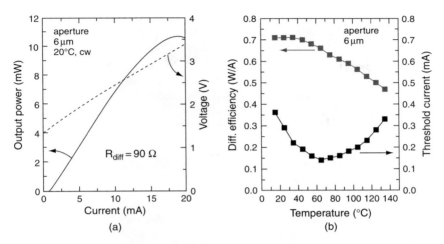

Figure 2.15 Characteristics of a multimode SML-QD-VCSEL: (a) L-U-I and (b) differential efficiency and threshold current vs temperature (this figure may be seen in color on the included CD-ROM).

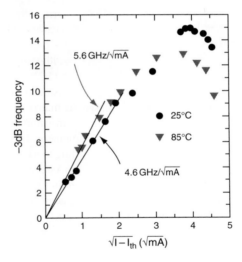

Figure 2.16 Small signal modulation bandwidth of a 6-μm SML QD-VCSEL for 25°C and 85°C. The maximum bandwidth is 15 and 13 GHz, respectively (this figure may be seen in color on the included CD-ROM).

currents, the modulation efficiency here is higher. In Figure 2.17, eye diagrams for 20 Gb/s NRZ 2^7-1 PRBS modulation at 25°C and 85°C are shown. A 25-GHz Discovery Semiconductors MM detector was used. The bias current and modulation voltage were kept constant at 13 mA and 0.8 V_{p-p} for this comparison. Both eyes are clearly open. The signal-to-noise ratio (S/N) changed only from 5.9 at 25°C to 4.3 at 85°C, the extinction ratio was above 4.0 dB. Figure 2.18 shows

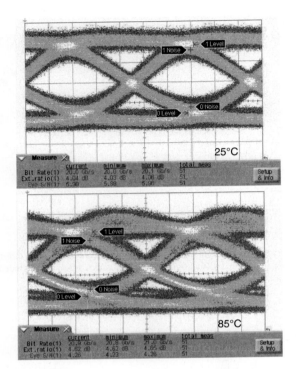

Figure 2.17 Eye diagram for 20 Gb/s modulation of a 6-μm SML-QD-VCSEL at 25°C and 85°C without change of the bias current and modulation voltage (this figure may be seen in color on the included CD-ROM).

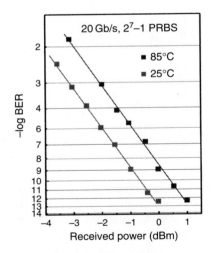

Figure 2.18 Bit-error-rate at 20 Gb/s with $2^7 - 1$ PRBS at 25°C and 85°C for the same bias current (this figure may be seen in color on the included CD-ROM).

the BER also at 25°C and 85°C. Except for the modulation voltage at 25°C of 1.2 Vp-p, all conditions were identical to the eye measurements in Figure 2.17. The device operates error free with a BER $< 10^{12}$ and no error floor even for 85°C. The penalty at 85°C is only 1 dB compared to the back-to-back error rate at 25°C.

2.10 ARE QDs A HYPE?

Some among the scientific community believe it. In actual fact, we are very far from having identified all advantages of semiconductor nanostructures for devices and having disclosed the differences between zero-dimensional and three-dimensional semiconductors. What do we know about electron–acoustic–phonon–interaction in QDs? What are the piezoelectric constants in QDs? Until now, model systems have been investigated. Zero-dimensionality or self-organization, however, are universal. Other zero-dimensional material systems than those investigated until now will show different properties, have surprising physical properties, and will allow applications different from those anticipated presently. We made the first steps on a road without seeing its end.

ACKNOWLEDGMENTS

It is great pleasure to thank my former and present co-workers, as well as colleagues and friends, who within several cooperations decisively contributed to the implementation of a new research field.

Paola Borri, Jürgen Christen, Sabine Dommers, Gerrit Fiol, Martin Geller, Marius Grundmann, Frank Heinrichsdorff, Axel Hoffmann, Friedhelm Hopfer, Leonid Karachinsky, Nils Kirstaedter, Matthias Kuntz, Matthias Lämmlin, Nicolai Ledentsov, Anatol Lochmann, Andreas Marent, Udo Pohl, Konstantin Pötschke, Sven Rodt, Kathrin Schatke, Andrei Schliwa, Robert Seguin, Vitali Shchukin, Oliver Stier, Erik Stock, André Strittmatter, Viktor Temyov, Victor Ustinov, Till Warming, Ulrike Woggon, Roland Zimmermann, and many others, who hopefully forgive me for not mentioning all who deserve it.

The work was funded by SANDiE Network of Excellence of the European Commission No. NoE-NMP-CT-2004–5001 01, ProFIT-Monopic, ProFIT-Optidot, ProFIT-NanoFlash, and EU-IST (TRIUMPH).

REFERENCES

[1] R. Dingle and C. Henry, "Quantum effects in heterostructure lasers," U.S. Patent No. 3982207, 1976.
[2] Y. Arakawa and H. Sakaki, "Multidimensional quantum well laser and temperature dependence of its threshold current," *Appl. Phys. Lett.*, 40, 939, 1982.
[3] M. Asada, Y. Miyamoto, and Y. Suematsu, "Gain and the threshold of three-dimensional quantum-box lasers," *IEEE J. Quantum Electron.*, 22, 1915, 1986.

[4] D. Bimberg, M. Grundmann, and N. N. Ledentsov, *Quantum Dot Heterostructures*, UK, John Wiley and Sons, 1998.
[5] H. Hirayama, K. Matsunaga, M. Asada, and Y. Suematsu, "Lasing action of $Ga_{0.67}In_{0.33}As$/GaInAsP/InP tensile-strained quantum-box laser," *Electron. Lett.*, 30, 142, 1994.
[6] H. Benisty, C. M. Sotomayor-Torrès, and C. Weisbuch, "Intrinsic mechanism for the poor luminescence properties of quantum-box systems," *Phys. Rev. B*, 44, 10945, 1991.
[7] V. A. Shchukin, N. N. Ledentsov, and D. Bimberg, *Epitaxy of Nanostructures*, Heidelberg, Germany, Springer 2003.
[8] L. Goldstein, F. Glas, J. Y. Marzin et al., "Growth by molecular beam epitaxy and characterization of InAs/GaAs strained-layer superlattices," *Appl. Phys. Lett.*, 47, 1099, 1985.
[9] S. Guha, A. Madhukar, and K. C. Rajkumar, "Onset of incoherency and defect introduction in the initial stages of molecular beam epitaxical growth of highly strained $In_xGa_{1-x}As$ on GaAs(100)," *Appl. Phys. Lett.*, 57, 2110, 1990.
[10] M. Tabuchi, S. Noda, and A. Sasaki, "Mesoscopic structure in lattice-mismatched heteroepitaxial interface layers," in *Science and Technology of Mesoscopic Structures* (S. Namba, C. Hamaguchi and T. Ando, eds), Tokyo, Japan, Springer, 1992, p. 379.
[11] M. Grundmann, J. Christen, N. N. Ledentsov et al., "Ultranarrow Luminescence Lines from Single Quantum Dots," *Phys. Rev. Lett.*, 74, 4043, 1995; N. N. Ledentsov, M. Grundmann, N. Kirstaedter et al., "Luminescence and structural properties of (In,Ga)As/GaAs quantum dots," in Proc. 22nd *ICPS*, Vancouver (1994).
[12] J. M. Moison, F. Houzay, F. Barthe et al., "Self-organized growth of regular nanometer-scale InAs dots on GaAs," *Appl. Phys. Lett.*, 64, 196, 1994.
[13] D. Leonard, M. Krishnamurthy, C. M. Reaves et al., "Direct formation of quantum-sized dots from uniform coherent islands of InGaAs on GaAs surfaces," *Appl. Phys. Lett.*, 63, 3203, 1993.
[14] V. A. Shchukin, N. N. Ledentsov, P. S. Kop'ev, and D. Bimberg, "Spontaneous Ordering of Arrays of Coherent Strained Islands," *Phys. Rev. Lett.*, 75, 2968, 1995.
[15] F. Heinrichsdorff, Ch. Ribbat, M. Grundmann, and D. Bimberg, "High-power quantum-dot lasers at 1100 nm," *Appl. Phys. Lett.*, 76, 556, 2000.
[16] E. Pehlke, N. Moll, A. Kley, and M. Scheffler, *Appl. Phys. A*, 65, 525, 1997; M. Meixner, E. Schöll, V. A. Shchukin, and D. Bimberg, "Self-Assembled Quantum Dots: Crossover from Kinetically Controlled to Thermodynamically Limited Growth," *Phys. Rev. Lett.*, 87, 236101, 2001; V. A. Shchukin, D. Bimberg, T. P. Munt, and D. E. Jesson, "Metastability of Ultradense Arrays of Quantum Dots," *Phys. Rev. Lett.*, 90, 076102, 2003.
[17] R. Heitz, A. Kalburge, Q. Xie et al., "Excited states and energy relaxation in stacked InAs/GaAs quantum dots," *Phys. Rev. B*, 57, 9050, 1998.
[18] N. Kirstaedter, N. N. Ledentsov, M. Grundmann et al., "Low threshold, large T_0 injection laser emission from (InGa)As quantum dots," *Electron. Lett.*, 30, 1416, 1994.
[19] M. Grundmann, R. Heitz, D. Bimberg et al., "Carrier Dynamics in Quantum Dots: Modeling with Master Equations for the Transitions between Micro-States," *phys. stat. sol. (b)*, 203, 121, 1997; M. Grundmann and D. Bimberg, "Theory of random population for quantum dots," *Phys. Rev. B*, 55, 9740, 1997.
[20] K. Scheerschmidt and P. Werner, in *Nano-Optoelectronics* (M. Grundmann, ed.), Heidelberg, Germany, Springer, 2002.
[21] O. Stier, M. Grundmann, and D. Bimberg, "Electronic and optical properties of strained quantum dots modeled by 8-band k.p theory," *Phys. Rev. B*, 59, 5688, 1999.
[22] S. Ruvimov, P. Werner, K. Scheerschmidt et al., "Structural characterization of (In,Ga)As quantum dots in a GaAs matrix," *Phys. Rev. B*, 51, 14766, 1995; A. Lenz, R. Timm, H. Eisele et al., "Reversed truncated cone composition distribution of $In_{0.8}Ga_{0.2}As$ quantum dots overgrown by an $In_{0.1}Ga_{0.9}As$ layer in a GaAs matrix," *Appl. Phys. Lett.*, 81, 5150, 2002.
[23] A. Schliwa, M. Winkelnkemper, and D. Bimberg, "Impact of size, shape, and composition on piezoelectric effects and electronic properties of In(Ga)As/GaAs quantum dots," *Phys. Rev. B*, 76, 205324, 2007.

2. Semiconductor Quantum Dots

[24] V. M. Ustinov, A. E. Zhukov, A. Yu. Egorov et al., "Low threshold quantum dot injection laser emitting at 1.9 µm," *Electron. Lett.*, 34, 670, 1998; N. N. Ledentsov, A. R. Kovsh, A. E. Zhukov et al., "High performance quantum dot lasers on GaAs substrates operating in 1.5 µm range," *Electron. Lett.*, 39, 1126, 2003.

[25] P. Borri, S. Schneider, W. Langbein, and D. Bimberg, "Ultrafast carrier dynamics in InGaAs quantum dot materials and devices," *J. Opt. A: Pure Appl. Opt.*, 8, S33, 2006; P. Borri, W. Langbein, S. Schneider et al., "Ultralong Dephasing Time in InGaAs Quantum Dots," *Phys. Rev. Lett.*, 87, 157401, 2001.

[26] R. Seguin, A. Schliwa, S. Rodt et al., "Size-Dependent Fine-Structure Splitting in Self-Organized InAs/GaAs Quantum Dots," *Phys. Rev. Lett.*, 95, 257402, 2005; R. Seguin, A. Schliwa, T. D. Germann et al., "Control of fine-structure splitting and excitonic binding energies in selected individual InAs/GaAs quantum dots," *Appl. Phys. Lett.*, 89, 263109, 2006.

[27] *The Quantum Information Science and Technology Roadmap*, http://qist.lanl.gov

[28] A. Lochmann, E. Stock, O. Schulz et al., "Electrically driven single quantum dot polarised single photon emitter," *Electron. Lett.*, 42, 774, 2006; A. Lochmann, E. Stock, O. Schulz et al., "Electrically driven quantum dot single photon source," *phys. stat. sol. (c)*, 4, 547, 2007.

[29] *The International Technology Roadmap for Semiconductors*, http://public.itrs.net/

[30] A. Marent, M. Geller, A. Schliwa et al., "10^6 years extrapolated hole storage time in GaSb/AlAs quantum dots," *Appl. Phys. Lett. in print.*

[31] D. Bimberg, "Quantum dots for lasers, amplifiers and computing," *J. Phys. D*, 38, 2055, 2005; F. Hopfer, A. Mutig, G. Fiol et al., "High speed performance of 980 nm VCSELs based on submonolayer quantum dots," in Proc. CLEO 2006. Techn. Digest of CLEO/QELS and PhAST 2006, Long Beach, USA, CPDB2, *postdeadline* (2006).

[32] V. M. Ustinov, A. E. Zhukov, N. A. Maleev et al., "1.3 µm InAs/GaAs quantum dot lasers and VCSELs grown by molecular beam epitaxy," *J. Cryst. Growth*, 227, 1155–1161, 2001.

[33] J. A. Lott, N. N. Ledentsov, V. M. Ustinov et al., "InAs-InGaAs quantum dot VCSELs on GaAs substrates emitting at 1.3 µm," *Electron. Lett.*, 36(16), 1384–1385, 2000.

[34] L. Ya. Karachinsky, T. Kettler, N. Yu. Gordeev et al., "High-power singlemode CW operation of 1.5µm-range quantum dot GaAs-based laser," *Electron. Lett.*, 41(8), 478–480, 2005.

[35] R. L. Sellin, Ch. Ribbat, D. Bimberg et al., "High-reliability MOCVD-grown quantum dot laser," *Electron. Lett.*, 38(16), 883–884, 2002.

[36] R. L. Sellin, Ch. Ribbat, M. Grundmann et al., "Close-to-ideal device characteristics of high-power InGaAs/GaAs quantum dot lasers," *Appl. Phys. Lett.*, 78(9), 1207–1209, 2001.

[37] D. O'Brien, S. P. Hegarty, G. Huyet et al., "Feedback sensitivity of 1.3 µm InAs/GaAs quantum dot lasers," *Electron. Lett.*, 39(25), 1819–1820, 2003.

[38] G. Huyet, D. O'Brien, S. P. Hegarty et al., "Quantum dot semiconductor lasers with optical feedback," *phys. stat. sol. (a)*, 201(2), 345–352, 2004.

[39] P. M. Varangis, H. Li, G. T. Liu et al., "Low-threshold quantum dot lasers with 201 nm tuning range," *Electron. Lett.*, 36(18), 1544–1545, 2000.

[40] Ch. Ribbat, R. Sellin, M. Grundmann et al., "Enhanced radiation hardness of quantum dot lasers to high energy proton irradiation," *Electron. Lett.*, 37(3), 174–175, 2001.

[41] D. Bimberg, M. Grundmann, N. N. Ledentsov et al., "Novel Infrared Quantum Dot Lasers: Theory and Reality," *phys. stat. sol. (b)*, 224(3), 787–796, 2001.

[42] Ch. Ribbat, R. L. Sellin, I. Kaiander et al., "Complete suppression of filamentation and superior beam quality in quantum-dot lasers," *Appl. Phys. Lett.*, 82(6), 952–954, 2003.

[43] M. Kuntz, G. Fiol, M. Lämmlin et al., "Direct modulation and mode locking of 1.3 µm quantum dot lasers," *New J. Phys.*, 6, 181, 2004.

[44] S. M. Kim, Y. Wang, M. Keever, and J. S. Harris, "High-frequency modulation characteristics of 1.3-µm InGaAs quantum dot lasers," *IEEE Photonics Technol. Lett.*, 16(2), 377–379, 2004.

[45] M. Kuntz, G. Fiol, M. Lämmlin et al., "10 Gbit/s data modulation using 1.3 µm InGaAs quantum dot lasers," *Electron. Lett.*, 41(5), 244–245, 2005.

[46] K. T. Tan, C. Marinelli, M. G. Thompson et al., "High bit rate and elevated temperature data transmission using InGaAs quantum-dot lasers," *IEEE Photonics Technol. Lett.*, 16(5), 1415–1417, 2004.
[47] M. Lämmlin, G. Fiol, C. Meuer et al., "Distortion-free optical amplification of 20–80 GHz modelocked laser pulses at 1.3 μm using quantum dots," *Electron. Lett.*, 42(12), 41, 2006.
[48] D. Bimberg, N. N. Ledentsov, and J. A. Lott, "Quantum-dot vertical-cavity surface-emitting lasers," *MRS Bull.*, 27(7), 531–537, 2002.
[49] T. Mano, R. Notzel, G. J. Hamhuis et al., "Formation of InAs quantum dot arrays on GaAs (100) by self-organized anisotropic strain engineering of a (In,Ga)As superlattice template," *Appl. Phys. Lett.*, 81(9), 1705–1707, 2002.
[50] A. R. Kovsh, A. E. Zhukov, N. A. Maleev et al., "Lasing at a wavelength close to 1.3 μm in InAs quantum-dot structures," *Semiconductors*, 33(8), 929–932, 1999.
[51] O. G. Schmidt, N. Kirstaedter, D. Bimberg et al., "Prevention of gain saturation by multi-layer quantum dot lasers," *Electron. Lett.*, 32(14), 1302–1304, 1996.
[52] Zh. I. Alferov, N. A. Bert, A. Yu. Egorov et al., "An injection heterojunction laser based on arrays of vertically coupled InAs quantum dots in a GaAs matrix," *Semiconductors*, 30(2), 194–196, 1996.
[53] N. N. Ledentsov and D. Bimberg, "Growth of self-organized quantum dots for optoelectronics applications: nanostructures, nanoepitaxy, defect engineering," *J. Cryst. Growth*, 255(1–2), 68–80, 2003.
[54] N. Nuntawong, S. Birudavolu, C. P. Hains et al., "Effect of strain-compensation in stacked 1.3 μm InAs/GaAs quantum dot active regions grown by metalorganic chemical vapor deposition," *Appl. Phys. Lett.*, 85(15), 3050–3052, 2004.
[55] D. G. Deppe, H. Huang, and O. B. Shchekin, "Modulation characteristics of quantum-dot lasers: the influence of p-type doping and the electronic density of states on obtaining high speed," *IEEE J. Quant. Electron.*, 38(12), 1587–1593, 2002.
[56] P. Bhattacharya and S. Ghosh, "Tunnel injection $In_{0.4}Ga_{0.6}As$/GaAs quantum dot lasers with 15 GHz modulation bandwidth at room temperature," *Appl. Phys. Lett.*, 80(19), 3482–3484, 2002.
[57] D. Bimberg, G. Fiol, M. Kuntz et al., "High speed nanophotonic devices based on quantum dots," *phys. stat. sol. (a)*, 203(14), 3523–3532, 2006.
[58] A. R. Kovsh, N. A. Maleev, A. E. Zhukov et al., "InAs/InGaAs/GaAs quantum dot lasers of 1.3 μm range with enhanced optical gain," *J. Cryst. Growth*, 251(1–4), 729–736, 2003.
[59] D. Ouyang, N. N. Ledentsov, D. Bimberg et al., "High performance narrow stripe quantum-dot lasers with etched waveguide," *Semicond. Sci. Technol.*, 18(12), L53–L54, 2003.
[60] K. Otsubo, N. Hatori, M. Ishida et al., "Temperature-insensitive eye-opening under 10-Gb/s modulation of 1.3-μm p-doped quantum-dot lasers without current adjustments," *Jpn. J. Appl. Phys. Lett.*, 43(8B), L1124–L1126, 2004.
[61] A. Gubenko, D. Livshits, I. Krestnikov et al., "High-power monolithic passively modelocked quantum-dot laser," *Electron. Lett.*, 41(20), 1124–1125, 2005.
[62] A. Martinez, A. Lemaitre, K. Merghem et al., "Static and dynamic measurements of the alpha-factor of five-quantum-dot-layer single-mode lasers emitting at 1.3 μm on GaAs," *Appl. Phys. Lett.*, 86(21), 211115, 2005.
[63] A. A. Lagatsky, E. U. Rafailov, W. Sibbett et al., "Quantum-dot-based saturable absorber with p-n junction for mode-locking of solid-state lasers," *IEEE Photonics Technol. Lett.*, 17(2), 294–296, 2005.
[64] P. Borri, W. Langbein, J. M. Hvam et al., "Spectral hole-burning and carrier-heating dynamics in InGaAs quantum-dot amplifiers," *IEEE J. Sel. Top. Quantum Electron.*, 6(3), 544–551, 2000.
[65] M. Kuntz, N. N. Ledentsov, D. Bimberg et al., "Spectrotemporal response of 1.3 μm quantum-dot lasers," *Appl. Phys. Lett.*, 81(20), 3846–3848, 2002.
[66] H. Dery and G. Eisenstein, "The impact of energy band diagram and inhomogeneous broadening on the optical differential gain in nanostructure lasers," *IEEE J. Quantum Electron.*, 41(1), 26–35, 2005.

2. Semiconductor Quantum Dots

[67] W. W. Chow and S. W. Koch, "Theory of semiconductor quantum-dot laser dynamics," *IEEE J. Quantum Electron.*, 41(4), 495–505, 2005.

[68] Z. Mi, P. Bhattacharya, and S. Fathpour, "High-speed 1.3 μm tunnel injection quantum-dot lasers," *Appl. Phys. Lett.*, 86(15), 153109, 2005.

[69] H. Su and L. F. Lester, J. Phys. D, "Dynamic properties of quantum dot distributed feedback lasers: high speed, linewidth and chirp," *Appl. Phys.*, 38(13), 2112–2118, 2005.

[70] F. Hopfer, A. Mutig, M. Kuntz et al., "Single-mode submonolayer quantum-dot vertical-cavity surface-emitting lasers with high modulation bandwidth," *Appl. Phys. Lett.*, 89, 141106, 2006.

[71] J. A. Kash, F. E. Doany, L. Schares et al., "Chip-to-chip optical interconnects," in Proc. OFC, OFA3, 2006. Conference: OFCNFOEC 2006. 2006 Optical Fiber Communication Conference and National Fiber Optic Engineers Conference, Anaheim, CA, USA, 5–10 March 2006.

[72] N. Suzuki, H. Hatakeyama, K. Fukatsu et al., "25-Gbps operation of 1.1-μm-range InGaAs VCSELs for high-speed optical interconnections," in Proc. OFC, OFA4, 2006. Conference: OFCNFOEC 2006. 2006 Optical Fiber Communication Conference and National Fiber Optic Engineers Conference, Anaheim, CA, USA, 5–10 March 2006.

[73] A. E. Zhukov, A. R. Kovsh, S. S. Mikhrin et al., "3.9 W cw power from sub-monolayer quantum dot diode laser," *Electron. Lett.*, 35, 1845–1847, 1999.

[74] S. S. Mikhrin, A. E. Zhukov, A. R. Kovsh et al., "0.94 μm diode lasers based on Stranski-Krastanow and sub-monolayer quantum dots," *Semicond. Sci. Technol.*, 15, 1061–1064, 2000.

[75] N. Ledentsov and V. Shchukin, "Optoelectronic device based on an antiwaveguiding cavity," United States Application 20050226294.

[76] S. Fathpour, Z. Mi, and P. Bhattacharya, "High-speed quantum dot lasers," *J. Phys. D: Appl. Phys.*, 38, 2103–2111, 2005.

3

High-speed low-chirp semiconductor lasers

Shun Lien Chuang, Guobin Liu, and Piotr Konrad Kondratko

Department of Electrical and Computer Engineering, University of Illinois, Urbana, IL, USA

This chapter provides a comparison of InAlGaAs with InGaAsP long-wavelength quantum well (QW) lasers in terms of high-speed performance and extraction of important parameters such as gain, differential gain, photon lifetime, and chirp. Both DC measurements and high-speed direct modulation of QW lasers are presented and comparison with theoretical model will be made. We also provide insight to novel quantum dot (QD) lasers for high-speed operation including the ideas of p-type doping and tunneling injection for reduced chirping and broad bandwidth operation.

3.1 INTRODUCTION

In the optical component market for telecom and datacom applications, cost reduction is a driving force for technology innovations. As a result, directly modulated semiconductor laser operating in multi-Gb/s optical communication systems has become the technology of choice for local area and metropolitan networks [1, 2]. It reduces component complexity and implementation cost, because it eliminates the need for a modulator. However, directly modulated laser tends to have wider spectral width under high-speed modulation due to chirping. Due to chromatic dispersion effects in the optical fiber, it presents a transmission limitation for directly modulated laser at 1.55-μm communication systems. Such systems will continue to use the extensive existing network of standard optical fiber, which is optimized for zero dispersion at 1.31 μm. This effect degrades the signal and limits the system bit rate and the fiber link length. The latest development to improve the modulation bandwidth and reduce the chirp of the directly modulated laser by using strained quantum well (QW) [3, 4] and quantum dot (QD) tunneling injection structures [5, 6] has been demonstrated.

For 1.55-μm long wavelength telecommunication semiconductor lasers, two major material systems have been utilized. $In_{1-x}Ga_xAs_yP_{1-y}$ system has been investigated by many groups, and high performance in terms of low threshold and high-modulation bandwidth has been demonstrated [7, 8]. However, it is a challenging task to design and fabricate uncooled $In_{1-x}Ga_xAs_yP_{1-y}$/InP QW lasers. The reasons are partly due to poor electron confinement in the conduction band of QW and partly due to Auger recombination process in this material system at high temperatures. The electron overflow will also deteriorate the optical transition efficiency and limit the modulation bandwidth. On the other hand, $In_{1-x-y}Ga_xAl_yAs$ material system has emerged recently as a candidate [1, 9, 10]. As it has a large conduction band offset ($0.7\Delta E_g$) compared to ($0.4\Delta E_g$) for $In_{1-x}Ga_xAs_yP_{1-y}$ systems, a better confinement of electrons for this system results in a higher temperature stability. More symmetrical injection of the holes and electrons to the QW region will improve the laser performance.

Previous reports have been given on the model comparing these two systems in terms of band structures [11], threshold current density and differential gain [12], and the temperature dependence of the threshold current [13]. There have been measurements of the contributions of the various recombination mechanisms for each system [14]. We have also performed detailed studies comparing the gain and recombination mechanisms of these two systems and comparing the existing many-body optical gain models with experimental measurements [10]. However, little work has been done to directly compare the high-speed modulation characteristics of these two material systems and their dependencies on the steady-state characteristics such as gain and differential gain.

In this chapter, we review our recent work on the high-speed modulation responses of one Fabry–Perot (FP) $In_{1-x}Ga_xAs_yP_{1-y}$ QW laser and two $In_{1-x-y}Ga_xAl_yAs$ QW lasers operating at 1.55 μm [3, 4]. We present experimental data for the high-speed modulation responses of both material systems, with the high-speed parameters directly measured from the fundamental optical gain spectrum setup. We then compare and discuss fundamental parameters for determining the high-speed modulation characteristics of these lasers.

Finally, high-speed operation and linewidth enhancement factor (LEF) of QD lasers is presented [5, 6, 15]. A novel structural engineering is presented for reducing chirp and increasing modulation bandwidth in these lasers. The p-doped and tunnel injection designs are discussed with their effect on gain, LEF, and direct high-speed modulation.

3.2 FUNDAMENTAL DC PROPERTIES OF LONG-WAVELENGTH QW LASERS

The theories used for studying strained QW lasers are summarized here. The detailed derivation can be found in Refs [16] and [17] for small-signal microwave modulation theory and in Refs [10] and [18] for optical gain model with many-body effects.

3.2.1 Theory of Optical Gain, Refractive Index Change, and LEF

To model the optical gain of QW lasers, the band structures of the QWs are calculated using a block-diagonalized 3×3 Hamiltonian based on the **k.p** method for valence subbands [19] and a simple isotropic parabolic band model for conduction subbands. Once the band structure is known, the optical gain $g(\omega)$ and the corresponding refractive index change $\delta n_{\mathrm{e}}^{\mathrm{int}}(\omega)$ due to the interband transition based on a non-Markovian gain model using a spontaneous-emission transformation method are given by Refs [10, 18, 20]

$$g(\omega) = \left[1 - \exp\left(\frac{\hbar\omega - \Delta F}{kT}\right)\right] \sqrt{\frac{\mu_0}{\epsilon}} \left(\frac{e^2}{m_0^2 \omega}\right)$$

$$\times \sum_{\sigma} \sum_{m} \sum_{l} \int_0^\infty dk_t \frac{k_t}{\pi L_z} |M_{lm}^\sigma|^2 f_l^c(k_t)(1 - f_m^v(k_t))$$

$$\times (1 - \mathrm{Re} q_{k_t}) \frac{\mathrm{Re} L(E_{lm}(k_t)) - \mathrm{Im} q_{k_t} \mathrm{Im} L(E_{lm}(k_t))}{(1 - \mathrm{Re} q_{k_t})^2 + (\mathrm{Im} q_{k_t})^2} \quad (3.1)$$

$$\delta n_{\mathrm{e}}^{\mathrm{int}}(\omega) = \sqrt{\frac{\mu_0}{\epsilon}} \left(\frac{e^2 c}{2 m_0^2 \omega^2}\right) \sum_{\sigma} \sum_{m} \sum_{l} \int_0^\infty dk_t \frac{k_t}{\pi L_z} |M_{lm}^\sigma|^2 (f_l^c(k_t) - f_m^\sigma(k_t))$$

$$\times \left[-\frac{(1 - \mathrm{Re} q_{k_t})\mathrm{Im} L + \mathrm{Im} q_{k_t} \mathrm{Re} L}{(1 - \mathrm{Re} q_{k_t})^2 + (\mathrm{Im} q_{k_t})^2}\right], \quad (3.2)$$

where ω is the angular optical frequency, ΔF is the quasi-Fermi level separation, μ_0 is the vacuum permeability, and ϵ is the dielectric constant. The term $|M_{lm}^\sigma|^2$ is the momentum matrix element in the QW; $f_l^c(k_t)$ and $f_m^v(k_t)$ are the Fermi functions for the lth conduction and the mth valence subbands, respectively. Also, $E_{lm}(k_t) = E_l^c(k_t) - E_m^v(k_t) + E_g + \Delta E_{\mathrm{SX}} + \Delta E_{\mathrm{CH}} - \hbar\omega$ is the renormalized transition energy between electron and hole subbands where E_g is the bandgap of the material, and ΔE_{SX} and ΔE_{CH} are the screened exchange and Coulomb-hole contribution to the bandgap renormalization. The factor q_{k_t} accounts for the excitonic or Coulomb enhancement of the interband transition matrix element [20]. The line shape function is Gaussian for the simplest non-Markovian quantum kinetics and is given by

$$\mathrm{Re} L(E_{lm}(k_t)) = \sqrt{\frac{\pi \tau_{\mathrm{in}}(k_t) \tau_c}{2\hbar^2}} \exp\left(-\frac{\tau_{\mathrm{in}}(k_t) \tau_c}{2\hbar^2} E_{lm}^2(k_t)\right) \quad (3.3)$$

and

$$\mathrm{Im} L(E_{lm}(k_t)) = \frac{\tau_c}{\hbar} \int_0^\infty \exp\left(-\frac{\tau_c}{2\tau_{\mathrm{in}}(k_t)} t^2\right) \sin\left(\frac{\tau_c E_{lm}(k_t)}{\hbar} t\right) dt, \quad (3.4)$$

where the intraband relaxation time τ_{in} and correlation time τ_c are the fitting parameters in the calculation. This gain model is used to fit our measured optical gain spectrum to extract the gain vs carrier density relation and therefore, the differential gain.

In addition to Eqn (3.2), there is a contribution from the free carrier plasma effect [21] for TE polarization:

$$\delta n_e^{plasma}(\omega) = -\frac{e^2 N}{2m_r^* \omega^2 n_r \epsilon_0}, \quad (3.5)$$

where N is the carrier concentration, n_r is the refractive index of the QW layer, and m_r^* is the reduced effective mass for electrons and holes. The LEF due to the interband transition is then obtained by

$$\alpha_e = -\frac{4\pi}{\lambda} \frac{\partial \delta n_e/\partial N}{\partial g/\partial N}, \quad (3.6)$$

where $\partial g/\partial N$ is the differential gain and $\partial \delta n_e/\partial N$ is the incremental change in the refractive index due to carrier injection, and the total refractive index change is

$$\delta n_e = \delta n_e^{int} + \delta n_e^{plasma} \quad (3.7)$$

for the TE mode and

$$\delta n_e = \delta n_e^{int} \quad (3.8)$$

for the TM mode. For the compressively strained and lattice-matched QW lasers, the TE mode is the dominant optical mode as verified by the experiment [3].

3.2.2 DC Measurements of Optical Gain, Refractive Index Change, and LEF

Three FP lasers grown on InP substrate are used in our experiment: a buried heterostructure laser with seven -0.9% compressively strained QWs made of the $In_{1-x}Ga_xAs_yP_{1-y}$ material system (sample A), a buried heterostructure laser with five -0.78% compressively strained QWs made of the $In_{1-x-y}Ga_xAl_yAs$ material system (sample B), as well as a buried heterostructure laser with five lattice-matched QWs of the $In_{1-x-y}Ga_xAl_yAs$ material system (sample C). Detailed information for the structure of each active region of three lasers is listed in Table 3.1. Parameters for samples B and C are also listed in Ref. [10]. All three lasers are bounded p-side up on high-speed transmission lines. During the experiment, the lasers are placed on a heat-sink mount controlled by an ILX Lightwave temperature controller with an environmental temperature variation less than $0.1°C$.

3. High-Speed Low-Chirp Semiconductor Lasers

Table 3.1

High-speed laser parameters and extracted values at 25°C from microwave modulation responses. After Ref. [3]

Parameter	Symbol	Sample A	Sample B	Sample C
Structural				
Well material		InGaAsP	InGaAlAs	InGaAlAs
Well strain		−0.9%	−0.78%	0.0%
Well width (Å)	L_w	80	54	86
Well PL wavelength (µm)		1.58	1.56	1.56
Barrier material (zero strain)		InGaAsP	InGaAlAs	InGaAlAs
Stripe width (µm)	w	1.2	1.5	1.46
Barrier width (Å)	L_b	87	57	50
Barrier PL wavelength (µm)		1.255	1.21	1.21
SCH width (Å)	L_{sch}	700	1000	600
Number of wells	N_w	7	5	5
Active volume ($\times 10^{-11}$ cm^3)	V	1.9	3.6	4.0
Extracted				
Optical confinement factor	Γ	0.10±0.02	0.06±0.02	0.11±0.02
Group velocity ($\times 10^9$ cm/s)	v_g	9.1±0.3	9.1±0.3	9.1±0.3
Photon lifetime (ps)	τ_p	1.8±0.2	2.5±0.2	3.1±0.2
Intrinsic loss (cm^{-1})	α_i	17±2	30±2	25±2
Mirror loss (cm^{-1})	$\bar{\alpha}$	44±5	14±2	11±2
Carrier lifetime (ns)	τ_w	0.11±0.04	0.12±0.04	0.22±0.04
Effective differential gain ($\times 10^{-16}$ cm^2)	g'/χ	8.9±0.4	9.2±0.4	6.9±0.4
K factor (ns)	K	0.19±0.04	0.46±0.08	0.96±0.10
Nonlinear gain suppression coefficient ($\times 10^{-17}$ cm^3)	ϵ	2.4±0.5	7.7±1.0	13.4±1.0

It is necessary to express those parameters in Eq. (3.16) in terms of the measurable parameters. Using a procedure described in Ref. [10], the optical gain spectra are measured for three lasers biased below threshold. The facet emission spectrum of each laser was measured for various currents below lasing threshold at $T = 25°C$. The facet output was collected using an E-TEK Dynamics laser optical fiber interface (LOFI) with a built-in polarizer to control the polarization of the measured emission. An optical isolator was used to prevent optical feedback effects, and the light was monitored using an HP optical spectrum analyzer with a spectral resolution of 0.08 nm. The net modal gain G_{net} for these devices is obtained using the well-known Hakki–Paoli method [22], in which the ratio of the maximum I_{max} and minimum I_{min} of the amplified spontaneous emission (ASE) spectrum yields the net modal gain G_{net} by the following relations:

$$G_{net} = \Gamma g - \alpha_i = \frac{1}{L}\ln\left(\frac{\sqrt{I_{max}} - \sqrt{I_{min}}}{\sqrt{I_{max}} + \sqrt{I_{min}}}\right) + \frac{1}{2L}\ln\left(\frac{1}{R_1 R_2}\right), \quad (3.9)$$

where g is the material gain for one well, Γ ($= N_w\Gamma_w$) is the optical confinement factor per well (Γ_w) multiplied by N_w (the number of wells), α_i is the intrinsic loss, L is the cavity length, and R_1 and R_2 are the facet power reflectivities. At threshold

$$\Gamma g_{th} = \alpha_i + \alpha_m, \qquad (3.10)$$

the mirror loss is

$$\alpha_m = \frac{1}{2L}\ln\left(\frac{1}{R_1 R_2}\right). \qquad (3.11)$$

The photon lifetime τ_p is obtained by

$$\tau_p = \frac{1}{v_g(\alpha_i + \alpha_m)}, \qquad (3.12)$$

where v_g is the group velocity, α_i is the intrinsic loss obtained from fitting the laser net model gain, and α_m is the mirror loss. The photon density S_0 can be expressed in terms of the optical power P_{out}

$$S_0 = \frac{\Gamma}{V v_g \hbar \omega \alpha_m \eta_c} P_{out}, \qquad (3.13)$$

where V is the active volume of the laser and η_c is the coupling efficiency. The optical power is obtained by measuring the $L-I$ curves of the laser. A microscope objective with a numerical aperture of 0.4 is used to ensure that all the output light is coupled to the detector; therefore, η_c is almost unity and is ignored in the analysis. Using this procedure, the photon density S_0 is directly linked to the modulation response at different injection current. This allows us to extract the intrinsic parameters based on the microwave modulation response.

To verify the above results, we also perform measurement of the LEF using the ASE spectroscopy. The ASE spectrum of the test laser is measured for two close currents below lasing threshold at $T = 25°C$. The peak wavelength of each FP mode in the ASE spectrum is a function of the effective refractive index $n_e(x)$. The change in the refractive index can be calculated from Ref. [23]

$$\Delta n_e = \frac{\lambda}{2L}\frac{\Delta \lambda_l}{\Delta \lambda_m}, \qquad (3.14)$$

where $\Delta \lambda_l$ is the wavelength shift of a FP mode due to a change in injected carriers at two bias currents. The term $\Delta \lambda_m$ is the mode spacing of two adjacent FP peaks. The LEF or α_e can then be extracted using

$$\alpha_e = -\frac{4\pi}{\lambda}\frac{\Delta n_e}{\Delta G_{net}}, \qquad (3.15)$$

where ΔG_{net} is change of the net modal gain [10].

3.2.3 Comparison Between Theory and DC-Measured Experimental Data

To understand the dependence of the intrinsic parameters for high-speed modulation of three lasers, steady-state characteristics of these lasers are studied. The net modal gain spectra for three lasers are measured for several injection currents up to the threshold and are plotted in Figure 3.1(a)–(c) in dashed lines. Also presented in Figure 3.1(a)–(c) are our calculated (solid curves) optical gain spectra using the many-body gain model for the three material systems. Through fitting the gain spectra, we obtain the parameters, including the intraband relaxation time τ_{in}, the correlation time τ_c, and a parameter C used in the formulation of the reference [20], which contributes to the many-body bandgap renormalization. Each of these parameters affect the peak gain position and overall shape of the spectra. They were selected to yield the best overall fit. The intrinsic loss is obtained from the plateau on the long-wavelength side of the gain spectrum [16]. Note that the optical gain using a conventional Lorentzian line shape function will give an anomalous absorption region below the bandgap, and the slow convergence of the Lorentzian leads to a very long tail of the gain spectra into the bandgap. An excellent fit to the experimental data using our Gaussian line shape theory provides those parameters for the analysis. The fitting of the gain spectra of these three material systems also allows us to obtain the density of carriers in the QWs corresponding to the measured injection currents.

Using these parameters, we plot the peak optical gain as a function of the carrier density in Figure 3.2(a). The product of the peak optical gain g vs the surface carrier density is plotted in Figure 3.2(b) for three lasers. The solid line is for sample A, the dashed line is for sample B, and the dotted line is for sample C. The symbols correspond to the experimentally extracted data from Figure 3.1. Above the threshold, the carrier densities are pinned at the threshold conditions. When the injection current increases, the optical gain will be pinned at the threshold. The gain-well width product is proportional to the laser modal gain.

The differential gain g' at the peak gain wavelength is plotted as functions of the carrier density in Figure 3.3. The arrows in the figure indicate the threshold conditions for three lasers. At the threshold, the differential gain at the peak optical gain wavelength is $9.76 \times 10^{-16}\,\text{cm}^2$ for sample A, $10.11 \times 10^{-16}\,\text{cm}^2$ for sample B, and $7.13 \times 10^{-16}\,\text{cm}^2$ for sample C. The values of the transport parameter χ (Section 3.3) for high-speed model are calculated to be 1.10, 1.10, and 1.03, respectively, for three lasers. The differential gain for three lasers obtained from high-speed modulation experiment of Section 3.3 (Table 3.1) agrees well with the values obtained from the steady-state experiment. The results extracted from the steady-state experiment explain the high-speed characteristics of the QW lasers.

From Figure 3.2(b), the curves for two compressively strained QW lasers (samples A and B) have a similar trend, implying that the lasers designed using the QW structures for samples A and B would have a similar steady-state behavior

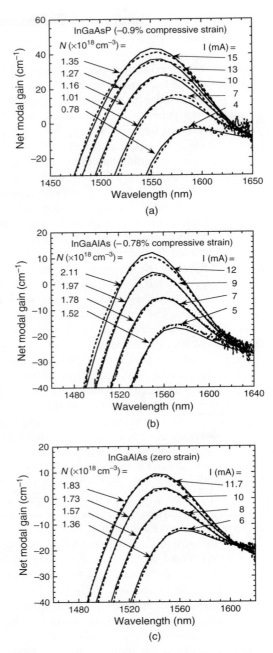

Figure 3.1 The measured modal gain (dashed) and calculated (solid) spectra from the electronic band structure model for (a) $In_{1-x}Ga_xAs_yP_{1-y}$ QW laser with -0.9% compressive strain (sample A), (b) $In_{1-x-y}Ga_xAl_yAs$ QW laser with -0.78% compressive strain (sample B), and (c) $In_{1-x-y}Ga_xAl_yAs$ QW laser with a zero strain (sample C) in the wells. After Ref. [3] (this figure may be seen in color on the included CD-ROM).

3. High-Speed Low-Chirp Semiconductor Lasers

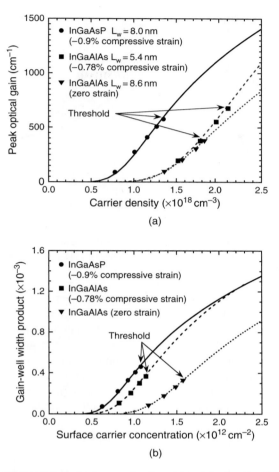

Figure 3.2 (a) Calculated peak optical gain g as a function of the carrier density n for three lasers (solid, dashed, and dotted lines). (b) Calculated peak optical gain–well width product (Γg_{th}) as a function of the surface carrier density (nL_w) for three lasers. The symbols correspond to the experimental data. After Ref. [3] (this figure may be seen in color on the included CD-ROM).

if the nonradiative process is ignored. From Figure 3.3, we can see that samples A and B can achieve almost the same amount of the peak differential gain, but sample A reaches the highest differential gain at a much lower carrier density. Considering the fact that both samples have similar amount of compressive strain (-0.9% vs -0.78%), but the QW width of sample A is 80 Å while the QW width of sample B is 54 Å, we can see that the $In_{1-x}Ga_xAs_yP_{1-y}$ material system has a higher optical transition efficiency compared to the $In_{1-x-y}Ga_xAl_yAs$ material system. However, this advantage is offset by a much more severe nonradiative Auger recombination process present in $In_{1-x}Ga_xAs_yP_{1-y}$ material system. As a

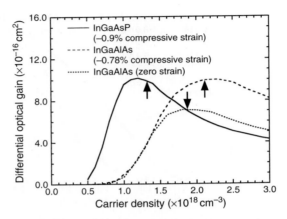

Figure 3.3 The calculated differential gain g' at the peak gain wavelength as a function of the carrier density for three lasers. The solid line is for sample A, the dashed line is for sample B, and the dotted line is for sample C. The arrows indicate the threshold values for three lasers. After Ref. [3] (this figure may be seen in color on the included CD-ROM).

result, the threshold condition is achieved at a much larger injection current density for $In_{1-x}Ga_xAs_yP_{1-y}$ material system. The characteristic temperature T_0 measured from the shift in threshold over a temperature range from 20°C to 50°C is 33.1 K for sample A, 49.3 K for sample B, and 48.6 K for sample C. These relative values indicate that Auger recombination is a major factor reducing T_0 for these laser systems [24, 25]. Shown in Figure 3.4, the experimental net modal gain vs the total current density for each laser system is plotted. In the $In_{1-x}Ga_xAs_yP_{1-y}$ system, due to an Auger coefficient about three to ten times

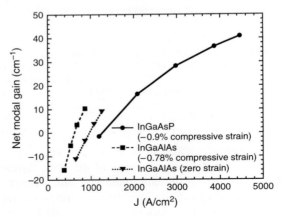

Figure 3.4 The experimental net modal gain ($\Gamma_g - \alpha_i$) at peak gain vs the total current density. The solid dots are for sample A, the squares are for sample B, and the triangles are for sample C. After Ref. [3] (this figure may be seen in color on the included CD-ROM).

larger than that of $In_{1-x-y}Ga_xAl_yAs$ lasers as reported in Ref. [10], the Auger recombination dominates at a much lower point and becomes much larger while approaching the threshold. This leads to a much larger threshold current density and a much lower slope efficiency of the optical gain with respect to the injection current density. The QW lasers based on the $In_{1-x-y}Ga_xAl_yAs$ system can be designed to be more efficient to lower the threshold current density and to reduce the temperature sensitivity.

3.2.4 Comparison of LEFs of InAlGaAs and InGaAsP Systems

We have fitted the optical gain for this laser previously. Using the same fitting parameters, the change of the refractive index due to the interband transition is calculated. In Figure 3.5, the LEF due to the interband transition (short dashed line), the plasma effect (long dashed line), and their sum is plotted. The theoretical curve (interband plus plasma effects) agrees with the experiment very well. Both the interband transition and the plasma effect contribute to the total LEF significantly for the laser studied. At the laser wavelength of 1550 nm, the LEF is 3.0. This value agrees with typical value of long-wavelength-strained QW lasers. The experimental uncertainty from both experiments is about 15%. Our theory will be very useful for designing and optimizing the QW lasers, especially to meet specification of high-speed performance in terms of high-differential gain and low chirp.

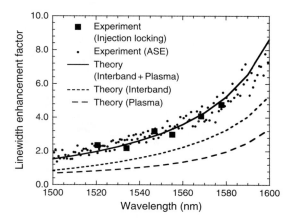

Figure 3.5 The linewidth enhancement factor is plotted as a function of the wavelength for the test laser. The solid squares are the experimental results using the injection-locking technique, the circles are the experimental data using the amplified spontaneous emission (ASE) spectroscopy. The theoretical linewidth enhancement factor is plotted for the interband contribution (short dashed line), the plasma effect contribution (long dashed line), and their total (solid line). After Ref. [4].

Our directly measured results using ASE are plotted as circles in Figure 3.5, and they are in excellent agreement with our results using the injection-locking technique (squares). When using the ASE spectroscopy method, the LEF is measured near threshold condition. On the other hand, using the injection-locking method, the LEF is measured above threshold and lasing condition. For high-speed modulation applications, the LEF measured at the lasing condition will be more important. The results show that the LEF does not change too much near threshold and at lasing conditions.

The ASE spectra for three strained QW lasers used above are processed. The effective refractive index differences are obtained for three strained QW lasers using Eqn (3.14). Also, the theoretical curves for the effective refractive index differences are calculated using the same parameters for fitting the net modal gain spectra with the plasma effect taken into account (Eqn (3.7)) for TE modes in the compressively strained and lattice-matched lasers. In Figure 3.6, the LEF obtained from the experiment and theory is plotted as a function of wavelength. As the refractive index change is roughly a constant over the wavelength range shown, the LEF spectrum is mainly determined by the gain difference profile. LEF tends to decrease at higher energies because of the increased differential gain caused by the band-filling effects. For the compressively strained and the lattice-matched samples, the theoretical results agree with the experimental data very well for all injection currents. The contribution from the interband transition and the plasma effect are both significant for the total induced refractive index change. At the lasing wavelength, the LEFs are 2.8 for sample A (1.55 μm), 3.0 for sample B (1.55 μm), and 5.3 for sample C (1.54 μm). These results show that two

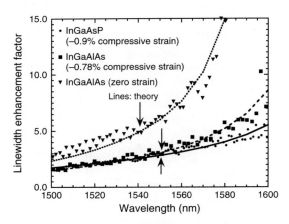

Figure 3.6 The measured linewidth enhancement factors (symbols) and calculated (lines) values from the electronic band structure for three strained QW lasers. The arrows indicate the linewidth enhancement factor at the lasing wavelength for each laser. The round symbols and the solid line are for sample A, the square symbols and the short dashed line are for sample B, and the triangle symbols and the short dashed lines are for sample C. (this figure may be seen in color on the included CD-ROM).

compressively strained QW lasers have a similar trend for LEF and are better than the lattice-matched laser in terms of low chirp for high-speed modulation applications. LEF factor can be decreased by adding strain in QW lasers and can be further decreased by adding p-doping in the strained active region [26–29].

3.3 HIGH-SPEED DIRECT MODULATION OF STRAINED QW LASERS

3.3.1 Theory for High-Speed Small-Signal Modulation Response

Based on the rate equation model, for a small signal electrical injection, $I(t) = I_0 + i(\omega)e^{j\omega t}$, the normalized modulation response for the output light is given by [16, 17]

$$\frac{M(\omega)}{M(0)} = \frac{\omega_r^2}{(1 + j\omega\tau_{bw})(\omega_r^2 - \omega^2 + j\omega\gamma)} \tag{3.16}$$

$$\omega_r^2 = \frac{(v_g g'/\chi)S_0}{\tau_p(1 + \epsilon S_0)}\left(1 + \frac{\epsilon}{v_g g'\tau_w}\right) \tag{3.17}$$

$$\gamma = K f_r^2 + \frac{1}{\chi\tau_w}, \tag{3.18}$$

where τ_{bw} is a constant that contributes to low-frequency rolloff, S_0 is the photon density, τ_p is the photon lifetime, ϵ is the nonlinear gain suppression coefficient, $\omega_r = 2\pi f_r$ is the relaxation frequency, γ is the damping factor, τ_w is the carrier lifetime, $\chi = 1 + \tau_{bw}/\tau_{wb}$ is a transport factor, defined by carrier transport time constants from quantum barrier to well τ_{bw} and from quantum well to barrier τ_{wb} and K is an important factor in high-speed modulation, which is defined as the slope of the damping factor vs relaxation frequency squared and is given by

$$K = 4\pi^2\left(\tau_p + \frac{\epsilon}{v_g g'/\chi}\right) \tag{3.19}$$

The modulation bandwidth of a QW laser is determined by the intrinsic response characterized by the relaxation frequency f_r and the damping factor γ and by the low-frequency rolloff effect characterized by the parameter τ_{bw}. A smaller value of the K factor will reduce the damping factor when laser operates at a large relaxation frequency. The maximum intrinsic relaxation frequency of the semiconductor laser is inversely proportional to the K factor and is expressed as $f_{max} = 2\pi\sqrt{2}/K$ under certain approximations. The intrinsic modulation response will set an upper limit for the overall modulation bandwidth of the laser. To design a semiconductor laser with higher modulation response, the K factor should be as small as possible.

Conventionally, most groups fit the modulation responses to extract the parameters such as g', χ, ϵ, and τ_p. Here, we determine some of these parameters based on separate measurements. As shown in Section 3.2, we measure the optical gain spectrum as a function of driving current from below threshold to threshold to extract the gain spectrum, $g(\lambda)$, and the peak gain as a function of bias current density, and the threshold gain, Γg_{th}. We then use the many-body optical gain model to extract the peak gain vs carrier density relation, $g_p = g(n)$, to obtain $g' = dg/dn$. By the optical gain measurement, we also extract the intrinsic loss α_i and therefore, the photon lifetime τ_p at the threshold gain condition.

3.3.2 Microwave Modulation Experiment

The experimental setup of small-signal microwave modulation experiment is shown in Figure 3.7. The HP 8510B network analyzer provides a small signal at frequencies swept from 45 MHz to 15 GHz, which is used to couple the test laser electrodes through a high-speed probe. The small signal response is converted to an electrical signal using a Newport high-speed photodetector with a 3-dB bandwidth of 29 GHz and a New Focus 18-dB microwave amplifier. The network analyzer measures the magnitude of the modulation response $|M(\omega)|^2$. An optical isolator is used to prevent reflection. The modulation response functions are taken for various currents for each of the three lasers.

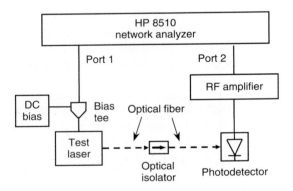

Figure 3.7 The experimental setup for the high-speed microwave modulation measurement. After Ref. [3].

3.3.3 Experimental Results of InAlGaAs and InGaAsP Systems and Comparison with Theory

The measured modulation response curves for several injection currents are shown as symbols in Figure 3.8(a) for sample A (-0.9% compressive strain, InGaAsP), Figure 3.8(b) for sample B (-0.78% compressive strain, InGaAlAs), and

3. High-Speed Low-Chirp Semiconductor Lasers

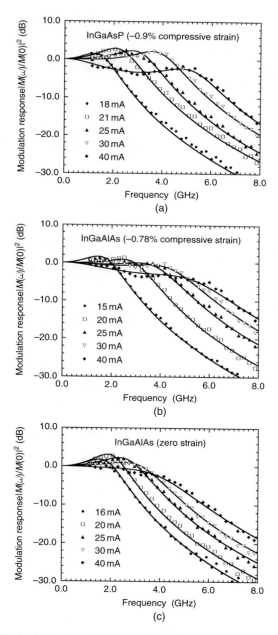

Figure 3.8 Normalized microwave modulation response for different injection currents for (a) $In_{1-x}Ga_xAs_yP_{1-y}$ with -0.9% compressive strain (sample A), (b) $In_{1-x-y}Ga_xAl_yAs$ with -0.78% compressive strain (sample B), and (c) $In_{1-x-y}Ga_xAl_yAs$ with zero strain (sample C). The symbols are experimental data and the lines are theoretical fits. After Ref. [3] (this figure may be seen in color on the included CD-ROM).

Figure 3.8(c) for sample C (zero strain, InGaAlAs). Each curve is fitted using Eqn (3.16) to obtain the relaxation frequency f_r, the damping factor γ, and the rolloff frequency $f_p = 1/(2\pi\tau_{bw})$. Note that the rolloff frequency consists of the contributions from the transport effect and the parasitic RC rolloff effect. However, these two effects are not distinguishable and thus, are characterized by one single parameter, f_p. The modulation responses calculated using the fitting parameters are also plotted in Figure 3.8 as solid curves.

With the results from DC measurement, the relaxation frequency squared, f_r^2, from the fitting is plotted as a function of the photon density for sample A (circles), sample B (squares), and sample C (inverted triangles) in Figure 3.9(a). The slopes

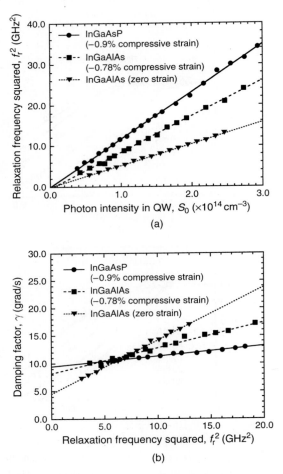

Figure 3.9 (a) The relaxation frequency squared vs photon intensity and (b) the damping factor vs the relaxation frequency squared at 25°C. The lines are the least-square fit to the data to extract the slopes. After Ref. [3] (this figure may be seen in color on the included CD-ROM).

of these curves are determined from Eqn (3.17) and are proportional to the effective differential gain g'/χ and the reciprocal of the photon lifetime τ_p. In Figure 3.9(b), the damping factor γ is plotted as a function of the relaxation frequency square f_r^2 for each laser. The intercepts of these curves with the γ axis (the vertical axis) are determined by the reciprocals of the carrier lifetimes, and the slopes of the curves are determined by the K factor. Evaluating and summarizing the results from Figure 3.9(a) and Figure 3.9(b), we obtained the parameters that characterize the high-speed performances of three lasers. These parameters are listed in Table 3.1.

As listed in Table 3.1, the effective differential gain, g'/χ, can be extracted to be $8.9 \times 10^{-16}\,\text{cm}^2$ for sample A, $9.2 \times 10^{-16}\,\text{cm}^2$ for sample B, and $6.9 \times 10^{-16}\,\text{cm}^2$ for sample C. However, in Figure 3.9(a), sample A has the highest slope compared to the other two samples. This is because the slope is also determined by the photon lifetime. Sample A (InGaAsP) is a short-cavity laser, which has a considerably shorter photon lifetime than the other two InGaAlAs lasers. Other factors for determining the slopes of the three curves in Figure 3.9(a), which involve the nonlinear gain suppression coefficient ϵ, will only affect the slope by no more than 10% for all three lasers. Noting that χ is close to unity for a well-confined QW, the slope of the relaxation frequency squared, vs the photon intensity is mainly affected by the differential gain and the photon lifetime. The differential gain plays a central role in determining the fundamental frequency response of semiconductor lasers. The laser with a high slope will reach higher relaxation frequency faster for a given photon intensity, which will give a higher intrinsic 3-dB modulation bandwidth.

In Figure 3.9(b), the damping factor γ is plotted as a function of the relaxation frequency squared. The K factors can be extracted from the slopes of the curves and are 0.19 ns for sample A, 0.46 ns for sample B, and 0.96 ns for sample C, respectively. The K factor value of 0.19 ns for sample A agrees with the results from the laser based on the similar $\text{In}_{1-x}\text{Ga}_x\text{As}_y\text{P}_{1-y}$ material system [16]. The sample A has the largest maximum 3-dB bandwidth among these three samples. According to Eqn (3.19), the K factor is determined from the photon lifetime and $\epsilon/(v_g g'/\chi)$. While the nonlinear gain suppression coefficient can be designed to be small enough to increase the modulation bandwidth, the photon lifetime will set a low limit for the K factor. Among the three tested lasers, samples B and C using the $\text{In}_{1-x-y}\text{Ga}_x\text{Al}_y\text{As}$ material system have a much larger nonlinear gain suppression coefficient ϵ than that of sample A. The K factors for these two samples are mainly determined by the quantity $\epsilon/(v_g g'/\chi)$ involving the nonlinear gain suppression coefficient, whereas for sample A, both this quantity and the photon lifetime contribute significantly to determine the K factor. A low value of $\epsilon/(v_g g'/\chi)$ will give a smaller K factor and a larger modulation bandwidth. As the QW structures for the laser samples B and C have not been optimized, and as the nonlinear gain suppression coefficient can be reduced by optimizing the strain in the QW, the number of the QW, and the cavity width [30, 31], it is expected that the performance of the $\text{In}_{1-x-y}\text{Ga}_x\text{Al}_y\text{As}$ lasers can be improved.

3.4 QUANTUM DOT LASERS

Room-temperature QD lasers are expected to exhibit high performance with large gain, differential gain, and modulation bandwidth. Moreover, the QDs have demonstrated low threshold current density and reduced temperature dependence of the threshold current density [32]. These performance features are directly the result of zero-dimensional carrier confinement and discrete density of states of QDs. For instance, the symmetric gain shape of QD lasers leads (through Kramers–Krönig relation) to low index of refraction change at the peak gain wavelength, therefore, QD lasers exhibit near zero LEF. As shown previously, LEF for InGaAs- and InGaAsP-strained QW lasers (see Figure 3.6) are on the order of 2 or higher. Recently, it has been demonstrated that QD lasers show reduced or near zero LEF [6, 33].

Recent experimental and theoretical studies proposing novel QD designs have demonstrated an improvement in direct laser modulation and LEF. The improvements were attributed to the use of p-type doping and tuneling injection within the active region of the QDs [6, 33, 34]. In the following sections, we discuss these QD laser improvements.

3.4.1 Optical Gain and LEF p-Doped and Tunneling Injection QDs

A theoretical and experimental study of the optical gain, refractive index change, and LEF of a p-doped QD laser has recently been reported [5]. These studies reveal that p-doped QD lasers show an increase in high-speed operation and reduced LEF as compared to undoped QDs [15].

Figure 3.10 shows the p-doped QD laser structure grown by molecular beam epitaxy. A 10-stack QD active region is sandwiched between 1.5-μm-thick $Al_{0.35}Ga_{0.65}As$ cladding layers. The QD array is formed by deposition of 2.5–3 monolayers of InAs and covered with 5-nm-thick $In_{0.15}Ga_{0.85}As$ QW cap layer. The surface density of QDs is estimated to be $5.0 \times 10^{10}\,cm^{-2}$ per QD layer. Each QD layer is separated by a 33-nm-thick GaAs barrier. Within the GaAs barriers, the central 10-nm-thick region doped with carbon is grown 9 nm before the next QD layer. The doping density of p-type acceptors is $5.0 \times 10^{17}\,cm^{-8}$ per layer (10 acceptors per QD). The structure was processed into a single-mode 3-μm-wide ridge waveguide.

Figure 3.11(a) shows the calculated net modal gain spectra of an undoped (dashed) and a p-doped (solid) QD FP laser. For the coherent comparison of the two lasers, surface electron densities per QD layer are adjusted such that the peak modal gains are equal for both p-doped and undoped QD FP laser. This is necessary because the band-filling behavior of the undoped QDs is different from that of the p-doped QDs. The p-doped QD FP laser requires less surface electron density to

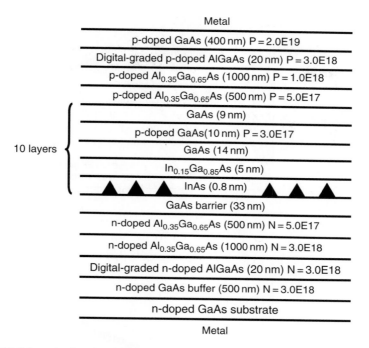

Figure 3.10 Schematic of a p-doped QD laser structure grown by molecular beam epitaxy. After Ref. [5].

achieve the same peak modal gain than the undoped QD FP laser does due to the hole supply from the p-type doping.

Figure 3.11(b) shows the calculated LEF spectra of the p-doped (solid) and the undoped (dashed) QD FP laser based on the calculated change in modal gain (Figure 3.11(a)) and refractive index. Because the p-doped QD laser has a smaller carrier density in the excited states of QDs near threshold, the LEF of the p-doped QD FP laser is smaller by a value of about 1 over the spectral range of 50 nm than that of the undoped QD FP laser.

The experimental gain and LEF data of p-doped QD were obtained considering pulsed operation (to reduce thermal heating effects of the device). In Figure 3.11(b), the measured LEF by pulsed current injection (symbols) is matched with the theoretical calculation. This measured and calculated LEF without the thermal effect is about 1.0 at the lasing wavelength of 1288 nm.

Experimentally, many improvements in operation of conventional QDs are not observed. This is mainly due to inhomogeneity of QD size variation, poor carrier collection (due to low QD growth densities), and charge redistribution within the QD layer [36]. The recent proposal of a QD laser employing an auxiliary QW is shown to have better performance in reducing the threshold current density as compared to a stand-alone QD layer within a laser [37]. Moreover, the QD

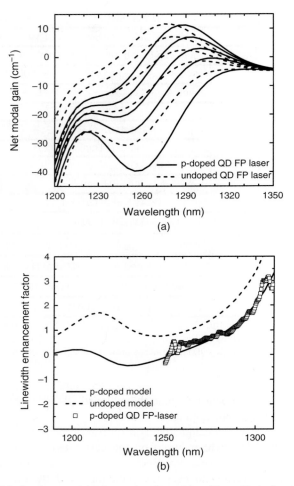

Figure 3.11 (a) Calculated net modal gain spectra of an undoped laser (dashed) compared with the calculated net modal gain spectra of the p-doped quantum dot (QD) Fabry–Perot (FP) QD laser (solid). (b) Calculated LEF spectra of the p-doped (solid) and the undoped (dashed) QD FP laser near threshold. The experimental LEF (scatter) of p-doped QD laser is obtained under pulsed laser operation to isolate thermal effects. The LEF is extracted from the modal gain spectra at two pulsed currents of 19 and 21 mA, which is matched with the theoretical prediction. After Ref. [5].

tunneling injection structure improves self-organization during growth, which in turn achieves higher density of self-assembled QDs with significant improvement in QD size distributions [38, 39]. The recent studies of tunneling injection QD lasers show symmetrical gain shape, low threshold, and therefore near zero LEF [6, 32, 40, 41].

The GaAs–InGaAs–CInAs 1.1-μm tunneling injection QD laser heterostructure is shown in Figure 3.12. The active medium was grown using low-pressure

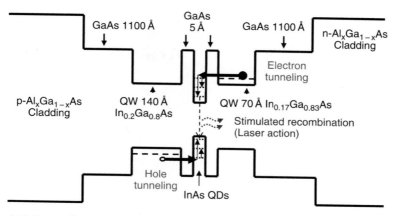

Figure 3.12 Energy diagram of an InAs – laser. The auxiliary QWs serve as collectors of electron and holes which tunnel to a QD for stimulated quantum dot (QD) quantum well (QW) recombination. After Ref. [6] Reprinted with permission from American Institute of Physics. (this figure may be seen in color on the included CD-ROM).

metalorganic chemical vapor deposition (LP-MOCVD) on an n-type GaAs substrate. The 70-Å QW on the n-side serves as an electron collector. Similarly, a 140-Å QW on the p-side of the dot layer serves as the hole collector. The carriers are then subject to tunneling through a thin barrier into the QD layer where they produce stimulated recombination for laser action. The tunneling lifetime has been studied theoretically for this QW–QD structure [42]. For a small barrier separation, the tunneling time is very fast, on the order of picoseconds. The QW–QD laser device growth methodology, structure, and operation are further detailed in Refs [37–39].

The optical gain spectra with increased injection current is plotted in Figure 3.13(a), showing symmetrical gain shape. The heating effects of the active medium with cw injection current is disregarded by operating the laser sample in pulsed mode. These thermal effects introduce a sign change in the magnitude of the refractive index change spectrum and can result in a zero and even negative LEF for the measurement of this material system under cw operation [40].

The tunneling injection QD laser is pulsed (2% duty cycle) to reduce heating effects. Subsequently, from the data at different current injection, the spectrum of refractive index change (using Eqn 3.14) and the LEF (using Eqn 3.15) are obtained. An increase in the bias current at a constant interval of 1 mA reveals the reduction in the magnitude of the refractive index. However, as the laser approaches threshold, the differential gain is also observed to decrease. The corresponding LEF is shown in Figure 3.13(b). A flat spectrum of the LEF over a large range of wavelength is observed. This is attributed to the broadened gain spectra of the dot ground states and contribution of dot excited states and the well states. The value of LEF at the lasing wavelength (spectra shown on the plot) is found to be 0.15.

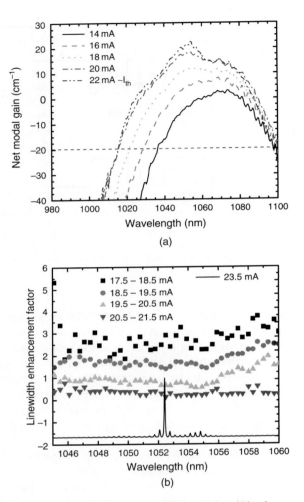

Figure 3.13 (a) DC modal gain spectra at increased current and (b) linewidth enhancement factor spectra obtained for the tunneling injection QD InAs tunneling injection laser. After Ref. [6] Reprinted with permission from American Institute of Physics. (this figure may be seen in color on the included CD-ROM).

3.4.2 High-Speed QD Lasers: p-type Doping and Tunneling Injection

Tunnel injection QD lasers, with their reduced carrier modulation of the refractive index in the active volume, have also demonstrated an improvement in direct modulation bandwidths. The QD laser employing tunnel injection and p-type doping in active region has demonstrated $\simeq 25$ GHz modulation response [15, 43, 44]. In comparison, the conventional QD lasers have shown about 6–9 GHz bandwidth [45].

3. High-Speed Low-Chirp Semiconductor Lasers

The limitations in direct modulation of conventional QDs are onset by hot carrier effects arising from the occupation of barrier and wetting layer states by the electrons injected into the active region at room temperature. The occupation of these unwanted states results in the gain compression, which reduces the modulation bandwidth. Moreover, thermal hole broadening in the valence band of QDs is another limiting factor in the modulation of these lasers. The hole broadening results in many states with small energy separations in QDs, therefore, a large injection hole density is required for a large gain of the ground state. The hot carrier effects and hole broadening directly relate to a decrease in the gain and decreased differential gain of QDs, therefore reducing the direct modulation bandwidth. To alleviate these limitations, tunneling injection has been proposed to reduce hot carrier effects [37, 38, 43, 46]. Whereas, p-doping has been proposed as solution to hole state broadening at room temperature [47, 48].

The p-doped and tunnel injection QD laser studies have reported that main advantage of p-doping lies in enhancing threshold [47] and linewidth enhancement characteristics (discussed in previous section). It is found that tunnel injection scheme for collection and cooling of carriers appears to be a better approach in achieving high-speed direct modulation QD laser [15].

Modulation response of QD lasers consisting of p-doped, tunnel injection, and p-doped tunnel injection active region was investigated in Ref. [15]. Figure 3.14(a)–(c) shows the direct modulation response of p-doped, tunnel injection, and p-doped with tunnel injection QD single-model 1.1-μm laser, respectively. The p-type doping is observed to slightly increase the bandwidth performance.

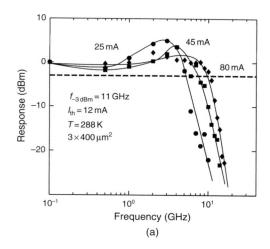

Figure 3.14 (a) Modulation response of single-mode 1.1-μm laser with p-doping in the active region of QD SCH laser at different bias currents showing a maximum 3-dB bandwidth of ≃11 GHz. (b) A tunneling injection design of the QD laser improves the modulation response of undoped QD lasers to 22 GHz. (c) When both tunnel injection and p-type doping are utilized in the active medium, the 3-dB frequency of 24.5-GHz direct modulation is achieved. After Ref. [15].

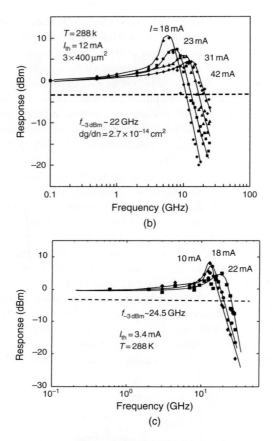

Figure 3.14 (Continued)

This is shown by the comparison of 9-GHz modulation of undoped QD-SCH laser to 11 GHz (Figure 3.14(a)). Moreover, the tunnel injection laser with undoped QDs response of 22 GHz is compared to 24.5 GHz for the tunnel injection laser with p-doping. The effect of tunnel injection on modulation bandwidth is more pronounced. The 3-dB modulation response is found to be more than double for p-doped and undoped QD lasers utilizing tunnel injection. For the modulation response of QD lasers presented in Figure 3.14(a)–(c), the differential gains are $6.9 \times 10^{-15}\,\text{cm}^2$, $2.7 \times 10^{-14}\,\text{cm}^2$, and $3.0 \times 10^{-14}\,\text{cm}^2$, respectively.

3.5 DISCUSSIONS

In conclusion, we reviewed high-speed small-signal modulation experiments of $In_{1-x}Ga_xAs_yP_{1-y}$ and $In_{1-x-y}Ga_xAl_yAs$ FP QW lasers. We discussed the intrinsic parameters that affect the laser steady-state and high-speed modulation performance. The topics such as the degradation of Al material system or temperature dependence, which are also important for designing high-performance telecommunication lasers, were not covered. The results show that the laser based on compressively strained $In_{1-x}Ga_xAs_yP_{1-y}$ QWs has a better performance in terms of higher modulation bandwidth due to a smaller K factor. However, the characteristics of $In_{1-x}Ga_xAs_yP_{1-y}$ lasers are severely limited by nonradiative Auger recombination processes [10, 14, 49–51]. The K factor determines the maximum intrinsic modulation bandwidth. The smaller the K factor, the larger the intrinsic modulation bandwidth. The lower limit of the K factor is set by $4\pi^2\tau_p$. The $In_{1-x}Ga_xAs_yP_{1-y}$ laser has a much smaller K factor than the other two samples. The large difference of the K factors comes from the large nonlinear gain suppression coefficients ϵ from two $In_{1-x-y}Ga_xAl_yAs$ lasers. For the $In_{1-x}Ga_xAs_yP_{1-y}$ laser, the contributions to the K factor from both the τ_p and ϵ are significant. It makes sense to talk about decreasing the photon lifetime to obtain broader bandwidth. For two $In_{1-x-y}Ga_xAl_yAs$ lasers, the K factors are mainly determined by large values of ϵ. Improvement on the nonlinear gain suppression ratio should be made to increase the modulation bandwidth of lasers based on $In_{1-x-y}Ga_xAl_yAs$ material system. The nonlinear gain suppression ratio is dependent on the intraband relaxation time for carriers and optical matrix element. Note that the structures of our tested $In_{1-x-y}Ga_xAl_yAs$ lasers are not optimized. The design parameters such as strain, QW width, number of QWs, and laser cavity structure need to be optimized to improve the performance of lasers based on $In_{1-x-y}Ga_x Al_yAs$ material systems. Discussion on this subject can be found, for example, in Refs [30, 31].

Moreover, discussion of QD lasers with increased differential gain and large symmetric gain present reduced LEF and therefore chirp. The p-doping within the QD active medium reduces the thermal broadening in the valance band, hence, increases the attainable gain and differential gain. The differential gain and LEF of a p-doped QD laser are shown to be superior to those of an undoped QD laser due to the reduced transparency carrier density caused by p-type doping. This p-doped QD lasers also demonstrate increased modulation response as compared to undoped QDs. The effect of tunnel injection design to reduce hot carrier effects and occupation of barrier and wetting layer states in QD lasers have shown to substantially increase the modulation response. Moreover, these novel QD active regions exhibit near zero LEF.

ACKNOWLEDGMENT

The authors thank S. H. Park, J. Minch, T. Keating, X. Jin, and J. Kim for many contributions during the past years of research and the publications related to this work.

REFERENCES

[1] C. E. Zah, R. Bhat, B. Pathak et al., "High performance uncooled 1.3 μm AlGaInAs/InP strained-layer quantum well lasers for subscriber loop applications," *IEEE J. Quantum Electron.*, 30, 511–523, 1994.

[2] S. Shirai, Y. Tatsuoka, C. Watatani et al., "120°C Uncooled Operation of Direct Modulated 1.3 μm AlGaInAs-MQWDFB Laser Diodes for 10-Gb/s Telecom Applications," in Proc. *OFC*, ThD5, Los Angeles, CA, 2004.

[3] G. Liu and S. L. Chuang, "High-speed modulation of long-wavelength InGaAsP and InAlGaAs strained quantum-well lasers," *IEEE J. Quantum Electron.*, 37, 1283–1291, 2001.

[4] G. Liu, X. Jin, and S. L. Chuang, "Novel techniques for measurement of linewidth enhancement factors of strained QW lasers using injection locking," *IEEE Photon. Technol. Lett.*, 13, 430–432, 2001.

[5] J. Kim and S. L. Chuang, "Theoretical and experimental study of optical gain, refractive index change, and linewidth enhancement factor of p-doped quantum-dot lasers," *IEEE J. Quantum Electron.*, 42(9), 942–952, September 2006.

[6] P. K. Kondratko, S.-L. Chuang, G. Walter et al., "Observations of near-zero linewidth enhancement factor in a quantum-well coupled quantum-dot laser," *Appl. Phys. Lett.*, 83(23), 4818, December 2003.

[7] P. J. A. Thijs, L. F. Tiemijer, P. I. Kuindersma et al., "High performance of 1.5 μm wavelength InGaAs-InGaAsP strained quantum-well lasers and amplifiers," *IEEE J. Quantum Electron.*, 27, 1426–1438, 1991.

[8] P. J. A. Thijs, L. F. Tiemijer, J. J. M. Binsma, and T. van Dongen, "Progress in long-wavelength strained-layer InGaAs(P) quantum-well semiconductor lasers and amplifiers," *IEEE J. Quantum Electron.*, 30, 477–499, 1994.

[9] M. Allovon and M. Quillec, "Interest in AlGaInAs on InP for optoelectronic applications," *IEE Proc. J. Optoelectronics*, 139, 148–152, 1992.

[10] J. Minch, S. H. Park, T. Keating, and S. L. Chuang, "Theory and experiment of $In_{1-x}Ga_xAs_yP_{1-y}$ and $In_{1-x-y}Ga_xAl_yAs$ long-wavelength strained quantum-well lasers," *IEEE J. Quantum Electron.*, 35, 771–782, 1999.

[11] T. Ishikawa and J. Bowers, "Band lineup and in-plane effective mass of InGaAsP or InGaAlAs on InP strained-layer quantum well," *J. Quantum Electron.*, 30, 562–569, 1994.

[12] O. Issanchou, J. Barrou, E. Idiart-Alhor, and M. Quillec, "Theoretical comparison of GaInAs/GaAlInAs and GaInAs/GaInAsP quantum-well lasers," *J. Appl. Phys.*, 78, 3925–3930, 1995.

[13] J. Pan and J.-I. Chyi, "Theoretical study of the temperature dependence of 1.3 μm AlGaInAs-InP multiple-quantum-well lasers," *J. Quantum Electron.*, 32, 2133–2138, 1996.

[14] M. C. Wang, K. Kash, C. Zah et al., "Measurement of non radiative Auger and radiative recombination rates in strained-layer quantum-well systems," *Appl. Phys. Lett.*, 62, 166–168, 1993.

[15] S. Fathpour, Z. Mi, and P. Bhattacharya, "High-speed quantum dot lasers," *J. Phys. D. Appl. Phys.*, 38, 2103–2111, June 2005.

[16] T. Keating, X. Jin, S. L. Chuang, and K. Hess, "Temperature dependence of electrical and optical modulation response of quantum-well lasers," *IEEE J. Quantum Electron.*, 35, 1526–1534, 1999.

[17] R. Nagarajan, T. Fukushima, S. W. Corzine, and J. E. Bowers, "Effects of carrier transport on high-speed quantum well lasers," *Appl. Phys. Lett.*, 59, 1835–1837, 1991.

[18] S. H. Park, D. Ahn, and S. L. Chuang, "Intraband relaxation time effects on non-Markovian gain with many-body effects and comparison with experiment," *Semicond. Sci. Technol.*, 15, 203–208, 2000.

[19] C. Y.-P. Chao and S. L. Chuang, "Spin-orbit-coupling effects on the valence-band structure of strained semiconductor quantum wells," *Phys. Rev. B*, 46, 4110–4122, 1992.

[20] W. W. Chow, S. W. Koch, and M. Sergent, III, *Semiconductor-Laser Physics*. Berlin, Springer, 1994.

[21] Y. Huang, S. Arai, and K. Komori, "Theoretical linewidth enhancement factor alpha of GaInAs/GaInAsP/InP strained-quantum-well structures," *IEEE Photon. Technol. Lett.*, 5, 142–145, 1993.

[22] B. W. Hakki and T. L. Paoli, "Gain spectra in GaAs double heterostructure injection lasers," *J. Appl. Phys.*, 46, 1299–1306, 1975.
[23] C. S. Chang, S. L. Chuang, J. R. Minch et al., "Amplified spontaneous emission spectroscopy in strained quantum-well lasers," *IEEE J. Set Top. Quantum Electron.*, 1, 1100–1107, 1995.
[24] S. J. Sweeney, A. F. Phillips, A. R. Adams et al., "The effect of temperature dependent processes on the performance of 1.5 μm compressively strained InGaAs(P) MQW semiconductor diode lasers," *IEEE Photon. Technol. Lett.*, 10, 1076–1078, 1998.
[25] E. P. O'Reilly and M. Silver, "Temperature sensitivity and high temperature operation of long wavelength semiconductor lasers," *Appl. Phys. Lett.*, 63, 3318–3320, 1993.
[26] T. Yamanaka, Y. Yoshikuni, K. Yokoyama et al., "Theoretical study on enhanced differential gain and extremely reduced linewidth enhancement factor in quantum-well lasers," *IEEE J. Quantum Electron.*, 29, 1609–1616, 1993.
[27] F. Kano, T. Yamanaka, N. Yamamoto et al., "Reduction of linewidth enhancement factor in InGaAsP-InP modulation-doped strained multiple-quantum-well lasers," *IEEE J. Quantum Electron.*, 29, 1553–1559, 1993.
[28] F. Kano, T. Yamanaka, N. Yamamoto et al., "Linewidth enhancement factor in InGaAsP/InP modulation-doped strained multiple-quantum-well lasers," *IEEE J. Quantum Electron.*, 30, 533–537, 1994.
[29] A. Schonfelder, S. Weisser, J. D. Ralston, and J. Rosenzweig, "Differential gain, refractive index, and linewidth enhancement factor in high-speed GaAs-based MQW lasers: Influence of strain and p-doping," *IEEE Photon. Technol. Lett.*, 6, 891–893, 1994.
[30] S. Seki, P. Sotirelis, K. Kess et al., "Theoretical analysis of gain saturation coefficients in InGaAs/AlGaAs strained layer quantum well lasers," *Appl. Phys. Lett.*, 61, 2147–2149, 1992.
[31] Y. Matsui, H. Murai, S. Arahira et al., "Enhanced modulation bandwidth for strain-compensated InGaAsAs-InGaAsP MQW lasers," *IEEE J. Quantum Electron.*, 34, 1970–1978, 1998.
[32] L. V. Asryan and S. Luryi, "Tunneling-injection quantum-dot laser: Ultrahigh temperature stability," *IEEE J. Quantum Electron.*, 37(7), 905–910, July 2001.
[33] T. C. Newell, D. J. Bossert, A. Stintz et al., "Gain and linewidth enhancement factor in InAs quantum-dot laser diodes," *IEEE Photon. Technol. Lett.*, 11(12), 1527, December 1999.
[34] J. Kim and S. L. Chuang, "Theoretical and experimental study of optical gain, refractive index change, and linewidth enhancement factor of p-doped quantum-dot lasers," *IEEE J. Quantum Electron.*, 42, 942–952, 2006.
[35] P. Bhattacharya, K. Kamath, J. Singh et al., "In(Ga)As/GaAs self-organized quantum dot lasers: DC and small-signal modulation properties," *IEEE Trans. Electron Devices*, 46(5), 871, May 1999.
[36] J. H. Ryou, R. D. Dupuis, G. Walter et al., "Properties of InP self-assembled quantum dots embedded in $In_{0.49}(Al_xGa_{1-x})_{0.51}P$ or visible light emitting laser applications grown by metalorganic chemical vapor deposition," *J. Appl. Phys.*, 91(8), 5313, April 2002.
[37] G. Walter, T. Chung, and N. Holonyak, "High-gain coupled InGaAs quantum well InAs quantum dot AlGaAs-GaAs-InGaAs-InAs heterojunction diode laser operation," *Appl. Phys. Lett.*, 80(7), 1126, February 2002.
[38] T. Chung, G. Walter, and N. Holonyak, "Coupled strained-layer InGaAs quantum-well improvement of an InAs quantum dot AlGaAsC-GaAsC-InGaAsC-InAs heterostructure laser," *Appl. Phys. Lett.*, 79(27), 4500, December 2001.
[39] G. Walter, T. Chung, and N. Holonyak, "Coupled-stripe quantum-well-assisted AlGaAs-GaAs-InGaAs-InAs quantum-dot laser," *Appl. Phys. Lett.*, 80(17), 3045, April 2002.
[40] P. Kondratko, S. L. Chuang, G. Walter et al., "A novel tunneling injection quantum-well-dot laser with extremely small chirp parameter," *Conf. Lasers and Electro-Optics*, 1523 (CthI1), 1487–1489, June 2003.
[41] P. Kondratko, S.-L. Chuang, G. Walter et al., "Gain narrowing and output behavior of InP-InGaAlP tunneling injection quantum-dot-well laser," *IEEE Photon. Technol. Lett.*, 17(5), 938–940, May 2005.
[42] S. L. Chuang and N. Holonyak, "Efficient quantum well to quantum dot tunneling: Analytical solutions," *Appl. Phys. Lett.*, 80(7), 1270, February 2002.

[43] P. Bhattacharya and S. Ghosh, "Tunnel injection Ino.4Gao.6As/GaAs quantum dot lasers with 15 GHz modulation bandwidth at room temperature," *Appl. Phys. Lett.*, 80(19), 3482, May 2002.

[44] S. Ghosh, S. Pradhan, and P. Bhattacharya, "Dynamic characteristics of high-speed Ino.4Gao.6As/GaAs self-organized quantum dot lasers at room temperature," *Appl. Phys. Lett*, 81(16), 3055, October 2002.

[45] K. Kamath, J. Phillips, H. Jiang et al., "Small-signal modulation and differential gain of single-mode self-organized Ino.4Gao.6As/GaAs quantum dot lasers," *Appl. Phys. Lett.*, 70(22), 2952–2953, June 1997.

[46] G. Walter, N. Holonyak, J. H. Ryou, and R. D. Dupuis, "Room-temperature continuous photo-pumped laser operation of coupled InP quantum dot and InGaP quantum well InPC-InGaP-CIn(AlGa)PC-InAlP heterostructures," *Appl. Phys. Lett.*, 79(13), 1956, September 2001.

[47] L. V. Asryan and S. Luryi, "Tunneling-injection quantum-dot laser: Ultrahigh temperature stability," *IEEE J. Quantum Electron.*, 37(7), 905, 2001.

[48] O. B. Shchekin and D. G. Deppe, "The role of p-type doping and the density of states on the modulation response of quantum dot lasers," *Appl. Phys. Lett.*, 80(15), 2758–2760, April 2002.

[49] R. Olshansky, C. B. Su, J. Manning, and W. Powazinik, "Temperature sensitivity and high temperature operation of long wavelength semiconductor lasers," *IEEE J. Quantum Electron.*, 20, 838–854, 1984.

[50] Y. Zou, J. S. Osinski, P. Grodzinski et al., "Experimental study of Auger recombination, gain, and temperature sensitivity of 1.5 μm compressively strained semiconductor lasers," *IEEE J. Quantum Electron.*, 29, 1565–1575, 1993.

[51] G. Agrawal and N. K. Dutta, *Semiconductor Lasers* (2nd edn), New York, Van Nostrand Reinhold, 1993, Ch. 3.

4

Recent advances in surface-emitting lasers

Fumio Koyama

Microsystem Research Center, P&I Lab, Tokyo Institute of Technology, Nagatsuta, Midori-ku, Yokohama, Japan

4.1 INTRODUCTION

A vertical-cavity surface-emitting laser (VCSEL) was invented in 1977 [1, 2]. The schematic structure is illustrated in Figure 4.1. A number of unique features have been proven:

(1) low-power consumption
(2) high-speed modulation with low driving current
(3) large scale two-dimensional (2D) array
(4) narrow circular beam for direct fiber coupling
(5) low-cost and small-packaging capability
(6) single longitudinal mode operation with vertical microcavity
(7) on-wafer wavelength control
(8) continuous wavelength (CW) tuning with electromechanical system
(9) wafer level testing for low-cost manufacturing.

The first lasing operation [3] and low threshold operation with a microcavity [4] were demonstrated in 1979 and 1987, respectively. The early long-term VCSEL research triggered worldwide research on VCSELs and formed the research community of VCSEL-based photonics [1]. The research history and principal references on VCSELs in wide spectral regions can be seen in Ref. [1]. After the room-temperature cw operation of GaAs VCSELs [5] and the demonstration of sub-mA microlasers [6], VCSEL research was accelerated for reducing threshold current and increasing efficiencies [7]. In particular, significant advances were achieved in oxide-confined VCSELs [8], which incorporate buried oxide apertures [9] with electrical and optical confinement. Oxide-confined VCSELs gave us a significant threshold reduction [8, 10] and high-power-conversion efficiencies of more than 50% [11, 12].

Figure 4.1 Schematic structure of VCSEL array (this figure may be seen in color on the included CD-ROM).

Gigabit Ethernet and Fiber Channel are currently major markets for VCSELs. Also, VCSEL-based 10G Ethernet modules are ready for practical systems. Commercial 850-nm GaAs VCSELs have been well established for these short-reach applications. However, long wavelength VCSELs emitting at 1200–1330 nm are currently attracting much interest for use in single-mode fiber (SMF) metropolitan area and wide-area networks. Various materials have been proposed and demonstrated for long-wavelength VCSELs, which include highly strained GaInAs/GaAs quantum wells (QWs) [13–19], GaInAs quantum dots (QDs) [20, 21], GaAsSb QW [22–24], GaInNAs [25–36], and GaAlInAs/InP QWs [37–47]. Low-cost and high-performance VCSELs emitting at 1.3 µm may drive significant cost reduction in high-speed links of over several kilometers with single-mode fibers.

High-speed modulation operations have also been demonstrated [48–50] and the highest modulation speed reached at 30 Gb/s for GaInAs VCSELs [51]. High-data throughput and high-interconnection densities are becoming important for high-end computing systems. Optical interconnection is a good candidate for avoiding bottlenecks in next-generation supercomputers. The development of high-speed VCSELs has been involved in the Japanese government project on the next-generation supercomputer "KEISOKU."

For higher data rates beyond 10 Gb/s, a wavelength-division multiplexing (WDM) technology would be an alternative even in very short-reach systems. There are various ways to control the lasing wavelength in VCSELs, and a multiple-wavelength integrated VCSEL array is a good candidate for short-reach WDM applications. A 2D multiple-wavelength VCSEL array was first realized using an inherent beam flux gradient in molecular beam epitaxy (MBE) growth [52]. When we use metalorganic chemical vapor deposition (MOCVD) on a patterned substrate, the local gradient of the chemical species in the gas phase changes the growth rate. The epitaxial growth of GaAs and AlGaAs on a nonplanar

GaAs substrate has been developed and an MOCVD-grown multiple-wavelength VCSEL array was demonstrated based on this technique [53]. Several reports on multiple-wavelength VCSELs emitting at 850 and 980 nm have been presented [54–56]. Ultrahigh capacity chip-to-chip optical interconnect based on WDM VCSEL technologies has been developed in the Terabus project [57]. The data rate is up to 20 Gb/s with 48 channels (4 wavelengths × 12 parallel channels), leading to a bit rate approaching 1 Tb/s.

Microelectromechanical system (MEMS)-based tunable filters and tunable VCSELs [58–62] are attracting much interest because of their unique features, such as wide continuous tuning, polarization-insensitive operation, and 2D array integration. Wide continuous wavelength tuning has been demonstrated when electrostatic force is used for moving a cantilever structure in micromachined filters and VCSELs.

However, all-optical signal processing using VCSELs has recently been investigated for future photonic networks. The optical injection locking is very useful for reducing chirp and for extending the modulation bandwidth [63]. The injection locking of VCSELs has been examined theoretically and experimentally [64–66]. Various interesting nonlinear behaviors were observed, which was dependent on injection power and frequency detuning. All-optical format conversion was demonstrated using a polarization switching in a VCSEL [67]. In this case, a complex configuration is required to control the polarization state of the input signal. Transverse-mode switching properties in an 850 nm VCSEL induced by optical injection were investigated experimentally [68]. An all-optical inverter was demonstrated using a two-mode VCSEL with the fundamental and first-high-order modes [69]. Also, a new function of high-speed polarization controller was proposed [70]. Another interesting application of a VCSEL structure for optical signal processing is to control the phase and the group delay of light, including tunable ultraslow light in 1550-nm VCSEL amplifier [71]. Also, a novel optical phase shifter was proposed and demonstrated [72] and, the intensity-dependent phase shift is induced by photo-carriers in a resonant cavity absorber, which would enable us to control the transient chirp and to compensate fiber nonlinearities in optical domain.

In this chapter, the recent advance of VCSEL photonics will be described. We present the wavelength engineering of VCSELs for use in high-speed short-reach applications, which include the advances of long-wavelength VCSELs, their wavelength integration and wavelength stabilization based on fully monolithic technologies. The potential and challenges for new functions of VCSEL photonics will be discussed.

4.2 LONG-WAVELENGTH VCSELs

Various materials have been proposed and demonstrated for long-wavelength VCSELs as shown in Figure 4.2, which include highly strained GaInAs/GaAs QWs [13–19], GaInAs QDs [20, 21], GaAsSb QW [22–24], GaInNAs [25–36], and GaAlInAs/InP QWs [37–47]. Highly strained GaInAs QW VCSELs have been developed, covering a wide wavelength window of 1.0–1.3 μm [13–19]. The

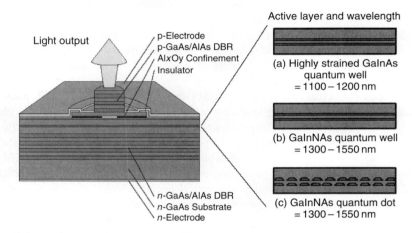

Figure 4.2 Schematic structure of long wavelength VCSELs (this figure may be seen in color on the included CD-ROM).

wavelength extension of highly strained GaInAs QWs grown by MOCVD was carried out up to 1.2 μm and the gain-offset enables lasing wavelengths of 1.3 μm [18, 19]. An advantage for highly strained GaInAs QW lasers is no noticeable penalty in crystal quality and their good temperature characteristics [73–77]. The low threshold current density of highly strained GaInAs/GaAs QW lasers emitting at nearly 1.2 μm wavelength was demonstrated [76–78]. The PL wavelength of grown GaInAs QWs could be extended over 1.2 μm without any degradation in crystal qualities.

The schematic structure of long wavelength VCSELs on a GaAs substrate is shown in Figure 4.3 [78]. The active region consists of 8-nm-thick $Ga_{0.66}In_{0.34}As$/GaAs triple QWs surrounded by a GaAs spacer layer to form a λ cavity. The n-type distributed Bragg reflector (DBR) consists of 35 pair $Al_{0.8}Ga_{0.2}As$/GaAs doped with Se grown at 695°C. The p-type DBR consists of 22-pair C-doped $Al_{0.8}Ga_{0.2}As$/GaAs. The grown samples were dry-etched by inductively coupled plasma (ICP) etching to form the 30 μm × 30 μm square mesas. We formed 4 μm × 4 μm oxide apertures by AlAs wet oxidation process. We fabricated highly strained GaInAs QW VCSELs either on (100) or (311)B GaAs substrates [15, 78]. We could expect excellent temperature characteristics due to the deep potential well of this material system. A characteristic temperature T_0 of GaInAs/GaAs QW edge emitting lasers is more than 200 K, which is the highest at 1.2–1.3 μm wavelength band. A low threshold current of below 1 mA, high-temperature operation of up to 450 K, and high reliability of >2000 hours were demonstrated [15]. This device was grown on a GaAs (311)B substrate, showing large orthogonal polarization suppression ratio of 30 dB, whereas VCSELs on a (100) substrate shows polarization switching. The maximum single-mode output power is over 3 mW. The threshold and slope

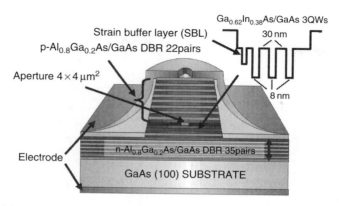

Figure 4.3 Schematic structure of GaInAs/GaAs VCSEL [78] (this figure may be seen in color on the included CD-ROM).

efficiency are almost unchanged up to 75°C as shown in Figure 4.4 [78]. The extension of the emission wavelength up to 1.2 µm enables high-speed data transmission in single-mode fibers. Although there is a difficulty using installed SMFs because of their standardization, a high-performance 1.2 µm VCSEL would be a good candidate for newly installed short-reach systems. The negative dispersion of a fiber is helpful for short-pulse transmission with the frequency chirp of a VCSEL [79]. The result shows a potential of highly strained GaInAs VCSELs for use in high-capacity networks beyond 10 Gb/s. Another important aspect on GaInAs/GaAs QW VCSEL is a prospect for reliability. The Al-free GaInAs/GaAs QW would be preferable for an active layer due to its indium-suppressed dislocation motion [51].

An important question arises: can we avoid an optical isolator in single-mode VCSEL modules? The feedback sensitivity for single-mode VCSELs is particularly important while no noticeable optical feedback noise can be seen for multimode VCSELs. It is needed to realize isolator-free operations for avoiding an

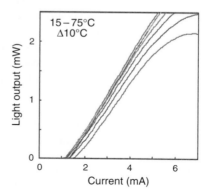

Figure 4.4 Temperature dependences of L/I characteristic of 1.13-mm GaInAs VCSELs [78] (this figure may be seen in color on the included CD-ROM).

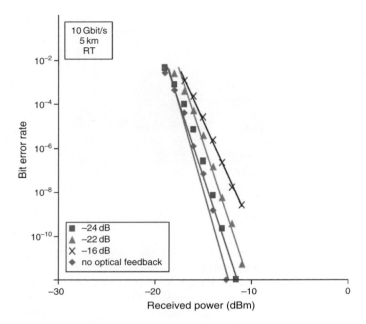

Figure 4.5 Bit error rates of 10 Gb/s, 5-km transmission with and without optical feedback for 1.1 μm VCSEL [81].

optical isolator in low-cost modules. There have been several reports on optical feedback sensitivity for long-wavelength VCSELs [80–84]. Figure 4.5 shows the BER for 10 Gb/s data transmission over 5 km of a standard SMF with and without optical feedback [81]. We could see no noticeable power penalty due to optical feedback effect. A power penalty is below 1 dB for a feedback level of −24 dB. While further increase in optical feedback up to −16 dB results in difficulties of error-free data transmission, we pointed out that it is very effective for improving the feedback sensitivity to increase the relaxation oscillation frequency [82]. A 10 Gb/s error-free SMF transmission was demonstrated using a 1.3 μm InP-based VCSEL under strong optical reflection (−13 dB) and without an optical isolator by increasing the relaxation oscillation frequency over 10 GHz [84].

For further elogation of lasing wavelength, GaInAs QDs [20, 21], GaAsSb QW [22–24], GaInNAs [25–36], and GaAlInAs/InP QWs [37–47] have been developed. Low-cost and high-performance VCSELs emitting at 1.3 μm may drive significant cost reduction in high-speed links. Although the first lasing operation of VCSELs was demonstrated using InP-based materials in the long-wavelength region in 1979 [3], progress on short-wavelength GaAs-based VCSELs was much faster, resulting in their commercialization for LAN market. This is because good semiconductor mirrors of GaAlAs systems are available on GaAs substrates and GaAs or InGaAs QWs show higher gain and better temperature characteristics than InP-based QWs. In recent years, drastic improvements have been achieved

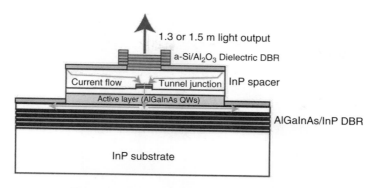

Figure 4.6 InP-based long-wavelength VCSELs [41].

on 1.55-μm InP-based VCSELs. The introduction of a tunnel junction allows the replacement of p-type layers with n-type layers as the tunnel junction can convert electrons to holes. p-type layers have to be located only near the active region and the total optical absorption can be reduced. In addition, with the help of the higher mobility of electrons, the injection current of VCSELs is much more uniform laterally when the current is injected through n-type layer, which avoids current crowding. Figure 4.6 shows the schematic structure of InP-based tunnel junction VCSELs emitting at 1.3 or 1.55 μm [41]. Drastic improvements such as cw operation at high temperature and uncooled high-speed operation have been demonstrated for both 1.3 and 1.55 μm InP-based VCSELs.

4.3 WAVELENGTH INTEGRATION AND CONTROL

The on-wafer wavelength control can be realized by grading the epitaxial layer thickness of a single-mode VCSEL structure. 2D multiple-wavelength VCSEL arrays were realized using an inherent beam flux gradient in MBE growth [52]. When we use MOCVD on a patterned substrate, the local gradient of the chemical species in the gas phase changes the growth rate [85]. The MOCVD-grown multiple-wavelength VCSEL array was demonstrated based on this technique [53]. Several reports on multiple-wavelength VCSELs emitting at 850 and 980 nm have been presented [54–56]. Yang et al. [56] reported the maximum wavelength span of 57 nm in a 980-nm band VCSEL array [56].

The schematic structure of a multiple-wavelength VCSEL array on a patterned substrate is illustrated in Figure 4.7 [86]. The spatial modulation on growth rates in MOCVD enables the wavelength control of VCSEL arrays. In the case of non-planar growth of MOCVD, the local gradient of the chemical species in the gas phase increases (decreases) the growth rate on mesas (in channels). Thus, the resonant wavelength is longer (shorter) on mesas (in channels). Also, the gain peak wavelength of QWs also becomes longer (shorter) simultaneously due to the well thickness variation. In addition, ternary GaInAs QWs enable wider wavelength

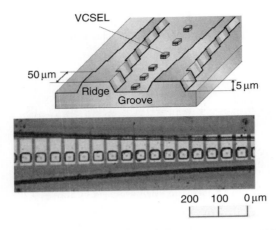

Figure 4.7 Densely integrated multiple-wavelength GaInAs/GaAs VCSEL array [86].

variation than GaAs QWs, because the composition modulation takes place by the difference in a diffusion coefficient between gallium and indium source in the gas phase.

A suitable tapered pattern shape was designed to give us densely integrated multiple wavelength VCSEL arrays [86]. A patterned substrate was prepared by standard photolithography followed by wet etching prior to the growth. The VCSEL structure is the same as in Figure 4.3 except a patterned substrate. The resonant wavelength was linearly graded with changing the widths of grooves and ridges at the same time. This linear array consists of 20 channels with a 50 μm pitch. Figure 4.7 shows the photograph of a fabricated multiple-wavelength VCSEL array. The threshold current is 0.52 ± 0.1 mA. All VCSELs exhibit an output power of over 1 mW and show uniform characteristics. Figure 4.8 shows the lasing spectra of the fabricated multiple-wavelength VCSEL array. All VCSELs in a linear array exhibit single-mode operation with a side-mode suppression ratio (SMSR) of >30 dB. The average wavelength separation is 0.8 nm.

For realizing a multiple-wavelength VCSEL in a wider wavelength span, it is necessary to control the gain peak shift and resonant wavelength shift at the same time. The gain peak shift of GaInAs/GaAs QWs is originated from the change of the QW thickness and the In composition modulation. However, the resonant wavelength shift is proportional to the thickness variation of GaAs/AlAs DBRs. It is a challenge to extend the wavelength span of multiple-wavelength VCSEL array. A main limiting factor in expanding the wavelength span in arrays is the offset between the gain peak and the resonant wavelength in an array. A growth pressure control in epitaxial growth on patterned substrates was proposed and demonstrated for further extension of lasing wavelength span in arrays. The growth pressure of MOCVD influences a diffusion constant of chemical species in gas phase [87]. The diffusion constant is inversely proportional to the growth pressure

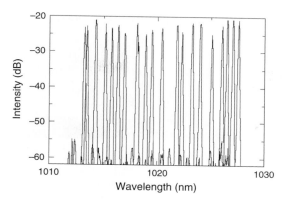

Figure 4.8 Lasing spectra of 20-channel multiple-wavelength VCSEL array on a patterned substrate [86].

[88]. Therefore, a growth rate enhancement and reduction on a patterned substrate can be controlled by changing the growth pressure. The enhancement of growth rates is larger under higher pressures. To expand the wavelength span of arrays with decreasing the gain-resonance mismatching, it is helpful to grow QWs under a higher pressure and to grow DBRs under a lower pressure. A 0.96- to 1.16-μm multiple-wavelength VCSEL array with highly strained GaInAs/GaAs QWs exhibits a record wavelength span of 192 nm [89].

There are still issues to be concerned in multiple-wavelength VCSEL arrays, which are the scalability of multiple-wavelength arrays and the precision control of wavelength interval. A 1.2-μm densely packed VCSEL array was demonstrated with an electrothermal tuning contact for precise wavelength control, exhibiting continuous wavelength tuning of 3 nm [90]. We also carried out large scale integration of a multiple-wavelength VCSEL array. A 110-channel VCSEL linear array was demonstrated with a mesa size of 15×15 μm and a pitch of 20 μm [91]. The device exhibits single-mode operation with a SMSR of over 38 dB. The thermal crosstalk is an important issue, which is discussed in Ref. [92]. Figure 4.9 shows the lasing spectra of the 110-channel high-density VCSEL array under testing one by one. The spatial grading of grown layers was partly used to form a multiple-wavelength array. The wavelength spacing could be precisely controlled to be 0.1 nm by adjusting the bias current for each element. A total of 110 VCSELs could be densely integrated in a 2.2-mm long bar. The integration density in space would be 100 times larger than conventional edge-emitting lasers, showing a benefit of small footprint in VCSELs. A 1.1 to 1.3-μm densely integrated multiple-wavelength VCSEL array can be expected to be used in ultrahigh-capacity short-reach applications.

MEMS-based VCSELs are attracting much interest for functioning as widely tunable sources. Wide continuous wavelength tuning has been demonstrated when electrostatic force is used for moving a cantilever structure in micromachined VCSELs [58]. A micromachined VCSEL and filter with a thermally actuated mirror was also proposed [93]. Various unique features can be expected in tunable

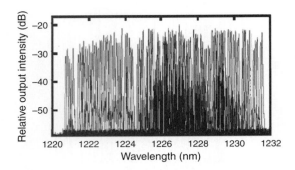

Figure 4.9 Lasing spectra of 110-channel VCSEL array [91] (this figure may be seen in color on the included CD-ROM).

VCSELs, which include wavelength temperature-insensitive operation, multiple-wavelength array integration, thermal wavelength tuning with a low tuning voltage, and so on [94]. Figure 4.10 shows the schematic structure and the operating principle of an "athermal" VCSEL with a thermally actuated cantilever [94]. The upper GaAlAs/GaAs or InP/GaInAsP DBR is freely suspended above the substrate. An air gap is formed between the upper and the lower DBRs. The structure is

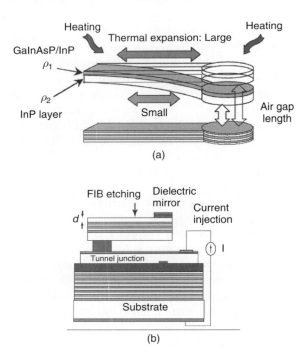

Figure 4.10 (a) Principle and (b) schematic structure of athermal VCSEL with a micromachined cantilever structure [94, 96] (this figure may be seen in color on the included CD-ROM).

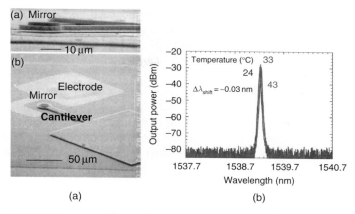

Figure 4.11 (a) SEM image and (b) lasing spectra of athermal InP-based VCSEL with a thermally actuated cantilever structure [96] (this figure may be seen in color on the included CD-ROM).

similar to that developed for tunable wavelength VCSELs with a micromachined cantilever [58]. A difference in the device is one additional thermal strain control layer on the upper DBR with a smaller or larger thermal expansion coefficient than the average expansion coefficient of the DBR. When temperature increases, a micromachined cantilever moves up if the top thermal strain control layer has a smaller thermal expansion, whereas it moves down if the top thermal strain control layer has a larger thermal expansion. This gives us either blue shift or red shift of wavelength. This thermally tunable cavity enables us to realize a temperature-insensitive VCSEL. When temperature increases, a micromachined cantilever moves down. This results in athermal operations of VCSELs. We can freely control the temperature dependence of the resonant wavelength of a micromachined vertical cavity. An extremely low-temperature dependence of below 0.0001 nm/K was predicted [95]. This is 700 times smaller than the temperature dependence of conventional VCSELs. Recently, an athermal InP-based VCSEL was realized with a thermally actuated cantilever structure [96]. The temperature dependence is as low as 0.0016 nm/K, which is 50 times smaller than that of conventional single-mode laser diodes (LDs) as shown in Figure 4.11.

4.4 PLASMONIC VCSELs

High-density optical data storages with terabytes capacity have been attracting much interest. The storage density of conventional optical memories such as CD and DVD is determined by the spot size of light, which is limited by optical diffraction. An optical near-field technology is one of the candidates to make a breakthrough for future optical storages [97, 98]. A storage density of terabit/inch^2 is expected when we reduce the spot size to be in the range of 10 nm. A high-density optical disk system using a VCSEL array was proposed [99] and a metal nanoaperture VCSEL was demonstrated for producing optical near-field localized at the metal nanoaperture [100, 101]. The voltage change induced by scattering in a nanoaperture enables us to use the same nanoaperture VCSEL chirp for optical near-field probing [102].

The evanescent wave is emitted from a metal nanoaperture formed on a VCSEL surface to an optical disk. The refractive index change on the disk may affect the threshold condition of a VCSEL, resulting in the change of an operating voltage under a constant current operation. We can use the same optical head for writing and reading out the data in this simple setup. A high spatial resolution can be expected by reducing the physical size of the metal aperture. The optical near-field intensity through a metal aperture is decreased with decreasing the diameter of the aperture, which is a common difficulty in near-field optics. It is a challenge to enhance the optical near-field from nanoaperture VCSELs. An interesting approach for increasing an optical near-field is to use surface plasmon in metallic nanostructures [103, 104]. Surface plasmon excited on a nanoparticle results in the significant increase of near-field intensity.

The schematic structure of a fabricated metal-aperture VCSEL is shown in Figure 4.12 [102]. The device includes a Au nanoparticle in the center of the metal aperture for plasmon enhancement of optical near-fields. The diameters of the metal aperture and a Au particle are 200 and 100 nm, respectively. The number of the p-type DBR pair was designed to be about a half of a standard design for increasing the near-field intensity through the nanoaperture. The structure except the top mirror design and the metal nanoaperture is the same as conventional GaAs

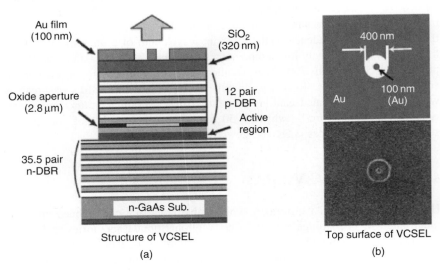

Figure 4.12 (a) Schematic structure and (b) top view of a fabricated nanoaperture VCSEL with a Au particle [102] (this figure may be seen in color on the included CD-ROM).

4. Recent Advances in Surface-Emitting Lasers

VCSELs. The lasing wavelength was 850 nm. The detailed structure is described in Ref. [52]. Here, we reduced the diameter of an oxide aperture for increasing the power density in a VCSEL cavity. The diameter of an oxide aperture is as small as 1.7 µm. A small oxide aperture is also helpful for transverse mode control. The threshold is as low as 300 µA.

Optical near-field measurement was carried out. The measurement system is based on a commercially available SNOM head (Seiko Instruments Inc., model SPA300). An aluminum-covered, sharpened fiber probe was scanned in the x, y direction just above the nanoaperture of the fabricated VCSEL. The distance between the fiber probe and the VCSEL was controlled to be constant and to be less than 20 nm by feedback control in the noncontact AFM mode of SNOM system. When the fiber probe is approaching the nanoaperture, the scattering from the fiber probe results in threshold changes of the VCSEL. This causes the change in a diode voltage under operating at a constant current. The spatial resolution of the SNOM is 130 nm. This setup also enables us to measure the topography of the device surface at the same time.

Figures 4.13(a) and (b) show the measured optical near-field intensity and voltage change of the nanoaperture VCSEL with a Au nanoparticle, respectively [102]. The spot size of optical near-field measured by SNOM shown in Figure 4.12 is estimated to be 100 nm. The power density of optical near-field localized at the nanoaperture is as large as 0.84 MW/cm^2, which is almost 10 times larger than that of conventional VCSELs, which results from strong optical confinement in the oxide microcavity and plasmon enhancement. The maximum power density is currently limited by thermal roll-over and the appearance of high-order transverse modes. The proposed near-field VCSEL may provide us ultrahigh power density at low power consumption, which may be useful for future optical storages and optical sensing applications.

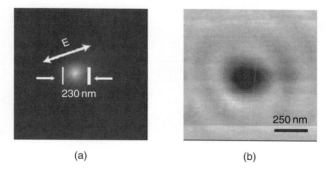

Figure 4.13 Measured near-field intensity (a) and voltage signal (b) of 200-nm metal-aperture VCSEL [102] (this figure may be seen in color on the included CD-ROM).

4.5 OPTICAL SIGNAL PROCESSING BASED ON VCSEL TECHNOLOGIES

In contrast to the optoelectronic regenerators, all-optical regenerators have a potential for low power consumption, as well as simple and cost–effective configuration. All-optical inverter based on transverse mode switching in a two-mode VCSEL was proposed, which has attractive features of low power consumption, dense packaging, and polarization insensitivity [69]. The transverse mode switching is induced when a first-high-order mode was injection-locked by a signal light. Figure 4.14 illustrates the principle of an optical inverter using a VCSEL. External light (signal light), with a wavelength slightly longer than that of a high-order transverse mode, is injected into the VCSEL. The dominant lasing mode switches from a fundamental mode to the high-order mode due to injection locking. If we look at the output power of the fundamental mode as a function of the input power, an optical inverter function with abrupt switching is obtained.

An external light from a tunable laser diode was injected through a standard single-mode fiber into a 1.55-μm InP-based VCSEL with a 7-μm circular tunnel junction aperture [41]. The dominant lasing mode at the bias current was the fundamental mode. When a high-order mode is injection-locked, an optical inverter function was obtained as shown in Figure 4.15. An abrupt modal switching with a large extinction ratio of >25 dB was obtained. The difference in threshold input power between the two orthogonal polarization inputs was as low as ~13%. Because the LP_{02} mode is circularly symmetric, we could obtain polarization-insensitive injection locking. The inverter operated even for the input signal with random polarization, showing a possibility of polarization-insensitive operation. The present device needs a few milliwatt of input power and thus an optical amplifier is needed in front of the inverter.

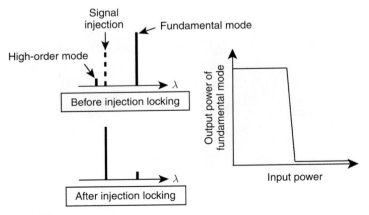

Figure 4.14 Principle of an optical inverter using an injection-locked two-mode VCSEL [69].

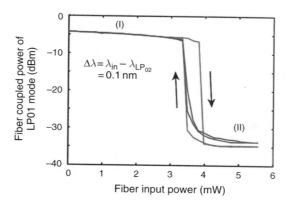

Figure 4.15 Optical input/output characteristics of the VCSEL with external light injected into the LP_{02} mode. The polarizations of the input light are (a) parallel and (b) orthogonal to that of the dominant lasing mode [69] (this figure may be seen in color on the included CD-ROM).

The obtained step-like transfer function would be useful for optical regeneration. A distorted signal of 1 Gb/s nonreturn-to-zero (NRZ) pseudorandom bit sequence with a word length of $2^7 - 1$ bit is injected into the VCSEL. The distorted input signal is successfully regenerated. The measured back-to-back BER before and after regeneration shows a receiver-sensitivity improvement of 1.2 dB at BER of 10^{-9} after the optical regeneration. The output waveform has been degraded by the relaxation oscillation of the VCSEL. Therefore, the switching speed of the VCSEL-based inverter is partly limited by the modulation bandwidth of the VCSEL. It could be improved to be beyond 10 Gb/s by using a VCSEL with over a 10 Gb/s modulation bandwidth, which was already demonstrated [41]. The VCSEL structure may provide us a possibility of low power consumption and polarization-insensitive operations for all-optical signal processing.

Fiber nonlinearities are dominant limiting factors for high-speed transmission systems of 40 Gb/s or beyond. Waveform distortion is induced by various fiber nonlinear effects such as SPM (self-phase modulation), XPM (cross-phase modulation), and FWM (four-wave mixing). The fiber nonlinear effects cannot be compensated by linear optical circuits in optical domain. If we realize a nonlinear optical compensator, which gives us a negative phase shift in an opposite sign of optical Kerr effect in fibers, we are able to compensate fiber nonlinearities by inserting the device. An optical nonlinear phase shifter based on a VCSEL structure with a saturable absorber was demonstrated [72].

Figure 4.16 shows the schematic structure and the operating principle of the optical phase shifter based on 1.55 μm VCSEL structure [41]. This device consists of dielectric multilayer structure (top DBR) and 40-pair InGaAlAs/InP (bottom DBR) and InGaAlAs QWs functioning as an absorber. An intensity-dependent negative refractive index change appears with an opposite sign of optical Kerr effect in the absorber, which is enhanced by a resonant vertical cavity. The

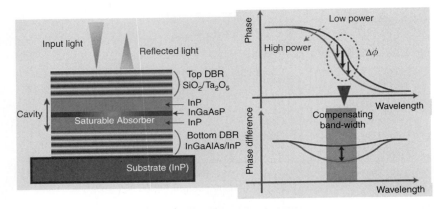

Figure 4.16 Schematic structure and operating principle of nonlinear-effect optical compensator using InGaAlAs vertical cavity absorber [41] (this figure may be seen in color on the included CD-ROM).

modeling result shows that either positive or negative phase shifts of reflected light can be obtained with 1.55-μm VCSEL, depending on the cavity Q-value [105]. Both positive and negative phase shifts are useful for the compensation of laser chirp and fiber nonlinearities in optical domain, respectively.

There is the trade-off between the phase shift and optical bandwidth. Therefore, we have to choose suitable design of mirrors for optimizing bandwidth and phase shift. An InGaAlAs absorber is sandwiched by the two mirrors. If the input light power coupled to this device increases, the phase difference is induced by the refractive index change in the absorber. The fabricated device consists of 3-pair SiO_2/Ta_2O_5 (top DBR), InGaAlAs/InP (bottom DBR with a reflectivity of >99.9%), and InGaAlAs QWs functioning as an absorber. The reflectivity is estimated to be 70%. The fabricated structure is not optimized in terms of the top mirror reflectivity. Also, no reverse bias gives slow recovery time of 1 ns range. The reflectivity and the group delay dependence on input power was measured by using an optical component analyzer (Advantest Q7761) with a tunable laser source. The phase shift was estimated from performing spectral domain integration of the measured group delay. Figure 4.17 shows the nonlinear phase shift from the data of 0.10 mW input power. The solid lines show the calculation. Here, we assume that saturation coefficient is 2 kW/cm², which corresponds to the case of $\tau = 1$ ns recovery time for an absorber without reverse bias. A positive group delay and negative phase shifts were observed as predicted in theory. We obtained a large negative phase shift of a −0.4 rad for input power of 0.42 mW, which is large enough for compensating 100-km-long fiber nonlinearities [105]. The reduction of absorption recovery time below 10 ps with reverse bias enables us to use the compensator for high bit-rate signals. The addition of 1.2 V reverse bias showed the transient response of nonlinear phase shifts even for 7 ps input pulses [41]. The novel nonlinear-effect compensator shows a large nonlinear negative phase shift depending on input

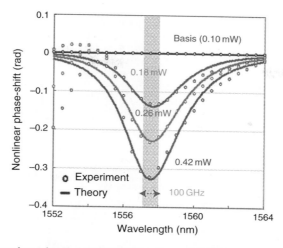

Figure 4.17 Measured wavelength dependence of nonlinear phase shift for different input power. The measured device is 1.55 μm InGaAlAs VCSEL without reverse bias [105] (this figure may be seen in color on the included CD-ROM).

power levels. The proposed concept may open up a novel technology for compensating fiber nonlinearities in optical domain.

4.6 VCSEL-BASED SLOW LIGHT DEVICES

The manipulation of the speed of light has been attracting much interest in recent years. In particular, slow light observed in photonic crystals, semiconductor amplifiers, and microresonators has been studied for optical buffer memories, optical delay lines, and so on [106–108]. Also, the slow group velocity of light dramatically reduces the size of various optical devices such as optical amplifiers, optical switches, and nonlinear optical devices [109, 110]. We have observed large waveguide dispersion and slow light [111] in Bragg waveguides where light is confined with highly reflective Bragg reflectors, [112]. We recently proposed a slow light modulator [8] with a Bragg waveguide, which predicts the possibility of low-modulation voltage (<1 V) even for ultracompact waveguide modulators (<20 μm). We expect the high-speed modulation of such an ultracompact waveguide modulator but reducing its parasitic capacitance.

We demonstrated an electroabsorption modulator consisting of Bragg waveguides with slow light enhancement as shown in Figure 4.18 [114]. The base structure is similar to that of a conventional InP-based VCSEL without tunnel junction [41]. The bottom mirror is AlGaInAs quarter-wavelength stack mirror. At first, 1.5-pairs Si/SiO_2 dielectric mirror was deposited over the entire surface except top electrodes and then a 5-pair Si/SiO_2 dielectric mirror was partly deposited to form a 20-μm-long Bragg waveguides. The role of 1.5-pair Si/SiO_2

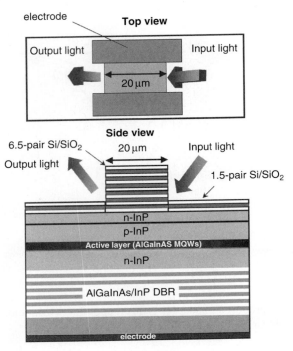

Figure 4.18 Schematic structure of a slow light AlGaInAs MQW electroabsorption modulator [104] (this figure may be seen in color on the included CD-ROM).

dielectric mirror is the efficient excitation of slow light in a Bragg waveguide. The absorption layer consists of AlGaInAs MQWs in a 1.5-µm wavelength band. By applying reverse bias voltages in its p–n junction, an electroabsorption takes place.

The group velocity decreases with increasing the waveguide dispersion when the wavelength approaches the cutoff wavelength. The slow-down factor, which is defined as the ratio of the group velocity of slow light vs that in conventional semiconductor waveguides, is over 10 in the wavelength range of 1550–1560 nm. Thus, the electroabsorption effect is enhanced by a factor of more than 10 in this wavelength range and we are able to reduce the size. Even for an ultracompact modulator, we expect an extinction ratio of 7 dB over 1550 nm, which will be large enough for short-reach optical links. We also expect low-polarization dependence, which is very difficult for that of photonic crystal slab waveguides. The important issue is how to couple with slow light in a Bragg waveguide. We proposed a simple and practical method of a tilt-coupling scheme as shown in Figure 4.19 [41]. The input beam is off from the vertical axis and the tilt angle is typically 30°. The coupling loss is less than 1.5 dB for TE and TM modes with a 4-µm–spot-size Gaussian beam input. This coupling scheme enables us to excite slow light propagating in a Bragg waveguide, where light is confined by Bragg mirrors. An extinction ratio of 6 dB and an insertion loss of 2 dB were realized for a

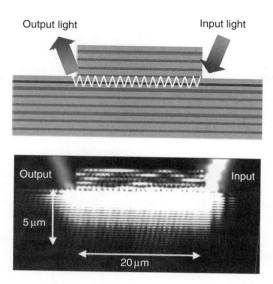

Figure 4.19 Model and calculated intensity distribution with tilt light input for efficient excitation of slow light [104] (this figure may be seen in color on the included CD-ROM).

20-μm-long compact waveguide modulator. The proposed structure can be monolithically integrated with VCSELs. The proposed modulator would be useful for ultrahigh speed short-reach optical links. In addition, simple coupling scheme with slow light in Bragg waveguides would also be useful for slow light photonic circuits involving optical switches, amplifiers, lasers, and so on.

4.7 CONCLUSION

Recent advances on VCSEL photonics were reviewed, including the wavelength engineering and new functions of VCSELs. The small footprint of VCSELs allowed us to form a densely packed VCSEL array both in space and in wavelength. The wavelength engineering of VCSELs may open up ultrahigh capacity networking. In addition, new functions of VCSELs for optical signal processing were addressed, which include optical regenerators based on an injection-locked VCSEL. Also, a novel nonlinear optical phase shifter enables us to compensate fiber nonlinearities in optical domain.

It has already been 30 years since a VCSEL was invented. The emission wavelength is from ultraviolet to infrared spectral regions. We have seen various potential applications including datacom, sensors, optical interconnects, spectroscopy, optical storages, printers, laser displays, laser radar, atomic clock, optical signal processing, and so on. On the other hand, VCSELs with external cavities, so-called VECSELs [115], allow a large emitting area of the device with a single-transverse mode and thus much high power of over 1 W can be obtained. The external cavity configuration also allows intracavity frequency doubling, which are currently attracting much interest for green light emitters of laser display applications [116]. These high-power applications are also challenging for future optoelectronics. Another interesting application is the spontaneous emission in microcavities. Microcavity VCSELs with QDs would be one of the good candidates for single-photon emitters [117].

REFERENCES

[1] K. Iga, "Surface-emitting laser-its birth and generation of new optoelectronics field," *IEEE J. Select. Topics Quantum Electron.*, 6(6), 1201–1215, December 2000.
[2] K. Iga, "Surface emitting laser," *Trans. IEICE*, C-I, J81-C-1(9), 483–493, September 1998.
[3] H. Soda, K. Iga, C. Kitahara, and Y. Suematsu, "GaInAsP/InP surface emitting injection lasers," *Jpn. J. Appl. Phys.*, 18, 2329–2330, December 1979.
[4] K. Iga, S. Kinoshita, and F. Koyama, "Microcavity GaAlAs/GaAs surface-emitting laser with I = 6 mA," *Electron. Lett.*, 23(3), 134–136, January 1987.
[5] F. Koyama, S. Kinoshita, and K. Iga, "Room-temperature continuouswave lasing characteristics of GaAs vertical cavity surface-emitting laser," *Appl. Phys. Lett.*, 55(3), 221–222, July 1989.
[6] J. L. Jewell, A. Scherer, S. L. McCall et al., "Low-threshold electrically pumped vertical-cavity surface-emitting microlasers," *Electron. Lett.*, 25(17), 1123–1124, August 1989.
[7] R. S. Geels, S. W. Corzine, and L. A. Coldren, "InGaAs vertical-cavity surface-emitting lasers," *IEEE J. Quantum Electron.*, 27(6), 1359–1367, June 1991.
[8] D. L. Huffaker, D. G. Deppe, K. Kumar, and T. J. Rogers, "Native-oxide defined ring contact for low threshold vertical-cavity lasers," *Appl. Phys. Lett.*, 65, 97–99, 1994.
[9] K. D. Choquette, K. M. Geib, C. I. H. Ashby et al., "Advances in selective wet oxidation of AlGaAs alloys," *IEEE J. Sel. Top. Quantum Electron*, 3(3), 916–926, June 1997.
[10] Y. Hayashi, T. Mukaihara, N. Hatori et al., "Lasing characteristics of low-threshold oxide confinement InGaAs-GaAlAs vertical-cavity surface-emitting lasers," *IEEE Photon. Technol. Lett.*, 7(11), 1234–1236, November 1995.
[11] K. L. Lear, K. D. Choquette, R. P. Schneider, Jr, et al., "Selectively oxidized vertical-cavity surface emitting lasers with 50% power conversion efficiency," *Electron. Lett.*, 31, 208–209, 1995.
[12] R. Jüger, M. Grabherr, C. Jung et al., "57%wallplug efficiency oxide-confined 850 nm wavelength GaAs VCSELs," *Electron. Lett.*, 33, 330–331, 1997.
[13] D. Schlenker, T. Miyamoto, Z. Chen et al., "1.17 µm highly strained GaInAs-GaAs quantum-well laser," *IEEE Photon. Technol. Lett.*, 11(8), 946–948, August 1999.
[14] F. Koyama, D. Schlenker, T. Miyamoto et al., "1.2 mm highly strained GaInAs/GaAs quantum well lasers for singlemode fiber datalink," *Electron. Lett.*, 35(13), 1079–1081, June 1999.
[15] N. Nishiyama, M. Arai, S. Shinada et al., "Highly strained GaInAs/GaAs quantum well verticalcavity surface-emitting laser on GaAs (311)B substrate for stable polarization operation," *IEEE J. Sel. Top. Quantum Electron.*, 7(2), 242–248, May 2001.
[16] C. Asplund, P. Sundgren, S. Mogg et al., "1260 nm InGaAs vertical-cavity lasers," *Electron. Lett.*, 38(13), 635–636, 2002.
[17] H. C. Kuo, Y. H. Chang, H. H. Yao et al., "High-speed modulation of InGaAs: Sb-GaAs-GaAsP quantum-well vertical-cavity surface-emitting lasers with 1.27 µm emission wavelength," 17(3), 528–530, 2004.
[18] R. M. von Wotemberg, P. Sundgren, J. Berggren et al., "1.3 µm InGaAs vertical-cavity surfaceemitting lasers with mode filter for single mode operation," *Appl. Phys. Lett.*, 85, 4851–4853, 2004.

[19] E. Pougeoise, P. Gilet, P. Grosse et al., "Strained InGaAs quantum well vertical cavity surface emitting lasers emitting at 1.3 μm," *Electron. Lett.*, 42(10), 584–586, 2006.
[20] R. P. Mirin, J. P. Ibbetson, K. Nishi et al., "1.3 μm photoluminescence from InGaAs quantum dots on GaAs," *Appl. Phys. Lett.*, 67(25), 3795–3797, 1995.
[21] J. A. Lott, N. N. Ledentsov, V. M. Ustinov et al., "InAs–InGaAs quantum dot VCSEL's on GaAs substrates emitting at 1.3 μm," *Electron. Lett.*, 36, 1384–1385, 2000.
[22] T. Anan, M. Yamada, K. Nishi et al., "Continuous-wave operation of 1.30 μm GaAsSb/GaAs VCSELs," *Electron. Lett.*, 37(9), 566–567, 2001.
[23] D. C. Kilper, F. Quochi, J. E. Cunningham, and M. Dinu, "High-speed dynamics of GaAsSb vertical-cavity lasers," *IEEE Photon. Technol. Lett.*, 14(4), 438–440, 2002.
[24] P. Dowd, S. R. Johnson, S. A. Feld et al., "Long wavelength GaAsP/GaAs/GaAsSb VCSELs on GaAs substrates for communications applications," *Electron. Lett.*, 39(13), 987–988, 2003.
[25] M. Kondow, K. Uomi, A. Niwa et al., "GaInNAs: A novel material for long-wavelength-range laser diodes with excellent high-temperature performance," *Jpn. J. Appl. Phys.*, 35, 1273–1275, 1996.
[26] K. D. Choquette, J. F. Klem, A. J. Fischer et al., "Room temperature continuous wave InGaAsN quantum well vertical-cavity lasers emitting at 1.3 μm," *Electron. Lett.*, 36, 1388–1390, 2000.
[27] N. Nishiyama, S. Sato, T. Miyamoto et al., "First CW operation of 1.26 μm electrically pumped MOCVD grown GaInNAs/GaAs VCSEL," presented at *IEEE ISLC 2000*, Postdeadline paper PD-1, Monterey, CA, 2000.
[28] B. Borchert, A. Y. Egorov, S. Illek et al., "1.29 μm GaInNAs multiple quantum-well ridge-waveguide laser diodes with improved performance," *Electron. Lett.*, 35, 2204–2206, 1999.
[29] T. Kageyama, T. Miyamoto, S. Makino et al., "High-temperature operation up to 170°C of GaInNAs-GaAs quantum-well lasers grown by chemical beam epitaxy," *IEEE Photon. Technol. Lett.*, 12, 10–12, 2000.
[30] A. W. Jackson, R. L. Naone, M. J. Dalberth et al., "OC-48 capable InGaAsN vertical cavity lasers," 37(6), 355–356, 2001.
[31] A. Ramakrishnan, G. Steinle, D. Supper et al., "Electrically pumped 10 Gbit/s MOVPE-grown monolithic 1.3 μm VCSEL with GaInNAs active region," *Electron. Lett.*, 38(7), 322–324, 2002.
[32] T. Takeuchi, Y.-L. Chang, M. Leary et al., "1.3 μm InGaAsN vertical cavity surface emitting lasers grown by MOCVD," *Electron. Lett.*, 38(23), 1438–1440, 2002.
[33] R.-S. Hsiao, J.-S. Wang, K.-F. Lin et al., "Single mode 1.3 μm InGaAsN/GaAs quantum well vertical cavity surface emitting lasers grown by molecular beam epitaxy," *Jpn J. Appl. Phys.*, 43(12A), L1555–L1557, 2004.
[34] T. Nishida, M. Takaya, S. Kakinuma, and T. Kaneko, "4.2-mW GaInNAs long-wavelength VCSEL grown by metalorganic chemical vapor deposition," *IEEE J. Sel. Top. Quantum Electron.*, 11(5), 958–961, 2005.
[35] J. Jewell, L. Graham, M. Crom et al., "1310 nm VCSELs in 1–10 Gb/s Commercial Applications," in Proc. *SPIE*, 6132, 2006, p. 613204.
[36] M. A. Wistey, S. R. Bank, H. P. Bae et al., "GaInNAsSb/GaAs vertical cavity surface emitting lasers at 1534 nm," *Electron. Lett.*, 42(5), 282–283, 2006.
[37] A. Karim, P. Abraham, D. Lofgreen et al., "Wafer bonded 1.55 μm vertical-cavity lasers with continuous-wave operation up to 105°C," *Appl. Phys. Lett.*, 78, 2632–2634, 2001.
[38] V. Jayaraman, M. Mehta, A. W. Jackson et al., "High-power 1320 nm wafer bonded VCSELs with tunnel junction," *IEEE Photon. Technol. Lett.*, 15(11), 1495–1497, November 2003.
[39] C.-K. Lin, D. Bour, J. Zhu et al., "High temperature continuous-wave operation of 1.3–1.55 μm VCSELs with InP/air-gap DBRs," *IEEE J. Sel. Top. Quantum Electron.*, 9(5), 1415–1421, September/October 2003.
[40] M. Ortsiefer, S. Baydar, K. Windhorn et al., "Long-wavelength monolithic VCSEL arrays with high optical output power," 41(14), 807–808, 2005.

[41] N. Nishiyama, C. Caneau, B. Hall et al., "Long wavelength vertical cavity surface emitting lasers on InP with lattice matched AlGaInAs/InP DBR grown by MOCVD," *IEEE J. Sel. Top. Quantum Electron.*, 11(5), 990–998, 2005.

[42] J. Cheng, C.-L. Shieh, X. Huang et al., "Efficient CW lasing and high-speed modulation of 1.3 μm AlGaInAs VCSELs with good high temperature lasing performance," *IEEE Photon. Technol. Lett.*, 17(1), 7–9, 2005.

[43] M. V. R. Murty, X. D. Huang, G. L. Liu et al., "Long-wavelength VCSEL-based CWDM scheme for 10-GbE links," *IEEE Photon. Technol. Lett.*, 17(6), 1286–1288, 2005.

[44] V. Iakovlev, G. Suruceanu, A. Caliman et al., "High-performance single-mode VCSELs in the 1310-nm waveband," *IEEE Photon. Technol. Lett.*, 17(5), 947–949, 2005.

[45] D. Feezell, D. A. Buell, and L. A. Coldren, "InP-based 1.3–1.6 μm VCSELs with selectively etched tunnel-junction apertures on a wavelength flexible platform," *IEEE Photon. Technol. Lett.*, 17(10), 2017–2019, 2005.

[46] J. Boucart, G. Suruceanu, P. Royo et al., "3.125-Gb/s modulation up to 70 deg/C using 1.3 μm VCSELs fabricated with localized wafer fusion for 10GBASE LX4 applications," *IEEE Photon. Technol. Lett.*, 18(4), 571–573, 2006.

[47] W. Hofmann, N. H. Zhu, M. Ortsiefer et al., "10-Gb/s data transmission using BCB passivated 1.55 μm InGaAlAs-InP VCSELs," *IEEE Photon. Technol. Lett*, 18(2), 424–426, 2006.

[48] A. N. Al-Omari and K. L. Lear, "Polyimide-planarized vertical-cavity surface-emitting lasers with 17.0-GHz bandwidth," *IEEE Photon. Technol. Lett.*, 16(4), 969–971, April 2004.

[49] P. Pepeljugoski, D. Kuchta, Y. Kwark et al., "15.6-Gb/s transmission over 1 km of next generation multimode fiber," *IEEE Photon. Technol. Lett.*, 14(5), 717–719, May 2004.

[50] D. M. Kuchta, P. Pepeljugoski, and Y. Kwark, "VCSEL Modulation at 20 Gb/s over 200 m of Multimode Fiber using a 3.3 V SiGe Laser Driver IC," *LEOS Summer Topical Meeting 2001*, paper WA1.2, Copper Mountain, CO, July/August 2001.

[51] K. Fukatsu, K. Shiba, Y. Suzuki et al., "30-Gbps Transmission over 100 m-MMFs using 1.1 μm-range VCSELs and Receivers," 19th *IPRM*, WeB3-4, Matsue, May 2007.

[52] C. J. Chang-Hasnain, J. P. Harbison, C. E. Zah et al., "Multiple wavelength tunable surface-emitting laser arrays," *IEEE J. Quantum Electron.*, 27(6), 1368–1376, 1991.

[53] F. Koyama, T. Mukaihara, Y. Hayashi et al., "Wavelength control of vertical cavity surface-emitting lasers by using nonplanar MOCVD," *IEEE Photon. Tech. Lett.*, 7(1), 10–12, 1995.

[54] L. E. Eng, K. Bacher, Y. Wupen et al., "Multiple-wavelength vertical cavity laser arrays on patterned substrates," *IEEE J. Sel. Top. Quantum Electron.*, 1(1), 624–628, 1995.

[55] G. G. Oritz, S. Q. Luong, S. Z. Sun et al., "Monolithic, multiple-wavelength vertical-cavity surface-emitting laser arrays by surface-controlled MOCVD growth rate enhancement and reduction," *IEEE Photon. Technol. Lett.*, 9(8), 1069–1071, 1997.

[56] K. Yang, Y. Zhou, X. D. Huang et al., "Monolithic oxide-confined multiple-wavelength vertical-cavity surface-emitting laser arrays with a 57-nm wavelength grading range using an oxidized upper Bragg mirror," *IEEE Photon. Technol. Lett.*, 10(4), 377–379, 2000.

[57] J. K. Kash, F. E. Doany, L. Schares et al., "Chip-to-Chip Optical Interconnects," in Proc. *OFC 2006*, paper OFA3, 2006.

[58] C. J. Chang-Hasnain, "Tunable VCSEL," *IEEE Sel. Top. Quantum Electron.*, 6, 978–987, 2000.

[59] M. C. Larson, B. Pezeshki, and J. S. Harris, Jr, "Broadly-tunable resonant-cavity light-emitting diode," *IEEE Photon. Technol. Lett.*, 7(11), 382–384, 1995.

[60] M. S. Wu, E. C. Vail, G. S. Li et al., "Tunable micromachined vertical cavity surface emitting laser," *Electron. Lett.*, 31, 1671–1672, 1995.

[61] A. Syrbu, V. Iakovlev, G. Suruceanu et al., "1.55 μm optically pumped wafer-fused tunable VCSELs with 32-nm tuning range," *IEEE Photon Technol. Lett.*, 16(9), 1991–1993, 2004.

[62] M. Maute, B. Kogel, G. Bohm et al., "MEMS-tunable 1.55 μm VCSEL with extended tuning range incorporating a buried tunnel junction," *IEEE Photon Technol. Lett.*, 18(5), 688–690, 2006.

[63] C. H. Chang, L. Chrostowski, and C. J. Chang-Hasnain, "Injection locking of VCSELs," *IEEE J. Sel. Top. Quantum Electron.*, 9, 1386–1393, September/October 2003.

[64] J. Y. Law, G.H.M. van Tartwijk, and G. P. Agrawal, "Effects of transverse-mode competition on the injection dynamics of vertical-cavity surface-emitting lasers," *Quantum Semiclassic. Opt.*, 9, 737–747, October 1997.
[65] Y. Hong, P. S. Spencer, P. Rees, and K. A. Shore, "Optical injection dynamics of two-mode vertical cavity surface-emitting semiconductor lasers," *IEEE J. Quantum Electron.*, 38, 274–278, March 2002.
[66] A. Valle, L. Pesquera, S. I. Turovets, and J. M. López, "Nonlinear dynamics of current-modulated vertical-cavity surface-emitting lasers," *Opt. Commun.*, 208, 173–182, July 2002.
[67] H. Kawaguchi, Y. Yamayoshi, and K. Tamura, "All-Optical Format Conversion using an Ultrafast Polarization Bistable Vertical-Cavity Surface-Emitting Laser," in Proc. *CLEO 2000*, CWU2, 379–380, 2000.
[68] F. Brown de Colstoun, G. Khitrova, A. V. Fedorov et al., "Transverse mode, vortices and vertical-cavity surface-emitting lasers," *Chaos Solitons Fractals*, 4, 1575–1596, August/September 1994.
[69] Y. Onishi, N. Nishiyama, C. Caneau et al., "Dynamic behavior of an all-optical inverter using transverse-mode switching in 1.55-μm vertical-cavity surface-emitting lasers," *IEEE Photon. Technol. Lett.*, 16, 1236–1238, May 2004.
[70] K. Hasebe, F. Koyama, N. Nishiyama et al., "All-Optical Polarization Controller Using Elliptical-Apertured 1.5 μm VCSEL," *CLEO 2006*, paper CWP1, May 2006.
[71] X. Zhao, P. Palinginis, B. Pesala et al., "Room Temperature Tunable Ultraslow Light in 1550 nm VCSEL Amplifier," in Proc. *ECOC 2005*, postdeadline paper, Th4.3.6, 2005.
[72] S. Suda, F. Koyama, N. Nishiyama et.al., "Optical Phase shifter using Vertical Microcavity with Saturable Absorber," *CLEO 2006*, paper CWK3, May 2006.
[73] Z. Pan, T. Miyamoto, D. Schlenker et al., "Low temperature growth of GaInNAs/GaAs quantum wells by metalorganic chemical vapor deposition using tertiarybutylarsine," *J. Appl. Phys.*, 84(11), 6409–6411, December 1998.
[74] D. Schlenker, T. Miyamoto, Z. Chen et al., "Growth of highly strained GaInAs/GaAs quantum wells for 1.2 μm wavelength lasers," *J. Cryst. Growth*, 209, 27–36, January 2000.
[75] N. Nishiyama, M. Arai, S. Shinada et al., "Growth and optical properties of highly strained GaInAs/GaAs quantum wells on (311)B GaAs by MOCVD," *J. Cryst. Growth*, 221, 530–534, 2000.
[76] Z. Chen, D. Schlenker, T. Miyamoto et al., "High temperature characteristics of near 1.2 μm InGaAs/AlGaAs lasers," *Jpn. J. Appl. Phys.*, 38(10B), L1178–1179, October 1999.
[77] T. Kondo, M. Arai, M. Azuchi et al., "Low threshold current density operation of 1.16 μm highly strained GaInAs/GaAs vertical cavity surface emitting lasers on (100) GaAs substrate," *Jpn. J. Appl. Phys.*, 41(5B), L562–L564, May 2002.
[78] T. Kondo, M. Arai, T. Miyamoto, and F. Koyama, "Higly strained GaInAs/GaAs 1.13 μm vertical cavity surface emitting laser with uncooled single mode operation, "The 16th Annual Meeting of the IEEE LEOS 2003," Tucson, USA, WD2, October 25–30, 2003.
[79] T. Kondo, M. Arai, M. Azuchi et al., "Singlemode fibre transmission using 1.2 μm band GaInAs/GaAs surface emitting laser," *Electron. Lett.*, 38(16), 901–903, August 2002.
[80] J. Geske, V. Jayaraman, T. Goodwin et al., "2.5-Gb/s transmission over 50 km with a 1.3-m vertical-cavity surface-emitting laser," *IEEE Photon. Technol. Lett.*, 12(12), 1707–1709, December 2000.
[81] T. Kondo, M. Arai, T. Miyamoto, and F. Koyama, "Isolator-free 10Gb/s singlemode fiber date transmission using 1.1 μm GaInAs/GaAs vertical cavity surface emitting laser," *Electron. Lett.*, 40(1), 65–66, 2004.
[82] M. Ariga, M. Arai, T. Kageyama et al., "Noise characteristics of GaInNAsSb 1300-nm-range VCSEL with optical feedback for isolator-free module," *IEEE J. Sel. Top. Quantum Electron.*, 11(5), 1074–1078, September/October 2005.
[83] L. Chrostowski, P. B. Subrahmanyam, Y. Zhou, and C. J. Chang-Hasnain, "VCSEL Tolerance to Optical Feedback for Inter-Chip Optical Interconnects," *IEEE ISLC.*, paper FA3, Shimane, Japan, 2004.

[84] N. Nishiyama, C. Caneau, S. Tsuda et al., "10-Gb/s error-free transmission under optical reflection using isolator-free 1.3 μm InP-based vertical-cavity surface-emitting lasers," *IEEE Photon. Technol. Lett.*, 17(8), 1605–1607, 2005.

[85] L. Buydens, P. Demeester, M. V. Ackere et al., "Thickness variations during MOVPE growth on patterned substrates," *J. Electron. Mater.*, 19, 317–321, 1990.

[86] A. Onomura, M. Arai, T. Kondo etal., "Densely integrated multiple-wavelength vertical-cavity surface-emitting laser array," *Jpn. J. Appl. Phys.*, 42(5B), L529–L531, May 2003.

[87] T. Fujii and M. Ekawa, "Origin of compositional modulation of InGaAs in selective area metalorganic vapor phase epitaxy," *J. Appl. Phys*, 78(9), November 1995.

[88] G. B. Stringfellow, *Organometallic Vapor-Phase Epitaxy*, New York, Academic Press, 1989.

[89] M. Arai, T. Kondo, A. Onumura et al., "Multiple-wavelength GaInAs/GaAs vertical cavity surface emitting laser array with extended wavelength span," *IEEE J. Sel. Top. Quantum Electron.*, 9(5), 1367–1373, September/October 2003.

[90] Y. Uchiyama, T. Kondo, K. Takeda et al., "Electro-Thermal Wavelength Tuning of 1.2 μm GaInAs/GaAs Vertical Cavity Surface Emitting Laser Array," The 18th Annual Meeting of the *IEEE* in LEOS 2005, TuS2, Sydeny, Australia, October 23–27, 2005, pp. 326–327.

[91] Y. Uchiyama, T. Kondo, K. Takeda et al., "1.2 μm band GaInAs/GaAs high-density multiple-wavelength vertical cavity surface emitting laser array," *Jpn. J. Appl. Phys.*, 44(6), L214–L215, February 2005.

[92] Y. Uchiyama, T. Kondo, K. Takeda et al., "Thermal cross-talk evaluation of densely integrated vertical cavity surface emitting laser array," *IEICE Electronics Express*, 1(17), 545–550, 2004.

[93] F. Koyama and K. Iga, "Wavelength Stabilization and Trimming Technologies for Vertical-Cavity Surface Emitting Lasers," *Quantum Optoelectronics of 1997 OSA Spring Topical Meeting, OSA Spring Topical '97*, 9, QTh-14, 1997, pp. 90–92.

[94] F. Koyama, T. Amano, N. Furukawa et al., "Micromachined semiconductor vertical cavity for temperature insensitive surface emitting lasers and optical filters," *Jpn. J. Appl. Phys.*, 39(3B), 1542–1545, 2000.

[95] T. Amano, F. Koyama, M. Arai, and A. Matsutani, "Micromachined GaAs/AlGaAs resonant-cavity light emitter with small temperature dependence of emission wavelength," *Jpn. J. Appl. Phys.*, 42(11B), L1377–L1379, November 2003.

[96] W. Janto, K. Hasebe, N. Nishiyama et al., "Athermal Operation of 1.55 μm InP-Based VCSEL with Thermally-Actuated Cantilever Structure," *IEEE ISLC*, PD1.1, Hawaii, 2006.

[97] M. Ohtsu, *Near-Field Nano/Atom Optics and Technology*, Tokyo, Springer-Verlag, 1998.

[98] A. Partovi, D. Peale, M. Wuttig et al., "High-power laser light source for near-field optics and its application to high-density optical data storage," *Appl. Phys. Lett.*, 75, 1515–1517, 1999.

[99] K. Goto, "Proposal of ultrahigh density optical disk system using a vertical cavity surface emitting laser array," *Jpn. J. Appl. Phys.*, 37, 2274–2278, 1998.

[100] S. Shinada, F. Koyama, N. Nishiyama et al., "Analysis and fabrication of micro-aperture GaAs/GaAlAs surface emitting laser for near field optical data storage," *IEEE J. Sel. Top. Quantum Electron.*, 7(2), 365–370, 2001.

[101] J. Hashizume and F. Koyama, "Plasmon-enhancement of optical near-field of metal nano-aperture surface emitting laser," *Appl. Phys. Lett.*, 84(17), 3226–3228, 2004.

[102] J. Hashizume and F. Koyama, "Plasmon enhanced optical near-field probing of metal nano-aperture surface emitting laser," *Opt. Express*, 12(25), 6391–6396, 2004.

[103] T. Thio, H. F. Ghaemi, H. J. Lezec et al., "Surface-plasmon-enhanced transmission through hole arrays in Cr films," *J. Opt. Soc. Am. B*, 16, 1743–1748, 1999.

[104] T. Thio, K. M. Pellerin, R. A. Linke et al., "Enhanced light transmission through a single subwavelength aperture," *Opt. Lett.*, 26, 1972–1974, 2001.

[105] S. Suda, F. Koyama, N. Nishiyama et al., "Optical Nonlinear-Effect Compensator Based on Vertical-cavity Saturable Absorber," in Proc. *ECOC 2006*, 2006.

[106] M. Notomi, K. Yamada, A. Shinya et al., "Extremely large group-velocity dispersion of line-defect waveguides in photonic crystal slabs," *Phys. Rev. Lett.*, 87, 235902, 2001.

[107] A. Yariv, Y. Xu, R. K. Lee, and A. Scherer, "Coupled-resonator optical waveguide: a proposal and analysis," *Opt. Lett.*, 24, 711713, 1999.
[108] X. Zhao, P. Palinginis, B. Pesala et al., "Room Temperature Tunable Ultraslow Light in 1550 nm VCSEL Amplifier," in Proc. *ECOC 2005*, postdeadline paper, Th4.3.6, 2005.
[109] M. Soljacic, S. G. Johnson, S. Fan et al., "Photonic-crystal slow-light enhancement of nonlinear phase sensitivity," *J. Opt. Soc. Am. B*, 19, 2052, 2002.
[110] E. Mizuta, H. Watanabe, and T. Baba, "All semiconductor low-photonic crystal waveguide for semiconductor optical amplifier," *Jpn. J. Appl. Phys.*, 45(8A), 6116–6120, 2006.
[111] Y. Sakurai and F. Koyama, "Control of group delay and chromatic dispersion in tunable hollow waveguide with highly reflective mirrors," *Jpn. J. Appl. Phys.*, 43(8B), 5828–5831, 2004.
[112] P. Yeh and A. Yariv, "Theory of Bragg fiber," *J. Opt. Soc. Am*, 68(9), 1196–1201, 1978.
[113] K. Kuroki and F. Koyama, "Propasal and Modeling of miniature Slow Light Modulator Integrated with VCSEL," 12th *MOC*, J-1, Seoul (Korea), 2006, pp. 222–223.
[114] G. Hirano, F. Koyama, K. Hasebe et al., "Slow Light Modulator with Bragg Reflector Waveguide," in proc. *OFC 2007*, postdeadline paper, PDP34, Anaheim, 2007.
[115] M. Kuznetsov, F. Hakimi, R. Sprague, and A. Mooradian, "Design and characteristics of high-power (>0.5-W CW) diode-pumped vertical-external-cavity surface-emitting semiconductor lasers with circular TEM00Beams," *IEEE J. Sel. Top. Quantum Electron.*, 5(3), 561–573, 1999.
[116] V. Shchegrov et al., "532-nm Laser Sources Based on Intracavity Frequency Doubling of Extended-Cavity Surface-Emitting Diode Lasers," in Proc. *SPIE*, 5332, 151–158, 2004.
[117] Y. Yamamoto and R. E. Slusher, "Optical processes in microcavities," *Phys. Today*, 46(6), 66–73, 1993.

5

Pump diode lasers

Christoph Harder

HPP, Etzelstrasse 58, Schindellegi, Switzerland

5.1 INTRODUCTION

The basis of optical fiber amplifiers is the pump diode laser, their optical power supply. The availability of cheap and reliable pump diode lasers enabled optical amplifiers, in the form of erbium-doped fiber amplifiers (EDFA), to conquer, the long-distance communication market [1].

Pump diode lasers are also enablers of fiber lasers (FL) [2] for material processing, ytterbium/erbium co-doped fiber amplifiers (YEDFA) for TV and fiber to the home (FTTH) distribution, and, more recently, direct diodes (DD) for printing and material processing. Pump lasers are also serving as versatile optical power supplies for active optics.

Nonlinear processes require high-power diode lasers, like pump diode lasers, but with severe additional specific requirements on the amplitude and phase noise. High-power laser diodes are enabling fiber amplifiers based on the Raman effect [3] as well as visible light generation in periodically poled lithium niobate (PPLN) waveguides and crystals [4].

5.1.1 Optical Power Supply

Pump diode lasers, being optical power supplies, have to perform with respect to transferred power efficiency and reliability. Thus, the pump diode has to be matched to the etendue (E) and absorption spectrum of the optical load, has to minimize the (parasitic) resistance to the electrical power supply, and has to allow for efficient removal of the dissipated heat. To meet the required low-failure rate in the field, an exact understanding of the reliability is also essential.

Pump diode lasers serve many different markets and are based on a variety of technologies. Here, we concentrate on fiber-coupled pump lasers in the 9xx nm (900–1100 nm) and the 14xx nm (1400–1500 nm) band. We also include fiber-coupled 14xx nm high-power diode lasers for Raman amplification and 9xx nm high-power diode lasers for consumer application.

There is a huge literature, even within this limited scope, of fiber-coupled devices in the two wavelength bands. We concentrate on referring to the origin of innovations and progress in the last few years, quoting articles as well as patents.

5.1.2 Telecom Optical Amplifiers

Until 5 years ago, the development of high-power pump and diode lasers was mainly driven by the demands of the telecommunication industry. Since the last review in this series of books [5], investments from the telecom industry have been dramatically reduced. The limited resources have been concentrated on scaling down fabs (while not losing the recipe) and on device cost reduction rather than performance improvement. Consequently, progress has slowed down; the field of pump diodes has matured. Section 5.2 is dedicated to 980-nm pump lasers and Section 5.3 to 1480-nm pump lasers and 14xx nm high-power diode lasers.

5.1.3 Power Photonics

Even before telecom applications, pump diode lasers have been used for diode-pumped solid-state lasers, (DPSSLs), mostly for 808-nm pumping of Nd-YAG 1060 nm rod lasers. The 808 nm pump diodes for low-power densities and low cost were developed to compete against the established pump technology (tungsten lamp) and to adapt to the system characteristics (high-pump photon-defect heating, nonhermetic, large etendue pump input).

With the demonstration of a diffraction-limited FL exceeding 100 W [6], based on "all fiber" telecom technology, the ground was prepared for converging telecom and industrial pump technologies. This technology, often called "Power Photonics," is all fiber based, all solid state, highly efficient, reliable, and completely free of open surfaces. The development has been spectacular, as demonstrated with the commercial availability in summer 2007 of a continuous-wave (cw) 3-kW diffraction-limited FL, all fiber based and powered up with multimode (MM)-pigtailed pumps. This progress is driven by *push* from telecom technology, looking to find new markets, and by *pull* from the industrial side and technology programs (BRIOLAS) [7], SHEDS and ADHELS [8].

In Section 5.4 we review the status of MM pigtailed pumps, as they are used for cladding pumping FLs, YEDFAs, and for DD applications. Section 5.5 is dedicated to the topic of increasing radiance by combining diode lasers, and Section 5.6 is a short review on pump vertical-cavity surface-emitting laser diodes (VCSELs).

5.2 SINGLE-MODE FIBER 980-NM PUMPS

Today, C and L band EDFAs are preferably pumped at 980 nm, resulting in lower noise figure, lower overall power consumption [9], and providing ample power since the availability of reliable high-power, Peltier-cooled 980-nm pump modules.

Undersea intercontinental links, based on dense wavelength-division multiplexing (DWDM), impose the most demanding requirements on single-mode (SM) fiber 980-nm pump diodes. For the undersea application, highly efficient and reliable pump modules operating over a wide temperature range (without Peltier temperature stabilization) had to be developed (Figure 5.1).

5.2.1 Materials for 980-nm Pump Diodes

Initially, 980-nm diode lasers for EDFAs were not available and DWDM systems were using 1480-nm pumps. No lattice-matched material on GaAs or InP substrate

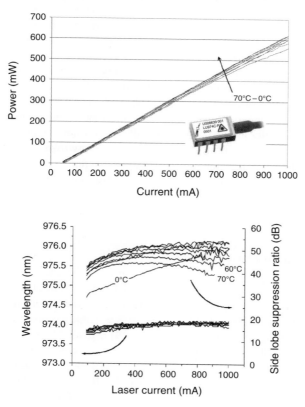

Figure 5.1 Performance of an uncooled 980 nm MiniDIL for undersea systems (courtesy Bookham) (this figure may be seen in color on the included CD-ROM).

with the corresponding bandgap is available. In the early 1980s, strain was considered to lead to unreliable devices. Still, work was done on strained InGaAs quantum wells (QWs) on GaAs substrate to investigate the possibility of threshold reduction through strain-induced changes in the density of states. It was soon discovered that adding indium improved the stability of GaAs [10–13]. Today, the benefits of strained InGaAs QW for high-power diodes from 900 to 1100 nm (often categorized as "9xx nm range"), fabricated by either metalorganic vapour phase epitaxy (MOVPE) or molecular-beam epitaxy (MBE), are well understood. In addition, different technologies have been developed to suppress the catastrophic degradation mode at the facets.

The impressive performance of pump diode lasers is based on the gain characteristics resulting from the steep density of states of the strained InGaAs QWs [14, 15] and the low-optical confinement factor in a low-loss large optical cavity (LOC). An active layer with quantum dots (QDs) should have an even steeper density of states and thus improve the performance. After 20 years of work on QD fabrication technology, one is still searching for a technology to produce InGaAs QD with the required uniformity. Unfortunately, due to inhomogeneous broadening, no improved device performance with respect to high-power lasers can be observed [16, 17].

The power conversion efficiency of a pump diode laser depends on many factors and can be improved by a clever design, but it is ultimately limited by material parameters such as the ratio of mobility and free-carrier absorption of holes in the p-type waveguide and, in addition, by the thermal resistance of the p-type waveguide and of the cladding. To look for the best material system, InGaAsP on GaAs substrate was investigated, and 980-nm pump diodes were fabricated with excellent characteristics [18, 19]. The InGaAsP system is also less reactive to oxygen. Surface passivation is eased (but still necessary), and overgrown structures [20] can be produced. Detailed investigations [21] have shown that the AlGaAs material system is superior with respect to mobility and thermal resistance in p-type waveguide and cladding (free-carrier absorption limits are not known yet).

5.2.2 Optical Beam

Pump modules have either a SM (HI 1060, HI 980) or polarization-maintaining (PM) (Panda PM 980 family) fiber pigtail with a numerical aperture (NA) between 0.12 and 0.2. No optical isolator is required; it is possible to couple directly into the fiber. A large variety of lensed fiber tips [22–24] is available to match various SM laser beams to a SM fiber. However, it is much more cost effective to design the diode laser beam for coupling into a simple wedge-shaped fiber as developed more than 30 years ago [25, 26] and refined later [27–29]. The diode laser fundamental mode needs to be designed to have the same slow-axis NA (or, equivalently, mode field diameter) as the fiber. Beam aspect ratio or fast-axis NA is of no concern, as the wedge-shaped fiber tip can adapt for fast-axis divergence.

Narrow Stripe Technology

To match the slow-axis NA of the diode to the NA of the fiber, a weak lateral waveguide is necessary. The required index difference, given by the well-known approximation $\Delta n = \mathrm{NA}^2/(2 \times n)$, is small: $\Delta n = 2 \times 10^{-3}$ and $\Delta n = 5 \times 10^{-3}$ for fiber NAs of 0.12 and 0.2, respectively. Such weak waveguides can be fabricated in various technologies: impurity disordered [30, 31], buried stripe [32, 33], ridge waveguide [34], and more recently by slab-coupled waveguide [35, 36] and, possibly, ARROW [37] (Figure 5.2).

Any deviation from the fundamental mode results in severe coupling issues into the fiber, known as kinks. It is common practice to suppress lasing of higher order modes by increasing their threshold gain. This is achieved most easily by introducing mode-selective losses [38–41].

In the early 1990s, pump diode manufacturers were baffled by kinks based on systematic beam steering (therefore sometimes called shift kink) which appeared periodically with pump current. It was soon discovered that high-power

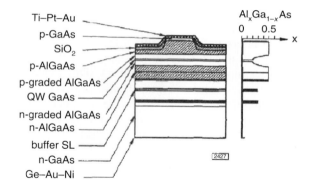

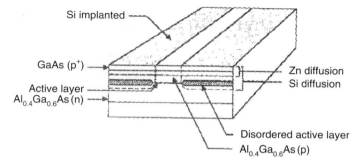

Figure 5.2 Cross-section of a ridge waveguide [34] and a silicon-implanted impurity disordered waveguide [31] laser diode on GaAs substrate.

fundamental mode operation of weak waveguides is limited by resonant coherent coupling of the fundamental mode into the first-order mode [42]. The modes of the passive waveguide are orthogonal and, without any asymmetric element, no coupling occurs. In practical diodes, slight asymmetries at the front mirror (e.g., angular misalignment, mirror coating, and heating) are a likely source for coupling. Power reflected from the fiber Bragg grating (FBG) in the pigtail also adds asymmetry (as the fiber is not perfectly aligned). The shift kink cannot be suppressed by the standard method of increasing the losses of the higher order mode as the power is resonantly coupled from the lasing fundamental mode to the first-order mode (the first-order mode can still be below threshold). Kink power level is controlled by controlling the beat length of the two modes, either by an individual adjustment of the cavity length [43] or by controlling the difference of propagation constants of the two modes by exactly controlling the ridge process [44] and by eliminating asymmetries as much as possible. Today's pump diodes display reduced shift kink issues, most likely due to long cavities, low front mirror reflectivities and low waveguide asymmetries.

Mode Filters

Broad-area (BA) diode lasers with mode filters to enforce fundamental mode operation, were thoroughly investigated. Best known are the master-oscillator power-amplifier (MOPA), alpha-distributed feedback (DFB) lasers laser, multimode interference (MMI) waveguide, and taper laser [45–51]. These structures are not in use for telecommunication, as they have not been able to meet the stringent requirement for power conversion efficiency, reliability, and ease of coupling. For special applications, the taper laser, preferably in an external grating feedback configuration, generating up to 3 W of nearly diffraction-limited power, has found industrial application [50].

5.2.3 Output Power Scaling

It has been possible to increase the pump diode rollover power by more than an order of magnitude (Figure 5.3). This was achieved by making the laser cavity longer, thus improving the cooling of the pump diode while scaling the other diode parameters appropriately.

Length Scaling of Laser Diode

Four key figures of a pump diode are (1) number of carriers in the QW (which is given by the required roundtrip gain G), (2) external quantum efficiency η, (3) photon lifetime in the cavity τ_{ph}, and (4) asymmetry of the laser cavity given by the ratio of power behind the front and back mirrors Pr.

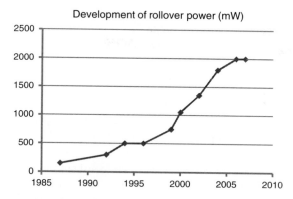

Figure 5.3 Development of rollover power of 980 nm single lateral mode pump diodes (cw, junction up, at room temperature) (this figure may be seen in color on the included CD-ROM).

For a back mirror reflectivity equal to 1, the four key figures, as function of the laser parameters, optical confinement factor Γ, optical loss α, and front mirror reflectivity R, are:

$$G = \left(\alpha + \frac{1}{2L} \cdot \ln\left(\frac{1}{R}\right)\right)/\Gamma, \quad \eta = \left(\frac{1}{2L} \cdot \ln\left(\frac{1}{R}\right)\right)\bigg/\left(\alpha + \frac{1}{2L} \cdot \ln\left(\frac{1}{R}\right)\right)$$

$$\tau_{\text{ph}} = 1\bigg/\left(v_{\text{gr}} \cdot \left(\alpha + \frac{1}{2L} \cdot \ln\left(\frac{1}{R}\right)\right)\right), \quad Pr = (1+R)/(2 \cdot \sqrt{R})$$

Ideally, the laser parameters (Γ, α, and R) are adapted to keep all four key figures (G, η, τ_{ph}, and Pr) constant while increasing at the same time the cavity length L to improve the thermally limited rollover power. This is not possible. Thus, gain G and external efficiency η are kept fixed, while photon lifetime OR power ratio are adapted, depending on the approach. For obvious reasons, the two cases are called "constant photon lifetime scaling" and "constant power ratio scaling."

For the constant photon lifetime scaling, we obtain the following scaling rules:

$$\Gamma(L) = \Gamma(L_0), \quad \alpha(L) = \alpha(L_0)$$

$$R(L) = R(L_0)^{\frac{L}{L_0}}$$

This method was used in the initial phase of increasing power; one can just cleave longer cavities from the same material and reduce the front mirror reflectivity exponentially, according to the exponential scaling rule, and still maintain efficiency. The power ratio becomes also exponentially larger with increasing cavity length. Due to this increased power ratio, longitudinal spatial hole burning is observed. Hole burning can be reduced by the introduction of a slightly flared adiabatic waveguide [52, 53]. Driving this scaling method too far (i.e., mirror reflectivity well below 1%) causes amplified spontaneous emission (ASE) and external FBG frequency stabilization issues.

Thus, for long cavities, constant power ratio scaling is favored with the following scaling rules:

$$R(L) = R(L_0)$$

$$\Gamma(L) = \frac{L_0}{L} \cdot \Gamma(L_0), \qquad \alpha(L) = \frac{L_0}{L} \cdot \alpha(L_0)$$

As a consequence of keeping the power ratio constant, the mirror reflectivity is also constant; therefore, this method is often called constant mirror reflectivity scaling. Requirements on confinement and optical losses are demanding. Both have to be reduced linearly with increased cavity length. The constant mirror reflectivity scaling has the advantage of making, to zeroth order, average optical power in the QW, pump current density, heat generation density, and ASE and spectral stability independent of length. A careful design of the vertical epitaxial structure to reduce the waveguide loss α and confinement Γ is necessary for long cavities.

Vertical Epitaxial Structure

The optical confinement Γ can be decreased by expanding the beam and by asymmetric placement of the QW in the waveguide [54, 55]. The waveguide loss α can be decreased by reducing the various contributions, such as scattering and leaking losses of the passive waveguide. Free-carrier absorption is reduced by reducing the carriers within the optical mode field and by careful choice of the material composition and fabrication details. Waveguides with losses $\alpha < 1\,\text{cm}^{-1}$ are routinely fabricated, and it is not known yet what the physical and engineering limits are. This is an important topic since the maximum power that one obtains from a longer pump laser (i.e., constant power ratio scaling) is ultimately limited by the abilities to scale the losses α in the waveguide. (Losses by the QW are scaled by the simultaneous scaling of the optical confinement.) The waveguide loss α has to be reduced while keeping the joule heating (series resistivity), the heat removal (thermal conductivity), and temperature sensitivity (T_0) under control. In addition, single lateral mode emission has to be ensured. 2 W of roll over power has been achieved by using such scaling rules (Figure 5.4).

Challenges for length-scaling BA pumps are very similar. In recent years, much progress has been made toward optimizing the vertical structure, documented mainly within the SHEDS (super high-efficiency diode source) program [8]. Of the many different vertical waveguide structures [57–60], the asymmetric LOC which expands the beam to a divergence below 30° [full width at half maximum FWHM], as shown in Figure 5.5, is favored. To reduce leaky losses for the fundamental mode, such LOC structures have an NA high enough to guide a

5. Pump Diode Lasers

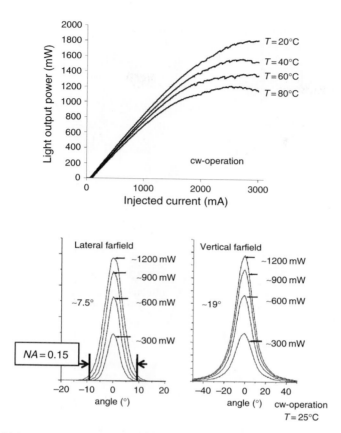

Figure 5.4 Light output vs injected current of ridge waveguide laser at various heat-sink temperatures and farfields at various power levels [56].

few modes. By placement of the QW and by introducing mode-selective losses, only the fundamental mode is brought above threshold gain.

For 1480-nm pumps, due to different material parameters and growth challenges, optical trap waveguides [61] have shown excellent performance.

5.2.4 Spectral Stability

The emission spectrum of the solitary Fabry–Perot 980-nm laser diodes is not stabilized well and, early on, displayed complex behavior for some (e.g., impurity disordered) waveguide structures [62]. The spectrum of the index-guided ridge waveguide laser diodes is very stable with the usual average shift with changing temperature (0.3 nm/K), but mode jumps are observed due to a modulated loss

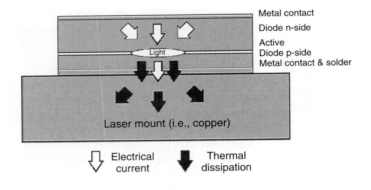

Figure 5.5 Cross-section of "junction-down" mounted BA diode laser and example of asymmetric LOC waveguide [21].

spectrum. This substructure is believed to be due to radiation from the leaky waveguide into the substrate made possible by the GaAs substrate, which is transparent and has a higher index of refraction than the waveguide [63].

Early on, it was realized [64, 65] that by placing a weak FBG into the pigtail, the spectrum of the Fabry–Perot laser could be stabilized effectively. The temperature drift of the spectrum can thus be reduced to 7 pm/K, nearly two orders of magnitude lower than temperature drifts of the solitary Fabry–Perot laser. Initial shortcomings were noise, cost, and reliability concerns due to the complexity the FBG-stabilized pump package. However, the gain flatness requirement of DWDM EDFAs required pump sources with a very stable emission spectrum, and thus, these issues had to be and were solved [66–68].

Polarization mode effects in the pigtail under FBG stabilization have to be controlled, either by a strict rule for fiber lay in the EDFA module [69] or, much more repeatable, by using a PM fiber pigtail. To reduce mode hopping noise, FBG-stabilized pumps are designed to operate in coherence collapse. Unfortunately, laser diodes can have the tendency to switch from time to time from coherence collapse into a coherent state, thus causing amplitude noise [70–74]. This switching noise can be effectively suppressed by a dither or a double FBG [75] beyond a critical distance.

A FBG-stabilized diode laser can also be designed to have a coherence length of a few centimeters and low noise. This is a useful source for generating visible light by frequency doubling in PPLN waveguides [76].

5.2.5 Packaging

The basis of a modern 980-nm pump package is a monolithic planar AlN substrate which acts as the optical bench. The laser chip is soldered p-side up directly to the metalized AlN substrate, and the fiber, tipped with a low-cost wedge lens, is attached in front of the chip on the AlN substrate.

The platform is placed in the MiniDIL housing and hermetically sealed. Package-induced failure (PIF) mode is suppressed, by filling the package with a mixture of 20% oxygen and 80% nitrogen, as it was used by some manufacturers in the 1980s for short-wave lasers and later on applied to 980-nm pumps [77, 78]. PIF is based on a photo-thermal decomposition of trace hydrocarbons in the atmosphere on a surface with a high-optical power density. Under oxygen, it is being *burned-off* faster than deposited [79]. Obviously, this phenomenon is not restricted to the laser diode but manifests itself on any surface in the laser beam. The presence of oxygen in a hermetic package, however, favors the long-term evolution of humidity, as a by-product of the burning of the hydrocarbon traces and hydrogen. A long-term humidity concentration below 5000 ppm (dew point at 0°C) has to be ensured by carefully cleaning the package of organic contamination, desiccation and dehydrogenation processes. Otherwise, a humidity getter has to be included inside the package [80].

The stability of this platform is outstanding as is evident from the results obtained in stress testing of an earlier generation (Figure 5.6) and ensures operation at 400 mW from the fiber pigtail in the very demanding undersea application.

It is standard practice to increase the output power by cooling the pump laser diode through a Peltier heat pump [81]. The chip can be stabilized at different temperatures (standard is to stabilize at 25°C), trading off power dissipation and maximum available optical power [66].

Pump powers in excess of 1 W from a SM fiber have been demonstrated already a few years ago [82–84], and products in the 700 mW range, corresponding to 70 MW/(cm^{2}sr), with outstanding reliability, are commercially available.

5.2.6 Reliability

Today's reliable high-power 980-nm pump diodes for the demanding telecom market are based on a fundamental understanding which the industry has. It has

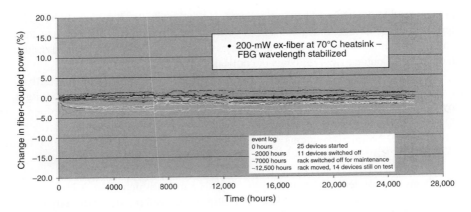

Figure 5.6 Stability of 25 uncooled FBG-stabilized 980-nm pump diode MiniDIL over 3 years of high temperature and maximum power operation (courtesy Bookham).

been possible to eliminate the gradual degradation completely, and today, 980-nm pump laser diode reliability is limited by a low rate of random sudden failures [79] due to point defects. Only little of this know how has been published, and the details are very complex and involved. Here, we just give a brief summary on the most important aspects of reliability of 980-nm pump laser diodes.

Failure Modes

To gain a very rough understanding of the failure modes of a 980-nm pump laser leading to sudden failures, three major topics should be considered:

(1) Catastrophic junction meltdown (CMD): A hot spot with thermal runaway resulting in CMD, mostly occurring at the front facet (thermally most exposed spot).
(2) Heating through nonradiative recombination (NRR), initiating a hot spot due to NRR of electrons and holes through mid-gap states at defects.
(3) Thermally accelerated decomposition (TAD) at the facets with even trace amounts of oxygen acting as catalyst, generating defects and NRR centers.

Catastrophic Junction Meltdown

A 9xx-nm laser diode, even without any defects but uneven cooling, can develop at high currents a hot spot with a thermal runaway. In a hot spot, the bandgap is lowered, resulting in higher carrier injection, both from the contacts (current) and the laser beam (absorption). It is assumed that above a critical temperature (temperature difference), this hot spot is leading to catastrophic thermal runaway with junction meltdown (CMD). Experimentally, this critical temperature was measured by initiating thermal runaway with an external heating source (argon laser spot) [85], and a value (averaged over a 1.5 μm spot size) of 120–140 K above room temperature was measured.

This CMD most often happens at the facets, as they are not well cooled, have the highest current density, highest optical field [86, 87], and usually some parasitic additional heat source (defects, mirror coating) which triggers the thermal runaway. Traditionally, the whole mirror was blown off, stopping the laser to work because of the damaged optical cavity, therefore called catastrophic optic mirror damage (COMD). Because of an improved thermal design, the junction can often melt only microscopically, without any visible defect under the optical microscope [88]. Nevertheless, the diode laser stops working since the molten and re-crystallized junction is electronically dead, that is, it provides no gain and has a high NRR current [89].

No consistent hot-spot model leading to CMD is available. Only some of the required material constants have been measured [90]. Effective countermeasures against CMD are blocking current injection into critical areas which leads to reduced temperatures [91] and uniform heat sinking through appropriate heat spreaders and junction up mounting [81].

NRR Heating

The major additional heating source of aged laser diodes is heating due to NRR [15]. NRR is the product of electron, holes, and NRR center density and thus, can be suppressed by suppressing *any* of the three terms. Usually minority carriers or NRR centers are suppressed. The minority carrier density can be kept low by a so-called window laser, and the density of NRR can be kept low by an appropriate facet passivation.

Within the window is a larger bandgap to avoid generation of minority carriers by absorption [therefore sometimes also called nonabsorbing mirror (NAM)] and to reduce the number of thermalized minority carriers. In addition, there is usually no p–n junction to avoid minority carrier injection. Such window lasers have been known for a long time and were initially realized by Zn diffusion [92]. Later on, window lasers were implemented through epitaxial growth through the active region [93]. The quality of this regrown interface is very important (and challenging). For very high-power lasers, the Si-doped and disordered window laser has proven itself [94–96] over many years and is in heavy use today throughout the industry. Recently, also good progress has been made with vacancy-induced disordering [97–99] to form a window region.

TAD at the Facets

It has been known that GaAs lasers degrade at the facets under high-power operation [100]. This effect is also seen at a slower rate in the so-called aluminum-free active-area laser diodes [101] and accelerated in AlGaAs. It was shown in the late 1980s/early 1990s in a series of papers that this degradation is TAD [102–105].

Only trace amounts (a fraction of a monolayer) of oxygen are enough to start breaking the bonds, decomposing GaAs around the p–n junction. This leaves behind atomic Ga and As and defects, which then act as nonradiating recombination centers. Oxygen does not form a stable oxide but acts as catalyst to decompose the facets [106]. Over time, NRR centers are building up, until the laser goes into CMD (or even COMD). Temperature accelerates decomposition, and an operation power dependency of the time to COMD [102] is observed.

Temperature is an important measure of the NRR rate. Sophisticated tools have been developed to measure temperature directly on the small facet with absolute temperature scale, by Raman [107, 108], with high-spatial resolution by electron beam-induced current (EBIC) charge thermography [109], and high-temperature resolution by thermoreflectance [110].

Decomposition at the facets can be completely stopped if the surface is sealed without any damage and without trace amounts of oxygen. This process is called passivation of the facet. One distinguishes in situ passivation [breaking the wafer in an ultra-high vacuum (UHV) system, followed by covering it with a passivation layer] and ex situ passivation: cleaving the wafer in air, then transferring it into a vacuum system followed by a gentle (to minimize structural damage to the laser facet) cleaning-off of the oxygen followed by a passivation layer.

There exist today essentially three in situ processes. All in situ processes are based on a complex UHV system, with load locks to transfer the wafer, and a bar cleaving mechanism operated inside the UHV system. One in situ process is based on the high-temperature growth of a single-crystal InGaAsP passivation layer [111]. Because of the high temperatures involved, the metallic contacts can only be applied after passivation at the individual bar level. Another one is based on a low-temperature ZnSe [112–114]. The most successful and widely used in situ passivation is based on low temperature and low-energy deposition of silicon [115, 116].

Numerous ex situ processes have been tried to reduce cost of passivation. It is suspected that an ex situ process will never be able to produce a completely defect-free, stochimetric surface. Therefore, the time to C(O)MD is expected to be finite for ex situ passivation, but this time could be well beyond the required useful life. Today, it is understood that passivation through sulphation [117] and nitridation [118, 119] has proven to be difficult to manufacture reproducibly with very high quality, most likely as it does not chemically bind any remaining oxygen completely. Most successful passivation techniques seem to be the ones which include the deposition of silicon (or ZnSe) after removing the oxygen either with a low-energy hydrogen plasma [120–122] or a low-energy ion beam, such as in the I3 processes [123, 124].

5.2.7 Failure Rate

Unlike other characteristics of a pump laser package, lifetime cannot be measured readily for each laser package but has to be estimated from understanding the physics of failure, massive testing, and controls on manufacturing stability to reproduce the lifetime. The industry has developed detailed knowledge, especially for the undersea application. Here, we just give a brief summary of the most important aspects of lifetime estimation of 980-nm pump laser packages.

Unfortunately, there are essentially no ways to accelerate aging for the overall package. Packages are qualified according to norms, such as GR-468, which guarantees robustness, but which does not allow for predicting reliable operation.

Damage levels are determined by driving it into destruction by overstressing with the various relevant parameters. These damage levels should be at a safe distance from operating conditions, and from an understanding of the physics of failures, one can attempt at predicting use failure rates.

The design of the FBG stabilization has to be robust so that the spectrum remains stable, even under worst case conditions and fully aged characteristics of the different elements. For this, it is important to investigate in detail the aging of the various critical parameters of the involved components [125].

A carefully designed and manufactured 980-nm pump diodes has very low-failure rates with respect to gradual wear out as well as wear out sudden failures [126]. But even if these two failure modes are completely suppressed, one is left with low rate of sudden failures, best described by a constant failure rate, that is, random sudden failures of the pump diode. As it is difficult to find a physical explanation for a random failure rate, the sudden failures are sometimes also modeled by a lognormal distribution [127] with a large σ (indicating a wide activation energy distribution of underlying physical processes).

It is expected that the failures of the chip can be accelerated by increasing the driving forces behind the degradation, that is, carrier density in the active region, local temperature in the active region, and so on. Unfortunately, these parameters are not directly accessible. Thus, the failure rate is accelerated by increasing operating temperature and drive current and power. Of course, one has to ensure by physical failure mode investigation that the diode running under accelerated conditions produces the same failure modes as observed under use conditions [128], in other words, the laser diode has to be designed to work not only properly at use condition but also at accelerated conditions, that is, design for use and testing.

The failure rate at use condition is usually extrapolated from the failure rates from blocs of laser diodes (stress cells) with each cell running at different accelerated conditions. Traditionally, the following heuristic functional dependence of failure rate on junction temperature (T_j), current density (j), and power density (p) is assumed:

$$\textit{Failure rate}\,(j, p, T_j) = FR_0 \cdot j^x \cdot p^y \cdot \exp(Ea/kT_j)$$

A maximum likelihood calculation is used to find the values x, y and Ea, for which the failure rate in the stress cells is most likely to be observed [129, 130]. From these most likely parameters, the most likely failure rate at use condition can be estimated. It has become customary to calculate upper confidence limits for use failure rates, based on Ea, x, and y, as determined by maximum likelihood,

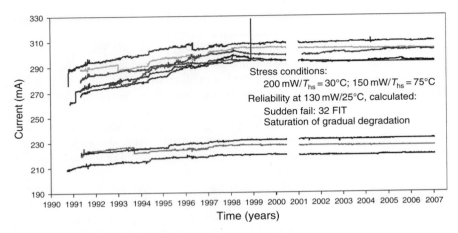

Figure 5.7 Sixteen years of constant power operation of 980-nm pump diodes at temperature and power stress (courtesy Bookham).

neglecting the confidence range of the maximum likelihood estimates. This is inconsistent and leads to very optimistic upper confidence limits. It has also been pointed out that the assumed functional dependence has a big impact on failure rate estimation at use conditions [131]. It must also be ensured that manufactured quality corresponds to the qualification sample, which is routinely done by defining and controlling the so-called critical process parameters.

Field returns and long-life tests in the laboratory are the test of the failure rate estimations. Both terrestrial field returns (<20 FIT on more than 20 billion device hours are estimated by the industry) as well as undersea field returns (<5 FIT with 60% confidence for redundant pump pairs [132, 133]) are testimony of this fact.

Long-life tests on 980-nm pumps (operated for more than 15 years at accelerated conditions) are shown in Figure 5.7. The gradual increase saturates after a while and the lasers run very stable with one random fail being the only observation. This observation is consistent with failure rate predictions for these accelerated test conditions, thus increasing confidence in the "art" of accelerated stress cell testing (Figure 5.7).

5.3 1480-nm PUMPS AND 14XX-nm HIGH-POWER LASERS

5.3.1 1480-nm Pumps

The stronghold of 1480-nm pumps clearly is remote pumping of preamplifier EDFAs [9], which requires very high 1480-nm power. It is an advantage that 1480-nm pump diodes can be produced on the basis of the established 1550-nm

5. Pump Diode Lasers

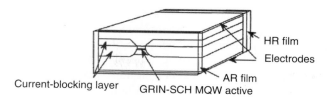

Figure 5.8 Schematic structure of a 14xx-nm diode laser chip [135].

signal laser technology, using the well-known material system of compressively strained multiquantum well (MQW) InGaAsP layer on InP substrate. The 1480-nm pumps were initially (and sometimes still are) used for pumping power booster EDFAs. Similar considerations as for 980-nm pump optimization have to be applied to optimize the output of 1480-nm pumps. Due to inherently higher series resistance, Auger losses, intervalence band absorption, and an inherently stronger temperature sensitivity of efficiency and of threshold current, it is very challenging to design and fabricate a high-power 1480-nm pump. Losses in the InGaAsP waveguide at 1480-nm have been reduced steadily over the last few years and, with an optimized asymmetric waveguide [134], losses in the range of $3\,\mathrm{cm}^{-1}$ have been obtained. Nevertheless, multiple QWs are necessary to provide enough modal gain to overcome the waveguide losses.

Because of the proximity of the pump wavelength to the signal band, reflections have to be suppressed by placing an optical isolator in the package. Such a package requires discrete lenses, which requires a round beam (for low-cost lenses, high-coupling efficiency). This can be achieved by a multigrowth step buried heterostructure [135], as shown in Figure 5.8. It has been shown that also a simple ridge laser diode, first demonstrated 30 years ago [136] on InP substrate, can be designed to have excellent characteristics, a rollover power of 1.2 W and a beam with an aspect ratio of around 2, using an optical super-lattice waveguide design [61, 137].

The reliability of 1480-nm pump lasers is given by gradual degradation and does not suffer from random sudden failures (like 980 nm). Reliability modeling as well as failure rate predictions are quite well understood. However, it is found that the failure rate accelerates quite strongly with temperature ($Ea = 0.6$ eV) [135], and thus, 1480-nm pump laser packages require a Peltier heat pump to cool the chip to 25°C. Due to the relatively inefficient chip, practical maximum power available from 1480-nm pump lasers is limited by the capability of the pump package to handle the dissipated power from the chip. Special Peltier elements and packages have been developed for 1480-nm pump lasers, and commercially, pigtail powers up to 400 mW are available today.

5.3.2 14xx-nm High-Power Diode Lasers

Research on Raman amplification in optical fibers was started in the early 1970s, but the required pump powers at the required wavelength 14xx nm were not readily available. Raman amplifiers have the big advantage that the gain is not limited to any specific spectral band and that amplification works also outside the EDFA C and L bands. With the availability of high-power 14xx-nm pump laser diodes and the requirement for ever higher capacity, transmission systems based on Raman were investigated in detail, and it was demonstrated that record capacity distance products are achieved with Raman amplification [3].

As a consequence of the Raman process, the requirements on the 14xx-nm pump diode with respect to power and noise are quite challenging [3], unlike 1480 or 980-nm pump lasers, where the noise is reduced by the long spontaneous lifetime of the erbium gain. Much progress has been made toward the required "quiet" 14xx-nm high-power diode laser by internal frequency stabilization and polarization scrambling [3, 138–141]. Raman gain blocks are available which contain usually a few 14xx-nm individual diode lasers with different wavelength [142] to establish flat gain over a few tens of nanometers. Diode laser modules with high-output powers as well as special fibers are available today, and spectacular all Raman amplified systems with record transmission performance have been obtained. However, progress for 14xx-nm pumps has slowed down as the industry seems to use EDFAs for cost-optimized C and L bands and the thirst for bandwidth extension has been small.

5.4 MULTIMODE FIBER-COUPLED 9xx nm PUMP LASERS

The recent application of the 9xx-nm telecom technology to the industrial markets, especially pumps for FL, disk and rod DPSSL, and also in the form of DD for material processing [143], has been very fruitful. Due to a concerted effort with programs for power conversion efficiency (SHEDS) and radiance (ADHELS and BRIOLAS) [7, 8, 144], very exciting results have been obtained in the last few years, especially in the field of MM fiber-coupled 9xx-nm pump lasers, one of the key enabling devices of the hermetic, all solid-state power photonics.

5.4.1 Classification of 9xx-nm MM pump Lasers

Available radiances of commercial 9xx-nm pump lasers are roughly 1 W in a SM fiber (which has the diffraction-limited etendue of $E = \lambda^2 = 1\,\mu m^2 sr$ for a 980-nm laser). A MM step index fiber with 200 μm/0.22 NA has $E = 5000\,\mu m^2 sr$. Thus, it should be possible to get 5 kW from a 200 μm/0.22-NA pump module, even without wavelength and polarization combiners.

Unfortunately, it is prohibitively expensive to manufacture such a MM pump due to two challenges: optical coupling and heat removal. Thus, we classify 9xx-nm MM pumps according to these two topics.

Heat Transport Classification

High-power MM diode lasers are thermally limited and need to be soldered *junction down* onto a heat sink. Best heat transport can be achieved by directly mounting the laser diode on a copper or, even better, diamond heat sink. Mismatch in thermal expansion material coefficients is managed by using a soft solder (e.g., indium) which accommodates the mechanical movements due to temperature changes during manufacturing (soldering) and use. Soft solders change their properties and shape, even under standard use conditions, causing a whole range of failure modes. Despite much work, only a few applications work satisfactorily within these failure modes. It is well known from telecom pump diode technology that reliable systems use hard solders (e.g., AuSn), together with expansion-matched substrates (e.g., CuW, electrically insulating AlN, or BeO, or layered Cu/Mo microchannel coolers [145]). Heat sinks need to be coated with a dedicated stack of metal layers (diffusion barrier, strain relief, heat spreader, and adhesion promoter) to enable a reproducible and reliable solder joint process. We classify heat sinking into (see Figure 5.9).

(1) passive cooling (also conductive cooling): two-dimensional heat transport.
(2) active cooling: one-dimensional heat transport.

In the case of passive cooling, the heat load is so small that it is possible to spread the heat flow in expansion-matched substrates. This package can then be pressed (by mechanical fixture) against cooling fins or a heat exchanger, which can have a different expansion coefficient (due to pressure contact and the small temperature variation at this removed interface). In the case of active cooling, the heat load is so big that it has to be removed in close proximity by a heat exchanger. Thus, for active cooling, the pump diode has to be attached with a hard solder onto

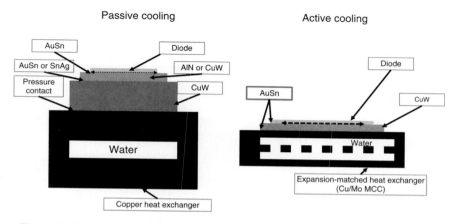

Figure 5.9 Schematic structure of a passive-cooled diode laser and an active-cooled diode laser.

an expansion-matched microchannel cooler, possibly with an expansion-matched submount in between to ease manufacturing. Experience has taught us that high-power and low-failure rate applications strictly require expansion-matched materials for solder joints and pressure contacts between nonexpansion-matched heat sinks.

Optical Classification

A complex optical system might be needed to couple a diode laser into a MM fiber, which has a tremendous impact on cost. Because of the many degrees of freedom of a general laser beam, it is difficult to systematically classify this topic. Fortunately, if we restrict ourselves to plain BA laser diodes (with no focal astigmatism) and step index MM fibers (both NAs limited systems), then emission width (diameter), NA, and power characterize the beam fully. The related optical characteristic parameters are given as follows: beam parameter product $BPP = NA \times diameter/2$, etendue $F = \pi^2 \times BPP^2$ (approximation for NA < 0.5), and *radiance = power/etendue* (Table 5.1).

Table 5.1

Beam parameter product (BPP) and etendue (E) for select pump diode lasers and high-power fibers.

Diode laser	Beam width (μm)	NA (rad)	Fast-axis BPP (μm rad)	Slow-axis BPP (μm rad)	Etendue (μm² sr)
Single-mode diode	5	0.12	0.3	0.3	1
Standard BA diode at low power	100	0.05	0.3	3	8
Standard BA diode	100	0.09	0.3	5	14
Low-NA wide-BA diode	200	0.09	0.3	9	28
Low-NA minibar	3,200	0.07	0.3	112	340

Fiber	Core diameter (μm)	NA (rad)	BPP (μm rad)	Etendue (μm² sr)
SM fiber	5	0.12	0.3	1
Input fiber for fiber combiners	105	0.15	8	610
Standard material processing delivery	200	0.22	22	4,800
High-power material processing delivery	400	0.22	44	19,000
Fiber of cladding pumped laser	400	0.46	92	84,000
High-power material-processing delivery	1,500	0.46	345	1,200,000

The pump diode is diffraction limited in the fast axis, and thus, it is trivial to couple it to any fiber. MM fibers even allow for extensive stacking in this direction. From slow-axis considerations, we distinguish three classes:

(1) Direct coupling: Pump laser width and NA in the slow axis are smaller than diameter and NA of the fiber. A simple fiber tip coupling, as with SM lasers, is used.
(2) Simple lens coupling: Pump laser slow-axis BPP is smaller than the BPP of the fiber. A simple focusing lens is used.
(3) Beam symmetrization optics coupling: Pump laser slow-axis BPP is larger than the BPP of the fiber. The beam has to be reshaped by symmetrization optics [146].

5.4.2 BA Pump Diode Laser

What has been discussed on SM diode laser above (Section 5.2) applies also to BA laser diodes with the exception of junction-down mounting, relaxed requirement on slow-axis BPP (lateral beam quality), and frequency stabilization. BA pump diodes have continued the progress on length scaling of SM diode lasers and the respective work is included in Section 5.2.

Slow-Axis BPP

To increase the output power of a diode laser, the emission width is made larger, up to 1 cm, the full width of today's available standard heat sinks. A simple wide contact stripe is sufficient, but sometimes a shallow ridge is etched to reduce lateral current spreading and to limit the lower slow-axis NA.

At large widths, the laser is wider than long and suffers from efficiency problems due to lateral ASE (or even lateral lasing). Lateral ASE is suppressed by limiting the width and by introducing optical isolation trenches on either side of the active stripe. As these trenches go through the active region, they have to be electrically insulated, that is, placed at a distance. They also serve as powerful mechanical insulation trenches (dislocations in the active-area layer cannot propagate beyond these insulation trenches). Usually, a subunit of such a divided wide-area diode laser is called BA diode, broad-area single-emitter (BASE), or MM single-emitter laser diode and has a width between 50 µm and a few hundred micrometers depending on desired lateral ASE suppression and NA optimization.

One could expect that an ideal BA diode is lasing in the fundamental lateral mode and that the BPP is independent of width (NA should decrease linearly with width). Unfortunately, measurements of the farfield (see figure 5.10) show that this is not the case. Already at low power, the slow-axis farfield has a top hat shape

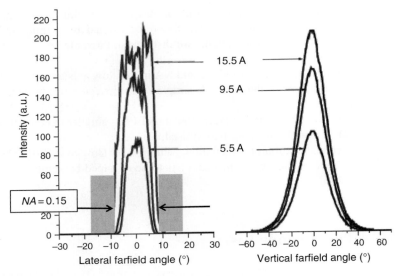

Figure 5.10 Slow (lateral)-and fast (vertical)-axis farfields of a 90-μm BA diode for various drive currents [147].

with a high BPP and the BPP even increases with power. Major causes for slow-axis NA degradation are as follows:

(1) Mode filling: Lasing onset of higher order lateral modes which milk the lateral gain profile more effectively than the fundamental mode, at high powers increased by spatial hole burning.
(2) Thermal waveguide.
(3) Gain guiding due to lateral gain/loss profile (resulting in a phase curvature) and carrier-induced index changes. Considered to be less important causes for low-loss waveguide and single QW gain active-region diode lasers.

Slow-axis NA degradation due to mode filling can be reduced by carefully tailoring the current injection, as proposed quite some time ago [148] and demonstrated by a 2.5-W nearly diffraction-limited beam from a BA laser [149] in the pulse mode with heating absent.

Thermal waveguiding is caused by the positive dependence of index of refraction on temperature. Due to the cooling geometry, the temperature raises in the center by ΔT more than at the edges of the BA laser, causing a lateral index difference of Δn. According to the effective index approach, the NA of such an index-guided beam is

$$NA = \sqrt{(n + \Delta n)^2 - n^2} \approx \sqrt{2 \cdot n \cdot \frac{\Delta n}{\Delta T} \cdot \Delta T}$$

A slow-axis $NA = 0.07$ results already for a small $\Delta T = 2$ K (using $\Delta n/\Delta T = 3 \times 10^{-4} \, \text{K}^{-1}$, $n = 3.6$, [150]). The lateral temperature difference can be reduced by reducing the improving efficiency and heat extraction from the active region (asymmetric waveguide, see also Section 5.2) and by improved heat sinking. In pursuit for higher powers, some of these parameters have been improved over the last few years and so have the slow-axis NA. Thermal waveguiding is still the major cause for NA degradation, and the remedy, an athermal waveguide (i.e., modal index does not depend on temperature), is still waiting to be developed.

As with narrow stripe laser diodes, techniques have been investigated to improve the slow-axis BPP by mode filters, most notably by taper lasers. The bothersome destabilization of the slow-axis focal plane with resulting sensitivities to drive current, reflections, and, possibly, aging makes it difficult to couple such devices even to MM fibers. Mode-filtered diode lasers are still awaiting a breakthrough in focal plane stability and in efficiency.

Frequency Stabilization

Spectral stabilization through FBG, as used for SM pump lasers (see Section 5.2.4.), is, due the MM nature of fiber, not effective. The aluminum-free InGaAsP material system offers the opportunity for a conventional regrown DFB stabilization [151]. A partial DFB with very good characteristics has been fabricated in the InGaAlAs [152]. Frequency can also be stabilized effectively through external elements, such as volume holographic grating [153].

5.4.3 Passive-Cooled 9xx-nm MM Pumps

Passive-cooled 9xx-nm pump packages with a MM pigtail are easily installed by simply attaching electrical wires, splicing a fiber to the pigtail, and by clamping them mechanically against a heat sink. Such a heat sink can either be forced air-cooled fins or an isothermal plate, temperature stabilized by a Peltier or directly by a heat-exchanging liquid. We distinguish, according to the classification above, between different optical-coupling arrangements.

Direct Coupling

The 9xx-nm pump with the largest volume and the most pigtailed power shipped every year (a few MW per year) is the direct-coupled package. A BA diode laser is directly coupled into a wedge-lensed (matching fast-axis NA) MM step index fiber, very much as it is done for telecom pump lasers.

The development of the BA diode lasers has moved in parallel with single-mode narrow stripe lasers, however, soldered junction-down on a heat sink. cw room temperature rollover powers from a 100-µm BA laser are in the range of 20 W at 25 A, that is, 19 W at 25 A, [152], 18.5 W at 23 A, [154], 19 W at 24 A [155], and the chips have been designed to be temperature insensitive (Figure 5.11).

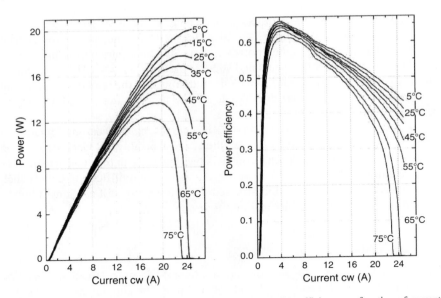

Figure 5.11 Typical rollover characteristics and power conversion efficiency as function of current (at various heat-sink temperatures) for a 90-μm wide BA diode laser [155].

Packaging is based on the telecom packaging technology but made more difficult by the high-power densities for heat removal as well as for optical coupling. Standard fibers for fused couplers are 105/125-μm MM step index fibers with $NA = 0.12$, 0.15, or 0.22. To ease alignment, the BA diode laser is usually only 95-μm wide, and it is beneficial for radiance to pick a fiber with an NA matched to the BA slow-axis NA. Due to the thin cladding of the 105/125-μm fiber and the high powers involved, not only the core coupling (which should be as large as possible) but also the coupling to the cladding (which has to be kept small) have to be optimized, otherwise one runs into issues with burning the protective fiber coating. Markets are cladding pumping of YEDFAs for TV and fiber to the premises (FTTP) boosters (940 nm), cladding pumping of FL (920, 940, 960, and 976 nm), and DD applications (9xx nm). The printing application, requiring high brightness but limited power, usually uses 50/125-μm fiber with an $NA = 0.12$.

Power requirements for pumping FL are very demanding, and the BA diode is operating at the reliability limit. Requirements for failures under use conditions are usually specified in "time to 5% cumulative failures," TT 5% (1 out of 20 failures) with requirements for TT 5% being larger than 10 or 30 kh, depending on application. Failure mode statistics and understanding is very much based on SM 9xx-nm pump diode laser experience, nevertheless there are fine but important differences. Testing for failure rate is done with multicell test [156], and scaled failure statistics with a Weibull distribution is obtained (Figure 5.12).

5. Pump Diode Lasers

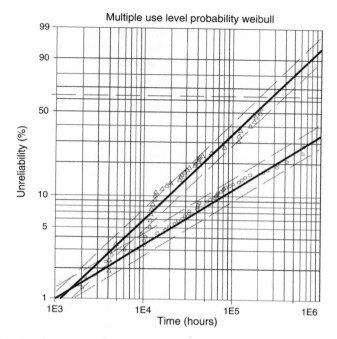

Figure 5.12 Projected percentage of failures as function of time, based on multicell test, for use conditions of 8 W, 30°C for two types of 100-μm BA diode lasers. Projected time to 5% failures are for the two types of BA diode lasers 8000 and 20,000 hours, respectively [156].

Reliability requirements are higher for YEDFA applications, and thus, such pumps are run at a derated power level.

Simple Lens Coupling

With simple lens coupling, the slow-axis NA of the BA diode is matched to the NA of the fiber (fast-axis NA can be matched by fiber tip or a separate simple lens). Because of the simplicity and because the typical NA of the diode is smaller than the NA of the fiber, a high-power, wide-stripe BA diode can be used. One recent example is the 16 W at 20 A in a 105/0.22 fiber from a 200-μm wide BA laser [154].

Beam Symmetrization Coupling Optics

Already a 50-μm $NA = 0.22$-μm fiber carries 600 lateral modes (both polarizations). It is therefore theoretically possible to couple 600 single lateral mode narrow stripe diode lasers, each with 1 W of power, into such a fiber. To ease this tedious task, one resorts to array techniques, that is, one uses a linear monolithic "single lateral mode emitter array laser" called single emitter array laser (SEAL) [157] and an array of lenses [158]. Pumps with 50 W from a 50 µm, 0.22 NA fiber are available [159], one order of magnitude away from the limits, partially due to derating of the SEAL vs the individual laser, but mostly to accommodate for alignment tolerances.

It becomes impractical to individually couple single lateral mode diode lasers into larger fibers (i.e., 40,000 modes in a 400 µm/0.22 NA fiber), and thus, arrays of BA diode lasers on one bar are used for larger fibers. Limited by the passive cooling, these bars have a low-packing density of around 10–30% (therefore often called low-fill factor (LFF) bars). Powers in the range of 400 W in a 200 µm/0.22 NA fiber, are commercially feasible [159], without wavelength and polarization multiplexing.

The 9xx-nm pump packages with a 100 µm/0.12 NA pigtail, based only on a few BA diodes, matched in BPP in the slow direction, and stacked in the fast axis, called multiple BA pumps, are predicted to become the next generation workhorse for passive hermetic fiber pumps.

The loss in radiance by coupling pump diode lasers to MM fibers is painful and limited by engineering limits which are still at least one order of magnitude away from the limits imposed by physics. Numerous beam-matching devices have been investigated, but the search for a breakthrough device is going on.

5.4.4 Active-Cooled Pump Diode Packages

In active cooling, the heat is removed by a liquid in close proximity. Among the different possibilities (immersion, micro-, or macrochannel coolers in different materials), the expansion-matched (Cu/Mo) microchannel cooler, run with pure water, is presently seen as the best compromise [145] between all the requirements. The requirement for high power and high reliability can only be accommodated by hard-soldering the bar on an expansion-matched microchannel cooler, possibly with an expansion-matched submount in between, for ease of manufacturing. Routine measurement methods are required to control during production thermal and mechanical integrity of this arrangement such as thermal wavelength shifts [160] and routine stress measurements by monitoring changes in bow. Active cooling systems are expensive. Therefore, they are used only for highest powers, which require beam symmetrization optics.

Beam Symmetrization Coupling Optics

It has been a standard for a long time that high-power bars, microchannel coolers, and lenses are designed for a 1-cm wide emission stripe. For ASE and NA optimization, the abovementioned optical isolation trenches are necessary to subdivide the 1-cm wide laser diode. Such a subdivided 1-cm wide laser can also be looked at as closely packed BA laser diodes and is therefore often also called high fill factor bar. For drive current optimization (trading off threshold current vs power density), the passive space between BA diodes can be increased, thereby reducing the fill factor. In the last few years cw hero powers have been increased from 300 W to 1 kW from a 1-cm aperture, demonstrating the raw capability of a 1-cm diode [161–164]. At such currents (up to 1000 A), one runs into ohmic resistance loss issues in the current supply cables, and as a consequence, the calculated intrinsic (ohmic losses subtracted) power conversion efficiency becomes much less relevant than the measured overall (with or without chiller power included) "wallplug efficiency."

As solution, a narrower standard for bars called Maxichip (also "3-mm bar," "1/3 bar," and "minibar") has been proposed, and products with high radiance (80 W from 3200 μm/0.08 NA) are available [165]. Maxichips, with their compact form factor (3600-μm wide and for now 3600-μm long, projected to become even longer very soon), might very well be the beginning of the end of the 1-cm legacy width for high-power bars.

5.4.5 Pulse Operation

High-power diode lasers are thermally limited in roll over power as well as in NA. If the diode laser is pulsed with a length shorter than the respective time constant [166], higher powers can be obtained. At low enough duty factors, heat sinking can be reduced allowing for monolithic vertical stacking [167]. At short pulses, the NA and radiance of a simple BA diode laser can be improved [149].

5.4.6 Combined Power

Power photonics would not be possible without the availability of all fiber MM-fused fiber combiners (with or without signal feed-through), Bragg gratings, and mode field adapters [168]. Key for this technology is their low-insertion loss, radiance conservation as well as the capability to handle high powers [169].

5.5 HIGH-RADIANCE DIODE LASER TECHNOLOGIES

Wavelength and polarization division multiplexing to combine beams of the same spacial mode is routinely used to increase the radiance. By wavelength combining a SEAL, a nearly diffraction-limited beam with 30 W cw power (radiance of 781 MW/(cm^{2}sr)) or pulsed beam with 50 W of power (3600 MW/(cm^{2}sr)) was demonstrated [170].

The holy grail of high-radiance diode lasers is coherent coupling of an array of lasers, locked by evanescent fields [171]. Progress to reproducibly lock the phases over a useful power and temperature range has been exceedingly difficult [172, 173], a robust design has not been found yet. With diode laser technology improving, the possibility arises for coherent combination of diodes with individually controlled frequency and phase [174].

5.6 VCSEL PUMP AND HIGH-POWER DIODE LASERS

Pump diode lasers are limited by thermal and by optical power density, and for this reason, a large area VCSEL should be an excellent pump source. A VCSEL has a large optical mode size with an already built-in epitaxial surface passivation. Low-power VCSELs on GaAs substrates are established high-volume technologies for datacom (850 nm MM) and sensing (optical PC mouse sensor). Due to the small active volume, such VCSELs have a high efficiency and high speed, already at very low currents and powers. In addition, the low NA symmetric beam lends itself to low-cost coupling.

In an exemplary effort, the concept of low-power VCSELs was adapted to high-power operation at 980 nm [175]. To control the beam quality at high power, an external optical cavity is added, either with an external mirror or, for lower powers, through an extended cavity. With the external mirror pump modules with 400 mW at 1.5 A drive current and with the extended cavity, an uncooled pumplet, designed for 50 mW, was demonstrated. Output power, power conversion efficiency, and reliability understanding fell short with respect to the competing edge emitters in telecom market.

As we understand from edge emitters, the thermal limits of a pump diode laser are ultimately limited by heat generation, given by joule heating and free-carrier absorption in the volume overlapping with the mode and the heat removal, given by the thermal resistance between the active region and the heat sink, from the active region.

The cooling area for a high-power narrow stripe edge-emitting device (5 μm × 4000 μm) and for a typical high-power NECSL with 150 μm diameter are comparable. It is a distinct characteristic (disadvantage) of a VCSEL configuration that carrier injection, optical mode reflection, and heat removal are all concentrated in the bottom-distributed Bragg reflector mirror, a GaAs/AlGaAs superstructure with relatively poor material constants with respect to high-power operation. A large research effort, for both, GaAs and InP devices, has been directed to ease this heating issue through clever designs and material choices. However, design space and available material impose limits.

Arrays of GaAs-based NECSL are ideally suited for intracavity frequency doubling in PPLN to generate light in the visible spectrum in the Watt regime as RGB light sources for laser projection displays (Figure 5.13).

5. Pump Diode Lasers

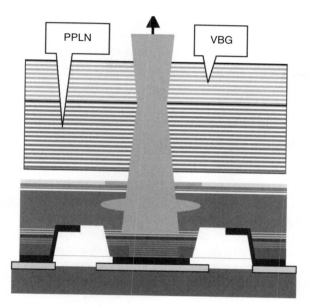

Figure 5.13 Schematic structure of a novalux extended cavity laser (NECL) array element of a visible light source by frequency doubling [176].

5.7 STATUS, TRENDS, AND OPPORTUNITIES

5.7.1 Status

In the 1980s, many different technologies, such as time-division multiplexing, coherent detection, soliton, and WDM, were investigated for cost-efficient exploitation of the fiber transmission capacity. All these technologies had to carefully manage the "loss budget." After the pioneering demonstrations on how to get rid of the loss concerns through Raman amplifiers [177], or EDFA amplifiers [178–180], a concerted and successful effort was started in the late 1980s to develop pump diode lasers, key enablers to power up these disruptive technologies. This development was fueled by the deregulation of the telecom industry, the need for long haul communication capacity, and the "new economy" financial experiment. Today, after a painful technology and company shake-out, pump diode technology has matured and only a few companies are supplying telecom pump lasers, but in large volume, "powering up the Internet," as initially strived for.

Based on the status of the diode laser technology, the optical power supplies of choice are 980-nm pump lasers and 1480-nm remote pump laser for C and L band EDFAs [9] and 14xx-nm high-power diode lasers for C and L band extension by Raman amplification. Best for distribution of TV and FTTH signals is the cladding-pumped YEDFA, powered up with an uncooled 940-nm MM pump module.

Power photonics, triggered through the demonstration by Gapontsev in 2002 [6] of a diffraction-limited 135-W cw *all fiber* laser, has since established itself in the printing and material-processing market.

5.7.2 Trends and Opportunities

After the shake-out of telecom pump technologies, it is fair to assume that the surviving pump lasers represent already the fittest solution, and no new opportunities will arise for a while. The trend to further reduce cost by gradually increasing power and manufacturing volume will continue.

In this next phase of power photonics, the trend to improve 9xx-nm MM fiber-coupled pumps will continue, mainly based on continuation of the successful length scaling. We do not know today what the limits to length scaling are: material parameters (mobility/free-carrier absorption) or financial limits? Instead of the calculated power conversion efficiency, wallplug efficiency (from wallplug to fiber) will be targeted with the goal to top today's 50% across all power ranges. Lowering cost will be achieved through standardization of parts and increased manufacturing volume.

In this new field of power photonics, there are many opportunities for pump lasers, just to name a few:

(1) BA laser with athermal waveguide to reduce high-power slow-axis NA.
(2) Semiconductor with better ratio of mobility/free carrier absorption and thermal properties.
(3) Pump diode waveguide with loss $\alpha < 0.2$ cm^{-1}.
(4) Expansion-matched heat sinks with higher thermal conductivity.
(5) Low-cost beam matcher between pump diode and fiber.
(6) High-power laser with controllable frequency and phase for coherent combination.
(7) Standardization of products to increase volume and reduce cost.

ACKNOWLEDGMENTS

I thank my colleagues for the exciting time of discoveries, which led to this disruptive technology, and rewarding experience for bringing this technology to the market, Alan Willner and Ivan Kaminov for their persuasive encouragement, Beni Muller and Hartwig Thomas for their expert help to write this chapter, and Yvonne for her loving support.

REFERENCES

[1] N. S. Bergano, "Wavelength division multiplexing in long-haul transoceanic transmission systems," *J. Lightw. Technol.*, 23(12), 4125–4139, 2005.
[2] J. Limpert, F. Roser, S. Klingebiel et al., "The rising power of fiber lasers and smplifiers," *IEEE. J. Sel. Top. Quantum Electron.*, 13(3), 537–545, 2007.
[3] J. Bromage, "Raman amplification for fiber communications systems," *J. Lightw. Technol.*, 22(1), 79–93, 2004.
[4] A. Mooradian, S. Antikichev, B. Cantos et al., "High Power Extended Vertical Cavity Surface Emitting Diode Lasers and Arrays and Their Applications," in Proc. *MOC*, Tokyo, 2005, pp. 1–4.
[5] B. Schmidt, S. Mohrdiek, and C. S. Harder, "Pump laser diodes," in *Optical Fiber Telecommunications* IV-A, San Diego: Academic Press, 2002.
[6] N. S. Platonov, D. V. Gapontsev, V. P. Gapontsev, and V. Shumilin, "135 W CW Fiber Laser with Perfect Single Mode Output," in Proc. *CLEO 2002*, CPDC3, Long Beach, USA, 2002, pp. 1–4.
[7] F. Bachmann, "Goals and Status of the German National Research Initiative," in Proc. *SPIE*, 6456, 2007, pp. 1–12.
[8] C. M. Stickley, M. E. Filipkowski, E. Parra, and E. E. Hach, III, "Overview of Progress in Super High Efficiency Diodes for Pumping High Energy Lasers," in Proc. *SPIE 2006*, 6104, pp. 42–51.
[9] D. G. Foursa, A. Lucero, C. R. Davidson et al., "2 Tb/s (200×10 Gb/s) Data Transmission Over 7,300 km Using 150 km Spaced Repeaters Enabled by ROPA Technology," in Proc. *OFC 2007*, PDP 25, 2007, pp.1–3.
[10] K. Fukagai, S. Ishikawa, K. Endo, and T. Yuasa, "Current density dependence for dark-line defect growth velocity in strained InGaAs/AlGaAs quantum well laser diodes," *Jpn. J. Appl. Phys.*, 30(3a), L371–L373, 1991.
[11] R. G. Waters, D. P. Bour, S. L. Yellen, and N. F. Ruggieri, "Inhibited dark-line defect formation in strained InGaAs/AlGaAs quantum well lasers," *IEEE Photon. Technol. Lett.*, 2(8), 531–533, 1990.
[12] S. E. Fischer, R. G. Waters, D. Fekete et al., "Long-lived InGaAs quantum well lasers," *Appl. Phys. Lett.*, 54(19), 1861–1862, 1989.
[13] W. T. Tsang, "Extension of lasing wavelengths beyond 0.87 µm in GaAs/AlxGa1-xAs double-heterostructure lasers by In incorporation in the GaAs active layers during molecular beam epitaxy," *Appl. Phys. Lett.*, 38(9), 661–663, 1981.
[14] J. Stohs, D. J. Bossert, D. J. Gallant, and S. R. J. Brueck, "Gain, refractive index change, and linewidth enhancement factor in broad-area GaAs and InGaAs quantum-well lasers," *IEEE J. Quantum Electron.*, 37(11), 1449–1459, 2001.
[15] L. A. Coldren, and S. W. Corzine, *Diode Lasers and Photonic Integrated Circuits*, New York, John Wiley & Sons, Wiley-Interscience, 1995.
[16] D. Rodríguez, I. Esquivias, S. Deubert et al., "Gain, index variation and linewidth-enhancement factor in 980-nm quantum-well and quantum-dot lasers," *IEEE J. Quantum Electron.*, 41(2), 117–126, 2005.
[17] S. C. Auzanneau, N. Michel, M. Calligaro et al., "High Brightness (1 W with $M^2 = 2.9$) GaInAs/(Al)GaAs Index Guided Quantum-Dot Tapered Lasers at 980 nm with a High Wavelength Stability," in Proc. *SPIE 2004*, 5452, 2004, pp. 1–13.
[18] M. Pessa, J. Nappi, A. Ovtchinnikov et al., "Aluminium-Free 980-nm Laser Diodes for Er-Doped Optical Fiber Amplifiers," in Proc. *SPIE 1995*, 2397, 1995, pp. 333–341.
[19] M. Krakowski, M. Calligaro, C. Larat et al., "High-Brightness Tapered Laser Diode Bars and Optical Modules with Al-Free Active Region ($\lambda = 980$ nm)," in Proc. *SPIE 2004*, 5620, 2004, pp. 128–136.

[20] H. Wenzel, A. Klehr, M. Braun et al., "Design and Realization of High-Power DFB Lasers," in Proc. *SPIE 2004*, 5594, 2004, pp. 110–123.
[21] M. Peters, V. Rossin, and B. Acklin, "High Efficiency, High Reliability Laser Diodes at JDS Uniphase," in Proc. *SPIE 2005*, 5711, 2005, pp. 142–151.
[22] S. M. Yeh, S. Y. Huang, and W. H. Cheng, "A new scheme of conical-wedge-shaped fiber endface for coupling between high-power laser diodes and single-mode fibers," *J. Lightw. Technol.*, 23(4), 1781–1786, 2005.
[23] C. A. Edwards, H. M. Presby, and C. Dragone, "Ideal microlenses for laser to fiber coupling," *J. Lightw. Technol.*, 11(2), 252–257, 1993.
[24] H. Yoda, and K. Shiraishi, "A new scheme of a lensed fiber employing a wedge-shaped graded-index fiber tip for the coupling between high-power laser diodes and single-mode fibers," *J. Lightw. Technol.*, 19(12), 1910–1917, 2001.
[25] K. Kurokawa, and E. E. Becker, "Laser fiber coupling with a hyperbolic lens," *IEEE Trans. Microw. Theory Tech.*, 23(3), 309–311, 1975.
[26] E. E. Becker and K. Kurokawa, "Hyperbolic type optical fiber lens coupler for coupling the fiber to an optical line source," U.S. Patent No. 3910677, 1975.
[27] V. S. Shah, L. Curtis, R. S. Vodhanel et al., "Efficient power coupling from a 980-nm, broad-area laser to a single-mode fiber using a wedge-shaped fiber endface," *J. Lightw. Technol.*, 8(9), 1313–1318, 1990.
[28] T. Takuji and S. Michitomo, "Optical fiber with lens and method of manufacturing the same," U.S. Patent No. 5845024, 1998.
[29] J. Xu, S. Kenji, and A. Takayuki, "Lensed optical fiber having high coupling efficiency, process of production and apparatus for production of same, and laser diode module," U.S. Patent No. 6597835, 2003.
[30] R. L. Thornton, R. D. Burnham, T. L. Paoli et al., "Low threshold planar buried heterostructure lasers fabricated by impurity-induced disordering," *Appl. Phys. Lett.*, 47(12), 1239–1241, 1985.
[31] D. F. Welch, D. R. Scifres, P. S. Cross, and W. Streifer, "Buried heterostructure lasers by silicon implanted, impurity induced disordering," *Appl. Phys. Lett.*, 51(18), 1401–1403, 1987.
[32] H. Horie, N. Arai, Y. Mitsuishi et al., "Greater than 500-mW CW kink-free single transverse-mode operation of weakly index guided buried-stripe type 980-nm laser diodes," *IEEE Photon. Technol. Lett.*, 12(10), 1304–1306, 2000.
[33] H. Horie, S. Nagao, K. Shimoyama, and T. Fujimori, "Weakly index guided buried-stripe type 980 nm laser diodes grown by a combination of gas source molecular beam epitaxy and metalorganic vapor phase epitaxy with an AlGaAs/InGaP/GaAs double etch stop structure," *Jpn. J. Appl. Phys.*, 38(10), 5888–5897, 1999.
[34] C. S. Harder, P. Buchmann, and H. P. Meier, "High-power ridge-waveguide AlGaAs GRIN-SCH laser diode," *Electron. Lett.*, 22(20), 1081–1082, 1986.
[35] J. N. Walpole, "Slab-coupled optical waveguide lasers: a review," in Proc. *SPIE 2004*, 5365, 2004, pp. 124–132.
[36] J. P. Donnelly, R. K. Huang, J. N. Walpole et al., "AlGaAs–InGaAs slab-coupled optical waveguide lasers," *IEEE J. Quantum Electron.*, 39(2), 289–298, 2003.
[37] J. C. Chang, J. J. Lee, A. Al-Muhanna et al., "Comprehensive above-threshold analysis of large-aperture (8–10 mm) antiresonant reflecting optical waveguide diode lasers," *Appl. Phys. Lett.*, 81(26), 4901–4903, 2002.
[38] S. Pawlik, S. Traut, A. Thies et al., "Ultra-High Power RWG Laser Diodes with Lateral Absorber Region," in Proc. *ISLC 2002*, Conference Digest, 2002, pp. 163–164.
[39] M. Yuda, T. Hirono, A. Kozen, and C. Amano, "Improvement of kink-free output power by using highly resistive regions in both sides of the ridge stripe for 980-nm laser diodes," *IEEE J. Quantum Electron.*, 40(9), 1203–1207, 2004.
[40] M. Buda, H. H. Tan, L. Fu et al., "Improvement of the kink-free operation in ridge-waveguide laser diodes due to coupling of the optical field to the metal layers outside the ridge," *IEEE Photon. Technol. Lett.*, 15(12), 1686–1688, 2003.

[41] D. P. Bowler, "Semiconductor laser with kink suppression layer," U.S. Patent No. 6366595, 2002.
[42] J. Guthrie, G. L. Tan, M. Ohkubo et al., "Beam Instability in 980-nm Power Lasers: Experiment and Analysis," *IEEE Photon. Technol. Lett.*, 6(12), 1409–1411, 1994.
[43] M. F. Schemmann, C. J. van der Poel, B. A. van Bakel et al., "Kink power in weakly index guided semiconductor lasers," *Appl. Phys. Lett.*, 66(8), 920–922, 1995.
[44] M. Achtenhagen, A. A. Hardy, and C. S. Harder, "Coherent kinks in high-power ridge waveguide laser diodes," *J. Lightw. Technol.*, 24(5), 2225–2232, 2006.
[45] D. F. Welch, D. G. Mehuys, D. R. Scifres, "Semiconductor gain medium with multimode and single mode regions," U.S. Patent No. RE37051, 2001.
[46] S. O'Brien, D. F. Welch, R. A. Parke et al., "Operating characteristics of a high-power monolithically integrated flared amplifier master oscillator power amplifier," *IEEE J. Quantum Electron.*, 29(6), 2052–2057, 1993.
[47] R. J. Lang, K. Dzurko, A. A. Hardy et al., "Theory of grating-confined broad-area lasers," *IEEE J. Quantum Electron.*, 34(11), 1998, 2196–2210.
[48] A. Guermache, V. Voiriot, and J. Jacquet, "Investigations of 14xx-nm Pump Lasers Formed by Active MMI Waveguide," in Proc. *SPIE 2006*, 6115, 2006, pp. 1–10.
[49] S. O'Brien, A. Schoenfelder, and R. J. Lang, "5-W CW diffraction-limited InGaAs broad-area flared amplifier at 970 nm," *IEEE Photon. Technol. Lett.*, 9(9), 1217–1219, 1997.
[50] J. Weber, M. T. Kelemen, S. Moritz, and M. Mikulla, "High-Power High-Brightness 3 W Tapered Amplifiers Tunable from 940 nm to 980 nm," in Proc. *SPIE 2006*, 6104, 1–7, 2006.
[51] K. Paschke, B. Sumpf, F. Dittmar et al., "Nearly-Diffraction Limited 980 nm Tapered Diode Lasers with an Output Power of 6.7 W," in Proc. *ISLC 2004*, Conference Digest, Piscataway, NJ, 2004, pp. 43–44.
[52] A. Guermache, V. Voiriot, D. Locatelli et al., "Experimental demonstration of spatial hole burning reduction leading to 1480-nm pump lasers output power improvement," *IEEE Photon. Technol. Lett.*, 17(10), 2023–2025, 2006.
[53] B. Schmidt, S. Pawlik, and N. Lichtenstein, "High power semiconductor laser diode and method for making such a diode," U.S. Patent No. 6798815, 2004.
[54] K. Shigihara, K. Kawasaki, Y. Yoshida et al., "High-power 980-nm ridge waveguide laser diodes including an asymmetrically expanded optical field normal to the active layer," *IEEE J. Quantum Electron.*, 38(8), 1081–1088, 2002.
[55] B. C. Qiu, S. D. McDougall, X. F. Liu et al., "Design and fabrication of low beam divergence and high kink-free power lasers," *IEEE J. Quantum Electron.*, 41(9), 1124–1130, 2005.
[56] B. Schmidt, N. Lichtenstein, B. Sverdlov et al., "Further Development of High Power Pump Laser Diodes," in Proc. *SPIE 2003*, 5248, 2003, pp. 43–54.
[57] M. V. Maximov, Y. M. Shernyakov, I. I. Novikov et al., "Longitudinal Photonic Bandgap Crystal Laser Diodes with Ultra-Narrow Vertical Beam Divergence," in Proc. *SPIE 2006*, 6115, 2006, pp. 1–13.
[58] A. Małąg, A. Jasik, M. Teodorczyk et al., "High-power low vertical beam divergence 800-nm-band double-barrier-SCH GaAsP–(AlGa)As laser diodes," *IEEE Photon. Technol. Lett.*, 18(15), 1582–1584, 2006.
[59] G. Lin, S. T. Yen, C. P. Lee, and D. C. Liu, "Extremely small vertical far-field Angle of InGaAs-AlGaAs quantum-well lasers with specially designed cladding structure," *IEEE Photon. Technol. Lett.*, 8(12), 1588–1590, 1996.
[60] P. Crump, W. Dong, M. Grimshaw et al., "100-W+ Diode Laser Bars Show > 71% Power Conversion from 790-nm to 1000-nm and Have Clear Route to > 85%," in Proc. *SPIE 2007*, 6456, 2007, pp. 1–11.
[61] N. Lichtenstein, A. C. Fily, and B. Reid, "High power semiconductor laser with a large optical superlattice waveguide," U.S. Patent No. 7085299, 2006.
[62] C. R. Giles, T. Erdogan, and V. Mizrahi, "Reflection-induced changes in the optical spectra of 980-nm QW lasers," *IEEE Photon. Technol. Lett.*, 6(8), 903–906, 1994.

[63] I. A. Avrutsky, R. Gordon, R. Clayton, and J. M. Xu, "Investigations of the spectral characteristics of 980-nm InGaAs–GaAs–AlGaAs lasers," *IEEE J. Quantum Electron.*, 33(10), 1801–1809, 1997.

[64] B. F. Ventrudo and G. Rogers, "Fibre-grating-stabilized diode laser," U.S. Patent No. 5485481, 1996.

[65] B. F. Ventrudo, G. A. Rogers, G. S. Lick et al., "Wavelength and intensity stabilisation of 980 nm diode lasers coupled to fibre Bragg gratings," *Electron. Lett.*, 30(25), 2147–2149, 1994.

[66] T. Pliška, S. Arlt, R. Battig et al., "Wavelength stabilized 980 nm uncooled pump laser modules for erbium-doped fiber amplifiers," *Opt. Laser Eng.*, 43, 271–289, 2005.

[67] M. K. Davis, G. Ghislotti, S. Balsamo et al., "Grating stabilization design for high-power 980-nm semiconductor pump lasers," *IEEE. J. Sel. Top. Quantum Electron.*, 11(5), 1197–1208, 2005.

[68] N. Matuschek, S. Mohrdiek, and T. Pliska, "Laser source with high relative feedback and method for making such a laser source," U.S. Patent No. 7099361, 2006.

[69] T. Pliška, N. Matuschek, S. Mohrdiek et al., "External feedback optimization by means of polarization control in fiber Bragg grating stabilized 980-nm pump lasers," *IEEE Photon. Technol. Lett.*, 13(10), 1061–1063, 2001.

[70] M. Achtenhagen, S. Mohrdiek, T. Pliška et al., "L–I characteristics of fiber Bragg grating stabilized 980-nm pump lasers," *IEEE Photon. Technol. Lett.*, 13(5), 415–417, 2001.

[71] R. Badii, N. Matuschek, T. Pliška et al., "Dynamics of multimode diode lasers with strong, frequency-selective optical feedback," *Phys. Rev. E*, 68(3), id. 036605, 1–12, 2003.

[72] J. Wang, and D. T. Cassidy, "Investigation of partially coherent interaction in fiber Bragg grating stabilized 980-nm pump modules," *IEEE J. Quantum Electron.*, 40(6), 673–681, 2004.

[73] A. Ferrari, G. Ghislotti, S. Balsamo et al., "Subkilohertz fluctuations and mode hopping in high-power grating-stabilized 980-nm pumps," *J. Lightw. Technol.*, 20(3), 515–518, 2002.

[74] I. Kostko, and R. Kashyap, "Modeling of Self-Organised Coherence-Collapse and Enhanced Regime Semiconductor Fibre Grating Reflector Lasers," in Proc. *SPIE 2004*, 5579, 2004, pp. 367–374.

[75] S. Mohrdiek, "Stabilized laser source," U.S. Patent No. 6771687, 2004.

[76] T. Pliška, N. Matuschek, J. Troger et al., "A Compact, Narrow Band, and Low Noise 800 mW Laser Source at 980 nm," in Proc. *SPIE 2005*, 5738, 2005, pp. 380–387.

[77] J. A. Sharps, "Reliability Of Hermetically Packaged 980 nm Diode Lasers," in Proc. *LEOS 1994*, 2, 1994, pp. 35–36.

[78] D. W. Hall, P. A. Jakobson, J. A. Sharps, and R. F. Bartholomew, "Packaging of high power semiconductor lasers," U.S. Patent No. 5770473, 1998.

[79] A. Oosenbrug, "Reliability Aspects of 980-nm Pump Lasers in EDFA Applications," in Proc. *SPIE 1998*, 3284, 1998, pp. 20–27.

[80] R. Battig, H. U. Pfeiffer, N. Matuschek, and B. Valk, "Reliability Issues in Pump Laser Packaging," in Proc. *LEOS 2002*, 2, 2002, pp. 542–543.

[81] X. Liu, M. H. Hu, C. G. Caneau et al., "Thermal Management Strategies for High Power Semiconductor Pump Lasers," in Proc. *ITHERM 2004*, 2, 2004, pp. 493–500.

[82] M. A. Bettiati, C. Starck, F. Laruelle et al., "Very High Power Operation of 980 nm Single-Mode InGaAs/AlGaAs Pump Lasers," in Proc. *SPIE 2006*, 6104, 2006, pp. 130–139.

[83] B. Sverdlov, B. Schmidt, S. Pawlik et al., "1 W 980 nm Pump Modules with Very High Efficiency," in Proc. *ECOC 2002*, 5, 2002, pp. 1–2.

[84] R Nagarajan,, V. V. Rossin, H. Ransom et al., "Grating Stabilized 0.5 W, 980 nm Pump Modules," in Proc. *ISLC 2000*, 2000, pp. 27–28.

[85] W. C. Tang, H. J. Rosen, P. Vettiger, and D. J. Webb, "Raman microprobe study of the time development of AlGaAs single quantum well laser facet temperature on route to catastrophic breakdown," *Appl. Phys. Lett.*, 58(6), 557–559, 1991.

[86] H. P. Dietrich, M. Gasser, A. Jakubowicz et al., "Semiconductor lasers and method for making the same," U.S. Patent No. 5940424, 1999.

[87] A. Baliga, D. C. Flanders, and R. Salvatore, "Long, high-power semiconductor laser with shifted-wave and passivated output facet," U.S. Patent No. 6519272, 2003.

[88] A. K. Chin, Z. Wang, K. Luo et al., "Failure-Mode Analysis of High-Power, Single-Mode, 980 nm, Pump Laser-Diodes," in Proc. *SPIE 2003*, 4993, 2003, pp. 84–90.

[89] R. E. Mallard, R. Clayton, D. Mayer, and L. Hobbs, "Failure analysis of high power GaAs-based lasers using electron beam induced current analysis and transmission electron microscopy," *J. Vac. Sci. Tech. A*, 16(2), 825–829, 1998.

[90] F. Beffa, H. Jackel, M. Achtenhagen et al., "High-temperature optical gain of 980 nm InGaAs/AlGaAs quantum-well lasers," *Appl. Phys. Lett.*, 77(15), 2301–2303, 2000.

[91] C. Hanke, F. U. Herrmann, and S. Beeck, "Investigation of the Mirror Temperature of Narrow-stripe GaAs/AlGaAs Quantum Well Lasers with Segmented Contacts," in Proc. *ISLC 1990*, Conference Digest, 1990, pp. 156–157.

[92] H. O. Yonezu, M. Ueno, T. Kamejima, and I. Hayashi, "An AlGaAs window structure laser," *IEEE J. Quantum Electron.*, 15(8), 775–781, 1979.

[93] J. Ungar, N. Bar-Chaim, and I. Ury, "High-power GaAlAs window lasers," *Electron. Lett.*, 22(5), 279–280, 1986.

[94] R. L. Thornton, D. F. Welch, R. D. Burnham et al., "High power (2.1 W) 10-stripe AlGaAs laser arrays with Si disordered facet windows," *Appl. Phys. Lett.*, 49(23), 1572–1574, 1986.

[95] D. F. Welch and D. R. Scifres, "Window laser with high power reduced divergence output," U.S. Patent No. 4845725, 1989.

[96] S. Yamamura, K. Kawasaki, K. Shigihara et al., "Highly Reliable Ridge Waveguide 980 nm Pump Lasers Suitable for Submarine and Metro Application," in Proc. *OFC 2003*, 1, 2003, pp. 398–399.

[97] D. A. Yanson, J. H. Marsh, S. Najda et al., "High-Power, High-Brightness, High-Reliability Laser Diodes Emitting at 800–1000 nm," in Proc. *SPIE 2007*, 6456, 2007, pp. 1–8.

[98] J. H. Marsh and C. J. Hamilton, "Semiconductor laser," U.S. Patent No. 6760355, 2004.

[99] Y. Yamada, Y. Yamada, T. Fujimoto, and K. Uchida, "High Power and Highly Reliable 980 nm Lasers With Window Structure Using Impurity Free Vacancy Disordering," in Proc. *SPIE 2005*, 5738, 2005, pp. 40–46.

[100] K. Isshiki, T. Kamizato, A. Takami et al., "High-power 780 nm window diffusion stripe laser diodes fabricated by an open-tube two-step diffusion technique," *IEEE J. Quantum Electron.*, 26(5), 837–842, 1990.

[101] K. H. Park, J. K. Lee, D. H. Jang et al., "Characterization of catastrophic optical damage in Al-free InGaAs/InGaP 0.98 μm high-power lasers," *Appl. Phys. Lett.*, 73(18), 2567–2569, 1998.

[102] A. Moser, E. E. Latta, and D. J. Webb, "Thermodynamics approach to catastrophic optical mirror damage af AlGaAs single quantum well lasers," *Appl. Phys. Lett.*, 55(12), 1152–1154, 1989.

[103] A. Moser, "Thermodynamics of facet damage in cleaved AlGaAs lasers," *Appl. Phys. Lett.*, 59(5), 522–524, 1991.

[104] A. Moser, A. Oosenbrug, E. Latta et al., "High-power operation of strained InGaAs/AlGaAs single quantum well lasers," *Appl. Phys. Lett.*, 59(21), 2642–2644, 1991.

[105] A. Moser, and E. E. Latta, "Arrhenius parameters for the rate process leading to catastrophic damage of AlGaAs-GaAs laser facets," *J. Appl. Phys.*, 71(10), 4848–4853, 1992.

[106] F. A. Houle, D. L. Neiman, W. C. Tang, and H. J. Rosen, "Chemical changes accompanying facet degradation of AlGaAs quantum well lasers," *J. Appl. Phys.*, 72(9), 3884–3896, 1992.

[107] H. Brugger, and P. W. Epperlein, "Mapping of local temperatures on mirrors of GaAs/AlGaAs laser diodes," *Appl. Phys. Lett.*, 56(11), 1049–1051, 1990.

[108] P. W. Epperlein, "Temperature, Stress, Disorder and Crystallization Effects in Laser Diodes: Measurements and Impacts," in Proc. *SPIE 1997*, 3001, 1997, pp. 13–28.

[109] A. Jakubowicz, "Microcharacterization of Semiconductor Laser Diodes – Materials and Devices," in Proc. *SPIE 1994*, 2780, 1994, pp. 344–354.

[110] E. Schaub, "Optical absorption rate determination, on the front facet of high-power GaAs laser diodes, by means of thermoreflectance technique," *Jpn. J. Appl. Phys.*, 40, 2752–2756, 2001.

[111] K. Hausler and N. Kirstaedter, "Method and device for passivation of the resonator end faces of semiconductor lasers based on III-V semiconductor material," U.S. Patent No. 7033852, 2006.

[112] M. McElhinney and P. Colombo, "Semiconductor lasers having single crystal mirror layers grown directly on facet," U.S. Patent No. 6590920, 2003.

[113] D. Crawford, M. McElhinney, R. McGowan et al., "Design and Performance of 980 nm Pump Laser Modules Exhibiting Greater than 400 mW Kink-Free Fiber-Coupled Power," in Proc. *OFC 2002*, 2002, pp. 482–483.

[114] N. Chand, "Passivated faceted article comprising a semiconductor laser," U.S. Patent No. 5665637, 1997.

[115] M. Gasser and E. E. Latta, "Method for mirror passivation of semiconductor laser diodes," U.S. Patent No. 5063173, 1991.

[116] L. W. Tu, E. F. Schubert, M. Hong, and G. J. Zydzik, "In-vacuum cleaving and coating of semiconductor laser facets using thin silicon and a dielectric," *J. Appl. Phys.*, 80(11), 6448–6451, 1996.

[117] H. Kawanishi, T. Morimoto, S. Kaneiwa et al., "Semiconductor laser device with a sulfur-containing film provided between the facet and the protective film," U.S. Patent No. 5208468, 1993.

[118] L. K. Lindstrom, N. P. Blixt, S. H. Soderholm et al., "Method to obtain contamination free laser mirrors and passivation of these," U.S. Patent No. 6812152, 2004.

[119] R. W. Lambert, T. Ayling, A. F. Hendry et al., "Facet-passivation processes for the improvement of Al-containing semiconductor laser diodes," *J. Lightw. Technol.*, 24(2), 956–961, 2006.

[120] M. Hu, L. D. Kinney, E. C. Onyiriuka et al., "Passivation of semiconductor laser facets," U.S. Patent No. 6618409, 2003.

[121] P. Ressel, G. Erbert, U. Zeimer et al., "Novel passivation process for the mirror facets of Al-free active-region high-power semiconductor diode lasers," *IEEE Photon. Technol. Lett.*, 17(5), 962–964, 2005.

[122] P. Ressel and G. Erbert, "Method for the passivation of the mirror-faces surfaces of optical semiconductor elements," U.S. Patent No. 20050287693, 2005.

[123] H. Horie, H. Ohta, and T. Fujimori, "Reliability improvement of 980-nm laser diodes with a new facet passivation process," *IEEE. J. Sel. Top. Quantum Electron.*, 5(3), 832–838, 1999.

[124] H. Horie, H. Ohta, T. Fujimori, "Compound semiconductor light emitting device and method of fabricating the same," U.S. Patent No. 6323052, 2001.

[125] G. Gelly, "Design and Technology for Highly Reliable 980 nm Submerged Pump Modules," in Proc. *SubOptic 2004*, paper We 11.2, 2004, pp. 1–3.

[126] M. Usami, N. Edagawa, Y. Matsushima et al., "First Undersea-Qualified 980 nm Pump Laser Diode Module evaluated with Massive Life Test," in Proc. *OFC/IOOC 1999*, vol. suppl., PD39, 1999, pp. 1–3.

[127] G. Ghislotti, and F. Fantini, "Design and screening of highly reliable 980-nm pump lasers," *IEEE Trans. Device and Mater. Rel.*, 2(2), 26–29, 2002.

[128] S. Arlt„ H. U. Pfeiffer, I. D. Jung et al., "Reliability Proving of 980 nm Pump Lasers for Metro Applications," in Proc. *ISLC 2002*, Conference Digest, 2002, pp. 167–168.

[129] H. U. Pfeiffer, S. Arlt, M. Jacob et al., "Reliability of 980 nm Pump Lasers for Submarine, Long-Haul Terrestrial and Low Cost Metro Applications," in Proc. *OFC 2002*, 2002, pp. 483–484.

[130] G. Yang, G. M. Smith, M. K. Davis et al., "Highly reliable high-power 980-nm pump laser," *IEEE Photon. Technol. Lett.*, 16(11), 2403–2405, 2004.

[131] J. Gardner, B. Dean, and A. Shadagopan, "The 980 nm Pump Laser Experience: Prediction and Practice," in Proc. *SubOptic 2004*, paper Tu B 2.4, 2004, pp. 1–3.

[132] B. Dean, and J. Gardner, "Reliability by design – in practice and in the field," *Subtelforum 2003*, 11, 2003, pp. 20–25.

[133] B. Dean, and P. Laverty, "Managing Reliability in a Changing Supplier Market," in Proc. *SubOptic 2004*, paper We 11.5, 2004, pp. 1–3.

[134] Y. Nagashima, S. Onuki, Y. Shimose et al., "1480-nm Pump Laser with Asymmetric Quaternary Cladding Structure Achieving High Output Power of > 1.2 W with Low Power Consumption," in Proc. *ISLC 2004*, Conference Digest, 2004, pp. 47–48.

[135] N. Tsukiji, J. Yoshida, T. Kimura et al., "Recent Progress of High Power 14XX nm Pump Lasers," in Proc. *SPIE 2001*, 4532, 2001, pp. 349–360.

[136] I. P. Kaminow, R. E. Nahory, M. A. Pollack et al., "Single-mode c.w. ridge waveguide laser emitting at 1.55 mm," *Electron. Lett.*, 15(23), 763–765, 1979.

[137] N. Lichtenstein, A. Fily et al., "1 Watt 14xy InGaAsP/InP Ridge Waveguide Pump Laser Diodes with Low Vertical Farfield and High Efficiency," in Proc. *OFC 2003*, 1, 2003, pp. 396–397.

[138] T. Fukushima, N. Tsukiji, J. Yoshida et al., "High Power 14XX nm Pump Lasers for Next Generation," in Proc. *SPIE 2002*, 4905, 2002, pp. 47–61.

[139] K. L. Bacher, C. Fuchs, J. S. Paslaski et al., "High Power 14xx nm Pump Laser Modules for Optical Amplification Systems," in Proc. *SPIE 2001*, 4533, 2001, pp. 47–59.

[140] R. P. Espindola, K. L. Bacher, K. Kojima et al., "High Power, Low RIN, Spectrally-Broadened 14xx DFB Pump for Application in Co-Pumped Raman Amplification," in Proc. *ECOC 2001*, 6, 2001, pp. 36–37.

[141] N. Tsukiji, "Advances in Diode Laser Pumps for Raman Amplification," in Proc. *OFC 2004*, 1, 2004.

[142] D. Garbuzov, A. Komissarov, I. Kudryashov et al., "High Power Raman Pumps Based on Ridge Waveguide InGaAsP/InP Diode Lasers," in Proc. *OFC 2003*, 1, 2003, pp. 394–396.

[143] N. Lichtenstein, B. Schmidt, A. Fily et al., "DPSSL and FL Pumps Based on 980nm-Telecom Pump Laser Technology: Changing the Industry," in Proc. *SPIE 2004*, 5336, 2004, pp. 77–83.

[144] C. M. Stickley, M. E. Filipkowski, E. Parra, and M. Sandrock, "High Power Laser Diodes and Applications to Direct Diode HELs," in Proc. *LEOS 2006*, 2006, pp. 468–469.

[145] M. Leers, "Microchannel heat sink with adjusted thermal expansion," http://www.ilt.fraunhofer.de/eng/100904.html

[146] F. Bachmann, P. Loosen, and R. Poprawe (eds), *High Power Diode Lasers. Technology and Applications*, Springer Series in Optical Sciences, Berlin, Heidelberg: Springer Verlag, 2007.

[147] S. Pawlik, B. Sverdlov, R. Battig et al., "9xx High Power Pump Modules," in Proc. *SPIE 2006*, 6104, 2006, pp. 1–8.

[148] C. P. Lindsey and A. Yariv, "Method for tailoring the two-dimensional spatial gain distribution in optoelectronic devices and its application to tailored gain broad area semiconductor lasers capable of high power operation with very narrow single lobed farfield patterns," U.S. Patent No. 4791646, 1988.

[149] J. R. O'Callaghan, J. Houlihan, V. Voignier et al., "Focusing properties of high brightness gain tailored broad-area semiconductor lasers," *IEEE Photon. Technol. Lett.*, 14(1), 9–11, 2002.

[150] S. Gehrsitz, F. K. Reinhart, C. Gourgon et al., "The refractive index of $Al(x)Ga(1-x)As$ below the band gap: Accurate determination and empirical modeling," *J. Appl. Phys.*, 87(11), 7825–7837, 2000.

[151] M. Kanskar, J. Cai, C. Galstad et al., "High Power Conversion Efficiency and Wavelength Stabilized, Narrow Bandwidth 975 nm Diode Laser Pumps," in Proc. *SPIE 2006*, 6216, 2006, pp. 1–7.

[152] M. Peters, V. Rossin, M. Everett, and E. Zucker, "High Power, High Efficiency Laser Diodes at JDSU," in Proc. *SPIE 2007*, 6456, 2007, pp. 1–11.

[153] C. Schnitzler, S. Hambuecker, O. Ruebenach et al., "Wavelength Stabilization of HPDL Array – Fast-Axis Collimation Optic with integrated VHG," in Proc. *SPIE 2007*, 6456, 2007, pp. 1–7.

[154] S. Pawlik, B. Sverdlov, J. Muller et al., "Broad Area Single Emitter (BASE) Modules with Improved Brightness," in Proc. *CLEO Europe 2007*, CB14-1-THU, 2007, pp. 1.

[155] V. Gapontsev, I. Berishev, G. Ellis et al., "9xx nm Single Emitter Pumps for Multi-kW Systems," in Proc. *SPIE 2006*, 6104, 2006, pp. 1–11.

[156] V. Rossin, M. Peters, E. Zucker, and B. Acklin, "Highly Reliable High-Power Broad Area Laser Diodes," in Proc. *SPIE 2006*, 6104, 2006, pp. 61–70.
[157] N. Lichtenstein, Y. Manz, P. Mauron et al., "High-Brightness 9xx and 14xx Single-Mode Emitter Array Laser Bars," in Proc. *SPIE 2005*, 5711, 2005, pp. 101–108.
[158] V. Lissotschenko, A. Mikhailov, I. Mikliaev, and M. Darsht, "Method and device for influencing light," U.S. Patent No. 20070127132, 2007.
[159] C. Naumer, N. Lichtenstein, B. Schmidt et al., "Brightness Required! More Efficient Laser Diodes with Refractive Micro-Optics," *Laser + Photonik*, 3, 26–29, 2007.
[160] J. Troger, M. Schwarz, and A. Jakubowicz, "Measurement systems for the investigation of soldering quality in high-power diode laser bars," *J. Lightw. Technol.*, 23(11), 3889–3892, 2005.
[161] N. Lichtenstein, Y. Manz, P. Mauron et al., "325 Watt from 1-cm Wide 9xx Laser Bars for DPSSL- and FL-Applications," in Proc. *SPIE 2005*, 5711, 2005, pp. 1–11.
[162] P. Crump, J. Wang, T. Crum et al., " >360 W and >70% Efficient GaAs-Based Diode Lasers," in Proc. *SPIE 2005*, 5711, 2005, pp. 21–29.
[163] J. Sebastian, H. Schulze, R. Hulsewede et al., "High Brightness – High Power 9xx nm Diode Laser Bars: Developments at JENOPTIK Diode Lab," in Proc. *SPIE 2007*, 6456, 2007, pp. 1–11.
[164] H. Li, T. Towe, I. Chyr et al., "Near 1 kW of continuous-wave power from a single high-efficiency diode-laser bar," *IEEE Photon. Technol. Lett.*, 19(13), pp. 960–962, 2007.
[165] Y. M. Manz, M. Krejci, S. Weiss et al., "Brightness Scaling of High Power Laser Diode Bars," in Proc. *Cleo Europe 2007*, CB-37-WED, 2007, 1.
[166] D. Schleuning, M. Griffin, P. James et al., "Robust Hard-Solder Packaging of Conduction Cooled Laser Diode Bars," in Proc. *SPIE 2007*, 6456, 2007, pp. 1–11.
[167] M. Behringer, F. Eberhard, G. Herrmann et al., "High Power Diode Lasers Technology and Application in Europe," in Proc. *SPIE 2003*, 4831, 2003, pp. 4–13.
[168] F. Gonthier, L. Martineau, N. Azami et al., "High-Power All-Fiber Components: The Missing Link for High Power Fiber Lasers," in Proc. *SPIE 2004*, 5335, 2004, pp. 266–276.
[169] A. Wetter, M. Faucher, M. Lovelady, and F. Seguin, "Tapered Fused-Bundle Splitter Capable of 1 kW CW Operation," in Proc. *SPIE 2007*, 6453, 2007, pp. 1–10.
[170] R. K. Huang, B. Chann, L. J. Missaggia et al., "High-brightness wavelength beam combined semiconductor laser diode arrays," *IEEE Photon. Technol. Lett.*, 19(4), 209–211, 2007.
[171] D. Scifres, W. Streifer, and R. Burnham, "Experimental and analytic studies of coupled multiple stripe diode lasers," *IEEE J. Quantum Electron.*, 15(9), 917–922, 1979.
[172] D. Botez, "High-Power, High Brightness Semiconductor Lasers," in Proc. *SPIE 2005*, 5624, 2005, pp. 203–212.
[173] F. Causa, and D. Masanotti, "Observation and analysis of phase-locking in parabolic bow-tie laser arrays," *IEEE J. Quantum Electron.*, 42(10), 1016–1022, 2006.
[174] A. Kewitsch, G. Rakuljic, and A. Yariv, "Semiconductor lasers in optical phase-locked loops," U.S. Patent No. 20060239312, 2006.
[175] E.M. Strzelecka, J.G. McInerney, A. Mooradian et al., "High Power, High Brightness 980 nm Lasers Based on the Extended Cavity Surface Emitting Lasers Concept," in Proc. *SPIE 2003*, 4993, pp. 57–67, 2003.
[176] G. T. Niven, and A. Mooradian, "Trends in Laser Light Sources for Projection Display," *IDW*, 1936–1939, 2006.
[177] J. W. Hicks, "Optical communication systems using Raman repeaters and components therefor," U.S. Patent No. 4616898, 1986.
[178] S. B. Poole, D. N. Payne, and M. E. Fermann, "Fabrication of low-loss optical fibres containing rare-earth ions," *Electron. Lett.*, 21(17), 737–738, 1985.
[179] R. J. Mears, L. Reekie, I. M. Jauncey, and D. N. Payne, "Low-noise erbium-doped fibre amplifier operating at 1.54 μm," *Electron. Lett.*, 23(19), 1026–1028, 1987.
[180] E. B. Desurvire, J. R. Simpson, and P. C. Becker, "High-gain erbium-doped traveling-wave fiber amplifier," *Opt. Lett.*, 12(11), 888–890, 1987.

6

Ultrahigh-speed laser modulation by injection locking

Connie J. Chang-Hasnain and Xiaoxue Zhao

Department of Electrical Engineering and Computer Sciences, University of California, Berkeley, CA, USA

6.1 INTRODUCTION

Optical injection locking (OIL) of semiconductor lasers has been shown to be an efficient and robust technique to improve the spectral and dynamic performance of a directly modulated diode laser. Injection locking refers to a state when the frequency and phase of an oscillator, referred to as the slave oscillator, are locked through direct coupling or injection of another oscillator, referred to as the master oscillator. The concept of injection locking was first observed and described by Huygens in 1655 [1]. He recorded the phenomenon that two pendulum clocks, hanging on the same wall, could synchronize with each other, even when they started with different frequencies and phases. Van der Pol [2] used this phenomenon to create forced oscillator circuit in 1927. Adler [3] further developed the technique in electronics and communications in 1945. After the invention of lasers in 1960s, Stover and Steier [4] studied injection locking on He–Ne lasers in 1966 using optical oscillators. Up to this point, the parameters of interests were oscillator frequency and phase. With the emergence of semiconductor diode lasers, injection locking entered a new era—additional parameters are found to be strongly altered by injection locking, particularly the dynamic modulation characteristics of the slave lasers.

In the late 1970s, the emerging of semiconductor lasers started the revolution of lightwave telecommunication. Numerous investigations were carried out to push the performance of diode lasers. Kobayashi and Kimura [5] reported the first demonstration of injection locking in 1980 on an AlGaAs semiconductor laser, which demonstrated frequency stabilization by injection locking. Iwashita and Nakagawa [6] subsequently demonstrated side-mode suppression of a Fabry–Perot (FP) laser diode by injection locking, turning a multimode laser into a single-mode laser with much lower mode partition noise. Lang [7] published the first theory paper on injection locking of semiconductor lasers, which established the theoretical framework and became the foundation of various subsequent theoretical predictions and studies. Over the next two decades, significant progress was made with numerical simulations and experimental demonstrations of chirp reduction [8–10], modulation bandwidth enhancement [11–14], and broadband noise reduction [11, 15, 16].

In this chapter, the physical origin and theoretical and experimental results of injection-locked lasers are reviewed with a particular emphasis on the dynamic performance of an injection-locked laser. Recent advances of vertical-cavity surface-emitting lasers (VCSELs) under ultrahigh injection-locking conditions are discussed. Some of the performance improvements include $\sim$20 dB increase of both spur-free dynamic range (SFDR) and radio frequency (RF) link gain, one order of magnitude increase in resonance frequency, as well as $\sim$20 dB reduction in laser noise. A record-high modulation bandwidth of 66 GHz is also attained by injection-locking a 10-GHz VCSEL [17]. These properties can potentially lead to an increase of the transmission distance or bandwidth. Finally, new applications of injection locking in wavelength-division multiplexed (WDM) systems are discussed for metropolitan area networks (MANs) and passive optical networks (PONs) applications.

6.2 BASIC PRINCIPLE OF OIL

OIL refers to a technique of injecting light from one laser (master) to another laser (slave) to result in an apparent locking condition of coherent oscillation of the two lasers. Usually, the master is kept under continuous-wave (CW) operation. As the roles of the two lasers are clearly defined, an isolation component is usually used in the setup to achieve unidirectional locking. Figure 6.1 shows the typical experimental configuration for either an edge-emitting laser or a surface-emitting laser as a slave laser. An optical isolator or circulator will be used accordingly.

The dynamics of injection-locked slave laser can be described by injection-locking rate equations. It was established by modifying the laser master equation within the framework of the semiconductor laser theory developed by Lamb [18] in 1964.

Figure 6.1 Typical OIL experimental configurations for (a) edge-emitting lasers (b) surface-emitting lasers (this figure may be seen in color on the included CD-ROM).

6. Ultrahigh-Speed Laser Modulation by Injection Locking

For free-running lasers, based on the laser master equation, the laser field equation can be written as

$$\frac{dE(t)}{dt} = \frac{1}{2}(G-\gamma)E(t) + j\omega E(t). \tag{6.1}$$

$E(t)$ is the laser field, G is the gain from the active material inside the laser cavity, γ is the loss including both material loss and mirror loss, which is equal to the inverse of the photon life time, $1/\tau_p$, and ω is the cavity resonance frequency. The laser field can be written in a complex form. The amplitude equation together with the carrier conservation equation for electrically injected diode lasers forms the well-known laser rate equations.

If $E(t) = |E(t)|e^{j\phi(t)}$, then

$$\frac{d|E(t)|}{dt} = \frac{1}{2}(G-\gamma)|E(t)|$$

$$\frac{dN(t)}{dt} = \frac{I}{qV} - \frac{N}{\tau_N} - G|E(t)|^2,$$

where N is the carrier density, I is the injected current, q is the electron charge, V is the active area volume, and τ_N is the carrier lifetime.

The first thorough theoretical study on OIL of semiconductor lasers was done by Roy Lang [7] in 1982. By adding in the external light injection term, the master equation changes to

$$\frac{dE_s(t)}{dt} = \frac{1}{2}\left[G(N) - \frac{1}{\tau_p}\right]E_s(t) + j\omega(N)E_s(t) + \kappa E_{\text{inj}}(t), \tag{6.2}$$

where $E_s(t) = E(t)e^{j[\omega_{FR}t + \phi_s(t)]}$ and $E_{\text{inj}}(t) = E_{\text{inj}}\, e^{j(\omega_{\text{inj}}t + \phi_{\text{inj}})}$ are the complex fields of the slave and master lasers.

In addition, as the external field enhances the stimulated emission inside the cavity, which will reduce the carrier density, the index of refraction, hence the cavity resonance will be red shifted. It will be seen in a later section that this cavity resonance shift plays a very important role in the resonance frequency enhancement of an injection-locked laser under direct modulation.

Plugging the complex form of both the injection field and the slave laser field into the modified master equation and separating the real and imaginary parts, a set of three equations can be derived and they are the well-known rate equations for injection-locked laser [19].

$$\frac{dE(t)}{dt} = \frac{1}{2}g \cdot v_g[N(t) - N_{\text{th}}] \cdot E(t) + \kappa E_{\text{inj}} \cdot \cos\phi(t) \tag{6.3a}$$

$$\frac{d\phi(t)}{dt} = \frac{\alpha}{2}g \cdot v_g[N(t) - N_{\text{th}}] - \Delta\omega - \kappa \frac{E_{\text{inj}}}{E(t)}\sin\phi(t) \tag{6.3b}$$

$$\frac{dN(t)}{dt} = \frac{I}{qV} - \frac{N(t)}{\tau_N} - \left\{\frac{1}{\tau_p} + g \cdot v_g[N(t) - N_{th}]\right\} E^2(t), \quad (6.3c)$$

where g is the differential gain of the slave laser, v_g is the group velocity of the field inside the slave laser cavity, N_{th} is the threshold carrier density and can be expressed as $g v_g N_{th} = 1/\tau_p$, $\Delta\omega = \omega_{inj} - \omega_{FR}$ is the frequency difference between the master and the slave lasers, often referred to as frequency detuning; $\phi(t) = \phi_s(t) - \Delta\omega t - \phi_{inj}$ is the relative phase between the master laser field and the slave laser field, α is the linewidth enhancement factor of the slave laser, and $\kappa = (v_g/2L)\sqrt{1-R}$ is the coupling coefficient depending on the group velocity v_g, the cavity length L, and the laser mirror reflectivity R. Spontaneous emission and the noise terms are neglected, as the steady state solution is of major interest here. Note that the phase equation, which does not show up in the free-running case, is equally important as the field and the carrier equations, indicating that OIL is a coherent process.

The steady state solution can be obtained by setting the time derivatives of the three rate equations to be zero. In the amplitude equation, by setting the derivative of the amplitude to be zero, carrier density change can be related to the steady state phase difference as

$$\Delta N = -\frac{2\kappa}{v_g g} \frac{E_{inj}}{E_0} \cos\phi_0. \quad (6.4)$$

Physically, the carrier density is decreased due to the enhanced stimulated emission caused by the external light injection. Therefore, the condition that $\Delta N < 0$ confines the phase term to be $0 < \cos\phi_0 < 1$. In the phase equation, by setting the derivative of the phase to be zero and utilizing the equation for (ΔN), an expression of frequency detuning is derived given by

$$\Delta\omega = -\kappa \frac{E_{inj}}{E_0}(\alpha\cos\phi_0 + \sin\phi_0). \quad (6.5)$$

The boundaries of injection locking are given by the maximum and minimum of the frequency detuning. By applying the constraints of the trigonometric functions on the above equation, it is straightforward to get

$$(\alpha\cos\phi_0 + \sin\phi_0)_{max} = \sqrt{1+\alpha^2}$$
$$(\alpha\cos\phi_0 + \sin\phi_0)_{min} = -1.$$

Therefore, the range within which the slave laser can be locked by the master laser is

$$-\kappa\frac{E_{inj}}{E_0}\sqrt{1+\alpha^2} \leq \Delta\omega \leq \kappa\frac{E_{inj}}{E_0}. \quad (6.6)$$

And the corresponding phase value is $-\pi/2 \leq \phi_0 \leq 0$.

Note the experimental detuning values often are presented in the units of wavelength, and the power ratio of the master laser and the slave laser is often referred as the injection ratio. The ratio of the injection field to the slave laser field

6. Ultrahigh-Speed Laser Modulation by Injection Locking

can be expressed as the square root of the power ratio of the two quantities. Therefore, the locking range can be rewritten in terms of wavelength as

$$-\frac{\kappa\lambda^2}{2\pi c}\sqrt{\frac{P_{inj}}{P_0}} \leq \Delta\lambda \leq \frac{\kappa\lambda^2}{2\pi c}\sqrt{\frac{P_{inj}}{P_0}}\sqrt{1+\alpha^2}, \tag{6.7}$$

where $\Delta\lambda = \lambda_{master} - \lambda_{slave}$, c is the speed of light, and P_{inj}/P_0 is the injection ratio.

Notice that the locking range is asymmetric due to the linewidth enhancement factor. The physical origin of this asymmetry can be explained qualitatively as follows. The external light injection reduces the carrier density which will in turn reduce the gain of the slave laser. As the linewidth enhancement factor couples the gain and the phase inside the slave cavity, the gain change through α gives a red shift of the slave-cavity wavelength. Therefore, the slave laser tends to be locked to a wavelength longer than its lasing wavelength (red side).

A two-dimensional stability plot can be generated with wavelength detuning as the vertical axis and injection ratio as the horizontal axis, which is usually used to determine the locking condition, including injection regime and locking range. Figure 6.2 shows such a plot both analytically and experimentally by plotting Eqn (6.7) and measuring the injection ratio and locking range, respectively. Locking regimes can be empirically separated into weak, moderate, strong and ultrahigh injection with injection ratio in a range that is below −20 dB, between −20 and −10 dB, greater than, −10 dB, and greater than 0 dB, respectively. The stronger the injection ratio is, the larger the locking range is. This shows the increase of the robustness of the technique with increase of the injection ratio, as can be seen in Figure 6.2. In addition, as will be shown in the following sections, the dramatic dynamic performance improvement of the slave laser occurs in the strong or ultrahigh-injection regimes. Hence, these are the regimes where the experiments were performed unless elsewhere stated.

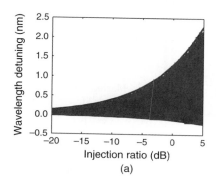
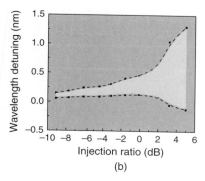

Figure 6.2 Locking range (a) calculated from Eqn (6.7) and (b) measured from experiment (this figure may be seen in color on the included CD-ROM).

If we use control theory to solve the stability of the injection-locked rate equations, there exist unstable and stable locking regions within the locking range shown in Eqns (6.6) and (6.7) [20]. In the unstable region, various nonlinear phenomena could take place, such as four-wave mixing, pulsation, and chaos. These effects have been studied by a number of groups [21–24]. However, to achieve performance improvement, usually the injection-locked laser is operated in the stable locking region.

Experimentally, the locking condition is measured and monitored by an optical spectrum analyzer (OSA). A stream of locking spectra is shown in Figure 6.3 at a particular injection ratio but with various detuning values. It is noted that in the ultrahigh-injection regime, even though the slave laser is injection locked by the master laser and is lasing at the master wavelength, the slave laser mode does not completely disappear. The slave laser cavity-resonance mode can still be seen using a high-resolution OSA. A rule of thumb to judge the locking range in such a case is that the slave-cavity mode is at least 30 dB below the master mode.

An illustrative model to understand the injection-locking dynamics is the phasor model developed by Henry et al. [25] in 1985. This model gives an intuitive

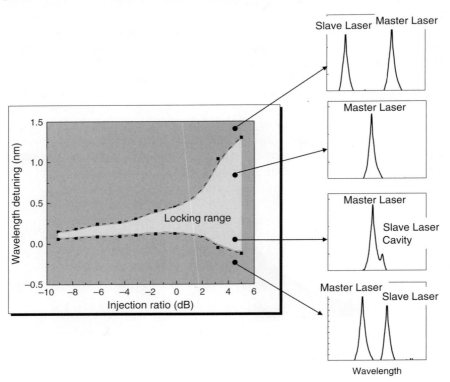

Figure 6.3 Experimentally measured optical spectra of injection locking at a fixed injection ratio but with various detuning values over the locking range (this figure may be seen in color on the included CD-ROM).

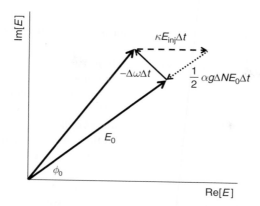

Figure 6.4 Phasor diagram showing injection locking dynamics.

description of the coherence interaction between the photons of the two lasers. It also explains the fact that the slave laser will lase at the master laser frequency under injection locking even though the master laser frequency is detuned from the actual cavity resonance frequency of the slave laser.

A phasor diagram of a statically injection-locked laser field is shown in Figure 6.4. The free-running slave laser field is labeled by E_0. When the slave laser is injection locked by a master laser field, the locked slave field vector rotates by $\Delta\omega\Delta t$ during each time interval Δt due to the frequency difference between the master and the slave laser, $\Delta\omega$, shown by the black arrow in Figure 6.4. In the addition, the external light field contributes to the field as shown by the dashed vector in Figure 6.4 (the coordinate is chosen such that the real part of the electrical field is aligned with the external field merely for calculation simplicity). However, due to the injected external stimulated emission, the slave laser gain is reduced, thus its field amplitude. This introduces the dotted vector shown in the phasor diagram. Therefore, the slave laser field phasor returns to its original position, indicating a steady state injection-locking condition. It should be noted that the phasor diagram is consistent with the first two OIL rate equations, 6.3a and 6.3b shown previously. It is also worth mentioning that due to the linewidth enhancement factor and the gain reduction, the slave laser cavity mode will be red shifted under OIL. This will be discussed in detail in the next section.

6.3 MODULATION PROPERTIES OF OIL VCSELs

6.3.1 Small-Signal Modulation Response

One of the predictions by the theory is that OIL can increase the resonance frequency of a directly modulated laser, which determines the data rate that a directly modulated laser transmitter can handle in an optical link. To verify the enhancement, small-signal frequency response is typically measured by a vector network analyzer experimentally.

OIL was initially studied on FP lasers early in the 1980s as a technique to attain single-mode lasers [6] for improved transmission performance in single-mode fibers (SMFs). As single-mode lasers, namely distributed feedback (DFB) lasers, became commercially available, OIL-DFB was demonstrated with various dynamic performance improvements, including resonance frequency [26], relative intensity noise (RIN) [15], SFDR which is defined as the signal-to-noise ratio at the input RF power for which the system noise floor equals the largest distortion spurious power [27].

High-speed VCSELs are known to have excellent single-mode behavior due to the short cavity length. In addition, VCSELs are promising because of their simple and low-cost fabrication and testing process. Hence, VCSELs are of great interest for many applications, such as MANs, PONs, and analog fiber-optic links. Although the modulation speed of VCSELs is usually limited by the laser parasitics, it has been demonstrated that the resonance frequency can be drastically improved by applying strong OIL [14].

The small-signal frequency response can be obtained by performing standard procedures to linearize the OIL rate equations. Numerical simulation on the linearized rate equations showing resonance frequency enhancement was published by a number of groups independently [11, 19, 25, 28].

Systematic experimental study has been done on 1.55-μm VCSELs [29, 30] with ultrahigh-light injection. As mentioned in the previous section, there are two parameters, injection ratio and detuning, to control the locking condition, so small-signal frequency response needs to be examined as a function of both parameters [14].

If the wavelength detuning, $\Delta\lambda$, is kept at a relative constant value, by varying the injection ratio within the strong or ultrahigh-locking regime, different response curves are shown in Figure 6.5(a) with the stability plot as the inset indicating the locking conditions. The response shown here is the intrinsic response of OIL VCSEL after de-embedding of the laser and system parasitics (including laser packaging, cable loss, detector response, etc.), in which way the features caused by OIL can be revealed clearly. As injection ratio goes high, the relaxation oscillation frequency increases accordingly up to 50 GHz. A steady-state solution with approximation in the strong locking regime shows that

$$\omega_r^2 \approx \omega_{r,\text{FR}}^2 + \kappa^2 \frac{P_{\text{inj}}}{P_0} \qquad (6.8)$$

This means that an OIL laser resonance frequency is dominated by the injection ratio and continues to increase under strong injection. However, the ultimate limit of the resonance frequency remains an interesting, important, and unanswered question.

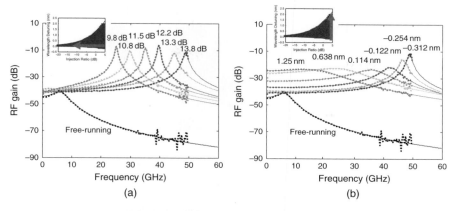

Figure 6.5 Experimental small-signal frequency response of an OIL VCSEL showing (a) injection ratio dependency and (b) detuning dependency [14] (this figure may be seen in color on the included CD-ROM).

For a fixed injection ratio within the strong or ultrahigh-locking regime, the frequency response as a function of wavelength detuning values is shown in Figure 6.5(b), with the stability plot as the inset indicating the locking condition. Figure 6.5(b) shows de-embedded frequency responses of an OIL VCSEL for various $\Delta\lambda$ under a fixed injection ratio of 13.8 dB. As can be seen from the figure, with a large blue $\Delta\lambda$ (master mode toward the short wavelength side of the free-running slave mode), a sharp resonance peak with RF gain of $\sim$20 dB is observed at 50 GHz, which is $\sim$10 times that of free-running VCSEL. As $\Delta\lambda$ increases (master mode moves toward the red side of the free-running slave mode), the response is more damped out; the resonance peak reduces and the modulation efficiency at lower frequencies increases by $\sim$20 dB to that of the free-running VCSEL.

For design purpose, the trends addressed so far can be shown together in the stability plot as in Figure 6.6. Clearly, to achieve high-resonance frequency, the locking condition should be in a high-injection ratio regime with a blue detuning value. Furthermore, by adjusting injection ratio and detuning independently, one can customize the frequency response to have a desired shape.

Notice that the actual 3-dB bandwidth is the determining factor for the modulation data rate in a digital fiber-optic link though the de-embedded response shows a very high-resonance frequency. However, this can be improved by engineering the laser and the system parasitics or deploying a cascaded optical injection locking (COIL) structure, which will be discussed later in the section.

Although the detuning-dependent behavior shown in Figure 6.5(b) was repeatedly observed, not only on VCSELs but also on other types of lasers [13, 26], a physically intuitive explanation that is critical to further the understanding was lacking. To provide an intuitive explanation, a simple model was established [31] based on the experimental observations of OIL VCSELs. However, it is also

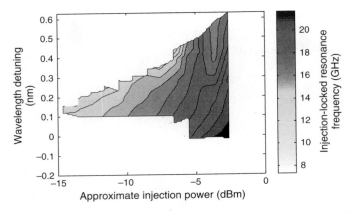

Figure 6.6 Contour plot of resonance frequency as a function of both wavelength detuning and injection power (this figure may be seen in color on the included CD-ROM).

applicable to other types of lasers under strong OIL [32]. Figure 6.7(a) shows the optical spectrum of an OIL VCSEL modulated at a single tone of 10 GHz. The modulation sidebands can be seen on two sides of the carrier wavelength 10 GHz away. Note the sideband on the longer wavelength (red) side has substantially greater amplitude. On the shorter sides, a small peak is observed corresponding to the amplified spontaneous emission (ASE) spectrum of the substantially red shifted VCSEL cavity due to the carrier density reduction by external optical injection. These features change as $\Delta\lambda$ increases, as shown in Figure 6.7(b). Both the asymmetry of the sidebands and the cavity resonance peak are more pronounced

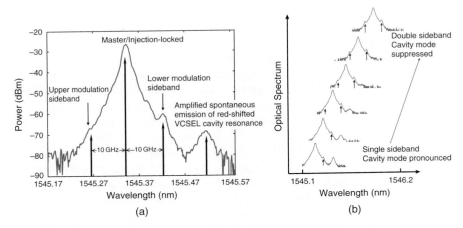

Figure 6.7 Optical spectra of a single-tone (10 GHz) modulated OIL VCSEL. (a) A zoom-in spectrum with major features labeled. (b) A series of spectra at various detuning values showing the gradual change of the features with the detuning (this figure may be seen in color on the included CD-ROM).

6. Ultrahigh-Speed Laser Modulation by Injection Locking

at the most negative $\Delta\lambda$ value. As $\Delta\lambda$ increases, that is, master wavelength gets to the red side, the two sidebands are more equal in value while the cavity resonance peak is more suppressed and closer to the injection-locked lasing mode.

All these experimental observations result from the dynamic coupling between photons and carriers as well as their density change under external light injection. It can be described by the OIL rate equations. From the steady-state solutions derived in Section 6.2, the carrier density change of the slave laser and the steady-state phase difference between the injection light field and the free-running field can be expressed as

$$\Delta N = -\frac{2\kappa}{v_g g}\frac{E_{inj}}{E_0}\cos\varphi_0 \tag{6.10}$$

$$\varphi_0 = \arcsin\left(\frac{-\Delta\omega}{\kappa\sqrt{1+\alpha^2}}\frac{E_0}{E_{inj}}\right) - \tan^{-1}\alpha$$
$$= \arcsin\left(\frac{2\pi c}{\lambda_0^2}\frac{\Delta\lambda}{\kappa\sqrt{1+\alpha^2}}\frac{E_0}{E_{inj}}\right) - \tan^{-1}\alpha \tag{6.11}$$

A Simple inspection on the above equations shows that the carrier density of the slave laser under external light injection is reduced. The underlying physics is that the stimulated emission introduced by the external field depletes the carriers of the slave laser. Therefore, the gain of the injection-locked slave laser is lower than its threshold value, which makes the locked slave laser act essentially like a gain-clamped optical amplifier. Furthermore, the gain level of the OIL laser amplifier can be tuned by the wavelength detuning, $\Delta\lambda$, because the amount of carrier reduction increases with $\Delta\lambda$. However, the stimulated emission from the master laser compensates the gain reduction in the slave laser, and overall the injection-locked slave laser is lasing at the master wavelength. Through the linewidth enhancement factor α, this gain reduction changes the refractive index of the slave laser, thus the cavity resonance frequency. The cavity shift was originally included in the rate equations and clearly pointed out in ref. [19]. The amount of the cavity resonance shift in wavelength can be derived as

$$\Delta\lambda_{cav}(N) = -\frac{\lambda_0^2}{2\pi c}\alpha v_g g\frac{\Delta N}{2} \tag{6.12}$$

As ΔN is always negative, the cavity resonance will be red shifted as observed in Figure 6.7(a). The steady state phase difference φ_0 can be estimated roughly in the range $(-\pi/2, 0)$ across the locking range of $\Delta\lambda$ according to Eqn (6.11). Therefore, on the blue $\Delta\lambda$ side, the carrier density only has a small amount of reduction from the threshold value, whereas on the red side, the reduction is larger according to Eqn (6.10). This trend can also be understood by examining the gain reduction change compensated by the same amount of stimulated emission from the master laser at different $\Delta\lambda$. In essence, the injection-locked VCSEL cavity works as an amplifier with a detuning-controlled gain level, thus a detuning-controlled Q factor, which can greatly enhance the RF response when the red side (lower frequency) modulation sideband coincides with the red shifted cavity resonance.

Based on the above understanding, a model of a Fabry–Perot (F–P) structure with detuning-controlled gain in the cavity can be established to describe an OIL laser acting as a variable-Q amplifier, thus resulting in the drastic resonance frequency enhancement and detuning-dependent RF response [34]. As shown in Figure 6.8(a), when the frequency response is being measured, the modulation sidebands, ω_m, scan over a certain frequency range beside the master mode. The sidebands, which work like probes, replicate the features of the VCSEL amplifier gain profile and represent it in the RF domain after detection. Therefore, main features including the gain of the resonance peak and the frequency at which the resonance peak is located in the RF domain, as functions of the detuning, should follow those of the cavity mode in the optical domain. To verify these strong correlations, hence the physical origin of the resonance enhancement, optical spectra and the frequency response of an OIL VCSEL are simulated using an F–P amplifier model. The modeling approach is as follows: (i) solve the OIL rate equations at steady-state operation, (ii) extract important parameters such as the carrier density reduction (Eqn 6.10) and the cavity resonance shift (Eqn 6.12) from these solutions, and (iii) simulate a conventional F–P structure using these parameters obtained in (ii) to determine the optical spectrum as well as the modulation response. A simple F–P amplifier structure is shown in Figure 6.8(b), which consists of two mirrors and active media in between. A reflection spectrum can be obtained from multiple reflections between the front and back mirror:

$$\frac{E_r}{E_i} = \frac{(1 - r_1^2) r_2 e^{(\gamma - \alpha_i)L} e^{-j2kL}}{1 - r_1 r_2 e^{(\gamma - \alpha_i)L} e^{-j2kL}} - r_1 \qquad (6.13)$$

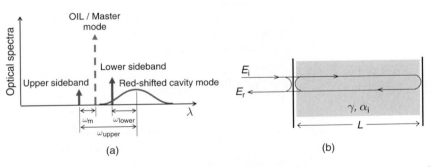

Figure 6.8 (a) Schematic optical spectrum of a modulated OIL laser. The modulation frequency ω_m is being scanned when frequency response measurement is performed. (b) A simple F–P amplifier model used to simulate the OIL VCSEL. E_i is the incident light field, E_r is the reflected light field, L is the cavity length, r_1 and r_2 are the mirror field reflectivity, γ and α_i are the gain and the distributed loss of the active medium [34] (this figure may be seen in color on the included CD-ROM).

6. Ultrahigh-Speed Laser Modulation by Injection Locking

The optical spectrum which measures the intensity of the field is the norm square of Eqn (6.13). The frequency response in the RF domain, including contributions from both the lower and the upper sidebands, can be expressed as

$$\text{amplitude response (dB)} = 10 \, \text{Log} \left| \left(\frac{E_r}{E_i}\right)_{\text{lower}} + \left(\frac{E_r}{E_i}\right)_{\text{upper}} \right|^2 \quad (6.14)$$

The frequency, hence the phase ($2kL$), information of each sideband is included in k, the wave vector.

For a particular injection-locking condition, with a known injection ratio and a detuning value, one can calculate the carrier density reduction from Eqn (6.10). This determines the gain of the slave laser through $\gamma = g(N_{\text{th}} + \Delta N)$, where g is the differential gain, and the threshold carrier density, N_{th}, can be obtained from the threshold gain $\gamma_{\text{th}} = \alpha_i - (1/L)\ln r_1 r_2$. The cavity resonance shift is attained through Eqn (6.12). This needs to be taken into account when computing the frequency of the modulation sidebands ω_{lower} and ω_{upper} relative to the cavity resonance as labeled in Figure 6.8(a), thus obtaining RF response from interactions of both sidebands with the cavity mode. Therefore, the phase of the two sidebands needed in Eqn (6.13) are

$$\phi_{\text{lower}} = 2\left(\frac{\omega_{\text{lower}}}{v_g}\right)L, \text{ and } \phi_{\text{upper}} = 2\left(\frac{\omega_{\text{upper}}}{v_g}\right)L \quad (6.15)$$

Simulation can be performed based on this modeling framework, using Eqn (6.14). Figure 6.9(a) shows the simulated optical spectra as well as the RF frequency response of an OIL laser. As detuning value increases the carrier density reduces further, resulting in a more and more suppressed amplifier peak. This reduced amplifier peak manifests itself in the RF domain as a more damped

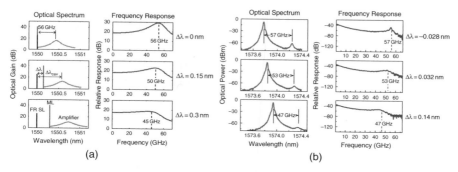

Figure 6.9 Optical spectra and frequency response of an OIL VCSEL at three detuning conditions (a) Simulated at detuning values 0, 0.15 and 3 nm. (FR: free-running, SL: slave laser, ML: master laser) (b) Measured at detuning values −0.028, 0.032 and 0.14 nm. The frequency response is raw data without removing cable loss and VCSEL parasitics [34] (this figure may be seen in color on the included CD-ROM).

resonance peak. Hence, strong correlations between the VCSEL cavity quality factor (Q), the bandwidth of the ASE spectrum, and the damping of the RF response are clearly illustrated. In addition, the carrier density reduction further red-shifts the slave cavity. However, the overall spacing between the master and the slave laser cavity mode is reduced. This spacing in fact determines the resonance frequency. And the resonance frequency is reduced as $\Delta\lambda$ is increased. Figure 6.9(b) shows typical experimental optical spectra and their corresponding modulation response of an OIL VCSEL with an injection ratio of 8 dB. The experimental results in Figure 6.9(b) agree qualitatively well with simulations in Figure 6.9(a), except the overall efficiency declination of the RF response, which is due to the parasitic response of the VCSEL with bandwidth ~10 GHz.

In addition to OIL VCSELs achieving high resonance and bandwidths, DFB lasers under ultra-strong injection condition show similar RF performance enhancements. A record resonance frequency at 72 GHz and 3-dB bandwidth of 44 GHz was reported by Lau et al. [33] on a DFB laser shown in Figure 6.10.

Although we have demonstrated the enhancement of the resonance frequency by OIL so far, to make an injection-locked laser a useful transmitter in a high-speed digital link, a broadband response is highly desired. For both VCSELs and DFB lasers under OIL, a significant reduction in the modulation efficiency is seen between low frequencies and the resonance frequency, as shown in Figure 6.10(b). As a result, the 3-dB bandwidth is typically much lower than that of the resonance frequency. However, by leveraging the bandwidth enhancement property of a directly modulated laser under OIL, a novel cascaded optical injection-locking (COIL) configuration is proposed and experimented to attain a tailorable broadband modulation response [17].

The COIL idea is schematically illustrated in Figure 6.11. The first slave laser is injection-locked and directly modulated, whereas the second slave laser is kept under CW operation and injection-locked by the output of the first stage. The slave laser is lasing at the master mode when it is injection-locked, while its cavity

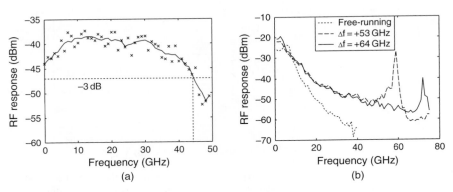

Figure 6.10 Experimental results showing (a) highest 3-dB bandwidth (b) highest resonance frequency of a DFB laser [33].

6. Ultrahigh-Speed Laser Modulation by Injection Locking

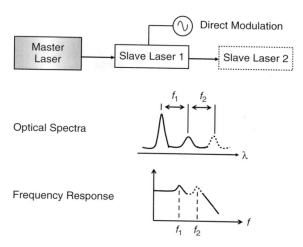

Figure 6.11 A schematic explaining the idea of cascaded optical injection locking (COIL) [17] (this figure may be seen in color on the included CD-ROM).

resonance is red-shifted and exhibits ASE in the optical spectrum, shown in solid and dotted for the first stage and the second stage, respectively. When the modulation lower sideband (with lower frequency) scans over the cavity mode and experiences the amplification, a resonance peak shows up in the frequency response at the frequency corresponding to the spacing between the OIL mode and the cavity mode as labeled in the Figure 6.11. Since the resonance peak is due to the amplification of the sideband by the slave laser cavity, a second resonance peak exhibits even though the second slave laser is not modulated. Therefore, by repeatedly utilizing the cavity effect based on the understanding of OIL dynamics, it is possible to achieve high-speed broadband devices using COIL.

Figure 6.12 shows the measured frequency response of COIL VCSELs. Dotted line is the response from the first OIL stage only, red-detuned and with an injection ratio of $\sim$14 dB. Then the second VCSEL is injection-locked with an injection ratio of $\sim$16 dB. As expected, the response of two-stage OIL is ameliorated as shown in the solid black line in Figure 6.12, and a 3-dB bandwidth of 66 GHz is achieved.

In addition, this system has the scaling-up potential to eventually reach ultra-wide band modulation (>100 GHz) by cascading more than two slave lasers in a daisy chain structure. The modulation signal can also be applied to the master laser instead of directly modulating the slave laser. Similar bandwidth enhancement can be obtained in a master-modulated configuration [35]. Figure 6.13 shows the frequency response of a COIL system, in which the master laser is externally modulated by an electro-optic modulator (EOM), whereas the two slave lasers are kept in CW operation. The dotted light line shows the link response without OIL (essentially the response of the EOM), having a 3-dB bandwidth of 25 GHz. When the first OIL stage is turned on and injection-locked by the modulated master light, the modulation response is increased and bandwidth increases to 36 GHz, shown in

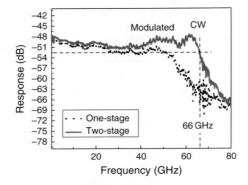

Figure 6.12 Frequency response of COIL. The first slave laser is directly modulated. 3-dB bandwidth of 66 GHz is obtained [17] (this figure may be seen in color on the included CD-ROM).

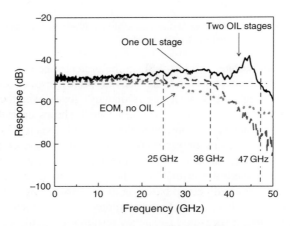

Figure 6.13 Frequency response of a master-modulated COIL configuration. Bandwidth is enhanced from 25 GHz to 47 GHz using two cascaded OIL VCSELs [35].

the dashed line. By adding a second OIL-VCSEL, the bandwidth can be further increased to 47 GHz, shown in the solid line. The shape of the resonance peaks and total response can be tailored by adjusting the injection ratio as well as the detuning values of the two OIL stages.

Therefore, COIL is a promising approach to achieve wide-bandwidth direct-modulated laser transmitters using relatively low-cost and low-speed devices.

6.3.2 Large Signal Modulation

Since the advent of fiber communication, reducing laser linewidth and chirp have been the research topics of great interest. Laser chirp reduces the bandwidth–distance product of the fiber by broadening the pulse after fiber propagation. This problem became more important after the appearance of erbium-doped fiber amplifier (EDFA), which effectively makes most of the communication systems lossless and dispersion limited.

To mitigate the dispersion effect in a typical SMF, the very first step is to use a single-mode laser as the transmitter. With their high power and high side-mode suppression ratio, DFB lasers dominate the long-haul communication system even though the lasers are expensive. Many research activities were conducted to further reduce the linewidth enhancement factor, α, of a DFB laser by either exploiting various high-speed active region design (as the linewidth enhancement factor is inversely proportional to differential gain) or better waveguide structures. Also due to the presence of spontaneous emission and gain compression, carrier density inside the cavity is not perfectly clamped at its threshold value, even when the laser operates above its threshold. In fact, the carrier density never reaches the ideal threshold carrier density. Furthermore, when the laser is modulated, the carrier density fluctuates with the modulation current. This fluctuation of the carrier density results in gain and refractive index variation of the cavity, which, in turn, manifests itself as the time-varying laser wavelength or chirp. The amount of chirp depends on the linewidth enhancement factor, modulation pattern, the laser output power, and some intrinsic parameters of the active region design.

Injection locking was demonstrated to reduce the frequency chirp for a direct amplitude-modulated laser [36]. This technique provides another method to minimize the dispersion penalty without complicating the laser design. In addition, both the master and the slave lasers can be driven with similar electronics without high-power consumption. This greatly simplifies the design of the circuits and hence reduces the cost of the transmitter modules.

From the injection-locked rate equations, the frequency variation, or chirp, of the injection-locked slave laser can be derived. Here, we use photon numbers instead of the electric field. The nonlinear gain is also considered as described by the gain compression factor ϵ. After some derivation, the chirp is expressed as

$$\frac{d\phi}{dt} = \frac{\alpha \epsilon S}{2 \tau_p} + \frac{\alpha}{2} \frac{1}{S} \frac{dS}{dt} + k_c \sqrt{\frac{S_{inj}}{S}}[\sin \phi(t) - \alpha \cos \phi(t)] - \Delta\omega,$$

where the first term is the adiabatic chirp, corresponding to the change of frequency for different current densities. Because of the presence of the gain compression factor, ϵ, the carrier density will not clamp at the threshold level. Whenever the laser is biased differently for various photon output, the carrier density changes and the wavelength follows adiabatically. The second term is the transient chirp. It dictates the wavelength variation based on how the photon is modulated. For a drastic photon number change in a short time, such as the case in digital modulation, the transient chirp is the dominating term of all. Even though the transient chirp can be reduced by increasing the photon density inside the cavity, this might not be feasible most of the time. Transient chirp remains one of the most challenging issues in deploying directly modulated lasers in a high-bit rate system. The last terms are the effect of injection locking, which offer a possibility to cancel both the adiabatic and the transient chirp in a modulated laser. The detuning contributes to a constant wavelength shift and can be either positive or negative.

At the first glance, it might seem plausible to have a positive detuning as it always reduces the chirp. However, as $\Delta\phi$ is time varying when the laser is modulated, the sine and cosine terms can be the more important factors if the injection intensity is high enough. In particular, when the detuning is at the red edge of the locking range, $\Delta\omega < 0$ and $\phi(t) \sim -\pi/2$. Therefore, $\sin\phi(t) - \alpha\cos\phi(t) \sim -1$, the chrip value can be positive, negative, or zero chirp. However, if detuning is at the blue edge of the locking range, $\Delta\omega > 0$ and $\phi(t) \sim 0$. Therefore, $\sin\phi(t) - \alpha\cos\phi(t) \sim -\alpha$, a zero, positive, or negative chirp may result. The trigonometric terms have the possibility to follow both the adiabatic and transient chirp closely but with a different sign.

The chirp reduction has been demonstrated on OIL VCSESLs by both experiment and numerical simulation [37, 38]. Figure 6.14(a) and (b) shows the

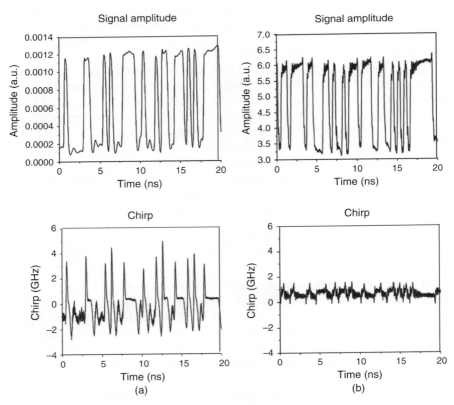

Figure 6.14 Output power and wavelength variation of (a) free-running VCSEL and (b) OIL VCSEL under large signal modulation.

6. Ultrahigh-Speed Laser Modulation by Injection Locking

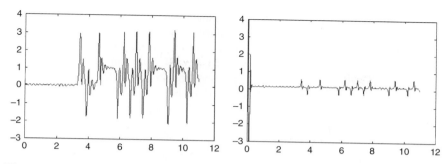

Figure 6.15 Simulated chirp of (a) free-running VCSEL (b) OIL VCSEL (this figure may be seen in color on the included CD-ROM).

experimentally measured signal amplitude and chirp before (free running) and after injection locking. An Advantest dynamic chirp measurement system was used to measure the real-time wavelength information of the signal. From the similarity of the waveform of both the amplitude and the chirp, it is obvious that chirp is modulation waveform dependent. The adiabatic chirp, defined as the steady state wavelength difference between the high and low level, is due to carrier density difference between the two levels. The transient chirp, which occurs at the transient period of the level change, can be understood as the wavelength variation due to carrier overshoot when the levels changed. Both adiabatic and transient chirp will result in dispersion after fiber propagation and will increase the power penalty at the receiver end. In this free-running case, the peak-to-peak frequency variation is 7.8 GHz, which is mainly due to the transient chirp. A much-reduced peak-to-peak frequency variation, less than 1.8 GHz, is obtained when the VCSEL is injection locked. Compared to the free-running case in Figure 6.14(a), the transient chirp is effectively reduced by the external power injection. Note that the adiabatic chirp is also suppressed from 2.5 to 0.5 GHz.

Numerical simulation can be performed according to the rate equations. The simulated chirp before and after injection locking is shown in Figure 6.15. The large amount of chirp reduction agrees well with the experimental results.

The chirp reduction by OIL indeed helps to improve the performance of the VCSEL as a digital transmitter. The bit-error rate (BER) results will be shown in a later section.

6.4 RF LINK GAIN ENHANCEMENT OF OIL VCSELs

Frequency response curves in Figure 6.5 all reveal an interesting phenomenon for VCSELs under ultrahigh injection locking. Very large modulation efficiency (RF gain) enhancement is attained at lower frequencies for red wavelength detuning. This phenomenon was never predicted by theory. Experimentally, the redder $\Delta\lambda$ is, the larger the gain is. Typically, the modulation efficiency of a free-running 1.55-μm VCSEL used in Section 6.3 is 0.2–0.3 W/A [29]. For a large injection ratio (14 dB), the low-frequency RF gain enhancement is up to 20 dB at large positive detuning (1.25 nm showing in Figure 6.5(b)), resulting in a modulation efficiency of 2.2 W/A, or an equivalent 2.75 photons generated per electron–hole pair due to the stimulated emission from the external light injection.

Under a fixed injection ratio, the RF gain varies with detuning. The minimal value is at a blue detuning associated with frequency response showing a sharp resonance peak. On the other hand, the maxima RF gain is obtained at red detuning values accompanied by frequency response highly damped. Hence, there exists a tradeoff between the resonance frequency and RF gain. Figure 6.16 shows the low-RF gain as a function of wavelength detuning studied on different devices. The correlation between RF gain enhancement and wavelength detuning is clearly visible. Furthermore, the maximum RF gain enhancement obtained at the red edge of the locking range keeps increasing with increasing the injection ratio. Figure 6.17 shows this enhancement vs injection ratio.

If we combine the dependence on both the detuning value and injection ratio, the RF gain can also be plotted on the stability plot shown in Figure 6.18. It was measured for a small-signal input of −20 dBm at 1 GHz. A high-RF link gain at low frequencies can be obtained when operating the OIL laser in the strong injection regime with red detuning values.

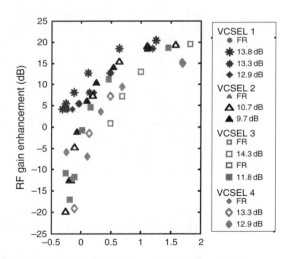

Figure 6.16 RF gain measured at 1 GHz for four vertical-cavity surface-emitting lasers (VCSELs) vs wavelength detuning. 20-dB RF gain is found for large detuning cases. (FR: free running, decibel values are injection ratios) (this figure may be seen in color on the included CD-ROM).

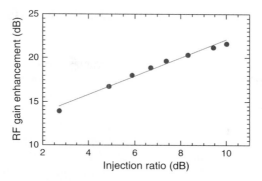

Figure 6.17 Maximum RF gain enhancement vs injection ratio for an OIL VCSEL (this figure may be seen in color on the included CD-ROM).

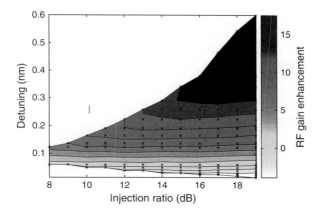

Figure 6.18 Contour plot of RF gain enhancement as a function of both wavelength detuning and injection ratio (this figure may be seen in color on the included CD-ROM).

At high frequencies, the RF gain comes primarily from the resonance peak. A high-resonance peak about 20 dB can be obtained at the blue detuning edge. This high-frequency narrow band RF gain is useful for some interesting RF applications such as optoelectronic oscillators (OEO) for ultra-low phase noise microwave signal generation [54, 60].

The RF gain enhancement, both low frequency and high frequency, can be qualitatively explained by the physical picture described in Section 6.3. As shown in Figure 6.7(b), at the blue detuning edge, the modulation sideband is amplified by the sharp slave-cavity resonance which is far from the carrier frequency (master mode), resulting in a high-frequency resonance peak. However, at the red detuning edge, the carrier frequency is close to the cavity resonance, thus mainly amplifying the low-frequency signals.

6.5 NONLINEARITY AND DYNAMIC RANGE OF OIL VCSELs

Linearity is an important figure-of-merit for analog fiber-optic links, as it represents the faithfulness of reproduction of the modulation signal. Typical optical link linearity is characterized by SFDR. It is defined as the signal-to-noise ratio at the input RF power for which the system noise floor equals the largest distortion spur power. Distortion occurs from several sources, including second- and third-order harmonics of the signal and the intermodulation distortion (IMD) between different channels. For device studies, the standard technique is to characterize the third-order intermodulation (IMD3) products due to two-tone modulation. The highest IMD3 SFDR reported for a direct-modulated diode laser at 1 GHz is 125 dB-Hz$^{2/3}$ for 1 GHz for a 1.3-μm DFB laser [39] and 113 dB-Hz$^{2/3}$ at 0.9 GHz for an 850-nm VCSEL [40].

Nonlinear distortion has been shown to be inversely proportional to the resonance frequency [41]. Hence, with the resonance frequency enhancement, injection-locked lasers are promising for exhibiting reduced nonlinear distortion. Experiments and simulations demonstrating decreased distortion and improved IMD3-limited SFDR using strong injection locking are shown below.

6.5.1 Experiments

To measure IMD3 SFDR, typically two RF synthesizers are used to directly modulate the slave laser, and the output signal is characterized using an RF spectrum analyzer. Figure 6.19 shows the SFDR improvement by injection-locked DFB laser [27]. The improved performance is mainly due to the

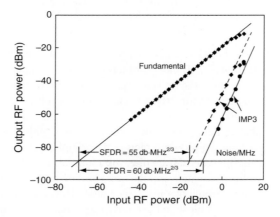

Figure 6.19 Measured RF fundamental tone and IMD3 for free-running and OIL-DFB lasers [27].

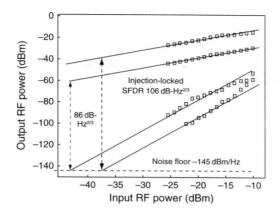

Figure 6.20 Two-tone spur-free dynamic range (SFDR) improvement at 50°C, 0.2-nm detuning and 1-GHz modulation by OIL. Inner lines are for free-running VCSEL [42].

resonance frequency increase in addition to the suppressed relaxation oscillation of the original free-running laser. Figure 6.20 shows that a significantly improved IMD3 SFDR of 106 dB-Hz$^{2/3}$ at 1 GHz [42] was obtained for an injection-locked 1.55-μm VCSEL. Due to both the fundamental tone enhancement and IMD3 reduction, a 20-dB improvement in the SFDR was attained by using strong OIL.

The fundamental tone and the IMD3 responses are shown together as a function of frequency in Figure 6.21. This shows that the large SFDR improvement can be attained for a wide range of modulation frequencies. The fundamental tone response is similar to the small-signal frequency response discussed in Section 6.3. The resonance frequency of the free-running VCSEL is about 2 GHz, and with injection locking, it is greatly increased leaving a flat response in the

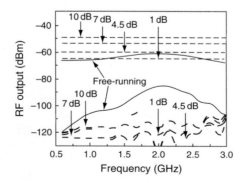

Figure 6.21 Experimental modulation response and two-tone IMD3 distortion vs modulation frequencies at different injection ratios [42].

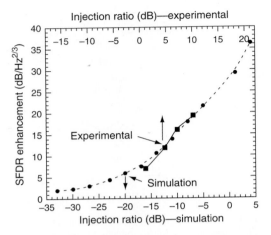

Figure 6.22 Experimental and simulated spur-free dynamic range (SFDR) enhancement vs. injection ratio at half of the resonance frequency [42].

frequency range shown here. For each injection ratio condition, wavelength detuning values were chosen to be close to the red edge where RF gain enhancement is more drastic to maximize the SFDR improvement. The IMD3 power, on the other hand, largest at the resonance peak, is reduced with the increasing of the resonance frequency. Therefore, the modulation efficiency enhancement with increasing injection ratio ranges from 0.5 to 16 dB, whereas the distortion is reduced from 11 to 19 dB. The SFDR enhancement is determined by the RF gain enhancement plus one-third of the distortion reduction; thus, the RF gain significantly contributes to the SFDR enhancement. If we pick the single frequency at 1 GHz and plot the SFDR improvement for various injection ratios as shown in Figure 6.22, the experiment demonstrates that an increasing injection ratio leads to an increasing SFDR. It should be mentioned that the measured distortion is source instrument limited, as evidenced by the similar oscillations observed in the measurement of the IMD of two synthesizers.

6.5.2 Simulations

Theoretical predictions for the distortion of an injection-locked laser have shown that a small reduction in distortion is possible for some injection conditions [43]. The injection conditions studied, however, were for a low-effective injection ratio, limited to typical edge-emitting laser injection-locking parameters. In the case of surface-emitting lasers, as the cavity length can be 300 times smaller than the edge emitters, even with a 99% reflectivity, the coupling coefficient κ can be 20 times larger. Numerical simulations were performed for injection-locked VCSELs under very strong injection conditions and demonstrate that a very large distortion

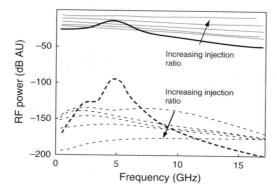

Figure 6.23 Numerical simulation for the modulation frequency response (fundamental tone) and the two-tone IMD3 [42].

reduction is possible [42]. The OIL laser rate equations are typically used for numerical simulations.

Figure 6.23 shows the simulated results for the signal and distortion frequency response. The thick curves are the fundamental tone (top solid) and IMD3 (bottom dotted) of the free-running laser, whereas the thin curves are those under various injection ratios. The wavelength detuning is chosen to be at the red edge of locking range. As can be seen, the modulation efficiency is increased as the injection ratio increases; meanwhile, the IMD3 is suppressed, which is consistent with the experimental results. For very high-injection ratios, substantial suppression of the IMD3 is observed.

The simulation results for the SFDR improvement vs injection ratio are also shown in Figure 6.20. They are plotted for the 2.4-GHz signal as a function of injection ratio for a wavelength detuning that is near the red edge of the locking range. For consistency, the comparison between the experimental and numerical results was done at half the resonance frequency.

The experimental and simulation results are comparable in trend. The simulation predicts that a very large dynamic range can be achieved for a high-injection ratio.

6.6 RELATIVE INTENSITY NOISE OF OIL VCSELs

RIN is defined as the dynamic range between the laser power and the laser intensity noise. For laser RIN measurement, the photodetector shot noise and the RF spectrum analyzer should be lower than the laser intensity noise. Otherwise, the system will be thermal or shot noise limited. High laser power would help increase the dynamic range of the measurement to prevent the system limit. Injection locking has been reported to reduce laser RIN for VCSELs and FP edge-emitting lasers [15, 16].

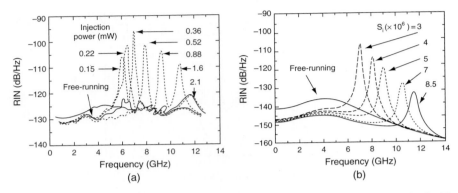

Figure 6.24 Experimental and simulated relative intensity noise (RIN) reduction due to OIL of a FP laser [15].

Figure 6.24 shows the RIN reduction of an injection-locked FP laser at different injection power levels [15] and the simulation results. For VCSELs, however, due to the low emission power (~ 1 mW) and fiber-coupling loss, the optical signal was too weak to be detected. The solution to this issue is to add an EDFA followed by an optical filter to amplify the noise signal [42]. An HP 71400C lightwave signal analyzer was then used to measure the RIN of the OIL VCSEL. The RIN spectra of the injection-locked VCSEL at different injection conditions are shown in Figure 6.25. The RIN spectrum of the free-running VCSEL is also shown as the reference at a 1-mA bias (2 × threshold for this device). The RIN peaks of both free-running and injection-locked conditions are in agreement with resonance frequencies in the small-signal modulation response of the laser. For an injection ratio of 8 dB, RIN data for various detuning values are shown, which has a similar behavior as the frequency response. A large RIN reduction at the low-frequency regime (0–13 GHz) is observed due to increased resonance peak. However, it is relatively constant over a large frequency span, even at the very low-frequency region. This is attributed to the limit from the noise of the EDFA. At a higher injection ratio (10 dB), the resonance frequency increases beyond the instrument limit, and the noise is reduced over nearly the entire visible frequency band.

The RIN reduction can be intuitively understood as the following: when the laser is injection locked, fewer carriers are needed to achieve lasing threshold. The spontaneous emission of the laser and, therefore, its noise is reduced. However, the increased resonance frequency enhancement is the most important reason that the noise value is decreased at the low-frequency regime as the noise peak is moved to higher frequencies. The combination of these two factors results in the RIN reduction of an injection-locked laser.

To model the OIL laser RIN, usually Langevin noise terms are applied [44]. Therefore, the complete OIL rate equations including the noise terms are presented here, which are used for RIN simulations.

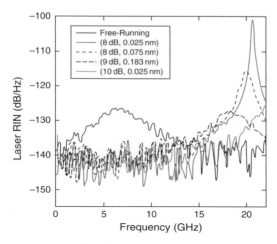

Figure 6.25 Experimental demonstration of relative intensity noise (RIN) reduction of OIL VCSEL at various detuning values compared with free-running VCSEL [42].

$$\frac{dS}{dt} = \frac{G_0(N - N_0)}{1 + \varepsilon S} \cdot S - \frac{S}{\tau_p} + 2k_c\sqrt{S \cdot S_{\text{inj}}} \cdot \cos(\phi(t) - \phi_{\text{inj}}) + R_{sp} + F_s$$

$$\frac{d\phi}{dt} = \frac{\alpha}{2} G_0(N - N_{\text{th}}) - 2\pi \cdot \Delta f - k_c\sqrt{\frac{S_{\text{inj}}}{S}} \cdot \sin(\phi(t) - \phi_{\text{inj}}) + F_\phi$$

$$\frac{dN}{dt} = \frac{I}{q} - \frac{N}{\tau_s} - \frac{G_0(N - N_0)}{1 + \varepsilon S} \cdot S + F_n$$

$$F_s = \sqrt{\frac{2S \cdot R}{\Delta t}} x_e$$

$$F_\phi = \sqrt{\frac{R}{2S \cdot \Delta t}} x_\phi$$

$$F_n = -F_s + \sqrt{\frac{2N}{\tau_s \cdot \Delta t}} x_n$$

As the RIN peak corresponds to the relaxation oscillation frequency in the small-signal frequency response, it should follow the same trend as the frequency response at various injection ratios and detuning values, as shown in Figures 6.22 and 6.23.

6.7 APPLICATIONS

6.7.1 Passive Optical Networks

The WDM-PON has long been recognized as an attractive upgrade solution for current access networks due to its large capacity, ease of upgradeability, and security guarantee. In that respect, laser transmitters in WDM-PONs are required to emit transmission wavelengths fixed to the WDM grid. This key requirement is to ensure minimal crosstalk with other wavelengths and transmission loss at the wavelength multiplexers and demultiplexers, for example, arrayed waveguide gratings (AWGs) [45]. Wavelength-specific DFB lasers, distributed Bragg reflector (DBR) lasers, and tunable lasers are considered the most expensive types of optical network unit (ONU) upstream transmitters. Active temperature control and wavelength feedback monitoring, for example, using FP etalons and Bragg filters, are necessary to maintain emission wavelengths at their designated wavelengths, thus leading to large-power consumption, complex system configuration, and more importantly, high cost.

A promising alternative to achieving low-cost wavelength-specific sources in a WDM-PON is to use VCSELs. VCSELs are well known for their excellent single-mode behavior and potential for low-cost manufacturing and electronics integration. A novel scheme that exploits the use of OIL VCSELs for operation as stable, directly modulated, and potentially uncooled ONU transmitters has been proposed and experimentally demonstrated. OIL allows the VCSEL wavelength to be locked to that of the master laser and dense wavelength-division multiplexing (DWDM) grid within a certain detuning range and thus, is locked onto the specific AWG port provided by the central office (CO) without additional wavelength locking or stabilizing elements [46]. Therefore, using injection-locked VCSELs as upstream transmitters expands the wavelength tolerance of the ONU and compatibility with various vendors and systems with slightly different DWDM grid. The proposed scheme does not require external injection-locking sources and external upstream modulators. At the CO, DFB lasers that carry downstream signals also serve a second function as master lasers to injection-lock ONU slave VCSELs onto the WDM grid. Both downstream and upstream data are intensity modulated and directly detected using standard components. The slave VCSEL is shown to respond only strongly to the wavelength but not the data from the master DFB laser, with good upstream transmission performance obtained over 25-km fiber. Thus, while the downstream laser's carrier wavelength is used to lock the upstream VCSEL, its data does not influence the upstream information. A similar work was proposed in Ref. [47] but used downstream signals in differential phase shift keying (DPSK) format to injection lock FP laser diodes (LDs), which brought extra expense of an external phase modulator for each wavelength channel at the CO and a DPSK demodulator at each ONU.

Figure 6.26 shows the proposed WDM-PON implementing OIL VCSELs. At the CO, master DFB lasers, either directly or externally modulated with downstream data, are temperature-tuned to emit distinct wavelengths on the DWDM grid. At each ONU, an optical splitter divides the optical power of the demultiplexed downstream signal to feed a downstream receiver and to injection lock a slave VCSEL. The splitting ratio must be carefully chosen to satisfy error-free detection at the downstream receiver while maintaining stable injection locking of the VCSEL. The OIL VCSEL is then directly modulated with upstream data and

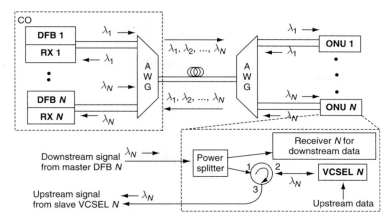

Figure 6.26 Proposed scheme of WDM-PON using directly modulated OIL VCSELs as upstream transmitters. Each slave VCSEL is injection locked by modulated downstream signal transmitted from a master DFB laser located at the CO [49].

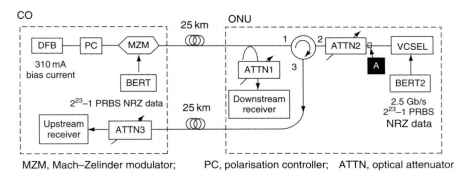

Figure 6.27 Experimental setup for single-channel demonstration [49].

transmits them back to the CO. For colorless operation, identical tunable VCSELs may be placed at each ONU [48].

The characterization and evaluation of the OIL VCSEL scheme were performed with a *non*-polarization-maintaining (*non*-PM) single-channel experiment with the setup shown in Figure 6.27. Detailed description about the experiment can be found in Ref. [49].

The BER of 2.5-Gb/s upstream signal from the VCSEL, injection locked by the master DFB laser with different line-rates and injection power levels, was evaluated. Figure 6.28 plots the BER curves measured for two injection power levels, -20 and -23 dBm (with coupling loss calibrated), and master laser operated at CW, 1.25-, and 2.5-Gb/s line-rates. BER measurements from a 2.5-Gb/s modulated free-running VCSEL are also plotted in Figure 6.28 as a performance comparison.

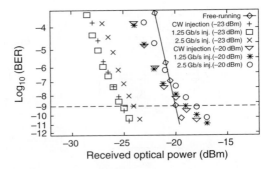

Figure 6.28 Upstream bit-error rate (BER) of free-running VCSEL and VCSEL injection locked by continuous wave (CW), 1.25- and 2.5-Gb/s master DFB laser with −23 and −20-dBm absolute injection powers [49].

These results were obtained with the upstream and downstream optical extinction ratio (ER) fixed to an identical and arbitrarily chosen value of 4.5 dB, enabling the OIL VCSEL upstream performance to be studied only as a function of optical injection power and downstream bit rate. Even though results in Figure 6.28 show the feasibility of transmitting upstream data from the free-running VCSEL, the setup is a single-channel experiment which neglected the potential high-transmission losses and out-of-band crosstalk arising from misalignment of the free-running VCSEL wavelength with the AWG passband. With injection locking at an absolute power of −20 dBm, an improvement in upstream transmission performance over the free-running case is observed but only at high-BER measurements, a feature applicable to networks implementing forward-error correction. As for the case of −23-dBm injection-locking power, a major improvement of 4–5 dB in BER was obtained over the free-running case, allowing an increased power margin in the network. With both injection-locking powers, the performance degrades with increasing master line-rate. These results, therefore, indicate that this scheme benefits from stable injection locking with low-injection power, which also benefits the detection of downstream signals at the ONU. The power splitter ratio can then be easily designed to maximize the performance at both the downstream receiver and the OIL VCSEL.

The experiment described above employed unidirectional fibers to evaluate the injection-locked behavior and transmission performance of the OIL VCSEL. In practice, bidirectional optical networks are more preferred owing to reduced management and fiber costs. However, with OIL, the master (downstream) and slave (upstream) wavelengths are identical, resulting in upstream performance degradation at the receiver in the CO from Rayleigh backscattered-induced intensity noise of the master DFB laser [50]. Compounding the issue is that the master light is not CW but modulated with downstream data. Therefore, the study of the effects of Rayleigh backscattered light together with the ER of the downstream signal on the proposed scheme is necessary.

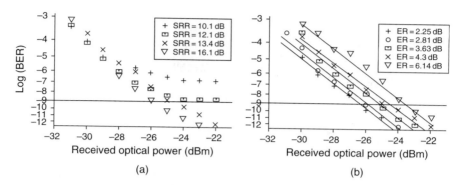

Figure 6.29 Upstream bit-error rate (BER) curves for different values of upstream signals to Rayleigh backscattering ratio (SRR) at (a) constant downstream extinction ratio (ER) = 4.5 dB and (b) downstream ER at constant SRR = 13.4 dB.

The performance dependence of OIL VCSELs on Rayleigh backscattering effects in a 25 km bidirectional optical link and on the modulated downstream data of the master DFB laser was studied. The setup is similar as shown in Figure 6.27 and described in detail in Ref. [51]. The BER of the modulated upstream signal against (a) the upstream signal to Rayleigh backscattering ratio (SRR) and (b) the optical ER of the modulated downstream signal at a constant injection power is shown in Figure 6.29.

Rayleigh backscattering effects are analyzed by adjusting the downstream and upstream ERs to a constant 4.5 dB while varying the SRR. Figure 6.29(a) shows BER measurements of 2.5 Gb/s upstream signals from the VCSEL optically injection-locked by 2.5 Gb/s downstream signals for different values of SRR. The upstream BER degrades with decreasing SRR, inducing power penalties and error floors. Nonetheless, results show that the influence of Rayleigh backscattering can be drastically reduced to achieve error-free transmission (at BER $<10^{-9}$) if the SRR is higher than 13.4 dB. This is easily achievable within the design limits of a practical system using lensed-fiber-pigtailed VCSELs with less than 1 dB coupling loss, thereby significantly increasing upstream signal power at the CO. Maintaining the SRR and upstream ER at 13.4 and 4.5 dB, respectively, the downstream ER was then varied with the corresponding upstream BER curves shown in Figure 6.29(b). While error-free transmission is achieved in all cases, the increase in downstream ER degrades upstream performance. In the worst case, an increase in downstream ER from 2.25 to 6.14 dB incurs ~3 dB degradation on the upstream BER.

Figure 6.30 plots the minimum power required to achieve BER $<10^{-9}$ against downstream ER for both upstream and downstream signals. The solid and dash lines represent the relative optical ER dependencies of both upstream and downstream signals in a bidirectional network and unidirectional back-to-back setup (i.e. no Rayleigh backscattering or transmission penalty), respectively. Results show a power penalty of less than 2 dB from 25 km of transmission and Rayleigh backscattering, but more importantly, a linear and small dependence of the

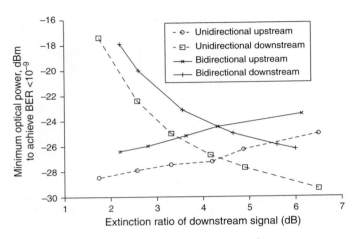

Figure 6.30 Received optical power at bit-error rate (BER) = 10^{-9} against downstream extinction ratio. —— Bidirectional 25 km transmission with Rayleigh backscattering effects – – – Unidirectional back-to-back transmission.

upstream signal on the downstream ER compared to high-exponential dependence of the downstream signal on the downstream ER.

As the WDM-PON is expected to be first deployed by business customers with high-capacity information in both upstream and downstream directions, real-time knowledge of a fiber fault and the location of the fault will ensure rapid rectification and restoration of transmission. However, despite all its advantages, a typical WDM-PON architecture does not allow for easy diagnostic of fiber failures due to its tree and branch topology. Conventional optical time-domain reflectometry (OTDR) based on a single wavelength source at the CO is not applicable in a WDM-PON due to the wavelength-routing characteristics of the AWG in the RN [52]. Aside from OTDR methods, other solutions require a broadband source with high-output power as well an optical spectrum analyser or a tunable filter with an integrated power meter, which increase the system cost substantially. More importantly, to install the fault monitoring and localization feature in an already deployed WDM-PON requires an upgrade of all ONUs whereby a wideband optical reflector or a wavelength-dependent filter is required to be added. However, the WDM-PON using OIL VCSEL proposed in this section can provide fault monitoring and localization without requiring any changes to be made at the ONU side [53].

The monitor is formed by the inherent mirror structure of the VCSEL located in each ONU, which is injection-locked by the downstream data as well as serving as the upstream transmitter. The feasibility of this scheme is demonstrated with characterizations of the fault monitor functioning reliably with high sensitivity (~ -67 dBm), low output power (~ -7 dBm) and low bandwidth (~ 2 kHz) requirements. It is also shown that the addition of the monitoring scheme to an existing infrastructure incurs a negligible penalty (~ 0.5 dB) on the upstream transmissions of the WDM-PON [53].

6.7.2 Metropolitan Area Networks

As shown in the previous sections, some of the performance improvements, such as bandwidth enhancement and chirp reduction, are very attractive for digital data transmission applications. A 2.5-Gb/s transmission has been demonstrated using OIL VCSEL but with another VCSEL as the master laser instead of a high-power DFB laser [38]. Hence, even with intrinsically large linewidth and low power, VCSEL can still serve as the master laser to lock another VCSEL.

Figure 6.31 shows the BER after transmission through 50-km SMF-28 optical fiber. The 2-dB power penalty reduction in BER measurement at 2.5 Gb/s after 50 km of SMF-28 fiber is concrete evidence of the improvement of injection locking for digital modulation application. This reduction of the slope is likely due to the interaction of fiber dispersion (17 ps/nm · km for SMF-28 fiber) and the chirp of the directly modulated VCSEL. Dispersion-related degradation manifests itself in a BER curve with a shallower slope.

6.7.3 Analog Optical Communications

For analog fiber-optic transmission, a faithful reproduction of the signal is desired, with negligible distortion, low noise, and a high-RF link gain. In addition, a large modulation bandwidth is desirable. External optical modulators are typically used for high-performance links. This solution, though effective, is costly and bulky. In contrast, directly modulated lasers are more desirable on these two accounts. In

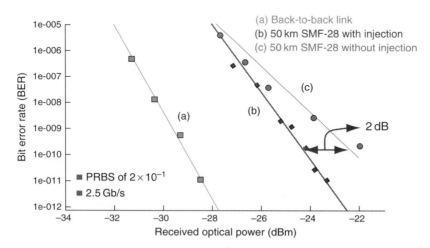

Figure 6.31 BER vs. received power for VCSEL-VCSEL OIL at different conditions. 2-dB power penalty reduction is achieved with injection locking after 50-km fiber transmission (this figure may be seen in color on the included CD-ROM).

particular, VCSELs are very promising for applications where low cost, small size, and low-power dissipation are required as much as high-signal fidelity. However, directly modulated lasers, including VCSELs, suffer from high distortion near the resonance frequency and can, therefore, only be used at very low-RF frequencies. The distortion physically originates from the nonlinear characteristics of the laser, dominated by the carrier–photon interaction. To avoid performance degradation due to this phenomenon, it is desirable to have the laser resonance frequency greatly exceed the highest RF frequency of use.

As shown in the previous sections, the use of an ultrahigh-injection ratio results in a high-resonance frequency, which leads to significant performance improvements including enhancements in bandwidth, RF-link gain, dynamic range, and reduction in laser noise. The performance and enhancements of OIL VCSELs is summarized in Table 6.1. All enhancements, except the large sharp resonance frequency, were obtained for a red wavelength detuning, for high-injection ratio, and are observable simultaneously. For the red detuning values, the resonance frequency is damped but still enhanced up to 30–40 GHz.

Very high-resonance frequency is achieved for several devices with all devices showing an initial free-running resonance frequency below 10 GHz. It is shown that the resonance frequency scales with increasing injection ratio and is expected to increase for even higher injection ratios. Under the ultrastrong injection conditions, we attain up to 20-dB RF link gain enhancement. This corresponds to a modulation efficiency enhancement from 0.22 to 2.2 W/A. The modulation efficiency increases with increasing injection ratios as well. Furthermore, injection locking is effective in reducing distortion, thereby improving the SFDR of a directly modulated laser, both experimentally and numerically. An OIL VCSEL with an IMD3 SFDR of 106 dB-Hz$^{2/3}$ at 1 GHz was achieved, representing an SFDR increase of 20 dB. A large dynamic range was observed as well over a wider frequency band of 0.6–3.0 GHz. The SFDR improvement is also shown to linearly increase with injection ratio. Similarly, the SFDR enhancement increases with

Table 6.1.

Microwave performance of OIL VCSELs for analog transmission.

Performance metric	OIL-VCSELs	Enhancement
Resonance frequency	50 GHz*	5–10 X
Intrinsic modulation bandwidth	>50 GHz	>5 X
Modulation efficiency	2.2 W/A	10 X
RF link gain	−19 dB	20 dB
Spur-free dynamic range (SFDR)	106 dB-Hz$^{2/3}$	20 dB
Relative intensity noise (RIN)	−140 dB/Hz*	15 dB

*Limited by available instrumentation.

increasing resonance frequency and RF gain. Due to the resonance frequency enhancement, OIL can reduce laser noise too.

These results suggest that OIL VCSELs are a promising high-performance solution for microwave photonic applications. OIL VCSELs may find applications in low-cost radio-over-fiber (RoF) distribution systems such as cellular telephone signals (global system for mobile communications (GSM) requires 90 dB Hz dynamic range) and wireless local area networks (WLANs) (802.11 ×). For injection locking to offer the enhancements described in this study, the resonance frequency must be increased to a very high value, which is achieved by using an ultrahigh-injection ratio.

6.8 CONCLUSION

In this chapter, the technique and performance of injection-locked lasers are discussed. The characteristics of a directly modulated laser can be fundamentally changed when it is locked to a master laser through coherent nonlinear interactions, resulting in superior device performance for various applications. There is no apparent restriction on the type of lasers to be used as slave or master lasers. Due to the rapid development and application needs of high-speed lasers, our main focus has been on the performance improvement of directly modulated semiconductor lasers using OIL for both digital and analog fiber-optic links. The demonstrated improvement includes chirp and noise reduction, modulation bandwidth, RF link gain, and SFDR enhancement.

Based on the superior performance of injection-locked diode lasers, they are utilized in various systems. We have discussed OIL VCSELs used as laser transmitter in PONs and MANs. Furthermore, it is also appealing to transmit analog signals. In addition to the applications we have presented here, OIL has been demonstrated to have advantages in many other microwave photonic applications. For example, optical injection phase-locked loop (OIPLL) combining phase-locking technique and injection-locking technique has been shown to improve the phase noise performance [55]. All optical regeneration, as a critical function in high-speed photonic networks, has been widely investigated by using an OIL-DFB laser [56] and a two-mode OIL FP laser [57]. Injection locking has also been proposed and demonstrated as one of the techniques for optical generation of millimeter-wave signals with low phase noise [54, 58–60], which would be crucial in a RoF system.

Despite all these intriguing device performance and system applications, there are many new ideas and novel applications requiring further investigation. To ensure the commercial use of OIL for future optical communication systems, low-cost integrated device is one important direction to pursue [26]. We believe that the OIL lasers will become one of the key technologies in the next generation high-speed fiber-optic networks.

ACKNOWLEDGMENTS

The authors thank Drs Chih-Hao Chang and Lukas Chrostowski for initializing and carrying out the early OIL VCSEL work at University of California at Berkeley. We thank Professor Markus-C. Amann, Werner Hofmann at Technical University of Munich and VERTILAS GmbH, Germany; Professor Ming C. Wu, Drs Erwin K. Lau and Hyuk-Kee Sung at University of California at Berkeley; Professor Rod S. Tucker and Dr Elaine Wong at University of Melbourne, Australia; and Dr. Stephen Pappert of DARPA for many stimulating discussions and fruitful collaborations. Finally, we thank DARPA for funding support.

REFERENCES

[1] A. Pikovsky, M. Rosenblum, and J. Kurths, Synchronization: A Universal Concept in Nonlinear Sciences, *Cambridge Nonlinear Science Series 12*. Cambridge, U.K.: Cambridge University Press, 2001.
[2] B. van der Pol, "Forced oscillations in a circuit with non-nonlinear resistance," *Philos. Mag.*, iii, 65–80, 1927.
[3] R. Adler, "A study of locking phenomena in oscillators," in Proc. *IRE*, 34, 351–357, 1946.
[4] H. L. Stover and W. H. Steier, "Locking of laser oscillators by light injection," *Appl. Phys. Lett.*, 8, 91–93, 1966.
[5] S. Kobayashi and T. Kimura, "Coherence on injection phase-locked AlGaAs semiconductor laser," *Electron. Lett.*, 16, 668–670, 1980.
[6] K. Iwashita and K. Nakagawa, "Suppression of mode partition noise by laser diode light injection," *IEEE Trans. Microw. Theory Tech.*, 30, 1657–1662, 1982.
[7] R. Lang, "Injection locking properties of a semiconductor laser," *IEEE J. Quantum Electron.*, QE-18, 976–983, 1982.
[8] C. Lin and F. Mengel, "Reduction of frequency chirping and dynamic linewidth in high-speed directly modulated semiconductor lasers by injection locking," *Electron. Lett.*, 20, 1073–1075, 1984.
[9] N. A. Olsson, H. Temkin, R. A. Logan et al., "Chirp-free transmission over 82.5 km of single mode fibers at 2 Gbit/s with injection locked DFB semiconductor lasers," *IEEE J. Lightw. Technol.*, LT-3(1), 63–66, 1985.
[10] C.-H. Chang, L. Chrostowski, and C. J. Chang-Hasnain, "Injection locking of VCSELs," *IEEE J. Sel. Top. Quantum Electron.*, 9(5), 1386–1393, 2003.
[11] T. B. Simpson, J. M. Liu, and A. Gavrielides, "Bandwidth enhancement and broadband noise reduction in injection-locked semiconductor lasers," *IEEE Photon. Technol. Lett.*, 7, 709–711, 1995.
[12] J. M. Liu, H. F. Chen, X. J. Meng, and T. B. Simpson, "Modulation bandwidth, noise, and stability of a semiconductor laser subject to strong injection locking," *IEEE Photon. Technol. Lett.*, 9, 1325–1327, 1997.
[13] H.-K. Sung, T. Jung, M. C. Wu et al., "Modulation bandwidth enhancement and nonlinear distortion suppression in directly modulated monolithic injection-locked DFB lasers," in *Int. Topical Microw. Photon. Meeting*, 2003, pp. 27–30.
[14] L. Chrostowski, X. Zhao, C. J. Chang-Hasnain et al., "50-GHz Optically Injection-Locked1.55-μm VCSELs," *IEEE Photon. Technol. Lett*, 18, 367–369, 2006.
[15] X. Jin and S. L. Chuang, "Relative intensity noise characteristics of injection-locked semiconductor lasers," *Appl. Phys. Lett.*, 77(9), 1250–1252, 2000.
[16] L. Chrostowski, C. H. Chang, and C. J. Chang-Hasnain, "Reduction of Relative Intensity Noise and Improvement of Spur-Free Dynamic Range of an Injection Locked VCSEL," in Proc. *LEOS*, 2, October 27–28, 2003. pp. 65–80.

[17] X. Zhao, D. Parekh, E. K. Lau et al., "Novel cascaded injection-locked 1.55-µm VCSELs with 66 GHz modulation bandwidth," *Opt. Express*, 15, 14810–14816, 2007.
[18] W. E. Lamb, "Theory of an optical maser," *Phys. Rev.*, 134(6A), 1429–1450, 1964.
[19] A. Murakami, K. Kawashima, and K. Atsuki, "Cavity resonance shift and bandwidth enhancement in semiconductor lasers with strong light injection," *IEEE J. Quantum Electron.*, 39, 1196–1204, 2003.
[20] F. Mogensen, H. Olesen, and G. Jacobsen, "Locking conditions and stability properties for a semiconductor laser with external light injection," *IEEE J. Quantum Electron.*, QE-21, 784–793, 1985.
[21] V. Annovazzi-Lodi, S. Donati, and M. Manna, "Chaos and locking in a semiconductor laser due to external injection," *IEEE J. Quantum Electron.*, 30, 1537–1541, 1994.
[22] V. Kovanis, A. Gavrielides, T. B. Simpson et al., "Chaos, Period-Doubling and Reverse Bifurcations in an Optically Injected Semiconductor Laser," in Proc. *IEEE Nonlinear Optics: Materials, Fundamentals, and Applications* (Cat. No. 94CH3370-4), New York, NY, USA, 1994, pp. 30–32.
[23] T. B. Simpson, J. M. Liu, A. Gavrielides et al., "Period-doubling route to chaos in a semiconductor laser subject to optical injection," *Appl. Phys. Lett.*, 64, 3539–3541, 1994.
[24] T. B. Simpson, J. M. Liu, K. F. Huang, and K. Tai, "Nonlinear dynamics induced by external optical injection in semiconductor lasers," *Quantum & Semiclassical Optics*, 9, 765–784, 1997.
[25] C. H. Henry, N. A. Olsson, and N. K. Dutta, "Locking range and stability of injection locked 1.54 µm InGaAsP semiconductor lasers," *IEEE J. Quantum Electron.*, QE-21, 1152–1156, 1985.
[26] X. J. Meng, C. Tai, and M. C. Wu, "Experimental demonstration of modulation bandwidth enhancement in distributed feedback lasers with external light injection," *Electron. Lett.*, 34, 2031, 1998.
[27] X. J. Meng, T. Chau, and M. C. Wu, "Improved intrinsic dynamic distortions in directly modulated semiconductor lasers by optical injection locking," *IEEE Trans. Microw. Theory Tech.*, 47, 1172–1176, 1999.
[28] J. Wang, M. K. Haldar, L. Li, and F.V.C. Mendis, "Enhancement of modulation bandwidth of laser diode by injection locking," *IEEE Poton. Technol. Lett.*, 8, 34–36, 1996.
[29] M. Ortsiefer, R. Shau, F. Mederer et al., "High-speed modulation up to 10 Gbit/s with 1.55 µm wavelength InGaAlAs VCSELs," *Electron. Lett.*, 38(20), 1180–1181, 2002.
[30] M. Ortsiefer, M. Furfanger, J. Rosskopf et al., "Singlemode 1.55-µm VCSELs with low threshold and high output power," *IEE Electron. Lett.*, 39, 1731–1732, 2003.
[31] X. Zhao, C. J. Chang-Hasnain, W. Hofmann, et al., "Modulation Efficiency Enhancement of 1.55-µm Injection-Locked VCSELs," presented at *International Semiconductor Laser Conference (ISLC)*, Big Island, Hawaii, USA, September, 2006.
[32] H.-K. Sung, E. K. L. Lau, and M. C. Wu, "Near-Single Sideband Modulation in Strong Optical Injection-Locked Semiconductor Lasers," in PROC. *OFC*, Anaheim, USA, March, 2006.
[33] E. K. Lau, H. K. Sung, and M. C. Wu, "Ultra-High, 72 GHz Resonance Frequency and 44 GHz Bandwidth of Injection-Locked 1.55-µm DFB Lasers," in Proc. *OFC*, Anaheim, USA, March, 2006.
[34] X. Zhao and C. J. Chang-Hasnain, "A New Amplifier Model for Resonance Enhancement of Optically-Injection-Locked Lasers," *IEEE Photon. Technol. Lett.*, to be published, January, 2008.
[35] X. Zhao, D. Parekh, E. K. Lau et al., "Cascaded Injection-Locked 1.55-µm VCSELs for High Speed Transmission," in Proc.*CLEO*, Postdeadline session, Baltimore, USA, May, 2007.
[36] C. Lin and F. Mengel, "Reduction of frequency chirping and dynamic linewidth in high-speed directly modulated semiconductor lasers by injection locking," *Electron. Lett.*, 20, 1073–1075, 1984.
[37] C. H. Chang, L. Chrostowski, and C. Chang-Hasnain, "Enhanced VCSEL performance by optical injection locking for analog and digital applications," in Proc. *IEEE LEOS*, invited talk, Tucson, AZ, 2003.
[38] C. H. Chang, L. Chrostowski, C. J. Chang-Hasnain, and W. W. Chow, "Study of long-wavelength VCSEL-VCSEL injection locking for 2.5-Gb/s transmission," *IEEE Photon. Technol. Lett.*, 14, 1635–1637, 2002.
[39] S. A. Pappert, C. K. Sun, R. J. Orazi, and T. E. Weiner, "Microwave Fiber Optic Links for Shipboard Antenna Applications," in Proc. *IEEE Int. Phased Array Syst. Technol. Conf.*, Piscataway, NJ, 2000, pp. 345–348.

[40] H. L. T. Lee, R. V. Dalai, R. J. Ram, and K. D. Choquette, "Dynamic range of vertical-cavity surface-emitting lasers in multimode links," *IEEE Photon. Technol. Lett.*, 11(11), 1473–1475, November 1999.

[41] K. Y. Lau and A. Yariv, "Intermodulation distortion in a directly modulated semiconductor injection laser," *Appl. Phys. Lett.*, 45(10), 1034–1036, 1984.

[42] L. Chrostowski, X. Zhao, and C. J. Chang-Hasnain, "Microwave performance of optically injection-locked VCSELs," *IEEE Trans. Microw. Theory Tech.*, 54, 788–796, 2006.

[43] G. Yabre and J. Le Bihan, "Reduction of nonlinear distortion in directly modulated semiconductor lasers by coherent light injection," *IEEE J. Quantum Electron.*, 33(7), 1132–1140, July 1997.

[44] N. Schunk and K. Petermann, "Noise analysis of injection-locked semiconductor injection lasers," *IEEE J. Quantum Electron.*, QE-22, 642–650, 1986.

[45] A. Banerjee, Y. Park, F. Clarke et al., "WDM-PON technologies for broadband access: A review," *OSA J. Opt. Networking*, 4, 737–758, November 2005.

[46] L. Chrostowski, C. Chang, and C. J. Chang-Hasnain, "Uncooled injection-locked 1.55μm tunable VCSELs as DWDM transmitter," in Proc. *OFC*, 2, Atlanta, GA, March 2003, pp. 753–754.

[47] W. Hung, C.-K. Chan, L.-K. Chen, and F. Tong, "An optical network unit for WDM access networks with downstream DPSK and upstream remodulated OOK data using injection-locked FP laser," *IEEE Photon. Technol. Lett.*, 15(10), 1476–1478, October 2003.

[48] W. Yuen, G. S. Li, R. F. Nabiev et al., "Electrically-Pumped Directly-Modulated Tunable VCSEL for Metro DWDM Applications," in *GaAS IC Symp. Tech. Dig.*, October 2001, pp. 51–52.

[49] E. Wong, X. Zhao, C. J. Chang-Hasnain et al., "Optically injection-locked 1.55-μm VCSELs as upstream transmitters in WDM-PONs," *IEEE Photon. Technol. Lett*, 18, 2371–2373, 2006.

[50] W. S. Jang, H. C. Kwon, and S. K. Han, "Suppression of Rayleigh backscattering in a bidirectional WDM optical link using clipped direct modulation," *IEE Proc., Optoelectron.*, 151(4), 219–222, 2004.

[51] E. Wong, X. Zhao, C. J. Chang-Hasnain et al., "Rayleigh backscattering and extinction ratio study of optically injection-locked 1.55 μm VCSELs," *Electron. Lett.*, 43(3), 182–183, 2007.

[52] C.L. Lin, *Broadband Optical access and FTTH* (Wiley & Sons, Ltd, 2006), Chapter 12, pp. 288.

[53] E. Wong, X. Zhao, and C. J. Chang-Hasnain, "Novel Fault Monitoring and Localization Scheme in WDM-PONs with Upstream VCSEL transmitters," in Proc. *OFC*, Anaheim, USA, March, 2007.

[54] X. Zhao, D. Parekh, H.-K. Sung et al., "Optoelectronic Oscillator using Injection-Locked VCSELs," *IEEE LEOS*, annual conference, Lake Buena Vista, Florida, USA, October, 2007.

[55] A. C. Bordonalli, C. Walton, and A. J. Seeds, "High-performance phase locking of wide linewidth semiconductor lasers by combined use of optical injection locking and optical phase-lock loop," *J. Lightw. Technol.*, 17, 328–342, 1999.

[56] S. Yamashita and D. Matsumoto, "Waveform reshaping based on injection locking of a distributed-feedback semiconductor laser," *IEEE Photon. Technol. Lett.*, 12, 1388–1390, 2000.

[57] S. Yamashita and J. Suzuki, "All-optical 2R regeneration using a two-mode injection-locked Fabry-Perot laser diode," *IEEE Photon. Technol. Lett.*, 16, 1176–1178, 2004.

[58] L. Goldberg, H. F. Taylor, J. F. Weller, and D. M. Bloom, "Microwave signal generation with injection-locked laser diodes," *Electron. Lett.*, 19, 491–493, 1983.

[59] R. P. Braun, G. Grosskopf, D. Rohde, and F. Schmidt, "Low-phase-noise millimeter-wave generation at 64 GHz and data transmission using optical sideband injection locking," *IEEE Photon. Technol. Lett.*, 10, 728–730, 1998.

[60] H.-K. Sung, E. K. Lau, X. Zhao et al., "Optically Injection-Locked Optoelectronic Oscillators with Low RF Threshold Gain," in Proc. *CLEO*, Baltimore, USA, May 2007.

7

Recent developments in high-speed optical modulators

Lars Thylén, Urban Westergren, Petter Holmström, Richard Schatz, and Peter Jänes[*]

Department of Microelectronics and Applied Physics, Royal Institute of Technology (KTH), Kista, Sweden
[*]Proximion Fiber Systems AB, Kista, Sweden

Abstract

The principles and basics of optical modulators are described, with emphasis on high-speed intensity modulators and on modulators where the modulating signal is electronic (as opposed to photonic modulation). The principles of operation of various modulators are elucidated, and the mechanisms employed are discussed as are the corresponding materials. State-of-the-art results for traveling-wave (TW) electroabsorption intensity modulators are described, and the development of some novel types of modulators (EAM) and future prospects in the field are elaborated on.

7.1 INTRODUCTION

With the rapid development in high data rate communications, brought about by the ever increasing demands from the Internet and in recent years also by the requirements for very high-speed computer interconnects, the optical single-mode waveguide-based modulator has emerged as a critical component. The modulator is sometimes called external modulator to emphasize that it is separated from the light source. This growing importance of the modulator has taken place because the external modulator is the only known device that can cope with time division multiplexed (TDM) signals at 100 Gb/s and beyond, rates that are increasingly required. The reason for the near total focusing on single-mode waveguide devices is their compatibility with the prevalent single-mode fiber but also the vastly superior performance in drive power as well as speed in comparison with multimode or even bulk-type modulators [1]. The main and ubiquitous

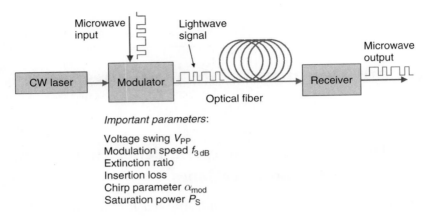

Figure 7.1 Principle of external modulation, where light from a continuous wave (CW) source is coupled (directly or through a fiber) to the modulator waveguide, where it is subsequently processed. In telecom applications, the modulated light is coupled to an optical fiber (this figure may be seen in color on the included CD-ROM).

alternative, direct modulation, where the injection current is used to control the light output power or wavelength of a laser diode, can be shown to be generally limited to below say 100-GHz bandwidth [2]. Figure 7.1 shows the principle of external modulation, where light from a continuous wave (CW) source is coupled (directly or via a fiber) to the modulator waveguide, where it is subsequently processed.

In the external modulator, an electric control signal (in general a voltage) is used to control the properties of the light output from the modulator, given a CW input signal from a laser source, instead of changing the light output properties by changing the injection current to the laser source. The hitherto prevalent modulation format is on-off keying (OOK), that is, a simple intensity modulation. However, the ever increasing bandwidth demand will probably change this in favor of more complex modulations, such as those employed in mobile communications, dictated by the shortage of available bandwidth. This chapter focuses nearly exclusively on OOK.

In addition to the sheer data rate advantage of the modulator, there are other advantages: The chirp or dynamic wavelength change can to a degree be engineered, alleviating the intrinsic dynamic wavelength changes due to electron carrier dynamics effects in directly modulated laser diodes and implying reduced dispersion penalty. Further, a high extinction ratio for large modulated optical power with constant electric drive power is possible, as is nonlinear pulse shaping. Phase modulation is easily accessible.

But there are also disadvantages such as added complexity, at least for discrete devices, added insertion loss, and in many cases high electric drive power.

The chapter is organized as follows: Section 7.2 gives a review of the principles and mechanisms of optical modulation, followed by a section on high-speed modulation, in many ways the raison d'être for the external modulator, and some associated transmission issues. Then, the basics of modulators based on phase and absorption changes and the associated physical mechanisms are treated, followed by an in depth discussion of state-of-the-art traveling-wave (TW) electroabsorption modulators (EAMs) based on indium phosphide (InP)-related materials. Novel or emerging modulator types are the subject of Section 7.7, notably intersubband (IS) as well as silicon-based modulators. Section 7.8 gives a summary and outlook.

7.2 PRINCIPLES AND MECHANISMS OF EXTERNAL OPTICAL MODULATION

In general, amplitude (intensity), phase, and polarization modulation are readily implemented with external modulators, thus they can be seen as basic light-field transformation devices. An input field

$$\hat{E} = \hat{x} A_x \sin(\omega t) \tag{7.1}$$

polarized in the x-direction with amplitude A_x and angular frequency ω can be transformed into

$$\hat{E} = \hat{x} A_x(t) \sin(\omega(t)t + \varphi_1(t)) + \hat{y} A_y(t) \sin(\omega(t)t + \varphi_2(t)), \tag{7.2}$$

where amplitude, frequency, and phase modulation are present. Depending on the physical mechanism used, these different quantities are more or less related to one another; obviously, frequency and phase are related through a time derivative. This whole transformation operation in essence rests on using an electronic signal to controllably change the complex refractive index $\tilde{n} = n' + in''$ of the modulator waveguide structure, where $\tilde{n} = \sqrt{\tilde{\varepsilon}\mu}$ and $\tilde{\varepsilon}$ is the complex dielectric constant. The real and imaginary parts of this are connected by the Kramers–Krönig relation [3]. The magnetic permeability μ is taken as 1 for the cases treated below. The change of the complex refractive index can be done in a wide variety of materials and employing a plethora of physical effects, as shown in the Appendix. Variation of the imaginary part of the refractive index gives intensity modulation, absorption, or amplification (often accompanied by an undesired phase modulation), whereas modulation of the real part of the refractive index (phase modulation) in principle can be employed to modulate all the parameters above. Thus, by changing the phase in interferometer-type modulators, such as Mach–Zehnder modulators, or directional coupler structures (Section 7.4), one can modulate amplitude or intensity. By changing the imaginary part of the refractive index, absorption changes result, due to the Franz–Keldysh or the quantum-confined Stark effects (QCSE) (Section 7.5). This can also be achieved by gain changes in semiconductor optical amplifiers. Polarization modulation is obtained by utilizing birefringent waveguides, with controllable introduction of off diagonal element in the dielectric tensor [1]. Frequency modulation is more readily achieved by direct modulation of lasers, since different feedback conditions such as a varying refractive index will vary the emitted wavelength, employed in tunable laser diodes [4].

With reference to polarization, it should be mentioned that basically all modulators will only operate efficiently for a single polarization. In LiNbO$_3$ modulators, based on employing the electrooptic tensor, the polarization dependence is due to the anisotropy of the material. In semiconductor modulators, the matrix elements of the involved transitions are different for different polarizations. The restriction to single-polarization operation is not overly severe, because in most cases the modulator is close to or integrated with a light source. Polarization-independent modulators will require special designs which will in general compromise performance.

Key performance data for external modulators are as follows:

- insertion loss and extinction ratio: should generally be lower than a few dB and larger than 10 dB, respectively, though strongly depending on application
- drive power (electrical drive power): should be lower than order of 100 mW into a 50 Ω load
- optical power-handling capability (saturation characteristics): strongly depending on the application, but in general a modulator should be able to handle several tens of mW input power
- chirping: should be adapted to the application in question, for example, not important for short-haul interconnect but decisive for long-haul back bone. In general, negative chirp parameters are favored.
- bandwidth (phase and amplitude): depends on application. Baseband applications require flat amplitude and phase response in contrast to carrier frequency-type application
- return loss (electrical and optical)
- microwave impedance (TW modulators): should in general be (close to) 50 Ω
- wavelength range: it is an advantage if the modulator can operate over a reasonably large wavelength range with no other modification than the drive voltage. The latter is true for LiNbO$_3$ modulators but not for EAM.
- physical size: in line with shrinking dimensions in photonics, the size should be smaller than a few hundred microns (cf. polymer and LiNbO$_3$ modulators with lengths of millimeters to centimeters)
- integratability with light sources: can be supplied by using III–V compounds

In this chapter, we focus on the high-speed issues and further limit ourselves to modulators, driven by electronic signals, that is, we do not treat the large field of optically controlled modulators, that is, modulators based on nonlinear optical effects. One reason is that information is generally generated and processed in the electrical domain. The so-called all optical modulators can generally achieve much higher frequencies (> THz) than optoelectronic ones though the limits of the latter are still a research item [5]. The higher bandwidth in the all optical case is achieved by using in general much higher drive powers, being a consequence of the prevalent weak optical nonlinearities.

7.3 HIGH-SPEED MODULATION

7.3.1 Basic Issues, Principles, and Limitations

The speed of the modulation of light is limited by several factors which may be divided into two groups:

(1) limitations inherent to the modulator due to charging times, charge carrier transit times, frequency-dependent loss, velocity mismatch of electrical and optical fields in distributed structures, and so on
(2) limitations caused by interaction with the drive electronics due to mismatch in impedances and interconnect parasitics.

The first group of limitations depends on the chosen modulator structure and will be different for different types of modulators. For example, velocity mismatch of electrical and optical fields may be an issue in relatively long distributed structures such as Mach–Zehnder (MZ) modulators while comparatively short traveling-wave EAMs (TWEAMs) are primarily limited by frequency-dependent loss in the electrical field. These inherent limitations will be discussed in more detail later in this chapter.

The second group of frequency limitations due to interactions with the drive electronics may appear not to be as important as the first as the causes are not fundamental from the point-of-view of device physics. However, to use a modulator in a fiber optics transmission system, it has to be connected to electronics which cannot be designed to handle any load impedance imposed by a modulator, and the interconnect technology itself will cause limitations regarding which impedances can be used without pulse distortion. When two semiconductor chips are connected to each other, typically by wire or flip-chip bonding, there will be capacitive and inductive interconnect parasitics which limit the maximum frequency and bitrate. Monolithic integration in the so-called optoelectronic integrated circuits (OEICs) can be used to circumvent these problems, but this will only become a viable solution if the yield of OEIC chips is improved in the future. The main solution in practice today is hybrid integration on a dielectric carrier of different modulator and electronic driver chips, manufactured in semiconductor technologies separately optimized for their respective functions.

Both high-speed electronics and interconnect technologies for high frequencies are developed for a standard systems impedance level of $50\,\Omega$, which has been chosen as a standard value as it falls in a range where microwave transmission line loss has a minimum, for example, in typical coaxial cables and microstrip technologies for printed circuit boards. If a device such as a modulator has an impedance which deviates substantially from $50\,\Omega$, it will cause a reflection of the electrical signal according to basic microwave theory. This reflection may in turn cause pulse distortion in digital modulation if the distance between the electronics and the modulator corresponds to delays comparable to the pulse length and the result from this pulse distortion may be errors in the data transmission. A device impedance much higher than 50 Ω will increase the impact of parallel capacitances caused by typical interconnect fringing fields, whereas a device impedance lower than 50 Ω will increase the frequency limitation caused by typical bonding series inductances. The latter will be discussed in some more detail as this is an important practical limitation for contemporary hybrid integrated systems in the range 10–100 Gb/s.

A typical interconnect in hybrid integration is a bond wire of gold or aluminum with about 25 μm diameter and a length of 0.5–1 mm, causing a series inductance of the order of 0.5–1 nH. This type of interconnect has acceptable parasitics for relatively low speeds, but when approaching 10 Gb/s, the reactance of the wire in a system with an impedance level of 50 Ω will cause pulse distortion due to the limitation of the bandwidth. A more detailed analysis of the parasitics of a bond wire interconnect shows that it can be modeled up to relatively high frequencies by a series inductance and capacitances caused by the electrical fields at the ends of each respective metal conductor on the chips (see Figure 7.2).

The problem with the bond-wire inductance can be reduced by shortening the wire, but there is a practical limit to the length if there is a height difference between two chips or between a chip and a carrier. Adding more wires in parallel is only effective if the wires are at an angle and/or separated, which is often not possible for reasons of geometry. A more efficient method to reduce the inductance is using flip-chip mounting where the bond wire is typically replaced by a ball of metal ("bump"), for example, indium (see Figure 7.2). This reduces the inductance to typically 0.05–0.1 nH compared to 0.5–1 nH for an ordinary bond wire as the length of the conductor between the two chips is reduced from typically 0.5–1 mm to 50–100 μm.

A simple equivalent circuit for the interconnect parasitics of both wire and flip-chip bonding is a π-type filter between an electronic driver circuit with the output impedance R_G and a modulator with the device impedance R_L, as shown in Figure 7.3.

If the capacitor values can be optimized by design of the chip geometry, the capacitances can be used to increase the maximum useful bitrate compared to a case with only series inductance and no capacitance, which has a 3-dB bandwidth

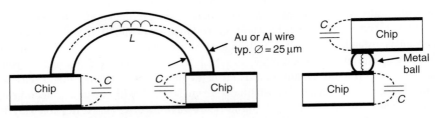

Figure 7.2 Bonding parasitics for wire bonding (left) and flip-chip bonding (right).

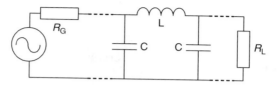

Figure 7.3 Equivalent circuit for bonding parasitics.

$(R_G + R_L)/2\pi L$. As a circuit designer's rule of thumb, it may be shown that the maximum useful bit rate in b/s with acceptable pulse distortion is equal to the value $f_{3\,dB}$ in hertz as approximately given by

$$f_{3\,dB} \approx \frac{5}{4} \cdot \frac{R_G + R_L}{2\pi L} \tag{7.3}$$

A typical flip-chip bonding may have an inductance of the order of 50 pH resulting in a ~400-GHz bandwidth for a typical 50 Ω system ($R_G = R_L = 50\,\Omega$), but it should be noted that the roll-off at high frequencies of the interconnect has to be treated together with the inherent frequency limitations of the modulator itself and may still contribute to a reduced total bandwidth. The interconnect bandwidth may also be reduced to values where applications such as 100 Gb/s transmission are seriously affected if the device impedance is below 50 Ω or if wire bonding is used with distances larger than about 100 μm. A high-speed modulator, such as a traveling-wave electroabsorption modulator designed for 100 Gb/s, will typically have an input impedance which is close to real and of the order of 30 Ω over a large frequency range, but closer to the 3-dBe bandwidth, the real part of the impedance will be transformed to higher values if a segmented structure is used (see Section 7.6), which means that a model for the complex device impedance as a function of frequency should be used when designing the exact interconnect network. However, the Eqn (7.3) will give a good indication of the limitation imposed by the interconnect also in a practical case. It may still be concluded that flip-chip bonding is expected to be used successfully when connecting a modulator to the drive electronics at bitrates of 100 Gb/s or more while wire bonding becomes increasingly difficult to use when exceeding 10 Gb/s.

For modulators with an impedance which is primarily capacitive, such as electrooptic and electroabsorption modulators based on a reverse-biased pin-diode structure, the charging time of the modulator can be reduced either by reducing the output impedance of the drive electronics or by creating a distributed structure where the parallel capacitance is divided into several sections separated by series inductances. The principle is shown in Figure 7.4 where the top circuit shows a lumped modulator with a single section of capacitance C, with a load impedance R introduced across the modulator to avoid microwave reflections when the generator and the modulator are physically separated. In the middle circuit, the capacitance has been divided into a number N of shorter sections where each is connected to the series inductance L/N. When increasing the number of

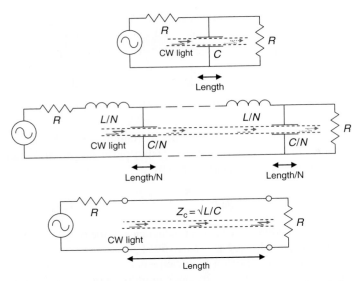

Figure 7.4 Lumped modulator (top), traveling-wave modulator (bottom), and equivalent traveling-wave structure with lumped elements (middle) (this figure may be seen in color on the included CD-ROM).

sections N, the structure eventually becomes distributed and is then often referred to as a "traveling-wave" structure (even though the total length of interaction between the electrical and optical fields is for some modulators much shorter than the wavelengths involved). According to basic microwave theory [6], the middle structure may be viewed as an equivalent circuit of a microwave transmission line when the number of sections N is increased. This transmission line has a characteristic impedance Z_c, and the charging time of the capacitance is not a limitation for the bandwidth of this structure as it is completely balanced by the series inductance. More details regarding traveling-wave modulators will be given in later sections of this chapter.

7.3.2 Chirp Issues in Modulators

The intensity modulation, $P(t) = P_0 + \Delta P(t)$, from a modulator is generally accompanied with phase fluctuations and hence frequency fluctuations, the so-called chirp, proportional to the relative modulation depth.

$$\Delta \varphi = \frac{\alpha_{\text{mod}}}{2} \frac{\Delta P}{P}, \tag{7.4}$$

where α_{mod} is the chirp factor of the modulator [7]. If the chirp factor is positive, the phase modulation will give a blue shift of the optical frequency during the rising edge of the optical pulse and a red shift during the falling edge of the optical pulse. The phase modulation is most often due to the coupling between the absorption, $\Delta\alpha$, and refractive index, Δn, of the optical waveguide material described by the Kramers–Krönig relation mentioned above:

$$\Delta n(\lambda) = \frac{\lambda^2}{2\pi^2} PV \int_0^\infty \frac{\Delta\alpha(\lambda')}{\lambda^2 - \lambda'^2} d\lambda', \qquad (7.5)$$

where PV denotes the Cauchy principal value of the integral. A spectrally asymmetric absorption change, $\Delta\alpha$, will hence in general be accompanied by a refractive index change, Δn, that, for small-signal modulation at the operational wavelength λ_{op}, can be described with the material chirp factor, α_{mat}.

$$\Delta n(\lambda_{op}) = \alpha_{mat}(\lambda_{op}) \frac{\lambda_{op}}{4\pi} \Delta\alpha(\lambda_{op}) \qquad (7.6)$$

For a TW loss-type modulator where the wavelength dependence is solely determined by the waveguide material, the modulator chirp factor in Eqn (7.4) equals the material chirp factor in Eqn (7.6). For modulators where the waveguide structure itself is wavelength selective, for example, MZ type, the modulator chirp will depend both on the waveguide structure and material chirp factor. In an ideal push–pull MZ modulator, the chirp is zero.

The induced phase fluctuations give a spectral broadening of the modulated signal which will, in combination with fiber dispersion, limit the transmission distance of the signal. To understand this, we can study the case of small-signal sinusoidal modulation.

$$P = P_0(1 + m\cos(2\pi f_m t)), \qquad (7.7)$$

where m is the modulation index and f_m the modulation frequency. The sinusoidal modulation will create new spectral components (modulation sidebands) around the optical carrier, spaced with the modulation frequency. Linear-intensity modulation will, in contrast to linear amplitude modulation, generate higher order sidebands, but for simplicity, we limit the discussion to the two first-order side bands on either side of the optical carrier. The fiber dispersion will cause the sidebands to propagate with different speed through the fiber. On detection with an intensity-dependent detector, each sideband will beat against the carrier and regenerate the sinusoidal signal. However, due to the initial-phase modulation and fiber dispersion, the two detected sideband components will be phase shifted with each other. The total amplitude of the received sinusoidal component will depend on this phase shift and can be written as [8]

$$P_{det,f_m} = m\sqrt{1 + \alpha_{mod}^2} \left| \cos\left(\frac{\pi \lambda^2 D L f_m^2}{c} + \arctan(\alpha_{mod}) \right) \right|. \qquad (7.8)$$

Here, D is the dispersion constant of the fiber and L is the fiber length. The detected signal component will be zero when the two sideband components are out-of-phase and cancel out. For a fixed fiber length, L, this cancellation will cause dips in the transfer function of the fiber at certain modulation frequencies, $f_{m,n}$, where n is an integer. This occurs when

$$\frac{\pi \lambda^2 DL f_{m,n}^2}{c} = \frac{\pi}{2} - \arctan(\alpha_{\text{mod}}) + n \cdot \pi \quad (7.9)$$

At these frequencies, the original modulation has, due to the fiber dispersion, been converted to pure phase modulation and is hence not detected by the intensity detector.

Equation (7.9) can be used to estimate the maximum transmission distance of the signal as a function of modulation frequency and chirp factor. For a modulation frequency of 10 GHz at 1.55-μm wavelength, one obtains a maximum transmission distance over standard fiber ($D = 17$ ps/nm/km) of 37 km with $\alpha_{\text{mod}} = 0$, which is reduced to half if $\alpha_{\text{mod}} = 1$ and approximately a third with $\alpha_{\text{mod}} = 2$. On the other hand, a negative α_{mod} is able to extend the maximum transmission distance up to a factor of 2 compared to the case $\alpha_{\text{mod}} = 0$. The transmission distance decreases with the inverse square of the modulation frequency, so at 40 GHz, the transmission distance is only around 1 km if $\alpha_{\text{mod}} = 1$. This is because frequency separation, and hence the differential time delay the sidebands experience due to dispersion, increases linearly with the modulation frequency. This leads to a quadratic dependence of the phase delay with respect to modulation frequency.

Equation (7.9) can also be used to measure the chirp parameter of the modulator and dispersion of the fiber of known length [8]. The dip frequencies, $f_{m,n}$, where the detected modulation signal is almost zero are measured and $f_{m,n}^2 L$ is plotted as a function of the dip number, n. This is according to Eqn (7.9), a linear relation and the dispersion, D, of the fiber can then be extracted from the slope and the chirp factor of the modulator, α_{mod}, from the crossing point with the abscissa.

The above discussion is strictly valid only for small-signal modulation. The chirp factor in electroabsorption modulators is dependent on the modulation voltage, that is, the index change as a function of absorption change is a nonlinear function. It normally decreases with reversed bias and is often negative at high bias (see Figure 7.5). For digital systems, the transmission length is mainly governed by the chirp factor close to the optical on-state. An effective chirp factor can be defined as the average value of α_{mod} within a 3-dB region from the on-state [9]. This means that in electroabsorption modulators, there is normally a trade-off between low-chirp factor and low-insertion loss. The chirp factor increases rapidly at longer wavelengths than the bandgap wavelength. Hence, an electroabsorption modulator will at long wavelengths function as a low-loss phase modulator, and this can be used for realization of semiconductor MZ modulators.

7. Recent Developments in High-Speed Optical Modulators

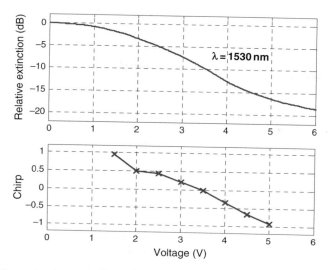

Figure 7.5 Measured voltage dependence of the transmission and chirp factor of a traveling-wave electroabsorption modulator fabricated at the Royal Institute of Technology (KTH), Stockholm (this figure may be seen in color on the included CD-ROM).

7.4 MODULATORS BASED ON PHASE CHANGES AND INTERFERENCE

Many of these types of modulators are based on the linear Pockels effect [1] in ferroelectric materials, such as $LiNbO_3$ [10]. The linear Pockels effect gives a change Δn in refractive index, linear in an applied electric field E:

$$\Delta n \propto rE, \qquad (7.10)$$

where r is the electrooptic coefficient (in reality a tensor).

This effect was discovered in 1893 and has subsequently been heavily used in bulk as well as integrated optics-type modulators. The effect has a response time much lower than 1 ps, has a low dispersion but is rather weak. However, as noted in Appendix, there are also several other ways of achieving refractive index changes by applying an electric field or current, but in view of the striking progress shown by the electrooptic $LiNbO_3$ modulators over the decades (the field of $LiNbO_3$ and high-speed modulators is quite old as evidenced by Refs [11, 12]) and their important role in systems, we will give a brief review of high-speed $LiNbO_3$ modulators.

A simple phase modulator is one where an applied electric field changes the refractive index (but nothing else). This is in apparent contradiction to the Kramers–Krönig relations [3] but can be achieved in materials such as lithium niobate, where the operating wavelengths are far removed from material resonances. The index change impresses a phase change $\Delta\varphi$ on the optical carrier:

$$\Delta\varphi = \Delta n k_0 L, \qquad (7.11)$$

where k_0 is the wavenumber in vacuum and L is the modulator length.

Such a modulator can obviously be used as a phase modulator in its own right for (differential) phase-shift keying (D)PSK, and most of the methods applied to accomplish high-speed amplitude modulators can also be applied to such a modulator.

- Push–pull operation: two voltage sources operated in antiphase
- Single-sided operation: only one voltage source

We will as noted focus on the so far prevalent amplitude or intensity modulation. This can be achieved by using interference between two arms, in a Mach–Zehnder configuration (Figure 7.6). By introducing a differential phase shift in the two arms, light in the input arm can be either transmitted to the output arm or radiated such that it escapes from the output waveguide, thus effecting an intensity modulation. The operation of this type of device rests on the input and output being single-mode waveguides, whereas the intermediate section, that is, the section between the two Y junctions, comprising two waveguides, is two moded. A zero (or multiple of 2π) relative phase shift between the two interferometer arms will correspond to an even mode incident on and propagating

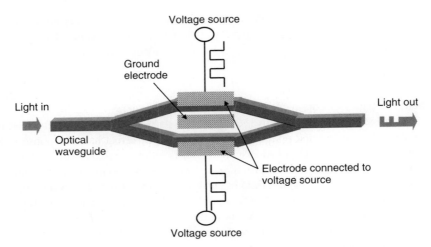

Figure 7.6 Structure of a Mach–Zehnder modulator, with voltage sources changing the effective refractive indices in the two middle waveguides, accomplishing a relative phase shift between the waveguides. This can be achieved through the Pockels electrooptic effect (this figure may be seen in color on the included CD-ROM).

in the single-mode output waveguide, whereas an odd multiple of π relative phase shift will produce an odd mode to be incident on the Y junction at the output port. Such a mode is, however, the next higher order mode that is not guided in the output single-mode waveguide and thus will escape from this and radiate away, hence achieving amplitude modulation. From Eqn (7.11), we note in passing that the required phase shift of π is easier to achieve at shorter wavelengths, everything else being equal.

In Section 7.3, the basics of high-speed modulation was reviewed, and here we quote some references on some high-speed $LiNbO_3$ modulators.

In Ref. [13] the current status of $LiNbO_3$ external-modulator technology is reviewed with emphasis on design, fabrication, system requirements, performance and reliability. The $LiNbO_3$ technology meets the requirements of current 2.5-, 10-, and 40-Gb/s digital communication systems and can be used for even higher data rates, normally at the expense of longer device structures and/or higher drive voltages. The reference shows the impressive results of $LiNbO_3$ development in terms of versatility and increased levels of integration.

A still more recent review, but now more geared to complex modulation formats, is given in Ref. [14]. Here, one can indeed, by employing structures with Mach–Zehnder structures inside Mach–Zehnder structures, implement (in addition to the easily obtained OOK and PSK) single-sideband modulation, SSB, frequency-shift keying, FSK, differential quaternary phase-shift keying, DQPSK and this at data rates approaching 100 Gb/s.

It appears probable that $LiNbO_3$ modulators will be further developed and a complement to semiconductor-based modulators.

7.5 INTENSITY MODULATORS BASED ON ABSORPTION CHANGES

Electroabsorption (EA), that is, the change of absorption spectrum of the active material by an applied electric field, is an efficient and intuitive mechanism for an optical modulator. In this section, the mechanisms for EA in bulk and quantum-well (QW) semiconductor structures, including the novel concept of using IS transitions, are introduced.

In semiconductors, absorption can be increased at energies below the unbiased bandgap by applying an electric field. The effect was predicted and theoretically estimated for bulk materials independently by Franz [15] and Keldysh [16] in 1958. Hence, EA in bulk materials is called the Franz–Keldysh (FK) effect. The absorption below the bandgap with an applied electric field arises as the electrons and holes have a finite penetration into the bandgap, that is, the absorption process can be viewed as photon-assisted tunneling of an electron from the valence band to the conduction band. The FK effect was subsequently demonstrated in a number of materials, and in 1964, an EA modulator structure was demonstrated, see Ref. [17] and references therein. Following the development of QWs in the 1970s and the 1980s, more efficient EA could be achieved by applying a field perpendicularly to the QWs. QW confinement of carriers in a well with a thickness on the order of the bulk exciton (a bound electron–hole pair) Bohr radius ($\sim$10 nm) increases the exciton-binding energy allowing observation of exciton absorption peaks at room temperature. And, importantly, the QW confinement also allows a substantial Stark red shift of the excitonic transitions by an applied electric field, without (completely) ionizing the excitons. EA in QWs is called the QCSE and was first demonstrated in 1984 [18]. About two times higher modulation speed vs drive voltage was found in demonstrated QCSE-based modulators compared to those based on the FK effect [19]. A recent experimental comparison of the merits of EA in bulk and QW-based materials is also available [20]. The drawbacks of the QCSE is that upon absorption of photons, the generated holes tend to pile up in the QWs yielding a rather low-absorption saturation power and that it in general gives a positive effective chirp parameter [21]. In the MZ modulator described in the previous section, the chirp can be engineered by the way in which the voltage is applied to the two arms. In contrast, in an EA modulator, the chirp is given by the induced change of the absorption spectrum as was discussed in Section 7.3.2 of this chapter. In general, a smaller chirp parameter is obtained by operating closer to the absorption edge. Thus, with the QCSE, a reduced chirp can be obtained at the expense of an increased transmission loss. This is also the case for the FK effect [22]. In QWs, the degeneracy of the light and heavy holes is broken so that in general the QCSE is dependent on light polarization [23]. If desired, the polarization dependence can be minimized by using QWs with tensile strain that pushes up the light-hole levels. An absorption modulation $\Delta\alpha \sim 2000\,\text{cm}^{-1}$ is typically achieved by the QCSE in the MQW material, enabling short active modulator lengths $\sim$100 µm, with low applied voltages. Thus, EA modulators have advantages over MZ modulators for applications in future optical-integrated circuits due to a much smaller size, lower drive power, and also the possibility for monolithic integration with semiconductor lasers [24].

By using material combinations with large conduction band offsets of around 1.5 eV or more, it is possible to achieve IS transitions in QWs at telecommunication wavelengths. In recent years, IS absorption at $\lambda = 1.55$ µm has been demonstrated in a number of QW materials, notably InGaN/AlAs/AlAsSb, CdS/ZnSe/BeTe, and (In)GaN/Al(In,Ga)N. IS absorption is peak shaped and can be stronger than interband absorption at similar wavelengths. This is because the nearly parallel subbands of the QW conduction band states give a peak-shaped joint density of states for IS transitions. IS-based EA modulation can be primarily due to modulation of the IS oscillator strengths [25] or primarily due to Stark shifting of the IS transition energy [26]. A proposed GaN/AlGaN/AlN step QW structure that achieves EA modulation by Stark shifting the IS transition is shown in Figure 7.7. IS transitions are characterized by very fast IS

7. Recent Developments in High-Speed Optical Modulators

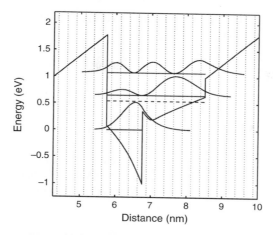

Figure 7.7 GaN/AlGaN/AlN step QW for intersubband-based quantum-confined Stark effect. The dashed line is the Fermi level. (Reproduced with permission from Ref. [26] © IEEE 2006.)

relaxation times of ~ 1 ps that renders them rather insensitive to optical saturation. A desirable small negative chirp parameter is readily achieved in an IS Stark-shift modulator, as the operating wavelength is preferably on the high-energy side of the IS resonance. A quick comparison of some properties of EA in bulk and QW materials is given in Table 7.1. The performance of IS modulators is predicted to be very much material dependent, and the success of them is subject to a continued development of IS materials. The prospects for IS-based modulators are further discussed in Section 7.7 of this chapter.

Table 7.1

Comparison of some properties of the Franz-Keldysh (FK) effect, the quantum confined stark effect (QCSE) and intersubband (IS)-based electroabsorption.

Properties	FK	QCSE	IS
Saturation intensity	High	Moderate	High
Chirp	Lower chirp at expense of insertion loss	Lower chirp at expense of insertion loss	Small negative
Polarization dependence	Small	Depends on QW design	Only TM
Materials for $\lambda = 1.55\,\mu m$	InGaAsP, SiGe	InGaAsP/InGaAsP, InGaAs/AlInAs, Ge/SiGe	InGaAs/AlAs/AlAsSb, GaN/AlGaN/AlN
Material quality requirement	Moderate	Moderate	High

7.6 TRAVELING-WAVE ELECTROABSORPTION MODULATORS (EAMs)

7.6.1 Introduction

To create a viable technology for practical implementation in high-speed fiber optics communications system, light-intensity modulators have to be designed to fulfill a large number of boundary conditions. They have to provide a high-frequency interface to the electronic circuits with a drive voltage low enough to comply with the breakdown limitations imposed by high-speed electronics together with an impedance level near the standard 50-Ω characteristic impedance of microwave transmission lines to avoid large electrical reflections. The electrical-to-optical signal conversion has to be efficient to provide a useful optical extinction ratio for typical light wavelengths of interest for fiber-optics communications, usually around 1550 nm.

EAMs have been shown to be suitable for use in compact fiberoptics transmitters with high speed and efficiency. The epitaxial structure allows monolithic integration with CW lasers, and they can be designed for voltages and currents that can be provided by electronics based on transistor technologies suitable for high speeds. Distributed EAMs with traveling-wave (TW) electrodes avoid the RC limitation due to the capacitance of a lumped EAM structure [27–31]. However, the characteristic impedance of the electrodes of efficient high-speed TWEAMs is typically lower than the standard high-frequency impedance of 50 Ω, and this results in relatively large electrical reflections when connected to high-speed circuits. A narrower waveguide increases the impedance but reduces the optical confinement and the modulation efficiency, while a thinner active layer can be used to increase modulation efficiency but this causes a reduction of the impedance and consequently an increase in the electrical reflection. Microwave periodic structures for the TW electrodes of EAMs have been demonstrated to increase the impedance and decrease electrical reflections with limited degradation of the efficiency [32].

State-of-the-art results for TWEAM include structures suitable for 40 Gb/s transmission with a required modulation voltage of the order of 1 V_{p-p} [33–35]. Modulation bandwidths of 100 GHz (−3 dBe, where dBe = dB change for electrical signal) have been demonstrated with electrical reflections lower than −10 dB in a 50 Ω system [36]. Transmission at 80 Gb/s with nonreturn-to-zero (NRZ) code has been demonstrated [36] for InP-based TWEAMs using electronic time-domain multiplexing (ETDM), indicating the possibility of reaching speeds of 100 Gb/s and beyond.

7.6.2 Ridge-Waveguide EAM Structure

Figure 7.8 shows a simplified cross section of a ridge-waveguide EAM structure which allows single-optical mode operation and which has been developed for

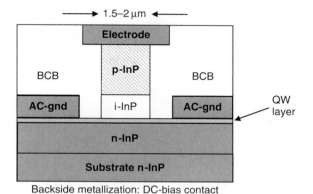

Figure 7.8 Vertical cross section of a typical ridge-waveguide TWEAM (not to scale). The PIN structure is reverse biased.

bitrates of the order of 100 Gb/s and higher [32]. A n-InP buffer layer (1.5 μm) is grown over a n-type InP substrate, and the active material contains InGaAsP/ InGaAsP QWs. The upper cladding consists of n^--InP where active EAM segments have been p-doped by Zn-diffusion almost down to the active layer. Selective wet etching, stopping above the QW layer, is used to form the optical ridge waveguide with a typical width in the range 1.5–2 μm. Metal planes are placed on either side of the mesa to provide ground for alternating currents (AC), and a metal contact on the back side of the conducting substrate is used for direct current (DC) connection.

The capacitance of the pin-diode structure is determined by the choice of both the thickness of the active QW layer, where absorption of light takes place, and the thickness of the i-layer directly on top of the QW layer. Decreasing the i-layer thickness will increase the electrical field across the QW layer and thus the optical absorption for any given voltage applied to the modulator, but this will also increase the capacitance across the pin-diode and thus reduce the impedance of the traveling-wave structure. The result is then a reflection of the applied high-frequency modulation signal, which in turn reduces the efficiency of the modulator in terms of optical extinction compared to applied voltage.

The positioning of the optical field relative to the QW absorption layer can be optimized for maximum overlap to improve modulation efficiency [36]. Figure 7.9 shows the result of an optical simulation which is part of the optimization process. The choice of the absorption layer thickness, including the choice of the number of QWs, is very important not only for the capacitance across the junction and thus for the speed but also for the modulation efficiency.

In the simulation in Figure 7.9, a circular Gaussian beam with a beam width of 4 μm was applied at the beginning of the waveguide. The optical field was calculated for the waveguide structure of the EAM with a 3-D beam-propagation method (BPM) program.

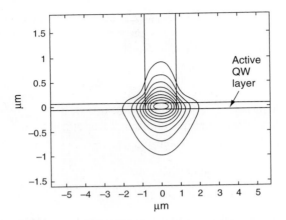

Figure 7.9 Fundamental-mode field for the waveguide structure in a EAM calculated with a 3-D BPM program. The circles correspond to optical intensities of 90 to 10% in 10% steps of the peak intensity. The input Gaussian beam with 1540 nm wavelength has a FWHM of 4 μm. Both vertical and horizontal scales are in micrometers. (Figure reproduced with permission from IEEE [36].)

7.6.3 TWEAM Structures for Increased Impedance

Traveling-wave modulator structures have been demonstrated in LiNbO$_3$-based intensity Mach–Zehnder interferometer (MZI) modulators with bandwidths of up to 70 GHz [37]. EAM with traveling-wave electrode structures have been shown to provide very efficient modulation at 40 Gb/s [33–35] as well as the potential for much higher modulation speeds [32]. However, there is a compromise between high efficiency and a device impedance comparable to the standard 50 Ω level as described above, and therefore, methods of increasing the impedance without significantly reducing the efficiency (in a limited frequency range) have been investigated.

Modeling of the microwave properties of a TWEAM may be used to optimize the design, and TWEAM models have been treated in several publications [38–40]. The microwave properties of a TWEAM can be estimated from the simplified equivalent circuit in Figure 7.10 [38]. In this circuit, R_C is the conductor

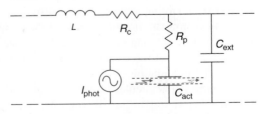

Figure 7.10 Equivalent circuit model of an incremental length of a TWEAM (this figure may be seen in color on the included CD-ROM).

resistance, L is the inductance, C_{act} is the active layer capacitance, C_{ext} is the fringing field capacitance, and R_p is the semiconductor layer series resistance usually dominated by the p-type semiconductor in the mesa, and I_{phot} is the induced photocurrent from the optical absorption. The circuit may be analyzed analytically [41], but it is also easily implemented in commonly available electronic circuit simulation software to allow for more flexibility, as may be required for more complex structures than a TWEAM with constant cross section.

An analysis of possible frequency limitations of a TWEAM shows [41] the following:

- The velocity mismatch between optical and electrical waves along the waveguide structure is not an important limitation as typical TWEAMs are short compared to the wavelength of the highest electrical frequencies of interest. For example, a typical TWEAM designed for 100 Gb/s may be of the order of 200-µm long while the wavelength on chip at 100 GHz is more than 0.5 mm. The term "traveling wave" is therefore somewhat misleading but has become more commonly used than more appropriate terms as, for example, "distributed" or "transmission-line" modulator.
- The transit time for the electrons and holes generated by absorbed photons corresponds to 3 dBe bandwidth of the order of 60–80 GHz for typical total intrinsic layer thicknesses around 0.4 µm. This will appear as a delay in the generated photocurrent I_{phot} which produces a voltage counteracting the applied modulation voltage. This results in a distortion of the pulses at high bitrates but happens only when the modulator starts to saturate due to the absorbed optical power. It may thus be argued that transit time limitations of the bandwidth can be disregarded as they occur under circumstances when the modulator is already less useful due to saturation.
- The frequency-dependent transmission line loss is usually the dominating bandwidth limitation of a TWEAM. For devices based on the InP material system, an InGaAs contact layer is usually required on top of the waveguide for a good contact with low series resistance. This contact layer can cause substantial optical absorption unless it is sufficiently separated from the active optical waveguide by a cladding layer which is normally thicker than 1.5 µm and creates series resistance. An analysis of the equivalent circuit in Figure 7.10 shows that a dominating contribution to the resulting attenuation at high frequencies is related to the device capacitance, and it may also be concluded that a high-bandwidth TWEAM design results in a low-modulator waveguide impedance.
- A TWEAM should be terminated by a resistive load impedance to reduce reflections caused by the end of the transmission line. If this termination resistance has a higher value than the characteristic impedance of the transmission line, the reflection will typically be out-of-phase with the incident wave from the drive electronics at high frequencies, causing a reduction of the effective modulation voltage at the TWEAM input. This

leads to reduced modulation at high frequencies and thus a lower bandwidth compared to a matched termination. On the other hand, if the termination resistance is smaller than the characteristic impedance, the reflected voltage will be in-phase with the incident voltage at high frequencies at the TWEAM input, causing increased modulation and thus a higher bandwidth. However, the conclusion that the termination resistance thus should be chosen lower than the characteristic impedance should be treated with care as a too large reflection will peak the optoelectronic transfer function at high frequencies and cause overshoot in the optical output pulses, which can cause errors in the receiver due to pulse distortion. The effect can thus be used to increase the bandwidth, but the design has to be carefully performed.

- For frequencies typically exceeding 100 GHz, the charging time implied by the combination of the p-InP layer resistance R_p and the active layer capacitance C_{act} will start to affect the modulator bandwidth. This charging time is obviously determined by the choice of total intrinsic layer thickness and the possibility to increase p-doping to reduce resistance.

It may be concluded that the choice of the thickness of the intrinsic layer is very important for the modulator performance. A complete analysis is complicated, but some principles for the design of a TWEAM may be easily derived. A thicker intrinsic layer is beneficial due to decreased capacitance, which leads to decreased attenuation, increased impedance, and decreased internal charging time, but the optical absorption is reduced for a given voltage. A characteristic impedance of 20–25 Ω may be used as a reasonable compromise between microwave reflections and optoelectronic modulation efficiency, but the resulting electrical reflections are then still higher than typical high-frequency circuit specifications of a reflection of −10 dB or less in a 50 Ω system.

A method to combine high-modulation efficiency with relatively high-device impedance using a segmented lateral layout structure has been developed [32]. Active EAM segments are separated by passive light waveguides and electrical transmission lines to form a microwave periodic structure which allows transformation of an active EAM characteristic impedance of 20–25 Ω into a higher device impedance Z_{dev} in the range 30–50 Ω at the input of the TWEAM in a limited frequency range. Figure 7.11 shows a schematic description of an example with two active and three passive segments where the active sections have a ~22 Ω characteristic impedance which is transformed to ~35-Ω device impedance Z_{dev}, thereby reducing the microwave reflection coefficient Γ in a 50 Ω system at the input to the left in the Figure. The magnitude of Z_{dev} will increase at high frequencies, but the design principle can be proven to be useful in practice up to at least 100 GHz with available technologies.

To obtain a relatively high characteristic impedance for the passive transmission line segments, a benzocyclobutene (BCB) layer is used to separate the electrode from the ground plane. Devices designed for a device impedance of

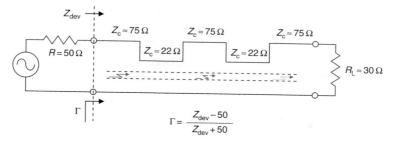

Figure 7.11 Schematic description of a segmented microwave structure with two active and three passive segments where the active sections have a ~22-Ω characteristic impedance which is transformed to ~35 Ω device impedance Z_{dev} (this figure may be seen in color on the included CD-ROM).

35 Ω are terminated with an integrated NiCr thin-film resistor placed on the BCB layer with vias to the metal-ground plane. The lateral device layout consists of five cascaded transmission line segments where two are active ($Z \approx 22\,\Omega$) and three are passive transmission lines ($Z \approx 75\,\Omega$) in parallel with passive optical waveguides. The total lateral active length of the two active segments is about 160 μm in the example in Figure 7.12, and the two active segments have the same length. The contact pads for the top electrode were designed for probed measurements with a width of 20 μm to minimize capacitive parasitics.

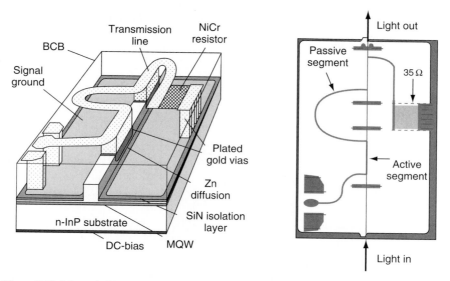

Figure 7.12 Schematic 3D device layout (left) and mask layout (right) of a segmented EAM optimized for a 35-Ω load. (Figure reproduced with permission from IEEE [36]) (this figure may be seen in color on the included CD-ROM).

The design described above has been fabricated and demonstrated to be capable of reaching performance which should be suitable for communication up to 100 Gb/s. The Figure 7.13 below shows the optical small-signal transfer function response (S_{21}) and the reflection ($\Gamma = S_{11}$) of the device. The optical response of a 35 Ω device has dropped by 1 dBe at 60 GHz, and model extrapolation results in a 3-dBe bandwidth of the order of 100 GHz [32]. The return loss is less than -12 dB up to 60 GHz. Optical DC characteristics are also plotted to the right in the figure.

The DC and high-frequency small-signal responses shown in Figure 7.13 are important for the design and evaluation of devices, but the ultimate test of the usefulness of a device intended for fiber optic communications is in a transmission test. To allow this, a fiber optic transmitter for ETDM signals with NRZ code optical output was assembled with a TWEAM, a 40-Gb/s bit-error rate test set (BERTS) and a 80-Gb/s multiplexer [36]. The transmitter showed open eye diagrams (Figure 7.14) with transition times limited by nonoptimum electrical amplification and connections. A 80-Gb/s electronic multiplexer [42] based on a SiGe bipolar technology with an f_T and f_{max} of greater than 200 GHz was used together with a III–V-based electronic driver to provide the input electrical signal to the TWEAM which was connected to the electronics by a high-frequency on-wafer probe. The transition time (20%–80%) of the drive signal was ~6 ps. The dynamic extinction ratio of the TWEAM measured at 50 Gb/s was 10.7 dB at the optical wavelength $\lambda = 1540$ nm for ~3 V_{p-p} incident drive voltage.

The transition time of the optical signal was limited by nonoptimum electrical connections, and it may be expected to be improved without changing the technologies for the modulator or the electronics. An analysis of the devices in the transmitter indicates that further optimization of the modulator and the driver may be possible in order to approach 100 Gb/s.

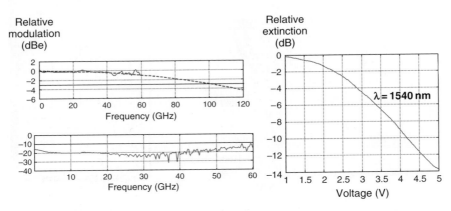

Figure 7.13 The top left figure shows the measured (solid) and simulated (dashed) optical small-signal frequency response for the 35-Ω device. The bottom left figure shows the measured return-loss for the device. The right figure shows the optical DC extinction of the 35-Ω device at $\lambda = 1540$ nm. (Figure reproduced with permission from IEEE [36]) (this figure may be seen in color on the included CD-ROM).

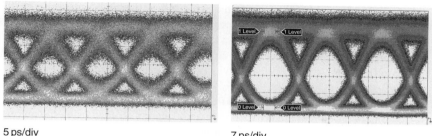

5 ps/div   7 ps/div

Figure 7.14 Optical eye diagram back-to-back measured at 80 Gb/s (left figure) and 50 Gb/s (right figure). Light parts show high occurrence in the sweep and dark parts low occurrence. (Figure reproduced with permission from IEEE [36]) (this figure may be seen in color on the included CD-ROM).

7.6.4 High-Efficiency Modulators for 100 Gb/s and Beyond

An analysis of the experimental results described above shows that it is possible to improve the efficiency of the TWEAM for a given bandwidth by changing two properties of the device design. The optical waveguide structure and the active layer design can be optimized to increase the coupling of the microwave and the optical fields. The microwave design can be improved by increasing the active modulator length with constant bandwidth through optimization of the layout structure with active segments and passive transmission lines. These two optimizations can be made independently, and the combined result is an expected substantial reduction of the drive voltage compared to the results reported earlier [36].

In electroabsorption materials, the transmitted light intensity is given by the absorption coefficient α and the incident light intensity as follows:

$$\frac{P_{\text{out}}}{P_{\text{in}}} = \exp(-\Gamma \alpha(V) L). \quad (7.12)$$

Γ, $\alpha(V)$, and L are the optical confinement factor, absorption coefficient change as a function of applied voltage V, and the modulator's effective length, respectively. Therefore, the extinction ratio (on/off ratio) in dB can be given as

$$[\text{on/off}] = 0.434 \cdot \Gamma \Delta \alpha(V) L \quad (7.13)$$

From (7.13), we find that there are three approaches to improve the modulation efficiency:

(1) Increase $\Delta\alpha(V)$: The value is limited by the inherent property of the QW. When the QW material is fixed, the ratio between $\Delta\alpha$ and V is determined.

(2) Increase L: This will increase the optical insertion loss and decrease the bandwidth unless tuning of the microwave response is introduced, which is discussed below.
(3) Increase Γ: This will not cause any disadvantages and can be achieved by careful design of the optical waveguide in the modulator.

The QW region in the design described in the previous section [36] had eight QWs and an optical-confinement factor estimated to be about 0.2. Increasing the number of QWs to 12 and optimizing the vertical position of the active layer results in an increase of the optical confinement factor to over 0.3 according to BPM simulations. This causes an expected improvement of modulating efficiency and a reduction of the required drive voltage for the EAM by $\sim$30%.

Further improvement in the efficiency can be obtained by tuning the microwave properties. The lateral layout of the TWEAM has earlier been designed with a constant periodic structure to minimize electrical reflections [32]. The resulting return-loss is typically less than -15 dB up to more than half of the -3-dBe bandwidth but increases toward the -3-dBe bandwidth limit to values larger than -10 dB. Allowing different lengths of the active segments in the design provides the possibility to numerically optimize the bandwidth with the condition of less than -10 dB electrical reflection throughout the -3-dBe bandwidth of the device. Combining this new degree of freedom in the lateral design with a reduction of the thickness of the intrinsic region to increase the electric field results in a reduction of the characteristic impedance of the active segments from 22 to about 20 Ω and makes it possible to increase the total active length to about 250 μm. The expected reduction of the drive voltage is of the order of 30%, and the added optical loss due to the increased length is limited to $\sim$1 dB. An example of a typical layout of an optimized TWEAM with three active modulator sections is shown in Figure 7.15

As the optimizations of the optical waveguide and the microwave design are independent of each other, the expected total reduction of the drive voltage will be $\sim$50%. The drive voltage for 10-dB optical extinction ratio at 100 Gb/s is then reduced from $3\,V_{p-p}$ to $<2\,V_{p-p}$. These values are based on simulations using models derived from measurements of previous structures. Figure 7.15 shows an example of the layout of the new designs with a chip size of $0.6 \times 0.8\,mm^2$ excluding CW laser. The layout also has larger pads than previous designs with 60 μm diameter to allow wire or flip-chip bonding. Realistic-bonding parasitics for flip-chip bonding have been included in the simulations of the designs when establishing small-signal bandwidth and large-signal transition times.

Figure 7.16 shows a comparison between experimental results at KTH and simulations of the optimized designs described above. The simulations indicate also that bandwidths suitable for 160 Gb/s should be possible to reach with this technology if interconnect parasitics can be minimized. However, substantial improvements beyond what these simulations show are difficult to envisage without a relatively radical change such as a different absorption mechanism.

7. Recent Developments in High-Speed Optical Modulators

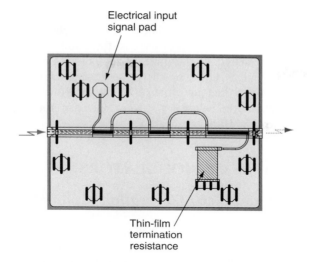

Figure 7.15 Layout of TWEAM with 100-GHz bandwidth and less than 2 V expected voltage swing for 10 dB ER. The size is ∼0.6 × 0.8 mm² (this figure may be seen in color on the included CD-ROM).

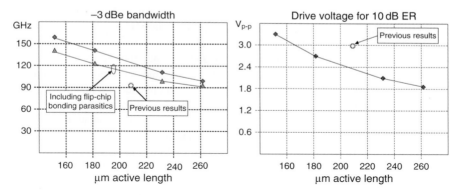

Figure 7.16 Measured ("previous results") and simulated bandwidths (left) and peak-to-peak drive voltages for optimized segmented TWEAM with total active lengths ranging from 150 to 260 µm (this figure may be seen in color on the included CD-ROM).

7.6.5 Future Development

The demonstrated TWEAMs have so far been optimized with a MQW active layer design similar to that of commercial devices with EAMs monolithically integrated with distributed feedback (DFB) CW lasers. Eventually, the efficiency of a modulator will depend on the chosen method of electroabsorption in the active region, and future modulators may rely on IS transitions in QWs in the conduction band. IS-modulator technology is being investigated to further improve modulation efficiency and/or reduce the length of high-speed light-intensity modulators, and the experimental work has been focused on the quality of the heterostructure interfaces as this will determine the limits of the possible reduction in required drive voltage. Calculations based on the present status of the development of epitaxial growth indicate that the voltage swing for 10-dB optical extinction ratio may be expected to be below 1 V_{p-p} for a bitrate of 100 Gb/s. IS is treated in more detail in Section 7.7 of this chapter.

7.7 NOVEL TYPES OF MODULATORS

7.7.1 Intersubband Electroabsorption Modulators

The QCSE-based interband EAM described in Sections 7.5 and 7.6 constitute a relatively mature technology. To improve the active material considerably, one must probably resort to some new concept. One promising avenue is offered by intersubband (IS) transitions in QWs. Primarily with the target to reduce the driving voltage and increase bandwidth, we have investigated IS transitions within the conduction band of QWs for high-speed EAMs [25, 26, 43, 44]. Thus, the optical modulation principle is still electroabsorption, but the active layer is now unipolar, only involving electrons, which has several interesting consequences. Using IS transitions, the subbands have a similar curvature giving a peak-shaped joint density of states, and thus, the absorption can be stronger, enabling a shorter and thus faster modulator. In contrast, QW interband transitions occur between subbands of opposite curvature. It is also interesting to note that the rapid LO phonon-mediated IS relaxation time on the order of 1 ps makes IS-based absorption insensitive to absorption saturation. The short relaxation time means that excited electrons can rapidly relax within the QW and be "recycled." Thus, the moderate optical saturation power afforded by interband QCSE, due to hole-pileup, can be considerably improved. Additionally, as no p-doping is necessary in an IS-based modulator, it will not suffer from the problems of the series resistance of the p-doped layer.

However, in the conventional III–V QW materials such as InGaAs/InAlAs and InGaAsP/InGaAsP, the IS resonance energy is relatively small, corresponding to wavelengths of 4 µm or longer. To reach the telecommunication wavelength of 1.55 µm (0.8 eV), a material combination with a large conduction band offset ΔE_c of about 1.5 eV or more is required. In recent years, IS absorption at 1.55 µm has been demonstrated in several QW materials InGaAs/AlAs/AlAsSb, GaN/Al(Ga)N, and CdS/ZnSe/BeTe [45–47]. We have performed simulations of RC-limited EAMs at $\lambda = 1.55$ µm based on IS transitions in GaN/AlGaN/AlN step QWs [26, 44] and in InGaAs/InAlAs/AlAsSb-coupled QWs. The high-speed performance of such modulators will ultimately be determined by the IS absorption linewidth $\Gamma_{\text{linewidth}}$ that can be achieved. A small linewidth is essential. It can be shown that the capacitance of an IS-based modulator depends on the linewidth as

$C \sim (\Gamma_{\text{linewidth}})^3$ [43]. At the small well-width required for IS transitions at 1.55 µm, the absorption linewidth is generally limited by the material quality due to interface roughness, well-width fluctuations, and is also affected by doping.

For comparison to the interband-based segmented TWEAM [32] that is described in Section 7.6 of this chapter, we here focus on the InGaAs/InAlAs/AlAsSb material of reference [25]. It is interesting as it can be grown on InP making only small alterations of our currently well-established TWEAM process necessary. It consists of strain-balanced coupled QWs separated by lattice-matched AlAsSb barriers, so that the MQW will as a whole be lattice matched to InP. Figure 7.17(a) shows one period of the MQW structure. To have IS absorption, the QW ground state needs to be populated. To this end, the barriers are n-type δ-doped with $n_D = 5 \times 10^{12}\,\text{cm}^{-2}$. Increasing the doping further would begin to populate bound states in the satellite X- and L-valleys of the conduction band. As these valleys have multiple equivalent minima and much heavier masses, an increased doping would primarily populate these satellite valleys instead of the QW ground state in the zone center Γ-valley, that is, not improving the performance of the modulator. Absorption modulation is achieved by a combination of two effects. First, there is a Stark shift, that is, the energy difference between the populated ground state and the first excited state is dependent on the applied electric field. This is seen in Figure 7.17(b) as a shift of the absorption peak. Second, the oscillator strength of this transition is dependent on applied electric field, due to the coupling of the excited states in the QW. In Figure 7.17(b), this is seen as the height of the peak varies with applied field. Further details on the simulation of this material in an RC-limited EAM are available [25].

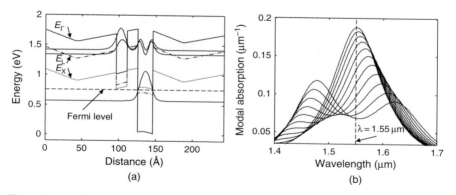

Figure 7.17 (a) Potential profile of one coupled quantum well, with moduli squared of the wave functions, at no applied electric field. The absorption at 1.55 µm is primarily due to the ground state to first excited state transition. The profiles of the Γ (zone center) and X,L (satellite valleys) conduction band edges are indicated. The dotted line is the Fermi energy. (b) Modal absorption spectra for different applied electric fields. Absorption modulation at 1.55 µm is achieved by a combination of Stark shift and modulation of the oscillator strengths. The difference in electric field between two curves is 26 kV/cm.

Comparing the experimental results of the fabricated TWEAMs with simulated results on IS modulators should be done with caution, as results depend considerably on assumed IS resonance linewidths and as the structures also concerning waveguide and cladding layers are differing. As a simple first comparison, we replace the interband QCSE active material in the segmented TWEAM with the IS-based InGaAs/InAlAs/AlAsSb material of reference [25], using the same active layer thickness 100 nm (six IS-based coupled QWs) as the interband QCSE material has. Hence, we obtain the calculated modal absorption in the active segments shown in Figure 7.17(b). Comparing the modal absorption change vs the applied electric field, we find that this nonoptimized IS-based structure offers an improvement by a factor of 3 compared to the measured values in the segmented TWEAM (see Section 7.6). This improvement would allow a reduced voltage swing and/or a shorter active modulator length.

Another way to compare is a direct comparison of the $f_{3\,\text{dB}}$ modulation bandwidth of the segmented TWEAM with $f_{3\,\text{dB}}$ of the short RC-limited IS-based modulator structures that we have simulated for 1.55 µm. To compare on an equal footing, we assume a 35-Ω termination resistor (note that a termination or shunt resistor was not employed in the original simulations [25, 26]). This gives $f_{3\,\text{dB}} = 220$ GHz for the InGaAs/InAlAs/AlAsSb structure (assuming a linewidth $\Gamma = 60$ meV) with 2.0 $V_{\text{p-p}}$ for 10-dB optical extinction ratio for a 25 µm modulator length and $f_{3\,\text{dB}} = 120$ GHz for the GaN/AlGaN/AlN structure (assuming a linewidth $\Gamma = 100$ meV) with 2.8 $V_{\text{p-p}}$ for a 13-µm modulator length [26, 44]. We note that the higher modulation speed obtained in the former structure can be explained by the different linewidths that were used, even when considering the higher voltage swing in the latter structure. It should also be noted that no microwave tuning of the IS devices has been considered yet, and this should increase the bandwidth substantially for any given device capacitance. IS-based material thus has a considerable potential for improvement over interband QCSE. However, both Sb-based and nitride materials are difficult to grow with high quality, and the main challenge in future work is to experimentally demonstrate relatively narrow IS linewidths at the high-doping conditions required for strong IS absorption.

An important characteristic of IS-based electroabsorption is that a negative and small-chirp parameter is readily obtained [26]. A small negative chirp parameter can substantially extend the usually dispersion-limited transmission distance of high-speed optical pulses on standard single-mode optical fiber [7, 48] as described in Section 7.3.2 of this chapter. In contrast, interband QCSE-based modulators in general give a positive chirp to generated pulses [49]. This is a consequence of employing electroabsorption on the low-energy side of the interband absorption step at the band edge. The chirp can be decreased by operating closer to the band edge, giving a trade-off between low chirp and insertion loss [7]. Such a trade-off is not necessary in an IS modulator, however. In Figure 7.18 is shown the calculated modal absorption, chirp parameter, and transmittance of an electroabsorption modulator for $\lambda = 1.55$ µm based on IS transitions in GaN/AlGaN/AlN step QWs (the step QW is shown in Figure 7.7 in this chapter) [26, 44].

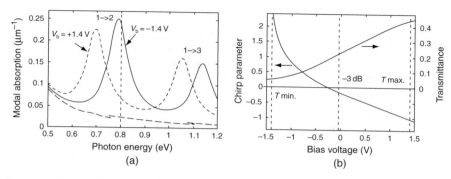

Figure 7.18 (a) Simulated modal absorption spectrum at the applied bias voltages $V_b = -1.4$ V (solid) and $V_b = +1.4$ V (dashed) in an electroabsorption modulator for $\lambda = 1.55\,\mu$m based on intersubband transitions in GaN/AlGaN/AlN step QWs. The vertical dashed line indicates the photon energy $\hbar\omega = 0.8$ eV ($\lambda = 1.55\,\mu$m). (b) Chirp parameter α and transmittance T of the modulator vs the applied voltage. The dashed lines indicate the applied voltage swing with minimum and maximum transmittance T at $V_b = -1.4$ V and $+1.4$ V, respectively, and where T has dropped by 3 dB. (Figures reproduced with permission from Ref. [26] © IEEE 2006.)

The negative chirp and high-optical transmission is achieved in the same part of the applied voltage swing, that is, when the IS resonance is tuned away from the photon energy. Hence, there is no trade-off between a negative chirp and low-insertion loss, by virtue of using the high-energy side of the IS resonance. To estimate the effective chirp parameter, it has been suggested to use the average of the chirp parameter during the top 3 dB of the transmittance [32]. Using that suggestion, an effective chirp parameter of $\alpha_{\text{eff}} = -0.6$ is derived from Figure 7.18. This is near optimal for standard single-mode optical fiber.

7.7.2 Silicon-Based Modulators

Among the optoelectronic materials, Si is attracting a strongly increasing interest. This is driven by a wish to exploit the very well developed and cheap Si-processing technologies and to integrate optical devices with electronics in a cost-efficient way. Fueled by investments from the electronics industry in recent years, substantial progress has been made regarding Si photonic-integrated circuits and systems for the telecommunications wavelength 1.55 μm [50, 51]. The other main target for Si photonics is optical interconnects, which, however, are not a priori bound to a specific wavelength [52, 53].

Achieving high-speed Si-based optical modulators is challenging due to the indirect bandgaps of Si and Ge, the low-carrier mobilities, and the absence of a linear electrooptic Pockels effect due to inversion symmetry of the crystals. Two excellent reviews of Si-based optical modulators and modulation mechanisms are available that cover developments until 2004 [54, 55]. We note also that a Si IS-based electroabsorption modulator has been proposed earlier utilizing the large conduction band offset in Si/CaF$_2$ [56]. However, recently, two significant and promising breakthroughs regarding Si-based modulators have been reported. One is the demonstration of QCSE electroabsorption due to the *direct* interband transitions in Ge/SiGe QWs [57]. The absorption change is of the same size as that in the conventional III–V-based QCSE and thus may pave the way for equally compact Si-based modulators. The other breakthrough is demonstration of high-speed (10 Gb/s) MZ modulators with a refractive index change based on carrier density modulation (plasma effect) in reverse-biased Si pn junctions [58–60], see Figure 7.19 and Chapter 11 by Gunn and Koch, and in MOS capacitors [61]. Phase modulation due to the plasma effect has been the traditional mechanism used in Si-based modulators. The new designs, using reverse-biased pn junctions or MOS capacitors, by relying on majority carriers circumvents the carrier lifetime switching speed limitation of earlier forward-biased pin-diode structures. But, in return, they require longer modulator lengths of 2–10 mm and higher drive voltages. An intrinsic switching time of 7 ps was found in a theoretical analysis of MZ modulators utilizing a Si pn junction at an applied voltage of 5 V [58]. Phase modulation by reverse biasing a pn junction has earlier been investigated in detail and found to be efficient in GaAs [62], where, however, a large contribution was due to the Pockels effect which as mentioned is absent in Si. Recently, 30-Gb/s data transmission was demonstrated with a traveling-wave MZ modulator utilizing a reverse-biased Si pn junction (Figure 7.19) [60]. Although the extinction ratio was low in the short 1-mm device, this confirmed the intrinsic high-speed capability of the Si pn junction.

In 2006, strong electroabsorption was demonstrated utilizing the direct transitions in Ge/SiGe QWs. This implies an intriguing possibility to achieve very compact and importantly low-power-consumption modulators in a process that is compatible with complementary metal–oxide–semiconductor (CMOS). An important reason that

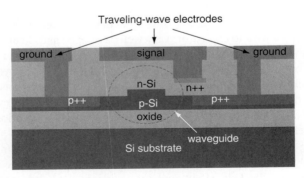

Figure 7.19 Schematic cross-sectional view of one arm in a Mach–Zehnder traveling-wave modulator including a silicon pn junction in a SOI waveguide for phase modulation. The transmission line has a high-frequency characteristic impedance of $\sim$20 Ω. (Reproduced with permission from reference [60] © OSA 2007) (this figure may be seen in color on the included CD-ROM).

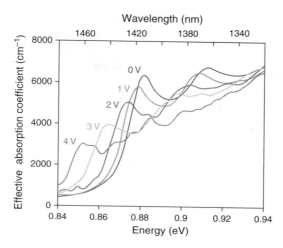

Figure 7.20 Effective absorption coefficient spectra due to the direct transitions in a Ge/SiGe MQW for different reverse bias voltages. The absorption was estimated from a photocurrent measurement. (Reprinted with permission from Macmillan publishers Ltd: Nature [57] © 2005) (this figure may be seen in color on the included CD-ROM).

QW-direct transitions (i.e., to the conduction band Γ-valley) can be observed is that the Γ–L intervalley scattering time is slow enough that bound QW states can be formed by the Γ-valley states. The Ge intervalley scattering time at low temperature has been measured to 570 fs [63]. Due to the somewhat too large Ge bandgap, useful electroabsorption was achieved at ~1440–1470 nm in the first device as shown in Figure 7.20. By employing wider 12.5 nm Ge wells and also utilizing a red shift of the absorption edge at elevated temperatures, the absorption modulation was extended out to 1550 nm [64]. But much work still remains to achieve strong absorption modulation at this wavelength. It can include adjusting the modulator design within the present SiGe material system or adjusting the QW material to increase the operating wavelength. GeSn is a possible QW material with lower bandgap [65]. A FK-type electroabsorption modulator design utilizing the direct transitions in bulk SiGe has also been evaluated [66]. In bulk, the unstrained direct bandgap of Ge 0.80 eV is instead slightly too small, so that operation at 1.55 µm should be possible using SiGe with a low Si content.

7.8 SUMMARY AND FUTURE PROSPECTS

In this chapter, we have treated the principles of optical modulation with emphasis on amplitude or intensity modulation. Different physical mechanisms and various materials to accomplish this were discussed. The basics of high-speed modulation, in many ways the raison d'être for the external modulator, were reviewed. State-of-the-art TWEAMs based on InP was described, reporting 100-GHz bandwidth

Table 7.2

A semiquantitative comparison between today's dominating intensity modulators. MZI: Mach–Zehnder interferometer modulator, EAM: Electro absorption modulator.

	External modulators for fiber optic communications	
	LiNbO$_3$ MZI	InP EAM
Drive voltage	+ Medium to low (>5 V)	+ Low (<3 V)
Bandwidth	+ Very high (~100 GHz)	+ Very high (~100 GHz)
Chirp	+ Adjustable	− Yes (but can be engineered)
Insertion loss	+ Low	Moderate
Size	− Large (several centimeters)	+ Small (< or <<500 μm)
λ-sensitivity	+ Low	− High
Monolithic laser, SOA, and tandem integration	− No	+ Yes

and discussing proposals for even larger bandwidths at a drive voltage of a few volts. At the present time, the electrooptic Mach–Zehnder-type LiNbO$_3$ and electroabsorption modulators are dominating, and their relative merits are in a semiquantitative fashion described in Table 7.2.

Further, some novel or emerging modulator types were treated, concentrating on IS transition structures and Si-based modulators.

However, with optical systems gaining in sophistication and in some ways approaching the spectrally efficient modulation formats of mobile telephony systems [67], modulators other than intensity or amplitude ones have gained in importance. If phase modulation is sufficient, the implementation is rather straightforward (on the transmitter side); however, frequency and polarization modulation are more complex issues.

There has been a surge of interest in Si modulators recently (Section 7.7), fueled by visions of monolithic optoelectronic integration in this material. However, the implementation of Si modulators is most likely only worthwhile with a Si-based light source; this is still an elusive target and will maybe remain so, due to the indirect bandgap. In III–V semiconductors, such integration between light source and modulators is naturally accomplished, and maybe hybrid integration between Si and III–Vs is the only viable way to achieve the coveted optoelectronic integration.

We have in this chapter not discussed materials such as polymers. The reason is that they currently do not basically seem to offer much over and above what lithium niobate can. However, relatively novel phenomena such as electromagnetically induced transparency and plasmonics do indeed offer very intriguing prospects for modulation, given that room temperature operation and the optical loss issues can be mastered [68].

It appears that the ultimate limits regarding the operating frequency and other characteristics of optical modulators, employing different mechanisms and materials, are remote and to some extent unknown. These ultimate limits are dictated by the pertinent fundamental issues, the remoteness from the limits mostly originating form practical matters. Thus, the development of modulators of ever higher performance, integration as well as capability for complex modulation formats, will certainly remain a challenging research topic, especially experimentally, for a long time to come. The application scope is ubiquitous in the quest for THz modulation rates.

ACKNOWLEDGMENTS

The part of the work reported here which was performed at KTH was supported in part by VINNOVA (Swedish Agency for Innovation Systems) and the KAW foundation. The authors wish to acknowledge contributions from Stefan Irmscher, Robert Lewén, Urban Eriksson, and Yichuan Yu.

APPENDIX

Table A

A compilation of different mechanisms to achieve the changes in complex refractive index required for modulation of optical radiation.

Physical mechanism	Operation mode	Characteristics	Materials
Index change through Pockels effect (linear EO effect)	Voltage applied over dielectric material or reverse-biassed pin junction	Speed usually limited by microwave/optical wave walk-off or RC-constants. Small nonresonant effect. $dn/n \sim 0.0001$ typically, no associated absorption.	All materials lacking inversion symmetry: LN, III–V SC
Index change through nonresonant Kerr effect (quadratic EO effect)	See above	See above. Usually very small	All materials
Free-carrier, intra-band transitions ("plasma" effect). Index and absorption change	Carrier injection or depletion through injection and reverse biasing, respectively	Usually smaller than other carrier-induced effects. Nonresonant effect, increases with the square of the wavelength	All SC and other materials in which free carriers can be induced
Absorption/gain change through bandfilling	Unipolar through reverse biasing of pin junction (depletion), bipolar forward biased	Unipolar operation is fast (RC-limited) but cannot give gain, bipolar operation limited by carrier recombination. Strong λ-dependence.	III–V (and other SC)

(Continued)

Table A (Continued)

Physical mechanism	Operation mode	Characteristics	Materials
Index change through bandfilling, related to the above by the KK relation	See above	See above. Large index changes achievable of the order $dn/n \sim 0.001-0.01$ Strong λ-dependence	See above
Absorption change by quantum-confined Stark effect (QCSE)	Reverse biassed pin junctions	RC—or transport—time-limited response (fast, depends on number of wells and barrier heights). Strong λ-dependence	Quantum Well SC
Index change through QCSE, related to the above by the KK relation	See above	RC-limited response dn/n 0.001–0.01 Strong λ-dependence	See above
Absorption change by Franz–Keldysh (FK) effect	See above	RC—or transport-time-limited response. Strong λ dependence	All SC
Temperature-induced absorption and index changes for λ's close the bandgap	Change of external temperature, and heat dissipation associated with carrier recombination	Large effect that partially counteracts the index and absorption changes induced by electronic effect. Depends on heat sinking and the ratio between radiative and non-radiative recombination channels	All SC

EO: Electrooptic, SC: Semiconductor, LN: LiNbO$_3$, KK: Kramers–Krönig relations [3].

Some materials for guided wave modulators

- LiNbO$_3$, LiTaO$_3$, SBN, and so on all offer high linear refractive index change in response to an applied electric RF field
  - LiNbO$_3$: ferroelectric (spontaneous nonzero) polarization material, allows comparatively simple fabrication of low-loss (<0.2 dB/cm) waveguides
  - also useful for nonlinear optics
- InP, GaAs, in general III–V-type semiconductors lacking inversion symmetry
  - have, like the ferroelectrics above, a linear refractive index change in response to applied electric field, however, smaller than the above, but in addition
  - coupled absorption and index changes (connected by Kramers–Krönig relations) due to bandgap effects (Franz–Keldysh effect, quantum-confined Stark effect, and bandfilling)

- Si
  - quadratic refractive index change (Kerr effect) in response to an applied electric RF field
  - absorption and index changes due to free-carrier nonbandgap-related (plasma) effect
- Polymers
  - poled polymers: linear refractive index change in response to applied electric RF field

REFERENCES

[1] T. Tamir (ed.) *Topics in Applied physics: Integrated Optics,* Berlin, Heidelberg, New York, Springer, 1975.
[2] A. Karlsson, R. Schatz, and G. Björk, "On the modulation bandwidth of semiconductor microcavity lasers," *IEEE Photon. Technol. Lett.,* 6(11), 1312–1314, November 1994.
[3] A. Yariv, *Quantum Electronics,* Wiley, 2nd edn, app. 1. New York, London, Sydney, Toronto, 1975.
[4] J. Buus and E. J. Murphy, "Tunable lasers in optical networks," *J. Lightw. Technol.,* 24(1), 5, January 2006.
[5] E. Forsberg, B. Hessmo, and L. Thylen, "Limits to modulation rates of electroabsorption modulators," *IEEE J. Quantum Electron.,* 40(4), 400–405, 2004.
[6] R. E. Collin, *Foundations for Microwave Engineering,* 2nd ed, McGraw-Hill Inc., Singapore, ISBN 0-07-112569-8, 1992.
[7] F. Koyama and K. Iga, "Frequency chirping in external modulators," *J. Lightw. Technol.,* 6(1), 87–93, January 1988.
[8] F. Devaux, Y. Sorel, and J. F. Kerdiles, "Simple measurement of fiber dispersion and chirp parameter of intensity modulated light emitter," *J. Lightw. Technol.,* 11, 1937–1940, December 1993.
[9] F. Dorgeuille and F. Devaux, "On the transmission performances and the chirp parameter of a multiple-quantum-well electroabsorption modulator," *IEEE J. Quantum Electron.,* 30(11), 2565–2572, November 1994.
[10] I. P. Kaminow, L.W. Stulz, and E. H. Turner, "Efficient strip waveguide modulators," *Appl. Phys. Lett.,* 27, 555, 1975.
[11] R. C. Alferness, "Waveguide electro-optic modulators" *IEEE Trans. Microw. Theory Tech.,* 30(8), 1121–1137, August 1982.
[12] M. Izutsu, Y. Yamane, and T. Sueta, "Broad-band traveling-wave modulator using a LiNbO3 optical waveguide," *IEEE J. Quantum Electron.,* QE-13(4), 287–290, April 1977.
[13] E. L. Wooten, K. M. Kissa, A. Yi-Yan et al., "A review of lithium niobate modulators for fiber-optic communications systems," *IEEE J. Sel. Top. Quantum Electron.,* 6, 69–82, 2000.
[14] T. Kawanishi, T. Sakamoto, and M. Izutsu, "High-speed control of lightwave amplitude, phase, and frequency by use of electrooptic effect," *IEEE J. Sel. Top. Quantum Electron.,* 13, 79–91, 2007.
[15] W. Franz, "Einfluß eines elektrischen Feldes auf eine optische Absorptionskante," *Z. Naturforsch.,* 13a, 484, 1958.
[16] L. V. Keldysh, "The effect of a strong electric field on the optical properties of insulating crystals," *Zh. Eksp. Teor. Fiz.,* 34, 1138, 1958 *(Sov. Phys. JETP,* 7, 788, 1958).
[17] G. Racette, "Absorption Edge Modulator Utilizing a P-N Junction," in Proc. *IEEE,* 52, New York, 1964, p. 716.
[18] D. A. B. Miller, D. S. Chemla, T. C. Damen et al., "Band-edge electroabsorption in quantum well structures: the quantum-confined stark effect," *Phys. Rev. Lett.,* 53, 2173–2176, 1984.

[19] K. Wakita, *Semiconductor Optical Modulators*, Boston, Kluwer Academic Publishers, 1998, p. 169.
[20] G. B. Morrison, J. W. Raring, C. S. Wang et al., "Electroabsorption modulator performance predicted from band-edge absorption spectra of bulk, quantum-well, and quantum-well-intermixed InGaAsP structures," *Solid-State Electron.*, 51, 38–47, 2006.
[21] F. Koyama and K. Iga, "Frequency chirping in external modulators," *J. Lightw. Technol.*, 6(1), 87–93, 1988.
[22] O. Sahlén, "Optimization of DFB lasers integrated with Franz-Keldysh absorption modulators," *J. Lightw. Technol.*, 12(6), 969–976, 1994.
[23] S.-L. Chuang, *Physics of Optoelectronic Devices*, New York: Wiley Interscience, 1995.
[24] J. W. Raring, L. A. Johansson, E. J. Skogen et al., "40-Gb/s widely tunable low-drive-voltage electroabsorption-modulated transmitters," *J. Lightw. Technol.*, 25(1), 239–248, 2007.
[25] P. Jänes and P. Holmström, "High-Speed Optical Modulator Based on Intersubband Transitions in InGaAs/InAlAs/AlAsSb Coupled Quantum Wells," in Proc. of the 15th *IPRM 2003*, Santa Barbara, USA, 2003, pp. 308–311.
[26] P. Holmström, "Electroabsorption modulator using intersubband transitions in GaN/AlGaN/AlN step quantum wells," *IEEE J. Quantum Electron*, 42, 810–819, 2006.
[27] Y.-J. Chiu, H.-F. Chou, V. Kaman et al., "High extinction ratio and saturation power traveling-wave electroabsorption modulator," *IEEE Photon. Technol. Lett.*, 14(6), 792–794, June 2002.
[28] M. Shirai, H. Arimoto, K. Watanabe et al., "40 Gbit/s electroabsorption modulators with impedence-controlled electrodes," *Electron. Lett.*, 39, 734–735, 2003.
[29] Y. Akage, K. Kawano, S. Oku et al., "Wide bandwidth of over 50 GHz traveling wave electrode electroabsorption modulator integrated DFB lasers," *Electron. Lett.*, 37(5), 299–300, 2001.
[30] S. Irmscher, R. Lewén, and U. Eriksson, "InP/InGaAsP high-speed traveling-wave electro-absorption modulators with integrated termination resistors," *IEEE Photon. Technol. Lett.*, 14, 923–925, 2002.
[31] G. L. Li, S. A. Pappert, P. Mages et al., "High-saturation high-speed traveling-wave InGaAsP-InP electroabsorption modulator," *IEEE Photon. Technol. Lett.*, 13(10), 1076–1078, 2001.
[32] R. Lewén, S. Irmscher, U. Westergren et al., "Segmented transmission-line electroabsorption modulators," *J. Lightw. Technol.*, 22(1), 172–179, January 2004.
[33] H. Fukano, T. Yamanaka, M. Tamura, and Y. Kondo, "Very-low-driving-voltage electroabsorption modulators operating at 40 Gb/s," *J. Lightw. Technol.*, 24(5), 2219–2224, May 2006.
[34] H. Fukano, T. Yamanaka, M. Tamura et al., "40 Gbit/s electroabsorption modulators with 1.1 V driving voltage," *Electron. Lett.*, 40(18), 1144–1146, September 2, 2004.
[35] H. Fukano, T. Yamanaka, M. Tamura et al., "Very low driving-voltage InGaAlAs/InAlAs electroabsorption modulators operating at 40 Gbit/s," *Electron. Lett.*, 41(4), 211–212, Feburary 17, 2005.
[36] Y. Yu, R. Lewen, S. Irmscher et al., "80 Gb/s ETDM Transmitter with a Traveling-Wave Electroabsorption Modulator" in "Proc. *OFC 2005*," in OWE1, Anaheim, California, March 2005, pp. 6–11.
[37] K. Noguchi, O. Mitomi, and H. Miyazawa, "Millimeter-wave Ti:LiNbO3 optical modulators," *J. Lightw. Technol.*, 16(4), 615–619, April 1998.
[38] R. Lewén, S. Irmscher, and U. Eriksson, "Microwave CAD circuit modeling of a traveling-wave electroabsorption modulator," *IEEE Trans. Microw. Theory Tech.*, 51(4), 1117–1128, April 2003.
[39] F. Bertazzi, F. Cappelluti, F. Bonani et al., "A Novel Coupled Physics-Based Electromagnetic Model of Semiconductor Traveling-Wave Structures for RF and Optoelectronic Applications," *GaAs 2003*, Munich, October 2003, pp. 239–242.
[40] H. H. Liao, K. K. Loi, C. W. Tu et al., "Microwave Structures for Traveling-Wave MQW Electroabsorption Modulators for Wide Band 1.3 µm Photonic Links," Proc. *SPIE*, 3006, 1997, pp. 291–300.
[41] R. Lewén, S. Irmscher, U. Eriksson et al., "Traveling-Wave Electroabsorption Modulators", in *Encyclopedic Handbook of Integrated Optics* (Prof. Kenichi Iga, ed.), Marcel Dekker, October 2005, Chapter 52.

[42] M. Meghelli, "A 108 Gbps multiplexer in 0.13um SiGe-bipolar Technology," in Proc. *ISSCC Digest*, Paper 13.3, Feburary 2004.
[43] P. Holmström, "High-speed mid-IR modulator using Stark shift in step quantum wells," *IEEE J. Quantum Electron.*, 37, 1273–1282, 2001.
[44] P. Holmström, "Intersubband Electroabsorption Modulator," in *Nitride Semiconductor Devices: Principles and Simulation* (J. Piprek ed.), Berlin: Wiley-VCH, 2007, pp. 253–277.
[45] A. V. Gopal, H. Yoshida, T. Simoyama et al., "Room-temperature dephasing time of intersubband transitions in heavily-doped InGaAs/AlAs/AlAsSb coupled quantum wells," *Appl. Phys. Lett.*, 83, 1854, 2003.
[46] X. Y. Liu, P. Holmström, P. Jänes et al., "Intersubband absorption at 1.5–3.5 µm in GaN/AlN multiple quantum wells grown by molecular beam epitaxy on sapphire," *Phys. Stat. Sol.*, B244(8), 2892–2905, 2007.
[47] B. S. Li, R. Akimoto, K. Akita, and T. Hasama, "$\lambda \sim 1.49$–3.4 µm intersubband absorptions in (CdS/ZnSe)/BeTe quantum wells grown by molecular beam epitaxy," *Appl. Phys. Lett.*, 88(22), 221915, 2006.
[48] F. Dorgeuille and F. Devaux, "On the transmission performances and the chirp parameter of a multiple-quantum-well electroabsorption modulator," *IEEE J. Quantum Electron.*, 30, 2565–2572, 1994.
[49] Y. Miyazaki, H. Tada, S. Tokizaki et al., "Small-chirp {40-Gbps} electroabsorption modulator with novel tensile-strained asymmetric quantum-well absorption layer," *IEEE J. Quantum Electron.*, 39, 813–819, 2003.
[50] B. Jalali and S. Fathpour, "Silicon photonics," *J. Lightw. Technol.*, 24(12), 4600–4615, 2006.
[51] T. L. Koch, "Opportunities and challenges in silicon photonics," in Proc. *IEEE-LEOS 2006 Annual Meeting*, Montreal, Canada, 2006, pp. 677–678.
[52] D. A. B. Miller, "Optical interconnects to silicon," *IEEE J. Sel. Top. Quantum Electron.*, 6(6), 1312, 2000.
[53] L. Pavesi and G. Guillot (eds), *Optical Interconnects: The Silicon Approach*, Berlin Heidelberg: Springer-Verlag, 2006.
[54] A. Irace, G. Breglio, M. Iodice, and A. Cutolo, "Light modulation with silicon devices," in *Silicon photonics* (L. Pavesi and D. J. Lockwood, eds), *Topics Appl. Phys.*, 94, 361–391, Springer-Verlag, Berlin Heiderlberg, 2004.
[55] G. T. Reed and C. E. J. Png, "Silicon optical modulators," *Mater. Today*, 8(1), 40–50, 2005.
[56] L. Friedman and R. A. Soref, "A proposed electroabsorption modulator at 1.55 µm in silicon/silicon-germanium asymmetric quantum-well structures," *IEEE Photon. Technol. Lett.*, 5(10), 1200–1202, 1993.
[57] Y.-H. Kuo, Y. K. Lee, Y. Ge et al., "Strong quantum-confined Stark effect in germanium quantum well structures on silicon," *Nature*, 437, 1334–1336, 2005.
[58] F. Gardes, G. Reed, N. Emerson, and C. Png, "A sub-micron depletion-type photonic modulator in silicon on insulator," *Opt. Express*, 13, 8845–8854, 2005.
[59] A. Huang, C. Gunn, G.-L. Li et al., "A 10 Gb/s Photonic Modulator and WDM MUX/DEMUX Integrated with Electronics in 0.13 µm SOI CMOS," *2006 IEEE International Conference on Solid-State Circuits*, 2006, pp. 922–929.
[60] A. Liu, L. Liao, D. Rubin et al., "High-speed optical modulation based on carrier depletion in a silicon waveguide," *Opt. Express*, 15(2), 660–668, 2007.
[61] L. Liao, D. Samara-Rubio, A. Liu et al., "High-speed metal–oxide–semiconductor capacitor-based silicon optical modulators," *Jpn. J. Appl. Phys.*, 45(8B), 6603–6608, 2006.
[62] J. G. Mendoza-Alvarez, L. A. Coldren, A. Alping et al., "Analysis of depletion edge translation lightwave modulators," *J. Lightw. Technol.*, 6(6), 793–808, 1988.
[63] G. Mak and W. W. Rühle, "Femtosecond carrier dynamics in Ge measured by a luminescence up-conversion technique and near-band-edge infrared excitation," *Phys. Rev. B*, 52(16), R11584–R11587, 1995.
[64] Y.-H. Kuo, Y. K. Lee, Y. Ge et al., "Quantum-confined Stark effect in Ge/SiGe quantum wells on Si for optical modulators," *IEEE J. Sel. Top. Quantum Electron.*, 12, 1503–1513, 2006.

[65] V. R. D'Costa, C. S. Cook, A. G. Birdwell et al., "Optical critical points of thin-film $Ge_{1-y}Sn_y$ alloys: A comparative $Ge_{1-y}Sn_y/Ge_{1-x}Si_x$ study," *Phys. Rev. B*, 73, 125207, 2006.

[66] J. Liu, D. Pan, S. Jongthammanurak et al., "Design of monolithically integrated GeSi electro-absorption modulators and photodetectors on an SOI platform," *Opt. Express*, 15, 623–628, 2007.

[67] P. J. Winzer and R.-J. Essiambre, "Advanced Optical Modulation Formats," in Proc. *IEEE*, 94(5), May 2006, p. 952.

[68] B. Hessmo, P. Holmström, and L. Thylen, "Shaping Light Pulses Using Electromagnetically Induced Transparency," in Proc. *CPT*, Tokyo, Japan (2002), invited paper.

8

Advances in photodetectors

Joe Charles Campbell

*School of Engineering and Applied Science,
Department of Electrical and Computer Engineering
University of Virginia, Charlottesville, VA, USA*

The evolution of fiber optic digital transmission systems toward 40 Gb/s and beyond, the development of microwave photonics, and the emergence of multiple-component all-optical functions has been facilitated by numerous improvements and a few breakthroughs in photodetector and receiver technologies, the full breadth of which would make up another book. This chapter will focus on three primary topics: high-speed waveguide photodiodes, photodiodes with high saturation current, and recent advances in telecommunications avalanche photodiodes.

8.1 WAVEGUIDE PHOTODIODES

The transition from 10-Gb/s transmission systems to 40 Gb/s and higher has necessitated a transformation in the basic photodiode structure due to the inherent trade-off between the responsivity and the bandwidth of normal-incidence photodiodes. Figure 8.1(a) shows a generic normal-incidence PIN photodiode structure. For most III–V compounds the absorption coefficient $\alpha \approx 10^4 \, cm^{-1}$, which means that $\sim 2\,\mu m$ is needed to absorb 88% of the light. Once the electron-hole pairs have been generated, they drift through the absorbing region, which is fully depleted. The thicker the depletion region, the longer it takes to collect these carriers. This is the origin of the transit-time limit to the bandwidth. The bandwidth-efficiency product is $\sim 20\,GHz$ [1], which is more than sufficient for 10-Gb/s receivers, but not adequate for subsequent generations.

In order to simultaneously achieve good responsivity and high bandwidth, photodiodes with waveguide structures have become the photodetectors of choice. Waveguide photodiodes, as the name implies, utilize an optical waveguide with

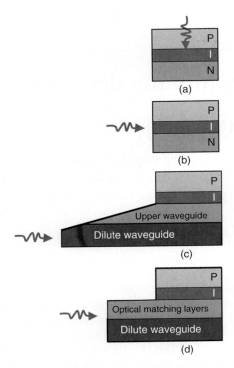

Figure 8.1 Illustrations of (a) normal-incidence, (b) edge-coupled waveguide, (c) tapered evanescently-coupled waveguide, and (d) multimode input waveguide evanescently coupled photodiodes (this figure may be seen in color on the included CD-ROM).

embedded or adjacent absorbing layers. The most straightforward approach, side illumination, is illustrated in Figure 8.1(b); the input signal is incident directly on the edge of the photodiode active region. Even though the confinement factor in the waveguide may be much less than unity, producing an effective absorption coefficient less than that of the bulk semiconductor, the interaction length (i.e. the length of the photodiode) is typically a few 10s of microns and thus the internal quantum efficiency can easily approach 100%. In contrast, the photogenerated carriers transit only the thin absorption/depletion region perpendicular to the epitaxial layers, which results in high bandwidths. This approach has successfully circumvented the bandwidth-responsivity trade-off because electrical and optical transports are not collinear.

8.1.1 Edge-Coupled Waveguide Photodiodes

The first side-illuminated waveguide photodiode consisted of an $In_{0.53}Ga_{0.47}As$ absorbing region sandwiched between p- and n-type InP layers [2]. The external quantum efficiency was 25%, and the bandwidth was 28 GHz. In order to improve the quantum efficiency and achieve higher bandwidths, K. Kato et al. [1] developed

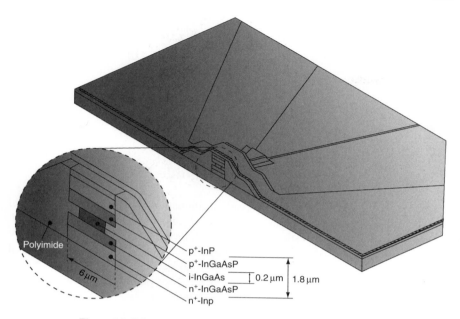

Figure 8.2 Schematic of mushroom-mesa waveguide photodiode [1].

the mushroom-mesa waveguide photodiode shown in Figure 8.2. A high external quantum efficiency of 50% was achieved by using a large optical cavity to increase the size of the guided mode to more closely match that of an optical fiber. The optical cavity consisted of a 0.2-μm thick $In_{0.53}Ga_{0.47}As$ absorber with 0.8 μm InGaAsP ($\lambda \sim 1.3$ μm) layers on each side. This "double core" was placed between InP-cladding layers. The thin depletion layers required to achieve the short transit times that enable very high bandwidths can result in high capacitance unless the device area is reduced to very small dimensions, which, in turn, can result in high contact resistance. The mushroom-mesa solved the RC time constant problem by maintaining a relatively large contact area to keep the resistance low (10 Ω) while minimizing the p–n junction area to achieve very low capacitance ($\sim$15 fF), which resulted in a bandwidth of 110 GHz.

Another approach to avoid the limitations of lump-element devices imposed by their resistance, capacitance, and inductance is to adopt traveling wave structures [3, 4]. The traveling wave photodiode is a "distributed" structure in that the photogenerated electrical signal propagates along the electrical contacts, which are designed as a transmission line with characteristic impedance matched to that of the external microwave circuit. Hence, the requirement to optimize the trade-off between the RLC bandwidth and the transit-time bandwidth can effectively be eliminated. Using this approach, Giboney et al. [5] achieved 172 GHz bandwidth and 76 GHz bandwidth-efficiency product. For traveling wave photodiodes, the bandwidth is determined by the transit-time bandwidth and the mismatch between the group velocity of the optical wave in the device and the electrical signal in the transmission line. The bandwidth is given by the expression [6]

$$f_{3dB} = \frac{f_{tt}}{\sqrt{1 + (f_{tt}/f_{vm})^2}}$$

where f_{tt} is the transit-time bandwidth and f_{vm} is the velocity-mismatch bandwidth.

8.1.2 Evanescently Coupled Waveguide Photodiodes

There are two negative aspects to the side-illumination approach. The first is that there is not much tolerance to lateral and vertical displacement of the input signal. Consequently, highly sophisticated packaging procedures are required to achieve low coupling losses, which degrades receiver sensitivity and adds to the total receiver cost. The second is that these photodiodes have limited optical power capabilities owing to the very high optical intensity on the input facet and the exponential decay profile of the photo-generated carriers [6]. Evanescently coupled photodiodes were developed to mitigate these drawbacks. It has been shown that evanescent coupling can achieve at least four times higher saturation current, i.e., the current at which the frequency response is compressed by 1 dB than side-illuminated photodiodes [7]. The simplest evanescently coupled structure consists of a photodiode located near or directly in contact with the waveguide core. Because the light couples evanescently from the waveguide to the photodiode, the absorption occurs relatively uniformly along the device length, which leads to an improved high-power capability [8]. However, this approach may be limited by a quantum efficiency–coupling efficiency trade-off. If the waveguide layer is thick in order to achieve good coupling efficiency, the evanescent field will be weak, which will result in low quantum efficiency for high-speed (i.e., short) devices. Thinner waveguide core layers can reduce the absorption length at the cost of higher insertion loss. It has been shown that the waveguide–photodiode coupling can be enhanced by grading the index of the underlying waveguide to "steer" light into the photodiode [9]. Responsivity of 0.96 A/W and 40 GHz bandwidth was reported. A more widely used approach exploits optical matching layers between the waveguide and the photodiode [10, 11]. Nevertheless, input coupling is limited by the fact that the waveguide mode tends to be much smaller than that of a standard single-mode fiber.

In order to improve the input coupling efficiency, several approaches to integrate a spot-size converter with evanescently coupled waveguide photodiodes have been reported. Mörl et al. [12] reported one of the first successful integrated spot-size converters. A side-view cross section is shown in Figure 8.1(c). It consists of a lower dilute waveguide and an upper single-layer core InGaAsP ($\lambda_g \sim 1.3\,\mu m$) waveguide. The dilute waveguide consists of thin InGaAsP layers in InP. The mode is not tightly confined and thus exhibits a spatial profile close to that of the input fiber. Recessed from the input and on top of the InGaAsP core is an InGaAs PIN photodiode. At the input, the thickness of the upper InGaAsP core is tapered so that it is too thin to support a guided mode. Hence, light is coupled into the underlying dilute waveguide, which consists of three 40-nm thick InGaAsP layers separated by 1-μm thick InP layers. The mode profile of the dilute waveguide is well matched to a single-mode optical fiber. As the thickness of the InGaAsP core of the upper waveguide increases, light from the dilute waveguide is adiabatically coupled to the upper waveguide whose dimensions are tailored for efficient coupling to the InGaAs photodiode. Tapering of the InGaAsP thickness is achieved by ion beam shadow etching [13]. Lateral optical confinement is has achieved with a rib waveguide structure. This evanescent coupling approach achieved 7 dB lower coupling loss for flat-end fibers and exhibited ±2 μm alignment tolerance for 1 dB excess loss.

In recent years, this structure has been refined and optimized to achieve maximum coupling between an optical fiber input and the photodiode and to achieve ultra-high bandwidths [14–16]. To increase the responsivity, the InGaAsP-tapered input waveguide was designed for efficient, uniform coupling to the InGaAs photodiode whereas the mode of the diluted waveguide was designed to match that of the fiber. Using this approach, fiber to chip coupling and absorption in the photodiode can be designed independently. Several structural enhancements and modifications that have been incorporated onto the chip have led to bandwidths in the 100-GHz region. These include the following: (1) integration of the bias tee on chip, (2) integration of a 50-Ω termination resistor, (3) fabrication of the complete waveguide stack from semi-insulating materials, (4) achieving efficient coupling to short, low-capacitance photodiodes, (5) utilization of thin depleted absorbing layers to minimize the carrier transit component of the bandwidth, and (6) air bridge connections to coplanar contact pads. Tapered-waveguide evanescently coupled photodiodes of this type, reported by Bach et al. [15], have achieved 100-GHz bandwidth, 53-GHz bandwidth-efficiency product, polarization dependent loss of 0.9 dB, and ±2 μm vertical and horizontal alignment tolerances. The 1-dB saturation point of the RF output power was +12 dBm RF at 100 GHz. Figure 8.3 shows (1) a schematic of this photodiode structure, (2) circuitry monolithically integrated on the chip, and (3) the relative frequency response.

Another approach to adiabatic spot size conversion in evanescently coupled photodiodes utilizes laterally tapered transition regions. Figure 8.4 shows a double taper implementation of this type of evanescently coupled photodiode [17]. Similar to the tapered-thickness photodiodes discussed above, a dilute waveguide (9 μm-wide) consisting of three thin InGaAsP ($\lambda_g \sim 1.05$ μm) layers embedded in a 0.7-μm thick layer of InP is used to match the mode of an optical fiber. To further facilitate input coupling, a slot for the optical fiber is etched at the input facet. An antireflection coating is deposited at wafer level and etched V-grooves (not shown) enable precise cleaving. In operation, light couples to a single-mode InGaAsP taper ($\lambda_g \sim 1.05$ μm). The light is then "steered" to the second InGaAsP ($\lambda_g \sim 1.4$ μm)

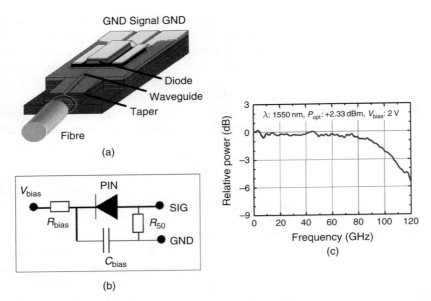

Figure 8.3 (a) Schematic of evanescently coupled waveguide photodiode with integrated spot size converter, (b) circuitry monolithically integrated on the chip, and (c) the relative frequency response [15] (this figure may be seen in color on the included CD-ROM).

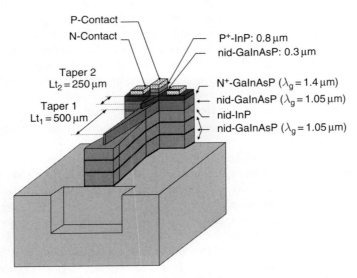

Figure 8.4 Schematic evanescently coupled waveguide photodiode with two lateral tapers [17] (this figure may be seen in color on the included CD-ROM).

multimode taper, which serves as the n-type contact layer and as an optical matching layer [18] to facilitate the transition into the photodiode. Responsivities of 0.37, 0.46, and 0.50 A/W were reported with less than 1-dB polarization dependence, on 10-, 20-, and 25-μm length diodes, respectively. The corresponding bandwidths were 50, 41, and 36 GHz. The saturation current was 7 mA at 50 GHz and −1 V reverse bias. Xia et al. [19] have reported a similar structure that utilizes a diluted large-fiber guide followed by transfer to a single low-loss lateral-taper coupler. The responsivity was 0.75 A/W with 0.4-dB polarization sensitivity and >40 GHz bandwidth.

Evanescently coupled photodiodes with vertically or laterally tapered couplers have achieved excellent performance, but precise control of the taper geometry is essential. For example, submicron lithography is often required to fabricate the laterally tapered couplers [19]. Recently, it has been shown that the critical taper technology can be replaced by exploiting mode-beating effects in multimode waveguides [10, 11, 20–24]. Figure 8.5 shows a schematic of an evanescently coupled photodiode that utilizes a short multimode input waveguide with optical matching layers [20]. The challenge for all high-speed waveguide photodiodes is to absorb as much light as possible over a short distance in order to minimize the area and, thus, the capacitance. The multimode waveguide consists of a diluted waveguide and two optical matching layers. The diluted waveguide is a stack of 10 periods of undoped InP/GaInAsP (1.1-μm band gap) layers. The number of periods was optimized to achieve high coupling efficiency with an input fiber and low

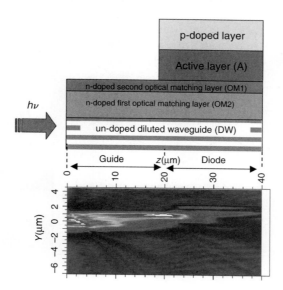

Figure 8.5 Schematic of evanescently coupled photodiode with short multimode input waveguide and two optical matching layers. The lower image is a simulation of the optical intensity as light propagates through the integrated waveguide/photodiode [20] (this figure may be seen in color on the included CD-ROM).

TE/TM polarization dependence. The two optical matching layers are n-doped GaInAsP with band gaps corresponding to 1.1 and 1.4 μm for the first and second optical matching layers, respectively. This provides a gradual increase of the optical refractive index from the diluted waveguide to the absorbing layer (as illustrated by the simulated optical intensity in the lower part of Figure 8.5), which results in a significant enhancement in the quantum efficiency. For this approach, since the waveguide to photodiode coupling is based on mode interference effects, the coupling efficiency oscillates along the propagation direction. The solid line in Figure 8.6 presents the responsivity simulation of 20-μm long photodiodes vs the input waveguide length. Oscillations related to inter-modal interferences are clearly shown in this figure. In agreement with the modeling, the optimal waveguide length was found to be 20 μm. 1.07 A/W responsivity was achieved at 1.543-μm wavelength (corresponding to a external quantum efficiency of 86%) with TE/TM polarization dependence less than 0.5 dB. It might be anticipated that this approach would lead to rapid variation of the quantum efficiency with wavelength; however, it was found that the efficiency changed <0.2 dB from 1500 to 1600 nm. The bandwidth of a 5 × 20-μm device as measured by heterodyning two DFB lasers was >50 GHz. Electro-optic sampling yielded a bandwidth of 65 GHz. As shown in Figure 8.5, the responsivity is a function of the length of the input multimode waveguide, which is determined by cleaving. It has been demonstrated that the use of etched V-grooves can control the input waveguide length to ±2 μm.

By using a dual-step coupling region that separates the input waveguide region from the coupling-waveguide region, the dependence of the responsivity on the length of the input waveguide has been essentially eliminated [22, 25]. Figure 8.7 shows a cross section and top view of this dual-step coupling device. High responsivity (~1.0 A/W) with a large cleaved tolerance (~50 μm) and bandwidths >40 GHz have been achieved [25]. Beling et al. [22] have adapted this

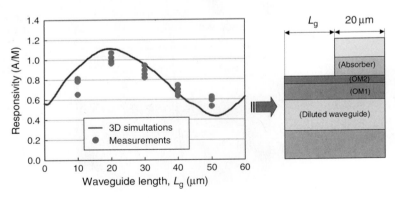

Figure 8.6 Responsivity of photodiode with planar multimode input waveguide vs guide length for a 20-μm long active region: the solid line is the simulated responsivity and the plotted circles are measurements at $\lambda = 1543$ nm [20] (this figure may be seen in color on the included CD-ROM).

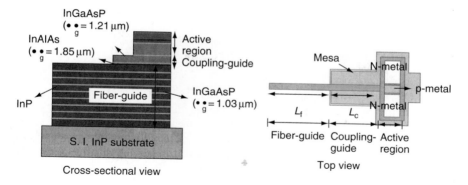

Figure 8.7 The cross-sectional view and top-view demonstrated of dual step input waveguide photodiode [25].

modification of the optical matching layers to achieve increased responsivity in waveguide photodiodes with very short absorber lengths. A schematic of the device structure is shown in Figure 8.8. Note that compared with the waveguide-integrated photodiode in Figure 8.5, the top matching layer protrudes a length L beyond the front facet of the photodiode. Figure 8.9 shows the calculated and measured responsivity vs the protrusion length, L, for two wafers with absorbing

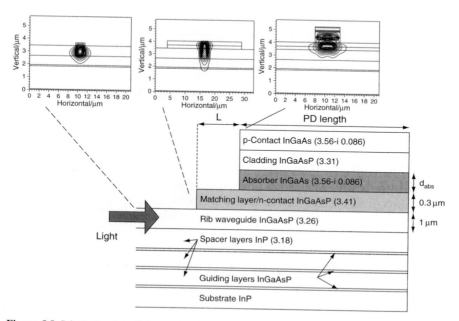

Figure 8.8 Schematic view of the most relevant layers of an evanescently coupled photodiode with extended optical matching layer. The numbers in parentheses are refractive indices of the various layers [22].

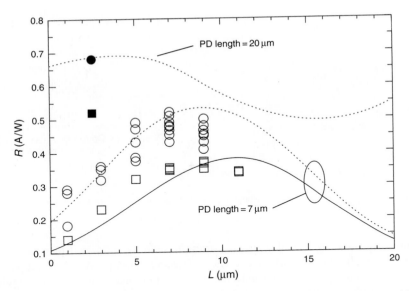

Figure 8.9 Responsivity vs the length of the matching layer protrusion. The solid and dotted lines represent simulated results for wafers having 430- and 350-nm thick absorbing layers, respectively. The circular and square data points represent the 430- and 350-nm-thick absorbing layers, respectively [22].

layer thickness of 430 and 350 nm and for two active areas, both 5-μm wide and lengths of 7 and 20 μm. For the 7-μm long devices, responsivities ~0.5 A/W were achieved for protrusion length, $L = 7$ μm. The reduced capacitance of these small area photodiodes enabled bandwidths up to 120 GHz. Wu et al. [24] have reported a similar structure with stepped matching layers and active layer length of 20 μm that achieved responsivity of 0.9 A/W and 60-GHz bandwidth.

8.2 BALANCED RECEIVERS

Coherent optical communications drove the initial development of balanced photodetector configurations, which consist of anti-parallel biased photodiodes. The advantages of coherent detection, as discussed in Chapters 3 and 4 of Volume B by K. Kikuchi and X. Liu et al., respectively, are increased receiver sensitivity, enabling electronic compensate for signal degradation such as dispersion, and the ability to process advanced modulation formats. For coherent communications, balanced receivers provide suppression of local oscillator intensity noise and they require less power than single photodiode configurations [26]. Because optimal performance is achieved when both receiver channels are perfectly matched electrically and optically, it is beneficial to monolithically integrate the receiver components, particularly the two photodiodes. Toward that end, there have been several demonstrations of monolithic integration of the different constituents of a typical coherent balanced heterodyne receiver. Deri et al. [27] reported integration of a 3-dB directional coupler with balanced interdigitated "phototransistor-like" detectors using mirror coupling through the substrate. Local oscillator intensity noise suppression of 33 dB was achieved. An integrated heterodyne receiver consisting of a tunable multiple-quantum-well distributed-Bragg-reflector laser, a 3-dB directional coupler mixer, and balanced multiple-quantum-well waveguide detectors achieved receiver sensitivity of −42.3 dBm and −39.7 dBm at 108 Mb/s and 200 Mb/s, respectively [28]. High-speed operation of balanced photodiodes was first achieved with twin back-illuminated PINs having integrated microlenses [29]. The external quantum efficiency was 80%, the bandwidth was 13 GHz, and intensity noise suppression of −30 dB was achieved up to 7 GHz. Chandrasekhar et al. [30] reported the first integration of balanced detectors with electronics. Two InGaAs p-i-n photodetectors were monolithically integrated, with a transimpedance preamplifier using InP-InGaAs heterojunction bipolar transistor technology. The receiver, with a bandwidth of 3 GHz and a common mode rejection of 25 dB, achieved a sensitivity of −49 dBm at 200 Mb/s.

Suppression of laser intensity noise is also important in analog optical links and radio frequency photonics applications such as cable television, phased array antennas, and photonic analog-to-digital converter systems. In addition, balanced receivers can also reduce amplified spontaneous emission noise from optical amplifiers [31, 32], which can ultimately lead to shot-noise limited performance at high optical powers [33]. However, at present, the primary driver for monolithic balanced receivers is the emergence of advanced modulation formats such as differential phase shift keying (DPSK) [34–37]. The evolution of high-bit-rate transmission systems to 40 Gb/s and beyond presents numerous technical challenges. For amplitude modulation formats (OOK), fiber-related impairments have been lessened by forward error correction and adaptive dispersion compensation. Alternative approaches include the adoption of phase modulation formats. DPSK using balanced detection has been shown to provide 3-dB sensitivity improvement relative to OOK [38, 39]. As an example of the performance that can be achieved, an RZ DPSK receiver sensitivity for 10^{-9}-bit error rate at 42.7 Gb/s of −37 dBm was achieved using balanced detection [40]. For RF photonics and DPSK, the critical parameters for balanced detectors are high symmetry of the two photodiodes, large bandwidth, high responsivity, and, in some instances, high power capability. The requirement for high symmetry favors monolithic balanced detectors.

Various approaches including normal incidence [41], waveguide photodiodes [42–44], and velocity-matched distributed configurations [45, 46] have been employed to fabricate high-speed monolithic balanced photodiodes. A balanced velocity-matched distributed MSM photodetector achieved a bandwidth of 13.8 GHz and 26-mA average current for small-signal modulation [45]; a balanced velocity-matched p-i-n photodetector achieved a DC linear photocurrent of 45 mA [46]. Li et al. [41] reported balanced photodetectors with top-illuminated, charge-compensated uni-traveling-carrier (UTC) photodiodes [47–50] designed for high-power operation. A 20-μm diameter photodiode balanced pair achieved a bandwidth of 10 GHz, and a large signal saturation current of 25 mA for each photodiode. A 10-μm diameter photodiode balanced pair achieved a bandwidth of 26 GHz, and a large-signal saturation current of 15 mA for each photodiode. A 46-dB common mode rejection ratio was achieved at 10 GHz. The normal-incidence structure restricted the responsivity to ~0.32 A/W.

In order to achieve higher responsivity, monolithic balanced detectors have been fabricated using the waveguide structures described in the previous section. Li et al. [44] used the planar multimode-waveguide input-coupling structure combined with conventional InGaAs PIN photodiodes and with partially depleted absorber (PDA) photodiodes [51, 52]. The PDA structure is similar to that of the UTC except the depleted drift region (collector) is the same material as the absorber (typically $In_{0.53}Ga_{0.47}As$) instead of a wide-bandgap material (typically InP). A balanced pair consisting of two $5 \times 20\,\mu m^2$ PIN (PDA) photodiodes exhibited a responsivity of 0.94 A/W (0.81 A/W), a bandwidth of 30 GHz (24 GHz), and a large-signal saturation current of 10 mA (24 mA) for each photodiode. The common mode rejection ratio was measured to be ~30 dB at 10 GHz. Agashe et al. [43] integrated a 3-dB multimode interference (MMI) coupler with a balanced detector fabricated with twin, single, laterally-tapered waveguide photodiodes. The bandwidth was 15 GHz, the responsivity was 0.32 A/W, and the common mode rejection ratio was 34 dB. Bias circuitry was integrated with the balanced detector reported by Beling et al. [42]. The vertically-tapered waveguide photodiodes with integrated spot size converters had active areas of $5 \times 16\,\mu m^2$ and 50-Ω load resistors were implemented on chip at the output transmission line. The chip circuit is illustrated in Figure 8.10 (left) and the frequency response of the individual photodiodes is shown on the right. The bandwidths are 70 GHz and the two frequency characteristics agree within 0.5 dB up to 50 GHz, indicating the symmetry of the photodiodes at high signal frequencies. The responsivity was 0.56 A/W for each photodiode with polarization dependence of less than 0.5 dB [53]. Up to 10 GHz the common mode rejection ratio was better than 25 dB, from 10 to 30 GHz the CMRR was to at least 20 dB.

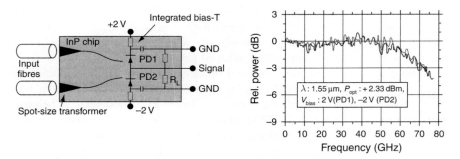

Figure 8.10 Left: Circuit of integrated balanced photodetector. Right: Relative frequency responses of the two photodiodes in the balanced photodetector [53] (this figure may be seen in color on the included CD-ROM).

Integration of an amplifier with the balanced photodetector can provide additional power gain while maintaining high symmetry and minimizing interconnection parasitics [54, 55]. Schramm et al. [55] have successfully demonstrated a fully packaged balanced photoreceiver designed for 40-Gb/s DPSK systems. Based on the vertically-tapered spot-size converter [12] two waveguide-integrated photodiodes were monolithically integrated with a distributed amplifier. Mounted in a butterfly package, a 3-dB cutoff frequency of 42 GHz was achieved. Compared to a balanced detector without integrated amplifier, an increase in sensitivity of 4 dB was achieved.

8.3 HIGH-POWER PHOTODETECTORS

Photodetectors with high bandwidths and high-power capability have become critical components for numerous analog and digital optical links. Analog applications include antenna remoting, optically fed, phased array antennas, photonic oscillators, and photonic analog-to-digital converter systems, all of which require high fidelity with large dynamic range [56 and references therein]. One means to accomplish this is to increase the optical carrier power [57], which means that the photodetectors in these systems must be able to operate at high photocurrent levels to minimize noise figures and at high speed while providing wide dynamic range. In the digital domain, the sensitivity of a conventional optical receiver, which typically consists of a PIN photodiode and a transimpedance amplifier (TIA), is frequently limited by the front-end noise of the electrical amplifier. Recently, high receiver sensitivity at 40 Gb/s has been achieved by increasing the incident optical power with Er-doped fiber amplifiers [58, 59] and semiconductor optical amplifiers [60] to the extent that the voltage swing at the output of the photodiode can directly drive digital logic circuits. This type of receiver eliminates the complications and expense of post-detection, wide-bandwidth, flat-phase RF amplifiers but requires photodiodes that can operate at high photocurrent levels without degradation of the bandwidth.

It is well known that photodiodes exhibit compression at high photocurrent levels [61–64]. In the frequency domain this is observed as a decrease in the bandwidth as illustrated in Figure 8.11(a), which shows the compression in the frequency response with increasing photocurrent. The saturation current of a photodiode is defined as the dc current or average current at which the frequency response decreases by 1 dB for small signal or large signal modulation, respectively. Figure 8.11(b) shows the pulse response of a PIN photodiode as the input pulse energy increases from 0.2 to 2 pJ/pulse [65]. The waveforms consist of two components, an initial fast component arising from electron transport and a tail due to the slower moving holes. We note that the electron component begins to saturate at relatively low energy and the pulse width increases slowly. The response of the holes degrades rapidly as energy increases.

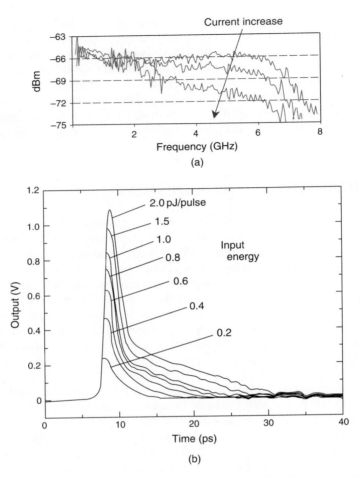

Figure 8.11 (a) Frequency response of a PIN photodiode with increasing photocurrent and (b) pulse response for input pulse energy in the range 0.2–pJ/pulse [65] (this figure may be seen in color on the included CD-ROM).

There are several physical mechanisms that impact saturation in photodiodes, including space-charge [63, 64, 66, 67], series impedance [68], and thermal effects [63, 67, 69, 70]. The thermal limit is determined by the heat dissipation characteristics of the constituent semiconductor layers, the photodiode geometry, and by the heat sink design. Joule heating can result in temperatures as high as 500°C in the depletion region [69, 70], which can cause device thermal and/or electrical failure. The space-charge effect has its origin in the spatial distribution of the photo-generated carriers as they transit the depletion layer. At high current densities, as electrons and holes travel in opposite directions, an internal space-charge field is generated that opposes the bias electric field. For sufficiently high optical input power levels, the space-charge-induced electric field can be strong enough to collapse the bias electric field, which will result in RF photocurrent compression [71]. When the electric field minimum becomes less than the field required for the carriers to reach their saturation velocities, the device speed is reduced. In addition to the space-charge effect and thermal considerations, for large signal modulation, when large RF power is delivered to the load, the voltage swing across the load and device series resistance can effectively remove available voltage bias from the depletion region and thus negatively impact the photodiode saturation current.

8.3.1 Uni-Traveling Carrier Photodiodes

In 1996, Davis et al. [72] reported a photodetector with a 3-μm thick undepleted InGaAs absorber and a 5-μm thick InP depletion region that was designed to decouple the device capacitance and quantum efficiency. The device had a RC-limited bandwidth of 295 MHz and a small-signal saturation current of 150 mA. The quantum efficiency at 1319 nm was close to 100%. In 1997, Ishibashi et al. [47] demonstrated that this kind of photodiode could achieve high speed and high saturation current operation, if the layer thicknesses were designed properly. The UTC photodiode has demonstrated exceptional power-bandwidth performance products. The band diagrams of the UTC-PD and a conventional PIN are schematically shown in Figures 8.12(a) and (b), respectively. The absorbing region of the UTC-PD is a narrow-gap (typically $In_{0.53}Ga_{0.47}As$) p-type layer adjacent to an undoped (or a lightly n-type doped) wide bandgap (typically InP) depleted drift layer. Since the absorbing layer is undepleted, the photo-generated excess hole density decays within the dielectric relaxation time. The electrons, on the other hand, move by both drift and diffusive forces through the $In_{0.53}Ga_{0.47}As$ absorber and are injected into the InP collector. Hence, the photocurrent is purely electron transport. This provides two distinct advantages compared with the PIN structure (Figure 8.12(b)), in which both electrons and holes are generated in and drift through the depletion layer. The first is

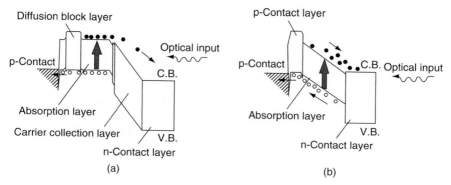

Figure 8.12 Band structure of (a) uni-traveling carrier and (b) conventional PIN photodiodes [65].

higher speed. Although both carriers contribute to the photocurrent in a PIN, since the hole velocity is much lower than that of electrons, it is the holes that limit the bandwidth and exacerbate the space-charge effect [73]. In addition, the frequency response of UTC photodiodes can be further enhanced by velocity overshoot of electrons [74, 75] in the depletion layer. Electron transport is frequently cited as one of the reasons why UTC devices exhibit good high-current performance; however, from a pure space-charge perspective, traditional PIN designs have been suggested to have the advantage of better space-charge balance in the depletion region [71]. It is noteworthy that the UTC structure can achieve high speed and high saturation output at low bias voltage [76], because electrons attain high velocity at relatively low electric fields [77].

The combination of high speed and high-output power has positioned UTC photodiodes as key components for high-speed photoreceivers, integrated optical gates for ultrafast signal processing, high-power millimeter (mm) wave generators for high-frequency measurement and sensing systems, and mm-wave transmitters for fiber radio communications systems [65, 78]. Figure 8.13 shows receiver configurations with (a) conventional electrical postamplification and (b) optical preamplification. The primary advantage of optical preamplifiers is that the

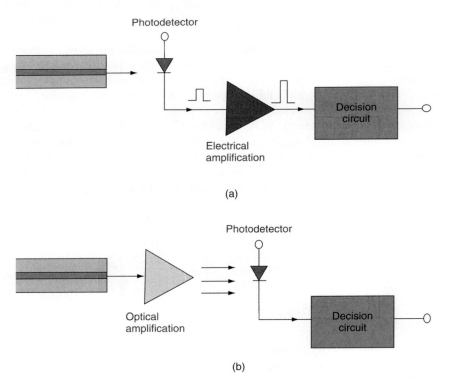

Figure 8.13 Optical receivers with (a) electrical post amplification and (b) optical preamplifier (this figure may be seen in color on the included CD-ROM).

8. Advances in Photodetectors

amplified stimulated emission noise overcomes the shot noise of the electrical amplifier, thereby approaching quantum noise-limited operation. The result is higher receiver sensitivity with simpler receivers, particularly for operation at 40 Gb/s and higher [79]. Implementation of optical preamplification, however, requires photodiodes that can deliver output voltages sufficient to drive digital circuits at high bandwidth, attributes of UTC photodiodes [80]. Miyamoto et al. [59] have demonstrated this type of receiver up to 80 Gb/s. The generated non-return-to-zero pseudo-random bit sequence (PRBS) signal at 80 Gb/s was transmitted through an 89-km optical fiber as one channel of 1.04 Tb/s WDM signal and amplified with an Er-doped optical amplifier. The UTC photodiode delivered a peak-to-peak voltage of 0.8 V. Error-free operation with an input sensitivity of less than −17 dBm was confirmed for all 10-Gb/s time-slots ($8 \times 13 = 104$ slots). Error-free operation of a UTC-photodiode receiver at 50 Gb/s with an output peak-to-peak voltage of 1.0 V for return-to-zero signal has also been demonstrated [79].

The high-output voltage of the UTC enables it to drive other optoelectronic components. The most successful implementation has been monolithic integration of UTC photodiodes with traveling wave electroabsorption modulators (PD-EAM). The PD-EAM has been used to demonstrate ultrafast optical gating [81], 320-Gb/s demultiplexing [82], 100-Gb/s wavelength conversion [82], 100-Gb/s error-free retiming [83], and 80-Gb/s optical sampling [84]. The PD-EAM was developed to circumvent speed limitations of conventional electronics. A schematic of the PD-EAM is shown in Figure 8.14. The UTC anode is

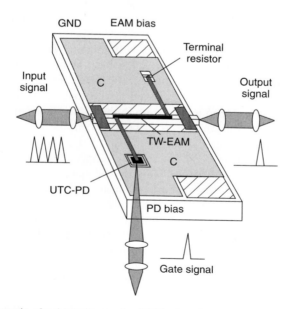

Figure 8.14 Schematic of uni-traveling-carrier (UTC) photodiode-traveling wave electroabsorption optical gate [65].

connected to the signal line (anode) of the EAM by a thin-film microstrip line. An optical input generates a positive drive voltage for the EAM. Thus, the PD-EAM functions as a transmission-type optical gate. A gate window of 2.3 ps and optical pulse gating at a corresponding data rate of 500 Gb/s were reported by S. Kodama et al. [85]. High-speed optical gating has been used to achieve 320–10-Gb/s demultiplexing [82]. The demultiplexer consisted of a PD-EAM, two Er-doped fiber amplifiers (one to generate a voltage pulse in the UTC photodiode and the other as the signal input to the EAM), and a variable optical delay line. For the demultiplexed 320-Gb/s input signal, the receiver sensitivity was −18 dBm at 10^{-9} bit error rate. The PD-EAM can also be used for wavelength conversion because the optical drive to the UTC (λ_{in}) is impressed on the signal through the EAM (λ_{out}). Since this is an optical-electrical-optical conversion method, the channel-to-channel crosstalk observed in cross gain modulation is avoided and operation over a broad range of input and output wavelengths can be achieved. Kodama et al. [82] have successfully demonstrated 100-Gb/s error-free wavelength conversion for $\lambda_{in} = 1545$–1570 nm with $\lambda_{out} = 1552.5$ nm and for $\lambda_{in} = 1552.5$ nm with $\lambda_{out} = 1545$–1565 nm. The power penalties were ∼2 dB.

Recently, modified thin depletion layer PDA and UTC structures have been developed to achieve higher saturation currents. Although these thin depletion layer devices help to suppress the space-charge effect, at sufficiently high photocurrent levels, the devices still exhibit voltage-dependent compression currents. The underlying cause for this behavior can be attributed to a combination of the space-charge effect with the potential for series impedance limitations [86]. Improvements to the series impedance are being pursued, but the lingering space-charge effects still remain. To address this, it has been suggested that the space-charge effect could be reduced further by a fixed distribution of background dopants [71, 87]. By doping the depletion region, the electric field can be preconditioned to be higher at the location where ultimately it will tend toward zero in the presence of high space charge. This type of charge compensation can possibly double the maximum photocurrent without the associated failure caused by increased thermal loading from higher bias voltages. To implement charge compensation in a UTC structure, a uniform n-type doping is used. The space charge in the depletion region, which arises from the electrons that are injected from the undepleted p-type absorption region, results in a uniform distribution of electrons under saturated carrier velocity conditions. However, a superior compensation design must take into account the position dependence of the carrier velocity in the collector, which is non-uniform when electric field-dependent carrier velocities or absorbing depletion layers are included. A complementary doping profile would be optimal; however, uniform doping has been easier to implement in these early demonstrations. For example, Shimizu et al. [88] showed that higher output voltage could be achieved by doping the InP collection layer with Si.

8.3.2 Charge-Compensated Uni-Traveling Carrier Photodiodes

Using a charge-compensated UTC (CC-UTC) structure consisting of a 450-nm p-type $In_{0.53}Ga_{0.47}As$ absorbing layer and collection region comprising a 200-nm n-doped InP layer ($N_d \sim 5 \times 10^{16}$ cm^{-3}), a 5-nm heavily n-doped InP charge layer, two 15-nm n-doped InGaAsP layers, and a 15-nm undoped $In_{0.53}Ga_{0.47}As$ layer. Li et al. [50] reported a 20-μm-diameter CC-UTC with 25 GHz bandwidth and large-signal 1-dB compression current greater than 90 mA; the RF output power at 20 GHz was 20 dBm. A smaller $\sim$100 μm^2 photodiode exhibited a bandwidth of 50 GHz and large-signal 1-dB compression current greater than 50 mA. The maximum RF output power at 40 GHz was 17 dBm. The responsivity was 0.45 A/W.

In the structures in which the space-charge effect is reduced, at very high-output power, the dynamic voltage developed across the load resistor reduces the bias on the photodiode. Hence, it is necessary to operate at relatively high bias to accommodate the load voltage swing. But, higher bias typically results in higher dark current and more joule heating, which ultimately leads to thermal failure. Thermal failure is closely tied to the increase in dark current with increasing temperature. The increased dark current causes an additional increase in the diode temperature, resulting in additional dark current, and so on (thermal runaway). For applications that require very high RF output power, device structures that can handle high bias and high thermal power are desired. One way to accomplish this is to increase the thickness of the depletion region, being mindful that thermal management with well-designed heat sink techniques will be required. Using a structure similar to that in Ref. [50] with a thicker InP collector (450 nm) and improved thermal management, Duan et al. [89] achieved saturation current as high as 250 mA and RF output power of 800 mW at 2 GHz. Table 8.1 summarizes the saturation currents and RF output powers for a range of device diameters.

Table 8.1

Saturation (or failure) current and RF output power of charge-compensated uni-traveling carrier photodiodes.

Diameter (μm)	Bandwidth (GHz)	Bias (V)	Saturation or failure current (mA)	RF output power
10[1]	50	4	50	17 dBm
20[1]	25	4	90	20 dBm
34[2]	10	6	140	24 dBm (250 mW)
40[2]	10	7	180	24.5 dBm (290 mW)
56[2]	6	8	180	25.3 dBm (340 mW)
100[2]	2	8	250	29 dBm (800 mW)

[1] Reference [50].
[2] Reference [89].

8.3.3 High-Power Partially Depleted Absorber Photodiodes

From thermal management and space charge design points of view, it is desirable to avoid thick InGaAs depletion layers due to the poor thermal conductivity of InGaAs [71]. Heat accumulates in the depletion layer, eventually leading to device failure at high optical powers/photocurrents. However, for normal-incidence detectors, thick (>2 μm) InGaAs layers insure high responsivity. Analysis [63] has predicted that the benefits of using thin depletion layers to obtain high current are substantial. This trade-off is frequently addressed by varying the thickness and ratio of InGaAs to InP in the drift layer of traditional PIN (100% InGaAs), dual depletion region (part InGaAs, part InP), and UTC (100% InP) photodiodes. This has been demonstrated by Li et al. [68, 90] where a thin InGaAs depletion layer was combined with an undepleted InGaAs absorber to increase optical responsivity. Devices of this structure have been labeled as Partially Depleted Absorber (PDA) photodiodes. In essence, the PDA photodiode is equivalent to the InP/InGaAs UTC with the InP collector replaced by an InGaAs collector. The homojunction aspect facilitates carrier transport because there are no band discontinuities at the p–i or i–n interfaces to impede carrier mobility. Relative to the UTC for the same p-doped InGaAs layer thickness, the PDA will achieve higher responsivity [91]; however, the lower bandgap of InGaAs may result in larger thermally-activated dark currents, which could yield heat-related failures sooner than in InP depletion layers, although this has not yet been verified.

PDA photodiodes having 1-μm thick p-type InGaAs absorbing layer and 700-nm depleted InGaAs collector layer illuminated through the substrate that achieved responsivity of 1.0 A/W. The large-signal modulation saturation currents and RF output powers for 28-, 34-, and 40-μm-diameter photodiodes were 70, 80, and 90 mA and 17, 18.5, and 19.5 dBm, respectively. Williams et al. [92] reported 1-dB small-signal compression currents for a 34-μm-diameter PDA photodiode of 700, 620, and 260 mA at 300 MHz, 1 GHz, and 6 GHz, respectively.

8.3.4 Modified Uni-Traveling Carrier Photodiodes

D.-H. Jun et al. [93] have reported that a modified UTC (MUTC), shown in Figure 8.15, formed by inserting an undoped $In_{0.53}Ga_{0.47}As$ layer between the InP drift layer and the p: $In_{0.53}Ga_{0.47}As$ absorption region of a UTC can achieve higher responsivity and higher bandwidth when the thickness of the added layer is optimized. Wang et al. [94] have shown that the MUTC can also achieve high saturation characteristics. Table 8.2 summarizes the large-signal modulation characteristics for various diameter devices. The responsivity at $\lambda = 1550$ nm was 0.75 A/W. At a bias of 6 V, the 28-μm-diameter MUTC exhibited 23-GHz bandwidth, 90-mA saturation current, and 83-mW RF output power, which yields a bandwidth-saturation current figure of merit of 2070 mA-GHz. This compares favorably with the value of 2500 mA-GHz reported for a $10 \times 10 \mu m^2$ CC-UTC [50].

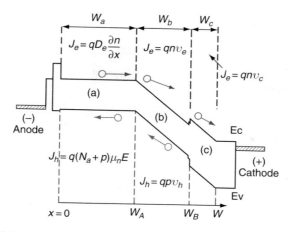

Figure 8.15 Band diagram of a modified uni-traveling-carrier photodiode with current components in each layer: v_e, v_h, and v_c are the electron and hole saturation velocity in the undoped InGaAs layer and the electron saturation velocity the InP layer, respectively: (a) p-type photo-absorption layer, (b) undoped photo-absorption layer, and (c) undoped InP layer [93] (this figure may be seen in color on the included CD-ROM).

Table 8.2

Bandwidth, saturation (or failure) current, and RF output power of modified uni-traveling carrier. Note that the responsivity of these photodiodes (0.75 A/W) is greater than twice that of the charge-compensated uni-traveling-carrier photodiodes in Table 8.1.

Diameter (μm)	Bias (V)	Bandwidth (GHz)	Saturation (mA)	RF output (dBm)
20	4	30	70	18.6
28	4	20	80	20
34	5	17	100	21.5
40	5	14	130	22.5

If responsivity is folded into the figure of merit, the bandwidth-saturation current-responsivity product is ~ 1.55 A^2-GHz/W compared with 1.13 A^2-GHz/W for the CC-UTC.

In order to increase the responsivity of UTC photodiodes while maintaining very high speed, an edge-illuminated refracting-facet structure (Figure 8.16) has been developed [95, 96]. Light is incident through an angled facet and refracted so as to achieve total internal reflection from the top surface, thus lengthening the optical absorption path length. Although the concept is similar to back-illumination and reflection from a metal electrode on the top surface, the refracting-facet approach avoids absorption in the metal and scattering at the rough metal–semiconductor interface. Responsivity as high as 1.0 A/W with less than 0.2 dB polarization dependence has been achieved; the bandwidth was 50 GHz with operating current of 20 mA [96].

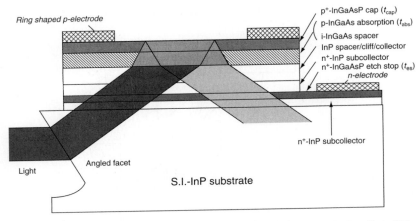

Figure 8.16 Schematic cross section of edge-coupled refracting-facet UTC photodiode [96].

8.3.5 High-Power Waveguide Photodiodes

The high-power UTC and PDA structures discussed above have been incorporated into waveguide photodiodes to improve their performance at high signal levels. In order to improve the high-power performance of evanescently coupled multimode-input waveguide photodiodes while maintaining high responsivity for bandwidths >40 GHz, UTC and PDA structures have been incorporated into the active photodiode regions. M. Achouche et al. [21] have reported waveguide UTC photodiodes with 0.76-A/W responsivity at 1.55 μm, polarization dependence <0.1 dB, >50 GHz bandwidth, and 22-mA saturation current. S. Demiguel et al. [52] first reported a PDA waveguide photodiode. These photodiodes exhibited 0.80-A/W responsivity with a fiber mode diameter of 6 μm. The polarization dependence was <0.5 ± 0.3 dB with −1-dB input coupling tolerances as high as ±2.0 and ±1.3 μm for horizontal and vertical displacements. The bandwidth was 50 GHz, and the −1-dB compression current at 40 GHz was 17 mA corresponding to +4.5 dBm RF output power. Y.-S. Wu et al. compared the performance of evanescently coupled waveguide PIN and PDA photodiodes with the same active areas (150 μm^2). For high photocurrent (∼5.5 mA) and low bias voltage (−1 V), the PDA photodiodes exhibited higher bandwidth, 26 GHz vs 8 GHz for the PIN photodiodes. At higher bias (−5 V), the saturation current was 23 mA and the bandwidth-saturation current product was 920 mA-GHz; the responsivity was 1.0 A/W.

Although incorporating high-power photodiodes into evanescently coupled waveguide structures has improved their power characteristics, it remains a fact that achieving higher and higher bandwidths necessitates reducing the active area in order to minimize the RC time constant. It follows that the current density also increases, to the detriment of high-power operation. One solution is to divide the optical signal between an array of photodiodes and to combine their photocurrents

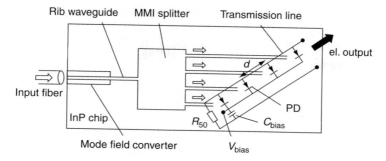

Figure 8.17 Schematic view of the traveling-wave photodiode array [98].

in phase with a transmission line [97]. This approach, illustrated in Figure 8.17, leads to scaling of the output power with the number of photodiodes in the array. Recently, A. Beling et al. [98] have successfully demonstrated the efficacy of this approach. A 1×4 MMI power splitter was employed to distribute the optical signal to four ($4 \times 7\,\mu m^2$) evanescently coupled traveling-wave waveguide photodiodes [13]. The InGaAs/InGaAsP heterostructure photodiodes with an intrinsic InGaAs absorption layer thickness of 200 nm were optimized to provide high responsivity [22]. A coplanar waveguide (CPW) transmission line was utilized to connect the photodiodes in parallel and to combine the four electrical output signals. The bandwidth of the traveling-wave photodiode array as determined with an optical heterodyne technique was 80 GHz. The high-power characteristics were measured at several fixed beat frequencies. At a signal frequency of 10 GHz, a maximum electrical output power of +10.3 dBm was achieved (Figure 8.18). Compared to a single photodiode from the same wafer with the same active area, this represents a 10-dB increase in output power and corresponds to an

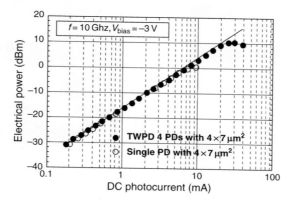

Figure 8.18 RF output power vs DC photocurrent at −3 V bias and small signal modulation at 10 GHz. The line indicates the ideal current–power relation at a 50 Ω load [98].

increase in the saturation current from 8 to 27 mA. A maximum electrical output power of −2.5 dBm was observed at 150 GHz. This is 7 dB higher than that of a single photodiode. At 200 GHz, the available power from the array was −9 dBm, and even at 400 GHz a power of −32 dBm was detected.

8.3.6 Thermal Considerations

The success of high-power photodiode structures in reducing the impact of space charge and the resulting increase in saturation currents and RF output power levels has heightened the importance of thermal management. Frequently, thermal failure is observed before current saturation. The better the heat transfers out of the photodiode depletion region, the higher the maximum dissipated power. InP is the lattice-matched substrate widely used for InGaAs photodetectors that operate at 1.3 and 1.55 μm. If the InP substrate can be replaced by materials with higher thermal conductivity such as Si, SiC, or diamond, the heat conduction through the substrate can be improved and higher power accommodated [63]. Direct semiconductor-to-semiconductor wafer bonding has been reported for transfering InGaAs high-power photodiodes onto Si substrates [99]. Li et al. [100] have employed a different approach, metal-to-metal bonding, to transfer InGaAs/InP photodiodes to an Si wafer. This is similar to the transferred-substrate technology that has been developed for InP/InGaAs heterojunction bipolar transistor circuits [101, 102] and resonant-cavity light-emitting diodes [103]. Figure 8.19 compares the two bonding procedures. The advantage of Au bonding is that the bonding

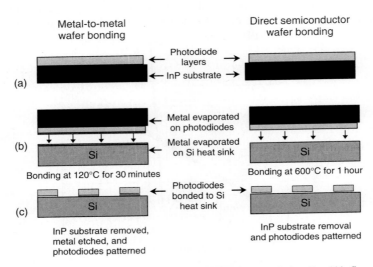

Figure 8.19 Illustration of metal-to-metal and direct semiconductor wafer bonding (this figure may be seen in color on the included CD-ROM).

temperature is much lower: 200°C compared with ~600°C. Hence, the stress induced by the thermal expansion mismatch between InP and Si is significantly reduced. For 20-μm-diameter photodiodes on InP, the maximum operating currents without saturation were 30 mA under 4 V bias, 35 mA under 5 V bias, and 40 mA under 6 V bias. Thermal failure occurred at 45 mA at 6 V. At bias voltages of 4, 5, and 6 V, the 20-μm photodiode bonded on Si showed very similar saturation behavior as the photodiode on InP. However, due to better thermal conductivity and larger specific heat capacity of Si, it could be operated at higher bias voltages of 7 and 8 V, which resulted in higher power dissipation. The highest operating current without saturation was ~50 mA at 8 V reverse bias. The maximum total dissipated power was more than 1.5 times higher than that of the same photodiode on InP.

8.4 AVALANCHE PHOTODIODES

Over the past five decades, avalanche photodiodes (APDs) have been utilized for a wide range of commercial, military, and research applications. In recent years, optical communications has been a primary driving force for research and development of APDs. It is well known that the internal gain of APDs provides higher sensitivity in optical receivers than PIN photodiodes [104–107], however, at the cost of more complex epitaxial wafer structures and bias circuits. APDs have been successfully deployed in optical receivers that operate up to 10 Gb/s and research on materials and device structures that will extend to higher-bit-rate applications is ongoing.

8.4.1 Separate Absorption, Charge, and Multiplication APDs

First-generation optical fiber communication systems, which operated in the wavelength range 800 to 900 nm, utilized Si PIN and APDs [108]. The evolution of transmission wavelengths to 1300 and 1550 nm to take advantage of the optimum windows for low dispersion and attenuation, motivated research on "long-wavelength" photodetectors. Tunneling at the high electric fields required for impact ionization in $In_{0.53}Ga_{0.47}As$ homojunctions [109, 110] led to the development of separate absorption and multiplication (SAM) APD structures [111]. In these APDs, the p–n junction and thus the high-field multiplication region is located in a wide bandgap semiconductor such as InP in which tunneling is insignificant and absorption occurs in an adjacent InGaAs layer. By properly controlling the charge density in the multiplication layer, it is possible to maintain a high enough electric field to achieve good avalanche gain while keeping the field low enough to minimize tunneling and impact ionization in the InGaAs absorber. However, the frequency response of SAM APDs, as originally implemented, was very poor owing to accumulation of photo-generated holes at the absorption/multiplication heterojunction interface [112]. Several methods to eliminate the slow release of trapped holes were reported; however, the approach that has been most widely adopted utilizes a transition region consisting of one or more lattice-matched, intermediate-bandgap $In_xGa_{1-x}As_{1-y}P_y$ layers [112–114]. A second modification to the original SAM APD structure has been the inclusion of a high–low doping profile in the multiplication region [115–117] similar to the reach-through structure that has been widely used for Si APDs [118]. In this structure, the wide-bandgap multiplication region consists of a lightly doped (usually unintentionally doped) layer in which the field is high and an adjacent, doped charge layer or field control region. This type of APD, which is frequently referred to as the SACM structure with the "C" representing the charge layer, decouples the thickness of the multiplication region from the charge density constraint in the SAM APD.

Most of the initial work on InP/InGaAsP/InGaAs SAM and SACM APDs utilized mesa structures because of their fabrication simplicity and reproducibility. However, the consensus that planar structures are more reliable than mesa-type photodiodes spurred the development of planar configurations. Some of the techniques that have been successfully demonstrated utilize a lateral extended guard ring [119–121], floating guard rings [122–125], pre-etched charge sheet with regrowth [126], etched diffusion well [127], or selective ion implantation of the charge region [116]. Each of these approaches has been successful in suppressing edge breakdown. Figure 8.20 shows a schematic cross section of an InP/InGaAsP/InGaAs SACM APD with a double-diffused floating guard ring [123]. The adjacent graph shows the electric field profile normal to the surface and illustrates how the charge layer is used to tailor the relative fields in the multiplication and absorption layers.

Although the InP/InGaAsP/InGaAs SACM APDs have achieved excellent receiver sensitivities up to 2.5 Gb/s [128–130], there are three factors that have

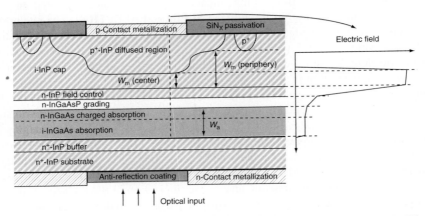

Figure 8.20 Schematic cross section of InP/InGaAsP/InGaAs SACM APD with double-diffused floating guard ring configuration [123].

limited their performance at higher bit rates. Because they operate under normal incidence and because the absorption coefficient of InGaAs at telecommunications wavelengths is $\sim 10^4$ cm^{-1} [131, 132], the absorption region must be approximately 2.5 μm thick in order to absorb >90% of the light that enters the detector. The associated transit times limit the bandwidth to 10 GHz at low gain. At higher gains, the relatively low gain-bandwidth product (<100 GHz) restricts the frequency response. As will be discussed below, the gain-bandwidth product and high excess noise are consequences of the reasonably unfavorable ionization coefficients of InP [133, 134]. Much of the recent work on APDs has focused on developing new structures and incorporating alternative materials that will yield lower noise and higher speed while maintaining optimal gain levels.

8.4.2 Low-Excess Noise APDs

The multiplication region of an APD plays a critical role in determining its performance, specifically the gain, the multiplication noise, and the gain-bandwidth product. According to McIntyre's local-field avalanche theory [135–137], both the noise and the gain-bandwidth product of APDs are determined by the electron, α, and hole, β, ionization coefficients of the semiconductor in the multiplication region, or more specifically, their ratio, $k = \beta/\alpha$. The noise power spectral density, φ, for mean gain, $<M>$, and mean photocurrent, $<I_{ph}>$, is given by the expression $\varphi = 2q<I_{ph}><M>^2 F(M)$. $F(M)$ is the excess noise factor, which arises from the random nature of impact ionization. Under the conditions of uniform electric fields and pure electron injection, the excess noise factor is

$$F(M) = <M^2>/<M>^2 = k<M> + (1-k)(2 - 1/<M>). \tag{8.1}$$

Equation (8.1) has been derived under the condition that the ionizations coefficients are in local equilibrium with the electric field, hence, the designation "local field" model. This model assumes that the ionization coefficients at a specific position are determined solely by the electric field at that position. It is clear from Eqn (8.1) that lower noise is achieved when $k \ll 1$. The gain-bandwidth product results from the time required for the avalanche process to build up or decay: the higher the gain, the higher the associated time constant and, thus, the lower the bandwidth. Emmons [138] has shown that the frequency-dependent gain can be approximated by the expression $M(\omega) = M_o/\sqrt{1 + (\omega M_o k \tau)^2}$, where M_o is the dc gain and τ is approximately (within a factor of ~ 2) the carrier transit time across the multiplication region. It follows from this expression that for $M_o > \alpha/\beta$ the frequency response is characterized by a constant gain-bandwidth product that increases as k decreases. There are three documented methods to achieve low excess noise in an APD. The best-known approach is to select a material such as Si [139–142] that has $k \ll 1$. The low-noise characteristics of Si are well documented;

however, Si photodiodes do not operate at the telecommunications wavelengths because the bandgap of Si is transparent at 1300 and 1550 nm.

Thin Multiplication Regions

Low excess noise and high gain-bandwidth product have also been achieved by submicron scaling of the thickness of the multiplication region, w_m. This is somewhat counterintuitive because it appears to contradict the local field model. As w_m is reduced, in order to maintain the same gain, the electric field intensity must increase in order to reduce the distance between ionization events. However, for high electric fields, the electron and hole ionization coefficients tend to merge so that k approaches unity. Consequently, based on the excess noise expression in Eqn (8.1), higher excess noise would be expected for the same gain. However, counter to the basic assumption of the local-field model, it is well known that impact ionization is non-local in that carriers injected into the high field region are "cool" and require a certain distance to attain sufficient energy to ionize [143]. This also applies to carriers immediately after ionization because their final states are typically near the band edge. The distance in which essentially no impact ionization occurs is frequently referred to as the "dead space," $d_{e(h)}$ for electrons (holes). If the multiplication region is thick, the dead space can be neglected and the local field model provides an accurate description of APD characteristics. However, for thin multiplication layers, the non-local nature of impact ionization has a profound impact. This can be explained as follows: As impact ionization is a stochastic process, it can best be described in terms of the probability distribution function (pdf), $p(x)$, which is the probability per unit length that a carrier ionizes a distance x from the injection point or the point where it was created by another impact ionization event. For the local-field theory, as shown in Figure 8.21(a), the pdf for electrons has the form $p_e(x) = \alpha^{-1}\exp(-\alpha x)$. At the high fields encountered in thin multiplication regions, the pdf must be modified to account for the fact that $p(x) \sim 0$ for $x <$ the dead space. Several analytical [144–148] and numerical

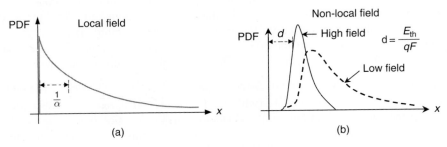

Figure 8.21 Probability distribution functions for (a) the local field model and (b) inclusion of the dead space for high field (solid line) and low field (dashed line) (this figure may be seen in color on the included CD-ROM).

models [149–151] have successfully been developed to accurately include the effect of the dead space. Although differing markedly in their approach, the physical picture that emerges from these models is consistent and matches up well with experimental measurements. Figure 8.21(b) illustrates qualitatively how incorporating the dead space alters the pdfs. First, it is clear that the dead-space length decreases with increasing field. As carriers transit the multiplication region, they continuously gain energy from the electric field and lose energy by optical phonon scattering. At the highest fields, phonon scattering becomes less significant because the phonon energy is small, a few tens of million electron volts. As a result, the carrier transport becomes "quasi-ballistic" and the dead space length is, to good approximation, equal to E_{th}/qF where E_{th} is the threshold energy for impact ionization, q is the electron charge, and F is the electric field strength. The decrease in dead-space length with increasing field tends to make it less significant at high fields. However, we note that the pdf also narrows significantly with increasing field. Because the width of the pdf decreases faster than the contraction in the dead space, the net result is that the ionization process becomes more deterministic, which reduces the variation in M. Figure 8.22 shows the calculated gain distributions for two $Al_{0.48}In_{0.52}As$ APDs with multiplication layer thickness of 1.0 μm (dashed line) and 0.1 μm (solid line) [152]. These APDs have the same average gain, $M \sim 20$, but the excess noise factors are 6.9 and 4.0 for the 1.0 and 0.1 μm APDs, respectively. The gain distribution of the 1.0-μm APD is broader than that of the 0.1-μm device, which gives rise to higher excess noise. This graph also shows that the thicker device has higher probabilities for both high gain ($M > 100$) and low gain ($M = 1$), whereas the

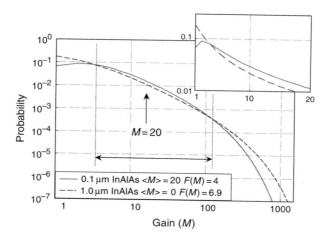

Figure 8.22 Comparison of the gain distribution curves for $Al_{0.48}In_{0.52}As$ APDs having multiplication region widths of 1.0 μm (dashed line) and 0.1 μm (solid line). The average gain for both APDs is $M \sim 20$ but the excess noise factors for the 1.0 and 0.1 μm APDs is 6.9 and 4, respectively [152] (this figure may be seen in color on the included CD-ROM).

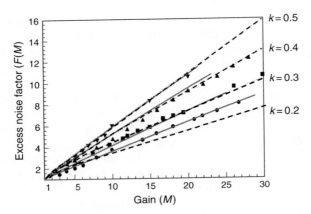

Figure 8.23 Comparison of calculated noise curves (solid lines) with experimental data for GaAs homojunction APD of different thickness 0.1 μm (●), 0.2 μm (■), 0.5 μm (▲), and 0.8 μm (▼) [146] (this figure may be seen in color on the included CD-ROM).

probabilities for the thin device are higher for gains in the range $2 < M < 100$. This is reasonable because they have different standard deviations in M while keeping $<M>$ the same. It is interesting to note that the 1.0 μm APD has a peak at $M = 1$, whereas the 0.1 μm APD has a peak at $M = 2$. This is consistent with the pdfs in Figure 8.21. The long tail in the distribution at low field, which is characteristic of thick multiplication regions, is indicative of a greater probability that a carrier will travel a longer distance, which in some cases can be the whole multiplication region, before ionizing. This has also been observed in Ref. [150].

Noise reduction in thin APDs has been demonstrated for a wide range of materials including InP [153–156], GaAs [150, 155–160], $Al_xIn_{1-x}As$ [155, 156, 161], Si [162, 163], $Al_xGa_{1-x}As$ [155, 156, 164–167], SiC [168], GaP [169], and GaInP [170]. Figure 8.23 shows the excess noise figure vs gain for GaAs APDs with w_m in the range 0.1–0.8 μm [146]. The dashed lines are plots of Eqn (8.1) for $k = 0.2$ to 0.5. These lines are not representative of the actual k values; they are presented solely for reference because the k value has become a widely used indirect figure of merit for excess noise. For constant gain, it is clear that the excess noise falls significantly with decreasing w_m.

AlInAs Multiplication Regions

Although shrinking, the multiplication region thickness is an effective approach to noise reduction, it should be noted that this is relative to the characteristic noise of the bulk (thick) material. Thus, it appears that lower noise can be achieved by beginning with "low-noise" semiconductors. For this reason, $In_{0.52}Al_{0.48}As$ is an attractive candidate for telecommunications APDs. Like $In_{0.53}Ga_{0.47}As$, $In_{0.52}Al_{0.48}As$ (referred to below as InAlAs) can be grown lattice-matched on InP substrates. Watanabe et al. [171] have measured the ionization coefficients for InAlAs and found that β/α was $\sim$0.3–0.4 for electric field in the range 400–650 kV/cm, which compares favorably with $\alpha/\beta \sim 0.4$–0.5 for InP. Lenox et al. [161] investigated the excess noise characteristics of PIN-structure InAlAs APDs; the excess noise was equivalent to $k = 0.2$ and 0.31 for $w_m = 200$ nm and 1600 nm, respectively. Thin layers of AlInAs have also been incorporated into the multiplication region of SACM APDs. Ning Li et al. [172] reported that mesa-structure undepleted-absorber InAlAs APDs with 180-nm thick multiplication regions exhibited excess noise equivalent to $k = 0.15$ and gain-bandwidth product of 160 GHz. Several planar InAlAs/InGaAs SACM APDs have also been developed. This has been more challenging than fabricating planar InP/InGaAs SACM APDs owing to the absence of a good n-type diffusant coupled with the requirement for electron injection. Watanabe et al. [173, 174] developed a quasi-planar structure with a InAlGaAs-InAlAs multiple quantum well multiplication region and Ti-implanted guard ring. These APDs exhibited dark current of 0.36 μA at a gain of 10, external quantum efficiency of 67%, 110 GHz gain-bandwidth product, and low-gain bandwidth of 15 GHz. Deployment of these APDs, however, has been limited by difficulties in optimizing Ti ion dosage and the Ti-activation anneal and the relatively high dark current [175]. Recently, an AlInAs/InGaAs planar SACM APD without a guard ring has been reported [175, 176]. Figure 8.24 shows a schematic cross section. The active region is defined by Zn diffusion through a transparent InP window layer to the InGaAs absorption region. In this case, the p–n junction and thus the high field region are located below the diffusion front at the interface between the p-type InAlAs charge layer and the thin (200 nm), unintentionally doped InAlAs multiplication region. These APDs have

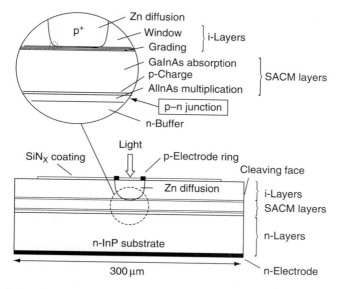

Figure 8.24 Schematic cross section of AlInAs/InGaAs planar SACM APD without a guard ring [175].

achieved gain >40, high external quantum efficiency (88%), 10 GHz low-gain bandwidth, and gain-bandwidth product of 120 GHz.

Heterojunction Multiplication Regions

Recently, it has been shown that the noise of APDs with thin multiplication regions can be reduced even further by incorporating new materials and impact ionization engineering (I^2E) with appropriately designed heterostructures [177–184]. Structurally, I^2E is similar to a truncated multiple quantum well (frequently mislabeled as "superlattice") APD [185, 186]; however, operationally, there is a fundamental difference in that these APDs do not invoke heterojunction band discontinuities. Their function relies instead on the differences in threshold energies for impact ionization between adjacent wide-bandgap and narrower bandgap materials. The structures that have achieved the lowest excess noise, to date, utilize multiplication regions in which electrons are injected from a wide-bandgap semiconductor into adjacent low-bandgap material. Initial work that demonstrated the efficacy of this approach utilized the $GaAs/Al_xGa_{1-x}As$ material system [177, 178, 180–182]. Excess noise equivalent to $k < 0.1$ has been demonstrated [180, 181]. Recently, InGaAlAs/InP implementations that operate at the telecommunications wavelengths have been reported. Using both a single-well structure and a pseudo-graded bandgap based on InAlAs/InGaAlAs materials, Wang et al. [183] demonstrated excess noise equivalent to $k \sim 0.12$ and dark current comparable to that of homojunction InAlAs APDs. Duan et al. [184] have incorporated a similar I^2E multiplication region into an MBE-grown InGaAlAs I^2E SACM APD. A cross section of the layer structure is shown in Figure 8.25. The compound I^2E multiplication region consisted of an unintentionally doped layer of $In_{0.52}Al_{0.48}As$ with a thickness of 80 nm, an unintentionally doped $In_{0.53}Ga_{0.17}Al_{0.3}As$ layer with a thickness of 80 nm, a p-type (Be, 2.2×10^{17} cm^{-3}) 120-nm thick $In_{0.53}Ga_{0.17}Al_{0.3}As$ and an 80-nm thick $In_{0.52}Al_{0.48}As$ layer with the same p-type doping level. The latter two layers also served as the field control or "charge" region. A 420-nm thick unintentionally doped $In_{0.53}Ga_{0.47}As$ layer was grown as the absorbing layer. Undoped InGaAlAs grading layers (50 nm) were inserted to reduce the barrier between $In_{0.52}Al_{0.48}As$ and $In_{0.53}Ga_{0.47}As$ in order to prevent carrier pile-up at the heterointerface. The absorber was slightly p-doped in order to suppress impact ionization in the absorption region. Ideally, the doping in the absorber would be graded to provide a slightly higher field in the direction of the multiplication region. This was approximated by step doping the absorber in two regions, one at 1×10^{16}/cm^3 and the other at 4×10^{16}/cm^3. Based on Monte Carlo simulations of similar $GaAs/Al_xGa_{1-x}As$ I^2E APDs [180], it can be inferred that there are relatively few ionization events in the $In_{0.52}Al_{0.48}As$ layer, owing to the combined effects of "dead space" and the higher threshold energy in $In_{0.52}Al_{0.48}As$. Figure 8.26 shows the excess noise factor, $F(M)$, vs gain. The dotted lines in Figure 8.26 are plots of $F(M)$ for $k = 0$–0.5. For $M \leq 4$, it appears that $k < 0$, which is unphysical and simply reflects the inapplicability of the local field model for this type of multiplication region. At higher gain, the excess noise is equivalent to a k

8. Advances in Photodetectors

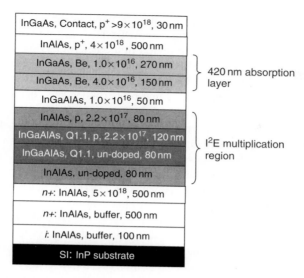

Figure 8.25 Layer structure of SACM InAlAs/InGaAlAs APD with I^2E multiplication region [184] (this figure may be seen in color on the included CD-ROM).

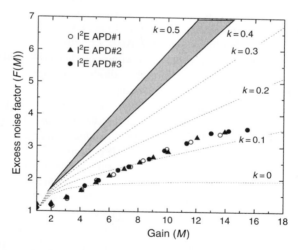

Figure 8.26 Excess noise factor, $F(M)$, vs gain for an SACM APD with I^2E $In_{0.52}Ga_{0.15}Al_{0.33}As/In_{0.52}Al_{0.48}As$ multiplication region [184] (this figure may be seen in color on the included CD-ROM).

value of ~0.12. For reference, the excess noise factor for InP/$In_{0.53}Ga_{0.47}As$ SACM APDs is shown as the shaded region in Figure 8.26.

An important question is whether the excess noise for these heterojunction APDs is lower than it would be in homojunction APDs having the same multiplication width and composed of either of the constituent materials. Because most of the impact ionization occurs in the narrow bandgap region, it might also be appropriate to compare to homojunction APDs having multiplication thickness equal to that of the narrow-bandgap region in the heterojunction devices. Groves et al. [179] studied avalanche multiplication and excess noise on a series of $Al_xGa_{1-x}As$–GaAs and GaAs–$Al_xGa_{1-x}As$ ($x = 0.3$, 0.45, and 0.6) single heterojunction P^+IN^+ diodes and concluded that properly designed heterojunctions can reduce the noise. They attribute this to two functions provided by the wide-bandgap layer. Electrons gain energy in the wide-bandgap layer but do not readily ionize owing to its high threshold energy. The hot electrons are then injected into the GaAs region, which has lower threshold energy, where they more readily ionize. The wide-bandgap layer also effectively suppresses hole ionization. Both of these effects reduce excess noise. Hayat et al. [187] have developed a modified dead-space multiplication theory (MDSMT) to describe injection of carriers with substantial kinetic energy into the multiplication region and have identified a mechanism, the "initial-energy effect" that reduces the excess noise. The energy buildup can occur through a sharp electric field gradient or, in the case of I^2E structures, in a wide-bandgap injector [188]. The initial energy of the injected carriers is linked to reduced noise through a reduction in the initial dead space associated with the injected avalanche-initiating carrier, i.e., "the strong localization of the first impact ionization event at the beginning of the multiplication region...is akin to having two injected carriers per absorbed photon" [187].

As was done for PIN photodiodes, in order to achieve high quantum efficiency without sacrificing bandwidth, waveguide APD structures have been developed. SACM APDs have been incorporated into edge-coupled [189–194] and evanescently coupled [195, 196] waveguide photodiodes. Nakata et al. [191] have reported an edge-coupled $In_{0.52}Al_{0.48}As/In_{0.53}Ga_{0.47}As$ APD that achieved 0.73 A/W responsivity, low-gain bandwidth of 35 GHz, and 140 GHz gain-bandwidth product. At 40 Gb/s, the receiver sensitivity was -19 dBm for 10^{-10}-bit error rate. Shiba et al. [192] have modified the edge-illuminated APD structure by tapering the width along the propagation direction in order to accommodate higher input power levels. A schematic of this structure, referred to as an asymmetric waveguide APD, is shown in Figure 8.27. Quantum efficiencies of 94 and 90% at 1.31 and 1.55 µm, respectively, were reported. The gain–bandwidth product was 110 GHz. At 10 Gb/s and 10^{-9} BER, the receiver sensitivity was -30.2 dBm. Demiguel et al. [196] have reported an evanescently coupled $In_{0.52}Al_{0.48}As/In_{0.53}Ga_{0.47}As$ SACM APD having a planar short multimode input waveguide. The input waveguide and coupling structure is similar to that reported by the same group for PIN [20] and PDA [52] photodiodes. A schematic cross section of this APD is shown in Figure 8.28. The photocurrent, dark current, and gain vs reverse bias are plotted in Figure 8.29(a). The breakdown occurred at ~ 18.5 V, and the dark current at 90% of the breakdown was in the range 100–500 nA. The responsivity was 0.62 A/W with TE/TM polarization dependence <0.5 dB. Figure 8.29(b) shows the bandwidth vs gain; at low gain, the maximum bandwidth was 35 GHz and the high-gain response exhibits a gain-bandwidth product of 160 GHz.

8. Advances in Photodetectors 255

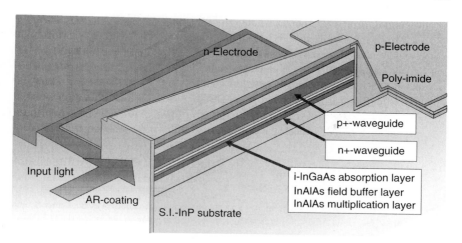

Figure 8.27 Schematic of edge-illuminated asymmetric waveguide avalanche photodiode [192] (this figure may be seen in color on the included CD-ROM).

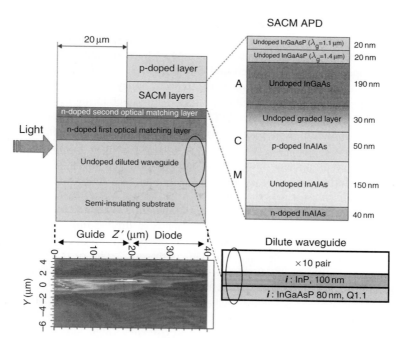

Figure 8.28 Schematic cross section of evanescently-coupled InAlAs/InAlAs waveguide APD [196] (this figure may be seen in color on the included CD-ROM).

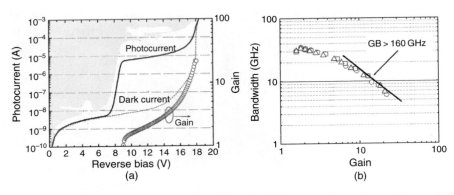

Figure 8.29 (a) Photocurrent, dark current, and gain and (b) bandwidth vs gain of evanescently-coupled InAlAs/InAlAs waveguide APD [196] (this figure may be seen in color on the included CD-ROM).

8.4.3 Single-Photon Avalanche Detectors

Quantum cryptography or quantum key distribution is a rapidly emerging field of optical fiber communications [197]. The goal is to provide a shared secret key to two authorized parties who desire to communicate securely, even if an eavesdropper has access to all of the message traffic. A key feature of quantum cryptography is that it provides the ultimate security based on the quantum mechanical properties of single photons. These systems are presently being studied and developed in laboratories around the world. A critical function for these systems is single-photon detection with high efficiency and minimal false positives. The photodetector of choice is the single-photon-counting avalanche photodetector (SPAD). Operation of an APD below breakdown voltage is referred to as linear mode operation. Operation above the breakdown voltage is fundamentally different. Above breakdown, an APD acts as a trigger element similar to a Geiger–Muller counter of nuclear radiation. Consequently, this mode of operation is frequently referred to as Geiger-mode operation. In the Geiger mode, an APD sustains the high bias across the depletion layer until a carrier is injected into the multiplication region. Once a carrier has initiated an avalanche process, owing to the high field, the current will continue to increase limited only by the external circuit. The net result is that a macroscopic current pulse is produced in response to a single carrier in the depletion layer. This is the "on" state of the Geiger-mode APD. The single carrier can be photogenerated, by which single-photon detection is achieved, or may have its origin in the dark current, giving rise to a dark count. Optimum performance is achieved when the single-photon detection efficiency is high and the dark count rate is low.

The first SPADs were Si, and these devices still achieve the best performance; a good review of the history of Si SPADs and state-of-the-art performance can be found in Ref. [198]. The development of fiber optic distribution systems has created a need for long-wavelength SPADs. To that end, InP/InGaAs SACM SPADs have been studied widely as single-photon counters. Although most of these studies have used commercially available APDs that were designed for fiber optic transmission, recently, there have been some efforts in designing InP-based SACM APDs for single-photon-counting applications [199–201]. Using gated quenching, Liu et al. [202] have reported high single-photon detection efficiency (~45%) and low dark count rate ($<10^4\,\text{s}^{-1}$) with a 40-μm diameter InP/InGaAs SACM APD similar in structure to that in Figure 8.20. Figure 8.30 shows the dark count rate vs the single-photon detection efficiency at 1.31 μm. Essentially, identical results were achieved at 1.55 μm. The reason that the dark count rate increases with increasing detection efficiency is that higher detection efficiencies are achieved by increasing the excess bias above breakdown, which results in higher dark counts. Although room temperature operation has been achieved, it is clear that cooling reduces the dark count rate by over two orders of magnitude. The single-photon detection efficiency is the product of the external quantum efficiency and the probability that a photogenerated carrier will initiate an avalanche breakdown. As a result, the external quantum efficiency (68% for these APDs) is the theoretical maximum for single-photon detection efficiency; hence, for these APDs, ~60% of the photons that are absorbed are counted. It is anticipated that future research on SPADs whose structures have been optimized for single-photon detection will achieve lower dark counts and higher detection efficiencies. One issue that will need to be addressed before high transmission capacities can be achieved is afterpulsing, i.e., the generation of false counts that arise from the emission of carriers that were trapped on deep levels in the multiplication region during earlier avalanche events. The more cooling that is required to achieve low dark counts, the longer the emission time and thus the more severe the effects of afterpulsing. At present, afterpulsing limits transmission rates at

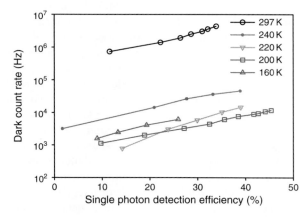

Figure 8.30 Dark count rate vs photon detection efficiency for 40 μm-diameter InP/InGaAs SPAD at 1310 nm [202] (this figure may be seen in color on the included CD-ROM).

telecommunication wavelengths to less than 10 Mb/s. Viable approaches to reduce afterpulsing include increasing the operating temperature as material quality improves to reduce the dark current and thus the dark counts, reducing the density of deep traps in the multiplication layer, and developing more sophisticated quenching circuits that limit the total charge flow and provide optimized hold-off times.

8.5 CONCLUSIONS

This chapter has focused on photodiodes for state-of-the-art and evolving advanced fiber optic systems. Before the ubiquitous deployment of Er-doped fiber amplifiers and the development of systems that operate at 40 Gb/s and higher bit rates, the dichotomy of photodetectors for optical receivers was rather simple, APDs for performance-driven systems and PIN photodiodes for low-cost systems. In recent years, however, the evolution of photodetectors for high-bit-rate receivers has been influenced by two factors. First, conventional normal-incidence photodiodes cannot achieve the requisite bandwidth-efficiency products for acceptable receiver sensitivities, hence, the move to waveguide photodiodes. Bandwidth-efficiency products of $\sim$60 GHz have been demonstrated. Another factor is that, to date, APDs have not achieved the sensitivity advantage relative to PIN photodiodes (typically $\sim$10 dB) at 40 Gb/s that they have demonstrated at lower bit rates. APDs with sufficient bandwidth for 40 Gb/s operation at low gain ($M < 10$) have been reported, but the gain-bandwidth products do not permit operation at the gain values required to achieve high receiver sensitivities. As a result, pending a breakthrough that produces APDs with gain-bandwidth products $\sim$400 GHz, 10 Gb/s appears to be a transition point where APDs still prove beneficial with a switch to high-speed waveguide PINs with optical amplification at higher bit rates. However, research on APDs may be transformed toward Geiger-mode operation with the emergence of interest in single-photon detection. The development of high-power photodiodes has occurred in response to a broad range of applications ranging from RF photonics to high-speed optoelectronic switches. In addition, moderately high saturation characteristics are needed for high-input-power optical amplification. To date, the highest power levels have been achieved with normal-incidence structures, but merging high-power approaches with high-speed waveguide structures is anticipated. Balanced receivers have been shown to be advantageous for coherent detection; however, they will become more prevalent as advanced modulation schemes such as DPSK become widely deployed. For these applications, monolithic structures will be required, and it is expected that higher levels of integration will become pervasive. All in all, photodetectors remain a critical component for optical communications, and the evolution of fiber optic transmission systems will continue to provide exciting research challenges and opportunities.

REFERENCES

[1] K. Kato, A. Kozen, Y. Muramoto et al., "110 GHz, 50% efficiency mushroom-mesa waveguide p-i-n photodiode for a 1.55-μm wavelength," *IEEE Photon. Technol. Lett.*, 6, 719–721, June 1994.
[2] J. E. Bowers and C. A. Burrus, "High-speed zero-bias waveguide photodetectors," *Electron. Lett.*, 22, 905–906, 1986.
[3] H. F. Taylor, O. Eknoyan, C. S. Park et al., "Traveling Wave Photodetectors," in Proc. *SPIE*, 1217, 1990, pp. 59–63.
[4] K. S. Giboney, M. W. J. Rodwell, and J. E. Bowers, "Traveling wave photodetectors," *IEEE Photon. Technol. Lett.*, 4(12), 1363–1365, 1992.
[5] K. S. Giboney, R. L. Nagarajan, T. E. Reynolds et al., "Traveling wave photodetectors with 172 GHz bandwidth and 76 GHz bandwidth-efficiency product," *IEEE Photon. Technol. Lett.*, 7(4), 412–414, 1995.
[6] K. Kato, "Ultrawide-band/high-frequency photodetectors," *IEEE Trans. Microw. Theory Tech.*, 47(7), 1265–1281, 1999.
[7] S. Demiguel, L. Giraudet, L. Joulaud et al., "Evanescently coupled photodiodes integrating a double stage taper for 40 Gb/s applications – compared performance with side-illuminated photodiodes," *J. Lightw. Technol.*, 12, 2004–2014, 2002.
[8] G. Unterborsch, D. Trommer, A. Umbach et al., "High-Power Performance of a High-Speed Photodetector," in Proc. *24th ECOC*, Heinrich-Hertz-Inst. fur Nachrichtentech. Berlin GmbH, Germany, 1998, pp. 67–68.
[9] G. Takeuchi, T. Nakata, K. Makita, and M. Yamaguchi, "High-speed, high-power, and high-efficiency photodiodes with evanescently coupled graded-index waveguide," *Electron. Lett.*, 36(11), 972–973, 2000.
[10] R. J. Hawkins, R. J. Deri, and O. Wada, "Optical power transfer in vertically integrated impedance-matched waveguide/photodetectors: Physics and implications for diode-length reduction," *Opt. Lett.*, 16(7), 470–472, 1991.
[11] R. J. Deri, W. Doldissen, R. J. Hawkins et al., "Efficient vertical coupling of photodiodes to InGaAsP rib waveguides," *Appl. Phys. Lett.*, 58(24), 2749–2751, 1991.
[12] L. Mörl, C. M. Weinert, F. Reier et al., "Uncladded InGaAsP/InP rib waveguides with integrated thickness tapers for efficient fibre-chip coupling," *Electron. Lett.*, 32(1), 36–38, 1996.
[13] D. Trommer, R. Steingrüber, R. Loeffler, and A. Umbach, "A Novel Flexible, Reliable and Easy to Use Technique for the Fabrication of Optical Spot Size Converters for InP Based PICs," in Proc. *11th IPRM'99*, postdeadline papers, Davos, Switzerland, May 16–20, 1999. p. 12.
[14] A. Umbach, "High-speed integrated photodetectors for 40 Gbit/s applications," *Proceedings of the SPIE – The International Society for Optical Engineering*, 5246(1), 434–442, 2003.
[15] H.-G. Bach, A. Beling, G. G. Mekonnen et al., "InP-based waveguide-integrated photodetector with 100 GHz bandwidth," *IEEE J. Sel. Top. Quantum Electron.*, 10(4), 668–672, July/August 2004.
[16] G. G. Mekonnen, H.-G. Bach, A. Beling et al., "80-Gb/s InP-based waveguide-integrated photoreceiver," *IEEE J. Sel. Top. Quantum Electron.*, 11(2), 356–360, March/April 2005.
[17] S. Demiguel, L. Giraudet, P. Pagnod-Rossiaux et al., "Low-cost, polarization independent, tapered photodiodes with bandwidth over 50 GHz," *Electron. Lett.*, 37(8), 516–517, 2001.
[18] L. Griaudet, F. Banfi, S. Demiguel, and G. Herve-gruyer, "Optical design of evanescently coupled, waveguide-fed photodiodes for ultra wide-band applications," *IEEE Photon. Technol. Lett.*, 11(1), 111–113, 1999.
[19] F. Xia, J. K. Thomson, M. R. Gokhale et al., "An asymmetric twin-waveguide high-bandwidth photodiode using a lateral taper coupler," *IEEE Photon. Technol. Lett.*, 13(8), 845–847, 2001.
[20] S. Demiguel, N. Li, X. Li et al., "Very high-responsivity evanescently-coupled photodiodes integrating a short planar multimode waveguide for high-speed applications," *IEEE Photon. Technol. Lett.*, 15, 1761–1763, 2003.

[21] M. Achouche, V. Magnin, J. Harari et al., "High performance evanescent edge coupled waveguide unitraveling-carrier photodiodes for >40 Gb/s optical receivers," *IEEE Photon. Technol. Lett.*, 16(2), 584–586, 2004.

[22] A. Beling, H.-G. Bach, G. G. Mekonnen et al., "Miniaturized waveguide-integrated p-i-n photodetector with 120 GHz bandwidth and high responsivity," *IEEE Photon. Technol. Lett.*, 17(10), 2152–2154, 2005.

[23] M. Achouche, V. Magnin, J. Harari et al., "Design and fabrication of a p-i-n photodiode with high responsivity and large alignment tolerances for 40 Gb/s applications," *IEEE Photon. Technol. Lett.*, 18(4), 556–558, 2006.

[24] Y.-S. Wu, P.-H. Chiu, and J.-W. Shi, "High-Speed and High-Power Performance of a Dual-Step Evanescently-Coupled Uni-Traveling-Carrier Photodiode at a 1.55 μm Wavelength," in Proc. *2007 OFC/NFOEC*, paper OthG-1, Anaheim, CA, 2007.

[25] Y.-S. Wu, P.-H. Chiu, and J.-W. Shi, "High-Speed and High-Power Performance of a Dual-Step Evanescently-Coupled Uni-Traveling-Carrier Photodiode at a 1.55 μm Wavelength," in Proc. *OFC 2007,* paper OThG1, Anaheim, CA, USA, March 2007.

[26] B. L. Kasper, C. A. Burrus, J. R. Talman, and K. L. Hall, "Balanced dual-detector receiver for optical heterodyne communications at Gbit/s rates," *Electron. Lett.*, 22, 413–415, 1986.

[27] R. J. Deri, T. Sanada, N. Yasuoka et al., "Low-loss monolithic integration of balanced twin-photodetectors with a 3 dB waveguide coupler for coherent lightwave receivers," *IEEE Photon. Technol. Lett.*, 2(8), 581–583, 1990.

[28] T. L. Koch, F. S. Choa, U. Koren et al., "Balanced operation of a GaInAs/GaInAsP multiple-quantum-well integrated heterodyne receiver," *IEEE Photon. Technol. Lett.*, 2, 577–580, 1990.

[29] M. Makiuchi, H. Hamaguchi, T. Kumai et al., "High-speed monolithic GaInAs twin-pin photodiode for balanced optical coherent receivers," *Electron. Lett.*, 25(17), 1145–1146, 1989.

[30] S. Chandrasekhar, B. Glance, A. G. Dentai et al., "Monolithic balanced p-i-n/HBT photoreceiver for coherent optical heterodyne communications," *IEEE Photon. Technol. Lett.*, 3(6), 537–539, 1991.

[31] E. Ackerman, S. Wanuga, D. Kasemset et al., "Maximum dynamic range operation of a microwave external modulation fiber-optic link," *IEEE Trans. Microw. Theory Tech.*, 41, 1299–1306, 1993.

[32] L. T. Nichols, K. J. Williams, and R. D. Esman, "Optimizing the ultrawide-band photonic link," *IEEE Trans. Microw. Theory Tech.*, 45, 1384–1389, August 1997.

[33] K. J. Williams and R. D. Esman, "Optically amplified downconverting link with shot-noise-limited performance," *IEEE Photon. Technol. Lett.*, 8(1), 148–150, January 1996.

[34] M. Rohde, C. Caspar, N. Heimes et al., "Robustness of DPSK direct detection transmission format in standard fibre WDM systems," *Electron. Lett.*, 36(17), 1483–1484, 2000.

[35] P. J. Winzer, S. Chandrasekhar, and K. Hoon, "Impact of filtering on RZ-DPSK reception," *IEEE Photon. Technol. Lett.*, 15(6), 840–842, 2003.

[36] P. J. Winzer and R.-J. Essiambre, "Advanced modulation formats for high-capacity optical transport networks," *J. Lightw. Technol.*, 24(12), 4711–4728, December 2006.

[37] Chapter 22 of this book.

[38] A. H. Gnauck, S. Chandrasekhar, J. Leuthold, and L. Stulz, "Demonstration of 42.7 Gb/s DPSK receiver with 45 photons/bit sensitivity," *IEEE Photon. Technol. Lett.*, 15(1), 99–101, January 2003.

[39] M. Mlejnek, "Balanced differential phase-shift keying detector performance: An analytical study," *Opt. Lett.*, 31(15), 2266–2268, 2006.

[40] J. H. Sinsky, A. Adamiecki, A. Gnauck et al., "RZ-DPSK Transmission using a 42.7 Gb/s integrated balanced optical front end with record sensitivity," *J. Lightw. Technol.*, 22(1), 180–185, January 2004.

[41] N. Li, H. Chen, S. Demiguel et al., "High-power charge-compensated unitraveling-carrier balanced photodetector," *IEEE Photon. Technol. Lett.*, 16(10), 2329–2331, 2004.

[42] A. Beling, H. G. Bach, D. Schmidt et al., "High-speed balanced photodetector module with 20dB broadband common-mode rejection ratio," in Proc. *OFC*, 1, 2003, pp. 339–340.

[43] S. S. Agashe, S. Datta, F. Xia, and S. R. Forrest, "A monolithically integrated long-wavelength balanced photodiode using asymmetric twin-waveguide technology," *IEEE Photon. Technol. Lett.*, 16(1), 236–238, 2004.

[44] N. Li, S. Demiguel, H. Chen et al., "Planar short-multimode waveguide evanescently-coupled balanced photodetectors with fully depleted and partially depleted absorber," in Proc. *OFC, Technical Digest*, Microelectron. Res. Center Texas/Univ. Austin, TX, USA; *OFC/NFOEC 2005, OFC, Technical Digest*, 2005, pp. 161–163.

[45] M. S. Islam, T. Jung, T. Itoh et al., "High power and highly linear monolithically integrated distributed balanced photo detectors," *J. Lightw. Technol.*, 20, 285–295, 2002.

[46] M. S. Islam, S. Murthy, T. Itoh et al., "Velocity-matched distributed photodetectors and balanced photodetectors with p-i-n photodiodes," *IEEE Trans. Microw. Theory Tech.*, 49, 1914–1920, 2001.

[47] T. Ishibashi, N. Shimizu, S. Kodama et al., "Uni-Traveling-Carrier Photodiodes," *Tech. Dig. Ultrafast Electronics and Optoelectronics*, 83–87, 1997.

[48] H. Ito, H. Fushimi, Y. Muramoto et al., "High-power photonic microwave generation at K- and Ka-bands using a uni-traveling-carrier photodiode," *J. Lightw. Technol.*, 20(8), 1500–1505, 2002.

[49] Y. Muramoto, K. Kato, M. Mitsuhara et al., "High-output-voltage, high speed, high efficiency uni-traveling-carrier waveguide photodiode," *Electron. Lett.*, 34, 122–123, 1998.

[50] N. Li, X. Li, S. Demiguel et al., "High-saturation-current charge-compensated InGaAs/InP uni-traveling-carrier photodiode," *IEEE Photon. Tech. Lett.*, 16(3), 864–866, 2004.

[51] X. Li, S. Demiguel, N. Li et al., "Backside illuminated high saturation current partially depleted absorber photodetectors," *Electron. Lett.*, 39, 1466–1467, 2003.

[52] S. Demiguel, X. Li, N. Li et al., "Analysis of partially depleted absorber waveguide photodiodes," *J. Lightw. Technol.*, 23(8), 2505–2512, 2005.

[53] H.-G. Bach, "Ultra-Broadband Photodiodes and Balanced Detectors Towards 100 Gbit/s and Beyond," in Proc. *SPIE – The International Society for Optical Engineering*, 6014(1), 2005, pp. 1–13.

[54] A. Beling, H.-G. Bach, G. G. Mekonnen et al., "Monolithically Integrated Balanced Photoreceiver OEIC Comprising a Distributed Amplifier for 40 Gbit/s Applications," in OSA Trends in Optics and Photonics Series, V95A, Optical Fiber Communication Conference (OFC), Postconference Digest, 2004, Los Angeles, CA, pp. 527–529.

[55] C. Schramm, H.-G. Bach, A. Beling et al., "High-bandwidth balanced photoreceivers suitable of 40 Gb/s RZ-DPSK modulation formats," *IEEE J. Sel. Top. Quantum Electron.*, 11(1), 127–134, 2005.

[56] V. J. Urick, M. S. Rogge, F. Bucholtz, and K. J. Williams, "Wideband (0.045–6.25 GHz) 40 km analogue fibre-optic link with ultra-high (>40 dB) all-photonic gain," *Electron. Lett.*, 42(9), 552–553, 2006.

[57] K. J. Williams, L. T. Nichols, and R. D. Esman, "Photodetector nonlinearity limitations on a high-dynamic range 3 GHz fiber optic link," *J. Lightw. Technol.*, 16(2), 192–199, 1998.

[58] Y. Miyamoto, M. Yoneyama, K. Hagimoto et al., "40 Gbit/s high sensitivity optical receiver with uni-travelling-carrier photodiode acting as decision IC driver," *Electron. Lett.*, 34(2), 214–215, 1998.

[59] Y. Miyamoto, K. Yonenaga, A. Hirano et al., "1.04-Tbit/s DWDM transmission experiment based on alternate-polarization 80-Gbit/s OTDM signals," in Proc. *24th ECOC*, 3, 1998, pp. 55–57.

[60] S. Takashima, H. Nakagawa, S. Kim et al., "40-Gbit/s Receiver with −21 dBm Sensitivity Employing Filterless Semiconductor Optical Amplifier," in Proc. *OFC*, 2, 2003, pp. 471–472.

[61] M. Dentan and B.D. de Cremoux, "Numerical simulation of a p-i-n photodiode under high illumination," *J. Lightw. Technol.*, 8(8), 1137–1144, August 1990.

[62] P.-L. Liu, K. J. Williams, M. Y. Frankel, and R. D. Esman, "Saturation characteristics of fast photodetectors," *IEEE Trans. Microw. Theory Tech.*, 47(7), 1297–1303, 1999.

[63] K. J. Williams and R. D. Esman, "Design considerations for high current photodetectors," *J. Lightw. Technol.*, 17, 1443–1454, 1999.

[64] K. J. Williams, R. D. Esman, and M. Dagenais, "Effects of high space-charge fields on the response of microwave photodetectors," *IEEE Photon. Technol. Lett.*, 6(5), 639–641, 1994.

[65] H. Ito and T. Nagatsuma, "High-speed and high-output-power unitraveling-carrier photodiodes," in Proc. *of the SPIE – The International Society for Optical Engineering*, 5246(1), 2003, pp. 465–479.

[66] M. S. Islam, A. Nespola, M. Yeahia et al., "Correlation Between the Failure Mechanism and Dark Currents of High Power Photodetectors," in Proc. *LEOS 2000 Annual Meeting*, 1, 2000, pp. 82–83.

[67] N. Shimizu, N. Watanabe, T. Furuta, and T. Ishibashi, "InPInGaAs Uni-traveling-carrier photodiode with improved 3-dB bandwidth of over 150 GHz," *IEEE Photon. Technol. Lett.*, 10(3), 412–414, 1998.

[68] X. Li, N. Li, S. Demiguel et al., "A comparison of front and backside-illuminated high-saturation power partially depleted absorber photodetectors," *IEEE J. Quantum Electron.*, 40(9), 1321–1325, 2004.

[69] J. Paslaski, P. C. Chen, J. S. Chen et al., "High-Power Microwave Photodiode for Improving Performance of RF Fiber Optic Links," in Proc. *SPIE, Photonics and Radio Frequency*, 2844, Denver, CO, 2006, pp. 110–119.

[70] N. Duan, X. Wang, N. Li et al., "Thermal analysis of high-power InGaAs-InP photodiodes," *IEEE J. Quantum Electron.*, 42(12), 1255–1258, 2006.

[71] K. J. Williams, "Comparisons between dual-depletion-region and uni-travelling-carrier p-i-n photodetectors," *IEE Proc.-Optoelectronics*, 149(4), 131–137, 2002.

[72] G. A. Davis, R. E. Weiss, R. A. LaRue et al., "A 920–1650 nm high current photodetector," *IEEE Photon. Technol. Lett.*, 8(10), 1373–1375, 1996.

[73] T. Furuta, H. Ito, and T. Ishibashi, "Photocurrent dynamics of uni-traveling-carrier and conventional pinphotodiodes," *Inst. Phys. Conf. Ser.*, 166, 419–422, 2000.

[74] T. Ishibashi, "High speed heterostructure devices," in *Semiconductors and Semimetals*, Vol. 41, San Diego: Academic Press, 1994, Chapter 5, p. 333.

[75] Y.-S. Wu, J.-W. Shi, and P.-H. Chiu, "Analytical modeling of a high-performance near-ballistic uni-traveling-carrier photodiode at a 1.55-µm wavelength," *IEEE Photon. Technol. Lett.*, 18(8), 938–940, 2006.

[76] H. Ito, T. Furuta, S. Kodama, and T. Ishibashi, "Zero-bias high-speed and high-output-voltage operation of cascade-twin uni-travelling-carrier photodiode," *Electron. Lett.*, 36, 2034–2036, 2000.

[77] B. Jalali and S. J. Pearton, *InP HBTs: Growth, Processing and Applications*, Boston: Artech House, 1995, p. 93.

[78] S. Kodama and H. Ito, "UTC-PD-based optoelectronic components for high-frequency and high-speed applications," *IEICE Trans. Electron.*, E90-C(2), 429–435, 2007.

[79] Y. Miyamoto, M. Yoneyama, T. Otsuji et al., "40-Gbit/s TDM transmission technologies based on high-speed ICs," *IEEE J. Solid-State Circuits*, 34, 1246–1253, 1999.

[80] K. Hagimoto, Y. Miyamoto, T. Kataoka et al., "Twenty-Gbit/s signal transmission using simple high-sensitivity optical receiver," *Tech. Dig. 16th OFC*, 1992, p. 48.

[81] S. Kodama, T. Ito, N. Watanabe et al., "2.3 picoseconds optical gate monolithically integrating photodiode and electroabsorption modulator," *Electron. Lett.*, 37(19), 1185–1186, September 2001.

[82] S. Kodama, T. Yoshimatsu, and H. Ito, "320 Gbit/s error-free demultiplexing using ultrafast optical gate monolithically integrating a photodiode and electroabsorption modulator," *Electron. Lett.*, 39(17), 1269–1270, August 2003.

[83] T. Yoshimatsu, S. Kodama, K. Yoshino, and H. Ito, "100-Gb/s errorfree wavelength conversion with a monolithic optical gate integrating a photodiode and electroabsorption modulator," *IEEE Photon. Technol. Lett.*, 17(11), 2367–2369, November 2005.

[84] S. Kodama, T. Shimizu, T. Yoshimatsu et al., "Ultrafast optical sampling gate monolithically integrating a photodiode and an electroabsorption modulator," *Electron. Lett.*, 40(11), 696–697, May 2004.

[85] S. Kodama, T. Yoshimatsu, and H. Ito, "500-Gb/s demultiplexing operation of a monolithic PD-EAM optical gate," *Electron. Lett.*, 40(9), 555–556, April 2004.
[86] Y. Muramoto and T. Ishibashi, "InP/InGaAs pin photodiode structure maximizing bandwidth and efficiency," *Electron. Lett.*, 39, 1749–1750, November 2003.
[87] K. J. Williams and R. D. Esman, "Design considerations for high-current photodetectors," *J. Lightw. Technol.*, 17(8), 1443–1454, 1999.
[88] N. Shimizu, N. Watanaba, T. Furuta, and T. Ishibashi, "InP-InGaAs uni-traveling-carrier photodiode with improved 3-dB bandwidth of over 150GHz," *IEEE Photon. Technol. Lett.*, 10, 412–414, 1998.
[89] N. Duan, X. Wang, N. Li et al., "Thermal analysis of high-power InGaAs-InP photodiodes," *IEEE J. Quantum Electron.*, 42(12), 1255–1258, 2006.
[90] X. Li, N. Li, X. Zheng et al., "High-saturation-current InP/InGaAs photodiode with partially depleted absorber," *IEEE Photon. Technol. Lett.*, 15, 1276–1278, 2003.
[91] Y. Muramoto and T. Ishibashi, "InP/InGaAs pin photodiode structure maximizing bandwidth and efficiency," *Electron. Lett.*, 39, 1749–1750, November 2003.
[92] K. J. Williams, D. A. Tulchinsky, J. B. Boos et al., "High-Power Photodiodes," *2006 Digest of the LEOS Summer Topical Meetings* (IEEE Cat. No. 06TH8863C), 2006, pp. 50–51.
[93] D.-H. Jun, J.-H. Jang, I. Adesida, and J.-I. Song, "Improved efficiency-bandwidth product of modified uni-traveling carrier photodiode structures using an undoped photo-absorption layer," *Jpn. J. Appl. Phys.*, 45(4B), 3475–3478, 2006.
[94] X. Wang, N. Duan, H. Chen, and J. C. Campbell, "InGaAs/InP photodiodes with high responsivity and high saturation power," *IEEE Photon. Technol. Lett.*, 19(16), 1272–1274, August 2007.
[95] H. Fukano, Y. Muramoto, and Y. Matsuoka, "High-speed and high output voltage edge-illuminated refracting-facet photodiode," *Electron. Lett.*, 35(18), 1581–1582, September 1999.
[96] Y. Muramoto, H. Fukano, and T. Furuta, "A polarization-independent refracting-facet uni-traveling-carrier photodiode with high efficiency and large bandwidth," *J. Lightw. Technol.*, 24(10), 3830–3834, 2006.
[97] C. L. Goldsmith, G. A. Magel, and R. J. Baca, "Principles and performance of traveling-wave photodetector arrays," *IEEE Trans. Microw. Theory Tech.*, 45(8), 1342–1350, 1997.
[98] A. Beling, H.-G. Bach, G. G. Mekonnen et al., "High-speed miniaturized photodiode and parallel-fed traveling-wave photodetectors based on InP," *IEEE J. Sel. Top. Quantum Electron.*, 13(1), 15–21, 2007.
[99] D. A. Tulchinsky, K. J. Williams, A. Pauchard et al., "High-power InGaAs-on-Si pin RF photodiodes," *Electron. Lett.*, 39(14), 1084–1086, 2003.
[100] N. Li, H. Chen, N. Duan et al., "High power photodiode wafer bonded to si using au with improved responsivity and output power," *IEEE Photon. Technol. Lett.*, 18(23), 2526–2528, 2006.
[101] D. Mensa, R. Pullela, Q. Lee et al., "48GHz digital IC's and 85-GHz baseband amplifiers using transferred-substrate HBT's," *IEEE J. Solid-State Circuits*, 34(9), 1196–1203, 1999.
[102] S. Lee, H. J. Kim, M. Utreaga et al., "Transferred-substrate InP/InGaAs/InP double heterojunction bipolar transistors with $f_{max} = 425$ GHz," *Electron. Lett.*, 37(17), 1096–1098, 2001.
[103] R.-H. Horng, W.-K. Wang, S.-Y. Huang, and D.-S. Wuu, "Effect of resonant cavity in waverbonded green InGaN LED with dielectric and silver mirrors," *IEEE Photon. Technol. Lett.*, 18(3), 457–459, 2006.
[104] S. D. Personick, "Receiver design for digital fiber-optic communication systems, Parts I and II," *Bell Syst. Tech. J.*, 52, 843–886, 1973.
[105] R. G. Smith and S. D. Personick, "Receiver design for optical fiber communications systems," in *Semiconductor Devices for Optical Communication*. New York: Springer-Verlag, 1980, Chapter 4.
[106] S. R. Forrest "Sensitivity of avalanche photodetector receivers for high-bit-rate long-wavelength optical communication systems," in *Semiconductors and Semimetals*, Vol. 22, *Lightwave Communications Technology*. Orlando, FL: Academic Press, 1985, Chapter 4.

[107] B. L. Kasper and J. C. Campbell, "Multigigabit-per-second Avalanche photodiode lightwave receivers," *J. Lightw. Technol.*, LT-5, 1351, 1987.

[108] H. Melchior, A. R. Hartman, D. P. Schinke, and T. E. Seidel, "Planar epitaxial silicon avalanche photodiode," *Bell Syst. Tech. J.*, 57, 1791–1807, 1978.

[109] S. R. Forrest, M. DiDomenico, Jr., R. G. Smith, and H. J. Stocker, "Evidence of tunneling in reverse-bias III-V photodetector diodes," *Appl. Phys. Lett.*, 36, 580–582, 1980.

[110] H. Ando, H. Kaaba, M. Ito, and T. Kaneda, "Tunneling current in InGaAsP and optimum design for InGaAs/InP avalanche photo-diodes," *Jpn. J. Appl. Phys.*, 19, 1277–1280, 1980.

[111] K. Nishida, K. Taguchi, and Y. Matsumoto, "InGaAsP heterojunction avalanche photodiodes with high avalanche gain," *Appl. Phys. Lett.*, 35, 251–253, 1979.

[112] S. R. Forrest, O. K. Kim, and R. G. Smith, "Optical response time of In$_{0.53}$Ga$_{0.47}$As avalanche photodiodes," *Appl. Phys. Lett.*, 41, 95–98, 1982.

[113] J. C. Campbell, A. G. Dentai, W. S. Holden, and B. L. Kasper, "High-performance avalanche photodiode with separate absorption, grading, and multiplication regions," *Electron. Lett.*, 18, 818–820, 1983.

[114] Y. Matsushima, A. Akiba, K. Sakai et al., "High-speed response InGaAs/InP heterostructure avalanche photodiode with InGaAsP buffer layers," *Electron. Lett.*, 18, 945–946, 1982.

[115] F. Capasso, A. Y. Cho, and P. W. Foy, " Low-dark-current low-voltage 1.3–1.6 μm avalanche photodiode with high-low electric field profile and separate absorption and multiplication regions by molecular beam epitaxy," *Electron. Lett.*, 20(15), 635–637, 1984.

[116] P. Webb, R. McIntyre, J. Scheibling, and M. Holunga, "A planar InGaAs APD fabricated using Si implantation and regrowth techniques," in Proc. *Tech. Digest of 1990 OFC*, New Orleans, 1988.

[117] L. E. Tarof, "Planar InP-InGaAs avalanche photodetectors with n-multiplication layer exhibiting a very high gain-bandwidth product," *IEEE Photon. Technol. Lett.*, 2, 643–645, 1990.

[118] H. W. Ruegg, "An optimized avalanche photodiode," *IEEE Trans. Electron Dev.*, ED-14, 239–251, 1966.

[119] K. Taguchi, T. Torikai, Y. Sugimoto et al., "Planar-structure InP/InGaAsP/InGaAs avalanche photodiodes with preferential lateral extended guard ring for 1.0–1.6 μm wavelength optical communication use," *J. Lightw. Technol.*, 6, 1643–1655, 1988.

[120] Y. Matsushima, Y. Noda, Y. Kushiro et al., "High sensitivity of VPE-grown InGaAs/InP-heterostructure APD with buffer layer and guard-ring structure," *Electron. Lett.*, 20(6), 235–236, 1984.

[121] J. N. Hollenhorst, D. T. Ekholm, J. M Geary et al., "High Frequency Performance of Planar InGaAs/InP APDs," *SPIE, High Frequency Analog Communications*, 995, 1988, pp. 53–60.

[122] Y. Liu, S. R. Forrest, J. Hladky et al., "A planar InP/InGaAs avalanche photodiode with floating guard ring and double diffused junction," *J. Lightw. Technol.*, 10(2), 182–192, 1992.

[123] M. A. Itzler, K. K. Loi, S. McCoy et al., "Manufacturable Planar Bulk-InP Avalanche Photodiodes for 10 Gb/s Applications," in Proc. *LEOS '99*, San Francisco, CA, November, 1999, pp. 748–749.

[124] S. R. Cho, S. K. Yang, J. S. Ma et al., "Suppression of avalanche multiplication at the periphery of diffused junction by floating guard rings in a planar In-GaAs–InP avalanche photodiode," *IEEE Photon. Technol. Lett.*, 12(5), 534–536, 2000.

[125] J. Wei, J. C. Dries, H. Wang et al., "Optimization of 10-Gb/s long-wavelength floating guard ring InGaAs-InP avalanche photodiodes," *IEEE Photon. Technol. Lett.*, 14(7), 977–979, 2002.

[126] L. E. Tarof, D. G. Knight, K. E. Fox et al., "Planar InPAnGaAs avalanche photodiodes with partial charge sheet in device periphery," *Appl. Phys. Lett.*, 57, 670–672, 1990; L. E. Tarof, "Planar InP/InGaAs avalanche photodetectors with π-multiplication layer exhibiting a very high gain-bandwidth product," *IEEE Photon. Technol. Lett.*, 2, 643–446, 1990.

[127] L. E. Tarof, R. Bruce, D. G. Knight et al., "Planar InP–InGaAs single growth avalanche photodiodes with no guard rings," *IEEE Photon. Technol. Lett.*, 7, 1330–1332, 1995.

[128] C. Y. Park, K. S. Hyun, S. K. Kang et al., "High-performance InGaAs/InP photodiode for 2.5 Gb/s optical receiver," *Opt. Quantum Elect.*, 24(5), 553–559, 1995.

[129] G. Hasnain, W. G. Bi, S. Song et al., "Buried-mesa Avalanche photodiodes," *IEEE J. Quantum Electron.*, 34(12), 1998.

[130] M. A. Itzler, C. S. Wang, S. McCoy et al., "Planar Bulk InP Avalanche Photodiode Design for 2.5 and 10 Gb/s Applications," in Proc. *ECOC '98 – 24th EC OC*, 1, Madrid, Spain, September 20–24, 1998, pp. 59–60.
[131] D. A. Humphreys and R. J. King, "Measurement of absorption coefficients of $Ga_{0.47}In_{0.53}As$ over the wavelength range 1.0–1.7μm," *Electron. Lett.*, 21(25/26), 1187–1189, 1985.
[132] F. R. Bacher, J. S. Blakemore, J. T. Ebner, and J. R. Arthur, "Optical-absorption coefficient of $In_{1-x}Ga_xAs/InP$," *Phys. Rev. B*, 37(5), 2551–2557, 1988.
[133] C. A. Amiento and S. H. Groves, "Impact ionization in (100)-, (110)-, and (111)-oriented InP avalanche photodiodes," *Appl. Phys. Lett.*, 43(2), 333–335, 1983.
[134] L. W. Cook, G. E. Bulman, and G. E. Stillman, "Electron and hole ionization coefficients in InP determined by photomultiplication measurements," *Appl. Phys. Lett.*, 40(7), 589–591, 1982.
[135] R. J. McIntyre, "Multiplication noise in uniform avalanche diodes," *IEEE Trans. Electron Dev.*, 13(1), 154–158, 1966.
[136] R. J. McIntyre, "The distribution of gains in uniformly multiplying avalanche photodiodes: Theory," *IEEE Trans. Electron Dev.*, ED-19, 703–713, 1972.
[137] R. J. McIntyre, "Factors affecting the ultimate capabilities of high speed avalanche photodiodes and a review of the state-of-the-art," *Tech. Dig. Int. Electron Dev. Mtg.*, 213–216, 1973.
[138] R. B. Emmons, "Avalanche-photodiode frequency response," *J. Appl. Phys.*, 38(9), 3705–3714, 1967.
[139] C. A. Lee, R. A. Logan, R. L. Batdorf et al., "Ionization rates of holes and electrons in silicon," *Phys. Rev.*, 134, A761–A773, 1964.
[140] J. Conradi, "The distributions of gains in uniformly multiplying avalanche photodiodes: experimental," *IEEE Trans. Electron Dev.*, ED-19(6), 713–718, 1972.
[141] W. N. Grant, "Electron and hole ionization rates in epitaxial silicon at high electric fields," *Solid-State Electron.*, 16, 1189–1203, 1973.
[142] T. Kaneda, H. Matsumoto, and T. Yamaoka, "A model for reach-through avalanche photodiodes (RAPD's)," *J. Appl. Phys.*, 47(7), 3135–3139, 1976.
[143] Y. Okuto and C. R. Crowell, "Ionization coefficients in semiconductors: a nonlocalized property," *Phys. Rev. B*, 10, 4284–4296, 1974.
[144] M. M. Hayat, B. A. E. Saleh, and M. C. Teich, "Effect of dead space on gain and noise of double-carrier multiplication avalanche photodiodes," *IEEE Trans. Electron Dev.*, 39(3), 546–552, 1992.
[145] R. J. McIntyre, "A new look at impact ionization – part 1: a theory of gain, noise, breakdown probability and frequency response," *IEEE Trans. Electron Dev.*, 48(8), 1623–1631, 1999.
[146] X. Li, X. Zheng, S. Wang et al., "Calculation of gain and noise with dead space for GaAs and $Al_xGa_{1-x}As$ avalanche photodiodes," *IEEE Trans. Electron Dev.*, 49, 1112–1117, 2002.
[147] B. Jacob, P. N. Robson, J. R. P. David, and G. J. Rees, "Fokker-Planck model for nonlocal impact ionization in semiconductors," *J. Appl. Phys.*, 90(3), 1314–1317, 2001.
[148] A. Spinelli and A. L. Lacaita, "Mean gain of avalanche photodiodes in a dead space model," *IEEE Trans. Electron Dev.*, 43(1), 23–30, 1996.
[149] G. M. Dunn, G. J. Rees, J. R. P. David et al., "Monte Carlo simulation of impact ionization and current multiplication in short GaAs pin diodes," *Semicond. Sci. Technol.*, 12, 111–120, 1997.
[150] D. S. Ong, K. F. Li, G. J. Rees et al., "A Monte Carlo investigation of multiplication noise in thin pin avalanche photodiodes," *IEEE Trans. Electron Dev.*, 45(8), 1998.
[151] S. A. Plimmer, J. R. P. David, D. S. Ong, and K. F. Li, "A simple model including the effects of dead space," *IEEE Trans. Electron Dev.*, 46(4), 769–775, 1999.
[152] J. C. Campbell, S. Demiguel, F. Ma et al., "Recent advances in Avalanche photodiodes," *J. Sel. Top. Quantum Electron.*, 10(4), 777–787, 2005.
[153] K. F. Li, S. A. Plimmer, J. R. P. David et al., "Low avalanche noise characteristics in thin InP p-i-n diodes with electron initiated multiplication," *IEEE Photon. Technol. Lett.*, 11, 364–366, 1999.
[154] J. C. Campbell, S. Chandrasekhar, W. T. Tsang et al., "Multiplication noise of wide-bandwidth InP/InGaAsP/InGaAs avalanche photodiodes.," *J. Lightw. Technol.*, 7(3), 473–477, 1989.

[155] P. Yuan, C. C. Hansing, K. A. Anselm et al., "Impact ionization characteristics of III-V semiconductors for a wide range of multiplication region thicknesses," *IEEE J. Quantum Electron.*, 36, 198–204, 2000.

[156] M. A. Saleh, M. M. Hayat, P. O. Sotirelis et al., "Impact-Ionization and noise characteristics of thin III-V Avalanche photodiodes," *IEEE Trans. Electron Dev.*, 48, 2722–2731, 2001.

[157] K. F. Li, D. S. Ong, J. R. P. David et al., "Avalanche noise characteristics of thin GaAs structures with distributed carrier generation," *IEEE Trans. Electron Dev.*, 47(5), 910–914, 2000.

[158] K. F. Li, D. S. Ong, J. R. P. David et al., "Avalanche multiplication noise characteristics in thin GaAs p-i-n diodes," *IEEE Trans. Electron Dev.*, 45(10), 2102–2107, 1998.

[159] C. Hu, K. A. Anselm, B. G. Streetman, and J. C. Campbell, "Noise characteristics of thin multiplication region GaAs avalanche photodiodes," *Appl. Phys. Lett.*, 69(24), 3734–3736, 1996.

[160] S. A. Plimmer, J. R. P. David, D. C. Herbert et al., "Investigation of impact ionization in thin GaAs diodes," *IEEE Trans. Electron Dev.*, 43(7), 1996.

[161] C. Lenox, P. Yuan, H. Nie et al., "Thin multiplication region InAlAs homojunction avalanche photodiodes," *Appl. Phys. Lett.*, 73, 783–784, 1998.

[162] C. H. Tan, J. C. Clark, J. R. P. David et al., "Avalanche noise measurements in thin Si p-i-n diodes," *Appl. Phys. Lett.*, 76(26), 3926–3928, 2000.

[163] C. H. Tan, J. R. P. David, J. Clark et al., "Avalanche Multiplication and Noise in Submicron Si p-i-n Diodes," in Proc. *SPIE, Silicon-based Optoelectronics II*, 3953, 95–102, 2000.

[164] S. A. Plimmer, J. R. P. David, G. J. Rees et al., "Impact ionization in thin $Al_xGa_{1-x}As$ ($x = 015 - 0.30$) p-i-n diodes," *J. Appl. Phys.*, 82(3), 1231–1235, 1997.

[165] B. K. Ng, J. R. P. David, G. J. Rees et al., "Avalanche multiplication and breakdown in $Al_xGa_{1-x}As$ ($x < 0.9$)," *IEEE Trans. Electron Dev.*, 49(12), 2349–2351, 2002.

[166] B. K. Ng, J. R. P. David, R. C. Tozer et al., "Excess noise characteristics of $Al_{0.8}Ga_{0.2}As$ avalanche photodiodes," *IEEE Trans. Elecront Dev.*, 48(10), 2198–2204, 2001.

[167] C. H. Tan, J. R. P. David, S. A. Plimmer et al., "Low multiplication noise thin $Al_{0.6}Ga_{0.4}As$ avalanche photodiodes," *IEEE Trans. Electron Dev.*, 48(7), 1310–1317, 2001.

[168] B. K. Ng, J. R. P. David, R. C. Tozer et al., "Nonlocal effects in thin 4H-SiC UV avalanche photodiodes," *IEEE Trans. Electron Dev.*, 50(8), 1724–1732, 2003.

[169] A. L. Beck, B. Yang, S. Wang et al., "Quasi-direct UV/blue GaP avalanche photodiodes," *IEEE J. Quantum Electron.*, 40(12), 1695–1699, 2004.

[170] C. H. Tan, R. Ghin, J. R. P. David et al., "The effect of dead space on gain and excess noise in $In_{0.48}Ga_{0.52}P$ pin diodes," *Semicon. Science and Technol.*, 18(8), 803–806, 2003.

[171] I. Watanabe, T. Torikai, K. Makita et al., "Impact ionization rates in (100) $Al_{0.48}In_{0.52}As$," *IEEE Electron Dev. Lett.*, 11(10), 437–439, 1990.

[172] N. Li, R. Sidhu, X. Li et al., "InGaAs/InAlAs avalanche photodiode with undepleted absorber," *Appl. Phys. Lett.*, 82, 2175–2177, 2003.

[173] I. Watanabe, M. Tsuji, K. Makita, and K. Taguchi, "A new planar-structure InAlGaAs-InAlAs superlattice avalanche photodiode with a Ti-implanted guard-ring," *IEEE Photon. Technol. Lett.*, 8(6), 827–829, 1996.

[174] I. Watanabe, T. Nakata, M. Tsuji et al., "High-reliability and low-dark-current 10-Gb/s planar superlattice avalanche photodiodes," *IEEE Photon. Technol. Lett.*, 9(12), 1619–1621, 1997.

[175] E. Yagyu, E. Ishimura, M. Nakaji et al., "Simple planar structure for high-performance AlInAs avalanche photodiodes," *IEEE Photon. Technol. Lett.*, 18(1), 76–78, 2006.

[176] E. Yagyu, E. Ishimura, M. Nakaji et al., "Investigation of guardring-free planar AlInAs avalanche photodiodes," *IEEE Photon. Technol. Lett.*, 18(11), 1264–1266, 2006.

[177] P. Yuan, S. Wang, X. Sun et al., "Avalanche photodiodes with an impact-ionization-engineered multiplication region," *IEEE Photon. Technol. Lett.*, 12, 1370–1372, 2000.

[178] O.-H. Kwon, M. M. Hayat, S. Wang et al., "Optimal excess noise reduction in thin heterojunction $Al_{0.6}Ga_{0.4}As$-GaAs avalanche photodiodes," *IEEE J. Quantum Electron.*, 39(10), 1287–1296, 2003.

[179] C. Groves, C. K. Chia, R. C. Tozer et al., "Avalanche noise characteristics of single $Al_xGa_{1-x}As(0.3 < x < 0.6)$–GaAs heterojunction APDs," *IEEE J. Quantum Electron.*, 41(1), 70–75, 2005.

[180] S. Wang, R. Sidhu, X. G. Zheng et al., "Low-noise avalanche photodiodes with graded impact-ionization-engineered multiplication region," *IEEE Photon. Technol. Lett.*, 13, 1346, 2001.

[181] S. Wang, F. Ma, X. Li et al., "Ultra-low noise avalanche photodiodes with a 'centered-well' multiplication region," *IEEE J. Quantum Electron.*, 39, 375–378, 2003.

[182] M. M. Hayat, O.-H. Kwon, S. Wang et al., "Boundary effects on multiplication noise in thin heterostructure avalanche photodiodes: theory and experiment," *IEEE Trans. Electron Dev.*, 49, 2114–2123, 2002.

[183] S. Wang, J. B. Hurst, F. Ma et al., "Low-noise impact-ionization-engineered avalanche photodiodes grown on InP substrates," *IEEE Photon. Technol. Lett.*, 14, 1722–1724, 2002.

[184] N. Duan, S. Wang, F. Ma et al., "High-speed and low-noise SACM avalanche photodiodes with an impact-ionization engineered multiplication region," *IEEE Photon. Technol. Lett.*, 17(8), 1719–1721, 2005.

[185] F. Capasso, W. T. Tsang, A. L. Hutchinson, and G. F. Williams, "Enhancement of electron impact ionization in a superlattice: a new avalanche photodiode with a large ionization rate ratio," *Appl. Phys. Lett.*, 40, 38–40, 1982.

[186] R. Chin, N. Holonyak, Jr, G. E. Stillman et al., "Impact ionization in multilayered heterojunction structures," *Electron. Lett.*, 16, 467–469, 1980.

[187] M. M. Hayat, O.-H. Kwon, S. Wang et al., "Boundary effects on multiplication noise in thin heterostructure avalanche photodiodes," *IEEE Trans. Electron Dev.*, 49, 2114–2123, 2002.

[188] O.-H. Kwon, M. M. Hayat, S. Wang et al., "Optimal excess noise reduction in thin heterojunction $Al_{0.6}Ga_{0.4}As$-GaAs avalanche photodiodes," *IEEE J. Quantum Electron.*, 39(10), 1287–1296, 2003.

[189] C. Cohen-Jonathan, L. Giraudet, A. Bonzo, and J. P. Praseuth, "Waveguide AlInAs/GaAlInAs avalanche photodiode with a gain-bandwidth product over 160 GHz," *Electron. Lett.*, 33(17), 1492–1493, 1997.

[190] T. Nakata, G. Takeuchi, I. Watanabe et al., "10Gbit/s high sensitivity, low-voltage-operation avalanche photodiodes with thin InAlAs multiplication layer and waveguide structure," *Electron. Lett.*, 36(24), 2033–2034, 2000.

[191] T. Nakata, T. Takeuchi, K. Maliita et al., "High-Sensitivity 40-Gbis Receiver with a Wideband InAlAs Waveguide Avalanche Photodiode," in Proc. ECOC '02, Photonic & Wireless Devices Res. Lab., NEC Corp., Ibaraki, Japan, Paper 10.5.1, 2002.

[192] K. Shiba, T. Nakata, T. Takeuchi et al., "High sensitivity asymmetric waveguide APD with over-30 dBm at 10 Gbit/s," *Electron. Lett.*, 42(20), 1177–1178, 2006.

[193] G. S. Kinsey, J. C. Campbell, and A. G. Dentai, "Waveguide avalanche photodiode operating at 1.55 µm with a gain-bandwidth product of 320 GHz," *IEEE Photon. Technol. Lett.*, 13(8), 842–844, 2001.

[194] T. Torikai, T. Nakata, T. Kato, and V. Makita, "40-Gbps Waveguide Avalanche photodiodes," *2005 OFC Technical Digest* (IEEE Cat. No. 05CH37672), Vol. 5, paper OFM3, 2005.

[195] J. Wei, F. Xia, and S. R. Forrest, "A high-responsivity high-bandwidth asymmetric twin-waveguide coupled InGaAs-InP-InAlAs avalanche photodiode," *IEEE Photon. Technol. Lett.*, 14(11), 1590–1592, 2002.

[196] S. Demiguel, X.-G. Zheng, N. Li et al., "High-responsivity and high-speed evanescently-coupled avalanche photodiodes," *Electron. Lett.*, 39(25), 1848–1849, 2003.

[197] W. P. Risk and D. S. Bethune, "Quantum cryptography," *Opt. Photon. News*, 13(7), 26–32, 2002.

[198] S. Cova, M. Ghioni, A. Lotito et al., "Evolution and prospects for single-photon avalanche diodes and quenching circuits," *J. Modern Opt.*, 51(9–10), 1267–1288, 2004.

[199] J. K. Forsyth and J. C. Dries, "Variations in the Photon-Counting Performance of InGaAs/InP Avalanche Photodiodes," in Proc. *IEEE LEOS Annual Conference*, Sensors Unlimited Inc., Princeton, NJ, USA, 2003, p. 777.

[200] K. A. McIntosh, J. P. Donnelly, D. C. Oakley et al., "InGaAsP/InP avalanche photodiodes for photon counting at 1.06 μm," *Appl. Phys. Lett.*, 81(14), 2505–2507, 2002.
[201] M. Itzler, R. Ben-Michael, C.-F. Hsu et al., *Single-Photon Workshop (SPW) 2005: Sources, Detectors, Applications and Measurement Methods*, 24–26 October 2005, Teddington, UK.
[202] M. Liu, X. Bai, C. Hu et al. "Low Dark Count Rate and High Single Photon Detection Efficiency Avalanche Photodiode in Geiger-Mode Operation," in Proc. *64 DRC*, paper II.A-4, Penn State Univ, University Park, PA, June 2006.

9

Planar lightwave circuits in fiber-optic communications

Christopher R. Doerr[*] and Katsunari Okamoto[†]

[*]Alcatel-Lucent, Holmdel, NJ, USA
[†]Okamoto Laboratory, Mito-shi, Ibaraki-ken, Japan

9.1 INTRODUCTION

Non-technical overview: Planar lightwave circuits are optical "chips" that are usually connected to optical fibers and perform various functions such as wavelength filtering, optical switching, and optical channel power control. They are used mostly in fiber-optic communication networks. They are sometimes also called "photonic integrated circuits," "integrated optics," "optical benches," or "waveguides." They can be made of several different materials, the most common being glass waveguides on top of a silicon wafer or InGaAsP waveguides on an InP wafer.

Planar lightwave circuits (PLCs) are a planar arrangement of waveguides on a substrate. PLCs exist in many different material systems, including SiO_2, SiON, polymer, Si, GaAs, InP, and $LiNbO_3$. In this chapter, we mainly focus on the lower refractive index materials, SiO_2, SiON, and polymer, mainly because we can cover only so much in one chapter, and also because another chapter in this volume covers InP PLCs.

Compared to electronic integrated circuits (EICs), photonic integrated circuits (PICs) (another name for PLCs) are simple. Whereas EICs commonly integrate millions of components, today's PICs integrate at most hundreds of components. In fact, the PIC market is still mainly dominated by simple optical couplers/splitters for fiber-to-the-home applications. One of the main reasons is that it is easier to achieve a strong nonlinear response with electrons than with photons. Another reason is that the wavelength of an electron is much smaller than the wavelength of a photon, facilitating much more dense integration of EICs than PICs.

However, whereas electrons win over photons in the particle picture, in the wave picture optics win over electronics. In the wave picture, optics and electronics are not so different, optics simply being electromagnetic waves oscillating at

Optical Fiber Telecommunications V A: Components and Subsystems
Copyright © 2008, Elsevier Inc. All rights reserved.
ISBN: 978-0-12-374171-4

terahertz frequencies and higher, whereas electronics are usually considered as electromagnetic waves oscillating at gigahertz frequencies and lower. Using optics as a carrier, extremely high signal bandwidths can be transported. The optical fiber provides a low-loss medium to do this, and optical amplifiers can re-amplify the signal with nearly ideal noise levels.

This chapter focuses on PLCs used with signal-transporting optical fibers. There are two main ways to achieve high bandwidth on an optical fiber, and PLCs are instrumental in both. One is to increase the number of "channels" on a fiber, each channel being at a different carrier frequency (or equivalently wavelength). This is called wavelength-division multiplexing (WDM), and inter-channel control PLCs play a large role here, serving mainly as channel multiplexers and demultiplexers. The other method to increase bandwidth is to increase the signal rate itself, and intra-channel control PLCs are beginning to play a significant role here, as well, serving, for example, as optical dispersion compensators and general optical equalizers.

9.2 BASIC WAVEGUIDE THEORY AND MATERIALS

9.2.1 Index Contrast

Probably the most important parameter in PLCs is the refractive index difference between the core and the cladding, often referred to as "delta":

$$\Delta = \frac{n_{core}^2 - n_{cladding}^2}{2n_{core}^2} \approx \frac{n_{core} - n_{cladding}}{n_{core}} \tag{9.1}$$

For silica (SiO_2) PLCs with buried waveguides, a typical $\Delta = 0.008$, usually called "0.8%" and can be as high as 0.015 [1]. For SiON PLCs with buried waveguides, Δ can be significantly higher, as much as 0.14 [2]. For semiconductor PLCs, such as InP and Si, Δ can range from typically 0.01 for buried waveguides to as high as 0.45 for air-clad waveguides. The largest possible value of Δ is 0.5.

Δ determines how many modes a certain waveguide size will guide, how tight a curve can be, and how much coupling there will be between neighboring waveguides. A useful number derived from Δ is the V-number. For slab waveguides,

$$V = \frac{\pi n_{core} w}{\lambda_0} \sqrt{2\Delta}, \tag{9.2}$$

where w is the waveguide width and λ_0 is the free-space wavelength. If the waveguide has a finite height, then Δ should be an effective Δ, found using the effective index method. The number of modes guided by the waveguide is $2V/\pi$.

It is commonly accepted that if $\Delta < 0.007$, it is "low delta," if $0.007 < \Delta < 0.009$, it is "medium delta," and if $\Delta > 0.009$ it is "high delta."

9.2.2 Optical Couplers/Splitters

Non-technical overview: An optical splitter splits an optical signal on one waveguide into multiple copies of the signal on different waveguides. An optical coupler (also called a combiner) is an optical splitter operated in reverse.

All couplers are actually $N \times N$ couplers, where N is an integer, because of the conservation of the number of modes in a closed, static system. However, some of the input or output modes might be radiation modes (which must be treated carefully, because radiation modes are a continuum). For example, a 1×2 coupler is really a 2×2 coupler with one "port" being a radiation mode. Also, in optics, most couplers are directional couplers, in that there are distinct inputs and distinct outputs. For example, if light enters one port of a 2×2 optical coupler, it exits at most only two ports of the coupler. However, historically, the term "directional coupler" has been reserved for evanescent couplers only.

Coupler action can be divided into two main categories: power splitting or mode separation (sometimes called mode evolution) [3] (or equivalently, power combining and mode combining). Of course, all couplers are really power splitters (or combiners), but in this terminology, power splitting means that energy that started in one local normal mode of the coupler at the coupler input is distributed among a plurality of local normal modes at some point in the coupler. A good example is the multimode interference (MMI) coupler. In contrast, mode separation means that energy that started in one local normal mode of the coupler at the coupler input always stays in one local normal mode. It originates from the concept of demultiplexing multiple modes in a single wide waveguide into multiple narrow waveguides, one mode in each. A good example is the adiabatic directional coupler.

However, we do not recommend to be strict about this categorization. This is because when waveguides are far apart, it is ambiguous whether the local normal mode is the mode in each individual waveguide or the mode of the entire structure. For example, for two narrow waveguides that are far apart, there are two choices for the set of two local normal modes. One is that the local normal modes are the mode of each individual waveguide; the other is that the local normal modes are the even and odd modes of the two waveguides viewed as a larger structure. Another reason is that, as pointed out in Ref. [4], practical mode-separation couplers are not truly adiabatic in the same sense of the definition as is used in other branches of physics, such as thermodynamics or quantum mechanics. This is because the phase slip between modes in an adiabatic coupler is usually not many π, otherwise the coupler would be far too long.

1×2 Couplers

The two most popular optical 1×2 couplers are the Y-branch coupler and the 1×2 MMI coupler, shown in Figures 9.1 and 9.2. The Y-branch is best for use in low Δ materials, and the MMI is best for use in high Δ materials. This is because the Y-branch requires a very narrow gap, whereas an MMI has much wider gaps.

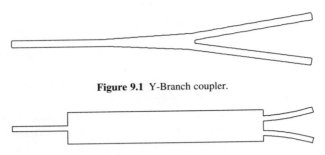

Figure 9.1 Y-Branch coupler.

Figure 9.2 1 × 2 MMI coupler.

Also, the principle of operation for an MMI coupler requires high Δ. However, the MMI waveguide width must be accurate to avoid excess insertion loss. Both have a robust coupling magnitude and phase, the magnitude squared being 50/50 and the relative phase being 0°. This is because of the symmetry in their layouts.

A 1 × 2 coupler is actually a 2 × 2 coupler with one input/output being a radiation mode. For example, if two signals enter the Y-branch coupler from the right side and they are out of phase, then the combined energy radiates away on both sides of the single waveguide, none of the energy ending up in the waveguide.

2 × 2 Couplers

There are many choices for 2 × 2 couplers. A 2 × 2 coupler is much more challenging to make than a 1 × 2 coupler because there is no longer symmetry about the input waveguide. In other words, because the two output waveguides are placed symmetrically about the input waveguide in a 1 × 2 coupler, the splitting magnitude squared is always 50/50, and the splitting phase is always 0. In a 2 × 2 coupler, it must be designed just right to get a 50/50 splitting magnitude squared and/or a desired splitting phase. Wavelength, polarization, and fabrication (WPF) changes can all affect the splitting characteristics. The action of a 2 × 2 coupler is conveniently written as a rotation matrix $R(\theta)$. θ is $\pi/4$ for a 50/50 coupler.

2 × 2 couplers are useful for switches and, as we will see later, for making optical filters with flat-top passbands with low insertion loss.

One type is the "directional coupler," shown in Figure 9.3, also called the evanescent coupler. It consists of two waveguides brought close enough together to couple. It has the lowest insertion loss of any 2 × 2 coupler. If the directional coupler is symmetric about a horizontal line, then the splitting phase is always

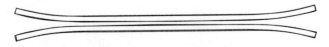

Figure 9.3 Directional coupler.

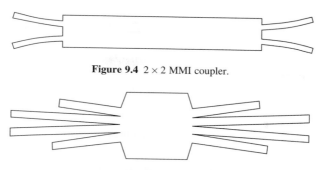

Figure 9.4 2 × 2 MMI coupler.

Figure 9.5 2 × 2 star coupler.

precisely 90°. The splitting magnitude varies with the length of the coupler. The length must be chosen just right to get a 50/50 splitting magnitude squared ratio, making the directional coupler highly sensitive to WPF changes.

Another type is the 2 × 2 MMI coupler (Figure 9.4). The MMI coupler is basically the same as the directional coupler with the gap filled in. Whereas the directional coupler is best used in low Δ waveguides, the MMI coupler is best used in high Δ waveguides. Whereas the directional coupler theoretically has zero excess loss, the MMI coupler has some intrinsic excess loss. The MMI coupler is less sensitive than the directional coupler to WPF changes, but it is still quite sensitive.

Another choice is the 2 × 2 star coupler (Figure 9.5). The star coupler is well known from its use in the arrayed waveguide grating (AWG). The star coupler is relatively robust to WPF changes. However, whereas an $N \times N$ star coupler, where N is large, can have very low loss, a 2 × 2 star coupler has high excess loss, typically at least 1 dB. This is because the waveguides converge and diverge too rapidly for a reasonable adiabatic transition. Nevertheless, the 2 × 2 star coupler is extremely valuable for avoiding polarization crosstalk and other fabrication sensitivities because each waveguide always sees a symmetric environment, unlike all the other 2 × 2 couplers discussed in this section, and is one of the reasons why the AWG is so successful.

The rest of the choices are all variations of the directional coupler, designed so as to mitigate the splitting magnitude ratio sensitivity to WPF changes. There exists a basic uncertainty principle in that as the coupler is desensitized in splitting magnitude, it is sensitized to splitting phase.

The first such coupler we describe is the asymmetric directional coupler. This is simply a directional coupler with one waveguide much wider than the other. In a symmetric directional coupler, as the coupler length increases the splitting magnitude squared goes all the way from 0/100 to 100/0. In an asymmetric directional coupler (Figure 9.6), the coupling never reaches 100/0 because the propagation speed in the two waveguides is different. The coupler is designed such that the maximum power transfer reaches at most 50% before the coupled light becomes out of phase with the incoming coupled light. Thus, the splitting magnitude is

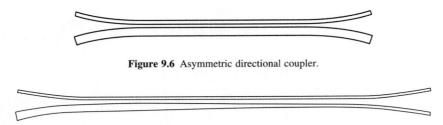

Figure 9.6 Asymmetric directional coupler.

Figure 9.7 Adiabatic directional coupler.

insensitive to wave length changes to first order. However, the splitting ratio is highly sensitive to changes in the waveguide width, and thus, the asymmetric directional coupler is rarely used in practical devices.

An elegant 2×2 coupler is the adiabatic directional coupler (Figure 9.7). This directional coupler simply starts with two waveguides of different widths and transitions gradually to two waveguides of the same width. If the transition is slow enough to not exchange energy between local normal modes, i.e., adiabatic, then the splitting magnitude squared ratio is precisely 50/50. This is because the mode in the wider waveguide on the left becomes the symmetric mode of the two waveguides on the right, and the mode in the narrow waveguide on the left becomes the anti-symmetric mode on the right. The two drawbacks to this coupler are that it is very long and that the splitting phase is highly uncertain in the right-to-left propagation direction (precisely 0 in the left-to-right propagation direction).

Probably the most practical WPF-desensitized directional coupler is the multi-section directional coupler. Designs with two [5], three [6], and four [7] sections have been demonstrated. Figure 9.8 shows a two-section example. One example is a two-section coupler with the first and second sections having a nominal 50/50 and 100/0 splitting magnitude squared ratios, respectively, with a relative path-length difference of 120° between the two sections. For a three-section coupler, a good 50/50 design is to make all three sections have the same length and nominal 50/50, and have 0° phase between the first and the second section and 120° phase between the second and the third sections.

The final WPF-desensitized directional coupler we discuss is the curved directional coupler (Figure 9.9) [8]. This operates on the same principle as the asymmetric directional coupler in that the bending results in the coupled light becoming out of phase before full coupling occurs. A significant advantage over the asymmetric coupler is that both waveguides have the same width, and thus, it is not sensitive to waveguide width

Figure 9.8 Multi-section directional coupler.

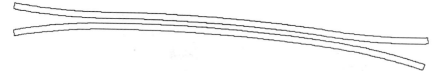

Figure 9.9 Curved directional coupler.

changes like the asymmetric directional coupler is. This coupler is the most compact WPF-desensitized directional coupler. Unfortunately, this coupler appears to exhibit significant polarization crosstalk when implemented in silica waveguides, and research is ongoing to mitigate this.

9.2.3 Polarization Effects

Non-technical overview: The oscillating electric field that makes up the optical signal can be oscillating predominantly either parallel to the PLC surface or perpendicular to it and in the direction of propagation. Which direction the electric field is oscillating determines its polarization. The light in PLCs usually travels faster or slower depending on its polarization due to birefringence. Birefringence can result from the molecular structure of the material in the waveguide, strain in the waveguide, or the waveguide shape.

Because PLCs are planar, they tend to have birefringence. Birefringence means that there are two different refractive indices for the two optical polarizations in the chip. The two polarizations are commonly called transverse electric (TE) and transverse magnetic (TM) polarizations, where TE has an electric field component parallel to the surface of the PLC, and TM has a magnetic field component parallel to the surface of the PLC. Note that eigenmodes of PLC waveguides are only truly TE and TM for slab waveguides. Modes of waveguides with side walls, i.e., most real PLC waveguides, cannot be perfect TE or TM modes and should be called quasi-TE and quasi-TM modes.

Birefringence results in a polarization-dependent wavelength (PDW) shift in PLC optical filters. It is important to realize that PDW shift is a function only of the birefringence. No matter what the channel spacing, bandwidth, etc. of a filter, it will always have the same PDW shift in a given birefringent PLC material.

The main sources of the birefringence are strain and geometry. In silica PLCs, the dominant source of birefringence is strain, whereas in InP PLCs, the dominant source of birefringence is geometry. Suppose we consider an optical filter that has a passband. If we define PDW = passband peak wavelength for TM − passband peak wavelength for TE, then silica typically has a PDW = +0.05 nm, and InP typically has a PDW = −3.0 nm.

The most common way to eliminate PDW in silica is to reduce the strain on the cores by doping the upper cladding glass so that its expansion coefficient matches that of the silicon substrate. The most common way to eliminate PDW in InP is to

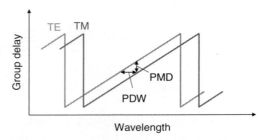

Figure 9.10 Group delay vs wavelength for a typical dispersion compensator, showing how PDW shift results in PMD (this figure may be seen in color on the included CD-ROM).

choose just the right waveguide height and width and layer composition. As the index contrast between core and cladding increases, it can become nearly impossible to achieve a low PDW because of the fabrication sensitivities. In such a case, other schemes, such as polarization diversity where two copies of the filter are made, one to handle each polarization, are employed.

If the optical filter has a nonlinear phase response, such as a dispersion compensator does, then PDW shift results in polarization-mode dispersion (PMD), as shown in Figure 9.10. PMD = |PDW × CD|, where CD is the chromatic dispersion.

Another effect that must be considered in PLCs is polarization crosstalk, also called polarization conversion. One might expect that because PLCs are planar, if a signal is launched as TE or TM into one end of the chip, it will exit in the same polarization. This is often untrue. If the signal ever experiences a condition where the local axis of strain or geometrical symmetry is not parallel or perpendicular to the PLC surface, polarization crosstalk can occur. Polarization crosstalk is especially prevalent in directional couplers [9] but can also be seen in bends [10] and even at the sidewalls of straight high-index-contrast straight waveguides [11].

Polarization crosstalk is problematic when trying to achieve polarization-insensitive destructive interference in an interferometer, such as is needed in variable optical attenuators (VOAs) and Mach–Zehnder delay interferometers (MZDIs) (see Figure 9.11). The PDW shift of an MZDI must typically be smaller than what can be achieved due to open-loop process birefringence control (for silica, 50 pm is possible, but <4 pm is often required). Thus, in the design of Figure 9.11, there is a thin half-wave plate inserted half-way through the MZDI. The waveplate swaps TE and TM half-way through the interferometer, thus achieving a net zero PDW shift for the interferometer, regardless of the waveguide birefringence. However, polarization crosstalk in the directional couplers prevents the half-wave plate from achieving a perfect zero net PDW shift. To minimize the polarization crosstalk, the design of Figure 9.11 uses "clamping," which consists of placing dummy waveguides around all the directional couplers in an effort to symmetrize the stress on the waveguides in the coupler.

However, the clamping solution can only partially equalize the stress on the upper and lower parts of the core. If the upper cladding is so heavily doped so as to make the

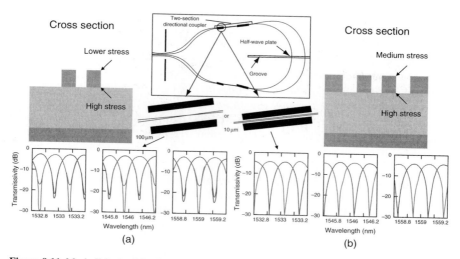

Figure 9.11 Mach–Zehnder delay interferometer made of multi-section directional couplers with coupler "clamping" to symmetrize stress on the coupled waveguides in order to mitigate polarization crosstalk (this figure may be seen in color on the included CD-ROM).

PDW shift zero or even negative (over compensation), we have found experimentally that significant polarization crosstalk in directional couplers occurs despite clamping.

A more robust approach is to use 2×2 star couplers instead of directional couplers (see Figure 9.5). This results in a perfectly symmetric environment for each waveguide, thus eliminating the polarization crosstalk. A Mach–Zehnder delay interferometer that uses star couplers to achieve low polarization crosstalk is described in Ref. [12]. The drawback is increased insertion loss due to the inefficiency of a 2×2 star coupler, and if the star coupler is not carefully designed, the phase difference between the two outputs when launching into one input of the star coupler will not be exactly 90°. The low polarization crosstalk of the star coupler is related to why AWGs usually do not exhibit polarization conversion—all the elements in a typical AWG provide a symmetric environment for all waveguides.

9.3 PASSIVE OPTICAL FILTERING, DEMODULATING, AND DEMULTIPLEXING DEVICES

This section covers passive devices that can filter and/or route wavelengths.

9.3.1 Mach–Zehnder Interferometers

A Mach–Zehnder interferometer (MZI) is an interferometer with two couplers and two arms. If one arm is longer than the other, then the MZI acts as an optical filter or optical demodulator.

Demodulators

Non-technical overview: A phase-shift keyed (PSK) signal has data encoded in the phase of the signal. Because a photodetector detects only optical power, the PSK signal must be converted to an amplitude-shift keyed (ASK) signal at the receiver. This can be done by interfering the signal with one or more delayed copies of itself or with a separate optical signal (i.e., a local oscillator).

Optical demodulators serve to convert PSK data to ASK data. The simplest example is a 1-bit delay, used to demodulate differential PSK (DPSK). It consists of an MZI with one arm longer by one bit length than the other; an example was shown in Figure 9.11.

A more complicated example is a differential quadrature PSK (DQPSK) demodulator. DQPSK consists of four phase levels. It can be demodulated using either two MZIs or one MZI with a 90° hybrid as the second coupler, as shown in Figure 9.12. A silica PLC implementation of the latter option, using a star coupler as the 90° hybrid, is shown in Figure 9.13.

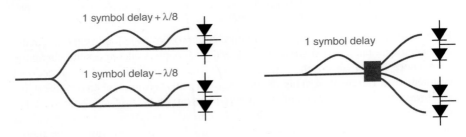

Figure 9.12 Two choices for a DQPSK demodulator (this figure may be seen in color on the included CD-ROM).

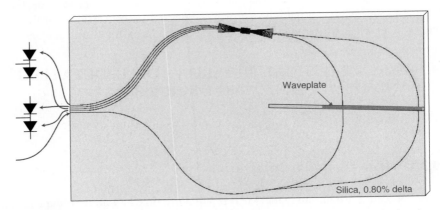

Figure 9.13 Silica PLC implementation of a DQPSK demodulator (this figure may be seen in color on the included CD-ROM).

Interleavers

Interleavers are 1×2 or 2×2 filters that are spectrally periodic. They can be made of concatenated MZIs and are discussed later on in this chapter.

9.3.2 Arrayed Waveguide Gratings

Non-technical overview: An AWG is somewhat analogous to a prism, in that it takes an input white light beam and separates it into many colored output light beams. This feature is called "demultiplexing" and is useful to separate out the signal channels at different wavelengths in an optical fiber. The device can also work in reverse, i.e., multiplex different colored light beams into one beam and is called "multiplexing." However, the prism analogy is imperfect, and AWGs can do much more, such as provide $N \times N$ wavelength routing and "colorless" routing.

The arrayed waveguide grating (AWG) is a generalized MZI. Other names for AWGs are PHASARs (phased arrays) and waveguide grating routers (WGRs). The basic AWG is shown in Figure 9.14. It consists of two $N \times N$ star couplers connected by an array of waveguides of linearly increasing path length.

An $N \times N$ AWG multiplexer is very attractive in optical WDM networks since it is capable of increasing the aggregate transmission capacity of a single strand optical fiber in a compact device that is relatively simple to manufacture [13–15]. The AWG consists of input/output waveguides, two focusing slab regions, and a phased array of multiple channel waveguides with a constant path-length difference ΔL between neighboring waveguides (Figure 9.14). In the first slab region, the input waveguide separation is D_1, the array waveguide separation is d_1, and the radius of curvature is f_1.

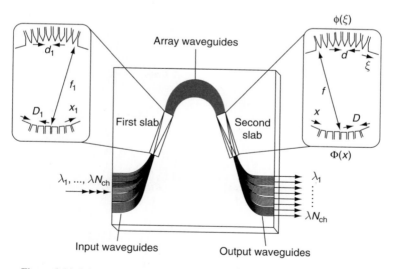

Figure 9.14 Schematic configuration of arrayed waveguide grating multiplexer.

Generally, the waveguide parameters in the first and the second slab regions may be different. Therefore, in the second slab region the output waveguide separation is D, the array waveguide separation is d, and the radius of curvature is f, respectively. The input light at the position of x_1 (x_1 is measured in a counter-clockwise direction from the center of input waveguides) is radiated to the first slab and then excites the arrayed waveguides. The excited electric field amplitude in each array waveguide is a_i ($i = 1 - N$) where N is the total number of array waveguides. The amplitude profile a_i is usually a Gaussian distribution. After traveling through the arrayed waveguides, the light beams constructively interfere into one focal point x (x is measured in a counterclockwise direction from the center of the output waveguides) in the second slab. The location of this focal point depends on the signal wavelength because the relative phase delay in each waveguide is given by $2\pi \Delta L / \lambda$. Figure 9.15 shows an

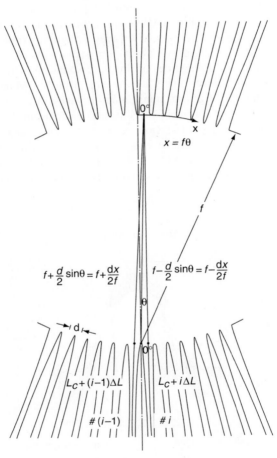

Figure 9.15 Enlarged view of the second slab region (rotated 180° from Figure 9.14).

enlarged view of the second slab region. Let us consider the phase retardations for the two light beams passing through the $(i-1)$-th and i-th array waveguides. The geometrical distances of two beams in the second slab region are approximated as shown in Figure 9.15. We have similar configurations in the first slab region as those in Figure 9.15. The difference of the total phase retardations for the two light beams passing through the $(i-1)$-th and i-th array waveguides must be an integer multiple of 2π in order that two beams constructively interfere at the focal point x. Therefore, we have the interference condition expressed by

$$\beta_s(\lambda_0)\left(f_1 - \frac{d_1 x_1}{2f_1}\right) + \beta_c(\lambda_0)[L_c + (i-1)\Delta L] + \beta_s(\lambda_0)\left(f + \frac{dx}{2f}\right)$$
$$= \beta_s(\lambda_0)\left(f_1 + \frac{d_1 x_1}{2f_1}\right) + \beta_c(\lambda_0)[L_c + i\Delta L] + \beta_s(\lambda_0)\left(f - \frac{dx}{2f}\right) - 2m\pi, \quad (9.3)$$

where β_s and β_c denote the propagation constants in slab region and array waveguide, m is an integer, λ_0 is the center wavelength of WDM system, and L_c is the minimum array waveguide length. Subtracting common terms from Eqn (9.3), we obtain

$$\beta_s(\lambda_0)\frac{d_1 x_1}{f_1} - \beta_s(\lambda_0)\frac{dx}{f} + \beta_c(\lambda_0)\Delta L = 2m\pi. \quad (9.4)$$

When the condition $\beta_c(\lambda_0)\Delta L = 2m\pi$ or

$$\lambda_0 = \frac{n_c \Delta L}{m} \quad (9.5)$$

is satisfied for λ_0, the light input position x_1 and the output position x should satisfy the condition

$$\frac{d_1 x_1}{f_1} = \frac{dx}{f}. \quad (9.6)$$

In Eqn (9.5), n_c is the effective phase index of the array waveguide ($2\pi n_c = \beta_c/k$, k: wavenumber in vacuum) and m is called the diffraction order. The above equation means that when light is coupled into the input position x_1, the output position x is determined by Eqn (9.6). Usually, the waveguide parameters in the first and the second slab regions are the same ($d_1 = d$ and $f_1 = f$). Therefore, input and output distances are equal as $x_1 = x$. The spatial dispersion of the focal position x with respect to the wavelength λ for the fixed light input position x_1 is given by differentiating Eqn (9.4) with respect to λ as

$$\frac{\Delta x}{\Delta \lambda} \approx -\frac{N_c f \Delta L}{n_s d \lambda_0}, \quad (9.7)$$

where n_s is the effective index in the slab region, and N_c is the effective *group* index of the effective index n_c of the array waveguide ($N_c = n_c - \lambda dn_c/d\lambda$). The spatial dispersion of the input-side position x_1 with respect to the wavelength λ for the fixed light output position x is given by

$$\frac{\Delta x_1}{\Delta \lambda} \approx \frac{N_c f_1 \Delta L}{n_s d_1 \lambda_0}. \tag{9.8}$$

The input and output waveguide separations are $|\Delta x_1| = D_1$ and $|\Delta x| = D$, respectively, when $\Delta \lambda$ is the channel spacing of the WDM signal. Putting these relations into Eqns (9.5) and (9.6), the wavelength spacing on the output side for a fixed light input position x_1 is given by

$$\Delta \lambda_{\text{out}} = \frac{n_s \, dD \lambda_0}{N_c f \Delta L}, \tag{9.9}$$

and the wavelength spacing on the input side for a fixed light output position x is given by

$$\Delta \lambda_{\text{in}} = \frac{n_s \, d_1 D_1 \lambda_0}{N_c f_1 \Delta L}. \tag{9.10}$$

Generally, the waveguide parameters in the first and the second slab regions are the same; they are $D_1 = D$, $d_1 = d$, and $f_1 = f$. Then the channel spacings are the same as $\Delta \lambda_{\text{in}} = \Delta \lambda_{\text{out}} \equiv \Delta \lambda$. The path-length difference ΔL is obtained from Eqns (9.9) or (9.10) as

$$\Delta L = \frac{n_s \, dD \lambda_0}{N_c f \Delta \lambda}. \tag{9.11}$$

The spatial separation of the mth and $(m+1)$th focused beams for the same wavelength is given from Eqn (9.4) as

$$X_{\text{FSR}} = x_m - x_{m+1} = \frac{\lambda_0 f}{n_s d}. \tag{9.12}$$

X_{FSR} represents the free *spatial* range of AWG, which is also called the Brillouin zone width of the star coupler. The number of available wavelength channels N_{ch} is given by dividing X_{FSR} with the output waveguide separation D as

$$N_{\text{ch}} = \frac{X_{\text{FSR}}}{D} = \frac{\lambda_0 f}{n_s dD}. \tag{9.13}$$

Figure 9.16(a) and (b) shows a beam propagation method (BPM) simulation of the light-focusing property in the second slab region for the (a) central wavelength λ_0 and (b) a shorter wavelength component $\lambda < \lambda_0$. For the signal component which converges into the off-center output port as in Figure 9.16(b), higher or lower order diffraction beams appear. Because one of the two diffraction beams in

9. Planar Lightwave Circuits in Fiber-Optic Communications

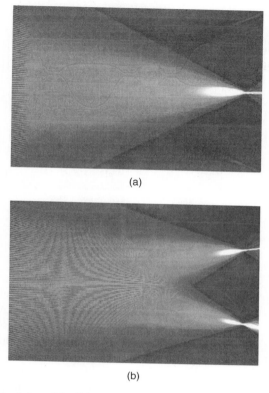

Figure 9.16 BPM simulation of the light-focusing property in the second slab region (rotated −90° from Figure 9.1) for the (a) central wavelength λ_0 and (b) shorter wavelength component $\lambda < \lambda_0$.

Figure 9.16(b) is usually thrown away, the insertion loss for the peripheral output port becomes $2 \sim 3$ dB higher than that for the central output port.

For further reading on AWG fundamentals see Refs [16–19].

Gaussian Spectral Response AWG

Non-technical overview: AWGs are optical filters with passbands that pass a certain spectral portion and block others. A "Gaussian AWG" means that the passband shape has a Gaussian shape. A Gaussian shape means that the passband transmission is of the form $\exp\{-[(f-f_0)/f_{bw}]^2\}$, where f is frequency, f_0 is the center frequency, and f_{bw} is the bandwidth. A real AWG passband is not precisely Gaussian-shaped and instead has sidelobes. Many applications do not favor Gaussian-shaped passbands because if a signal must pass through many such passbands in series, such as in a long optical link with many add–drop nodes, the net passband gets narrower and narrower, eventually distorting the signal.

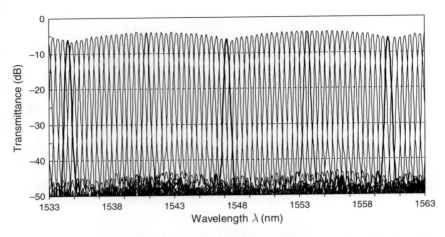

Figure 9.17 Demultiplexing properties of 32-ch 50-GHz spacing AWG over four diffraction orders.

A "Gaussian" AWG denotes that the frequency response has a Gaussian shape. Various kinds of multiplexers including a 50-nm-spacing 8-channel AWG and a 25-GHz spacing 400-channel AWG have been fabricated [20–22].

Figure 9.17 shows the measured transmission spectra of a 32-channel 50-GHz-spacing AWG over four diffraction orders. As explained by the BPM simulations in Figure 9.16(a) and (b), insertion losses for the peripheral output ports are $2.5 \sim 3$ dB higher than those for central output ports. In order to obtain uniform loss characteristics in WDM system applications, the total number of channels N_{ch} of AWG should be significantly larger than the channel number N_{system} of the system. The number of output ports N_{system} is usually $0.5 \sim 0.6 \times N_{ch}$ to guarantee a loss variation of less than 1 dB.

Figure 9.18 shows the demultiplexing properties of a 400-channel, 25-GHz-spacing AWG. Crosstalk levels of about −30 dB have been achieved. Super-high-Δ waveguides with a refractive-index core-cladding contrast of 1.5% were used in the 400-channel 25-GHz AWG so as to minimize the area of the array waveguide region. Figure 9.19 shows the demultiplexing properties of a 32-channel AWG with a very narrow channel spacing of 10 GHz without using post processing for the phase-error (non-uniformity in the optical path-length difference $N_c \Delta L$) compensation. Here, a high-Δ waveguide with refractive-index difference of 0.75% was used. Adjacent channel crosstalk of approximately −30 dB has been achieved even in the 10-GHz spacing AWG. Postfabrication phase-error compensation techniques can further improve the crosstalk of AWGs.

AWG Crosstalk The crosstalk of an AWG is mainly attributed to the phase (optical path length) fluctuation in the arrayed waveguides [17]. If there were no phase (or amplitude) errors, we could obtain about a −60-dB level in the example

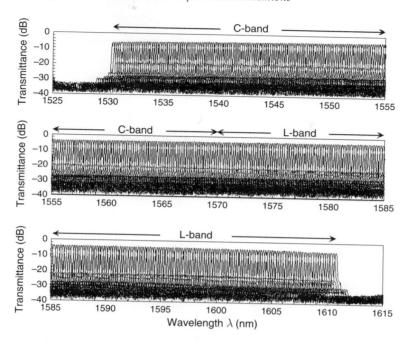

Figure 9.18 Demultiplexing properties of 400-ch 25-GHz-spacing AWG.

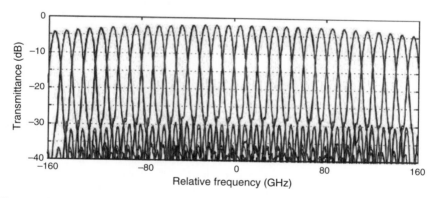

Figure 9.19 Demultiplexing properties of 32-ch 10-GHz-spacing AWG (this figure may be seen in color on the included CD-ROM).

with $\delta n = 0$ in Figure 9.20. The reason why we cannot achieve such a low crosstalk level in reality is that the envelope of the electric field at the second slab interface does not have a smooth Gaussian shape, because of phase errors in the AWG arms. Phase errors are caused by effective-index non-uniformities in the arrayed waveguide region; that is, refractive index fluctuations and core width and thickness non-uniformities. Excitation of higher order waveguide modes in the grating arms

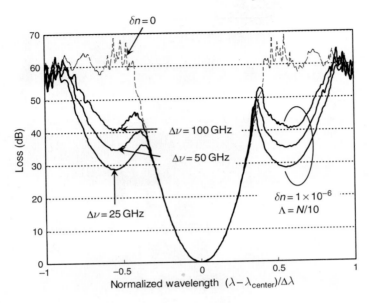

Figure 9.20 Theoretical demultiplexing properties of AWGs having sinusoidal phase fluctuation of $\delta n = 1 \times 10^{-6}$ with $\Lambda = N/10$ for three kinds of AWGs with 100-, 50-, and 25-GHz channel spacing.

can also cause phase errors. Figure 9.20 shows the demultiplexing properties of AWGs having sinusoidal phase fluctuations of $\delta n = 1 \times 10^{-6}$ with $\Lambda = N/10$ for three kinds of AWGs with 100-, 50-, and 25-GHz channel spacing, respectively. The spatial sinusoidal phase fluctuation is expressed by

$$\tilde{n}_c(\ell) = n_c(\ell) + \delta n(s) \cdot \sin\left[\ell \frac{\pi}{\Lambda(s)}\right] \qquad (9.14)$$

where ℓ is an integer from 1 to N, and s denotes array number and order of spatial frequency in fluctuation, $n_c(\ell)$ is the ideal effective-index distribution, δn is an amplitude of fluctuation, and Λ is a half period of index fluctuation, respectively. It is known empirically that crosstalk degrades by $\sim$5 dB when the channel spacing is cut in half. It is also known that crosstalk degrades by $\sim$10 dB when δn becomes about 3.2 ($\sqrt{10}$) times larger.

AWG Dispersion Phase errors cause not only crosstalk but also chromatic dispersion in the AWG [23]. Figure 9.21 shows theoretical dispersion properties for 64-channel 100-GHz spacing to 64-channel 25-GHz spacing AWGs having spatially sinusoidal phase fluctuations given by Eqn (9.14). It is shown that a spatially low-frequency phase fluctuation (small s) causes dispersion inside the passband of AWG. Here, dispersion values are fixed to be -15 ps/nm in Figure 9.21(a)–(c) and

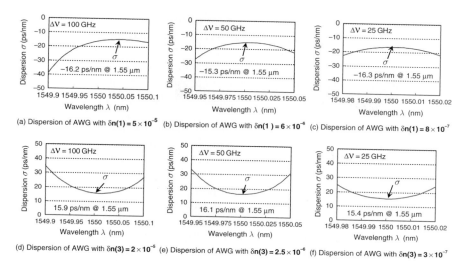

Figure 9.21 Theoretical dispersion characteristics of three kinds of AWGs with 100-, 50-, and 25-GHz channel spacing for various kinds of index fluctuation δn (this figure may be seen in color on the included CD-ROM).

+15 ps/nm in Figure 9.21(d)–(f), respectively. Then the index fluctuation δn that generates the specified dispersion is calculated. It is known empirically from several calculations including Figure 9.21 that the dependence of dispersion σ on index fluctuation δn and channel spacing $\Delta \nu$ is approximately obtained as $\sigma \propto \delta n / \Delta \nu^3$. Because σ is inversely proportional to $\Delta \nu^3$, the dispersion of an AWG with a narrow channel spacing is highly sensitive to phase errors.

Flat Spectral-Response AWGs

Non-technical overview: When multiplexing WDM channels together or demultiplexing them apart, one usually wants the optical passbands of the (de)multiplexer to have a magnitude vs wavelength that looks like a rectangle and a phase vs wavelength that is a straight line. This puts no distortion on a signal that is completely within the rectangle and filters out all noise outside of the rectangle. Such passbands are usually called "flat" passbands, and (de)multiplexers with such passbands are often called "wide-band" (de)multiplexers.

Because the rate of displacement of the focal position x with respect to the wavelength λ is constant [Eqn (9.7)], the transmission loss of a Gaussian AWG monotonically increases around the center wavelength of each channel. This places tight restrictions on the wavelength tolerance of laser diodes and requires accurate temperature control for both the AWGs and the laser diodes. Moreover, because optical signals may be transmitted through several filters in the WDM ring/bus networks, the cumulative passband width of each channel becomes much narrower than that of a single-stage AWG filter. One solution is to make the

Gaussian AWG passband very wide compared to the channel spacing, as discussed in Section "Gaussian Spectral Response AWG." This may be acceptable when the AWG is used as a multiplexer. However, it is not acceptable when the AWG is used as a demultiplexer, because of high crosstalk due to overlapping adjacent passbands. Therefore, flattened and broadened spectral responses with steep spectral skirts (i.e., a rectangular passband) are often required for AWG demultiplexers.

Several approaches have been proposed to flatten the passbands of AWGs [24–29]. One is to create a flat electric field distribution at the input waveguide. Because the AWG is an imaging device, the flat electric field is reproduced at the output plane. The overlap integral of a flat field with a Gaussian local normal mode gives a flat spectral response. Parabolic waveguide horns [25] or 1 × 2 MMI couplers [26] are used to create a flat electric field distribution at the input waveguide. The second method is to engineer array waveguide design to create a flattened electric field at the output plane. In this case, the input field is a normal Gaussian distribution. There are mainly two kinds of methods. One is to make a sinc-like electric field envelope in the array waveguides [24]. Because the focused electric field profile is a Fourier transformed image of the electric field in the array waveguides, we can generate a flattened field distribution at the output plane. The other is to make two focal spots at the output plane [27]. Here, the light-focusing direction of the array waveguide is alternately changed to two separate focal positions. In this case, the flattened field profile consists of two Gaussian beams.

All the above techniques pay a penalty in insertion loss by flattening the passband. There are some other flattening techniques employing concatenated interferometers that theoretically do not pay any loss penalty [28, 29]. In these techniques, the input position of the beam moves synchronously with the wavelength change of the signal, and they are discussed in Section "AWGs with Periodic Stationary Imaging." Then the output beam lies in a fixed position independent of wavelength within one channel span.

One way to create a flat spectral-response AWG is to place a non-adiabatic parabolic horn taper on the input waveguide at the first star coupler. The width of the parabolic horn along the propagation direction z is given by [30]

$$W(z) = \sqrt{\frac{2\alpha\lambda}{n_c} z + D^2}$$

where α is a constant less than unity and D is the core width of the channel waveguide. At the proper horn length $z = \ell$ less than the collimator length, a slightly double-peaked intensity distribution can be obtained as shown in Figure 9.22. A broadened and sharp-falling optical intensity profile is obtainable by the parabolic waveguide horn, which is quite advantageous for achieving a wide passband without deteriorating the nearest neighbor crosstalk characteristics. The broadened and double-peaked field is imaged onto the entrance of an output waveguide having a normal core width. The overlap integral of the focused field

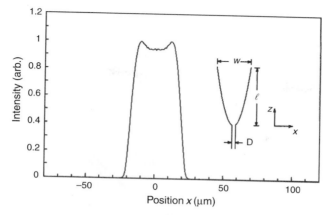

Figure 9.22 Intensity profile calculated by the beam propagation method in the parabolic input waveguide. Inset shows the schematic configuration of parabolic waveguide.

with the local normal mode of the output waveguide gives a flattened spectral response for the AWG. Figure 9.23 shows the demultiplexing properties of a 32-channel 100-GHz-spacing AWG having parabolic horns with $W = 26.1\,\mu\text{m}$ and $\ell = 270\,\mu\text{m}$.

In a parabola-type flat AWG, the double-peaked electric field distribution is created by the interference of the fundamental mode and the second order mode. Generally, phase retardations of the fundamental mode and second order mode are different. Therefore, the total phase at the end of the parabolic waveguide horn is not a uniform phase distribution. A non-uniform phase distribution in the parabola

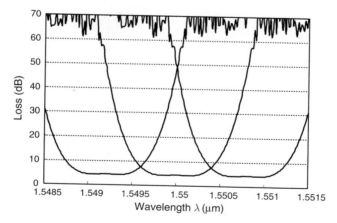

Figure 9.23 Demultiplexing properties of 32-ch 100-GHz-spacing AWG having parabolic horns with $W = 26.1\,\mu\text{m}$ and $\ell = 270\,\mu\text{m}$.

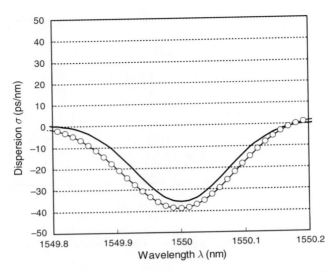

Figure 9.24 (a) Theoretical and (b) experimental dispersion characteristics of 32-ch 100-GHz parabola-type flat AWG.

input waveguide causes non-uniformity of phase in the array waveguides. It is known that the non-uniform phase is similar to the effective-index fluctuations in the array waveguides and thus can cause chromatic dispersion as described in the previous section. Figure 9.24 shows theoretical and experimental dispersion characteristics of a 32-channel 100-GHz parabola-type flat AWG.

Phase retardations between the fundamental and second order modes at the end of the parabolic waveguide horn can be adjusted by adding a straight multimode waveguide as shown in Figure 9.25 [31]. At the proper multimode waveguide length, phase retardations of the fundamental and second order modes are almost equalized. The total phase distribution becomes uniform while the double-peaked electric field distribution is maintained. Theoretical and experimental dispersion characteristics of 32-ch 100-GHz parabola-type flat AWG with $W = 26.1$ μm, $\ell = 270$ μm, and $L_{multi} = 85$ μm are shown in Figure 9.26. Chromatic dispersion has been reduced to a negligible value.

AWGs with Periodic Stationary Imaging *Non-technical overview: This section covers the concatenation of two or more interferometers, such as an MZI connected to an AWG or two AWGs connected together. By connecting multiple interferometers together, one can achieve a flat-top passband theoretically without introducing any extra loss.*

Achieving Multiple Zero-Loss Maxima How rectangular a passband is can be quantified by the number of maxima in the passband. For instance, a standard AWG has a Gaussian-shaped passband, which has one maximum in the passband.

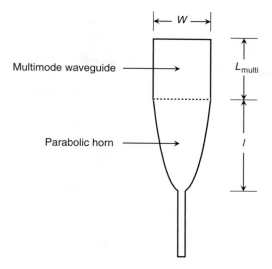

Figure 9.25 Parabolic waveguide having a straight multimode waveguide.

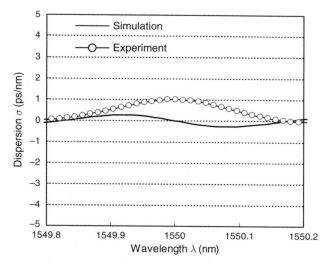

Figure 9.26 Theoretical and experimental dispersion characteristics of 32-ch 100-GHz parabola-type flat AWG with $W = 26.1\,\mu\text{m}$, $\ell = 270\,\mu\text{m}$ and $L_{\text{multi}} = 85\,\mu\text{m}$, respectively.

Such a passband is far from being rectangular. However, if we have two maxima in the passband, then the passband starts to look rectangular. The maxima in the passband can be at different frequencies or at the same frequency. When many of the same optical filter are placed in series, the net cascaded passband is simply a stretching of the scale of the passband, if it is plotted in dBs (in actuality, small variation from filter to filter complicates this). For example, the 0.3-dB down point on the passband becomes the 3-dB down point after 10 cascades. The passband shape is stretched downward around the maxima. Thus, the more maxima (provided the maxima are at the same magnitude), the more the passband is "held open."

As shown in Section "Parabola-Type AWG," one way to add more maxima to an AWG passband is image mismatching, i.e., the image from the input waveguide does not match the image that an output waveguide would make. Such a method is compact and robust. However, image mismatching adds optical loss. This section addresses adding maxima without adding loss.

Any optical filter can be represented in the time domain by its impulse response, $u(t)$, and in the frequency domain by the Fourier transform of its impulse response, $\tilde{u}(f)$ (f is now frequency rather than star coupler length). For the discussion here, we will limit ourselves to directional optical filters that are periodic, with a free-spectral range (FSR) equal to Δf. Actually, any optical filter that is based on interference of different paths has a finite FSR. The time-domain impulses of such a filter are spaced by $1/\Delta f$.

$$u(t) = \sum_{k=1}^{K} c_k \delta\left(t - \frac{k}{\Delta f}\right) \tag{9.16}$$

$$\tilde{u}(f) = \sum_{k=1}^{K} c_k \exp\left(j2\pi k \frac{f}{\Delta f}\right) \tag{9.17}$$

where K is the number of impulses in the impulse response. If a filter has one zero-loss maximum per FSR, then the sum of the magnitudes of the impulses (in optical amplitude, not power) is equal to unity. Each impulse rotates in phase as the optical carrier frequency is changed, and they are all in phase once per FSR.

$$\sum_{k=1}^{K} |c_k| = 1 \quad \text{for one zero-loss maximum per FSR.} \tag{9.18}$$

If a filter has more than one zero-loss maximum per FSR, then this sum must be greater than unity. Otherwise, all the impulses must be in phase to achieve unity, and this can happen only once per FSR.

$$\sum_{k=1}^{K} |c_k| > 1 \quad \text{for multiple zero-loss maxima per FSR.} \tag{9.19}$$

Consider an arbitrary optical filter structure consisting of couplers and path lengths. If it has multiple zero-loss maxima per FSR, then the sum of the impulse magnitudes must be greater than unity, as we just discussed. Now suppose we adjust the phases of all the various path lengths in any manner we choose. It must be impossible for us to put all the impulses in phase at any frequency, or else, we will violate the conservation of energy. Consider a case in which all the couplers in this optical filter structure are only $1 \times N$ and $N \times 1$ couplers, where N is any integer and can vary from coupler to coupler. If so, then at all the gathering couplers (i.e., all the $N \times 1$ couplers), we can adjust the phase of the input waveguides to these couplers such that the impulses exiting the coupler are all in phase. If the input to a gathering coupler is a sequence of impulses generated by other couplers, then all the impulses in that sequence have been put in phase by the preceding couplers. Subsequently, all the impulses exiting the filter can be adjusted to be in phase. Thus, an optical filter containing only $1 \times N$ and $N \times 1$ couplers cannot have more than one zero-loss maxima per FSR. Thus, we have proven the following theorem:

Flat-Top Filter Construction Theorem: a filter comprising couplers and connecting paths can have more than one zero-loss maximum per FSR only if at least one coupler is an effective $M \times N$ coupler, with M and $N > 1$.

By effective, we mean that coupler ports that are not used in a connection between a specified input and a specified output of the filter are disregarded. For example, for an MZI consisting of a 1×2 and a 2×2 coupler, the 2×2 coupler is effectively a 2×1 coupler, because we look at only one output port of the 2×2 coupler at a time. Also, if a coupler is connected to another one with equal path lengths, then those two couplers can be viewed as a single effective coupler. For example, if a 1×2 coupler is connected with equal path lengths to a $2 \times N$ coupler, then that structure can be viewed as a $1 \times N$ coupler.

Figure 9.27(a) and (b) shows examples of filters that can have at most one zero-loss maxima per FSR, and Figure 9.27(c) and (d) shows some that *may* have multiple zero-loss maxima per FSR. One will notice that Figure 9.27(a) is a conventional AWG.

A practical example of a filter with multiple zero-loss maxima per FSR is the Fourier filter interleaver, shown in Figure 9.28. This filter uses effective 2×2 couplers. One stage, shown as Circuit 1, has a nonlinear phase response, though. This can be overcome by cascading two identical filters in series, in which one connects one port of the first filter to the complementary port of the second filter [32]. This works because in a 2×2 filter, the amplitude transmissivity from one input port to one output port is equal to the complex conjugate of the amplitude transmissivity between the other input port to the other output port. The next section shows a multiple zero-loss maxima filter that has a linear phase response without needing two stages.

AWG with Two-Arm-Interferometer Input Based on the theorem just proven, we realize that a conventional AWG can have only one zero-loss maximum per FSR. In this Section, we create an AWG with two zero-loss maxima per FSR. To do this, we will have to connect the device input to more than one input of

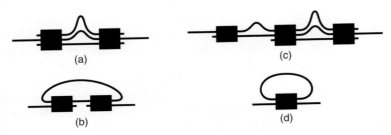

Figure 9.27 Example of optical filter designs. (a) and (b) cannot have multiple zero-loss maxima per FSR, because they contain only effective $1 \times N$ or $N \times 1$ couplers. (c) and (d) may have multiple zero-loss maxima [(d) actually has an infinite number of zero-loss maxima], because (c) contains an effective 2×3 coupler, and (d) contains an effective 2×2 coupler.

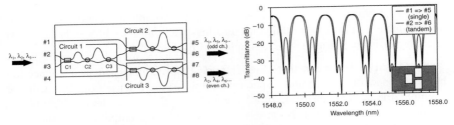

Figure 9.28 Fourier filter-based interleaver and concept for achieving zero dispersion with 2×2 filters. As one can see, the port-pair combination in the left stage is complementary to the port-pair combination in the right stage.

the first AWG star coupler (or equivalently connect the device output to more than one output from the second AWG star coupler) with different path lengths in the connections.

To figure out what to connect between the device input and the two star coupler inputs note that the problem with a conventional AWG is that the image in the second star coupler is always moving as wavelength increases. To create two zero-loss maxima per FSR, we need to have the image focus on the output waveguide in the second star coupler twice per FSR. To do this, we need to move the image in the first star coupler as wavelength changes. We can accomplish this by having a length-imbalanced MZI connected to the two inputs of the first star coupler, where the light moves from one input to the second as wavelength changes such that both images focus on the same output waveguide [29, 33]. If we then match the FSR of the MZI to the channel spacing of the AWG, then one MZI can create dual zero-loss maxima for all channels.

Figure 9.29 shows a plot of the optical power from the MZI at the MZI—star coupler junction as the wavelength changes. The light switches from one output to the other. In between these states, the light is in both outputs simultaneously. One can view the "center of mass" of the optical distribution as moving continuously

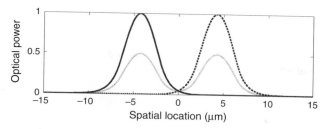

Figure 9.29 Change in optical field at end of a length-imbalanced MZI with a directional 2 × 2 coupler as the optical frequency is changed.

between the two states. Because this is imaged by the AWG onto the output waveguide, one can view the MZI as making the center of mass of the image stationary on the output waveguide. This is what creates the desired low-loss flat-top passband.

Two possible ways of laying out the MZI–AWG are shown in Figure 9.30. Figure 9.30(a) shows the MZI and AWG having the same sign of "curvature." This is necessary for the synchronization of the MZI and AWG at their interconnection point. Figure 9.30(b) shows a switching of the sign of curvature. This is possible by having a sort of waveguide crossing in the MZI, effected by using a 100/0 coupler. It is found that a relative phase shift of 180° between the waveguides between the 100/0 coupler and the 50/50 coupler minimizes the performance sensitivity to wavelength, fabrication, and polarization changes.

Figure 9.31 shows an actual waveguide layout and the resulting measured transmissivity. Theoretically, the chromatic dispersion in the passband is zero.

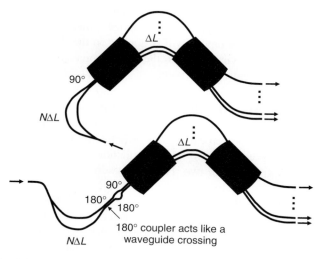

Figure 9.30 Possible layouts for a rectangular passband demultiplexer consisting of an MZI directly connected to an AWG (this figure may be seen in color on the included CD-ROM).

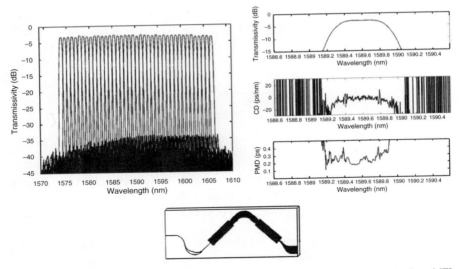

Figure 9.31 Waveguide layout and measured results from a demultiplexer consisting of an MZI directly connected to an AWG.

However, if the 50/50 coupler has a splitting ratio error, then the device will exhibit dispersion. This can be understood by following a signal passing from right to left through the device. Suppose the 50/50 coupler is too weak. Then, when the signal from the AWG is imaged onto the upper waveguide of the 50/50 coupler, which is the short-wavelength side of the passband, then instead of the signal being divided evenly between the two MZI arms, more light will be coupled to the longer MZI arm. The effective path length through an interferometer is equal to the average of the path lengths weighted by the optical power in each arm. Thus, the total path length through the device will be longer on the shorter wavelength side of the passband. It will be vice versa for the longer wavelength portion. Thus, the passband will exhibit negative dispersion.

AWG with Three-Arm Interferometer Input We just showed how to create a passband with two zero-loss maxima by connecting a two-arm interferometer to an AWG. One can extend this principle by using a larger number of arms in the input interferometer. However, some challenges arise once the number of arms is greater than two. Here, we address them for an arm number of three [34, 35].

The main challenge with going from two to three arms is the directional coupler that connects to the AWG. As we will show, a three-arm directional coupler is not trivial, and it has an undesirable phase relationship between the outputs. One might think why not turn the input interferometer into a small AWG and use a star coupler? However, a star coupler with a low port count (approximately <8) has a very low efficiency, undoing our intent of making a low-loss flat-top passband.

9. Planar Lightwave Circuits in Fiber-Optic Communications

A star coupler with a low port count has a low efficiency because the waveguides converge too quickly to convert the plane waves of the free-space region to only first-order Bloch modes of the waveguide array.

The problem with the three-waveguide directional coupler is that the difference of the phase between adjacent output waveguides from the coupler when illuminating one input waveguide of the coupler is always an integer multiple of 90°, whereas we would like it to be an integer multiple of 120°. We would like it to be 120° because this is the phase difference encountered in an ideal star coupler, and it is a star coupler that creates a linearly moving image. The phase between port m_1 on one side and m_2 on the other side of an ideal $M \times M$ star coupler is

$$\phi = \frac{2\pi}{M}\left(m_1 - \frac{M+1}{2}\right)\left(m_2 - \frac{M+1}{2}\right) \qquad (9.20)$$

For $M = 3$, one can see that this gives either 0° or 120° phase difference. The 90° phase difference encountered in the directional coupler means that the three passbands per FSR from the input interferometer are unequally spaced in frequency, and the resulting interferometer-plus-AWG device exhibits mostly a two zero-loss maxima passband. However, using a multi-section three-waveguide directional coupler, one can achieve the desired phase difference and achieve a three zero-loss maxima passband. The design and results are shown in Figure 9.32.

If we wish to use more than three arms and not use a star coupler for the second coupler of the input interferometer, then we have a difficult problem with the design of the input interferometer. As shown in Ref. [36], if one makes a demultiplexer using an MMI coupler (and likewise with a directional coupler), the arm lengths no longer follow a linear progression in path length when the arm number is greater than three.

AWGs with High Spectral Sampling

Non-technical overview: An AWG demultiplexer has a passband for each output port, the passband shifted in wavelength from port to port. If the passbands are very narrow compared with their spacing, then the spectral sampling of the AWG is low. If the passbands are very wide compared with their spacing (they can even be wider than the spacing in some designs), then the spectral sampling is high. If the passbands are about the same width as their spacing, then the spectral sampling is about 1.0.

Suppose we have a conventional AWG with one input and N outputs, and we launch a signal into the input. The N outputs take discrete "samples" of the spectrum of the input signal, spaced in spectrum by a certain amount and "filtered" by a

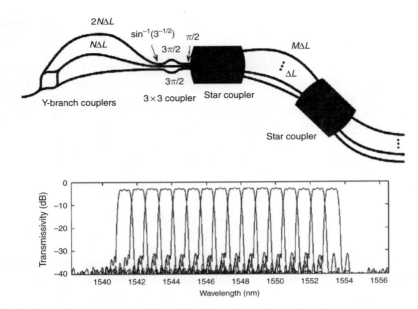

Figure 9.32 16-Channel demultiplexer consisting of a three-arm interferometer directly connected to an AWG. (a) schematic and (b) measured passbands.

certain "bandwidth" [37]. This is analogous to sampling in the time domain, in which a waveform is sampled by discrete-time sample points spaced in time by a certain amount and filtered by a certain "bandwidth."

Sampling Theorem For the time-domain case, if a spectrum is zero for frequency $|f| > f_B/2$, and the signal is sampled in time by spacing T, then we can define a sampling coefficient

$$s = \frac{1}{Tf_B}. \tag{9.21}$$

The sampling theorem states that if $s \geq 1$, then the signal is uniquely determined by its samples.

We can derive a similar relation for the AWG in the spectral domain. The field along one edge of a star coupler is approximately the spatial spectrum (i.e., Fourier transform) of the field along the other edge. The output waveguides of the AWG are thus sampling the spatial spectrum of the field from the grating arms. If we make this field from the grating arms zero outside of a spatial width w, and if the output waveguides are spaced closely enough, then the samples in the output waveguides can uniquely determine the field along the grating arm edge of the output star coupler in the AWG. This field is determined by the input signal spectrum. Thus, we can uniquely determine the input spectrum by these discrete samples. To make the field zero outside of width w, we can simply use a limited number of grating arms in the AWG. $w = Ma$, where M is the number of grating arms, and a is the grating arm pitch at the connection to the output star coupler. Note that the star coupler can still have dummy waveguides outside of width w, to which the field can couple, because the dummy waveguides are necessary for the star coupler to perform an accurate Fourier transform.

If the output waveguides have pitch b at their connection to the output star coupler, then the sampling coefficient is

$$s = \frac{R\lambda}{Mab}, \quad (9.22)$$

where R is the star coupler radius (called f in earlier sections in this chapter), and λ is the wavelength in the slab (i.e., not the free-space wavelength). Thus, if $s \geq 1$, the information in the output waveguides is sufficient to uniquely determine the AWG input signal. Note that because there are a finite number of output waveguides, the signal time-frequency spectrum must be band-limited for this to be true. It is analogous to the time-domain case, in which the signal must be limited in its time extent because only a finite number of time-sampling points can be recorded.

Interestingly, when $s = 1$, the output star coupler is an $M \times M$ star coupler, although it is possible that less than M output waveguides are actually used.

One use for an AWG with high sampling is as a multiplexer with very wide passbands and no concern for crosstalk level. Other uses including making a dynamic gain equalization filter (DGEF) and a band de/multiplexer, described later in this section.

Aberrations A conventional Gaussian-passband AWG demultiplexer has $s \approx 0.5$. As s is increased toward one and beyond, a significant problem in the AWG usually develops: the output waveguides become so close that there is significant mutual coupling between them. If left uncorrected, this results in aberrations that give high loss and large passband shoulders. To mitigate the mutual coupling, one could increase b. However, this results in a longer R [see Eqn (9.22)] and a larger device size. An alternative solution is to move the aiming point of the grating waveguides into the output waveguides and then slightly modify the grating arm path lengths [38].

To calculate how much to adjust the grating arm lengths, one must calculate the transmissivity from each grating waveguide through the second star coupler to each output waveguide and use the phase of the transmissivity to adjust the grating arm lengths. When there is mutual coupling between the waveguides, a useful tool is BPM. Many types of BPM exist, such as Fourier transform BPM (FT-BPM) and finite-difference BPM (FD-BPM). FT-BPM can handle wide angles, but is slow and cannot handle large index steps. FD-BPM is much faster and can handle large index steps, but cannot handle wide angles. There is another type called sinc-BPM [39], which is based on FT-BPM, but stays in the angular spectrum domain. It is fast and accurate for periodic waveguide arrays, found, for example, in star couplers.

To calculate the transmissivities mentioned earlier, we first calculate the waveguide mode of a grating waveguide at a point that is far enough away from the slab region as to be uncoupled to other grating waveguides. Then we use BPM to propagate it until it reaches the slab boundary. Call this field u_1. Then we likewise calculate the waveguide mode in an output waveguide at a point far from the slab region and use BPM to propagate it until it reaches the slab boundary. Call this field u_2. The transmissivity through the second star coupler in an AWG from a grating waveguide at angle θ_1 to an output waveguide at angle θ_2 is

$$t(\theta_1, \theta_2) = \frac{\int_{-\infty}^{\infty} u_1^{\text{prop}}(x) u_2(x) \exp[jkx(\theta_2 - \angle AB)] dx}{\sqrt{\int_{-\infty}^{\infty} |u_1(x)|^2 dx \int_{-\infty}^{\infty} |u_2(x)|^2 dx}} \quad (9.23)$$

$$AB = R\left(e^{-j\theta_1} - 1 + e^{j\theta_2}\right) \quad (9.24)$$

$$u_1^{\text{prop}}(x) = F^{-1}\left\{F\{u_1(x) \exp[jkx(\theta_1 - \angle AB)]\} \exp\left(j\sqrt{k^2 - k_x^2}|AB|\right)\right\} \quad (9.25)$$

$$F\{u(x)\} = \int_{-\infty}^{\infty} u(x) e^{jxk_x} dx \quad (9.26)$$

If there is no mutual coupling among the waveguides connected to the slab region, then $\angle t(\theta_1, \theta_2)$ is a linear function of θ_1 and θ_2. If there is mutual coupling, then the path lengths of the waveguides should be adjusted so as to make $\angle t(\theta_1, \theta_2)$ a linear function of θ_1 and θ_2.

In general, when designing a star coupler with mutual coupling between waveguides, it is best to aim for making u_1 and u_2 such that the power coupled to each of the two adjacent waveguides is $\sim$4 dB less than the power in the launched waveguide. Increasing the mutual coupling beyond this tends to make the device too sensitive to fabrication details.

Just as segmentation [40], vertical tapering [41], or selective UV exposure [42] (see Figure 9.33) improve the AWG grating efficiency, these techniques can also improve the coupling efficiency to highly sampled output waveguides. They work by reducing the abruptness of the transition from the waveguide array to the slab region. Segmentation is the easiest to implement because it does not require any extra fabrication steps. In segmentation, one simply adds sections (15 is a typical number) perpendicular to the waveguides. The section center-to-center

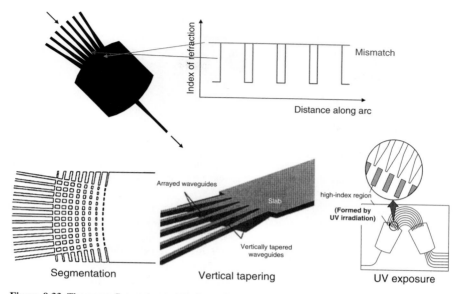

Figure 9.33 The upper figure shows the mismatch in refractive index distribution at the slab-to-array transition in a star coupler, which is a significant source of insertion loss in AWGs. The lower figures show various demonstrated schemes for reducing this loss (this figure may be seen in color on the included CD-ROM).

spacing is constant, but their widths gradually decrease as their distance from the slab increases.

Dynamic Gain Equalization Filter In Section "Achieving Multiple Zero-Loss Maxima," we proved that we need $N \times M$ couplers, where $N, M > 1$ to create a passband with more than one zero-loss maximum. In Sections "AWG with Two-Arm-Interferometer Input" and "AWG with Three-Arm Interferometer Input," we then showed how to make a two-maxima passband and a three-maxima passband by connecting two-arm and three-arm interferometers, respectively, directly to an AWG. Here, we connect one AWG to another AWG through an arrayed waveguide lens (AWL) to create a passband with numerous zero-loss maxima. By inserting either phase shifters or VOAs into the AWL arms, we create a dynamic filter that can control the channel powers in a WDM link, often called a DGEF [43].

An AWL is an AWG with a grating order of zero, i.e., the path lengths of all the arms in the "grating" are equal. In order to lay out such a structure without waveguides running into each other, the conventional "U"-shaped structure of an AWG cannot be used. Instead, it is best to use a "W"-shaped structure.

We would like the DGEF to be able to pass all the WDM channels in a given spectral region without any distortion. In order to do this, the AWL ports must

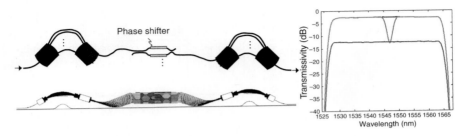

Figure 9.34 DGEF consisting of two AWGs connected by an AWL. Upper left is the concept, lower left is the actual waveguide layout (has 16 waveguides in the AWL), and the right is the measured response, fiber-to-fiber (this figure may be seen in color on the included CD-ROM).

perfectly or over sample the spectra from the AWGs, i.e., $s \geq 1$ for all AWL ports and all λ in the spectral region of interest.

A layout for a DGEF is shown in Figure 9.34. There is an MZI with thermooptic phase shifters in each arm. Each MZI must be driven in a push–pull fashion so as to maintain a constant phase difference between adjacent AWL arms. Using push–pull also reduces the polarization dependence, because it reduces the maximum required phase shift, and a typical silica thermooptic phase shifter shifts TM-polarized light 4% more efficiently than TE-polarized light. Furthermore, push–pull reduces the worst-case power consumption and keeps the total heat dissipation constant [43]. A measured spectral response through the DGEF is shown in Figure 9.34(b) when all VOAs are set for minimum attenuation, when one VOA is set for increased attenuation, and when all VOAs are set for increased attenuation. One can see that the spectral response is smooth and ripple-free, although the spectrum is broken into discrete sections by the AWGs.

There is another DGEF arrangement which has a lower insertion loss, which is shown in Figure 9.35. It consists of a large interferometer with one arm containing

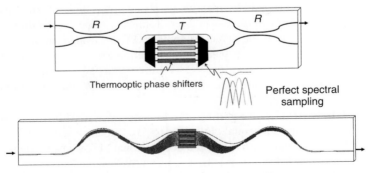

Figure 9.35 Lower loss DGEF using a large interferometer. Concept (upper) and actual waveguide layout (lower) (this figure may be seen in color on the included CD-ROM).

an AWG–AWL–AWG series [44]. In this DGEF, each AWL arm contains just a phase shifter rather than an MZI. The couplers of the interferometer do not have to have a 50/50 coupling ratio. The overall insertion loss is lower than the insertion loss of the AWG–AWL–AWG series because only a portion of the light passes through the AWG–AWL–AWG series. The main drawback to this design is that because of the large interferometer with different elements in each arm, it is difficult to make it insensitive to the environment and the input light polarization. One way to solve the polarization-dependence difficulty is to use polarization diversity.

Band Multiplexer With appropriate settings of the phase shifters, the DGEF can act as a band filter, a filter with a very wide passband and very steep sidewalls. With a small modification, one can change it into a band de/multiplexer (we will call it just a multiplexer from here on) [45]. The concept is shown in Figure 9.36. In this case, the AWL is split into groups at one end. Because each group enters the right-hand AWG at a different location, the spectral portions carried by each group are sent to different outputs. Note that in this design it is not necessary that the AWL arm path lengths be the same from group to group.

It is important to note that this band multiplexer does not exhibit any chromatic dispersion. This is because all path lengths are the same for all wavelengths (provided the aberrations have been compensated for). This is possible because AWG filters are non-minimum phase filters [46]. Thus, AWGs need not obey the Kramers–Kronig relations and so can have rectangular shaped passbands and linear phase simultaneously. This is unlike conventional band filters which are made from thin-film technology. Those band filters are made from a series of resonances and are infinite-impulse response filters. In transmission, thin-film filters are minimum phase filters and thus obey the Kramers–Kronig relations. These filters have significant dispersion at sharp passband corners, resulting in signal distortion.

A waveguide layout and measured results from a 5-band 8-skip-0 multiplexer are shown in Figure 9.37 [47]. A-skip-B means that each band carries A number of channels (typically on a 100-GHz grid), and there are dead zones of B channels between them. As expected, the chromatic dispersion is negligible, even at the passband corners.

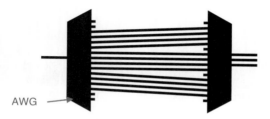

Figure 9.36 Concept of band de/multiplexer using AWGs (this figure may be seen in color on the included CD-ROM).

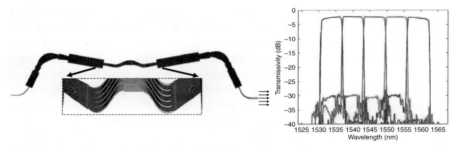

Figure 9.37 8-skip-0 5-band de/multiplexer (this figure may be seen in color on the included CD-ROM).

There is a type of WDM called coarse WDM (CWDM), which has a 20-nm channel spacing. The channel spacing is so wide because it is designed for non-temperature-controlled laser transmitters. The passbands for CWDM typically must be wide and flat with a bandwidth greater than 14 nm. The band multiplexer is a good way to make a CWDM de/multiplexer. Figure 9.38 shows a waveguide a layout for an 8-channel CWDM multiplexer and the measured response [48]. Because the channel spacing is so wide, the AWG grating orders are very small and having unequal path lengths in the AWL adds a negligible amount of

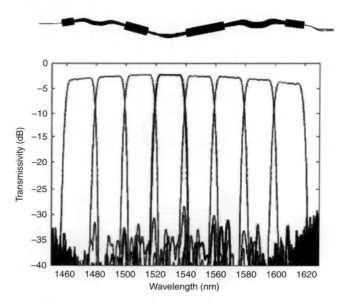

Figure 9.38 8-channel CWDM (de)multiplexer. Waveguide layout (upper) and measured passbands (lower).

chromatic dispersion (allowing one to use a "U" shape for the AWL, making it more compact). Thus, the AWGs take on a "W" shape, and the AWL takes on a "U" shape, the opposite of the case in Figure 9.37.

A highly compact silica waveguide PLC 4-channel CWDM multiplexer made using a completely different technique, a two-dimensional reflective hologram, is shown in Section 9.3.3.

Athermal (Temperature-Insensitive) AWGs

Non-technical overview: The refractive index of PLCs changes with temperature. This causes the filter response of PLCs to shift wavelength when their temperature is changed. The amount of shift depends only on the refractive index change with temperature and not the filter design itself. Silica PLCs shift approximately $+0.01$ nm/°C, InP and Si PLCs shift approximately $+0.1$ nm/°C, and polymer PLCs shift approximately -0.3 nm/°C. The shift is undesirable in DWDM systems. To avoid this, often the PLC is temperature controlled, using either a thermo-electrical cooler or a heater. Another approach is to make the PLC "athermal," the topic of this section.

The temperature sensitivity of the passband center wavelength (frequency) in the silica-based AWG is about $d\lambda/dT = 1.2 \times 10^{-2}$(nm/deg) [$d\nu/dT = -1.5$(GHz/deg)], which is mainly determined by the temperature dependence of silica glass itself [$d_{nc}/dT = 1.1 \times 10^{-5}$(deg^{-1})]. The AWG multiplexer should be temperature controlled with a heater or a Peltier cooler to stabilize the channel wavelengths. This requires a constant power consumption of a few Watts and significant equipment for the temperature control. Various kinds of AWG configurations to achieve athermal operation have been proposed [49–52].

Figure 9.39 shows a schematic configuration of an athermal AWG in which temperature-dependent optical path-length change is compensated by movement

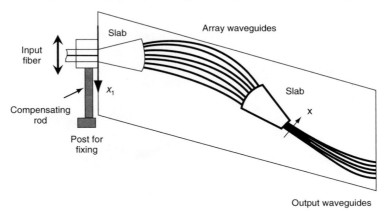

Figure 9.39 Configuration of an athermal AWG with a temperature compensating input position: From Ref. [49].

of the input fiber position. The input coupling device consists of three parts: namely one part holding the input fiber, an additional post to fix the whole coupling device to the chip, and a metal compensating rod between them. The compensating rod is made of a material with a high thermal expansion coefficient, aluminum. It changes its length with the ambient temperature T and shifts the input fiber along the endface of the slab waveguide to compensate for the thermal drift of the passband center wavelength in the AWG.

Figure 9.40 shows the shift of several channels' peak transmissions with temperature for a 200-GHz module. The temperature is varied from -35 to $+80°C$. The center wavelength of the AWG filter is nearly independent of the ambient temperature. This method may be applied to spectrographs and phased array filters realized in any material.

Figure 9.41 shows a schematic configuration of a completely solid-state athermal AWG. The temperature-dependent optical path difference in silica waveguides is compensated with a trapezoidal groove filled with silicone adhesive which has a negative thermooptic coefficient. Because the passband center wavelength is given by $\lambda_0 = n_c \Delta L / m$, the optical path-length difference $n_c \Delta L$

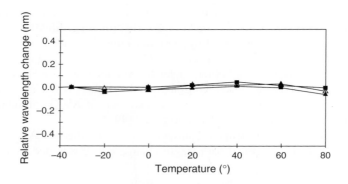

Figure 9.40 Temperature dependences of the passband center wavelengths.

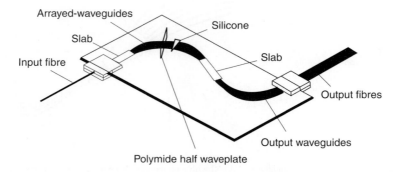

Figure 9.41 Schematic configuration of silicone-filled athermal AWG.

must be made insensitive to temperature. Therefore, the groove is designed to satisfy the following conditions

$$n_c \Delta L = n_c \Delta \ell + \hat{n}_c \Delta \hat{\ell}, \tag{9.27}$$

and

$$\frac{d(n_c \Delta L)}{dT} = \frac{dn_c}{dT} \Delta \ell + \frac{d\hat{n}_c}{dT} \Delta \hat{\ell} = 0, \tag{9.28}$$

where $\hat{n}$ is the refractive index of silicone and ΔL and $\Delta \hat{\ell}$ are the path-length differences of silica waveguides and silicone region, respectively. Equation (9.27) is a condition to satisfy the AWG specifications and Eqn (9.28) is the athermal condition, respectively. The temperature sensitivity of silicone is $d\hat{n}_c/dT = -37 \times 10^{-5} (\deg^{-1})$. Therefore, the path-length difference of silicone is $\Delta \hat{\ell} \cong \Delta \ell / 37$. Figure 9.42 shows temperature dependencies of passband center wavelengths in conventional and athermal silicone-groove-filled AWGs. The temperature-dependent wavelength change has been reduced from 0.95 to 0.05 nm in the $0 \sim 85°C$ range. The excess loss caused by the groove is about 2 dB, which is mainly due to diffraction loss in the groove. The insertion loss caused by diffraction loss can be reduced by segmenting a single trapezoidal silicone region into multiple groove regions [53]. In the segmented groove regions, the light beam is periodically refocused and the insertion loss is reduced to about 0.4 dB.

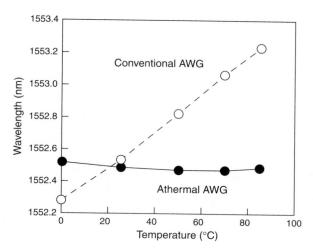

Figure 9.42 Temperature dependences of passband center wavelengths in a conventional AWG and a silicone-filled athermal AWG.

Tandem AWG Configuration

The maximum available wafer size of a PLC is limited by the fabrication apparatus such as the deposition machine, electric furnace, mask aligner, etc. Therefore, it is not easy to fabricate AWGs with very large channel counts. One possible way to increase the total number of channels is to use two kinds of AWGs, which are cascaded in series. These AWGs are connected with a single-mode connection and so have no relation to the periodic stationary imaging principle. Figure 9.43 shows the configuration of a 10-GHz-spaced 1010-channel WDM filter that covers both the C and the L bands [54]. It consists of a primary 1×10 flat-top AWG (AWG #k with $k = 1, 2, \ldots, 10$) with a 1-THz channel spacing and ten secondary 1×101 AWGs with 10-GHz spacing and 200 channels. Crosstalk levels of the secondary AWGs are around -32 dB, and the sidelobe levels in these passbands are less than -35 dB. The tandem configuration enables us to construct flexible WDM systems, that is, secondary AWGs can be added when bandwidth demand increases. Also, this configuration will be essential for the construction of hierarchical cross-connect (XC) systems such as fiber XC, band XC, and wavelength XC. Output port #k ($k = 1, 2, \ldots, 10$) of the primary AWG is connected to the input port of AWG #k through an optical fiber. Two conditions are imposed on these AWGs. First, the center wavelength of 200 channels of AWG #k should be designed to coincide with that of the flat-top passband #k from primary AWG output #k. Then, passband #k is sliced with the AWG #k without any noticeable loss. Second, the sidelobe components of flat-top passband #k are removed as shown in the inset of Figure 9.43. Therefore, one passband within the FSR is obtained from one output port of AWG #k. 101 wavelengths are selected from every AWG #k. Figure 9.44 shows the demultiplexing properties of all the channels of the tandem AWG filter.

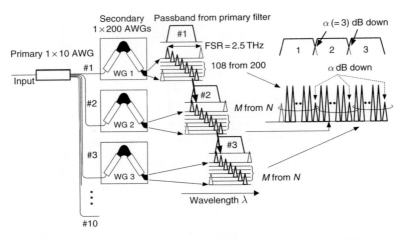

Figure 9.43 Configuration of 10 GHz-spaced 1010-channel tandem AWG.

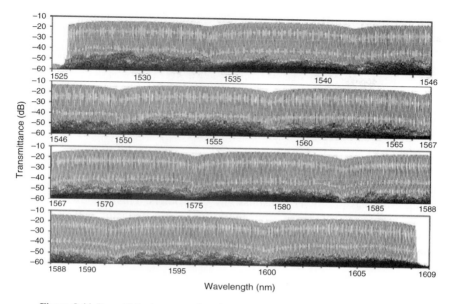

Figure 9.44 Demultiplexing properties of all the channels of the tandem AWG filter.

There are a total of 1010 channels and they are all aligned at 10 GHz intervals with no missing channels in the 1526- to 1608-nm wavelength range. The loss values ranged from 13 to 19 dB. The main origin of the 13 dB loss is the 10-dB intrinsic loss of the primary AWG. This could be reduced by about 9 dB when a 1×10 interference filter is used instead of a flat-top AWG or by using a flat-top AWG design that does not have intrinsic loss. A 4200-channel AWG with 5-GHz channel spacing has also been fabricated by using the tandem configuration [55–57].

Chirped AWGs

Non-technical overview: A bird's chirping is a change in pitch while the bird is uttering a sound. A chirped grating is a change in pitch across the grating. A chirped AWG means that the path-length difference between adjacent arms is not constant.

All the AWGs discussed so far have had a constant path-length difference between adjacent grating waveguides. If we use a variable path-length difference, we create a "chirped" AWG. We will discuss two main types of chirped AWGs. The first is where the arm path length vs arm number has a quadratic component. We will call this "linear chirping." The second is where the path-length difference between adjacent arms oscillates between two values. We will call this "interleave chirping."

Linear Chirping By adding a significant quadratic component to the arm length distribution, and then quantizing it to an integer multiple of a desired wavelength, one can have an imaging focal length in the AWG that changes significantly with wavelength [58, 59]. In other words,

$$L(i) = \text{round}\left\{m\left[i + \gamma\left(i - \frac{I+1}{2}\right)^2\right]\right\}\frac{\Delta L}{m} \qquad (9.29)$$

where γ is the amount of chirp, m is the operating grating diffraction order, and I is the total number of grating waveguides. "round" means round to the nearest integer. We call this linear chirping. Linear chirping can be used to make one grating order have a higher transmissivity than others. This was used to make a tunable laser consisting of a silica AWG and semiconductor optical amplifier oscillate in a desired grating order [60]. If one wishes to have multiple ports in focus simultaneously in a linearly chirped AWG, one needs to distort the output star coupler. One can tilt the focal line of the star coupler at the output ports, vary the grating port pitch linearly, or a combination of both.

Interleave Chirping An interleave-chirped AWG has interleaved sets of grating arm lengths with different properties. Two main types have been demonstrated. One is the different sets of waveguides have a small difference in their path-length difference and has been used to create flat-top passbands in AWGs [61], as mentioned in Section "Flat Spectral-Response AWGs." This technique necessarily adds insertion loss, however, in accordance with the Flat-Top Filter Construction Theorem. The other is the different sets of waveguides have approximately the same path-length difference, but there is a phase-shift difference between them, and has been used to create a wavelength-selective switch (WSS) [62].

Figure 9.45 shows the waveguide layout for the WSS. It consists of two interleave-chirped AWGs connected by an AWL with thermooptic phase shifters in each arm. The odd-numbered grating waveguides in the interleaver-chirped AWGs have a $\lambda/4$ extra path length as compared with the even-numbered grating waveguides. This causes two images to be generated by the AWG at each wavelength. The AWL is connected so as to pick up these two images plus one extra image from outside the central Brillouin zone. Thus, there are three images per wavelength channel. There are two input waveguides and two output waveguides for the WSS. By controlling the relative phases between the three images for each channel, each channel can be routed independently from either input waveguide to either output waveguide. The three phases help provide an improved extinction ratio. The lower part of Figure 9.45 shows the measured transmissivity for one switching state. Although this WSS has a low insertion loss and is compact, thermal crosstalk between phase shifters in the WSS makes it difficult to control.

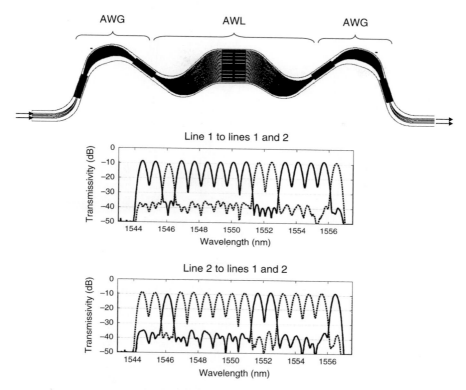

Figure 9.45 WSS made using two interleave-chirped AWGs and an AWL.

9.3.3 Planar Holographic Bragg Reflectors

Non-technical overview: This type of optical device has very little to do with real holograms, which are images where the magnitude and phase of the light is recorded, rather than just the intensity as in conventional photography, allowing for three-dimensional images. The device here is simply a two-dimensional array of weak reflectors. In one-dimension, it is simply a Bragg reflector. The reflections add constructively at various wavelengths and in various directions to create optical filters.

A planar holographic Bragg reflector consists of very fine etched features in part of the core (some of the core must remain in order to maintain vertical confinement) in concentric arcs. These act as two-dimensional Bragg reflectors. High-resolution lithography is required to fabricate such devices. This type of device is not a photonic crystal. By definition, a photonic crystal is made of high index contrast, so that it can exhibit a band gap over a large range of

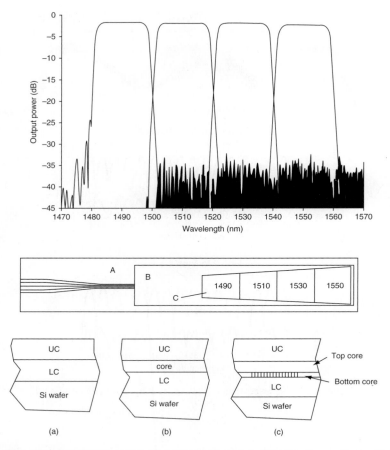

Figure 9.46 4-Channel CWDM multiplexer made using a planar holographic Bragg reflector. The two-dimensional Bragg reflector region is in area C. From Ref. [63].

propagation directions, whereas a planar holographic Bragg reflector is typically made of low index contrast material. An example of 4-channel CWDM multiplexer using this technology is shown in Figure 9.46 [63].

9.4 INTER-SIGNAL CONTROL DEVICES

Non-technical overview: By inter-signal control, we mean control of the overall spectrum of multiple wavelength channels in a WDM system. The devices in this section are dynamic devices that can power-regulate and route wavelength channels in WDM networks.

9.4.1 Reconfigurable Optical Add/Drop Multiplexers

A reconfigurable optical add/drop multiplexer (ROADM) is a device that gives simultaneous access to all wavelength channels in WDM communication systems. The first PLC integrated optic ROADM was fabricated and basic functions of individually routing 16 different wavelength channels with 100-GHz channel spacing were demonstrated in Ref. [64]. The waveguide configuration of the 16-channel optical ROADM is shown in Figure 9.47. It consists of four arrayed-waveguide gratings and 16 double-gate thermooptic switches (TOSWs). Four AWGs are allocated with crossing their slab regions with each other. These AWGs have the same grating parameters; they are the channel spacing of 100 GHz and the free spectral range of 3300 GHz (26.4 nm) in the 1.55-μm region. Equally spaced WDM signals, $\lambda_1, \lambda_2, \ldots, \lambda_{16}$, which are coupled to the main input port (add port) in Figure 9.47 are first demultiplexed by the AWG_1 (AWG_2) and then 16 signals are introduced into the left-hand-side arms (right-hand-side arms) of double-gate TOSWs. The crossing angle of the intersecting waveguides is designed to be larger than 30° so as to make the crosstalk and insertion loss negligible.

Here, the "off" state of a double-gate switch is defined as the switching condition in which a signal from a lower left input port (right input port) goes to an upper right output port (left output port) in Figure 9.47. The "on" state is then defined as the condition in which a signal from a lower left input port (right input

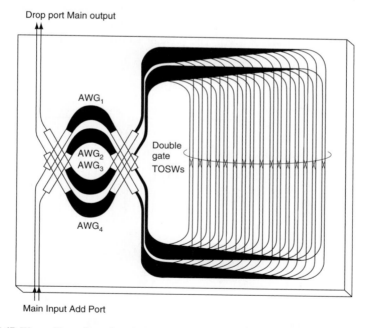

Figure 9.47 Waveguide configuration of 16ch ROADM with double-gate thermo-optic switches (TOSWs).

port) goes to a lower left output port (right output port). When a double-gate switch is "off," the demultiplexed light by AWG_1 (AWG_2) goes to the cross arm and is multiplexed again by AWG_3 (AWG_4). On the other hand, if a double-gate switch is "on" state, the demultiplexed light by AWG_1 (AWG_2) goes to the through arm and is multiplexed by AWG_4 (AWG_3). Therefore, any specific wavelength signal can be extracted from the main output port and led to the drop port by changing the corresponding switch condition. Signals at the same wavelength as that of the dropped component can be added to the main output port when it is coupled into the add port in Figure 9.47.

Figure 9.48 shows light transmission characteristics from main input port to main output port (solid line) and drop port (dotted line) when all TO switches are "off." The on-off crosstalk is smaller than $-33\,dB$ with the on-chip losses of $7.8 \sim 10.3\,dB$. When TO switches SW_2, SW_4, SW_6, SW_7, SW_9, SW_{12}, SW_{13}, and SW_{15}, for example, are turned to "on" the selected signals λ_2, λ_4, λ_6, λ_7, λ_9, λ_{12}, λ_{13}, and λ_{15} are extracted from main output port (solid line) and led to the drop port (dotted line) as shown in Figure 9.49. The on-off crosstalk is smaller than $-30\,dB$ with the on-chip losses of $8 \sim 10\,dB$.

Figure 9.50 is a waveguide configuration of an athermal 16-channel ROADM [65]. The silicone-type athermal technique is incorporated in this ROADM. Transmission spectra become essentially insensitive to temperature change, as shown in Figure 9.51.

Although the electric power necessary to drive a double-gate switch is two times larger than the conventional TO switch, the power consumption itself can be reduced from $\sim 1/5$ to $1/2$ when bridge-suspended phase shifters [66] or a trench/groove structure [67] are utilized. The present ROADM can transport all input

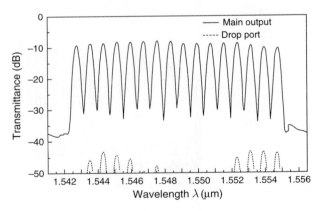

Figure 9.48 Transmission spectra from main input port to main output port (solid line) and drop port (dotted line) when all TO switches are "off."

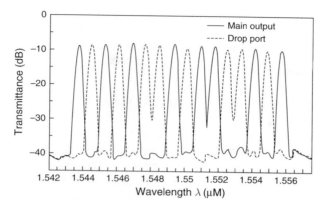

Figure 9.49 Transmission spectra from main input port to main output port and drop port when TO switches SW_2, SW_4, SW_6, SW_7, SW_9, SW_{12}, SW_{13} and SW_{15} are "on."

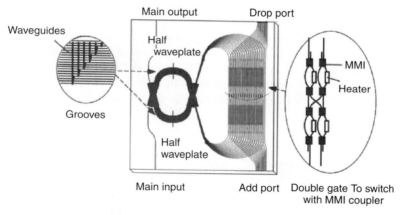

Figure 9.50 Configuration of athermal 16-ch ROADM (this figure may be seen in color on the included CD-ROM).

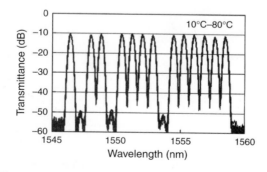

Figure 9.51 Transmission spectra of athermal 16-ch ROADM.

signals to the succeeding stages without inherent power losses. Therefore, these ROADMs are very attractive for all optical WDM-routing systems and allow the network to be transparent to signal formats and bit rates.

9.4.2 Large Channel Count ROADMs

The first demonstrated PLC ROADM was discussed above in Section 9.4.1. This 16-channel demonstration has numerous waveguide crossings, which can require significant wafer real estate because the crossings must occur at a steep angle to mitigate loss and crosstalk. To scale to a very large channel count, one approach is to use a reflective design with a striped mirror facet [68]. One such design had through passbands with two zero-loss maxima by employing two waveguides per channel. One design used perfect spectral sampling across the entire spectrum to achieve no gaps between through channels, shown in Figure 9.52. A difficulty with this design is that after fabrication, one had to manually trim the phases in all the waveguides using electrical hyperheating to phase-align all the spectral components. Another design that is easier to manufacture used perfect spectral sampling only in pairs [69]. In this case, only the phase alignment within each pair needed to be trimmed. However, the full-flat passband between adjacent through channels was lost. Yet another design used an MZI input to create a 64-channel colored ROADM with flat-top passbands (as described in Section "AWG with

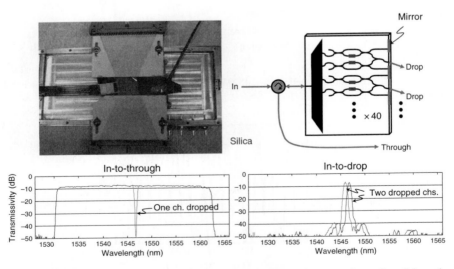

Figure 9.52 40-Channel reflective ROADM using an AWG with perfect spectral sampling. Schematic (upper right), photograph (upper left), measured in-to-through spectra (lower left), and measured in-to-drop spectra (lower right) (this figure may be seen in color on the included CD-ROM).

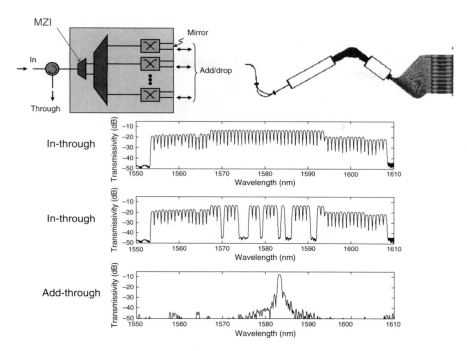

Figure 9.53 64-Channel reflective ROADM using an MZI input to the AWG (this figure may be seen in color on the included CD-ROM).

Two-Arm-Interferometer Input"), shown in Figure 9.53 [70]. In this case, only one phase alignment in the entire device, the MZI to the AWG, needed to be performed.

9.4.3 Wavelength-Selective Switches

A J × K WSS is a device that routes WDM signals from J inputs to K outputs and can be used to make a ROADM. There are many different types. We showed two examples of 2 × 2 WSSs in Sections "Interleave Chirping" and "Reconfigurable Optical Add/Drop Multiplexers." The most common deployed type is a 1 × K WSS. A 1 × K WSS can be used to create a colorless ROADM or an optical mesh node. Another type is a 1 × 1 WSS, which is often called a blocker. It can either block or pass each WDM channel that passes through it. Several PLC blocker demonstrations have been reported [71].

Figure 9.54 shows a waveguide layout of a 1 × 9 WSS [72]. It uses ten 200-GHz-spacing, 8-channel AWGs (1 as a demultiplexer and 9 as multiplexers) and a tree arrangement of 1 × 2 thermooptic switches and a final stage of shutters for improved switching extinction ratio and also to provide a VOA function. It also includes a star coupler with eight VOAs to act as an add-channel combiner.

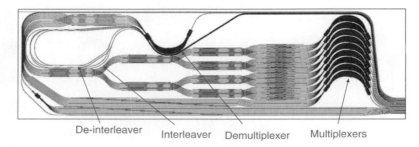

Figure 9.54 8-channel 1 × 9 WSS (this figure may be seen in color on the included CD-ROM).

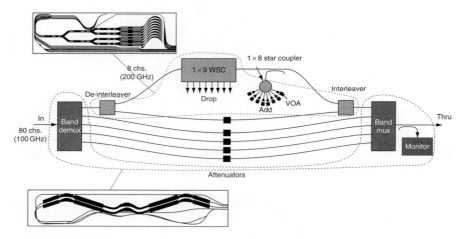

Figure 9.55 Modular 80-channel partially colorless ROADM (this figure may be seen in color on the included CD-ROM).

Furthermore, it includes a de-interleaver to divide an incoming stream of 100-GHz-spaced channels into two streams of 200-GHz-spaced channels and a corresponding interleaver.

This 1 × 9 WSS PLC, along with a band de/multiplexer PLC, were developed to create a modular 80-channel ROADM, shown in Figure 9.55 [73].

9.5 INTRA-SIGNAL CONTROL DEVICES

Non-technical overview: By intra-signal control, we mean control of the characteristics of the spectrum of a single wavelength channel. The devices in this section are dynamic devices that can modify the waveforms of optical signals, in order to create desired pulse shapes, overcome transmission impairments, etc.

9.5.1 Temporal Pulse Waveform Shapers

Non-technical overview: A temporal pulse waveform shaper is an arbitrary waveform generator (which unfortunately has the same acronym as AWG). It consists of taking a period pulse train, spectrally demultiplexing all the spectral components, controlling the amplitude and phase of each component, and then multiplexing them back together.

The shaping and encoding of optical pulse waveforms are important for a variety of applications in optical communications, optical radar, and picosecond and femtosecond spectroscopy. Control of the pulse temporal profile is achieved by spatially dispersing the optical frequency components, whose amplitude and phase are arbitrarily weighted and multiplexing them again into a single optical beam. Weiner et al. [74] first demonstrated a technique for optical pulse shaping using a grating pair as a dispersive element and masks for amplitude and phase filtering. Because they used a grating pair, the size of the experimental apparatus was of the order of 1 m². Also, the weighting functions for the amplitude and phase masks were fixed because they fabricated them by metal deposition and reactive-ion etching of the silica glass.

A schematic configuration of a fully integrated optic dynamic temporal pulse waveform shaper is shown in Figure 9.56 [75]. It consists of an AWG pair for demultiplexing (AWG_1) and multiplexing (AWG_2) the spectral components of mode-locked optical pulses and TOSWs and phase shifters for arbitrarily patterning the spectral components. The channel spacing and total number of channels of

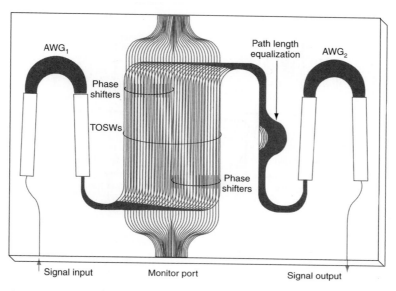

Figure 9.56 Schematic configuration of temporal pulse waveform shaper (this figure may be seen in color on the included CD-ROM).

the AWG are 40 and 80 GHz, respectively, which are centered at $\lambda_0 = 1.55\,\mu$m. About 32 channels among 80 channels of the AWG are used for spectral filtering. An array of 32 TOSWs and phase shifters are allocated between the AWG pair. All the optical path lengths from AWG_1 to AWG_2 are made equal by employing path-length adjustment waveguides. The switching ratio of each TO switch, which can be controlled essentially from 0 to 1, is measured by using the monitor port. Also, the amount of phase shift in each path is determined by comparing the relative phase difference with a reference arm. The fiber-to-fiber insertion loss is 12 dB, and the extinction ratio is $\sim$30 dB. The average electric power for each TO switch is $\sim$300 mW. The minimum controllability of the electric power is 1–2 mW. This enables us to obtain 0.1–0.2 dB amplitude controllability. The average electric power necessary to obtain π phase shift is also about 300 mW. According to this minimum controllability of electric power, the phase-shifter resolution is about $\pi/100$.

The temporal pulse waveform shaper can be utilized in a variety of applications for optical pulse multiplexing, pulse waveform shaping, frequency chirping compensation, and frequency-encoding code division multiplexing (FE-CDM). N times optical pulse multiplication can easily be accomplished by filtering the line spectral components of the mode-locked pulse in every Nth interval.

A square-shaped optical pulse is a particularly useful pulse shape with potential application to nonlinear optical metrology, coherent transient spectroscopy, and future all-optical switching and optical demultiplexing [76, 77]. Figure 9.57 shows

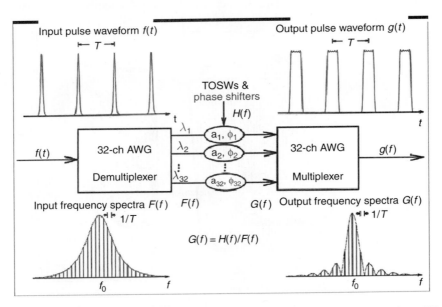

Figure 9.57 Schematic pulse waveforms and frequency spectra in square pulse generation scheme (this figure may be seen in color on the included CD-ROM).

the schematic temporal pulse waveforms and frequency spectra in the square pulse-generation scheme. The frequency spectrum of a mode-locked pulse train with the waveform $f(t) = A \cdot \mathrm{sech}(t/t_0)$ is given by

$$F(f) = \sum_{m=-\infty}^{\infty} A \cdot \mathrm{sech}\left[\pi^2 t_0 (f - f_0)\right] \delta\left(f - f_0 - \frac{m}{T}\right), \qquad (9.30)$$

where f_0, T, and t_0 denote center frequency, pulse interval, and pulse width (FWHM width $\tau = 2\cosh^{-1}\sqrt{2} \cdot t_0 \cong 1.763 t_0$), respectively. In order to generate square pulses with rise and fall times of $(t_2 - t_1)$, the corresponding sinc-like frequency spectra with the form

$$G(f) = \sum_{m=-\infty}^{\infty} \frac{\sin[\pi(t_2+t_1)(f-f_0)]}{\pi(t_2+t_1)(f-f_0)} \cdot \frac{\cos[\pi(t_2-t_1)(f-f_0)]}{1 - [2(t_2-t_1)(f-f_0)]^2} \cdot \delta\left(f - f_0 - \frac{m}{T}\right), \qquad (9.31)$$

should be synthesized. Then, the amplitude and phase-weighting function $H(f)$ of the pulse shaper for each spectral component is determined by $H(f) = G(f)/F(f)$. Here, we assume $T = 25\,\mathrm{ps}$, $\tau = 8\,\mathrm{ps}$, $t_1 = 3.3\,\mathrm{ps}$, $t_2 = 5.3\,\mathrm{ps}$, and $f_0 = 1.931\,\mathrm{THz}$.

Figure 9.58(a) and (b) shows the experimental original spectra and auto-correlated pulse waveforms [78]. The original pulse had an FWHM of $\tau = 0.9\,\mathrm{ps}$. Figure 9.59(a) and (b) shows synthesized spectra and corresponding cross-correlated pulse waveforms for designed pulse widths of $\tau = 11.9\,\mathrm{ps}$. Dotted and solid lines show designed and experimental values, respectively. The ripples in the flat-top pulse region are caused by the finite available bandwidth and are estimated to be 0.1 dB compared with the calculated values of 0.2 dB. The rise and fall time (10% to 90%) is 2.9 to

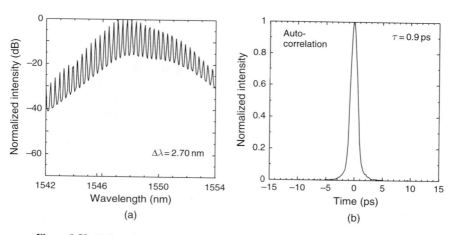

Figure 9.58 (a) Experimental original spectra and (b) auto-correlated pulse waveforms.

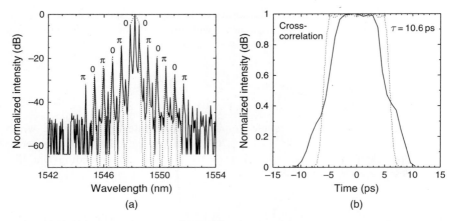

Figure 9.59 (a) Synthesized spectra and (b) corresponding cross-correlated pulse waveforms for designed pulse widths of $\tau = 11.9$ ps.

5.4 ps, while the designed values are both 1.5 ps. The deterioration in the rise and fall time is mainly brought about by phase-setting errors that originated in the thermal crosstalk among the phase shifters used for TO phase adjustment. Phase-setting errors could be reduced by using heat-insulating grooves.

9.5.2 General Optical Equalizers

Non-technical overview: An optical equalizer (at least a linear one) is simply an optical filter that serves to improve a signal that was degraded due to deterministic distortions. By "optical filter," we mean any optical element with some sort of spectral response, in amplitude, phase, or both. It does not have to filter anything, and in fact, it is not an equalizer's job to mitigate noise. A tunable optical dispersion compensator is a type of equalizer, and indeed, the first dispersion compensators were called equalizers.

An equalizer is a device that at least partially restores a signal corrupted by deterministic intersymbol interference (ISI). ISI is the mixing together of different time components in the signal. Example causes of ISI are limited transmitter modulator bandwidth, chromatic dispersion in optical fiber, narrow optical filters, and group delay ripple in optical filters. Equalizers successfully undo only deterministic ISI and not random processes, such as spontaneous emission from amplifiers. Equalizers in the electrical domain are a widely used technology. The electrical technology is called electrical equalization (EEQ). Here, we will focus on optical equalization (OEQ).

The simplest type of equalizer is a tapped delay line. A tapped delay line consists of a series of weighted taps and delays which are re-added to the signal. Adjusting the weights' amplitudes and phases allows one to create a desired

9. *Planar Lightwave Circuits in Fiber-Optic Communications* 323

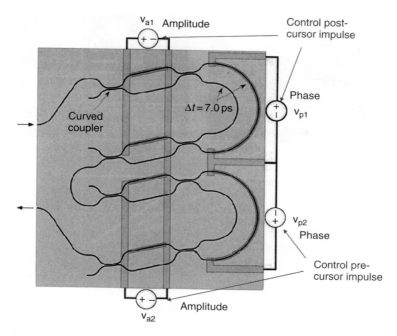

Figure 9.60 Two-tap general OEQ (this figure may be seen in color on the included CD-ROM).

impulse response. Ideally, this impulse response is the inverse of the impulse response that created the ISI, i.e., $h_{EQ}(t) = F^{-1}\{1/F\{h_{ISI}(t)\}\}$. To create a tapped delay line in optics, one can use MZIs. A simple one was demonstrated in [79], and a more complex one, using lattice filters, was demonstrated in [80].

Figure 9.60 shows a simple OEQ—a two-tap OEQ, with one pre-cursor tap and one post-cursor tap. For many ISI impairments, it is found that the best time tap spacing is ∼0.7 times the bit period of the digital data signal. Two-tap OEQs have been demonstrated for both 40- and 100-Gb/s signals, correcting modulation bandwidth limitations, extinction ratio limitations, chromatic dispersion, and signal distortion due to narrow optical filtering. It has also been shown to mitigate PMD impairments by signal shaping. By making the OEQ have an FSR that aligns with the WDM channel spacing, a plurality of WDM channels can be equalized simultaneously, provided the ISI on the channels is equal.

Equalization reshapes not only the signal but also the noise. For on-off keyed (OOK) formats, it is better to equalize the signal when the optical signal-to-noise ratio (OSNR) is high, but for PSK formats, initial indications are such that the equalizer is equally effective equalizing when the OSNR is high or low [81].

Such a two-tap OEQ, with a time tap spacing of 7 ps, was used to compensate ISI due to modulator bandwidth limitation and achieve a high-quality 107 Gb/s non-return-to-zero signal [82], as shown in Figure 9.61.

Figure 9.61 107-Gb/s NRZ eye diagram, (a) before and (b) after optical equalization, showing an example of equalization of a bandwidth-limited transmitter (this figure may be seen in color on the included CD-ROM).

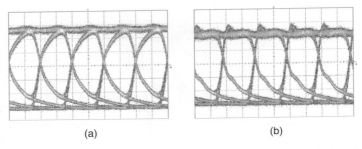

Figure 9.62 Electrical eye diagram of 10-Gb/s signal from a band-limited photodiode before (a) and after (b) optical equalization (this figure may be seen in color on the included CD-ROM).

Although OEQ is highly effective against most transmitter impairments, it is ineffective against most receiver impairments. This is because in order to correct optically for a band-limited direct-detection receiver, one needs to generate a negative optical power, which is impossible. For example, most photodiodes are limited in speed by either a resistor-capacitor (RC) time constant or carrier transport barriers, both of which have an asymmetric impulse response. Thus, a typical band-limited photodiode gives an electrical eye diagram as shown in Figure 9.62(a). An attempt to open the eye using a two-tap OEQ results in the eye diagram shown in Figure 9.62(b) (because the ISI impulse response is asymmetric, the best OEQ setting was also asymmetric). The OEQ can fix the top rail, but is unable to do anything about the bottom rail, because it cannot generate a negative optical power. Instead, one must use an electrical equalizer after the photodiode to fix a band-limited receiver.

A general OEQ optimally needs to have independent adjustment of the amplitude and phase of all its impulses. Some ISI impulse responses are symmetric, whereas others are asymmetric. For example, if a modulator is limited in bandwidth due to an RC effect, the ISI impulse response is asymmetric, whereas if the modulator bandwidth limitation is due to RF transmission line loss, the ISI impulse response may be symmetric. An even more general OEQ can independently control the impulses for each polarization, and a step toward such an OEQ was shown in Ref. [83].

Although the above OEQs used finite-impulse response (FIR) filters, an OEQ using an infinite-impulse filter was demonstrated using ring resonators in silicon waveguides [84]. Such an OEQ is not a general OEQ, because that OEQ has an asymmetric impulse response, and there is limited control over individual impulses. Such an OEQ is instead a specific equalizer, and the next section covers equalizers specific to optical dispersion compensation. Specific equalizers can compensate much larger distortions than general equalizers, because all the impulses do not need to be individually controlled.

9.5.3 Optical Chromatic Dispersion Compensators

An optical dispersion compensator (ODC) is an optical equalizer optimized to compensate chromatic dispersion. It is often used to compensate the chromatic dispersion encountered in optical fibers; standard single-mode fiber (SSMF) has a dispersion of approximately 17 ps/nm/km in the C-band. If the amount of ODC dispersion is adjustable, the device is called a tunable ODC (TODC). Figure 9.63 shows most of the TODC types that have been demonstrated. They are divided into two main categories: infinite impulse response (IIR) and finite impulse response (FIR) types.

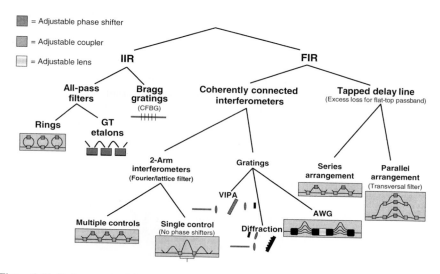

Figure 9.63 Various types of demonstrated TODCs (this figure may be seen in color on the included CD-ROM).

Infinite Impulse Response TODCs

An IIR filter is a filter with an infinite number of impulses and optically is created using resonators. Chromatic dispersion in optical fiber comes from both material effects and waveguiding effects. Both are resonant in nature and thus have an IIR. It is natural to equalize them with another IIR filter.

All-Pass Filters   A convenient way to make an IIR dispersion compensator is to use an all-pass filter [85]. An all-pass filter is an IIR filter with constant unity transmissivity but a non-linear phase response as a function of optical frequency.

Ring Resonators: A good example of an all-pass filter is a waveguide coupled to a ring resonator. If the waveguide, coupler, and ring resonator are loss-less, then the light does not experience any loss after passing through the bus waveguide. However, light that has an optical frequency that is resonant in the ring will take a longer time to pass through the bus waveguide. Thus, the group delay will exhibit one peak each ring FSR. By placing several rings in series, all coupled to the same waveguide, and controlling the ring coupling and length, one can create an approximately linear region of group delay, thus generating chromatic dispersion, which can be used to equalize the dispersion encountered in the optical fiber. Although one can imagine other ring configurations to create a TODC, the configuration of Figure 9.64 is the only one demonstrated to date.

A ring resonator coupled to a bus waveguide uses a 2×2 coupler and thus meets the criterion of the Flat-Top Filter Construction Theorem, and, interestingly, generates an infinite number of zero-loss maxima, which is why it is an all-pass filter. Demonstrations of such a TODC are shown in Refs [86, 87].

Gires–Tournois Etalons: A Gires–Tournois etalon consists of a cavity with a 100% reflector at one end and a partial reflector at the other end. If the etalon is illuminated at a slight angle, so that the beam exits at different angle than which it enters and thus can be separated out, then, neglecting beam walk-off, it operates on the same principle as ring resonators. Gires–Tournois etalons have been demonstrated only in bulk material to date (Figure 9.65).

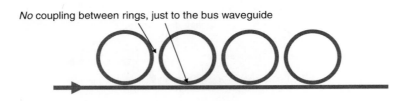

Figure 9.64   TODC using ring resonators in series. Each ring resonator path length is tuned, and usually the coupling to the waveguide bus is also tuned (this figure may be seen in color on the included CD-ROM).

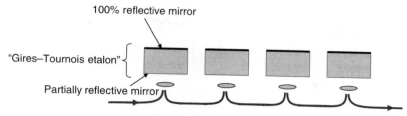

Figure 9.65 TODC using Gire–Tournois etalons (this figure may be seen in color on the included CD-ROM).

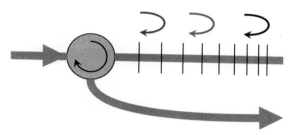

Figure 9.66 Chirped Bragg grating TODC (this figure may be seen in color on the included CD-ROM).

Bragg Gratings Bragg gratings consist of many reflectors in series. Wavelengths that match twice the spacing between reflectors divided by an integer are reflected back. If the Bragg grating has a constantly varying period, i.e., it is linearly chirped, then different wavelengths reflect at different points in the grating, creating chromatic dispersion (Figure 9.66). The dispersion is usually tuned by heating the grating non-uniformly [88]. Bragg gratings are most commonly made in optical fibers, but they have also been demonstrated in silica [89], polymer [90], and InP PLCs.

Finite Impulse Response TODCs

One can also implement dispersion compensation using an FIR filter [91]. An all-pass FIR filter is impossible, so the filter must be designed to have both a linear group delay and a constant transmissivity over the bandwidth of interest. Optically, FIR filters are constructed out of feed-forward interferometers, e.g., MZIs and AWGs.

Tapped Delay Lines We saw an example of an optical tapped delay line in the general OEQ section. Tapped delay line equalizers are commonly used in RF electronics.

Serial Arrangement: A simple tapped delay line filter can be constructed from a serial arrangement of single-mode connected MZIs, as we saw in the general

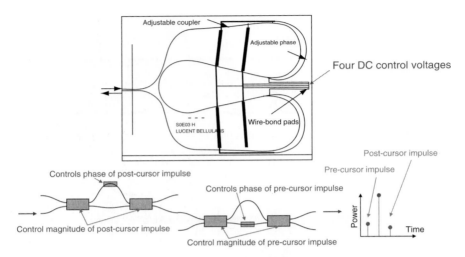

Figure 9.67 Two-tap optical equalizer using a serial arrangement of MZIs (this figure may be seen in color on the included CD-ROM).

OEQ section. Figure 9.67 shows more details of this OEQ. This design has only 1×2 and 2×1 couplers, and thus, according to the Flat-Top Filter Construction Theorem, it has intrinsic loss when used as a TODC.

Parallel Arrangement: Another way to construct a tapped delay line is to use a parallel arrangement of waveguides of different lengths and is called a transversal filter. An example in silica is shown in Figure 9.68 [92], and an example in InP is shown in Figure 9.69 [93]. Again, because transversal filters contain only $1 \times N$ or $N \times 1$ couplers, they have intrinsic loss when used as TODCs. The intrinsic loss

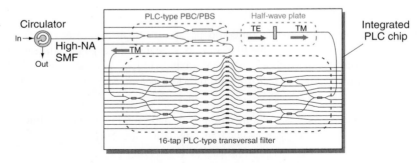

Figure 9.68 TODC using a transversal filter in silica waveguides. The left and right regions contain adjustable couplers, and the center region contains adjustable phase shifters. From H. Kawashima of Ref. [92] (this figure may be seen in color on the included CD-ROM).

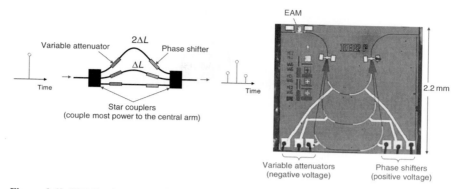

Figure 9.69 TODC using a transversal filter in InP waveguides. This TODC is integrated with an electroabsorption modulator to create a transmitter with integrated dispersion compensation. At left is the concept and at right is a photograph of the chip (this figure may be seen in color on the included CD-ROM).

increases the larger the desired TODC figure of merit (FOM) (defined later in this section).

Coherently Connected Interferometers The previous FIR-based TODCs had only one mode in connections between interferometers. By coherently connected inteferometers, we mean that there are multiple modes (i.e., multiple single-mode waveguides) in connections between interferometers. In other words, there are effective $N \times M$ couplers where N and $M > 1$. This is for the purpose making low intrinsic loss TODCs, according to The Flat-Top Filter Construction Theorem.

Two-Arm Interferometers
Multiple Controls: Figure 9.70 shows a TODC using a series arrangement of 2×2 couplers. Such an arrangement can construct any desired impulse response with N impulses with an arbitrary number of zero-loss maxima per FSR, thus theoretically allowing for an ideal TODC, provided there are at least $N-1$ length-imbalanced MZI stages. To tune the dispersion, all the couplers and all the phases in each MZI arm must be adjusted. Each control affects all the impulses, so the control is complicated.

Single Control: The previous section showed a TODC design constructed from 2×2 couplers with many independent controls. With certain arrangements, one can eliminate the phase controls in the MZI arms and control all the 2×2 couplers with a single control signal.

Figure 9.71 shows such a 3-stage TODC. In this example, the bandwidth was chosen very narrow so as to achieve a large dispersion range. Because of the narrow bandwidth, a filter was integrated with the TODC for having the TODC track the signal wavelength. A feedback loop adjusts the TODC temperature to keep it centered on the signal wavelength. The TODC FSR is chosen quite small so that the temperature change required to lock onto any wavelength is at most ±8°C [94].

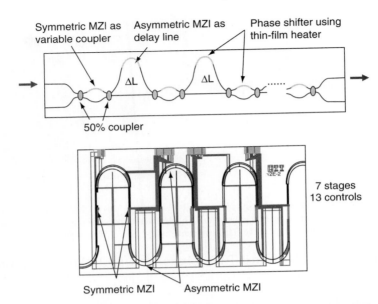

Figure 9.70 TODC using coherently coupled MZIs (this figure may be seen in color on the included CD-ROM).

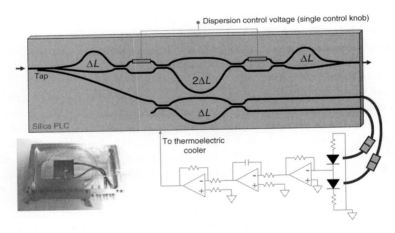

Figure 9.71 3-Stage, 2-arm TODC with a single control for tuning the dispersion. It includes an integrated wavelength-locking monitor (this figure may be seen in color on the included CD-ROM).

A 4-stage design with a single control is shown in Figure 9.72. A 4-stage design has a larger TODC FOM than a 3-stage design. An implementation in silica is shown in Figure 9.73 with an FSR of 100 GHz, which is suitable for 40-Gb/s applications [95]. Results from a 33.3-GHz FSR version are shown in Figure 9.74, which is suitable for 10 Gb/s applications.

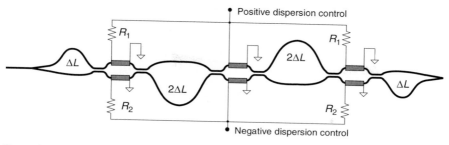

Figure 9.72 4-Stage, 2-arm TODC with a single control for tuning the dispersion (this figure may be seen in color on the included CD-ROM).

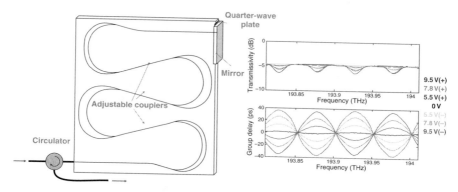

Figure 9.73 4-Stage version realized in a silica PLC. FSR = 100 GHz (this figure may be seen in color on the included CD-ROM).

Gratings: Another way to construct a TODC is to use a diffraction grating, such as a ruled grating, a virtually imaged phased array (VIPA), or an AWG. The basic concept is to spread the spectrum of the signal over an element than can provide a controllable parabolic phase profile (dispersion means a linear group delay which means a parabolic phase vs frequency) and then recombine the spectrum.

Virtually Imaged Phased Array: The VIPA (Figure 9.75) [96] acts like an echelle diffraction grating with extremely high dispersion constructed using a glass or silicon plate with a 100% reflection coating on one side and a partial reflection coating on the other. The plate is tilted such that multiple reflections in the plate create multiple images.

Diffraction Grating: A TODC using a diffraction grating and thermally deformable mirror was demonstrated in Ref. [97]. It used four passes on the grating to achieve a wide passband with low loss at high dispersion settings. No PLC diffraction grating-based TODCs have been reported, to our knowledge.

Arrayed Waveguide Grating: This TODC design consists of two AWGs connected at their second star coupler outer boundaries (where output waveguides would normally be connected). The frequency response measured from input to

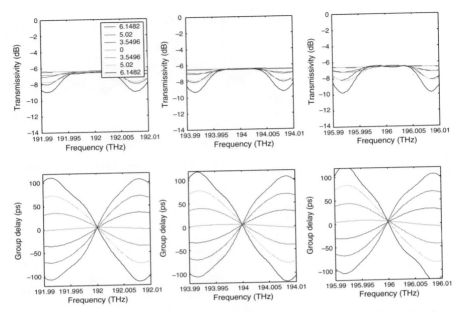

Figure 9.74 4-Stage version in silica PLC with an FSR of 33.3 GHz. The legend shows the voltage setting (this figure may be seen in color on the included CD-ROM).

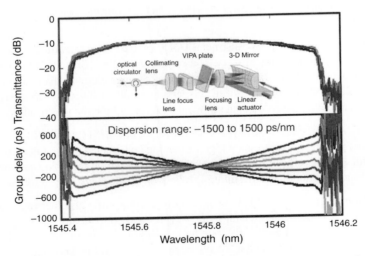

Figure 9.75 TODC using a VIPA. From Ref. [96] (this figure may be seen in color on the included CD-ROM).

output has a Gaussian-shaped passband with a constant negative chromatic dispersion, the dispersion given by

$$D_0 = -\frac{2Rf}{n(a\Delta f)^2}, \qquad (9.32)$$

where R is the star coupler radius, n is the refractive index in the slab, a is the grating waveguide pitch at the star coupler, f is the optical frequency (not focal length, as it was used for earlier), and Δf is the AWG FSR. It is usually best to make Δf equal to the WDM channel spacing. Because this device contains an $N \times N$ coupler, it can act as a TODC with low intrinsic loss.

To understand why the device has a negative dispersion, consider the lower part of Figure 9.76. The center lines of the Gaussian beams for two different wavelengths within an FSR are shown. From the left-hand AWG, the wavelengths are focused onto the middle of the double star coupler. Wavelengths longer than the channel center wavelength are imaged below, shorter wavelengths above. After passing the focal line, these beams continue in a straight line, and when they reach the grating arms of the second AWG they are no longer centered. Because the effective path length through an AWG is equal to the average of the path lengths weighted by the power in each path, longer wavelengths end up traveling a shorter path than shorter wavelengths. This is negative dispersion. Note that the off-center beams end up coupling to the output waveguide with a reduced efficiency, and thus, the net passband has a rounded, single-maxima shape.

To tune the dispersion amount, one needs a lensing action as shown at the lower right of Figure 9.76. For example, if a lens with a focal length equal to $R/2$ is placed in the double star center, all the beams are recentered on the grating arms of the second AWG, and the device exhibits zero dispersion. In such a case, the passband is ideally perfect flat, having numerous zero-loss maxima.

One way to create the adjustable lens is to use thermooptics. Because all angles are much less than 1 rad, a lens is just a quadratic index distribution. One way to

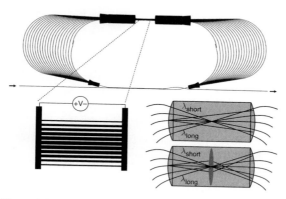

Figure 9.76 AWG-based TODC using a silica thermooptic lens.

create a quadratic index distribution is to drive an array of closely spaced heaters in parallel with a quadratic variation in width. Figure 9.76 shows such a device [98].

However, the electrical power consumption of such a lens when using silica on a silicon substrate can be quite high. In the case above, the power consumption was ~7 W to tune the entire range.

One way to reduce the power consumption is to cut the design in half and make it reflective. This cuts the power consumption in half. To further reduce it, one can use either a deformable mirror [99] or a polymer thermooptic lens, as shown in Figures 9.77 and 9.78 [100]. Polymer gives a large power reduction because the magnitude of its index change with temperature is ~30 times that of silica, and the thermal conductivity of polymer is ~8 times less than that of silica.

Figure of Merit

Let us define an FOM for a TODC, which is usable bandwidth, B (in frequency), squared multiplied by the dispersion adjustment range, ΔD. This FOM is non-dimensional, and signal bandwidth squared multiplied by dispersion is an approximate number of bit slots that each bit is spread across by the dispersion. This FOM describes the complexity of the TODC. For example, it is linearly related to the number of ring resonators, the number of MZI stages, or the number of AWG arms in the TODC.

$$\text{TODC FOM} = B^2 \Delta D \frac{\lambda_0^2}{c_0} \tag{9.33}$$

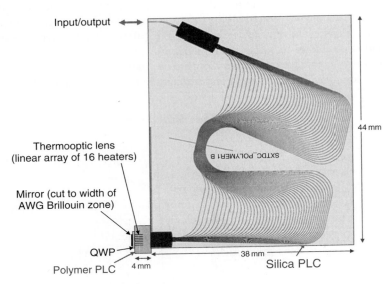

Figure 9.77 AWG-based TODC using a polymer thermooptic lens, in a reflective configuration (this figure may be seen in color on the included CD-ROM).

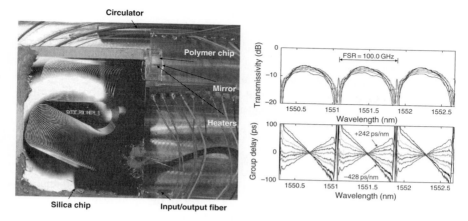

Figure 9.78 Photograph and measured response of AWG-based TODC using a polymer thermo-optic lens (this figure may be seen in color on the included CD-ROM).

Table 9.1
Figure of merit for various demonstrated TODCs.

Type	Reference	ΔD (ps/nm)	B (GHz)	FOM
VIPA	[96]	3000	68	111
CFBG	[101]	800	80	41
AWG*	[82]	1300	39	16
		560	57	15
Etalon	[102]	3400	18	9
Ring*	[87]	3000	19	9
MZI*	[95]	4200	14	7
		420	45	7
Transversal filter*	[92]	200	60	6
Ring*	[86]	900	25	5
MZI*	[91]	300	40	4

* = PLC TODC

Table 9.1 shows approximate FOMs of some currently demonstrated TODCs.
There is actually another FOM for dispersion compensators, but it applies only to dispersion compensating fiber [103]:

$$\text{DCF FOM} = D/L, \qquad (9.34)$$

where D is the dispersion and L is the loss. A typical DCF FOM is ∼300 ps/(nm-dB). However, the DCF FOM does not apply well to PLC TODCs, because one can easily devise a PLC TODC with arbitrarily large dispersion, but arbitrarily small bandwidth, without significantly affecting the insertion loss and thus even the simplest PLC TODC could exhibit an arbitrarily large DCF FOM.

9.6 CONCLUSION

This chapter gave an overview of many PLC devices that have been demonstrated. We regret that there are many more demonstrated PLC devices that we did not have room to include. We concentrated mostly on silica-based PLCs, but did present some SiON, polymer, and InP PLCs. Si PLCs are currently being vigorously investigated and hold the promise of integration with low-cost electronics. Also, InP PLC research is experiencing a resurgence, because of increasing total data rates. A 10-channel × 10-Gb/s InP PLC transmitter and InP PLC receiver are now being widely sold commercially [104], and InP PLCs that can produce multilevel formats are currently of great interest.

We focused exclusively on PLCs for fiber-optic communications; and telecommunications is where PLCs were born, and nearly all the significant PLC applications are. However, we believe that for PLC research and development to advance significantly, PLCs must find applications in other areas. There have been small niche non-telecom applications for PLCs, such as fiber-optic gyroscopes; but these are not large enough to have a significant effect on PLC development. Some possible significant alternative PLC applications include radio-frequency (RF) signal processing, biophotonics, displays, and beam steerers/scanners.

It is important to realize that the PLC device with the largest deployed volume today is a simple $1 \times N$ splitter for fiber-to-the-home networks. A key question for PLCs is will photonic circuits more complex than a simple splitter succeed in the *mass market*, like electronic circuits have.

REFERENCES

[1] S. Sohma, T. Goh, H. Okazaki et al., "Low switching power silica-based super high delta thermo-optic switch with heat insulating grooves," *Electron. Lett.*, 38, 127–128, 2002.

[2] B. E. Little, S. T. Chu, P. P. Absil et al., "Very high-order microring resonator filters for WDM applications," *IEEE Photon. Technol. Lett.*, 16, 2263–2265, October 2004.

[3] W. Burns and A. F. Milton, "Mode conversion in planar-dielectric separating waveguides," *IEEE J. Quantum Electron.*, QE-11(1), 32–39, January 1975.

[4] H. Ishikawa, "Fully adiabatic design of waveguide branches," *J. Lightw. Technol.*, 25, 1832–1840, July 2007.

[5] K. Jinguji, N. Takato, A. Sugita, and M. Kawachi, "Mach-Zehnder interferometer type optical waveguide coupler with wavelength-flattened coupling ratio," *Electron. Lett.*, 26, 1326–1327, August 16, 1990.

[6] C. R. Doerr, L. W. Stulz, D. S. Levy et al., "Cross-Connect-Type Wavelength Add-Drop Node with Integrated Band Muxes, Interleavers, and Monitor," in Proc. *OFC* Conference Digest, postdeadline paper PD33, 2003.

[7] M. Oguma, T. Kitoh, Y. Inoue et al., "Compactly Folded Waveguide-Type Interleave Filter With Stabilized Couplers," in Proc. *OFC* Conference Digest, paper TuK3, 2002, pp. 70–72.

[8] C. R. Doerr, M. Cappuzzo, E. Chen et al., "Bending of a planar lightwave circuit 2 × 2 coupler to desensitize it to wavelength, polarization, and fabrication changes," *IEEE Photon. Technol. Lett.*, 17, 1211–1213, June 2005.

[9] R. Narevich, G. Heise, E. Narevicius et al., "Novel Wide-Band Low-PDL Integrated Variable Optical Attenuator in Silica-on-Silicon," in Proc. *OFC*, paper OthV6, 2005.
[10] K. Takada and S. Mitachi, "Polarization crosstalk dependence on length in silica-based waveguides measured by using optical low coherence interference," *J. Lightw. Technol.*, 16, 1413–1422, August 1998.
[11] M. A. Webster, R. M. Pafchek, A. Mitchell, and T. L. Koch, "Width dependence of inherent TM-mode lateral leakage loss in silicon-on-insulator ridge waveguides," *IEEE Photon. Technol. Lett.*, 19, 429–431, March 2007.
[12] C. R. Doerr, M. A. Cappuzzo, E. Y. Chen et al., "Polarization-insensitive planar lightwave circuit dual-rate Mach-Zehnder delay-interferometer," *IEEE Photon. Technol. Lett.*, 18, 1708–1710, August 2006.
[13] M. K. Smit, "New focusing and dispersive planar component based on an optical phased array," *Electron. Lett.*, 24, 385–386, 1988.
[14] H. Takahashi, S. Suzuki, K. Kato, and I. Nishi, "Arrayed-waveguide grating for wavelength division multi/demultiplexer with nanometer resolution," *Electron. Lett.*, 26, 87–88, 1990.
[15] C. Dragone, C. A. Edwards, and R. C. Kistler, "Integrated optics N × N multiplexer on silicon," *IEEE Photon. Tech. Lett.*, 3, 896–899, 1991.
[16] M. K. Smit and C. Van Dam, "PHASAR-based WDM-devices: Principles, design, and applications," *IEEE J. Sel. Top. Quantum Electron.*, 2, 236–250, June 1996.
[17] K. Okamoto, *Fundamentals of Optical Waveguides* (2nd edn), Academic Press, 2006, Chapter 9.
[18] Y. P. Li and C. H. Henry, "Silicon optical bench waveguide technology," *in Optical Fiber Telecommunications IIIB*, New York, Academic Press, 1997, Chapter 8.
[19] P. Munoz, D. Pastor, and J. Capmany, "Modeling and design of arrayed waveguide gratings," *J. Lightw. Technol.*, 20, 661–674, April 2002.
[20] K. Okamoto, K. Moriwaki, and S. Suzuki, "Fabrication of 64 × 64 arrayed-waveguide grating multiplexer on silicon," *Electron. Lett.*, 31, 184–185, 1995.
[21] K. Okamoto, K. Syuto, H. Takahashi, and Y. Ohmori, "Fabrication of 128-channel arrayed-waveguide grating multiplexer with a 25-GHz channel spacing," *Electron. Lett.*, 32, 1474–1476, 1996.
[22] Y. Hida, Y. Hibino, T. Kitoh et al., "400-channel arrayed-waveguide grating with 25 GHz spacing using 1.5%-Δ waveguides on 6-inch Si wafer," *Electron. Lett.*, 37, 576–577, 2001.
[23] K. Okamoto, "AWG Technologies: Design and their Applications," in Proc. *OFC 2002*, SC113, Baltimore CA, 2002.
[24] K. Okamoto and H. Yamada, "Arrayed-waveguide grating multiplexer with flat spectral response," *Opt. Lett.*, 20, 43–45, 1995.
[25] K. Okamoto and A. Sugita, "Flat spectral response arrayed-waveguide grating multiplexer with parabolic waveguide horns," *Electron. Lett.*, 32, 1661–1662, 1996.
[26] M. R. Amersfoort, J. B. D. Soole, H. P. LeBlanc et al., "Passband broadening of integrated arrayed waveguide filters using multimode interference couplers," *Electron. Lett.*, 32, 449–451, 1996.
[27] D. Trouchet, A. Beguin, H. Boek et al., "Passband Flattening of PHASAR WDM Using Input and Output Star Couplers Designed with Two Focal Points," in Proc. *OFC '97*, ThM7, Dallas, Texas, 1997.
[28] G. H. B. Thompson, R. Epworth, C. Rogers et al., "An Original Low-Loss and Pass-Band Flattened SiO_2 on Si Planar Wavelength Demultiplexer," in Proc. *OFC '98*, TuN1, San Jose, CA, 1998.
[29] C. R. Doerr, L. W. Stulz, and R. Pafchek, "Compact and low-loss integrated box-like passband multiplexer," *IEEE Photon. Technol. Lett.*, 15, 918–920, 2003.
[30] W. K. Burns, A. F. Milton, and A. B. Lee, "Optical waveguide parabolic coupling horns," *Appl. Phys. Lett.*, 30, 28–30, 1977.
[31] T. Kitoh, Y. Inoue, M. Itoh, and Y. Hibino, "Low chromatic-dispersion flat-top arrayed waveguide grating filter," *Electron. Lett.*, 39, 1116–1118, 2003.

[32] T. Chiba, H. Arai, K. Ohira et al., "Chromatic Dispersion Free Fourier Transform-Based Wavelength Splitters for D-WDM," in Proc. *OECC*, paper 13B2-2, 2000.

[33] C. Dragone, "Frequency routing device having a wide and substantially flat passband," U.S. Patent 5488680, 1996.

[34] C. R. Doerr, M. A. Cappuzzo, E. Y. Chen et al., "Low-loss rectangular-passband multiplexer consisting of a waveguide grating router synchronized to a three-arm interferometer," *IEEE Photon. Technol. Lett.*, 17, 2334–2336, November 2005.

[35] C. R. Doerr, M. A. Cappuzzo, E. Y. Chen et al., "Wide-band arrayed waveguide grating with three low-loss maxima per passband," *IEEE Photon. Technol. Lett.*, 18, 2308–2310, November 2006.

[36] P. A. Besse, M. Bachmann, C. Nadler, and H. Melchior, "The integrated prism interpretation of multileg Mach-Zehnder interferometers based on multimode interference couplers," *Opt. Quantum Electron.*, 27, 909–920, 1995.

[37] C. R. Doerr, M. Cappuzzo, E. Laskowski et al., "Dynamic wavelength equalizer in silica using the single-filtered-arm interferometer," *IEEE Photon. Technol. Lett.*, 11, 581–583, May 1999.

[38] C. Dragone, "An $N \times N$ optical multiplexer using a planar arrangement of two star couplers," *IEEE Photon. Technol. Lett.*, 3, 812–815, 1991.

[39] C. R. Doerr, "Beam propagation method tailored for step-index waveguides," *IEEE Photon. Technol. Lett.*, 13, 130–132, February 2001.

[40] Y. P. Li, "Optical device having low insertion loss," U.S. Patent No. 5745618, April 28, 1998.

[41] A. Sugita, A. Kaneko, K. Okamoto et al., "Very low insertion loss arrayed-waveguide grating with vertically tapered waveguides," *IEEE Photon. Technol. Lett.*, 12, 1180–1182, September 2000.

[42] K. Maru, T. Chiba, M. Okawa et al., "Low-loss arrayed-waveguide grating with high index regions at slab-to-array interface," *Electron. Lett.*, 37, 1287–1289, 2001.

[43] C. R. Doerr, R. Pafchek, and L. W. Stulz, "16-band integrated dynamic gain equalization filter with less than 2.8-dB insertion loss," *IEEE Photon. Technol. Lett.*, 14, 334–336, March 2002.

[44] C. R. Doerr, C. H. Joyner, and L. W. Stulz, "Integrated WDM dynamic power equalizer with potentially low insertion loss," *IEEE Photon. Technol. Lett.*, 10, 1443–1445, October 1998.

[45] C. R. Doerr, R. Pafchek, and L. W. Stulz, "Integrated band demultiplexer using waveguide grating routers," *IEEE Photon. Technol. Lett.*, 15, 1088–1090, August 2003.

[46] G. Lenz, B. J. Eggleton, C. K. Madsen et al., "Optical dispersion of optical filters for WDM systems," *IEEE Photon. Technol. Lett.*, 10, 567–569, April 1998.

[47] S. Chandrasekhar, C. R. Doerr, and L. L. Buhl, "Flexible waveband optical networking without guard bands using novel 8-skip-0 banding filters," *IEEE Photon. Technol. Lett.*, 17, 579–581, March 2005.

[48] C. R. Doerr, M. Cappuzzo, L. Gomez et al., "Planar ligthwave circuit eight-channel CWDM multiplexer with <3.9-dB insertion loss," *J. Lightw. Technol.*, 23, 62–65, January 2005.

[49] G. Heise, H. W. Schneider, and P. C. Clemens, "Optical Phased Array Filter Module with Passively Compensated Temperature Dependence," in Proc. *ECOC '98*, Madrid, Spain, September 20–24, 1998, pp. 319–320.

[50] Y. Inoue, A. Kaneko, F. Hanawa et al., "Athermal silica-based arrayed-waveguide grating multiplexer," *Electron. Lett.*, 33, 1945–1946, 1997.

[51] R. Gao, K. Takayama, A. Yeniay, and A. F. Garito, "Low-Insertion Loss Athermal AWG Multi/Demultiplexer Based on Perfluorinated Polymers," in Proc. *ECOC '02*, 6.2.2, Copenhagen, Denmark, September 8–12, 2002.

[52] J. B. D. Soole, M. Schlax, C. Narayanan, and R. Pafchek, "Athermalisation of silica arrayed waveguide grating multiplexers," *Electron. Lett.*, 39, 1182–1184, 2003.

[53] A. Kaneko, S. Kamei, Y. Inoue et al., "Athermal Silica-Based Arrayed-Waveguide Grating (AWG) Multiplexers with New Low Loss Groove Design," in Proc. *OFC-IOOC '99*, TuO1, San Diego, CA, 1999, pp. 204–206.

[54] K. Takada, M. Abe, T. Shibata, and K. Okamoto, "10 GHz-spaced 1010-channel Tandem AWG filter consisting of one primary and ten secondary AWGs," *IEEE Photon. Technol. Lett.*, 13, 577–578, 2001.

[55] K. Takada, M. Abe, T. Shibata, and K. Okamoto, "Three-stage ultra-high-density multi/demultiplexer covering low-loss fiber transmission window 1.26–1.63 µm," *Electron. Lett.*, 38, 405–406, 2002.

[56] K. Takada, M. Abe, T. Shibata, and K. Okamoto, "A 25-GHz-spaced 1080-channel tandem multi/demultiplexer covering the S-, C-, and L-bands using an arrayed-waveguide grating with Gaussian passbands as a primary filter," *IEEE Photon. Technol. Lett.*, 14, 648–650, 2002.

[57] K. Takada, M. Abe, T. Shibata, and K. Okamoto, "5 GHz-spaced 4200-channel two-stage tandem demultiplexer for ultra-multi-wavelength light source using supercontinuum generation," *Electron. Lett.*, 38, 572–573, 2005.

[58] C. R. Doerr, M. Shirasaki, and C. H. Joyner, "Chromatic focal plane displacement in the parabolic chirped waveguide grating router," *IEEE Photon. Technol. Lett.*, 9, 625–627, May 1997.

[59] C. R. Doerr and C. H. Joyner, "Double-chirping of the waveguide grating router," *IEEE Photon. Technol. Lett.*, 9, 776–778, June 1997.

[60] C. R. Doerr, L. W. Stulz, R. Pafchek et al., "Potentially low-cost widely tunable laser consisting of a semiconductor optical amplifier connected directly to a silica waveguide grating router," *IEEE Photon. Technol. Lett.*, 15, 1446–1448, October 2003.

[61] A. Rigny, A. Bruno, and H. Sik, "Multigrating method for flattened spectral response wavelength multi/demultiplexer," *Electron. Lett.*, 33, 1701–1702, 1997.

[62] C. R. Doerr, L. W. Stulz, J. Gates et al., "Arrayed waveguide lens wavelength add-drop in silica," *IEEE Photon. Technol. Lett.*, 11, 557–559, May 1999.

[63] D. Iazikov, C. M. Greiner, and T. W. Mossberg, "Integrated holographic filters for flat passband optical multiplexers," *Opt. Express*, 14, 3497–3502, 2006.

[64] K. Okamoto, M. Okuno, A. Himeno, and Y. Ohmori, "16-channel optical add/drop multiplexer consisting of arrayed-waveguide gratings and double-gate switches," *Electron. Lett.*, 32, 1471–1472, 1996.

[65] T. Saida, A. Kaneko, T. Goh et al., "Athermal silica-based optical add/drop multiplexer consisting of arrayed waveguide gratings and double gate thermo-optical switches," *Electron. Lett.*, 36, 528–529, 2000.

[66] A. Sugita, K. Jinguji, N. Takato et al., "Bridge-Suspended Thermo-Optic Phase Shifter and its Application to Silica-Waveguide Optical Switch," in Proc. *IOOC '89*, paper 18D1-4, 1989, p. 58.

[67] R. Kasahara, M. Yanagisawa, A. Sugita et al., "Low-power consumption silica-based 2×2 thermooptic switch using trenched silicon substrate," *IEEE Photon. Technol. Lett.*, 11, 1132–1134, 1999.

[68] C. R. Doerr, L. W. Stulz, M. Cappuzzo et al., "40-wavelength add-drop filter," *IEEE Photon. Technol. Lett.*, 11, 1437–1439, November 1999.

[69] C. R. Doerr, L. W. Stulz, R. Monnard et al., "40-wavelength planar channel-dropping filter with improved crosstalk," *IEEE Photon. Technol. Lett.*, 13, 1008–1010, September 2001.

[70] C. R. Doerr, L. W. Stulz, R. Pafchek, and S. Shunk, "Compact and low-loss manner of waveguide grating router passband flattening and demonstration in a 64-channel blocker/multiplexer," *IEEE Photon. Technol. Lett.*, 14, 56–58, January 2002.

[71] C. R. Doerr, L. W. Stulz, M. Cappuzzo et al., "2×2 wavelength-selective cross connect capable of switching 128 channels in sets of eight," *IEEE Photon. Technol. Lett.*, 14, 387–389, March 2002.

[72] C. R. Doerr, L. W. Stulz, D. S. Levy et al., "Eight-wavelength add-drop filter with true reconfigurability," *IEEE Photon. Technol. Lett.*, 15, 138–140, January 2003.

[73] C. R. Doerr, L. W. Stulz, D. S. Levy et al., "Wavelength add-drop node using silica waveguide integration," *J. Lightw. Technol.*, 22, 2755–2757, December 2004.

[74] A. M. Weiner, J. P. Heritage, and E. M. Kirshner, "High-resolution femtosecond pulse shaping," *J. Opt. Soc. Am. B*, 5, 1563–1572, 1988.

[75] K. Okamoto, T. Kominato, H. Yamada, and T. Goh, "Fabrication of frequency spectrum synthesiser consisting of arrayed-waveguide grating pair and thermo-optic amplitude and phase controllers," *Electron. Lett.*, 35, 733–734, 1999.

[76] A. M. Weiner, J. P. Heritage, and R. N. Thurston, "Synthesis of phase coherent, picosecond optical square pulses," *Opt. Lett.*, 11, 153–155, 1986.

[77] S. Kawanishi, H. Takara, T. Morioka et al., "200 Gbit/s, 100 km time-division-multiplexed optical transmission using supercontinuum pulses with prescaled PLL timing extraction and all-optical demultiplexing," *Electron. Lett.*, 31, 816–817, 1995.

[78] K. Takiguchi, K. Okamoto, T. Kominato et al., "Flexible pulse waveform generation using a silica waveguide-based spectrum synthesis circuit," *Electron. Lett.*, 40, 537–538, 2004.

[79] C. R. Doerr, S. Chandrasekhar, P. J. Winzer et al., "Simple multichannel optical equalizer mitigating intersymbol interference for 40-Gb/s nonreturn-to-zero signals," *J. Lightw. Technol.*, 22, 249–251, January 2004.

[80] M. Bohn, G. Mohs, C. Scheerer et al., "An Adaptive Optical Equalizer Concept for Single Channel Distortion Compensation," in Proc. *ECOC*, Mo.F.2.3, 2001.

[81] A. H. Gnauck, G. Charlet, P. Tran et al., "25.6- Tb/s C+L-Band Transmission of Polarization-Multiplexed RZ-DQPSK Signals," in Proc. *OFC*, PDP19, 2007.

[82] C. R. Doerr, P. J. Winzer, G. Raybon et al., "A Single-Chip Optical Equalizer Enabling 107-Gb/s Optical Non-Return-to-Zero Signal Generation," in Proc. *ECOC*, post deadline paper, 2005.

[83] M. Bohn, W. Rosenkranz, P. M. Krummrich et al., "Experimental Verification of Combined Adaptive PMD and GVD Compensation in a 40 Gb/s Transmission Using Integrated Optical FIR-Filters and Spectrum Monitoring," in Proc. *OFC*, TuG3, 2004.

[84] D. M. Gill, M. S. Rasras, X. Liu et al., "CMOS Compatible Guided-Wave Tunable Optical Equalizer," in Proc. *OFC* Anaheim, OTuM6, 2007.

[85] C. K. Madsen and J. H. Zhao, *Optical Filter Design and Analysis*, John Wiley and Sons, 1999.

[86] F. Hosrt, C. Berendsen, R. Beyeler et al., "Tunable Ring Resonator Dispersion Compensators Realized in High-Refractive-Index Contrast SiON Technology," in Proc. *ECOC*, PD paper 2.2, 2000.

[87] W. Chen, S. Chu, B. Little et al., "Compact, Full C-Band, Widely Tunable Optical Dynamic Dispersion Compensators" in Proc. *OFC*, paper PDP8, 2006.

[88] B. J. Eggleton, A. Ahuja, P. S. Westbrook et al., "Integrated tunable fiber grating for dispersion management in high-bit rate systems," *J. Lightw. Technol.*, 18, 1418–1432, October 2000.

[89] Y. Hibino, T. Kitagawa, K. O. Hill et al., "Wavelength division multiplexer with photoinduced Bragg gratings fabricated in a planar-lightwave-circuit-type asymmetric Mach-Zehnder interferometer on Si," *IEEE Photon. Technol. Lett.*, 8, 84–86, January 1996.

[90] L. Eldada and L. W. Shacklette, "Advances in polymer integrated optics," *IEEE J. Sel. Top. Quantum Electron.*, 6, 54–68, 2000.

[91] K. Takiguchi, K. Jinguji, K. Okamoto, and Y. Ohmori, "Variable group-delay dispersion equalizer using lattice-form programmable optical filter on planar lightwave circuit," *IEEE J. Sel. Top. Quantum Electron.*, 2(2), 270–276, June 1996.

[92] H. Kawashima, N. Matsubara, and K. Nara, "Tunable dispersion compensator using PLC type optical transversal filter," Furukawa Review, no. 29, 13–18, 2006.

[93] C. R. Doerr, L. Zhang, L. L. Buhl et al., "40-Gb/s Modulator With Monolithically Integrated Tunable Dispersion Compensator," in Proc. *OFC*, PDP45, 2007.

[94] C. R. Doerr, S. Chandrasekhar, L. L. Buhl et al., "Optical dispersion compensator suitable for use with non-wavelength-locked transmitters," *J. Lightw. Technol.*, 24, 166–170, January 2006.

[95] C. R. Doerr, S. Chandrasekhar, M. A. Cappuzzo et al., "Four-stage Mach-Zehnder-type tunable optical dispersion compensator with single-knob control," *IEEE Photon. Technol. Lett.*, 17, 2637–2639, December 2005.

[96] Y. Yamauchi, H. Sonoda, H. Furukawa, and Y. Kubota, "Variable Dispersion Compensation Using a VIPA With an Extended Bandwidth," in Proc. *ECOC*, paper Th 1.5.3, 2005.

[97] D. T. Neilson, R. Ryf, F. Pardo, V. A. Aksyuk, M.-E. Simon, D. O. Lopez, D. M. Marom, S. Chandrasekhar, "MEMS-based channelized dispersion compensator with flat passbands," *J. Lightw. Technol.*, 22, 101–105, January 2004.

[98] C. R. Doerr, L. W. Stulz, S. Chandrasekhar, and R. Pafchek, "Colorless tunable dispersion compensator with 400-ps/nm range integrated with a tunable noise filter," *IEEE Photon. Technol. Lett.*, 15, 1258–1260, September 2003.

[99] D. M. Marom, C. R. Doerr, M. A. Cappuzzo et al., "Compact colorless tunable dispersion compensator with 1000-ps/nm tuning range for 40-Gb/s data rates," *J. Lightw. Technol.*, 24, 237–241, January 2006.

[100] C. R. Doerr, R. Blum, L. L. Buhl et al., "Colorless tunable optical dispersion compensator based on a silica arrayed-waveguide grating and a polymer thermooptic lens," *IEEE Photon. Technol. Lett.*, 18, 1222–1224, June 2006.

[101] Y. Painchaud, A. Maillux, H. Chotard et al., "Multi-Channel Fiber Bragg Gratings for Dispersion and Slope Compensation," in Proc. *OFC*, ThAA5, 2002.

[102] G. Shabtay, D. Mendlovic, and Y. Itzhar, "Optical Single Channel Dispersion Compensation Devices and Their Application," in Proc. *ECOC*, We1.2.1, 2005.

[103] L. Gruner-Nielsen, M. Wandel, P. Kristensen et al., "Dispersion-compensating fibers," *J. Lightw. Technol.*, 23, 3566–3579, November 2005.

[104] R. Nagarajan et al., "Large-scale photonic integrated circuits," *IEEE J. Sel. Top. Quantum Electron.*, 11, 50–65, 2005.

10

III–V photonic integrated circuits and their impact on optical network architectures

Dave Welch, Chuck Joyner, Damien Lambert, Peter W. Evans, and Maura Raburn

Infinera Inc., Sunnyvale, CA, USA

10.1 INTRODUCTION

When Miller and Chynoweth [1] published the first volume in this series of books in 1979, few communications experts were convinced that fiberoptic-based telecommunication systems would become the backbone of the communications industry. While confidence in optical technology increased slowly through the 1980s, it was not until the introduction of the erbium-doped fiber amplifier (EDFA) in 1992 and the subsequent growth of dense wavelength-division multiplexed (DWDM) signals that fiberoptic-based communications became the technology of choice for a high-bandwidth communications infrastructure. In the decade since its widespread adoption, optical transport has seen several major technological advancements. In particular, the introduction of 10 Gb/s transmission optics and the more recent advancement of reconfigurable optical add/drop multiplexers (ROADMs) have continued to strengthen the capabilities of the fiberoptic network. Until recently, the network has remained essentially "all optical." Information traveled in optical format from source to final destination as shown in the upper half of Figure 10.1. One of the inherent shortcomings in management of this type of communications network is the high cost of gaining access to the bits themselves. In response, the conventional optical network has been designed to minimize the need for management of the signal, with the result that it has been burdened instead with the need for "analog management" and forced to use proxies

Optical Fiber Telecommunications V A: Components and Subsystems
Copyright © 2008, Elsevier Inc. All rights reserved.
ISBN: 978-0-12-374171-4

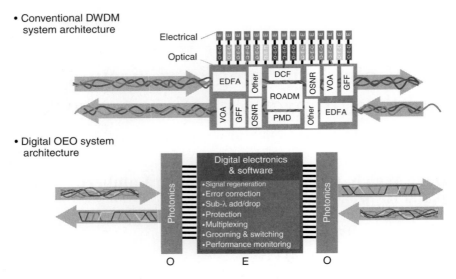

Figure 10.1 A comparison of a transport node in an "all optical" or analog transport system (upper) vs a digital transport system (lower). VOA, variable optical attenuator; GFF, gain flattening filter; DCF, dispersion compensating fiber; PMD, polarization mode distortion; OEO, optical-to-electrical-to-optical (this figure may be seen in color on the included CD-ROM).

such as optical signal to noise ratio (OSNR) to infer the bit-error rate (BER) of the system.

This strategy has indeed improved the capacity and cost/Gb of today's fiber-optic networks compared to those operating circa 1990, but at the cost of a reduction in network functionality, fidelity, and flexibility.

In a digital optical network (DON), as shown in the lower half of Figure 10.1, the information in the fiber is transformed from the optical domain to the electrical domain before optical retransmission at every node. This format allows full access to both the data in the network and optically transmitted network control signals at every node. Signals undergo complete conventional regeneration (retiming, reshaping, and reamplification) with error correction as well. Information may be added or dropped at sub-wavelength granularity. Information can be groomed, switched, and routed in multiple directions at greater depth within the network. And complete performance monitoring at the BER level is possible at every node.

If one had to construct a digital network entirely from discrete, commercially available components, it would be technically impractical, due in large part to the number of fiber connections involved. It would be economically prohibitive, due to the number of expensive individual hermetic packages that would be required. Similar concerns existed in the electronics industry circa 1975 when Robert Noyce [2] and others made the case for larger scale electronic integration. In Figure 10.2, the individual chip cost is expected to increase with increased function count as, given a constant defect density, the chance of failure increases exponentially with

10. III–V Photonic Integrated Circuits

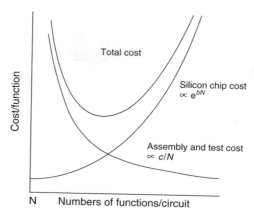

Figure 10.2 Conceptual plot showing the trade-off between chip cost vs assembly plus testing, where b and c are constants to be determined by fitting the curves.

chip size. Test and assembly costs drop as $1/N$ because the number of parts assembled and tested becomes smaller. The logic that prevailed then was simple. *The economic value derived from the integrated component must outweigh the cost of the integration itself.* Regretfully for photonics, overcoming the entrenched industry practices to choose the appropriate technology platform, and accurately assessing the cost of integration as well as the economic benefit derived, has not been as simple as Noyce's argument might suggest.

Recent advances in InP substrate quality, commercial availability of high-uniformity epitaxial growth and fabrication equipment, and process-line management tools pioneered by the silicon electronics industry combine to make large-scale monolithic integration on InP a practical reality.

In this chapter, we will look at the current state of InP-based photonic integration; recent manufacturing advances in the fabrication of III–V devices; and the impact that this is having on optical network architectures. Finally, we will comment on the possible directions future VLSI photonics devices may take optical networks.

10.2 PHOTONIC MATERIAL INTEGRATION METHODS

Individual component devices that make up an integrated photonic circuit are generally best made from materials of different bandgaps and junction placements. For example, laser material must enable efficient light emission under forward bias, modulator material should allow low and high attenuation in the "on" and "off" states under reverse bias, and passive waveguides require a material with minimal absorption. A multitude of successful material integration methods exist.

One of the most popular approaches, butt-joint regrowth [3, 4], involves etching away part of the initial epitaxial layer structure and regrowing a different material in the newly opened space, Figure 10.3(a). The initial epitaxial layer is often protected with the dielectric mask used to perform this etch.

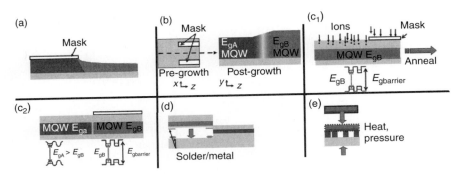

Figure 10.3 Materials integration methods. (a) Regrowth, (b) selective area growth, quantum well intermixing showing (c_1) ion implant and (c_2) post-anneal results, (d) flip-chip bonding, (e) direct-contact wafer bonding (this figure may be seen in color on the included CD-ROM).

Selective area growth (SAG), shown in Figure 10.3(b), enables the simultaneous growth of regions of varying bandgap in a single plane [5–8]. Due to the difference in gas phase diffusion coefficients of In and Ga precursors during metalorganic vapor phase epitaxy (MOVPE), dielectric patterns with masked vs open areas lead to quantum wells that are thicker and more In-rich when the masked areas are wide or the open gaps narrow. Through the quantum-size effect and alloy composition change, multiple quantum well (MQW)-region bandgap red shifting by well over 200 nm is possible [6].

Quantum well intermixing (QWI) involves the exchange of group III and V species between wells and barriers to create changes of well width and barrier height in selected areas post-growth [9–12], as shown in Figure 10.3(c). Bandgap blue shifting by over 100 nm has been achieved using QWI in the InP system [10, 12].

Flip-chip bonding is achieved through placing two substrates or chips in contact, face-to-face, and securing them with molten solder or other adhesive material (Figure 10.3(d)) [13]. Polymer or an epoxy can also be used to indirectly bond a patterned waveguide structure face-down onto a host substrate for subsequent growth substrate removal and further waveguide fabrication for double-side-processed vertically coupled waveguide devices [14].

Direct-contact wafer bonding involves bringing two very clean substrates, epitaxial layers, or processed devices into contact and annealing under pressure to create a strong adhesive-free bond. Successful bonds can have mass transport at the bonded interface such that no voids or gaps exist, enabling good optical properties and, in some cases, current flow across the interface. The materials can be dissimilar, such as InP and Si, InP and GaAs, Si and SiO_2, or InP and SiO_2. They can also be patterned on the contact surface, as in Figure 10.3(e). Direct-contact wafer bonding of materials has been used to create devices of superior performance, not possible with a single material system, [15–18], and unusual device geometries that cannot be reasonably created by any other means [19].

10.3 III–V PHOTONIC INTEGRATED CIRCUIT SMALL-SCALE INTEGRATION

10.3.1 Single-Channel Multi-Component Chips

Recent photonic integrated circuit (PIC) transmitter (TX) advances include the development of uncooled TX optical sub-assemblies (TOSAs). Devices which do not require a thermoelectric cooler (TEC), or work at a high temperature though temperature controlled, enable a lower electrical power budget and reduced production costs. Uncooled 1.55-μm 10 Gb/s TOSAs operate over temperature ranges as great as 80°C, with low dispersion, reasonable output powers, and stable extinction ratios [20–22]. Distributed feedback (DFB) lasers designed for high-temperature operation require detuning of the Bragg grating wavelength from the gain peak at room temperature such that they will become aligned at the higher operating temperature. In general, the semiconductor gain peak and the Bragg grating wavelength shift at different rates with temperature, as shown in Figure 10.4(a). The intended operating temperature range must be predetermined to be able to calculate the required grating offset. In Figure 10.4(a), a 25-nm offset of the grating pitch was required to allow an operating temperature around 65°C. Such high-temperature DFBs have enabled 10 Gb/s transmission at 90°C [23].

Figure 10.4(b) shows the power of the simulation tools available for high-temperature laser design. Here, the solid curves are calculated while the points are experimental data for the measured parameters. The characteristic temperature, T_0, is an indication of the laser threshold current sensitivity to temperature, as defined by $I_{th} = I_0 \, e^{(T/T_0)}$.

Another notable commercialization is that of widely tunable 1.55 μm-based lasers and electroabsorption-modulated lasers (EMLs). One tuning approach is

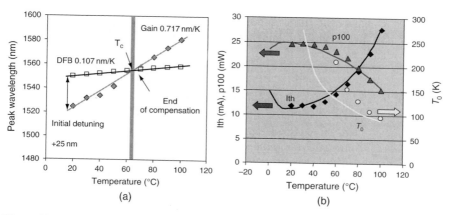

Figure 10.4 Temperature dependence of distributed feedback (DFB) with +25-nm Bragg grating detuning from the gain peak. (a) Thermal dependence of DFB (grating) and gain peak wavelengths and (b) simulated and experimental laser thresholds, power at 100 mA, and T_0 thermal behavior [23].

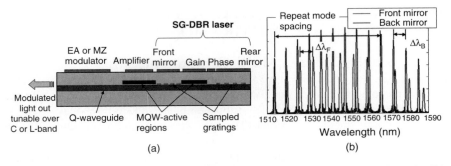

Figure 10.5 (a) Schematic of a wavelength-tunable SG-DBR laser integrated with an SOA and an EA modulator. (b) Reflectivity spectrum of the sampled grating mirrors showing the multiple mirror peaks used to cover the tuning range [24].

with a sampled grating as shown in the sampled-grating distributed Bragg reflector (SG-DBR) laser in Figure 10.5(a) [24]. These periodically blanked gratings produce reflection "combs" in wavelength (Figure 10.5(b)). The differing comb pitches of the front and back mirrors lead to large changes, relative to standard DBRs, in lasing wavelength with tuning current. The device in Figure 10.5 features integration with an electroabsorption modulator (EAM) and semiconductor optical amplifier (SOA) through an offset-quantum well etch-and-regrow approach. Sampled gratings producing reflection combs can also be used with multi-contact-chirped gratings for widely tunable lasers [25]. Although SG-DBRs can span lasing wavelengths of many tens of nanometers with reasonable powers, wavelength stability and mode-hop-free tuning of such lasers can be complex. Tuning with simpler wavelength stability can be achieved with a multi-wavelength DFB array and a micro-electrical mechanical systems (MEMS) tilt mirror [26].

10.3.2 Multi-Channel Interference-Based Active Devices

Interference-based devices allow manipulation of individual channels with multiplexing and demultiplexing functionality. Butt-joint regrowth, SAG, and QWI have enabled multiple-bandgap active interference-based devices. The 64-channel dual-arrayed waveguide grating (AWG) optical-code division multiple access (O-CDMA) encoder/decoder chip is shown in Figure 10.6. Fabricated through an offset quantum well etch-and-regrowth technique, it is one of the most compact of its kind [27, 28]. Multi-channel wavelength-division multiplexed (WDM) modulators consisting of AWG, EAM, and SOA arrays have been fabricated using butt-joint selective growth techniques [29]. Butt-joint regrowth has also been used to realize integrated InP AWG-based multi-wavelength single state [30] and dual state [31] SOA ring lasers. SAG has been used to make single-growth SOA-based Mach–Zehnder interferometer switches [32] suitable for multi-channel integration.

10. III–V Photonic Integrated Circuits

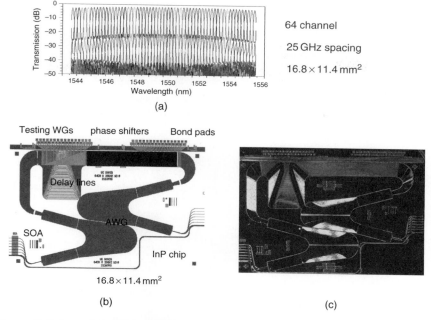

Figure 10.6 Dense InP AWG (25 GHz)-based 64-channel O-CDMA encoder/decoder chip [27]. (a) Transmission as a function of wavelength, (b) schematic of device layout, and (c) fabricated InP device (this figure may be seen in color on the included CD-ROM).

Multi-Channel Ring Resonator Devices

Micro-ring and micro-disk resonators enable very compact and powerful multiplexing, demultiplexing, and filtering devices. Although small-bend radii can have high-optical losses, the smallness of the disks or rings enable the dense channel spacing in the spatial domain. Fabrication of such devices from III–V materials enables active switching and the addition of gain or loss. Vertically coupled rings made through techniques such as wafer bonding, as shown in Figure 10.7 [33], have less stringent waveguide spacing tolerances than horizontally coupled rings with air gaps between coupled guides [14]. Also, vertical waveguide spacing set by epitaxial growth thicknesses is typically easier to control than a spacing set by etched grooves, as is the case with horizontally coupled guides.

Novel Devices and Integration Methods

One noteworthy integration technique, wafer bonding, has enabled the realization of hybrid silicon–InP lasers and detectors [34, 35]. As silicon has a higher refractive index than 1.55-μm bandgap InGaAsP, the optical mode in Figure 10.8(a) is largely confined to the silicon. However, the part of the mode residing in the InP experiences an optical gain sufficient for the mode to lase. A scanning electron micrograph

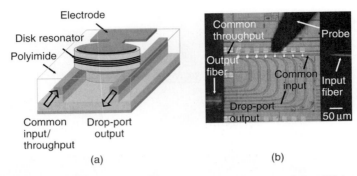

Figure 10.7 (a) Schematic of active micro-disk resonator vertically coupled to I/O bus lines and (b) micrograph showing top view of a fabricated eight-channel active micro-disk demultiplexer [33].

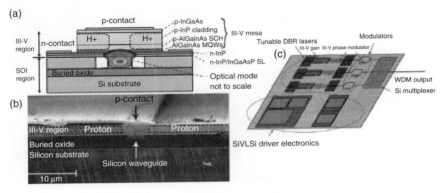

Figure 10.8 (a) Schematic of the hybrid laser structure with the optical mode superimposed. (b) SEM cross-sectional image of fabricated hybrid AlGaInAs–Si evanescent laser. (c) Proposed structure of Si electronics wavelength-division multiplexed (WDM) transmitter [34, 35].

(SEM) of the cross-section of the laser is shown in Figure 10.8(b). An idea for a WDM TX is shown in Figure 10.8(c), with the active optical devices (lasers, amplifiers, and modulators) made from InP/silicon hybrid structures and the passive optical devices (multiplexer) and electronics made from silicon only.

Optical isolators allow light propagation in one direction while greatly inhibiting propagation in the opposite direction. In general, the principle of Faraday rotation is used in conjunction with polarizers to produce this effect on both TE and TM optical modes. On-chip integration of optical isolators would allow more robust PIC layouts involving back-reflection-sensitive devices, such as lasers, where optical feedback may cause mode hoping with accompanying unwanted frequency shifts. So far, fabrication attempts based on Faraday rotation have yielded insufficient forward-to-reverse isolation ratios in InP-based devices. A new approach is shown in Figure 10.9. As most laser sources on InP use compressively strained quantum wells, the dominant emission is in the TE mode. Therefore, as a practical matter, an effective isolator only

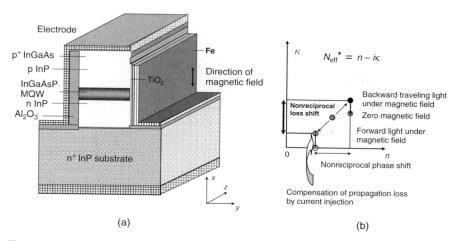

Figure 10.9 (a) Device structure of the TE mode waveguide optical isolator. Fe layer is placed on one sidewall of the high-mesa SOA waveguide. (b) Schematic of operational principle of waveguide optical isolator based on the nonreciprocal loss shift [36] (this figure may be seen in color on the included CD-ROM).

needs to achieve nonreciprocal propagation for the TE mode. The isolator of Figure 10.9 is built around a SOA. A thin-film Fe layer is deposited on only one side of the SOA waveguide and poled so that the magnetization vector of the magnetooptic material (Fe) is aligned parallel to the magnetic field vector of the TE mode light, perpendicular to both the waveguide and the substrate. TE light traveling in the forward direction is pushed laterally to the low-loss side of the SOA. TE light traveling in the reverse direction is drawn laterally toward the higher loss side of the waveguide where the magnet is located. A TE isolation ratio of nearly 15 dB [36] is achieved. In addition, device loss is compensated by the built-in SOA.

Polarization-independent monolithically integrated waveguide optical circulators have also been proposed using InP half-wave plates and Faraday rotators in waveguide form [37].

10.3.3 Large-Scale III–V Photonic Integration

Since 2004, large-scale photonic circuits based on InP have been commercially available [38]. These monolithically integrated transmitter (TX) and receiver (RX) PICs have been employed to form a DON as described above and in Section 10.5 below. A schematic of the first generation of TX large-scale photonic integrated circuits (LS-PICs) is shown in Figure 10.10. The optical signals in each of the 10 channels originate in an active section and are multiplexed into a single output channel exiting the chip. The active train of each channel includes a tunable EML—a tunable DFB laser integrated with an EAM operating at a data rate of 10 Gb/s. The EMLs are individually controlled with direct current (DC) bias on the DFBs and 10 Gb/s input

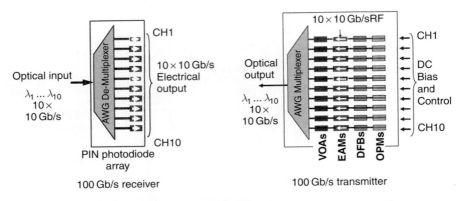

Figure 10.10 Schematic diagrams of 100-Gb/s transmitter and receiver PICs.

on the EAMs. The wavelength is chirped across the monolithically integrated DFB array to yield 10 distinct and highly controllable wavelengths that can each be fine-tuned to the International Telecommunications Union (ITU) frequency grid. A photodiode is monolithically integrated into the back of each channel to monitor the output power of the DFB over the lifetime of the chip. Control of the output power profile across all channels is enabled with a monolithically integrated variable optical attenuator (VOA), essentially an absorber in the active train light path of each channel. Multiplexing the 10 wavelengths into a single output channel is accomplished with an AWG router monolithically integrated in the passive section of the chip. The single output channel is terminated in a spot-size converter for optimized fiber coupling. AWGs are ideally suited for LS-PICs, given their integrability in InP, high-channel count, and low-insertion loss characteristics. This chip represents the first realization of the combined functionality—10-Gb/s modulated source integrated with continuous tunability, AWG frequency-selective multiplexing, and power profiling through the use of VOAs, all operating at performance levels required by commercial class carrier networks.

Some of the key performance metrics of the TX PIC are shown in Figure 10.11.

In Figure 10.11(a), a typical light vs current/voltage (L–I–V) plot of the DFB laser array is shown with all 10 channels superimposed. The data were taken using the per channel power monitor shown in Figure 10.1 as the detector. The lasing threshold (I_{th}) for all channels ranges from 20 to 28 mA. A typical voltage at turn on is 1.2 V. The devices exhibit a forward resistance of 5.5 (±0.3) ohms and operate in a current range of 60–80 mA, typically.

In Figure 10.11(b), the DFB spectrum is superposed with the AWG multiplexer transmission function. This is achieved by forward biasing the EAMs to produce an ASE emission source to map the AWG passbands. The center position of the AWG passband comb is tuned using the TEC in the packaged module to match the ITU grid. Each DFB is subsequently individually thermally tuned onto the ITU

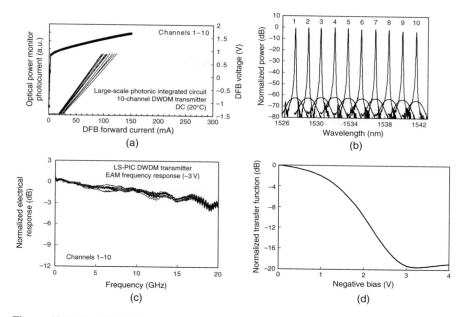

Figure 10.11 Key TX PIC performance metrics: (a) the LIV curves for all 10-DFB lasers. (b) Superposition of the 10-channel DFB spectrum with the AWG multiplexer transmission function for all 10 channels. (c) The small-signal frequency response for the 10 EAMs of a transmitter superimposed. (d) The power transfer function of a VOA.

grid as well. The tunable DFBs are further employed to account for small frequency drifts to maintain the frequency of each channel to within ±3 GHz over life. The tuning range of the DFBs exceeds 300 GHz (data not shown). Control of the relative channel spacing aperiodicity on the 200 GHz grid for the AWG passbands is better than ±4 GHz (one sigma of a normal distribution). The loss penalty due to misalignment of any AWG passband to the ITU grid for the TX is less than 0.1 dB.

The response vs frequency curves for the modulators of all 10 channels have been superimposed and are shown in Figure 10.11(c). The −3-dB bandwidth is better than 17 GHz for the worst channel. The variation in frequency response from channel to channel is on the order of 0.5 dB and largely due to variation in error of measurement with probe placement. The ripple on the plots is a combination of small electrical reflections from the PIC layout and reflections internal to the probe. This data indicates that the uniformity of the modulator capacitance and the impedance matching of the bond pad configuration across the array are very high.

The VOAs are used to control the shape of the 10-channel output power spectrum envelope. They are the last element in the TX signal chain before multiplexing the signals from all channels through the AWG. This capability is critical to achieving the optimal TX power spectrum before it is launched into the fiber. In addition, the VOAs can be used to compensate for any small differences in output power between

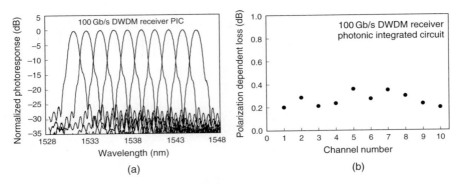

Figure 10.12 Key DC performance characteristics of the Receiver (RX) PIC. (a) Normalized photoresponse of the 100 Gb/s (10 × 10 Gb/s) dense wavelength-division multiplexed (DWDM) RX PIC. (b) Polarization-dependent loss as a function of channel number for the 100 Gb/s (10 × 10 Gb/s) DWDM RX PIC.

the DFBs over their lifetime. The DC transfer function for a typical VOA element is shown in Figure 10.11(d). An attenuation range of 5.5 dB is possible by employing between 0 and −2 V reverse bias.

Two of the DC performance metrics for the RX PIC, depicted schematically in Figure 10.10, are shown in Figure 10.12. Figure 10.12(a) shows the normalized photoresponse of the RX. The input to the device is a tunable laser. The output is measured as the DC photocurrent at the PIN array. The nominal fiber-coupled responsivity of the integrated device is sufficient to effectively transmit data across all long-haul links encountered in real-world networks and compares favorably with currently available discrete combinations of commercial components.

Unlike the TX PIC, the integrated RX PIC chip needs to be polarization independent. To determine the polarization-dependent loss (PDL) and polarization-dependent wavelength shift (PDWS) of the RX PIC, the maximum and minimum photoresponse is measured as the polarization state of the input light is varied. Figure 10.12 shows the nominal PDL of the RX as a function of channel number. The per-channel PDL is less than 0.5 dB for all 10 channels with a median of 0.3 dB. For the single polarization spectrum shown in Figure 10.11, the flatness (which affects the dynamic range of the RX) is better than 0.4 dB. The median PDWS for all 10 channels is 14 GHz (the minimum value is 12 GHz and the maximum value is 17 GHz).

PICs are not reach limited and do not require an advanced modulation format to achieve ultralong-haul transmission. Typical TX and RX line cards were used to perform the systems experiment shown in Figure 10.13.

This test bed configuration has 26 spans of 75–100 km fiber each with a hybrid Raman/EDFA after every span (Figure 10.13(c)). A gain flattening element is used between nodes 13 and 14 to adjust the power spectrum midway through the 2000-km link. Figure 10.13(b) shows the Q measurement vs wavelength for zero dispersion across 2000 km. The data are plotted as $20\log(Q)$ as a function of OSNR. An excellent

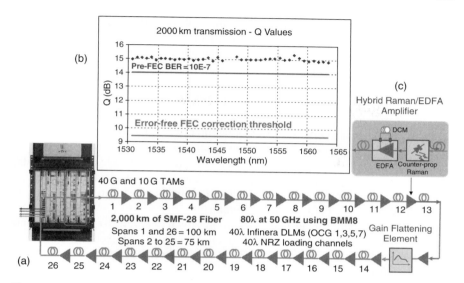

Figure 10.13 (a) Schematic of a nonreturn-to-zero (NRZ) system test bed utilized for evaluating the performance of the DWDM transceiver PIC-based line cards over multiple fiber spans. (b) Q vs wavelength of a DWDM PIC-based transmitter–receiver module pair on a system line card. (c) The hybrid Raman/EDFA amplifier, containing a dispersion compensation module (DCM), is used at each skipped transport node (this figure may be seen in color on the included CD-ROM).

discussion on the definition of Q, its practical applications, and its experimental determination may be found in a publication by Bergano et al. [39]. In addition, the same set up was used to show error-free transmission for 65 ps of added differential group delay (DGD) at a data rate of 40 Gb/s.

Until just a few years ago, forward error correction (FEC) was utilized solely in high-end submarine transmission systems. There are several FEC schemes that have been demonstrated to date, and most of them are proprietary. Recently, ITU has approved G.975.1 (02/2004), the implementation of FEC in DWDM systems.

The on-PIC VOA allows one to correct, within limits, for a large part, the imperfections in gain flatness of an EDFA chain. The demonstrated uniformity in C-band transmission performance over 26 fiber segments is only possible with a very tight control of the chirp and other modulation properties of the EAM. This, is turn, is enabled by precise controls realized during PIC fabrication.

10.4 MANUFACTURING ADVANCES FOR III–V FABRICATION IMPLYING SCALABILITY

The rapid growth of the DON would not have been possible without the achievement of low-cost optical-to-electrical-to-optical (OEO) conversion. Such an economically compelling optical transport network would not have been possible without the integration of multiple optoelectrical discrete devices within one single pair of chips as described above. In this section, we will review many of the methodologies, which leveraged the InP integration experience from the silicon-based VLSI industry and enabled LS-PIC technology [38].

10.4.1 Design for Manufacturability

Some of the major challenges to the production of LS-PICs were addressed by the implementation of a robust *design for manufacturability*. The fabrication of LS-PICs leveraged the experience developed in the silicon-based VLSI industry by employing similar design techniques, process equipment, and manufacturing technology. Breaking the traditions of the III–V industry, designs were screened for process capability before implementation.

Consider that the gaussian-like distribution shown in Figure 10.14(a) represents the frequency of values (y-axis) of resist thickness (x-axis) obtained in a well-tuned wafer coating process as measured on many samples run over a period of time. The *control limits* of the process are set at this 6σ range. As long as any subsequent thickness measurement falls within this 6σ distribution, the tool is "in control." The fabrication process, however, must be designed with wider *specification limits* to accept resist thicknesses both thicker and thinner than this range by a certain margin. Process capability (C_p) is defined in Figure 10.14 where the upper specification limit (USL) and the lower specification limit (LSL) bound the performance values required for a component to pass the product specification, for that figure of merit. The 6σ in the denominator refers to the full distribution range of values (resist thicknesses, in this example) for the figure of merit normally achieved in fabricating or testing the component. C_p values greater than 1 describe a capable process.

C_{pk}, the capability index, is a further description of process health, indicating how far the mean (μ) of the achievable distribution is inside the closest specification limit.

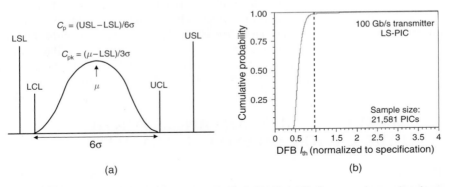

Figure 10.14 Process capability (C_p) and the capability index (C_{pk}) (this figure may be seen in color on the included CD-ROM).

For a process to be robustly capable, the C_{pk} should be greater than 2. A process with a C_{pk} greater than three may be considered "good by design" and require only occasional monitoring. In the example of Figure 10.14(a) the distribution has "walked" too close to the LSL and is due to be "recentered" to achieve a better C_{pk}.

If a process reaches the upper control limit (UCL) or the lower control limit (LCL), then corrective action is required. Maintaining *statistical process control* (SPC) for each significant step in a process insures that the specification limits will seldom be reached.

A complex manufacturing flow should be partitioned into simple process steps obeying well-defined, robust design rules. The resulting step-level design simplifications greatly enhance overall process capabilities. In a few instances, this means that any individual component may suffer a small reduction in some figure of merit for noncritical parameters as determined by transmission system design analysis. However, the advantages of having all of the components on a single vertically integrated platform, with no fiber couplings to connect them, more than offsets any small individual performance loss. In addition, the reliability for each component of a complex system is multiplicative. The monolithic integration of multiple discrete devices within one single chip addresses both manufacturing cost and system reliability issues.

Figure 10.14(b) shows a cumulative probability plot for the maximum threshold current (I_{th}) figure of merit for monolithically integrated DFBs on more than 20,000 TX PICs. Here, the x-axis shows the DFB maximum I_{th} value normalized to 1. The y-axis shows the percentage population of all threshold values. The portion of the distribution to the right of the dotted line at "1" is the population that fails. In this example, less than 2% of all lasers tested fail to meet the maximum I_{th} specification. By using SPC on all the contributing factors to threshold performance, it is possible to reduce the number of fails to an arbitrarily small percentage. As a practical matter, the cost of such a highly capable process at all steps may be prohibitive. The decision regarding where or when to invest in equipment or engineering effort is made clear and objective by having the SPC data along with a working economic model for the entire production process.

During the development of a new process, modern methods of process experimentation such as the design of experiments (DOE) [40] or the *Taguchi* method should be employed to maximize gained information while minimizing all required resources. Through the proper design of a manufacturing step, the process can be made insensitive to variations, thus avoiding the costly eventualities of rejection and/or rework.

Factory *automation* addresses the manufacturing challenges generated by the need to mange complexity while improving repeatability and reproducibility. Automation allows managers to optimize a production line subject to changing requirements and demands for increasing scalability of production volumes. The synchronized automation of factory operations, production equipment, material handling systems, factory database systems, and facility management systems provides the ability to set polices and procedures to better control the production within the factory. Similarly, the automation of both process and metrology equipment enhances the ability to extract real-time information and to quickly measure process impact on performance. The reduction of manual handling of wafers and chips considerably reduces wafer breakage and defect generation, which in turn significantly improves yields.

To achieve high-capacity ramp rates in the semiconductor wafer fabrication process, strict attention must be paid to the minimization of *cycle time* through each phase of product development. Rapid turns through a process are also the most powerful way to move up the learning curve to higher yields. Initially, a form of cycle time reduction may be obtained by the implementation of *short loops* which only investigate variations on restricted parts of a process. Short loops provide rapid process feedback for down selection among many competing options in a process.

Process monitor (PM) wafers are another help to reduce in-line errors and requalify tools after process maintenance. PM wafers are not for product and designed specifically to test a narrow range of critical metrics on a process tool to confirm that it is working properly for a particular task.

Keeping the work-in-progress (WIP) restricted only to the wafers viable to yield for product is another way to reduce cycle time and keep wasted resources to a minimum. In the next section, we will discuss several techniques to accomplish in-line monitoring for early rejection of wafers that will not yield well.

10.4.2 In-Line Testing

The highly serialized nature of wafer processing for LS-PICs requires metrology in between the various processing steps. Wafer test metrology equipment must be extensively used to frequently verify that the wafers in process meet all product quality requirements. In-line testing enables early detection of process drifts and screens poor quality wafers out of the production line [41]. Most metrology measurements are made on special in-line wafer test structures which are in cells on the wafer devoted entirely to process quality management. Additional in-line wafer testing is performed on dropped-in test structures inserted in the layout on product die, which provide yield estimation and reliability prediction based on the outcome of the measurements [42]. If the number of failed chips on a wafer exceeds a predetermined threshold, the wafer is scrapped rather than being invested into further processing. As a result, material cost, equipment utilization, and labor utilization are minimized for low-yielding wafers. In addition, drifting pieces of equipment or deviant processes are identified in a real-time manner enabling corrective action plans to take effect immediately and restricting the volume of wafers at risk for low yields.

Upon completion of the wafer level fabrication steps, wafers are subject to a variety of computer-controlled electrical tests to determine whether the LS-PICs each function properly. Modern methods of pattern recognition are used to enable full automation of the alignment of the chips from a wafer to probe cards of an electronic tester. During wafer probing, all individual dice are tested for functional defects on every component within a dice. Defects are cataloged as to component, function, and position. Once all tests pass for a specific die, its position is electronically recorded for later use during visual inspection and singulation. Wafer maps provide an efficient screen that enables assembly operations to concentrate die-fab resources only on devices performing to specifications. Once the assembly of a LS-PIC onto a carrier is complete, test patterns similar to those employed during wafer probing are used to perform more advanced tests and characterization. These procedures assess both performance and reliability of the chip before passing final outgoing quality assurance gates.

10.4.3 Yield Management Methodologies

Similar to what has been proven in the silicon-based IC industry, one of the highest leverage activities for improving the profitability of manufacturing LS-PICs is probe yield improvement. Integrated yield management (IYM) is one of the most powerful methods to achieve this goal. The technique was developed at IBM by Mel Effron and presented at Silicon Systems Incorporated in 1994.

IYM methodologies quantify random vs systematic defects, identify all sources of yield loss, isolate the root cause of yield loss, and control the cause of yield loss. A simplified version of an IYM triangle, shown in Figure 10.15, illustrates the overall sequential yield analysis methodology [43]. The IYM triangle is partitioned into six levels, starting from the general cluster analysis at the top to the more detailed analysis toward the bottom and finally the in-line and in situ process controls at the base.

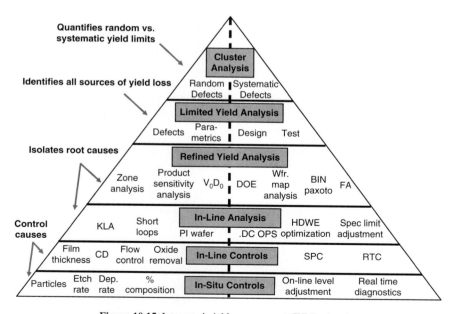

Figure 10.15 Integrated yield management (IYM) triangle.

Cluster Analysis Using the Negative Binomial Model

For a given product line, both random defect yield limits and systematic yield limits can be extracted using cluster analysis [44]. Reorganization of wafer probe maps grouping adjacent devices into cells containing 1, 2, 3, 4 (2 × 2), 6 (2 × 3), and 9 (3 × 3) chips enables an analyst to emulate and calculate the percent yield for dice with various area sizes. All chips of a given cell must pass all tests for a grouped cell to pass and qualify for production. Using a least square method, the six yield points obtained for the various areas are fitted to an equation representing the defect distribution. When the cluster factor α is unknown for the defect density probability, the negative binomial model has been recognized to be the most mathematically rigorous random defect distribution model:

$$Y = Y_S \cdot Y_D \quad \text{with} \quad Y_D = \frac{1}{(1+(AD/\alpha))^\alpha}, \tag{10.1}$$

where Y_S is the systematic yield limit, Y_D is the random defect yield limit, D is the density of random fatal defects in units of defects per die, A is the area of the cell, and α is the *cluster factor*, which measures the tendency for defects to cluster or depart from total randomness. Small values for α indicate a high degree of clustering and a greater probability for killer defects to be located on the same or adjacent chips. By contrast, values of α approaching infinity indicate that the yield probability approaches a Poisson distribution probability, and the defects are distributed randomly at the surface of the wafers.

Limited Yield Analysis

While the cluster analysis accurately evaluates the relative contribution of the yield limits Y_S and Y_D in the final yield, it does not provide any information about the sub-components of Y_S nor does it help find the root causes of the yield loss. The systematic yield limit Y_S can be broken down into its sub-components. These sub-components are represented by the design yield limits Y_P, the test yield limits Y_T, and the systematic process limits Y_{SP}:

$$Y_S = \prod_i Y_{P\hat{i}} \cdot \prod_j Y_{T\hat{j}} \cdot \prod_k Y_{SP\hat{k}} \tag{10.2}$$

with

$$Y_{SP\hat{k}} = \int F(SP_k) \cdot Y(SP_k) \cdot dSP_k \tag{10.3}$$

where $F(SP_k)$ is frequency distribution for parameter SP_k and $Y(SP_k)$ is the yield as a function of that parameter. Root causes for each of the yield limits Y_D, Y_P, Y_T, and Y_{SP} can be further investigated using the refined yield analysis.

Refined Yield Analysis

Zone analysis identifies specific locations on the surface of wafers that contain systematic defects contributing to Y_{SP}. In this method, wafer yields are partitioned by zones on wafers. Yield improvement efforts are then concentrated on specific regions of the wafers that are responsible for the majority of the yield loss. The *product sensitivity analysis* (PSA) compares and correlates in-line electrical or physical parameters to performance parameters tested at wafer probe. This method highlights design or process deficiencies contributing to Y_P and Y_{SP} yield limits. It offers significant advantages compared to other techniques because it requires relatively few wafers, it can be implemented early in a development cycle, and it quickly elucidates possible causes of yield loss. One method to investigate the systematic process limits Y_{SP} consists of performing a DOE analysis. Typically, well-constructed designs attempt to test and statistically manage corners of the process parameters space. Combining extremes of the fabrication steps by intentionally varying multiple process parameters reveals how they interact and how they can impact a number of sensitive performance parameters. The information extracted from the DOE analysis can be leveraged to update designs or process specification limits.

In-Line Analysis

The *defect budgeting* methodology offers a means to limit the search for sources of random defects. This technique isolates short subsets of process sequence steps in a complex manufacturing flow and introduces strategic in-line visual inspections. Identified defects or particles are totaled during the inspections and correlated to the most probable sources within the process flow.

Wafer maps built from multiple in-line *visual inspections* provide important knowledge regarding visible killer defects and how the defects correlate with performance parameters at probe test. The study is typically performed on a sample of random wafers at multiple key processing steps. The work performed on process yield improvement is prioritized based on the random yield limits calculated from the overlay of the defect maps in correlation with the probe maps.

10.4.4 Defect Density and Functionality Per Chip

Scaling DWDM transmitter and receiver communication rates with a cost-effective manufacturing business model entails increasing the amount of integrated components in a single LS-PIC. Typically, this requirement results in a steady increase of chip size. In order for the economics of manufacturing large-size devices to remain compelling, random defect density must be constantly improved. Indeed, reducing defect density ensures that the manufacturing yields remain high and that the allowable chip size scales up with device integration demand.

Recently, Infinera has demonstrated the ability to manufacture LS-PICs with yields commensurate for cost-effective commercialization [45]. In part, this was achieved by a constant effort to reduce both visual and electrical defect densities on the surface of wafers through strict manufacturing process controls and robust component designs. A cluster analysis employing the negative binomial distribution model discussed in Section 10.4.3 was applied on yield data from wafer-level probe testers. Random yield limit, systematic yield limit, fatal defect density, and cluster factors were extracted from the calculations. Excellent data fits are obtained with $D = 0.86\,\text{cm}^{-2}$ and $\alpha = 4.7$ for the second half of 2006. Similar good fits were obtained for the first half of 2006 and the second half of 2005. As displayed in Figure 10.16, the random defect density on currently manufactured PICs displays continued reduction every 6 months. A defect density of $0.86\,\text{cm}^{-2}$ and a cluster factor of 4.7 are evidence of the high-yield capability rapidly achieved in the manufacturing of LS-PICs. World-leading IC foundries measure cluster factors between 8 and 100.

In terms of random defect density control, Figure 10.17 exhibits that current LS-PIC technology is comparable to what was achievable for the silicon IC industry in the early 1990s [46]. At that time, Intel was manufacturing 486^{TM} processors integrating over one million transistors in a single chip. Recent development work extending the level of integration from the device displayed in Figure 10.10 has led to the demonstration of a single 40-channel LS-PIC TX chip, which is capable of operating at an aggregate data rate of 1.6 Tb/s [47]. The L–I–V curves for a single chip integrating over 200 optoelectronic components is displayed in Figure 10.18(a). The output power of the DFBs was extracted from the linear photocurrent measured by the integrated power monitor located at the rear of the lasers. The true output power of the DFBs was computed taking into account the insertion loss between the laser and the detector as well as the responsivity of the detector.

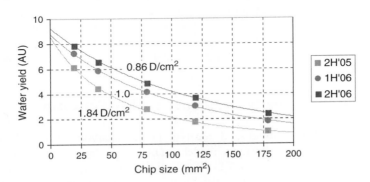

Figure 10.16 Progression of wafers yields vs chip size for various time periods of the LS-PIC production. A low-defect density of $0.86\,\text{cm}^{-2}$ combined with a high-cluster parameter of 4.7 indicate a high-production yield process capability. A cluster analysis with a negative binomial distribution model was employed to fit test probe yield data (this figure may be seen in color on the included CD-ROM).

10. III–V Photonic Integrated Circuits

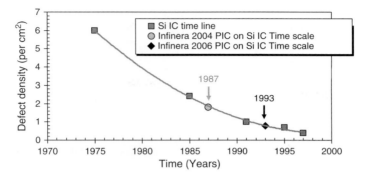

Figure 10.17 Position of PIC technology relative to the VLSI evolutionary time scale. LS-PIC defect densities during 2004 and 2006 are compared to the Si IC defect density through a period of 23 years. This plot provides an estimate of the LS-PIC technology's maturity with respect to the VLSI evolution.

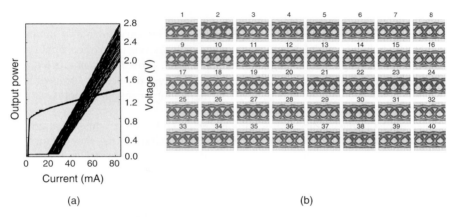

Figure 10.18 (a) The L–I–V characteristics of the DFB array on a 40-channel LS-PIC show a tight distribution on all channels. (b) The eye diagrams for all channels operating at 40 Gb/s (this figure may be seen in color on the included CD-ROM).

Figure 10.18(b) shows the eye diagrams for each channel in the 1.6-Tb/s TX PIC operated independently at 40 Gb/s. The normalized fiber-coupled output power spectrum of the 1.6 Tb/s LS-PIC at a temperature stabilized around 25°C is shown in Figure 10.19.

10.4.5 Scalability: Chip Size Versus Cost as a Function of Manufacturing Metrics

In 1961, the first commercialized IC had only one transistor, three resistors, and one capacitor. Subsequently, the commercial scaling of ICs has never ceased to expand, progressing from single components during the early 1950s to circuits

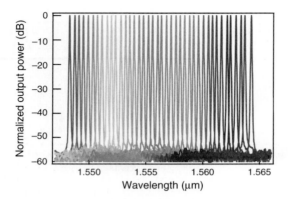

Figure 10.19 The normalized output power distributions on a 40-channel LS-PIC reveal a precise wavelength control of the DFB array (this figure may be seen in color on the included CD-ROM).

containing over 50 components in 1965 [48] and recently to Intel's Itanium® processors, which contained in 2002 over 220 million transistors on a single chip. Since the very beginning, the geometric integration growth was nourished by a need for an ever-increasing computing power. Without the benefit of integration, the cost of interconnecting and packaging single components in complex systems would have been colossal.

In contrast to the commercial scaling experienced by the IC industry, optoelectronic integration remained nonexistent from the creation of the first commercial laser diode in 1975 until the early 1990s. As displayed in Figure 10.20, integration

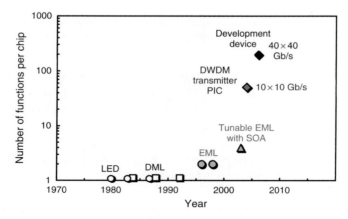

Figure 10.20 Progression chart representing the scaling of the number of functions per chip for InP-based transmitter devices utilized in commercially deployed networks.

of optoelectronic components remained marginal until the commercialization of the first LS-PIC by Infinera in 2004. The major factor preventing the scaling rate for PICs to initially follow a geometric growth has been the immaturity of existing networks. Indeed, for a long time, system requirements for data capacity have remained low. As a result, the lack of demand for added capacity did not support higher manufacturing costs induced by integration. Recently, however, network capacity has been consumed as fast as installed or made available. The intensification of Internet usage and the development of increasingly more complex applications or contents have generated a demand for high-speed Internet access that continues surging year after year. Photonic integration constitutes the natural evolution resulting from the recent changes in the landscape of telecommunication networks.

As both Moore [48] and Noyce [2] discovered, the optimal level of integration auto-regulates itself in such a way that the cost added per function integrated on a chip balances the cost saved per function in assembly work. As a result, an increased demand for functions needed to satisfy more complex applications drives assembly costs upward, which in turn are compensated by continued integration. Photonic integration enables continued cost-effective scaling of network capabilities in response to a demand for increased data capacity; however, for photonic networks, it is not only the packaging/assembly cost which drives component integration, it is the network assembly cost and the systems operational advantages that act as additional powerful drivers for further integration. As evident in Figure 10.21, the scaling of transmission data rate per chip has been growing exponentially at a pace doubling every 26 months. Such geometric progression is reminiscent of the integration trend for ICs expressed by Moore's law, which stipulated in 1965 that the number of transistors on an integrated circuit for minimum component cost should double every 18 months.

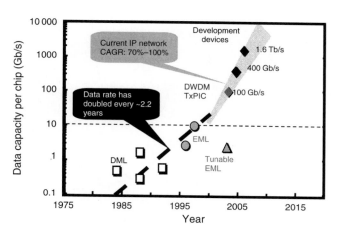

Figure 10.21 Progression chart representing the scaling of the data capacity for InP-based transmitter chips utilized in commercialized telecommunication networks. Over the past 25 years, the data capacity per chip has doubled every 2.2 years on average. CAGR, cumulative average growth rate.

10.5 NETWORK ARCHITECTURE IMPACT OF LSI PICs

10.5.1 Component Consolidation Advantages

The 100 Gb/s Tx LSI module integrates 10 channels of DFB lasers, modulators, power monitors, VOAs, and wavelength-locking elements multiplexed with a low-loss AWG. It is mounted in a Tx module, complete with digital drive electronics, ADC, wave locking, FEC, power monitoring, TEC temperature control functions, and a single fiber output (Figure 10.22).

The 100-Gb/s Rx PIC integrates a polarization-insensitive de-multiplexing AWG with 10 channels of photodiodes. It is packaged in a separate module with a single fiber input. DFB laser thresholds, output power, wavelength accuracy, and linewidth all perform to industry standards [38]. Each digital line module (DLM) contains an Rx and Tx module pair, deploying 100 Gb/s of capacity per card. DLM wavelengths from eight PICs spanning different optical channel groups at 100 Gb/s each are combined with a band mux module (BMM) to give up to 800 Gb/s transmission and reception over 80 wavelengths on a fiber pair. Through large-scale integration of optical components, this 800 Gb/s system reduces the number of intrasystem fiber couplings from 260 (conventional) to less than 10 (LSI-based PIC).

Testing and screening of integrated parts is made easier in that fewer parts need to be moved physically and are electrically tested in parallel. Clustered defect patterns that may be otherwise lost in a sparser use of real estate become apparent and help to screen weak parts and provide feedback for continuous improvement of line yields.

Integration of 50 or more components on a PIC directly improves network-level functionality, reliability, and cost. Bandwidth deployment is modular and comes in

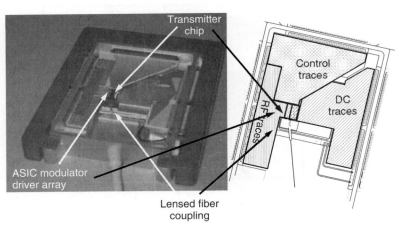

Figure 10.22 Photograph and schematic of the 100 Gb/s DWDM LSI-PIC transmitter module with the hermetic seal removed (this figure may be seen in color on the included CD-ROM).

increments of 10 10-Gb/s channels. As 10 channels of optical components are physically located on a fine pitch, integrated electronics are also deployed 10 channels at a time. This greatly reduces the chip cost, packaging cost, footprint, and thermal expense of the digital electronics and the system, compared to like technology based on discrete optical components. The dynamic range that needs to be supported by EDFA-based "skip" sites is reduced because 10 channels are the minimum number deployed on the system. As each channel runs at 10 Gb/s, the system can deploy with adjunct optical system technologies, such as EDFA and Raman-based line amplifiers, dispersion-compensating fibers, grating-based devices, and add/drop technologies such as fixed filters and conventional ROADMs. With 5–25 × fewer fiber couplings, the PIC-based system is inherently more reliable than traditional DWDM systems based on discrete or SSI optical components, as fiber-coupling failure accounts for 60%–70% of device failures. By June 2007, LSI PICs deployed in the field have cumulatively delivered 21.7 million field hours of data reliably. This corresponds to 42 per PIC pair failures in time (FIT) at a 60% confidence level (CL). This represents greater than 1.2 billion PIC element hours corresponding to less than 1 FIT per PIC element (at 60% CL) or 2 FIT per PIC element (at 90% CL). This compares to field data from discrete commercial 980-nm pump diodes at 5 FIT at 60% CL [G.Yang, JDSU, OFC 2007, paper JWA30].

Compact, integrated optics and electronics result in a smaller, less-expensive system. Prolific deployment of OEO in the PIC-based DON system produces systems that are readily, fully, and remotely reconfigurable (through GMPLS). The PIC-based optical transport system is therefore flexible, maintainable, economical, efficient, and compact.

10.5.2 The DON Architecture

The first generation DON is deployed in increments of 100 Gb/s bandwidth through highly integrated photonic devices and highly integrated electronic drive circuitry working in concert to enable multiple digital functions available to the network supervisor through generalized multi-protocol label switching (GMPLS). Each card contains an electronic cross point switch operating at a 2.5-Gb/s data rate granularity. The GMPLS on the network control board allows management of a wide variety of client side inputs that can go over this 100 Gb/s capacity per line card. Just as the optical components are integrated 10 channels per chip, so are the electronic drivers. The 10-channel FEC function is located within the same module as the Tx PIC and Tx driver.

At each node, the 10 × 10 Gb/s signals are returned from the optical domain to the electrical domain to correct errors, retime the signals, and monitor the system performance. A variety of tributary adapter module (TAM) interfaces are available to operate at tributary speeds from 1 to 40 Gb/s. Slower signals are aggregated by routers onto higher speed formats prior to entering the digital transport system (Figure 10.23).

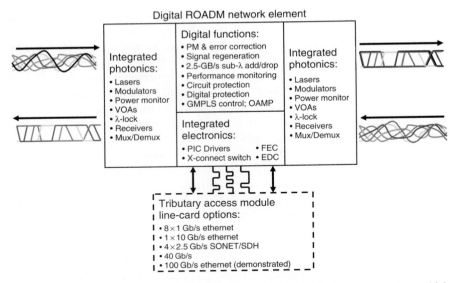

Figure 10.23 Digital ROADMs use integrated optics and integrated electronics to execute multiple digital functions at network nodes (this figure may be seen in color on the included CD-ROM).

While router speeds and the sheer volume of geometrically growing Internet traffic push the transport backbone to higher speeds, there remain and will continue to remain needs for sub-wavelength bandwidth provisioning. Otherwise, bandwidth is underutilized when a full channel is dedicated to carry a slower signal. Having full access to the bits, as one does in a digital ROADM-based network, ensures full access to the bandwidth. This also ensures that there can be no channel blocking as traffic approaches the capacity of the system. The electronic regeneration of bits also enables broadcast functions, as required by cable TV operation and other applications. Recently, high-bandwidth applications are developing which require multiple 10G wavelengths to support them. Edge-router speeds have already passed 100 Gb/s and are expected to keep rising, so that optical transport networks will come under increasing pressure to develop efficient solutions for delivering that amount of bandwidth and more [49, 50]. At the SC06 International Conference in November 2006, Infinera teamed with Finisar, Level3 Communications, Internet2, and the University of California at Santa Cruz (UCSC) to demonstrate the first successful 100-gigabit ethernet (100 GbE) transmission through Level3's live production network over a link spanning Florida to Texas. Digital grooming of the 10×10 Gb/s channels enabled this feat that otherwise would not have worked because of dispersion impairments. Beyond access to the bandwidth, access to the bits enables BER checks and therefore digital protection. Degrading circuits can be accurately detected and alternate routes configured prior to loss of service. This is key for broadcast video, video on demand (VoD), and especially intra-LATA and inter-LATA traffic [51].

10.5.3 Q-Improvement Cost

Digital networks efficiently use integrated electronics to maintain Q as defined by the equation below.

$$Q = \sqrt{2}\ \mathrm{erfc}^{-1}\ (2\ \mathrm{BER})$$

The BER is the probability of error per bit. Another way to think of Q is as the difference between the means of the signal 1s and 0s divided by the sum of their standard deviations, assuming that standard deviations are gaussian in distribution. As a rule of thumb, a BER of 10^{-9} corresponds to a Q of ~6. High Q-transmission originates from clean, powerful signals. Chromatic dispersion, attenuation, and power splitting are just some of the mechanisms that degrade Q. In the optical domain, attenuation is compensated with amplification, and dispersion is typically compensated with specially designed fiber. In the electronic domain, FEC is used to correct errors that come as the result of weak signal and high noise, resulting in Q restoration of up to 8 dB. Another critical technology for low-cost long-reach applications is electronic dispersion compensation (EDC). Multiple commercial vendors offer EDC for reaches of 2400 ps/nm in a tunable manner with adequate margins at 2.5 and 10.7 Gb/s data rates [52]. Q cost is cheaper in the electronic domain than the optical, and the integration cost is also less [53]. This is shown graphically in Figure 10.24.

Digital networks efficiently leverage integrated electronics to maintain link Q.

10.5.4 Data Ingress/Egress in a Digital ROADM

Optical networks built with ROADMs (based on PICs) give the network operator full access to the link bandwidth with swift, remote reconfigurability. As OEO is executed at each node, the network manager has full access to the deployed *bandwidth*, not just the wavelengths. Sonet/synchronus digital hierarchy (SDH) runs at the sub-lambda 2.5 Gb/s data rate, so that if they must be switched, it is done electronically in a "digital ROADM." As digital networks are not limited to full wavelength routing, there is no stranded bandwidth resulting from a mismatch between the data rate and the wavelength discretization.

In all optical networks, when wavelengths are exhausted faster, then more tributary and transport OEO are required to support other bandwidth demands. Also, when conventional (all optical) ROADMs exhaust their switching capacity, it forces more circuitous paths for additional bandwidth to be deployed. This, in turn, necessitates further deployment of intermediate OEO for the purpose of grooming the degrading signal.

DONs, executing full OEO and switching as needed at each digital ROADM, only require extra OEO at the points of data ingress and egress where more

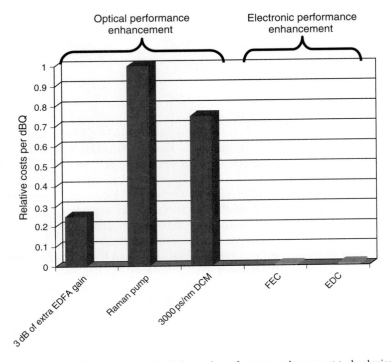

Figure 10.24 Relative Q-cost for optical and electronic performance enhancement technologies (this figure may be seen in color on the included CD-ROM).

bandwidth is added, until the full system bandwidth is exhausted. As no excess intermediate OEO is required, the DON is fully reconfigurable regardless of the traffic pattern and may harness the full bandwidth of the system without physical reconfiguration.

In a network simulation of the number of OEOs deployed vs number of 2.5G demands, researchers found that conventional (all optical) ROADMs require up to twice as many OEOs for service termination, signal regeneration, and wavelength conversion as a digital ROADM (Figure 10.25) [49]. This flexibility makes networks based on digital ROADMs simpler to plan, cheaper to operate, quicker to reconfigure, and easier to predict.

Optical networks built with digital ROADMs have full access to the system bandwidth and are concurrently suitable for sub-lambda, lambda, and super-lambda applications.

10.5.5 Network Management Advantages

Harnessing the power of highly integrated optics, electronics, and digital functions simplifies digital network management. After physically installing a new module

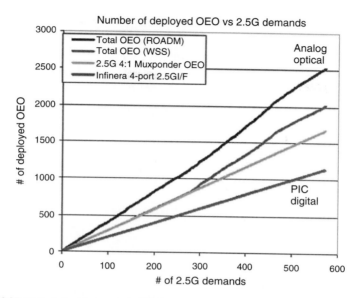

Figure 10.25 Optical-electrical-optical (OEO) consumption vs 2.5G (Sonet/SDH) demands in a simulated optical network (this figure may be seen in color on the included CD-ROM).

on the digital network, an auto-discovery feature recognizes that the correct part has been installed and updates the topological map of the network. This same feature enables fault detection and rerouting. In a ring or mesh network, this also results in a backhoe-proof system. New bandwidth, supplied in increments of 100 Gb/s on the transport side and 1.25, 2.5, 10, and 40 Gb/s on the tributary access side, is immediately available for provisioning. When traffic is not fully utilizing all channels, diagnostic BER tests are run to assess the health of the system and enable proactive servicing if needed. Access to the bits and the electronic switching fabric means that bandwidth is not stranded by channel blocking or by wave-centric switching constraints. Fewer, yet more highly integrated, electrical and optical components mean far fewer delicate fiber splices, electrical boards, connections, heaters, chillers, and fans, which directly results in a highly reliable optical transport system. It also yields high bandwidth per square foot in network hubs that tend to be highly deployed in expensive locations. Fewer, yet more highly functional, network elements make network planning, provisioning, sparing, and growth simpler, quicker, and robust.

Wide area network (WAN) speeds have recently exceeded the typical single-wavelength optical transport speed of 10 Gb/s. Multiple wavelengths are required to transport the data efficiently. "Super-lambda" links, where several optical carriers are bundled into a larger "virtual data stream," can readily support this need in digital architectures, as there is no wavelength blocking. Digital architectures also seamlessly maintain the freedom to support super-lambda, lambda, and

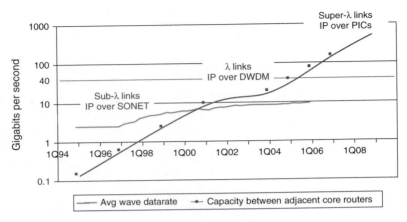

Figure 10.26 Historical progression of router-to-router bandwidth compared to WAN connectivity speeds (this figure may be seen in color on the included CD-ROM).

sub-lambda traffic diversity. Figure 10.26 shows how super-lambda links will be needed in the future to support IP traffic that exceeds the bandwidth of any individual carrier frequency.

If one thing can be predicted about tomorrow's network, it is that it will grow unpredictably. Digital networks uniquely deliver reliable bandwidth on top of an existing DWDM infrastructure in sub-, full-, or super-wave speeds, which is at once economical and easy to operate.

10.6 THE FUTURE OF OEO NETWORKS ENABLED BY III–V VLSI

It is now widely accepted that the world-wide web and Internet service availability is key to the global economy and society, providing real-time connectivity, business services, entertainment, and communications in a ubiquitous and accessible manner. These benefits are being continually expanded and consequently are driving continued scaling of the Internet in terms of applications, users, and connectivity and leading to significant ongoing growth in IP network capacity required to sustain these developments. Enabling and sustaining the growth of Internet network capacity will be critical to the roll-out of new services, including high-resolution video, Internet protocol television (IPTV), application-based software computing, grid computing, and other bandwidth-intensive applications. The growth in IP core network capacity has the consequence of imposing important scaling requirements on the underlying optical transport network used to carry this traffic. Network infrastructures carrying multi-terabit information capacity per fiber will be essential in the future.

10. III–V Photonic Integrated Circuits

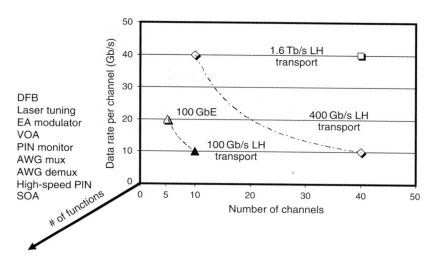

Figure 10.27 Possible future directions for large-scale photonic circuits.

Future PICs may increase network bandwidth by expanding in many directions as depicted in Figure 10.27. One option is to increase the number of channels while narrowing the channel spacing. Another is to increase the data rate per channel. It seems quite likely, at the time of this writing, that both options may be required. The current level of PIC manufacturing is capable of increasing channel count and data rate simultaneously for future generations. The most powerful option that photonic integration offers, however, is to increase the number of functions allowed on the optical chip. Just as with electronics, as time progresses, more and more of the entire network function will take place on the photonic circuit. This will not only allow a continued cost reduction as the number of discrete packages continues to dwindle, but new capabilities will emerge.

10.6.1 On Chip Amplifiers Offer Additional Bandwidth

Consider, as an example, the fact that current transmission windows used in all optical networks are largely dictated by EDFA gain spectrum. Ironically, the same technology credited with the fostering of optical networks in 1992 has now become a growth limiting factor to expanded bandwidth usage. Polarization-independent (PI) SOAs may be tailored in bandgap to amplify at any wavelength in the optical fiber transmission spectrum. Figure 10.28 shows a recently developed PIC in which PI-SOAs have been integrated into the input waveguide of a 40-channel RX. The on-chip gain is greater than 22 dB with a 50 nm optical bandwidth. The polarization-dependent gain (PDG) of the entire circuit is less

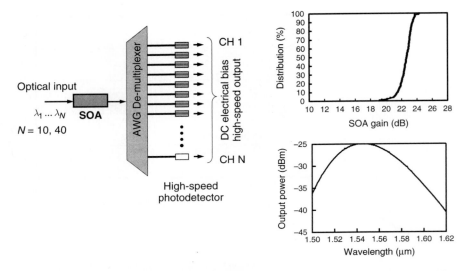

Figure 10.28 A receiver PIC with integrated PI-SOA. A gain of 22 db over 50 nm is achieved with a total PDG of 0.8 db over 40 channels (this figure may be seen in color on the included CD-ROM).

than 0.8 dB over all channels, [54]. PIC-based architectures allow one to cost effectively escape the artificial constraint of fiber-based amplifiers.

With the introduction of quantum dot-based active regions, SOAs of the future may yield a very broad gain bandwidth, a large signal gain, and high-saturation power while providing multi-wavelength amplification with no cross talk at high-bit rates [55].

Future DONs may be able to use upward of 80 nm more of the total available fiber spectrum to transport information than conventional all optical networks.

10.6.2 Power Consumption and the Thermal Bottleneck Challenge

Increased bandwidth demands require increased power consumption for all currently commercially deployed network hardware designs. For terabit networks of the future to be viable, reducing the total wall plug power required per bit will be essential. As on-chip functions continue to be added to PICs as described in Sections 10.4.4 and 10.4.5, a further challenge to designers will be to reduce the module power requirements simultaneously with increasing function. If this is not done, then the rate at which the highly compacted components generate heat will exceed the rate at which it can be removed from chassis. One high-power consumer is the TEC found in modules to maintain temperatures near 20°C. To maintain such a low temperature on the component side against an ambient of

65°C is a power-intensive task. All else being equal, to create a 40° temperature difference by cooling with a TEC requires four times the power that using a resistive heater would require to create the same temperature delta by heating. In Section 10.3.1, we have seen EML designs with improved high-temperature performance. Constructing components that perform well, even though working at elevated temperatures, is a very effective way of combating runaway heat generation.

The smaller the temperature difference from ambient required, the less the power consumed for temperature control.

Figure 10.29 shows the early results of a 10-channel transmitter PIC constructed to work at elevated temperature. In Figure 10.29(a), the output power shows only small degradation up to 50°C. In Figure 10.29(b), the Q-factor is shown at an OSNR of 23 dB and 0 ps/nm of dispersion. Figure 10.29(c) displays the eye diagrams for all 10 channels from 20°C to 85°C. By running hot, future PICs will reduce the overall wall plug power required for their operation by more than a factor of 2.

10.6.3 Mobile Applications for Optical Communications

Aerospace applications have severe limitations on equipment size, weight, and power (SWaP) because the equipment deployed must be lifted to an altitude to perform, and renewable sources of electrical power are much more expensive for

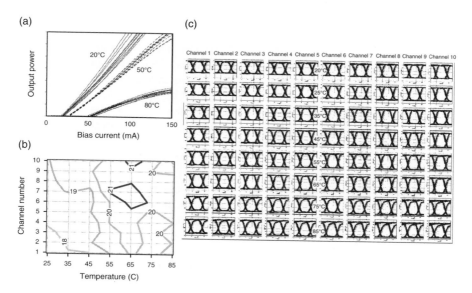

Figure 10.29 (a) LI curves for all 10 channels at 20°C, 50°C, and 80°C. (b) Q-factor as a function of temperature. (c) 10 Gb/s eye diagrams from 20°C to 85°C.

mobile platforms than on the ground. By reducing the number of individual modules required for a given function, PICs reduce weight by a factor of 3 and both power and size by a factor of 5 over conventional communications hardware solutions performing the same function. In addition to the SWaP advantages that PICs provide, the reduction of FIT (Section 10.5.1) by a factor of 5 over conventional hardware created from discrete components is another factor making PIC-based systems the aerospace method of choice for communications. Mobile applications range from internal communications systems on planes and naval vessels to any air-to-air or air-to-ground application including free space satellite communications networks.

The same SwaP advantages that make PIC-based networks enabling to mobile applications are becoming increasingly important to land-based service providers as bandwidth demands increase simultaneously with the expense of central office real estate.

10.7 CONCLUSION

In summary, we have reviewed the many integration methodologies that exist for III–V photonic circuits and their current status. The manufacturing methodologies that make LS-PICs continue to scale with high reliability have been highlighted. We have pointed out the ways in which digital networks differ from conventional analog networks in superior service functionality and we have taken a look at some of the future directions that OEO networks may take to meet increasing bandwidth demands.

DONs have now been deployed since 2004. The highly integrated nature of the photonic circuits composing these networks makes them economically compelling. This alone would be enough to explain their expanding use in the world-wide communications infrastructure. However, it may be that in the future, service providers will come to see their ease of deployment, ease of maintenance, and configuration flexibility as stronger reasons to deploy OEO networks rather than price alone. In partial summary, DONs offer

- full digital OAM&P at each digital node
- optical link and impairment management isolation between nodes
- an optical service layer that is independent of the optical transmission layer
- the elimination of all constraints on end-to-end service path routing
- an enhanced access to PM and OAM&P data at all nodes
- the simplification of network planning, system engineering, and service turn-up
- the ability to implement digital reconfigurable optical add/drop services at every node
- the ability to configure "super wavelength" virtual data streams to obtain maximum use of the available network bandwidth.

As the cost of OEO continues to drop, we will see these digital networks deployed closer and closer to the end user. The ultimate impact of digital transmission on the world-wide communications infrastructure may be comparable to the replacement of the electrical technology format with the first optical technology format over 20 years ago.

REFERENCES

[1] S. E. Miller and A. G. Chynoweth, *Optical Fiber Telecommunications*, NY, Academic Press, 1979.
[2] R. N. Noyce, "Large-scale integration: what is yet to come?," *Science*, 195, 1102–1106, March 1977.
[3] Y. Kawamura, K. Wakita, Y. Yoshikuni et al., "Monolithic integration of a DFB laser and an MQW modulator in the 1.5 pm wavelength range," *IEEE J. Quantum Electron.*, QE-23, 915–918, 1987.
[4] J. Binsma, P. Thijs, T. VanDongen et al., "Characterization of butt-joint InGaAsP waveguides and their application to 1310 nm DBR-Type MQW ganin-clamped semiconductor optical amplifiers," *IEICE Trans. Electron.*, E80-C, 675–681, 1997.
[5] M. Aoki, H. Sano, M. Suzuki et al., "Novel structure MQW electroabsorption modulator/DFB-laser integrated device fabricated by selective area MOCVD growth," *Electron. Lett.*, 27, 2138–2140, 1991.
[6] C. H. Joyner, S. Chandrasekhar, J. W. Sulhoff, and A. G. Dentai, "Extremely large band gap shifts for MQW structures by selective epitaxy on SiO_2 masked substrates," *IEEE Photon. Technol. Lett.*, 4(9), 1006–1009, 1992.
[7] Y. D. Galeuchet and P. Roentgen, "Selective area MOVPE of GaInAs/InP heterostructures on masked and nonplanar (100) and (111) substrates," *J. Cryst. Growth*, 107, 147–150, 1991.
[8] M. Aoki, M. Suzuki, H. Sano et al., "InGaAs/InGaAsP MQW electroabsorption modulator integrated with a DFB laser fabricated by bandgap energy control selective area MOCVD," *IEEE J. Quantum. Electron.*, 27, 2281–2295, 1993.
[9] J. H. Marsh, "Quantum well intermixing," *Semicond. Sci. Technol.*, 8, 1136–1155, 1993.
[10] S. McDougall, O. Kowalski, C. Hamilton et al., "Monolithic integration via a universal damage enhanced quantum-well intermixing technique," *IEEE J. Sel. Top. Quantum Electron.*, 4(4), 636–646, 1998.
[11] S. Charbonneau, E. Kotels, P. Poole et al., "Photonic integrated circuits fabricated using ion implantation," *IEEE J. Sel. Top. Quantum Electron.*, 4(4), 772–793, 1998.
[12] E. Skogen, J. Barton, S. DenBaars, and L. Coldren, "A quantum-well intermixing process for wavelength-agile photonic integrated circuits," *IEEE J. Sel. Top. Quantum Electron.*, 8(4), 863–869, 2002.
[13] J. Weiland, H. Melchior, M. Q. Kearley et al., "Optical receiver array in silicon bipolar technology with selfaligned, low parasitic III/V detectors for DC-1 Gbit/s parallel links," *Electron. Lett.*, 27, 2211, 1991.
[14] I. Christiaens, G. Roelkens, K. De Mesel et al., "Thin-film devices fabricated with benzocyclobutene adhesive wafer bonding," *J. Lightw. Technol.*, 23(2), 517–523, 2005.
[15] D. I. Babic, J. Piprek, K. Streubel et al., "Design and analysis of double-fused 1.55-μm vertical-cavity lasers," *IEEE J. Quantum Electron.*, 33(8), 1369–1383, 1997.
[16] A. R. Hawkins, T. E. Reynolds, D. R. England et al., "Silicon hetero-interface photodetector," *Appl. Phys. Lett.*, 68, 3692–3694, 1996.
[17] H. Park, A. W. Fang, O. Cohen et al., "Design and fabrication of optically pumped hybrid silicon-AlGaInAs evanescent lasers," *IEEE J. Sel. Top. Quantum Electron.*, 12, Part 2, 1657–1663, 2006.

[18] Kurt W. Eisenbeiser and J. Ramdani, "Structure and method for fabricating semiconductor structures and devices not lattice matched to the substrate," US Patent No. 20020030246, filed 25 July 2001, published March 14, 2002.
[19] M. Raburn, B. Liu, P. Abraham, and J. E. Bowers, "Double-bonded InP-InGaAsP vertical coupler 1:8 beam splitter," *IEEE Photon. Technol. Lett.*, 12(12), 1639–1641, 2000.
[20] N. C. Frateschi, J. Zhang, R. Jambunathan et al., "Uncooled long-reach performance of 10 Gb/s laser-integrated modulator modules with InGaAlAs/InP and InGaAsP/InP MQW electro-absorption modulators monolithically integrated with semiconductor amplifiers," *IEEE Photon. Technol. Lett.*, 17(7), 1378–1380, 2005.
[21] S. Makino, K. Shinoda, T. Shiota et al., "Wide Temperature (15°C to 95°C), 80-km SMF Transmission of a 1.55-μm, 10-Gb/s InGaAlAs Electroabsorption Modulator Integrated DFB Laser," in Proc. *OFC*, OMS 1, Anaheim, CA, March 26–30, 2007.
[22] J.-R. Burie, G. Glastre, S. Fabre et al., "10Gb/s 100km Transmission up to 80C over Single Mode Fiber at 1.55 μm with an Integrated Electro-Absorption Modulator Laser," in Proc. *OFC 2007*, OMS 2, Anaheim, CA, March 26–30, 2007.
[23] C. Jany, J. Decobert, F. Alexandre et al., "10 Gb/s 1.55um 25 km transmission at 90C with New Self Thermally Compensated AlGaInAs Directly Modulated Laser", in Proc. *OFC*, JWA32, Anaheim, CA, March 26–30, 2007.
[24] Y. A. Akulova, G. A. Fish, P.-C. Koh et al., "Widely tunable electroabsorption-modulated sampled-grating DBR laser transmitter," *IEEE J. Sel. Top. Quantum Electron.*, 8(6), 1349–1357, 2002.
[25] A. J. Ward, D. J. Robbins, G. Busico et al., "Widely tunable DS-DBR laser with monolithically integrated SOA: design and performance," *IEEE J. Sel. Top. Quantum. Electron.*, 11(1), 149–156, 2005.
[26] B. Pezeshki, E. Vail, J. Kubicky et al., "12 element multi-wavelength DFB arrays for widely tunable laser modules," in *Proc. OFC 2002*, ThGG71, Anaheim, CA, 2002, pp. 711–712.
[27] S. J. B. Yoo, "Next Generation Networking Systems Laboratory UCDavis," Projects, Optical CDMA, 2006. February 2007, http://sierra.ece.ucdavis.edu/, and private communication, March 2007.
[28] J. Chen, R. G. Broeke, Y. Du et al., "Monolithically integrated InP-based photonic chip development for O-CDMA systems," *IEEE J. Sel. Top. Quantum Electron.*, 11(1), 66–77, 2005.
[29] Y. Suzaki, H. Yasaka, H. Mawatari et al., "Integrated Eight-Channel WDM Modulator Module with 25-GHz Channel Spacing," in Proc. *OFC 2003*, 1, 2003, pp. 186–187.
[30] E. A. J. M. Bente, Y. Barbarin, J. H. den Besten et al., "Wavelength selection in an integrated multiwavelength ring laser," *IEEE J. Quantum Electron.*, 40(9), 1208–1216, 2004.
[31] M. T. Hill, T. de Vries, H. J. S. Dorren et al., "Integrated two-state AWG-based multiwavelength laser," *IEEE Photon. Technol. Lett.*, 17(5), 956–958, 2005.
[32] X. Song, N. Futakuchi, F. C. Yit et al., "28-ps switching window with a selective area MOVPE all-optical MZI switch," *IEEE Photon. Technol. Lett.*, 17(7), 1480–1482, 2005.
[33] S. J. Choi, Z. Peng, Q. Yang et al., "An eight-channel demultiplexing switch array using vertically coupled active semiconductor microdisk resonators," *IEEE Photon. Technol. Lett.*, 16(11), 2517–2519, 2004.
[34] A. W. Fang, H. Park, O. Cohen et al., "Electrically pumped hybrid AlGaInAs-silicon evanescent laser," *Opt. Express*, 14, 9203–9210, 2006.
[35] H. Park, A. Fang, S. Kodama et al., "Hybrid silicon evanescent laser fabricated with a silicon waveguide and III-V offset quantum wells," *Opt. Express*, 13(23), 9460–9464, 2005.
[36] H. Shimizu and Y. Nakano, "Fabrication and characterization of an InGaAsP/InP active waveguide optical isolator with 14.7 dB/mm TE mode nonreciprocal attenuation," *J. Lightw. Technol.*, 24(1), 38–43, 2006.
[37] T. R. Zaman, X. Guo, and R. J. Ram, "Proposal for a polarization-independent integrated optical circulator," *IEEE Photon. Technol. Lett.*, 18(12), 1359–1361, 2006.
[38] R. Nagarajan et al., "Large-scale photonic integrated circuits," *IEEE J. Sel. Top. Quantum Electron.*, 11(1), January–February 2005.

[39] N. S. Bergano, F. W. Kerfoot, and C. R. Davidsion, "Margin measurements in optical amplifier system," *IEEE Photon. Technol. Lett.*, 5(3), 304–306, 1993.

[40] A. R. Alvarez, B. L. Abdi, D. L. Young et al., "Application of statistical design and response surface methods to computer-aided VLSI device design," *IEEE Trans. Computer-Aided Design*, CAD-7(2), 272–288, February 1988.

[41] R. G. Cosway, M. G. Ridens, M. Peters et al., "Use of On-Product Measurements for Process Control for Improved Manufacturing Efficiency and Reduced Costs," *IEEE/SEMI Advanced Semiconductor Manufacturing Conference and Workshop*, November 1996, pp. 370–374 Cambridge, MA, USA.

[42] C. K. Hansen, "Effectiveness of yield-estimation and reliability-prediction based on wafer test-chip measurements," in Annual Proc. *RAMS*, January 1997, pp. 142–148.

[43] R. Ross and N. Atchison, "Yield analysis," *Texas Instruments Tech. J.*, 58–103, October–December 1998.

[44] C. H. Stapper and R. J. Rosner, "Integrated circuit yield management and yield analysis: development and implementation," *IEEE Trans. Semicond. Manuf.*, 8(2), 95–102, May 1995.

[45] C. H. Joyner, J. L. Pleumeekers, A. Mathur et al., "Large-Scale DWDM Photonic Integrated Circuits: A manufacturable and scalable integration platform," in Proc. *IEEE LEOS 2005*, October 2005, pp. 344–345.

[46] D. Potter, R. Bowman, and L. Peters, "Integrated Circuit Engineering Corporation," *Cost Effective IC Manufacturing*, ISBN-10: 1877750603, January 1997.

[47] R. Nagarajan, M. Kato, J. Pleumeekers et al., "Single-chip 40-channel InP transmitter photonic integrated circuit capable of aggregate data rate of 1.6 Tbit/s," *IEEE Electron. Lett.*, 42(13), 771–773.

[48] G. E. Moore, "Cramming more components onto integrated circuits," *Electronics*, 38(8), 82–85, January 1965.

[49] S. Melle, "Next-Generation Higher Speed Ethernet: 100GB/s Ethernet and Beyond," 2007 white paper available on Infinera website.

[50] S. Melle and V. Vusirikala, "Network Planning and Architecture Analysis of Wavelength Blocking in Optical and Digital ROADM Networks," in Proc. *OFC/NFOEC*, NTuC, Anaheim, CA, March 25–29, 2007.

[51] Vusirikala and Melle, "Digital Protection for Ethernet and Video Transport Oriented Metro Optical Networks," in Proc. *OFC/NFOEC*, NThD, Anaheim, CA, March 25–29, 2007.

[52] A. Ghiasi, A. Momtaz, A. Dastur et al., "Experimental results of EDC based receivers for 2400 ps/nm at 10.7 Gb/s for emerging telecom standards," in Proc. *OFC 2006*, OTuE3, Anaheim, CA, March 5–10, 2006.

[53] O. Welch et al., "The realization of large scale photonic integrated circuits and the associated impact on fiberoptic communication systems," *J. Lightw. Technol.* Special Issue on Optoelectronics, 24(12), 4674–4682, December 2006.

[54] Nagarajan PDP32 *OFC* 2007.

[55] Eisenstein OWP5 *OFC* 2007.

11

Silicon photonics

Cary Gunn[*] **and Thomas L. Koch**[†]

[*] *Chief Technology Officer, Luxtera, Inc., Carlsbad, CA, USA*
[†] *Center for Optical Technologies Sinclair Laboratory, Lehigh University, Bethlehem, PA, USA*

11.1 INTRODUCTION

Silicon photonics refers to the implementation of integrated optics and photonic integrated circuit (PIC) technologies in silicon. Beyond the advantageous use of silicon substrates, it refers specifically to the use of silicon itself as an optical or waveguide medium, as well as an active medium for modulation, tuning, or switching. Beginning with the seminal work of Soref and coworkers in the 1980s, recent years have seen a dramatic increase in the sophistication, performance, and complexity of devices and circuits implemented with silicon photonics [1–3]. This chapter will outline the foundations and current status of silicon photonics.

The unprecedented maturity of silicon materials, processing, device and circuit design technologies, stemming from worldwide R&D investments now exceeding $40 billion annually [4] (Figure 11.1), clearly offers an enticing opportunity for leveraging into photonics applications and markets. To fully leverage these investments, significant work has been recently focused on true Complementary Metal Oxide Semiconductor (CMOS) process compatibility and foundry manufacture. The field has also grown to include the implementation of active devices in other materials enabling photodetection and modulation, such as Ge and SiGe alloys that have already been embraced by mainstream CMOS-integrated circuit manufacturing, as well as the exploration of III–V and rare-earths for emitters which are not included in today's commercial CMOS process.

Most silicon photonics is implemented on Silicon-On-Insulator (SOI), which provides many attractive design attributes for photonic circuits, including a very high index contrast for ultra-small bend radius index-guided passive circuits and

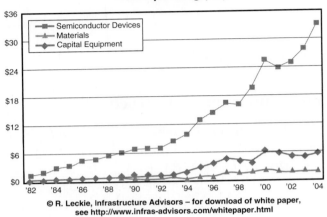

Figure 11.1 Annual R&D spend for devices, materials, and capital equipment in the silicon microelectronics industry. Even back in 2004, this had reached a level of over $41 billion per year (this figure may be seen in color on the included CD-ROM). (*Sources*: S & P, SIA, SEMI, INFRASTRUCTURE ADVISORS.)

devices, as well as a variety of high-performance photonic crystal concepts. While silicon has historically not been viewed as a high-performance *active* optical material, recent work has now demonstrated that Mach–Zehnder devices based on plasma index mechanisms can be quite efficient in field-effect configurations, effectively bypassing the previous bandwidth limitations of thermo-optic devices or forward-injection designs. Ge-based waveguide photodetectors, implemented with commercial SiGe very large scale integration (VLSI) epitaxy technologies, have also demonstrated high-speed operation and acceptable leakage currents for many applications. The result of these studies is a nearly full suite of active and passive photonic devices that can be implemented with high density and yield using commercial CMOS foundry technologies, with well-documented low costs per unit wafer area.

An obvious additional benefit of designing into a CMOS foundry is the ability to integrate optical functions with standard VLSI circuitry functions on a single silicon chip. For optical fiber telecom and optical interconnect applications, in addition to the potential for voltage scaling and low power consumption through reduced dimensions in field-effect devices, intimate integration with electronics can relax impedance matching design issues relative to conventional multi-chip configurations, providing paths for further potential power reduction. For markets outside of telecom and data links, such as sensors and signal-processing applications, the ability to integrate optical functions with advanced preamplification, digital signal processor (DSP), and controller functions on a single low-cost chip could be dramatically enabling.

11.2 SOI WAFER TECHNOLOGY

Silicon photonic devices rely heavily on SOI technology, and as a result, the more comprehensive development of silicon photonics began in earnest only after SOI became a mainstream CMOS technology. SOI began as a boutique military and aerospace technology, driven by the radiation tolerance benefits of a thin transistor body. For several decades, this boutique technology existed as silicon epitaxially grown on sapphire wafers, known as Silicon-On-Sapphire. While the crystalline quality and achievable film thickness of the technique is satisfactory for optical devices, it has failed to become mainstream and migrate to deep ultra-violet (DUV) fabrication—another fundamental requirement of modern silicon photonics.

Eventually, microprocessor designers began to take notice of the simplified processing and reduced parasitics present in an SOI transistor and chose to exploit these benefits by constructing CMOS circuitry on SOI wafers. The technique first used to create a very uniform, wafer-scale, thin film in mainstream production was called separation by implantation of oxygen (SIMOX) [5]. In this technique, oxygen is ionized and implanted at high dose into a silicon wafer to provide the stoichiometry for a buried SiO_2 layer. Low defect densities can be achieved despite the high implant dose due to a subsequent thermal anneal that acts to repair the top silicon surface while creating a clean boundary between the oxide and the silicon layers. While there has been some interesting multilayer silicon photonics work done using this technique to define optical devices [6], the technique suffers from the presence of silicon crystals in the oxide layer that appear as defects in semiconductor inspection equipment and also poses serious challenges in achieving the thinner films required to scale SOI transistors.

The technology which has become a winner in the marketplace is SOI created by the SMARTCUT technique, developed at CEA/LETI in France, and commercialized by Soitec [7]. This technique, illustrated in Figure 11.2, works by implanting hydrogen in a first substrate and growing oxide on a second substrate. After the two substrates are wafer bonded, a rapid annealing step expands the hydrogen, creating an accurate cleave plane and leaving a uniform silicon film on top of the oxide.

In 2001, IBM, Freescale, and AMD were all manufacturing microprocessors on SOI wafers. Today, nearly all major semiconductor fabs have an SOI process, and perhaps the most telling fact regarding the health of the SOI market is that all major gaming systems are built around an SOI microprocessor.

The benefit of the commercial success of SMARTCUT SOI is that high-quality SOI wafers are now available to the silicon photonics community in sizes up to 12 in. (300 mm), and there is an increasing awareness of how to handle and process SOI material. There are, however, two major looming challenges to the silicon photonics community. First, the CMOS SOI roadmap, driven by the desire to fully deplete the transistor body of carriers, is moving toward ever-thinner silicon films. The optimal thickness for a strip silicon waveguide is in the order of 200 nm (see Section 11.3), and that thickness corresponded to the 0.18-μm node, which is becoming obsolete. Second, due to evanescent leakage into the substrate, SOI for silicon photonics

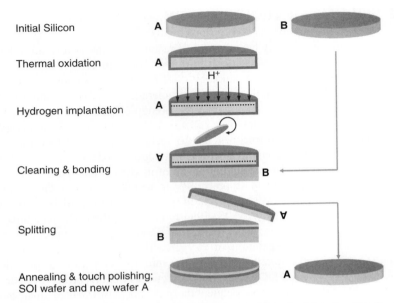

Figure 11.2 Processing sequence for SMARTCUT technology SOI wafer fabrication (this figure may be seen in color on the included CD-ROM).

requires a buried oxide (BOX) layer thickness on the order of 1 μm or more, which is substantially thicker than required for CMOS transistors. Thus, optically capable SOI also requires an extended oxidation period during manufacturing, which adds cost to the substrate as well as requiring nonstandard processing.

11.3 HIGH-INDEX-CONTRAST WAVEGUIDE TYPES AND PERFORMANCE ON SOI

Waveguide types in SOI technology can be divided into three classes, as illustrated in Figure 11.3.

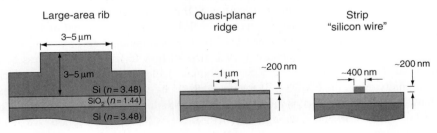

Figure 11.3 Three common waveguide variants for SOI technology, each with its distinct, application-dependent advantages (this figure may be seen in color on the included CD-ROM).

11. Silicon Photonics

Historically, early silicon-based waveguides were demonstrated without SOI, utilizing the plasma index depression from highly doped substrates, with losses as low as 1.3 dB/cm [8, 9]. On SOI, most early work on high-performance silicon photonic waveguides focused on the "large-area rib" designs. While the wavelength of 1.55 μm light in silicon is <0.5 μm, it is remarkable that these guides can display effectively single-mode behavior even with very large core dimensions. This results because modes that are higher order in the vertical direction in the core have lower effective indices than the fundamental modes in the lateral slab, and are thus highly leaky [10]. Such guides have demonstrated losses as low as 0.1 dB/cm in structures with direct butt-coupling losses to single-mode fiber as low as 0.3 dB [11]. While such structures indeed provide inherently lower losses due to a lower fraction of the optical power at the waveguide sides which often have fabrication-induced interface roughness, the large modal size of this guide design also brings a low tolerance to waveguide bends. In addition to the large device sizes implied by this feature, the scale of the vertical morphology also presents compatibility problems with modern CMOS process and device trends.

The other extreme is the silicon strip or "wire" waveguide, with no high-index lateral slab regions. Such guides require lateral dimensions typically below 0.5 μm to guarantee single-mode operation, and as a result provide very small, highly confined optical modes. The high lateral index contrast allows for very tight micron-scale bend radii, as illustrated experimentally in Figure 11.4, allowing for ultra-compact passive devices and routing waveguides. The high field amplitude at the etched boundaries tends to produce somewhat higher losses from scattering, although current loss values for "wire" waveguides are now at the 2- to 3-dB/cm level [12].

Intermediate between the large-area rib and the silicon "wire" is the ridge waveguide, which can be designed as single-mode for a variety of widths depending upon the etch depth of the ridge. The bending performance is also scalable with etch depth as illustrated in Figure 11.6, and shallow etching also tends to yield

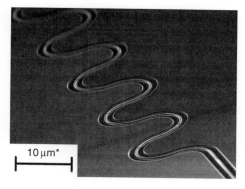

Figure 11.4 Vlasov et al., IBM: Example of losses for high-index-contrast "wire" waveguides: For 6.5 μm radius bends, losses are 0.0043 dB per 180° turn.

lower scattering losses. Because this guide is closest to the slab waveguide, it provides the tightest vertical confinement for devices designs targeting field overlap with a planar active region. Perhaps, most importantly, this configuration allows for conductive lateral electrical access to drive and control active devices. This configuration has also been shown to produce losses in the range of 0.3 dB/cm or less when optimized ridge fabrication processes are employed [13–15]. While the ridge guide offers ample design freedom for transverse electric (TE) mode operation, care must be exercised when designing for transverse magnetic (TM) mode operation because the high-index contrast in the SOI system can lead to significant lateral radiation loss unless precise waveguide widths are selected [16].

One interesting attribute of high-index-contrast waveguides is the distinctive polarization characteristics that can be achieved. A particularly striking example is the "slot waveguide" which can provide unusually large field amplitudes, or modal confinement, in suitably configured low-index dielectric layers or air [17]. This results simply from advantageous use of the simple boundary condition for continuity of the normal component of the displacement field D, which for TM modes is substantial, at a dielectric boundary. This is best illustrated in a simple two-dimensional slab of Si with a layer of SiO_2 embedded in the middle. This continuity requires that the electric field amplitude (which is primarily normal to the interface) increases by a factor of $n_{Si}^2/n_{SiO_2}^2$ ~5.8 in the embedded SiO_2 relative to its amplitude in the Si "core," while the tangential component of magnetic field H is continuous. The intensity, or component of the Poynting vector down the waveguide axis $(E \times H) \cdot \hat{z}$, thus increases by this factor of ~5.8 as illustrated by the modal intensity profile in the example of Figure 11.5, where a 44.3% confinement is achieved in an oxide layer only 50 nm thick.

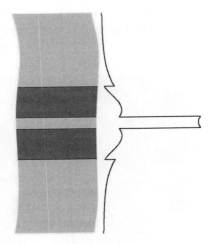

Figure 11.5 Numerically calculated intensity profile for a "slot" waveguide comprising two 0.14 μm-thick slabs of Si embedded in SiO_2. Although the slot layer is only 50 nm thick, it has an exceptionally high optical confinement factor of 44.3% (this figure may be seen in color on the included CD-ROM).

Loss mechanisms in SOI guides include the aforementioned bending losses, as well as radiation leakage to the high-index substrate, defect-induced losses in the silicon or BOX layers, fabrication-induced scattering losses, scattering and absorption losses in amorphous or polycrystalline silicon if used in the guide, doping and impurity losses, and losses that may result from surface chemistry. Bending losses are often computed using either numerical techniques, or modeling based on analytic expressions that can be derived using cylindrical field solutions or using conformal mapping techniques to convert the curved waveguide into a straight guide [18–20]. For example, the simple approximate expression derived by Marcuse predicts an exponential dependence of losses on curvature radius R in the form $\alpha = K e^{-C \cdot R}$ where $C \approx \frac{2}{3}\beta \left((n_{\text{eff}}^2 - n_{\text{clad}}^2)/n_{\text{eff}} \right)^{3/2} \approx \frac{2}{3}\beta \left(2\Delta n_{\text{eff}}/n_{\text{eff}} \right)^{3/2}$ and the constant K depends on the core and cladding indices and thicknesses. The approximate $(\Delta n_{\text{eff}}/n_{\text{eff}})^{3/2}$ dependence clearly indicates why the high-index contrast between the effective index and the cladding medium, or cladding slab modes in an equivalent two-dimensional effective index picture, is so important in achieving small bend radius. Also note that the lateral effective index steps are often markedly different for TE and TM mode leading to sometimes dramatic differences in bending losses. Examples illustrating the TE mode bending loss dependence on ridge waveguide design parameters are shown in Figure 11.6 [13].

To preclude excessive leakage to the high-index Si substrate, the BOX layer must also be sufficiently thick, with the actual value depending on both the thickness of the Si core or "device" layer, and the polarization, with the slower vertical evanescent decay of TM modes requiring thicker BOX layers. As a simple example, for a slab

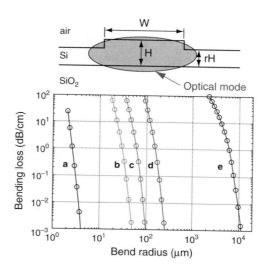

Figure 11.6 Bending losses for various guide configurations [13]. (a) "wire" guide with $H = 0.22\,\mu\text{m}$, $W = 0.4\,\mu\text{m}$, and single-mode ridges with $H = 0.205\,\mu\text{m}$ and (b) $r = 0.72$, $W = 0.65\,\mu\text{m}$, (c) $r = 0.8$, $W = 0.8\,\mu\text{m}$, and (d) $r = 0.92$, $W = 1.36\,\mu\text{m}$, and large-area rib with $H = 4.3\,\mu\text{m}$, $r = 0.6$, and $W = 3.7\,\mu\text{m}$ (this figure may be seen in color on the included CD-ROM).

waveguide with a 0.2-μm thick device layer with air cladding on top, a BOX layer of 0.75 μm is required to keep substrate leakage below 0.1 dB/cm for TE modes, while the TM mode requires a 1.91-μm BOX layer thickness to achieve this loss. These values must be additionally increased when lateral guiding is introduced.

Scattering losses can be evaluated by extending the treatment of Lacey and Payne [21], which highlights the important role of both the RMS surface roughness and the correlation length for perturbations. For example, for a symmetric slab guide of thickness t and index n_1 embedded in a medium of index n_2, one obtains for a mode of propagation constant β a loss of

$$\alpha = \frac{(n_2^2 - n_1^2)^2}{4\pi n_1} \frac{\omega^3}{c^3} |\varepsilon(t/2)|^2 \int_0^\pi \tilde{R}(\beta - n_2 \frac{\omega}{c} \cos\theta) d\theta, \quad (11.1)$$

where $\varepsilon(t/2)$ is the amplitude of the modal electric field at the boundary normalized such that $\int_{-\infty}^{\infty} |\varepsilon(t/2)|^2 dy = 1$, and $\tilde{R}(\Omega)$ is the transform of the autocorrelation function of the surface roughness. For example, if it is assumed to be exponential with a correlation length L_c, yielding $R(u) = \sigma^2 e^{-|u|/L_c}$ then we have $\tilde{R}(\Omega) = \frac{2\sigma^2 L_c}{1 + L_c^2 \Omega^2}$.

Methods to reduce surface roughness include oxidation smoothing of the waveguide boundaries following etching [22], hydrogen-induced surface reflow following fabrication [23], and forming the waveguide using a local oxidation process rather than etching [16], yielding surfaces with root mean square (RMS) roughness values below $\sigma \sim 0.3$ nm, close to that of a pristine SOI wafer.

Another key loss mechanisms result from free carriers provided by doping or injection, where the data of Soref and Bennett provide the relationship [24]

$$\Delta \alpha = 8.5 \times 10^{-18} \Delta N + 6.0 \times 10^{-18} \Delta P \quad (11.2)$$

This indicates that losses of 0.1 dB/cm require doping concentrations below 2.7×10^{15} cm^{-3} N-type or 3.8×10^{15} cm^{-3} P-type material. For active devices, requiring electrical access through doped semiconductors, or utilizing free-carrier modulation, these losses have significant design implications.

Finally, losses from polycrystalline or amorphous Si scattering are quite complex, and depend in great detail on the annealing conditions, the size of polycrystalline grains, and the state of dangling bonds at grain boundaries or in the amorphous material. Losses in amorphous Si can range from several hundred dB/cm down to values of 1.6 dB/cm [25] depending on processing conditions.

11.4 INPUT–OUTPUT COUPLING

While the high index contrast available to silicon photonics allows for remarkably compact waveguide structures, signals must be transported to or from the chip in most telecommunications applications. Additionally, in spite of progress to be discussed later in developing on-chip laser sources in silicon photonics, at present commercially practical approaches require a source laser to be coupled with optimized efficiency onto chips with modulators, filters, and signal-processing elements. Because the waveguides are often even more compact than those in III–V semiconductor lasers and PICs, this coupling poses a significant engineering challenge. The two most viable approaches used to date include mode transformers or tapers to increase mode sizes for input–output coupling, or grating-based solutions for distributed coupling from waveguides into beams that emerge normal to the waveguide plane.

11.4.1 Tapers

Silicon photonics borrows from the extensive research in III–V PIC technology, where the idea of a reverse taper was introduced to form "expanded-beam" or "spot-size converted" lasers that are more tolerant to the limited precision achievable with passive alignment to waveguides or fibers [26, 27]. In such structures, rather than increase the waveguide core to scale the mode size, the core is decreased to lose confinement and increase mode size. While this reduction in core size can be accomplished by a number of means, the most common and practical technique in silicon photonics utilizes readily available high-resolution lithography to form wedge-shaped or triangular tapers, as seen in a plan view, using reactive ion etching. Such structures, whether in III–V or silicon, are often accompanied by a larger, low index contrast waveguide so that the fundamental waveguide mode is adiabatically transformed to the larger waveguide as it loses confinement in the original core. The larger core guide can be in a dielectric or polymer, for example, and can be engineered to provide a good match to optical fibers, other waveguides, or laser sources.

The design challenges in such a structure include avoiding reflections due to effective index discontinuities and making the taper gradual enough at all points to optimize the adiabatic nature of the mode transition for low loss. Modeling of such structures is typically done with beam propagation methods. Such structures have demonstrated coupling losses to smaller core single-mode fibers as low as 0.5 dB [28], with the micron-scale alignment tolerances expected from the $\sim 8\,\mu m$ mode field size of a fiber. Figure 11.7 shows an example of such a taper from the literature, in which the final effective index of the propagating mode achieves not only a good size match but a good effective index match to fiber due to the predominant confinement in air and SiO_2 [29].

11.4.2 Gratings

A second technique commonly employed for coupling into a silicon wire waveguide is a grating coupler. This device has a rich history and was reasonably well understood by the early 1970s [30]. However, the devices developed for silicon

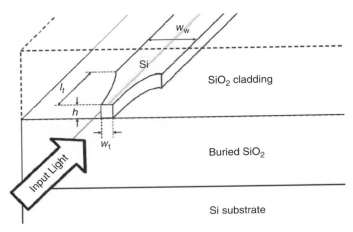

Figure 11.7 Schematic of waveguide with nano-taper coupler [28].

photonics [31, 32] have a number of unique traits—a relatively small number of grating teeth, a desire to mode-match to the Gaussian output of a single-mode fiber, and a remarkably wide optical bandwidth.

A silicon-grating coupler exploits the high index contrast between silicon and silicon dioxide, as well as the sub-wavelength patterning capabilities of a modern DUV lithography process, to create a grating capable of creating a well-controlled optical mode with lateral dimensions equivalent to the core of an optical fiber, or ∼10 μm in length. An example of a grating coupler is shown in Figure 11.8.

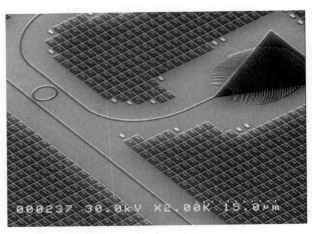

Figure 11.8 Oblique view of a planar grating coupler structure employing curved gratings, non-uniform pitch, and non-uniform periodicity. This particular device is formed using curved polysilicon gratings formed on top of a planar silicon slab which is tapered down to a single mode silicon waveguide. Image courtesy Luxtera, Inc.

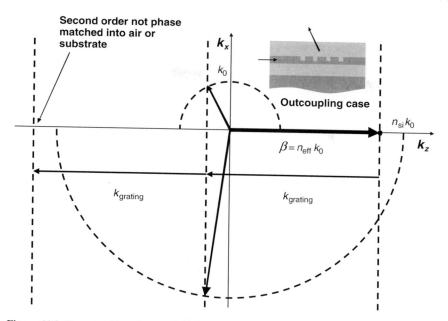

Figure 11.9 Phasematching diagram for backward scattering grating outcoupler, in which only first order is phasematched to radiate. Upper portion shows phasematching to air, while lower portion shows phasematching into silicon substrate (this figure may be seen in color on the included CD-ROM).

Waveguide-grating couplers in silicon function similarly to those in other material systems, with one exception being the increased scattering strength of the etched grooves. Using rigorous vector Finite Difference Time Domain (FDTD) modeling, it is relatively straightforward to analyze coupling from typical 200 nm silicon guide layers with grating grooves, revealing that the strong index contrast allows for efficient extraction of light in only 20 or so grooves. An additional difference with silicon-grating couplers is the magnitude of k-space that must be traversed with the grating.

Figure 11.9 shows a k-space diagram illustrating the coupling condition between a mode propagating in the silicon waveguide and the external free-space mode for fiber coupling. In this discussion, we consider only the outcoupled mode and rely on reciprocity to guarantee incoupling. The forward propagating slab waveguide mode has a propagation constant, β, which is related to the effective index n_{eff} of the propagating mode by

$$\beta = n_{\text{eff}} k_0 = \frac{2\pi n_{\text{eff}}}{\lambda_0}. \tag{11.3}$$

The allowed propagation directions of the diffracted light are determined by the horizontal component $k_{z,m}$ of the diffraction orders as described with the grating equation

$$k_{z,m} = \beta + m\left(\frac{2\pi}{\Lambda}\right) \qquad m = 0, \pm 1, \pm 2, \ldots \tag{11.4}$$

where Λ represents the physical grating pitch. The horizontal diffraction orders of a properly designed backward scattering grating coupler are shown as vertical dashed lines in Figure 11.9.

The total wavevector $\bar{k}$ for each material with respective refractive index n_{mat} must satisfy

$$\bar{k}^2 = k_x^2 + k_{z,m}^2 = (n_{mat}k_0)^2 = \left(\frac{2\pi n_{mat}}{\lambda_0}\right)^2 \tag{11.5}$$

This constraint is illustrated by the semi-circles in k-space as shown in the figure, in which the circle is only shown in the portions of the diagram where the final propagating region is present. The constraint for diffraction into air with index $n_{air} = 1$ is shown as the small semicircle in the upper quadrants, and the constraint for material radiated into the silicon substrate with $n_{si} \sim 3.5$ is shown as the circle in the bottom quadrants. Note that in this configuration only two diffracted orders are thus allowed, one near surface normal into air, and another into the substrate. Radiation into the substrate can be reduced by designing the gratings to scatter preferentially in the upward direction and by placing a reflector that returns this light to the upward direction. By incorporating these design elements, it is theoretically possible to design a grating coupler that is close to 100% efficient.

In practice, these devices have proven to be quite robust, manufacturable, and highly efficient. Record devices with coupling below 1 dB have been demonstrated [32]. Experimental results from a typical device are shown in Figure 11.10.

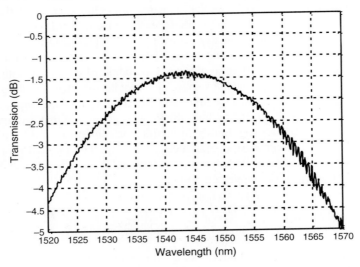

Figure 11.10 Experimentally achieved fiber-coupling efficiency of an SOI grating coupler as a function of wavelength. Data courtesy of Luxtera, Inc.

11.5 PASSIVE WAVEGUIDE DEVICES AND RESONATORS

11.5.1 Splitters and Couplers

As noted earlier, the high index contrast and low bending loss available in the SOI system suggests that optical components can be significantly miniaturized compared with more common integrated optic materials systems. Examples of passive devices, and thermo-optic active devices, that have been demonstrated include the full suite of splitters, directional and multimode interference (MMI) couplers, thermo-optic switches based on these components, and for wavelength division multiplexing (WDM) functions, arrayed waveguide grating filters, grating-based filters and add–drops, and ring-based filters and add–drops.

Figure 11.11 illustrates an example of some highly compact splitters demonstrated by Vlasov and coworkers at IBM [33]. The adiabatic splitter shown at left has demonstrated a flat 3-dB split over a wavelength range from 1450 to 1700 nm, while other approaches, utilizing resonant concepts as illustrated at right, can be even more compact with footprints of only $4\,\mu m^2$. Splitters with fanouts as large as 1×16 have also been demonstrated based on MMI couplers [34, 35].

Silicon photonic waveguides are readily incorporated into directional couplers, which in turn have been combined with splitters and thermo-optics to make more complex device architectures. The example in Figure 11.12 from Yamada and coworkers [36] illustrates a 1×2 thermo-optic switch utilizing the change in index of refraction with temperature to be discussed later.

11.5.2 Filters

Filters obviously are key to WDM functionality, signal conditioning, advanced transmit and receive functions, as well as amplified spontaneous emission (ASE)

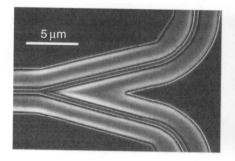
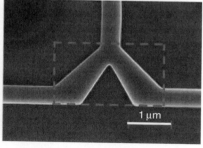

Figure 11.11 Highly compact splitters realized in the SOI system. Left: Near adiabatic split, and right: even more compact, resonant splitter. Photos courtesy of IBM (this figure may be seen in color on the included CD-ROM).

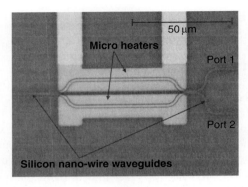

Figure 11.12 Highly compact 1 × 2 interferometric thermo-optic switch utilizing a splitter, and an output directional coupler.

noise filtering in systems employing amplification. Filters are almost exclusively realized using interferometric concepts, either using gratings, multipath waveguide configurations, or resonators. As an example of a grating-based WDM functionality, an add–drop filter has been demonstrated by incorporating lateral corrugation gratings along the waveguide and directional couplers in suitable configurations [37]. The arrayed waveguide grating, which has become a key component in WDM systems, has also been realized by a number of groups in silicon. Larger core waveguides, with less ability to tolerate sharp bends, have demonstrated excellent flat spectral characteristics over the C-band with crosstalk levels of approximately −35 dB [38]. However, it has been challenging to achieve this level of performance while also taking full advantage of the ultra-compact silicon wire waveguides. Figure 11.13 illustrates an example with 4 channels spaced at 11 nm demonstrating −14 dB crosstalk, but realized in a highly compact footprint of only 42 × 51 μm [39].

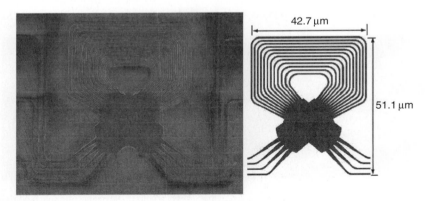

Figure 11.13 Ultra-compact 4-channel arrayed-waveguide grating filter realized in SOI using deposited waveguides (this figure may be seen in color on the included CD-ROM).

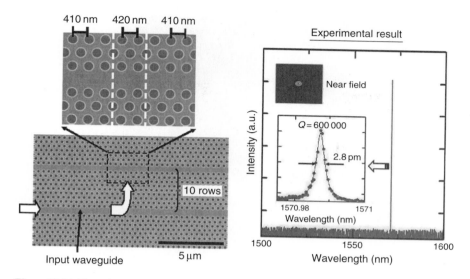

Figure 11.14 Photonic crystal add–drop filter with $Q \sim 600{,}000$ to achieve a 340 MHz bandwidth (this figure may be seen in color on the included CD-ROM).

11.5.3 Photonic Crystal Resonators

High finesse filters usually rely on resonant phenomena, and resonant structures have been demonstrated based on photonic crystal designs as well as ring-based designs [40]. Figure 11.14 illustrates the remarkably narrow channel drop filter by Noda and coworkers using a photonic crystal design realized in a suspended device layer [41]. This technology is described in detail elsewhere in this book. On the basis of lateral coupling into a resonator achieved by introducing a local shift or "defect" in a photonic crystal waveguide, a remarkable bandwidth of 340 MHz was achieved in a highly compact device.

11.5.4 Ring Resonators

Very low-loss resonators are also readily achievable utilizing low-loss waveguides directly in a ring configuration. Their simplicity and process compatibility, combined with their small size compared with other filter designs, makes a compelling case for system architectures based predominantly on rings. Additionally, the use of rings for measurement of waveguide loss and effective index makes them useful diagnostic tools to be employed in a small area within a wafer. For this reason, we will go into a bit more detail on the ring configuration.

When used as a filter, a ring resonator is typically deployed between two waveguides as shown in the example by Luxtera in Figure 11.15. This configuration has a well-known intensity transmission characteristic given by [42]

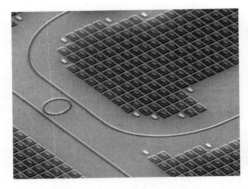

Figure 11.15 Ring resonator configured as an add/drop filter. The ring diameter is 5 μm. Photo courtesy of Luxtera, Inc.

$$T = \frac{\left(1 - |t_1|^2\right)\left(1 - |t_2|^2\right)e^{-\alpha \pi r}}{\left|1 - t_1 t_2 e^{\left(i\frac{\omega}{c} n_{\text{eff}} - \alpha/2\right)2\pi r}\right|^2}. \tag{11.6}$$

Here, the couplers are assumed loss-free, and t_1 and t_1 are the amplitude transmission coefficients for each of the two couplers for light staying in its original path, r is the ring radius, n_{eff} is the effective index of the waveguide mode. For low values of the total distributed loss in the ring, $\gamma = \alpha + \frac{1}{\pi r}\ln|t_1 t_2|$, including propagation loss α as well as the effective distributed input and output coupling losses, the transmission will display a near-Lorentzian response near the transmission peaks. If in addition the waveguide propagation losses are low, $\alpha \ll \frac{1}{\pi r}\ln|t_1 t_2|$, and the two transmission coefficients t_1 and t_1 are equal, then nearly 100% intensity transmission is achieved with a functional form near the peaks approximated by

$$T \approx \frac{1}{1 + 4(\omega - \omega_N)^2 \tau_{\text{ph}}^2}, \tag{11.7}$$

where $\tau_{\text{ph}} = (\gamma c/n_g)^{-1}$ is the photon lifetime, n_g is the group effective index, and the successive resonances are $\omega_N = \frac{c}{n_{\text{eff}} r} \cdot N$, with N integral. Here, we have ignored any frequency offset from the fixed phase of the transmission coefficients. The free spectral range (FSR) and Q of a cavity are given by $\text{FSR} = \frac{c}{n_g 2\pi r}$ and $Q = \omega \tau_{\text{ph}}$. To have a single ring capable of isolating an individual wavelength over the C-band requires a radius of less than ~3.75 μm, and while these values are exceptionally small, they are still well within the technical capabilities of the high-index-contrasts silicon waveguide technologies discussed above if exceptionally large Q values are not required.

Ring Add–Drop Filters

Single ring filters can provide good channel discrimination. Figure 11.16 illustrates experimental fiber-to-fiber results for a packaged device with a 15-μm radius

11. Silicon Photonics

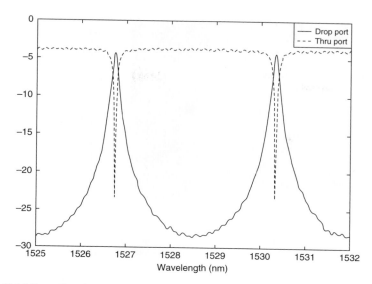

Figure 11.16 Data taken from the drop and transmission ports of a fiber-coupled ring filter. Data courtesy of Luxtera, Inc.

ring with $Q \sim 10\,900$, an FSR of 450 GHz, and a passband width of 18 GHz, residual filter insertion loss of <0.5 dB, and transmission bandwidth that is nearly adequate for 10 Gb/s bandwidth. This devices is capable of handling 4 channels separated by 100 GHz while maintaining crosstalk below −20 dB. However, the Lorentzian roll-off characteristic of a single ring does pose challenges in keeping adequate filter bandwidth to minimize a link intersymbol interference (ISI) penalty while keeping adjacent channel crosstalk at a low value.

Higher order, cascaded ring filters have been known to provide flat-top filters with sharp roll-off characteristics for significantly higher efficiency in usable bandwidth [43]. Figure 11.17 shows both experimental and simulated response for highly compact cascaded-ring-optical-waveguide (CROW) filters realized in the SOI system with 1, 2, 4, 8, and 16 elements [44]. While these designs illustrate the power of cascading, they do require strict manufacturing uniformity and operational temperature uniformity to ensure that all rings have identical resonance frequencies, or some form of active control.

11.6 ACTIVE MODULATION SILICON PHOTONICS

11.6.1 Modulation Effects

As can be seen in the previous section, modern fabrication techniques have led to highly compact, low-loss passive components. Full functionality in silicon

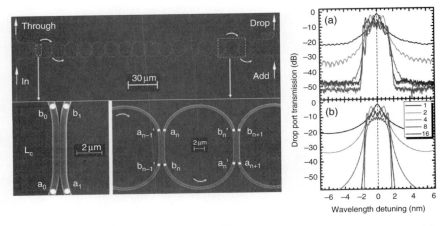

Figure 11.17 (a) Measured and (b) simulated transmission spectra at drop ports of CROWs containing 1, 2, 4, 8, and 16 racetrack resonators (this figure may be seen in color on the included CD-ROM).

photonics requires active components such as modulators, tunable filters, switches, as well as emitters and detectors which convert between electrical and optical signals. Many of the former functions can be accomplished with interferometric structures relying on active index manipulation in silicon.

Thermo-optic devices rely on index change with temperature. For example, near room temperature, the index of refraction of Si changes with temperature at 1.55 µm according to [45]

$$\Delta n \approx 1.84 \times 10^{-4} \times \Delta T, \qquad (11.8)$$

and this effect can be implemented in devices, for example, with thin-film Pt surface heaters near waveguide sections. While this effect can be used to tune, switch, or trim, resonant or phase-sensitive optical device, it is typically too slow to be useful as a modulation technique for information coding, with typical response times of 100 µs [36].

For higher speed operation, the early work of Soref and Bennett [24] examined electrooptical effects in silicon, including contributions to real index and loss resulting from (a) free carrier changes that may result from injection, depletion, or doping, (b) electro-refractive effects from the Franz–Keldysh effects and Kerr effects. While the combined electro-refractive effects produced index changes increasing more than quadratically with field, at 1.55 µm they only resulted in small index changes of $\Delta n = 1.6 \times 10^{-6}$ even at fields of 10^5 V/cm.

However, free-carrier effects are quite large, with contributions dominated by free-carrier plasma contributions, but including in principle weak effects from band-filling effects and band-gap shrinkage due to exchange correlation interactions in the indirect transition absorption edge. The loss mechanism noted earlier from free-carrier absorption in silicon,

$$\Delta\alpha = 8.5 \times 10^{-18} \times \Delta N + 6.0 \times 10^{-18} \times \Delta P \tag{11.9}$$

can be used as a modulation technique, for example, by increasing free carriers in a signal path using P–N junction in forward bias. However, it is typically more effective to use interferometric techniques and harness the changes in the real index of refraction.

The simple Drude model of real index change yields

$$\Delta n = -\frac{e^2 \lambda^2}{8\pi^2 c^2 \varepsilon_0 n} \left(\frac{\Delta N}{m_e^*} + \frac{\Delta P}{m_h^*} \right) \tag{11.10}$$

but in Soref and Bennett's work it was found that the contributions to real index change from holes were not quite linear as predicted by the Drude model, and the total real index change was instead best fit by [24]

$$\Delta n = -8.8 \times 10^{-22} \times \Delta N - 8.5 \times 10^{-18} \times (\Delta P)^{0.8} \tag{11.11}$$

A linearization of the hole-density-dependence at $P = 1 \times 10^{17} \text{cm}^{-3}$ reveals that the effect of a differential changing in hole concentration on the index is actually three times that of electrons, again in contrast with the Drude model. The index change from these free-carrier plasma effects is by far the most commonly used modulation technique in silicon photonics, and it can be realized dynamically in a device configuration by forward injection in a P–N junction, by depleting a P–N junction, or by accumulation or inversion in field-effect devices using insulating gate dielectrics.

The index-changing mechanisms described above are based on the plasma index change. In the III–V semiconductor arena, the use of the Franz–Keldysh and Quantum-Confined Stark Effect (QCSE) modulators have been known for many years to provide highly compact, efficient modulation. Both silicon and germanium are indirect-gap materials, and this is generally viewed as precluding lasing because any excited carrier density will populate the indirect valley. However, the direct gap of Ge or SiGe alloys can have values in the range of interest for near-IR optical communications, and will provide strong absorption. This absorption can certainly be modulated by the Franz–Keldysh effect [24, 46], and experimentally, the scattering time to the indirect valley is comparable to, or longer than, the exciton dephasing time, enabling suitably designed quantum well structures to also exhibit a very well-defined QCSE as illustrated in Figure 11.18 [47].

11.6.2 P–N Junction in Depletion

Assuming for simplicity an abrupt junction with donor and acceptor concentrations of N_D and N_A as illustrated in Figure 11.19, Gauss' law provides for a linear

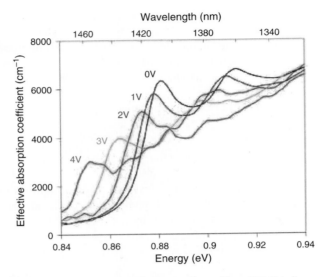

Figure 11.18 QCSE modulation in Ge/SiGe quantum wells on silicon [47] (this figure may be seen in color on the included CD-ROM).

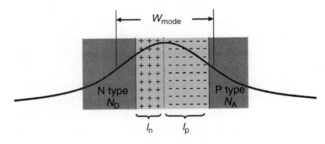

Figure 11.19 Simple analysis of abrupt P-N junction being depleted within an optical waveguide mode (this figure may be seen in color on the included CD-ROM).

increase in electric field to a maximum value E_{max} at the junction with the respective widths of the depleted p-type and n-type regions, l_p and l_n:

$$\nabla \cdot D = \rho \Rightarrow \frac{dE}{dx} = \frac{-eN_A}{\varepsilon_r\varepsilon_0} \Rightarrow E_{max} = \frac{-eN_A}{\varepsilon_r\varepsilon_0} \cdot l_p \quad (11.12)$$

and proceeding similarly for l_n we have

$$l_p = \frac{|E_{max}|\varepsilon_r\varepsilon_0}{eN_A} \quad \text{and} \quad l_n = \frac{|E_{max}|\varepsilon_r\varepsilon_0}{eN_D} \quad (11.13)$$

11. Silicon Photonics

By integrating the fields, E_{max} can be expressed in terms of the voltage V applied to the P terminal with the N terminal grounded. For $V < V_0$, where V_0 is the built-in voltage of the junction, this yields

$$|E_{max}| = \left[\frac{2e}{\varepsilon_r\varepsilon_0}\left(\frac{N_D N_A}{N_D + N_A}\right) \cdot (V_0 - V)\right]^{\frac{1}{2}} \quad (11.14)$$

If we consider applying a voltage ranging from $\sim V_0$ through zero to negative values, swinging the width of the depletion region from near zero to larger values, we know that this will change the index of refraction in the region spanned by the largest depletion width due to the free-carrier plasma effect. As illustrated in the simple two-dimensional picture shown in Figure 11.19, we can envision an optical mode with width W_{mode} that has significant overlap with this depletion region. If we linearize the index change with carrier density from the expression presented earlier, we have

$$\Delta n_{real} = \left(\frac{\partial n_{real}}{\partial N}\Delta N\right) \cdot \Gamma_N + \left(\frac{\partial n_{real}}{\partial P}\Delta P\right) \cdot \Gamma_P$$

where Γ_N and Γ_P are the modal confinement factors for the regions whose carrier density changes. Because here $\Delta N = N_D$ and $\Delta P = N_A$, and if we assume that the optical mode is well-positioned on the depletion region, we will have

$$\Gamma_N \cong \frac{l_n}{W_{mode}} \quad \text{and} \quad \Gamma_P \cong \frac{l_p}{W_{mode}}.$$

If we then substitute our expressions for l_n and l_p derived earlier, we have the simple result

$$\Delta n_{real} = \left(\frac{\partial n_{real}}{\partial N} + \frac{\partial n_{real}}{\partial P}\right)\frac{\varepsilon_r\varepsilon_0 E_{max}}{e W_{mode}} \quad (11.15)$$

Clearly, this is maximized by making the optical mode small, but also by increasing E_{max}. To see the design limits, we assume that the applied voltage is increased to the breakdown field in the medium, E_{BD}, and that the doping level is raised to a high-enough value such that the depletion width at this breakdown field level is essentially contained within the optical mode. Higher values may not be desirable due to the residual free-carrier losses within the mode. If this index change is going to be used in a push–pull Mach–Zehnder interferometer, then we would require that the available differential index change satisfy

$$\frac{2\pi \Delta n_{real} L}{\lambda} = \frac{\pi}{2} \quad (11.16)$$

and substituting the expression above with $E_{max} = E_{BD}$, we finally have

$$\frac{L}{W_{mode}} = \frac{\lambda e}{4\varepsilon_r\varepsilon_0|E_{BD}|\left(\frac{\partial n_{real}}{\partial N} + \frac{\partial n_{real}}{\partial P}\right)} \quad (11.17)$$

In silicon, we have $E_{BD} \approx 5 \times 10^5$ V/cm at a doping level of $P \cong 10^{17}$ cm^{-3} [48], and from the expression of Soref, expanding about $P \cong 10^{17}$ cm^{-3}, we have

$\left(\frac{\partial n_{\text{real}}}{\partial N} + \frac{\partial n_{\text{real}}}{\partial P}\right) \approx -3.6 \times 10^{-21} \text{cm}^{-3}$. This gives the first-order result, roughly independent of the relative doping strength on each side of the junction, of

$$\frac{L}{W_{\text{mode}}} \geq 3.3 \times 10^3 \qquad (11.18)$$

For typical silicon photonic compact ridge waveguide modes of $W_{\text{mode}} \sim 0.5\,\mu\text{m}$, this indicates that a Mach–Zehnder device made using this approach will be at least ~ 1.6 mm long, and using larger mode sizes inherently will scale the device to be longer. Detailed extinction ratio and insertion loss analysis also requires evaluation of the loss induced by the same change in carrier density. Again using the linearized expression for index change vs carrier density, the plasma index change required for a $\pm \pi/2$ push–pull phase swing in a Mach–Zehnder will also carry with it a differential field amplitude change of

$$e^{\pm\frac{\Delta\alpha}{2}L} = \exp\left[\frac{\pm\left(\frac{\partial\alpha}{\partial N} + \frac{\partial\alpha}{\partial P}\right)}{\left(\frac{\partial n_{\text{real}}}{\partial N} + \frac{\partial n_{\text{real}}}{\partial P}\right)} \times \frac{\lambda}{8}\right] = e^{\pm 0.078} \qquad (11.19)$$

This will provide an extinction ratio of 22 dB, and <0.7 dB inherent insertion loss in the on-state. The extinction ratio can be readily improved by introducing very slight asymmetry in the Mach–Zehnder splitters if critical to the application.

The speed of a reverse-biased device will be limited primarily by RC times for practical designs, because the charge redistribution for the majority carrier device is typically governed by the faster dielectric relaxation times. The lateral junction, with little additional parasitic capacitance, is desirable in this respect, although the doping distribution and contact placement for lateral electrical access must be optimized to reduce resistance while minimizing unnecessary doping-induced optical losses or exposure of the optical mode to metallic contacts.

11.6.3 P–N Junction in Forward Bias

P–N junctions can also be driven into forward bias to inject minority carriers into each region for net plasma index increase. However, in this case, the response time is generally limited by the minority carrier recombination time τ with a typical single-pole filter response function of the form

$$R(\omega) \sim \left|\frac{1}{1+i\omega\tau}\right|^2 \qquad (11.20)$$

If τ is short, either from extrinsic, Auger, or from surface recombination in small fabricated structures, the response can be near-GHz level. Equalization techniques have been applied, and additionally, if the modulation swing includes both forward and reverse bias, and the reverse bias depletion width is comparable to a diffusion length to fully extract all injected carriers, then speeds comparable to reverse-biased operation can be obtained, and some enhancement in dynamic range can be obtained relative to a purely reverse-biased device [49].

11.6.4 Carrier Accumulation on Gate Oxides

Another approach is to use charge accumulation and/or depletion, not in a P–N junction, but at the interface between silicon and a dielectric using designs that are analogous to gate structures in CMOS field-effect transistors [50]. In this instance, we have the picture shown in Figure 11.20.

This structure, with an embedded gate oxide sandwiched by P-type and N-type silicon layers, can be viewed as a simple capacitor with the enhanced charge density on each side being given respectively by

$$\Delta N = \frac{\varepsilon_0 \varepsilon_r}{e\, t_{ox} t_{ch-N}} [V - V_0] \quad \text{and} \quad \Delta P = \frac{\varepsilon_0 \varepsilon_r}{e\, t_{ox} t_{ch-P}} [V - V_0] \qquad (11.21)$$

where t_{ox} is the oxide thickness and t_{ch-N} and t_{ch-P} are the charge accumulation thicknesses (approximately the Debeye lengths) on the N and P sides, and V_0 is the flat-band voltage. Because the confinement factor will be given by

$$\Gamma_N \cong \frac{t_{ch-N}}{W_{mode}} \quad \text{and} \quad \Gamma_P \cong \frac{t_{ch-P}}{W_{mode}},$$

we again have (with a linearized differential index change) the simple expression

$$\begin{aligned}\Delta n_{real} &= \left(\frac{\partial n_{real}}{\partial N}\Delta N\right)\cdot \Gamma_N + \left(\frac{\partial n_{real}}{\partial P}\Delta P\right)\cdot \Gamma_P \\ &= \left(\frac{\partial n_{real}}{\partial N} + \frac{\partial n_{real}}{\partial P}\right)\frac{\varepsilon_0 \varepsilon_r}{e\, t_{ox} W_{mode}}[V - V_0]\end{aligned} \qquad (11.22)$$

where in this case the ε_r is that of gate dielectric. Noting that

$$C \approx \frac{\varepsilon_0 \varepsilon_r A}{t_{ox}} \qquad (11.23)$$

is approximately the capacitance of a region of area A, we see that the index change is really just proportional to the charge density per unit area induced on the capacitor and inversely proportional to the width of the mode in the direction perpendicular to the interface. If we compare this to the previous expression for the P–N junction, we see that it is very similar, except here the field E_{max} is the field across the gate dielectric, which is typically SiO_2 with $\varepsilon_r \approx 3.9$. Because the breakdown field of SiO_2 is larger than that of Si, typically $>10^7$ V/cm, despite

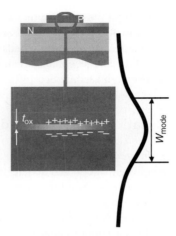

Figure 11.20 Charge accumulation at a CMOS-like gate dielectric, contained in a waveguide within an optical mode (this figure may be seen in color on the included CD-ROM).

the lower dielectric constant one can achieve in principle significantly shorter devices than in the case of the P–N junction. This is further enhanced by the fact that typically the modal dimension in the vertical direction is relatively easier to design at a small value, perhaps as narrow as 0.2 µm. Together, these effects suggest that the field effect device could be an order of magnitude shorter in length than the P–N junction devices.

Below we will see that there are additional approaches, such as resonant modulators, which can reduce the required size even more dramatically, but at the expense of fixed-wavelength operation.

11.6.5 Modulator Devices

Silicon photonic devices based on the free carrier index and loss mechanisms described above have been realized and are commercially available. One example is the variable attenuator design illustrated in Figure 11.21, which is based on the

Figure 11.21 Cross-section of variable attenuator waveguide from Kotura, Inc., illustrating optical mode overlap with laterally injected carriers to provide free-carrier-induced loss [51] (this figure may be seen in color on the included CD-ROM).

large-area rib waveguide design to improve fiber-coupling efficiency, and exhibiting dynamic range in attenuation of >40 dB with 60 mA current drive and <1 μs response times [51]. These devices have also been fabricated in highly uniform eight element arrays with integrated directional coupler taps and flip-chip bonded monitor photodiodes for use in power management for WDM optical networking applications.

To realize a Mach–Zehnder modulator based on depleting a P–N junction, a typical exemplary design is illustrated in Figure 11.22. Figure 11.23 illustrates high-performance devices that have been realized based on a lateral depletion

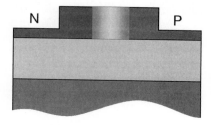

Figure 11.22 Illustrative example of a laterally depleted ridge waveguide as element in Mach–Zehnder modulator (this figure may be seen in color on the included CD-ROM).

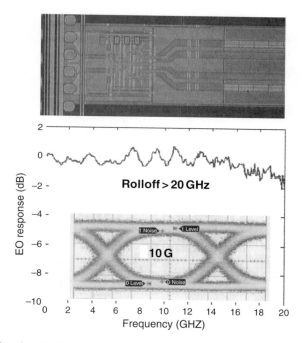

Figure 11.23 Plan view of Mach Zehnder modulator fabricated in fully CMOS compliant foundry by Luxtera, Inc., exhibiting >20 GHz bandwidth and excellent 10 Gb/s eye [52] (this figure may be seen in color on the included CD-ROM).

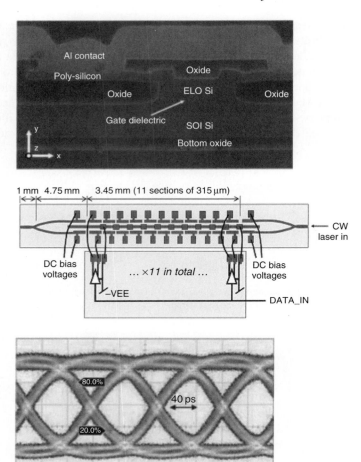

Figure 11.24 Scanning electron micrograph of large-area rib design with embedded gate dielectric from Intel Corp. Plan view illustrates segmented electrode in push–pull configuration, and bandwidth is high enough for 10 Gb/s operation (this figure may be seen in color on the included CD-ROM).

concept in a fully CMOS compliant fabrication process [52]. These devices have 3 dB modulation bandwidths exceeding 20 GHz, and exhibit excellent 10 Gb/s eye diagrams.

Mach–Zehnder modulators based on the CMOS gate configuration as illustrated in Figure 11.24, have also been realized [50]. In this example, the large mode size realized by using the large-area rib design precluded scaling the device length down to the levels discussed earlier, but ∼10 Gb/s operation was realized in a CMOS compliant process.

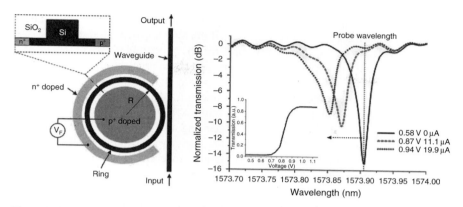

Figure 11.25 Left: Ring modulator with lateral P–N junction for current injection. Right: Transmission spectrum of the ring under increasing injection, which both shifts the resonance to shorter wavelength and reduces the Q due to increased loss [53] (this figure may be seen in color on the included CD-ROM).

Another technique that has been introduced for reducing the size of modulators is to employ resonant structures. This concept essentially re-circulates the optical pathlength in the modulated medium to get a larger accumulated phase shift, but this comes at the expense of narrow-band operation. A representative device structure is illustrated schematically in Figure 11.25 [53] in which a ring that is critically coupled to a input/output waveguide (i.e., the coupling to ring is roughly the same as the round-trip loss of the ring).

This gives high absorption at the resonance of the ring within a bandwidth of $\Delta \nu_{\text{FWHM}} \approx \frac{1}{2\pi\tau_{\text{ph}}}$, where τ_{ph} is the photon lifetime, or $1/e$ energy decay, of the loaded ring. The ring is fabricated from a ridge guide as shown earlier, and lateral minority carrier injection reduces the index and shifts the resonance of the ring to shorter wavelength according to

$$\frac{\Delta\lambda}{\lambda} \approx \frac{\Delta n}{n_g}, \qquad (11.24)$$

where n_g is the group index of the guide. If a continuous input beam is initially on resonance, it is extinguished until the modulation shifts the resonance to the side, as illustrated in the figure, thereby allowing transmission.

It is tempting to think that a resonant modulator offers fundamentally improved sensitivity due to the sharp, resonant nature of the transmission, and also that it may have higher speed due to lower parasitics, but neither of these are necessarily true. To provide good extinction the ring modulator must shift its transmission peak by an amount $\Delta \nu \approx -\frac{1}{2\pi\tau_{\text{ph}}}$ and, with $\frac{\Delta \nu}{\nu} \approx -\frac{\Delta n}{n_g}$, this requires and index shift of

$$\Delta n \approx \frac{n_g}{2\pi\nu\tau_{\text{ph}}} = \frac{n_g}{Q} \qquad (11.25)$$

If we compare this to a single-arm drive Mach-Zehnder requiring an index shift of $\Delta n = \frac{\lambda}{2L}$, it can achieve identical sensitivity if the length is increased to $L = Q \cdot \frac{\lambda}{2n_g}$, which is typically not prohibitively long for the moderate Qs required to give acceptable modulation bandwidth of the ring. To examine the relative speeds of these two modulators, first note that the lumped-element Mach–Zehnder modulator will be transit-time limited by the optical transit delay of $T_d = \frac{n_g L}{c}$. This produces a sinc^2 modulation response with a bandwidth given roughly by $f_{3dB} \approx \frac{1}{2T_d} = \frac{c}{2n_g L} = \frac{\nu}{Q}$. However, the resonant ring modulator also has a bandwidth limitation precisely due to the resonant character, i.e, it cannot be shifted faster than the photon lifetime or the cumulative interference that gives rise to the narrow resonance is not effective. This cumulative interference is, in fact, just a "folded" version of the equivalent-sensitivity Mach–Zehnder device discussed above.

While a full analysis of a ring composed of a temporally modulated medium is non-trivial, some basic observations can illustrate the speed limitations of this concept. Consider the response of the ring transmission to a step-function change in index, from a state that is initially not aligned in frequency with a CW input beam, and is thus in a transmitting state, to a state that is aligned and nontransmitting. There is no significant light in the ring in the initial state, and the intensity in the ring upon switching to the non-transmitting state then exponentially builds up with a photon lifetime τ_{ph}. This build-up, which is required to produce the interferometric cancellation of the transmitted beam and provide extinction, thus requires a photon lifetime, and the response will appear to be "RC-like" and have a 3-dB bandwidth given by $f_{3dB} \approx \frac{1}{2\pi \tau_{ph}} = \frac{\nu}{Q}$ which is indeed the same as the equivalent-sensitivity Mach–Zehnder. The opposite case, with a step function from a transmitting state to a nontransmitting state, can also be analyzed as an instantaneous parametric frequency conversion of the resonant light in the ring, and ultimately yields a similar response time. Thus, the ring and the lumped-element MZ of comparable bandwidth will have comparable sensitivity, and it is then clear that a *traveling-wave* MZ, well-known to be an improvement on the lumped-element device, will actually have better sensitivity than the ring also if microwave transmission line losses do not pose a serious limitation.

Similarly, while the capacitance may be reduced by a ratio of the lengths of the ring to the Mach–Zehnder length, the resistance is increased by the same proportion so the lumped-element RC bandwidth will be the same using the same guide technology. If the device resistance is not the limiting resistance, then some benefit can of course be realized by the smaller resonant modulator.

The two distinct and potentially important advantages the resonant modulator has are its small very small size and its low capacitance. If the power supplied is due exclusively to charging and discharging the capacitance, its power consumption will be $P \sim \frac{1}{2} C V^2 \times f_{mod}$, and the low capacitance is an obvious benefit. However, these benefits come at the operational complexity of fixed wavelength operation, and maintaining alignment to the optical signal may require large-scale tuning, for example, using the thermo-optic effect discussed earlier, which will also likely consume undesirable power. Figure 11.26 illustrates a ring-based

11. Silicon Photonics

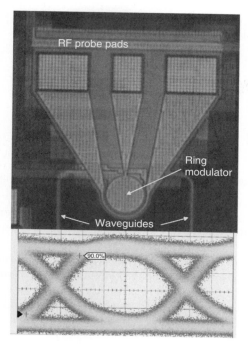

Figure 11.26 Top: Compact resonant ring modulator from Luxtera, Inc. exhibiting open eye diagrams at 10 Gb/s (Bottom) (this figure may be seen in color on the included CD-ROM).

resonant modulator fabricated in fully compliant CMOS foundry manufacture, with bandwidths sufficient for 10 Gb/s operation [54].

11.7 GERMANIUM PHOTODETECTORS AND PHOTORECEIVERS FOR INTEGRATED SILICON PHOTONICS

The ability to convert optical signals to electrical signals is a key function that must be achieved in nearly every optical system. However, while an SOI CMOS process offers convenient materials for optical waveguides and a variety of modulation techniques, it is missing a means to create efficient photodetectors. Thus, there is a broad interest in integration of germanium photodetectors into a CMOS process. This section will first focus on the theory and rational for such integration, and then will show results from integration work at Luxtera in Freescale Semiconductor's 0.13 μm SOI process.

Many silicon photonic devices do not perform as well as their discrete counterparts. However, when Germanium photodetectors are integrated in immediate

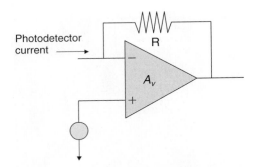

Figure 11.27 Simplified conceptual schematic of transimpedance amplifier configuration.

proximity to their amplifier transistors, there are substantial performance benefits that promise to make these devices among the highest performance devices achievable.

Figure 11.27 shows a greatly simplified schematic of a transimpedance amplifier (TIA), which converts an input photocurrent to an output voltage. Maximizing gain suggests that R be set as high as possible to achieve the largest voltage for a given input current. However, this adversely impacts bandwidth, and for a given bandwidth, B, input capacitance, C_{in}, and amplifier voltage gain A_v, the upper limit is given by

$$R < \frac{A_v}{2\pi C_{in} B} \qquad (11.26)$$

In practical systems, the bandwidth is set to accommodate the bit rate, and the gain of amplifiers is set by the CMOS process in which integration is occurring. Thus, reduction of input capacitance is the dominant variable left to address. Fortunately, when integrated into a CMOS process, Germanium detectors can be placed immediately adjacent to the TIA and connected with the CMOS metallization. This achieves an incredibly low capacitance as compared with any technique used to package hybrid devices. The highest performance hybrid technology is where the photodetector is flip-chipped directly on top of the TIA, utilizing bond pads on both die and a conductive epoxy or solder between the two.

A convenient approximation for the sensitivity of a fast TIA photoreceiver, optimized according to Eqn (11.26) with input capacitance assumed to be dominated by the photodetector capacitance C, and including thermal noise and shot noise from the detector dark current I_d, is given by

$$OMA = \frac{SNR \sqrt{\frac{16\pi kTCB^2}{A_v} + 2qI_d B + \frac{2\pi CV_{th}B}{A_v}}}{Rsp} \qquad (11.27)$$

11. Silicon Photonics

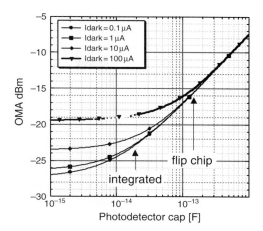

Figure 11.28 Required optical modulation amplitude as a function of photodetector capacitance for different values of dark current. Values of parameters used are $B = 9$ GHz, SNR = 14.1, $A_v = 5$, $V_{th} = 10$ mV, and Rsp = 0.8 A/W (this figure may be seen in color on the included CD-ROM).

where OMA is the optical intensity modulation amplitude, B is the bandwidth, SNR is the signal-to-noise ratio required to achieve 10^{-12} bit error rate, A_v is the voltage gain of the amplifier input stage, T is temperature, k is Boltzman's constant, and Rsp is the detector responsivity. This formula has been augmented to include the effects of the threshold voltage sensitivity, V_{th}, of the follow-on limiting amplifier which is not shown in Figure 11.27. Figure 11.28 shows a comparison between the capacitance, and thus sensitivity achievable using flip-chip or fully integrated photodetectors and using the assumptions contained in the discussion above. The OMA is shown as a function of capacitance for a variety of dark currents, and assuming bandwidth adequate for a 10-Gb/s signal. In practice, it is possible to achieve a capacitance of ∼0.2 pF for flip-chip detectors, while the first integrated devices have been demonstrated at one order of magnitude lower [55]. Note that the performance of the device can exceed that of flip-chip technology even with relatively high dark currents. Substantial benefits are realized with dark currents below 10 μA, while decreasing returns are achieved below 1 μA. This is of great importance because the dark current of Germanium grown on silicon has proven to be higher than bulk material. However, by controlling the size of the integrated detectors, 1 μA is readily achieved. Thus, using a very reasonable set of assumptions and an easily achieved input capacitance, it is easy to see the large improvement possible with integrated detectors. Additional improvements can be achieved with Germanium detectors because they can be constructed in waveguide configuration, which improves responsivity, and they can be operated in avalanche mode with gain as high as 10 dB already demonstrated.

These striking technical benefits, along with the cost, scalability, and form factor benefits make Germanium detectors a compelling technology, but there are significant hurdles to realize such a device. First, Germanium has a 4% lattice mismatch with silicon, which complicates growth of crystalline germanium material on a silicon substrate. However, this has been overcome through use of a two-step temperature process, which has demonstrated the ability to grow fully strain-relaxed germanium films [56–58]. The absorption coefficients of this material displays a substantial redshift of the absorption edge, making use across the C-band practical. In contrast, if an SiGe alloy is used, it is a substantial challenge to find a wavelength regime in which both high passive waveguide transparency and efficient photocollection can be achieved. Germanium devices have been realized as stand-alone components by a number of researchers [59–63] and, finally, have been integrated into a CMOS process [64, 65].

A further challenge for integration involves the type of germanium detector to be grown, surface illuminated vs waveguide configuration. Normal incidence detectors require several microns of Ge material to be grown. However, due to the presence of numerous chemical-mechanical planarization steps, this is incompatible with standard CMOS processing. Thus, there are benefits to pursing waveguide photodetectors, which can be quite thin while maintaining an appropriately long interaction length. Through the use of selective epitaxy on silicon using a silicon oxide/nitride growth mask, patterned films with thicknesses of 200 nm are readily realized.

Yet, another architectural decision involves which type of detector to construct. Three primary options are shown in Figure 11.29. In the single heterojunction device, contact is made to both the silicon and the germanium layers, and photocurrent is passed across the heterointerface. While this configuration has been proven to have high photocurrent collection efficiency, the main drawback is the need to develop a CMOS-compatible germanium contact module. The homojunction photodetector involves only contact to germanium implants. While the field confinement of this configuration achieves a good collection efficiency, and no current crosses the heterointerface, it has the drawback that both n- and p-type contacts must be developed, and n-type contacts have proven, to date, to have high leakage current. The final device, which has contacts only to silicon, has the benefit of using standard contact recipes, but at the price of requiring very high bias voltages in order to extract the carriers across the two interfaces.

A key input to the type of photodetector to pursue involves the behavior of the heterointerface, and the choice of which material is to be the anode or cathode. Comparisons of germanium heterojunction devices of both configurations are

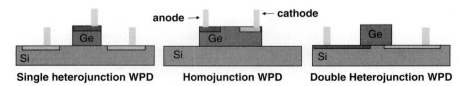

Figure 11.29 Schematics of different contacting schemes for Ge waveguide photodetectors (WPDs) (this figure may be seen in color on the included CD-ROM).

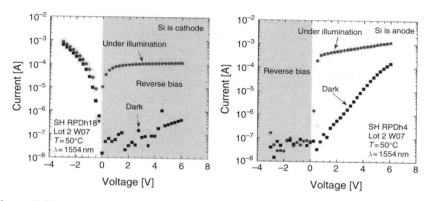

Figure 11.30 Dark currents for Ge single heterojunction photodetectors. Left: Device doping configured for silicon as cathode. Right: Device doping configured for silicon as anode (this figure may be seen in color on the included CD-ROM).

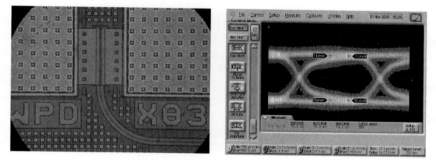

Figure 11.31 Left: Plan view photograph of Ge waveguide photodiode realized in CMOS foundry. Right: 10 Gb/s eye diagram of device implanted to provide avalanche gain of 4 dB (this figure may be seen in color on the included CD-ROM).

shown in Figure 11.30. Note that when the silicon is employed as the cathode, there is a substantial improvement in the dark current under bias. Due to the poor performance of the n-type Germanium, detectors have been constructed with p-type Ge contacts, and a typical configuration illustrating the curved input waveguide is shown in Figure 11.31. This device achieved a responsivity of 0.5 A/W and bandwidth of 14 GHz and a capacitance of only 30 fF.

In addition, this silicon implant profile of the single heterojunction device can be modified to allow formation of an avalanche photodetector. Up to 10 dB of gain has been achieved, though with high dark current. The eye diagram resulting from avalanche gain of 4 dB using 5 V is shown in Figure 11.31. There has been a substantial amount of work done to validate the reliability of these devices, with promising results. Though they have yet to be demonstrated in commercial products, these devices have great potential and are reportedly under development in a number of corporations, though few details have been made public. It is quite likely that these components will be integrated into products in the near future.

11.8 CMOS INTEGRATION AND INTEGRATED SILICON PHOTONICS

11.8.1 SOI Transistor Description

Silicon photonics is fundamentally interesting due to the ability to leverage the massive investment in CMOS-processing infrastructure and processes development. However, it is far too easy to create a silicon photonics device that is not realizable using these standard processes. Here, we provide a brief description of a standard SOI process, and additional detail may be found in the Refs [66, 67].

An SOI transistor is formed by first etching a blanket silicon film to define the extents of the silicon transistor body. The resulting silicon sidewalls are then treated to prevent leakage, and the trenches are filled with dielectric material. The transistor bodies are then implanted to define n- or p-type transistors. Subsequently, the polysilicon gates and dielectric spacers are formed. Next, the contacts are formed by salicidation of the gate and source/drain regions of the transistor, and this is covered with a number of dielectric films.

Contact holes are then etched through the dielectric stack and tungsten "plugs" are plated into the void. On top of this stack, a copper and low-k dielectric backend is then formed, often containing 8–10 levels of metallization and the associated vias.

Figure 11.32 shows a cross-section of a submicron SOI transistor fabricated at Freescale Semiconductor, while Figure 11.33 shows a larger scale view illustrating one plane of metallization with contact vias. Note the relatively large number of films and materials in immediate proximity to the silicon: polysilicon, gate oxide, gate liner, gate spacers, salicidation, contact etch stop, interlevel dielectric spacer, BOX, substrate, trench fill oxidation, trench liner, contact metallization, Cu diffusion barrier, and the lowest metal layer. Each of these films has the potential to impact the flow of light in the silicon and must be considered in any design where photonics are incorporated with transistors. Furthermore, if the light is to enter and exit the chip from the top, such as through a grating coupler, the remainder of the dielectric and passivation films in the process must be considered as a thin film stack in the path of the light. Thus, the optical properties of the numerous films must be understood in order to evaluate suitability for integration into any baseline process.

A remaining challenge for integration of photonics into an SOI process is compatibility with the numerous process modifications that are slated for the

11. Silicon Photonics

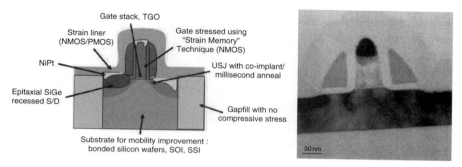

Figure 11.32 A cutting edge SOI transistor schematic and scanning electron micrograph (SEM) detail (image provided courtesy of Freescale) (this figure may be seen in color on the included CD-ROM).

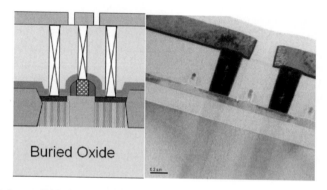

Figure 11.33 Larger field view of metalization vias extending down to device layer in SOI wafer, image provided courtesy of Freescale (this figure may be seen in color on the included CD-ROM).

coming process nodes. First, strain engineering in silicon has already become widely adopted for cutting edge SOI devices. Strain is built into the starting substrate and is enhanced during processing with the addition of SiGe to the source and drain regions. Additionally, the trench gap-filling dielectric and the conformal passivation dielectric have engineered strain properties. Due to the dependence of silicon's refractive index and carrier mobility on strain, this has an uncertain impact on optical devices and a detailed understanding of how strain impacts optical devices such as waveguides and modulators has yet to be fully understood. Second, newer transistors are also implemented in silicon films that are widely understood to be far too thin for silicon photonic devices. Third, the motivation to reduce capacitance in the backend metallization layers has led to the development and usage of porous low-k dielectric materials, which now become the cladding of the silicon waveguides. Fourth, it is widely discussed that rare-earth materials will be incorporated in the next generation high-k gate dielectric. These numerous changes pose both a threat and an opportunity to silicon photonics designers, and it is not clear that photonic integration in future generations of CMOS will be achieved similarly to current developments in the 0.13-μm node.

One benefit to the enormous changes underway in the CMOS fab is that foundry managers and process engineers, once reputed to be stalwart opponents to introduction of new materials, now embrace a wide variety of materials and have developed processes for introduction of new material in their process lines. This open-minded approach is essential for future development of photonic devices, particularly those related to light sources.

11.8.2 Examples of Integrated Silicon Photonics Chips

The ultimate measure of success of any new technology is often judged in the commercial marketplace. In many people's minds, silicon photonics has the promise to offer optical communications at the cost structure of CMOS, while others are focused on the ability to leverage integration to develop new capabilities beyond the reach of traditional optical component technology. In this section, we will discuss example transceivers where silicon photonics are integrated into a production CMOS process—a path that promises to broaden the marketplace for optical communications.

A 20 Gb/s Dual XFP Transceiver

Figure 11.34 shows a die photograph of the first silicon photonics transceiver fully integrated into CMOS [52]. This die has the same functionality as two independent 10 Gb/s XFP transceivers (XFP is a small form factor protocol-independent

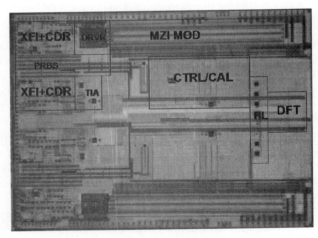

Figure 11.34 Two independent 10 Gb transceivers on a single die (contains complete PHY circuitry). Photograph provided by Luxtera, Inc. (this figure may be seen in color on the included CD-ROM).

pluggable transceiver standard). Light is supplied by two laser diodes that are flip-chip bonded to the die using the grating coupling scheme described earlier. Modulation is achieved using two silicon MZ modulators driven by on-chip drivers. Light is transmitted through grating couplers down two separate fibers. On the receive side, light enters the chip through grating couplers and travels to photodetectors that are flip-chip bonded over integrated CMOS TIAs. PHY (Physical Layer) circuitry is included in this device and is capable of retiming the input and output electrical signals according to the 10-Gb/s XFI interface standard.

A 40-Gb/s WDM Transceiver

Manufacture of silicon photonics in a CMOS process opens up the possibility to add numerous optical components to a single die at nearly no cost. This is a compelling concept for silicon photonics because it can scale in bandwidth while maintaining a minimum number of fiber connections to the die—thus controlling the cost of packaging, a major cost driver for any optical communications device.

In this transceiver, shown in Figure 11.35, 4 wavelengths on a 200-GHz grid are employed and optical mux/demux is performed on the die using interleavers. Modulation is performed on each of the four wavelengths using MZ modulators with integrated drivers. The four wavelengths are then multiplexed using a tunable, two-stage interleaver. Light is then transmitted and received down a fiber using grating couplers. At the receiver, the light is demultiplexed using a similar two-stage interleaver and sent to separate flip-chip photodetectors with integrated TIAs [68]. Details of the transmitter section and the receiver section are shown in Figures 11.36 and 11.37.

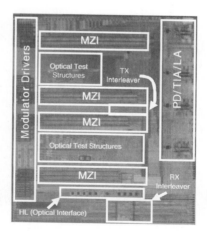

Figure 11.35 Die photograph of a 40 Gb/s WDM transceiver containing PMD electronics and optical components. Photograph provided by Luxtera, Inc. (this figure may be seen in color on the included CD-ROM).

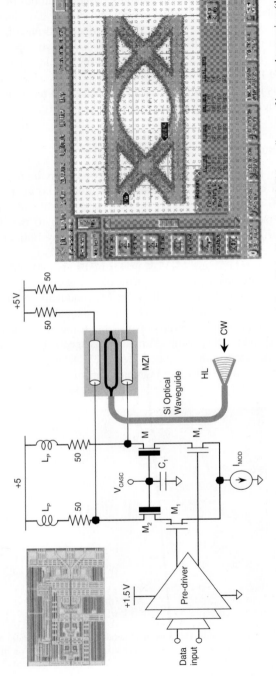

Figure 11.36 Left: Circuit schematic and photograph of integrated driver and MZ modulator. Right: Transmitted 10 Gb/s eye diagram of integrated transceiver (this figure may be seen in color on the included CD-ROM).

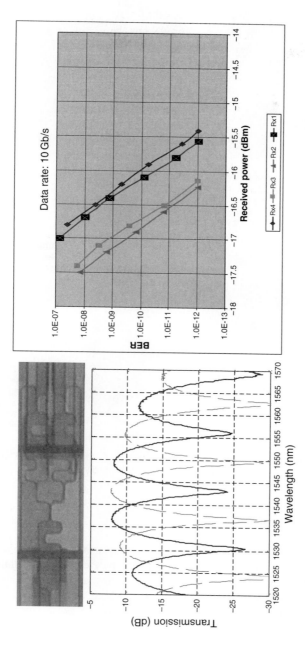

Figure 11.37 Left: Photograph of the two-stage silicon photonics interleaver and demuxing performance of a single stage. Right: Receiver performance for all four channels of transceiver, achieving −15 dBm for 10^{-12} BER (this figure may be seen in color on the included CD-ROM).

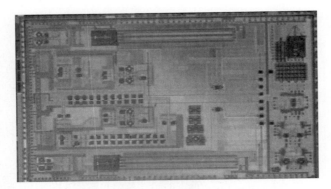

Figure 11.38 Plan view of integrated chip providing two ultra-low phase noise 10 GHz RF oscillators (this figure may be seen in color on the included CD-ROM).

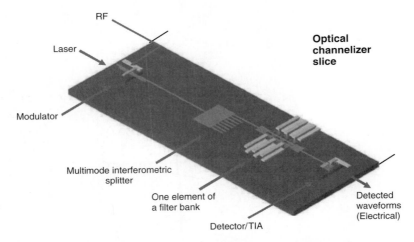

Figure 11.39 One frequency channel slice of real-time optical "fast fourier transform" μwave signal spectral analysis (this figure may be seen in color on the included CD-ROM).

11.8.3 Microwave OE oscillators

Figure 11.38 shows an implementation of an optoelectronic oscillator in Luxtera's CMOS photonics technology [69]. Microwave optoelectronic oscillators insert long, low-loss optical link delays or resonators, using a transmitter and receiver to impose the microwave modulation onto the optical signal, into a microwave electronic gain loop, producing low microwave phase noise due to the long optically generated delay [70]. In this example, the long optical delay is provided with off-chip fibers, and the die contains silicon photonic components for two 10-GHz microwave oscillators, including modulators, modulator drivers, optical splitters and taps, polarization filtering, TIAs, radio frequency (RF) bandpass filters, 10 GHz amplifiers, modulator drivers, and fiber-coupling grating couplers.

11.8.4 Microwave Spectrum Analyzer

Figure 11.39 gives another example of an integrated CMOS silicon photonic circuit architecture under investigation for optical signal processing [71]. The chip modulates an optical signal with microwave RF data, and the resulting optical spectrum is then divided using a splitter and a set of high-performance, narrow-band optical filter banks. The resulting filtered signal is then detected with Ge-based detectors and TIA preamps to provide the functionality of a microwave spectrum analyzer, with an aim to achieve nearly two orders of magnitude reduction in each metric of size, weight, and power.

11.9 NONLINEAR EFFECTS

Due to the exceptionally small size and high confinement of silicon photonic waveguides, nonlinear effects can be greatly magnified. The principal phenomena of interest include two-photon and related nonlinear absorption, Raman amplification, and Kerr nonlinearities that can induce self-phase modulation, 4-wave mixing, parametric amplification, and other nonlinear propagation such as soliton formation.

Two-photon absorption is typically characterized by a coefficient β which would appear in the equation governing the attenuation of intensity

$$\frac{dI}{dz} = -\alpha I - \beta I^2 \qquad (11.28)$$

where α is the usual linear attenuation coefficient in cm^{-1} and the two-photon absorption coefficient has typical experimentally determined values in silicon of $\beta \approx 4.5 \times 10^{-10}$ cm/W, or ~0.45 cm/GW at 1.54 μm [72]. Similarly, the Kerr coefficient appears in the nonlinear index $n = n + n_2 I$, as a value of $n_2 \approx 6 \times 10^{-14}$ cm^2/W. These values appear small, but a 50-mW signal in a 0.2×0.6 μm cross section will have an intensity of ~40 MW/cm^2, leading to a -3-dB decay length, and nonlinear π phase shift lengths, both on the order of ~35 cm. Because many silicon photonic configurations are much smaller than this, this might suggest that nonlinearities are not severe. However, two-photon absorption creates free carriers, and the integrated, cumulative effect can lead to very significant free-carrier-induced losses. With a free-carrier volume generation rate given by

$$\frac{1}{2h\nu}\frac{dI}{dz} = \frac{\beta}{2h\nu}I^2,$$

a carrier lifetime of τ will yield free-carrier losses of

$$\Delta \alpha = 14.5 \times 10^{-18} \cdot \frac{\beta \tau}{2h\nu} I^2.$$

With the same 50 mW signal, and an assumed carrier lifetime of 100 ns, we have a -3-dB loss in only ~170 μm. In many cases, the carrier lifetime can be shorter due to surface recombination effects, and the optical power levels may be much lower, but this underscores the potentially deleterious effects of two-photon absorption through free-carrier absorption.

Perhaps the most dramatic demonstration of nonlinear effects has been the advantageous use of Raman amplification. The Raman gain coefficient of $g_R = 2.9 \times 10^{-8}$ cm/W is $\sim 10^4$ higher than values in typical optical fibers [73]. When combined with the ultra-small dimensions achievable in silicon waveguides, this suggests that practical amplification may be possible in highly compact structures. One challenge in realizing useful Raman gain has been the aforementioned accumulation of two-photon-induced free-carrier losses. By interrogating the sample with pulses, to preclude accumulation of free carriers, Raman gain and Raman lasing has been demonstrated based on the gain peak of ~ 105 GHz bandwidth stokes shifted from the pump by 15.7 THz [74, 75]. More practical structures have also been implemented with reverse-biased active structures similar to the modulators shown earlier, sweeping out two-photon-induced free carriers to eliminate this loss mechanism. Using these structures, optically pumped continuous wave (CW) lasing has been demonstrated in silicon photonic waveguides

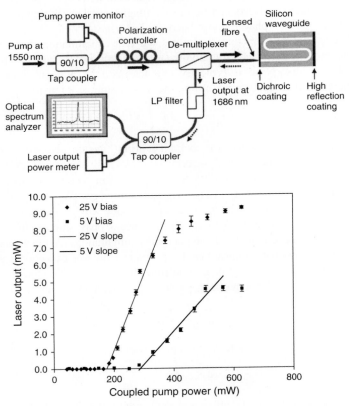

Figure 11.40 Upper: Configuration for optically pumped CW silicon laser using Raman gain. Lower: Laser output vs pump laser input illustrating impact of bias on removing free-carrier losses.

based on Raman amplification as shown in Figure 11.40 [76]. Because the gain bandwidth typically captures only a mode or two of a multi-longitudinal mode laser, such structures are typically driven with a high-power single-mode laser which could have been modulated in the first place rather than frequency-converted. Nevertheless, Raman amplification may be useful for boosting signals, generating light in new wavelength regimes, or generating light with particular spectral characteristics.

Additional recent work has shown that the enhancement of the Kerr nonlinearity, due to the high intensities encountered in silicon photonic waveguides, can enable highly compact parametric amplifiers [77] and even the formation of solitons in very short distances [78].

11.10 TOWARD A SILICON LASER

The examples given above clearly indicate that the active and passive elements available in today's integration technology offer a powerful suite of tools for addressing a host of telecommunications and signal-processing applications. However, it can be argued that for any application requiring a laser source, the hybrid coupling employed thus far represents only a shift in what function gets performed in silicon (previously just electronics) vs III–V compounds because the latter is still required in the module. It would certainly be desirable if laser sources could also be integrated into a silicon or even CMOS-compatible platform. To this end significant research is underway worldwide to re-examine this possibility. While light emission has been seen from porous silicon and silicon nanocrystals, the work discussed here will limit attention to light with photon energies below the Si bandgap for compatibility with active silicon photonics discussed thus far.

Three general approaches are emerging to achieve an electrically pumped laser for silicon photonics. The first involves hybrid wafer-bonded III–V epitaxial structures to make active waveguides using the well-established active properties of III–V materials such as InP and related quantum-well gain media. This approach, while not truly CMOS compatible, does allow for some level of wafer-scale or local wafer batch processing of lasers using lithography, thereby eliminating costly packaging and alignments of input laser sources. Figure 11.41

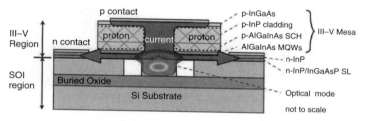

Figure 11.41 Schematic of wafer-bonded hybrid laser. Optical mode resides primarily in silicon waveguide but extends into AlGaInAs MQW stack electrically pumped in III–V bonded superstructure (this figure may be seen in color on the included CD-ROM).

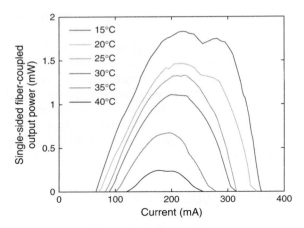

Figure 11.42 CW light-current curves at various temperatures for hybrid wafer-bonded laser (this figure may be seen in color on the included CD-ROM).

illustrates an example of such a structure, in which the active laser waveguide is primarily confined in a silicon waveguide, but some portion of the waveguide mode extends into the electrically activated InP/AlInGaAs quantum-well stack for gain [79]. Figure 11.42 illustrates the rapid progress made using this approach, where room temperature CW operation at milliwatt levels has already been achieved. It is anticipated that such a process might be compatible with back-end CMOS processing in which no extreme temperature excursions are utilized, and contamination of the CMOS device layers could be avoided.

A different approach envisions the use of "extrinsic" gain, i.e., resulting from the inclusion of some species such as Erbium or other rare earths, lead sulfide quantum dots, silicon nanocrystals, nanoengineered defect states, or combinations of these materials to enable excitation. Optically pumped lasing has been observed at low temperature based on defects states [80] and relatively efficient electrically pumped Er-based luminescence at 1.5 μm has been observed in CMOS gate structures [81, 82]. However, to date the latter have employed Fowler–Nordheim tunneling or hot electron injection, which tend to produce rapid degradation of oxides and lead to unreliable operation. Nevertheless, these studies are highly intriguing, and when coupled with waveguides such as the slot configuration discussed earlier for high confinement in thin oxide layers, may point toward a path for laser action.

A third approach focuses on epitaxial approaches such as III–V materials, or even trying to make advantageous use of strained SiSn and GeSn layers, or Ge itself, to provide direct gap behavior. While epitaxial growth of III–V materials on silicon have been pursued for decades, it has been plagued by serious lattice mismatch problems and defects that preclude reliable minority-carrier devices such as lasers. These can be remedied to some degree with thick buffer layers, but recent work has illustrated that in some circumstances a precise misfit dislocation array can be formed which is confined just to the monolayer at the interface [83].

Such structures may provide a means to high-quality selective-area epitaxy of III–V compounds just where required for active gain media, or even ultra-high performance gates for FETs. In the case of strained Ge and Si alloys, studies have shown that $Ge_{1-y}Sn_y$ becomes direct gap for $y > 0.09$, and SiSn on GeSn can be formed with direct bandgaps corresponding to the 1.3- to 1.55-μm range [84]. Other studies have projected direct bandgap gain from tensile strained Ge when high n-type doping is used to fill the indirect bandgap L valleys [85]. These approaches are intriguing in that they could provide high-quality epitaxial interfaces for reliable, high-yield devices and a simplified true wafer-scale processing similar to today's CMOS electronics processing.

11.11 FUTURE TRENDS AND APPLICATIONS

Silicon photonics offers an opportunity for highly integrated photonics to leverage the continuing progress of VLSI silicon investments and technology. Moreover, the high index contrast offers an unusual opportunity for a higher level of miniaturization and integration, at low cost structure, than had previously been envisioned. Finally, as has been illustrated in work already performed to date, relatively sophisticated CMOS circuitry can be readily integrated monolithically with the photonics.

One aspect that always confronts the proposition of integration is the actual cost savings involved. Since photonic integration is fundamentally different than digital electronic integration and its massively repetitive interconnected logic cells, this value proposition needs to be examined carefully.

Photonic integration derives its strong value from the ability to monolithically interconnect waveguide components using lithography, thereby eliminating expensive alignments and exotic packaging materials. Further benefits are realized by the increased performance and reliability that can be realized with stable, efficient, low-reflection serial connections between optical components.

Silicon photonics, in the absence of high-performance monolithic lasers, still requires an efficient laser connection for most applications. As long as module cost is dominated by packaging costs rather than die costs, integration will offer its strongest benefit when larger numbers of optical components are interconnected monolithically than are connected using conventional packaging. Single element transceivers, for example, will need to realize other advantages such as power or size reduction to provide a compelling impetus for silicon photonics.

However, in applications in which there is significant parallel or serial integration, silicon photonics may be compelling from a pure cost perspective. Examples could be arrays of transceivers driven by single laser for parallel ribbon fiber linecards in datacom or telecom. WDM offers an outstanding example where both parallel and serial integrations plays a key role, and silicon photonics may offer a pathway for low-cost WDM interconnection. These interconnections may play a role in medium and short-reach telecommunications, but they are also being carefully examined for datacom applications at the backplane level, the board level, and even for possible intra-chip applications as multi-core microprocessor architectures look into optical solutions to enable scaling.

In other markets, the combination of optical modulators, detectors and filters, together with CMOS control, AD/DA converters, DSPs, and memory may make a compelling case for a variety of low-cost sensor and signal processing applications for biomedicine, defense, automotive, and industrial applications. Silicon photonics could even enable consumer applications for things that are not today associated with photonics because it has historically been too expensive.

While the progress in the field of silicon photonics is rapid and impressive, it is important to note that the highest performance applications are still likely to be in the domain of III–V devices which offer nearly uncompromised performance. The InP material system has also recently made great strides in integration, with notable examples described elsewhere in this volume such as 1.6 Tb/s WDM transmitters including over 100 components [86]. Until a silicon compatible laser is developed, modules for telecommunications will contain hybrid combinations of silicon and III–V photonics, and the question posed by the recent advances in silicon photonics suggest that the boundary may move more in the direction of silicon than was previously thought.

REFERENCES

[1] R. Soref, "The past, present, and future of silicon photonics," *IEEE J. Quant. Electron.*, QE-12, 1678–1687, 2006.
[2] G. T. Reed and A. P. Knights, *"Silicon Photonics: An Introduction,"* John Wiley & Sons, Ltd., West Sussex, England, 2004.
[3] L. Pavesi and D. L. Lockwood, Eds., *"Silicon Photonics,"* Springer-Verlag, Berlin Heidelberg, 2004.
[4] R. Leckie, White Paper, "Funding the Future," Infrastructure Advisors, commissioned by SEMI, available at http://www.infras-advisors.com/whitepaper.html.
[5] J.-P. Colinge, *Silicon-On-Insulator Technology: Materials to VLSI*, 2nd Edition, Section 2.7, "SIMOX," Kluwer Academic Publishers, 1997.
[6] P. Koonath, T. Indukuri, and B. Jalali, "Vertically-coupled micro-resonators realized using three-dimensional sculpting in silicon," *Appl. Phys. Lett.*, 85, 1018–1020, 2004.
[7] M. Bruel, "Silicon on insulator material technology," *Electron. Lett.*, 31, 1201–1202, 1995.
[8] R. A. Soref and J. P. Lorenzo, "All-silicon active and passive guidedwave components for $\lambda = 1.3$ and 1.6 µm," *IEEE J. Quant. Electron.*, 2(6), 873–879, 1986.
[9] A. Splett and K. Petermann, "Low loss single-mode optical waveguides with large cross-section in standard epitaxial silicon," *IEEE Photon. Technol. Lett.*, 6(3), 425–427, 1994.
[10] R. A. Soref, J. Schmidtchen, and K. Petermann, "Large single-mode rib waveguides in GeSi-Si and Si-on-SO$_2$," *IEEE J. Quantum Electron.*, 27(8), 1971–1974, 1991.
[11] U. Fischer, T. Zinke, J.-R. Kropp et al., "0.1 dB/cm waveguide losses in single-mode SOI Rib waveguides," *IEEE Photon. Technol. Lett.*, 8(5), 647–648, 1996.
[12] S. Itabashi, H. Fukuda, T. Tsuchizawa et al., "Silicon Wire Waveguides and Silicon Microphotonic Devices," *NTT Technical Review*, 4(3), 48–56, March 2006.
[13] M. A. Webster, R. M. Pafchek, G. Sukumaran, and T. L. Koch, "Low-Loss Thin SOI Waveguides and High-Q Ring Resonators," in Proc. *Optical Society of America Annual Meeting*, paper FTuV3, Tucson, October 18, 2005.

[14] M. A. Webster, R. M. Pafchek, G. Sukuraman, and T. L. Koch, "Low-loss quasi-planar ridge waveguides formed on thin silicon-on-insulator," *Appl. Phys. Lett.*, 87, 231108–231109, 2005.
[15] C. Gunn, "CMOS Photonics Technology," presented at NSF Workshop on Very Large Scale Photonic Integration, Arlington, VA, March 19–29, 2007 *(unpublished)*.
[16] M. A. Webster, R. M. Pafchek, A. Mitchell, and T. L. Koch, "Width dependence of inherent TM-mode lateral leakage loss in silicon-on-insulator ridge waveguides," *IEEE Photon. Technol. Lett.*, 19(6), 429–431, 2007.
[17] V. Almeida, Q. Xu, C. Barrios, and M. Lipson, "Guiding and confining light in void nanostructure," *Opt. Lett.*, 29(11), 1209, 2004.
[18] E. A. J. Marcatili, "Bends in optical dielectric guides," *Bell Syst. Tech. J.*, 48, 2103–2132, 1969.
[19] D. Marcuse, "Bending losses of the asymmetric slab waveguide," *Bell Syst. Tech. J.*, 50, 2551–2563, 1971.
[20] M. Heiblum and J. H. Harris, "Analysis of curved waveguides by conformal transformation," *IEEE J. Quant. Electron.*, QE-11, 75–83, 1975.
[21] J. R. P. Lacey and F. P. Payne, "Radiation loss from planar waveguides with random wall imperfections," *IEE Proc. Pt-J.*, 137, 282, 1990.
[22] K. K. Lee, D. R. Lim, and L. C. Kimerling, "Fabrication of ultralow-loss Si/SiO_2 waveguides by roughness reduction," *Opt. Lett.*, 26(23), 1888–1890, 2001.
[23] M.-C. M. Lee and M. C. Wu, "Thermal Annealing in Hydrogen for 3-D Profile Transformation on Silicon-on-Insulator and Sidewall Roughness Reduction," *J. Microelectromechanical Sys.*, 15(2), 338–343, April 2006.
[24] R. A. Soref and B. R. Bennett, "Electrooptical effects in silicon," *IEEE J. Quant. Electron.*, QE-23(1), 123–129, 1987.
[25] A. Harke, M. Krause, and J. Mueller, "Low-loss singlemode amorphous silicon waveguides," *Electron. Lett.*, 41(25), 1377–1379, 2005.
[26] T. L. Koch, U. Koren, G. Eisenstein et al., "Tapered waveguide InGaAs/InGaAsP multiple-quantum-well lasers," *IEEE Photon. Technol. Lett.*, 2, 88–90, February 1990.
[27] I. Moerman, P. P. Van Daele, and P. M. Demeester, "A review on fabrication technologies for the monolithic integration of tapers with III-V semiconductor devices," *IEEE J. Sel. Top Quant. Electron.*, 6, 1308–1320, 1997.
[28] T. Shoji, T. Tsuchizawa, T. Watanabe et al., "Low loss mode size converter from 0.3 micron square Si wire waveguides to singlemode fibers," *Electron. Lett.*, 38, 1669–1670, 2002.
[29] V. R. Almeida, R. Panepucci, and M. Lipson, "Nanotaper for Compact Mode Conversion," *Opt. Lett.*, 28(15), 1302–1304, August 2003.
[30] R. G. Hunsberger, in *Integrated Optics: Theory and Technology* (T. Tamir, ed.), 3rd Edition, Springer Series in Optical Sciences, Vol. 33, Springer-Verlag, 1991.
[31] F. Van Laere, G. Roelkens, M. Ayre et al., "Compact and highly efficient grating couplers between optical fiber and nanophotonic waveguides," *J. Lightw. Technol.*, 25, 151–156, 2007.
[32] C. Gunn, "Silicon photonics: poised to invade local area networks," *Photonics Spectra*, 40(3), 62–68, March 2006.
[33] L. Sekaric, S. J. McNab, and Yu. A. Vlasov, "Y-splitters in Photonic Wires and Photonic Crystal Waveguides," presentation at *PECSVI 2005*, available at www.research.ibm.com/photonic.
[34] P. D. Trinh, S. Yegnanarayanan, F. Coppinger, and B. Jalali, "Compact Multimode Interference Couplers in Silicon-On-Insulator Technology," in Proceedings of Conference on Lasers and Electro-Optics, *CLEO 1997*, Optical Society of America, 11, 441, 18–23 May, 1997.
[35] K. K. Lee, D. R. Lim, A. Agarwal et al., "Performance of Polycrystalline Silicon Waveguide Devices for Compact On-Chip Optical Interconnection," in Proc. *SPIE Photonics East Symposium*, Boston, MA, 120–125, September 1999.
[36] T. Chu, H. Yamada, S. Ishida, and Y. Arakawa, "Compact $1 \times N$ thermo-optic switches based on silicon photonic wire waveguides," *Opt. Express*, 13, 10109–10114, 2005.
[37] H. Yamada, T. Chu, S. Ishida, and Y. Arakawa, "Optical add-drop multiplexers based on Si-wire waveguides," *Appl. Phys. Lett.*, 86, 191107, 2005.

[38] B. T. Smith, D. Feng, H. Lei et al., "Fundamentals of Silicon Photonic Devices," available at http://www.kotura.com/pdf/.
[39] D. Dai, L. Liu, L. Wosinski, and S. He, "Design and fabrication of ultra-small overlapped AWG demultiplexer based on a-Si nanowire waveguides," *Electron. Lett.*, 42, 400–402, 2005.
[40] Y. Akahane, T. Asano, B. S. Song, and S. Noda, "High-Q photonic nanocavity in a two-dimensional photonic crystal," *Nature*, 425, 944–947, 2003.
[41] B. S. Song, S. Noda, T. Asano, and Y. Akahane, "Ultra-high-Q photonic double-heterostructure nanocavity," *Nature Materials*, 4, 207–210, 2005.
[42] J. M. Choi, R. K. Lee, and A. Yariv, "Ring fiber resonators based on fused-fiber grating add–drop filters: application to resonator coupling," *Opt. Lett.*, 27, 1598–1600, 2002.
[43] B. E. Little, S. T. Chu, H. A. Haus et al., "Microring resonator channel dropping filters," *J. Lightw. Technol.*, 15, 998–1005, 1997.
[44] F. Xia, L. Sekaric, M. O'Boyle, and Y. Vlasov, "Coupled resonator optical waveguides based on silicon-on-insulator photonic wires," *Appl. Phys. Lett.*, 89, 041122, 2006.
[45] B. J. Frey, D. B. Leviton, and T. J. Madison, "Temperature-dependent refractive index of silicon and germanium," in *Optomechanical Technologies for Astronomy* (E. Atad-Ettedgui, J. Antebi, D. Lemke eds) Proceedings of the *SPIE*, 6273, 62732J, 2006.
[46] J. Liu, D. Pan, S. Jongthammanurak et al., "Design of monolithically integrated GeSi electro-absorption modulators and photodetectors on a SOI platform," *Opt. Express*, 15, 623–628, 2007.
[47] Y.-H. Kuo, Y.-K. Lee, Y. Ge et al., "Strong quantum-confined Stark effect in germanium quantum-well structures on silicon," *Nature*, 437, 1334–1336, 2005.
[48] S. M. Sze, *Physics of Semiconductor Devices*, John Wiley and Sons, N.Y., 1981.
[49] Q. Xu, S. Manipatruni, B. Schmidt et al., "12.5 Gbit/s carrier-injection-based silicon microring silicon modulators," *Opt. Express*, 15, 430436, 2007.
[50] L. Liao, D. Samara-Rubio, M. Morse et al., "High speed silicon Mach-Zehnder modulator," *Opt. Express*, 13, 3129–3135, 2005.
[51] UltraVOA Array Application Note: Transient Suppression and Channel Identification and Control, available at http://www.kotura.com/pdf/.
[52] A. Huang, C. Gunn, G.-L. Li et al., "A 10 Gb/s Photonic Modulator and WDM MUX/DEMUX Integrated with Electronics in 0.13 µm SOI CMOS," in Proc. *ISSCC 2006*, 922–929, 2006.
[53] Q. Xu, B. Schmidt, S. Pradhan, and M. Lipson, "Micrometre-scale silicon electro-optic modulator," *Nature*, 435, 325–327, 19 May 2005.
[54] C. Gunn, "CMOS Photonics," in Proc. 2nd International Conference *GFP 2005*, paper WC5, Antwerp, 2005.
[55] L. C. Gunn, G. Masini, J. Witzens, and G. Capellini, "CMOS photonics using germanium photodetectors," *ECS Tran.*, 3, 17, 2006.
[56] M. Lee and E. A. Fitzgerald, "Optimized Strained Si/Strained Ge Dual-Channel Heterostructures for High Mobility p- and n-MOSFETs," *IEDM Tech. Digest*, 429, 2003.
[57] G. Masini, L. Colace, and G. Assanto, "Germanium thin films on silicon for detection of near-infrared light," in *Handbook of Thin Films Materials* (H. S. Nalwa, ed.), Academic Press, San Diego, 2001.
[58] J. M. Hartmann, A. Abbadie, A. M. Papon et al., Reduced pressure–chemical vapor deposition of Ge thick layers on Si(001) for 1.3–1.55 µm photodetection, *J. Appl. Phys.*, 95, 5905–5907, 2004.
[59] L. Colace, M. Balbi, G. Masini et al., "Ge on Si p-I-n photdiodes operating at 10 Gbit/s," *Appl. Phys. Lett.*, 88, 101111, 2006.
[60] G. Dehlinger, S. J. Koester, J. D. Schaub et al., "High-speed Germanium-on-SOI lateral PIN photodiodes," *IEEE Photon. Technol. Lett.*, PTL, 16, 2547–2549, 2004.
[61] O. I. Dosunmu, D. D. Cannon, M. K. Emsley et al., "Resonant cavity enhanced Ge photodetectors for 1550 nm operation on reflecting si substrates," *IEEE J. Sel. Top Quant. Electron.*, 10, 694, 2004.
[62] L. Colace, G. Masini, G. Assanto et al., "Metal-semiconductor-metal near-infrared light detector based on epitaxial Ge/Si," *Appl. Phys. Lett.*, 72, 317, 1998.
[63] G. Masini, L. Colace, G. Assanto et al., "High performance pin Si photodetectors for the near infrared: From model to demonstration," *IEEE Trans. Electron. Dev.*, 48(6), 1092, 2001.

11. Silicon Photonics

[64] L. C. Gunn, "Method of Incorporating Germanium in a CMOS Process," US Patent No. 6,887,773, March 2005.

[65] L. C. Gunn, G. Masini, J. Witzens, and G. Capellini, "CMOS Photonics Using Germanium Photodetectors," *ECS Tran.*, 3, 17, 2006.

[66] See, for example, J.-P. Colinge, *Silicon-On-Insulator Technology: Materials to VLSI*, 2nd Edition, Kluwer Academic Publishers, Boston, 1997.

[67] See, for example, Proceedings of the 2006 *IEEE International SOI Conference*, Niagara Falls, October 2–5 (IEEE, Piscataway, NJ) (2006).

[68] A. Narasimha, B. Analui, Y. Liang, et al., "A Fully Integrated 4 x 10Gb/s DWDM Optoelectronic Transceiver in a standard 0.13μm CMOS SOI," in proceedings of the 2007 IEEE International Solid State Circuits Conference, ISSCC 2007 (IEEE, Piscataway, NJ) February 11–15, San Francisco, 2007, pp. 42–586.

[69] C. Gunn, D. Guckenberger, T. Pinguet, et al., "A Low Phase Noise 10 GHz Optoelectronic RF Oscillator Implemented Using CMOS Photonics," in proceedings of the 2007 IEEE International Solid State Circuits Conference, ISSCC 2007 (IEEE, Piscataway, NJ) February 11–15, San Francisco, 2007, pp. 570–622.

[70] X. S. Yao and L. Maleki, "Optoelectronic microwave oscillator," *J. Opt. Soc. Am. B.*, 13(8), 1725–1735, August 1996.

[71] L. C. Kimerling, D. Ahn, A. B. Apsel et al., "Electronic-Photonic Integrated Circuits on the CMOS Platform," in *Proc. SPIE*, 6125, 612502, 2006.

[72] H. K. Tsang, C. S. Wong, T. K. Liang et al., "Optical dispersion, two-photon absorption and self-phase modulation in silicon waveguides at 1.5 μm wavelength," *Appl. Phys. Lett.*, 80, 416, 2002.

[73] R. L. Espinola, J. I. Dadap, R. M. Osgood, Jr et al., "Raman amplification in ultrasmall silicon-on-insulator wire waveguides," *Opt. Express.*, 12, 3713–3718, 2004.

[74] O. Boyraz and B. Jalali, "Demonstration of a silicon Raman laser," *Optics Express*, 12(21), 5269–5273, 2004.

[75] Q. Xu, V. R. Almeida, and M. Lipson, "Time-resolved study of Raman gain in highly confined silicon-on-insulator waveguides," *Opt. Express*, 12, 4437–4442, 2004.

[76] H. Rong, R. Jones, A. Liu et al., "A continuous-wave Raman silicon laser," *Nature*, 433, 725–728, 2005.

[77] M. A. Foster, A. C. Turner, J. E. Sharping et al., "Broad-band optical parametric gain on a silicon photonic chip," *Nature*, 441, 960–963, 2006.

[78] J. Zhang, Q. Lin, G. Piredda et al., "Optical solitons in a silicon waveguide," *Opt. Express*, 15, 7682–7688, 2007.

[79] A. W. Fang, H. Park, O. Cohen et al., "Electrically pumped hybrid AlGaInAs-silicon evanescent laser," *Opt. Express*, 14, 9203–9210, 2006.

[80] S. G. Cloutier, P. A. Kossyrev, and J. Xu, "Optical gain and stimulated emission in periodic nanopatterned crystalline silicon," *Nature Materials*, 4, 887–891, 2005.

[81] M. E. Castagna, S. Coffa, L. Caristia, et al. "Quantum Dot Materials and Devices for Light Emission in Silicon," Proceedings of the 32nd European Solid State Devices Research Converence, ESSDERC 2002 (IEEE, Piscataway, NJ), September 24–26, Bologna, pp. 439–442, 2002.

[82] J. M. Sun, W. Skorupa, T. Dekorsy et al., "Efficient electroluminescence from rare-earth implanted SiO_2 metal-oxide-semiconductor structures," in *Proc. GFP 2005*, paper ThB3, Antwerp, 48–50, 2005.

[83] D. L. Huffaker, G. Balakrishnan, S. Huang et al., "Monolithic Integration of Sb-based Photo-pumped Lasers on Si," in *Proc. GFP 2005*, paper WA1, Antwerp, 149–150, 2005.

[84] R. Soref, J. Kouvetakis, and J. Menendez, "Advances in SiGeSn/Ge Technology," *Mater. Res. Soc. Symp. Proc.*, 958, paper 0958-L01–08, 2006.

[85] J. Liu, Sun, D. Pan, et al., "Tensile-strained, n-type Ge as a gain medium for monolithic laser integration on Si," *Optics Express*, 15(18), 11272–11277, 2007.

[86] R. Nagarajan, C. Joyner, R. Schneider et al., "Large-scale photonic integrated circuits," *IEEE J. Sel. Top Quant. Electron.*, 11, 50–65, 2005.

12

Photonic crystal theory: Temporal coupled-mode formalism

Shanhui Fan

Ginzton Laboratory, Department of Electrical Engineering, Stanford, CA, USA

Abstract

Photonic crystal has been of great interest for optical information-processing applications, in part because it provides a common platform to miniaturize a large number of optical components on-chip down to single wavelength scale. For this purpose, many devices are designed starting from a crystal with a photonic band gap, followed by introducing line and point defect states into the gap. Different functionalities, such as filters, switches, modulators, delay lines, and buffers, can then be created by controlling the coupling between these defect states. In this chapter, we review the temporal coupled-mode theory formalism that provides the theoretical foundation of many of these devices.

12.1 INTRODUCTION

Photonic crystals [1–6] have been of great interest for optical information-processing applications, in part because these crystals provide a common platform to miniaturize a large number of optical components on-chip down to single wavelength scales [7, 8]. For this purpose, at least conceptually, many devices are conceived by starting from a perfect crystal with a photonic band gap, i.e., a frequency range where light is prohibited to propagate inside the crystal. Appropriate line and point defect geometries are then designed to introduce defect states in the gap that serves either as waveguides or as micro-cavities (Figure 12.1). Afterward, different functionalities are synthesized by coupling of these defect states together.

Designing photonic crystal devices based on controlled coupling between waveguides and cavities has been a very fruitful area of study. A few examples of devices that are particularly relevant in optical communications include channel add/drop

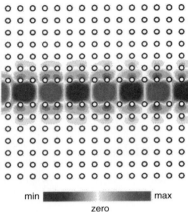

Figure 12.1 Point and line defect states in a photonic crystal (this figure may be seen in color on the included CD-ROM).

filters [9–17] for wavelength division multiplexing (WDM), optical switches [18–24], isolators [25–27], static delay lines [28–32], and dynamic optical buffers [33–38].

In this chapter, we review the temporal coupled-mode theory formalism that provides the theoretical foundation for many of these devices. As examples of applications of this formalism, we then discuss nonlinear bistable optical switch and dynamic optical buffers.

12.2 TEMPORAL COUPLED-MODE THEORY FOR OPTICAL RESONATORS

Temporal coupled-mode theory [39] is a powerful theoretical framework for understanding a variety of effects that arise from the coupling between waveguides and resonators. This theory is applicable when the quality factor of the resonator is sufficiently high, such that the amplitude inside the resonator can be assumed to decay exponentially in time in the absence of external stimulus. On the basis of this assumption, a complete theoretical formalism can then be systematically built up using very general arguments such as energy conservation and time-reversal symmetry. The beauty of the theory is that aside from a few parameters representing resonant frequencies and coupling constants, the details of the resonators and the waveguides are irrelevant. Hence, such a theory is particularly attractive for photonic crystal device structures, which tend to have complicated geometry.

As an illustration of temporal coupled-mode theory, we will provide a unified treatment of three different device structures (Figure 12.2). (1) A directly coupled resonator configuration, in which two truncated waveguides are coupled together through a resonator; (2) a side-coupled resonator configuration, in which a single waveguide

12. Photonic Crystal Theory: Temporal Coupled-Mode Formalism 433

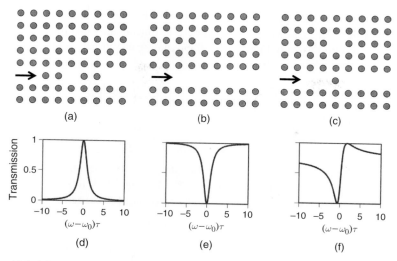

Figure 12.2 A few structures formed by coupling point and line defect together in a photonic crystal. (a) A transmission resonator configuration. (b) A side-coupled resonator configuration. (c) A side-coupled resonator configuration that exihibits asymmetric Fano lineshape. (d–f) The corresponding transmission spectrum for each structure are shown in (a–c), respectively.

is side-coupled to a standing wave resonator; and (3) a Fano-interference configuration, which is similar to (2), except an additional scatterer is placed inside the waveguide. These devices provide a number of useful filtering and switching functionalities.

We develop our theory by adopting the formalism in Ref. [39]. For illustration purposes, we use the theoretical model, schematically shown in Figure 12.3, that consists of a single-mode optical resonator coupled with two ports. Figure 12.3 is plotted based on Figure 12.2c, but here we use it to illustrate the general features in all three structures in Figure 12.2. In general, for light to go through from one port to the other, it can either go through a *resonant pathway* by coupling through the resonator (black lines in Figure 12.3), or it can go through a *direct pathway* that bypasses the

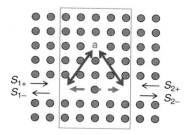

Figure 12.3 Graphic illustration of the various parts of the temporal coupled-mode theory when applied to a photonic crystal structure. The structural region inside the box is the resonator system. The gray and black arrows indicate the direct and the resonant pathways for light to go through this system (this figure may be seen in color on the included CD-ROM).

resonator (gray line in Figure 12.3). As we will show below, by controlling the relative strength of these two pathways, a wide variety of functionality can be synthesized.

For the system shown in Figure 12.3, the dynamic equations for the amplitude a of the resonance mode can be written as

$$\frac{da}{dt} = \left(j\omega_0 - \frac{1}{\tau}\right)a + (\kappa_1 \quad \kappa_2)\begin{pmatrix} s_{1+} \\ s_{2+} \end{pmatrix}, \tag{12.1}$$

$$\begin{pmatrix} s_{1-} \\ s_{2-} \end{pmatrix} = C\begin{pmatrix} s_{1+} \\ s_{2+} \end{pmatrix} + a\begin{pmatrix} d_1 \\ d_2 \end{pmatrix}, \tag{12.2}$$

where ω_0 is the center frequency of the resonance, τ is the lifetime of the resonance and is related to the quality factor Q by $Q = \frac{\omega_0 \tau}{2}$. The amplitude a is normalized such that $|a|^2$ corresponds to the energy inside the resonator [39]. The resonant mode is excited by the incoming waves $\begin{pmatrix} s_{1+} \\ s_{2+} \end{pmatrix}$ from the two ports, respectively, with the amplitude-coupling coefficients ($\kappa_1 \quad \kappa_2$). The resonant mode, once excited, couples with the outgoing waves $\begin{pmatrix} s_{1-} \\ s_{2-} \end{pmatrix}$ at the two ports with the coupling constants $\begin{pmatrix} d_1 \\ d_2 \end{pmatrix}$. In addition to the resonance-assisted coupling between the ports, the incoming and outgoing waves in the ports can also couple through a direct pathway (Figure 12.3), as described by a scattering matrix C. In order to provide a unified treatment of all three structures shown in Figure 12.2, the matrix C here is taken to be an arbitrary scattering matrix, i.e., any unitary and symmetric matrix.

Equations (12.1) and (12.2) generalize the standard temporal coupled-mode theory [39], in which C is a diagonal matrix. Also, while for illustration purposes we have only considered the case of two ports and one resonance, the formalism developed here can be readily adopted to incorporate multiple ports [40] and multiple resonances [41].

The coefficients κ, d, and C are not independent; rather, they are related by energy conservation and time-reversal symmetry constraints. Below, we will exploit the consequence of these constraints to develop a minimal set of parameters that completely characterize the system. First, for externally incident excitations $\begin{pmatrix} s_{1+} \\ s_{2+} \end{pmatrix}$ at a frequency ω, we can write the scattering matrix S for the system described by Eqns (12.1) and (12.2), as:

$$\begin{pmatrix} s_{1-} \\ s_{2-} \end{pmatrix} \equiv S\begin{pmatrix} s_{1+} \\ s_{2+} \end{pmatrix} = \left[C + \frac{1}{j(\omega - \omega_0) + 1/\tau} \times \begin{pmatrix} d_1 \\ d_2 \end{pmatrix}(\kappa_1 \quad \kappa_2)\right]\begin{pmatrix} s_{1+} \\ s_{2+} \end{pmatrix}$$
$$= \left[C + \frac{1}{j(\omega - \omega_0) + 1/\tau} \cdot \begin{pmatrix} d_1\kappa_1 & d_1\kappa_2 \\ d_2\kappa_1 & d_2\kappa_2 \end{pmatrix}\right]\begin{pmatrix} s_{1+} \\ s_{2+} \end{pmatrix} \tag{12.3}$$

Because the scattering matrix S has to be symmetric due to time-reversal symmetry, we have

$$d_1\kappa_2 = d_2\kappa_1 \tag{12.4}$$

12. Photonic Crystal Theory: Temporal Coupled-Mode Formalism

Also, with incoming wave amplitudes $\begin{pmatrix} s_{1+} \\ s_{2+} \end{pmatrix}$, the amplitude of the resonant mode is

$$a = \frac{1}{j(\omega - \omega_0) + 1/\tau}(\kappa_1 s_{1+} + \kappa_2 s_{2+}) \tag{12.5}$$

Instead of considering the case where the resonator is excited by externally incident waves $\begin{pmatrix} s_{1+} \\ s_{2+} \end{pmatrix}$, let us now consider an alternative scenario, in which the external incident wave is absent, i.e., $\begin{pmatrix} s_{1+} \\ s_{2+} \end{pmatrix} = 0$, and at $t=0$, there is a finite amplitude of the resonance. At $t > 0$, the resonant mode shall decay exponentially into the two ports, as

$$\frac{d|a|^2}{dt} = -\left(\frac{2}{\tau}\right)|a|^2 = -\left(|s_{1-}|^2 + |s_{2-}|^2\right) = -|a|^2\left(|d_1|^2 + |d_2|^2\right), \tag{12.6}$$

which requires that

$$|d_1|^2 + |d_2|^2 = 2/\tau. \tag{12.7}$$

Now, let us perform a time-reversal transformation for the exponential decay process as described by Eqn (12.6). The time-reversed case is represented by feeding the resonator with exponentially growing waves at a complex frequency $\omega = \omega_0 - j \cdot 1/\tau$, with incident amplitudes at the two ports when $t=0$ equal to $\begin{pmatrix} s_{1-}^* \\ s_{2-}^* \end{pmatrix}$. Such excitations cause a resonance amplitude a^* at $t=0$ to grow exponentially in time. Using Eqn (12.1) at the complex frequency $\omega = \omega_0 - j \cdot 1/\tau$, we have $a^* = \frac{\tau}{2}(\kappa_1 s_{1-}^* + \kappa_2 s_{2-}^*) = \frac{\tau}{2}(\kappa_1 d_1^* + \kappa_2 d_2^*)a^*$, and therefore

$$\kappa_1 d_1^* + \kappa_2 d_2^* = \frac{2}{\tau} \tag{12.8}$$

Combining Eqns (12.4), (12.7) and (12.8), we are led to an important conclusion:

$$\kappa_1 = d_1, \kappa_2 = d_2. \tag{12.9}$$

The time-reversed excitation $\begin{pmatrix} s_{1-}^* \\ s_{2-}^* \end{pmatrix}$ also has to satisfy the condition that no outgoing wave shall occur upon such excitations, i.e.,

$$0 = C \begin{pmatrix} s_{1-}^* \\ s_{2-}^* \end{pmatrix} + a^* \begin{pmatrix} d_1 \\ d_2 \end{pmatrix} = a^* C \begin{pmatrix} d_1 \\ d_2 \end{pmatrix}^* + a^* \begin{pmatrix} d_1 \\ d_2 \end{pmatrix}. \tag{12.10}$$

Thus, the coupling constants $|d\rangle$ have to satisfy a further condition:

$$C\begin{pmatrix} d_1^* \\ d_2^* \end{pmatrix} = -\begin{pmatrix} d_1 \\ d_2 \end{pmatrix}. \quad (12.11)$$

Hence, the coupling constants in general cannot be arbitrary but are instead related to the scattering matrix of the direct process.

To check that the constraints, Eqns (12.7) and (12.9) and (12.11), indeed produce a self-consistent temporal coupled-mode theory, we need to ensure that the scattering matrix S, as defined by Eqn (12.3), is unitary. For this purpose, we note that

$$S^+ = C^+ + \frac{1}{-j(\omega-\omega_0)+1/\tau} \cdot \begin{pmatrix} d_1 \\ d_2 \end{pmatrix}^* (d_1 \quad d_2)^* \quad (12.12)$$

and therefore

$$SS^+ = CC^+ + \frac{(2/\tau)}{(\omega-\omega_0)^2+(1/\tau)^2} \begin{pmatrix} d_1 \\ d_2 \end{pmatrix} (d_1^* \quad d_2^*)$$

$$+ \frac{1}{-j(\omega-\omega_0)+(1/\tau)} C \begin{pmatrix} d_1^* \\ d_2^* \end{pmatrix}(d_1^* \quad d_2^*) + \frac{1}{j(\omega-\omega_0)+(1/\tau)} \begin{pmatrix} d_1 \\ d_2 \end{pmatrix} (d_1 \quad d_2)C^+ \quad (12.13)$$

Taking advantage of Eqn (12.11) and its complex conjugate,

$$(d_1 \quad d_2)C^+ = -(d_1 \quad d_2)^*, \quad (12.14)$$

we can indeed prove the unitary property of the matrix S:

$$SS^+ = CC^+ +$$

$$\left[\frac{(2/\tau)}{(\omega-\omega_0)^2+(1/\tau)^2} - \frac{1}{-j(\omega-\omega_0)+(1/\tau)} - \frac{1}{j(\omega-\omega_0)+(1/\tau)} \right] \begin{pmatrix} d_1 \\ d_2 \end{pmatrix}(d_1^* \quad d_2^*)$$

$$= CC^+ = I \quad (12.15)$$

Below, we apply the general formalism to the structures shown in Figure 12.2. In general, once the magnitudes of the coupling constants d_1 and d_2 are fixed, the phases of the coupling constant can be determined from the scattering matrix C of the direct process. For the structures shown in Figure 12.2, however, the theory can be further simplified, because they all possess mirror symmetry. For these structures, if we further assume that the resonance mode is even with respect to the mirror that maps the two ports, we should have $d_1 = d_2 = d$ if the reference planes are symmetrically placed on each side of the mirror plane. With such a symmetric choice of reference plane location, the scattering matrix for the direct transport process also acquires a special form [39]:

$$C = e^{j\phi} \begin{pmatrix} r & jt \\ jt & r \end{pmatrix}, \qquad (12.16)$$

where r, t, and ϕ are real constants with $r^2 + t^2 = 1$. Using Eqns (12.7) and (12.11), we can determine d_1 and d_2 as

$$d_1 = d_2 = je^{\frac{j\phi}{2}} \sqrt{r + jt} \cdot \sqrt{\frac{1}{\tau}} \qquad (12.17)$$

and consequently the scattering matrix S for the overall system as

$$S = e^{j\phi} \left\{ \begin{pmatrix} r & jt \\ jt & r \end{pmatrix} - \frac{1/\tau}{j(\omega - \omega_0) + 1/\tau} (r + jt) \begin{pmatrix} 1 & 1 \\ 1 & 1 \end{pmatrix} \right\}. \qquad (12.18)$$

From Eqn (12.18), the power transmission coefficient T is therefore

$$T = \frac{t^2(\omega - \omega_0)^2 + r^2(1/\tau)^2 + 2rt(\omega - \omega_0)(1/\tau)}{(\omega - \omega_0)^2 + (1/\tau)^2} \qquad (12.19)$$

Using Eqn (12.19), we now discuss the three configurations as shown in Figure 12.2:

(a) Directly coupled resonator configuration (Figure 12.2a). For this structure, in the absence of the resonator, a photonic crystal region that acts as a tunneling barrier separates the two waveguides. Therefore $r \approx 1$ and $t \approx 0$. The power transmission spectrum

$$T = \frac{(1/\tau)^2}{(\omega - \omega_0)^2 + (1/\tau)^2} \qquad (12.20)$$

exhibits a Lorentzian peak (Figure 12.2(d)). Such a configuration therefore functions as a band-pass filter.

(b) Side-coupled resonator configuration (Figure 12.2(b)). For this structure, in the absence of the resonator, light propagates in the waveguide unperturbed. If one chooses the reference plane to be exactly where the resonator is, one has $C = \begin{pmatrix} 0 & 1 \\ 1 & 0 \end{pmatrix}$. Therefore, in Eqn (12.16) one should set $r = 0$, $t = 1$, and $\phi = -\frac{\pi}{2}$.

Using Eqn (12.17), we then have $d_1 = d_2 = j\sqrt{\frac{1}{\tau}}$. Hence, the power transmission spectrum becomes

$$T = \frac{(\omega - \omega_0)^2}{(\omega - \omega_0)^2 + (1/\tau)^2} \quad (12.21)$$

At resonance, the destructive interference between the direct transmission and the resonant pathway produces complete reflection (Figure 12.2(e)). The structure therefore functions as a band-rejection filter.

(c) Fano interference configuration (Figure 12.2(c)). By introducing additional scatterer inside the waveguide, one can alter the direct pathway such that neither r nor t is close to zero. In such a case, the transmission spectrum exhibits a sharp asymmetric Fano lineshape [42], in which the transmission coefficient rapidly varies from 0 to 100% within a narrow frequency range (Figure 12.2(f)). The width of such a frequency range can be smaller than $1/\tau$. Consequently, the use of such sharp asymmetric lineshapes provides an opportunity for enhancing the sensitivity in sensors and in reducing the power requirements for optical switches [43].

These examples quoted above highlight the general theory that describes waveguide–cavity interaction in photonic crystals.

12.3 USING TEMPORAL COUPLED-MODE THEORY TO PREDICT OPTICAL SWITCHING

In addition to its application in linear devices, as outlined above, the temporal coupled-mode theory can also be used to design nonlinear switching devices. Optical bistable devices are of great importance for all-optical information-processing applications [44]. Photonic crystal switches can be constructed by introducing Kerr nonlinear material into the resonator for all the three structures shown in Figure 12.2. Compared with conventional devices, the use of photonic crystal resonator, which possesses high-quality factor and small modal volume, results in greatly reduced power requirements [18–24].

A judicious choice of waveguide-cavity coupling schemes further allows one to optimize the device characteristics that are important for practical applications. For example, in integrated two-port bistable devices, an important consideration is the contrast ratio in the transmission between the two bistable states. A high contrast ratio is beneficial for maximum immunity to noise and detection error, and for fan-out considerations.

Here, as an application of the temporal coupled-mode theory, we compare the nonlinear behaviors of the structures shown in Figure 12.2a and 12.2b, when a Kerr nonlinear material is introduced into the cavity region. (The general Fano interference case corresponding to Figure 12.2(c) is more complicated and can

12. Photonic Crystal Theory: Temporal Coupled-Mode Formalism

be found in Refs [45–48].) A Kerr nonlinear material has an intensity-dependent index $n = n_0 + n_2 I$. Assuming light is injected into the structure from port 1 only, i.e., $S_{2+} = 0$, from Eqn (12.5), the energy in the cavity therefore in both cases become

$$|a|^2 = \frac{1/\tau}{(\omega - \omega_{\text{res}})^2 + (1/\tau)^2} |s_{1+}|^2 \qquad (12.22)$$

(This equation in fact holds true for the general Fano interference case as well.) For the nonlinear system, the resonant frequency ω_{res} becomes dependent upon the total energy inside the cavity by

$$\omega_{\text{res}} = \omega_0 - \eta |a|^2 \qquad (12.23)$$

where ω_0 is the resonant frequency of the linear cavity, and η is a constant that relates to the Kerr coefficient, n_2, and is inversely proportional to the cavity modal volume V. [19]

For the directly coupled resonator geometry, the transmitted power $P_{\text{trans}} \equiv |S_{2-}|^2$ is proportional to the energy inside the cavity by $P_{\text{trans}} = \frac{1}{\tau}|a|^2$. Consequently, Eqn (12.23) can be converted into

$$\frac{P_{\text{trans}}}{P_{\text{in}}} = \frac{1}{1 + (P_{\text{trans}}/P_0 + \delta)^2} \qquad (12.24)$$

where $\delta = (\omega - \omega_0)\tau$ is the normalized detuning between the signal frequency with the resonant frequency of the cavity itself in the low-intensity limit, and $P_0 = \frac{1}{\eta \tau^2}$ defines the scale of the power, because $\eta \propto 1/V$ and $\tau \propto Q$. The power scale is reduced by a factor Q^2/V when a resonator is used. The photonic crystal resonator, with its high-quality factor and small modal volume therefore provides an important opportunity in general for reducing the power requirement in nonlinear optical devices.

A plot of Eqn (12.24), at $\delta = -2\sqrt{3}$, reveals a bistable dependency of output power on the input power (Figure 12.4a). The contrast ratio in the transmission, however, is limited: In general, due to weakness of nonlinearity, it is necessary to choose the operating frequency to be in the vicinity of the resonant frequency in order to reduce the incident power requirement. ($\delta = -\sqrt{3}$ is the minimum frequency detuning that still permits bi-stability. [44]) Doing so, however, also decreases the ratio of the optical energy of the two states inside the resonator.

The contrast ratio in the transmission can be drastically enhanced with the use of the side-coupled resonator geometry as shown in Figure 12.2b. Similar to the direct-coupled resonator geometry, the optical energy inside the cavity can exhibit

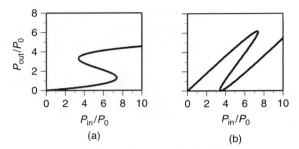

Figure 12.4 Transmitted vs input power, for the two structures shown in Figure 12.2a and 12.2b, when a Kerr nonlinear material is introduced into the resonator.

bistable dependence on the incident power level and can switch between two states with either low or high optical energy. However, in this case, one could take advantage of the interference between the propagating wave inside the waveguide and the decaying wave from the cavity to greatly enhance achievable contrast ratio in the transmission between the two bistable states.

Again, starting with Eqs (12.22) and (12.23), but noting in the side-coupled case, the reflected power $P_{\text{ref}} \equiv |s_{1-}|^2 = \frac{1}{\tau}|a|^2 = P_{\text{in}} - P_{\text{trans}}$, the transmitted power ratio T for a nonlinear side-coupled resonator can be written as

$$T \equiv \frac{P_{\text{trans}}}{P_{\text{in}}} = \frac{(P_{\text{ref}}/P_0 + \delta)^2}{1 + (P_{\text{ref}}/P_0 + \delta)^2} \quad (12.25)$$

For the same frequency detuning $\delta = -2\sqrt{3}$, the output power vs input power for this system is plotted in Figure 12.4b. We note that one of the bistable states can possess near-zero transmission coefficient, and thus, the contrast ratio can be infinitely high. This occurs when there is sufficient energy inside the cavity such that the resonance frequency of the cavity is shifted to coincide with that of the incident field.

As a physical implementation of the theoretical idea above, we consider the photonic crystal structure shown in Figure 12.5a and Figure 12.5b. The crystal consists of a square lattice of high dielectric rods ($n = 3.5$) with a radius of 0.2 l (l is the lattice constant) embedded in air ($n = 1$). We introduce the waveguide into the crystal by removing a line of rods and create a side-coupled cavity that supports a single resonant state by introducing a point defect with an elliptical dielectric rod, with the long and short axis lengths of l and 0.2 l, respectively. The defect region possesses instantaneous nonlinear Kerr response with a Kerr coefficient of $n_2 = 1.5 \times 10^{-17}$ W/m^2, which is achievable using nearly instantaneous nonlinearity in many semiconductors [49]. The use of the elliptical rod generates a single-mode cavity and also enhances the field localization in the nonlinear region.

12. Photonic Crystal Theory: Temporal Coupled-Mode Formalism

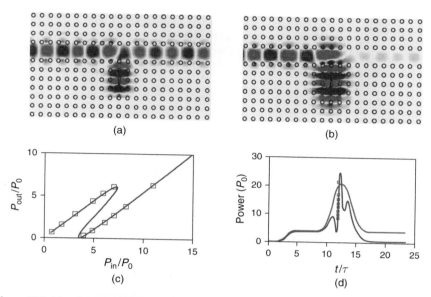

Figure 12.5 (a) and (b) The high- and low-transmission states for a nonlinear photonic crystal switch. (c) and (d) Comparison between the temporal coupled-mode theory formalism and the exact nonlinear finite difference time domain simulation. (c) Output vs input power. The dots are simulations and the line is theory. (d) Input and output power as a function of time. The light gray curve is the input power, the dark gray curve is the output as obtained from theory, and the dots are the output power as obtained from simulation (this figure may be seen in color on the included CD-ROM).

We perform nonlinear finite-difference time-domain (FDTD) simulations [50] for the TM case with electric field parallel to the rod axis for this photonic crystal system. The simulations incorporate a perfectly matched layer (PML) boundary condition specifically designed for photonic crystal waveguide simulations [51]. At a low incident power level where the structure behaves linearly, we determine that the cavity has a resonant frequency of $\omega_{res} = 0.371$ $(2\pi c/l)$, which falls within the band gap of the photonic crystal, a quality factor of $Q = 4494$. Using these parameters, the theory predicts a characteristic power level of $P_0 = 4.4$ mW/μm for 1.55 μm wavelength used in our simulations. For a three-dimensional structure, with the optical mode confined in the third dimension to a width about half a wavelength, the characteristic power is only on the order of a few milliwatt.

To study the nonlinear switching behavior, we excite an incident continuous wave (CW) in the waveguide detuned by $\delta = -2\sqrt{3}$ from the cavity resonance. We vary the input power and measure the output power at steady state. In particular, we observe a bistable region between $3.39 P_0$ and $7.40 P_0$. In fact, the FDTD results fit almost perfectly with the theoretical prediction (Figure 12.5(c)). (Note that on the theory curve, the region where there are no FDTD data points is unstable.) The contrast ratio between the upper and the lower branches approaches infinity as transmission drops to zero in the lower branch in transmission.

Figure 12.5(a) and 12.5(b) show the field patterns for the two bistable states for *the same input CW power* level of $3.95P_0$. Figure 12.5(a) corresponds to the high transmission state. In this state, the cavity is off resonance with the excitation. The field inside the cavity is low and thus the decaying field amplitude from the cavity is negligible. Figure 12.5b corresponds to the low transmission state. Here, the field intensity inside the cavity is much higher, pulling the cavity resonance frequency down to the excitation frequency of the incident field. The decaying field amplitude from the cavity is significant, and it interferes destructively with the incoming field. Thus, it is indeed the interference between the wave propagating in the waveguide and the decaying amplitude from the cavity that result in the high contrast ratio in transmission.

The FDTD analysis also reveals that the transmission can be switched to the lower branch from the upper branch with a pulse. Figure 12.5(d) shows the peak power in each optical period in the waveguide as a function of time, as we switch the system between the two bistable states shown in Figures 12.5(a) and 12.5(b). As the input is initially increased to the CW power level of $3.95P_0$, the system evolves into a high transmission state, with the transmitted power of $3.65P_0$. The switching then occurs after a pulse, which possesses a peak power $20.85P_0$, the same carrier frequency as that of CW, and a rise time and a width equal to the cavity lifetime, is superimposed upon the CW excitation. The pulse pushes the stored optical energy inside the cavity above the bistable threshold. After the pulse has passed through the cavity, the system switches to the bistable state with low transmission power of $0.25P_0$.

The switching dynamics, as revealed by the FDTD analysis, can also in fact be completely accounted for with temporal coupled-mode theory developed in this chapter. The coupled-mode equations (Eqn (12.1)) can be adopted to solve the reflected amplitude s_{1-} in terms of the input amplitude s_{1+} as

$$\frac{ds_{1-}}{dt} = i\omega_0 \left(1 - \frac{1}{2Q}\frac{|s_{1-}|^2}{P_0}\right)s_{1-} - \frac{1}{\tau}s_{1-} - \frac{1}{\tau}s_{1+}, \qquad (12.26)$$

from which the transmitted power can then be obtained as $|s_{1+}|^2 - |s^2_{1-}|$. It is important to note that the FDTD analysis takes into account the full effects of the nonlinearity. The coupled-mode theory, on the other hand, neglects higher harmonics of the carrier frequency generated by the nonlinearity. Nevertheless, because the switching and the cavity decay time scales are far larger than the optical period, the agreements between the coupled-mode theory and FDTD simulations are excellent as shown in Figure 12.5(d). Thus, we show that the nonlinear dynamics in photonic crystal structures can be completely accounted for using coupled-mode theory, which provides a rigorous and convenient framework for analyzing complex nonlinear processes and devices.

12.4 STOPPING LIGHT IN DYNAMIC PHOTONIC CRYSTALS

As a last example of this chapter, we show that one could use temporal coupled-mode theory as a guide for designing dynamic photonic crystals. The idea of dynamic photonic crystal is to modulate the property of a crystal while an optical pulse is inside the crystal. In doing so, the spectrum of the pulse can be molded almost arbitrarily with a small refractive index modulation, leading to highly nontrivial information-processing capabilities on chip. As examples of such capabilities, we show that the bandwidth of a light pulse can be compressed to zero, resulting in all-optical stopping and storage of light [33–38].

12.4.1 Tuning the Spectrum of Light

Here, we first provide a simple example to show how the spectrum of electromagnetic wave can be modified by a dynamic photonic structure. Consider a linearly polarized electromagnetic wave in one-dimension, the wave equation for the electric field is

$$\frac{\partial^2 E}{\partial x^2} - (\varepsilon_0 + \varepsilon(t))\mu_0 \frac{\partial^2 E}{\partial t^2} = 0 \qquad (12.27)$$

Here, $\varepsilon(t)$ represents the modulation and ε_0 is the background dielectric constant. We assume that both ε_0 and $\varepsilon(t)$ are independent of position. Hence, different wavevector components do not mix in the modulation process. For a specific wavevector component at k_0, with electric field described by $E(t) = f(t)e^{i(\omega_0 t - k_0 x)}$, where $\omega_0 = \frac{k_0}{\sqrt{\mu_0 \varepsilon_0}}$, we have

$$-k_0^2 f - [\varepsilon_0 + \varepsilon(t)]\mu_0 \left[\frac{\partial^2 f}{\partial t^2} + 2i\omega_0 \frac{\partial f}{\partial t} - \omega_0^2 f\right] = 0 \qquad (12.28)$$

By using a slowly varying envelope approximation, i.e., ignoring the $\frac{\partial^2 f}{\partial t^2}$ term, and by further assuming that the index modulations are weak, i.e., $\varepsilon(t) \ll \varepsilon_0$, Eqn (12.28) can be simplified as

$$i\frac{\partial f}{\partial t} = \frac{\varepsilon(t)\omega_0}{2[\varepsilon(t) + \varepsilon_0]}f \approx \frac{\varepsilon(t)\omega_0}{2\varepsilon_0}f \qquad (12.29)$$

which has an exact analytic solution:

$$f(t) = f(t_0) \exp\left[-i\omega_0 \int_{t_0}^{t} \frac{\varepsilon(t')}{2\varepsilon_0} dt'\right], \qquad (12.30)$$

where t_0 is the starting time of the modulation. Thus, the "instantaneous frequency" of the electric field for this wavevector component is

$$\omega(t) = \omega_0 \left(1 - \frac{\varepsilon(t)}{2\varepsilon_0}\right) \quad (12.31)$$

We note that frequency change is proportional to the magnitude of the refractive index shift alone. Thus, the process defined here differs in a fundamental way from traditional nonlinear optical processes. For example, in a conventional sum frequency conversion process, in order to convert the frequency of light from ω_1 to ω_2, modulations at a frequency $\omega_2 - \omega_1$ need to be provided. In contrast, in the process described here, regardless of how slow the modulation is, as long as light is in the system, the frequency shift can always be accomplished [52]. Below, we will demonstrate some very spectacular consequence of such frequency shift in the dynamic photonic crystal, in its application for stopping a light pulse all-optically.

12.4.2 General Conditions for Stopping Light

By stopping light, we aim to reduce the group velocity of a light pulse to zero, while completely preserving all the coherent information encoded in the pulse. Such capability holds the key to the ultimate control of light and has profound implications for optical communications and quantum information processing [53, 54].

There has been extensive work attempting to control the speed of light using optical resonances in static photonic crystal structures. Group velocities as low as 10^{-2} c for pulse propagation with negligible distortion have been experimentally observed in waveguide band edges or with coupled resonator optical waveguides (CROW) [29, 55–56]. Nevertheless, such structures are fundamentally limited by the delay-bandwidth product constraint—the group delay from an optical resonance is inversely proportional to the bandwidth within which the delay occurs [32, 57]. Therefore, for a given optical pulse with a certain temporal duration and corresponding frequency bandwidth, the minimum group velocity achievable is limited. In a CROW waveguide structure, for example, the minimum group velocity that can be accomplished for pulses at 10 Gbit/s rate with a wavelength of 1.55 μm is no smaller than 10^{-2} c. For this reason, static photonic structures could not be used to stop light.

To stop light, it is therefore necessary to use a dynamic system. The general condition for stopping light [33] is illustrated in Figure 12.6. Imagine a dynamic photonic crystal system, with an initial band structure possessing a sufficiently wide bandwidth. Such a state is used to accommodate an incident pulse, for which each frequency component occupies a unique wavevector component. After the pulse has entered the system, one can then stop the pulse by flattening the dispersion relation of the crystal adiabatically, while preserving the translational invariance.

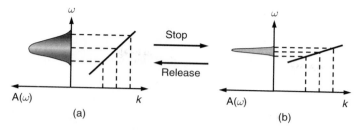

Figure 12.6 The general conditions for stopping a light pulse. (a) The large-bandwidth state that is used to accommodate an incident light pulse; (b) The narrow-bandwidth state that is used to hold the light pulse. An adiabatic transition between these two states stops a light pulse (this figure may be seen in color on the included CD-ROM).

In doing so, the spectrum of the pulse is compressed, and its group velocity is reduced. In the meantime, because the translational symmetry is still preserved, the wavevector components of the pulse remain unchanged, and thus, one actually preserves the dimensionality of the phase space. This is crucial in preserving all the coherent information encoded in the original pulse during the dynamic process.

12.4.3 Tunable Fano Resonance

To create a dynamic photonic crystal, one needs to adjust its properties as a function of time. This can be accomplished by modulating the refractive index, either with electro-optic or with nonlinear optic means. However, the amount of refractive index tuning that can be accomplished with standard optoelectronics technology is generally quite small, with a fractional change typically on the order of $\frac{\delta n}{n} : 10^{-4}$. Therefore, we employ Fano interference schemes in which a small refractive index modulation leads to a very large change of the bandwidth of the system. The essence of Fano interference scheme is the presence of multipath interference, in which at least one of the paths includes a resonant tunneling process [42]. As we have argued in Section 12.2, such interference can be used to greatly enhance the sensitivity of resonant devices to small refractive index modulation [43].

Here, we consider a waveguide side-coupled to two cavities [41]. The cavities have resonant frequencies $\omega_{A,B} \equiv \omega_0 \pm \frac{\delta\omega}{2}$ respectively. (This system represents an all-optical analogue of atomic systems exhibiting Electromagnetically Induced Transparency (EIT) [58]. Each optical resonance here is analogous to the polarization between the energy levels in the EIT system [59].) Consider a mode in the waveguide passing through the cavities; from Eqn (12.18), the transmission and reflection coefficients ($t_{A,B}$ and $r_{A,B}$ respectively) with a single side cavity is

$$t_{A,B} = \frac{j(\omega - \omega_{A,B})}{j(\omega - \omega_{A,B}) + 1/\tau} \quad (12.32)$$

$$r_{A,B} = \frac{1/\tau}{j(\omega - \omega_{A,B}) + 1/\tau} \qquad (12.33)$$

When two cavities are cascaded together, the transmission spectrum can be derived as

$$T = \left(\frac{|t_A t_B|}{1 - |r_A r_B|}\right)^2 \frac{1}{1 + 4\left(\frac{\sqrt{|r_A r_B|}}{1 - |r_A r_B|}\right)^2 \sin^2 \theta} \qquad (12.34)$$

θ is one-half the round trip phase accumulated in the waveguides: $\theta = \frac{1}{2} Arg\left[r_A r_B e^{-2j\beta(\omega)L_1}\right]$, where $\beta(\omega)$ is the waveguide dispersion relationship and L_1 is the spacing between the cavities.

The transmission spectra of one- and two-cavity structures, calculated using Eqn (12.32–12.34), are plotted in Figure 12.7. In the case of one-cavity structure, the transmission features a dip in the vicinity of the resonant frequency, with the width of the dip controlled by the strength of waveguide-cavity coupling (Figure 12.7a). With two cavities, when the condition

$$2\beta(\omega_0)L = 2n\pi \qquad (12.35)$$

is satisfied, the transmission spectrum features a peak centered at ω_0. The width of the peak is highly sensitive to the frequency spacing between the resonances $\delta\omega$. When the cavities are lossless, the center peak can be tuned from a wide peak when $\delta\omega$ is large (Figure 12.7b) to a peak that is arbitrarily narrow with $\delta\omega \to 0$ (Figure 12.7c). The two-cavity structure, appropriately designed, therefore

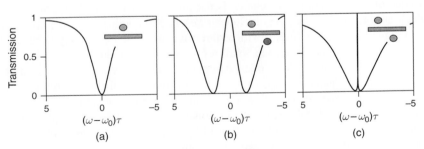

Figure 12.7 The insets show the structures considered. The bar is the waveguide and the circles are cavities. (a) Transmissions spectrum through a waveguide side-coupled to a single-mode cavity. (b) and (c) Transmission spectra through a waveguide side coupled to two cavities. The spectra are calculated using Eqs (12.32–12.34), The parameters for the cavities are $\omega_0 = \frac{2\pi c}{L_1}, \frac{1}{\tau} = 0.05\omega_0$. And the waveguide satisfies a dispersion relation $\beta(\omega) = \frac{\omega}{c}$, where c is the speed of light in the waveguide and L_1 is the distance between the cavities. In (b) $\omega_{a,b} = \omega_0 \pm 1.5\frac{1}{\tau}$. In (c) $\omega_{a,b} = \omega_0 \pm 0.2\frac{1}{\tau}$.

behaves as a tunable bandwidth filter (as well as a tunable delay element), in which the bandwidth can be in principle adjusted by any order of magnitude with very small refractive index modulation.

12.4.4 From Tunable Bandwidth Filter to Light-Stopping System

By cascading the tunable bandwidth filter structure as described in the previous section, one can construct a structure that is capable of stopping light (Figure 12.8a). In such a light-stopping structure, the photonic band diagram becomes highly sensitive to small refractive index modulation.

The photonic bands for the structure in Figure 12.8a can be calculated using a transfer matrix method. For the two-port systems defined in Figure 12.2, the transfer matrix T is defined as

$$\begin{pmatrix} s_{2+} \\ s_{2-} \end{pmatrix} = T \begin{pmatrix} s_{1+} \\ s_{1-} \end{pmatrix} \tag{12.36}$$

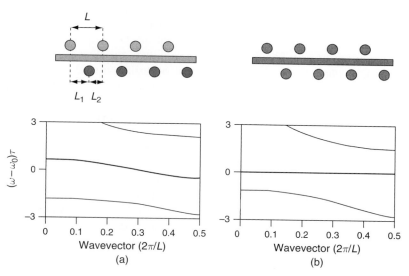

Figure 12.8 The top panels in each figure are the schematic of the structure used to stop light. The bar is the waveguide, and the circles are cavities. The bottom panels are the photonic band structures, calculated using the same waveguide and cavity parameters as in Figure 12.7(b) and 12.7(c), with the additional parameter $L_2 = 0.7 L_1$. The thicker lines highlight the middle band that can be used to stop a light pulse.

For a waveguide side coupled to a single resonator with resonance frequency ω_i can be calculated from the scattering matrix S (Eqn (12.3)) as [60]:

$$T_{c_i} = \begin{pmatrix} 1 + j/(\omega - \omega_i)\tau & j/(\omega - \omega_i)\tau \\ -j/(\omega - \omega_i)\tau & 1 - j/(\omega - \omega_i)\tau \end{pmatrix} \quad (12.37)$$

The transfer matrix through an entire unit cell in Figure 12.8 can then be determined as

$$T = T_{c_1} T_{l_1} T_{c_2} T_{l_2}, \quad (12.38)$$

where $T_{l_i} = \begin{pmatrix} e^{-j\beta L_i} & 0 \\ 0 & e^{j\beta L_i} \end{pmatrix}$ is the transmission matrix for a waveguide section of length L_i. Here, β is the wavevector of the waveguide at a given frequency ω.

Because $\det(T) = 1$, the eigenvalues of T can be represented as e^{ikl}, e^{-ikl}, where $L = L_1 + L_2$ is the length of the unit cell, and k (when it is real) corresponds to the Bloch wavevector of the entire system. Therefore, we obtain the band diagram of the system as [35]

$$\frac{1}{2} Tr(T) = \cos(kL) = f(\omega) \equiv \cos(\beta L) + \frac{C_+}{(\omega - \omega_A)} + \frac{C_-}{(\omega - \omega_B)}, \quad (12.39)$$

where $C_\pm = \frac{1}{\tau}\sin(\beta L) \pm \frac{2\sin(\beta L_1)\sin(\beta L_2)}{(\omega_A - \omega_B)\tau^2}$. In the frequency range where $|f(\omega)| < 1$, the system supports propagating modes, while $|f(\omega)| > 1$ corresponds to the frequency ranges of the photonic band gaps.

The band diagrams thus calculated are shown in Figure 12.8, in which the waveguide and cavity parameters are the same as those used to generate the transmission spectrum in Figure 12.7. In the vicinity of the resonances, the system supports three photonic bands, with two gaps occurring around ω_A and ω_B. The width of the middle band depends strongly on the resonant frequencies ω_A, ω_B. By modulating the frequency spacing between the cavities, one goes from a system with a large bandwidth (Figure 12.8(a)) to a system with a very narrow bandwidth (Figure 12.8b). In fact, it can be analytically proved that the system can support a band that is completely flat in the entire first Brillouin zone [35], allowing a light pulse to be frozen inside the structure with the group velocity reduced to zero. Moreover, the gaps surrounding the middle band have sizes on the order of the cavity-waveguide coupling rate, $1/\tau$, and are approximately independent of the slope of the middle band. Thus, by increasing the waveguide-cavity coupling rate, this gap can be made large, which is important for preserving the coherent information during the dynamic bandwidth compression process [33].

12.4.5 Numerical Demonstration in a Photonic Crystal

The system presented above can be implemented in a photonic crystal of a square lattice of dielectric rods ($n = 3.5$) with a radius of $0.2a$ (a is the lattice constant)

12. Photonic Crystal Theory: Temporal Coupled-Mode Formalism

embedded in air ($n = 1$) [35] (Figure 12.9). The photonic crystal possesses a band gap for TM modes with electric field parallel to the rod axis. Removing one row of rods along the pulse propagation direction generates a single-mode waveguide. Decreasing the radius of a rod to $0.1a$ and the dielectric constant to $n = 2.24$

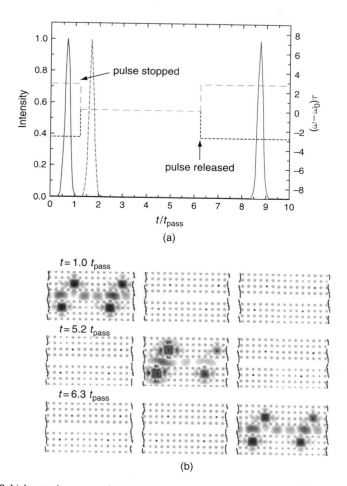

Figure 12.9 Light-stopping process in a photonic crystal simulated using finite-difference time-domain methods. The crystal consists of a waveguide side coupled to 100 cavity pairs. Fragments of the photonic crystal are shown in part b. The three fragments correspond to unit cells 12–13, 55–56, 97–98. The dots indicate the positions of the dielectric rods. The black dots represent the cavities. (a) The straight lines represent the variation of ω_A and ω_B as a function of time. Also shown in the figure are the intensity of the incident pulse as recorded at the beginning of the waveguide (left-most Gaussian pulse form), and the intensity of the output pulse at the end of the waveguide, in the absence (middle Gaussian pulse form) and the presence (right-most Gaussian pulse form) of modulation. t_{pass} is the passage time of the pulse in the absence of modulation. (b) Snapshots of the electric field distributions in the photonic crystal at the indicated times. The gray regions represent areas with significant field (this figure may be seen in color on the included CD-ROM).

provides a single-mode cavity with resonance frequency at $\omega_c = 0.357 \times (2\pi c/a)$. The nearest neighbor cavities are separated by a distance of $l_1 = 2a$ along the propagation direction, and the unit cell periodicity is $l = 8a$. The waveguide-cavity coupling occurs through barrier of one rod, with a coupling rate of $\frac{1}{\tau} = \omega_c/235.8$. The resonant frequencies of the cavities are tuned by refractive index modulation of the cavity rods.

We simulate the entire process of stopping light for $N = 100$ pairs of cavities with FDTD method, which solves Maxwell's equations without approximation. The dynamic process for stopping light is shown in Figure 12.9. We generate a Gaussian pulse in the waveguide (The process is independent of the pulse shape). The excitation reaches its peak at $t = 0.8 t_{\text{pass}}$, where t_{pass} is the traversal time of the pulse through the static structure. During the pulse generation, the cavities have a large frequency separation. The field is concentrated in both the waveguide and the cavities (Figure 12.9b, $t = 1.0 t_{\text{pass}}$), and the pulse propagates at a relatively high speed of $v_g = 0.082$ c. After the pulse is generated, we gradually reduce the frequency separation of the two cavities. During this process, the speed of light is drastically reduced to zero. As the bandwidth of the pulse is reduced, the field concentrates in the cavities (Figure 12.9b, $t = 5.2 t_{\text{pass}}$). When zero group velocity is reached, the photon pulse can be kept in the system as a stationary waveform for any time duration. In this simulation, we store the pulse for a time delay of $5.0 t_{\text{pass}}$, and then release the pulse by repeating the same index modulation in reverse (Figure 12.9(b), $t = 6.3 t_{\text{pass}}$). The pulse intensity as a function of time at the right end of the waveguide is plotted in Figure 12.9a, and shows the same temporal shape as both the pulse that propagates through the unmodulated system and the initial pulse recorded at the left end of the waveguide. Thus, the pulse is perfectly recovered without distortion after the intended delay.

12.4.6 Future Prospects of Dynamic Photonic Crystal System

In the all-optical light-stopping scheme presented above, for a small refractive index shift of $\delta n/n = 10^{-4}$ achievable in practical optoelectronic devices, and assuming a carrier frequency of approximately 200 THz, as used in optical communications, the achievable bandwidths are on the order of 20 GHz, which is comparable to the bandwidth of a single-wavelength channel in high-speed optical systems. The storage times are limited only by the cavity lifetimes, which may eventually approach millisecond time scales as limited by residual loss in transparent materials. The loss in optical resonator systems might be further counteracted with the use of gain media in the cavities, or with external amplification. With such performance, the capabilities for on-chip stopping light should have important implications for optical communication systems. As an important step toward its eventual experimental demonstration, the required EIT-like two-cavity interference effects have recently been observed in a micro-ring cavity system on a silicon chip [61]. The general concept of introducing dynamics into photonic crystal systems could also be very promising for creating new optical signal-processing functionalities far beyond the capabilities of static systems.

12.5 CONCLUDING REMARKS

In this chapter, we provide a brief review of the basis of the temporal coupled-mode theory, as well as its applications in nonlinear and dynamic photonic crystals, drawing examples from our own recent work in this area. These developments highlight two general trends in the theoretical work in this field. On one hand, using computational electromagnetic techniques such as the FDTD methods [50] in combination with modern large-scale computing architectures, almost any complex optical processes in photonic crystal can now be simulated through exact numerical solutions of Maxwell's equations. On the other hand, with the band structures and modal properties of passive dielectric photonic structures largely mapped out, one can now create analytic models with only a few dynamic variables based on these modal properties, in order to describe the essential physics of optical processes in photonic crystals. These developments in both theory and simulations, in the context of very rapid progress in experimental fabrications of photonic crystals, are leading to ways of controlling light that are truly unprecedented.

ACKNOWLEDGMENT

This work is supported in part by NSF and DARPA. The author acknowledges Professor Mehmet Fatih Yanik, Professor Marin Soljacic, Professor John Joannopoulos, Dr Wonjoo Suh, Dr Zheng Wang, Dr Michelle Povinelli, and Sunil Sandhu for contributions to all aspects of this work.

REFERENCES

[1] E. Yablonovitch, "Inhibited spontaneous emission in solid state physics and electronics," *Phys. Rev. Lett.*, 58, 2059–2062, 1987.
[2] S. John, "Strong localization of photons in certain disordered dielectric superlattices," *Phys. Rev. Lett.*, 58, 2486–2489, 1987.
[3] J. D. Joannopoulos, R. D. Meade, and J. N. Winn, *Photonic Crystals: Molding the Flow of Light*, Princeton University Press, Princeton, NJ, 1995.
[4] C. Soukoulis, ed., *Photonic Crystals and Light Localization in the 21st Century*, The Netherlands, NATO ASI Series, Kluwer Academic Publisher, 2001.
[5] S. G. Johnson and J. D. Joannopoulos, *Photonic Crystals: The Road from Theory to Practice*, Kluwer Academic Publisher, Boston, 2002.
[6] K. Inoue and K. Ohtaka, *Photonic Cryst.*, Berlin, Springer-Verlag, 2004.
[7] J. D. Joannopoulos, P. R. Villeneuve, and S. Fan, "Photonic crystals: Putting a new twist on light," *Nature*, 386, 143–147, 1997.

[8] A. Mekis, J. C. Chen, I. Kurland et al., "High transmission through sharp bends in photonic crystal waveguides," *Phys. Rev. Lett.*, 77, 3787–3790, 1996.

[9] S. Fan, P. R. Villeneuve, J. D. Joannopoulos, and H. A. Haus, "Channel drop tunneling through localized states," *Phys. Rev. Lett.*, 80, 960–963, 1998.

[10] S. Fan, P. R. Villeneuve, J. D. Joannopoulos, and H. A. Haus, "Channel drop filters in a photonic crystal," *Opt. Express.*, 3, 4–11, 1998.

[11] C. Manolatou, M. J. Khan, S. Fan et al., "Coupling of modes analysis of resonant channel add-drop filters," *IEEE J. Quant. Electron.*, 35, 1322–1331, 1999.

[12] S. Noda, A. Chutinan, and M. Imada, "Trapping and emission of photons by a single defect in a photonic band gap structures," *Nature*, 407, 608–610, 2000.

[13] C. Jin, S. Fan, S. Han, and D. Zhang, "Reflectionless multi-channel wavelength demultiplexer in a transmission resonator configuration," *IEEE J. Quant. Electron.*, 39, 160–165, 2003.

[14] Y. Akahane, T. Asano, B.-S. Song, and S. Noda, "Investigation of high-Q channel drop filters using donor-type defects in two-dimensional photonic crystal slabs," *Appl. Phys. Lett.*, 83, 1513–1515, 2003.

[15] K. H. Hwang and G. H. Song, "Design of a high-Q channel add/drop multiplexer based on the two-dimensional photonic crystal membrane structure," *Opt. Express.*, 13, 1948–1957, 2005.

[16] Z. Zhang and M. Qiu, "Coupled-mode analysis of a resonant channel drop filter using waveguides with mirror boundaries," *J. Opt. Soc. Am. B.*, 23, 104–113, 2006.

[17] A. Shinya, S. Mitsugi, E. Kuramochi, and M. Notomi, "Ultrasmall multi-port channel drop filter in two-dimensional photonic crystal on silicon-on-insulator substrate," *Opt. Express.*, 14, 12394–12400, 2006.

[18] S. F. Mingaleev and Y. S. Kivshar, "Nonlinear transmission and light localization in photonic-crystal waveguides," *J. Opt. Soc. Am. B*, 19, 2241–2249, 2002.

[19] M. Soljacic, M. Ibanescu, S. G. Johnson et al., "Optimal bistable switching in nonlinear photonic crystals," *Phys. Rev. E*, 66, 055601(R), 2002.

[20] M. Soljacic, C. Luo, J. D. Joannopoulos, and S. Fan, "Nonlinear photonic microdevices for optical integrations," *Opt. Lett.*, 28, 637–639, 2003.

[21] M. F. Yanik, S. Fan, and M. Soljacic, "High-contrast all-optical bistable switching in photonic crystal microcavities," *Appl. Phys. Lett.*, 83, 2741–3, 2003.

[22] M. F. Yanik, S. Fan, M. Soljacic, and J. D. Joannopoulos, "All-optical transistor action with bistable switching in photonic crystal cross-waveguide geometry," *Opt. Lett.*, 28, 2506–2508, 2003.

[23] M. Notomi, A. Shinya, S. Mitsugi et al., "Optical bistable switching action of Si high-Q photonic crystal nanocavities," *Opt. Express.*, 13, 2678–2687, 2005.

[24] Y. Tanaka, H. Kawashima, N. Ikeda et al., "Optical bistable operations in AlGaAs-based photonic crystal slab microcavity at telecommunication wavelengths," *IEEE Photon. Technol. Lett.*, 18, 1996–1998, 2006.

[25] Z. Wang and S. Fan, "Magneto-optical defects in two-dimensional photonic crystals," *Appl. Phys. B*, 81, 369–375, 2005.

[26] Z. Wang and S. Fan, "Optical circulators in two-dimensional magneto-optical photonic crystals," *Opt. Lett.*, 30, 1989–1991, 2005.

[27] Z. Wang and S. Fan, "Add-drop filter in two-dimensional magneto-optical photonic crystals and suppression of disorder effects by time-reversal breaking," *Photonics and Nanostructures: Fundamentals and Applications*, 18, 1996–1998, 2006.

[28] N. Stefanou and A. Modinos, "Impurity bands in photonic insulators," *Phys. Rev. B*, 57, 12127–12133, 1998.

[29] A. Yariv, Y. Xu, R. K. Lee, and A. Scherer, "Coupled-resonator optical waveguide: a proposal and analysis," *Opt. Lett.*, 24, 711–713, 1999.

[30] M. Bayindir, B. Temelkuran, and E. Ozbay, "Tight-binding description of the coupled defect modes in three-dimensional photonic crystals," *Phys.l Rev. Lett.*, 84, 2140–2143, 2000.

[31] Y. Xu, Y. Li, R. K. Lee, and A. Yariv, "Scattering-theory analysis of waveguide resonator coupling," *Phys. Rev. E*, 62, 7389–7404, 2000.

[32] Z. Wang and S. Fan, "Compact all-pass filters in photonic crystals as the building block for high capacity optical delay lines," *Phys. Rev. E*, 68, Art. No. 066616, 2003.
[33] M. F. Yanik and S. Fan, "Stopping light all-optically," *Phys. Rev. Lett.*, 92, Art. No. 083901, 2004.
[34] M. F. Yanik and S. Fan, "Time-reversal of light with linear optics and modulators," *Phys. Rev. Lett.*, 93, Art. No. 173903, 2004.
[35] M. F. Yanik, W. Suh, Z. Wang, and S. Fan, "Stopping light in a waveguide with an all-optical analogue of electromagnetically induced transparency," *Phys. Rev. Lett.*, 93, Art. No. 233903, 2004.
[36] M. F. Yanik and S. Fan, "Stopping and storing light coherently," *Phys. Rev. A*, 71, Art. No. 013803, 2005.
[37] M. F. Yanik and S. Fan, "Dynamic photonic structures: Stopping, storage, and time-reversal of light," *Studies in Applied Mathematics*, 115, 233–254, 2005.
[38] S. Sandhu, M. L. Povinelli, M. F. Yanik, and S. Fan, "Dynamically-tuned coupled resonator delay lines can be nearly dispersion free," *Opt. Lett.*, 31, 1981–1983, 2006.
[39] H. A. Haus, *Waves and Fields in Optoelectronics*, New Jersey, Englewood Cliffs: Prentice-Hall, 1984.
[40] S. Fan, W. Suh, and J. D. Joannopoulos, "Temporal coupled mode theory for Fano resonances in optical resonators," *J. Opt. Soc. Am. A*, 20, 569–573, 2003.
[41] W. Suh, Z. Wang, and S. Fan, "Temporal coupled-mode theory and the presence of non-orthogonal modes in lossless multimode cavities," *IEEE J. Quant. Electron.*, 40, 1511–1518, 2004.
[42] U. Fano, "Effects of configuration interaction on intensities and phase shifts," *Phys. Rev.*, 124, 1866–1878, 1961.
[43] S. Fan, "Sharp asymmetric lineshapes in side-coupled waveguide-cavity systems," *Appl. Phys. Lett.*, 80, 908–910, 2002.
[44] H. M. Gibbs, *Optical Bistability: Controlling Light with Light*, Orlando, Academic Press, 1985.
[45] A. R. Cowan and J. F. Young, "Optical bistability involving photonic crystal microcavities and Fano lineshapes," *Phys. Rev. E*, 68, Art. No. 046606, 2003.
[46] V. Lousse and J. P. Vigeron, "Use of Fano resonances for bistable optical transfer through photonic crystal films," *Phys. Rev. B*, 69, Art. No. 155106, 2004.
[47] A. E. Miroshnichenko and Y. S. Kivshar, "Engineering Fano resonance in discrete arrays," *Phys. Rev. E*, 72, Art. No. 056611, 2005.
[48] S. F. Mingaleev, A. E. Miroshnichenko, Y. S. Kivshar, and K. Busch, "All-optical switching, bistability, and slow-light transmission in photonic crystal waveguide-resonator systems," *Phys. Rev. E*, 74, Art. No. 046603, 2006.
[49] M. Sheik-Bahae, D. C. Hutchings, D. J. Hagan, and E. W. Van Stryland, "Dispersion of bound electronic nonlinear refraction in solids," *IEEE J. Quant. Electron.*, 27, 1296–1309, 1991.
[50] A. Taflove and S. C. Hagness, *Computational Electrodynamics*, Norwood MA, Artech House, 2000.
[51] M. Koshiba and Y. Tsuji, "High-performance absorbing boundary conditions for photonic crystal waveguide simulations," *IEEE Microwave and Wireless Compon. Lett.*, 11, 152–154, 2001.
[52] E. J. Reed, M. Soljacic, and J. D. Joannopoulos, "Color of shock waves in photonic crystals," *Phys. Rev. Lett.*, 91, Art. No. 133901, 2003.
[53] C. Liu, Z. Dutton, C. H. Behroozi, and L. V. Hau, "Observation of coherent optical information storage in an atomic medium using halted light pulses," *Nature*, 409, 490–493, 2001.
[54] D. F. Phillips, A. Fleischhauer, A. Mair et al., "Storage of light in atomic vapors," *Phys. Rev. Lett.*, 86, 783–786, 2001.
[55] M. Notomi, K. Yamada, A. Shinya et al., "Extremely large group-velocity dispersion of line-defect waveguides in photonic crystal slabs," *Phys. Rev. Lett.*, 87, Art. No. 253902, 2001.
[56] Y. A. Vlasov, M. O'Boyle, H. F. Harmann, and S. J. McNab, "Active control of slow light on a chip with photonic crystal waveguides," *Nature*, 438, 65–69, 2005.

[57] G. Lenz, B. J. Eggleton, C. K. Madsen, and R. E. Slusher, "Optical delay lines based on optical filters," *IEEE J. Quant. Electron.*, 37, 525–532, 2001.
[58] S. E. Harris, "Electromagnetically induced transparency," *Phys. Today.*, 50, 36–42, 1997.
[59] L. Maleki, A. B. Matsko, A. A. Savchenkov, and V. S. Ilchenko, "Tunable delay line with interacting whispering-gallery-mode resonators," *Opt. Lett.*, 29, 626–628, 2004.
[60] S. Fan, P. R. Villeneuve, J. D. Joannopoulos et al., "Theoretical investigation of channel drop tunneling processes," *Phys. Rev. B.*, 59, 15882–15892, 1999.
[61] Q. Xu, S. Sandhu, M. L. Povinelli et al., "Experimental realization of an on-chip all-optical analogue to electromagnetically induced transparency," *Phys. Rev. Lett.*, 96, Art. No. 123901, 2006.

13

Photonic crystal technologies: Experiment

Susumu Noda

Department of Electronic Science and Engineering, Kyoto University, Kyoto, Japan

Abstract

Photonic crystals, in which the refractive index changes periodically, provide an exciting new tool for the manipulation of photons and have received keen interest from a variety of fields. This manuscripts review the recent progresses and future prospects of photonic crystals and their applications to photonic nanostructure devices.

13.1 INTRODUCTION

Photonic crystals are a kind of nanostructures for light, the refractive indices of which change in a periodic fashion. They are characterized by the formation of band structures with respect to photon energy [1, 2] from the analogy to those with respect to electron energy in solid-state electronics. Examples of two- and three-dimensional (2D and 3D) photonic crystals and their band structures are shown in Figure 13.1. In 3D photonic crystals [1–4, 43] shown in Figure 13.1(b), a complete photonic band gap is formed; the transmission (and even the presence) of light with frequencies lying in the band gap is not allowed. By applying various types of engineering described below to photonic crystals, with particular focus on the band gap, the band edge, and the transmission band in the band structures, the manipulation of photons in a variety of ways becomes possible:

(1) **Band gap/defect engineering:** This type of engineering focuses on band gaps that prohibit the presence of light. In structures with complete photonic band gaps, light with certain frequencies is blocked from the crystals. However, by introducing an artificial periodic disturbance, or "defect," into the crystal light can be controlled in various ways. For example, by introducing a

Optical Fiber Telecommunications V A: Components and Subsystems
Copyright © 2008, Elsevier Inc. All rights reserved.
ISBN: 978-0-12-374171-4

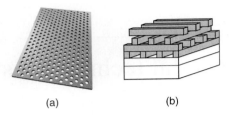

Figure 13.1 Schematic representations of (a) a 2D photonic crystal and (b) a 3D photonic crystal.

line-shaped defect, it is possible to form an ultrasmall waveguide that permits the transmission of light only along the defect. Light can be trapped at certain points by introducing point defects, thereby forming photonic nanocavities. By combining these line and point defects, it is possible to form ultrasmall photonic circuits (or chips) with various functions. Furthermore, the band gap itself makes it possible to suppress spontaneous emission, which is a fundamental factor limiting the performance of various photonic devices.

(2) *Band edge engineering:* This type of engineering focuses on the band edge, where the group velocity of light becomes zero. At the band edge, light being propagating in various directions is combined by Bragg reflection to form a standing wave. By using this standing wave as a cavity mode, for example, a laser that enables coherent oscillation over a large 2D area can be realized. The creation of various types of novel nonlinear optical phenomena might also be possible.

(3) *Band engineering:* This type of engineering focuses on the transmission bands that allow the propagation of light. Various types of light control become possible when the unique dispersion relation obtained from the band structure can be manipulated. It is anticipated that by considerably slowing the propagation speed and by altering the propagation direction of the light in a photonic crystal, or by making negative refraction possible, a variety of novel applications and photonic devices will be developed.

In addition to the various types of engineering described above, the recent introduction of a concept of "photonic heterostructures" accelerates the progresses of the field of photonic crystals. On the basis of the engineering described above, we would like to introduce some recent developments concerning 2D and 3D photonic crystals.

13.2 BAND GAP/DEFECT ENGINEERING

13.2.1 Two-Dimensional Photonic Crystal Slabs

Two-dimensional photonic crystal research initially targeted structures with periodicity in two dimensions and a third dimension assumed to be of infinite length [5], but recently the focus of research has been slab structures [2, 6, 7], such as the one shown in Figure 13.1(a), that have a thickness of the order of the wavelength of the light. The photonic crystal in this figure has a triangular lattice structure. The slab material is assumed to be a high refractive index medium, such as Si and III–V semiconductors, and the lattice points are assumed to be comprising a low refractive index medium, such as air. In 2D photonic crystal slabs, the confinement of light occurs in the in-plane direction due to a photonic band gap effect, and light is confined in the perpendicular direction by total internal reflection due to a difference in refractive indices. Pseudo-3D light control becomes possible as a result. It is, of course, important to optimize the structure so that leakage of light in the direction perpendicular to the slab is minimized. Recently, as described below, 2D photonic crystal slabs have been subjected to various types of band gap/defect engineering, and research with the aim of realizing ultrasmall photonic circuits has been advancing steadily.

Control of Light by Combined Line/Point-Defect Systems

As shown in Figure 13.2(a), when a line defect is introduced into a 2D photonic crystal slab, it will act as an ultrasmall optical waveguide. To operate the waveguide efficiently, it is necessary to devise a way in which the leakage of light in the direction perpendicular to the slab can be nullified. In 1999, an optical waveguide experiment was reported [8], but the perpendicular leakage of light was not taken into consideration and the propagation loss was hence extremely large, exceeding 70 dB/mm. A year later in 2000, detailed results of an examination of the effects of the slab thickness and the ratio of the low refractive index [the medium (air) that forms the lattice points] to the high refractive index (the slab material) were reported, and a theoretical zero-loss waveguide structure was proposed for the first time [9]. In 2001, based on these design guidelines, a new waveguide experiment was conducted in which a low propagation loss of 7 dB/mm was obtained [10]. At the same time, discussions were taking place regarding the effect of modification of the waveguide width on various waveguide characteristics. It was also pointed out that an increase in propagation loss was caused by vertical asymmetry in the slab [11]. Regarding this, the importance of using crystals with a 2D in-plane full band gap was pointed out [12]. Recently, an extremely

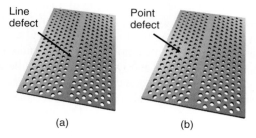

Figure 13.2 Schematic representations of (a) a line-defect waveguide introduced into a 2D photonic crystal slab and (b) a combined point-defect resonator and line-defect waveguide system.

low waveguide loss less than 0.7 dB/mm, one-tenth of the previous best value, has been achieved [13–15], and it should be possible to fabricate waveguides with propagation losses of less than 1 dB/cm in the near future. In view of the merits of the miniaturization enabled by the use of photonic crystals, these recent developments show that the minimization of waveguide loss in 2D crystal slabs has advanced to the point where it can almost be ignored.

In conjunction with studies of straight-line waveguides, detailed investigations on waveguide bends have also been carried out. It was initially pointed out that the bending of a waveguide was possible in purely 2D crystals, in regions of very broad-wavelength transmission [5, 16], but that, in slab structures, the effect of light scattering to the outside of the slab meant that the low-loss band region in the bend part of the waveguide could not be enlarged to a great extent [9]. Thereafter, the importance of controlling lattice point shapes positioned in the curved part was highlighted [17], and the enlargement and shifting of the band area in the curved region were then reported [18, 19]. In 2004, it was also shown that low-loss curvature of a waveguide at a very broadband area might be possible even for a bend of 120° [20]. In addition, there have been some studies of low-loss connections between line-defect waveguides and external optical systems [21–24]. Empirical studies have demonstrated that such waveguides can be connected with optical fibers with losses of less than a few dB [23, 24].

When a point defect is introduced into a photonic crystal, it becomes possible for this to act as a photonic nanocavity; in other words, as a trap of photons [6, 7]. In photonic crystals, a point defect and a line defect can integrate quite spontaneously, thus point defects have been extensively discussed with respect to their combination with line defects. High-Q nanocavities and fusion with nonlinear and/or active media will be discussed later. Figure 13.2(b) shows such a combined system of defects [7, 25]. Here, the point defect is formed by filling three lattice points with a high-refractive index medium. If the resonant frequency of the point defect is defined as f_i and light with various wavelengths is introduced from the waveguide, light that has a frequency of f_i is trapped by the point defect. The trapped light resonates in the point-defect cavity and is emitted perpendicular to the slab surface due to a breakdown in the total internal reflection conditions. Such optical behavior in a combined line/point-defect system could be used, for example, as a surface emitting-type ultracompact channel add/drop functional device. It would also be possible to apply it as an ultracompact device for sensing and/or trapping of nanomaterials.

In combined line/point-defect systems, a maximum of 50% of the light introduced into the waveguide is trapped by the point defect and emitted to free space [26], which can be deduced by coupled mode theory. This maximum efficiency is obtained when $Q_{in} = Q_v$, where Q_{in} and Q_v are defined as the quality factors of point-defect cavity for in-plane and vertical directions, respectively. More concretely, Q_{in} is determined by the optical coupling between the point-defect cavity and line-defect waveguide, and Q_v is determined by the total internal reflection conditions of the point defect in the direction perpendicular to the slab. The overall Q factor is expressed as $1/Q = 1/Q_{in} + 1/Q_v$ and determines the wavelength resolution. Figure 13.3 shows an electron microscope image of a sample comprising a

13. Photonic Crystal Technologies: Experiment

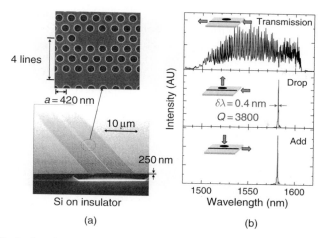

Figure 13.3 Device based on the photonic crystal slab design in Figure 13.2(b) and its characteristics: (a) electron microscope image; (b) transmission, drop, and add spectra (this figure may be seen in color on the included CD-ROM).

combined point and line defect, and its transmission, drop, and add spectra [25]. The device was created using Si/SiO$_2$/Si, that is, an SOI system. A photonic crystal is formed in Si in the top layer. The SiO$_2$ layer beneath was removed. In the transmission spectrum shown in Figure 13.3, the many sharp resonance peaks observed are attributed to Fabry–Perot resonance at both edges of the waveguide. This phenomenon can be eliminated by applying an antireflection coating at the waveguide surfaces. An important point to note is the sharp decrease in transmission seen in the vicinity of 1.58 μm, which was attributed to the emission of light through the point defect in the perpendicular direction. Figure 13.3 also shows the add spectra obtained when light was introduced from free space into the point defect. It can be clearly seen that light is effectively entering and exiting through an extremely small point-defect cavity. The resolution determined from these drop (or add) spectrum was estimated to be 0.4 nm, and a relatively large Q value of 3800 was obtained. By comparison with theoretical calculations, a high optical drop (or add) efficiency of up to 50% was obtained.

As outlined above, the use of 2D photonic crystal slabs has allowed the investigations of the basic behavior of line and point defects, individually and in combination, and a basis has been established for the realization of advanced functional photonic circuits.

The Effect of Introducing Heterostructures

Along with the advances in band gap/defect engineering described above, new concepts for controlling light with a greater degree of freedom are also emerging.

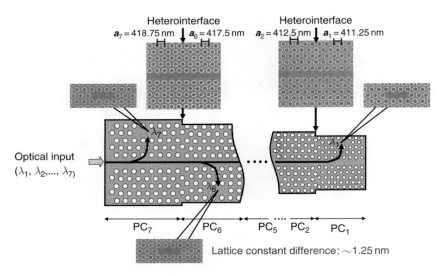

Figure 13.4 Example of an in-plane heterostructure: The insets show electron microscope images of a fabricated sample. The schematic drawing highlights the differences in lattice constant, which is too small (1.25 nm) to see in the microscope images (this figure may be seen in color on the included CD-ROM).

One of these is the fabrication of photonic crystal heterostructures [27, 28]. In the field of semiconductor electronics, the fundamental technology involving heterostructures is concerned with optical/electronic devices such as laser diodes and transistors, the importance of which does not need elaboration here. However, in addition to their functions in electronic systems, heterostructures are also playing important, albeit different, roles in photonic crystal research. The purpose of developing heterostructures in photonic crystals is explained below using an in-plane heterostructure [27] in a 2D crystal slab.

Figure 13.4 shows an example of an in-plane heterostructure. This photonic crystal has the same basic structure as that shown in Figure 13.3, except that the lattice constant has been slightly varied to give seven different values. Figure 13.4 also shows an electron microscope image of each part of the sample actually created. The schematic drawing emphasizes the variation in lattice constant, but in the fabricated sample the lattice constant in adjacent areas differed by only 1.25 nm, too small to be discerned from the electron microscope image. One of the promising properties of this type of in-plane heterostructure is the possibility of obtaining an optimized multiwavelength behavior while maintaining the high values of Q (3800) and high drop efficiency (50%) for the configuration shown in Figure 13.3. As described in the previous section, the overall point-defect Q value is determined by a combination of the in-plane value Q_{in} and the vertical value Q_v, and the radiation efficiency is highest when $Q_{in} = Q_v$. If the size of only one point defect is changed, in turn changing the operational wavelength, the effective refractive index of the defect will also be altered, resulting in a change in Q_v. However, Q_{in} will be relatively unaffected; therefore, the overall Q value will be altered and the drop efficiency will change together with the wavelength. The difference in the overall Q value will also cause a change in the wavelength resolution. When a heterostructure such as that in Figure 13.4 is used to vary the structure of the overall crystal proportionally, although the operating wavelength will change with the lattice constant, the Q value and drop efficiency, which are dimensionless quantities, will be maintained at constant values. In other words, multiwavelength operations will be possible as long as the wavelength resolution and drop efficiency are kept constant. Figure 13.5 shows the results of such an experiment. It shows the light emitted from the point defect in each region as the irradiated wavelength is varied. Figure 13.6 shows the corresponding drop spectra. The spectrum for λ_7 has been omitted due to light scattering effects at the waveguide edge. The relationship between the transmission and drop spectra are also shown. It can be seen that the drop wavelength interval was almost constant at $\sim$5 nm, as per the design, and the Q value was maintained at 3800 in all wavelength regions. It is also demonstrated that the drop efficiency was essentially constant for λ_1 to λ_4. Detailed calculations showed that in these wavelength areas a high drop efficiency of $\sim$40% was obtained. In contrast, although the Q values for λ_5 and λ_6 were similar to those for λ_1 to λ_4, it is clear that far greater drop efficiencies than the theoretical maximum of 50% were obtained. The reason for this unexpected observation is concealed in the relationship between the transmission spectrum and drop spectrum in the inset to Figure 13.6. It can be seen that λ_5 and λ_6 are outside the transmission area, that is, the cut-off area. Figure 13.7(a) illustrates this mechanism schematically, where a heterostructure comprising two photonic crystals, PC_m and PC_n, that have different lattice constants is shown: a part of the light introduced into the waveguide in PC_m is transmitted (T) without being trapped by

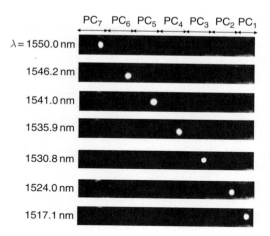

Figure 13.5 Multichannel drop operation of a device with an in-plane heterostructure.

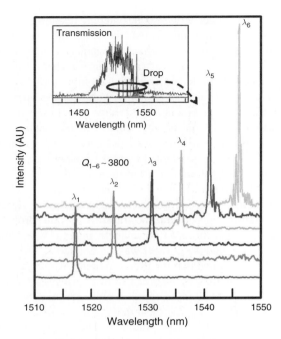

Figure 13.6 Drop spectra of a device with an in-plane heterostructure: the inset shows the relation between transmission and drop spectra (this figure may be seen in color on the included CD-ROM).

the point defect. A different part of the light is trapped by the point defect, but it is reflected (R) to the entrance side of the crystal without being emitted in the upward direction. The rest of the light is emitted upward from the point defect, that is, it drops (D). As described previously, based on coupled mode theory it can be deduced that light D will have a maximum radiation efficiency of 50% and that $T = R = 25\%$. When there is a heterointerface and the emitted light cannot enter PC_n; in other words, a cutoff condition is reached, the transmitted light T will be reflected by the interface. When the reflected light T interferes with R in an out-of-phase fashion, T and R will cancel each other out and, as a result, the efficiency of light D becomes 100%. Figure 13.7(b) shows the calculated radiation efficiency when reflection at the interface is incorporated. The phase difference (θ) between light T reflected by the heterointerface and light R reflected by the defect is shown on the horizontal axis, and a high D efficiency can be obtained due to the broad phase conditions. If θ is selected carefully, it is possible to obtain a light radiation efficiency of 100%. The experimental results for λ_5 and λ_6 in Figure 13.6 shows this effect clearly.

As described above, important effects can be expected from the use of heterostructures, such as dramatically improving the degrees of freedom available in the design of devices, and improving the operation efficiency. As described later, reflection at a heterointerface also plays an important role in the realization of

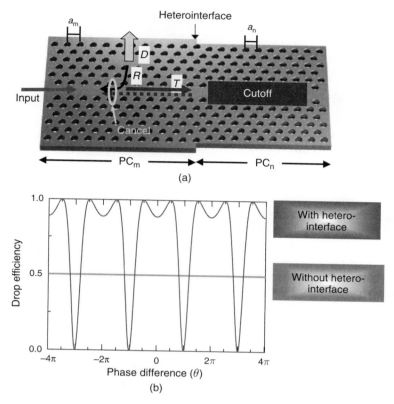

Figure 13.7 Schematic drawing showing the effects of reflection at a heterointerface (a) and the calculated properties of these effects (b). In (a), T, R, and D represent the transmitted light (transmission rate), reflected light (reflection rate), and drop light (drop efficiency), respectively (this figure may be seen in color on the included CD-ROM).

ultrahigh-Q nanocavities. Furthermore, chirped photonic crystal waveguides could be made feasible through heterostructure deformations, and it might also be possible to use such heterostructures in a broad sense for controlling light propagation [29]. In the future, it is expected that heterostructures will continue to gain in importance as a technology for the realization of photonic crystal devices.

High-Q Nanocavities

The devices shown in Figures 13.3 and 13.4 utilize photons that have been emitted to free space from a point-defect cavity. However, if the Q-factor of the point defect could be increased sufficiently, the emission of photons outside the slab plane would be inhibited, even if photons were trapped by the point defect. An in-plane light circuit can then be composed [30].

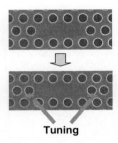

Figure 13.8 Method of enlarging the Q-value of a donor-type defect in which three lattice points are filled (see Figure 13.3) (this figure may be seen in color on the included CD-ROM).

It is thus expected that the development of high-Q nanocavities with very small modal volumes (V) will advance to enable the development of in-plane photonic circuits and to play an extremely important role in a broad range of other applications. Examples include the creation of strong coupling system between photon and electron systems for quantum computing, single photon source for quantum communication, future photonic memory, and supersensitive sensing. The applications envisaged will be based on the extreme strengthening of the interaction between light and matter by strongly trapping light in a very small area over an extended timescale. In the following, discussions on recent advances in high-Q nanocavities are described.

The Q-factor of the point-defect cavity shown in Figure 13.3 was 3800, but recently, as shown in Figure 13.8, a phenomenon has been discovered whereby Q can be increased by at least a factor of 10 by simply tuning the lattice points (air holes) at both edges of the point-defect cavity [31]. Before this phenomenon was found, it was revealed that a major cause of the leakage was light within the cavity being emitted in the direction perpendicular to the slab, due to abrupt reflection at the ends of the cavity along the long axis. To prevent such leakage, it has been found that the prevention of steep changes in the electric field gradient at the edge of the defect is important: this should have a gradual (ideally Gaussian) distribution [31]. The tuning of the lattice points, as depicted in Figure 13.8, corresponds to a change of the Bragg reflection condition at the cavity edge. The Bragg reflection is determined by a summation of partial reflections at a series of lattice points near the cavity. When lattice points near the cavity edges are tuned, the Bragg reflection condition should be modified. Because the phases of partial reflections at the tuned rods are changed, the resultant phase-mismatch weakens the magnitude of Bragg reflection. To compensate for the reduction of the reflection, light is considered to penetrate further inside the Bragg reflection mirror and be reflected perfectly. It means that the electric field profile at the cavity edge becomes gentler. On the basis of the above ideas, nanocavities with various lattice point shifts were created, as shown in Figure 13.9, and the corresponding Q-factors were measured. The highest value of Q (45,000) was obtained when the lattice point was shifted by $\sim 0.15a$ (60 nm), where a is lattice constant. When the modal

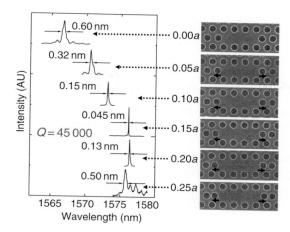

Figure 13.9 Electron microscope images of a nanocavity with various lattice point shifts and corresponding resonance spectra (this figure may be seen in color on the included CD-ROM).

volume was standardized, this value corresponded to a record light confinement effect.

Thereafter, tuning was performed not only for lattice points directly adjacent to the nanocavity, but also for those that were two and three points away. As a result, Q-value of 100,000 was attained [32]. Furthermore, very recently, it was discovered that a photonic double heterostructure (Figure 13.10) that utilizes reflection based on mode gap effects at in-plane heterointerfaces, discussed previously, is able to achieve even higher Q-values of $\sim$1,000,000 [33].

It can be concluded from the above results that, in accordance with the guidelines derived from Gaussian confinement, the use of even simple structures such as

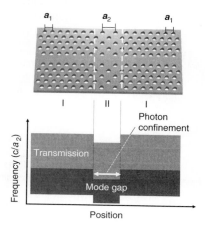

Figure 13.10 Photonic double heterostructure and its photon-confinement mechanism (this figure may be seen in color on the included CD-ROM).

a 2D slab, together with lattice point tuning and the concept of a photonic double heterostructure, mean that the realization of a cavity that does not leak is no longer just a dream.

Introduction of NonLinear and/or Active Functionality

As described above, various advances have been made in 2D photonic crystal slab structures. All of these have occurred based on the use of material with a constant refractive index that forms the photonic crystals, the material used being transparent. As ever, more advanced functionality is demanded, dynamical control of the photonic crystals becomes important. Great effort is currently being applied to realize such control. For example, in Figure 13.3 the resonance (drop) wavelength is fixed, but if the refractive index of the defect is dynamically varied, it becomes possible to tune the resonance wavelength. Changes in the refractive index can be induced by thermal effects and/or carrier plasma effects [34, 35]. Figure 13.11 shows an example of the outcome of such tuning. By introducing light onto the point defect and thermally changing the refractive index of the defect, tuning of the resonance wavelength by up to 5 nm is achieved. This device is extremely compact, so that even the use of a thermal effect allows a dynamic speed to be obtained in a matter of microseconds. A high-speed, nonlinear effect has recently been tried to be implemented. There have also been recent reports of the fabrication of a bistable switch that is operated at a low-input intensity [36].

Advances have also been made in the realization of extremely small active devices by the introduction of active media into a 2D photonic crystal slab. 2D photonic crystal point-defect lasers were reported first in 1999 [6]. Thereafter, various improvements to the structure of the resonator were made, improving the performance. In 2004, a current injection operation into a single point-defect device was carried out [37]. Research of line-defect lasers was continuing at the same time, and, in 2002, the first oscillations from a line-defect laser were

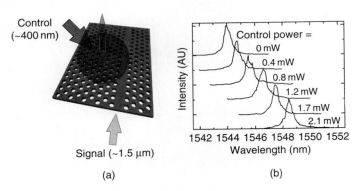

Figure 13.11 Tuning of resonance spectra: (a) schematic representation of light incident on a point defect; (b) changes in resonance spectra for various incident control-light intensities (this figure may be seen in color on the included CD-ROM).

reported [38]. This showed that substantial wavelength tuning was possible by varying the width of the waveguide [39]. Recently, an attempt was made to introduce quantum dots [40, 41] into the high-Q nanocavity shown in Figure 13.8 [42]. Further research and development related to single-photon sources and optical quantum information devices are expected using methods that introduce an active medium.

13.2.2 Three-Dimensional Photonic Crystals

As discussed above, band gap/defect engineering of slab structures in 2D photonic crystals has advanced remarkably. The key to developing slab structures lies in the confinement of light in the direction perpendicular to the slab. However, in 3D photonic crystals, light could also, in principle, be trapped in the direction perpendicular to the crystal using the photonic band gap effect. A 3D photonic crystal is thus the ideal structure and promises the capacity to combine various scientifically interesting effects, including the ultimate control of spontaneous emission. However, because more sophisticated fabrication technology is required compared to that used for 2D photonic crystals, a major focus of current research is on methods of fabrication. The next section briefly touches upon the current state of fabrication technology for 3D photonic crystals, and then introduces some results that prove that it is finally possible to control natural emission, a concern that has been present ever since the beginning of photonic crystal research. The following section also summarizes the current situation regarding light propagation control.

Current State of 3D Photonic Crystal Fabrication Technology

Performing the ultimate band gap/defect engineering necessitates the development of 3D photonic crystals possessing a complete band gap in all directions. The structure shown in Figure 13.1(b) is one of the most common of such structures, and is one of the types being investigated most actively. Crystals displaying most ideal band gap effect in the optical wavelength region were fabricated in 1999 [43]. Figure 13.12 shows an electron microscope image of the structure. This 3D crystal has a so-called woodpile or stacked-stripe structures in which semiconductor stripes (width 200 nm, thickness 200 nm, and period 700 nm) mutually overlap. It has a clear band gap in the wavelength region of 1.55 µm.

Various attempts at developing 3D crystals continued thereafter, including the use of a micromanipulation technique [44]. Crystals with a band gap in the wavelength region 3–4 µm were realized. The development of 3D crystal templates based on direct laser beam delineation has also advanced [45–47]. Using surface micromachining technology, crystals are being developed in the region of optical transmission [48].

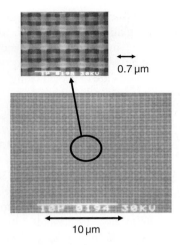

Figure 13.12 Scanning electron microscope image of a 3D photonic crystal operating at the optical transmission wavelengths.

Control of Spontaneous Emission

When the concept of photonic crystals was first introduced, the vision that attracted the most attention was the possibility of fundamentally controlling a light-emitting phenomenon whereby "even if a material that can emit light is introduced into a three-dimensional crystal, the light emission will be restrained from its source by a complete photonic band gap effect. Conversely, when the periodicity of the crystal is artificially disturbed, a strong emission of light from the material will be possible in that disturbance or, in other words, in the defect part" [1, 49]. This vision was the key factor in attracting worldwide interest in the nascent field of photonics. Photonic crystal research advanced in various forms thereafter, but this scientifically important and significant hypothesis remained unproven for a considerable period of time. The following discussion outlines the results [50] that empirically verified the possibility of ultimately controlling emitted light.

Figure 13.13 shows a schematic drawing and an electron microscope image of a 3D photonic crystal that was developed to control emitted light. The basic structure of the crystal is the same as that in Figure 13.1, but it has been designed such that it possesses a photonic band gap in the optical transmission region (1.5 μm wavelength). Moreover, a light-emitting material has been added in the center of the crystal. The light emitter is an InP/InGaAsP quantum well structure generically used for the transmission of light in the 1.5 μm region, corresponding to the photonic band gap. Adjacent to this crystal, as a reference, is a region containing only an emission layer that does not have a photonic crystal structure. The samples that were prepared had striped stacks of five and nine layers in total, including the emission layer.

13. Photonic Crystal Technologies: Experiment

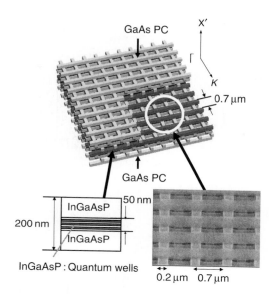

Figure 13.13 Schematic representation and scanning electron microscope image of a 3D photonic crystal developed for the control of light emission. An emission layer has been introduced so that periodicity of the crystal is not disturbed. Two types of samples were fabricated, one with five layers and one with nine layers (this figure may be seen in color on the included CD-ROM).

When the light-emitting substance was first introduced into the center of the 3D photonic crystal, excited by external light and thus induced into a light-emitting state, the photonic crystal was examined to determine what effect it had on that material. Photonic crystal effects were examined at that time by measuring the emission spectra in various directions from the light emitter trapped inside the photonic crystal, being normalized using the emitted spectra of the reference layer. The results are shown in Figure 13.14(a) for the five-layer stacked crystal, and in Figure 13.14(b) for the nine-layer stacked crystal. Even when the number of stacked layers was relatively low at five, light emission was inhibited over a broad wavelength area. When the number of layers was increased to nine, the suppression of the light emission was dramatically increased. For the 1.45–1.6 µm region in particular, a maximum attenuation of 20 dB was achieved. It was clear that this wavelength region corresponded to a complete photonic band gap area of the crystal; the extent of the suppression of light emission very closely matched the calculated values. These results indicated that light emission from material could be inhibited by a photonic band gap effect.

Next, as shown in Figure 13.15(a), an artificial defect of various sizes (i)–(vii) was introduced into the 3D crystal. Figure 13.15(b) shows the emitted light spectra from each defect and for the entire crystal with no defect. It can be seen that there is strong light emission from the defect in comparison with the complete crystal. When the defect was large, as shown in (i), emission with a broad spectrum

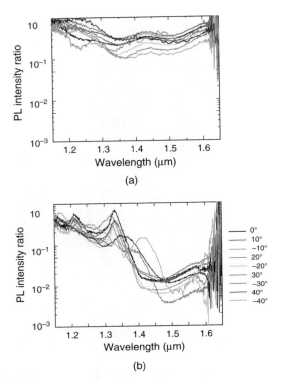

Figure 13.14 Inhibition of light emission from a complete 3D photonic crystal: emission spectra measured at various angles from light emitters introduced into (a) a five-layer stacked crystal and (b) a nine-layer stacked crystal (this figure may be seen in color on the included CD-ROM).

was seen, but as the defect became smaller, the spectral width of the emission narrowed. For the smallest defect (vii), a narrow spectrum was observable. This indicates that, by introducing a defect, a state is generated in the band gap whereby light can be emitted in accordance with the dimensions of the defect, and that the emission from the substance will be strong.

To demonstrate the above phenomenon more directly, Figure 13.16 shows the results of mapping the emission at each site of the photonic crystal, centred on the defect, using the defect in Figure 13.15(a) as an example. Figure 13.16(a) and (b) shows the results when the probe wavelength was set to 1.3 µm (outside the complete band gap region) and 1.55 µm (inside the complete band gap). It is clear that light with a wavelength outside the complete band gap region was emitted from everywhere, no difference being seen between the complete crystal and the defect. However, it is evident that light with a wavelength inside the photonic band gap was emitted only from the defect and was inhibited in the complete crystal. This clearly indicates that light emission was inhibited by the complete 3D photonic crystal, but that light emission was enabled by introducing an artificial defect.

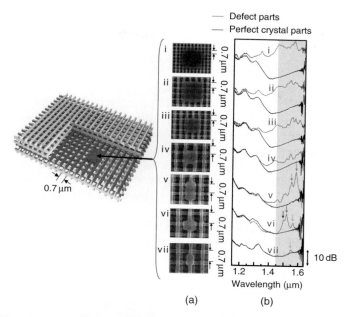

Figure 13.15 Artificial defects introduced into a complete 3D photonic crystal and light emission from them: (a) schematic representations and electron microscope images of various shapes of defects; (b) corresponding emission spectra from the defects (gray line). The black line shows the emission spectra from the complete crystal (this figure may be seen in color on the included CD-ROM).

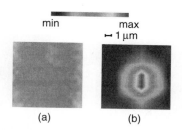

Figure 13.16 Micro PL mapping of emission at different sites of a photonic crystal, with the defect in the center, using defect (v) in Figure 13.15(b), for (a) probe-wavelength set at 1.3 μm (outside the band gap); (b) probe-wavelength set at 1.5 μm (inside the band gap) (this figure may be seen in color on the included CD-ROM).

Here, we should note that very recently, the use of 2D photonic-crystal slab, which can make a quasi-3D confinement of light, has also been carried out [51] for the control of spontaneous emission. A very unique control of spontaneous emission reflecting the 2D nature of the slab has been demonstrated, where the spontaneous emission rate has been suppressed significantly (by a factor of 5 to 15) by the 2D PBG effect, and simultaneously the stored excited carriers have been utilized to emit light to the direction normal to the photonic-crystal plane where the PBG effect does not exist.

Control of Light Propagation

As discussed in "Control of Light by Combined Line/Point-Defect Systems" research on line-defect waveguides in 2D photonic crystal slabs has advanced remarkably. However, light propagation [52] has recently been observed also in 3D photonic crystals. Figure 13.17(a) shows a schematic drawing of a fabricated defect waveguide, created by thickening a single stripe in a 3D crystal. A similar 3D crystal structure to that discussed previously was used and, for the defect part, the single stripe width was thicker than its surroundings. Figure 13.17(b) shows a scanning electron microscope image of a five-layer state in which one GaAs striped layer including the waveguide was fused with a four-layer photonic crystal. Figure 13.17(c) shows a scanning electron microscope image of a cross section of the crystal after its completion. It is clearly shown that a 3D photonic crystal waveguide has been successfully formed. A light propagation experiment on the fabricated samples verified that light can be propagated in a wavelength band of

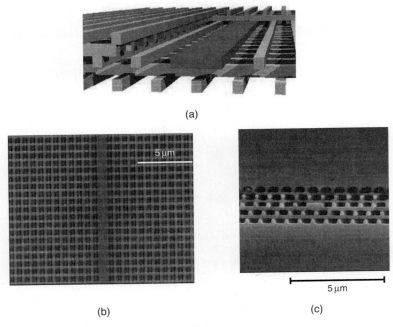

Figure 13.17 3D photonic-crystal waveguide: (a) schematic representation of a fabricated waveguide; (b) electron microscope image of the surface of a five-layer configuration in which a four-layer photonic crystal was combined with a single GaAs striped layer containing a waveguide; (c) scanning electron microscope image of a cross section of the 3D waveguide after fabrication (this figure may be seen in color on the included CD-ROM).

1.5 μm [53]. This experiment represented the first successful propagation of light in a fiber optic wavelength region in 3D crystals. Currently, the number of layers that have been stacked is about nine, so the propagation loss is still relatively large, but it is expected that an increase in the number of layers will cause the propagation loss to decrease exponentially.

13.3 BAND EDGE ENGINEERING

13.3.1 Outline of Resonance Action at the Band Edge

As is well known, distributed feedback semiconductor lasers have a 1D periodic lattice. The propagating waves undergo Bragg diffraction from this lattice and are diffracted in the opposite direction. As a result, the propagating waves merge with the diffracted waves, which in turn causes a standing wave to be set up and a resonator to be formed. This is equivalent to providing a state in which the resonator loss is minimized at the band edge of 1D photonic crystal band gaps, which enables oscillation [54]. By incorporating this idea into the development of photonic crystals with 2D lattices and utilizing the combination of several light waves by Bragg diffraction within the 2D plane, it becomes possible to compose a stable wave state that extends throughout the entire plane [55]. As a result, it is possible to obtain an oscillation mode in which the electromagnetic distribution is fully controlled at each lattice point of the 2D crystal. In other words, it would be possible in a 2D crystal to regulate, not only the longitudinal mode for the laser oscillation, but also the beam pattern—the so-called transverse mode. It would thus be possible to create a laser that could oscillate in single longitudinal and transverse modes at any time, irrespective of the size of the area, and would be possible to create a completely new laser that goes beyond conventional concepts.

Oscillation using current injection based on these principles was first realized in 1999 [55]. Research into the laser action of light pumping using organic dyes has also advanced [56, 57]. Recently, resonators based on this principle have been used in the field of quantum cascade lasers [58]. This shows promise as a powerful laser resonator system in 2D photonic slabs with an introduced active medium [59, 60]. Another feature of devices based on this principle is that the formation of band gaps is unnecessary: such devices can therefore be envisaged for various types of active media and crystal structures.

13.3.2 Broad Area Coherent Laser Action

Figure 13.18 shows an example of a laser structure based on band edge engineering [60–63]. This laser comprises two wafers, A and B. Wafer A includes an active layer, on top of which a photonic crystal has been introduced. By joining wafer A with wafer B, the photonic crystal is enclosed, thereby completing the device.

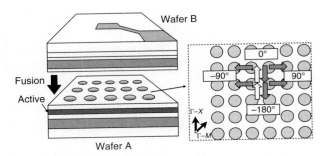

Figure 13.18 Laser structure constructed by band edge engineering (this figure may be seen in color on the included CD-ROM).

The actual material used is an InGaAs/GaAs semiconductor. As can be seen in the inset to Figure 13.18, the photonic crystals used have a square lattice structure and are set up such that the lattice period in the Γ–X direction matches the emitted wavelength in the active layer. Light propagated in a specific Γ–X direction will be diffracted both in the opposite direction (−180°) and at angles of −90° and 90°. The resulting four equivalent light waves will be combined, causing a 2D resonator to form. Figure 13.19(a) shows the corresponding band structure. The resonance

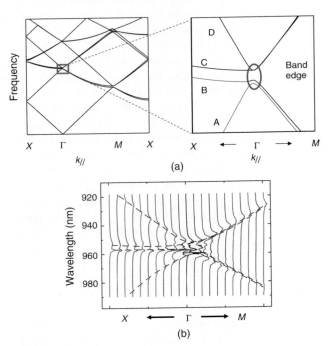

Figure 13.19 (a) Calculated and (b) measured band structures for 2D photonic crystal laser (this figure may be seen in color on the included CD-ROM).

13. Photonic Crystal Technologies: Experiment

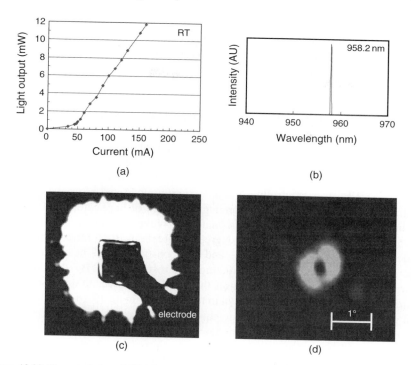

Figure 13.20 Characteristics of 2D photonic crystal laser: (a) current–light output characteristics; (b) resonance spectrum; (c) near-field image; (d) far-field image (this figure may be seen in color on the included CD-ROM).

point is found at the Γ point for the four bands A, B, C, and D in the magnified figure on the right. Of these four band edges, detailed analysis showed that the Γ point in band A has the highest Q-factor, suggesting that it is prone to oscillation [62]. In this device the diffraction effect takes place simultaneously in a direction perpendicular to the surface, thus it could act as a surface-emitting device.

Figure 13.19(b) shows the measured band structure for a device that has actually been fabricated [63]. It is clear that this result matches the band structure shown on the right of Figure 13.19(a) very closely. Figure 13.20(a)–(c) shows the current–light output characteristics, oscillation spectrum, and near-field image of the device, respectively. One can see that despite the size of the electrode being as large as $50 \times 50\,\mu m^2$, single wavelength operation is obtained throughout the device. A comparison of the obtained oscillation wavelength and the band structure in Figure 13.19(b) reveals that oscillation occurred at the Γ point of band A. Figure 13.20(d) shows the far-field pattern measured perpendicular to the surface; the beam pattern has a small divergence angle $<1°–2°$ and a donut shape (the light is polarized in a tangential direction). This beam pattern was in close agreement with one calculated from the electromagnetic field distribution during oscillation at the Γ point of band A.

The investigations described above show that devices can be fabricated that operate at the band edge of 2D photonic band structures and that operate in stable single mode over large areas. The continuous room-temperature oscillation of such a device has been realized [61].

13.3.3 Lattice Point Control and Polarization Mode Control

A donut-shaped far-field pattern has been obtained using a photonic crystal laser, as shown in Figure 13.20(d). This type of beam shape is important for optical tweezers [64] and other applications, such as a light source for omniguide photonic-crystal fibers [65]. However, controlling the polarization mode and beam patterns would be important steps for further expanding the range of applications for this laser. In the following, some of the efforts made to achieve this are described.

Photonic crystals have various degrees of freedom, such as the shape of the lattice points, the lattice intervals, and the construction of heterostructures as discussed in "The Effect of Introducing Heterostructures." By designing these characteristics, it is possible to control the oscillation mode. In the case of lattice points with a circular shape, as shown in the inset to Figure 13.18, four light waves that propagate in a Γ–X direction are equivalent; in a state of resonance, a highly symmetrical (i.e., rotationally symmetrical) electromagnetic field distribution is thus achieved. Consequently, beams that are emitted in a direction perpendicular to the surface possess the different polarized light components equivalently; the polarized light does not become unidirectional. In the center of the pattern, the electromagnetic field is cancelled out and the beam becomes donut-shaped. Here let us consider that lattice point shape changes from round to elliptical. The inset to Figure 13.18 shows that the coupled state of the waves progressing in the 0° and 180° directions and the coupled state of the waves progressing in the +90° and −90° directions are modulated, which causes the symmetry of the in-plane electromagnetic field to break down. As a result, the unidirectional polarized light state is enhanced and the polarization mode becomes controllable [66]. Moreover, it has currently been demonstrated that it is possible to generate a range of beam patterns based on more advanced engineering of photonic crystal structures [67].

Although it has not been discussed here, band edge engineering can not only enhance the characteristics of lasers as mentioned above, but also a range of nonlinear optical effects over large areas. Various types of interesting advances are therefore likely in the near future.

13.4 BAND ENGINEERING

This section discusses control of light by utilizing the dispersion relations in the transmission bands of photonic crystals, where the utilization of the dispersion relations in transmission bands of line-defect waveguides shown in Figure 13.2(a) is included here.

One of the uses of the specific dispersion characteristics of photonic crystals is the super prism phenomenon reported in 1998 [68]. The phenomena mean that the direction of propagating light incident to photonic crystals changes dramatically as a sensitive response to slight changes in the wavelength and incident angle. In addition, when the refractive index contrast between high and low index of the medium that comprises the photonic crystal becomes sufficiently large, the phenomenon of negative refraction will be evident [69], which implies that positive and negative refraction phenomena can be controlled artificially.

It is also expected that the specific dispersion relations of photonic crystals could be used for controlling the propagation of optical pulses, in other words, for the control of pulse speed and shape. In the following, the focus is on the control of group velocity of light.

13.4.1 Control of Group Velocity of Light

Studies on an optical pulse propagation in 3D photonic crystals were first performed in 2000, where bulk 3D photonic crystals were used [70]. Analogous experiments were then performed using 2D photonic crystals with the same bulk shape, which indicated that the group velocity could be controlled [71]. Research then progressed to include the control of group velocity using line-defect waveguides in 2D photonic crystal slabs, as shown in Figure 13.2(a). Figure 13.21(a) shows the dispersion relation of the waveguide in Figure 13.2(a) [9]. A line-defect waveguide mode is formed in the band gap. The area shown in gray is the frequency zone where propagation is in principle possible without radiation loss. In this zone, the slope of the waveguide mode becomes more gradual as the frequency is lowered. It can be seen that the mode edge with a wave number of 0.5 $(2/a)$ is the point at which a standing wave, also discussed in Section 13.3, is constructed. However, at sites of greater frequency than this mode edge, the group velocity of the light became very slow within the condition that it does not reach zero. In 2001, the Fabry–Perot interference present at cleaved facets on both edges of a waveguide was examined, and the group refractive index was estimated based on the wavelength intervals in the interference spectrum [72]. This indicated the possibility that the group velocity could become very slow at the mode edge. By using an ultrashort optical pulse injected into the waveguide with subpicosecond range, it became possible to directly measure the changes in group velocity by varying the frequency of the incident light [73]. Figure 13.21(b) shows the results of these direct measurements. The group velocity was evaluated by measuring the time interval between the pulse that was directly transmitted from the waveguide and the pulse emitted from the waveguide for a second time after it had been reflected by the surface of the waveguide edge and had traveled back and forth within the waveguide. It can clearly be seen from Figure 13.21(b) that the group velocity slows with decreasing distance to the waveguide mode edge.

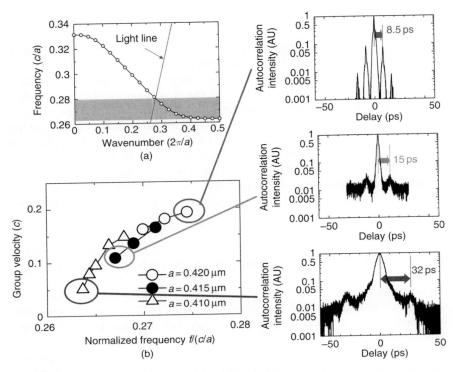

Figure 13.21 Control of group velocity of a light pulse: (a) dispersion relation of the line-defect waveguide in Figure 13.2(a). The right-hand line represents the boundary of the leakage of light that propagates through the waveguide. There is no light leakage at frequencies equal to or lower than this line. The value c on the longitudinal axis represents the speed of the light, and a represents the lattice constant. (b) measured group velocity calculated using an autocorrelation method (this figure may be seen in color on the included CD-ROM).

As mentioned above, this phenomenon is expected to provide basic data relevant to various applications, including pulse shaping and optical delay line function.

13.4.2 Angular Control of Light Propagation

It is easy to see that photonic crystals can play a role as diffraction gratings. By actively utilizing this diffraction grating action, it becomes possible, for example, to improve the light extraction efficiency of light-emitting diodes. From this perspective, great effort is being invested in improving the efficiency of light-emitting diodes using photonic crystal structures. Current interest is especially focused on applications [74–77] for GaN system light-emitting diodes and organic EL elements. The light-emitting efficiency of both of these instruments was reportedly improved by a factor of ~1.5 when photonic crystals were introduced. For such applications, it is important to develop technology that allows photonic crystals to be easily formed. Integration with nanoprocessing techniques such as nanoimprinting is also an extremely interesting development [78, 79].

13.5 SUMMARY AND FUTURE PROSPECTS

Recent advances in photonic crystals have been discussed in terms of (i) band gap and defect engineering, (ii) band edge engineering, and (iii) band engineering. Although many of the explanations given may have been insufficient due to space limitations, it is hoped that the reader has come away with a sense of the steady advances made in the last few years in fundamental technology for operating and controlling light in extremely small volumes, which was the initial expectation of photonic crystals.

It is expected that, over the next 10 years, nanoprocessing technology will dramatically advance, and that more reliable and precise devices will continue to be developed. In the case of 2D photonic crystal slabs, there is promise of remarkable advances in Si-based systems, together with progress in integration with electronic circuits. Further advances can be expected in combined optical and electronic circuits equipped with features such as optical switching, tuning, and delay functionality. It is expected that the main components of such circuits will be optical, and optical/electronic chips will be developed. There is no doubt that the size and power consumption of such devices will be more than tens to hundreds of times smaller than they are now. Advances in a large number of applications can also be expected, including developments in next generation miniaturized multiple-wavelength light sources based on the addition of active functionality, supersensitive sensors, optical memory functions that require a high Q-value, and single photon light sources.

It is expected that band edge lasers will enable surface emission over a large area with a single wavelength, single light polarization, and single spot, and that by controlling the photonic crystal structure the beam pattern will also be fully controllable. It is likely that band edge lasers will be applied in future to a wide range of fields including information processing, communications, and bio-related fields. It is believed that progress in integrating photonic crystals with organic EL and blue LED devices will also be made, and that important advances will occur in highly efficient emission diodes and display technology.

It is expected that the technology available for the fabrication of 3D photonic crystals, which are considered to be more difficult to fabricate than 2D crystals at the present time, will also advance over the coming decade, and that 3D crystals will witness completely new levels of light control that will also enable intricate "complete control of fields."

ACKNOWLEDGMENT

This work was supported by the Core Research for Evolutional Science and Technology (CREST) program of the Japan Science and Technology Agency, by a Grant-in-Aid for Scientific Research of the Ministry of Education, Culture, Sports, Science and Technology of Japan, by the Global COE program of Kyoto University.

REFERENCES

[1] E. Yablonovitch, "Applied physics - How to be truly photonic," *Science*, 289, 557–559, 2000.
[2] S. Noda and T. Baba (eds), *Roadmap on Photonic Crystals*, Kluwer Academic Publishers, Springer, 2003.
[3] K. M. Ho, C. T. Chan, and C. M. Soukoulis, "Existence of a photonic gap in periodic dielectric structures," *Phys. Rev. Lett.*, 65, 3152–3155, 1990.
[4] S. Noda, N. Yamamoto, and A. Sasaki, "New realization method for three-dimensional photonic crystal in optical wavelength region," *Jpn. J. Appl. Phys.*, 35, L909–L912, 1996.
[5] J. D. Joannopoulos, P. R. Villeneuve, and S. Fan, "Photonic crystals: Putting a new twist on light," *Nature*, 386, 143–149, 1997.
[6] O. Painter, R. K. Lee, A. Scherer et al., "Two-dimensional photonic band-gap defect mode laser," *Science*, 284, 1819–1821, 1999.
[7] S. Noda, A. Chutinan, and M. Imada, "Trapping and emission of photons by a single defect in a photonic bandgap structure," *Nature*, 407, 608–610, 2000.
[8] T. Baba, N. Fukaya, and J. Yonekura, "Observation of light propagation in photonic crystal optical waveguides with bends," *Electron. Lett.*, 35, 654–655, 1999.
[9] A. Chutinan and S. Noda, "Waveguides and waveguide bends in two-dimensional photonic crystal slabs," *Phys. Rev. B*, 62, 4488–4492, 2000.
[10] M. Notomi, A. Shinya, K. Yamada et al., "Singlemode transmission within photonic bandgap of width-varied single-line-defect photonic crystal waveguides on SOI substrates," *Electron. Lett.*, 37, 293–295, 2001.
[11] Y. Tanaka, T. Asano, Y. Akahane et al., "Theoretical investigation of a two-dimensional photonic crystal slab with truncated cone air holes," *Appl. Phys. Lett.*, 82, 1661–1663, 2003.
[12] S. Takayama, H. Kitagawa, T. Asano et al., "Experimental demonstration of complete photonic band gap in two-dimensional photonic crystal slabs," *Appl. Phys. Lett.*, 87, Art. No. 061107, 2005.
[13] S. J. McNab, N. Moll, and Y. A. Vlasov, "Ultra-low loss photonic integrated circuit with membrane-type photonic crystal waveguides," *Opt. Express*, 11, 2927–2939, 2003.
[14] Y. Sugimoto, Y. Tanaka, N. Ikeda et al., "Low propagation loss of 0.76 dB/mm in GaAs-based single-line-defect two-dimensional photonic crystal slab waveguides up to 1 cm in length," *Opt. Express*, 12, 1090–1096, 2004.
[15] E. Kuramochi, M. Notomi, S. Hughes et al., "Disorder-induced scattering loss of line-defect waveguides in photonic crystal slabs," *Phys. Rev. B*, 72, Art. No. 161318, 2005.
[16] A. Mekis, J. C. Chen, I. Kurland et al., "High transmission through sharp bends in photonic crystal waveguides," *Phys. Rev. Lett.*, 77, 3787–3790, 1996.
[17] A. Chutinan, M. Okano, and S. Noda, "Wider bandwidth with high transmission through waveguide bends in two-dimensional photonic crystal slabs," *Appl. Phys. Lett.*, 80, 1698–1700, 2002.
[18] Y. Sugimoto, Y. Tanaka, N. Ikeda et al., "Two dimensional semiconductor-based photonic crystal slab waveguides for ultra-fast optical signal processing devices," *IEICE Trans. Electron.*, 87C, 316–327, 2004.
[19] H. Takano, Y. Akahane, T. Asano, and S. Noda, "In-plane-type channel drop filter in a two-dimensional photonic crystal slab," *Appl. Phys. Lett.*, 84, 2226–2228, 2004.
[20] P. I. Borel, A. Harpoth, L. H. Frandsen et al., "Topology optimization and fabrication of photonic crystal structures," *Opt. Express*, 12, 1996–2001, 2004.
[21] E. Miyai, M. Okano, M. Mochizuki, and S. Noda, "Analysis of coupling between two-dimensional photonic crystal waveguide and external waveguide," *Appl. Phys. Lett.*, 81, 3729–3731, 2002;

E. Miyai and S. Noda, "Structural dependence of coupling between a two-dimensional photonic crystal waveguide and a wire waveguide," *J. Opt. Soc. Am. B*, 21, 67–72, 2004.

[22] A. Gomyo, J. Ushida, M. Shirane et al., "Low optical loss connection for photonic crystal slab waveguides," *IEICE Trans. Electron.*, E87C, 328–335, 2004.

[23] T. Shoji, T. Tsuchizawa, T. Watanabe et al., "Low loss mode size converter from 0.3 µm square Si wire waveguides to singlemode fibres," *Electron. Lett.*, 38, 1669–1670, 2002.

[24] M. Notomi, A. Shinya, S. Mitsugi et al., "Waveguides, resonators and their coupled elements in photonic crystal slabs," *Opt. Express*, 12, 1551–1561, 2004.

[25] Y. Akahane, T. Asano, B. S. Song, and S. Noda, "Investigation of high-Q channel drop filters using donor-type defects in two-dimensional photonic crystal slabs," *Appl. Phys. Lett.*, 83, 1512–1514, 2003.

[26] A. Chutinan, M. Mochizuki, M. Imada, and S. Noda, "Surface-emitting channel drop filters using single defects in two-dimensional photonic crystal slabs," *Appl. Phys. Lett.*, 79, 2690–2692, 2001.

[27] B. S. Song, S. Noda, and T. Asano, "Photonic devices based on in-plane hetero photonic crystals," *Science*, 300, 1537, 2003.

[28] S. Kawakami, T. Sato, K. Miura et al., "3-D photonic-crystal hetero structures: Fabrication and in-line resonator," *IEEE Photon. Technol. Lett.*, 15, 816–818, 2003.

[29] D. Mori and T. Baba, "Dispersion-controlled optical group delay device by chirped photonic crystal waveguides," *Appl. Phys. Lett.*, 85, 1101–1103, 2004.

[30] H. Takano, Y. Akahane, T. Asano, and S. Noda, "In-plane-type channel drop filter in a two-dimensional photonic crystal slab," *Appl. Phys. Lett.*, 84, 2226–2228, 2004; H. Takano, B. S. Song, T. Asano, and S. Noda, "Highly efficient multi-channel drop filter in a two-dimensional hetero photonic crystal," *Opt. Express*, 14, 3492–3496, 2006.

[31] Y. Akahane, T. Asano, B. S. Song, and S. Noda, "High-Q photonic nanocavity in a two-dimensional photonic crystal," *Nature*, 425, 944–947, 2003.

[32] Y. Akahane, T. Asano, B. S. Song, and S. Noda, "Fine-tuned high-Q photonic-crystal nanocavity," *Opt. Express*, 13, 1202–1214, 2005.

[33] B.S. Song, S. Noda, and T. Asano, "Ultra-high-Q photonic double-heterostructure nanocavity," *Nat. Mater.*, 4, 207–210, 2005; T. Asano, B. S. Song, and S. Noda, "Analysis of the experimental Q factors (similar to 1 million) of photonic crystal nanocavities," *Opt. Express*, 14, 1996–2002, 2006.

[34] T. Asano, W. Kunishi, M. Nakamura et al., "Dynamic wavelength tuning of channel-drop device in two-dimensional photonic crystal slab," *Electron. Lett.*, 41, 37–38, 2005.

[35] T. Baba, M. Shiga, K. Inoshita, and F. Koyama, "Carrier plasma shift in GaInAsP photonic crystal point defect cavity," *Electron. Lett.*, 39, 1516–1518, 2003.

[36] M. Notomi, A. Shinya, S. Mitsugi et al., "Optical bistable switching action of Si high-Q photonic-crystal nanocavities," *Opt. Express*, 13, 2678–2687, 2005.

[37] H. G. Park, S. H. Kim, S. H. Kwon et al., "Electrically driven single-cell photonic crystal laser," *Science*, 305, 1444–1447, 2004.

[38] A. Sugitatsu and S. Noda, "Room temperature operation of 2D photonic crystal slab defect-waveguide laser with optical pump," *Electron. Lett.*, 39, 123–125, 2003.

[39] A. Sugitatsu, T. Asano, and S. Noda, "Characterization of line-defect-waveguide lasers in two-dimensional photonic-crystal slabs," *Appl. Phys. Lett.*, 84, 5395–5397, 2004.

[40] Y. Arakawa and H. Sakaki, "Multidimensional quantum well laser and temperature-dependence of its threshold current," *Appl. Phys. Lett.*, 40, 939–941, 1982.

[41] M. Tabuchi, S. Noda, and A. Sasaki, "Mesoscopic structure in lattice-mismatched heteroepitaxial interface layers," *Science and Technology of Mesoscopic Structures* edited by Nanba et al., 1992, pp. 379–384; Y. Nabetani, T. Ishikawa, S. Noda, and A. Sasaki, "Initial growth stage and optical-properties of a 3-dimensional inas structure on gaas," *J. Appl. Phys.*, 76, 347–351, 1994.

[42] T. Yoshie, A. Scherer, J. Hendrickson et al., "Vacuum Rabi splitting with a single quantum dot in a photonic crystal nanocavity," *Nature*, 432, 200–203, 2004.

[43] S. Noda, K. Tomoda, N. Yamamoto, and A. Chutinan, "Full three-dimensional photonic bandgap crystals at near-infrared wavelengths," *Science*, 289, 604–606, 2000.

[44] K. Aoki, H. T. Miyazaki, H. Hirayama et al., "Microassembly of semiconductor three-dimensional photonic crystals," *Nat. Mater.*, 2, 117–121, 2003.

[45] M. Campbell, D. N. Sharp, M. T. Harrison et al., "Fabrication of photonic crystals for the visible spectrum by holographic lithography," *Nature*, 404, 53–56, 2000.

[46] V. Mizeikis, K. K. Seet, S. Juodkazis, and H. Misawa, "Three-dimensional woodpile photonic crystal templates for the infrared spectral range," *Opt. Lett.*, 29, 2061–2063, 2004.

[47] M. Deubel, G. V. Freymann, M. Wegener et al., "Direct laser writing of three-dimensional photonic-crystal templates for telecommunications," *Nat. Mater.*, 3, 444–447, 2004.

[48] M. Qi, E. Lidorikis, P. T. Rakich et al., "A three-dimensional optical photonic crystal with designed point defects," *Nature*, 429, 538–542, 2004.

[49] E. Yablonovitch, "Inhibited spontaneous emission in solid-state physics and electronics," *Phys. Rev. Lett.*, 58, 2059–2062, 1987.

[50] S. Ogawa, M. Imada, S. Yoshimoto et al., "Control of light emission by 3D photonic crystals," *Science*, 305, 227–229, 2004 (Published online 3 June 2004 (10.1126/science.1097968)).

[51] M. Fujita, S. Takahashi, Y. Tanaka et al., "Simultaneous inhibition and redistribution of spontaneous light emission in photonic crystals," *Science*, 308, 1296–1298, 2005; K. Kounoike, M. Yamaguchi, M. Fujita et al., "Investigation of spontaneous emission from quantum dots embedded in two-dimensional photonic-crystal slab," *Electron. Lett.*, 41, 1402–1403, 2005.

[52] A. Chutinan and S. Noda, "Highly confined waveguides and waveguide bends in three-dimensional photonic crystal," *Appl. Phys. Lett.*, 75, 3739–3741, 1999; M. Okano, S. Kako, and S. Noda, "Coupling between a point-defect cavity and a line-defect waveguide in three-dimensional photonic crystal," *Phys. Rev. B*, 68, Art. No. 235110, 2003.

[53] M. Imada, L. H. Lee, M. Okano et al., "Development of three-dimensional photonic-crystal waveguides at optical-communication wavelengths," *Appl. Phys. Lett.*, 88, Art. No. 171107, 2006.

[54] H. Kogelnik and C. V. Shank, "Coupled-wave theory of distributed feedback lasers," *J. Appl. Phys.*, 43, 2327–2335, 1972.

[55] M. Imada, S. Noda, A. Chutinan et al., "Coherent two-dimensional lasing action in surface-emitting laser with triangular-lattice photonic crystal structure," *Appl. Phys. Lett.*, 75, 316–318, 1999; M. Imada, A. Chutinan, S. Noda, and M. Mochizuki, "Multidirectionally distributed feedback photonic crystal lasers," *Phys. Rev. B*, 65, Art. No. 195306, 2002.

[56] M. Meier, A. Mekis, A. Dodabalapur et al., "Laser action from two-dimensional distributed feedback in photonic crystals," *Appl. Phys. Lett.*, 74, 7–9, 1999.

[57] M. Notomi, H. Suzuki, and T. Tamamura, "Directional lasing oscillation of two-dimensional organic photonic crystal lasers at several photonic band gaps," *Appl. Phys. Lett.*, 78, 1325–1327, 2001.

[58] R. Colombelli, K. Srinivasan, M. Troccoli et al., "Quantum cascade surface-emitting photonic crystal laser," *Science*, 302, 1374–1377, 2003.

[59] S. H. Kwon, S. H. Kim, S. K. Kim et al., "Small, low-loss heterogeneous photonic bandedge laser," *Opt. Express*, 12, 5356–5361, 2004.

[60] D. Ohnishi, K. Sakai, M. Imada, and S. Noda, "Continuous wave operation of surface emitting two-dimensional photonic crystal laser," *Electron. Lett.*, 39, 612–614, 2003.

[61] D. Ohnishi, T. Okano, M. Imada, and S. Noda, "Room temperature continuous wave operation of a surface-emitting two-dimensional photonic crystal diode laser," *Opt. Express*, 12, 1562–1568, 2004.

[62] M. Yokoyama and S. Noda, "Finite-difference time-domain simulation of two-dimensional photonic crystal surface-emitting laser having a square-lattice slab structure," *IEICE Trans. Electron.*, E87C, 386–392, 2004.

[63] K. Sakai, D. Ohnishi, T. Okano et al., "Lasing band-edge identification for a surface-emitting photonic crystal laser," *IEEE J. Sel. Area Commun.*, 23, 1335–1340, 2005.

[64] A. Ashkin, "History of optical trapping and manipulation of small-neutral particle, atoms, and molecules," *IEEE J. Sel. Top. Quantum Electron.*, 6, 841–856, 2000.

[65] Y. Fink, D. J. Ripin, S. H. Fan et al., "Guiding optical light in air using an all-dielectric structure," *J. Lightw. Technol.*, 17, 2039–2041, 1999.

[66] S. Noda, M. Yokoyama, M. Imada et al., "Polarization mode control of two-dimensional photonic crystal laser by unit cell structure design," *Science*, 293, 1123–1125, 2001.
[67] E. Miyai1, K. Sakai1, T. Okano et al., "Lasers producing tailored beams," *Nature*, 441, 946, 2006.
[68] H. Kosaka, T. Kawashima, A. Tomita et al., "Superprism phenomena in photonic crystals," *Phys. Rev. B*, 58, R10096–R10099, 1998.
[69] M. Notomi, "Theory of light propagation in strongly modulated photonic crystals: Refractionlike behavior in the vicinity of the photonic band gap," *Phys. Rev. B*, 62, 10696–10705, 2000.
[70] T. Tanaka, S. Noda, A. Chutinan et al., "Ultra-short pulse propagation in 3D GaAs photonic crystals," *Technical Digest of International Workshop on Photonic and Electromagnetic Crystal Structure (PESC-II)*, W4-14, 2000; *Opt. Quantum Electron.*, 34, 37–43, 2002.
[71] K. Inoue, N. Kawai, Y. Sugimoto et al., "Observation of small group velocity in two-dimensional AlGaAs-based photonic crystal slabs," *Phys. Rev. B*, 65, Art. No. 121308, 2002.
[72] M. Notomi, K. Yamada, A. Shinya et al., "Extremely large group-velocity dispersion of line-defect waveguides in photonic crystal slabs," *Phys. Rev. Lett.*, 87, Art. No. 253902, 2001.
[73] T. Asano, K. Kiyota, D. Kumamoto et al., "Time-domain measurement of picosecond light-pulse propagation in a two-dimensional photonic crystal-slab waveguide," *Appl. Phys. Lett.*, 84, 4690–4692, 2004.
[74] K. Orita, S. Tamura, T. Takizawa et al., "High-extraction-efficiency blue light-emitting diode using extended-pitch photonic crystal," *Jpn. J. Appl. Phys.*, 43, 5809–5813, 2004.
[75] M. Fujita, T. Ueno, T. Asano et al., "Organic light-emitting diode with ITO/organic photonic crystal," *Electron. Lett.*, 39, 1750–1752, 2003.
[76] Y. R. Do, Y. C. Kim, Y. W. Song et al., "Enhanced light extraction from organic light-emitting diodes with 2D SiO2/SiNx photonic crystals," *Adv. Mater.*, 15, 1214–1218, 2003.
[77] M. Fujita, T. Ueno, K. Ishihara et al., "Reduction of operating voltage in organic light-emitting diode by corrugated photonic crystal structure," *Appl. Phys. Lett.*, 85, 5769–5771, 2004.
[78] A. Yokoo, H. Suzuki, and M. Notomi, "Organic photonic crystal band edge laser fabricated by direct nanoprinting," *Jpn. J. Appl. Phys.*, 43, 4009–4011, 2004.
[79] K. Ishihara, M. Fujita, I. Matsubara et al., "Direct fabrication of photonic crystal on glass substrate by nanoimprint lithography," *Jpn. J. Appl. Phys.*, 45, L210–L212, 2006.

14

Photonic crystal fibers: Basics and applications

Philip St John Russell

Max-Planck Research Group (IOIP)
University of Erlangen-Nuremberg, Erlangen, Germany

14.1 INTRODUCTION

Photonic crystal fibers (PCFs)—fibers with a periodic transverse microstructure—first emerged as practical low-loss waveguides in early 1996 [1, 2]. The initial demonstration took 4 years of technological development, and since then the fabrication techniques have become more and more sophisticated. It is now possible to manufacture the microstructure in air–glass PCF to accuracies of 10 nm on the scale of 1 μm, which allows remarkable control of key optical properties such as dispersion, birefringence, nonlinearity, and the position and width of the photonic band gaps (PBGs) in the periodic "photonic crystal" cladding. PCF has in this way extended the range of possibilities in optical fibers, both by improving well-established properties and by introducing new features such as low-loss guidance in a hollow core.

In this chapter, the properties of the various types of PCF (the main classes of PCF structure are illustrated in Figure 14.1) are first introduced, followed by a detailed discussion of their established or emerging applications. An account of the basic characteristics of PCF is available in [3].

14.2 FABRICATION TECHNIQUES

Photonic crystal fiber structures are currently produced in many laboratories worldwide using a variety of different techniques (see Figure 14.2 for some example structures). The first stage is to produce a "preform"—a macroscopic version of the planned microstructure in the drawn PCF. There are many ways to do this, including stacking of capillaries and rods [4], extrusion [5–8], sol-gel

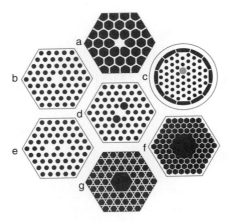

Figure 14.1 Representative sketches of different types of PCF. The black regions are hollow, the white regions are pure glass, and the grey regions are doped glass. (a) highly nonlinear PCF (high-air-filling fraction, small core); (b) endlessly single-mode solid-core PCF; (c) double-clad PCF with offset doped lasing core and high-numerical aperture inner cladding for pumping (the photonic crystal cladding is held in place by thin webs of glass); (d) birefringent PCF; (e) dual-solid-core PCF; (f) 19-cell hollow-core PCF; (g) hollow-core PCF with a Kagomé cladding lattice. PBG-guiding all-solid glass versions of (b) have been made in which the hollow channels are replaced with raised-index doped glass strands.

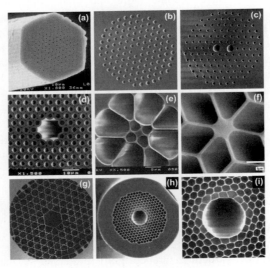

Figure 14.2 Selection of scanning electron micrographs of PCF structures. (a) The first working PCF—the solid glass core is surrounded by a triangular array of 300-nm-diameter air channels, spaced 2.3 μm apart [4]; (b) endlessly single-mode solid-core PCF (interhole spacing ∼2 μm; Erlangen); (c) polarization-preserving solid-core PCF (Erlangen); (d) the first hollow-core PCF [11]; (e) a PCF extruded from Schott SF6 glass with a core ∼2 μm in diameter [7]; (f) highly nonlinear PCF (core diameter 800 nm [12]); (g) hollow-core PCF with Kagomé lattice (Erlangen); (h) hollow-core PCF guiding by a PBG (core diameter 20 μm; BlazePhotonics Limited); (i) detail of (h).

casting [9], injection molding, and drilling. The materials used ranged from silica to compound glasses, chalcogenide glasses, and polymers [10].

The most widely used technique is stacking of circular capillaries. Typically, meter-length capillaries with an outer diameter of ~1 mm are drawn from a starting tube of high-purity synthetic silica with a diameter of ~20 mm. The inner/outer diameter of the starting tube, which typically lies in the range from 0.3 up to beyond 0.9, largely determines the d/Λ (hole diameter to spacing ratio) value in the drawn fiber. The capillaries are stacked horizontally, in a suitably shaped jig, to form the desired crystalline arrangement. The stack is bound with wire before being inserted into a jacketing tube, and the whole assembly is then mounted in the preform feed unit for drawing down to fiber. Judicious use of pressure and vacuum during the draw allows some limited control over the final structural parameters, for example, the d/Λ value.

14.2.1 Design Approach

The successful design of a PCF for a particular application is not simply a matter of using numerical modeling to calculate the parameters of a structure that yields the required performance. This is because the fiber-drawing process is not lithographic, but introduces its own (highly reproducible) types of distortion through the effects of viscous flow, surface tension, and pressure. As a result, even if the initial preform stack precisely mimics the theoretically required structure, several modeling and fabrication iterations are usually needed before a successful design can be reached.

14.3 CHARACTERISTICS OF PHOTONIC CRYSTAL CLADDING

The simplest photonic crystal cladding is a biaxially periodic, defect-free, composite material with its own well-defined dispersion and band structure. These properties determine the behavior of the guided modes that form at cores (or "structural defects" in the jargon of photonic crystals). A convenient graphical tool is the propagation diagram—a map of the ranges of frequency and axial wavevector component β, where light is evanescent in all transverse directions regardless of its polarization state (Figure 14.3) [13]. The vertical axis is the normalized frequency $k\Lambda$ (vacuum wavevector k), and the horizontal axis is the normalized axial wavevector component $\beta\Lambda$. Light is unconditionally cutoff from propagating (due to either total internal reflection (TIR) or PBG) in the black regions.

In any sub region of isotropic material (glass or air) at fixed optical frequency, the maximum possible value of $\beta\Lambda$ is given by $k\Lambda n$, where n is the refractive index (at that frequency) of the region under consideration. For $\beta < kn$ light is free to propagate, for $\beta > kn$ it is evanescent, and at $\beta = kn$ the critical angle is reached—denoting the onset of TIR for light incident from a medium of index larger than n.

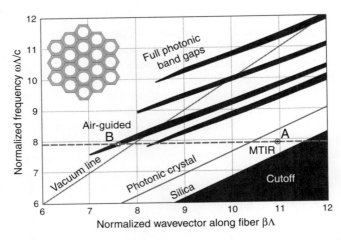

Figure 14.3 Propagation diagram for a PCF with 45% air-filling fraction (after Ref. [13]). Note the different regions of propagation, separated by the slanting straight lines. On the left-hand side of each line, light is able to propagate freely in the relevant region of the fiber structure. On the right-hand side of the silica line, light is completely cutoff from propagating in the structure. The black "fingers" indicate the positions of full 2D photonic band gaps. At point A, guidance can occur in a solid-glass core by a modified form of TIR. At point B, a guided mode can form in a hollow core provided its area is sufficiently large.

The slanted guidelines (Figure 14.3) denote the transitions from propagation to evanescence for air, the photonic crystal, and glass. At fixed optical frequency for $\beta < k$, light propagates freely in every subregion of the structure. For $k < \beta < kn_g$ (n_g is the index of the glass), light propagates in the glass subregions and is evanescent in the hollow regions. Under these conditions, the "tight binding" picture approximately holds, and the structure may be viewed as an array of coupled glass waveguides.

14.3.1 Maximum Refractive Index

The maximum axial refractive index $n_{max} = \beta_{max}/k$ in the photonic crystal cladding lies in the range $k < \beta < kn_g$ as expected of a composite glass–air material. Its value depends strongly on frequency, even when neither the air nor the glass is dispersive; microstructuring itself creates dispersion, through a balance between transverse energy storage and energy flow that is highly dependent upon frequency. By averaging the square of the refractive index in the photonic crystal cladding, it is simple to show that

$$n_{max} \rightarrow n_{max}^{\infty} = \sqrt{(1-F)n_g^2 - Fn_a^2} \qquad (14.1)$$

14. Photonic Crystal Fibers: Basics and Applications

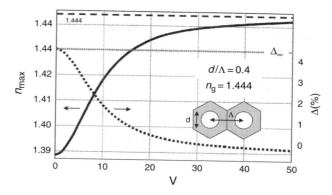

Figure 14.4 Maximum axial refractive index in the photonic crystal cladding as a function of the frequency parameter $v = k\Lambda \sqrt{n_g^2 - 1}$ for $d/\Lambda = 0.4$ and $n_g = 1.444$ (the index of silica at 1550 nm). For this filling fraction of air (14.5%), the value at long wavelengths ($V \to 0$) is $n_{max} = 1.388$, in agreement with Eqn (14.1). The values of Δ are also plotted (dotted curve).

in the long-wavelength limit $k\Lambda \to 0$ for a scalar approximation, where F is the air-filling fraction and n_a is the index in the holes (which we take to be 1 in what follows).

As the wavelength of the light falls, the optical fields are better able to distinguish between the glass regions and the air. The light piles up more and more in the glass, causing the effective n_{max} "seen" by it to change. In the limit of small wavelength $k\Lambda \to \infty$, light is strongly excluded from the airholes by TIR, and the field profile "freezes" into a shape that is independent of wavelength. The resulting frequency dependence of the maximum cladding index in a particular case is shown in Figure 14.4 (calculated using the approximate analysis in Ref. [14]).

14.3.2 Transverse Effective Wavelength

The transverse effective wavelength in the *i*th material is defined as follows:

$$\lambda_{\text{eff}}^i = \frac{2\pi}{\sqrt{k^2 n_i^2 - \beta^2}} \tag{14.2}$$

where n_i is its refractive index. This wavelength can be many times the vacuum value, tending to infinity at the critical angle $\beta \to kn_i$, and being imaginary when $\beta > kn_i$. It is a measure of whether the light is likely to be resonant within a particular feature of the structure, for example, a hole or a strand of glass, and defines PCF as a wavelength-scale structure.

14.3.3 Photonic Band Gaps

Full 2D PBGs exist in the black finger-shaped regions in Figure 14.3. Some of these extend into the region $\beta < k$, where light is free to propagate in vacuum, confirming the feasibility of trapping light within a hollow core.

14.4 CHARACTERISTICS OF GUIDANCE

In single-mode fiber (SMF), guided modes form within the range of axial refractive indices $n_{cl} < n_z < n_{co}$, when light is evanescent in the cladding ($n_z = \beta/k$; core and cladding indices are n_{co} and n_{cl}). In PCF, three distinct guidance mechanisms exist: a modified form of TIR [14, 15], PBG guidance [11, 16], and a leaky mechanism based on a low density of photonic states in the cladding [17]. In the following subsections, we explore the role of resonance and antiresonance, and discuss chromatic dispersion, attenuation mechanisms, and guidance in cores with refractive indices raised and lowered relative to the "mean" cladding value.

14.4.1 Classification of Guidance

A useful classification of the different types of PCF is based on the sign of the core-cladding refractive index difference $\Delta = (n_{co} - n_{cl})/n_{cl}$, which takes the value Δ_∞ in the long-wavelength limit [using Eqn (14.1) to estimate the core and cladding indices]. When Δ_∞ is positive, TIR operates, at least at longer wavelengths, and when Δ_∞ is negative a PBG is the only viable low-loss confinement mechanism. Of course, PBGs can also appear, over selected wavelength ranges, for positive or even zero values of Δ_∞.

14.4.2 Positive Core-Cladding Index Difference $\Delta_\infty > 0$

This type of PCF may be defined as one where the mean cladding refractive index in the long-wavelength limit $k \to 0$ (Eqn (14.5)) is lower than the core index (in the same limit). Under the correct conditions, PBG guidance may also occur in this case, although experimentally the TIR-guided modes will dominate. A striking feature of solid-core PCFs with $d/\Lambda < 0.43$ is that they are "endlessly single mode" (ESM), that is, the core never becomes multimode, no matter how short the wavelength of the light [14]. Although the guidance in some respects resembles conventional TIR, it turns out to have some interesting and unique features that distinguish it markedly from step-index fiber. These are due to the piecewise discontinuous nature of the core boundary—sections where (for $n_z > 1$) airholes strongly block the escape of light are interspersed with regions of barrier-free glass. In fact, the cladding operates in a regime where the transverse effective wavelength in silica (Eqn (14.8)) is comparable with the size of the glass substructures in the cladding. The zone of operation in Figure 14.3 is $n_{\max} < n_z < n_g$ (point A).

ESM behavior can be understood by viewing the array of holes as a modal filter or "sieve." The fundamental mode in the glass core has a transverse effective wavelength $\lambda_{\text{eff}}^g \approx 4\Lambda$. It is thus unable to "squeeze through" the glass channels between the holes, which are $\Lambda - d$ wide and thus below the Rayleigh resolution limit $\approx \lambda_{\text{eff}}^g/2 = 2\Lambda$. Provided the relative hole size d/Λ is small enough, higher order modes are able to escape—their transverse effective wavelength is shorter so they have higher resolving power. As the holes are made larger, successive higher order modes become trapped.

It may also be explained by the strong dispersion of the refractive index in the photonic crystal cladding, which forces the core-cladding index step $\Delta(\lambda)$ to fall as the wavelength gets shorter (Figure 14.4) [14, 15]. This counteracts the usual trend toward increasingly multimode behavior at shorter wavelengths. In the limit of very short wavelength, the light strikes the glass–air interfaces at glancing incidence, and is strongly rejected from the air holes. In this regime, the transverse single-mode profile freezes to a constant shape, independent of wavelength. As a consequence, the angular divergence (roughly twice the numerical aperture) of the emerging light is proportional to wavelength—in SMFs it is approximately constant owing to the appearance of more and more higher order guided modes as the frequency increases.

Thus, the refractive index of the photonic crystal cladding increases with optical frequency, tending toward the index of silica glass in the short-wavelength limit. If the core is made from a glass of refractive index lower than that of silica (e.g., fluorine-doped silica), guidance is lost at wavelengths shorter than a certain threshold value [18]. Such fibers have the unique ability to prevent transmission of short-wavelength light—in contrast to conventional fibers which guide more and more modes as the wavelength falls.

Ultralarge Area Single Mode

The modal filtering in ESM-PCF is controlled only by the geometry (d/Λ for a triangular lattice). A corollary is that the behavior is quite independent of the absolute size of the structure, permitting single-mode fiber (SMF) cores with arbitrarily large areas. A single-mode PCF with a core diameter of 22 μm at 458 nm was reported in 1998 [19]. In conventional step-index fibers, where $V < 2.405$ for single-mode operation, this would require uniformity of core refractive index to ~1 part in 10^5—very difficult to achieve if modified chemical vapor-deposition (MCVD) is used to form the doped core.

Fibers with Multiple Cores

The stacking procedure makes the production of multicore fiber quite straightforward. A preform stack is built up with a desired number of solid (or hollow) cores and drawn down to fiber in the usual manner [20]. The coupling strength between the cores depends on the sites chosen, because the evanescent decay rate of the fields changes with azimuthal direction. Multicore structures have many potential applications, from high-power fiber lasers (see Section 14.6.2) to imaging [21] and curvature sensing [22].

14.4.3 Negative Core-Cladding Index Difference $\Delta_\infty < 0$

Because TIR cannot operate under these circumstances, low-loss waveguiding is only possible if a PBG exists in the range $\beta < kn_{\text{core}}$.

Hollow-Core Silica/air

In silica–air PCF, larger air-filling fractions and small inter-hole spacings are necessary to achieve PBGs in the region $\beta < k$. The relevant operating region in Figure 14.3 is to the left of the vacuum line, inside one of the band gap fingers (point B). These conditions ensure that light is free to propagate, and form guided modes, within the hollow core while being unable to escape into the cladding. The number N of such modes is controlled by the depth and width of the refractive index "potential well" and is approximately given by

$$N \approx k^2 \rho^2 (n_{\text{high}}^2 - n_{\text{low}}^2)/2 \qquad (14.3)$$

where n_{high} and n_{low} are the refractive indices at the edges of the PBG at fixed frequency and ρ is the core radius. Because the band gaps are quite narrow ($n_{\text{high}}^2 - n_{\text{low}}^2$ is typically a few percent), the hollow core must be sufficiently large if a guided mode is to exist at all. In the first hollow-core PCF, reported in 1999 [11], the core was formed by omitting seven capillaries from the preform stack (Figure 14.2(d)). Electron micrographs of a more recent structure, with a hollow core made by removing 19 missing capillaries from the stack, are shown in Figures 14.2(h) and 14.2(i) [23].

Higher Refractive Index Glass

Achieving a band gap in higher refractive index glasses for $\beta < k$ presents at first glance a dilemma. Whereas a larger refractive index contrast generally yields wider band gaps, the higher "mean" refractive index seems likely to make it more difficult to achieve band gaps for incidence from vacuum. Although this argument holds in the scalar approximation, the result of calculations show that vector effects become important at higher levels of refractive index contrast (e.g.,

2:1 or higher), and a new species of band gap appears for smaller filling fractions of air than in silica-based structures. The appearance of this new type of gap means that it is actually easier to make hollow-core PCF from higher index glasses such as tellurites or chalcogenides [24].

Surface States on Core-Cladding Boundary

Similar guided modes are commonly seen in hollow-core PCF, where they form surface states (related to electronic surface states on semiconductor crystals) on the rim of the core, confined on the cladding side by PBG effects. These surface states become phase-matched to the air-guided mode at certain wavelengths, creating couplings (anticrossings on the frequency–wavevector diagram) that perturb the group velocity dispersion (GVD) and contribute additional attenuation (see the absorption and scattering section) [25–27].

All-Solid Structures

In all-solid PBG-guiding fibers, the core is made from low-index glass and is surrounded by an array of high-index glass strands [28–30]. Because the mean core-cladding index contrast is negative, TIR cannot operate, and PBG effects are the only possible guidance mechanism. These structures have some similarities with 1D "ARROW" structures, where antiresonance plays an important role [31].

When the cladding strands are antiresonant, light is confined to the central low-index core by a mechanism not dissimilar to the modal filtering picture in Section 14.4.2; the high-index cores act as the "bars of a cage," so that no features in the cladding are resonant with the core mode, resulting in a low-loss guided mode. Guidance is achieved over wavelength ranges that are punctuated with high-loss windows where the cladding 'bars' become resonant. Remarkably, it is possible to achieve PBG guidance by this mechanism even at index contrasts smaller than 1% [28, 32], with losses as low as 20 dB/km at 1550 nm [33].

Low Leakage Guidance

The transmission bands are greatly widened in hollow-core PCFs with a kagomé cladding lattice [17] [Figures 14.1(g) and 14.2(g)]. The best examples of such fibers show a minimum loss of $\sim$1 dB/m over a bandwidth of several hundred nanometers. Numerical simulations show that, although the cladding structure supports no band gaps, the density of states is greatly reduced near the vacuum line. The consequential poor overlap between the core states, together with the greatly reduced number of cladding states, appears to inhibit the leakage of light without completely preventing it. There is also a large number of anticrossings between the core mode and the multiple states in the cladding, causing rapid changes in dispersion over narrow wavelength bands within the transmission window. The precise mechanism of low-loss guidance in kagomé PCF is a continuing topic of debate.

14.4.4 Birefringence

The guided modes of a perfect sixfold symmetric core-plus-cladding structure are not birefringent [34]. Birefringence can result from residual strain following the drawing process, or slight accidental distortions in geometry, which can cause significant birefringence because of the large glass–air index difference. If, however, the core is deliberately distorted so as to become twofold symmetric, extremely high values of birefringence can be achieved. For example, by introducing capillaries with different wall thicknesses above and below a solid glass core [Figures 14.1(d) and 14.2(c)], values of birefringence some 10 times larger than in conventional fibers can be obtained [12]. Hollow-core PCF with moderate levels of birefringence ($\sim 10^{-4}$) can be realized either by forming an elliptical core or by adjusting the structural design of the core surround [35, 36]. By suitable design, it is even possible to realize a strictly single-polarization PCF in which one polarization state is cutoff from guidance over a certain range of wavelengths [37].

Experiments show that the birefringence in pure silica PCF is some 100 times less sensitive to temperature variations than in conventional fibers, a feature that is important in many applications [38]. This is because traditional "polarization maintaining" fibers (bow tie, elliptical core, or Panda) contain at least two different glasses, each with a different thermal expansion coefficient. In such structures, the resulting temperature-dependent stresses make the birefringence a strong function of temperature.

14.4.5 Group Velocity Dispersion

Group velocity dispersion is a factor crucial in the design of telecommunications systems and in all kinds of nonlinear optical experiments. PCF offers greatly enhanced control of the magnitude and sign of the GVD as a function of wavelength. In many ways, this represents an even greater opportunity than a mere enhancement of the effective nonlinear coefficient (which PCF can also deliver).

Solid Core

As the optical frequency increases, the GVD in SMF changes sign from anomalous ($D > 0$) to normal ($D < 0$) at ~ 1.3 µm. In solid-core PCF, as the holes get larger, the core becomes increasingly separated from the cladding, until it resembles an isolated strand of silica glass. If the whole structure is made very small (core diameters less than 1 µm have been made), the zero dispersion point of the fundamental guided mode can be shifted to wavelengths in the visible [39, 40]. For example, the PCF in Figure 14.2(f) has a dispersion zero at 560 nm.

By careful design, the wavelength dependence of the GVD can also be reduced in PCFs at much lower air-filling fractions. Figure 14.5 shows the flattened GVD profiles of three PCFs with cores several microns in diameter [41, 42]. These fibers

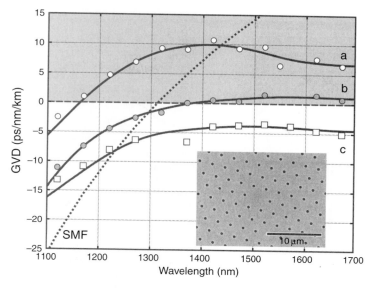

Figure 14.5 Group velocity dispersion profiles, against wavelength, for three different PCFs designed to have low-level ultraflattened GVD [41, 42]. The curve for Corning SMF-28 is included for comparison.

operate in the regime where SMF is multimoded. Although the fundamental modes in both SMF and PCF have similar dispersion profiles, the presence of higher order modes (not guided in the PCF, which is ESM) makes the use of SMF impractical.

A further degree of freedom in GVD design may be gained by working with multicomponent glasses, such as Schott SF6 (Figure 14.2(e)), where the intrinsic zero dispersion point occurs at $\sim 1.8\,\mu m$ [7]. In highly nonlinear small-core PCF, this shifts the whole dispersion landscape to longer wavelengths than in a silica-based PCF with the same core size and geometry.

Hollow Core

Hollow-core fiber behaves in many respects rather like a circular–cylindrical hollow metal waveguide, which has anomalous dispersion (the group velocity increases as the frequency increases). The main difference, however, is that the dispersion changes sign at the high-frequency edge, owing to the approach of the photonic band edge and the weakening of the confinement (Figure 14.6) [43].

14.4.6 Attenuation Mechanisms

An advantage common to all fibers is the very large extension ratio from preform to fiber, which has the effect of smoothing out imperfections, resulting in a

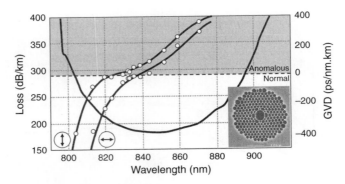

Figure 14.6 Measured attenuation and GVD spectra for a hollow-core PCF designed for 850 nm transmission [43]. The core is slightly elliptical (see scanning electron micrograph in inset), so the dispersion in each eigenstate of polarization is different.

transverse structure that is extremely invariant with distance along the fiber. This is the chief reason for the ultralow attenuation displayed by fibers compared with other waveguide structures. In PCF the losses are governed by two main parameters: the fraction of light in glass and roughness at the glass–air interfaces. The light-in-glass fraction can be controlled by judicious design, and ranges from close to 100% in solid-core fibers to less than 1% in the best hollow-core fibers.

Absorption and Scattering

The best reported loss in solid-core PCF, from a group in Japan, stands at 0.28 dB/km at 1550 nm, with a Rayleigh scattering coefficient of 0.85 dB/km/μm^4. A 100-km length of this fiber was used in the first PCF-based penalty-free dispersion-managed soliton transmission system at 10 Gb/s [44, 45]. The slightly higher attenuation compared with SMF is due to roughness at the glass–air interfaces [26].

It is hollow-core PCF, however, that has the greatest potential for extremely low loss, because the light travels predominantly in empty (or gas-filled) space. Although the best reported attenuation in hollow-core PCF stands at 1.2 dB/km [23], values below 0.2 dB/km or even lower seem feasible with further development of the technology. The prospect of improving on conventional fiber, at the same time greatly reducing the nonlinearities associated with a solid glass core, is intriguing. By using infrared (IR) glasses, transmission can be extended into the IR [46], and recent work shows that silica hollow-core PCF can even be used with acceptable levels of loss in the mid-IR [47], owing to the very low overlap between the optical field and the glass.

In the latest hollow-core silica PCF, with loss levels approaching 1 dB/km at 1550 nm, very small effects can contribute significantly to the attenuation floor. The ultimate loss limit in such fibers is determined by surface roughness caused by

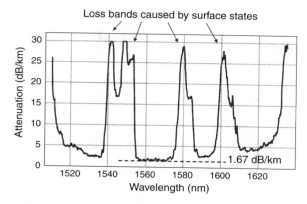

Figure 14.7 Attenuation spectrum of a typical ultralow loss hollow-core PCF designed for operation in the 1550 nm telecommunications band. [see scanning electron micrograph in Figure 14.2(h)].

thermally driven capillary waves, which are present at all length scales. These interface ripples freeze in when the fiber cools, introducing scattering losses for modes that are concentrated at the interfaces, such as surface modes guided on the edge of the core. The pure core mode does not itself "feel" the ripples very strongly, except at anticrossing wavelengths where it becomes phase-matched to surface modes, causing light to move to the surface and experience enhanced scattering.

The result is a transmission spectrum consisting of windows of high transparency punctuated with bands of high attenuation (Figure 14.7). This picture has been confirmed by measurements of the surface roughness in hollow-core PCFs, the angular distribution of the power scattered out of the core, and the wavelength dependence of the minimum loss of fibers drawn to different scales [23]. The thin glass shell surrounding the hollow core can be designed to be antiresonant with the core mode, permitting further exclusion of light from the glass [48].

Ignoring material dispersion, the whole transmission landscape shifts linearly in wavelength in proportion to the overall size of the structure (a consequence of Maxwell's equations). This means that the smallest loss at a given wavelength will be obtained by drawing a fiber to a particular diameter. The optical overlap with the surface roughness scales inversely with the size with the fiber, and the scattering itself may be regarded as being governed by the density of states into which scattering can occur, which in three-dimensions scales as λ^{-2}. Thus, the wavelength of minimum loss scales as λ^{-3}, in contrast to the λ^{-4} dependence of Rayleigh scattering in bulk glass.

Bend Loss

Conventional fibers suffer additional loss if bent beyond a certain critical radius R_{crit}, which depends on wavelength, core-cladding refractive index step, and—most notably—the third power of core radius, a^3 [49]. For wavelengths longer than a certain value (the "long-wavelength bend edge"), all guidance is effectively lost.

A starting point for understanding bend loss in solid-core ESM-PCF (perhaps the most interesting case) is the long-wavelength limit. ESM behavior occurs when $d/\Lambda < 0.43$, which sets the highest air-filling fraction at 16.8% and yields an area-averaged cladding refractive index of 1.388 (silica index of 1.444)—valid in the limit $k \to 0$. This index step is some 10 times higher than in Corning SMF-28, making ESM-PCF relatively much less susceptible to bend loss at long wavelengths. For a step-index fiber with a Ge-doped core, 40 mol% of GeO_2 would be needed to reach the same index step (assuming 0.0014 index change per mol% GeO_2 [50]). The result is that the long-wavelength bend edge in ESM-PCF is in the (IR) beyond the transparency window of silica glass, even when the core radius is large [51].

ESM-PCF also exhibits a *short*-wavelength bend edge, caused by bend-induced coupling from the fundamental to higher order modes, which of course leak out of the core [14]. The critical bend radius for this loss varies as:

$$R_c \sim \Lambda^3/\lambda^2 \tag{14.4}$$

compared with $R_c \sim \lambda$ for SMF. The reciprocal dependence on λ^2 makes it inevitable that a short-wavelength bend edge will appear in ESM-PCF. It turns out that a step-index fiber with core-cladding index difference Δ_∞ (i.e., the value for an ESM-PCF in the long-wavelength limit) is multimode over wide parameter ranges where ESM-PCF has negligible bend loss.

In contrast, hollow-core PCF is experimentally very insensitive to bend loss—in many cases no appreciable drop in transmission is observed until the fiber breaks. This is because the effective depth of "potential well" for the guided light (see Section "Hollow-Core Silica/Air"), given by the distance $\Delta\beta$ between the edges of the PBG, is substantially larger than in SMF.

Confinement Loss

The photonic crystal cladding in a realistic PCF is of course finite in extent. For a guided mode, the Bloch waves in the cladding are evanescent, just like the evanescent plane waves in the cladding of a conventional fiber. If the cladding is not thick enough, the evanescent field amplitudes at the cladding/coating boundary can be substantial, causing attenuation. In the solid-core case for small values of d/Λ, the resulting loss can be large unless a sufficiently large number of periods is used [41].

Very similar losses are observed in hollow-core fibers, where the "strength" of the PBG (closely related to its width in β) determines how many periods are needed to reduce confinement loss to acceptable levels. Numerical modeling is useful for giving an indication of how many periods are needed to reach a required loss level. The cladding field intensity in the ultralow loss PCF reported in Ref. [23] falls by $\sim$9 dB per period, reaching -63 dB at the edge of the photonic crystal region.

14.4.7 Kerr Nonlinearities

The ability to enhance or reduce the effective Kerr nonlinearity, and at the same time control the magnitude and wavelength dependence of the GVD, makes PCF a versatile vehicle for studies of effects such as four-wave mixing, self-phase modulation, modulation instability, soliton formation, and stimulated Raman scattering (SRS). For a composite PCF made from two or more homogeneous materials, the γ coefficient of fiber nonlinearity takes the form [52]

$$\gamma = k \sum_i \frac{n_2^i}{A_i} \qquad (14.5)$$

where n_2^i is the nonlinear refractive index of material i and A_i its nonlinear effective area:

$$A_i = \frac{n_m \left(\iint \psi^2(x,y) \, dx \, dy \right)^2}{\iint u_i(x,y) \, \psi^4(x,y) \, dx \, dy} \qquad (14.6)$$

where $u_i(x, y)$ equals n_i in regions made from material i and zero elsewhere. In this expression, n_m is the guided mode phase index and $\psi(x, y)$ is the scalar transverse field amplitude profile.

In air, the nonlinear refractive index is 2.9×10^{-23} m^2/W^1, in silica it is 2.5×10^{-20} m^2/W^1, and in multicomponent glasses it can be an order of magnitude or more higher. The highest nonlinearity available in conventional step-index fibers is $\gamma \simeq 20$ W/km^1 at 1550 nm [53]. By comparison, a solid-core PCF similar to the one in Figure 14.1(f) but with a core diameter 1 μm has a nonlinearity of $\gamma \simeq 240$ W^1/km^1 at 850 nm, and values as high as $\gamma = 550$ W/km^1 at 1550 nm have been measured for PCFs made from multicomponent glasses [54]. In complete contrast, hollow-core PCF has extremely low levels of nonlinearity, owing to the small overlap between the glass and the light. In a recent example, a fiber was reported with a nonlinear coefficient $\gamma = 0.023$ W/km^1 (some $\times 10^4$ smaller than in a typical highly nonlinear solid-core PCF) [55, 56].

Although the level of nonlinearity is clearly important, the actual nonlinear effects that appear in a particular case are also strongly dependent on the magnitude, sign, and wavelength dependence of the GVD (see Section 14.6.3) as well as on the characteristics of the laser pulses [57].

14.5 INTRA-FIBER DEVICES, CUTTING AND JOINING

As PCF becomes more widely used, there is an increasing need for effective cleaves, low-loss splices, multiport couplers, intra-fiber devices, and mode-area transformers. The airholes provide an opportunity not available in standard fibers: the creation of dramatic morphological changes by altering the hole size by collapse (under surface tension) or inflation (under internal over-pressure) when heating to the softening temperature of the glass. Thus, not only can the fiber be stretched locally to reduce its cross-sectional area, but the microstructure can itself be radically altered.

14.5.1 Cleaving and Splicing

PCF cleaves cleanly using standard tools, showing slight end-face distortion only when the core crystal is extremely small (inter-hole spacing $\sim 1\,\mu m$) and the air-filling fraction very high (>50%). Solid glass end-caps can be formed by collapsing the holes (or filling them with sol–gel glass) at the fiber end to form a core-less structure through which light can be launched into the fiber. Solid-core PCF can be fusion-spliced successfully both to itself and to step-index fiber using resistive heating elements (electric arcs do not typically allow sufficient control). The two fiber ends are placed in intimate contact and heated to softening point. With careful control, they fuse together without distortion. Provided the mode areas are well-matched, splice losses of <0.2 dB can normally be achieved except when the core is extremely small (less than $\sim 1.5\,\mu m$). Fusion splicing hollow-core fiber is feasible when there is a thick solid glass outer sheath [e.g., as depicted in Figure 14.2(h)], although very low splice losses can be obtained simply by placing identical fibers end-to-end and clamping them (the index-matching "fluid" for hollow-core PCF is vacuum). The ability to hermetically splice gas-filled hollow-core PCF to SMF has made it possible to produce in-line gas cells for (SRS) in hydrogen and frequency measurement and stabilization (using acetylene). These developments may lead for the first time to practical miniature gas laser devices that could even be coiled up inside a credit card [58].

14.5.2 Mode Transformers

In many applications, it is important to be able to change the mode area without losing light. This is done traditionally using miniature bulk optics—tiny lenses precisely designed to match to a desired numerical aperture and spot size. In PCF, an equivalent effect can be obtained by scanning a heat source (flame or carbon dioxide laser) along the fiber. This causes the holes to collapse, the degree of collapse depending on the dwell time of the heat. Drawing the two fiber ends apart at the same time provides additional control. Graded transitions can fairly easily be made—mode diameter reductions as high as 5:1 have been realized with low loss.

Ferrule methods have been developed for making low-loss interfaces between conventional SMFs and PCFs [59]. Adapted from the fabrication of PCF preforms from stacked tubes and rods, these techniques avoid splicing and are versatile enough to interface with virtually any type of index-guiding silica PCF. They are effective for coupling light into and out of all the individual cores of a multicore fiber without input or output crosstalk. The technique also creates another opportunity—the use of taper transitions to couple light between a multimode fiber and several SMFs. When the number of SMFs matches the number of spatial modes in the multimode fiber, the transition can have low loss in both directions. This means that the high performance of SMF devices can be reached in multimode systems, for example, a multimode fiber filter with the transmission spectrum of a SMF Bragg grating [60], a device that has applications in earth-based astronomy, where the high throughput of a multimode fiber can be retained while unwanted atmospheric emission lines are filtered out.

A further degree of freedom may be gained by pressurizing the holes during the taper process [61]. The resulting hole inflation permits radical changes in the guidance characteristics. It is possible, for example, to transform a PCF, with a relatively large core and small air-filling fraction, into a PCF with a very small core and a large air-filling fraction, the transitions having very low loss [62].

14.5.3 In-Fiber Devices

Precise use of heat and pressure induces large changes in the optical characteristics of PCF, giving rise to a whole family of new intra-fiber components. Micro-couplers can be made in a PCF with two optically isolated cores by collapsing the holes so as to allow the mode fields to expand and interact with each other, creating local coupling [63]. Long-period gratings, which scatter the core light into cladding modes within certain wavelength bands, can be made by periodic modulation of hole size [64]. By rocking a birefringent PCF to and fro, while scanning a carbon dioxide laser along it, so-called rocking filters can be made, which transfer power from one polarization state to the other within a narrow band of wavelengths [65]. All these components have one great advantage over equivalent devices made in conventional fiber: being formed by permanent changes in morphology, they are highly stable with temperature and over time.

Heating and stretching PCF can result in quite remarkable changes in the scale of the micro/nanostructures, without significant distortion. Recently, a solid-core PCF was reduced five times in linear scale, resulting in a core diameter of 500 nm (Figure 14.8). This permitted formation of a PCF with a zero dispersion wavelength that matched the 532 nm emission wavelength of a frequency-doubled Nd:YAG laser [66] (this is important for supercontinuum (SC) generation—see Section Supercontinuum Generation). A further compelling advantage of the tapering approach is that it neatly sidesteps the difficulty of launching light into submicron-sized cores; light is launched into the entry port (in this case with core diameter 2.5 μm) and adiabatically evolves, with negligible loss, into the mode of the 500 nm core.

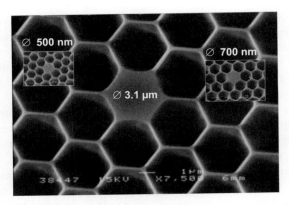

Figure 14.8 Scanning electron micrographs (depicted to the same scale) of the fiber cross sections produced by tapering a solid-core PCF. The structures are well preserved, even down to core diameters of 500 nm [66].

14.6 APPLICATIONS

The diversity of new or improved features, beyond conventional fibers, means that PCFs are finding an increasing number of applications in every widening area of science and technology.

14.6.1 High Power and Energy Transmission

Larger mode areas allow higher power to be carried before the onset of intensity-related nonlinearities or damage, with obvious benefits for delivery of high laser power, and for high-power fiber amplifiers and lasers. ESM-PCF's ability to remain single mode at all wavelengths where it guides, and for all scales of structure, means that the core area can be increased without the penalty of introducing higher order guided modes [19]. This also suggests that it should have superior power-handling properties, with applications in, for example, the field of laser machining. A key issue is bend loss, and as we have seen, it turns out that PCF offers a wider bandwidth of useful single-mode guidance than high-delta SMF, because it can operate in the multimode regime of SMF while remaining single mode (Section "Bend Loss"). This also shifts the long wavelength bend-edge to longer wavelengths than is possible in standard fibers.

Hollow-core PCF is also an excellent candidate for transmitting high continuous-wave power as well as ultrashort pulses with very high peak powers. Solitons have been reported at 800 nm using a Ti:sapphire laser [55] and at 1550 nm with durations of 100 fs and peak powers of 2 MW [67]. The soliton energy is of course determined by the effective value of γ (Section 14.4.7) and the magnitude of the anomalous GVD. As mentioned in Section "Hollow Core," the GVD changes sign

14. Photonic Crystal Fibers: Basics and Applications

across the band gap, permitting choice of normal or anomalous dispersion depending upon the application [43]. Further studies have explored the ultimate power-handling capacity of hollow-core PCF [25, 68, 69].

14.6.2 Fiber Lasers and Amplifiers

Photonic crystal fiber lasers can be straightforwardly produced by incorporating a rare-earth-doped cane in the preform stack. Many different designs can be realized, such as cores with ultralarge mode areas for high power [70], and structures with multiple lasing cores [71–75]. Cladding–pumping geometries for ultrahigh power can be fashioned by incorporating a second core (much larger and multimode) around a large off-center ESM lasing core. Using microstructuring techniques, this "inner cladding waveguide" can be suspended by connecting it to an outer glass tube with very thin webs of glass (see Figure 14.9) [76, 77]. This results in a very large effective index step and thus a high-numerical aperture (>0.9), making it easy to launch and guide light from high-power diode-bar pump lasers—which typically have poor beam quality. The multimode pump light is efficiently absorbed by the lasing core, and high-power single-mode operation can be achieved [78, 79, 80]. Microchip-laser-seeded Yb-doped PCF amplifiers, generating diffraction-limited 0.45 ns duration pulses with a peak power of 1.1 MW and a peak spectral brightness of greater than 10 kW/(cm^2.sr.Hz), have been reported [81–83].

Hollow-core PCF with its superior power-handling and designable GVD is ideal as the last compression stage in chirped-pulse amplification schemes. This permits operation at power densities that would destroy conventional glass-core fibers [84, 85].

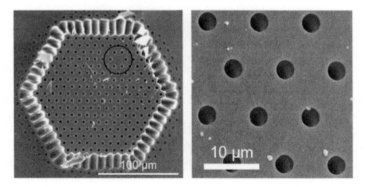

Figure 14.9 Scanning electron micrograph of an early air-clad lasing PCF (Yb-doped core circled). The structural parameters are hole-spacing 9.6 μm, $d/\Lambda = 0.4$, core diameter 15.2 μm, 54 webs of width 350 nm, inner cladding diameter 143 μm across flats, and 164 μm across corners [78].

14.6.3 Kerr-Related Nonlinear Effects

The nonlinear characteristics are determined by the relative values of the nonlinear length $L_{nl} = 1/(\gamma P_0)$ where P_0 is the peak power, the dispersion length $L_D = \tau^2/|\beta_2|$, where τ is the pulse duration, and the effective fiber length $L_{eff} = (1 - \exp(-\alpha L))/\alpha$, where α/m is the power attenuation coefficient [86]. For a solid-core PCF with $\gamma = 240/\text{W}^1/\text{km}^1$, a peak power of 10 kW yields $L_{nl} < 0.5$ mm. For typical values of loss (usually between 1 and 100 dB/km), $L_{eff} \gg L_{nl}$ and the nonlinearity dominates. For dispersion values in the range $-300 < \beta_2 < 300$ ps^2/km and pulse durations $\tau = 200$ fs, $L_D > 0.1$ m. Because both these lengths are much longer than the nonlinear length, it is easy to observe strong nonlinear effects.

Supercontinuum Generation

One of the most successful applications of nonlinear PCF is to SC generation from nanosecond, picosecond, and femtosecond laser pulses. When high-power pulses travel through a material, their frequency spectrum can be broadened by a range of interconnected nonlinear effects [87]. In bulk materials, the preferred pump laser is a regeneratively amplified Ti–sapphire system producing high (mJ)-energy femtosecond pulses at 800 nm wavelength and kHz repetition rate. Supercontinua have also previously been generated in SMF by pumping at 1064 or 1330 nm [88], the spectrum broadening out to longer wavelengths mainly due to stimulated SRS. Then in 2000, it was observed that highly nonlinear PCF, designed with zero GVD close to 800 nm, massively broadens the spectrum of low (few nJ) energy unamplified Ti–sapphire pulses launched into just a few centimeters of fiber [89–91]. Removal of the need for a power amplifier, the hugely increased ($\sim$100 MHz) repetition rate, and the spatial and temporal coherence of the light emerging from the core, makes this source unique. The broadening extends both to higher and to lower frequencies because four-wave mixing operates more efficiently than SRS when the dispersion profile is appropriately designed. This SC source has applications in optical coherence tomography [92, 93], frequency metrology [94, 95], and all kinds of spectroscopy. It is particularly useful as a bright low-coherence source in measurements of group delay dispersion based on a Mach–Zehnder interferometer.

A comparison of the bandwidth and spectrum available from different broadband light sources is shown in Figure 14.10; the advantages of PCF-based SC sources are evident. In addition to 800 nm pumping, supercontinua have been generated in different PCFs at 532 [66], 647 [96], 1064 [97], and 1550 nm [7]. Compact PCF-based SC sources, pumped by inexpensive microchip lasers at 1064 or 532 nm, are finding important applications in many diverse areas of science. A commercial ESM-PCF-based source, using as pump source a 10 W fiber laser delivering 5 ps pulses at 50 MHz repetition rate, produces an average spectral power density of $\sim$4.5 mW/nm in the range 450–800 nm [98]. The use of

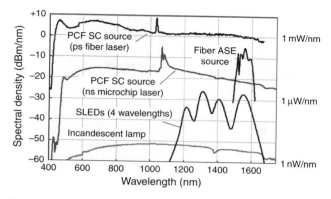

Figure 14.10 Comparison of the brightness of various broadband light sources (SLED, superluminescent light-emitting diode; ASE, amplified spontaneous emission; SC, supercontinuum). The microchip laser SC spectrum was obtained by pumping at 1064 nm with 600 ps pulses (updated version of a plot by H. Sabert).

multicomponent glasses such as Schott SF6 or tellurite glass allows additional adjustments in the balance between nonlinearity and dispersion, as well as offering extended transparency into the (IR) [99].

Parametric Amplifiers and Oscillators

In step-index SMFs, the performance of optical parametric oscillators and amplifiers is constrained by the limited scope for GVD engineering. In PCF, these constraints are lifted, permitting flattening of the dispersion profile and control of higher order dispersion terms. The new range of experimentally available GVD profiles has, for example, led to studies of ultrashort pulse propagation in the 1550 nm wavelength band with flattened dispersion [41, 42]. The effects of higher order dispersion in such PCFs are subtle [100, 101]. Parametric devices have been designed for pumping at 647, 1064, and 1550 nm, the small effective mode areas offering high gain for a given pump intensity, and PCF-based oscillators synchronously pumped by femtosecond and picosecond pump pulses have been demonstrated at relatively low power levels [102–105].

Correlated Photon Pairs

The use of self-phase modulation to generate bright sources of correlated photon pairs is unsuccessful in step-index fibers owing to high-Raman-related noise. This is because for $\beta_2 < 0$ and $\beta_n = 0$, $n > 2$, the modulational instability sidebands are situated very close to the pump frequency, within the Raman gain band of the glass. By flattening the GVD profile and making β_2 small, however, higher order GVD terms become important and gain bands can appear in the normal dispersion regime for $\beta_2 > 0$ and $\beta_4 < 0$ [42]. The relevant expression for the sideband gain (m^{-1}), including β_4 and β_6, is

$$\text{gain} = \text{Im}\left[\sqrt{Q(Q + 2\gamma P)}\right] \quad (14.7)$$
$$Q = \beta_2 \Omega^2/2 + \beta_4 \Omega^4/24 + \beta_6 \Omega^6/720$$

where γ is the nonlinear coefficient (Eqn (14.5)), P the pump power, and Ω the angular frequency offset from the pump frequency. This expression may be used to show that the sidebands can be widely spaced from the pump frequency, their position and width being controllable by engineering the even-order higher order dispersion terms (Figure 14.11). It is straightforward to arrange that these sidebands lie well beyond the Raman gain band, reducing the Raman noise and allowing PCF to be used as a compact, bright, tunable, single-mode source of pair-photons with wide applications in quantum communications [106, 107].

In a recent example, a PCF with zero dispersion at 715 nm was pumped by a Ti:sapphire laser at 708 nm (normal dispersion) [108]. Under these conditions, phase-matching is satisfied by signal and idler waves at 587 and 897 nm, and 10 million photon pairs per second were generated and delivered through single-mode fiber to Si avalanche detectors, producing $\sim 3.2 \times 10^5$ coincidences per second for a pump power of 0.5 mW. These results point the way to practical and efficient sources of entangled photon pairs that can be used as building blocks in future multiphoton interference experiments.

More complicated and potentially engineerable effects can be obtained by adding one or more pump lasers at different frequencies; this has the effect of altering the MI gain spectrum.

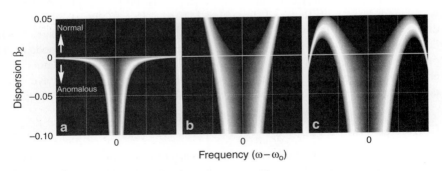

Figure 14.11 Density plots of the level of modulational instability gain as a function of frequency offset from a strong single-frequency pump laser and the level of β_2 at the pump frequency. The black to white regions represent gain on a linear scale of grey; gain is zero in the black regions. The nonlinear length $L_{nl} = 1/(\gamma P_0)$ is 0.1 in each case (see Section 14.6.3): (a) $\beta_4 = 0$, $\beta_2 = 0$; (b) $\beta_4 = -10^{-3}$, $\beta_2 = 0$; (c) $\beta_4 = -10^{-3}$, $\beta_2 = 1.24 \times 10^{-5}$. The units in each case are ps^n/km for β_n.

Soliton Self-Frequency Shift Cancellation

The ability to create PCFs with negative dispersion slope at the zero dispersion wavelength (in SMF the slope is positive, i.e., the dispersion becomes more anomalous as the wavelength increases) has made it possible to observe Čerenkov-like effects in which solitons (which form on the anomalous side of the dispersion zero) shed power into dispersive radiation at longer wavelengths on the normal side of the dispersion zero. This occurs because higher order dispersion causes the edges of the soliton spectrum to phase-match to linear waves. The result is stabilization of the soliton self-frequency shift, at the cost of gradual loss of soliton energy [109]. The behavior of solitons in the presence of wavelength-dependent dispersion is the subject of many recent studies, for example, Ref. [110]. Interactions between solitons have also been studied in a number of recent papers [111, 112].

14.6.4 Brillouin Scattering

The periodic micro/nanostructuring in ultrasmall core glass–air PCF strongly alters the acoustic properties compared with that in conventional SMF [113–116]. Sound can be guided in the core both as leaky and as tightly confined acoustic modes. In addition, the complex geometry and "hard" boundaries cause coupling between all three displacement components (radial, azimuthal, and axial), with the result that each acoustic mode has elements of both shear and longitudinal strain. This complex acoustic behavior strongly alters the characteristics of both forward and backward Brillouin scattering.

Backward Scattering

When a solid-core silica–air PCF has a core diameter of around 70% of the vacuum wavelength of the launched laser light, and the air-filling fraction in the cladding is very high, the spontaneous Brillouin signal displays multiple bands with Stokes frequency shifts in the 10 GHz range. These peaks are caused by discrete guided acoustic modes, each with different proportions of longitudinal and shear strain, strongly localized to the core [117]. At the same time, the threshold power for stimulated Brillouin scattering increases fivefold—a rather unexpected result, because conventionally one would assume that higher intensities yield lower nonlinear threshold powers. This occurs because the effective overlap between the tightly confined acoustic modes and the optical mode is actually smaller than in a conventional fiber core; the sound field contains a large proportion of shear strain, which does not contribute significantly to changes in refractive index. This is of direct practical relevance to parametric amplifiers (Section "Parametric Amplifiers and Oscillators"), which can be pumped five times harder before stimulated Brillouin scattering appears.

Forward Scattering

The very high air-filling fraction in small-core PCF also permits sound at frequencies of a few gigahertz to be trapped purely in the transverse plane by acoustic TIR [118] or phononic band gap effects [119]. The ability to confine acoustic energy at zero axial wavevector $\beta_{ac} = 0$ means that the ratio of frequency ω_{ac} to wavevector β_{ac} becomes arbitrarily large as $\beta_{ac} \to 0$, and thus can easily match the value for the light guided in the fiber, c/n. This permits phase-matched interactions between the acoustic mode and two spatially identical optical modes of different frequency. Under these circumstances, the acoustic mode has a well-defined cutoff frequency ω_{cutoff}, above which its dispersion curve—plotted on an (ω, β) diagram—is flat, similar to the dispersion curve for optical phonons in diatomic lattices (Figure 14.12). The result is a scattering process that is Raman-like (i.e., the participating phonons are optical-phonon-like), even though it makes use of acoustic phonons; Brillouin scattering is turned into Raman scattering, power being transferred into an optical mode of the same order, frequency-shifted from the pump frequency by the cutoff frequency. Used in stimulated mode, this effect may permit generation of combs of frequencies spaced by $\sim$2 GHz at 1550 nm wavelength.

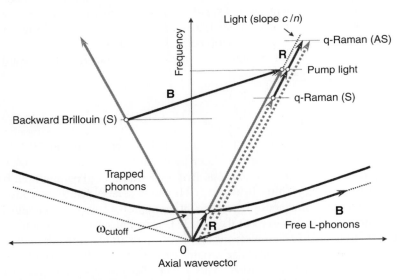

Figure 14.12 Illustrating the different phase-matching processes for sound–light interactions in PCF. Backward Brillouin scattering (B) proceeds by inelastic collisions with free phonons that travel predominantly along the fiber axis; in this case phase-matching is a strict condition, and the frequency shift is a strong function of pump frequency [117]. In quasi-Raman scattering (R), trapped acoustic phonons automatically phase-match to light at the acoustic cutoff frequency, and scattering proceeds in the forward direction. In contrast to Brillouin scattering, the frequency shift is insensitive to the frequency of the pump light [119].

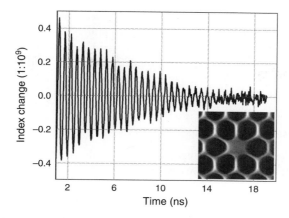

Figure 14.13 Quasi-Raman impulse response of a PCF with small solid core (1.1 μm in diameter) and high-air-filling fraction. Scanning electron micrograph of the core region of the PCF is shown in the inset.

These transversely trapped resonances can be electrostrictively excited by means of 100 ps pulses, and the resulting resonances monitored using interferometry (Figure 14.13). The creation and control of high-quality acoustic resonances may ultimately lead to efficient acoustooptic modulators built into the PCF structure.

14.6.5 Gas-Based Nonlinear Optics

It is a hundred years since Lord Rayleigh first explained the relationship between depth of focus and focal spot size. For a Gaussian beam focused to a spot of radius a, the depth of focus is $2\pi a^2/\lambda$, which means that the product of intensity and interaction length is independent of a. This consequence of the diffraction of light beams in free space presents an apparently insuperable barrier to achieving efficient nonlinear interactions between laser light and low-density media such as gases. The requirements of high-intensity, long interaction length, and good-quality (preferably single-mode) transverse beam profiles simply cannot be met. A structure conceptually capable of delivering all these requirements simultaneously would be a perfectly guiding hollow-core waveguide supporting a single transverse mode with low attenuation losses. Although theoretically this could be realized using a perfect metal, the attenuation in real metals at optical frequencies is much too high, especially when the bore is small enough to yield single-mode operation. A number of conventional approaches have been used to circumvent this problem, including focusing a laser beam into a gas with suitable optics, using a ~100-μm bore fiber capillary to confine the gas and provide some degree of guidance for the light [120], and employing a gas-filled high-finesse Fabry–Pérot cavity to increase the interaction length [121]. None of these approaches comes close, however, to the performance offered by hollow-core PCF [122]. At a bore diameter of 10 μm, for example, a focused free-space laser beam is marginally preferable to a capillary, whereas a hollow-core PCF with 1.2 dB/km attenuation is × 1,000,000 more effective. Such huge enhancements are rare in any field and are leading to dramatic improvements in all sorts of nonlinear laser–gas interactions.

Stimulated Raman Scattering

In 2002, SRS was reported in a hydrogen-filled hollow-core PCF at threshold pulse energies ~100× lower than previously possible [122]. In hollow-core PCF, guidance can only occur when a PBG coincides with a core resonance. This means that only restricted bands of wavelength are guided. This feature can be very useful for suppressing parasitic transitions by filtering away the unwanted wavelengths, as has been done in the case of rotational Raman scattering in hydrogen, where the normally dominant vibrational Raman signal can be very effectively suppressed using the PCF with the transmission characteristic shown in Figure 14.14. This resulted in a 1 million fold reduction in the threshold power for rotational Raman scattering in a single-pass hydrogen cell 30 m long, and near-perfect quantum efficiency in a cell 3 m long [123]. Gas cells of this type have been hermetically spliced to all-solid glass SMF for 1550 nm operation [58], making their use in standard optical fiber communication systems a realistic possibility.

High-Harmonic Generation

Hollow-core PCF is likely to have a major impact in other areas of nonlinear optics, such as X-ray generation in noble gases pumped by femtosecond

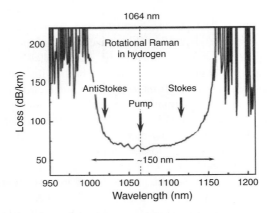

Figure 14.14 Attenuation spectrum of a hollow-core PCF designed for low-loss transmission of 1064 nm light. The pump, Stokes, and antiStokes Raman frequencies for rotational scattering in hydrogen are marked in. The vibrational Raman band lies well outside the low-loss window.

Ti:sapphire laser pulses [124]. The conversion efficiency of this process can be further enhanced by modulating the bore diameter of the core so as to phase-match the light and the X-rays [125]. This could be implemented in a hollow-core PCF, for example, by heat treatment with carbon dioxide laser light.

Electromagnetically Induced Transparency

Hollow-core PCF filled with acetylene vapor at low pressure has recently been used to demonstrate 70% electromagnetically induced transparency (for both Λ-and V-type interactions) over several lines of the R-branch of the $v_1 + v_3$ ro-vibrational overtone band [126, 127]. The well-controlled single-mode environment provided by hollow-core PCF makes effects of this kind much easier to control, monitor, and potentially to engineer into a practical devices.

14.6.6 Telecommunications

There are many potential applications of PCF or PCF-based structures in telecommunications, although whether these will be adopted remains an open question. One application that seems quite close to being implemented is the use of solid-core PCF or "hole-assisted" SMF for fiber-to-the-home, where the lower bend-loss is the attractive additional advantage offered by the holey structure [128]. Other possibilities include dispersion-compensating fiber and hollow-core PCF for long-haul transmission. Additional opportunities exist in producing bright sources of correlated photon pairs for quantum cryptography, parametric amplifiers with improved characteristics, highly nonlinear fiber for all-optical switching and amplification, acetylene-filled hollow-core PCF for frequency stabilization at 1550 nm, and the use of sliced SC spectra as wavelength division multiplexing (WDM) channels. There are also many possibilities for ultra-stable in-line devices based on permanent morphological changes in the local holey structure, induced by heating, collapse, stretching, or inflation.

The best reported loss in solid-core PCF, from a group in Japan, stands at 0.28 dB/km at 1550 nm, with a Rayleigh scattering coefficient of 0.85 $dB/km^1/\mu m^4$. A 100-km length of this fiber was used in the first PCF-based penalty-free dispersion-managed soliton transmission system at 10 Gb/s [45]. The slightly higher attenuation compared with that in SMF is due to roughness at the glass–air interfaces [26].

A New Telecommunications Window?

Hollow-core PCF is radically different from solid-core SMF in many ways. This makes it difficult to predict whether it could be successfully used in long-haul telecommunications as a realistic competitor for SMF-28. The much lower Kerr nonlinearities mean that WDM channels can be much more tightly packed without nonlinear crosstalk, and the higher power-handling characteristics mean that more

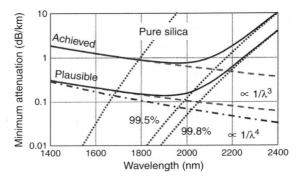

Figure 14.15 Achieved and plausible minimum attenuation spectra for hollow-core PCF. The scattering floor is proportional to λ^{-3}, its overall level depending on the amplitude of the surface roughness. Higher air-filling percentages push the infrared absorption edge out to longer wavelengths. As a result, the predicted low-loss window sits at 1900 nm, with an attenuation of 0.2 dB/km [23]. For comparison, the infrared absorption edge and the Rayleigh scattering floor (dot–dashed line) of pure silica glass are also indicated.

overall power can be transmitted. The effective absence of bend losses is also a significant advantage, particularly for short-haul applications. On the other hand, work still needs to be done to reduce the losses to 0.2 dB/km or lower, and to understand—and control—effects such as polarization mode dispersion, differential group delay, and multipath interference. It is interesting that the low-loss window of a plausible hollow-core PCF is centered at 1900 nm, because light travels predominantly in the hollow regions, completely changing the balance between scattering and infrared absorption (Figure 14.15) [23].

Dispersion Compensation

The large glass–air refractive index difference makes it possible to design and fabricate PCFs with high levels of group velocity dispersion. A PCF version of the classical W-profile dispersion-compensating fiber was reported in 2005, offering slope-matched dispersion compensation for SMF-28 fiber at least over the entire C-band (Figure 14.16). The dispersion levels achieved (−1200 ps/nm.km) indicate that only 1 km of fiber is needed to compensate for 80 km of SMF-28. The PCF was made deliberately birefringent to allow control of polarization mode dispersion [129].

14.6.7 Bragg Gratings in PCF

Bragg mirrors are normally written into the Ge-doped core of an SMF by ultraviolet (UV) light, using either two-beam interferometry or a zero-order nulled phase mask [130]. The mechanism for refractive index change is related to the UV

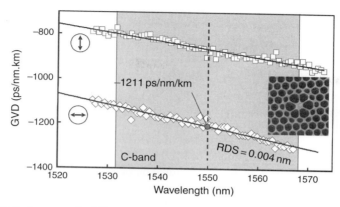

Figure 14.16 Performance of a PCF designed to provide slope-matched dispersion compensation for Corning SMF-28 over the C-band.

photosensitivity of the glass and is quite complex, with several different regimes of operation. In PCF, the presence of many holes in the cladding will inevitably scatter the UV light strongly and reduce (or enhance) the field amplitudes in the core. In addition, pure silica glass is only weakly photosensitive, requiring exposure to very intense light for formation of useful Bragg gratings. There have nevertheless been several reports of the inscription of Bragg gratings into pure-silica PCF, using 125 fs pulses at 800 nm wavelength [131] and multiphoton processes [132]. In the case of polymer PCF, low-power CW laser sources at 325 nm wavelength are sufficient to write Bragg gratings for 1570-nm operation [133].

14.6.8 Laser Tweezers in Hollow-Core PCF

A focused light beam produces both a longitudinal (accelerating) and a transverse (trapping) force on dielectric microparticles [134]. For maximum trapping force, the intensity gradient of the light must be as high as possible. This can be achieved by tight focusing of a laser beam using a high-numerical aperture lens. The Rayleigh length for such a tightly focused beam is rather short, and furthermore, it turns out that at high numerical apertures a particle also becomes trapped axially through of the effects of obliquely refracted rays, which produce forces much stronger than gravity. If one wishes to have a constant propulsive force, at the same time trapping the particle transversely, one would require a nondiffracting or guided beam—something that has only recently become feasible with the appearance of hollow core PCF. That did not prevent the demonstration of guidance of both solid particles [135] and atoms [136] along lengths of hollow capillary (channel diameter ~100 μm). Of course, the use of a capillary prevented full exploitation of the possibilities of a hollow waveguide. Small core sizes are needed for strong transverse confinement and single-mode operation, so that capillary losses are very high (they scale with the inverse cube of the radius), even on length scales of a few millimeters. The same is not true in hollow-core PCF, where core diameters of ~10 μm can be realized with losses as low as a few dB/km. Such small spot sizes give strong transverse trapping forces for a given longitudinal force, so that it becomes possible to envisage guiding particles even along tightly coiled fibers at reasonably high speeds. The first steps toward achieving this goal were taken in 2002 [137].

A further opportunity lies in the use of laser tweezering to manipulate biological cells and channel them into a PCF filled with a biological buffer solution. By controlling the power of the light, individual particles can be held stationary inside a hollow liquid-filled core and probed or excited optically using techniques such as UV-excited fluorescence. Microfluidic flow of chemicals through laser (or focused ion-beam) machined side-channels offers the possibility of adjusting the chemical microenvironment. By varying the power and controlling the polarization state of the light, the trapped particles can be moved up or down, or rotated. In effect, this represents a new kind of flow cytometry, a well-established technique where cells are passed single-file through capillaries with inner diameters of 50 to several 100 μm, laminar conditions being achieved by careful design of the flow at the entrance to the capillary.

14.6.9 Optical Sensors

Sensing is so far a relatively unexplored area for PCFs, although the opportunities are myriad, spanning many fields including environmental monitoring, biomedical sensing, and structural monitoring [138]. Solid-core PCF has been used in hydrostatic pressure sensing [139]. Multicore PCF has been used in bend and shape sensing [22, 140] and Doppler difference velocimetry [141]. Fibers with a central single-mode core surrounded by a highly multimode cladding waveguide (e.g., Figure 14.9) are useful in applications such as two-photon fluorescence sensing [142], where short pulses are delivered to the sample by the central single-mode core, and the resulting multiphoton fluorescence efficiently collected by inner cladding waveguide, which has a large numerical aperture. Sensitivity enhancements of 20 times have been reported. Given the high level of interest, and large amount of effort internationally, it seems very likely that many more important sensing applications of PCF will emerge over the next few years.

14.7 FINAL REMARKS

Photonic crystal fiber has improved the efficiency of nonlinear gas–laser interactions by a factor approaching 10 million. It has led to relatively inexpensive and compact white light sources that are 10,000 times brighter than incandescent lamps. It allows the Kerr nonlinearity of the glass to be adjusted over four orders of magnitude by manipulating the proportion of light in glass. It allows the realization of polarization-maintaining fiber, and in-fiber devices 100 times more stable than in conventional step-index fibers. In-fiber devices based on morphological changes in the holey structure are intrinsically extremely stable (over time) compared with those made by photorefractive effects. Such devices permit the realization of low-loss transitions and mode transformers between radically different core shapes and sizes. Remarkable adjustments in the shape of the dispersion landscape are possible by exquisite control of the PCF nanostructure. The behavior of GHz acoustic waves is strongly affected by the microstructuring, suggesting a new generation of highly efficient acoustooptic devices. These are just a few examples of the impact that PCF has, and is having, in many areas of optics and photonics.

By moving away from constraints of conventional fiber optics, PCF has created new opportunities in diverse areas of science and technology. Already, they are being adopted in many close-to-commercial fields, including compact SC sources, frequency comb systems for frequency metrology, constrained photo- or bio chemistry and microfluidics (making use of the hollow channels [143, 144]), biophotonic and biomedical devices, medical imaging, astronomy, particle delivery, high-power fiber lasers, fiber delivery of high-power laser light in manufacturing, and gas-based fiber devices.

GLOSSARY

A_i nonlinear effective area of subregion i
α intensity decay rate of guided mode, m^{-1}
β component of wavevector along the fiber axis
β_m axial wavevector of guided mode
β_{max} maximum possible axial component of wavevector in the PCF cladding
β_n nth-order dispersion in units of ps^n/km
c velocity of light in vacuum
d hole diameter
D $\left(=\frac{\partial}{\partial \lambda}\frac{1}{v_g}=\frac{\partial^2 \beta_m}{\partial \lambda \partial \omega}\right)$, the group velocity dispersion of guided mode in engineering units (ps/nm.km)
Δ $(n_{co}-n_{cl})/n_{cl}$, where n_{co} and n_{cl} are respectively the core and cladding refractive indices of a conventional step-index fiber
Δ_∞ value of Δ in the long-wavelength limit
γ W^{-1}/km^1 nonlinear coefficient of an optical fiber
k vacuum wavevector $2\pi/\lambda = \omega/c$
L_D $(=\tau^2/|\beta_2|)$, dispersion length, where τ is the pulse duration and β_2 is the GVD in ps^2/km
L_{eff} $[=(1-\exp(-\alpha L))/\alpha]$, effective fiber length before losses become significant

L_{nl} $(=1/(\gamma P_0))$, nonlinear length, where P_0 is the peak power
λ vacuum wavelength
λ_{eff}^i $(=2\pi/\sqrt{k^2 n_i^2 - \beta^2})$, the effective transverse wavelength in subregion i
Λ inter-hole spacing, period, or pitch
n_2^i nonlinear refractive index of subregion i
n_z axial or z-component of refractive index
n_m phase index of guided mode
n_{max} maximum axial index supported by the PCF cladding (fundamental space-filling mode)
n_q^∞ $\left(=\sqrt{\sum_i n_i^2 a_i / \sum_i a_i}\right)$, the area-averaged refractive index of an arbitrary region q of a microstructured fiber, where n_i and a_i are respectively the refractive index and area of subregion i
ω angular frequency of light
ν $\left(=k\Lambda\sqrt{n_g^2-1}\right)$, normalized frequency parameter used in the evaluation of Figure (14.4)
ρ core radius

LIST OF ACRONYMS

AS   anti-Stokes
ASE   amplified spontaneous emission
ESM   endlessly single mode
GVD   group velocity dispersion
IR   infrared
MCVD   modified chemical vapor deposition
PBG   photonic band gap
PCF   photonic crystal fiber
S   Stokes
SC   supercontinuum
SMF   single-mode fiber
SRS   stimulated Raman scattering
TIR   total internal reflection
WDM   wavelength division multiplexing

REFERENCES

[1] J. C. Knight, T. A. Birks, P. St J. Russell, and D. M. Atkin, "Pure Silica Single-Mode Fiber with Hexagonal Photonic Crystal Cladding," in Proc. *OFC 1996*, San Jose, California, Optical Society of America, 1996.
[2] P. St J. Russell, "Photonic crystal fibers," *Science*, 299, 358–362, January 2003.
[3] P. St J. Russell, "Photonic-crystal fibers," *J. Lightw. Technol.*, 24, 4729–4749, 2006.

[4] J. C. Knight, T. A. Birks, P. St J. Russell, and D. M. Atkin, "All-silica single-mode optical fiber with photonic crystal cladding," *Opt. Lett.*, 21, 1547–1549, October 1996.

[5] D. C. Allan, J. A. West, J. C. Fajardo et al., "Photonic crystal fibers: Effective index and bandgap guidance," in *Photonic Crystals and Light Localisation in the 21st Century*, (C. M. Soukoulis, ed.), Kluwer, Dordrecht, 2001, pp. 305–320.

[6] V. V. R. K. Kumar, A. K. George, J. C. Knight, and P. St J. Russell, "Tellurite photonic crystal fiber," *Opt. Express*, 11, 2641–2645, October 2003.

[7] V. V. R. K. Kumar, A. K. George, W. H. Reeves et al., "Extruded soft glass photonic crystal fiber for ultrabroad supercontinuum generation," *Opt. Express*, 10, 1520–1525, December 2002.

[8] K. M. Kiang, K. Frampton, M. Monro et al., "Extruded single-mode non-silica glass holey optical fibers," *Electron. Lett.*, 38, 546–547, 2002.

[9] R. Bise and D. J. Trevor, "Sol-Gel Derived Microstructured Fiber: Fabrication and Characterization," in *Proc. OFC 2005*, Anaheim, 2005.

[10] M. C. J. Large, A. Argyros, F. Cox et al., "Microstructured polymer optical fibres: New opportunities and challenges," *Mol. Cryst. Liq. Cryst.*, 446, 219–231, 2006.

[11] R. F. Cregan, B. J. Mangan, J. C. Knight et al., "Single-mode photonic band gap guidance of light in air," *Science*, 285, 1537–1539, September 1999.

[12] A. Ortigosa-Blanch, J. C. Knight, W. J. Wadsworth et al., "Highly birefringent photonic crystal fibers," *Opt. Lett.*, 25, 1325–1327, September 2000.

[13] T. A. Birks, P. J. Roberts, P. St J. Russell, D. M. Atkin, and T. J. Shepherd, "Full 2-D photonic band gaps in silica/air structures," *Electron. Lett.*, 31, 1941–1942, 1995.

[14] T. A. Birks, J. C. Knight, and P. St J. Russell, "Endlessly single-mode photonic crystal fiber," *Opt. Lett.*, 22, 961–963, July 1997.

[15] J. C. Knight, T. A. Birks, P. St J. Russell, and J. P. de Sandro, "Properties of photonic crystal fiber and the effective index model," *J. Opt. Soc. Am. A – Opt. Image Sci. Vis.*, 15, 748–752, March 1998.

[16] J. C. Knight, J. Broeng, T. A. Birks, and P. St J. Russell, "Photonic band cap guidance in optical fibers," *Science*, 282, 1476–1478, November 1998.

[17] F. Couny, F. Benabid, and P. S. Light, "Large-pitch kagome-structured hollow-core photonic crystal fiber," *Opt. Lett.*, 31, 3574–3576, December 2006.

[18] B. J. Mangan, J. Arriaga, T. A. Birks et al., "Fundamental-mode cutoff in a photonic crystal fiber with a depressed-index core," *Opt. Lett.*, 26, 1469–1471, October 2001.

[19] J. C. Knight, T. A. Birks, R. F. Cregan et al., "Large mode area photonic crystal fibre," *Electron. Lett.*, 34, 1347–1348, June 1998.

[20] B. J. Mangan, J. C. Knight, T. A. Birks et al., "Experimental study of dual-core photonic crystal fibre," *Electron. Lett.*, 36, 1358–1359, August 2000.

[21] M. A. van Eijkelenborg, "Imaging with microstructured polymer fibre," *Opt. Express*, 12, 342–346, January 2004.

[22] W. N. MacPherson, M. J. Gander, R. McBride et al., "Remotely addressed optical fibre curvature sensor using multicore photonic crystal fibre," *Opt. Commun.*, 193, 97–104, June 2001.

[23] P. J. Roberts, F. Couny, H. Sabert et al., "Ultimate low loss of hollow-core photonic crystal fibres," *Opt. Express*, 13, 236–244, January 2005.

[24] J. M. Pottage, D. M. Bird, T. D. Hedley et al., "Robust photonic band gaps for hollow core guidance in PCF made from high index glass," *Opt. Express*, 11, 2854–2861, November 2003.

[25] G. Humbert, J. C. Knight, G. Bouwmans et al., "Hollow core photonic crystal fibers for beam delivery," *Opt. Express*, 12, 1477–1484, April 2004.

[26] P. J. Roberts, F. Couny, H. Sabert et al., "Loss in solid-core photonic crystal fibers due to interface roughness scattering," *Opt. Express*, 13, 7779–7793, October 2005.

[27] J. A. West, C. Smith, N. F. Borrelli et al., "Surface modes in air-core photonic band-gap fibers," *Opt. Express*, vol. 12, 1485–1496, 2004.

[28] A. Argyros, T. A. Birks, S. G. Leon-Saval et al., "Photonic bandgap with an index step of one percent," *Opt. Express*, 13, 309–314, January 2005.

[29] J. C. Knight, F. Luan, G. J. Pearce et al., "Solid photonic bandgap fibres and applications," *Jpn. J. Appl. Phys. Part 1 – Regular Pap Short Notes & Review Papers*, 45, 6059–6063, August 2006.
[30] F. Luan, A. K. George, T. D. Hedley et al., "All-solid photonic bandgap fiber," *Opt. Lett.*, 29, 2369–2371, October 2004.
[31] N. M. Litchinitser, S. C. Dunn, B. Usner et al., "Resonances in microstructured optical waveguides," *Opt. Express*, 11, 1243–1251, 2003.
[32] A. Argyros, T. A. Birks, S. G. Leon-Saval et al., "Guidance properties of low-contrast photonic bandgap fibres," *Opt. Express*, 13, 2503–2511, April 2005.
[33] G. Bouwmans, L. Bigot, Y. Quiquempois et al., "Fabrication and characterization of an all-solid 2D photonic bandgap fiber with a low-loss region (<20 dB/km) around 1550 nm," *Opt. Express*, 13, 8452–8459, 2005.
[34] M. J. Steel, T. P. White, C. M. De Sterke et al., "Symmetry and degeneracy in microstructured optical fibers," *Opt. Lett.*, 26, 488–490, 2001.
[35] P. J. Roberts, D. P. Williams, H. Sabert et al., "Design of low-loss and highly birefringent hollow-core photonic crystal fiber," *Opt. Express*, 14, 7329–7341, August 2006.
[36] D. R. Chen and L. F. Shen, "Ultrahigh birefringent photonic crystal fiber with ultralow confinement loss," *IEEE Photon. Technol. Lett.*, 19, 185–187, January–February 2007.
[37] H. Kubota, S. Kawanishi, S. Koyanagi et al., "Absolutely single polarization photonic crystal fiber," *IEEE Photon. Technol. Lett.*, 16, 182–184, January 2004.
[38] D. H. Kim and J. U. Kang, "Sagnac loop interferometer based on polarization maintaining photonic crystal fiber with reduced temperature sensitivity," *Opt. Express*, 12, 4490–4495, September 2004.
[39] J. C. Knight, J. Arriaga, T. A. Birks et al., "Anomalous dispersion in photonic crystal fiber," *IEEE Photon. Technol. Lett.*, 12, 807–809, July 2000.
[40] D. Mogilevtsev, T. A. Birks, and P. St J. Russell, "Group-velocity dispersion in photonic crystal fibers," *Opt. Lett.*, 23, 1662–1664, November 1998.
[41] W. H. Reeves, J. C. Knight, P. St J. Russell, and P. J. Roberts, "Demonstration of ultra-flattened dispersion in photonic crystal fibers," *Opt. Express*, 10, 609–613, July 2002.
[42] W. H. Reeves, D. V. Skryabin, F. Biancalana et al., "Transformation and control of ultra-short pulses in dispersion-engineered photonic crystal fibres," *Nature*, 424, 511–515, July 2003.
[43] G. Bouwmans, F. Luan, J. C. Knight et al., "Properties of a hollow-core photonic bandgap fiber at 850 nm wavelength," *Opt. Express*, 11, 1613–1620, July 2003.
[44] K. Kurokawa, K. Tajima, and K. Nakajima, "10-GHz 0.5-ps pulse generation in 1000-nm band in PCF for high-speed optical communication," *J. Lightw. Technol.*, 25, 75–78, January 2007.
[45] K. Kurokawa, K. Tajima, K. Tsujikawa, and K. Nakagawa, "Penalty-free dispersion-managed soliton transmission over a 100-km low-loss PCF," *J. Lightw. Technol.*, 24, 32–37, 2006.
[46] G. J. Pearce, J. M. Pottage, D. M. Bird et al., "Hollow-core PCF for guidance in the mid to far infra-red," *Opt. Express*, 13, 6937–6946, September 2005.
[47] J. D. Shephard, W. N. MacPherson, R. R. J. Maier et al., "Single-mode mid-IR guidance in a hollow-core photonic crystal fiber," *Opt. Express*, 13, 7139–7144, September 2005.
[48] P. J. Roberts, D. P. Williams, B. J. Mangan et al., "Realizing low loss air core photonic crystal fibers by exploiting an antiresonant core surround," *Opt. Express*, 13, 8277–8285, October 2005.
[49] A. W. Snyder and J. D. Love, *Optical Waveguide Theory*, London, Chapman & Hall, 1983.
[50] E. M. Dianov and V. M. Mashinksy, "Germania-based core optical fibers," *J. Lightw. Technol.*, 23, 3500–3508, 2005.
[51] M. D. Nielsen, N. A. Mortensen, M. Albertsen et al., "Predicting macrobending loss for large-mode area photonic crystal fibers," *Opt. Express*, 12, 1775–1779, 2004.
[52] J. Laegsgaard, N. A. Mortensen, J. Riishede, and A. Bjarklev, "Material effects in air-guiding photonic bandgap fibers," *J. Opt. Soc. Am. B – Opt. Phys.*, 20, 2046–2051, 2003.
[53] M. Onishi, T. Okuno, T. Kashiwada, et al., "Highly nonlinear dispersion-shifted fibers and their application to broadband wavelength converter," *Opt. Fib. Technol.*, 4, 204–214, 1998.

[54] P. Petropoulos, H. Ebendorff-Heidepriem, V. Finazzi, et al., "Highly nonlinear and anomalously dispersive lead silicate glass holey fibers," *Opt. Express*, 11, 3568–3573, 2003.

[55] F. Luan, J. C. Knight, P. St J. Russell, et al., "Femtosecond soliton pulse delivery at 800 nm wavelength in hollow-core photonic bandgap fibers," *Opt. Express*, 12, 835–840, March 2004.

[56] C. J. Hensley, D. G. Ouzounov, A. L. Gaeta et al., "Silica-glass contribution to the effective nonlinearity of hollow-core photonic band-gap fibers," *Opt. Express*, 15, 3507–3512, March 2007.

[57] A. Efimov, A. J. Taylor, F. G. Omenetto et al., "Time-spectrally-resolved ultrafast nonlinear dynamics in small-core photonic crystal fibers: Experiment and modelling," *Opt. Express*, 12, 6498–6507, December 2004.

[58] F. Benabid, F. Couny, J. C. Knight et al., "Compact, stable and efficient all-fibre gas cells using hollow-core photonic crystal fibres," *Nature*, 434, 488–491, March 2005.

[59] S. G. Leon-Saval, T. A. Birks, N. Y. Joly et al., "Splice-free interfacing of photonic crystal fibers," *Opt. Lett.*, 30, 1629–1631, July 2005.

[60] S. G. Leon-Saval, T. A. Birks, J. Bland-Hawthorn, and M. Englund, "Multimode fiber devices with single-mode performance," *Opt. Lett.*, 30, 2545–2547, 2005.

[61] T. A. Birks, G. Kakarantzas, P. St J. Russell, and D. F. Murphy, "Photonic Crystal Fiber Devices," in Proceedings of the Society of Photo-Instrumentation Engineers, 4943, 2002, pp. 142–151.

[62] W. J. Wadsworth, A. Witkowska, S. Leon-Saval, and T. A. Birks, "Hole inflation and tapering of stock photonic crystal fibers," *Opt. Express*, 13, 6541–6549, 2005.

[63] G. Kakarantzas, T. E. Dimmick, T. A. Birks et al., "Miniature all-fiber devices based on CO_2 laser microstructuring of tapered fibers," *Opt. Lett.*, 26, 1137–1139, August 2001.

[64] G. Kakarantzas, T. A. Birks, and P. St J. Russell, "Structural long-period gratings in photonic crystal fibers," *Opt. Lett.*, 27, 1013–1015, June 2002.

[65] G. Kakarantzas, A. Ortigosa-Blanch, T. A. Birks et al., "Structural rocking filters in highly birefringent photonic crystal fiber," *Opt. Lett.*, 28, 158–160, February 2003.

[66] S. G. Leon-Saval, T. A. Birks, W. J. Wadsworth et al., "Supercontinuum generation in submicron fibre waveguides," *Opt. Express*, 12, 2864–2869, June 2004.

[67] D. G. Ouzounov, F. R. Ahmad, D. Muller et al., "Generation of MW optical solitons in hollow-core photonic band-gap fibers," *Science*, 301, 1702–1704, 2003.

[68] J. D. Shephard, F. Couny, P. St J. Russell et al., "Improved hollow-core photonic crystal fiber design for delivery of nanosecond pulses in laser micromachining applications," *Appl. Opt.*, 44, 4582–4588, July 2005.

[69] J. D. Shephard, J. D. C. Jones, D. P. Hand et al., "High energy nanosecond laser pulses delivered single-mode through hollow-core PBG fibers," *Opt. Express*, 12, 717–723, February 2004.

[70] X. Peng and L. Dong, "Fundamental-mode operation in polarization-maintaining ytterbium-doped fiber with an effective area of 1400 μm^2," *Opt. Lett.*, 32, 358–360, February 2007.

[71] R. J. Beach, M. D. Feit, S. C. Mitchell et al., "Phase-locked antiguided multiple-core ribbon fiber," *IEEE Photon. Technol. Lett.*, 15, 670–672, May 2003.

[72] P. L. Cheo, A. Liu, and G. G. King, "A high-brightness laser beam from a phase-locked multicore Yb-doped fiber laser array," *IEEE Photon. Technol. Lett.*, 13, 439–441, May 2001.

[73] L. J. Cooper, P. Wang, R. B. Williams et al., "High-power Yb-doped multicore ribbon fiber laser," *Opt. Lett.*, 30, 2906–2908, November 2005.

[74] L. Michaille, C. R. Bennett, D. M. Taylor et al., "Phase locking and supermode selection in multicore photonic crystal fiber lasers with a large doped area," *Opt. Lett.*, 30, 1668–1670, July 2005.

[75] J. Nilsson, W. A. Clarkson, R. Selvas et al., "High-power wavelength-tunable cladding-pumped rare-earth-doped silica fiber lasers," *Opt. Fiber Technol.*, 10, 5–30, January 2004.

[76] G. Bouwmans, R. M. Percival, W. J. Wadsworth et al., "High-power Er:Yb fiber laser with very high numerical aperture pump-cladding waveguide," *Appl. Phys. Lett.*, 83, 817–818, August 2003.

[77] W. J. Wadsworth, R. M. Percival, G. Bouwmans et al., "Very high numerical aperture fibers," *IEEE Photon. Technol. Lett.*, 16, 843–845, March 2004.

[78] W. J. Wadsworth, R. M. Percival, G. Bouwmans et al., "High power air-clad photonic crystal fibre laser," *Opt. Express*, 11, 48–53, January 2003.

[79] A. Tunnermann, S. Hofer, S. Liem et al., "Power scaling of high-power fiber lasers and amplifiers," *Laser Phys.*, 15, 107–117, 2005.

[80] J. Limpert, T. Schreiber, S. Nolte et al., "High-power air-clad large-mode-area photonic crystal fiber laser," *Opt. Express*, 11, 818–823, April 2003.

[81] F. Di Teodoro and C. D. Brooks, "1.1 MW peak-power, 7 W average-power, high-spectral-brightness, diffraction-limited pulses from a photonic crystal fiber amplifier," *Opt. Lett.*, 30, 2694–2696, October 2005.

[82] C. D. Brooks and F. Di Teodoro, "Multimegawatt peak-power, single-transverse-mode operation of a 100 mu m core diameter, Yb-doped rodlike photonic crystal fiber amplifier," *Appl. Phys. Lett.*, 89, September 2006.

[83] C. D. Brooks and F. Di Teodoro, "1-mJ energy, 1-MW peak-power, 10-W average-power, spectrally narrow, diffraction-limited pulses from a photonic-crystal fiber amplifier," *Opt. Express*, 13, 8999–9002, October 2005.

[84] C. J. S. de Matos, S. V. Popov, A. B. Rulkov et al., "All-fiber format compression of frequency chirped pulses in air-guiding photonic crystal fibers," *Phys. Rev. Lett.*, 93, September 2004.

[85] J. Limpert, T. Schreiber, S. Nolte et al., "All fiber chirped-pulse amplification system based on compression in air-guiding photonic band gap fiber," *Opt. Express*, 11, 3332–3337, 2003.

[86] G. P. Agrawal, *Nonlinear Fiber Optics*, Academic Press, San Diego, London, 2001.

[87] R. R. Alfano, *The Supercontinuum Laser Source*, New York, Springer-Verlag, 1989.

[88] S. V. Chernikov, Y. Zhu, J. R. Taylor, and V. P. Gapontsev, "Supercontinuum self-Q-switched ytterbium fiber laser," *Opt. Lett.*, 22, 298–300, 1997.

[89] J. M. Dudley, G. Genty, and S. Coen, "Supercontinuum generation in photonic crystal fiber," *Rev. Mod. Phys.*, 78, 1135–1184, October–December 2006.

[90] W. J. Wadsworth, A. Ortigosa-Blanch, J. C. Knight et al., "Supercontinuum generation in photonic crystal fibers and optical fiber tapers: a novel light source," *J. Opt. Soc. Am. B – Opt. Physics*, 19, 2148–2155, September 2002.

[91] J. K. Ranka, R. S. Windeler, and A. J. Stentz, "Visible continuum generation in air–silica microstructure optical fibers with anomalous dispersion at 800 nm," *Opt. Lett.*, 25, 25–27, 2000.

[92] G. Humbert, W. J. Wadsworth, S. G. Leon-Saval et al., "Supercontinuum generation system for optical coherence tomography based on tapered photonic crystal fibre," *Opt. Express*, 14, 1596–1603, February 2006.

[93] I. Hartl, X. D. Li, C. Chudoba et al., "Ultrahigh-resolution optical coherence tomography using continuum generation in an air-silica microstructure optical fiber," *Opt. Lett.*, 26, 608–610, May 2001.

[94] H. Hundertmark, D. Kracht, D. Wandt, et al., "Supercontinuum generation with 200 pJ laser pulses in an extruded SF6 fiber at 1560 nm," *Opt. Express*, 11, 3196–3201, December 2003.

[95] R. Holzwarth, T. Udem, T. W. Haensch, et al., "Optical frequency synthesizer for precision spectroscopy," *Phys. Rev. Lett.*, 85, 2264–2267, September 2000.

[96] S. Coen, A. H. L. Chau, R. Leonhardt, et al., "Supercontinuum generation by stimulated Raman scattering and parametric four-wave mixing in photonic crystal fibers," *J. Opt. Soc. Am. B – Opt. Phys.*, 19, 753–764, April 2002.

[97] W. J. Wadsworth, N. Joly, J. C. Knight et al., "Supercontinuum and four-wave mixing with Q-switched pulses in endlessly single-mode photonic crystal fibres," *Opt. Express*, 12, 299–309, January 2004.

[98] Fianium Limited, www.fianium.com.

[99] F. G. Omenetto, N. A. Wolchover, M. R. Wehner et al., "Spectrally smooth supercontinuum from 350 nm to 3 um in sub-centimeter lengths of soft-glass photonic crystal fibers," *Opt. Express*, 14, 4928–4934, May 2006.

[100] M. Yu, C. J. McKinstrie, and G. P. Agrawal, "Modulational instabilities in dispersion-flattened fibers," *Phys. Rev. E*, 52, 1072–1080, 1995.

[101] A. Y. H. Chen, G. K. L. Wong, S. G. Murdoch et al., "Widely tunable optical parametric generation in a photonic crystal fiber," *Opt. Lett.*, 30, 762–764, April 2005.

[102] Y. J. Deng, Q. Lin, F. Lu et al., "Broadly tunable femtosecond parametric oscillator using a photonic crystal fiber," *Opt. Lett.*, 30, 1234–1236, May 2005.

[103] J. Lasri, P. Devgan, R. Y. Tang et al., "A microstructure-fiber-based 10-GHz synchronized tunable optical parametric oscillator in the 1550-nm regime," *IEEE Photon. Technol. Lett.*, 15, 1058–1060, August 2003.

[104] J. E. Sharping, M. Fiorentino, P. Kumar, and R. S. Windeler, "Optical parametric oscillator based on four-wave mixing in microstructure fiber," *Opt. Lett.*, 27, 1675–1677, October 2002.

[105] J. D. Harvey, R. Leonhardt, S. Coen et al., "Scalar modulation instability in the normal dispersion regime by use of a photonic crystal fiber," *Opt. Lett.*, 28, 2225–2227, November 2003.

[106] J. Fan and A. Migdall, "A broadband high spectral brightness fiber-based two-photon source," *Opt. Express*, 15, 2915–2920, March 2007.

[107] J. G. Rarity, J. Fulconis, J. Duligall et al., "Photonic crystal fiber source of correlated photon pairs," *Opt. Express*, 13, 534–544, January 2005.

[108] J. Fulconis, O. Alibart, W. J. Wadsworth et al., "High brightness single mode source of correlated photon pairs using a photonic crystal fiber," *Opt. Express*, 13, 7572–7582, September 2005.

[109] D. V. Skryabin, F. Luan, J. C. Knight, and P. St J. Russell, "Soliton self-frequency shift cancellation in photonic crystal fibers," *Science*, 301, 1705–1708, September 2003.

[110] N. Y. Joly, F. G. Omenetto, A. Efimov et al., "Competition between spectral splitting and Raman frequency shift in negative-dispersion slope photonic crystal fiber," *Opt. Commun.*, 248, 281–285, April 2005.

[111] F. Luan, D. V. Skryabin, A. V. Yulin, and J. C. Knight, "Energy exchange between colliding solitons in photonic crystal fibers," *Opt. Express*, 14, 9844–9853, October 2006.

[112] A. Podlipensky, P. Szarniak, N. Y. Joly et al., "Bound soliton pairs in photonic crystal fiber," *Opt. Express*, 15, 1653–1662, Feburary 2007.

[113] V. Laude, A. Khelif, S. Benchbane et al., "Phononic band-gap guidance of acoustic modes in photonic crystal fibers," *Phys. Rev. B*, 71, 045107, 2005.

[114] S. Guenneau and A. B. Movchan, "Analysis of elastic band structures for oblique incidence," *Arch. Rational Mech. Anal.*, 171, 129–150, 2004.

[115] P. St J. Russell, E. Marin, A. Diez et al., "Sonic band gaps in PCF preforms: enhancing the interaction of sound and light," *Opt. Express*, 11, 2555–2560, October 2003.

[116] P. St J. Russell, "Light in a tight space: enhancing matter-light interactions using photonic crystals," in Proc. NOC 2002, Hawaii, 2002, pp. 377–379.

[117] P. Dainese, P. St J. Russell, N. Joly et al., "Stimulated Brillouin scattering from multi-GHz-guided acoustic phonons in nanostructured photonic crystal fibres," *Nat. Phy.*, 2, 388–392, June 2006.

[118] J. C. Beugnot, T. Sylvestre, H. Maillotte et al., "Guided acoustic wave Brillouin scattering in photonic crystal fibers," *Opt. Lett.*, 32, 17–19, January 2007.

[119] P. Dainese, P. St J. Russell, G. S. Wiederhecker et al., "Raman-like light scattering from acoustic phonons in photonic crystal fiber," *Opt. Express*, 14, 4141–4150, May 2006.

[120] P. Rabinowitz, A. Stein, R. Brickman, and A. Kaldor, "Efficient tunable hydrogen Raman laser," *Appl. Phys. Lett.*, 35, 739–741, 1979.

[121] L. S. Meng, K. S. Repasky, P. A. Roos, and J. L. Carlsten, "Widely tunable continuous-wave Raman laser in diatomic hydrogen pumped by an external-cavity diode laser," *Opt. Lett.*, 25, 472–474, 2000.

[122] F. Benabid, J. C. Knight, G. Antonopoulos, and P. St J. Russell, "Stimulated Raman scattering in hydrogen-filled hollow-core photonic crystal fiber," *Science*, 298, 399–402, October 2002.

[123] F. Benabid, G. Bouwmans, J. C. Knight, et al., "Ultrahigh efficiency laser wavelength conversion in a gas-filled hollow core photonic crystal fiber by pure stimulated rotational Raman scattering in molecular hydrogen," *Phys. Rev. Lett.*, 93, September 2004.

[124] T. Brabec and F. Krausz, "Intense few-cycle laser fields: Frontiers of nonlinear optics," *Rev. Mod. Phys.*, 72, 545–591, 2000.
[125] A. Paul, R. A. Bartels, R. Tobey et al., "Quasi-phase-matched generation of coherent extreme-ultraviolet light," *Nature*, 421, 51–54, 2003.
[126] F. Couny, P. S. Light, F. Benabid, and P. St J. Russell, "Electromagnetically induced transparency and saturable absorption in all-fiber devices based on $^{12}C_2H_2$-filled hollow-core photonic crystal fiber," *Opt. Commun.*, 263, 28–31, July 2006.
[127] S. Ghosh, J. H. Sharp, D. G. Ouzounov, and A. L. Gaeta, "Resonant optical interactions with molecules confined in photonic band-gap fibers," *Phys. Rev. Lett.*, 94, 093902, 2005.
[128] K. Nakajima, K. Hogari, J. Zhou et al., "Hole-assisted fiber design for small bending and splice losses," *IEEE Photon. Technol. Lett.*, 15, 1737–1739, 2003.
[129] P. J. Roberts, B. Mangan, H. Sabert et al., "Control of dispersion in photonic crystal fibers," *J. Opt. Fib. Commun. Rep.*, 2, 435–461, 2005.
[130] R. Kashyap, *Fiber Bragg Gratings*, Academic Press, San Diego, 1999.
[131] S. J. Mihailov, D. Grobnic, H. M. Ding et al., "Femtosecond IR laser fabrication of Bragg gratings in photonic crystal fibers and tapers," *IEEE Photon. Technol. Lett.*, 18, 1837–1839, September–October 2006.
[132] D. N. Nikogosyan, "Multi-photon high-excitation-energy approach to fibre grating inscription," *Meas. Sci. Technol.*, 18, R1–R29, January 2007.
[133] H. Dobb, D. J. Webb, K. Kalli et al., "Continuous wave ultraviolet light-induced fiber Bragg gratings in few- and single-mode microstructured polymer optical fibers," *Opt. Lett.*, 30, 3296–3298, December 2005.
[134] A. Ashkin, "Acceleration and trapping of particles by radiation pressure," *Phys. Rev. Lett.*, 24, 156–159, 1970.
[135] M. J. Renn, R. Pastel, and H. J. Lewandowski, "Laser guidance and trapping of mesoscale particles in hollow-core optical fibers," *Phys. Rev. Lett.*, 82, 1574–1577, 1999.
[136] M. J. Renn, D. Montgomery, O. Vdovin et al., "Laser-guided atoms in hollow-core optical fiber," *Phys. Rev. Lett.*, 75, 3253–3256, 1995.
[137] F. Benabid, J. C. Knight, and P. St J. Russell, "Particle levitation and guidance in hollow-core photonic crystal fiber," *Opt. Express*, 10, 1195–1203, October 2002.
[138] T. M. Monro, W. Belardi, K. Furusawa et al., "Sensing with microstructured optical fibres," *Meas. Sci. Technol.*, 12, 854–858, 2001.
[139] W. N. MacPherson, E. J. Rigg, J. D. C. Jones et al., "Finite-element analysis and experimental results for a microstructured fiber with enhanced hydrostatic pressure sensitivity," *J. Lightw. Technol.*, 23, 1227–1231, March 2005.
[140] P. M. Blanchard, J. G. Burnett, G. R. G. Erry et al., "Two-dimensional bend sensing with a single, multi-core optical fibre," *Smart Mater. Struct.*, 9, 132–140, April 2000.
[141] W. N. MacPherson, J. D. C. Jones, B. J. Mangan et al., "Two-core photonic crystal fibre for Doppler difference velocimetry," *Opt. Commun.*, 223, 375–380, August 2003.
[142] M. T. Myaing, J. Y. Ye, T. B. Norris et al., "Enhanced two-photon biosensing with double-clad photonic crystal fibers," *Opt. Lett.*, 28, 1224–1226, July 2003.
[143] C. E. Finlayson, A. Amezcua-Correa, P. J. A. Sazio et al., "Electrical and Raman characterization of silicon and germanium-filled microstructured optical fibers," *Appl. Phys. Lett.*, 90, March 2007.
[144] P. J. A. Sazio, A. Amezcua-Correa, C. E. Finlayson et al., "Microstructured optical fibers as high-pressure microfluidic reactors," *Science*, 311, 1583–1586, 2006.

15

Specialty fibers for optical communication systems

Ming-Jun Li, Xin Chen, Daniel A. Nolan, Ji Wang, James A. West, and Karl W. Koch

Corning Inc., Corning, NY, USA

15.1 INTRODUCTION

Optical fiber communications have changed our lives in many ways over the last 40 years. There is no doubt that low-loss optical transmission fibers have been critical to the enormous success of optical communications technology. It is less well known however, that fiber-based components have also played a critical role in this success. Initially, fiber optic transmission systems were point to point systems, with lengths significantly less than 100 km. Then in the 1980s, rapid progress was made on the research and understanding of optical components including fiber components. Many of these fiber components found commercial applications in optical sensor technology such as in fiber gyroscopes and other optical sensor devices. Simple components such as power splitters, polarization controllers, multiplexing components, interferometric devices, and other optical components proved to be very useful. A significant number of these components were fabricated from polarization maintaining fibers (PMFs). Although not a large market, optical fiber sensor applications spurred research into the fabrication of new components such as polarization multiplexers for the optical gyroscope, for example. Apart from these polarization maintaining components, other components such as power splitters were fabricated from standard multimode (MM) or single-mode telecommunication fiber. In the telecommunication sector, the so-called passive optical network was proposed for the already envisioned fiber-to-the-home (FTTH) network. This network relied heavily on the use of passive optical splitters. These splitters were fabricated from standard single-mode fibers (SMFs). Although FTTH, at a large scale, did not occur until decades later, research into the use of components for telecommunications applications continued.

Optical Fiber Telecommunications V A: Components and Subsystems
Copyright © 2008, Elsevier Inc. All rights reserved.
ISBN: 978-0-12-374171-4

The commercial introduction of the fiber optic amplifier in the early 1990s revolutionized optical fiber transmissions. With amplification, optical signals could travel hundreds of kilometers without regeneration. This had major technical as well as commercial implications. Rapidly, new fiber optic components were introduced to enable better amplifiers and to enhance these transmission systems. Special fibers were required for the amplifier, for example, erbium-doped fibers. The design of high-performance amplifier fibers required special considerations of mode field diameter, overlap of the optical field with the fiber active core, core composition, and use of novel dopants. Designs radically different from those of conventional transmission fiber have evolved to optimize amplifier performance for specific applications. The introduction of wavelength division multiplexing (WDM) technology put even greater demands on fiber design and composition to achieve wider bandwidth and flat gain. Efforts to extend the bandwidth of erbium-doped fibers and develop amplifiers at other wavelength such as 1300 nm have spurred development of other dopants. Codoping with ytterbium (Yb) allows pumping from 900 to 1090 nm using solid-state lasers or Nd and Yb fiber lasers. Of recent interest is the ability to pump Er/Yb fibers in a double-clad geometry with high power sources at 920 or 975 nm. Double-clad fibers are also being used to produce fiber lasers using Yb and Nd.

Besides the amplification fiber, the erbium-doped amplifier requires a number of optical components for its operation. These include wavelength multiplexing and polarization multiplexing devices for the pump and signal wavelengths. Filters for gain flattening, power attenuators, and taps for power monitoring among other optical components are required for module performance. Also, because the amplifier-enabled transmission distances of hundreds of kilometers without regeneration, other propagation properties became important. These properties include chromatic dispersion, polarization dispersion, and nonlinearities such as four-wave mixing (FWM), self- and cross-phase modulation, and Raman and Brillouin scattering. Dispersion compensating fibers were introduced in order to deal with wavelength dispersion. Broadband compensation was possible with specially designed fibers. However, coupling losses between the transmission and the compensating fibers was an issue. Specially designed mode conversion or bridge fibers enabled low-loss splicing among these three fibers, making low insertion loss dispersion compensators possible. Fiber components as well as microoptic or in some instances planar optical components can be fabricated to provide for these applications. Generally speaking, but not always, fiber components enable the lowest insertion loss per device. A number of these fiber devices can be fabricated using standard SMF, but often special fibers are required.

Specialty fibers are designed by changing fiber glass composition, refractive index profile, or coating to achieve certain unique properties and functionalities. In addition to applications in optical communications, specialty fibers find a wide range of applications in other fields, such as industrial sensors, biomedical power delivery and imaging systems, military fiber gyroscope, high-power lasers, to

15. Specialty Fibers for Optical Communication Systems

name just a few. There are so many kinds of specialty fibers for different applications. Some of the common specialty fibers include the following:

- Active fibers: These fibers are doped with a rare earth element such as Er, Nd, Yb or another active element. The fibers are used for optical amplifiers and lasers. Erbium doped fiber amplifiers are a good example of fiber components using an active fiber. Semiconductor and nanoparticle doped fibers are becoming an interesting research topic.
- Polarization control fibers: These fibers have high birefringence that can maintain the polarization state for a long length of fiber. The high birefringence is introduced either by asymmetric stresses such as in Panda, and bow-tie designs, or by asymmetric geometry such as elliptical core, and dual airhole designs. If both polarization modes are available in the fiber, the fiber is called PMF. If only one polarization mode propagates in the fiber while the other polarization mode is cutoff, the fiber is called single polarization fiber.
- Dispersion compensation fibers: Fibers have opposite chromatic dispersion to that of transmission fibers such as standard SMFs and nonzero dispersion-shifted fibers (NZDSFs). The fibers are used to make dispersion compensation modules for mitigating dispersion effects in a fiber transmission system.
- Highly nonlinear optical fibers: Fibers have high nonlinear coefficient for use in optical signal processing and sensing using optical nonlinear effects such as the optical Kerr effect, Brillouin scattering, and Raman scattering.
- Coupling fibers or bridge fibers: Fibers have mode field diameter between the standard SMF and a specialty fiber. The fiber serves as an intermediate coupling element to reduce the high coupling loss between the standard SMF and the specialty fiber.
- Photo-sensitive fibers: Fibers whose refractive index is sensitive to ultraviolet (UV) light. This type of fiber is used to produce fiber gratings by UV light exposure.
- High numerical aperture (NA) fibers: Fibers with NA higher than 0.3. These fibers are used for power delivery and for short distance communication applications.
- Special SMFs: This category includes standard SMF with reduced cladding for improved bending performance, and specially designed SMF for short-wavelength applications.
- Specially coated fibers: Fibers with special coatings such as hermitic coating for preventing hydrogen and water penetration, metal coating for high-temperature applications.
- Mid-infrared fibers: Non-silica glass-based fibers for applications between 2 and 10 μm.
- Photonic crystal fibers (PCFs): Fibers with periodic structures to achieve fiber properties that are not available with conventional fiber structures.

This is by no means an exhaustive list. It is practically impossible to cover all specialty fiber in one chapter or even in one book. For optical communications-related specialty fibers, some of them have been discussed in great detail in previous editions of Optical Fiber Telecommunications. For example, erbium-doped fibers and amplifiers are discussed in Chapter 2 of the third edition and Chapter 4 of the fourth edition. Fiber gratings and devices are covered in Chapter 7 of the third edition and Chapter 10 of the fourth edition.

In this chapter, we select a few specialty fiber topics for discussion. Section 2 discusses dispersion compensation fibers. Section 15.3 deals with PMFs and single polarization fibers (SPFs). Section 15.4 focuses on nonlinear fibers. Section 15.5 describes double-clad fiber for high-power lasers and amplifiers. And the last section, Section 15.6 covers the topic of photonic crystal fibers.

15.2 DISPERSION COMPENSATION FIBERS

15.2.1 Dispersion Compensation Technologies

High-capacity WDM long-haul transmission systems with large channel counts ($\geq$40 channels) and high channel speeds ($\geq$10 Gb/s) are being proposed and deployed to meet the increasing demand for bandwidth [1–4]. In such systems, chromatic dispersion is one of the primary limits [5–7]. There are two approaches to combating the effects of dispersion. One is to design new fibers with reduced dispersion, and the other is to compensate the dispersion with dispersion compensation modules. Standard SMF with a dispersion of around 17 ps/nm/km at 1550 nm limits the transmission distance to less than 100 km for a system with 10 Gbit/s bit rate or higher. To increase the transmission distance, NZDSFs with reduced dispersion were developed for WDM systems. However, even with NZDSFs, some dispersion compensation at the terminals is needed for a 10-Gb/s WDM system with 500-km distance. For a 10-Gbit/s WDM system with a longer reach or a 40 Gb/s WDM system, both dispersion and dispersion slope compensation are required.

Different technologies have been proposed to achieve dispersion and dispersion slope compensation. They can be classified into four categories:

- Dispersion compensation fibers
- Higher order mode dispersion compensation fibers
- Fiber Bragg grating-based devices
- Virtual image-phased array (VIPA)-based devices
- Planar waveguide-based devices

Among the dispersion compensation technologies listed above, dispersion compensation fiber [8–21] is currently the best choice because it has good overall performance and has been proven to be a reliable solution. Furthermore, large-scale production processes are available. Although the other technologies possess some unique features and potentials, they have issues that need to be addressed.

This section discusses dispersion compensation fibers using the fundamental mode.

15.2.2 Principle of Dispersion Compensating Fibers

Fiber Dispersion

When an optical pulse propagates in an optical fiber, different frequency components travel with different velocities. The phase velocity is given by $v_p = \omega/\beta$, where ω is the frequency, and β is the propagation constant. The phase velocity describes the propagation of the phase front. For an optical pulse, the center of the pulse travels with a speed called group velocity, which is defined as $v_g = d\omega/d\beta$. When a pulse travels though an optical fiber, the time delay per unit length due to the group velocity is called group delay: $\tau_g = 1/v_g = d\beta/d\omega$. The change of the group delay with frequency results in fiber dispersion, which is defined as

$$D = \frac{d\tau_g}{d\lambda} = -\frac{2\pi c}{\lambda} \frac{d^2\beta}{d\omega^2} \approx \frac{\lambda}{c} \frac{d^2 n_{eff}}{d\lambda^2} \tag{15.1}$$

where λ is the wavelength, c is the speed of light in the vacuum, and the n_{eff} is the effective index of the mode in the fiber. The fiber dispersion causes pulse broadening that limits the transmission data rate.

The fiber dispersion consists of two parts: material dispersion and waveguide dispersion. For silica-based fiber material doped with Germania (GeO_2) or Fluorine, the material dispersion changes slightly with the dopant level and its value is positive in the wavelength range of interest between 1300 and 1650 nm. The waveguide dispersion can be changed by fiber profile design and its value can be either positive or negative.

Fibers that have been installed or are being installed in long-haul systems can be classified into three categories: standard SMF, dispersion-shifted fiber (DSF), and NZDSF. The fiber dispersion for these three types of fiber is shown in Figure 15.1, together with the fiber attenuation curve. These fibers have different dispersion and dispersion slope values in the 1550-nm window. Table 15.1 compares the values of dispersion, dispersion slope, and dispersion/dispersion slope ratio of different types of fibers at 1550 nm. It is common to use the κ value, that is, the ratio of dispersion to dispersion slope, when describing fiber dispersion characteristics. The fibers listed in Table 15.1 have κ values ranging from 50 to 300 nm, which means they need different amounts of dispersion and dispersion slope compensation.

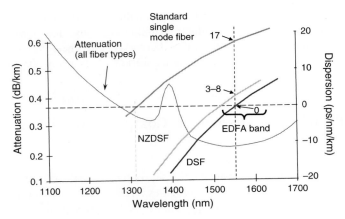

Figure 15.1 Dispersion curves of different single mode fibers (this figure may be seen in color on the included CD-ROM).

Table 15.1
Different transmission fiber types.

Fiber type	Dispersion (ps/nm.km)	Dispersion slope (ps/nm².km)	κ (nm)
Standard single mode fiber	17	0.058	298
NZDSF Type 1	4.2	0.085	50
NZDSF Type 2	4.5	0.045	100
NZDSF Type 3	8	0.057	140

Dispersion Compensation

The objective of dispersion compensation is to use a device that has opposite dispersion to that of the transmission fiber to achieve a total low residual dispersion for a transmission link. The total residual dispersion D_r of a transmission fiber span with a dispersion compensation fiber is defined as

$$D_r = D_{Tr} L_{Tr} + D_{DC} L_{DC} \quad (15.2)$$

where D_{Tr} and D_{DC} are the dispersion of the transmission and the dispersion compensation fibers, respectively. L_{Tr} and L_{DC} are the lengths of the transmission and dispersion compensation fibers, respectively. The dispersion change with wavelength of any type of fiber in a transmission window can be described to a good approximation by a three-term Taylor expansion around the center wavelength λ_0:

$$D = D_0 + S(\lambda - \lambda_0) + C(\lambda - \lambda_0)^2 \quad (15.3)$$

where $S = dD/d\lambda$ is the dispersion slope and $C = (1/2)d^2D/d\lambda^2$ is the dispersion curvature. If dispersion compensation is needed only for a single wavelength, only the dispersion at that wavelength needs to be matched and the dispersion slope and curvature need not be considered. We call this type of dispersion compensation fiber, single wavelength dispersion compensation. The first deployment of dispersion compensation fibers was for upgrading single wavelength 1310-nm transmission systems to 1550 nm. As system designs changed from single wavelength to multiple wavelengths using WDM technology, the need arose for simultaneous compensation of dispersion for all the wavelengths in the transmission window. For WDM systems, both the dispersion slope and the curvature must be considered. For most transmission fibers, the dispersion can be represented by a linear function because the dispersion curvature is small and can be neglected. For a linear dispersion function, we define a parameter κ, which is the ratio of the dispersion D to the dispersion slope S:

$$\kappa = \frac{D}{S} \quad (15.4)$$

A simultaneous compensation of the dispersion and dispersion slope of the transmission fiber implies

$$\kappa_{Tr} = \kappa_{DC} \quad (15.5)$$

We call this type of dispersion compensation fiber, dispersion slope compensation fiber. For a dispersion slope compensation fiber with a large κ value (>200 nm), the dispersion has a linear dispersion function along with negative dispersion slope that can be designed to match the dispersion and dispersion slope of transmission fibers resulting in very low residual dispersion. If more accurate dispersion compensation is needed, it is necessary to consider the dispersion curvature of the dispersion slope compensation fiber. This can be accomplished by adding a second compensation fiber to compensate the dispersion curvature. We call this type of dispersion compensation fiber, dispersion curvature compensation fiber. If the residual dispersion after the dispersion slope compensation fiber has a residual curvature of C_r, the dispersion curvature compensation fiber needs to have a curvature with the opposite sign, that is

$$C_{CC} = -C_r \quad (15.6)$$

In addition to matching fiber dispersion characteristics of transmission fibers, there are several other factors that must be considered. First, a dispersion compensation fiber needs to have high negative dispersion value and low attenuation. It is common to use a figure of merit to describe the efficiency of a dispersion compensation fiber, defined as the dispersion value D divided by the fiber attenuation α:

$$F = \frac{D}{\alpha} \quad (15.7)$$

The figure of merit is related to the attenuation increase in a link due to the dispersion compensation fiber. The higher the figure of merit is, the lower the incremental attenuation that is added to the system. The fiber attenuation is affected by many factors. For an optical fiber, the total attenuation can be expressed by

$$\alpha = \alpha_{\text{absorption}} + \alpha_{\text{scattering}} + \alpha_{\text{imperfection}} + \alpha_{\text{bending}} \tag{15.8}$$

In Eqn 15.8, $\alpha_{\text{absorption}}$ is the attenuation caused by material absorptions at UV and infrared wavelengths and color centers. $\alpha_{\text{scattering}}$ is the attenuation due to linear scattering that is due to fluctuations in material density and composition. This type of scattering is also called "Rayleigh scattering." In addition to linear scattering, nonlinear scattering can occur, if the power is sufficiently high, which increases fiber attenuation. The nonlinear scattering includes stimulated Brillouin scattering (SBS) due to acoustic phonons and stimulated Raman scattering (SRS) due to optical phonons. The third term is the attenuation due to waveguide imperfections such as defects and stresses. The last term, α_{bending} is the attenuation due to fiber bending including both the macro- and the microbending. The absorption loss does not change much with fiber design for Ge-doped silica fibers. On the other hand, different profile designs can change the other three terms significantly.

Second, a dispersion compensation fiber needs to have a sufficiently large effective area to minimize the nonlinear effects introduced to the system. If the effective area is too small, the nonlinear effects in the dispersion compensation fiber can introduce system penalty and increase splice losses. In the nonlinear regime, a higher figure of merit as defined in Eqn 15.7 does not mean that the system will have better performance. From a nonlinear perspective, a different figure of merit, the so-called nonlinear figure of merit, is necessary to take into account the impact of nonlinear effects [22]. The nonlinear figure of merit takes into consideration the dispersion D, fiber attenuation α, effective area A, and nonlinear coefficient γ of both the dispersion compensation fiber and the transmission fiber, and is defined as follows

$$F_{\text{nl}} = \frac{\gamma_{\text{Tr}}}{\gamma_{\text{DC}}} \frac{A_{\text{DC}}}{A_{\text{Tr}}} \frac{\alpha_{\text{DC}}}{\alpha_{\text{Tr}}} \frac{(l_{\text{Tr}} - 1)^2}{l_{\text{Tr}}} \frac{l_{\text{DC}}}{(l_{\text{DC}} - 1)^2} \tag{15.9}$$

where $l_{\text{Tr}} = e^{\alpha_{\text{Tr}} L_{\text{Tr}}}$, $l_{\text{DC}} = e^{\alpha_{\text{DC}} L_{\text{DC}}} = e^{\alpha_{\text{DC}} L_{\text{Tr}} D_{\text{Tr}}/D_{\text{DC}}}$. The nonlinear figure of merit is closely related to the system performance. A higher nonlinear figure of merit results in a lower nonlinear signal degradation. The design of dispersion compensation fiber should aim for increasing the dispersion and effective area while reducing the attenuation. Because the dispersion and the dispersion slope of dispersion compensation fiber are controlled by waveguide dispersion, the effective area tends to become smaller when the dispersion and the dispersion slope become more negative. It is a challenging task to design a dispersion compensation fiber with a high figure of merit and a large effective area.

15. Specialty Fibers for Optical Communication Systems

Third, a dispersion compensation fiber must introduce low multiple path interference (MPI) to the system when splicing to transmissions fibers. MPI happens when more than one mode propagates in the fiber [23], and the power coupling between the fundamental mode and the higher-order modes causes the transmitted power to fluctuate with wavelength. For single-mode dispersion compensation fibers, this implies the effective cutoff wavelength must be below the operating window.

Fourth, the polarization mode dispersion (PMD) of a dispersion compensation fiber must be low. Because of the high delta and small core designs for dispersion compensation fibers, their PMD values are normally much higher than that of transmission fibers. Low PMD requires more careful control of the fiber geometry and stresses during the fiber-manufacturing process. In addition, fiber-spinning techniques can be used to mitigate the PMD of dispersion compensation fibers [24, 25].

15.2.3 Dispersion Compensation Fibers

Fiber Profile Designs

In designing dispersion compensation fibers, one needs to manipulate the fiber refractive index profile to achieve negative waveguide dispersion and dispersion slope. The profile designs for dispersion compensation fibers can be grouped into four design types, as shown in Figure 15.2 [8–21]. Design (a) is a simple step core with a single cladding that was used in the early days of dispersion compensation fiber development. The design has two parameters only: the relative core refractive index change Δ and the core radius r. Total negative dispersion is achieved by creating more negative waveguide dispersion to move the zero-dispersion wavelength to a wavelength longer than the operating wavelength such as 1550 nm. This is done by adjusting the core delta and core radius. Negative waveguide dispersion results from the rapid change of power from the core to the cladding with wavelength. As the wavelength increases, the mode power extends further

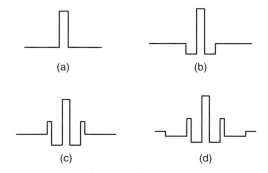

Figure 15.2 Profile types for dispersion compensation fibers.

into the cladding. Because the cladding has a lower refractive index than that of the core, the group index decreases and the group velocity increases as more and more power moves into the cladding. Thus, the group delay, which is inversely proportional to the group velocity, decreases with wavelength. As a result, the waveguide dispersion is negative. If the rate of power change is accelerated when increasing the wavelength, the slope of waveguide dispersion also becomes negative. While this design can achieve high negative dispersion in the 1550-nm window, it cannot achieve large negative dispersion slope to compensate the dispersion slope of transmission fibers.

To design a fiber with negative dispersion slope, it is necessary to increase the rate of power change. This can be done by designing a fiber in which the fundamental mode has a cutoff wavelength. When the mode is close to the cutoff wavelength, the power will move rapidly from the core to the cladding, which enables more negative waveguide dispersion and negative waveguide dispersion slope. To create a cutoff for the fundamental mode, one can add a depressed cladding layer to form a W-shape profile as a shown in design (b) of Figure 15.2. This design has four profile parameters to control: core delta Δ_1, core radius r_1, depressed cladding delta Δ_2, and depressed cladding radius r_2. The addition of a depressed cladding offers more degree of freedom to the profile design. By choosing the right profile parameters, both negative dispersion and negative dispersion slope can be achieved. However, in the high negative dispersion and slope wavelength region, the fiber is more bend-sensitive because the fundamental mode is near the cutoff wavelength.

To increase the cutoff wavelength, a ring can be added to the W-profile to form a W-ring profile as shown in design (c) in Figure 15.2. This type of design has six parameters that can be changed: core delta Δ_1, core radius r_1, depressed cladding delta Δ_2, depressed cladding radius r_2, ring delta Δ_3, and ring radius r_3. The ring raises the effective index of the fundamental mode. As a result, the cutoff wavelength of the fundamental mode is increased and the bending loss is reduced.

But for some fiber designs, to get the desired dispersion properties and bending performance, the fiber cutoff of higher order modes has to be higher than the operating window. The high cutoff may cause the MPI problem as discussed in the earlier section. To reduce the higher order mode cutoff wavelength, design (d) in Figure 15.2 can be used to solve this problem. In this design, a depressed cladding is introduced after the ring. If the index and dimension of this depressed cladding are chosen properly, it can strip the higher order modes without affecting the fundamental mode significantly.

Design Examples

In the following paragraphs, design examples using the W-ring type of profile for compensating standard SMF and NSDSF will be discussed. Figure 15.3 shows an example for dispersion and slope compensation for a standard SMF. The dispersion of the standard SMF as a function of wavelength in the C band is plotted in the upper

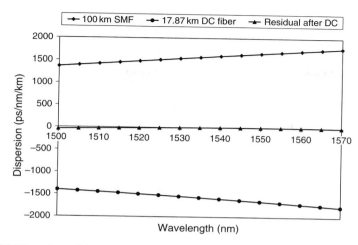

Figure 15.3 Dispersion and slope compensation of 100 km of standard single mode fiber in the C-band.

curve in Figure 15.3. In this example, the κ-value is about 300 at 1550 nm. To compensate the dispersion and the slope of this fiber, a fiber using the W-ring profile is designed that has a negative dispersion and slope that match that of the standard SMF. The dispersion curve of 17.87 km of the dispersion compensation fiber is plotted in the lower curve. By putting the two fibers together, the total dispersion becomes very small as shown by the middle curve in Figure 15.3. The residual dispersion is less than ± 0.04 ps/nm/km in the whole wavelength range of C band.

The second design example is for a NZDSF whose dispersion in 100 km in the C-band is shown in Figure 15.4. Compared to the standard SMF case, the

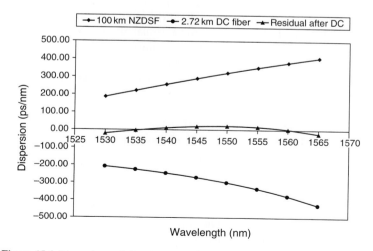

Figure 15.4 Dispersion and slope compensation of 100 km of NZDSF in the C band.

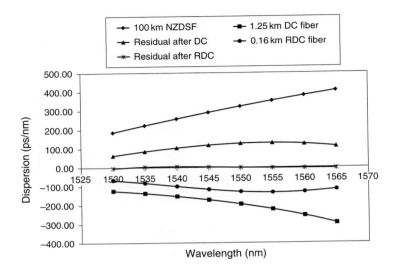

Figure 15.5 Modeled dispersion and dispersion slope compensation of 100 km NZDSF fiber in the C band using dispersion compensation and residual dispersion correction fibers.

dispersion of the NZDSF is much lower. The κ-value is about 50 at 1550 nm, much smaller than that of the standard SMF in the first example. A dispersion compensation fiber is designed for this fiber using also a W-ring design. The total dispersion of 2.72 km of the dispersion compensation fiber is shown in the lower curve in Figure 15.4. The total residual dispersion is ±0.2 ps/nm/km.

In the NZDSF example, the residual dispersion is still too large for 40 Gb/s applications, because the dispersion curvature of the two fibers are not well-matched. One way to further reduce the residual dispersion is to add another fiber to correct the dispersion curvature. To illustrate this, a design example is shown in Figure 15.5 for the same NZDSF. Now, instead of using only one dispersion compensation fiber, a residual dispersion correction fiber is added after the dispersion compensation fiber. The dispersion compensation fiber of 1.25 km long is used to compensate most of the dispersion and the slope. The residual dispersion correction fiber of 0.16 km is employed to correct residual dispersion. As can be see in Figure 15.5, modeling results show that the residual dispersion is reduced below ±0.04 ps/nm/km.

Experimental Results

Dispersion compensation fibers using different designs have been manufactured and tested by several companies. Table 15.2 summarizes experimental results on different types of dispersion and slope compensation fibers for the transmission

15. Specialty Fibers for Optical Communication Systems

Table 15.2
Experimental Results on different types of dispersion and slope compensation fibers.

Profile	Wavelength or band	Dispersion	Slope	κ (nm)	MFD (μm)	A_{eff}	Fiber attenuation (dB/km)	PMD	FOM	Residual dispersion (ps/nm/km)	Reference
W-ring	1550	−45	0.07				0.4		75		8
W-ring	1550	−90	0.00				0.65		140		8
W-ring	1550	−77	−0.02				0.5		130		8
W-ring	1550	−100	0.07				0.65		155		8
W-ring	1550	−120	−0.2	600			0.45		267		11
W-ring	1550	−37	−0.11	336			0.28		132		11
W	C	−125.2	−0.410	305				0.14			9
W-ring	C	−105	−0.33	318			0.60	0.10	210	0.1	10
W-ring	C	−105.4	−0.33	307						0.087	10
W-ring	C	−168	−1.59	106	4.9		0.75		224	0.09	10
W-ring	L	−98	−0.63	156	4.5		0.72		136		13
W-ring	C	−144	−1.35	106	4.4	15	0.55	0.058	261	0.11	13
W	L	−11.8	−0.105	112	5.5	23	0.434	0.05	27	0.04	18
W	L	−8.2	−0.066	124	5.4	22	0.25	0.05	33		14
W-ring	C	−37	−0.26	142			0.32	0.10	116	0.1	14
W-ring	C	−100	−0.220	455	5.2		0.5		200	0.08	17
W-ring	C	−98	−0.333	286	5.1		0.5		190	0.08	16
W-ring	C	−100	−0.67	149	4.5		0.68		150	0.08	16
W	C	−87	−0.71	123		19	0.3		290		19
W	C	−145	−1.34	108		15	0.5		290		19
W-ring	C			50	5.5		0.60			0.1	20
W-ring	L			108	4.4		0.60			0.1	20
W-ring	C			298	4.9		0.59			0.1	20
W-ring	C									0.005	21
W-ring	L									0.02	21

fiber types in Figure 15.1. Key characteristics of the fibers are discussed as follows:

Dispersion and Dispersion Slope Table 15.2 shows that dispersion from -8 to -178 ps/nm/km and the dispersion slope from 0.07 to -1.59 ps/nm^2/km have been demonstrated. A wide range of κ-values from 50 to above 300 has been realized, which is suitable for dispersion and slope compensation for different transmission fibers.

Attenuation Higher negative dispersion is realized at the expense of higher attenuation. This can be seen in Figure 15.6, where the attenuation is plotted against the fiber dispersion. The data are taken from Table 15.2. Although the fibers are from different companies, there is a clear trend in which the attenuation increases when the dispersion becomes more negative. This attenuation increase is primarily due to bending loss increase, which sets the limit on how large a negative dispersion can be realized in a practical fiber. To reduce the loss due to bending in a dispersion compensation module, loosely coiled fiber has been proposed.

Mode Field Diameter and Effective Area The mode field diameter of the fibers in Table 15.2 ranges from 4.4 to 5.5 μm, which corresponds to an effective area of less than 23 μm^2. A useful formula to calculate the effective area A_eff from the measured mode field diameter w is

$$A_\text{eff} = kw^2,$$

where k is a constant. It has been found that the k for dispersion compensation fiber is 0.76. The effective area achievable is limited by the acceptable bend

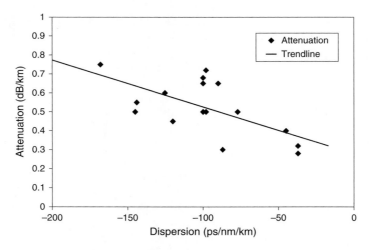

Figure 15.6 Fiber attenuation vs dispersion.

sensitivity. Because the high dispersion acts to rapidly increase the effective area as a function of wavelength, the trade-off between effective area and bend loss we observe in any fiber design is intensified. In order to achieve a fiber with good bend characteristics, the effective area must be reduced. The result is dispersion compensation fibers with effective areas less than 20 μm^2, although advanced coating formulations can be leveraged to increase this figure.

Nonlinear Properties Because the nonlinear refractive index n_2 of dispersion compensation fiber is only slightly higher than conventional fiber, the major factor affecting the nonlinear properties is the effective area. The common perception is that small effective areas of dispersion compensation fiber are unacceptable because propagation nonlinearities will be increased. Self-phase modulation (SPM) and cross-phase modulation (XPM) generated in the dispersion compensation fiber are the nonlinear impairments most likely to affect transmission. SPM is a significant impairment when the power launched per channel exceeds several milliwatts in a typical dispersion compensation module. XPM and FWM penalties are observed to be very small compared with transmission fiber. This is because (1) the fiber has high dispersion and (2) the power launched into the dispersion compensation fiber is not proportionally higher. SPM, on the other hand, is a single-channel impairment, which depends more on the fiber nonlinearity (n_2/A_{eff}) rather than on local dispersion. The nonlinear response of a typical dispersion compensation module (consisting of several kilometers of small effective area fiber) is often negligible in long-haul systems as long as the dispersion compensation module is deployed where the power per channel is not extremely large. We should note that with the proliferation of Raman amplification, smaller effective areas in dispersion compensation fibers may actually prove beneficial.

PMD As discussed earlier, it is a challenge to make dispersion compensation fibers with low PMD because dispersion compensation fibers are more sensitive to core deformation than conventional fibers due to the high core delta. To reduce the PMD in dispersion compensation fibers, extreme care must be taken to reduce core asymmetry due to geometry and stresses. In addition, a common technique to improve PMD is to use fiber spinning during the fiber draw. It has been reported that fiber spinning reduces the effective fiber PMD distribution. The average PMD is lowered from 0.21 to 0.091 ps/km$^{1/2}$, and the standard deviation is reduced from 0.19 to 0.065 ps/km$^{1/2}$.

PMD is sensitive to fiber deployment. For dispersion compensation fiber that is coiled in a dispersion compensation module, the coil diameter has a significant impact on fiber PMD. In a coiled fiber, fiber bending introduces stress birefringence. This stress-induced birefringence causes a significant increase in fiber PMD because dispersion compensation fibers are very sensitive to fiber bending. One way to reduce the stress-induced birefringence is to reduce the fiber diameter from the standard 125 μm, for example, to 80 μm. However, a fiber with reduced fiber diameter is more susceptible to microbending loss. In addition, reducing the cladding diameter does nothing to improve bent fiber form birefringence, another potentially significant source of birefringence. Both PMD and microbending loss must be considered in dispersion compensation fiber design. With proper designs and advanced manufacturing technologies, PMD below $0.1\,\text{ps/km}^{1/2}$ has been demonstrated, as summarized in Table 15.2.

Splicing Loss Dispersion compensation fibers have a mode field diameter of about 5 μm, which is much smaller than the 10.5 μm mode field diameter of standard SMF. Theoretical calculations show that the coupling loss due to the mode field mismatch is as high as 2.2 dB. However, practical fusion splice loss between dispersion compensation fiber and standard SMF is much lower than this value. This is because doping materials in the core regions will diffuse when the fiber is heated during the fusion-splicing process, which creates a tapered region with increased mode field diameter. By optimizing the fusion splice parameters, the splice loss between dispersion compensation fiber and standard SMF is reduced to less than 0.25 dB.

15.2.4 Summary

Dispersion compensation fibers were discussed in this section. Design concepts and physics for achieving negative dispersion and dispersion slope are explained. Compensation fiber designs for standard SMF and NZDSF fibers are described in detail.

Results on single mode dispersion compensation fibers from different companies were reviewed. They show that dispersion compensation fibers with different dispersion and slope values are possible and can be used for different types of transmission fibers. With the continuous development of high-data-rate, long-distance WDM optical networks, it is expected that dispersion will remain an important issue. As in the past, transmission fiber technologies and dispersion compensation technologies will work together to mitigate dispersion effects. Future developments are anticipated in the following areas:

- Dispersion-managed fiber cable systems. In these systems, the cable is made of matched positive and negative dispersion fibers. By managing the lengths of the two fibers, optimum system performance can be achieved. Because the positive and negative dispersion fibers compensate each other, this kind of system can have very low residual dispersion that can be easily compensated, if further compensation is needed.
- Continuous improvements in performance of dispersion compensation fibers. These improvements include increasing the fiber's effective area, reducing residual dispersion, and reducing bending and splice losses.

- Dispersion compensation fiber amplifiers. This type of dispersion compensation fiber devices incorporate SRS to compensate both dispersion and power loss.
- Development of dynamic tunable dispersion compensation devices. For systems with a bit rate greater than 40 Gb/s, very accurate dispersion control is required. Such systems will necessitate dynamic tunable dispersion devices on a channel by channel basis.
- Better understanding of electronic dispersion compensation (EDC) technique in terms of its capability and limitations. This will enable designing new transmission and dispersion compensation fibers to work with EDC.

15.3 POLARIZATION MAINTAINING AND SINGLE POLARIZATION FIBERS

15.3.1 Polarization Effects in Optical Fibers

Although SMFs are often misunderstood to be fibers that can only transmit a "single" mode by simply taking the words literally, SMFs can actually support two polarization modes. By one definition [26], SMFs are optical fibers which support only a single mode per polarization direction at a given wavelength. Each polarization mode has its own effective propagation constant, which can take different values resulting in polarization effects. Extensive study of polarization effects in SMF began several decades ago [27]. In the past decade, they received increasing attention as they can induce significant impairment in higher bit-rate optical communication systems.

In practice, it is very difficult to make fibers with perfect circular symmetry so that the two orthogonally polarized HE_{11} modes have the same propagation constant. The origin of polarization effects in SMFs is related to the asymmetry in either the fiber geometry or the stress distribution in the fibers. The latter results from imperfections in the preform-making process, which could be modified chemical vapor deposition (MCVD), outside vapor deposition (OVD), or other processes. The distortion of the geometry of the fiber core leads directly to different propagation constants or phase velocities in the two polarization modes. At different radial positions in the fiber, the fiber is formed with different chemical composition with distinct thermal expansion coefficients. The distortion in geometry can cause the stress field in the core to become asymmetric as the fiber is drawn and cooled down from a high temperature. Through the stress–optic effect, the asymmetric stress is translated into an anisotropic refractive index change across the fiber. Even after the fiber is made, the polarization properties of SMFs can be altered through external factors such as bending, twisting, or applying lateral load to the fibers when the fibers are deployed for actual use.

The fundamental polarization property of a fiber is characterized by birefringence B, which is defined as the difference of the effective refractive indices in the two polarization modes,

$$B = \Delta n_{\text{eff}} = n_{\text{eff}}^x - n_{\text{eff}}^y \qquad (15.10)$$

Birefringence is directly related to the difference of the propagation constants between the two local polarization modes $\Delta\beta (= \beta_x - \beta_y = \frac{\omega \Delta n_{\text{eff}}}{c})$. The birefringence affects the polarization evolution of fields propagating in the fiber, which results in a periodic evolution of the state of polarization along the fiber. The characteristic length or period of the polarization evolution is often referred to as the beat length. The beat length, L_B, is directly related to the birefringence $\Delta\beta$ and the difference of the refractive index Δn_{eff},

$$L_B = \frac{2\pi}{\Delta\beta} = \frac{\lambda}{B} \qquad (15.11)$$

The concept described in Eqn (15.10) is based on the phase velocity difference. However, in the telecommunication systems, the light is often transmitted in the form of pulses, which can be better described by the local difference of the group velocity as related to pulse broadening. Therefore, we can further define the concept of group birefringence and group beat length. In order to define the group birefringence, we first introduce the group index calculated from the phase index,

$$N_{\text{eff}} = n_{\text{eff}} + \omega \frac{dn_{\text{eff}}}{d\omega} = n_{\text{eff}} - \lambda \frac{dn_{\text{eff}}}{d\lambda} \qquad (15.12)$$

The group birefringence B_g and group beat length L_B^g can thus be defined in a similar manner to those in Eqns (15.10–15.11),

$$B_g = N_{\text{eff}}^x - N_{\text{eff}}^y = B - \lambda \frac{dB}{d\lambda}, \qquad (15.13)$$

$$L_B^g = \frac{\lambda}{\Delta N_{\text{eff}}} \qquad (15.14)$$

In contrast, the birefringence and the beat length defined in Eqns (15.10–15.11) are also referred to as phase or modal birefringence and phase beat length, respectively. In conventional fibers including transmission fibers and specialty fibers such as PMFs, the difference between the phase and the group birefringence is often negligible. In those situations, people often ignore the difference in their definitions and use the concepts interchangeably.

The role of polarization effects in telecommunication systems can be detrimental as it can cause pulse broadening and impair the system performance. In many devices where a linear state of polarization is desired, the birefringence of the fiber can alter the state of polarization at the fiber output. The fiber polarization properties can be affected by many factors such as bending, lateral loading, spinning, and twisting. A detailed understanding of the polarization evolution in birefringent fibers is needed. The Jones matrix formalism can be used to describe various effects in polarization evolutions [28, 29]. In the local coordinate system and linear polarization representation, the evolution of the amplitudes associated with each electric field can be described as follows:

$$\begin{pmatrix} \frac{da_x(z)}{dz} \\ \frac{da_y(z)}{dz} \end{pmatrix} = \begin{pmatrix} i\frac{\text{Re}(\Delta\beta)}{2} & \alpha(z) + i\frac{\text{Im}(\Delta\beta)}{2} \\ -\alpha(z) + i\frac{\text{Im}(\Delta\beta)}{2} & -i\frac{\text{Re}(\Delta\beta)}{2} \end{pmatrix} \cdot \begin{pmatrix} a_x(z) \\ a_y(z) \end{pmatrix} \qquad (15.15)$$

where $\Delta\beta$ is the superimposed birefringence taking into account original (or intrinsic) fiber birefringence $\Delta\beta_0$, the induced birefringence $\Delta\beta_i$ due to either bending or lateral load, and the relative orientation between the two sources of the birefringence,

$$\Delta\beta = \Delta\beta_0 + \Delta\beta_i \exp(-2i \int_0^z \alpha(z')dz'), \qquad (15.16)$$

where $\alpha(z)$ is the rate of change of the original birefringent axis over position. The Jones matrix which describes the full details of the polarization evolution can be calculated from the above equation, and an additional transformation can be made to recover the full information in the fixed coordinate system [29].

In practice, the efforts of mitigating the undesired polarization effects, however, diverge in two directions. For optical fibers used for transmission purpose, the polarization effects are primarily described using PMD. Although extensive studies have been conducted to understand PMD behaviors in terms of its origin, measurement, and statistical features, etc., a critical issue is to make fibers with low PMD [25]. This can be achieved by improving the fiber-manufacturing process so that the fiber geometry is consistently as circular as possible. Another widely adopted approach is to spin the fiber so that the linear birefringence and therefore the PMD can be significantly reduced even when the local birefringence cannot be made with arbitrarily low values. On the other hand, for many devices, a linear state of polarization from one part of the device to another need strictly be maintained. In these cases, an opposite approach is taken. PMFs or SPFs (sometimes also referred to as polarizing fibers), which intrinsically have significant asymmetric structure or high birefringence, have been used to achieve such goals.

The PMF and SPF as specialty fibers is the subject of the remaining part of this section. In the following subsections, we will review the progress in these fields both with historic approaches and with recent advances.

15.3.2 Polarization-Maintaining Fibers

Maintaining the state of polarization in SMFs can be achieved by making fibers with very high birefringence or very low birefringence. In a fiber with very low birefringence, the polarization state is very sensitive to external perturbation. A slight perturbation, such as bending or pressing the fiber, will induce sufficient birefringence to alter the state of the polarization. In practice, people typically utilize fibers with very high birefringence on the order of 10^{-4} and align the linear input state of polarization with one of the birefringence axes of the PMF to achieve a polarization-maintaining operation. Birefringence induced by external factors, which are several orders of magnitude smaller, is not sufficient to alter the state of polarization in these fibers. The ability of the PMF to maintain the state of polarization is characterized by the mode-coupling parameter h determined from the measured crosstalk [30],

$$CT = 10\log(P_y/P_x) = 10\log[\tanh(hl)], \qquad (15.17)$$

where P_x and P_y are the powers of the excited mode and the coupled mode after going through a fiber with length l.

In the early 1980s, PMFs were widely studied and became commercially available [30]. The mechanisms of achieving high birefringence in PMFs fall into two categories. In one category, the birefringence is induced with stress. The cross-section views of several such fibers, including Panda-type [31], Bow-Tie-type [32], and Elliptical Cladding [33] PMFs are shown in Figure 15.7. Among many designs, Panda-type and Bow-Tie-type PMFs are most widely used. The stress members can be chosen from materials, such as borosilicate, which has high mismatch in thermal

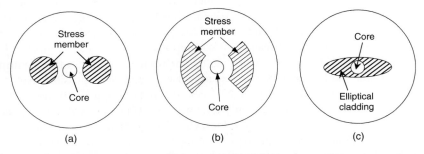

Figure 15.7 Cross-section view of several polarization-maintaining fibers with birefringence mechanism based on the stress effect. (a) Panda-type PMF, (b) Bow-tie-type PMF, (c) Elliptical cladding PMF.

expansion coefficient compared with pure silica surrounding the stress members. The birefringence of these fibers typically ranges from 3×10^{-4} to 7.2×10^{-4}. Other kinds of PMF based on the use of geometrical effects were also widely studied, which include elliptical core fibers [34] and side tunnel fibers [35]. However, these PMFs typically exhibit high loss due to the large refractive-index difference between the core and the cladding and the imperfections at the interfaces. The stress-based PMFs are more widely used.

In recent years, the field of PMFs has seen renewed interest due to the intensive study of microstructured fibers, which are also referred to as PCFs. PCFs have been explored to achieve high birefringence because they can have much higher refractive index contrast than conventional fibers, and complex microstructures can be implemented by the stack-and-draw fabrication process. In one category of PCFs, the light-guiding mechanism is known to be through index guiding, similar to conventional fibers. High birefringence can be achieved either by having different airhole diameters along the two orthogonal axes or by asymmetric core design, which induces high form (modal) birefringence [36–38]. Various configurations have been theoretically and experimentally studied. The modal birefringence of these fibers is in the order of 10^{-3}, which is one order of magnitude higher than conventional PMFs based on stress effects and also results in better polarization-maintaining properties. The high birefringence in index-guiding PCFs can also occur by adding stress rods outside of several layers of airholes which form the photonic crystal structure. These fibers enjoy the additional benefit of having large-mode area (LMA) suitable for high-power fiber laser applications [39].

Another category of PCFs provides the light guidance through photonic bandgap (PBG) effect [40, 41]. A highly birefringent PBG fiber was experimentally demonstrated [41]. A cross-section view of the fiber is shown in Figure 15.8. In Ref. [41], the group birefringence as high as 0.025 was measured. It was later clarified that the large group birefringence is due to a large negative slope of the

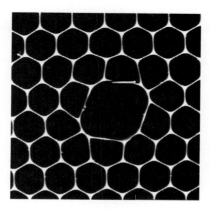

Figure 15.8 Cross section of the hollow-core PBGF. The dark regions are air, while the white regions are glass.

modal birefringence with wavelength, which contributes to the second term of group birefringence in Eqn (15.13) [42]. The modal birefringence of this type of fiber is 10^{-3}, one order of magnitude lower than the group birefringence, which is quite different from conventional fibers. It is also notable that despite the high birefringence, the polarization-maintaining properties of such fibers have not yet been demonstrated because of the involvement of surface modes [43, 44]. However, with a recently developed technique based on a resonant coupling mechanism [45], the higher-order modes may be suppressed to achieve high extinction ratio and polarization-maintaining PBG fibers.

15.3.3 Single-Polarization Fibers

Polarization-maintaining fibers guide two polarization modes. In order to maintain the linear state of polarization, one must align the input polarization state with one of the birefringent axes. This method suffers from the drawback that polarization coupling degrades performance over a long fiber length. Unlike PM fibers, SPFs guide only one polarization mode and thus avoid problems related to PMD and polarization-dependent loss.

One can achieve SPFs by increasing the attenuation of one polarization mode while keeping the other unaffected over a specific wavelength range. To meet this goal, large birefringence is required for separating the two polarization modes, and a fundamental mode cutoff mechanism is needed to eliminate the unwanted polarization mode. In combination of the two effects, one polarization mode will reach the cutoff sooner or at a lower wavelength than the other polarization mode. Single-polarization performance can be achieved within the wavelength window between the two cutoff wavelengths.

Different approaches have been adopted to achieve single polarization in fibers. One approach, which relies on thermal stress-induced birefringence from boron-doped elliptical cladding region, was proposed and implemented with a single-polarization bandwidth of around 45 nm [46]. The bandwidth was further improved to 100 nm [47].

In the past few years, hole-assisted SPFs were implemented and commercialized [48, 49]. A schematic cross-section view of this fiber is shown in Figure 15.9.

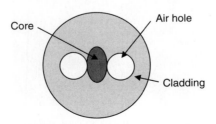

Figure 15.9 Cross section of a dual airhole fiber as SPF and PMF.

It has a pair of airholes placed next to an elliptical core. The relatively high refractive index contrast between the fiber core and the surrounding airholes provides high birefringence. The birefringence is largely due to the geometric effect. The low-index airholes also provide a mechanism for fundamental mode cutoff. The core dimension along the direction determined by the centers of the two airholes can be adjusted to yield the single-polarization behavior at the desired operating wavelengths. Figure 15.10 shows measured single-polarization operating windows of three SPFs achieved in the wavelength windows of interest around 1060, 1310, and 1550 nm. SPFs based on the dual airhole fiber structure have reached single-polarization bandwidth exceeding 55 nm [48]. The fibers exhibit very high extinction ratio of around 60 dB for a fiber length less than 30 cm in the single-polarization operating region.

In SPFs based on the dual airhole structure, airholes play the dual role of inducing high birefringence and fundamental cutoff. To have sufficiently high birefringence, the refractive index of the core should be high enough, which limits the core size of the fiber for single-mode operation. A recent design illustrated in Figure 15.11 utilizes both airholes and stress rods aligned perpendicular to one another [50]. Different from the airholes, the stress rods contribute birefringence through the stress–optic effect. Although the nature of the birefringence contributed from the airholes and the stress members is different, the birefringence from the two sources is additive. The fact that large birefringence can be achieved from stress birefringence eases the requirement for high-core refractive index. Therefore, this design offers the flexibility in choosing lower doping level in the core to allow for larger core dimension. Numerical modeling based on this scheme shows that single polarization over much larger operating windows and with larger effective area

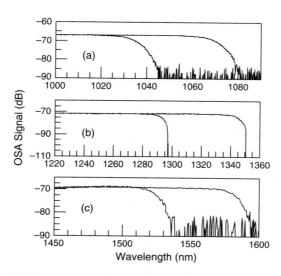

Figure 15.10 Measured cutoff wavelengths of the two polarization modes.

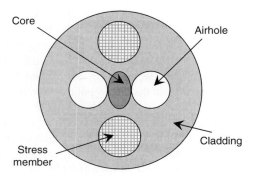

Figure 15.11 A single-polarization fiber design utilizing both airholes and stress members.

can be achieved. For example, one can achieve greater than 150 nm single-polarization operating window with effective areas greater than 120 μm^2 around 1550 nm.

SPFs can also be realized with photonic crystal structures. SPF utilizing two or more larger diameters of airholes asymmetrically aligned near the core region have been both theoretically and experimentally studied [51, 52]. The PCF structure allows the extinction of light propagation above or below certain wavelengths, providing two cutoff mechanisms to achieving single-polarization operation. In Refs [51, 52], the birefringence mechanism is from geometric effects. Alternative approaches have also been proposed by placing large stress members outside symmetric photonic crystal structures [52] or embedding stress rods within a photonic crystal structure by replacing some of the airholes with stress rods [53].

15.4 NONLINEAR FIBERS

Nonlinear effects in optical fibers such as SRS, SBS, and optical Kerr effect have many useful applications in telecommunications and in optical signal processing [54, 55]. The optical Kerr effect, in which the refractive index changes with optical power leads to various secondary effects, such as SPM, XPM, FWM, and modulation instability. Applications using the Kerr effect include optical parametric amplification and frequency conversion [56], optical phase conjugation [57], pulse compression [58] and regeneration [59], optical soliton propagation [60], etc. Each of the applications requires properly designed SMFs with high nonlinearity, appropriate dispersion properties, and low attenuation.

The nonlinear coefficient of optical fiber is determined by two factors [54]. The first one is the nonlinear refractive index n_2, which depends on the fiber material. The second one is the effective area, which is related to the fiber design. To increase the nonlinear coefficient, one can either increase the nonlinear refractive index n_2 or decrease the effective area. The most common material used to make optical fiber is Ge-doped silica whose n_2 is 3.2×10^{-16} cm/W. Although materials with n_2 higher than Ge-doped silica glass exist (e.g. chalcogenide glass), it is difficult to make an optical fiber with low loss using these materials. For Ge-doped silica fiber, because the increase of n_2 is not very significant with Ge concentration, the main player for increasing the nonlinearity is to decrease the effective area. However, the effective area is only one of the many fiber parameters that need to be considered. Other parameters such as dispersion, cutoff wavelength, and attenuation are also important for nonlinear applications. Therefore, an optimum nonlinear fiber design depends on the tradeoffs among the different parameters.

This section discusses nonlinear fiber designs for applications using optical Kerr effects.

15.4.1 Nonlinear Fiber Requirements

In designing a nonlinear fiber, several requirements need to be considered. The first one is that the fiber needs to have high nonlinearity to achieve an efficient nonlinear interaction. The fiber nonlinearity can be described by the nonlinear coefficient γ, which is defined by [54]

$$\gamma = \frac{2\pi}{\lambda} \frac{n_2}{A_{\text{eff}}} \tag{15.18}$$

where n_2 is the fiber nonlinear refractive index, and A_{eff} is the effective area of the mode, and λ is the wavelength of the light. For Ge-doped fibers, n_2 does not change significantly with Ge-doping level. Therefore, the effective area has to be small to increase the nonlinear coefficient.

The second requirement is that the fiber needs to have low attenuation to increase the effective interaction length, which is defined as

$$L_{\text{eff}} = \frac{1 - e^{-\alpha L}}{\alpha} \tag{15.19}$$

where L is the fiber length. For a long fiber, L_{eff} is approximately $1/\alpha$. Lowering the fiber attenuation increases the effective interaction length. The four factors that contribute to total attenuation are described by Eqn (15.8). Some of these factors can be changed by profile designs.

If the loss due to the nonlinear scattering is not significant, it is common to use a figure of merit to describe the nonlinear efficiency:

$$F_0 = \frac{\gamma}{\alpha} \tag{15.20}$$

When the pump power reaches the threshold of nonlinear scattering, the loss due to the nonlinear scattering becomes dominant. However, in the nonlinear scattering regime where the loss due to nonlinear scattering increases with the pump power, the conventional figure of merit definition is no longer appropriate to describe the nonlinear efficiency. A new figure of merit definition was introduced by taking the pump power P_0 into account:

$$F = \frac{P_0 \gamma}{\alpha} = \frac{L_{\text{eff}}}{L_{\text{nl}}} \quad (15.21)$$

Note that $1/\alpha$ is the effective length and $1/P_0\gamma$ is the nonlinear length. The physical meaning of the nonlinear figure of merit is that the nonlinear effect is enhanced by increasing the effective length or decreasing the nonlinear length. Because the threshold of SRS is normally much smaller than that of SBS, the pump power that can be used is mainly limited by SBS. Therefore, in the nonlinear scattering regime, the pump power loss is approximately proportional to the average gain coefficient of SBS $\bar{g}_B$ and the pump power:

$$\alpha = \bar{g}_B P_0 \quad (15.22)$$

The figure of merit in Eqn (15.21) becomes:

$$F = \frac{\gamma}{\bar{g}_B} \quad (15.23)$$

This new definition is convenient to compare fibers when the SBS limits the amount of useable power. High figure of merit means more pump power can be used to produce the desired nonlinear effect.

The third requirement is that the fiber must have the right amount of dispersion, which depends on the application. For example, for FWM, the dispersion needs to be near zero to achieve the phase-matching condition. For optical soliton effects, a certain amount of anomalous ($D > 0$) dispersion is required depending on the pump power level. Furthermore, it is advantageous to design a dispersion-decreasing fiber (DDF) to adapt the power decrease along the fiber for this type of application [61].

Another requirement is that the fiber must have low PMD. There are two ways to reduce fiber PMD [25]. The first one is to minimize asymmetries in the index profile and stress profile [62]. The second method for reducing fiber PMD is to introduce controlled polarization-mode coupling by fiber spinning [24, 63].

15.4.2 Fiber Designs

Fiber profile design plays an important role in meeting the three above-mentioned requirements. For designing a nonlinear fiber, the simplest design is the step-index

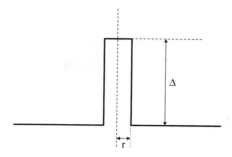

Figure 15.12 Simple step-index profile.

profile [64] shown in Figure 15.12. It has a core surrounded by a cladding. The design has only two parameters: the core refractive index change Δn and the core radius r. It is more convenient to use the relative refractive index change (delta) to describe the refractive index change, which is defined as

$$\Delta = \frac{n_1^2 - n_2^2}{2n_1^2} \approx \frac{n_1 - n_2}{n_1} \quad (15.24)$$

where n_1 is the refractive index of the core, and n_2 is the refractive index of the cladding.

To achieve a small effective area, the core Δ needs to be high and the core radius needs to be small. Figure 15.13 plots the effective area as a function of core

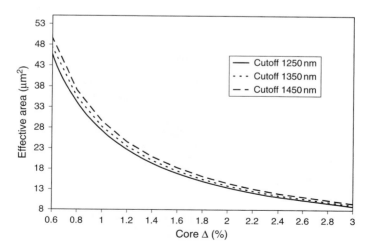

Figure 15.13 Effective area as a function of core delta for step-index profiles with different cutoff wavelengths.

Δ for three different cutoff wavelengths. To maintain the same cutoff wavelength, the core radius was adjusted in the calculation. It can be seen that the effective area decreases with increasing core Δ. For the same core delta, a fiber with lower cutoff wavelength has smaller effective area because the core radius is smaller. But the effective area change does not significantly change with cutoff wavelength. For a 200 nm shift in cutoff wavelength, the effective area change is less than 10%.

One drawback of increasing the core Δ is that the fiber attenuation also increases. This is because higher Ge-doping level increases both the loss due to Rayleigh scattering and the loss due to the waveguide imperfection. The effect of core Δ on fiber attenuation is shown in Figure 15.14. The attenuation increases almost linearly with the increase in core Δ. We notice that the rate of change with core Δ for the attenuation is less than that for the effective area. Therefore, the overall figure of merit is improved by increasing the core Δ. However, the core Δ is limited by the doped material properties and the process capability. For practical Ge-doped fibers, the maximum core Δ that can be made is about 3%, and the smallest effective area is limited to about $10\,\mu m^2$.

The effective area also depends on fiber chromatic dispersion. Figure 15.15 shows the chromatic dispersion change with core Δ for three cutoff wavelengths. It is clear that higher-core Δ reduces the dispersion value in the positive dispersion region and increases the dispersion value in the negative dispersion region. The dispersion puts a limit on how small an effective area can be realized. For example, if a positive dispersion of 5 ps/nm/km is required, the smallest effective area that can be achieved is about $20\,\mu m^2$. We also notice that high positive dispersion with small effective area is difficult to realize with a step-index design.

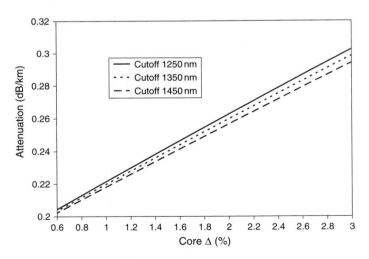

Figure 15.14 Attenuation as a function of core delta for step-index profiles with different cutoff wavelengths.

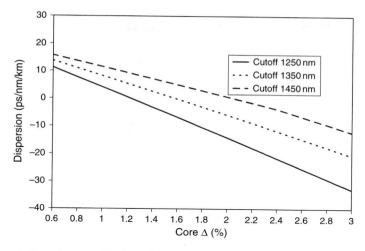

Figure 15.15 Dispersion as a function of core delta for step-index profiles with different cutoff wavelengths.

A more flexible design is to use the W-profile [65] as shown in Figure 15.16. In this design, a lower-index inner cladding or depressed cladding is added to the step-index core. The design has four parameters to control: core refractive index Δ_1, core radius r_1, depressed cladding refractive index Δ_2, and the depressed cladding radius r_2. The addition of a depressed cladding offers more degrees of freedom to the fiber design. With this design, different dispersion values can be achieved. Especially, high positive dispersion is possible. Furthermore, the effective area can be kept small.

To show the effects of the depressed cladding on nonlinear fiber properties, we calculated the effective area change with refractive index in the depressed

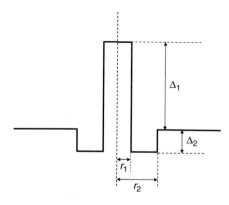

Figure 15.16 W-profile design.

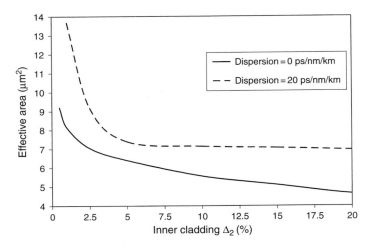

Figure 15.17 Effective area change with inner cladding delta Δ_2 for two dispersion values.

cladding. Figure 15.17 plots the effective area as a function of delta in the depressed cladding for two different dispersion values, 0 and 20 ps/nm/km. In the calculation, the core delta Δ_1 was kept to a constant value of 3% for the zero-dispersion curve, while the core delta was 2% for the 20 ps/nm/km dispersion curve. For each depressed delta Δ_2, the core radius and the depressed cladding radius were adjusted to achieve the desired dispersion value. In Figure 15.17, for both dispersion values, the effective area decreases very rapidly with the depressed cladding delta up to about 5%. Then the change becomes much slower. For the dispersion of 20 ps/nm/km, effective areas as small as 7 μm^2 can be achieved with this structure. For the dispersion value of 0 ps/nm/km, the effective area is even smaller because of the high-core delta.

We have seen earlier that, with the simple step design, it is impossible to design a fiber with a large positive dispersion and a small effective area simultaneously. This feature can be realized by using the W-profile design as shown in Figure 15.18. In generating Figure 15.18, only the depressed cladding delta was changed, while the other three parameters were kept constant: the core delta was 2%, the core radius was 1.9 μm, and the depressed cladding radius was 6 μm. In Figure 15.18, the dispersion increases while the effective area decreases with the increase of depression level in the inner cladding. Figure 15.18 demonstrates that a fiber dispersion as high as 70 ps/nm/km and an effective area as small as 7 μm^2 is possible with a Δ_2 value of 20%.

Using conventional fiber-making technologies such as MCVD, OVD, and PCVD, the maximum delta in the inner cladding that can be made is less than 2%. To increase the depression level, one can use hole-assisted W-fiber design [66] as shown in Figure 15.19. In the hole-assisted design, a ring of airholes are placed next to the core. The airholes lower the average index in the inner cladding,

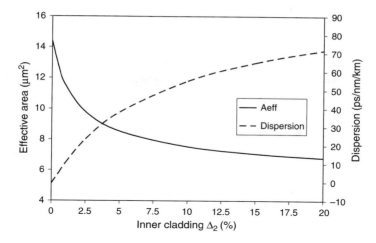

Figure 15.18 Dispersion and effective area change with inner cladding delta Δ_2.

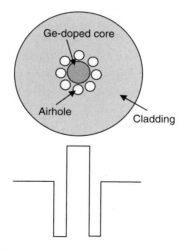

Figure 15.19 Hole-assisted W-profile design.

creating an effective depressed cladding structure. In this design, the depression level in the inner cladding depends on the air fraction. For an average delta of −20%, the air-fill fraction is about 65%, which can be realized in a practical hole-assisted fiber. Using a hole-assisted design, a fiber with dispersion of 43.8 ps/nm/km and effective area of 23 μm^2 has been demonstrated [66].

For applications in optical signal processing such as adiabatic pulse compression using optical soliton effects, DDFs are desired to achieve effective amplification through the change of the dispersion. There are two ways to make DDFs.

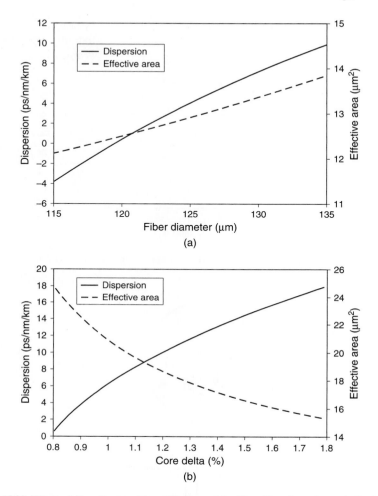

Figure 15.20 Effects of fiber diameter (a) and fiber core delta (b) on fiber dispersion and effective area.

The first one is to change the core diameter, and the second one is to change the core delta. The core diameter can be changed easily by changing the fiber diameter during the fiber draw process. Figure 15.20(a) shows an example of dispersion change with the fiber diameter. The profile in Figure 15.20(a) is a W-profile with the following parameters when the fiber diameter is 125 μm: $\Delta_1 = 2\%$, $r_1 = 1.89$ μm, $\Delta_2 = -0.4\%$, and $r_2 = 7$ μm. By drawing the fiber diameter from 135 to 115 μm, the dispersion is decreased by about 14 ps/nm/km. Meanwhile, the effective area is also decreased. The core delta change can be realized by changing the dopant level in the core during the preform-making process. An example of dispersion change with core delta is shown in Figure 15.20(b). The profile in

Figure 15.20(b) is also a W-profile. By reducing the core delta from 1.8 to 0.8%, the dispersion is decreased by about 16 ps/nm/km. Unlike the case of fiber diameter change, the effective area increases with decreasing dispersion. One advantage of changing the core delta is that SBS effects can be reduced, while changing the core diameter has very little impact on SBS.

15.4.3 Typical Fiber Performance

Silica-based nonlinear fibers can be manufactured by different manufacturing processes such as OVD, MCVD, and PCVD. For nonlinear fibers made by the OVD process, typical fiber performance is listed in Table 15.3. It can be seen that the effective area is in the range between 15 and 25 μm^2, depending on the dispersion and dispersion slope of the fiber. Dispersion values can be controlled from 0 to 20 ps/nm/km. Dispersion slope is less than 0.06 ps/nm^2/km. Flat dispersion around 1550 nm can also be achieved. The cutoff wavelength is less than 1550 nm, which is suitable for applications at the 1550-nm wavelength window. Typical attenuation is about 0.5 dB/km.

Table 15.3
Typical properties of nonlinear fibers.

Parameter	Typical value
Effective area	15–25 μm^2
Dispersion	0–20 ps/nm/km
Dispersion slope	0–0.06 ps/nm^2/km
Cutoff wavelength	1200–1500 nm
Attenuation	0.4–0.7 dB/km

15.5 DOUBLE-CLAD FIBERS FOR FIBER LASERS AND AMPLIFIERS BY OVD

15.5.1 Introduction

The concept of double-clad fibers first emerged in the late 1980s at Polaroid Corporation [67] as a means to boost power-scaling properties in fiber lasers and amplifiers. Fiber lasers/amplifiers had been invented in the 1960s by Snitzer [68] at the then American Optical, not long after the first demonstration of solid-state lasers [69]. However, it was not until the 1980s that these fiber-based devices began to be exploited in industrial applications. There were two key-enabling drivers developed during that period. The first was the development of semiconductor laser diode, which enabled a practical and efficient pumping mechanism for fiber-based lasers and amplifiers. The second was achievement of low-loss, rare-earth-doped optical fiber [70] as the gain media in the fiber devices. Early work on single-transverse-mode fiber lasers focused on devices that could be pumped with single-mode laser diodes [71]. Then, it became apparent later in the 1980s that scaling to higher powers was limited by the single-mode pump access. To overcome the limitation, the group at Polaroid fabricated the first double-clad fiber, also called "cladding-pumped fiber," as a way to explore the use of higher power capable, MM semiconductor laser diodes/bars to generate higher power fiber laser devices [72]. The concept generated intense interest worldwide on the development of practical high-power fiber laser oscillators and amplifiers with superior beam-quality to other conventional solid-state lasers.

Attractive features of fiber lasers are many that include small footprint, light weight, low operation/maintenance cost, minimal thermal-effect, excellent quantum-efficiency and reliability, superior beam-quality (often near diffraction-limited), flexible beam delivery, and capable of a variety of operating wavelengths for both CW- and pulsed-mode operation. In recent years, high-power fiber lasers have clearly emerged as a new generation of lasers. They are highly competitive for many important industrial applications such as in material processing, biomedical, printing, microfabrication, military, coherent, and space communications systems. Most typically, they compete in those applications traditionally served by diode- or lamp-pumped solid-state YAG lasers, and potentially also in areas covered by high-power CO_2 gas lasers among others. These recognitions have enabled fiber lasers to steadily gain ground on other laser markets, especially after several recent impressive demonstrations of near-diffraction-limited beam qualities at kilowatt output power levels from fiber laser sources [73].

New glass-composition fiber design and advanced fiber fabrication have, and will continue, to play key roles as technological enablers for further scaling the output power of fiber laser oscillator or amplifier with high beam quality [74]. This section will briefly review the fundamental working principle of double-clad optical fibers, and then introduce some newly established fiber design and fabrication method by OVD for the development of advanced double-clad optical fibers that can mitigate, or eliminate, various detrimental nonlinear optical effects, such as SBS, SRS, and other Kerr-type nonlinear effects in higher power applications.

15.5.2 Double-Clad Fiber 101

Different from conventional "single-clad" optical fiber, double-clad optical fiber has a core with two cladding layers, the inner- and outer clad, as schematically shown in Figure 15.21. The inner-clad is optically used as a "pump-core" to allow the launched light coupling into the fiber from MM semiconductor laser diodes/bars, while the outer clad accordingly serves as a "pump-clad" to secure the pump guiding and coupling from the MM laser-diodes or laser-bars. Hence the reason in the early days, these fibers were often referred to as "cladding-pumped fibers."

15. Specialty Fibers for Optical Communication Systems

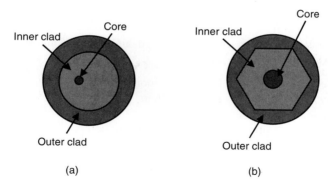

Figure 15.21 Schematic of double-clad optical fibers (this figure may be seen in color on the included CD-ROM).

The pump coupling is conventionally launched through either or both fiber-end(s). However, there are novel methods developed by pumping through the sides of double-clad fibers [75]. The core is preferably Yb-doped in the double-clad fiber for high-power applications. This is because the Yb^{3+} has a very simple energy-level scheme as shown in Figure 15.22. It has key characteristics like low quantum-defect (~10% when pumping at 976 nm and lasing at 1080 nm), and high quantum-yield (no ESA or upconversion transitions). Plus, Yb^{3+} has high absorption cross sections at 976, 940, and 915 nm (also shown in Figure 15.22), where powerful semiconductor laser diodes or bars are readily available in hundreds of kWs commercially. These make Yb an ideal dopant, and well suited, for high-power applications. Other typical rare-earth dopants for active fibers include Er^{3+}, Tm^{3+}, Nd^{3+}, Pr^{3+}, Ho^{3+}, Tb^{3+}, and Sm^{3+}. The doped core is preferably centered in the fiber but having a noncircular inner-clad perimeter shape; otherwise, an off-centered core is necessary [76] as given in Figure 15.21(a). This is to allow good overlap of the MM pump-light with the actively doped core largely through

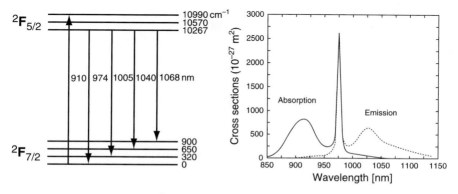

Figure 15.22 Yb^{3+} energy levels and spectral properties in silica.

mode-mixing in a shortest fiber-length possible, which will be discussed later in more detail.

The basic requirements for a double-clad high-power fiber are generally to satisfy the following:

- large-core effective-area for minimal nonlinear impairment;
- high coupling efficiency from the pump;
- good pump absorption with the core;
- highest dopant concentration possible without compromising Quantum Efficiency (QE).

Accordingly, when designing double-clad optical fibers, the preferred large effective area core, or LMA, needs a rather low NA to secure a high-brightness fiber laser output. Figure 15.23 shows the effective area change with core size for different NAs for step-index fiber. Compromise is required as to how low an NA is practical. Ideally, the fiber is designed such that the high-order modes (HOMs) are stripped off through macro-bend loss mechanisms, while the fundamental mode is left intact to realize good beam quality. For a Yb-doped high-power fiber laser operating around 1 μm with a step-index core, an NA of 0.07 or lower is typical and core effective area up to $1000\,\mu m^2$ or greater is not uncommon. These fibers are slightly MM as shown in Figure 15.24. Hence, suppression of HOMs through bending or coiling is necessary to achieve high beam quality.

Bending is effective for removing HOMs as shown in Figure 15.25, where the bending losses of LP_{01} and LP_{11} modes are plotted against the bend radius for 20- and 30-μm core fibers. The NA value for both fibers is 0.06. To create a bend loss of several dB/m, the bend radius is about 10 cm for a 20-μm core fiber and 5 cm for a 30-μm core fiber. For a fiber with a core larger than 30 μm, the bend radius is even smaller than 5 cm. However, as the bend radius decreases, two primary

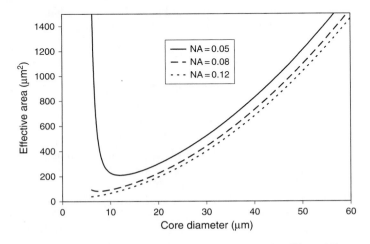

Figure 15.23 Effective area changes with core diameter for different NAs.

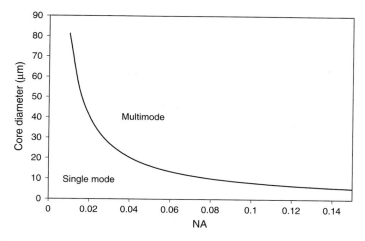

Figure 15.24 Core diameter as a function of NA for cutoff wavelength of 1060 nm.

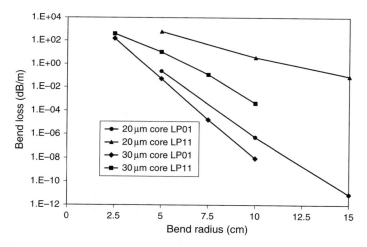

Figure 15.25 Bending losses of LP01 and LP11 modes for the fibers with 20- and 30-μm core diameters.

concerns arise. The first is the mechanical durability/reliability optical fiber due to bending a rather large OD double-clad fiber in a small bend-radius. The other is the mode-field deformation and the effective area reduction resulting from a small bend-radius, which is shown in Figure 15.26. For the fiber with a 20-μm core, the effective area change is negligible for bend-radii down to 5 cm. But, for the fiber with a 30-μm core, the effective area drops rather quickly as bend radius decreases. For a bend radius of 5 cm, the effective area is reduced by more than 25%. For a core size larger than 30 μm, the effective area drops even faster and the benefit of

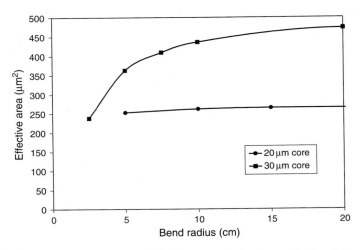

Figure 15.26 Effective area changes with the bend radius for the fibers with 20- and 30-μm core.

larger core size disappears. Therefore, more advanced alternatives or additional fiber-design measures are required for extra LMA (e-LMA) fiber design. Known examples in this regard include helical-core double-clad fibers [77] and HOM suppression/conversion methods [78].

To allow the high-power MM pump light from laser diode bars to be efficiently coupled into double-clad fiber, a high NA for the inner-clad "pump-core" is critical. This is achieved through the use of a low-refractive-index glass preferably, or polymer as the outer-clad material. In high-power applications, the polymer near the inner–outer clad interface can easily burn or gradually degrade during fiber laser operation. This degradation is caused by moisture, stress, or absorption of radiation, particularly in the case of bent fibers [79] that rely on differential modal loss to achieve diffraction-limited output. Polymers, however, are still mostly used in reality due to limited or very poor commercial availability of low-index glasses that are compatible with silica [80]. The inner-clad NA of 0.3–0.5 and dimension in the range of about 200–450 μm are most typical. A center-positioned core is often preferred for the purpose of practical alignment for coupling/splicing, etc. Much work has been done over the years on the inner-clad perimeter shape with core absorption. Noncircular inner-clad perimeter shapes are generally much better than circular inner-clad shapes for core-absorption efficiency [81], while an off-centered core in a circular inner cladding also provides some improvement as shown in Figure 15.27. The differences are found to be small with different noncircular inner-clad perimeter shapes with regard to effective core absorption similarly launched from the MM-pump [82]. Figure 15.27 shows the core-absorption efficiency with different inner-clad perimeter shape and core position in double-clad fiber by Müller et al. [82]. Practically, at the moment, a symmetric but noncircular inner-clad perimeter shape is the most

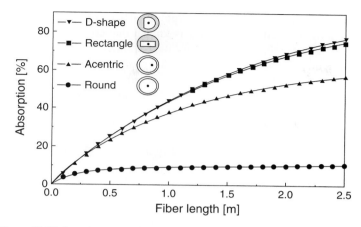

Figure 15.27 Core absorption efficiency with inner-clad shape or core position [83].

convenient for good pump absorption by the core. Ideally, the scenario of a circular inner-clad perimeter shape with a centered core remains highly desirable if the core absorption issue can be resolved. The asymmetric clad or core not only poses significant coupling issues but also causes considerable complications, for example, in fiber manufacture and grating impregnation.

Silica is practically the sole choice as fiber host material for high-power (>100 W) fiber lasers, primarily due to its low-loss characteristics and superior thermal/mechanical stability. Multicomponent (soft) glasses have been demonstrated as fiber host materials in fiber lasers at lower powers [84] and also have been shown to offer many desirable properties for a variety of application-specific uses for both active and passive fiber-based devices [85]. However, an efficient means to purify soft glasses for low-loss fiber characteristics remains a significant challenge. This challenge remains even after decades of intensive worldwide efforts. As a result, soft-glass fibers generally have much higher passive losses, ranging from ∼0.1–1 dB/m typically, compared with that of ∼0.2 dB/km in silica-based fibers. This effectively eliminates soft-glass fibers from use in further power scaling. The high losses per unit length exhibited in them lead to excessive heating and eventual destruction. Fluoride fibers have been thoroughly investigated for their ability to emit laser radiation in the visible or mid-IR spectral range [86]. Because of the lower phonon energy exhibited in fluoride glass, it makes nonradiative decay from the upper lasing levels by the multiphonon processes much less probable. But, fluoride fibers are prone to defect formation at higher power levels and are costly in manufacturing and nonpractical in handling.

It is preferable to have the Yb^{3+}-dopant concentration as high as possible for higher power scaling in fiber lasers, as long as it does not compromise Yb^{3+} quantum yield and fiber background loss. An alternative longer fiber length promotes the detrimental nonlinear effects under high-power operation.

These include SBS, SRS, and Kerr-type nonlinear effects depending upon the actual fiber-device configuration. It is known that codoping large amount of Al^{3+} is necessary in order to realize an effective high Yb-doping concentration benefit [87]. However, both aluminum and Yb have a rather high-index contribution when added to silica [88]. This makes the already difficult fiber-making of low-NA core, LMA double-clad fiber by conventional MCVD fiber process even harder.

15.5.3 All-Glass, Double-Clad Fibers by OVD [89]

Vapor-doping OVD Process

To overcome the limitations in conventional MCVD process coupled with solution-doping techniques [90] for making double-clad fiber, a robust all-glass, high-NA, Yb-doped double-clad fiber-making method has been reported by Corning [89] through a Yb/Al vapor-doping OVD process. The schematic of the OVD process flow is shown in Figure 15.28.

For the vapor-doped OVD double-clad fiber-making process, a precise, well-controlled gas-phase Yb- and Al-precursor generation and delivery is critical and was achieved [91]. The predetermined amount of various vapor-phase materials for each core or clad stage is delivered along with oxygen to react in the burner flame where the desired glass-soot particles are formed. The soot particles are then deposited onto a rotating bait rod for the designed soot preform of the Yb-doped core and germanium up-doped inner cladding for high inner-clad NA. After removing the bait rod, the soot preform is consolidated into a solid glass preform

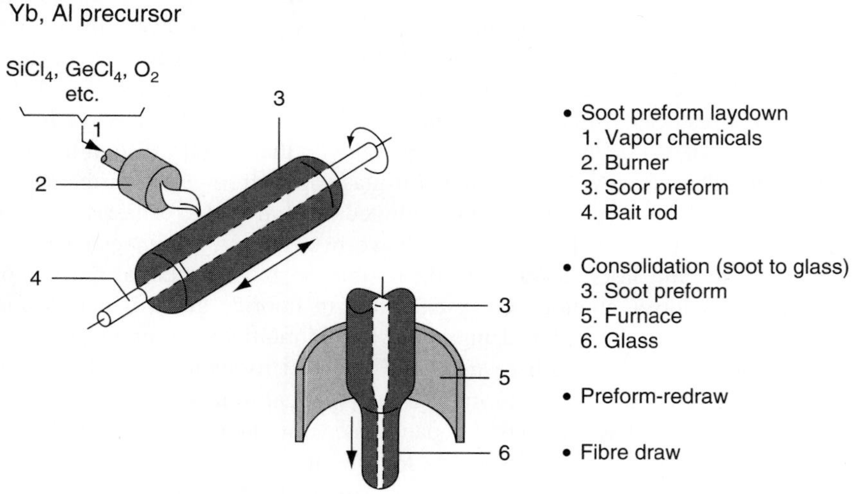

- Soot preform laydown
  1. Vapor chemicals
  2. Burner
  3. Soor preform
  4. Bait rod

- Consolidation (soot to glass)
  3. Soot preform
  5. Furnace
  6. Glass

- Preform-redraw

- Fibre draw

Figure 15.28 Schematic of the OVD process.

in a high-temperature furnace. A tailored consolidation condition was used taking into consideration the Al/Yb-dopant effects in order to achieve a crystal-free glass preform. It was then machined to have the desired outer perimeter shape. The same OVD soot-to-glass process with B/F down-doped silica was applied to the machined preform to form the outer cladding. The fiber drawing step is conventional. The all-glass double-clad fiber has the following core/inner clad/outer clad compositional format: $Yb_2O_3 \cdot Al_2O_3 \cdot SiO_2 \cdot GeO_2 \cdot F/SiO_2 \cdot GeO_2/B_2O_3 \cdot F \cdot SiO_2$. Each component is eventually optimized to ensure high device efficiency of the double-clad fiber laser. The inner clad in particular has a noncircular geometric shape to break the circular symmetry, thus maximizing the pump absorption efficiency as mentioned before.

Advantages of Vapor-Doping OVD Process

The vapor-doping OVD fiber-making process offers three distinct major advantages over the solution-doping MCVD process for double-clad fiber fabrication. Firstly, the OVD process is a totally synthetic, true all-vapor (including Yb- and Al-dopant precursor delivery) process. Consequently, it produces low-loss fiber characteristics as shown in Figure 15.29 for both the core (~1 dB/km at 1310 nm) and the inner cladding (~3 dB/km at 1310 nm) with desirable waveguide-quality materials for all the fiber sections without exception. In the MCVD process, the Yb/Al-solution-doped core often renders compromised loss from the solution-doping process leaving behind impurities from the starter-solution used, in addition a somewhat low reproducibility is observed. Moreover, the inner-clad glass quality is heavily dependent upon that of the starting silica tube used, commonly not a waveguide-quality material.

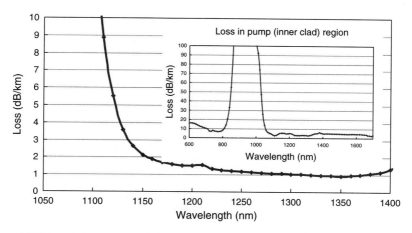

Figure 15.29 Low-loss characteristics in typical Yb-doped double-clad fiber by vapor-doping OVD (this figure may be seen in color on the included CD-ROM).

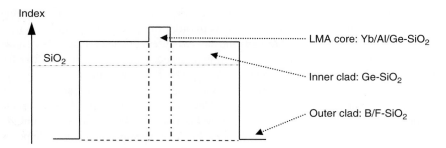

Figure 15.30 Flexible index composition with OVD double-clad fiber method.

The second is that the OVD process offers considerable compositional flexibility with regard to both the core and the cladding enabling desired index profiles, whereas in MCVD is much more limited by the starting silica tube used for the inner clad, and the solution-doping for the core. The OVD compositional flexibility conveniently allows not only low-NA core LMA fiber, but also the realization of high-NA (>0.3), all-glass, double-clad fiber structures for the first time through a B/F-codoped silica composition process for the outer-clad glass layer [92]. Figure 15.30 shows the typical index profile capable by the OVD method. As can be seen, it allows considerable flexibility for making LMA, all-glass, double-clad fibers with various low-core NAs and high inner-clad NAs through compositional designs.

Thirdly, the OVD vapor-doping allows high Yb-doping concentration, ~1.2 wt% Yb_2O_3, to be made regularly in a double-clad fiber without compromising the Yb quantum-efficiency as shown in Figure 15.31. Figure 15.31 shows the

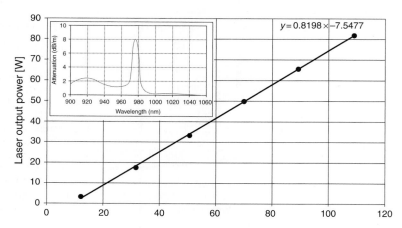

Figure 15.31 Yb laser slope-efficiency and absorption spectra in a high Yb-doped OVD double-clad fiber (this figure may be seen in color on the included CD-ROM).

Yb^{3+} absorption spectra measured for power launched through the MM inner-clad pump region (insert in the figure) and the laser-slope efficiency. The fiber has a 30-μm/350-μm core/inner-clad ratio, and the doped core has ~1.2 wt% Yb$_2$O$_3$. The Yb peak absorption of about ~8 dB/m at 976 nm, is the highest to date, to our knowledge, with an uncompromised Yb^{3+} laser slope-efficiency of ~82%.

15.5.4 High-Performance Yb-Doped Double-Clad Fibers

Introduction

In this section, we give an example of using the above-described "vapor-doping OVD" process to the development of a high-performance Yb-doped double-clad fiber, which mitigates detrimental nonlinear SBS. As fiber laser technology continues to progress with higher power output into the kilowatt range and beyond [93], one important way to further high-power scaling is through coherent beam-combining technology for 10s and 100s of kWs. Fiber lasers suitable for such applications require not only high power and good beam-quality but also narrow linewidth. High-power, narrow-linewidth fiber lasers are generally achieved by using a master-oscillator-power-amplifier (MOPA) configuration. Typical fiber cavity lengths of several meters make the discrimination of closely spaced longitudinal modes difficult in fiber lasers. The narrow-linewidth requirement of the fiber MOPA means that SBS must be effectively managed for high-power operation, as it becomes the limiting factor in the maximal power achievable. SBS in a fiber laser can be dealt with through geometric [94]/compositional effect(s) in the fiber profile design and/or through the distributed temperature [95] and/or stress/strain [96] effect(s) in the fiber laser-packaging design. Either way, raising the SBS threshold in a high-power fiber laser represents a considerable challenge.

SBS Effect in Optical Fiber

The origin of SBS lies in the backscattering of light by acoustic waves generated in the fiber material through the electrostrictive effect [97]. This backscattered light has a characteristically narrow linewidth (~50 MHz in silica) and is down-shifted in frequency by the Brillouin shift, which depends on the fiber material [98]. For common silica-based fibers, the shift is about 11 GHz at ~1.55 μm (the third telecomm window) and about 16 GHz at ~1 μm (the Yb-fiber laser wavelength). The narrow linewidth aspect of SBS makes it particularly harmful for narrow-linewidth high-power fiber laser operation. The backscattered light in a fiber interferes with the signal light and reinforces the density grating formed by the electrostrictive effect. This reinforced grating then *stimulates* more of the forward signal light to be backscattered. When the gain of the backscattered light is greater than the losses due to fiber attenuation we reach the *SBS-threshold*, after which the amount of backscattered light increases very rapidly with increasing input power until nearly all the input power is reflected. The transmitted power at the fiber output saturates at a level that barely increases with increasing input power. For SMF of lengths about 10 km, the SBS threshold can lie in the range of 8–10 dBm. In general, larger core size, shorter fiber length, higher signal light frequency, and broader signal linewidth all help to raise the nonlinear SBS threshold [99].

A thorough and comprehensive analysis of the SBS threshold was conducted [100] recently with regard to the fiber-related parameters. An expression for the threshold in terms of fiber and material parameters was derived, yielding

$$P_{th} \propto \frac{KA_{eff}\alpha_u}{g(\nu_{max})L_{eff}\bar{I}_u^{ao}}, \qquad (15.25)$$

where $g(\nu_{max})$ is the effective gain coefficient at peak frequency, α_u is the acoustic attenuation coefficient for the acoustic mode of order u, A_{eff} is the optical effective area, K is the polarization factor, L_{eff} is the effective interaction length, and $\bar{I}_u^{ao}$ is the normalized overlap integral between the optic and acoustic modes. The above equation indicates that the SBS threshold can be increased by increasing the effective area, polarization factor, and acoustic loss or by decreasing the overlap integral and the maximum gain coefficient. Among all the factors that affect SBS, the effective area and overlap integral can be controlled by fiber refractive index profile design and acoustic velocity profile design. The acoustic loss can be controlled by glass composition. The gain coefficient can also be lowered by creating an axially nonuniform Brillouin gain spectrum through distributed glass dopant or stress concentrations along the fiber. And, finally, the polarization factor can be increased through polarization control.

SBS-Managed Fiber Composition Design

As seen from Eqn (15.25), one way to raise the SBS threshold is by increasing the optical effective area, A_{eff}. Indeed, this was how the now well-known LMA fiber design concept initially came about [94]. As a matter of fact, the LMA core increases the power thresholds for all nonlinear effects such as SRS, SBS, and all optical Kerr-type effects in the fiber. This is simply because of the geometrically reduced power density in such core design.

By taking the double-clad fiber requirement into account, we have also considered feasible compositional means to reduce SBS.

Table 15.4 lists the dopants that can be used for making silica-based fibers and their effects on relevant optical and acoustic properties [101]. Among all the dopants listed, alumina remarkably, is the only dopant known that lowers the acoustic index in silica. It has been well known that boron and fluorine are the only dopants lowering the optical index in silica. The nature and mechanism as to why/how the addition of alumina lowers the acoustic index in silica, while nothing else does, is an extremely complex phenomenon.

Two compositional approaches can be used for reducing the SBS effect. One method is to design a fiber structure using different dopants that guides the optical

Table 15.4
Effect on optical and acoustic properties of various dopants in silica.

Dopant	GeO_2	P_2O_5	TiO_2	B_2O_3	F	Al_2O_3
Optical refractive index	↑	↑	↑	↓	↓	↑
Acoustic refractive index	↑	↑	↑	↑	↑	↓

wave but antiguides the acoustic wave. This is accomplished, for example, by choosing a dopant in the core such as Al_2O_3 to increase the optical index but decrease the acoustic index [102] or by choosing a dopant in the cladding such as F to decrease the optical index but increase the acoustic index [103]. The resulting optical and acoustic refractive index profiles are shown schematically in Figure 15.32(a). Because the acoustic wave is not guided in the core region, the interaction between the optical and the acoustic waves is reduced. Increased SBS threshold using this approach has been experimentally demonstrated [102, 103].

The second approach to suppressing SBS effects is to use different dopants within the core by changing the optical and acoustic field distribution to reduce the overlap-integral, $\bar{I}_u^{ao}$ between the optical and the acoustic field as indicated from the Eqn (15.25). This overlap effect is ignored in conventional SBS theory [104]. For an SMF with a step-index profile made using a single dopant, this overlap integral is approximately unity. Thus, ignoring this factor does not affect the SBS evaluation. For all other cases however, this overlap term can be significantly smaller than unity and cannot be neglected as demonstrated in Ref. [100]. An example of two doping regions in the core is shown in Figure 15.32(b). In this case, the first core region is doped with GeO_2 and the second core region is doped with Al_2O_3. The optical refractive index profile is selected to be a step, but the acoustic index profile becomes a W-shape. The optical field is confined in both the core regions while the acoustic field is confined in the first core region only. As a result, the overlap integral is reduced significantly. Numerical modeling indicates that the SBS threshold increase of 3–6 dB is possible using this approach.

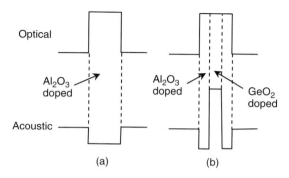

Figure 15.32 Optical and acoustic profiles in fiber design.

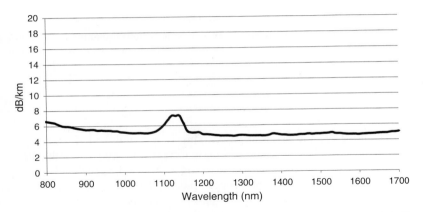

Figure 15.33 Wavelength-independent high scattering loss in a two-layer core single-mode fiber.

However, an ideal match in optical index between the different compositional sections in the core region is in practice difficult to accomplish. When an optical interface is generated within the core, as a result of using this type of core structure, strong light scattering occurring at the interface induces very high background loss as presented in Figure 15.33. This figure shows a typical wavelength-independent spectral loss curve characteristically due to strong-scattering effect of an SMF fabricated from a two-layer core. In fact, all such fibers made with this process have presented the same characteristically wavelength-independent high-scattering loss of about 4–5 dB/km as shown, regardless of the improvement made toward the flat optical index profile. This loss value is more than an order of magnitude higher than those of the typical one-section, Ge-doped core SMF, ~0.18 dB/km, and Al-doped core SMF, 0.35 dB/km [102], respectively.

This layer-scattering effect can be expected to be much more intense in double-clad fibers due primarily to the low-NA core requirement, in addition to the conventional relatively high Yb-doping-induced-loss effect [105]. Furthermore, intensely scattered light is particularly detrimental to high-power fiber laser operation as it directs part of the scattered energy to the outer polymer coating of the double-clad fiber. This can cause additional thermal problems, an extra burden to bear, which has already posed considerable concern in the operation and reliability of high-power fiber lasers in general. To overcome this difficulty of achieving low optical-loss SBS-suppressed fiber through the same "reduced optical/acoustic overlap" concept, we have then proposed a new "interface-free" core composition design as described in the following section.

Al/Ge Counter-Graded Composition Design and Fabrication

The key issue in our earlier multilayer core approach lies essentially in the manufacturing difficulty of realizing a true optical flat index profile among the different core composition layers. With the low-NA (~0.06) core requirement for the overall double-clad fiber design, it makes the task even harder in practice. To avoid the optical layer interface, while still managing a "reduced-overlap" between the acoustic and the optical field distributions in the fiber-core profile design, a new "interface-free" Al/Ge counter-graded fiber-core composition profile has been proposed [106] as shown in Figure 15.34.

Figure 15.34 shows the schematic of the new Al/Ge counter-graded core profile design applied to the Yb-doped all-glass double-clad fiber. This Al/Ge-counter-graded composition profile design eliminates the physical interface within the core, thus overcoming the limitation of our initially attempted multilayer fiber-core composition approach. In fiber fabrication, the flat optical-index profile can be conveniently achieved by up-doping germania content and at the same time down-doping alumina concentration in a manner that they compensate one another in their optical refractive index contribution. The slope of the germania up-doping is about ~30% greater than that of alumina down-doping due to the latter's higher refractive index contribution in silica. The linear dependence of optical index on the Al- or Ge-doping concentration makes this a much easier approach in practice. On the other hand, linearly down-grading alumina and up-grading germania from the center of the core causes the acoustic wave to be primarily guided or distributed toward the edge of the core. Contrary to a flat optical-index profile, the acoustic-index profile in the core is a V-shaped one for the new core-composition profile design. As a result, the overlap between the optical and the acoustic fields has been reduced. Modeling efforts in the step-index case have indicated an increased SBS threshold of ~3 dB according to Eqn (15.25).

A low-loss SMF with this Al/Ge counter-graded core-composition profile design has been successfully fabricated. Without a discontinuous interface in the design, the fiber exhibits satisfactory low-loss characteristics as shown in Figure 15.35. The minimal loss of ~0.4 dB/km@1550 nm has been measured. At the same time, an SBS threshold increase of 2.5 dB has also been experimentally demonstrated in this SMF. Through the verification of low-loss characteristic and high-SBS threshold in

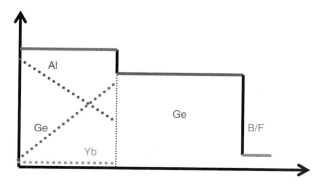

Figure 15.34 Schematic of composition design of SBS-managed, Yb-doped double-clad fiber (this figure may be seen in color on the included CD-ROM).

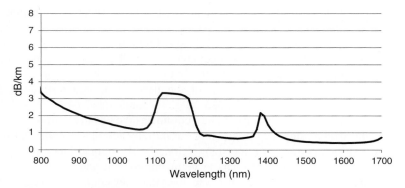

Figure 15.35 Low-loss (0.4 dB/km@1550 nm) characteristic of an Al/Ge counter-graded SBS SM fiber.

the Al/Ge counter-graded SMF, we have implemented this core profile design in the making of the Yb-doped, SBS-managed LMA double-clad all-glass fibers.

Characteristics of SBS-Managed LMA Double-Clad Fibers

The SBS-managed, Yb-doped double-clad fiber has been successfully made by implementing the Al/Ge counter-graded core composition design, schematically shown in Figure 15.34 as described earlier. To validate the SBS improvement in the LMA fiber, one other fiber was also made with a standard step-index profile using only germania as the dopant. The properties of the two fibers are summarized in Table 15.5.

Both fibers had a 30-μm diameter core, a 300-μm flat-to-flat hexagonal inner cladding, and were doped with 0.5 wt% of ytterbia. The core- and inner-cladding NAs were 0.06 and 0.32, respectively. To verify the acoustic profile of the fiber, the acoustic velocity across the preform was measured using a scanning-acoustic-microscope (SAM). The result of this measurement is shown in Figure 15.36. This clearly shows the high acoustic velocity (low acoustic index) in the center of the fiber and the low acoustic velocity (high acoustic index) guiding region at the edge of the core.

Table 15.5
Properties of fibers for SBS threshold measurement.

Fiber type	Core diameter (μm)	Length (m)	Yb_2O_3-concentration (wt-%)
Standard step-index (LMA-1)	30	12	0.5
Graded Al/Ge step-index (LMA-2)	30	12	0.5

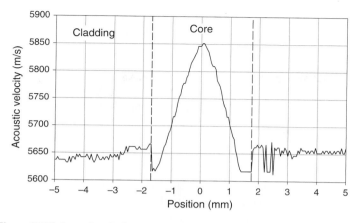

Figure 15.36 Acoustic velocity profile of fiber with reduced acousto–optic overlap.

The SBS threshold characteristics of the two LMA Yb-doped fibers were measured in the latter stage of a two-stage amplifier. The power amplifier stage was pumped by a maximum of 350 W at each end. The seed source was at 1064 nm with a 3-kHz linewidth. The 150-mW output was amplified to 4 W with the first amplification stage. Mid-stage isolation prevented parasitic oscillation and saturation of the first stage by the ASE produced in the second stage. A beamsplitter was placed between the two amplifier stages to monitor the power entering the second stage and the backward propagating power and spectrum. The SBS threshold is recorded as the forward power at which the backward propagating power dramatically increases. The results of the SBS threshold measurements are shown in Figure 15.37. The standard LMA fiber (LMA-1) exhibits an SBS threshold of approximately

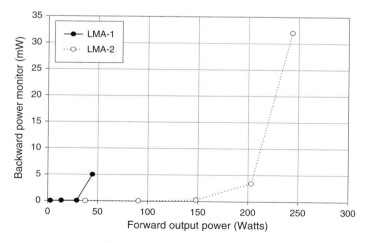

Figure 15.37 SBS threshold measurements.

40 W while the fiber with reduced acousto-optic overlap (LMA-2) showed a threshold of between 150 and 200 W. This measurement clearly demonstrates the feasibility of raising the SBS threshold through dopant profiling in the core of the fiber. Furthermore with the LMA-2 fiber, a fiber laser slope efficiency of over 80% has been measured, demonstrating that such profiling provides uncompromised laser performance. Record high-power, narrow-linewidth fiber amplifiers have also been demonstrated with this SBS-managed fiber design, and the details are given in Ref. [107].

15.5.5 Conclusions

Vapor-doping, all-glass, double-clad fiber processes have been established through the OVD method. The major benefits of such processes include ultralow loss, high Yb-doping concentration, and compositional flexibility. Experimental realization of double-clad fibers from this process provide superior power scaling and handling reliability for practical use due to its all-glass structure. Using OVD fiber-making advantages, recent work on the composition, design, and fabrication of SBS-managed, LMA-2 double-clad fiber lasers have been described. An aluminum/germanium (Al/Ge) counter-graded fiber-core composition profile has been proposed and demonstrated as a practical means for reducing the nonlinear SBS effect, through the reduced overlap between optical and acoustic fields in the fiber. Such Al/Ge counter-graded composition-profile designs overcome the limitation of multilayer fiber-core compositional approach in low-loss fiber fabrication. The new compositionally SBS-managed, LMA-2 Yb-doped double-clad fiber laser has been demonstrated with low-loss characteristics and high laser efficiency.

15.6 MICROSTRUCTURED OPTICAL FIBERS

15.6.1 Introduction

Microstructured optical fibers represent a new class of optical fibers that has reinvigorated the field of specialty fibers and has redefined our understanding of optical waveguides. As with any new advance, we can see precursors of the technology dating back for decades before reaching a critical mass that defined the field in its own right. In the case of microstructured fibers, the early roots of the technology are in evidence in the holey fibers of Kaiser et al. in 1974 [108], in the Bragg fibers from CalTech in 1978 [109], and even in the segmented core designs of dispersion-compensating fiber and NZDSF fibers of the late 1980s and 1990s. The turning point in the technology came in the early 1990s with the development of the formal ideas surrounding the PBGs produced by materials with periodic variations of refractive index [110]. It was at this time that a new class of optical fiber was proposed based on photonic crystal effects [111], but the decade would nearly end before a true PCF would be realized [112].

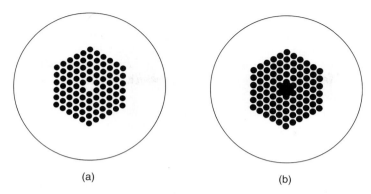

Figure 15.38 Representative geometries for the two major class of air-silica microstructured optical fibers: (a) EI-MOF and (b) PBGF. The dark areas represent airholes or voids and the light areas represent solid glass or other optically transparent material.

The scope of microstructured fibers is very large and cannot be covered in its entirety in this article. Rather than surveying the myriad structures and applications, we choose to focus on air/silica-microstructured fibers to convey the general principles and most important applications of this technology. We begin with a discussion of the fundamental principles behind waveguiding in these unusual waveguides and then follow with more detailed discussions of the properties of the two natural categories of microstructured optical fibers: effective index-microstructured optical fibers (EI-MOFs) and photonic band-gap fibers (PBGFs) as illustrated in Figure 15.38.

15.6.2 Waveguiding in Microstructured Optical Fibers

Until the introduction of photonic crystals it was generally accepted that waveguides required total internal reflection (TIR) to define guided optical modes. Incomplete reflections such as those provided by metallic or dielectric surfaces can define relatively low-loss cavities but true modes of conventional waveguides require a high-index core that produces TIR at the core/cladding interface. This is of course a simplified explanation motivated by a ray picture of guiding light, but it is a conceptual picture that has had a powerful influence on the way we understand waveguides. A deeper explanation of the physics behind waveguiding ultimately resorts to a full application of Maxwell's equations in which the rays are replaced by electromagnetic waves, and the index profile takes on the role of a potential. It is from this more complete electromagnetic description that the concepts of photonic crystals and PBGs take form, ultimately leading to a broader understanding of the requirements to observe optical waveguiding.

As we have described it, a conventional waveguide requires a high index core surrounded by a lower index cladding that is considered to extend to infinity.

In reality, many conventional waveguides and optical fibers have complex core geometries that may contain regions with refractive indices lower than that of the cladding, making it difficult to explain the waveguide in terms of TIR at a single interface. In anticipation of the PBG discussion, we approach conventional waveguiding from a different direction using concepts borrowed from photonic crystals [113].

An important yet often unappreciated result is that a homogenous material such as a block of glass has a region of frequencies for which propagation is forbidden. To understand this, we examine the relationship between the optical frequency ω and the propagation constant β of the optical field. In a homogeneous material of refractive index n_1, a propagating solution with a propagation constant β must have a frequency greater than $\beta c/n_1$. If the frequency is less than this value, the optical field decays and fails to propagate. This then defines a range of forbidden frequencies from zero to $\beta c/n_1$. If the refractive index is increased to n_2, the forbidden gap is smaller going from zero to $\beta c/n_2$, and frequencies between $\beta c/n_2$ and $\beta c/n_1$ that are forbidden in the region of refractive index n_1 are now allowed in the region of refractive index n_2. Thus, an embedded high-index region supports propagation for a range of frequencies at which propagation with propagation constant β is forbidden in the lower-index cladding. This is the basic principle behind conventional waveguides.

The above explanation still relies on two distinct macroscopic regions with a well-defined single interface. But the concept of forbidden frequencies is a general one, and even for more complicated waveguide geometries, there is still a well-defined cladding whose forbidden frequencies can be calculated. The core must then be designed to place modes at frequencies inside the forbidden range of the cladding in order to confine the light to the core. The precise design of the core determines the number of modes and their spatial and optical properties.

In conventional waveguides the forbidden frequency range of a homogeneous cladding is accessed by adding a core containing high-index regions as described above. The core may contain some regions with material indices lower than that of the cladding, but guided modes are obtained only if the modal frequencies lie within the forbidden gap of the cladding.

We have been implicitly assuming that the cladding must be homogeneous. But what if the cladding contains a well-defined structure such as those shown in Figure 15.38? Such a cladding also produces a forbidden frequency region from $\omega = 0$ to $\omega = \beta c/n_{\text{eff}}$, where n_{eff} is the effective cladding index with a strong dispersion related to the scale of the geometry. Once again, the core must be designed to push modes into the range of forbidden frequencies of this cladding. The most common periodic cladding uses a triangular lattice of airholes extending continuously throughout the length of the fiber. If we follow the approach used for conventional waveguides and introduce a core region with a higher index than the effective cladding index, we have an EI-MOF (as shown in Figure 15.38(a)). From this discussion, we see that there is no fundamental difference between conventional fibers and EI-MOFs.

It is often pointed out that the periodic cladding region is invariably truncated and the modes of microstructured fibers are not truly bound. Strictly speaking, this is true and in EI-MOFs with just a few rows of holes this may be a concern. But as more rows of holes are added, it becomes a very good approximation to assume that the cladding is infinite. One must remember that the situation is identical in conventional fiber because the finite cladding is jacketed by polymer materials with higher refractive indices than that of the cladding. In theory, light tunnels through the cladding even in conventional fibers, but this can be neglected if the cladding diameter is large enough.

If a periodic cladding only produced the forbidden zone from $\omega = 0$ to $\omega = \beta c/n_{\text{eff}}$, this chapter would include only EI-MOFs. However, the periodicity also produces other forbidden zones at higher frequencies for which propagation would be allowed if the material was truly homogeneous. We refer to these regions as band gaps, indicating that they are gaps between bands of allowed frequencies. To make a waveguide that supports propagation, we must once again introduce a core region into the cladding to provide modal frequencies in the band gaps. If we introduce high-index core regions, we access the lower forbidden region discussed previously for EI-MOFs. However, if we introduce core regions with indices lower than the effective index of the cladding, we can create allowed modes in the band gaps. A fiber that guides light using one of the band gaps is referred to as a PBGF and is shown in Figure 15.38(b). Although it may share many similar properties with EI-MOFs, a PBGF is a fundamentally different waveguide and must not be confused with the EI-MOFs.

In the following sections, we cover the categories of holey fibers, discussing design considerations, principles of operation, attributes, and example applications. Our objective is to present the reader with a clear understanding of the advantages and limitations so that full potential of this new class of fiber can be realized in applications that are appropriate to the strengths of the technology.

15.6.3 Effective-Index Microstructured Optical Fibers

Basic Concepts

Although the original goal was to produce a PBGF, the first of the modern EI-MOF to be realized experimentally [114] was the air-silica PCF. The dopant-free fiber preform was made by stacking hexagonal silica capillaries and then the resulting bundle of capillaries was drawn into fiber. This remains one of the techniques still used for making these fibers today.

The basic cross section of a PCF is shown in Figure 15.39(a). Airholes of diameter d in a glass or other dielectric are spaced with a lattice parameter or pitch given by Λ. In a PCF, the pitch Λ defines the overall scale of fiber including the core diameter and the lattice spacing. Ignoring the wavelength dependence of the refractive index of the glass, we find that the fiber properties are the same for fibers

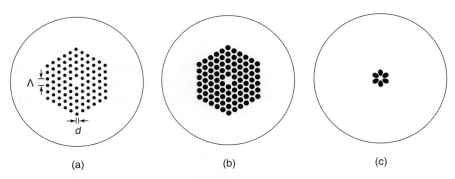

Figure 15.39 Examples of microstructured optical fibers: (a) Low-delta endlessly single-mode PCF, (b) high-delta PCF with a limited single-mode regime, and (c) small-core, air-clad fiber with noncircular airholes.

made from the same preform with the same ratio of Λ/λ. This behavior is a manifestation of the scale invariance of Maxwell's equations.

Most EI-MOFs have a regular pattern of microstructures in the cladding. The most typical configuration is the periodic triangular lattice shown in Figure 15.39, but there is no reason that EI-MOFs cannot be made with almost any lattice, the most common alternative being the square lattice [115]. Although the lattice structures in Figure 15.39 are not cylindrically symmetric, microstructured fibers are not birefringent if they have greater than twofold rotational symmetry [116].

An important parameter that determines many of the EI-MOF parameters is the air-filling fraction F which for circular holes is expressed as

$$\text{Triangular lattice: } F = \frac{\pi}{2\sqrt{3}} \left(\frac{d}{\Lambda}\right)^2,$$
$$\text{Square lattice: } F = \frac{\pi}{4} \left(\frac{d}{\Lambda}\right)^2. \tag{15.26}$$

In the limit that the airholes touch each other $(d = \Lambda)$, the triangular lattice produces a structure with $F \cong 90.7\%$ whereas the square lattice reaches only $F \cong 78.5\%$. This difference in ultimate air-filling fraction can play an important role in applications where effective index contrast is important. However, experimental EI-MOFs with noncircular airholes (see Figures 15.39(c) and 15.41) have air-filling fractions approaching unity, and in such fibers, the air-filling fraction is not given by the simple expressions in Eqns (15.26).

Endless Single Mode, Large Effective Area, and Coupling

In a conventional optical fiber (with core radius a, cladding index n_{clad}, and core index n_{core}), the number of guided modes at wavelength λ is related to the V-number given by

15. Specialty Fibers for Optical Communication Systems

$$V = \left(\frac{2\pi}{\lambda}a\right)\sqrt{n_{\text{core}}^2 - n_{\text{clad}}^2}. \qquad (15.27)$$

For conventional silica-based fibers, the difference in refractive index is a very slow function of wavelength so the *V*-number (and hence the number of guided modes) grows without limit as $a/\lambda \to \infty$ (either large core or short wavelength). However, in a PCF, the *V*-number is given by a slightly different expression from Eqn (15.27) with a replaced by Λ and n_{clad} replaced by the effective cladding index of the microstructured cladding. This effective cladding index no longer has the same wavelength dependence as the core index and asymptotically approaches the core index in the short wavelength limit. Because of this behavior, both the effective refractive index difference and the NA go to zero in this limit.

Surprisingly, in a PCF, the behavior of the effective cladding index cancels out the wavelength dependence of the *V*-number such that the *V*-number (and thus the number of modes) approaches a constant value as $\Lambda/\lambda \to \infty$. As a consequence, PCFs with $d/\Lambda < 0.4$ can be made *endlessly single-mode* (ESM) with no higher order modes at any wavelength [117]. Furthermore, because of the scale invariance of PCFs, the ESM behavior occurs not only for all wavelengths but also for all core diameters. Thus, in principle, one can make a single-mode PCF with a modal area $>1000\,\mu m^2$. However, in practice, the vanishing of the core-clad index difference leads to very large macrobend losses and large-core fibers exhibit not only the usual long-wavelength macrobend edge but also a short wavelength bend edge. Although this macrobend loss places a severe constraint on the utility of short-wavelength and large effective-area operation, the ESM effect has proved useful in high power laser applications in which short lengths and potentially straight deployments avoid bend issues [118].

In any EI-MOF, the airholes directly surrounding the core produce a modal profile that deviates from the typical circular Gaussian profile that approximates the fundamental mode in most conventional fibers. However, many microstructured fibers have fundamental modes that have excellent overlap with the modes of conventional fibers of comparable mode-field diameter. Because of this, coupling losses can be limited to less than 0.1 dB in many cases. One caveat is that, because of the wavelength dependence of the effective cladding index, coupling with microstructured fibers may vary with wavelength.

Small Modal Area and Nonlinearity

The preceding section dealt with the modal properties of PCFs with periodic claddings. However, an equally important class of EI-MOFs takes advantage of the high-index contrast offered by air and glass to produce fibers with enhanced properties. Typically, these fibers have solid cores surrounded by one or more rings of large airholes as shown in Figure 15.39(c). In effect, these fibers are simply air-clad silica rods suspended in a larger fiber superstructure. In the absence of dopants, the truncated holey cladding does not produce truly guided modes and can lead to high attenuation. However, in applications where low loss is more critical, additional rows of holes can be added to produce similar optical performance with improved attenuation.

The large index contrast in air-clad fibers requires a small core diameter to achieve single-mode operation. This combination of high-index contrast and small modal area can lead to high birefringence (in asymmetric designs), high dispersion, and high nonlinearity. In many applications, these properties are not desirable and small modal area can make efficient coupling difficult. However, in nonlinear device applications, the small modal area leads to significant nonlinear enhancement and the dopant-free design lends itself to other nonsilica glass systems with higher nonlinear coefficients. The resulting fibers [119] lead to enhanced nonlinear interactions in short device lengths with low latency. However, because nonsilica fibers have very high losses (>1 dB/m), nonlinear figures of merit that include attenuation may remain lower than those obtained using silica fibers.

Dispersion

Dispersion is an important property in many optical fiber applications, and although standard fiber designs offer a wide range of dispersion characteristics, EI-MOFs broaden and extend this range to limits which may be difficult to achieve with standard fiber-processing techniques [120]. The benefit comes from the large index contrast of the airholes and because of the wavelength dependence of the effective-cladding index described above.

Early work [121] concentrated on using air-clad EI-MOF designs that had large air-filling fractions and small pitches to produce a very large waveguide dispersion feature (see Figure 15.40). At long wavelengths, this feature produces a dispersion

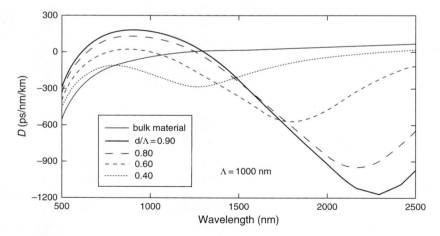

Figure 15.40 Modeled dispersion parameter for a standard air-silica PCF design with $\Lambda = 1000$ nm and for a range of air-filling fractions. The material refractive index is also plotted.

parameter D with a large negative component (less than–2000 ps/nm/km) that has potential application in dispersion compensation for telecommunication. However, such applications have been hampered by the nonlinearity and coupling issues resulting from the small core size. At short wavelengths, the dispersive feature produces a surprisingly large positive-D component (>100 ps/nm/km) that can be used to balance the large material dispersion at visible wavelengths leading to fibers with zero-dispersion wavelengths as low as 550 nm [122]. This zero-dispersion wavelength combined with the small core size enables efficient nonlinear processes that lead to dramatic effects such as broadband continuum generation [123].

More recently, EI-MOF designs have focused on slope compensation [120] and dispersion-flattened [124] designs. These designs, with more complex core geometries, have more reasonable effective areas ($>40\,\mu m^2$), but the performance improvement over conventional fiber has not been sufficient to drive commercial application.

Attenuation

Although early EI-MOF fibers had losses >100 dB/km, improvements in processing and design now yield silica-based EI-MOFs with attenuation as low as 0.28 dB/km [125]. It has been suggested that EI-MOFs might be the ultimate low-loss fiber because of the absence of all dopants. However, the large index contrast greatly enhances scattering at the air/silica interfaces and can lead to excess scattering losses, which are typically negligible in conventional fibers. The attenuation can be very dependant on the air-filling fraction, the pitch Λ, and the water content in the fiber [126], thus it is not possible to achieve low attenuation in every application, especially those in which the light is confined to a very small core diameter. However, in most applications, transmission losses can be expected to be on the order of 0.5–5 dB/km.

As mentioned in the discussion of ESM, macrobending losses can be very significant in EI-MOFs at long and short wavelengths. However, if designed correctly, holey fibers can achieve excellent bend loss especially when used in combination with more conventional fiber designs [127]. Such designs have been suggested for FTTH applications but have yet to achieve acceptance because of their nonstandard modal properties and higher production cost.

Microbending in EI-MOFs has received little attention, but it is predicted to be somewhat better than conventional fibers [128].

15.6.4 Hollow-Core PBGF

The second major class of microstructured optical fibers is the PBGF. The first demonstration of guiding light in air with a 2D dielectric PBGF was reported in 1999 [112]. While those initial demonstrations were in fibers only a few centimeters long, they clearly demonstrated that light could be confined to the hollow core of a dielectric waveguide. Since then, enormous strides in understanding, processing, and performing have taken place. In this section, we review the current understanding of how these fibers guide light and the optical properties of the hollow-core modes.

Origin of PBGs

In the introduction, we described a consistent picture of guidance in optical waveguides in which bound modes of a waveguide exist when fields are allowed to propagate in the core and prohibited from propagating in the cladding. In high air-filling fraction PBGFs, like that shown in Figure 15.41, each of the interstitial dielectric regions in the cladding can in theory support optical modes whose properties are determined primarily by their shape, size, and refractive index contrast. These interstitial regions are not isolated but are coupled to neighboring interstitial regions in such a way that the entire cladding region produces dense frequency bands of allowed cladding modes. For a periodic structure, there can be gaps between these bands of frequencies and it is these band gaps that can be used

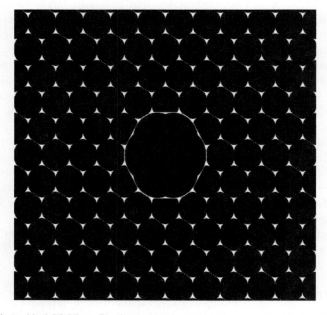

Figure 15.41 An ideal PBGF profile illustrating the lattice of dielectric resonators formed at the interstitial regions between adjacent airholes. The larger hole in the center of the profile is the core of the waveguide. The cladding has an air-filling fraction of >94% achievable only with noncircular holes.

to confine light to defects in the periodic structure. The defect serves as a core of the waveguide.

Most PBGFs have employed a triangular lattice, which is mechanically robust and resistant to perturbation. The practical benefit of other lattices has not been demonstrated for hollow-core fibers and many lattices, in fact, produce smaller more fragmented band gaps unsuitable for hollow-core guidance. The exception is a so-called Kagome lattice that has no band gaps to support hollow-core guidance but has found applications nonetheless in enhancing the interaction between optical fields and atoms and molecules filling the hollow core [129]. As in the case of EI-MOF, air fraction plays the key role in determining the band-gap width, as well as the dispersion, loss, and nonlinearity of the core modes. In addition, it also affects the position of the band gap. In air-silica PBGF, larger air fraction increases the band-gap width and decreases the center wavelength of the band gap.

The finite frequency range of these band gaps limits the bandwidth of operation for band-gap-guided modes. However, normalized band-gap widths, $\Delta\omega/\omega$ (the spectral width of the gap divided by the central frequency of the gap), as large as 0.3 have been demonstrated [130], which translates to a $\sim$500-nm bandwidth centered at 1500 nm. As mentioned previously, if material dispersion can be neglected, Maxwell's equations are scale invariant. Therefore, the center frequency of the band gap can be shifted to longer or shorter wavelengths by respectively increasing or decreasing the scale of the PBGF profile. If the geometry of the fiber profile is maintained through these scale changes, then the normalized band-gap width remains unchanged.

The core of a PBGF is defined by a defect formed in the photonic crystal lattice, typically through the removal of a series of adjacent capillaries. The removal of seven adjacent capillaries forms a hexagonal defect in the profile. This core has been demonstrated to support three to five different transverse mode profiles [130, 131]. Larger defects have also been demonstrated; removing the next row of capillaries results in a 19-cell core [131]. Similar to conventional optical fibers, large-core PBGFs support additional higher-order modes.

Dispersion in Hollow-Core PBGF

Two important attributes of fiber propagation are chromatic and modal dispersion. In this section we discuss chromatic and modal dispersion, as well as the transverse profiles of the bound modes associated with hollow-core PBGFs.

If we think of PBGFs as waveguides that prohibit propagating modes in the cladding, a simple analytical model can be used to approximate the boundary between the core and cladding, treating the PBGF as a perfectly reflecting hollow cylinder [113]. This approach is a very good approximation for estimating the dispersion and mode profiles in hollow-core PBGFs. More accurate descriptions of these and other optical properties require numerical modeling, which can be accomplished using a variety of techniques [132–135]. The modes and dispersion of a perfectly reflecting hollow cylinder have the simple relationship between the frequency, ω, and propagation constant, β, of the modes:

$$\frac{n^2\omega^2}{c^2} = \beta^2 + \left(\frac{j_{pq}}{R}\right)^2, \qquad (15.28)$$

where n is the refractive index of the core material, R is the radius of the core, c is the speed of light in a vacuum, and j_{pq} is the qth zero of the pth-order Bessel function of the first kind. It is convenient to recast Eqn (15.28) in terms of the effective index of the mode, $n_{\text{eff}} \equiv \beta c/\omega$:

$$n_{\text{eff}}^2 = n^2 - \left(\frac{j_{pq}c}{\omega R}\right)^2. \qquad (15.29)$$

Interestingly, the effective indices of these modes are less than the index of the core material, n. For a hollow core, with $n=1$, the mode effective index is less than one, which means the phase fronts propagate in the fiber with a velocity greater than the speed of light in vacuum. The group velocity of such modes is less than the speed of light in vacuum and the effective group index, n_g, is given by

$$n_g = n_{\text{eff}} + \omega \frac{dn_{\text{eff}}}{d\omega} = 1/n_{\text{eff}}. \qquad (15.30)$$

The fundamental mode of the hollow core is characterized by the first zero of the zeroth-order Bessel function, $j_{01} \cong 2.4048$. In Figure 15.42, we plot the band gaps associated with the PBGF profile shown in Figure 15.41, along with the dispersion of the hollow-core modes of Eqn (15.29). The calculated core modes of the profile are in close agreement with the modes of a perfectly reflecting hollow cylinder, except for the regions near the band edges, where coupling between the core and band modes causes the dispersion of the core mode to deviate from that of a perfectly reflecting hollow cylinder mode.

Unlike conventional optical fibers, some PBGF designs also support modes that exist on the boundary between the core and the periodic lattice [44, 136]. Due to their strong localization to small glass regions of the fiber profile, these surface modes are highly susceptible to structural variations in these regions. This results in significantly higher losses for these modes as compared with hollow-core modes, which have a much smaller overlap with the glass structure and thus experience a smaller impact due to structural variations. The surface modes have very different dispersion from the core modes (see Figure 15.42), which results in a strong interaction between the modes that have the same propagation constant at a given frequency. When the symmetry of a surface mode is the same as that of a core mode, they can interact strongly and result in an anticrossing, as seen in Figure 15.42. This strong interaction and mixing of the modes lead to increased attenuation of core-guided modes at frequencies near these crossing regions.

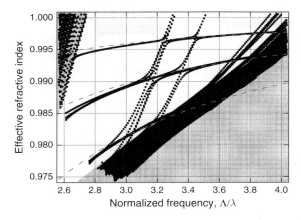

Figure 15.42 Effective refractive index, n_{eff} as a function of the normalized frequency for the fiber profile shown (Figure 7.4). The dashed lines of the perfectly reflecting hollow cylinder model closely approximate the dispersion of the core modes. The two sets of steep curves represent the dispersion of the surface modes. The interaction between the core and surface modes can be seen in their avoided crossing behavior. The shaded areas denote the bands of allowed frequencies of the photonic crystal cladding (this figure may be seen in color on the included CD-ROM).

As discussed earlier in this chapter, dispersion of the waveguide modes is an important property of an optical waveguide. There are two main contributors to the dispersion in hollow-core modes of PBGF: the dispersion associated with that of a perfectly reflecting hollow cylinder and the dispersion from the band edges of the PBG. Dispersion at the band edges results from the frequency dependence of the amplitude transmission of the band gap as required through the Kramers–Kronig relationship. The long-wavelength band edge produces a large positive dispersion parameter, and a symmetric large negative dispersion parameter appears at the short-wavelength band edge. If this was the only contributing factor, there would be a zero-dispersion point in the middle of the band gap, giving anomalous dispersion on the long-wavelength half of the band gap and normal dispersion on the short-wavelength half. However, the additional dispersion from the hollow cylinder mode, which is always anomalous and increases with increasing wavelength, shifts the zero-dispersion point to shorter wavelengths [137]. The large dispersion at the band edges is not significantly altered by the influence of the PRHC dispersion. Interestingly, the contribution from material dispersion for the hollow-core PBGF is insignificant, due primarily to the fact that the hollow-core modes have little overlap with the glass in the profile. This can be seen as an advantage, but also serves as a disadvantage, since the lack of material dispersion removes a knob by which the dispersion could be adjusted. This dispersion has been characterized in a real PBGF using a cross-correlation technique of the transmitted pulse with the incident pulse [137]. For surface modes, the material dispersion still contributes to the overall dispersion of these modes because of their larger overlap with the glass regions of the fiber.

Modal Properties of Hollow-Core PBGF

In Figure 15.43, we show the first four core modes corresponding to the fiber profile in Figure 15.41. Similar to an EI-MOF, the shape of the PBGF-mode profile is not a circular Gaussian. However, coupling to Gaussian modes from other structures can be efficient, because the center of the mode profile, containing a majority of the energy, is well approximated by a circular Gaussian beam. The structure of the fiber is overlayed in the modal images in the Figure 15.43 to indicate the relationship of the mode profile to the fiber profile. The overlap between the mode and the glass structure of the fiber profile is seen to be extremely small (<0.4% for the fundamental mode). This small overlap provides many of the distinguishing features of hollow-core PBGF. For example, the effective optical nonlinearity of the mode can be greatly reduced with respect to the nonlinearity of the glass.

The optical nonlinearity of a composite fiber profile can be modeled as resulting from contributions from the various materials in the profile [138]. Summing the intensity-dependent contributions of the various components, one arrives at an effective nonlinear response for the composite. Experiments to characterize the nonlinear response of hollow-core PBGF were carried out by measuring the spectral broadening of short pulses propagating through the fiber near the zero-dispersion wavelength [137]. The nonlinear phase shift was measured as a function of power, and the results enabled the researchers to estimate the effective nonlinearity of the optical mode. The results showed an air-filled PBGF had nonlinearity nearly equal to that of the nonlinearity of air itself, which is about a factor of 1000 times smaller [139] than the nonlinearity of standard SMF [140]. When the hollow volume of the fiber profile was evacuated, the effective nonlinearity of the fiber was measured to be 10,000 times smaller than that of standard SMF [141].

Initially, losses in hollow-core PBGF waveguides were quite high, on the order of 100 to 1000 dB/m, but improved to values of ~1 dB/m [142] until Corning Incorporated presented results [143] showing that the losses could be substantially reduced. Currently, losses of 13 dB/km for a so-called 7-cell core [130, 143] and 1.2 dB/km for a so-called 19-cell core [144] are the state of the art. There are a number of contributing factors that have been suggested that would limit the losses

Figure 15.43 First five transverse mode intensity profiles of a hollow-core PBGF. The structure of the fiber profile is superimposed on the intensity profiles of the modes. Each of these transverse profiles has two degenerate polarization states (this figure may be seen in color on the included CD-ROM).

in such fibers: material losses, tunneling through a finite cladding, surface-mode coupling, and surface roughness [131, 136]. When the same synthetic processes as those used to make low-loss telecommunications glass are employed to make the glass in a PBGF, the material losses are predicted to be less than 0.0015 dB/km for a mode profile with a 1% overlap with the glass. The tunneling loss of a leaky mode through the finite number of cladding-hole layers is predicted [135] to be less than 0.1 dB/km for 10 rows of holes with a hole diameter-to-pitch ratio of 0.95, typical of the number of rows for PBGF reported in the literature. The losses due to surface-mode coupling to the hollow-core modes can be greater than 100 dB/km near the resonance of the coupling, and is predicted to be on the order of that observed in the lowest loss fibers, far from the resonance. However, coupling between the hollow-core modes and fluctuations of the structure along the length of the fiber are believed to be responsible for a majority of the losses in the low-loss spectral regions of PBGF [131].

15.6.5 Concluding Remarks

The nature of microstructured optical fibers leads to an endless variety of fibers with varying geometries, materials, properties, and applications. In this section, we have only dealt with air-silica fibers, but there are a number of other examples of related fibers that possess unique optical properties. These fibers include the recent examples of the Bragg fibers [145], liquid-filled PCFs [146] and PBGFs [147], hybrid fibers that guide with TIR and PBG effects [148], and all-glass PCFs [113] and PBGFs [149]. While the physics of the broad class of microstructured optical fibers appears to be well-understood at this point, the adaptation of the technology to applications appears to be the main challenge for the future.

REFERENCES

[1] K. S. Jepsen, U. Gliese, B. R. Hemenway et al., "Network Demonstration of 32h × 10 Gb/S Across 6 Nodes Of 640 × 640 Wsxcs with 750 km Raman-Amplified Fiber," in Proc. *OFC 2000*, paper PD35, Baltimore, Maryland, March, 2000.

[2] T. N. Nielsen, A. J. Stentz, K. Rottwitt et al., "3.28-Tbh (82~40 Gbh) Transmission Over 3 × 100 km Nonzero-Dispersion Fiber Using Dual C- and L-Band Hybrid Raman/Erbium-Doped Inline Amplifiers," in Proc. *OFC 2000*, paper PD23, Baltimore, Maryland, March, 2000.

[3] G. Vareille, F. Pitel, and J.F. Marcerou, "3 Tbit/s (300 × 1 l. GGbit/s) Transmission Over 7380 km Using C + L Band with 25 GHz Channel Spacing and NRZ Format," in Proc. *OFC 2001*, paper PD 22, Anaheim, California, March, 2001.

[4] B. Zhu, L. E. Nelson, S. Stulz et al., "6.4-Tb/s (160 × 42.7 Gb/s) Transmission with 0.8 bit/s/Hz Spectral Efficiency Over 32 × 100 km of Fiber Using CSRZ-DPSK Format," in Proc. *OFC 2003*, paper PD 19, Atlanta, Georgia, March 2003.

[5] A. H. Gnauck and R. M. Jopson, "Dispersion compensation for optical fiber systems," Optical fiber Telecommunications, vol. IIIA, Chapter 7, Academic Press, 1977, pp. 162–195.

[6] B. Jopson and A. Gnauck, *IEEE Commun. Mag.*, 96–102, 1995.

[7] V. Srikant, "Broadband Dispersion and Dispersion Slope Compensation in High Bit Rate and Ultra Long Haul Systems," in Proc. *OFC 2001*, paper TuH1, Anaheim, California, 2001.

[8] A. J. Antos and D. K. Smith, "Design and characterization of dispersion compensating fiber based on the LP_{01} mode," *J. Lightw. Technol.*, 12, 1739–1745, 1994.

[9] T. Kashiwada, S. Ishikawa, T. Okuno et al., "Broadband Dispersion Compensating Module Considering its Attenuation Spectrum Behavior for Wdm System," in Proc. *OFC 1999*, paper WM12, San Diego, California, 1999.

[10] L. Gruner-Nielsen, S. N. Knudsen, T. Veng et al., "Design and Manufacture of Dispersion Compensating Fiber for Simultaneous Compensation of Dispersion and Dispersion Slope," in Proc. *OFC 1999*, paper WM13, San Diego, California, 1999.

[11] G. E. Berkey and M. R. Sozanski, "Negative Slope Dispersion Compensating Fibers," in Proc. *OFC 1999*, paper WM14, San Diego, California, 1999.

[12] S. N. Knudsen and T. Veng, "Large Effective Area Dispersion Compensating Fiber for Cabled Compensation of Standard Single Mode Fiber," in Proc. *OFC 2000*, paper TuG5, Baltimore, Maryland, 2000.

[13] L. Gruner-Nielsen, T. Veng, S. N. Knudsen et al., "New Dispersion Compensating Fibers for Simultaneous Compensation of Dispersion and Dispersion Slope of Non-Zero Dispersion Shifted Fibers in the C or L Band," in Proc. *OFC 2000*, paper TuG6, Baltimore, Maryland, 2000.

[14] R. Sugizagi, "Dispersion Slope Compensating Fibers for L-Band Wdm Systems Using NZDSF," in Proc. *OFC 2000*, paper TuG4, Baltimore, Maryland, 2000.

[15] A. H. Gnauck, L. D. Garrett, Y. Danziger et al., "Dispersion and Dispersion-Slope Compensation of Nzdsf For 40-Gb/S Operation Over the Entire C Band," in Proc. *OFC 2000*, Paper PD8, Baltimore, Maryland, 2000.

[16] L. Gruner-Nielsen, S. N. Knudsen, B. Edvold et al., "Dispersion Compensating Fibers and Perspectives for Future Developments," in Proc. *ECOC 2000*, Munich, Germany, 1, 2000, pp. 91–93.

[17] M. Hirano, T. Kato, K. Fukuda et al., "Novel Dispersion Flattened Link Consisting of New Nz-Dsf and Dispersion Compensating Fibers Module," in Proc. *ECOC 2000*, Munich, Germany, 1, 2000, pp. 99–100.

[18] N. T.Quang Le T. Veng, L. Gruner-Nielsen, "New Dispersion Compensating Module," in Proc. *OFC 2001*, paper TuH5, Anaheim, California, 2001.

[19] T. Kato, M. Hirano, K. Fukuda et al., "Design Optimization of Dispersion Compensating Fiber for Nz-Dsf Considering Nonlinearity and Packaging Performance," in Proc. *OFC 2001*, paper TuS6, Anaheim, California, 2001.

[20] M. J. Li, "Recent Progress in Fiber Dispersion Compensators," in Proc. *ECOC 2001*, Amsterdam, The Netherlands, September–October, 4, 2001, pp. 486–489.

[21] S. R. Bickham, P. Diep, S. Challa et al., "Ultra-Wide Band Fiber Pairs with Accurate Slope Compensation Over an Extended Wavelength Range," in Proc. *ECOC 2002*, Copenhagen, Denmark, September, paper 5.1.4, 2, 2002, pp. 8–12.

[22] F. Forghieri, R. W. Tkach, and A. R. Chraplyvy, "Dispersion compensation fiber: Is there merit in the figure of merit?" *Photon. Technol. Lett.*, 9, 970–972, 1997.

[23] M. J. Li, and C. Saravanos, "Optical fiber designs for field mountable connectors," *J. Lightw. Technol.*, 8(3), 314–319, 2000.

[24] M. J. Li, X. Chen, and D. A. Nolan, "Ultra Low PMD Fibers by Fiber Spinning," in Proc. *OFC'2004*, paper FA1, Los Angeles, California, March, 2004.

[25] D. A. Nolan, X. Chen, and M. J. Li, "Fibers with low polarization-mode dispersion," *J. Lightw. Technol.*, 22(4), 1060–1077, 2004.

[26] Single mode fiber definition at Encyclopedia of Laser Physics and Technology, http://www.rp-photonics.com/single_mode_fibers.html.

[27] I. P. Kaminow, "Polarization in optical fibers," *IEEE J. Quant. Electron.*, QE-17, 15–22, 1981.

[28] M. J. Li, A. F. Evans, D.W. Allen, and D. A. Nolan, "Effects of lateral load and external twist on polarization-mode dispersion of spun and unspun fibers," *Opt. Lett.*, 24, 1325, 1999.

[29] X. Chen, M.-J. Li, and D. A. Nolan, "Polarization Evolution in Spun Fibers," in Proc. *SPIE*, Shanghai, China, 6019, 2005, pp. 60192J.
[30] J. Noda, K. Okamoto, and Y. Sasaki, "Polarization-Maintaining Fibers and Their Applications," *J. Lightw. Technol.*, LT-4, 1071–1089, 1986.
[31] T. Hosaka, K. Okamoto, T. Miya et al., "Low-loss single-polarization fibers with asymmetrical strain birefringence," *Electron. Lett.*, 17, 530–531, 1981.
[32] M. P. Varnham, D. N. Payne, R. D. Birch, and E. J. Tarbox, "Single-polarization operation of highly birefringent bow-tie optical fibers," *Electron. Lett.*, 19, 246–247, 1983.
[33] T. Katsuyama, H. Matsumura, and T. Suganuma, "Propagation characteristics of single polarization fibers," *Appl. Opt.*, 22, 1748–1753, 1983.
[34] R. B. Dyott, J. R. Cozens, and D. G. Morris, "Preservation of polarization in optical-fiber waveguides with elliptical cores," *Electron. Lett.*, 15, 380–382, 1979.
[35] T. Okoshi, K. Oyamada, M. Nishimura, and H. Yokota, "Side-tunnel Fiber: An approach to polarization-maintaining optical waveguiding scheme," *Electron. Lett.*, 18, 824–826, 1982.
[36] A. Ortigosa-Blanch, J. C. Knight, W. J. Wadsworth et al., "Highly birefringent photonic crystal fibers," *Opt. Lett.*, 25, 1325–1327, 2000.
[37] T. P. Hansen, J. Broeng, S. E. B. Libori et al., "Highly birefrigent index-guiding photonic crystal fibers," *IEEE Photon. Technol. Lett.*, 13, 588–590, 2001.
[38] K. Suzuki, H. Kubota, S. Kawanishi et al., "Optical properties of a low-loss polarization-maintaining photonic crystal fiber," *Opt. Express*, 9, 676–680, 2001.
[39] J. R. Folkenberg, M. D. Nielsen, N. A. Mortensen et al., "Polarization maintaining large mode area photonic crystal fiber," *Opt. Express*, 12, 956–960, 2004.
[40] K. Saitoh and M. Koshiba, "Photonic bandgap fibers with high birefringence," *IEEE Photon. Technol. Lett.*, 14, 1291–1293, 2002.
[41] X. Chen, M.-J. Li, N. Venkataraman et al., "Highly birefringent hollow-core photonic bandgap fiber," *Opt. Express*, 12, 3888–3893, 2004.
[42] M. S. Alam, K. Saitoh, and M. Koshiba, "High group birefringence in air-core photonic bandgap fibers," *Opt. Lett.*, 30, 824–826, 2005.
[43] K. Saitoh, N. A. Mortensen, and M. Koshiba, "Air-core photonic band-gap fibers: the impact of surface modes," *Opt. Express*, 12, 394, 2004.
[44] J. A. West, C. M. Smith, N. F. Borrelli et al., "Surface modes in air-core photonic band-gap fibers," *Opt. Express.*, 12, 1485, 2004.
[45] K. Saitoh, N. Florous, T. Murao, and M. Koshiba, "Design of Large Hollow-Core Photonic Band-Gap Fibers with Suppressed Higher-Order Modes," in Proc. *OFC/ENFOEC CD-ROM 2007*, paper OML2, Washington, DC, Optical Society of America, 2007.
[46] J. R. Simpson, R. H. Stolen, F. M. Sears et al., "A Single polarization fiber," *J. Lightw. Technol.*, LT-1, 370–373, 1983.
[47] M. J. Messerly, J. R. Onstott, and R. C. Mikkelson, "A broad-band single polarization optical fiber," *J. Lightw. Technol.*, 9, 817, 1991.
[48] D. A. Nolan, G. E. Berkey, M.-J. Li et al., "Single-polarization fiber with a high extinction ratio," *Opt. Lett.*, 29, 1855, 2004.
[49] D. A. Nolan, M.-J. Li, X. Chen, and J. Koh, "Single Polarization Fibers and Applications," in Proc. *OFC/ENFOEC CD-ROM 2006*, paper OWA1, Washington, DC, Optical Society of America, 2006.
[50] M.-J. Li, X. Chen, J. Wang et al., "Fiber Designs for Higher Power Lasers," in Proc. *SPIE*, San Jose, California, 6469, 2007, pp. 64690H.
[51] H. Kubota, S. Kawanishi, S. Koyanagi et al., "Absolutely Single Polarization Photonic Crystal Fiber," *IEEE Photon. Technol. Lett.*, 16, 182, 2004.
[52] J. R. Folkenberg, M. D. Nielsen, and C. Jakobsen, "Broadband single-polarization photonic crystal fiber," *Opt. Lett.*, 30, 1446–1448, 2005.
[53] T. Schreiber, F. Roser, O. Schmidt et al., "Stress-induced single-polarization single-transverse mode photonic crystal fiber with low nonlinearity," *Opt. Express*, 13, 7621–7630, 2005.

[54] G. P. Agrawal, *Nonlinear Fiber Optics*, San Diego, CA, Academic Press, 1989.
[55] T. Okuno, M. Onishi, T. Kashiwada et al., "Silica-base functional fibers with enhanced nonlinearity and their applications," *IEEE J. Sel. Top. Quant. Electron.*, 5, 1385–1391, September 1999.
[56] R. Stolen and J. Bjorkholm, "Parametric amplification and frequency conversion in optical fibers," *IEEE J. Quant. Electron.*, 18, 1062–1072, July 1982.
[57] S. Wen, "Optical phase conjugation of multiwavelength signals in a dispersion-shifted fiber," *J. Lightw. Technol.*, 15, 1061–1070, July 1997.
[58] M. Sauer, D. Nolan, M. Li, and G. Berkey, "Simultaneous Multichannel Pulse Compression for Broadband Dynamic Dispersion Compensation," in Proc. *OFC 2003*, Atlanta, Georgia, March, 1, 2003, pp. 23–28, 298–300.
[59] S. Li, M. Sauer, Z. D. Gaeta et al., "Broad-band dynamic dispersion compensation in nonlinear fiber-based device," *J. Lightw. Technol.*, 22, 29–38, January 2004.
[60] H. A. Haus, "Optical Fiber Solitons, Their Properties and Uses," in Proc. *IEEE*, 81, July 1993, pp. 970–983.
[61] Ming-Jun Li, Shenping Li, and Daniel A. Nolan, "Nonlinear Fibers for Signal Processing Using Optical Kerr Effects," *J. Lightw. Technol.*, 23(11), 3606–3614, 2005.
[62] D. Q. Chowdhury and D. A. Nolan, "Perturbation model for computingoptical fiber birefringence from two-dimensional refractive index profile," *Opt. Lett.*, 20, 1973–1975, 1995.
[63] M. J. Li and D. A. Nolan, "Fiber spin-profile designs for producing fibers with low polarization mode dispersion," *Opt. Lett.*, 23, 1659–1661, 1998.
[64] M. J. Adams, *An Introduction to Optical Waveguides*, Chichester, John Wiley & Sons Ltd., 1981.
[65] B. J. Ainslie and C.R. Day, "A review of single-mode fibers with modified dispersion characteristics," *J. Lightw. Technol.*, 4, 967–979, August 1986.
[66] T. Hasegawa, E. Sasaoka, M. Onishi, and M. Nishimura, "Hole-assisted lightguide fiber for large anomalous dispersion and low optical loss," *Opt. Express.*, 9, 681–686, December 2001.
[67] E. Snitzer, H. Po, F. Hakimi et al., "Double Clad Offset Core Nd Fiber Laser," in Proc. *OFSTM*, paper PD 5, New Orleans, Louisiana/USA, 1988.
[68] E. Snitzer, "Proposed Fiber Cavities for optical masers," *J. Appl. Phys.*, 32(1), 36, 1961.
[69] T. H. Mainman, "Optical and microwave optical experiments in Ruby," *Phys. Rev. Lett.*, 4(11), 564–566, 1960.
[70] S. B. Poole, D. N. Payne, and M. E. Fermann, "Fabrication of low loss optical fibers containing rare-earth ions," *Electron. Lett.*, 21, 737–738, 1985.
[71] R. J. Mears, L. Reekie, S. B. Poole, and D. N. Payne, "Neodymium-doped silica single-mode fibre laser," *Electron. Lett.*, 21, 738–740, 1985.
[72] H. Po, E. Snitzer, R. Tumminelli et al., "Double Clad High Brightness Nd Fiber Laser Pumped by GaAlAs Phased Array," in Proc. *OFC*, paper PDP7, Houston, TX/USA, 1989.
[73] A. Liem, J. Limpert, H. Zellmer et al., "1.3 kW Yb-Doped Fiber Laser with Excellent Beam Quality," in Proc. *CLEO 2004*, post-deadline paper CPDD2, San Francisco, USA, May, 2004, pp. 16–21.
[74] J. Kirchhof, S. Unger, and A. Schwuchow, "Fiber Lasers: Materials, Structures and Technologies," in *Proc. SPIE*, 4957, 2003, pp. 1–15.
[75] J. P. Koplow, S. W. Moore, and D. A. V. Kliner, "A new method for side pumping of double-clad fiber sources," *J. Quant. Elelectron.*, 39(4), 529–540, 2003.
[76] E. Snitzer, "Rare-earth fiber laser," *J. LessCommon Met.*, 148, 45–58, 1989.
[77] P. Wang, L. J. Cooper, J. K. Sahu, and W. A. Clarkson, "Efficient single-mode operation of a cladding-pumped ytterbium-doped helical-core fiber laser," *Opt. Lett.*, 31, 226–228, 2006.
[78] J. M. Fini, M. D. Mermelstein, M. F. Yan et al., "Fibers with Resonant Mode Suppression," in Proc. *SPIE*, 6453, 2007, p. 64530F.
[79] S. L. Logunov and M. E. DeRosa, "Effect of coating heating by high power in optical fibres at small bend diameters," *Electron. Lett.*, 39, 897–898, 2003.
[80] K. -F. Klein, H. S. Eckhardt, C. Vincze et al., "High NA-Fibers: Silica-Based Fibers for New Applications," in Proc. *SPIE*, San Jose, California, 5691, 2005, pp. 30–41.

[81] A. Liu and K. Ueda, "The absorption characteristics of circular, offset, and rectangular double-clad fibers," *Opt. Commun.*, 132, 511–518, 1996.

[82] A. S. Kurkov, A. Yu. Laptev, E. M. Dianov et al., "Yb^{3+}-Doped Double-Clad Fibers and Lasers," in Proc. *SPIE*, Moscow, Russia, 4083, 2000, pp. 118–126.

[83] H.-R. Müller, J. Kirchhof, V. Reichel, and S. Unger "Fibers for high-power lasers and amplifiers," *C. R. Physique*, 7, 154–162, 2006.

[84] J. D. Minelly, E. R. Taylor, K. P. Jedrzejewski et al., "Laser-Diode-Pumped Nd-Doped Fiber Laser with Output Power > 1W," in Proc. *CLEO 1992*, Anaheim, CA, 1995, p. 246.

[85] E. R. Taylor, D. Taylor et al. "Application-Specific Optical Fibers Manufactured from Multi-component Glasses," in Proc. *MRS*, 172, 1990, pp. 321–327.

[86] J. Schneider, "Properties of a fluoride fiber laser operating at 3.9 µm," in Proc. *SPIE*, Denver, Colorado, 2841, 1996, pp. 230–236.

[87] J. Wang, W. S. Brocklesby, J. R. Lincoln et al., "Local structures of rare-earth ions in glasses: the 'crystal-chemistry' approach," *J. Non-Cryst. Solids*, 163, 261–267, 1993.

[88] F. Gan, *Calculations of Physical Properties of Inorganic Glasses*, Shanghai science press, 1978.

[89] J. Wang, D. T. Walton, and L.A. Zenteno, "All-glass, high NA, Yb-doped double-clad laser fibers made by outside-vapor deposition", *Electron. Lett.*, 40(10), 590–592, 2004.

[90] J. E. Townsend, S. B. Poole, and D. N. Payne, "Solution-doping technique for fabrication of rare-earth-doped optical fibers," *Electron. Lett.*, 23, 329–331, 1987.

[91] J. Wang, R. Kimball, R. Knowlton et al., "Optical fiber and method for making such fiber," Patent Application WO2005082801A2, 2005.

[92] J. Wang, D. T. Walton, M. J. Li et al., "Recent Specialty Fiber Research at Corning Towards High-Power and High-Brightness Fiber Lasers", in Proc. *SPIE*, 6028, 2005, pp. 6028021–6028026.

[93] V. Reichel, K. W. Moerl, S. Unger et al., "Fiber-Laser Power Scaling Beyond the 1-Kilowatt Level by Nd:Yb Co-Doping," in Proc. *SPIE*, 5777, 2005, pp. 404–407.

[94] D. Taverner, D. J. Richardson et al., "158-mJ pulses from a single-transverse-mode, large-mode-area erbium-doped fiber amplifier," *Opt. Lett.*, 22, 378, 1997.

[95] J. Hansryd, F. Dross, M. Westlund, and P. A. Andrekson, "Increase of the SBS threshold in a short highly nonlinear fiber by applying a temperature distribution," *J. Lightw. Technol.*, 19(11), 1691, 2001.

[96] A.Wada, T. Nozawa, D. Tanaka, and R. Yamauchi, "Suppression of SBS by Intentionally Induced Periodic Residual-Strain in Single-Mode Optical Fibers," in Proc. *17th ECOC*, paper No.B1.1, Paris, France, 1991.

[97] R. G. Smith, "Optical power handling capacity of low-loss optical fiber as determined by SRS and SBS," *Appl. Opt.*, 11, 2489–2494, 1972.

[98] C. L. Tang, "Saturation and spectral characteristics of the stokes emission in the stimulated Brillouin process," *J. Appl. Phys.*, 37, 2945, 1966.

[99] X. P. Mao, R. W. Tkach, A. R. Chraplyvy et al., "Stimulated Brillouin threshold dependence on fiber type and uniformity," *IEEE Photon. Technol. Lett.*, 4, 66–69, 1992.

[100] M. -J. Li, X. Chen, J. Wang et al., "Fiber Designs for Reducing Stimulated Brillouin Scattering," in Proc. *OFC*, paper# OTuA4, Anaheim, California, 2006.

[101] C. K. Jen, J. E. B. Oliveira, N. Goto, and K. Abe, "Role of guided acoustic wave properties in single-mode optical fiber design," *Electron. Lett.*, 24, 1419–1420, 1988.

[102] Ji Wang, Ming-Jun, Li, and Dan Nolan, "The Lowest Loss of 0.35 dB/km in an Al-Doped SM Fiber", in Proc. *OFC 2006*, paper OThA1, Anaheim, CA, USA, 2006.

[103] P. D. Draglc, C. -H. Liu, G. C. Papen, and Almantas Galvanauskas, "Optical Fiber with an Acoustic Guiding Layer for Stimulated Brillouin Scattering Suppression," in Proc. *CLEO 2005*, paper CThZ3, San Jose, California, 2005.

[104] G. P. Agrawal, *Nonlinear Fiber Optics*, San Diego, CA, Academic Press, 1989.

[105] J. Kirchhof, S. Unger, V. Reichel, and St. Grimm, "Drawing-Dependent Losses in Rare-Earth and Heavy Metal Doped Silica Optical Fibers," in Proc. *OFC 1997*, paper WL21, 1997, Dallas, TX/USA, 1997.

[106] J. Wang, S. Gray, D. Walton et al., "High performance Yb-doped double-clad optical fibers for high-power, narrow-linewidth fiber laser applications," in Proc. *SPIE*, 6351, 2006, p. 635109.

[107] S. Gray, A. Liu, D. T. Walton, J. Wang et al., "502 Watt, single transverse mode, narrow linewidth, bidirectionally pumped Yb-doped fiber amplifier," *Opt. Express*, 15, 17044–17050, 2007.

[108] P. V. Kaiser and H. W. Astle, "Low-loss single-material fibers made from pure fused silica," *Bell System Technical Journal*, 53, 1021–1039, 1974.

[109] P. Yeh and A. Yariv, "Bragg Reflection Waveguides," *Opt. Commun.*, 19, 427, 1976; P. Yeh, A. Yariv, and E. Marom, "Theory of Bragg fiber," *J. Opt. Soc. Am.*, 68, 1196–1201, 1978.

[110] S. John, *Phys. Rev. Lett.*, 53, 2169, 1984; ibid 58, 2486, 1987; E. Yablonovitch, *Phys. Rev. Lett.*, 58, 2059, 1987.

[111] T. A. Birks, P. J. Roberts, P. St. J. Russell et al., "Full 2-D photonic bandgaps in silica/air structures," *Electron. Lett.*, 31, 1941–1942, 1995.

[112] R. F. Cregan, B. J. Mangan, J. C. Knight et al., "Single-mode photonic band gap guidance of light in air," *Science*, 285, 1537–9, 1999.

[113] D. C. Allan, J. A. West, J. C. Fajardo et al., "Photonic crystal fibers: Effective-index and band-gap guidance" in *Photonic Crystals and Light Localization in the 21st Century* (C. M. Soukoulis, ed.), The Netherlands, Kluwer Academic Press, 2001, pp. 305–320.

[114] J. C. Knight, T. A. Birks, P. St. J. Russell, and D. M. Atkin, "All-silica single-mode optical fiber with photonic crystal cladding," *Opt. Lett.*, 21, 1547–1549, 1996.

[115] E. Marin, A. Diez, and P. St. J. Russell, "Optical Measurement of Trapped Acoustic Mode at Defect in Square-Lattice Photonic Crystal Fiber Preform," in Proc. *QELS*, paper JTuC5, Baltimore, Maryland, 2001.

[116] M. J. Steel, T. P. White, C. M. de Sterke et al., "Symmetry and degeneracy in microstructured optical fibers," *Opt. Lett.*, 26, 488–490, 2001.

[117] T. A. Birks, J. C. Knight, and P. S. J. Russell, "Endlessly single-mode photonic crystal fiber," *Opt. Lett.*, 22, 961–963, 1997.

[118] J. C. Knight, T. A. Birks, R. F. Cregan et al., "Large-mode-area photonic crystal fibre," *Electron. Lett.*, 34, 1347–1348, 1998; J. Limpert, T. Schreiber, S. Nolte et al., "High-power air-clad large-mode-area photonic crystal fiber laser," *Opt. Express*, 11, 818–823, 2003.

[119] X. Feng, A. K. Mairaj, D. W. Hewak, and T. M. Monro, "Nonsilica Glasses for Holey Fibers," *J. Lightw. Technol.*, 23, 2046–2054, 2005.

[120] J. Laegsgaard, S. E. Barkou Libori, K. Hougaard et al., "Dispersion Properties of Photonic Crystal Fibers-Issues and Opportunities," in Proc. *MRS*, 797, paper W7.1.1, Boston, MA, 2004.

[121] T. A. Birks, D. Mogilevtsev, J.C. Knight, and P.St.J. Russell, "Single Material Fibres for Dispersion Compensation," in Proc. *OFC 1999*, paper FG2–1, San Diego, California, 1999.

[122] J. C. Knight, J. Arriaga, T. A. Birks et al., "Anomalous dispersion in photonic crystal fiber," *IEEE Photon. Technol. Lett.*, 12, 807–809, 2000.

[123] J. K. Ranka, R. S. Windeler, and A. J. Stentz, "Visible continuum generation in air silica microstructure optical fibers with anomalous dispersion at 800 nm," *Opt. Lett.*, 25, 25–27, 2000.

[124] W. Reeves, J. Knight, P. Russell, and P. Roberts, "Demonstration of ultra-flattened dispersion in photonic crystal fibers," *Opt. Express*, 10, 609–613, 2002.

[125] K. Tajima, J. Zhou, K. Nakajima, and K. Sato, "Ultra Low Loss and Long Length Photonic Crystal Fiber," in Proc. *OFC 2003*, paperPD1, Atlanta, Georgia, 2003.

[126] L. Farr, J. C. Knight, B. J. Mangan, and P. J. Roberts, "Low Loss Photonic Crystal Fibre," in Proc. *28th ECOC*, paper PD1.3, Copenhagen, Denmark, 2002.

[127] K. Saitoh, Y. Tsuchida, and M. Koshiba, "Bending-insensitive single-mode hole-assisted fibers with reduced splice loss," *Opt. Lett.*, 30, 1779–1781, 2005.

[128] A. Bjarklev, T. P. Hansen, K. Hougaard et al., "Microbending in Photonic Crystal Fibres – An Ultimate Loss Limit?" in Proc. *27th ECOC*, paper WE.L.2.4, Amsterdam, The Netherlands, 2001.

[129] F. Couny, F. Benabid, and P. S. Light, "Large-pitch kagome-structured hollow-core photonic crystal fiber," *Opt. Lett.*, 31, 3574–3576. 2006.

[130] C. M. Smith, N. Venkataraman, M. T. Gallagher et al., "Low-loss hollow-core silica/air photonic bandgap fibre," *Nature*, 424, 657–9, 2003.

[131] P. J. Roberts, F. Couny, H. Sabert et al., "Ultimate low loss of hollow-core photonic crystal Fibres," *Opt. Express*, 13, 236–44, 2004.

[132] S. G. Johnson and J. D. Joannopoulos, "Block-iterative frequency-domain methods for Maxwell's equations in a planewave basis," *Opt. Express*, 8, 173–190, 2001.

[133] T. M. Monro, D. J. Richardson, N. G. R. Broderick, and P. J. Bennett, "Holey optical fibers: An efficient modal model," *J. Lightw. Technol.*, 17, 1093–1102, 1999.

[134] T. P. White, B. T. Kuhlmey, R. C. McPhedran et al., "Multipole method for microstructured optical fibers. I. Formulation," *J. Opt. Soc. Am. B*, 19, 2322–2330, 2002; B. T. Kuhlmey, T. P. White, G. Renversez et al., "Multipole method for microstructured optical fibers. II. Implementation and results," *J. Opt. Soc. Am. B*, 19, 2331–2340, 2002.

[135] K. Saitoh and M. Koshiba, "Leakage loss and group velocity dispersion in air-core photonic bandgap fibers," *Opt. Express*, 11, 3100–3109, 2003.

[136] M. J. F. Digonnet, H. K. Kim, J. Shin et al., "Simple geometric criterion to predict the existence of surface modes in air-core photonic-bandgap fibers," *Opt. Express*, 12, 1864–1872, 2004.

[137] D. G. Ouzounov, F. R. Ahmad, D. Müller et al., "Generation of megawatt optical solitons in hollow-core photonic band-gap fibers," *Science*, 301, 1702–1704, 2003.

[138] J. Lægsgaard, N. A. Mortensen, and A. Bjarklev, "Mode areas and field-energy distribution in honeycomb photonic bandgap fibers," *J. Opt. Soc. Am. B*, 20, 2037–45, 2003.

[139] E. T. J. Nibbering, G. Grillon, M. A. Franco et al., "Determination of the inertial contribution to the nonlinear refractive index of air, N2, and O2 by use of unfocused high-intensity femtosecond laser pulses," *J. Opt. Soc. Am. B*, 14, 650, 1997.

[140] A. Boskovic, S. V. Chernikov, J. R. Taylor et al., "Direct continuous-wave measurement of n2 in various types of telecommunication fiber at 1.55 mm," *J. Opt. Soc. Am. B*, 21, 1966–1968, 1996.

[141] C. J. Hensley, D. G. Ouzounov, A. L. Gaeta et al., "Silica-glass contribution to the effective nonlinearity of hollow-core photonic band-gap fibers," *Opt. Express*, 15, 3507–3512, 2007.

[142] J. A. West, "Demonstration of an IR-Optimized Air-Core Photonic Band-Gap Fibre," in Proc. *ECOC 2001*, paper Th.A.2, Amsterdam, 2001.

[143] N. Venkataraman, M. T. Gallagher, D. Müller et al., "Low-Loss (13 dB/km) Air-Core Photonic Band-Gap Fibre", in Proc. *ECOC 2002*, paper PD1.1, Copenhagen, Denmark, 2002.

[144] G. Humbert, J. Knight, G. Bouwmans et al., "Hollow core photonic crystal fibers for beam delivery," *Opt. Express*, 12, 1477–1484, 2004.

[145] S. Johnson, M. Ibanescu, M. Skorobogatiy et al., "Low-loss asymptotically single-mode propagation in large-core OmniGuide fibers," *Opt. Express*, 9, 748–779, 2001.

[146] P. Mach, M. Dolinski, K. W. Baldwin et al., "Tunable microfluidic optical fiber," *Appl. Phys. Lett.*, 80, 4294–4296, 2002.

[147] T. Larsen, A. Bjarklev, D. Hermann, and J. Broeng, "Optical devices based on liquid crystal photonic bandgap fibres," *Opt. Express*, 11, 2589–2596, 2003.

[148] A. Cerqueira, S. Jr, F. Luan, C. M. B. Cordeiro et al., "Hybrid photonic crystal fiber," *Opt. Express*, 14, 926–931, 2006.

[149] A. Argyros, T. Birks, S. Leon-Saval et al., "Guidance properties of low-contrast photonic bandgap fibres," *Opt. Express*, 13, 2503–2511, 2005.

[150] K. Saitoh, and M. Koshiba, "Single-polarization single-mode photonic crystal fibers," *IEEE Photon. Technol. Lett.*, 15, 1384–1386, 2003.

[151] E. A. Golovchenko, E. M. Dianov, A. S. Kurkov et al., "Generation of Fundamental Solitons at High Repetition Rate and other Applications of Fibers with Varying Dispersion," in Proc. *IOEC 1988*, paper PD32, July 1988, pp. 18–21.

16

Plastic optical fibers: Technologies and communication links

Yasuhiro Koike[*] and Satoshi Takahashi[†]

[*]*Keio University ERATO Koike Photonics Polymer Project, Yokohama, Japan*
[†]*The Application Group, Shin-Kawasaki Town Campus, Keio University, Kawasaki, Japan*

16.1 INTRODUCTION

Plastic optical fiber (POF) consists of a plastic core and plastic cladding of a refractive index lower than that of core.

POFs have very large core diameters compared to those of glass optical fibers (Figure 16.1).

Even when the diameter has a thickness of between 0.5 and 1 mm in diameter, POFs are flexible. These features enable easy installation and safe handling. There is no fear for POFs to stick into human skin (Figure 16.2). Owing to the large diameter, high-positioning accuracy is not required for connections so that cost of connecting devices can be reduced.

POFs have been used extensively in short-distance datacom applications, such as digital audio interface (Figure 16.3). POFs are also used for data transmission equipment, control signal transmission for numerical control machine tools and railway rolling stocks, and so on. During the late 1990s, POF has come to be used for optical data bus in automobiles.

As high-speed communication will be installed in homes in the near future, POF will be a promising candidate for network wiring because of its advantageous properties for consumer use.

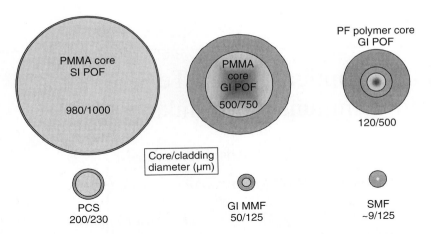

Figure 16.1 Comparison of cross-sectional views of various types of optical fibers: polymethylmethacrylate (PMMA) core step index (SI) POF, PMMA core graded index (GI) POF, perfluorinated (PF) polymer core GI POF, plastic clad silica (PCS) optical fiber, GI multimode glass optical fiber (MMF), and single-mode optical fiber (SMF). Core diameter for SMF denotes a typical mode field diameter at 1310 nm (this figure may be seen in color on the included CD-ROM).

Figure 16.2 POF enables easy and safe handling suitable to consumer use (this figure may be seen in colour on the included CD-ROM)

Figure 16.3 Plugs for optical digital audio interfaces (this figure may be seen in color on the included CD-ROM).

Attenuation and bandwidth of graded index (GI) POFs have been improving in the past 10 years. This chapter describes a brief history of the development, salient features, and datacom applications of POFs.

16.2 DEVELOPMENT OF POFs

16.2.1 Development in Attenuation of POF

The first POF was demonstrated in the mid-1960s by DuPont. Named "Crofon," it had a step index (SI) polymethylmethacrylate (PMMA) core. Mitsubishi Rayon commercialized SI POF for the first time with trade name of "Eska" in 1975. Other major POF manufacturers, Asahi Chemical and Toray, followed in 1980s.

Experimental work on loss reduction of PMMA-core SI POF was conducted in 1980s by Kaino et al [1]. They also reported low-loss SI POF experimentally obtained by employing perdeuterated PMMA [2]. Impressive results were obtained during the end of 1980s [3, 4].

Among GI-type POFs, the first report of PMMA-core GI POFs was presented by Keio University in 1976 [5]. Near-parabolic refractive-index profile in the first GI POF was formed by copolymerizing methylmethacrylate (MMA) monomer, whereas the second monomer had a refractive index higher than MMA. Actually, the attenuation first measured for GI POF composed of MMA and vinyl benzoate was 1000 dB/km, which was ~10 times higher than that of SI POF [6]. However, it was reported that the resulting copolymer composition is mainly divided into two compositions, which cause large increase of the inherent excess scattering loss [7].

To decrease such an excess scattering loss caused by the difference of monomer reactivity, a new interfacial copolymerization process based on random copolymerization was developed [8, 9]. The attenuation of MMA-benzyl methacrylate copolymer GI POF by this random copolymerization process was greatly decreased to about 200 dB/km from 1000 dB/km. An excess scattering loss of about 100 dB/km due to the heterogeneous structure in the copolymer still remained. Fundamental research on the relationship between scattering loss and heterogeneous structure in polymer materials finally made a breakthrough on the high-attenuation problem mentioned above [10, 11]. The process of doping low-molecular weight compound was then invented, as a replacement for the copolymerization process [12]. Refractive-index profile of the GI POF was formed by the radial concentration distribution of the dopant. By designing the dopant structure to be compatible with PMMA, the size of the heterogeneous structure in the polymer was decreased, resulting in reducing the attenuation to that of PMMA-core SI POF [13].

The doping method triggered the research and development of low-attenuation polymer materials. General aliphatic polymer has high absorption loss due to carbon–hydrogen stretching vibration. However, by substituting fluorine for all hydrogen bonding in polymer molecules, very low attenuation was achieved in a perfluorinated (PF) polymer-based GI POF, at wavelength of 1300 nm. The first PF polymer-based GI POF was reported in 1994 [14]. In 2000, an amorphous PF polymer-based GI POF was commercialized by Asahi Glass Co., Ltd. with the trade name of "Lucina."

Typical attenuation spectra of GI POFs with PMMA-based polymer, deuterated PMMA, and PF polymer are shown in Figure 16.4.

The history of attenuation reduction of SI POFs and G POFs is summarized in Figure 16.5.

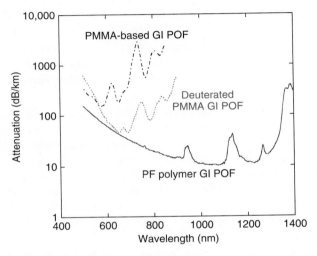

Figure 16.4 Typical attenuation spectra of graded index plastic optical fibers (GI POFs) of different core materials (this figure may be seen in color on the included CD-ROM).

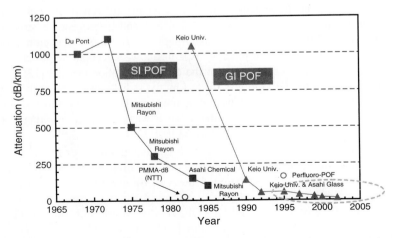

Figure 16.5 History of attenuation reduction of plastic optical fibers (POFs) (this figure may be seen in color on the included CD-ROM).

16.2.2 Development in High-Speed Transmission by POF

Several attempts to employ POF (mainly SI-type) for high-speed communication were conducted by researchers in many countries during the 1990s [15, 16]. However, the high-bandwidth characteristic of GI POFs was experimentally verified in 1990 for the first time [17]. The 3-dB bandwidth was 17.3 GHz for a 15-m length at 670-nm wavelength. Technological breakthrough was demonstrated in 1994, following the successive development of the semiconductor edge-emitting red laser [18] and vertical-cavity surface-emitting laser (VCSEL) emitting at 670-nm wavelength [19].

After 1994, gigabit per second (Gbps) transmission experiments employing GI POF began to be actively pursued worldwide using one of these light sources. The first 2.5-Gbps transmission demonstration using a PMMA-core GI POF 100-m long was reported as a cooperative work with NEC in 1994 [20]. The successful experimental transmission of 11 Gbps. 100 m in 1999 by Asahi Glass Co., Ltd. and Bell Laboratories in the United States was an extremely significant result, which demonstrated the high-bandwidth performance surpassing that of silica glass multimode optic fibers.

Because the wavelength dependency of the refractive index of the PF polymer is smaller than those for PMMA and silica, the material dispersion, which is one of factors limiting bandwidth, is small [21]. Figure 16.6 shows calculated values for the PF polymer-based GI POF and silica multimode optical fiber for fiber lengths of 100 m.

The figure shows bandwidth as a function of wavelength where refractive index distributions of both fibers are optimized to give maxima at a wavelength of 850 nm. The bandwidth of PF polymer GI POF surpasses that for silica multimode glass optical fiber (MMF), and PF polymer-based GI POF retains broadband features for a wide range of wavelengths covering from the visible to the 1.4-μm wavelength region.

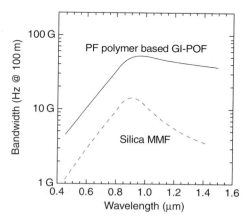

Figure 16.6 Calculated bandwidth of perfluorinated (PF) polymer-based (GI POF) graded index plastic optical fiber compared with that of Silica multimode glass optical fiber (MMF) (this figure may be seen in color on the included CD-ROM).

A 1-Gbps campus LAN utilizing PF GI POF was deployed at Keio University in 2000. Subsequently, GI POF has been used in Tokyo in housing complexes, hospitals, medical conference halls, and so on.

16.3 VARIETIES OF POFs, POF Cords, and Cables

16.3.1 Product-Specifications of POFs

Eight major types of POFs are specified as category A4 multimode fibers in IEC 60793-2-40. Table 16.1 summarizes their attributes.

The IEC standard specifies material as "plastic"; however, commercially available POFs A4a through A4e are made of PMMA-based resins, and A4f through A4h are PF polymer-based POFs.

Table 16.1
Attributes of POFs specified in IEC 60793-2-40.

Category	Structure			Characteristics		
	Diameter (μm)			Maximum attenuation	Minimum bandwidth	Operating wavelength
	Cladding	Core	NA	(dB/100 m)	(MHz at 100 m)	(nm)
A4a	1000		0.50	30[*]	10	650
A4b	750	15–35 smaller than cladding	0.50	30[*]	10	650
A4c	500		0.50	30[*]	10	650
A4d	1000		0.30	18[**]	100[**]	650
A4e	750	Min. 500	0.25	18[*]	200[*]	650
A4f	490	200	0.19	4[***]	1500[***]	650, 850, 1300
A4g	490	120	0.19	3.3[***]	1880[***]	650, 850, 1300
A4h	245	62.5	0.19	3.3[***]	1880[***]	650, 850, 1300

* Measured at 650 nm, under equilibrium launch condition.
** Measured at 650 nm, under launch NA of 0.3.
*** Measured at 850 nm.

16.3.2 Typical Construction of PMMA POF Cords

PMMA-based POFs are often used as buffered fibers or tight-buffered cords, which have no strength member. Typical constructions of simplex and duplex cords are shown in Figure 16.7.

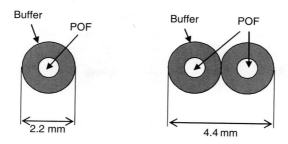

Figure 16.7 Typical constructions of simplex and duplex PMMA POF cords.

16.3.3 Typical Constructions of Cables with PF Polymer-Based POFs

Examples of some types of cable construction for PF polymer-based POFs are shown in Figure 16.8. Typical constructions are loose tube cables with reinforcing materials.

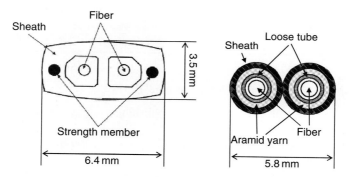

Figure 16.8 Typical Constructions of Cables with PF Polymer-based POFs (this figure may be seen in color on the included CD-ROM).

16.4 PASSIVE AND ACTIVE COMPONENTS FOR POFs

16.4.1 Connectors and Termination Methods for PMMA-based POFs

Connectors for PMMA-Based POFs

A majority of connectors for PMMA-based POFs are shown in Table 16.2.

The thick diameter of POFs and the simple structure of POF cords enable simple and inexpensive connector structures. These connectors are made by resin molding and consist of only two or three parts.

Table 16.2
Major connectors for PMMA-based POFs.

Name	Standard	Appearances
Type F05	IEC 60874-17	
Type PN	IEC 61754-16	
Type SMI	IEC 61754-21	
Versatile Link	AVAGO original	

Termination Methods for PMMA-Based POFs

PMMA-based POFs can be terminated by polishing, cutting, end milling, or using a hot-plate method.

Hot-plate termination is a unique and quick termination method applicable to PMMA-based POFs (Figure 16.9).

This termination procedure takes only 40 seconds per plug and does not require special skill.

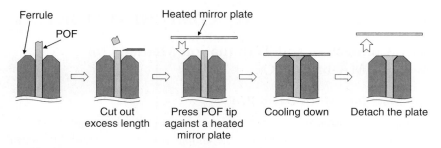

Figure 16.9 Schematic diagram of hot-plate termination (this figure may be seen in color on the included CD-ROM).

16.4.2 Connectors and Termination Methods for PF Polymer POFs

Connectors for PF Polymer POFs

Some connectors used for silica optical fibers, such as SC, LC, and MT-RJ, are applicable to PF polymer-based POFs. Figure 16.10 is a commercially available clip-on SC connector for PF polymer POF that enables easy termination.

Figure 16.10 Clip-on SC connector (this figure may be seen in color on the included CD-ROM).

Termination Methods for PF Polymer POFs

Perfluorinated polymer POFs are usually terminated by polishing.

16.4.3 Transceivers for POFs

Optical transceivers for PMMA-based POFs operate at wavelengths of visible region, ($\sim$650 nm). Transceivers for PMMA-based POFs standardized in IEC are shown in Table 16.3.

For operation with PF polymer-based POF, conventional 850- and 1310-nm transceivers designed for silica fibers are applicable.

Table 16.3
IEC standards for transceivers for PMMA-based POFs.

Title	Standard	Appearances
PN 1 × 9 plastic optical fiber transceivers	IEC 62148-4	
650-nm 250-Mbit/s plastic optical fiber transceivers	IEC 62149-6	

16.5 DATACOM APPLICATIONS WITH POFs

POFs are used for short-range data communication such as factory automation and digital audio interfaces. Market size of automotive LAN with POF has seen remarkable growth in the past several years. PF polymer POFs will open up a new market of customer premises network.

Table 16.4 shows major datacom standards with POFs.

Table 16.4
Standards for data communication with POFs.

Standard	Title
IEC 60958-3	Digital audio interface – Part 3: Consumer applications
IEC 62300	Consumer audio/video equipment digital interface with plastic optical fibre
The ATM Forum AF-PHY-POF155	155-Mb/s Plastic Optical Fiber and Hard Polymer Clad Fiber PMD Specification
IEEE 1394.b	High-Performance Serial Bus (High-Speed Supplement)
Media-Oriented Systems Transport	MOST specification Multimedia and Control Networking Technology

REFERENCES

[1] T. Kaino, M. Fujiki, S. Oikawa, and S. Nara, "Low-loss plastic optical fibers," *Appl. Opt.*, 20(17), 2886–2888, 1981.
[2] T. Kaino, K. Jinguji, and S. Nara, "Low loss poly (methyl methacrylate-d8) core optical fibers," *Appl. Phys. Lett.*, 42(7), 567–569, 1983.
[3] T. Kaino, M. Fujiki, K. Jinguji, and S. Nara, "Preparation of plastic optical fibers," *Rev. Electron. Commun. Lab.*, 32(3), 478–488, 1984.
[4] T. Kaino, "Influence of water absorption on plastic optical fibers," *Appl. Opt.*, 24(23), 4192–4195, 1985.
[5] Y. Ohtsuka and Y. Hatanaka, "Preparation of ligh-focusing plastic fiber by heat-drawing process," *Appl. Phys. Lett.*, 29(11), 735–737, 1976.
[6] Y. Koike, Y. Kimoto, and Y. Ohtsuka, "Studies on the light-focusing plastic rod. 12: The GRIN fiber lens of methyl methacrylate-vinyl phenyl acetate copolymer," *Appl. Opt.*, 21(6), 1057–1062, 1982.
[7] Y. Koike, "High-bandwidth graded-index polymer optical fibre," *Polymer*, 32(10), 1737–1745, 1991.
[8] Y. Koike and E. Nihei, "Method of manufacturing optical transmission medium from synthetic resin," JP Patent 3005808: US Patent 5,382,448: EU Patent 0497984, 1990.
[9] Y. Koike and E. Nihei, "Method of manufacturing a graded optical transmission medium made of synthetic resin," JP Patent 3010369: US Patent 5,253,323: EU Patent 0496893, 1990.
[10] Y. Koike, N. Tanio, and Y. Ohtsuka, "Light scattering and heterogeneities in low-loss poly (methyl methacrylate) glasses," *Macromol.*, 22(3), 1367–1373, 1989.
[11] Y. Koike, S. Matsuoka, and H. E. Bair, "Origin of excess light scattering in poly (methyl methacrylate) glasses," *Macromol.*, 25(18), 4807–4815, 1992.
[12] Y. Koike, "Optical resin materials with distributed refractive index, process for producing the materials, and optical conductors using the materials," JP Patent 3332922: US Patent 5,541,247: EU Patent 0566744: KR Patent 170358: CA Patent 2098604, 1991.

[13] T. Ishigure, E. Nihei, and Y. Koike, "Graded-index polymer optical fiber for high speed data communication," *Appl. Opt.*, 33(19), 4261–4266, 1994.
[14] Y. Koike and M. Naritomi, "Gradede-refractive-index optical plastic material and method for its production," JP Patent 3719733: US Patent 5,783,636: EU Patent 0710855: KR Patent: 375581: CN Patent ZL951903152: TW Patent 090942, 1994.
[15] R. J. S. Bates, "Equalization and mode partition noise in all-plastic optical fiber data links," *IEEE Photon. Technol. Lett.*, 4(10), 1154–1157, 1992.
[16] S. D. Walker, and R. J. S. Bates, "Towards gigabit plastic optical fiber data links: present progress and future prospects," in Proc. 2nd *POF 1993*, The Netherlands, Den Haag, 1993, pp. 8–13.
[17] Y. Ohtsuka, E. Nihei, and Y. Koike, "Graded-index optical fibers of methyl methacrylate-vinyl benzoate copolymer with low loss and high bandwidth," *Appl. Phys. Lett.*, 57(2), 120–122, 1990.
[18] S. Yamazaki, H. Hotta, S. Nakaya et al., "2.5Gb/s 100m GRIN plastic optical fiber data link at 650nm wavelength," *20th ECOC 1994*, Post Deadline Paper, Oslo, Norway, 1994, pp. 1–4.
[19] D. M. Kuchta, J. A. Kash, P. Pepeljugoski et al., "High speed data communication using 670 nm vertical cavity surface emitting lasers and plastic optical fiber," in Proc. 3rd *POF 1994*, Yokohama, Japan, 1994, pp. 135–139.
[20] T. Ishigure, E. Nihei, S. Yamazaki et al., "2.5Gb/s 100m data transmission using graded-index polymer optical fiber and high-speed laser diode at 650nm wavelength," *Electron. Lett.*, 31(14), 467–468, 1995.
[21] Y. Koike, and T. Ishigure, "High-bandwidth plastic optical fiber for fiber to the display," *J. Lightw. Technol.*, 24(12), 4541–4553, 2006.

17

Polarization mode dispersion

Misha Brodsky[*], Nicholas J. Frigo[†], and Moshe Tur[‡]

[*]AT&T Labs – Research, Middletown, NJ, USA
[†]Department of Physics, U.S. Naval Academy, Annapolis, MD, USA
[‡]School of Electrical Engineering, Tel Aviv University, Ramat Aviv, Israel

17.1 INTRODUCTION

Advances in optical fiber telecommunications technology have been truly breathtaking: Since the introduction of the first fiber optic systems in the early 1980s, the throughput on optical fibers has grown by a factor of about 20,000, with laboratory results another factor of 20 above that. As the technology grew, limiting impairments (loss, chromatic dispersion, non-linearities) were faced and were solved, so that fiber optics is now ubiquitous and is the universal transmission medium for high-speed terrestrial and undersea communications.

In this chapter, we examine another impairment, polarization mode dispersion (PMD), which has loomed as a limiting impairment for nearly 20 years. In brief, PMD is an effect in which polarization-dependent propagation delays of optical pulses in fibers lead to distorted signals at the end of an optical link. As with the other impairments, it has yielded, so far, to efforts by fiber-optic researchers around the globe. But PMD has a complex and random nature that could frustrate full exploitation of optical communication's promise.

PMD was first described by Poole and Wagner [1], with the mathematical formulation [2] and statistical descriptions [3, 4] following shortly thereafter. The field rapidly matured, and several reviews [5, 6], a tutorial [7], and two books [8, 9] have appeared which provide comprehensive views of the subject. Since the turn of the century, interest has increasingly moved to more practical questions of how PMD actually behaves in the traffic-bearing fibers that are installed in operating networks. We have discussed earlier how results from field tests around the world have led us to modify the classical (theory-driven) viewpoint of PMD behavior in installed fibers [10]. In this chapter, we discuss the system implications of these results.

This chapter is organized as follows. In Section 17.2, we give a self-contained view of PMD concepts that are used in the remainder of the chapter. Section 17.3 describes the "hinge" model which we have found helpful in describing field test results. In Section 17.4, we briefly review results from a series of reported field tests and show how they can be understood in terms of the hinge model. Section 17.5 is devoted to penalties and system outages due to pure first-order PMD, which is very important from the practical point of view. Next, in Section 17.6, we review higher PMD orders and discuss their importance in today's systems. The ability to evaluate systems without the burdens of full field tests has motivated PMD emulation, and we describe the promises and current limitations in Section 17.7. Finally, Section 17.8 touches upon the subtle interaction between fiber nonlinear effects and PMD, with emphasis on nonlinear polarization rotation.

An editorial goal for this book was that the chapters be self contained, develop a coherent view, be comprehensive, and still be of manageable length. As noted earlier [6], the subject can no longer be reviewed in its entirety in less than book length: important topics, alternative viewpoints, illuminating discussions, and many references that were off our chosen theme had to be neglected. In particular, consideration of polarization-dependent loss/gain (PDL/G) (originating mainly from in-line components rather than from the fiber), although important in a full description of lightwave communications systems' is omitted.

17.1.1 Notation Convention

To ease the divide between geometrical and analytical descriptions, we use the convention that a vector's magnitude and unit vector have the same symbol as the vector itself, i.e., most notably, $\vec{\beta} = \beta\hat{\beta}$, rather than the earlier convention $\vec{\beta} = (\Delta\beta)\hat{\beta}$ [6], and similarly for $\vec{\tau}$. This convention encroaches on the use of β as the propagation constant, which we denote by "k." Consequently, the alert reader will note that some of our formulas agree with earlier formulas in content, but have fewer "Δ" symbols.

17.2 BACKGROUND

In this section, we give a self-contained but dense introduction to those PMD concepts we need in order to describe recent developments. The comprehensive reviews [6, 9] and tutorial [7] are indispensable for more detailed study.

17.2.1 Propagation

Optical transmission media, such as glass, are characterized by two optical velocities, phase and group, both of which play a role in PMD. After traversing a distance z, light in a homogeneous medium can be described as a superposition of plane waves, exp [i($\omega t - kz$)], in which each Fourier component's wave number, k, depends on frequency. After a first-order expansion of k about central frequency ω_0 in the signal's Fourier transform, it is easy to show that the field at z is related to the incident field by

$$\mathbf{E}(z,t) = \mathbf{E}\left(0, t - \frac{z}{v_g}\right) \exp\left\{i\omega_0\left(t - \frac{z}{v_p}\right)\right\} \quad (17.1)$$

This form describes the electric field as a plane wave at central frequency ω_0, propagating at phase velocity $v_p = [k(\omega_0)/\omega_0]^{-1}$, modulated by the input signal's envelope, $\mathbf{E}(0,t)$, traveling at the group velocity $v_g = [dk/d\omega|_{\omega_0}]^{-1}$, with higher orders revealing chromatic dispersion and its slope. Associated with the phase (group) velocity is a phase (group) index of refraction, and a phase (group) delay after propagation through a medium which, in a homogeneous medium, is independent of $\mathbf{E}'s$ polarization. As we show below, PMD is essentially a time domain effect (manifested in group velocity differences) that is most easily described in the frequency domain.

17.2.2 Birefringence

Perturbations, such as stresses, break the ideal symmetry of a homogeneous fiber, creating two preferred polarization states (the "fast" and "slow" eigenstates) such that a wave with an arbitrary state of polarization (SOP) can be resolved into components along these two orthogonal states. The velocity difference for these polarizations, the birefringence, then creates differing delays, and the arrival time differences between the *phases* of the two polarization components and the *signals* of the two components (after traveling a distance L in the fiber) are the differential phase delay and the differential group delay (DGD), respectively. For the simplest case of a fiber with uniform birefringence along its length, these are given by Brodsky et al. [10] as

$$\Delta \tau_p = \frac{L}{v_{ps}} - \frac{L}{v_{pf}} = \left[\frac{k_s - k_f}{\omega}\right]L = \frac{\Delta k}{\omega}L \equiv \frac{\beta}{\omega}L \quad \text{and} \quad (17.2)$$

$$\Delta \tau_g = \frac{L}{v_{gs}} - \frac{L}{v_{gf}} = \left[\frac{d(k_s - k_f)}{d\omega}\right]L = \left[\frac{d(\Delta k)}{d\omega}\right]L \equiv \frac{d\beta}{d\omega}L = \beta'L \quad (17.3)$$

respectively, with subscripts indicating the slow and fast components of the phase and group velocities. Conventionally, $\beta = \Delta k = \Delta n(\omega/c)$ is defined as the fiber's birefringence. (We remind the reader that some other notations call the birefringence "$\Delta\beta$.") The similarities of the values for phase and group velocities in glasses lead to similarities in their differential delays, so that β/ω and $\beta' = d\beta/d\omega$ are approximately equal. For uniform birefringence, the frequency and time domain pictures are similar, but for real fibers with non-uniform birefringence, the pictures diverge [11].

The most useful description of the effects of birefringence is through a geometrical representation based on the Poincare sphere and the closely related Stokes vectors: it has been described from various viewpoints [7, 6, 12, 13, 14]. This coordinate-free approach treats linear, elliptical, and circular birefringence on an equal footing: associated with each SOP is a unit vector (i.e., a point on the unit sphere) and associated with each section's *birefringence* is another vector, aligned in direction with the unit vectors of that birefringence's eigenstates.

A brief resume of the geometrical representation for monochromatic waves is that:

(1) each SOP is associated with a unique vector, $\vec{s}$, on the surface of the unit sphere;
(2) the projection of one SOP onto another is given by $\cos(\theta/2)$, where θ is the angle between the two vectors on the sphere, and because orthogonal SOPs are 180° apart, any two antipodal points can form a basis set for SOPs;
(3) the fiber birefringence is represented by a vector $\vec{\beta}$ pointing in the same direction as that birefringence's "slow" eigenstate, and whose magnitude is β as above;
(4) in propagating a distance L through a section of fiber with birefringence represented by $\vec{\beta}$, the vector, $\vec{s}$, representing the lightwave's SOP, precesses about vector $\vec{\beta}$ through an angle of βL radians in the sense of the right-hand rule. (This precession is equivalent to a rotation R about the axis formed by $\vec{\beta}$.) The rotation angle is proportional to the differential phase delay of the wave.

These points are illustrated in the upper right inset of Figure 17.1, showing the usual convention that linear SOPs are on the equator (i.e., $|x\rangle$ and $|+45\rangle$) while right circular polarization $|R\rangle$ is at the north pole of the representation. (Not shown are the three states orthogonal to these, namely $|y\rangle$, $|-45\rangle$, and $|L\rangle$, respectively, reflected through the origin.) The inset depicts SOP evolution from $\vec{s}_{in}$ to $\vec{s}_{out}$, precessing about birefringence $\vec{\beta}$ in a single section, according to the right-hand rule. As light progresses down a fiber, it will encounter successive birefringent sections, each with its own birefringence $\vec{\beta}$, each imposing its own precession on the SOP, so that the evolution of SOP $\vec{s}$ can be viewed as traversing a succession of arcs on the sphere.

The remainder of Figure 17.1 illustrates a system of two concatenated birefringent fiber sections, represented schematically in the lower part of the drawing. Light at frequency ω_0 enters a birefringent fiber section with SOP $\vec{s}_0$, exits that section with SOP $\vec{s}_1^0$, propagates through the second birefringent section, and finally exits with SOP $\vec{s}_2^0$. (In this development, we use *subscripts* to denote section number, and *superscripts* to denote frequency.) The equivalent situation in the geometrical representation is shown on the large sphere in Figure 17.1. The incoming SOP, $\vec{s}_0$, precesses about $\vec{\beta}_1^0$ (solid vector) tracing out a circle and evolving into state $\vec{s}_1^0$ (dark arc). At that point, the SOP becomes the input SOP to the second section, and it continues to evolve by precessing about $\vec{\beta}_2^0$ to $\vec{s}_2^0$.

17. Polarization Mode Dispersion

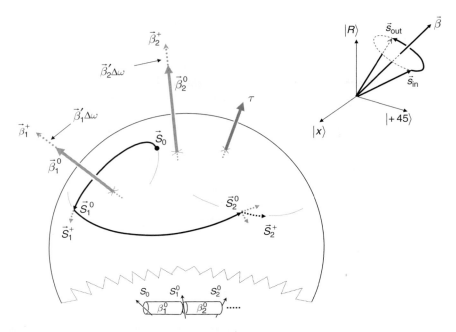

Figure 17.1 Geometrical representation of polarization evolution and PMD. SOPs $\vec{s}$ evolve as precessions about a birefringence vector $\vec{\beta}$ (inset). In a two-section fiber, light at ω_0 evolves along the solid curve under the influence of birefringence vectors $\vec{\beta}_1^0$ and $\vec{\beta}_2^0$ (solid vectors). At $\omega^+ = \omega_0 + \Delta\omega$, the birefringence vectors change slightly (dotted arrows), leading to slightly altered trajectories (dashed gray arcs), resulting in a net evolution in frequency (dashed black arc) for the output SOP that is described as a rotation about PMD vector $\vec{\tau}$ (this figure may be seen in color on the included CD-ROM).

The successive precessions could also have been written as $\vec{s}_2^0 = R_2 R_1 \vec{s}_0$, with the Rs representing appropriate rotations about the birefringence vectors.

17.2.3 Polarization Mode Dispersion

Now consider light at $\omega^+ = \omega_0 + \Delta\omega$, again launched with SOP $\vec{s}_0$. At this new frequency, the birefringence has changed slightly, expressed to first order as $\vec{\beta}_1^+ = \vec{\beta}_1^0 + \vec{\beta}_1' \Delta\omega$, where the first-order expansion term is shown as a dotted vector with the same direction as $\vec{\beta}_1^0$, the usual assumption [6]. (Because $\vec{\beta}' \Delta\omega \approx (\Delta\omega/\omega)\vec{\beta}$, as in Section 17.2.2, these dotted vectors are greatly exaggerated in scale.) At this new frequency, $\vec{s}_0$ progresses to $\vec{s}_1^+$, slightly farther than it did at ω_0, through an additional arc subtending $\beta_1' \Delta\omega z_1$ radians (shown in gray). We emphasize that there are two equivalent views of this angle. First, as $(\beta_1' \Delta\omega) z_1$, it describes the final SOP's extra rate of change with *position* at the offset frequency (our usual frequency domain view): in this view, the term in the parentheses is a differential birefringence. The second view of this arc is as

$(\beta'_1 z_1)\Delta\omega$, which describes the final SOP's rotation rate with *frequency*: in this view, the term in the parentheses is the DGD for that section, as in Eqn (17.2). Thus, at the end of a section we can identify the SOP rotation rate (with respect to frequency) as the DGD for that section. We define $\vec{\beta}'_1 z_1$ as the PMD vector for that section. The reason for the name is that light in the two polarization modes "disperses" in propagating down the fiber, by the DGD, as described below.

In a similar manner, the second section also rotates SOPs farther at ω^+ as light propagates down section length z_2 (i.e., after a total distance $z_1 + z_2$), and the second section will rotate $\vec{s}_1^+$ into state $\vec{s}_2^+$. But if $\vec{s}_1^+$ is viewed, geometrically, as the vector $\vec{s}_1^0$ plus the small gray vector arc to $\vec{s}_1^+$, one sees that the evolution of $\vec{s}_2^0$ to $\vec{s}_2^+$ (as frequency is increased by $\Delta\omega$) can be described as a vector sum of the two small gray arcs in Figure 17.1. These two gray arcs, to first order, consist of a composite of rotations: one about $\vec{\beta}'_2 \Delta\omega z_2$ and one about $\vec{\beta}'_1 \Delta\omega z_1$ *after it has been rotated about* $\vec{\beta}_2^0$. The resultant sum comprises a first-order rotation about a new vector, the total PMD $\vec{\tau}$ for the system. This is the essence of the PMD "concatenation rule" that an additional section rotates its "input" PMD by its birefringence and adds its own PMD [6].

$$\vec{\tau} = R_2 \vec{\tau}_1 + \vec{\tau}_2 \qquad (17.4)$$

This development can be extended recursively, each section rotating the previous PMD vector and adding its *own* PMD vector (i.e., $\vec{\beta}'_j z_j$) to the result. The total PMD vector is thus seen to be a sum of rotated PMD vectors.

Formally, the above can be expressed as a system of differential equations, using the fact that vector cross products generate rotations [2]:

$$\frac{\partial \vec{s}}{\partial z} = \vec{\beta} \times \vec{s} \qquad (17.5a)$$

$$\frac{\partial \vec{s}}{\partial \omega} = \vec{\tau} \times \vec{s} \qquad (17.5b)$$

$$\frac{\partial \vec{\tau}}{\partial z} = \vec{\beta}' + \vec{\beta} \times \vec{\tau} \qquad (17.5c)$$

Equation (17.5) has the same physical meaning as discussed in the earlier paragraphs. The first of these equations describes the rotation *with distance* of an SOP at a given frequency about the birefringence vector in propagating down a fiber. The second equation states that, at any given distance, the SOP vector rotates *with frequency* about the PMD vector at a rate determined by the DGD ($|\vec{\tau}|$) of the system. The final equation describes the PMD vector evolution as the differential form of the concatenation rule: in traveling a section of differential length, the PMD vector is rotated by the same birefringence that rotates the SOP in Eqn (17.5a), and that section's PMD is added to the result.

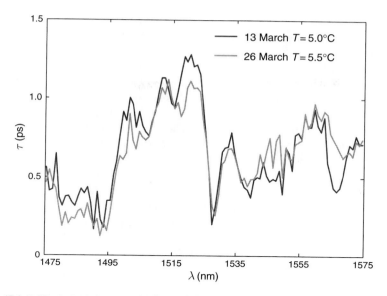

Figure 17.2 DGD of a buried cable with several exposed sections. Traces were taken at different times, separated by 2 weeks, for which the outside temperature was approximately the same (this figure may be seen in color on the included CD-ROM).

Generally speaking, the birefringence changes with frequency at each point along the fiber, as indicated in Figure 17.1, and so Eqn (17.5) must be integrated anew at each frequency. This leads to changes in PMD (and hence DGD) with optical frequency, and an experimental example of this is shown in Figure 17.2, for which the DGD of a fiber is plotted at two different times. We will return to this figure later, but note here that the DGD looks like a random function of frequency, with large relative variations. The randomness can be appreciated in view of Eqn (17.5), which imply that at different frequencies (or wavelengths), while the rotations in each section may change only slightly, after many concatenated sections the resultant PMD vector changes in a random fashion.

17.2.4 Properties of PMD in Fibers

In this section, we introduce the notions needed to view PMD as a system impairment. Considering our two-section system in Figure 17.1, it is clear that the rotations about $\vec{\beta}_1^0$ and $\vec{\beta}_2^0$ form a composite rotation, R_2R_1, taking the sphere into a rotated version of itself, $\vec{s}_0$ to $\vec{s}_2^{\,0}$ in particular. But, there must be *some* launch SOP $\vec{s}_0$ that is rotated into the particular final state $\vec{s}_2^{\,0}$ which is coincident with $\hat{\tau}$, a consequence of Euler's theorem [14]. Equation (17.5b) then indicates that to first order in frequency, i.e., over a narrow enough band, this output SOP remains stationary: $\vec{s}_2^{\,+} = \vec{s}_2^{\,0} = \hat{\tau}$. This special SOP and its orthogonal counterpart

$-\vec{\tau}$ are called the "Principal States of Polarization (PSP)" of the system [1], and light emerging in these two SOPs experience the maximal and minimal group delays, respectively, separated by the DGD, τ. Because the output PSPs are orthogonal, they can be used as a basis set for describing SOPs, and one may advantageously launch SOPs which evolve into these principal states. (Note that these PSPs are *not* the eigenstates of the birefringence. *Those* eigenstates are aligned to the axis of the composite rotation which took $\vec{s}_0$ to $\vec{s}_2^0$. See Ref. [15] for a fuller discussion.) Thus, because a narrowband signal's SOP can be resolved into the principal states, the output consists of two orthogonal replicas of the signal, separated by the DGD. Penalties associated with this distortion and higher order effects (PMD varying over the signal bandwidth) are discussed in later Sections 17.5 and 17.6, but here we turn to a few general properties of PMD in fibers.

Equation (17.5c) shows that the birefringence's derivative, $\vec{\beta}'$, determines a fiber's PMD, and there are two limiting cases. If $\vec{\beta}'$ is collinear with $\vec{\beta}$ and constant, then the randomization implied by rotations in Eqn (17.5c) does not take place: we have a uniform birefringent section and $\vec{\tau}$ grows linearly with distance. In contrast, if the direction of the driving term, $\vec{\beta}'$, varies randomly with distance, there can be only weak correlation between $\vec{\tau}$ and $\vec{\beta}'$ after some distance. The most obvious physical description of PMD incorporating this notion is the "retarder plate" model suggested in Figure 17.3. In this model, the fiber is viewed as a series of birefringent sections (retarder plates in Figure 17.3(a)) whose

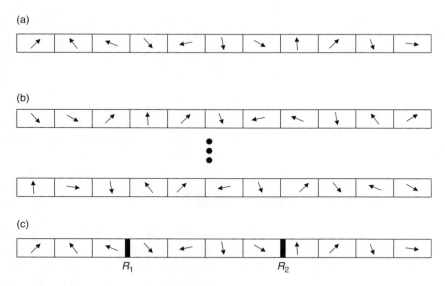

Figure 17.3 Fiber models. (a) Retarder plate model. Fiber is considered to be a concatenation of sections with randomly oriented birefringence vectors. (b) Fiber ensemble. Statistical analysis contemplates an ensemble of fibers, each a different realization drawn from the same sample space. (c) Hinge model. In addition to the fiber's concatenated sections, which are relatively inert, there are several active "hinges" which act as discrete polarization rotators.

orientation (and retardation) vary randomly from section to section, due to unavoidable variations and perturbations in the drawing process. Disregarding the rotation (second term) in Eqn (17.5c), $\vec{\tau}$ would be the sum of random vectors corresponding to the retarder plates, and the resultant vector would have an expected length given by the Maxwell distribution, in analogy to the average velocity of a classical gas. Adding rotations to the vectors in Eqn (17.5c) would not change the length of the vectors for each retarder plate, but would simply re-randomize their sum, leading to the same Maxwell distribution. Early in the formal development of the statistical theory of PMD, Curti et al. [3] developed a more rigorous analogy to Brownian motion, noting that the PMD vector's evolution would have random-walk properties. Independently, Foschini and Poole [4] introduced Stochastic Differential Equation techniques, launching a decade of analysis by groups across the world.

Analyses support the general intuitions developed above. Given a mean magnitude $\langle |\vec{\beta}'| \rangle$ over a fiber of total length L, consideration of Eqn (17.5c) and Figure 17.1 show us that each of the different "arc lengths" $\vec{\beta}'\mathrm{d}z$ of all the vectors that make up $\vec{\tau}$ will remain fixed in magnitude, regardless of rotations. To preclude a linear growth in PMD with fiber length, long lengths of uniform birefringent sections must therefore be avoided. If the birefringence correlation length (the distance over which $\vec{\beta}'$ is essentially constant) is L_c, then the PMD vector can be viewed as a composition of L/L_c random vectors, each of mean length $\beta' L_c$. Viewed as a random walk, the expected length of the PMD vector is then approximately $\beta'\sqrt{LL_c}$, showing that *ceteris paribus* we would like the correlation length to be as short as possible. (This is the motivation behind spun fibers [16], in which fiber is spun about the drawing axis as it is solidifying.) This correlation length-sets a scale for the retarder plate model: the fiber is viewed as concatenated sections, approximately L_c in length, over which the birefringence is approximately constant. Since one expects the average magnitude of $\vec{\beta}'$ and the average L_c to be a characteristic of a given fiber, one often lumps these together as the "PMD coefficient" $\beta'\sqrt{L_c}$, expressed in units of $ps\sqrt{\text{km}}$.

In addition to the statistics associated with the orientation of each section's PMD at ω_0, there are other statistical ensembles important for analyzing installed fiber plant. Unavoidable variations in the fiber-drawing process lead to the ensemble of fibers that are statistically equivalent to a given fiber, as shown in Figure 17.3(b). In this view, another fiber, whose sections are *statistically* identical to our chosen fiber, might have a radically different PMD due to the particular details of the orientations and rotations, even though they are drawn from the same Maxwellian distribution. That is, some realizations drawn from the ensemble might have a very high PMD. The second statistical set to consider is the same fiber, but viewed over such a wide range of wavelengths that successive rotations implied in Eqn (17.5c) significantly change the relative orientations of distinct sections, compared to the original fiber: again, the vector sum imaged at the output, the PMD, would be re-randomized. In an "ergodic" ensemble, taking either the frequency average of a single fiber, as in Figure 17.3(a), over a wide enough frequency or taking the sample average, Figure 17.3(b), should yield the same statistics. We will return to this point later.

Finally, we note a remarkable general scaling property of PMD analyzed by Shtaif and Mecozzi [17] as a general property of the Brownian motion nature of PMD in long fibers [4]: when normalized to the mean DGD, all PMD curves of long fibers are statistically equivalent in a certain sense. That is, if one stretched a 10-nm horizontal patch of data in Figure 17.2 into the full scale, and reduced the vertical scale by 10, the resulting curves would be valid representations of another fiber with a DGD 10 times smaller, a point that has significant system implications. The wide variations in Figure 17.2 lead one to expect that its DGD distribution reasonably approximates the "true" Maxwellian distribution and provides a good estimate for the mean DGD of this fiber. However, if the data had been limited to a 10-nm patch, it is clear that one would be unlikely to form a reasonable estimate for the mean. In fact, the standard deviation of the measurement when the bandwidth is restricted to B is given by Refs [17, 18]:

$$\sigma \approx \sqrt{\langle \tau \rangle / B}, \qquad (17.6)$$

which makes accurate estimates of low PMD links a difficult experimental challenge.

In the development above, we described general properties necessary to appreciate the system consequences of PMD. But, once the fiber is drawn, cabled, characterized, installed in a carrier's plant, tested, and certified for traffic, the carrier is no longer a consumer interested in the general properties of fiber: the issue becomes focused on properties of the very specific fiber at hand.

17.3 ELEMENTARY MODEL OF INSTALLED FIBER PLANT

With suitable parameter choices, the retarder plate model of Figure 17.3(a) represents any given fiber, and that might have been the end of the story. However, early anecdotal evidence with high PMD fibers gave the impression that the fiber was "alive" and varying fairly rapidly over time. In terms of Figure 17.3(a), the view was that the sections must be changing slightly under the influence of external stimuli and unknown internal processes. This gave rise to a third statistical ensemble: any fiber, itself, must also sample the ensemble over time. That is, perturbations force the model fiber in Figure 17.3(a) to visit other realizations of the ensemble as in Figure 17.3(b) over time. This leads to the troubling conclusion that every fiber, sooner or later, will visit the tails of the Maxwell distribution, i.e., every "good" fiber will someday go "bad."

This conclusion changes the carrier's focus to questions such as "How long will this particular fiber's PMD at ω_0 remain at its current level?" "Over what time scale is it likely that the PMD will remain at tolerable levels?" and "For how many other wavelengths will this fiber also be suitable, and for how long?" The most

17. Polarization Mode Dispersion

hopeful analytical outlook is what we have termed the "fast mixing assumption," which assumes that the entire Maxwell distribution is sampled over some time, perhaps on the order of several months. Then, for instance, if a fiber (drawn from an ensemble whose mean DGD is known) has a 10^{-5} probability of exceeding the PMD failure threshold, one expects this particular fiber to be unavailable for approximately 5 minutes per year. The drive to investigate this assumption motivated a number of experimental investigations, which we described earlier [10] and review below. Generally speaking, the results of a wide range of experiments do not support the fast mixing assumption: the PMD appears to change in a slower and more restricted manner. These outcomes led to the development of what we have called the "hinge model," depicted in Figure 17.3(c). We defer presenting supporting evidence for the model until the next section but describe its main features here.

Most of the long fiber links in backbone networks are buried underground in conduits containing cables with multiple fibers. Often, these conduits run along railroad lines, highways, and other rights of way, sometimes with other public utilities. The hinge model proposes that variations in PMD come from two sources: the PMD from the "retarder plates" in the fiber's buried sections, as in Figure 17.3(a), and several "hinges" which are short, exposed sections, such as R_1 and R_2 in Figure 17.3(c), which are vulnerable to external perturbations. These hinges are also "retarder plate" fiber sections as above, differing only in that they are more exposed to the environment than the buried sections.

Consider a fiber in a buried conduit forming a long (80 km) link along a highway. When the fiber route encounters bridges (e.g., at a river or another highway), its conduit emerges from underground, is led under the bridge, and re-enters the buried track on the other bank. The environmental stresses seen by the fiber in bridge crossings and in buried tracks are quite different. The buried fiber experiences few mechanical perturbations and only slow thermal diffusion processes which damp out temperature excursions and reduce their rapidity. In contrast, a rapidly moving cold front will be felt by fibers in the exposed conduit within times on the order of minutes.

A schematic representing these different levels of environmental susceptibility is shown in Figure 17.3(c), and it models hinges R_1 and R_2 as polarization rotator plates. In the language of Eqn (17.5c), a hinge is a short section dz which contributes very little in terms of overall PMD, but its birefringence rotates $\vec{\tau}$ in an environmentally sensitive manner. From the discussion in Section 17.2.2, one sees that the angle through which such a hinge rotates $\vec{\tau}$ is given in terms of the phase and group velocities as

$$\theta = \beta L \approx (\beta' L)\omega = DGD \times \omega \qquad (17.7)$$

so that an index difference of 10^{-7} over 10 m would rotate the PMD by π radians (Eqn (17.5c)) while contributing only a few femtoseconds of DGD. This illustrates the motivation for the word "hinge" in Figure 17.3(c). We view the link as comprising long spans of quiescent "dead" sections of fiber for which the PMD vectors are relatively stable, connected by livelier hinges which essentially rotate those large chains as the environment changes. This model does not preclude the "dead" sections from evolving over a long time scale as well, but on short time scales, the hinges will dominate PMD dynamics.

The hinge model predicts different dynamics than the "fast mixing" assumption, as can be appreciated by considering a long link with one active hinge. In the fast mixing model, all the sections are capable of reorienting, so the fiber samples the entire Maxwellian distribution. In this view, the hinge has no important bearing on the statistics, because it is merely another retarder plate. In the hinge model, however, the same fiber would be viewed as two fixed PMD sections that can only be re-oriented about that one point. If the hinge were to fully exercise all possible angles, the DGD would be bounded by the sum and difference of the DGDs for the two subspans. From a system perspective this would be quite good news: the resulting truncated statistical distribution implies that knowledge of today's PMD might be a better estimate of tomorrow's PMD, compared with the fast mixing model. Analytic results have been developed in analogy to a PMD emulator with freely rotating birefringence sections [19, 20]. Some of the system implications of these analytic results are promising [15, 20, 21], but have not yet been fully exploited on the system level.

A point to note is that the hinge model is a physical description, not an *ad hoc* computational device. Thus, there is no a priori reason to assume that any given hinge *will* exercise all possible angles. While it may be computationally expedient to assume hinges rotate isotropically, such behavior is a separate assumption. Simulations with hinges having a restricted range of motion have been shown to replicate experimentally observed features [10].

17.4 SURVEY OF FIELD TESTS

The recognition of PMD as an impairment to high-speed communication links motivated an international set of investigations into the actual dynamics of installed fiber. We use the machinery developed in Sections 17.2 and 17.3 to give a retrospective survey of these investigations and show that they form a coherent picture.

Early work on installed fiber by a Telecom Italia group [22] established that systematic and almost "reversible" variations in DGD could be traced to rapid fluctuations in the above-ground cabinets. Later work by that group verified the diurnal nature of the variations by examining the covariance between DGD spectra as a function of time: peaks at 24-hour intervals were superimposed on a more gradual week-long decay [23]. A very thorough experiment and analysis was done by Karlsson et al. [24] on two fibers in a buried cable, and they tracked the PMD vector itself over 36 days. This allowed them to demonstrate that SOPs drifted faster than the PMD in their stable system, and showed that there were correlations

17. Polarization Mode Dispersion

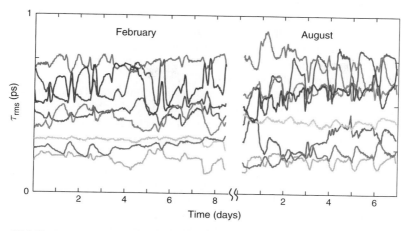

Figure 17.4 Field results for an installed cable. The rms DGD for eight fibers in a suburban environment is plotted for two spans of time (9 and 7 days) separated by 6 months. Diurnal variations are evident (this figure may be seen in color on the included CD-ROM).

between the two fibers. Contour plots of the DGD vs time and wavelength (we will show examples later) also bolstered insights into the system, unveiling persistent spectral features and system-wide time variations. Their system was relatively stable, and a drift analysis was performed which connected the data to their theoretical auto-correlation work. In a similar vein, Figure 17.4 [10] displays PMD measurements we made, displayed as the rms DGD taken over the optical spectrum. A set of eight fibers from a single cable were measured for a 9-day period and, 6 months later, another 7-day period. As above, diurnal variations are evident, but in this case, some fibers displayed greater variations than others.

These results are qualitatively encompassed by the description in Section 17.3. Environmental fluctuations change the details of the PMD vector (and hence the DGD) in two ways. The diurnal variations seem to be due to the hinges from exposed sections [25]. Because these sections are susceptible to fluctuations such as temperature, which can be nearly cyclic on a daily scale, it is not surprising to see nearly reversible effects that are strongly correlated on a daily basis, but lose correlation over longer periods of time. In contrast, in environmentally benign situations [24], the PMD can remain quite stable: there is no a priori reason to invoke the fast mixing assumption, since our viewing time may be much shorter than the mixing time. The mean DGD for a given fiber (i.e., averaged over frequency) should ideally remain constant: although the PMD spectra will drift, leading to changes in the shape of the DGD spectra (and thus its time covariance), the mean DGD should remain the same. However, finite measurement bandwidths, Eqn (17.6), imply that changing spectra lead to changing *measured* means, as observed in Figure 17.4. The fact that some fibers see large DGD changes and others do not is due to the random strength and orientation of those few hinges for that particular fiber at that particular time. Given enough hinges and enough variation, the variance of the measured DGD should approach that of Eqn (17.6), while smaller observed variances are indications of frustrated sampling of the ensemble. Fluctuations in Figure 17.4 show that fibers in that cable experienced different levels of ensemble sampling.

An illustration of both reversibility and different time scales between the hinges and the inert sections of fiber is shown in Figure 17.5 [25, 26]. Here, another field experiment was run, and spectrally resolved DGD curves taken over 21 days were sorted on the Weather Service's reported temperature at the time the curve was taken. The sorted DGDs at three wavelengths are presented, an early point at 0°C was chosen as a reference point for each wavelength, and for each trace, the DGD difference (compared to its reference point) is plotted vs the reported temperature at the time of the trace. While there is no discernible meaning to the shapes of the curves, what is clear is that each wavelength roughly follows some deterministic function. That is, as time progressed, the temperature ran up and down the horizontal axis, and each of the symbols initially traced out a curve [25] which gradually spread into a broad, but distinct, group of points. The segregation of the points indicates that something is changing rapidly with temperature (distinct shapes), but whatever it is, it is also somewhat deterministic (reversibility). We conclude that the hinges present a temperature-dependent birefringence to the fiber links, while the buried sections, although they differ with wavelength, are mostly stable.

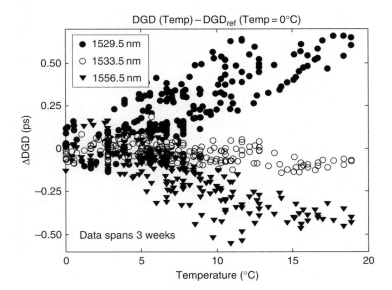

Figure 17.5 DGD changes vs outside temperature. DGD changes, referred to a reference point at 0°C, for three wavelengths tracked over a 3-week period. The data for each wavelength, plotted as a function of outside temperature, follow distinct curves (from Brodsky et al. [27]).

17. Polarization Mode Dispersion

Two traces with the widest temporal separation (but equal temperature) are shown in Figure 17.2. Even though the temperature and the DGD spectra had changed greatly over that time span, when the temperature returned to 5°C, the DGD spectrum looked similar to the earlier one. Our interpretation is that the fiber between the hinges is relatively stable over such times, while the hinges present temperature-dependent birefringences to the link. Slower variations in the fiber, gradually changing characteristics of the hinges, or non-uniform temperature distribution over the route may account for the gradual change in the shapes of the curves traced out by the three wavelengths.

After Karlsson et al. [24], contour plots of DGD, resolved temporally and spectrally, became the standard way of presenting long-term field data. An example is shown in Figure 17.6, which plots the DGD, normalized to the mean, for a concatenation of two 95-km fibers connected by an erbium-doped fiber amplifier (EDFA) in a hut [28]. This plot shows many of the features which characterize all such data reported to date. First, there is evidence of diurnal variations, marked by horizontal striations in the data with a period of 1 day. These data *also* show some faster time variations, but the diurnal structure is pronounced. Second, there is a marked persistence of spectral structure, shown as long vertical bands: regions with high DGD tend to remain high. Although they did not identify the detailed structure of their route, simulations were performed of a four-hinge link with limited motions gave rise to plots with a similar structure [10].

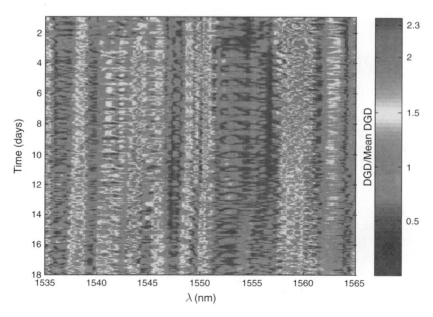

Figure 17.6 Experimental DGD (false color, normalized to mean DGD) over two concatenated 95-km links, plotted vs wavelength (horizontal) and time (vertical) axes (from Kondamuri et al. [29]) (this figure may be seen in color on the included CD-ROM).

Qualitatively, we can understand the above behavior in the following way. The temporal variations can be attributed to hinges with environmental sensitivity. In addition to fluctuations in outside temperature, however, we also showed that components *inside* huts and offices can act as polarization hinges [30], and we observed this effect in a long-haul field trial [10, 31], with shorter term fluctuations following the HVAC cycling in a field office. Spectral persistence can also be explained in terms of the hinge model of Figure 17.3(c). In that Figure, the birefringence vectors are represented at ω_0, and after they are rotated and added as in Eqns (17.5), each of the chains of sections separated by R_1 and R_2 can be viewed as a long, stable retarder plate section with a stable PMD. At some other optical frequency far from ω_0, the resultant sums for each chain may be quite different, due to the different rotations experienced by the vectors for each section. Consider the possibility that the PMD vector for the middle chain (i.e. between R_1 and R_2) has a much larger magnitude than the other two chains in a given wavelength region. Then, regardless of the relative orientations of R_1 and R_2, the total PMD vector will be large for those wavelengths. In another region of the optical spectrum, each chain may have a smaller PMD vector, so that the total PMD vector will always be more modest. Over long enough periods of time, perhaps the magnitude of the individual chains will evolve, which would destroy the vertical structure of plots such as Figure 17.6, but to our knowledge, there is no published experimental data to date that shows such behavior for buried fibers.

A field experiment on a 562-km commercial Raman-amplified 40 Gb/s system [31] gave us an opportunity to observe these notions from another viewpoint. The route passed through five huts in a loop-back configuration [10], and variations in the DGD at various optical frequencies were observed to be related to temperature variations in the huts [10], that is, the huts could be viewed as active hinges with known time dependence. Figure 17.7 shows spectral persistence in this case by plotting the observed distribution functions of the DGD for two different optical frequencies over a 60-hour period. Both frequencies exhibited wide variations in DGD, but evidently each obeyed different statistics, with distinctly different means (1.0 and 1.8 ps) [32, 33]. Apparently, between some set of hinges the system has longer PMD vectors at 186.65 THz.

Our discussion to this point has emphasized the optical foundations for PMD, its ability to impair signals, the basic model for PMD in installed fibers, and a brief survey of field results which led to the model. Before turning to more details of the system impact, however, we address a high-level question: Is it possible that advances in fiber-drawing techniques will result in fibers with PMD so low that PMD will no longer be a limitation to high speed transmission? For instance, will advances in spun fiber make PMD a moot point? The idea of spun fiber is that, as the preform is drawn, the fiber is rotated or rocked about the drawing axis. In terms of Figure 17.1, the motivation is clear: by rapidly re-orienting $\vec{\beta}'$ as the fiber is drawn, the birefringence correlation length L_c is reduced as in the retarder plate model [16]. However, the exact nature of limits to the improvement are still a matter for research [34]. We note that a more mundane limit, due to in-line optical components [31]

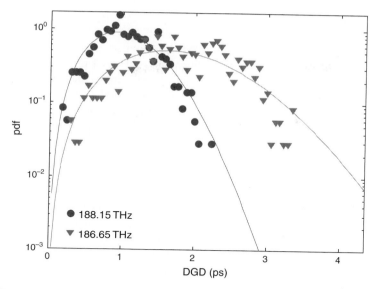

Figure 17.7 Experimental DGD probability density of observations (symbols); Maxwellian distribution with matched means (thin lines) (from Brodsky et al. [33]) (this figure may be seen in color on the included CD-ROM).

might impose more practical PMD limitations. In an effort to budget PMD for a 986-km system, we computed known PMDs due to fiber spans and DCMs, but found that an excess PMD of about 0.6 ps per span was required to explain the observed level of PMD. This unbudgeted PMD was apparently the result of other optical system components residing in offices and in huts. As optical technology improves, with optical add/drop multiplexers and other in-line devices becoming ubiquitous and link lengths lengthening, there will be an ever-increasing number of optical components between terminals. It seems more likely that an accumulation of apparently insignificant component PMDs (when viewed as "root-sum of squares" contributions) may need to be properly budgeted: what seem like reasonable optical component specifications today may need to be tightened in the future. In short, while it is foolish to bet against the ingenuity of researchers in solving the problems posed by PMD, it is a safer bet that such ingenuity will be required for the foreseeable future.

17.5 TRANSMISSION IMPAIRMENTS CAUSED BY THE FIRST-ORDER PMD

Recall (Section 17.2.4) that narrowband signals of arbitrary polarization can be resolved into the fast and slow principal states of the fiber, and that at the end of the system these two signals are separated in time by the DGD, τ. In this section, we discuss the transmission impairments caused by this first-order mechanism.

17.5.1 Poole's Approximation

Consider a pulse of any polarization at the input of a fiber, resolved into a superposition of two pulses in the orthogonally polarized principal states of the fiber. Because the two pulses have orthogonal polarizations, they can be viewed as independent: their energy adds up to the total energy of the original pulse, and the fraction of the energy in the slow principal state is denoted by the power splitting ratio γ. The remainder of the optical power, $1 - \gamma$, is in the fast principal state. The resulting electric field at the output of the fiber is then

$$E(t) = \sqrt{\gamma}E(t + \tau/2)\vec{s}_{\text{slow}} + \sqrt{(1-\gamma)}E(t - \tau/2)\vec{s}_{\text{fast}}, \qquad (17.8)$$

where $\vec{s}_{\text{slow}}$ and $\vec{s}_{\text{fast}}$ are the orthogonal Stokes vectors of the slow and fast PSP, τ is the DGD, and $E(t)$ is the electric field. Because conventional detectors are insensitive to polarization, the two terms on the right-hand side of Eqn (17.8) add when converted from optical to electrical power (each term is squared during the conversion), thus producing a distorted electrical pulse. Such distortions increase the bit-error ratio (BER) for transmission systems.

As an example of this impairment, Figure 17.8 presents concurrent measurements of the DGD and the system BER taken on a commercial Raman-amplified system deployed over a 1000-km-long fiber link with a mean DGD of about 4 ps [35]. The observed DGD variations over time were due to environmental changes,

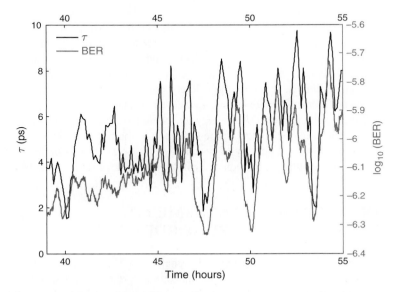

Figure 17.8 DGD (left axis) measured concurrently with BER (right axis) on a 1000-km operational system. Correlations are evident (from Boroditsky et al. [35]).

and the system BER generally follows the DGD [10, 32, 33]. Because the DGD measurement procedure required rapid and random polarization changes in the launch SOP and the BER reading was relatively slow, each point on the BER curve shown represents an average over several unknown input SOPs. Despite this ambiguity, good correlation between the DGD data and the BER data is apparent to the naked eye.

From a system-planning perspective, a metric different from the BER has proved to be more useful. This metric is the optical signal-to-noise ratio (OSNR) penalty. The OSNR penalty taken at a given BER level indicates by how much the OSNR needs to be improved (increased) to keep the BER level unchanged after introduction of a certain system's impairment. The first analytical estimation of the first-order PMD penalty was proposed by Poole et al. [36]:

$$\varepsilon(\mathrm{dB}) = A \left(\frac{\tau}{T}\right)^2 \gamma (1 - \gamma) \tag{17.9}$$

Here, T is the bit period (or inverse bit rate), and A is a receiver-specific parameter, which ranges from 20 to 40 for commercial receivers, being smaller for RZ formats. The derivation of Eqn (17.9) assumes that a small penalty is caused by and is equal to the pulse broadening, which was computed based on the moment of the received electrical pulse.

The penalty ε is represented by a grayscale surface as a function of τ/T and γ in Figure 17.9. Constant $A = 49.6$ was taken from Ref. [37]. Naturally, no DGD results in zero penalty. The penalty is also zero for $\gamma = 0$ and $\gamma = 1$, corresponding to the PSP launch condition in which all of the optical power is in the slow (for

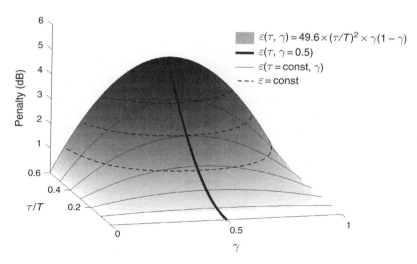

Figure 17.9 Penalty as a function of DGD (τ) and power splitting ratio (γ) for Poole's penalty formula. $A = 49.6$ as in Boroditsky et al. [37] (this figure may be seen in color on the included CD-ROM).

$\gamma = 0$) or fast ($\gamma = 1$) fiber axis and thus no distorting replicas are present. The largest effect is produced when an optical pulse is split equally between the two PSPs, i.e., the solid black line at $\gamma = 0.5$. For fixed τ, the penalty is a parabolic function of γ (thin lines). The contours of constant penalty (dashed lines) are used in the outage map concept [38] described below. For any fixed value of γ, the penalty is quadratic with both the DGD value τ and the bit rate $B = 1/T$, emphasizing the growing importance of PMD as technological developments enable increasing bit rates. The fairly steep quadratic increase of the PMD penalty with increasing DGD (τ) suggests the existence of a threshold level for a maximum tolerable τ. In fact, by considering two identical replicas of an NRZ signal arriving at the detector with a relative delay $\tau > T$ one sees that the upper (lower) rail goes through the middle of the optical eye for isolated ones (zeros). More rigorous numerical models, which account for optical beat noise and filtering details, show that various modern receivers reach 1 dB penalty near a threshold value much less than a bit period, at approximately $\tau_{th} \approx 0.3T - 0.45T$ [39]. The range in τ_{th}, which is modest, comes from different modulation formats [39, 40].

Equation (17.9), now widely used, was meant to approximate small signal distortions only, and penalties experienced by modern receivers deviate from this expression in one way or another. Still, the formula provides insight into the basic properties of PMD penalties and serves as a good qualitative tool. In particular, it allows a closed form expression for PMD outages, which we discuss below. As a foundation for our discussion of PMD outages, the next section highlights the similarities and differences between recently measured penalty data and Eqn (17.9).

17.5.2 First-order PMD Penalties from Real Receivers

The versatility of Poole's approximation, Eqn (17.9), comes at the expense of ambiguity in A. Generally, A is determined by the shape of the electrical pulse and thus depends on many factors. Recent numerical studies by Winzer et al. [41] demonstrated an intricate interplay between two eye closure mechanisms: one raising the lower rail and another lowering the upper rail of an optical eye. In practical terms the effect appears as a strong and non-monotonic dependence of A on the bandwidth of the electrical filtering for both NRZ and RZ modulation formats. Interestingly, while experimental values of A follow the trends predicted by simulations on average, the experimental uncertainty is large. This suggests an influence of other receiver parameters, besides the electrical bandwidth.

Figure 17.10 presents the results of a commercial receiver characterization, plotted in a fashion similar to Figure 17.9 [42]. The standard 7% FEC overhead is added to the 10-Gb/s data stream, so the receiver operates at 10.709 Gb/s. The penalty was measured at BER $= 10^{-3}$. There are two main differences between the data in Figure 17.10 and the analytical model in Figure 17.9. First, the actual receiver exhibits a much steeper rise in penalty with increasing τ. Second, the penalty is not a symmetric (parabolic) function of γ. Note, for example, that the

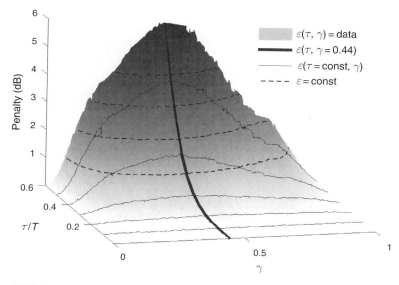

Figure 17.10 Penalty as a function of DGD (τ) and power splitting ratio (γ) for a commercial receiver [42] (this figure may be seen in color on the included CD-ROM).

worst launch condition is for $\gamma = 0.44$ rather than 0.5 as expected. In general, such asymmetry exists whenever the received electrical pulse replica (i.e., when free from PMD distortion) has an asymmetric shape [38]. Then, in the presence of PMD, the receiver will see a superposition of weak-advanced and strong-delayed replicas for launch condition $\gamma = \gamma_0$ and a time-reversed superposition for the orthogonal launch: $\gamma = 1 - \gamma_0$. The shape of the combined pulse differs in these two situations, thus resulting in different penalties. Asymmetries create the most dramatic deviations from the Poole's approximation because of their significance in the PMD outage calculations [38].

Finally, we consider deviations from Poole's formula for larger penalties. In the past years, two models suggested deterministic corrections for large first-order penalties ($\varepsilon > 2\,\text{dB}$). The first, based on numerically simulated penalties, suggested a "modified quadratic" expression for penalties: $\varepsilon_1 = A_1(\tau/T)^2\gamma(1-\gamma)/(1-\alpha(\tau/T)^2)$ [41]. The second correction came from fitting the experimental penalty data with "quartic" expression: $\varepsilon_2 = (A_2/4)L^2 + B_2 L^4$, where $L = 2(\tau/T)\sqrt{\gamma(1-\gamma)}$ [37]. In both the "quartic" experiment and a "modified quadratic" simulation [41] penalties were calculated at BER $= 10^{-9}$. These two functions are plotted in Figure 17.11 for $\gamma = 0.5$. The worst launch condition for the commercial receiver penalty data of Figure 17.10 (taken at BER $= 10^{-3}$) is plotted here as well for comparison (Squares). Interestingly, the commercial receiver is best fitted by yet a different function $\varepsilon_3 = B_3 L^4$ (thin black line), which reflects a steeper growth of penalty with DGD. The constants used in the plot are $A_1 = 51$, $\alpha = 0.41$; $A_2 = 27.5$, $B_2 = 33.4$; $B_3 = 46.6$. The quartic

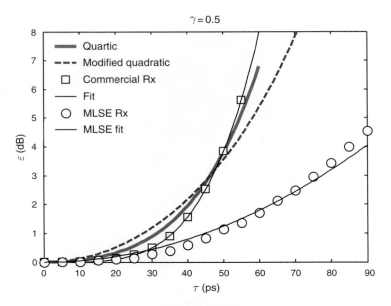

Figure 17.11 Penalty curves taken as a function of DGD (τ) for constant γ (worst launch) reported by various groups. Quartic and modified quadratic correspond to BER = 10^{-9}, the others to BER = 10^{-3} (this figure may be seen in color on the included CD-ROM).

corrections seem to capture the steeper (in comparison with Eqn (17.9)) rise seen in the commercial receiver. Note that $\tau_{1dB} = \tau(\varepsilon = 1\ \text{dB}) = 28, 31, 36$ ps are about a third of the bit period for all the models. Presumably, future technological advances will extend this limit. However, it is hard to envision a dramatic change in τ_{1dB} for systems without PMD compensation[*].

The emergence of electronic PMD mitigation techniques in recent years, such as multiple likelihood sequence estimation (MLSE) [43] has resulted in relatively more tolerant PMD receivers with $\tau_{1dB} = 48$ ps at 10 Gb/s [44, 45]. While the economic impact of such improvement is enormous, we will show later that the increase in τ_{1dB} is relatively too small to affect the applicability of the first-order approximation. Just for illustration, the data obtained from MLSE transponder for $\gamma = 0.5$ are plotted in the same figure as well as the Poole's fit to it. Quite surprisingly, despite sophisticated correction and decision processes in this MLSE receiver, its first-order performance can still be characterized by a single parameter A from Eqn (17.6) [45].

---

[*] It is worth noting that the variable L above represents a measurable quantity. For first-order PMD, it is the length of an arc that an SOP traces on the Poincaré sphere as the frequency is swept across the signal spectrum. This "string length" can be measured directly by a frequency resolved polarimeter (e.g., see Refs [46, 47]) and thus could serve as an in situ indicator of PMD penalties [37, 48].

To summarize, all of the penalty dependences reviewed above are somewhat similar in shape. Yet, existing differences indicate that there are numerous subtle parameters, which vary among receivers, that determine the exact functional dependence of penalty $\varepsilon(\tau, \gamma)$. Thus, reliable quantitative prediction of $\varepsilon(\tau, \gamma)$ might be impractical to achieve. Therefore, full PMD characterization for transponders before their deployment, that is the experimental measurements of the function $\varepsilon(\tau, \gamma)$, could be necessary.

17.5.3 Probability of PMD-Induced Outages

Reliable calculation of PMD-induced outages is perhaps the most difficult and yet the most important subject in the field. The difficulty arises from the necessity to make trustworthy assumptions about the statistics of several changing quantities, such as PMD and launch SOP. In contrast, calculated values of outage probability bear direct impact on the systems' deployment decision.

PMD differs from the other transmission impairments because of its dynamic nature: the instantaneous value seen by a channel changes in time, occasionally reaching large values. These large values, in turn, could cause system penalties—reductions of the system's OSNR margin. When a system's penalty is larger than an allocated threshold (usually, but not necessarily, $\varepsilon_{th} = 1$ dB) the system is said to experience an outage. As PMD and the launch SOP evolve in time, a given system gets into and out of outages, described statistically as the outage probability. If a system were operational for an infinite time (i.e., an interval much longer than the time-scale of PMD variations), the outage probability would be a ratio of the time in an outage state to the total time.

Discussions of outages and reliability have long historical roots in telephony practice and standards, and here, we take a short digression to alert the reader to potential pitfalls in the application of some of the concepts. The outage probability is sometimes described through its complement as $P_{out} = 1 - a$, where a is the system *availability*, the fraction of time a system is in a functioning condition, namely $a = \langle uptime \rangle / (\langle uptime \rangle + \langle downtime \rangle)$. (The brackets denote averaging over an infinite time.) Thus, an availability of "five nines," $a = 0.99999$, corresponds to an outage probability of $P_{out} = 10^{-5}$. Note that the system availability of a complex system a_Σ drops with the growing number of components. Indeed, the system's availability $a_\Sigma \approx 1 - \sum_i^N (1 - a_i)$ because low outage probabilities simply add to first order. Thus, if a system consists of ten components, each of *five nines* availability ($P_{out} = 10^{-5}$), the system's availability will only be *four nines* ($P_{out} = 10^{-4}$). To achieve high availability for a system comprising thousands of electronic components is a formidable task, requiring network operators to implement protection and restoration schemes. For unprotected systems, the availability is limited by the physical layer, part of which is the optical layer we are discussing. However, electronic failures limit the availability to the range of *three to four nines*. For more details, we refer the reader to another Chapter in this book-Commercial Optical Networks, Overlay Networks, and Services. From its telephony heritage, however, the PMD community considers *five nines* availability as a benchmark for PMD. While meeting such a benchmark has the advantage of removing PMD as a limitation to system availability, it has an economic cost and implementation hurdle, and might be profitably revisited in the future. Nonetheless, in light of the current practice, we continue to use this standard throughout the Chapter as changes in this number won't influence our conclusions significantly.

In its most general form, the outage probability is an integral of the joint probability function PDG(τ, γ) evaluated over the area in (τ, γ) space, for which the penalty exceeds a set threshold: $\varepsilon(\tau,\gamma) > \varepsilon_{\text{th}}$.

$$P_{\text{out}} = P[\varepsilon(\tau,\gamma) > \varepsilon_{\text{th}}] = \iint\limits_{\varepsilon(\tau,\gamma) > \varepsilon_{\text{th}}} PDF(\tau,\gamma)\mathrm{d}\tau\mathrm{d}\gamma \quad (17.10)$$

Equation (17.10) permits separation of the deterministic effect of a given PMD on a particular receiver from the PMD temporal statistics it suffers in its deployed environment, and is a basis for the two-step outage calculation method proposed by Winzer et al. [38]. When applied to the first-order PMD outage calculation, this method first draws an "outage map"—a set of iso-penalty contours [dashed lines in Figures 17.9 and 17.10] on the (τ, γ) plane and then calculates the integral over the area $\varepsilon(\tau, \gamma) > \varepsilon_{\text{th}}$. Although the outage maps were *calculated* in the original paper [38], they can also be *measured* directly by collecting the data similar to that shown in Figure 17.10 or they can be found *analytically* [6, 41, 49]. Once the outage maps are determined, any PMD temporal statistics can be applied to find the system outage. For instance, some recent calculations used Maxwellian PDF(τ) [38], while others [49, 50] utilized more advanced statistics [19] describing the hinge model [10]. As an example, one of the outage maps used in Ref. [38] is shown in the left panel of Figure 17.12.

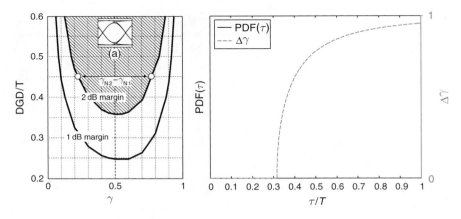

Figure 17.12 Left panel: Outage map from Winzer et al. [38] (© 2004 IEEE), shading indicates outage region. Right panel: probability density of τ is shown together with the outage weight $\Delta\gamma$ (this figure may be seen in color on the included CD-ROM).

Because τ depends on the fiber and γ depends on the launch SOP, they can be considered statistically independent, so their joint probability function is a product of two individual probability density functions, one of the DGD [i.e., PDF(τ)] and the other of γ [i.e., PDF(γ)]. For a uniform distribution in γ (corresponding to uniform Poincare Sphere coverage), this permits further simplification of Eqn (17.10) [38]:

$$P_{\text{out}} = P[\varepsilon(\tau,\gamma) > \varepsilon_{\text{th}}] = \iint\limits_{\varepsilon(\tau,\gamma)>\varepsilon_{\text{th}}} PDF(\tau)PDF(\gamma)\mathrm{d}\tau\mathrm{d}\gamma = \int_0^\infty PDF(\tau)\Delta\gamma(\tau)\mathrm{d}\tau$$

(17.11)

Here, the outage weight $\Delta\gamma(\tau)$ is the range of γ over which the penalty taken at a fixed τ exceeds the threshold $\varepsilon(\tau,\gamma) > \varepsilon_{\text{th}}$. The length of the arrow in the left panel of the Figure 17.12 indicates $\Delta\gamma$ for $\tau/T = 0.45$. A typical shape of the functional dependence of $\Delta\gamma(\tau)$ is shown on the right panel (right axis) of the Figure 17.12 together with a PDF(τ) (left axis). The cutoff value, below which $\Delta\gamma(\tau)$ is zero, corresponds to τ_{th}—the value of τ at which the penalty $\varepsilon(\tau, \gamma)$ reaches the maximum allocation $\varepsilon(\tau, \gamma) = \varepsilon_{\text{th}}$ for the worst γ. So, once the outage maps are determined for a particular receiver and thus $\Delta\gamma(\tau)$ is obtained, the outage probability is just an integral of the tail of PDF(τ) weighted by $\Delta\gamma(\tau)$.

Finally, an analytical expression of the penalty $\varepsilon(\tau,\gamma)$ could exist, such as the Poole's formula or various corrections to it described above [37, 41]. As long as the dependence of the penalty on γ remains the same as in the Poole's formula (Eqn 17.9), a closed form expression for the outage probability is possible [6, 41, 49]. Viewed from another angle, this assumption means that the penalty is a function of only the component of the PMD vector perpendicular to the launched SOP: $\tau_\perp = 2\tau\sqrt{\gamma(1-\gamma)}$. The penalty from a real receiver does not necessarily satisfy this condition as is obvious from Figure 17.10. In fact, the results of Ref. [38] demonstrate that the strong asymmetry in the γ dependence of the penalty function could affect the exact values of the outage probability. Although lacking exactness, the analytical expressions permit elegant qualitative conclusions, to a discussion of which we will now turn.

As an example of this procedure, for penalties governed by the Poole's approximation (Eqn (17.9)) and for the Maxwellian temporal PMD statistics, the closed form expression for the outage probability is derived in Refs [5, 6]:

$$P_{\text{out}} = P[\varepsilon(\tau,\gamma) > \varepsilon_{\text{th}}] = \exp\left(-\frac{\varepsilon_{\text{th}}}{\varepsilon_{\text{mean}}}\right),$$

(17.12)

where $\varepsilon_{\text{mean}}$ is the average penalty, $\varepsilon_{\text{mean}} = (\pi A/16T^2)\tau_{\text{mean}}^2$. There are two major consequences of the Eqn (17.12) that govern a carrier's deployment decision for a given PMD system. Usually, a system's PMD tolerance is *known* before deployment and a carrier needs to *choose* a fiber route with an appropriate value of τ_{mean}, so that the probability of a PMD outage is sufficiently low. The probability formula above allows one to connect the receiver tolerance and the fiber property. Indeed, it follows from the Eqns (17.12) and (17.9) that

$$\frac{\tau_{mean}}{T} = \frac{4\sqrt{\varepsilon_{th}}}{\sqrt{\pi A \ln\left(1/P_{out}\right)}} \quad (17.13a)$$

$$\frac{\tau_{mean}}{\tau_{th}} = \frac{2\sqrt{\varepsilon_{th}}}{\sqrt{\pi \ln\left(1/P_{out}\right)}} \quad (17.13b)$$

The first formula connects τ_{mean} to the bit period. By using it for a typical modern receiver ($A \approx 45$) and desired probability of $P_{out} = 10^{-5}$, one obtains the familiar relation $\tau_{mean} = 0.1T$. A more advanced MLSE receiver ($A \approx 20$) could be deployed over slightly worse fiber with $\tau_{mean} = 0.15T$.

The second formula connects τ_{mean} to the threshold value τ_{th}, i.e., the DGD at which system penalty reaches ε_{th} for the worst launch condition. This formula not only neatly relates the receiver property with that of a fiber but actually contains no parameters but P_{out}. The value τ_{th}, empirically found for ε_{th} of 1 or 2 dB at the worst launch condition, is currently adopted by the industry as *de facto* system's PMD specification. (Hopefully, the complete receiver characterization in connection with the outage maps will replace this standard eventually.) It follows from Eqn (17.13b) that in order to reach the outage probability of $P_{out} = 10^{-5}$, the mean DGD should be about a third of the threshold value. Note, however, that the dependence of τ_{mean}/τ_{th} on P_{out} is extremely weak. Indeed for $P_{out} = 10^{-6}, 10^{-5}, 10^{-4}$ the ratio τ_{mean}/τ_{th} remains nearly constant by taking values of 0.30, 0.33, and 0.37, correspondingly.

Unfortunately, due to the stochastic nature of PMD and the bandwidth limitation of modern PMD measurement instruments, measurement errors of at least 5–10% are unavoidable for fibers with τ_{mean} of up to 10 ps [10, 17, 27]. *Thus, the bad news is that it is nearly impossible to ensure an outage probability of exactly 10^{-5}!* Instead, a carrier has to tolerate a two order of magnitude wide range of the outage probabilities around 10^{-5}. In contrast, a somewhat lower choice of $\tau_{mean}/\tau_{th} = 0.23 - 0.26$ would correspond to outages almost never occurring: $P_{out} = 10^{-10} - 10^{-8}$. Such "mitigation by avoidance" could be done by either deploying a system over a better fiber (a standard practice nowadays) or, if better fiber is not available, halving the reach of a system thus reducing the total τ_{mean} of the link by about a factor of $\sqrt{2}$ (that is from $\tau_{mean} \approx 0.33\tau_{th}$ to $\tau_{mean} \approx 0.23\tau_{th}$).

All currently operating systems are deployed over a sufficiently low PMD fiber and, therefore, are not PMD limited. In other words, the probability of the PMD outage in these systems is practically zero. Let us repeat the condition for "sufficiently low" PMD. Despite the variety of existing modulation formats, modern telecom systems have similar PMD tolerances with 1 dB penalty corresponding to $\tau_{th} \approx 0.3T - 0.45T$ [39]. Taking $P_{out} = 10^{-5}$ as the threshold between PMD-limited and PMD-free systems, one can see that systems deployed over a fiber with $\tau_{mean} < 0.33\tau_{th}$ (or $\tau_{mean} < 0.1T - 0.15T$) are outage free. For higher PMD value of $\tau_{mean} > 0.1T - 0.15T$, the system experiences nonnegligible outages.

As a side note, we point out that quite often the PMD outage probability for various values of τ_{th}/τ_{mean} is approximated by estimating the area under the tail of Maxwellian pdf above τ_{th}—a well-tabulated function. From the outage map point of view, this corresponds to using a step-like outage weight $\Delta\gamma(\tau)$, which is zero below τ_{th} and one above, i.e., a hard failure at τ_{th}. Because essentially such an approximation ignores the distortion-free PSP launch condition, it overestimates the outage probability. Surprisingly, in a wide range of P_{out}, this simple method overestimates P_{out} by about a factor of 4. When this factor is taken into account, the method could serve as a legitimate approximation. Alternatively, to get the correct P_{out}, the area under the tail needs to be computed above $1.06 \times \tau_{th}$.

Finally, we point the reader to some unsolved issues regarding outage calculations. First, precise outage estimations need to take into account the properties of the particular receiver used in the system [38]; thus, the two-step approach including experimental characterization of the outage map might need to be adopted. Second, although the effect of the frequency dependence of the PMD vector on penalty might not be dominant (Section 17.6 below), additional margin allocation might be required. Last, but not least, is the concern that the assumption behind these outage calculations, namely that the system samples the entire ensemble described by Maxwellian statistics, is not necessarily appropriate for the time scales considered.

17.5.4 Implication of the Hinge Model on Probability of PMD-Induced Outages

The experimental evidence presented in Section 17.4 has changed the fundamental view of system vulnerability to PMD: the existence of channel-specific temporal DGD distributions suggest that instead of all channels being equally vulnerable at all times to PMD-induced outages, some channels in systems with hinges should be outage-free for long time periods, whereas the others should experience frequent outages [10, 20]. Two complementary measures, the Compliant Capacity Fraction [21] and the Non-compliant Capacity Ratio [50] were introduced to quantify the analysis. The first measure (CCF) is the fraction of channels with an outage probability less than a specified P_{spec}. For example, 90% of the channels have an outage probability less than 10^{-5} for a realistic case of 15 or fewer hinges and $\tau_{th} = 3\tau_{mean}$. Furthermore, it turns out that a significant fraction of channels are *guaranteed* to be outage-free for long periods of time, as long as the sections of the route between "hinges" do not change [51]. This last property follows directly from the pdf's truncation at a finite τ: the PMD vector cannot be longer than the sum of the (relatively fixed) buried section PMDs. Unfortunately, in a system with hinges, a small fraction of channels have a *higher* than specified outage probability. To account for this, the second measure (NCR) was introduced [50], which used the outage map approach rather than simply evaluating the probability for an instantaneous DGD to exceed a threshold [21]. Interestingly, the more rigorous analysis [50] exhibits similar qualitative behavior.

This new way of looking at outages highlights the utmost importance of in-service PMD monitoring techniques [37, 46, 48] and possibly opens a new paradigm in addressing the PMD impairment altogether. Indeed, if it were possible to know which channels are outage-free at any given time, these channels could be used for high availability services. Alternatively, increasing the tolerance toward PMD either by improving the system or by choosing a slightly better fiber should increase the fraction of outage-free channels. For example, as described in Ref. [21], this fraction can be tuned from 75 to 96% by changing the tolerance from $\tau_{max} = 2.5\tau_{mean}$ to $\tau_{max} = 3\tau_{mean}$ for a system with 10 hinges. The compliance calculations also hold open the possibility of new system strategies. For instance, if a route has 15 or fewer hinges, then by using tunable transponders (currently an emerging technology), the PMD outage problem could be solved by simply underutilizing the overall capacity by as little as 10% and using the 90% of "good" channels in the system. Over time, the "good" 90% of the channels may move, and the transponders would track them. Finally, because service level agreements are typically written in terms of the outage per month/year, it might make sense to artificially add extra degrees of freedom (say, several slow polarization scramblers mid-span) to force more predictable PMD dynamics closer to those described by a Maxwellian distribution over a desired timescale.

17.6 HIGH-ORDER EFFECTS

17.6.1 PMD Taylor Expansion

In this section, we describe some implications of the frequency dependence of the PMD vector $\vec{\tau}$. The elementary narrow-band picture of PMD penalties presented in Section 17.5 describes a PMD vector that is frequency independent. That is why historically the vector $\vec{\tau}$ is expanded into the Taylor series around the carrier frequency ω_0 and often only the lowest, frequency independent, term is taken into account.

$$\vec{\tau} = \sum_{n=0} \frac{1}{n!} \vec{\tau}^{(n)}(\omega_0) \Delta\omega^n = \vec{\tau}(\omega_0) + \vec{\tau}^{(1)}(\omega_0)\Delta\omega + \sum_{n=2} \frac{1}{n!} \vec{\tau}^{(n)}(\omega_0) \Delta\omega^n \quad (17.14)$$

Here, $\vec{\tau}^{(n)}(\omega_0) = (d^n\vec{\tau}/d\omega^n)|_{\omega=\omega_0}$ is the nth derivative of the PMD vector, which is usually called the $(n+1)$th order of the PMD, and $\Delta\omega = \omega - \omega_0$ is the frequency deviation from the carrier. Generally, the kth-order PMD approximation truncates the series by including only the first k PMD orders, and by leaving out the kth and higher frequency derivatives. First- and second-order PMD are the most studied PMD approximations, partially due to their relative simplicity.

The first (0th-order in frequency) term of this expansion corresponds to the most widely used first-order PMD approximation. The magnitude of the first term $\tau = |\vec{\tau}(\omega_0)|$ is the DGD, now familiar as the polarization-dependent pulse delay evaluated at the signal carrier frequency. Two optical pulses launched into the fiber's principal states (i.e., with polarizations parallel and antiparallel to the vector $\vec{\tau}(\omega_0)$ in the Stokes space) experience group delays that differ by τ.

The second-order PMD approximation adds the second term of the Taylor's expansion (first-order frequency derivative). The effect of the first derivative of the vector $\vec{\tau}$ (the second term in the Taylor expansion) is more complex. With changing frequency, the PMD vector can change either its direction or its magnitude. So, the frequency derivative of $\vec{\tau}$ can be broken into two parts: $d\vec{\tau}/d\omega = (d\hat{\tau}/d\omega)\tau + \hat{\tau}(d\tau/d\omega)$, where $\hat{\tau} = \vec{\tau}/\tau$ is the principal SOP (PSP) vector—a unit vector in the direction of the vector $\vec{\tau}$ and τ is the magnitude of the vector $\vec{\tau}$. The first part of the sum reflects how $\vec{\tau}$ changes in direction, whereas the second part describes changes in the magnitude of $\vec{\tau}$. This change in the magnitude of τ, $(d\tau/d\omega)|_{\omega=\omega_0}$, is a linear change in the group delay across the signal spectrum, i.e., chromatic dispersion. This additional chromatic dispersion is polarization dependent. Pulses launched into the two PSPs (which are parallel and anti-parallel to *the PMD vector*) $\vec{\tau}(\omega_0)$ and $-\vec{\tau}(\omega_0)$ see different amounts of chromatic dispersion, with the difference being proportional to $d\tau/d\omega$ [6].

Interestingly, there exists another somewhat similar case: the effect of the second-order PMD considered alone [52, 53]. However, there are subtle but important differences between the two situations. If the first-order PMD is compensated and the PMD orders higher than the second are absent, the PMD vector of a fiber can be represented by the second term of Eqn (17.14). The absence of the higher orders makes the magnitude of this SOPMD constant in frequency. So, again, this situation is equivalent to chromatic dispersion, but with different polarization dependence. First, the effect of the second order alone results in a differential chromatic dispersion accumulated by optical pulses parallel and anti-parallel to the *first derivative* of the PMD vector $\vec{\tau}^{(1)}(\omega_0)$. And second, the difference is proportional to the absolute value of the full derivative of the PMD vector: $|\vec{\tau}^{(1)}(\omega_0)|$. These results allow estimating the effect on system penalties caused by the second order alone, and we will return to them later.

17.6.2 Validity of PMD Taylor Expansion

The validity of any particular kth-order PMD approximation on system penalty is very hard to assess analytically. In contrast to the well-developed understanding of the first-order penalties (Section 17.5), there exists no model for the effects of high-order terms of Eqn (17.14) on PMD penalty. In fact, even the utility of considering the PMD frequency dependence in terms of discrete orders has been questioned [52]. Despite this, another interesting question can be answered—How many orders are required for faithful representation of the PMD vector itself? In other words, how fast is the PMD vector changing with frequency over a given bandwidth? Below, we consider some approaches to this question that were developed over the last several years and will show that, on average, the PMD vector does not change significantly over the bandwidth of interest (i.e., the signal modulation bandwidth) in a typical system.

The most rigorous approach is to consider the frequency autocorrelation function of the PMD vector. This function, when averaged over an ensemble of fiber realizations, shows the frequency range over which the PMD vector remains nearly constant and was derived analytically by two independent groups [54, 55]:

$$\langle \vec{\tau}(\omega') \bullet \vec{\tau}(\omega) \rangle = \frac{3}{\Delta\omega^2}\left[1 - \exp\left(-\frac{\pi\Delta\omega^2}{8}\tau_{\text{mean}}^2\right)\right] \qquad (17.15)$$

Here, $\Delta\omega = \omega - \omega'$ is the frequency difference and τ_{mean} is the average (over frequency) length of the PMD vector, i.e., the average DGD. Notice that the autocorrelation is defined solely by the first-order PMD, τ_{mean}, and the smaller the value of τ_{mean}, the wider is the correlation bandwidth. Thus, high-order PMD effects are relatively less important in low PMD fibers.

Various ways to quantify the correlation bandwidth have been proposed [54, 55]. Here, we compare two metrics and relate them to the widely used bandwidth of the PSP bandwidth: $\nu_{\text{PSP}} = 0.125/\tau_{\text{mean}}$ [6]. To avoid possible confusion in notation, we quote all the results below expressed in frequency ν, and *not* the angular frequency $\omega = 2\pi\nu$. The first metric is the equivalent rectangular width of the autocorrelation function: $\nu_c = (2\sqrt{2}/\pi)(1/\tau_{\text{mean}}) \approx 0.90/\tau_{\text{mean}}$ [54]. The second is the FWHM (3 dB bandwidth) of the autocorrelation peak: $f_{\text{3dB}} \approx 0.64/\tau_{\text{mean}}$ [55]. The two metrics are similar and when expressed in terms of PSP bandwidth, the former is about seven, $\nu_c \approx 7\nu_{\text{PSP}}$, and the latter is about five: $f_{\text{3dB}} \approx 5\nu_{\text{PSP}}$. So, a frequency range approximately equal to six PSP bandwidths $\nu_{\text{decor}} \approx 6\nu_{\text{PSP}}$ seems to be a reasonable scale over which the vector $\vec{\tau}$ decorrelates.

The autocorrelation analysis also indicates another important frequency range—that over which the PMD vector $\vec{\tau}$ remains nearly *constant*. It is approximately equal to the PSP bandwidth $\nu_{\vec{\tau}=\text{const}} \approx \nu_{\text{PSP}}$. Indeed, if a signal bandwidth is equal to ν_{PSP}, then the dot product of the vector $\vec{\tau}$ estimated at the center frequency and $\vec{\tau}$ estimated at the edge of the signal band is, on average, nearly one: $\langle \vec{\tau}(\nu_0) \bullet \vec{\tau}(\nu_0 \pm (\nu_{\text{PSP}}/2)) \rangle \approx 0.97$. Figure 17.13 shows the PMD vector autocorrelation as a function of frequency, expressed in terms of the PSP bandwidth. The metrics discussed are marked by symbols.

As an aside for the benefit of the curious reader, we list some additional results of autocorrelation studies. First, numerical simulations have shown that the

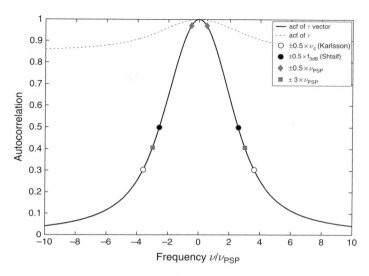

Figure 17.13 Normalized autocorrelations of the PMD vector and its magnitude. Markers show various ways to estimate the width of the peak (this figure may be seen in color on the included CD-ROM).

normalized autocorrelation of the full length PMD vectors $\langle \vec{\tau}(\nu') \bullet \vec{\tau}(\nu) \rangle / \langle \tau^2 \rangle$ is nearly identical to that of the PSP vectors $\langle \hat{\tau}(\nu') \bullet \hat{\tau}(\nu) \rangle$ [17]. Second, the correlation bandwidth of the DGD is close to that of the full PMD vector. Numerical simulations of the autocorrelation of the DGD (known techniques do not allow its analytical derivation) result in FWHM of the DGD autocorrelation peak to be $f_{3dB}^{DGD} \approx 0.82/\tau_{mean}$ [17]. So, the direction and magnitude of the PMD vector change in frequency with similar rates. As a final note, we mention that while a rigorous analytical expression for the DGD autocorrelation $\langle \tau(\nu')\tau(\nu) \rangle$ does not exist, an excellent approximation is available in an analytical form [17]:

$$\langle \tau(\omega')\tau(\omega) \rangle \approx \frac{3\pi - 8}{\pi} \left\{ \frac{2}{\Delta\omega^2} - \frac{16}{\pi\Delta\omega^4 \tau_{mean}^2} \left[1 - \exp\left(-\frac{\pi\Delta\omega^2 \tau_{mean}^2}{8} \right) \right] \right\} + \tau_{mean}^2$$

(17.16)

Equation (17.9) is based on the properly scaled autocorrelation of the squared DGD $\langle \tau^2(\nu')\tau^2(\nu) \rangle$. This function is plotted in Figure 17.13 as well. Again, it depends only on the fiber parameter τ_{mean}.

It is instructive to consider the relation between the autocorrelation function of Eqn (17.15) and the PSP bandwidth from another viewpoint as was proposed in Ref. [6]. The value of the autocorrelation between two vectors drops to from 1 to just 0.89 for a frequency separation equal to the PSP bandwidth. For a larger separation of six PSP bandwidths, the autocorrelation drops to 0.11. Thus, the empirically introduced PSP bandwidth is a good measure of how fast the PMD vector changes in frequency. First, $\vec{\tau}(\omega)$ is relatively *constant* over the PSP bandwidth and, second, two different values of $\vec{\tau}$ evaluated at two frequencies separated by more than six PSP bandwidths are nearly *statistically independent*. For frequency intervals smaller than the PSP bandwidth, the first term of Eqn (17.14) is sufficient for the description of the PMD vector. In other words, the first-order PMD approximation is applicable as long as the signal bandwidth does not exceed the PSP bandwidth: $B < \nu_{\text{PSP}}$.

We now consider a range of signal bandwidths reaching beyond the PSP bandwidth and follow an illustrative approach that was proposed in Refs [15, 56]. Using the results of Ref. [55], it is possible to calculate the rms value (over an ensemble of fibers) of each term in the Taylor series Eqn (17.14). This value, normalized by the zeroth term $T_0 = \tau_{\text{rms}}$, is given by

$$T_n = \frac{1}{\tau_{\text{rms}}} \left(\frac{1}{n!} \vec{\tau}^{(n)}(\omega_0) d\omega^n \right)_{\text{rms}} = \frac{\sqrt{(2n!)}}{n! \sqrt{(n+1)!}} \left[\frac{1}{2} \left(\frac{\pi}{4} \right)^3 \right]^{n/2} \left(\frac{\nu}{\nu_{\text{PSP}}} \right)^n \quad (17.17)$$

Figure 17.14 plots the results of Eqn (17.17) as a function of ν, the frequency deviation from center, expressed in terms of ν_{PSP}. For any given signal bandwidth B, the value of the function T_n taken at B/2 shows the average correction to $\vec{\tau}$ at the wings of that signal bandwidth arising from the $(n+1)$th PMD order. Note, that when the signal bandwidth $B = \nu_{\text{PSP}}$, the normalized rms value of the second PMD order at the edges of the signal spectrum, i.e. at $\nu/\nu_{\text{PSP}} = \pm 1/2$, is about 0.25

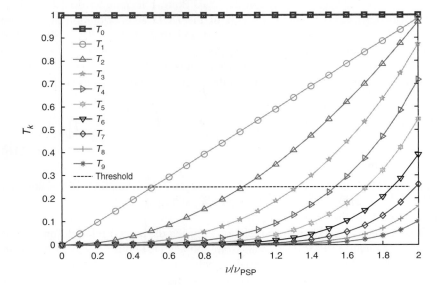

Figure 17.14 RMS Magnitudes of the normalized Taylor's PMD expansion terms estimated across the signal band (based on Refs [15, 56]) (this figure may be seen in color on the included CD-ROM).

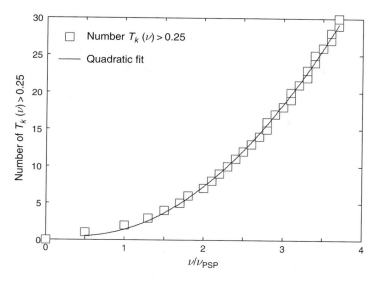

Figure 17.15 Number of terms above the threshold (black line in Figure 17.14) (based on Refs [15, 56]) (this figure may be seen in color on the included CD-ROM).

(circles). Recall that the autocorrelation analysis establishes this frequency band as the boundary of the first-order approximation.

By extending this criterion further, one can find the number of Taylor terms exceeding a certain threshold (horizontal line in the figure) as a function of frequency. For the threshold value of 0.25, we plot this number of "significant" terms in Figure 17.15. A quadratic fit is given as a guide to the eye. The shape of this curve does not depend strongly on the exact value of the threshold. This curve indicates how many terms of Eqn (17.14) are required for an accurate PMD description for any given bandwidth. The number grows practically quadratically with increasing signal bandwidth (or increase in the fiber's mean DGD $\tau_{mean} \propto 1/\nu_{PSP}$). The explosive growth of the number of significantly large terms with increasing bandwidth brings into question the utility of the PMD description in terms of truncated Taylor series. Indeed, even for $\nu = 2\nu_{PSP}$, the first *seven* terms become necessary. Although the truncation criterion was chosen somewhat arbitrarily, it has been shown that more rigorous conditions (based on the estimation of the remainder of the series) produce similar results [56].

17.6.3 Practical Consideration—When High PMD Orders are Negligible

The autocorrelation functions Eqns (17.15) and (17.16) determine how fast, on average, the PMD vector or its magnitude change with changing frequency.

The rate of change is inversely related to the mean DGD τ_{mean}. Consider a modulated telecom signal with optical bandwidth B. For sufficiently small bandwidth $B < \nu_{\text{PSP}}$, the PMD changes little over the spectral width, and the vector $\vec{\tau}(\omega)$ could be well approximated by its value at the central frequency $\vec{\tau}(\omega_0)$. In such cases, the first-order PMD approximation is sufficient. Let us stress here that the validity criterion for $B < \nu_{\text{PSP}}$ comes *NOT* from *receiver* properties but rather from intrinsic stochastic properties of the *fiber*. The first-order approximation is valid not because of the absence of high-order PMD penalties, but rather from the fact that high-order components themselves are essentially absent. In fact, the exact penalty mechanisms for higher PMD orders remain unknown. But, fiber properties ensure that the PMD vector $\vec{\tau}(\omega)$ stays almost constant over bandwidth B as long as $B \leq \nu_{\text{PSP}}$.

Let us examine the bandwidth limitation from the viewpoint of outage probabilities due to the *first-order* PMD penalty. Conventionally, system designers allocate a small OSNR margin for PMD. As described in Section 17.5, numerous experiments show that modern receivers reach a threshold penalty of $\varepsilon_{\text{th}} = 1$ dB (i.e., the penalty equal to the allocated PMD margin) at an instantaneous DGD threshold value of $\tau_{\text{th}} \approx 0.3T$–$0.45T$ [37, 41, 44, 45]. This ubiquitous threshold is determined by receiver properties, and arises from the detection of two delayed pulse replicas, and its value is relatively uniform across receiver types [39]. In fact, the highest value of $0.45T$ corresponds to the most advanced of the modern receivers utilizing MLSE technology.

In order to ensure that PMD outage events (i.e., penalty exceeding threshold ε_{th}) are sufficiently rare, transmission systems are deployed over fibers with average DGD τ_{mean} much smaller than the threshold value τ_{th} (Section 17.5). The exact relation between the two depends on many receiver parameters and is very sensitive to the details of the DGD temporal statistics models used for the outage calculations [21, 32, 41, 50]. Still, the relation $\tau_{\text{mean}} \approx \tau_{\text{th}}/3$ can be used as a good rule of thumb. Thus, modern receivers can operate outage free over fibers with $\tau_{\text{mean}} < 0.1$–$0.15T$, where T is the bit period. Rewriting this condition in terms of the signal bandwidth $B \approx 1/T$ and PSP bandwidth $\nu_{\text{PSP}} = 1/8\tau_{\text{mean}}$, we obtain $B < (0.8$–$1.2)\nu_{\text{PSP}} \approx \nu_{\text{PSP}}$.

The limit for the first-order approximation described above $B < \nu_{\text{PSP}}$ is nominally the same as the first-order outage-free condition $B < \nu_{\text{PSP}}$. This coincidental similarity between the two conditions causes confusion in the industry and we would like to clarify it here. The coincidence is astonishing because the latter comes from the ability of receivers to detect delayed pulse replicas whereas the former is purely due to properties of the fiber. An important conclusion can be derived from these two conditions. Indeed, in order for a system with a modulation band B to be PMD outage free, it has to be deployed over a fiber with $B < \nu_{\text{PSP}}$. But, when ν_{PSP} is larger than B, the PMD vector $\vec{\tau}(\omega)$ stays almost constant over any frequency interval equal to B. Thus, the fiber statistics dictate that it is highly improbable for a system operating with an outage probability of $P_{\text{out}} \leq 10^{-5}$ to experience large high-order effects.

Once again, it is instructive to employ Poole's formalism here to examine the potential implication of allocating different penalty margins ε_{th}, and using better receivers. Our rationale is independent of the modulation speed and format (as long as Poole's model applies). Indeed, by expressing Eqn (17.13a) in terms of bandwidth ($\nu_{PSP} = 1/8\tau_{mean}$ and $B = 1/T$) and then by employing $\varepsilon_{th} = A/4(\tau_{th}/T)^2$ we arrive at

$$\frac{B}{\nu_{PSP}} = \frac{16}{\sqrt{\pi}}\sqrt{\frac{\varepsilon_{th}}{A\ln(1/P_{out})}} = \frac{16}{\sqrt{\pi}}\frac{1}{\sqrt{\ln(1/P_{out})}} \times \frac{\tau_{th}}{T} \qquad (17.18)$$

Equation (17.18) compares the signal bandwidth B and ν_{PSP} needed to obtain the outage probability P_{out} for a receiver that can tolerate τ_{th}. Only when B becomes much larger than ν_{PSP} do high PMD orders come into play. Using $\tau_{th} \approx 0.3T - 0.45T$, we obtain B that is nearly equal to the PSP bandwidth $B = 0.8\nu_{PSP} - 1.2\nu_{PSP}$, and thus, the vector $\vec{\tau}$ is nearly constant over the signal bandwidth. Equation (17.18) permits a qualitative comparison of the effect of different receivers and different margin allocations. Smaller values of A (better receivers) and larger values of ε_{th} (higher PMD margins) result in an increase in the likelihood of occurrence of the high PMD orders. This may sound somewhat counterintuitive but recall that the formula is derived for a system *limited* by first order PMD—in fact, the relation $B < \nu_{PSP}$ was suggested as one of the first-order PMD outage criterion [6]. In other words, better receivers and higher margins permit increased fiber DGD while staying just at the boundary of the PMD-limited regime. Systems based on better receivers can tolerate more of the first-order PMD for the same signal bandwidth and, when deployed on high PMD fibers, these systems are more likely to experience high-order effects.

17.6.4 Experimental and Numerical Studies of High-Order Penalties

Finally, we highlight recent experimental and numerical attempts to quantify the high-order penalties [37, 48, 52, 57]. All of them have shown that the introduction of high-order PMD results in a small correction to the penalty caused by the first order only.

First-order PMD penalty depends only on two variables: $\varepsilon(\tau, \gamma)$. And, hence, it can be easily presented in a variety of ways—as a surface (such as in Figures 17.9 and 17.10) or as a set of one—dimensional functions of either variable τ or γ, with the second one as a parameter: $\varepsilon(\tau, \gamma = \text{const})$ or $\varepsilon(\tau = \text{const}, \gamma)$. For penalties symmetric in γ, that is, when the penalty is a function of the component of the PMD vector perpendicular to the launched SOP $\tau_\perp$ (see Section 17.5), the entire surface can even be collapsed into a single curve $\varepsilon(\tau_\perp) = \varepsilon(\tau \times 2\sqrt{\gamma(1-\gamma)})$. With the introduction of high orders, the penalty becomes dependent on a multitude of parameters. Thus, often the high-order penalties are considered statistically even by experimentalists. Usually, the data taken over an ensemble of PMD realizations and launched SOPs are represented as a scatter plot on the same axes as the first-order only penalty.

Figure 17.16 shows an example of such a plot, based on experiments in Refs [37, 48]. In these experiments the receiver was characterized for both polarization-maintaining (PM) fiber and high-order fiber. The solid line represents a fit to the first-order only (PM fiber) results, while the symbols correspond to the high-order data. The effect of the higher orders on penalty is two-fold. First, there is a slight increase in the OSNR penalty, and second, there is a 1-dB scatter in the data. The scatter reflects the random variation in the penalty for a fixed $\tau_\perp$ and occurs due to the presence of the high orders and variations in the launched SOP. The inset shows that the penalty spread (defined as the deviation from the lower bound) increases with the magnitude of the second-order PMD. Despite the high-order effects, the overall dependence on $\tau_\perp$ remains similar to that of the first order alone, suggesting the dominance of the latter. Small scatter indicates the lack of high PMD orders over the signal bandwidth. A numerical penalty simulation based on pulse broadening [57] and simulation of the eye-opening penalty [52] further substantiated these experimental findings. Similar results were obtained for MLSE receivers [44, 45]. Intriguingly, some papers reported that the presence of

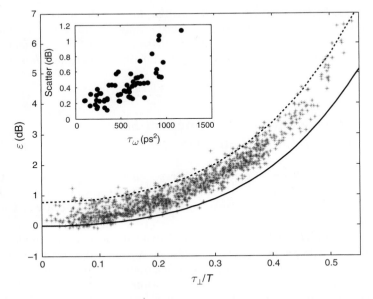

Figure 17.16 Penalty as a function of $\tau_\perp >$. Solid line is a fit to the first-order penalty. Stars represent higher order penalties. Spread (deviation from solid line) is correlated with a magnitude of the frequency derivative of PMD vector (from Ref. [48]) (this figure may be seen in color on the included CD-ROM).

high-order PMD can, at least for some input SOPs, lower the penalty in comparison with the first-order only case [45, 52]. Although the solid line in Figure 17.16 serves as a lower bound for the high-order data for this particular receiver, that does not seem be a universal rule. Summarizing the reviewed experiments and simulations, we conclude that the principal contribution to PMD-induced OSNR penalties comes from first-order PMD.

Although PMD-compensated systems seem to be far from the deployment stage, several groups have investigated PMD penalties and outages in such systems [52, 53, 58]. In particular, one concept was tested: Can PMD penalties in systems with compensated first order be described by the second order only? Such a hypothetical situation can be described as polarization-dependent chromatic dispersion (PDCD) [52, 53, and Section 17.5]. Both analytical and numerical results indicate that the exact second-order solution overestimates the PMD penalties, thus necessitating consideration of all higher order terms taken together. Still, this "all-order-induced" penalty is shown to be correlated with the magnitude of the second order [52] in accordance with experimental data in the inset of Figure 17.16. This suggests a strong correlation between all high PMD orders, which could also be seen in Figure 17.14. In this regard, Shtaif and Boroditsky posed an interesting question—could high-order effects on penalty be described by a small number of parameters instead of an infinite number of terms of the Taylor's series?

Optical PMD compensation has already been proven technically feasible in several high bit-rate field trials [59–63], though shortening system reach as described in Section 17.5 still seems to be most practical solution from a business perspective [64, 65]. But, eventually, the growth in bit rate will necessitate the deployment of PMD compensation. Then, the PMD outages will be solely due to high-order effects. Hopefully, by that time, the telecom community will have developed the theoretical tools needed to permit quantification of high-order penalties.

17.7 PMD EMULATION

Terminal equipment, designed to provide bounded error rates for transmission over a fiber link with a specified amount of PMD, must be tested to ensure proper operation in the presence of this important impairment. But, PMD is a statistical phenomenon and because the complete realization space of the distributed birefringence along a real fiber link may take many months to be fully covered (see Sections 17.3, 17.4), field testing of terminal equipment to ensure that PMD-induced outages do not exceed the design value, is time consuming, prohibitively expensive, and quite impractical. Instead, extensive efforts have been made to emulate PMD in the laboratory, not only for system testing but also for evaluating PMD compensators (PMDCs), as well as other novel polarization sensitive subsystems.

Unfortunately, proper emulation of PMD of a real fiber link is far from straightforward due to the random and complex nature of PMD: not only do the magnitude and direction of the PMD vector change randomly from one wavelength channel to another (and even within the bandwidth of a single channel), they also change over time for a given channel. Recent developments have taken two distinctly different approaches to laboratory generation of PMD. In the first approach, an effort is made to closely mimic the statistics of a real fiber, using a *PMD emulator* (PMDE). Typically, such emulators generate PMD in a time-varying, unpredictable and, preferably, fast manner. A PMDE can be useful for long-term tests of receivers and PMDCs in order to evaluate PMD-induced outage probabilities. However, the test results are not repeatable because the PMD states of the emulator itself are themselves not repeatable. Quite a different approach to testing is represented by the *PMD source* (PMDS), which attempts to produce PMD in a deterministically predictable and repeatable manner, allowing performance comparison among different systems, when subjected to the same PMD stress. This section discusses available emulation/source techniques, emphasizing progress made in recent years, and also addresses the impact of the hinge model described in Section 17.3. However, in order to understand the design goals of proper emulation/generation of PMD, we start with a brief summary of some of the properties of fiber PMD, relevant to both PMDEs and PMDSs.

17.7.1 First-Order Emulation and its Limitations

It is not too difficult to accurately emulate fiber PMD for a narrowband lightwave systems, i.e., where the PMD vector, $\vec{\tau}(\omega)$, does not change much over time and can be considered constant over the system's optical bandwidth. In such a case, a polarization-dependent variable delay structure, see Figure 17.17, can generate any given delay, τ, between the two linearly polarized components of the wave incident on its entrance polarization beam-splitter. In principle, to correctly emulate the vector nature of the PMD vector, the delay line should be sandwiched between two adjustable polarization controllers, where the first determines the polarization of the input principal states, while the second controls the polarization

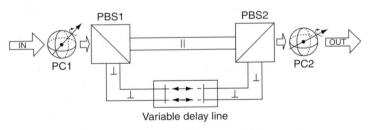

Figure 17.17 A PMD source/emulator, capable of generating a variable and arbitrary, though frequency-independent DGD. PBS1 is a linear polarization beamsplitter, decomposing the input polarization into its parallel and perpendicular components. PBS2 serves as a polarization beamcombiner. PC1 and PC2 are polarization controllers.

of the output principal states. By dialing-in the magnitude and direction of any required $\vec{\tau}$, one can easily implement a PMDS to test a narrowband receiver (Note that the PMD tolerance of a receiver is mainly determined by its handling of the PMD-induced distortion of the incoming bit stream, rather than by the SOP of the incident beam. Thus, the output polarization controller PC2 is practically redundant, unless the receiver includes a built-in optical PMD-mitigation elements, which may be sensitive to the incoming SOP). Alternatively, the same device can function as a PMDE (i.e., an emulator) by randomly changing the delay according to a given statistical distribution (e.g., Maxwellian [6] for the classical "fast mixing" models of the fiber, or a different one for the hinge model of Section 17.3 [19]). At the same time, the polarization controllers can be adjusted so that the input polarization to the linearly polarized delay line, as well as the polarization emerging from the second polarization controller uniformly cover the Poincare sphere. If system characteristics were so easily captured (i.e., a single narrowband optical channel over the fiber link), it would have been quite simple to use the device of Figure 17.17 to compensate for the fiber PMD, and PMD in lightwave systems would essentially be solved.

In practice, however, WDM technology populates many tens of optical channels over tens of nanometers of spectrum in the wide low-loss transmission window of the fiber, so the constant PMD offered by Figure 17.17 is inadequate for system emulation. Furthermore, the bandwidth of lightwave systems is steadily increasing, and thus, even for a single optical channel, $\vec{\tau}(\omega)$ may no longer be constant over the channel bandwidth of typical fiber links. To model this variation, it has become customary (for a different approach, see Ref. [66]) to represent the wavelength dependence of $\vec{\tau}(\omega)$ as a Taylor series around the system center frequency ($\hat{\tau}(\omega) \equiv \vec{\tau}(\omega)/|\vec{\tau}(\omega)|$) is a unit vector parallel to $\vec{\tau}(\omega)$, and all frequency derivatives are evaluated at ω_0) [6 and Section 17.6.1]:

$$\vec{\tau}(\omega_0 + \Delta\omega) = \vec{\tau}(\omega_0) + \frac{d\vec{\tau}}{d\omega} \cdot \Delta\omega + \cdots = \vec{\tau}(\omega_0)$$
$$+ \left(\frac{d\tau}{d\omega} \cdot \hat{\tau} + \tau \cdot \frac{d\hat{\tau}}{d\omega}\right) \cdot \Delta\omega + \cdots \quad (17.19)$$

$\vec{\tau}(\omega_0)$ is the first-order PMD, which is the only term of importance for narrowband systems, as discussed above. The coefficient of $\Delta\omega$ in the next term, the so-called SOPMD describes the higher order nature of PMD and comprises two contributions: (i) $d\tau/d\omega \cdot \hat{\tau}(\omega_0)$, whose magnitude measures the variation of DGD with frequency and, because it has the same units as chromatic dispersion, it is called PDCD; and (ii) $\tau(\omega_0) \cdot d\hat{\tau}/d\omega$, which describes the dependence of the PSP on frequency, also called the *depolarization* of $\vec{\tau}(\omega)$ with frequency. Note that the device of Figure 17.17 being a strictly first-order device, cannot emulate/generate PDCD, depolarization, or any other higher order PMD (see Ref. [67] for the inadequacy of first order alone to predict receiver performance).

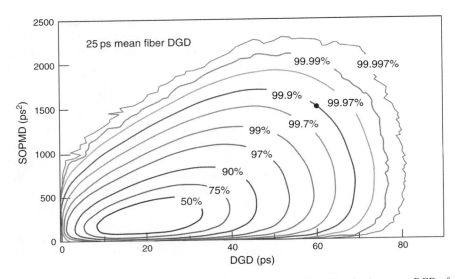

Figure 17.18 Joint probability function of the DGD and SOPMD for a fiber having mean DGD of $\tau_{mean} = 25$ ps. The percentages associated with each contour represent the probability that a (DGD, SOPMD) state will occur inside the contour (from [67] © 2006 IEEE) (this figure may be seen in color on the included CD-ROM).

Detailed expressions for probability distribution functions for SOPMD can be found in [6, Table 4.1], and all are characterized by the remarkable feature that they depend on, and therefore scale with the mean DGD (τ_{mean}) [17]. While those expressions were derived assuming perfectly random fiber birefringence, most PMDE designers see the reproduction of these ideal probability distributions as a quality criterion. An important distribution, not mentioned in Ref. [6], is the joint distribution of the DGD (τ) and the SOPMD, as measured by magnitude of the frequency derivative of the PMD vector, $|d\vec{\tau}(\omega)/d\omega|$. Obtained by numerical simulation techniques [67, 68], this distribution is shown in Figure 17.18 for a mean DGD of $\tau_{mean} = 25$ ps. However, this figure is actually universal because the DGD axis scales with τ_{mean}, and the SOPMD axis scales with τ_{mean}^2.

Thus, for example, consider the point (DGD = 60 ps, SOPMD = 1500 ps^2), which touches the 99.9% contour from its inside. In a fiber link with a mean DGD of 10 ps (rather than the 25 ps value of Figure 17.18), the same point (with respect to the universal contour) will have the coordinates [DGD = (10/25) × 60 = 24 ps, SOPMD = $10^2/(25^2)$ × 1500 = 240 ps^2]. The slightly slanted shape of this joint distribution function indicates that for a given mean DGD, the occurrence of a large value of the instantaneous DGD is statistically associated with a large value of the SOPMD, as well. This observation has directed recent work on PMDEs and sources toward putting more emphasis on proper reproduction of second-order effects.

Another all-order statistical measure of PMD is the spectral autocorrelation function of the PMD vectors at two optical frequencies [54, 55, and Eqn (17.14) of Section 17.4]. Again, this autocorrelation function is fully characterized by the mean DGD and tends to zero for a large enough frequency separation. As described earlier in this chapter, the classical modeling of the fiber, as being statistically homogeneous along its length, has been challenged by the observation that the long-buried sections of a long-haul link almost do not change in time and all statistical variations originate from "hinges", connecting the buried sections. The resulting PMD distributions are quite different from the Maxwellian model, as has been discussed in previous sections [19, 20, 21]. We are now well equipped to review current implementations of PMDEs and PMDSs.

17.7.2 PMD Emulators

In PMD *emulation*, one tries to imitate the actual time-varying statistics of real fibers. Much like the retarder plate models used for numerical simulations [6, Section 5.5], most implementations use a concatenation of linearly birefringent sections, whose birefringence axes have random angular orientations. Made of pieces of PM fiber or birefringent crystals, these sections have lengths determined by the required mean DGD of the emulator, τ_{mean}, and various methods to couple neighboring sections and to control the relative angles between them. For N sections with DGDs $\{\tau_i, i = 1, \ldots, N\}$, and large enough N, the mean DGD is given by $\tau_{\text{mean}} = \sqrt{(8/3\pi) \sum_{i=1}^{N} \tau_i^2}$. Normally, an average value for the section's PMD is determined from $\tau_i = \sqrt{3\pi/(8N)} \times \tau_{\text{mean}}$, and then, to avoid periodicity of PMD statistics in the frequency domain, the N values for $\{\tau_i\}$ are drawn from a random distribution, usually Gaussian [69], with the above-mentioned average. (Note that the introduced randomness in $\{\tau_i\}$ will result in a mean DGD for the whole emulator, which may slightly deviate from the design value.) Due to their similarity to real fibers, these PMDEs generate PMD of all orders in an inseparable and uncontrolled way. Some of the criteria for high-quality classical emulation include: (i) At any fixed frequency, the emulator should accurately reproduce the DGD and SOPMD statistics of the classical fiber model, when averaged over an ensemble of emulator realizations, including the generation of low-probability events; and (ii) the ensemble-averaged frequency autocorrelation function of these all-order emulators should follow the prediction of Eqn (17.15), tending to zero away from its central peak.

Most PMDE implementations use sections of fixed DGD and differ from one another in the way sections are coupled, see Figure 17.19. Methods for joining neighbor sections include (i) fixed splicing at 45° (or random) angle, in which different polarization behavior is achieved by controlled heating of the various sections, thereby controlling the evolution of the SOP of the optical wave as it propagates along the section; (ii) controlled mechanical rotation of the relative

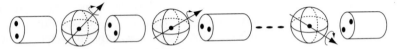

Figure 17.19 PMDEs based on the concatenation of birefringent sections of fixed, though not necessarily equal, DGD values, with polarization-controllers, or their equivalents, between neighboring sections.

angle; and (iii) mechanized polarization controllers between sections, providing additional degrees of freedom in the polarization transformation between sections. Different realizations are achieved by changing the appropriate control values according to a prescribed algorithm.

Detailed studies of PMDEs [69, 70] have indicated that proper emulation of real fibers requires tens of sections to achieve: (i) integrated deviations of less than 10^{-3} between the probability distribution functions of first-order PMD and SOPMD and their theoretical values; and (ii) background levels of the frequency autocorrelation function (i.e., away from the peak) of only a few percent of the peak value. The large number of required sections means that cost and complexity issues cannot be ignored, and not all proposed implementations are of the same practical importance.

A few recently reported realizations include (i) Fixed high birefringence sections, spliced at a fixed angle of 45°, but with an electrically controlled heater per section to provide adjustable polarization transformation between sections [69]. Here, the technology allows for quite a few sections, 30 in the experiment, but performance (filled diamonds in Figure 17.20) is inferior to the case in which polarization scattering between sections is fully randomized; and (ii) A polarization controller of a modified three-paddles Lefevre-design [71], which supports multiple windings of many high birefringence sections of different lengths [72, 73].

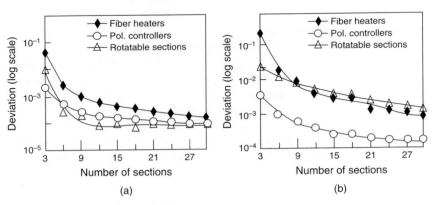

Figure 17.20 Deviation of DGD and SOPMD from their theoretical statistical distributions for three different emulators with the same average DGD, as a function of the number of emulator sections, as obtained from numerical simulations. The solid lines are polynomial fits to the results: (a) DGD deviation from Maxwellian distribution; (b) Second-order PMD magnitude deviation from theory (from [69] © 2004 IEEE).

In summary, dynamic PMDEs do mimic real fibers and can serve a variety of useful testing functions for lightwave systems. However, the evolution of the SOP through the emulator is adjustable but not fully controlled (e.g., the polarization transformation of each birefringent section depends on temperature). Thus, even for an emulator with many sections, it may require impractically large number of attempts to correctly reproduce the tails of the probability distribution, which involves large values of the DGD (i.e., many aligned section) and SOPMD.

17.7.3 PMD Sources

A properly designed PMDS should be (i) programmable and capable of generating prescribed PMD states (beyond first order) across the entire realization space of interest; (ii) repeatable; and (iii) stable with respect to both time and the environment to allow full system characterization with no change in the injected PMD state. In terms of implementing a model of Eqn (17.19), the variable delay line of Figure 17.17, while satisfying the above criteria, is only a first-order PMD source, because its PMD is represented by a constant, frequency-independent vector, $\vec{\tau}_0$. Various researchers [74–79] have proposed implementations of first-order and higher order effects by concatenating several such devices and controlling their degrees of freedom deterministically. But the task is not that simple: when two first-order stages, with PMD vectors, $\vec{\tau}_1$ and $\vec{\tau}_2$, are concatenated [74], the resultant PMD vector is given by Eqn (17.4): $\vec{\tau}_{(2\ \text{stages})}(\omega) = R_2(\omega)\vec{\tau}_1 + \vec{\tau}_2$, where $R_2(\omega)$ is the frequency-dependent polarization transformation of the second stage. Because $\vec{\tau}_1$ and $\vec{\tau}_2$ are frequency-independent, the resultant PMD vector, $\vec{\tau}_{(2\ \text{stages})}(\omega)$, precesses, as a function of frequency, around $\vec{\tau}_2$ with a *frequency-independent* length of $\tau_{(2\ \text{stages})} = \sqrt{\tau_1^2 + \tau_2^2 + 2\tau_1\tau_2\cos(\varphi)}$, where φ is the frequency-independent angle between $R_2(\omega)\vec{\tau}_1$ and $\vec{\tau}_2$ in Stokes space. Thus, while exhibiting a second-order effect, only depolarization (of magnitude $|d\vec{\tau}/d\omega| = |\tau \cdot (d\hat{\tau}/d\omega)| = \tau_1\tau_2\sin(\varphi)$) can be reproduced, but not PDCD. With three concatenated stages, PDCD is no longer zero, because $|\vec{\tau}_{(3\ \text{stages})}(\omega)|^2 = a_0 + a_1\cos(\omega\tau_2)$, where a_0 and a_1 are frequency-independent, and τ_2 is the differential delay of the middle stage. Clearly, to ensure stable and repeatable frequency dependence, τ_2 must be very precisely controlled (its value is multiplied by the optical frequency to produce a phase, which cannot fluctuate by more than a tiny fraction of 2π). Adding a fourth stage results in a richer Fourier spectrum for $|\vec{\tau}(\omega)|^2$ [80]

$$|\vec{\tau}_{(4\ \text{stages})}(\omega)|^2 = a_0 + a_1\cos(\omega\tau_2) + a_2\cos(\omega\tau_3) + a_4\cos(\omega(\tau_3 - \tau_2)) \\ + a_5\cos(\omega(\tau_3 + \tau_2)) \quad (17.20)$$

with all $\{a_i\}$ being frequency-independent. Note that the first and last stages do not contribute to the frequency-dependent terms. More stages give rise to additional

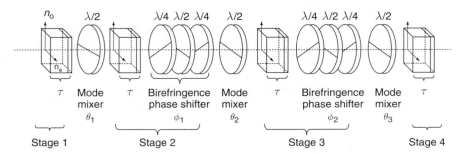

Figure 17.21 A block diagram of the Enhanced Coherent higher Order (ECHO) PMD source. There are four delay stages with intermediate mode mixers. Birefringent phase shifters are inserted in the center two stages. Below shows the location and alignment of the birefringent crystals and waveplates for the first two stages (from [81] © 2003 IEEE).

Fourier components, leading to a whole Fourier-based analysis of DGD developed in Ref. [80]. Again, having the optical frequency in the arguments of the various cosine functions means that optical phase control is mandatory to achieve repeatable and stable operation. A thoroughly investigated and useful special case of Eqn (17.20) is described in Figure 17.21 [81].

Here, all $\{\tau_i, i = 1, 4\}$ in Eqn (17.20) are made equal to τ_{stage} by using four identical uniaxial birefringent crystals, precisely cut to the same prescribed length (to achieve temperature stability, each delay element comprises two different crystal slabs in series, having aligned birefringence axes, but with opposite temperature dependence of their respective birefringence coefficients). But, mechanical precision is not good enough to ensure $\omega\tau_2 = \omega\tau_3$. Rather, this condition is met by a more precise control of the birefringent phase of the second and third sections, using the Evans phase shifters [82], whose rotation provides the required adjustment of the birefringent phase. Finally, all delay elements are housed with their birefringence axes parallel to one another, and half-wave plates are used for mode-mixing between neighboring stages. Under these coherent (i.e., full control of the birefringent phase) and harmonic (the arguments of the cosine functions are integer multiples of a basic unit) conditions, Eqn (17.20) reduces to

$$|\vec{\tau}_{(\text{coherent 4 stages})}(\omega)|^2 = b_0 + b_1 \cos(\omega\tau_{\text{stage}}) + b_2 \cos(2\omega\tau_{\text{stage}}) \quad (17.21)$$

As the angles of the mode mixers vary, with $\theta_1 = \theta_3$ (to reduce the number of controllable degrees of freedom), the following maximum values for first- and SOPMD can be achieved: $DGD_{\max} = 4\tau_{\text{stage}}$, $|d\vec{\tau}/d\omega|_{\max} \approx 4\tau_{\text{stage}}^2$. A wide range of values for the pair (DGD, SOPMD), between (0,0) and $(4\tau_{\text{stage}}, 4\tau_{\text{stage}}^2)$, can be programmed by using two degrees of freedom: θ_1 and θ_2, with $\theta_3 = \theta_2$. But, τ_{stage} cannot be arbitrarily increased to give more coverage of the (DGD, SOPMD) space: because all delays are equal, the device is periodic in frequency with a free spectral range of $FSR = 1/\tau_{\text{stage}}$, and, thus, to ensure that a whole optical

17. Polarization Mode Dispersion

(a)
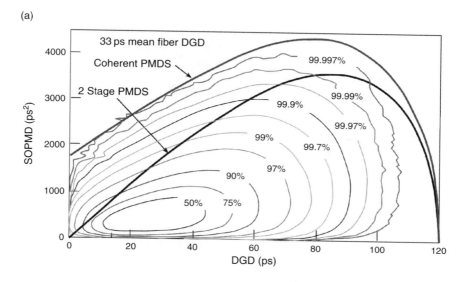

(b)
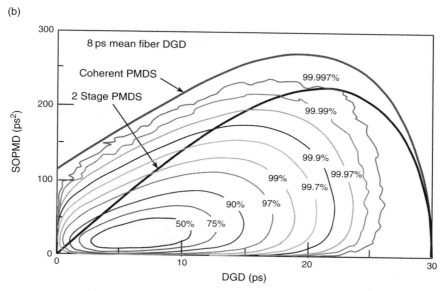

Figure 17.22 The axes and the surrounding highest smooth curve bound the PMD states accessible by (a) a 10 Gb/s and (b) 40 Gb/s coherent PMDS, with $\tau_{stage} = 33$ ps, FSR = 30 GHz, and 8 ps, FSR = 125 GHz, respectively. Over 99.997% of the (DGD, SOPMD) plane is covered. For comparison, the PMD states accessible by the corresponding two-stage PMDS are shown by the second-highest smooth curve (from [67] © 2006 IEEE) (this figure may be seen in color on the included CD-ROM).

channel can be conveniently located within the *FSR*, τ_{stage} cannot be too large. Figure 17.22 describes the good coverage of the (DGD, SOPMD) parameter space obtained by this technique for the interesting cases of 10 and 40 Gb/s [67], where at each bit rate, τ_{stage} was chosen such that the maximum possible delay τ_{stage} will exceed the bit duration by 20%.

With a stable and repeatable PMDS, receivers with or without PMDCs can be repeatedly tested for their BER (or Q) sensitivity to a dense enough array of values of (DGD, SOPMD) PMD states to ensure accurate evaluation of the system performance [83]. With the penalty measured for each PMD state (a quite tedious job), an estimate of the outage probability can be calculated by counting all those points in the tested PMD space for which the penalty *exceeded* the system allowance, but only after weighing each point by its theoretical joint probably distribution [67].

There are, though, a few points of caution about PMDSs:

- If the SOP entering the PMDS coincides with one of its PSPs, all first-order effects will vanish, despite the non-zero DGD of the device. To avoid such pitfalls, a range of input states of polarization must be scanned.
- An infinite number of PMD realizations share the same (DGD, SOPMD) values. Consider a point in Figure 17.18, with specified coordinates. Because only $|d\vec{\tau}(\omega)/d\omega|$ is specified, various combinations of depolarization and PDCD, which may have different effects on the device under test, will have the same SOPMD value. Even if a more complicated device is conceived, in which both the depolarization and the PDCD could be individually specified, still the PMD vector, $\vec{\tau}(\omega)$, of that device may have significant non-zero higher order terms. In conclusion: two different designs of PMDSs programmed to the same (DGD, SOPMD) state will most probably have different high-order PMD.
- There is theoretical evidence, [52], that for a high DGD fiber, the expansion implied in Eqn (17.19) may not converge rapidly, and many, if not all, high-order terms of the PMD are of the same importance, or at least as important as the SOPMD, see Section 17.6. Thus, specifying a large value of SOPMD implicitly implies large higher orders of PMD, whose emulated values cannot be independently specified. Even if repeatable, these high orders will probably be quite different than those of the real fiber, whose PMD effects are to be emulated.

Notwithstanding this, PMDSs of a specific design, made to tight tolerances, may provide a useful tool in alerting system engineers to vulnerabilities in comparing the behavior of different receivers.

17.7.4 Emulators with Variable DGD Sections

Adding DGD tunability to the individual sections of Figure 17.19 [84] opens new emulation horizons, at the expense of greater complexity. Thus, if one chooses

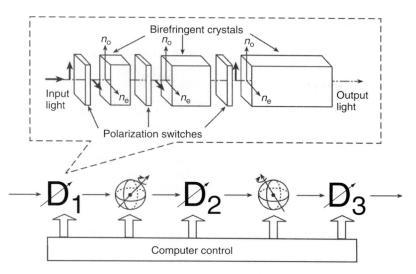

Figure 17.23 PMD emulator with three programmable DGD elements separated by two electrically driven polarization controllers [86] (© [2004] IEEE).

the individual sections' DGD values from a Maxwellian-distributed population, only a few sections are required for accurate emulation of the expected first-order (1 section) and second-order (15 sections) statistics. Rather than implementing this idea by replicating the polarization-sensitive delay line of Figure 17.17, researchers used concatenated birefringent crystals, having their birefringent axes parallel to one another, and surrounded by half-wave plates [85], or magneto-optic polarization switches [86], to build a discretely tunable DGD unit (Figure 17.23). In this implementation [86, 87], the lengths of the birefringent crystals were arranged in a binary power series, and each magneto-optic switch would either let its input beam go through or rotate it by 90°. Thus, if the DGD of the thinnest crystal is $\delta\tau$, a 6-bit module can be computer-programmed to generate DGD values from $-63\delta\tau$ to $+63\delta\tau$ in steps of $2\delta\tau$. With only three sections, separated by programmable, squeezer-based polarization controllers, good Maxwellian fits were achieved for different programmed values of the mean DGD, although SOPMD values were somewhat lower than expected for ideal fibers with the same mean DGD. Like other multi-stage architectures (cf. PMDSs above), it is not enough that the birefringent elements will be stable: the birefringent phase must be strictly controlled. In an attempt to add phase control to the structure of Figure 17.23, inter-stage state of polarization monitors, together with a feedback mechanism, were used to achieve overall stability and repeatability [88].

If proven to be stable and scalable, such devices can also emulate the "hinge" model of Section 17.3 [89]. Moreover, the same architecture, when properly stabilized, offers a practical way to implement the efficient emulation of rare events, using importance sampling [86, 90] and Brownian bridge [91] methods.

Finally, the functionality offered by the structure of Figure 17.23, and more, can be implemented in principle using cascaded integrated-optics variable delay lines [92].

In summary, while recent years have seen major progress in the generation/emulation of PMD, with fairly low values of insertion loss (~1 dB) and PDL (0.5 dB), a fully satisfying solution has not yet emerged. Multisection emulators have a single built-in mean DGD and are not repeatable; PMD sources stably emulate only a few low orders of PMD and thus do not faithfully represent higher orders of real fibers, and variable DGD implementations, while very flexible, need increases in scale (number of sections), stability, and repeatability, before becoming universal emulators. For a recent review of currently available practical solutions to PMD emulation, as well as PMD compensation, see Ref. [93 and also 94].

17.8 PMD AND OPTICAL NONLINEARITIES

17.8.1 Introduction

There is clear evidence that the combined effects of PMD with chromatic dispersion and optical nonlinearities can be of great importance to long fiber-optic links. It has been shown that in some cases the overall penalty of the nonlinear PMD-impaired system is lower than the sum of the individual penalties [95–98], whereas in other instances the opposite is true [99]. The reasons for the specific behavior of a given system are still only marginally understood, and the only proven tool for predicting the PMD tolerance in the nonlinear case has been numerical simulations. Nevertheless, some of the aspects of nonlinear propagation of light in randomly birefringent fibers can be elegantly explained in terms of tools developed in this chapter.

In this section, we review the equations governing the propagation of light in the presence of all these detrimental effects and concentrate on one particular aspect of PMD and nonlinear propagation that has recently received significant attention in the literature. This aspect concerns the way that different channels of a WDM system affect each other's polarization through cross-polarization modulation (XPolM).

17.8.2 The Coupled Nonlinear Schrodinger Equations

Let us follow an optical pulse at a typical operating wavelength of $\lambda = 1.55\,\mu m$, as it propagates through a communication-grade single-mode optical fiber, afflicted not only with birefringence (and PMD), but also with chromatic dispersion and Kerr-type, third-order nonlinearities [100]. In arriving at equations that describe this propagation, close attention is paid to the different important length scales, which characterize these distorting phenomena.

17. Polarization Mode Dispersion

Let us model the fiber as a smooth and continuous concatenation of locally constant birefringent sections of random lengths (on the order of the correlation length, $L_c \sim$ several to tens of meters, see Section 17.2.4). The orientation of the birefringence axes and the strength of the birefringence [as measured by β of Eqn (17.3) of Section 17.1 with $\Delta n \sim 10^{-6}$], are assumed to randomly change from one section to another (Section 17.1, Figure 17.3). We further assume, with some experimental support [101], that each section is *linearly* birefingent, so that the Stokes vector, $\vec{\beta}$, of Section 17.1, representing the local birefringence, lies in the equator of the Poincare sphere in the laboratory frame of reference. In the presence of chromatic dispersion and Kerr third-order nonlinearities, the propagation of light in a *single* linearly birefringent section is governed by the coupled nonlinear Schrodinger equation [we use $k(\omega)$ instead of the common $\beta(\omega)$ in order to be consistent with the notation of previous sections] [102]:

$$-j\frac{\partial A_x}{\partial z} = (k_x(\omega_0) - k_0)A_x + j\frac{dk_x}{d\omega}\bigg|_{\omega_0} \cdot \frac{\partial A_x}{\partial t} - \frac{1}{2}\frac{d^2 k_x}{d\omega^2}\bigg|_{\omega_0} \cdot \frac{\partial^2 A_x}{\partial t^2}$$
$$+ j\frac{\alpha}{2}A_x + \gamma\left(|A_x|^2 + 2|A_y|^2/3\right)A_x + \frac{\gamma A_y^2 A_x^*}{3}$$
$$-j\frac{\partial A_y}{\partial z} = (k_y(\omega_0) - k_0)A_y + j\frac{dk_y}{d\omega}\bigg|_{\omega_0} \cdot \frac{\partial A_y}{\partial t} - \frac{1}{2}\frac{d^2 k_y}{d\omega^2}\bigg|_{\omega_0} \cdot \frac{\partial^2 A_y}{\partial t^2}$$
$$+ j\frac{\alpha}{2}A_y + \gamma\left(|A_y|^2 + 2|A_x|^2/3\right)A_y + \frac{\gamma A_x^2 A_y^*}{3}$$

(17.22)

$A_x(z, t)$ and $A_y(z, t)$ are the complex envelopes of the x and y components of the electromagnetic field of the propagating mode, centered around ω_0 (Note that unlike the PMD community, see Section 17.2.1, literature on optical nonlinearities uses $\exp[j(k_0 z - \omega_0 t)]$, rather than $\exp[j(\omega_0 t - k_0 z)]$, see Eqn (17.1)):

$$E_x(z,t) = A_x(z,t)\exp[j(k_0 z - \omega_0 t)], \quad E_y(z,t) = A_y(z,t)\exp[j(k_0 z - \omega_0 t)]$$

The amplitudes $\{A_x, A_y\}$ are properly scaled so that $|A_x|^2 + |A_y|^2$ represents the optical power. k_0 is the average of the wavenumbers $k_x(\omega_0)$ and $k_y(\omega_0)$, which characterize the distinct phase velocities of the x and y polarizations, $\beta = k_x(\omega_0) - k_y(\omega_0)$ is the birefringence in the section (we assume x is the "slow axis"), $\beta' = dk_x/d\omega|_{\omega_0} - dk_y/d\omega|_{\omega_0}$ is the difference of group velocities (see Section 17.2.2), eventually responsible for the fiber PMD, and $k'' \equiv d^2 k_x/d\omega^2|_{\omega_0} \approx d^2 k_y/d\omega^2|_{\omega_0}$ is the fiber dispersion at ω_0, which we assume to be polarization independent (the * stands for complex conjugation). We have ignored higher order dispersion terms. α is the power loss per unit length. The terms preceded by the factor γ (~ 1–30/Watt/km [100], ~ 1.3/Watt/km for Corning SMF-28 single-mode fiber) represent the third-order (Kerr) nonlinear terms, and while contributing very weakly to the $\partial A_{x,y}/\partial z$, their integrated influence along thousands of kilometers of modern fiber links cannot be overlooked. It is clear from Eqn (17.22) that several of the involved effects require detailed understanding of the relative phases of the propagating signals, and so, in principle, the geometric representation of the previous sections might seem inadequate to provide a proper description. Instead, we study here the evolution of the Jones 2X1 vector, $[E_x(z,t), E_y(z,t)]^T$ (T stands for transpose). Each Jones vector (see Ref. [7] for a lucid review) has an image in the geometric representation, but an overall phase is lost. Nonetheless, as we show here, after suitable averaging and mathematical manipulations, elements of the non-linear optical interactions can be transferred to the geometrical description.

Each of the terms on the right-hand side of Eqn (17.22) has its own length scale. We have already seen in Section 17.2.2 that birefringence, represented by the first term: $(k_{x,y}(\omega_0) - k_0(\omega_0))A_{x,y}$, will rotate the SOP of the propagating light, completing a full circle around the s_1 axis of the Poincare sphere every beat-length L_b, where $L_b = 2\pi/\beta \approx 1$ m. This rotation can be removed from Eqn (17.22) by introducing a phase-shifted complex envelope, $\{B_x, B_y\}$, related to $\{A_x, A_y\}$ by a diagonal unitary transformation, $A_{x,y}(z,t) = B_{x,y}(z,t) \exp[\pm jkz/2]$(+ for x, − for y), to obtain:

$$-j\frac{\partial B_x}{\partial z} = j\frac{dk_x}{d\omega}\bigg|_{\omega_0} \cdot \frac{\partial B_x}{\partial t} - \frac{1}{2}k''\frac{\partial^2 B_x}{\partial t^2} + j\frac{\alpha}{2}B_x + \gamma\left(|B_x|^2 + 2|B_y|^2/3\right)B_x$$

$$-j\frac{\partial B_y}{\partial z} = j\frac{dk_y}{d\omega}\bigg|_{\omega_0} \cdot \frac{\partial B_y}{\partial t} - \frac{1}{2}k''\frac{\partial^2 B_y}{\partial t^2} + j\frac{\alpha}{2}B_y + \gamma\left(|B_y|^2 + 2|B_x|^2/3\right)B_y$$

(17.23)

In writing the *incoherent* version of the coupled nonlinear schrodinger equation, Eqn (17.23), we ignored the last terms of Eqn (17.22): $\gamma B_y^2 B_x^* \exp[-j2kz]$ and $\gamma B_x^2 B_y^* \exp[+j2kz]$. This is justified because the z-dependent exponential, $\exp[\pm j2kz]$, will average their contribution to zero over a single fiber section of only a few beat-lengths long, while, the effects of interest begin to be significant only after hundreds of meters.

We move now to the first term on the right-hand side of Eqn (17.23) [the second term in Eqn (17.22)]: it introduces a DGD of $\beta' L_c = \Delta n L_c/c$ (approximately a few hundredths of a picosecond) per section. Because the local birefringence axes change their orientation from section to section in a random way, as the pulse propagates through many such sections over the transmission link, the DGD will not grow linearly with distance, but rather as its square root, reaching values for the PMD coefficient of $0.1 \text{ ps}/\sqrt{\text{km}}$ for modern spun fibers [103]. Thus, in modern high-speed long-haul WDM systems, with bit periods, T, of tens of picoseconds (and a corresponding Fourier transform that spans a few tens of gigahertz), PMD becomes important only after hundreds of kilometers. We have so far ignored the significant bandwidth of the pulse. The chromatic dispersion of modern fibers used for long-haul transmission ($D = -2\pi c k''/\lambda^2 = 5-17 \text{ ps/nm/km}$)

will widen our pulse, regardless of its polarization, such that substantial and detrimental broadening will take place, but only after quite a few kilometers, thereby bringing another length scale, $T^2/|k''|$, into play. Another effect of dispersion is walk-off: in a WDM system, different channels have different wavelengths, so that pulse trains, nanometers apart, will slide in time with respect to each other, as they propagate along the fiber at different velocities, bringing another important length scale into play. The next term in Eqns (17.22) and (17.23) deals with fiber attenuation $\alpha > 0$ (or gain, $\alpha < 0$). Because fiber attenuation, assumed to be polarization independent, is very low, the accompanying scale is of order $1/\alpha \sim 22$ km. Finally, optical nonlinearities, represented by the last terms, will significantly distort the pulse, but not before a propagation distance, $(\gamma P_0)^{-1}$ (P_0 is the input power) of order of tens of kilometers. (Due to attenuation, nonlinear interaction becomes ineffective after a distance of order $L_{eff} = 1/\alpha$. But in-line amplifiers, distributed along the fiber, will let these effects accumulate throughout the whole link length). In practice, Eqn (17.23) can be numerically solved by the split-step algorithm [99, 100]: in each section, one solves for the contributions of the linear and nonlinear terms separately, and adds the results. Moving from one section to another is done by proper rotation transformation. However, more insight can be gained from a more analytical approach, which also leads to much faster numerical algorithms.

17.8.3 The Manakov-PMD Equation

Analytically, we can take advantage of the different length scales involved. In all practical scenarios, we are interested in propagation ranges (tens to hundreds of kilometers), which are much longer than L_c. Under these circumstances, the nonlinear terms, which depend on the pulse polarization [i.e., eqns (17.22) and (17.23)], can be mathematically treated in such a way that only those contributions that do not average with distance are retained. The result for an information-carrying optical channel centered at ω_0, is the *Manakov-PMD* equation [97, 104, 105]:

$$j\frac{\partial \mathbf{U}(z,t)}{\partial z} + j\alpha \mathbf{U}(z,t) - \frac{1}{2}k''\frac{\partial^2 \mathbf{U}(z,t)}{\partial t^2} + \frac{8}{9}\gamma |\mathbf{U}(z,t)|^2 \mathbf{U}(z,t)$$

$$= -j\beta'(z)\bar{\sigma}(z)\frac{\partial \mathbf{U}(z,t)}{\partial t} + \begin{bmatrix} A \text{ nonlinear PMD term} \\ \text{to be ignored} \end{bmatrix} \quad (17.24)$$

Here, $\mathbf{U}(z,t) = [U_x(z,t), U_y(z,t)]^T$ is a 2X1 Jones vector of the complex amplitudes of the field, measured along the axes of a rotating frame of reference, which rotates with the local birefringence axes of the fiber, with a further unitary transformation that ensures that a CW wave at the center frequency, ω_0, will have a distance-independent polarization, similar to the transformation used to transform Eqn (17.22) to Eqn (17.23). With proper scaling to account for the intensity distribution across the mode, $1/k'(\omega_0)|\mathbf{U}(z,t)|^2$ correctly represents the power of the wave. Moreover, the time parameter, t, in Eqn (17.24) is a retarded time: $t = t_{real} - k'(\omega_0)z$, where $1/k'(\omega_0)$ is the group velocity $k'(\omega_0) \equiv 0.5(dk_x/d\omega|_{\omega_0} + dk_y/d\omega|_{\omega_0})$. The equation allows for loss ($\alpha > 0$), or gain ($\alpha < 0$). k'' and $\beta'(z)$, represent, respectively, the chromatic dispersion, and local polarization group dispersion along the fiber (in the rotating frame), and $\bar{\sigma}(z)$ is a 2×2 unitary matrix, generated in the process of removing from $\mathbf{U}(z,t)$ its birefringence-related rotation at ω_0. The 8/9 factor, preceding the nonlinear term, originates from the averaging of the nonlinear interactions due to birefringence-induced rotation of the polarization vector [in the laboratory frame of reference, not in the rotating frame of Eqn (17.24), where it has been frozen at ω_0]. The first term on the right-hand side of the equation represents PMD. The unspecified second term, standing for nonlinear PMD, has not been spelled out due to both its complexity and the fact that it can be omitted as long as the correlation length of the birefringence is short compared with the nonlinear interaction length. This condition is well obeyed by current and future optical communication systems [106]. In sum, Eqn (17.24) can predict the propagation of optical signals over global distances, taking into account all important impairments (attenuation/gain, chromatic dispersion, PMD and nonlinearities), at the expense of averaging over phenomena that change over short scales (approximately tens of meters). Thus, when the system includes optical amplifiers, spaced a few tens of kilometers apart, the average loss is zero and the second term on the left-hand side is averaged out (but then the nonlinear term is modulated by the gain/loss factor).

Due to the large length scales involved in all of its terms, the Manakov-PMD equation can be numerically solved faster than Eqn (17.23), because larger distance steps can be implemented [107] and procedures have been developed for proper choice of $\beta'(z)$, and the 2×2 unitary matrix $\bar{\sigma}(z)$. When the PMD is set to zero, Eqn (17.24) is called the Manakov equation [108], and is quite useful in analyzing situations in which the PMD is much less important than dispersion and nonlinearities.

The Manakov-PMD and its simplified versions have been heavily used in recent years for analytical studies, and more frequently, for numerical simulations of multiwavelength, long- and ultralong-haul optical communication systems, producing experimentally confirmed results.

In the following subsections, we describe some of the interactions between PMD and third-order (Kerr) nonlinearities, relevant to fiber-optic communication systems, emphasizing recent developments.

17.8.4 Nonlinear Polarization Rotation and PMD

The many powerful wavelength channels of a modern WDM lightwave system may nonlinearly interact with one another. An important aspect of this interaction is the modulation of the SOP of a channel by power variations of its neighbors. Equation (17.24) has been used to describe this nonlinear polarization modulation

17. Polarization Mode Dispersion

in the Stokes space and to gain understanding into the important role of PMD in these interactions [109–111].

Consider the propagation of two polarized, intensity-modulated *signal* and *pump* waves, propagating at wavelengths λ_s and λ_p, with different group velocities, and let us ignore loss, intrasignal chromatic dispersion effects [the $\partial^2/\partial t^2$ term in Eqn (17.24)], and PMD. Substituting the sum of the Jones vector representations of these two signals, $\mathbf{U}(z,t) = \mathbf{P}(z,t) + \mathbf{S}(z,t)$, in Eqn (17.24), it is found that in the coordinate frame moving with the signal, and under the above-mentioned simplifying conditions (with $\bar{\gamma} = 8/9\gamma$, and the adjoint "+" superscript indicating conjugate and transpose operations of the relevant column 2×1 Jones vectors) [109]:

$$\frac{\partial \mathbf{S}(z,t)}{\partial z} = -j\bar{\gamma} \left[\underbrace{(\mathbf{S}^+\mathbf{S})\mathbf{S}}_{\text{SPM}} + \underbrace{(\mathbf{P}^+\mathbf{P})\mathbf{S} + (\mathbf{P}^+\mathbf{S})\mathbf{P}}_{\text{XPM}} \right] \quad (17.25\text{a})$$

$$\frac{\partial \mathbf{P}(z,t)}{\partial z} + d_{sp}\frac{\partial \mathbf{P}(z,t)}{\partial t} = -j\bar{\gamma} \left[\underbrace{(\mathbf{P}^+\mathbf{P})\mathbf{P}}_{\text{SPM}} + \underbrace{(\mathbf{S}^+\mathbf{S})\mathbf{P} + (\mathbf{S}^+\mathbf{P})\mathbf{S}}_{\text{XPM}} \right] \quad (17.25\text{b})$$

The second term on the left-hand side of the pump equation is required to account for the different group delays experienced by the two wavelengths, and d_{sp} is the difference between $k'(\omega_p) = 1/v_g(\omega_p)$ and $k'(\omega_s) = 1/v_g(\omega_s)$, Eqn (17.2). The self-phase modulation (SPM) terms reflect the process, in which the intensity of a propagating optical pulse slightly modifies the refractive index it sees, resulting in intensity-dependent phase modulation, pulse chirping (the peak of the pulse accumulates more nonlinear phase than its wings), and spectral broadening (which in the presence of chromatic dispersion converts the temporal phase changes into pulse distortion). The cross-phase modulation (XPM) terms have two parts. According to the first phase-matched part, e.g., $(\mathbf{P}^+\mathbf{P})\mathbf{S}$, the intensity $(\mathbf{P}^+\mathbf{P})$ of the modulated pump wave will imprint its bit pattern on the common refractive index, also seen by the signal, $\mathbf{S}$, whose phase will now be contaminated by the contents of the first channel. Again, in the presence of chromatic dispersion, the induced phase contamination will be transformed into intensity crosstalk. The second part of the XPM terms in Eqn (17.25), e.g., $(\mathbf{P}^+\mathbf{S})\mathbf{P}$, represents actual modulation of the signal polarization by the pump. This process is often called XPolM [112] (other names: nonlinear polarization scattering and XPM-induced polarization scattering (XPMIPS), [113]). Interestingly enough, Eqn (17.25) can be mathematically manipulated to be expressed in terms of the Stokes vectors of the signal and pump [109]:

$$\frac{\partial \vec{s}(z,t)}{\partial z} = \bar{\gamma}\vec{p}(z, t - d_{sp}z) \times \vec{s}(z,t) \quad (17.26\text{a})$$

$$\frac{\partial \vec{p}(z, t - d_{sp}z)}{\partial z} = \bar{\gamma}\vec{s}(z,t) \times \vec{p}(z, t - d_{sp}z), \tag{17.26b}$$

where $\vec{s}(z,t)$ and $\vec{p}(z, t - d_{sp}z)$ are (unnormalized) 3×1 Stokes vectors, representing the 2×1 Jones vectors **S** and **P**, and $\times$ stands for vector cross-product. Thus, each wave modifies the polarization of the other, unless the cross product is zero, i.e., unless the pump and signal are either copolarized or orthogonally polarized. In the CW case [no time dependence in Eqns (17.26)], it is found (by summing the two equations) that $\vec{m} = \vec{s}(z) + \vec{p}(z)$ is a constant, z-independent vector that serves as a pivot around which $\vec{s}(z,t)$ and $\vec{p}(z)$ rotate as they travel in the z direction [114]:

$$\begin{aligned} \frac{\partial \vec{s}(z)}{\partial z} &= \bar{\gamma}\vec{m} \times \vec{s}(z) \\ \frac{\partial \vec{p}(z)}{\partial z} &= \bar{\gamma}\vec{m} \times \vec{p}(z) \end{aligned} \tag{17.27}$$

The CW model will now be extended to the case of on-off-keying (OOK)-modulated signal and pump, but still *without* chromatic dispersion. According to Eqn (17.27), a pump "mark" bit will continuously rotate the polarization of the section of a signal "mark" bit that travels alongside (and together with) that "mark" bit. Zero pump bits will have no effect. Thus, XPolM will *depolarize* the signal wave in the sense that its output polarization is no longer constant but rather changes in time from bit to bit in a pattern dictated by the intensity modulation of the pump. The picture obviously becomes more complicated when chromatic dispersion is taken into account, introducing walk-off between the two bit streams. Each signal "mark" bit will now nonlinearly interact with many pump bits, as these two bit streams travel down the fiber with different speeds, and each and every "collison" between a signal "mark" bit and a pump "mark" bit will further alter the SOPs of the bits. Obviously, for a given link length, the accumulated rotation angle of the signal SOP will decrease as the walk-off length between the bit streams decreases, preventing the signal bit from a continuous cumulative interaction with an overlapping pump bit [109]. An approximate expression for the resulting degree of polarization (DOP) [6] in the case of two propagating channels (with no PMD) is given by Ref. [112]:

$$DOP \approx 1 - \frac{1}{3}\gamma^2 P_{total}^2 L_{walk-off}L; \quad \left(L_{walk-off} = \frac{\text{Pulse Width}}{d_{sp}|\lambda_s - \lambda_p|}\right), \tag{17.28}$$

where P_{total} is the total optical power, L is the total *effective* fiber length (e.g., the nonlinear effective length, ($L_{eff} \approx 20$ km) multiplied by the number of amplifier spans in the link), and $L_{walk-off} \sim 10$ km, is the walk-off length [100]. See also Refs [109, 115] for alternative quantifications of the XPolM effect.

The combined effects of PMD and XPM (including XPolM) are yet to be analyzed in a simple intuitive way (for numerical solutions, see for example Refs [97, 116–118]. Suffice it to say that in the presence of a sizable PMD [119], the different SOPs of the WDM channels become uncorrelated and change quickly over short distances, thereby averaging over the XPolM effects and decreasing their influence.

XPolM bears important consequences for WDM systems afflicted with PMD. In polarization-multiplexed WDM systems [6], nearby channels are launched with *orthogonal* polarizations. In the absence of PMD, and according to Eqn (17.27), XPolM poses no problem, because all channels are either co-polarized or orthogonally polarized. However, because WDM channels propagate along the fiber with different wavelengths, each channel sees slightly different birefringence, and this manifestation of PMD will spoil the parallel/orthogonal alignment of the channel polarizations, triggering nonlinear depolarization. At the receiver, each polarized WDM channel, after being spectrally selected, also goes through a polarizer, preceded by a polarization controller, in order to properly isolate it from crosstalk from its neighbors. Depolarization of the incoming bits will be translated by the polarizer into random intensity modulation of the received pattern, resulting in a BER penalty of up to 10 dB [120].

XPolM also modulates the DGD and PSP of the fiber. Consider the study of Boroditsky et al. [121], in which the (static) PMD of a 20-km spool of AllWave fiber was measured at 1537.5 nm, using a weak CW probe (~0 dBm; weak enough to eliminate nonlinear effects associated with the probe, while strong enough to dwarf pump leakage through the filter), in the presence of a nearby strong (up to 17 dBm) CW pump at 1543 nm. Rotation of the pump polarization gave rise to more than 30% variation of the DGD value at 1537.5 nm and caused the direction of the PMD vector to subtend a cone of 20° on the Poincare sphere around its no-pump value, see Figure 17.24.

In a dynamic environment [119], XPolM has a direct influence on the evolution of the SOP of every bit as it propagates along the fiber, changing this evolution in a random and fast (<1 ns) way, which depends on the bit contents of all interfering channels. Consequently, different trajectories of the SOP will cause the bits to experience different birefringence, and different PMD-induced delay and distortion, and different PSP [97]. In short, polarization-wise, the propagation of every bit can be simplistically viewed as governed by its own PMD vector, which changes very rapidly in time. Thus, it is to be expected that XPolM will also introduce time jitter in the received bit stream, as is indeed the case [115, 122]. Its effect on the operation of PMDCs is discussed next.

17.8.5 The Effect of Nonlinear Polarization Rotation on PMDCs

PMD is widely regarded as a limiting impairment in high-speed optical communication systems, whose severity grows with distance and bit rate. PMDCs [94, 123],

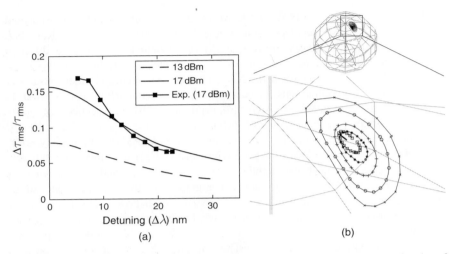

Figure 17.24 (a) RMS deviation of nonlinearity-induced DGD, normalized to its rms as a function of wavelength separation for different wavelength separations in a 20-km fiber spool with 0.35 ps mean DGD. The line with the solid squares is experimental, while the rest are theoretical [121]; (b) Example of the PSP evolution with pump SOP scanning the equator of Poincaré sphere for pump power of 5, 8, 11, 14, and 17 dBm, with the inner curve corresponding to the lowest power. The PSP deviations for the highest pump power (the outermost curve) are about 20°. The probe wavelength is fixed at 1537.5 nm (from [12] © 2006 IEEE) (this figure may be seen in color on the included CD-ROM).

attempting to mitigate PMD effects, are challenged by the complex dependence of the PMD vector on ω, as well as by its dynamic and random nature. Also, for proper operation, the PMDC must be controlled by a feedback signal, derived from a particular figure of merit of system performance, e.g., the system bit error rate (BER). Current compensators and their feedback mechanisms operate at rates much slower than transmission bit rates, but fast enough to follow the relatively slow (approximately kHz) dynamics of the PMD vector. In modern lightwave systems, which use WDM architectures, it is advantageous to apply compensation on a channel-by-channel basis, and modern compensators can indeed improve channel performance, at the expense of some penalties and cost [94]. Vital to successful channel-by-channel PMD compensation is the absence of crosstalk from neighboring channels. But the extremely fast (<1 ns) XPolM-induced depolarization should pose a significant problem to optically based PMDCs that try to instantaneously adjust their own DGD and PSP to compensate for the link PMD. Having control and feedback mechanisms that can respond with millisecond rates, a PMDC will not be able to follow the XPolM-induced fast changes in the fiber PMD vector. The reduction in efficiency of PMD compensation has been correlated with nonlinearly induced depolarization and also demonstarated in a number of experiments/ simulations [112, 115, 124–126], see Figure 17.25. Here, the upper left-hand curve indicates that under the assumed simulation conditions, [112], optical nonlinearities

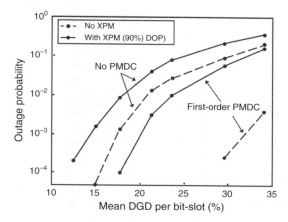

Figure 17.25 Outage probability vs average DGD, for no XPolM (dashed) and with XPolM (solid). The arrows indicate the influence of the XPolM, i.e., the increased Outage probabilty that is induced (from [112] © 2006 IEEE).

deteriorate the system performance even without PMDC. With no XPolM (the linear case), PMDC appears to be very effective in significantly reducing the outage probability, but its efficiency dramatically decreases in the presence of XPolM-induced depolarization [112]. Using the same simulations, it was found that the efficiency of optical active PMDCs can be preserved, if the DOP exceeds 97%. Using Eqn (17.28), this condition translates, [112], to:

$$\gamma P_{\text{tot}} \sqrt{L_{\text{wo}} L} < 0.3 \qquad (17.29)$$

Thus, no more than a few milliwatts per channel are allowed, if highly efficient PMD compensation is required in modern lightwave systems.

Because the DOP is less affected by short walk-off length, Eqn (17.28), it is to be expected that the efficiency reduction effect of XPolM on PMDCs will be less of a problem as the bit rate increases. Indeed, not only the bit period decreases but also the frequency separation between neighboring channels increases, both processes leading to a shorter $L_{\text{walk-off}}$. Thus, it has been shown in Ref. [115] that in an ultralong-haul, dense WDM system, XPolM will reduce the PMDC efficiency by 50% for a 10 Gb/s data rate, but only by 20% for a 40 Gb/s rate. Similarly, it has been noted [119] that nonnegligible PMD will mitigate the effect of XPolM on PMDC compensatoin.

We conclude this section by noting that other means to mitigate PMD effects may be less vulnerable to XPolM effects. These include different modulation formats (such as constant amplitude, phase-modulated techniques (e.g., DPSK), as well as modulation techniques that use the polarization degree of freedom [126]), the use of forward error correction, and electronic compensation [94], but more research is required before their robustness to nonlinear effects can be fully assessed.

17.8.6 Other Interactions Between Third-Order Nonlinearities and PMD

Here, we briefly mention some recent studies involving the interaction of PMD with Kerr nonlinearities, other than XPolM. For example, adding PMD to a system suffering from SPM-induced penalties may improve its performance [96, 97]. This improvement is the result of pulse compression, originating from the interaction between the induced nonlinear chirp [100] and the PDCD (discussed in Section 17.7), an SOPMD phenomenon. An example showing the detrimental combined effect of PMD and nonlinearities is discussed in Ref. [99]. In a numerical study of a long nonsolitonic system (60×100 km), it was found that in the first part of the system the PMD derails the propagating waveform from its unperturbed trajectory, inflicting serious penalties, which can be significantly reduced by invoking in-line optical filters. PMD can also increase the inaccuracies of various measurement procedures relying upon nonlinearities, e.g., the determination of the zero dispersion point using four wave mixing [127].

To summarize: chromatic dispersion, optical nonlinearities (including the Raman and Brillouin inelastic processes), PMD and amplified spontaneous emission of optical amplifiers, all interact in a very complex way in modern high bit rate, ultralong-haul WDM lightwave systems. While numerical techniques are the working horse of analysis, physical understanding of isolated processes always offers great help in developing an intuitive grasp of these complex but extremely important systems.

17.9 CONCLUSION

PMD has had an interesting evolution in the two decades since it was formally described. Initially, it was a laboratory curiosity, then an impairment only for delicate analog signals. As it became better understood and it loomed as a potential roadblock to the telecom expansion, PMD became the focus of an international research effort of major significance. In the intervening years, with dramatically improved fiber fabrication techniques, tremendous advances in our understanding of the physics and technology of the topic, and with inflated capacity projections popped after the telecom bubble, much of the millennial hysteria surrounding PMD has abated. So much so that, on one hand, there is a camp which perceives PMD as a solved problem.

On the other hand, as we look into the future, it is possible to discern forces which make such confidence seem slightly misplaced. Global increases in communications and the digital revolution support the inexorable logic of, and pull for, extension of the Ethernet hierarchy to 100 Gb/s which will place much greater demands on transmission fiber that is installed today. Rising complexity of network architectures will push more optical components into the network, and these will counteract the lower birefringence of deployed transmission fibers. The natural tendency to "cherry pick" better fibers on existing routes and economic constraints on new fiber deployment will stress the utility of existing fiber plants. In short, one can reasonably view the last several years as a hiatus, not as a permanent victory over PMD impairments.

Balanced against these potentially troubling forces are a mature understanding of the basic physical processes, and an increasingly comprehensive view of how fibers react in installed networks. Future research to attain more sophisticated understanding of the interplay between the various impairments, will undoubtedly result in better modeling, emulation, and mitigation of PMD-related impairments in next generation lightwave communication systems.

ACKNOWLEDGMENTS

The authors gratefully acknowledge numerous discussions with colleagues who contributed both to our research and to the completion of this manuscript. In particular, we deeply appreciate the generous contributions of C. Antonelli, M. Boroditsky, G. Carter, K. Cornick, S. Dods, A. Eyal, A. Galtarossa, R. Jopson, M. Karlsson, H. Kogelnik, P. Magill, B. Marks, A. Mecozzi, D. Menashe, C. Menyuk, J. Nagel, L. Nelson, L. Palmieri, M. Santagiustina, M. Schiano, M. Shtaif, A. Willner, P. Winzer, H. Yaffe and A. Zadok.

REFERENCES

[1] C. D. Poole and R. E. Wagner, "Phenomenological approach to polarization dispersion in long single-mode fibers," *Electron. Lett.*, 23, 1113–1114, 1986.
[2] C. D. Poole, J. H. Winters, and J. A. Nagel, "Dynamical equation for polarization dispersion," *Opt. Lett.*, 16(6), 372–374, March 1991.
[3] F. Curti, B. Daino, G. DeMarchis, and F. Matera, "Statistical treatment of the evolution of the principal states of polarization in single-mode fibers," *IEEE J. Lightw. Technol.*, LT-8, 1162–1166, 1990.
[4] G. J. Foschini and C. D. Poole, "Statistical theory of polarization dispersion in single mode fibers," *IEEE J. Lightw. Technol.*, LT-9, 1439–1456, 1991.
[5] C. D. Poole and J. A. Nagel, "Polarization effects in lightwave systems," in *Optical Fiber Telecommunications III A* (I. P. Kaminow and T. L. Koch, eds), San Diego, CA, Academic Press, 1997, pp. 114–161.
[6] H. Kogelnik, R. M. Jopson, and L. E. Nelson, "Polarization mode dispersion," in *Optical Fiber Telecommunications IV B* (I. P. Kaminow and T. Li, eds), San Diego, CA, Academic Press, 2002, pp. 725–861.
[7] J. P. Gordon and H. Kogelnik, "PMD Fundamentals: Polarization mode dispersion in optical fibers," *Proc. Natl. Acad. Sci. USA*, 97(9), 4541–4550, 2000.
[8] J. N. Damask, *Polarization Optics in Telecommunications*, New York, NY, Springer, 2004.
[9] A. Galtarossa and C. R. Menyuk, eds, *Polarization Mode Dispersion*, New York, NY, Springer, 2005.

[10] M. Brodsky, N. J. Frigo, M. Boroditsky, and M. Tur, "Polarization mode dispersion of installed fibers," *J. Lightw. Technol.*, 24(12), 4584–4599, 2006.
[11] N. Gisin and J. P. Pellaux, "Polarization mode dispersion: time vs frequency domains," *Opt. Commun.*, 89, 316–323, 1992.
[12] R. Ulrich, "Representation of codirectional coupled waves," *Opt. Lett.*, 5, 109–111, 1977.
[13] R. Ulrich and A. Simon, "Polarization optics of twisted single-mode fibers," *Appl. Opt.*, 18(13), 2241–2251, 1979.
[14] N. J. Frigo, "A generalized geometrical representation of coupled mode theory," *IEEE J. Quantum Electron.*, QE-22, 2121–2140, 1986.
[15] H. Kogelnik, L. E. Nelson, and J. P. Gordon, "Emulation and inversion of polarization mode dispersion," *J. Lightw. Technol.*, 21(2), February 2003.
[16] A. Galtarossa, P. Griggio, L. Palmieri, and A. Pizzinat, "Low-PMD spun fibers," *J. Opt. Fiber Commun. Rep.*, 1, 32–62, 2004.
[17] M. Shtaif and A. Mecozzi, "Study of the frequency autocorrelation of the differential group delay in fibers with polarization mode dispersion," *Opt. Lett.*, 25, 707–709, 2000.
[18] N. Gisin, B. Gisin, J. P. Von der Weid, and R. Passy, "How accurately can one measure a statistical quantity like polarization-mode dispersion," *IEEE Photon. Technol. Lett.*, 8, 1671–1673, 1996.
[19] C. Antonelli and A. Mecozzi, "Statistics of the DGD in PMD emulators," *IEEE Photon. Technol. Lett.*, 16(8), 1840–1842, 2004.
[20] C. Antonelli and A. Mecozzi, "Theoretical characterization and system impact of the hinge model of PMD," *IEEE J. Lightw. Technol.*, 24(11), 4064–4074, 2006.
[21] M. Boroditsky, M. Brodsky, P. D. Magill et al., "Outage probabilities for fiber routes with finite number of degrees of freedom," *IEEE Photon. Technol. Lett.*, 17(2), 345–347, 2005.
[22] C. De Angelis, A. Galtarossa, G. Gianello et al., "Time evolution of polarization mode dispersion in long terrestrial links," *IEEE J. Lightw. Technol.*, 10, 552–555, May 1992.
[23] R. Caponi, B. Riposati, A. Rossaro, and M. Schiano, "WDM Design Issues with Highly Correlated PMD Spectra of Buried Optical Cables," in Proc. *OFC*, 2002, pp. 453–455.
[24] M. Karlsson, M. J. Brentel, and P. Andrekson, "Long-term measurement of PMD and polarization drift in installed fibers," *J. Lightw. Technol.*, 18, 941–951, 2000.
[25] M. Brodsky, P. Magill, and N. J. Frigo, "Polarization mode dispersion of installed recent vintage fiber as a parametric function of temperature," *IEEE Photon. Technol. Lett.*, 16(1), 209–211, 2004.
[26] M. Brodsky, P. Magill, and N. J. Frigo, "Evidence for Parametric Dependence of PMD on Temperature in Installed 0.05 ps/km$^{1/2}$ Fiber," in Proc. *ECOC*, 4, paper 9.3.2, 2002.
[27] M. Brodsky, P. Magill, and N. J. Frigo, "Long-Term" PMD Characterization of Installed Fibers – How Much Time is Adequate," in Proc. *OFC*, paper FI5, 2004.
[28] P. K. Kondamuri, C. Allen, and D. L. Richards, "Variation of PMD-Induced Outage Rates and Durations with Link Length on Buried Standard Single-Mode Fibers," in Proc. *OFC*, paper OThX3, 2005.
[29] P. K. Kondamuri, C. Allen, and D. L. Richards, "Study of Variation of the Laplacian Parameter of DGD Time Derivative with Fiber Length Using Measured DGD Data," in Proc. *SOFM*, 2004, pp. 91–94.
[30] M. Brodsky, J. Martinez, N. J. Frigo, and A. Sirenko, "Dispersion Compensation Module as a Polarization Hinge," in Proc. *ECOC*, paper We 1.3.2, 2005, pp. 335–336.
[31] M. Brodsky, M. Boroditsky, P. D. Magill et al., "Field Measurements Through a Commercial, Raman-Amplified ULH Transmission System," in Proc. *LEOS PMD Summer Topical Meeting*, paper MB3.3, 2003, pp. 15–16.
[32] M. Brodsky, M. Boroditsky, P. Magill et al., "Channel-to-Channel Variation of Non-Maxwellian Statistics of DGD in a Field Installed System," in Proc. *ECOC*, 3, paper We 1.4.1, 2004, pp. 306–309.
[33] M. Brodsky, M. Boroditsky, P. Magill et al., "Persistence of spectral variations in DGD statistics," *Opt. Express*, 13, 4090–4095, May 2005.

[34] D. A. Nolan, X. Chen, and M.-J. Li, "Fibers with low polarization-mode dispersion," *J. Lightw. Technol.*, 22(4), 1066–1077, 2004.
[35] M. Boroditsky, M. Brodsky, N. J. Frigo et al., "In-service measurements of polarization mode dispersion and correlation to bit-error rate," *IEEE Photon. Technol. Lett.*, 15, 572–574, 2003.
[36] C. D. Poole, R. W. Tkach, A. R. Chraplyvy, and D. A. Fishman, "Fading in lightwave systems due to polarization-mode dispersion," *IEEE Photon. Technol. Lett.*, 3(1), 68–70, January 1991.
[37] M. Boroditsky, K. Cornick, C. Antonelli et al., "Comparison of system penalties from first- and multi-order polarization-mode dispersion," *IEEE Photon. Technol. Lett.*, 17, 1650–1652, August 2005; see also correction in 19(8), 628, April 2007.
[38] P. J. Winzer, H. Kogelnik, and K. Ramanan, "Precise outage specification for first-order PMD," *IEEE Photon. Technol. Lett.*, 16(2), February 2004.
[39] P. J. Winzer and R.-J. Essiambre, "Advanced Optical Modulation Formats," in Proc. *IEEE*, 94(5), May 2006.
[40] C. Xie, L. Moller, H. Haunstein, and S. Hunsche, "Polarization Mode Dispersion Induced Impairments in Different Modulation Formats," in Proc. *OFC*, 1, 2003, p. 258.
[41] P. J. Winzer, H. Kogelnik, C. H. Kim et al., "Reciever impact on first-order PMD outage," *IEEE Photon. Technol. Lett.*, 15(10), October 2003.
[42] M. D. Feuer and M. Brodsky, private communication.
[43] H. F. Haunstein, W. Sauer-Greff, A. Dittrich et al., "Principles for electronic equalization of polarization mode dispersion," *J. Lightw. Technol.*, 22(4), April 2004.
[44] J. M. Gene, P. J. Winzer, S. Chandrasekhar, and H. Kogelnik, "Simultaneous compensation of polarization mode dispersion and chromatic dispersion using electronic signal processing," *J. Lightw. Technol.*, 25(7), 1735–1741, July 2007.
[45] K. E. Cornick, M. Brodsky, M. Birk, and M. D. Feuer, "MLSE receiver tolerance to all-order polarization mode dispersion," *Opt. Express*, 15(24), 15999–16004, November 2007.
[46] S. X. Wang, A. M. Weiner, M. Boroditsky, and M. Brodsky, "Monitoring PMD-induced penalty and other system performance metrics via a high-speed spectral polarimeter," *IEEE Photon. Technol. Lett.*, 18, 1753–1755, August 2006.
[47] S. X. Wang and A. M. Weiner, "A complete spectral polarimeter design for lightwave communication systems," *J. Lightw. Technol.*, 24(11), 3982–3991, November 2006.
[48] K. Cornick, M. Boroditsky, N. J. Frigo et al., "Experimental Comparison of System Penalties Due to 1st Order and Multi-Order Polarization Mode Dispersion," in Proc *OFC*, OFF6, 2005.
[49] C. Antonelli, A. Mecozzi, K. Cornick et al., "PMD-induced penalty statistics in fiber links," *IEEE Photon. Technol. Lett.*, 17(5), 1013–1015, 2005.
[50] H. Kogelnik, H. P. J. Winzer, L. E. Nelson et al., "First-order PMD outage for the hinge model," *IEEE Photon. Technol. Lett.*, 17, 1208–1210, June 2005.
[51] A. Mecozzi, C. Antonelli, M. Boroditsky, and M. Brodsky, "Characterization of the time dependence of polarization mode dispersion," *Opt. Lett.*, 29, 2599–2601, November 2004.
[52] M. Shtaif and M. Boroditsky, "The effect of the frequency dependence of PMD on the performance of optical communications systems," *IEEE Photon. Technol. Lett.*, 15(10), 1369–1371, 2003.
[53] H. Kogelnik, L. E. Nelson, and P. J. Winzer, "Second-order PMD outage of first-order compensated fiber system," *IEEE Photon. Technol. Lett.*, 16(4), April 2004.
[54] M. Karlsson and J. Brentel, "Autocorrelation function of the polarization-mode dispersion vector," *Opt. Lett.*, 24(14), 939, July 1999.
[55] M. Shtaif, A. Mecozzi, and J. A. Nagel, "Mean-square magnitude of all orders of polarization mode dispersion and the relation with the bandwidth of the principal states," *IEEE Photon. Technol. Lett.*, 12(1), 53, January 2000.
[56] H. Chen, R. M. Jopson, and H. Kogelnik, "On the bandwidth of higher-order polarization-mode dispersion: the Taylor series expansion," *Opt. Express*, 11(11), 1270–1282, June 2003.

[57] C. Antonelli, A. Mecozzi, K. Cornick, and M. Boroditsky, "A Statistical Theory of PMD-Induced Power Penalty," in Proc. *ECOC*, We 1.3.3, 2005.

[58] E. Forestieri and G. Prati, "Exact analytical evaluation of second-order PMD impact on the outage probability for a compensated system," *J. Lightw. Technol.*, 22(4), April 2004.

[59] J. C. Rasmussen, A. Isomura, and G. Ishikawa, "Automatic compensation of polarization mode dispersion for 40 Gb/s transmission systems," *J. Lightw. Technol.*, 20(12), 2101–2109, December 2002.

[60] M. Schmidt, M. Witte, F. Buchali et al., "Adaptive PMD Compensation for 170 Gbit/s RZ Transmission Systems with Alternating Polarization," in Proc. *OFC 2005*, paper JWA38, 2005.

[61] R. Leppla, S. Vorbeck, M. Schmidt et al., "PMD Tolereance of 8×170 Gbit/s Field Transmission Experiment Over 430km of SSMF with and Without PMDC," in Proc. *OFC*, paper OFF2, 2005.

[62] S. Kiechbusch, S. Ferber, H. Rosenfeldt et al., "Automatic PMD compensator in a 160-Gb/s OTDM transmission over deployed fiber using RZ-DPSK modulation format," *J. Lightw. Technol.*, 23(1), 165–171, January 2005.

[63] M. Daikoku, T. Miyazaki, I. Morita et al., "8×160 Gb/s WDM field transmission experiment with single polarization RZ-DPSK signals and PMD compensator," *IEEE Photon. Technol. Lett.*, 18(2), January 2006.

[64] M. Eiselt, L. Garrett, J. Wiesenfeld, and R. Tkach, "Is PMD Compensation Really Useful," in Proc. *OFC*, 1, paper MF113, 2003, p. 142.

[65] M. Eiselt, "PMD Compensation: A System Perspective," in Proc. *OFC*, paper ThF4, 2004.

[66] A. Eyal, W. K. Marshall, M. Tur, and A. Yariv, "Representation of second-order polarisation mode dispersion," *Electron. Lett.*, 35(I9), 1658, 16 September 1999.

[67] H. H. Yaffe and D. L. Peterson, Jr, "Experimental determination of system outage probability due to first and second-order PMD," *J. Lightw. Technol.*, 24(11), November 2006.

[68] S. L. Fogal, G. Biondini, and W. L. Kath, "Multiple Importance Sampling for First- and Second-Order PMD," in Proc. *OFC*, Anaheim, CA, USA (ThA1), 2002, pp. 371–372.

[69] M. C. Hauer, Q. Yu, E. R. Lyons et al., "Electrically controllable all-fiber PMD emulator using a compact array of thin-film microheaters," *J. Lightw. Technol.*, 22(4), 1059–1065, 2004.

[70] I. T. Lima, Jr, R. Khosravani, P. Ebrahimi et al., "Comparison of polarization mode dispersion emulators," *J. Lightw. Technol.*, 19(12), 1872, December 2001.

[71] H. C. Lefevre, "Single-mode fractional wave devices and polarization controller," *Electron. Lett.*, 16, 778, 1980.

[72] Y. K. Lizé, L. Palmer, N. Godbout et al., "Scalable polarization-mode dispersion emulator with proper first- and second-order statistics," *IEEE Photon. Technol. Lett.*, 17(11), 2451–2453, 2005.

[73] Y. K. Lize, L. Palmer, M. Aubé et al., "Autocorrelation function of the single PC polarization-mode dispersion emulator," *IEEE Photon. Technol. Lett.*, 18(1), 217–219, 2006.

[74] M. Wegmuller, S. Demma, C. Vinegoni, and N. Gisin, "Emulator of first- and second-order polarization-mode dispersion," *IEEE Photon. Technol. Lett.*, 14(5), 630–632, 2002.

[75] P. B. Phua and H. A. Haus, "A deterministically controlled four-segment polarization-mode dispersion emulator," *J. Lightw. Technol.*, 20(7), 1132–1140, 2002.

[76] P. B. Phua and H. A. Haus, "Variable secondorder PMD module without first-order PMD," *J. Lightw. Technol.*, 20(11), 1951–1956, 2002.

[77] P. B. Phua, H. A. Haus, and E. P. Ippen, "Deterministic broad-band PMD emulator," *IEEE Photon. Technol. Lett.*, 16(6), 1486–1488, 2004.

[78] R. A. Chipman and R. Kinnera, "First- and second-order polarization mode dispersion generated by a two-stage emulator," *Appl. Opt.*, 43(33), 6074–6079, 2004.

[79] N. Y. Kim and N. Park, "Independently tunable first- and second-order polarization-mode dispersion emulator," *IEEE Photon. Technol. Lett.*, 17(3), 576–578, 2005.

[80] J. N. Damask, "Methods to construct programmable PMD Source. I. Technology and theory," *J. Lightw. Technol.*, 22(4), 997–1005, April 2004.

[81] J. N. Damask, P. R. Myers, A. Boschi, and G. J. Simer, "Demonstration of a coherent PMD source," *IEEE Photon. Technol. Lett.*, 15(11), 1612–1614, November 2003.

[82] J. W. Evans, "The birefringent filter," *J. Opt. Soc. Am.*, 39(3), March 1949.

[83] J. N. Damask, G. Gray, P. Leo et al., "Method to measure and estimate total outage probability for PMD-impaired systems," *IEEE Photon. Technol. Lett.*, 15(1), 48–50, January 2003.

[84] J. H. Lee, M. S. Kim, and Y. C. Chung, "Statistical PMD emulator using variable DGD elements," *IEEE Photon. Technol. Lett.*, 15(1), 54–56, 2003.

[85] J. N. Damask, G. J. Simer, K. B. Rochford, and P. R. Myers, "Demonstration of a programmable PMD source," *IEEE Photon. Technol. Lett.*, 15(2), 296–298, February 2003.

[86] L. Yan, M. C. Hauer, Y. Shi et al., "Polarization-mode-dispersion emulator using variable differential-group-delay (DGD) elements and its use for experimental importance sampling," *J. Lightw. Technol.*, 22(4), 1051–1057, 2004.

[87] L. Yan, C. Yeh, G. Yang et al., "Programmable group-delay module using binary polarization switching," *J. Lightw. Technol.*, 21(7), 1676, July 2003.

[88] L. S. Yan, M. Hauer, A. E. Willner et al., "High-Speed, Stable and Repeatable PMD Emulator wlth Tunable Statistics," in Proc. *OFC*, Atlanta, Georgia, USA (MF6), 2003, pp. 6–7.

[89] L.-S. Yan, X. S. Yao, and A. E. Willner, "Enabling 'Hinge' model in polarization-mode-dispersion statistics using variable differential-group-delay-based emulator," *IEEE Photon. Technol. Lett.*, 18(2), 427, January 2006.

[90] G. Biondini and W. L. Kath, "Polarization-mode dispersion emulation with maxwellian lengths and importance sampling," *IEEE Photon. Technol. Lett.*, 16(3), 789–791, 2004.

[91] M. Shtaif, "The brownian-bridge method for simulating polarization mode dispersion in optical communication systems," *IEEE Photon. Technol. Lett.*, 15(1), 51–53, 2003.

[92] C. K. Madsen, M. Cappuzzo, E. J. Laskowski et al., "Versatile integrated PMD emulation and compensation elements," *J. Lightw. Technol.*, 22(4), 1041–1050, 2004.

[93] L. Yan, S. Yao, M. Hauer, and A. Willner, "Practical solutions to polarization mode dispersion emulation and compensation," *J. Lightw. Technol.*, 24(11), November 2006.

[94] F. Buchali and H. Bülow, "Adaptive PMD compensation by electrical and optical techniques," *J. Lightw. Technol.*, 22(4), 1116, April 2004.

[95] L. Möller, Y. Su, G. Raybon et al., "Penalty interference of nonlinear intra-channel effects and PMD in ultra high-speed TDM systems," *Electron. Lett.*, 38(6), 281–283, 2002.

[96] B. S. Marks and C. R. Menyuk, "Polarization-Mode Dispersion Enhancement of Nonlinear Propagation," in Proc. *ECOC*, paper Mo3.7.1, 2003.

[97] C. R. Menyuk and B. S. Marks, "Interaction of Polarization Mode Dispersion and Nonlinearity in Optical Fiber Transmission Systems," *J. Lightw. Technol.*, 24(7), 2806–2826, 2006.

[98] G. Zhang, J. T. Stango, X. Zhang, and C. Xie, "Impact of fiber nonlinearity on PMD penalty in DWDM transmission systems," *IEEE Photon. Technol. Lett.*, 17(2), 501–503, 2005.

[99] E. Alperovich, A. Mecozzi, and M. Shtaif, "PMD penalties in long nonsoliton systems and the effect of inline filtering," *IEEE Photon. Technol. Lett.*, 8(10), 1179, May 2005.

[100] G. Agrawal, "Nonlinear Fiber Optics (Optics and Photonics)," 3rd edn, London, Academic Press, 2001.

[101] I. P. Kaminow, "Polarization in optical fibers," *IEEE J. Quantum Electron.*, QE-17(1), 15, January 1981.

[102] C. R. Menyuk, "Pulse propagation in an elliptically birefringent Kerr medium," *IEEE J. Quantum Electron.*, 25(12), 2674–2682, December 1989.

[103] M.-J. Li, X. Chen, and D. A. Nolan, "Ultra Low PMD Fibers by Fiber Spinning," in Proc. *OFC*, paper FA1, 2004.

[104] P. K. A. Wai and C. R. Menyuk, "Polarization mode dispersion, decorrelation, and diffusion in optical fibers with randomly varying birefringence," *J. Lightw. Technol.*, 14(2), 148, February 1996.

[105] C. R. Menyuk, "Application of multiple-length-scale methods to the study of optical fiber transmission," *J. Eng. Math.*, 36, 113–136, 1999.

[106] D. Marcuse, C. R. Menyuk, and P. K. A. Wai, "Application of the Manakov-PMD equation to studies of signal propagation in optical fibers with randomly varying birefringence," *J. Lightw. Technol.*, 15(9), 1735, September 1997.

[107] P. K. A. Wai, W. L. Kath, C. R. Menyuk, and J. W. Zhang, "Nonlinear polarization-mode dispersion in optical fibers with randomly varying birefringence," *J. Opt. Soc. Am. B*, 14(11), 2967, November 1997.

[108] S. V. Manakov, "On the theory of two-dimensional stationary self-focusing of electromagnetic waves," *Sov. Phys. JETP*, 38, 248–253, 1974.

[109] A. Bononi, A. Vannucci, A. Orlandini et al., "Degree of polarization degradation due to cross-phase modulation and its impact on polarization-mode dispersion compensators," *J. Lightw. Technol.*, 21(9), 1903–1913, 2003.

[110] D. Wang and C. R. Menyuk, "Reduced model of the evolution of the polarization states in wavelength-division-multiplexed channels," *Opt. Lett.*, 23, 1677–1679, 1998.

[111] D. Wang and C. R. Menyuk, "Polarization evolution due to Kerr nonlinearity and chromatic dispersion," *J. Lightw. Technol.*, 17, 2520–2529, 1999.

[112] M. Karlsson and H. Sunnerud, "Effects of nonlinearities on PMD-induced system impairments," *J. Lightw. Technol.*, 24(11), November 2006.

[113] M. R. Phillips, S. L. Woodward, and R. L. Smith, "Cross-polarization modulation: theory and measurement in subcarrier-modulated WDM systems," *J. Lightw. Technol.*, 24(11), November 2006.

[114] L. F. Mollenauer, J. P. Gordon, and F. Heismann, "Polarization scattering by soliton–soliton collisions," *Opt. Lett.*, 20(20), 2060, October 1995.

[115] C. Xie, L. Möller, D. C. Kilper, and L. F. Mollenauer, "Effect of cross-phase-modulation-induced polarization scattering on optical polarization mode dispersion compensation in wavelength-division-multiplexed systems," *Opt. Lett.*, 28(23), 2303, December 2003.

[116] Q. Lin, and G. P. Agrawal, "Effects of polarization-mode dispersion on cross-phase modulation in dispersion-managed wavelength-division-multiplexed systems," *J. Lightw. Technol.*, 22(4), 977–987, 2004.

[117] S. Pachnicke, E. Voges, E. De Man et al., "Experimental Investigation of XPM-Induced Birefringence in Mixed-Fiber Transparent Optical Networks," in Proc. *OFC*, Anaheim, California, USA, JThB9, 2006.

[118] N. Hanik, "Influence of Polarisation Mode Dispersion on the Effect of Cross Phase Modulation in Optical WDM Transmission," in Proc. *of Transparent Optical Networks*, 1, 2004, pp. 190–194.

[119] R. Khosravani, Y. Xie, L.-S. Yan et al., "Limitations to First-Order PMD Compensation in WDM Systems Due to XPM-Induced PSP Changes," in Proc. *OFC*, 3, 2001, WAA5-1–WAA5-3.

[120] B. C. Collings and L. Boivin, "Nonlinear polarization evolution induced by cross-phase modulation and its impact on transmission systems," *IEEE Photon. Technol. Lett.*, 12(11), 1582, November 2000.

[121] M. Boroditsky, M. Bourd, and M. Tur, "Effect of nonlinearities on PMD," *J. Lightw. Technol.*, 24(11), November 2006.

[122] R. Khosravani, "Timing-Jitter in High Bit-Rate WDM Communication Systems Due to PMD-Nonlinearity Interaction," in Proc. *OFC*, 5, 2005, p. 3.

[123] H. Sunnerud, C. Xie, M. Karlsson et al., "A comparison between different PMD compensation techniques," *J. Lightw. Technol.*, 20(3), 368–378, 2002.

[124] J. H. Lee, K. J. Park, C. H. Kim, and Y. C. Chung, "Effects of nonlinear crosstalk in optical PMD compensation," *IEEE Photon. Technol. Lett.*, 14(8), 1082, August 2002.

[125] E. Corbel, J. P. Thiéry, S. Lanne et al., "Experimental Statistical Assessment of XPM Impact on Optical PMD Compensator Efficiency," in Proc. *OFC*, paper ThJ2, 2003, pp. 499–501.

[126] Z. Pan, Q. Yu, Y. Arieli, and A. E. Willner, "The effects of XPM-induced fast polarization-state fluctuations on PMD compensated WDM systems," *IEEE Photon. Technol. Lett.*, 16(8), 1963–1965, 2004.

[127] Q. Lin and G. P. Agrawal, "Impact of polarization-mode dispersion on measurement of zero-dispersion wavelength through four-wave mixing," *IEEE Photon. Technol. Lett.*, 15(12), 1719, December 2003.

[128] L. Moller, L. Boivin, S. Chandrasekhar, and L. L. Buhl, "Setup for demonstration of cross channelinduced nonlinear PMD in WDM system," *Electron. Lett.*, 37(5), 306, March 2001.

18

Electronic signal processing for dispersion compensation and error mitigation in optical transmission networks

Abhijit Shanbhag, Qian Yu, and John Choma

Scintera Inc., Sunnyvale, CA, USA

Abstract

The use of smart electronic signal processing is emerging as a key technology to enable economical, high-performance transceivers for high-speed optical networks (10 Gb/s and beyond). This chapter provides an overview of adaptive equalization and other electronic compensation techniques and the associated broadband implementation techniques and challenges for optical transmission systems. Furthermore, we survey the role, scope, limitations, and challenges of this technology for enterprise, access, and core networks, to enable cost-effective optical transceivers and to reuse the installed fiber infrastructure in current and next-generation systems. Some of the state-of-the-art technology results in these areas will also be presented.

18.1 INTRODUCTION: ROLE OF ELECTRONIC SIGNAL PROCESSING IN OPTICAL NETWORKS

The worldwide increase in the number of wireline and wireless voice and broadband subscriber base together with the introduction and success of a large number of new broadband applications and services over the last few years have led to increasing demand for bandwidth within enterprise, access, storage, and core networks. This has resulted in a myriad of industry-wide activity to deliver the most cost-effective optical networking systems carrying data rates of 10 Gb/s and even beyond [1–5]. Data transmission at 10 Gb/s and beyond across the installed fiber infrastructure can result in intersymbol interference (ISI) within the fiber optic systems. An effective approach to mitigate such signal impairments uses adaptive electronic equalization at appropriately high frequencies (multi-GHz). The ability to electronically process faithfully input signals whose spectra contain very high signal frequencies and build high-speed electronic signal-processing architectures continues to be an underscored focus of signal-processing integrated circuit (IC) research and development.

This chapter provides a fairly broad overview, including the potential, feasibility, the current state-of-the-art and future trends, of electrical signal processing techniques that can be used to mitigate such ISI effects within a range of multi-Gb/s optical applications of current and future importance. Adaptive signal processing and, in particular, electronic equalization have found widespread applications in wireless communications, disc drives, cable and DSL modems, radar, sonar, etc. The discussion in this chapter will focus on practical engineering methodologies for applying some of these techniques in multi-Gb/s optical networks and will provide an update on the current industry-wide efforts which use such high-speed adaptive electronic processing for different optical networking applications. Portions of this chapter are taken from our earlier review paper [6] (© [2006] IEEE/OSA).

A significant effort has been expended in industry and academia to identify electronic signal processing as a cost-effective technique for upgrading data transmission to 10 Gb/s for various applications over installed fibers [7–15]. These applications include local area networks (LAN), storage area networks (SAN), metro-area networks, and long-haul systems. The dominant installed fiber infrastructure for LAN and SAN is multimode fiber (MMF) and for metro and long-haul is single-mode fiber (SMF). Such different installed media lead to very different engineering challenges due to different dispersion environment and introduce very different performance bounds. In addition to enabling upgrades to 10 Gb/s over installed fiber, there is also current activity in the industry in using plastic optic fiber (POF) with electronic signal processing as the most cost-effective and most power-efficient technique to enable 10-Gb/s transmission within data centers and smaller enterprises, as compared with 10GBASE-T over unshielded twisted copper pairs. Furthermore, there is also some effort on the use of electronic signal processing in 10 Gb/s Ethernet Passive Optical Networks (10GEPON) for Access Networks.

The electronic signal-processing techniques can be broadly classified as adaptive equalization at the receiver, predistortion at the transmitter, and electric-field domain signal processing. The electronic dispersion compensation (EDC) at the receiver can be most conveniently designed to be fully adaptive and, due to its ease of use and attractive economics, this chapter will emphasize this approach.

18.2 ELECTRONIC EQUALIZATION AND ADAPTATION TECHNIQUES

18.2.1 Overview

The equalization theory for linear time-invariant channels is well understood [16, 17]. A few important issues concerning application of this theory to different optical networks may be listed as follows:

(1) Implementation considerations: Because the target data rates are high (multi-Gb/s), the equalization architectures that can provide ideal performance could be infeasible or too complex to implement, especially in the classical form at such high speeds.
(2) Low bit-error rate (BER) application: The link using MMF, which is commonly installed in enterprise networks, is only statistically characterized and requires an uncoded BER of 10^{-12} or less. This requires a very robust equalizer adaptation that is carried out blind, that is, without a predetermined sequence of data symbols, and that can converge to a good state accurately for a wide range of channels.
(3) Nonclassical channel: The link using SMF in metro and long-haul networks is typically a nonlinear channel in the electrical domain due to the use of square-law intensity detection, and as such is nonclassical.

In the remaining subsections of this section, we will provide a quick introduction to some of the equalizer structures and adaptation techniques that are considered for different optical networking applications. The signal-processing models of EDC depend on the properties of optical sources, transmission fibers, and noise sources. Typically, an MMF link can be treated as a linear baseband channel with additive Gaussian noise. The channel model suitable for an optically amplified SMF transmission system will be explained in Section 18.4.

Throughout this section, we assume the received signal $r(t)$ in the electrical domain is given by

$$r(t) = \sum_{k=-\infty}^{\infty} a_k p(t - kT) + n(t), \qquad (18.1)$$

where t is time, a_k are transmitted symbols, $p(t)$ is the continuous-time pulse response of the channel, $1/T$ is the symbol rate, and $n(t)$ is additive Gaussian noise. Furthermore, we assume $p(t)$ is a rectangular pulse rec(t) (height = 1, width = T) passed through a baseband filter that represents the combined effects of transmitter shaping, fiber propagation, and receiver filtering. In optical systems using low-cost laser devices, the transmitter's relative intensity noise (RIN) is important, and the total received noise is not a white noise. We assume the double-side power spectrum of $n(t)$ is $N_0 S_n(f)$ in which $S_n(f)$ is normalized so that $S_n(0) = 1$, and the symbols $a_k = \pm a$ are independent identically distributed (i.i.d.) binary random variables.

18.2.2 Linear Equalizer

The ISI of a linear channel can be removed by a linear filter that inverts the channels' frequency response. When noise is taken into account, the optimum linear equalizer (LE) in theory is a continuous time-matched filter followed by a T-spaced sampled discrete-time filter. The Q-factor of an ideal LE (with infinite number of taps) output sample is given by Refs [16, 17].

$$Q^2(\text{ideal LE}) = \frac{a^2}{N_0} \left[\int_{-1/(2T)}^{1/(2T)} \frac{1}{S(f)} df \right]^{-1} \tag{18.2}$$

where

$$S(f) = \frac{N_0}{a^2 T} + \frac{1}{T^2} \sum_{k=-\infty}^{\infty} M\left(f + \frac{k}{T}\right) \tag{18.3}$$

$$M(f) = \left| \int_{-\infty}^{\infty} p(t) e^{i2\pi ft} dt \right|^2 / S_n(f) \tag{18.4}$$

Note that $N_0 S_n(f)$ is the power spectrum of noise, $M(f)$ is the frequency spectrum of the matched filter's output pulse, and $M(0) = T^2$. In order to achieve a BER $< 10^{-12}$, it is necessary to have $20 \log_{10} Q > 17$ dB.

The LE normally has a structure of transversal filter, which has a tapped delay line and sums the delayed copies of input signal with adjustable weights (Figure 18.1). It is conceptually similar to the finite-impulse-response (FIR) filter in digital signal processing (DSP). The following are some of the important considerations in selecting or designing an appropriate transversal filter for optical applications:

- Tap spacing: The tap spacing for an analog transversal filter in most optical applications is desired to be less than a symbol period to reduce noise enhancement, because the LE carries out matched filtering in addition to ISI suppression. For a digital or sampled implementation, the subsymbol tap spacing also provides more robustness particularly when the channel impulse response has significant energy beyond the Nyquist frequency. However, as the tap spacing is reduced, the adaptation can become unstable as the channel correlation matrix can become singular. Generally, a selection of the tap spacing between $0.3T$ and $0.6T$ provides for a reasonable trade off.
- Number of taps: Once the tap spacing is determined, the minimum number of taps needs to be selected so that the span of the LE exceeds the delay spread of the channel. However, as the taps are increased (in addition to added

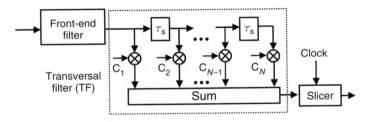

Figure 18.1 Schematic diagram of a linear equalizer.

implementation complexity and increased power consumption), there can be adaptation-related nonidealities such as slow convergence or even unstable behavior. A recommended value is about 25–50% more than is required to cover the delay spread. Again, this requires prudent engineering analysis to determine the number of taps in the different applications.
- Analog vs digital implementation: The analog implementation generally has significant power advantage, with some possible sacrifice in performance, as compared with digital implementation, especially at the process corners.
- Adaptation algorithm for the tap weights: This will be further discussed in Section 18.2.7.

18.2.3 Decision Feedback Equalizer

Decision feedback equalizer (DFE) is a nonlinear equalizer that may provide significantly better performance than an LE in the presence of severe ISI. The idea of decision feedback is that, if decisions on past symbols have been made, the past-symbol-induced ISI on the current symbol can be canceled before a decision is made on the current symbol. While the fractionally spaced LE in the earlier subsection compensates for the delay distortion, the DFE can provide significant performance gain in the presence of amplitude distortion and in relatively low signal-to-noise ratio (SNR) environment. Figure 18.2 shows a DFE structure that consists of a linear feed-forward equalizer (FFE) followed by a nonlinear feedback equalizer. The FFE is used to transform the received pulse to a new pulse with minimum ISI induced by future symbols. The feedback equalizer adjusts the slicer's input level symbol-by-symbol based on known past symbols. The optimum DFE can be found from the minimum-phase spectral factorization of $S(f)$ [16, 18]. The output Q-factor of the ideal DFE is given by Ref. [19]

$$Q^2(\text{ideal DFE}) = \frac{a^2 T}{N_0} \exp\left\{ T \int_{-1/(2T)}^{1/(2T)} \ln[S(f)] \mathrm{d}f \right\} \qquad (18.5)$$

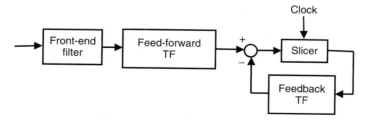

Figure 18.2 Schematic diagram of a decision feedback equalizer.

Throughout this chapter, the DFE is defined as the combination of feed-forward and feedback equalizers.

The following are some of the important considerations in selecting or designing an appropriate DFE for optical applications:

- Tap spacing within the FFE: This is again dictated by the trade off between the desire to reduce noise enhancement and correspondingly to select a smaller tap spacing, which results in SNR enhancement due to matched filtering, vs adaptation nonidealities. A selection of the tap spacing between $0.3T$ and $0.6T$ again provides for a reasonable trade off.
- Number of taps for the FFE: This is dictated mainly by the precursor delay spread within the channel vs implementation complexity.
- Number of taps in feedback: This is mostly dictated by the postcursor delay spread within the channel vs the implementation complexity. Commonly, the span of the feedback taps is less than the span of the FFE within the DFE.
- Analog continuous-time vs digital sampled implementation: The analog implementation generally can have significant power advantage, at the possible expense of performance, as compared with digital implementation, especially at the process corners. In particular, analog-sliced continuous-time feedback can provide an attractive low-power alternative to the more common sampled feedback approach. This will be further discussed in Section 18.3.3.
- Adaptation algorithm for the tap weights: This will be further discussed in Section 18.2.7.

18.2.4 Maximum Likelihood Sequence Estimation

The maximum likelihood sequence estimation (MLSE) makes decision to choose the most probable sequence of symbols among all sequences that could be transmitted [20]. The Viterbi algorithm is an efficient implementation of the MLSE for a discrete-time channel [21]. In order to apply the Viterbi algorithm, a linear filter is required to turn the continuous-time channel into an equivalent discrete-time channel whose output samples are statistically independent. The optimum prefilter in theory is referred to as the whitened matched filter (WMF) that can be found from the minimum-phase spectral factorization of $S(f)\ |_{N_0=0}$ [16]. The discrete-time samples y_k at the output of the WMF is given by

$$y_k = \sum_{j=0}^{L} d_j a_{k-j} + n_k, \qquad (18.6)$$

where $\{d_j\}$ is a minimum-phase FIR filter with $d_0 = 1$, and $\{n_k\}$ is white Gaussian noise. Given the observation sequence $\{y_k\}$, the MLSE chooses the symbol sequence that minimizes the squared Euclidean distance

$$\sum_{k=0}^{K} |y_k - \tilde{s}_k|^2 \text{ with } \tilde{s}_k = \sum_{j=k-L}^{k} d_{k-j}\tilde{a}_j, \quad (18.7)$$

among all trial sequences $\{\tilde{a}_j\}_{j=0}^{K}$ for sufficiently large K. The complexity of Viterbi algorithm is proportional to 2^L. Note that the squared Euclidean distance is proportional to the log-likelihood for Gaussian noise. A more general expression of the metric optimized by MLSE is given in Eqn (18.41).

The performance of MLSE cannot exceed the matched filter bound given by Refs [16, 17]

$$Q^2(\text{MF}) = \frac{a^2}{N_0} \int_{-\infty}^{+\infty} M(f) df, \quad (18.8)$$

which is the maximum SNR for detecting a single isolated pulse among noise. Although MLSE outperforms DFE, it is still difficult to equalize a channel that has spectral nulls or cutoff in the folded spectrum $S(f)\,|_{N_0=0}$ because the WMFs output noise power approaches infinity in this case.

The WMF is commonly implemented as a linear transversal filter. The span of this filter is again commonly selected to be approximately the delay spread of the channel and the tap-spacing at about $0.5T$. The sequence detector which follows the WMF is conceived mostly in the digital domain and would thus require a high-speed ($\geq 10\,\text{GSample/s}$) analog-to-digital converter (ADC).

18.2.5 Maximum A Posteriori Equalization

The optimal equalizer in terms of minimizing the uncoded BER is the *Maximum A Posteriori* (MAP) symbol detector. Again, as in the earlier subsection, if the discrete-time sample y_k at the output of the WMF is given by Eqn (18.6), the MAP equalizer chooses symbol-by-symbol detection, in the following form:

$$\hat{a}_n = \min{}_q \left\{ \sum_{\{\tilde{a}_j\}, \tilde{a}_n = q} \exp\left(\sum_{k=0}^{K} |y_k - \tilde{s}_k|^2 \sigma_n^{-2}\right) \right\}, \quad (18.9)$$

where the left side "Σ" sums over all trial sequences $\{\tilde{a}_j\}_{j=0}^{K}$ with fixed $\tilde{a}_n = q$, $q \in \{a, -a\}$, and σ_n^2 is the variance of noise n_k. An efficient implementation of the MAP equalizer uses the BCJR algorithm (also referred to as Forward–Backward algorithm) wherein, similar to the Viterbi algorithm, the complexity grows linearly rather than exponentially with the sequence size, but grows exponentially with the channel memory [22]. Unlike the Viterbi algorithm, which moves forward only along the trellis structure, the BCJR algorithm moves forward and backward along the trellis structure and requires the entire sequence of received samples to be available before decisions can be made [23].

The MAP equalizer provides the *a posteriori* error probability of each symbol. This information about the reliability of the decision can be used by a soft-decision forward error correction (FEC) code to substantially improve the system performance with FEC. However, for uncoded systems (or systems using hard-decision FEC), the gain of the MAP equalizer over the MLSE tends to be vanishing. Note that the MAP equalizer, like the MLSE, requires the channel estimate integrated within constructing the branch metrics, which further increases the complexity of developing an adaptive MAP equalizer for channels with unknown *a priori* characteristics.

18.2.6 Variants on the Theme

While the MAP- and MLSE-based equalizers provide the theoretical optimal performance for an uncoded system, their application may get somewhat limited as equalizers for uncoded 10 Gb/s and beyond, due to adaptation challenges down to low BER, high complexity leading to high power, and expected difficulty to produce in production environment. The performance with these techniques can fall quite steeply with implementation nonidealities. A number of other equalization architectures have also been considered which can potentially provide a more attractive performance-complexity trade offs. A few of the more promising architectures are listed below:

- Fixed Delay Tree Search: This is an equalization scheme which uses a sequence detector or a tree-search estimator replacing the slicer within the DFE-based architecture. While this scheme has been observed to provide performance close to the maximum-likelihood bound for optical channels at a lower complexity than MLSE, the challenge continues to be a robust adaptation algorithm in addition to the increased datapath complexity.
- DFE with precursor canceller: In this approach, just as the effect of the postcursor ISI is cancelled after the decision on the postcursor symbols are made, tentative decisions are made on the precursor symbols as well, and the precursor ISI is also cancelled. The performance of this scheme can be limited especially by error propagation from the tentative precursor symbols, and it is important to design the tentative decision block prudently.

18.2.7 Adaptation Techniques

Commonly, the channels for optical communications that need equalization or ISI compensation are not precisely known *a priori* and are frequently time-varying as well. Adaptation techniques are, thus, required to control the different weights within the range of equalizer architectures that have been described.

An adaptive filter is defined by the following aspects:

- The input and output signals within the filter;
- The filter structure which defines how the output signal is derived from the input;
- The adapting parameters within the filter structure;
- The figure of merit which needs to be optimized by the adaptation technique;
- The form of the adaptation algorithm which is commonly some kind of an optimization routine;
- The control parameters input into the adaptation algorithm;
- If the adaptation algorithm is assisted by an *a priori* known sequence/training sequence, or if it is blind (including possibly, decision-directed);
- The hardware–software and analog–digital partitioning of the adaptation technique.

The performance metrics for an adaptive filter or technique which guides some of the choices above include

- misadjustment from optimal point in steady-state performance;
- stability (both numerical and dynamical);
- initial convergence time;
- tracking error.

The figure of merit which needs to be optimized by the adaptive algorithm is the system BER. However, the BER surface as a function of the adapting coefficients can be highly multimodal. Furthermore, it can be quite complex to build hardware-based adaptation to minimize the BER and it is more feasible to develop a software-based approach. Thus, such techniques can suffer from slow adaptation speed for some applications.

A simple adaptation technique relies on a performance monitor such as the eye opening or BER to obtain the gradient information for tap weights adjustment. For example, if FEC is used in the system, the BER before error correction is an accurate performance monitor. FEC may also provide bit-pattern-dependent error probabilities that are useful for adaptation [24]. The eye opening can be directly measured from a synchronous histogram using a sample-and-hold (S/H) circuit followed by analog-to-digital (A/D) conversion. The sampling rate is a fraction of the bit rate and is slow enough to facilitate A/D conversion. The synchronous histogram also provides performance metrics like the mean-square error and the Q-factor. An alternative scheme of eye monitor requires only one variable-threshold decision circuit followed by an integrator [25]. A synchronous histogram monitor using this idea is shown in Figure 18.4. If the binary symbols ($+a$) are transmitted at equal probability, the histogram is measured from the variation of the decision circuit's output DC offset as a function of the decision threshold (see Figure 18.3). The adaptation

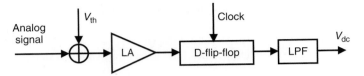

Figure 18.3 Schematic diagram of a synchronous histogram monitor using a variable-threshold limiting amplifier (LA) followed by a D-flip-flop. The low-pass filter (LPF) needs a narrow bandwidth, for example, 10 kHz, to extract the output DC offset. The histogram is measured from $\Delta V_{dc}/\Delta V_{th}$ vs V_{th} (after Ref. [6], © 2006 IEEE/OSA).

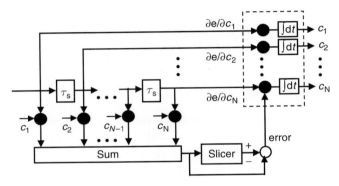

Figure 18.4 Schematic diagram of an LMS adaptive transversal filter (after Ref. [6], © 2006 IEEE/OSA).

of equalizers using a performance monitor is important for ≥ 40 Gb/s systems, because other adaptation techniques are hardly feasible in practice at such high speeds.

Due to the simplicity in hardware implementation, it is far more common to minimize the error signal power, where the error signal is the difference between the desired data signal and the equalizer output. This minimum mean-square-error (MMSE) criterion is quadratic in the adaptive filter coefficients within the LE and the DFE, in the presence of a known desired signal, and ensures a unique optimum when the corresponding channel correlation matrix is nonsingular. In the absence of an explicitly provided desired data signal, the error signal is the difference between the hypothesized data signal and the equalizer output before the decision device. In such a case of blind adaptation with decision-directed operation, it is known that the MSE surface, in the presence of severe ISI, can result in a multi-modal error surface.

A common adaptation technique for a number of equalizers including the LE and the DFE based on the MMSE criterion is the least-mean-squares (LMS) adaptation. The stochastic-gradient LMS algorithm requires the correlation of the error signal with the gradient of the error signal, which in the case of LE and

DFE is simply the delayed copies of the transversal filter's input signal. Mathematically,

$$\underline{c}_{n+1} = \underline{c}_n + \mu \cdot e_n \cdot \underline{X} \tag{18.10}$$

where $\underline{c}_n$ is the adapting coefficient vector in the nth iteration, e_n is the corresponding error signal and $\underline{X}$ includes the delayed copies of the input signal which are weighted by the coefficient vector $\underline{c}_n$. μ is an adaptation parameter which is set based on the desired adaptation speed, stability, and tolerable misadjustment for different classes of optical channels. There are considerable variants from this simple form to allow for one or more of further implementation simplicity, stability, fast convergence, and/or tracking.

An LE using LMS adaptation is shown in Figure 18.4, where we assume the multiplier, summer, and slicer have zero latency for simplicity. Timing alignment is important in the practical implementation of LMS adaptation in multi-Gb/s systems. The error signal for LMS adaptation is a combined effect of residual ISI and noise. When the required uncoded BER is very low ($<10^{-12}$) so the system operates at high SNR, the LMS adaptation tends to minimize the residual ISI at the cost of excessive noise enhancement and suboptimum BER. The LMS adaptation can also become rather suboptimum relative to adaptation approaches that minimize the BER, if the channel to be equalized has significant nonlinear ISI.

In certain applications, the transmitter can send a known sequence of symbols periodically, and the channel's pulse response can be estimated from the received waveform and known training sequence using an LMS algorithm. Adaptive maximum-likelihood sequence detection is achievable using the periodically updated channel estimates.

18.3 HIGH-SPEED ELECTRONIC IMPLEMENTATION: TECHNIQUES, ALTERNATIVES, AND CHALLENGES

18.3.1 Overview

There are several different dimensions relating to implementation of adaptive electronic signal-processing solutions at such high speeds, apart from the specific form of signal processor from different broadband design techniques to architectural choices. Successful implementation of such high-speed electronic signal processors requires careful codesign in the systems, circuits and device design paradigms, or abstraction levels.

The process technology choice is clearly a critical one which can constrain the architectural design choices. When the electronic signal processor is required to operate as high as a few tens of gigahertz, the natural inclination may be to select a lumped circuit-based design methodology and some fabrication process with unity-gain frequency that is more than five times the maximum frequency of interest and then Silicon–Germanium (SiGe), Gallium Arsenide and other III–V compound devices can be the candidates. However, from many standpoints, it is far more desirable to use CMOS as the process of choice, particularly for high-volume enterprise, storage, and possibly metro applications. The large number of foundries for CMOS and aggressive cost scalability across high-volume applications has led to CMOS to be the most mature and lowest cost solution for any high-volume application. The complicated multimask process for bipolar technologies may limit yield and, in addition, digital circuits implemented in bipolar technology tend to dissipate large amounts of power.

The reasonably high unity-gain bandwidth of deep-submicron MOS transistors, due to aggressive scaling to follow Moore's law which has increased the unity-gain bandwidth of MOS transistors by more than three orders of magnitude over the last three decades, coupled with pragmatic bandwidth-enhancing circuit design strategies, as discussed in this section, can result in adequate gain-bandwidth for each individual stage within different signal processor functions. The several metal layers provided within the deep submicron processes (e.g., 8 for a 0.13-μm CMOS process) can be used to design passives such as inductors and transmission lines, and capacitors, within these stages, are important components to help overcome the speed limitation with CMOS technology. Although MOS technology devices are exclusively addressed in the sections of material that follow, the concepts disclosed apply equally well to other monolithic device technologies, inclusive of conventional bipolar, SiGe heterostructure bipolar, and III–V compound transistors. Next, we provide an overview of the salient circuit concepts and theories that underlie the design of broadband amplifiers.

18.3.2 Broadbanding Techniques

Principles of Broadbanding Techniques

Virtually, all broadbanding strategies exploit one or both of two fundamental network theoretic principles. The first of these principles is that negative feedback applied around a dominant-pole open loop produces a closed loop that projects an improved frequency response. Indeed, the closed loop 3-dB bandwidth of such feedback architectures is larger than the 3-dB bandwidth of their open loop counterparts by a factor of one plus the zero frequency value of the loop gain. There are, however, inherent limitations to broadband feedback compensation measures. The primary limitation stems from the fact that the realization of a single-pole open loop amplifier becomes an increasingly more daunting design challenge as system performance specifications impose wider passband requirements. In particular, global feedback applied around a broadband active network generally manifests a diminished damping factor associated with the interaction of the presumably dominant pole and the unavoidable higher frequency poles implicit to the network. In addition to affording a bandwidth improvement that is less dramatic than that predicted for the case of true pole dominance, a smaller damping factor compromises the integrity of the frequency and time domain responses and portends of potential circuit instability.

The second principle underlying prudent broadbanding is that the high-frequency performance of dominant-pole amplifiers featuring no finite frequency zeros is inherently limited. When no finite frequency zeros are evidenced in a dominant-pole amplifier, broadbanding reduces to the problems of determining the designable parameters that affect the frequency of the dominant pole and then adjusting an appropriate subset of these variables to incur an increased dominant-pole frequency. But the frequencies of the presumably nondominant network poles are influenced (and invariably decreased) by the parametric changes adopted to effect an increased dominant-pole frequency. Even if these other frequencies are magically unaffected, moving the dominant pole to higher frequencies means that the originally dominant pole is displaced to a region of the complex frequency plane that is populated by higher order poles. The immediate effect of this repopulation of poles in the complex frequency plane is a nondominant pole amplifier whose bandwidth is difficult to predict reliably because of its vulnerability to numerous, often ill-controlled and nebulously defined, active device and circuit parameters.

An important dimension relating to foundations of circuit design relate to the design of analog wideband datapath, which can be classified as distributed vs lumped circuit topology. A distributed architecture uses multiple parallel or serial "lumped circuits" as the signal paths working harmoniously to achieve a certain function. An example of the distributed architecture includes the transmission line which has commonly been used in III–V technologies for microwave applications. A distributed circuit methodology allows better delay-bandwidth and gain-bandwidth trade offs. Distributed analog signal-processing approaches using deep submicron CMOS have been considered within the last few years for optical interconnects, amplifiers, oscillators, and equalizers. While significant benefits accrue with distributed design approaches at such high frequencies, it is imperative to have very careful attention to layout techniques, careful modeling of active and passive components and minimization of layout parasitics and noise, especially in the high-speed datapath across production corners.

RC Broadbanding

The common source stage is the workhorse of MOS technology systems requiring significant I/O gain. Figure 18.5(a) offers a common source topology in which a degeneration resistance, R_f, is inserted into the source lead of the transistor. This resistance acts as a negative feedback element to extend the

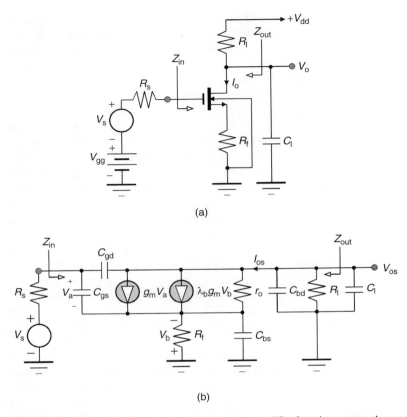

Figure 18.5 (a) Basic schematic diagram of a common source amplifier featuring a source degeneration resistance, R_f. (b) Small signal, high-frequency model of the amplifier in (a). The transistor is presumed biased in saturation (this figure may be seen in color on the included CD-ROM).

3-dB bandwidth of the basic common source stage at the expense of reducing the zero frequency gain of the stage. As is confirmed ultimately, the degeneration resistance also allows for the convenient introduction of a left half-plane zero that extends the degenerated bandwidth without incurring an additional gain penalty. Assuming transistor operation in its saturation regime, Figure 18.5(b) is the pertinent small signal model, in which resistance R_l is the drain load resistance, C_l is the capacitance loading the amplifier output port, and R_s is the Thévenin resistance of the signal source. The transistor is represented by its traditional small signal parameters. In particular, g_m is the forward transconductance, λ_b is the bulk transconductance factor, r_o is the small signal channel resistance, C_{gs} is the gate-source capacitance, C_{gd} is the gate-drain capacitance, C_{bd} is the bulk-drain capacitance, and finally, C_{bs} symbolizes the bulk-source capacitance.

If the amplifier in Figure 18.5(a) delivers a dominant pole frequency response, an analysis of the model in Figure 18.5(b) leads to a voltage gain, $A_v(s)$, of

$$A_v(s) \triangleq \frac{V_{os}}{V_s} = \frac{A_v(0)}{1 + (s/B_u)}, \quad (18.11)$$

where V_{os} is the small signal component of the net output voltage, V_o. In Eqn (18.11),

$$A_v(0) = -\frac{g_m(r_o \| R_l)}{1 + (R_f/(R_l + r_o))[1 + (1 + \lambda_b)g_m r_o]} \approx -\frac{g_m R_l}{1 + g_m R_f} \quad (18.12)$$

is the zero frequency gain of the amplifier, and the approximated result reflects the reasonable presumptions that $r_o \gg R_l$, $\lambda_b \ll 1$, and $g_m r_o \gg 1$. Moreover, B_u is the uncompensated bandwidth of the network; that is, the 3-dB radial bandwidth before invoking appropriate broadband compensation elements.

The uncompensated bandwidth is commonly estimated as the sum of the open circuit time constants associated with the capacitors of the small signal equivalent circuit [26]. In any active RC network, a necessary condition for pole dominance is that one of these open circuit time constants dominates all other computed time constants. In the model at hand, several reasons can be offered to justify the contention that the net capacitance prevailing at the network output port establishes the dominant open-circuit time constant. First, this net output capacitance is terminated in a reasonably large resistance, if the zero-frequency voltage gain is at least moderately large. Second, the effective resistance facing the net capacitance incident with the source terminal is of the order of $1/g_m$ and is therefore relatively small. Third, Miller multiplication of the gate-drain capacitance is likely to be inconsequential because the transconductance, g_m, of the transistor is attenuated by a factor of $(1 + g_m R_f)$. Finally, the resistance facing the gate-source capacitance, C_{gs}, is smaller than the signal source resistance, R_s, which is rarely large in a high-frequency circuit application. Thus, for large channel resistance and small bulk transconductance factor,

$$B_u \approx \frac{1}{R_l(C_l + C_{bd} + C_{gd})}. \quad (18.13)$$

Clearly, the bandwidth predicted by Eqn (18.13) can be extended through either a reduction of the net capacitance $(C_l + C_{bd} + C_{gd})$ or a reduction in the load resistance, R_l. Unfortunately, the circuit designer typically has little control over the net output port capacitance, while a reduction in load resistance R_l is consummated by a proportionate reduction in the zero-frequency gain.

A broadbanding clue derives from the observation that the net load capacitance reduces the load impedance from its zero-frequency value of R_l to a value that approaches zero at very high frequencies. Since the zero-frequency gain is the ratio of the load impedance to a linear function of resistance R_f, it is logical to consider replacing R_f in the amplifier by a shunt interconnection of R_f and an appropriate capacitance, say C_f, as shown in the schematic diagram of the proposed compensated amplifier in Figure 18.6(a). To the extent that the gain is directly dependent on the ratio of load impedance to feedback impedance, at least at low signal frequencies, a feedback impedance time constant matched to the time constant of the load impedance allows the feedback impedance to track the load impedance over frequency, whence a broadbanded amplifier frequency response.

The simplified small signal model for the structure in Figure 18.6(a) appears in Figure 18.6(b). The model is simplified in that the early resistance, r_o, and bulk transconductance factor, λ_b, are tacitly ignored, as are all device capacitances that do not contribute to the dominant time constant established at the amplifier output port. Moreover, capacitances C_{lo} and C_{fo} are given by

$$\left.\begin{array}{l} C_{fo} = C_f + C_{bs} \\ C_{lo} = C_l + C_{bd} + C_{gd} \end{array}\right). \tag{18.14}$$

An analysis of the subject model leads readily to a compensated gain, $A_{vc}(s)$, of

$$A_{vc}(s) = \frac{V_{os}}{V_s} = \frac{A_v(0)(1 + sR_fC_{fo})}{(1 + sR_fC_{fo}/(1 + g_mR_f))\left(1 + \frac{s}{B_u}\right)}, \tag{18.15}$$

where $A_v(0)$ is given by Eqn (18.12) and bandwidth B_u of the uncompensated amplifier is defined by Eqn (18.13). If the time constant, R_fC_{fo}, established by the

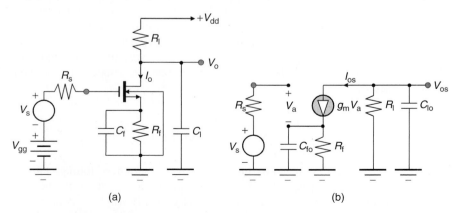

Figure 18.6 (a) Schematic diagram of compensated degenerative amplifier. (b) Approximate small-signal, high-frequency equivalent circuit of the amplifier in (a) (this figure may be seen in color on the included CD-ROM).

net compensation capacitance, C_{fo}, is selected so that it equates to the inverse frequency, $1/B_u$, of the uncompensated dominant pole, the resultant compensated bandwidth, B_c, is

$$B_c = \frac{1 + g_m R_f}{R_f C_{fo}} = (1 + g_m R_f) B_u \qquad (18.16)$$

The result indicates that the simple action of appending a capacitance across the feedback resistance in the source lead of the transistor can extend the uncompensated bandwidth by as much as one plus the product, $g_m R_f$, which is the zero-frequency loop gain of the amplifier.

Engineering care must be exercised when exploiting the foregoing result because the pole-zero cancellation on which it is premised is imperfect. In particular, Eqn (18.16) derives from Eqn (18.15), which presumes that pole dominance prevails in the original uncompensated amplifier. Imperfections also abound with respect to equating time constant $R_f C_{fo}$ to inverse B_u. For example, capacitance C_{fo} is a function of transistor bulk-source capacitance C_{bs}, which is rarely known accurately. Moreover, bandwidth B_u relies on capacitance C_{lo}, which in turn is functionally dependent on transistor bulk-drain and gate-drain capacitances C_{bd} and C_{gd}, respectively, whose precise numerical values, like that of C_{bs}, are elusive. When $B_u R_f C_{fo}$ is precisely one, idealized pole-zero cancellation is achieved, whereby Eqn (18.16) accurately reflects the factor by which the uncompensated bandwidth is widened. If $B_u R_f C_{fo}$ is larger than one, perhaps because of an inaccurate estimate of either B_u or C_{fo}, the uncompensated bandwidth is nonetheless extended, but only at the expense of potentially unacceptable gain peaking. Finally, $B_u R_f C_{fo} < 1$ results in a compensated bandwidth that can be significantly smaller than its uncompensated version. A pragmatic mitigation of these issues entails incorporating a capacitance adjustment capability in either C_l or C_f to fine-tune the ultimately realized frequency response.

Shunt-Peaked Broadbanding

The preceding section illustrates the positive impact exerted on the frequency response of a common source amplifier by a suitable left half-plane zero introduced into the amplifier transfer function. In the foregoing discussion, the requisite zero is forged through shunt RC impedance degeneration inserted into the transistor source terminal. Unfortunately, several costs accompany the laudable bandwidth enhancement afforded by source RC degeneration. The first of these costs is the increased power dissipation resulting from a drain-biasing current that necessarily flows through the resistance inserted in the transistor source lead. A second cost is decreased voltage gain. In particular, the closed-loop gain is smaller than the amplifier open-loop gain by an amount that roughly equals the bandwidth improvement factor afforded by the incorporated feedback subcircuit. In other words, the gain-bandwidth products of the uncompensated and the compensated amplifiers remain nominally the same, thereby implying that gain is traded for enhanced bandwidth. Finally, resistance in the transistor source lead degrades the noise characteristics of the amplifier by increasing the equivalent input noise voltage. This noise voltage is a critical amplifier metric in that it defines the minimum input signal level that can be captured reliably by the network.

Fortunately, a few alternative means of incorporating transmission zeros into the transfer function of a common source amplifier mitigate most of the foregoing shortfalls. One such method, known as shunt peaking, installs the requisite zero through use of an inductor inserted into the transistor drain circuit, as depicted in Figure 18.7. Usually, shunt peaking does not result in a bandwidth enhancement that is as dramatic as that boasted by appropriate source degeneration. But the bandwidth improvement is nonetheless substantial and is afforded without compromising the zero-frequency gain of the amplifier. Shunt peaking therefore delivers increased bandwidth by actually increasing the gain-bandwidth product of the uncompensated, common source configuration. A further advantage of shunt-peaked compensation is, ignoring the small winding resistance implicit to the realization of the inductor, that the introduced inductance incurs no increase in static circuit power dissipation. Although inductor winding resistance is detrimental to circuit power dissipation, the quality factor of the utilized inductor, which can be poor in silicon monolithic realizations, is inconsequential to the small signal dynamics of the shunt-peaked amplifier because the inductance is inserted in series with the drain load resistance.

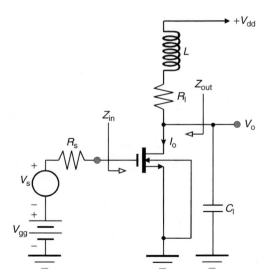

Figure 18.7 Basic schematic diagram of a shunt-peaked common source amplifier. The transistor operates in its saturation regime (this figure may be seen in color on the included CD-ROM).

Shunt peaking, as diagramed in Figure 18.7, is an effective broadbanding strategy, provided that the net capacitance incident at the output port, which includes the indicated load capacitance, C_l, and parasitic output port capacitances associated with the transistor and its layout, establishes the dominant time constant in the uncompensated (meaning $L=0$) amplifier. To this end, a potential concern is the Miller multiplication of the transistor gate-drain capacitance, for in deep submicron technologies, self-aligned gate processes do not eliminate entirely the gate oxide overlap with the drain region. When the dreaded Miller multiplication of the gate-drain capacitance in the common source transistor observably influences the 3-dB bandwidth of the uncompensated circuit, the common source–common gate cascode configuration diagramed in Figure 18.8 may prove efficacious, albeit at the expense of an increase in requisite static power dissipation. The circumvention of most of the Miller-multiplied capacitance problems stems from an effective load resistance seen by the common source transistor, M1, in Figure 18.8 that is nominally the inverse of the transconductance, g_{m2}, of the common gate device, M2. To the extent that $g_{m2}R_l \gg 1$, the cascode renders the time constant precipitated by the gate-drain capacitance of M1 inconsequential in comparison to the time constant observed at the amplifier output port. To be sure, the inclusion of transistor M2 manifests additional time constants that are not embodied by the simple shunt peak network in Figure 18.7. For example, the bulk-drain capacitance of M1, as well as the bulk-source, and gate-source

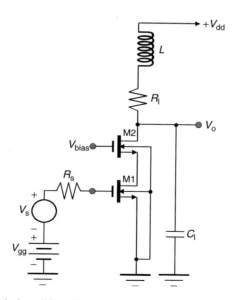

Figure 18.8 A shunt-peaked amplifier utilizing a common base cascode, M2, to mitigate the effects of Miller multiplication of the gate-drain capacitance of transistor, M1 (this figure may be seen in color on the included CD-ROM).

capacitances of M2, faces an effective resistance of roughly $1/g_{m2}$, which, as noted above, is presumably small. Nonetheless, the common gate current buffer is likely to improve the high-frequency response because of the potentially dramatic reduction of the Miller multiplication of the gate-drain capacitance in the common source driver.

In the shunt-peaked stage in Figure 18.7 or the cascode version shown in Figure 18.8, the amplifier driving the load comprising of capacitance C_1 in shunt with the series interconnection of resistance R_l and inductance L emulates a transconductor. If the resistively terminated transconductor functions as a dominant pole amplifier whose dominant pole frequency is determined by the net capacitance, C_{lo}, incident at the load port, the pertinent small-signal equivalent circuit is the structure suggested in Figure 18.9. An inspection of the subject diagram reveals a compensated voltage gain, $A_{vc}(s)$, of

$$A_{vc}(s) = \frac{V_{os}}{V_s} = -\frac{g_m R_l (1 + (sL/R_l))}{1 + sR_l C_{lo} + s^2 L C_{lo}}. \tag{18.17}$$

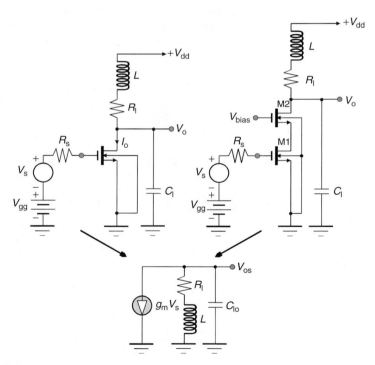

Figure 18.9 Approximate small-signal, high-frequency model of a common source or a common source–common gate cascode shunt-peaked amplifier. The indicated model presumes pole dominance established at the output port when inductance L is zero (this figure may be seen in color on the included CD-ROM).

Not surprisingly, the zero-frequency voltage gain is $A_{vc}(0) = -g_m R_1$. If gain $A_{vc}(s)$ is normalized to this zero-frequency value, Eqn (18.17) can be cast in the form,

$$A_n(s) \triangleq \frac{A_{vc}(s)}{A_{vc}(0)} = \frac{1 + (Qs/\omega_n)}{1 + (s/Q\omega_n) + (s/\omega_n)^2}, \qquad (18.18)$$

where

$$\omega_n = \frac{1}{\sqrt{LC_{lo}}} \qquad (18.19)$$

is the self-resonant frequency of the circuit and

$$Q = \frac{1}{\omega_n R_1 C_{lo}} = \frac{B_u}{\omega_n} = \frac{\sqrt{L/C_{lo}}}{R_1} \qquad (18.20)$$

is the circuit quality factor, which is seen to be the uncompensated bandwidth, B_u, normalized to the circuit self-resonant frequency. Equations (18.17) and (18.18) confirm that the insertion of the inductance in the drain load circuit of the uncompensated topology establishes a left half-plane zero that can be exploited for broadbanding purposes. But Eqn (18.18) also suggests that the compensated structure may exhibit undue frequency response peaking. The obvious source of this peaking is the left half-plane zero, which can lie at too low a frequency within the circuit passband if Q is too large and/or ω_n is too small. The second and more subtle cause of unacceptable peaking is that complex circuit poles, which arise for $Q > \frac{1}{2}$, conduce underdamped responses.

In order to circumvent issues related to underdamped responses, the shunt-peaked amplifier undergoing examination is generally designed to deliver a maximally flat magnitude (MFM) frequency response [27]. Such a response is attained when the circuit quality factor is set to a value, say Q_m, which is

$$Q_m = \sqrt{\sqrt{2} - 1} = 0.6436. \qquad (18.21)$$

For this value of circuit quality factor, the compensated bandwidth, B_c, which in general can be shown to derive from the relationship

$$\frac{B_c}{B_u} = \sqrt{\frac{(2Q^4 + 2Q^2 - 1) + \sqrt{(2Q^4 + 2Q^2 - 1)^2 + 4Q^4}}{2Q^4}}, \qquad (18.22)$$

is $B_c = 1.722 B_u$. Thus, shunt peaking implemented to ensure MFM response delivers a bandwidth that is 72.2% larger than the bandwidth of the uncompensated circuit. It is to be understood that this bandwidth enhancement is not the largest that can be achieved in a shunt-peaked amplifier. Instead, it is the largest achievable enhancement, subject to the constraint of no-frequency-response peaking, which is tantamount to requiring a monotonically decreasing magnitude response.

Series-Peaked Broadbanding

In a shunt-peaked amplifier, as diagramed in Figure 18.9, the inductive branch is placed in parallel with the amplifier output port to which the net capacitance serving to establish the dominant pole of the uncompensated amplifier is incident. In contrast, a series-peaked amplifier places the inductive branch in series with the load capacitance, C_l, that the amplifier is compelled to drive. As can be seen in the pertinent schematic diagram of Figure 18.10(a), the compensating inductance, L, separates the load capacitance from the amplifier output port capacitance, C_o, which is fundamentally the sum of the bulk-drain and gate-drain capacitances associated with the common gate cascode transistor, M2. The additional capacitance, C_x, appended to the amplifier output port proves indispensable with respect to achieving a broadbanded, MFM response in the compensated structure. The pertinent small signal model, given in Figure 18.10(b), is premised on the presumptions that the drain-source channel resistance of both transistors is sufficiently large to warrant its tacit neglect and bulk-induced threshold voltage modulation is negligible.

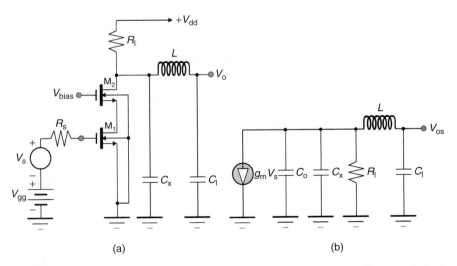

Figure 18.10 (a) Basic schematic diagram of a series-peaked amplifier. (b) Approximate small-signal, high-frequency model of the circuit in (a). Both transistors are presumed to operate in saturation, and pole dominance attributed to the net output port capacitance is presumed when $L=0$ (this figure may be seen in color on the included CD-ROM).

Additionally, with $L=0$, pole dominance attributed to the net output port capacitance is presumed. Accordingly, the uncompensated radial 3-dB bandwidth, B_u, is

$$B_u \approx \frac{1}{R_1(C_o + C_1)}. \tag{18.23}$$

Note that capacitance C_x is not included in this expression, for C_x is a required branch element in only the compensated, series-peaked topology.

A straightforward analysis of the model in Figure 18.10(b) sets forth a compensated voltage transfer function of

$$A_{vc}(s) = \frac{V_{os}}{V_s} = -\frac{g_m R_1}{1 + sR_1(C_o + C_x + C_1) + s^2 LC_1 + s^3 LC_1 R_1(C_o + C_x)}, \tag{18.24}$$

where, $-g_m R_1$ is confirmed as the zero-frequency value, $A_{vc}(0)$, of the voltage gain. A design goal of MFM compels the requirements,

$$\left.\begin{array}{l} \dfrac{2}{B_c} = R_1(C_o + C_x + C_1) \\[4pt] \dfrac{2}{B_c^2} = LC_1 \\[4pt] \dfrac{1}{B_c^3} = LC_1 R_1(C_o + C_x) \end{array}\right\}. \tag{18.25}$$

When combined with Eqn (18.23), these relationships imply a compensated circuit bandwidth, B_c, given by

$$\frac{B_c}{B_u} = \frac{2}{1 + (C_x/C_o + C_1)}, \tag{18.26}$$

which, assuming $C_x \ll (C_o + C_1)$, portends almost a doubling of the uncompensated circuit bandwidth. Equation (18.25) stipulates a requisite circuit inductance, L, and shunt output capacitance, C_x, of

$$\left.\begin{array}{l} L = \dfrac{2}{B_c^2 C_1} \\[4pt] C_x = \dfrac{C_1}{3} - C_o \end{array}\right\}. \tag{18.27}$$

Obviously, C_o is no smaller than zero, while from Eqn (18.27), C_o can be no larger than $C_1/3$. Accordingly, the factor by which the uncompensated circuit bandwidth can be improved through third-order, MFM compensation is between 1.5 and 2.0. In other words, third-order, series-peaked bandwidth compensation can conceivably exceed the bandwidth enhancement factor implicit to maximally flat shunt-peaked compensation.

Series Shunt-Peaked Broadbanding

An intriguing broadband circuit architecture that combines the attributes of series and shunt peaking and is capable of forging an MFM frequency response is the architecture depicted in Figure 18.11 [28, 29]. This configuration, which is an outgrowth of classic shunt and double series-peaking strategies, appends a bridging capacitance, C_f, across the two symmetrical coils to achieve a purely resistive load impedance, $Z_l(s)$, that terminates the drain of the common gate cascode transistor, M2. A resistive $Z_l(s)$ does not guarantee an acceptably flat frequency response for the small signal voltage gain, V_{os}/V_s. To this end, an additional design degree of freedom is forged by purposefully laying out the two coils to incur a controllable mutual inductance, M. This mutual inductance is characterized classically by the so-called coupling coefficient, k_c, between the two coils, such that

$$k_c = \frac{M}{L/2}. \qquad (18.28)$$

For $k_c = 0$, the inductors operate independently of one another (meaning that the electromagnetic fields surrounding one inductor do not envelope any portion of the coils implicit to the other inductor), while for $k_c = 1$, the inductors are maximally coupled, as in an ideal transformer. In silicon integrated circuits, spiral inductors can be laid out to yield coupling coefficients in the range of $k_c \leq 0.5$ [30]. However,

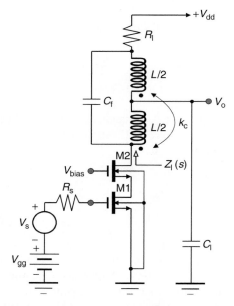

Figure 18.11 Series shunt-peaking architecture capable of realizing MFM frequency response. Observe the use of magnetically coupled inductors and the incorporation of a capacitance that bridges the coupled coils (this figure may be seen in color on the included CD-ROM).

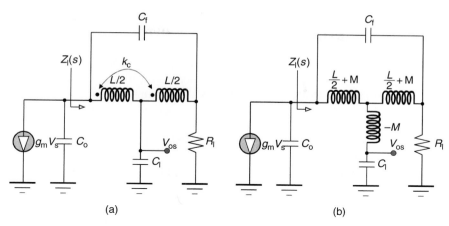

Figure 18.12 (a) Small signal, high-frequency equivalent circuit of the amplifier shown in Figure 18.11. (b) Alternative form of the model in (a) (this figure may be seen in color on the included CD-ROM).

proximately located bond wires can be positioned to achieve inductive coupling coefficients of $k_c \leq 0.7$.

The small-signal, high-frequency model of the amplifier in Figure 18.11 appears in Figure 18.12(a). In order to facilitate circuit analysis, the two coupled inductances can be represented electrically as the uncoupled tee inductor two-port network drawn in Figure 18.12(b). It is interesting to note that the second model suggests that the immediate effect of inductive coupling is to incur additional series peaking of the circuit load capacitance, C_1, thereby arguably leading to the possibility of improving the bandwidth established by conventional series or shunt peaking.

A conventional two-port analysis of the network appearing to the right of capacitance C_o in Figure 18.12(b) establishes the fact that the indicated impedance, $Z_l(s)$, is constant and identical to the load resistance, R_l, if capacitance C_f is chosen as

$$C_f = \frac{C_1}{4}\left(\frac{1-k_c}{1+k_c}\right) \qquad (18.29)$$

and inductor L is set in accordance with

$$L = \frac{R_l^2 C_1}{1+k_c}. \qquad (18.30)$$

Equations (18.29) and (18.30) comprise necessary and sufficient conditions underlying the realization of an effective load impedance, $Z_l(s)$, that is purely resistive and equal to the load resistance, R_l, for all signal frequencies. Note that this constant resistance characteristic can be achieved even when the two utilized inductors are uncoupled, that is, when $k_c = 0$. When the inductors are perfectly coupled in the sense of $k_c = 1$, Eqn (18.29) indicates that no bridging capacitance is required to achieve constant input resistance.

Using Eqns (18.29) and (18.30), the small signal voltage transfer function, $A_v(s) = V_{os}/V_s$, can be cast in a useful closed form format. This gain analysis is facilitated by the fact that Eqns (18.29) and (18.30) imply, $Z_l(s) \equiv R_l$ in Figure 18.12(b), which implies that the time constant associated with the transconductor output capacitance, C_o, is simply $R_l C_o$. If the inverse of this time constant is significantly larger than the desired compensated radial bandwidth, B_c, capacitance C_o can be ignored tacitly to simplify the analysis procedure. Then, it can be demonstrated that the compensated gain, normalized to its zero-frequency value, is

$$A_n(s) = \frac{1}{1 + (s/Q_c \omega_x) + (s/\omega_x)^2}, \qquad (18.31)$$

where

$$Q_c = \sqrt{\frac{1 - k_c}{1 + k_c}}, \qquad (18.32)$$

and frequency parameter, ω_x, is, recalling Eqn (18.23) for the uncompensated 3-dB bandwidth,

$$\omega_x = \frac{2}{\sqrt{LC_l(1 - k_c)}} = \frac{2}{Q_c R_l C_l} \approx \frac{2B_u}{Q_c}\left(1 + \frac{C_o}{C_l}\right). \qquad (18.33)$$

If circuit quality factor Q_c is set to the inverse of root two, the coupled inductor compensation scheme delivers a second-order, MFM frequency response whose radial 3-dB bandwidth is precisely ω_x. From Eqn (18.32), $Q_c = 1/\sqrt{2}$ is tantamount to the requirement, $k_c = 1/3$, which lies well within the coupling coefficient domain of monolithic inductor pairs. For this design constraint, Eqn (18.33) gives a resultant compensated bandwidth, B_c, of

$$B_c \equiv \omega_x \approx 2\sqrt{2}\left(1 + \frac{C_o}{C_l}\right) B_u, \qquad (18.34)$$

which suggests an impressive bandwidth enhancement factor approaching three.

It is important to stress that ω_x is the 3-dB bandwidth of the network represented by the transfer function in Eqn (18.31), only if Q_c is chosen to achieve MFM. In general, the compensated bandwidth, B_c, can be shown to be given by

$$B_c = \frac{2B_u}{Q_c}\left(1 + \frac{C_o}{C_l}\right) \sqrt{\sqrt{\left[\left(\frac{1}{2Q_c^2} - 1\right)^2 + 1\right]} - \left(\frac{1}{2Q_c^2} - 1\right)}. \qquad (18.35)$$

An assiduous investigation of Eqn (18.35) reveals an observation that is appealing from an integrated circuit design and layout perspective. In particular, the compensated bandwidth is not a particularly sensitive function of the coupling coefficient, whose precise numerical value is vulnerable to routinely encountered process vagaries and parametric uncertainties. For example, over the coefficient range, $0.1 \leq k_c \leq 0.5$, B_c/B_u varies from only 2.67 to its peak of 2.82 to 2.72.

18.3.3 Architectural Considerations

A fundamental architectural choice is the partitioning of the signal processing between the analog continuous-time, analog sampled, and digital domains. Conceptually, the achievable performance may be very close between these different computing paradigms. Analog continuous-time processing can generally lead to the lowest power solution with the smallest chip area at such high frequencies. However, the nonlinearities, noise, offsets with any bandwidth limitations, and other nonidealities need to be carefully considered for a robust design across process and temperature corners. Furthermore, because continuous-time processing is generally clockless, there may be need to accurately time-align different signal-processing data paths. Finally, the performance robustness and power considerations dictate the partitioning between continuous-time and discrete-time, analog and digital datapath.

Several high-performance equalizers such as the DFE and fixed delay tree-search equalizers, require low-latency feedback loops having latency of about a symbol interval. Realization of such low-latency feedback loops may be implemented in the continuous-time domain, where stability and accuracy may be concerns. Alternative approaches in the discrete-time convert the feedback to an equivalent feed-forward structure with two or more parallel paths [31]. While the latter structure can have benefits, adaptation can be a challenge.

Yet another dimension to carefully consider is analog vs digital adaptation. Analog adaptation uses analog signals to minimize a metric in the analog domain using steepest gradient or other optimization schemes. Digital adaptation uses discrete signals to minimize a metric in the analog domain or a metric that may be obtained only digitally. A commonly used metric is the mean-squared error, which can be obtained in the analog domain. The analog adaptation techniques may generally have faster time constants to adaptation but can be more vulnerable toward DC offsets [32], resulting in somewhat suboptimal adaptation. Digital adaptation typically would require larger IC area and may have slower convergence speeds. Again, some combination of analog and digital techniques for adaptation generally leads to an optimal solution.

All these different dimensions result in different architectural choices for each signal processor at multi-Gb/s data rates. The final selection may be guided as much by determinism in early design success as by power, form factor, and performance.

Classical implementations of adaptive equalizers utilize an ADC followed by DSP. Advanced equalization techniques like MLSE and MAP equalizers can only be implemented by this method, which requires ADC sampling at the symbol rate or twice of the symbol rate. For example, equalization of a 10-Gb/s optical channel needs a 10-GSample/s (or 20-GS/s) ADC with $\geq$ 4-bit resolution. At the time of writing, the IC technology can realize a 22-GS/s 5-bit flash ADC in SiGe BiCMOS process [33, 34], and a 20-GS/s 8-bit time-interleaved ADC in CMOS [35]. A time-interleaved ADC uses a number of ADCs operating in parallel and requires periodic calibration to correct for offset, gain, and timing mismatch among sub-ADCs [33, 36].

An analog signal-processing architecture without using high-speed ADC is suitable for the FFE/DFE. For example, an analog FFE that has a structure of transversal filter is shown in Figure 18.1. The DFE illustrated in Figure 18.2 needs clock recovery, bit-by-bit decision, and feedback from the decision slicer's output to the input. Typically, the electronic circuit of a clocked slicer consists of a limiting amplifier followed by a D-flip-flop. The time latency of the clocked slicer is usually more than one bit period, because the D-flip-flop itself has a latency of at least half the bit period. Therefore, in high bit-rate lightwave systems, it is difficult to realize decision feedback by means of analog feedback from a clocked slicer output to the slicer input. A nonlinear canceller using multiple detectors with different thresholds is a well-known technique that avoids analog feedback [7]. On the other hand, a limiting amplifier is a continuous-time unclocked slicer. Reducing the time latency of limiting amplifiers to less than a bit period is possible and allows for analog feedback from the amplifier output to the input. Figure 18.13 illustrates such an analog DFE with two feedback coefficients. The round-trip traveling times of the first- and second-stage feedback loops equal to T and $2T$, respectively. Using an S/H circuit that samples the FFE output at bit rate, the feedback equalizer still performs T-space-sampled signal processing. If the S/H circuit is further removed, we obtain an unclocked DFE that can provide the analog waveform of equalized signal without clock recovery [37]. Note that the latency of a high-speed limiting amplifier is proportional to its small signal gain, and a short-latency limiting amplifier is usually

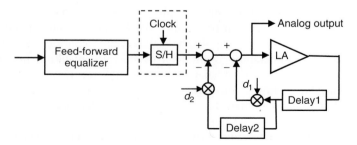

Figure 18.13 Block diagram of a continuous-time DFE using analog feedback. LA: limiting amplifier. Two possible structures: (i) clocked DFE with the sample-and-hold (S/H) block, (ii) unclocked DFE without the S/H block.

a "soft" slicer. Our computer simulation shows that a continuous-time DFE using a soft slicer has significantly reduced error propagation effect as compared with the traditional DFE using a hard slicer.

18.4 ELECTRONIC COMPENSATION FOR 10-GB/S APPLICATIONS

18.4.1 LAN: Transmission Over MMF

Overview

The IEEE 802.3aq 10-Gb/s Ethernet standard (10GBASE-LRM) defines the long-wavelength (1310 nm) links for LAN, which will achieve a distance of at least 220 m over legacy 62.5 and 50-μm MMFs. This standard defines a serial 10 Gb/s solution and will enable small form-factor transceivers at low-power dissipation enabling higher port densities. The 10GBASE-LRM modules are expected to be the dominant 10-Gb/s modules for LAN by 2008/2009.

The dispersive effect within the legacy 62.5-and 50-μm fiber arises due to the large number of mode groups within the fiber which arrive at the receiver with different delays, depending on the refractive index profile of the MMF core, and is referred to as differential modal dispersion (DMD). Furthermore, the power distribution across the mode groups depends on the transmit launch condition [38], and furthermore, mode-mixing effects due to the connectors in the link can substantially alter the DMD profile of the link. Thus, given the stochastic variations in the DMD across these links, the DMD over such population of links is characterized in a statistical sense. In the absence of compensation of DMD, the links are limited to less than 100 m with fairly high probability. With adaptive electronic equalization techniques, the link reach can exceed 220 m. Given that the link transfer function is only statistically characterized and, in addition, because these links exhibit significant time variance, preemphasis or precoding techniques at the transmitter provide little gain in effectiveness.

The 10GBASE-LRM standard subsumes the use of an adaptive electronic equalizer at the receiver within the link. While the initial 10GBASE-LRM optical modules will have the electronic equalizer at their receiver, there is currently an industry-wide initiative driven by enterprise and storage network requirements to further reduce the 10-Gb/s optical module form-factor. In this initiative, the direction is to move the equalizer from the optical module to the network switch or router. With this partitioning, the optical 10-Gb/s module, referred to as SFP+, has analog electrical linear interface at its receiver far-end and the optical module can be designed to have smaller form-factor and lower power dissipation. In this scenario, the electronic equalizer will need to compensate for the DMD within the MMF and also the ISI induced within the copper trace and the connector between the optical module and the switch or router, in the presence of the connector-induced cross talk.

Electronic Equalization for 10GBASE-LRM

The MMF link model for 10GBASE-LRM can be approximated as a linear baseband channel described in (18.1). Using a statistical MMF model developed by a research group from Cambridge University [39], our simulation of system performance shows that the DFE can provide good coverage of installed 62.5-μm FDDI grade fibers with up to 220-m length, and pure LE can hardly meet the requirement of 10GBASE-LRM [6].

As compared to DFE, the limited performance improvement of MLSE in theory may not justify the selection of MLSE for 10GBASE-LRM, which is highly sensitive to the cost and power consumption of the EDC solution. The practical implementation of MLSE at such high data rates has to meet the following challenges: (1) power dissipation associated with the high-speed A/D conversion followed with a DSP-based approach to carry out the trellis computations may get too high; (2) need for robust dynamic adaptation for the WMF and the trellis processing which requires accurate and continuously updated channel estimates; and (3) ensuring very low BER ($<10^{-12}$) without FEC.

As an instance of the capability of the current state of the art in CMOS technology, we consider a commercial EDC IC for 10.3-Gb/s MMF systems, developed by Scintera Networks. This device is a small-form factor (5 × 5 mm), low-power CMOS IC. It has an analog, continuous-time datapath, incorporating a feed-forward and nonlinear feedback path, and uses analog LMS adaptation [40]. Figure 18.14 shows a typical result of the BER vs the received optical power for three stressor channels selected by the IEEE standard draft of 10GBASE-LRM.

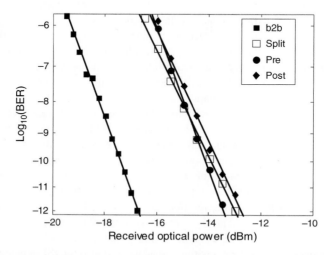

Figure 18.14 Measured BER vs average optical power for PRBS31 pattern at 10.3125 Gb/s. A commercial EDC integrated circuit is used to equalize three stressor channels that have symmetric split, precursor, and postcursor pulse responses, respectively, as defined by 10GBASE-LRM.

These stressor channels are generated from a direct modulation DFB-laser driven by a feed-forward transversal filter that can emulate channels of different pulse responses. The modulation extinction ratio is 6 dB. Using the above EDC, a power penalty of ~4 dBo can be achieved for the IEEE 10GBASE-LRM stressor channels.

18.4.2 Metro and Long-Haul Transmission over SMF

Overview

The early interest in applying EDC techniques in metro area networks at 10 Gb/s has been to enhance the link reach by compensating for the laser chirp and chromatic dispersion, use 2.5 G optical components for 10-Gb/s optical transmission, and improve the effective signal-to-noise ratio (due to the matched filtering gain) over proprietary or nonstandard-based links. In addition, receiver-based EDC techniques were also intended to improve the interoperability of different vendors' transponder modules. Given the cost and performance benefits of EDC, it was natural for different standards groups to initiate standardization of EDC-based links for these networks. The Optical Internetworking Forum and International Telecommunication Union have been coordinating efforts across different optical module, system, and component companies over the past few years to develop new application codes utilizing EDC to lower network cost and allow more flexible networks, and provide a seamless upgrade path from OC48 to OC192.

There is also strong interest in electronics methods for dispersion and error compensation in long-haul optical networks. While these networks tend to be less sensitive to standards-setting bodies, the use of receiver-based EDC techniques here was primarily to improve the link margin and relax the requirement on optical components. Receiver-based EDC can be employed to mitigate the degradation caused by polarization mode dispersion (PMD) and residual chromatic dispersion, whereas the bulk dispersion for these networks is still compensated using optical techniques such as dispersion management. However, it is technically feasible to compensate for the bulk chromatic dispersion in the E-field domain with high-speed electronic signal processing.

Modeling of Transmission Channel and Impairments

In the electrical domain, SMF systems using highly coherent transmitters and square-law direct detection are nonlinear channels. In order to analyze the impact of direct detection on the equalization of fiber dispersion, we assume the complex envelop of the received optical signal, represented by a 2×1 complex vector $\mathbf{A}(t)$, is written as

$$\mathbf{A}(t) = \sum_k b_k \mathbf{g}(t - kT), \quad \mathbf{g}(t) = \mathbf{h}_D(t) * g_0(t) \qquad (18.36)$$

where b_k are real-value data symbols, $g_0(t)$ is the complex envelop of back-to-back received optical pulse, $\mathbf{h}_D(t)$ represents the effect of fiber dispersion, and "*" denotes convolution. The Fourier transform of $\mathbf{h}_D(t)$ is given by

$$\int_{-\infty}^{+\infty} \mathbf{h}_D(t) \exp(i\Omega t) dt = \exp\left(\frac{i}{2}\beta_2 L\Omega^2\right) \exp[i\mathbf{B}(\Omega)]\mathbf{u} \quad (18.37)$$

where β_2 is the group-velocity dispersion (GVD) coefficient, L is the fiber length, Ω is the angular frequency deviation, $\exp[i\mathbf{B}(\Omega)]$ is a 2×2 unitary matrix representing all-order PMD (so $\mathbf{B}$ is a zero-trace Hermitian matrix), and $\mathbf{u}$ is a unit polarization vector. A direct detection O/E receiver detects the signal power,

$$s(t) = |\mathbf{A}(t)|^2 = \sum_k \sum_{m \geq 0} b_k b_{k+m} s_m(t - kT) \quad (18.38)$$

$$S_m(t) = \begin{cases} 2\text{Re}[\mathbf{g}(t) \cdot \mathbf{g}(t - mT)], & m > 0 \\ |\mathbf{g}(t)|^2, & m = 0 \end{cases} \quad (18.39)$$

Similar nonlinear channel models were proposed for non-return-to-zero (NRZ), on-off-keying (OOK) systems [41, 42], where the transmitted data $b_k \in \{0,1\}$ and the received data $a_k = b_k$. The model is also valid for many advanced modulation formats, for example, optical duobinary and differential-phase-shift-keying modulation [43]. For those two modulation formats, the transmitted data $b_k \in \{-1,1\}$; the received data $a_k = (b_{k-1}b_k + 1)/2$, and $a_k \in \{0,1\}$.

In order to have a statistical model of PMD, we assume

$$\exp[i\mathbf{B}(\Omega)] = \prod_{n=1}^{N} \exp\left(\frac{i}{2}\Omega \mathbf{B}_n\right), \mathbf{B}_n = \begin{bmatrix} \tau_{n1} & \tau_{n2} + i\tau_{n3} \\ \tau_{n2} - i\tau_{n3} & -\tau_{n1} \end{bmatrix} \quad (18.40)$$

where N is the number of cascaded elements, and τ_{nj} are i.i.d. Gaussian random variables with zero mean. As a result, the fiber's differential group delay (DGD or $\Delta\tau$) has an exact Maxwellian distribution for any number of N, and the outage probability of second-order PMD approaches the theoretical probability given by reference [44] as N increases to $N \geq 10$.

Nonlinear optical effects have significant impact on SMF systems [45]. In the case of single-wavelength transmission, ISI is induced by the combined effect of fiber dispersion and self-phase modulation. In multiwavelength systems, cross-phase modulation, and four-wave mixing degrade the system through interchannel cross talk. Although fiber nonlinearities are important in practical applications, we neglect them in the channel model to simplify the theoretical analysis of EDC.

In SMF systems, well-known noise sources include the optical ASE noise, the receiver's shot noise, and thermal noise [46]. Another important noise source that may significantly limit the performance of EDC is the transmitter's phase noise that can be converted to RIN by chromatic dispersion [47]. The received noise after direct detection is generally signal-dependent and non-Gaussian.

Electronic Equalization to Compensate Chromatic Dispersion

Fiber chromatic dispersion induces both linear and nonlinear ISI in an OOK system. This is evident from Eqn (18.38), which shows that the broadening of $s_0(t)$ pulse generates linear ISI and the broadening of $s_m(t)$ pulses generates nonlinear ISI. Assuming optical ASE is the dominant noise source and neglecting nonlinear optical effects, the chromatic dispersion limit of an SMF system is defined as the maximum GVD that can be tolerated with 3-dB optical signal-to-noise ratio (OSNR) penalty at 10^{-9} BER, denoted as $(\beta_2 L)_{3dB}$. Without any dispersion compensation, a chirp-free NRZ–OOK system has a chromatic dispersion limit of $(\beta_2 L)_{3dB} = 0.12T^2$. Using a DFE (FFE included), the dispersion limit can be improved to $(\beta_2 L)_{3dB} = 0.23T^2$ if the DFE is adapted to minimum BER [48]. Due to the impact of nonlinear ISI, a DFE using the LMS adaptation has significantly suboptimum BER.

The MLSE can provide reasonably higher tolerance to chromatic dispersion than the DFE in direct-detection OOK systems, to a large extent because of the "nonlinear, non-minimum-phase nature of the channel" [42]. In order to convert the continuous-time channel to a discrete-time channel, a simple approach is to sample the low-pass filtered output of the photodetector at the bit rate [36, 49]. Assuming the samples of signal plus noise, denoted by y_k, are independent, MLSE maximizes the log-likelihood function written as

$$\sum_{k=0}^{K} \log[pdf(y_k|\ \tilde{a}_k, \tilde{a}_{k-1}, \ldots, \tilde{a}_{k-L})] \tag{18.41}$$

for trial sequences $\{\tilde{a}_k\}_{k=0}^{K}$. This expression is also valid for non-Gaussian probability density functions (PDF). Using one-sample/bit MLSE, the dispersion limit of an NRZ–OOK system is $(\beta_2 L)_{3dB} = 0.32T^2$ [50]. The performance of bit-rate sampled MLSE is sensitive to the sampling clock phase, and the optimum sampling clock phase shifts half a bit when $\beta_2 L$ increases from 0 to $0.2T^2$ [50]. An over-sampling MLSE using two samples/bit can avoid this problem and significantly improve the GVD tolerance to $(\beta_2 L)_{3dB} = 0.98T^2$ [50]. Experimental demonstration of a two samples/bit MLSE is reported in Ref. [51].

There have been numerous experimental demonstrations of the DFE and MLSE for enhanced GVD tolerance, for example, [11, 52, 53]. The use of adaptive EDC techniques at the receiver is included in the ITU-T G.959.1 application code to extend SMF-28 reach at 10.7 Gb/s from 1600 to 2400 ps/nm, utilizing transmit optics specified to reach 1600 ps/nm. As part of the effort within the Optical Internetworking Forum, four semiconductor companies (AMCC, Broadcom, Scintera Networks, Vitesse) with EDC solutions experimentally demonstrated the feasibility of this application code [54], which also utilizes the negative-chirp optical modulation and the beneficial self-phase modulation effect to overcome chromatic dispersion.

Electronic Compensation of PMD

The first-order effect of PMD is known as a DGD between two principal states of polarizations and produces a notch in the frequency spectrum of the direct-detected signals. Electronic equalization compensates for the linear ISI arising from first-order PMD at the cost of noise enhancement and residual penalties [55]. Various electronic equalization techniques for PMD mitigation can be compared from the maximum instantaneous DGD that can be tolerated with 3-dB OSNR penalty at BER $= 10^{-9}$, denoted as $\Delta\tau_{3dB}$, for the case of equal power splitting. In an NRZ–OOK system without any PMD compensation, we have $\Delta\tau_{3dB} = 0.5T$. A DFE optimized for minimum BER improves $\Delta\tau_{3dB}$ to $0.7T$ [48]. The ultimate performance of electronic equalization is achievable using two-samples/bit MLSE, which can improve $\Delta\tau_{3dB}$ to $0.95T$ [50]. However, PMD is a stochastic degradation changing constantly with time, and the equalization of PMD requires robust adaptation, which could be a problem of the MLSE when training sequence is not available.

Frequency-dependent variation of the first-order PMD, that is the principal states of polarizations and the DGD, induces higher order PMD effect, which is important in systems in which the fiber's average DGD is $> 0.25T$ [56]. To improve the PMD tolerance, one may use a simple optical PMD compensator combined with electronic equalization to mitigate all-order PMD. We did computer simulation to investigate such a hybrid PMD compensation scheme. Let us consider an NRZ–OOK system with zero accumulated chromatic dispersion. Using the PMD model described in Eqn (18.40), we simulated 10^5 statistical samples of the fiber with an average $DGD \overline{\Delta\tau} = 0.4T$. The DGD of the fiber link at the center of the signal's spectrum is fully compensated for by an optical DGD element. Figure 18.15(a) shows the OSNR penalty vs the normalized second-order PMD in the case without electronic equalization. If we consider only second-order PMD that is modeled as a differential chromatic dispersion element, the penalty is determined by the magnitude of second-order PMD [44]. However, in the presence of all-order PMD, the result becomes a scatter plot. Figure 18.15(b) shows the performance improvement by a DFE with eight half-spaced FFE taps and two feedback taps for the worst 917 of 10^5 fibers, which have a $>$ 2-dB penalty without equalization. If the DFE tap weights are adjusted to minimize BER, the maximum OSNR penalty among 10^5 statistical samples is reduced from 7.2 to <3 dB.

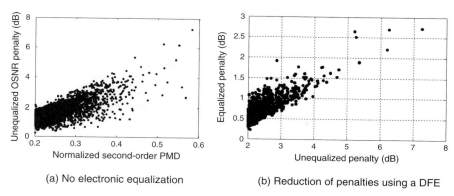

(a) No electronic equalization (b) Reduction of penalties using a DFE

Figure 18.15 OSNR penalties in a system in which PMD is compensated to first order by DGD nulling at the optical carrier frequency. Equation 18.40 was used as a statistical PMD model (average DGD equal to 40% the bit period).

E-Field Domain Signal Processing

The received signal after direct detection is proportional to the optical intensity, while the received signal after optical coherent detection is proportional to the electric field [57, 58]. Therefore, coherent detection allows for full compensation of chromatic dispersion and PMD through linear equalization in the *E*-field domain. Using optical heterodyne detection (please see Chapters 3 and 4 of Volume B for further treatments on this), a dispersive microwave stripline can compensate for a fixed value of chromatic dispersion [59]. A recent experiment using this approach combined with duobinary modulation demonstrated 10.7 Gb/s, 100- to 375-km transmission over standard SMF with $\leq$2-dB OSNR penalty at BER $= 10^{-3}$ [2]. The dispersion tolerance can be further improved to $\sim$10000 ps/nm with the addition of MLSE in the decision circuit [60]. Coherent detection requires beating of the optical signal with a local oscillator laser. If the optical fiber has negligible PMD so that the lightwave at fiber output is not depolarized, one can use an optical polarization controller to achieve polarization tracking. Employing polarization- and phase-diversity detection [61], a general scheme for optical coherent detection and electronic signal processing is shown in Figure 18.16. The in-phase and quadrature components of the electric field in each of the two orthogonal polarizations are detected [16]. Carrier recovery, coherent demodulation, and equalization can all be done by high-speed A/D conversion followed by DSP, and very high tolerance to PMD and chromatic dispersion can in principle be achieved. A recent experiment demonstrated the phase-diversity detection of this method using offline signal processing and a free running laser as the local oscillator [62].

Instead of coherent detection, another technique of *E*-field signal processing utilizes a vector electrooptical modulator that can modulate the real/imaginary

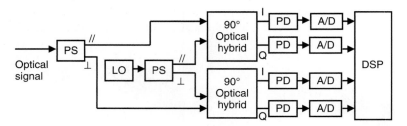

Figure 18.16 Coherent detection followed by electronic signal processing. PS: polarization beam splitter; PD: photo detector; LO: local oscillator laser; A/D: analog-to-digital conversion; DSP: digital signal processing (after Ref. [6], © 2006 IEEE/OSA).

part (or intensity/phase) of the optical signal independently, and precompensates the chromatic dispersion at the transmitter [63–65]. The vector modulator can be an optical E-field modulator with separate I/Q drives [64] or a near-ideal dual-drive Mach–Zehnder modulator [65]. The pros and cons of the two configurations of modulator are discussed in Ref. [66]. The modulator's driving signals are generated from linear and nonlinear digital filtering followed by high-speed digital-to-analog (D/A) conversion. Using an I/Q modulator, this technique is easily applied to modulation formats like duobinary and DPSK [66]. Electronic precompensation of ~60,000-ps/nm accumulated dispersion at 10 Gb/s has been demonstrated in a recirculating loop experiment over a distance of 3840 km [3]. For the purpose of adaptive dispersion compensation, this technique requires information feedback from the receiver to the transmitter.

Utilizing optical E-field modulation in the transmitter and coherent detection in the receiver, various coding, modulation, and signal-processing techniques developed for radio-frequency digital communications can be replicated in SMF systems. This approach has the potential to realize optical communication systems that have very high dispersion tolerance and have a communication performance or capacity approaching the theoretical limit.

18.4.3 Storage Area Networks

Fiber Channel is a dominant standard interface for storage area network applications, and T-11 is the corresponding standard-setting body. The dominant installed fiber infrastructure within this application is legacy 50 µm MMF. Currently 4Gb/s Fiber Channel (4G-FC) modules and systems are being installed, either as new installations or as an upgrade to 2Gb/s Fiber Channel (2G-FC) links.

Again, the DMD within the MMF resulted in a link reach of 300 m at 2G-FC, which would be reduced to 150 m with 4G-FC. The early agreement within the

T-11 committee in 2004 had been to allow for a variant, which uses EDC technology at 4G-FC to enhance the reach to 300 m, so that all 2G-FC links could be seamlessly upgraded to 4G-FC. However, most fiber channel links tend to be less than 100–150 m, and there has not been much momentum behind enhancing the reach of 4G-FC to 300 m. Currently, there is a significant effort in developing 8-Gb/s Fiber Channel (8G-FC) standards within the T-11 group and the optimal solution from the cost, power, and performance standpoint is under discussion at the time of writing of this chapter. It is expected that EDC will be adopted for at least some version of 8G-FC.

18.4.4 Data Center Applications

As earlier noted, 10GBASE-LRM is widely expected to be used across enterprise backbones, which use legacy 62.5- and 50-µm MMF. In addition, 10GBASE-LRM is also used in many other interrack and interswitch data center links. A large percentage of these links today run at 1 Gb/s over unshielded twisted pair (UTP), mostly CAT-5 cables. There is a very significant effort within IEEE, referred to as 10GBASE-T, to standardize 10 Gb/s over a new grade of UTP that requires new cable installation. Furthermore, it is expected that silicon solutions for 10GBASE-T will run at high power, initial estimates running as high as 10 W. This has kindled some interest in looking into more power-efficient solutions, which use a different but low-cost media.

One such media, which is being looked by a limited number of companies, includes the graded-index plastic optical fiber (GI-POF) (Chapter 6 of Volume A provides a detailed treatment on POF), which is specified at 300 MHz-km overfilled launch bandwidth. The POF is a low-cost medium and generally easier to use with better bending specification than other MMEs. To meet the requirements of a wide majority of data center links, the reach for these links needs to be close to 100 m. In the absence of any dispersion compensation, the GI-POF can achieve a distance close to 50 m. With adaptive electronic equalization at the receiver, the links using GI-POF can exceed 100 m. Furthermore, because the total power consumption of the GI-POF module can be less than 1.5 W, the GI-POF with EDC could emerge as an alternative for 10GBASE-T for data center links.

18.5 PROSPECTS AND TRENDS FOR NEXT-GENERATION SYSTEMS

This chapter provided a survey of the potential, feasibility, and the current state-of-the-art of electrical signal-processing techniques that can be used to mitigate ISI effects within a range of optical applications of commercial importance within industry and also within academia.

It is expected that as the need for increased bandwidth continues unabated, there will be need to upgrade transmission systems to 40 Gb/s or even 100 Gb/s while keeping the same installed base and also a need to enhance the reach and/or performance at 10 Gb/s. Efforts have already been initiated within the IEEE802 committee to define the next Ethernet technology at higher speeds, likely to be an aggregate of 100 Gb/s with serial data rate at 25 Gb/s or above. Analog equalizers (40-Gb/s) have been realized with SiGe ICs [67]. The steady progress of very high-speed A/D and D/A converters will enable DSP-based optical transmitters and receivers that will provide high-performance optical communication systems. As digital electronics continues to scale with Moore's law and the unity-gain bandwidth of the CMOS process increases, it can safely be predicted that the importance of electronic signal processing to optical communication will only continue to grow. Advances in optical communications will be significantly influenced by advances in electronics, including high-performance signal-processing systems.

REFERENCES

[1] T. Nielsen and S. Chandrasekhar, "OFC 2004 workshop on optical and electronic mitigation of impairments," *J. Lightw. Technol.*, 23(1), 131–142, January 2005.
[2] A. H. Gnauck, J. Sinsky, P. J. Winzer, and S. Chandrasekhar, "Linear Microwave Domain Dispersion Compensation of 10-Gb/s Signal Using Heterodynce Detection," in Proc. *OFC 2005*, Anaheim, CA, paper PDP31, 2005.
[3] D. McGhan, C. Laperle, A. Savchenko et al., "5120 km RZ-DPSK Transmission Over G652 Fiber at 10 Gb/s with no Optical Dispersion Compensation," in Proc. *OFC 2005*, Anaheim, CA, paper PDP27, 2005.
[4] N. Nishiyama, C. Caneau, J. D. Downie et al., "10-Gbps 1.3 and 1.55-μm InP-Based VCSELs: 85°C 10-km Error-Free Transmission and Room Temperature 40-km Transmission at 1.55-μm with EDC," in Proc. *OFC 2006*, Anaheim, CA, paper PDP23, 2006.
[5] P. Winzer, F. Fidler, M. J. Matthews et al., "10-Gb/s upgrade of bidirectional CWDM systems using electronic equalization and FEC," *J. Lightw. Technol.*, 23(2), 749–763, February 2005.
[6] Q. Yu and A. Shanbhag, "Electronic data processing for error and dispersion compensation," *J. Lightw. Technol.*, 24(12), 4514–4525, August 2006.
[7] J. H. Winters, R. D. Gitlin, and S. Kasturia, "Reducing the effects of transmission impairments in digital fiber optic systems," *IEEE Commun. Mag.*, 31(6), 69–76, June 1993.
[8] D. Schlump, B. Wedding, and H. Bülow, "Electronic Equalization of PMD and Chromatic Dispersion Induced Distortion After 100 km Standard Fiber at 10 Gbit/s," in Proc. *24th ECOC*, Madrid, Spain, 1, 1998, pp. 535–536.
[9] F. Buchali, H. Bülow, W. Baumert et al., "Reduction of the Chromatic Dispersion Penalty at 10 Gbit/s by Integrated Electronic Equalizers," in Proc. *OFC*, Baltimore, Maryland, 3, 2000, pp. 268–270.
[10] M. D. Feuer, S.-Y. Huang, S. L. Woodward et al., "Electronic dispersion compensation for a 10-Gb/s link using a directly modualted laser," *IEEE Photon. Technol. Lett.*, 15(12), 1788–1790, December 2003.
[11] A. Faerbert, "Application of Digital Equalization in Optical Transmission Systems," in Proc. *OFC 2006*, Anaheim, CA, paper OTuE5, 2006.
[12] H. Wu, J. Tierno, P. Pepeljugoski et al., "Integrated transversal equalizer in high-speed fiber-optic systems," *IEEE J. Solid-State Circuits*, 38(12), 2131–2137, December 2003.

18. Electronic Signal Processing in Optical Networks

[13] J. P. Weem, P. Kirkpatrick, and J. Verdiell, "Electronic Dispersion Compensation for 10 Gigabit Communication Links Over FDDI Legacy Multimode Fiber," in Proc. *OFC 2005*, Anahiem, CA, paper OFO4, 2005.

[14] N. Alic, G. Papen, R. Saperstein et al., "Experimental Demonstration of 10 Gb/s NRZ Extended Dispersion-Limited Reach Over 600 km-SMF Link Without Optical Dispersion Compensation," in Proc. *OFC 2006*, Anaheim, CA, paper OWB7, 2006.

[15] Y. Sun, M. E. Ali, K. Balemarthy et al., "10 Gb/s Transmission Over 300 m OM3 Fiber From 990–1080 nm with Electronic Dispersion Compensation," in Proc. *OFC 2006*, Anaheim, CA, paper OTuE2, 2006.

[16] E. A. Lee and D. G. Messerschmitt, *Digital Communication*, Boston, Kluwer Academic Publishers, 1988, pp. 289–407.

[17] J. G. Proakis, *Digital Communications*, New York, McGraw Hill Inc., 1995, pp. 601–626.

[18] J. F. Claerbout, *Foundamentals of Geophysical Data Processing*, chapter 3. Available: http://sep.stanford.edu/oldreports/fgdp2/

[19] J. Saltz, "Optimum mean-square decision feedback equalization," *Bell Syst. Tech. J.*, 52, 1341–1373, October 1973.

[20] G. D. Forney, Jr, "Maximum-likelihood sequence estimation of digital sequences in the presence of intersymbol interference," *IEEE Trans. Inf. Theory*, IT-18, 363–378, May 1972.

[21] A. J. Viterbi, "Error bounds for convolutional codes and an asymptotically optimum decoding algorithm," *IEEE Trans. Inf. Theory*, IT-13, 260–269, April 1967.

[22] L. Bahl, J. Cocke, F. Jelinek, and J. Raviv, "Optimal decoding of linear codes for minimizing symbol error rate," *IEEE Trans. Inf. Theory*, IT-20, 284–287, March 1974.

[23] M. Z. Win, J. H. Winters, and G. M. Vitetta, "Equalization techniques for mitigating transmission impairments," in *Optical Fiber Telecommunications IV B* (I. Kaminow and T. Li, eds), New York, Academic Press, 2002, pp. 985–991.

[24] H. F. Haunstein, W. Sauer-Greff, A. Dittrich et al., "Principles for electronic equalization of polarization-mode dispersion," *J. Lightw. Technol.*, 22(4), 1169–1182, April 2004.

[25] W. Baumert, "Eye monitor," U.S. Patent No. 6784653, August 2004.

[26] J. Choma, *Electrical Networks: Theory and Analysis*, New York, Wiley Interscience, 1985, pp. 424–429.

[27] W.-K. Chen, *Passive and Active Filters: Theory and Implementations*, New York, John Wiley & Sons, 1986, pp. 50–54.

[28] T. H. Lee, *The Design of CMOS Radio-Frequency Integrated Circuits*, Cambridge, UK, Cambridge University Press, 2004, pp. 279–282.

[29] J. Choma and W.-K. Chen, *Feedback Networks: Theory and Circuit Applications*, Singapore, World Scientific Press, 2007, pp. 771–785.

[30] B. M. Ballweber, R. Gupta, and D. J. Allstot, "A fully integrated 0.5–5.5 GHz CMOS distributed amplifier," *IEEE J. Solid-State Circuits*, 35, 231–239, February 2000.

[31] J. H. Winters and S. Kasturia, "Adaptive nonlinear cancellation for high-speed fiber-optic systems," *J. Lightw. Technol.*, 10(7), 971–977, July 1992.

[32] A. Shoval, D. A. Johns, and W. M. Snelgrove, "Comparison of DC offset effects in four LMS adaptive algorithms," *IEEE Trans. on Circuits and Systems II*, 42(3), 176–185, March 1995.

[33] P. Schvan, D. Pollex, S.-C. Wang et al., "A 22 GS/s 5b ADC in 0.13 µm SiGe BiCMOS," in Proc. *IEEE Internal Solid-State Circuits Conferences*, paper 31.4, San Francisco, CA, 2006.

[34] J. Lee, P. Roux, U. Koc et al., "A 5-b 10-GSample/s A/D converter for 10-Gb/s optical receivers," *IEEE J. Solid-State Circuits*, 39(10), 1671–1679, October 2004.

[35] K. Poulton, R. Neff, B. Setterberg et al., "A 20 GS/s 8b ADC with a 1MB memory in 0.18 µm CMOS," in Proc. *IEEE Internal Solid-State Circuits Conferences*, paper 18.1, San Francisco, CA, 2003.

[36] O. E. Agazzi, M. R. Hueda, H. S. Carrer, and D. E. Crivelli, "Maximum-likelihood sequence estimation in dispersive optical channels," *J. Lightw. Technol.*, 23(2), 749–763, February 2005.

[37] E. Ibragimov, Q. Yu, and P. Choudhary. "Decision feedback equalizer with dynamic feedback control," U.S. Patent No. 7120193, October 2006.

[38] M. Webster, L. Raddatz, I. H. White, and D. G. Cunningham, "A statistical analysis of conditioned launch for Gigabit Ethernet links using multimode fiber," *J. Lightw. Technol.*, 17(9), 1532–1541, September 1999.

[39] J. D. Ingham, R. V. Penty, and I. H. White, "Extension of Statistical Multimode Fiber Channel Model to Electronic Dispersion Compensation," in Proc. *IEEE 802 Interim Meeting: FDDI-grade MM Fiber Study Group*, Vancouver, Canada, Janaury 2004. Available: http://grouper.ieee.org/groups/802/3/10GMMFSG/public/jan04/penty_1_0104.pdf

[40] A. Phanse, A. Shanbhag, Q. Yu et al., "Adaptation structure and methods for analog continuous time equalizers," U.S. Patent No. 7016406, March 2006.

[41] B. E. Saleh and M. I. Irshid, "Coherence and intersymbol interference in digital fiber optic communication systems," *IEEE J. Quantum Electron.*, QE-18(6), 944–951, June 1982.

[42] O. E. Agazzi and V. Gopinathan, "The Impact of Nonlinearity on Electronic Dispersion Compensation of Optical Channels," in Proc. *OFC 2004*, Los Angeles, CA, paper TuG6, 2004.

[43] W. Kaiser, T. Wuth, M. Wichers, and W. Rosenkranz, "Reduced complexity optical duobinary 10-Gb/s transmission set resulting in an increased transmission distance," *IEEE Photon. Technol. Lett.*, 13(8), 884–886, August 2001.

[44] H. Kogelnik, L. E. Nelson, and P. J. Winzer, "Second-order PMD outage of first-order compensated fiber systems," *IEEE Photon. Technol. Lett.*, 16(4), 1053–1055, April 2004.

[45] G. P. Agrawal, *Applictions of Nonlinear Fiber Optics*, Boston, Academic Press, 2001, chapter 7.

[46] B. L. Kasper, O. Mizuhara, and Y.-K. Chen, "High bit-rate receivers, transmitters, and electronics," in *Optical Fiber Telecommunications IV A* (I. Kaminow and T. Li, eds), New York, Academic Press, 2002, pp. 794–800.

[47] S. Walklin and J. Conradi, "Multilevel signaling for increasing the reach of 10 Gb/s lightwave systems," *J. Lightw. Technol.*, 17(11), 2235–2248, November 1999.

[48] Q. Yu, "On the decision-feedback equalizer in optically-amplified direct-detection systems," *J. Lightw. Technol.*, 25(8), August 2007.

[49] A. J. Weiss, "On the performance of electrical equalization in optical fiber transmission systems," *IEEE Photon. Technol. Lett.*, 15(9), 1225–1227, September 2003.

[50] T. Foggi, E. Forestieri, G. Colavolpe, and G. Prati, "Maximum-likelihood sequence detection with closed-form metrics in OOK optical systems impaired by GVD and PMD," *J. Lightw. Technol.*, 24(8), 3073–3087, August 2006.

[51] J. Elbers, H. Wernz, H. Griesser et al., "Measurement of the Dispersion Tolerance of Optical Duobinary with an MLSE-Receiver at 10.7 Gb/s", in Proc. *OFC 2005*, Anaheim, CA, paper OThJ4, 2005.

[52] P. M. Watts, V. Mikhailov, S. Savory et al., "Performance of single-mode fiber links using electronic feed-forward and decision feedback equalizers," *IEEE Photon. Technol. Lett.*, 17(10), 2206–2208, October 2005.

[53] H. Bae, J. Ashbrook, J. Park et al., "An MLSE Receiver for Electronic Dispersion Compensation of OC-192 Fiber Links," in Proc. *IEEE Internal Solid-State Circuits Conferences*, paper 13.2, San Francisco, CA, 2006.

[54] A. Ghiasi, F. Chang, E. Ibragimov et al., "Experimental Results of EDC Based Receivers for 2400 ps/nm at 10.7 Gb/s for Emerging Telecom Standards," in Proc. *OFC 2006*, Anaheim, CA, paper OTuE3, 2006.

[55] L. Möller, A. Thiede, S. Chandrasekhar et al., "ISI mitigation using decision feedback loop demonstrated with PMD distorted 10 Gbit/s signals," *Electron. Lett.*, 35(24), 2092–2093, 1999.

[56] H. Bülow, F. Buchali, W. Baumert et al., "PMD mitigation at 10 Gbit/s using linear and nonlinear integrated electronic equalizer circuits," *Electron. Lett.*, 36(2), 163–164, 2000.

[57] L. Kazovsky, S. Benedetto, and A. Willner, *Optical Fiber Communication Systems*, Boston, Artech House, 1996, chapter 4.

[58] B. Spinnler, P. Krummrich, and E. Schmidt, "Chromatic Dispersion Tolerance of Coherent Optical Communication Systems with Electrical Equalization," in Proc. *OFC 2006*, Anaheim, CA, paper OWB2, 2006.

[59] N. Takachio and K. Iwashita, "Compensation of fiber chromatic dispersion in optical heterodyne systems," *Electron. Lett.*, 24(2), 108–109, 1988.
[60] A. H. Gnauck, S. Chandrasekhar, and P. J. Winzer, "Dispersion tolerant 10-Gb/s duobinary system employing heterodyne detection and MLSE," *IEEE Photon. Technol. Lett.*, 18(5), 697–699, May 2006.
[61] T. Okoshi and Y. H. Cheng, "Four-port homodyne receiver for optical fiber communications comprising phase and polarization diversities," *Electron. Lett.*, 23(8), 337–338, 1987.
[62] M. G. Taylor, "Coherent detection method using DSP for demodulation of signal and subsequent equalization of propagation impairments," *IEEE Photon. Technol. Lett.*, 16(2), 674–676, February 2004.
[63] M. M. El Said, J. Stich, and M. I. Elmasry, "An electrically pre-equalized 10-Gb/s duobinary transmisison systme," *J. Lightw. Technol.*, 23(1), 388–400, January 2005.
[64] J. McNicol, M. O'Sullivan, K. Roberts et al., "Electrical Domain Compensation of Optical Dispersion," in Proc. *OFC 2005*, Anaheim, CA, paper OThJ3, 2005.
[65] R. I. Killey, P. M. Watts, V. Mikhailov et al., "Electronic diserpsion compensation by signal predistortion using digital processing and a dual-drive Mach-Zehnder modulator," *IEEE Photon. Technol. Lett.*, 17(3), 714–716, March 2005.
[66] D. McGhan, "Electronic Dispersion Compensation," in Proc. *OFC 2006*, Anaheim, CA, tutorial OWK1, 2006.
[67] H. Jiang and R. Saunders, "Advances of SiGe ICs for 40 Gb/s Signal Equalization," in Proc. *OFC 2006*, Anaheim, CA, paper OTuE1, 2006.

19

Microelectromechanical systems for lightwave communication

Ming C. Wu[*], Olav Solgaard[†], and Joseph E. Ford[‡]

[*]Berkeley Sensor and Actuator Center (BSAC) and Electrical Engineering & Computer Science Department, University of California, Berkeley, CA, USA
[†]Department of Electrical Engineering, Edward L. Ginzton Laboratory, Stanford University, Standford, CA, USA
[‡]Department of Electrical and Computer Engineering, University of California, San Diego, CA, USA

19.1 INTRODUCTION

Nearly three decades ago, Petersen [2] published a paper on micromechanical spatial light modulators (SLMs) [1] and another on silicon torsion mirrors. Thirty years later, this has become a thriving field known as optical microelectromechanical systems (MEMS), sometimes also called microoptoelectromechanical systems (MOEMS), with several conferences dedicated to the field. It is a key enabling technology for *dynamic* processing of optical signals. The first market driver of optical MEMS was display technology [3, 4]. The digital micromirror devices (DMD) developed by Texas Instruments are one of the most successful MEMS products. They are now widely used in portable projectors, large-screen TVs, and digital cinemas [3]. Applications of optical MEMS in telecommunications started in the 1990s [5, 6]. The early efforts have focused on the development of optical MEMS devices and fabrication technologies [7–10]. The telecom boom in the late 1990s and early 2000s has accelerated maturation of the technology. A wide range of optical MEMS components were taken from laboratories to reliable products that meet Telcordia qualifications. Although not all commercialization endeavors were successful due to the market downturn, the technology developed is available for new applications in communications and other areas [11].

In this chapter, we will review the recent developments of Optical MEMS for communication applications. With the rapid expansion of the field and proliferation of literature, it is not possible to cover all published works. Instead, we will focus on a selected set of applications, and discuss the design trade-offs in MEMS devices and systems. Topics selected in this chapter include optical switches, filters, dispersion compensators, spectral equalizers, spectrometers, tunable lasers, other dense-wavelength-division-multiplexing (DWDM) devices such as wavelength add/drop multiplexers, wavelength-selective switches (WSS), and crossconnects (WXSC). Emerging technologies will be discussed at the end of this chapter.

19.2 OPTICAL SWITCHES AND CROSSCONNECTS

19.2.1 Two-Dimensional MEMS Switches

Protection switches are made of $1 \times N$ or small $N \times N$ switches. This can be realized by a two-dimensional (2D) array of vertical micromirrors commonly known as a 2D MEMS switch. Figure 19.1 shows the generic schematic of such a switch. The optical beams are collimated to reduce diffraction loss. The micromirrors are "digital," they either direct the optical beams to the orthogonal output ports or pass them to the drop ports. Generally, only one micromirror in a column or row is in the reflection position during operation.

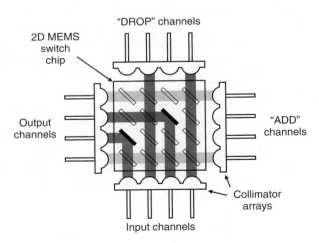

Figure 19.1 Schematic of 2D MEMS optical switches (this figure may be seen in color on the included CD-ROM).

The first MEMS 2D switch (2 × 2) was reported by Toshiyoshi and Fujita [12], and quickly followed by related work [13, 14]. For 2 × 2 switches, low insertion loss (0.6 dB) can be achieved without using collimators, especially when the micromirror is immersed in index-matching fluid [15]. Latchable 2 × 2 switches incorporating MEMS bistable structures were later commercialized [16, 17]. Larger switches require optical collimators to reduce diffraction loss. Switches with 8 × 8 and 16 × 16 ports were demonstrated [18, 19]. There are two basic approaches for the actuation of the micromirror: the first is based on out-of-plane rotation of the micromirror [12, 18, 20, 21], and the other moves micromirrors in and out of the optical paths without changing the mirror angle [13–15, 19, 22, 23]. The 2D switches have been realized using both bulk- [12–15] and surface-micromachining technologies [18–20, 22]. Electrostatic actuation is most commonly used [12–19, 22]. Magnetic actuation has also been demonstrated [13], some in conjunction with electrostatic clamping [20].

The port count of 2D switches is determined by several factors, including the mirror angle, size, fill factor (mirror width divided by unit cell width), and curvature [24, 25]: 16 × 16 switches have been realized and 32 × 32 switches are within the capability of today's technology. Figure 19.2(a) shows a scanning electron micrograph (SEM) of OMM's 2D switch [19]. A vertical mirror is attached at the tip of a cantilever. The tilted cantilever can be pulled down electrostatically. The mirror angle is maintained at 90° during switching. The switch is fabricated using a standard three-polysilicon-layer surface-micromachining process. The mirrors are assembled into vertical position with angular distribution of (90 ± 0.1)°. The hermetic switch package is shown in Figure 19.2(b) [26]. Maximum insertion losses of 1.7 and 3.1 dB have been obtained for 8 × 8 and 16 × 16 switches, respectively, and the crosstalk is less than −50 dB. The switching time is less than 7 ms. Packaging is critical to attain long-term reliability and satisfy Telcordia qualification for telecommunication applications [26].

Although the primary focus of this chapter is on MEMS technology, it should be mentioned that several other technologies are also serious contenders for lightwave communications applications. Silica or silicon planar lightwave circuits (PLCs) provide a guided-wave platform for integrating the switch fabric monolithically. The thermooptically switched PLC has been widely researched. Examples include 16 × 16 [27] and 1 × 128 [28] matrix switches. Switches using microfluidic actuation have also been employed to change total internal reflection (TIR) conditions in arrays of intersecting waveguides. These include Agilent's Champagne switch (also called "bubble" switch) [29] and NTT's OLIVE switches [30]. The Champagne switch used thermally generated bubbles to displace index-matching fluids at waveguide intersections, causing the light to bend by TIR. The OLIVE switch used thermal-capillary force to move trapped bubbles. One drawback of these approaches is the cumulative losses and crosstalk through multiple waveguide intersections. The maximum port counts achieved are 32 × 32 and 16 × 16 for the Champagne and the OLIVE switches, respectively.

(a)

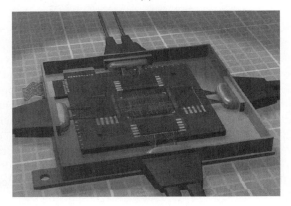

(b)

Figure 19.2 (a) SEM of OMM's 16 × 16 switch (reprinted from Fan et al. [19] with permission). (b) photograph of the packaged switch (reprinted from Dobbelaere et al. [26] with permission).

19.2.2 3D MEMS Switches

A transparent optical crossconnect (OXC) with large port count can be realized by 3D MEMS switches as illustrated in Figure 19.3. The input and output (I/O) fibers are arranged in 2D arrays. The optical beams are steered in three dimensions by two stages of dual-axis micromirrors, directing them toward the desired output ports. The 3D MEMS switch can implement OXC with a large port count [31–33]. Because the optical path lengths of all switching states are similar, it has a uniform insertion loss. The OXC was a subject of intense interest during the telecom boom around the turn of the century [34–38]. The early efforts (before 2002) focused on OXCs with port

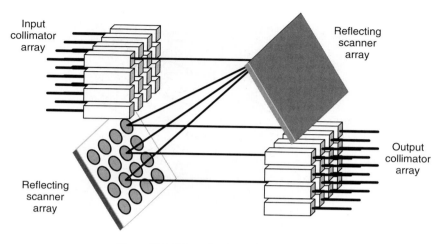

Figure 19.3 Schematic of a 3D MEMS switch.

count $\sim 1000 \times 1000$ [39, 40], driven by the explosion of Internet data transport. Recently, interest has shifted to applications in metropolitan area networks, including metro access and metro-core networks, which require OXC with medium port count ($\sim 100 \times 100$), with emphasis on low cost, low-power consumption, and small footprint [36, 41]. Our discussion here will focus on this trend.

Detailed design trade-offs of the OXC have been reported [34–38]. Two schemes have been proposed to reduce the size of the switch and the tilt angle of the micromirror. Lucent inserted a Fourier lens between the two micromirror chips with the focal length equal to the Rayleigh range of the optical beam (Figure 19.4) [42]. This reduces the required scan angle of the mirror. In addition, the mirrors can be placed at the beam waist, resulting in $\sqrt{2}$ times smaller optical beam. This permits the use of smaller mirrors and/or reduction of the crosstalk. Fujitsu used a "roof-top" mirror to connect two adjacent micromirror chips (photograph shown in Figure 19.5) [36]. The roof-top mirror shifts the optical beams laterally, reducing the tilt angle requirement. Folding of the optical beam also shrinks the footprint of the switch.

In the compact switch category, Lucent's 64×64 switch has a size of $100 \times 120 \times 20\,\mathrm{mm}^3$, which can be mounted on a standard circuit board [41]. The insertion loss is 1.9 dB. Fujitsu's 80×80 switch has a packaged size of $77 \times 87 \times 53\,\mathrm{mm}^3$ [36]. The average insertion loss is 2.6 dB. Impressively, the switch continues to operate under vibration or 50 G shock without any signal degradation. The total power consumption of Fujitsu's switch is only 8.5 W, thanks to the low operating voltage of the mirrors. NTT's 100×100 switch has a size of $80 \times 60 \times 35\,\mathrm{mm}^3$, with an insertion loss of 4 dB [35].

The two-axis micromirror array is the key enabling device of the 3D switch. Important parameters include the size, tilt angle, flatness, fill factor, and resonant frequency of the mirror. Additionally, the stability of the mirror plays a critical role

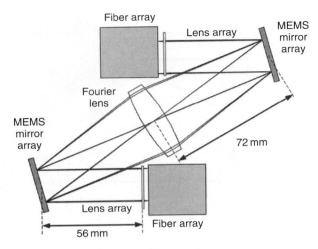

Figure 19.4 Lucent's optical system layout for OXC (reprinted from Aksyuk et al. [42] with permission). A Fourier lens is inserted between the two MEMS chip to reduce the required tilt of the mirror and beam size.

Figure 19.5 Photograph of Fujitsu's 80 × 80 OXC with a roof-top reflector connecting the two MEMS chips (reprinted from Yano et al. [36] with permission). The packaged size is 77 × 87 × 53 mm^3.

in the complexity of the control schemes. Early development focused on surface-micromachined two-axis scanners [43, 44]. The residual stress limits the mirror size to approximately $\lesssim$1 mm, and the different thermal expansion coefficients between the mirror and the metal coating also cause mirror curvature to change with temperature. Bulk-micromachined single-crystalline silicon micromirrors are often used in high-port-count OXCs that require larger mirror size [38, 45–48].

Electrostatic actuation is most commonly used because of low-power consumption and ease of control. Early devices used parallel-plate actuators, which have high actuation voltage and limited scan angle due to pull-in instability [49]. The mirror angle beyond ~1/3 of the maximum mechanical angle is not stable because the electrostatic force can no longer be balanced by the spring force. The mirror will continue to rotate until it hits a mechanical stop. Although the pull-in effect can be mitigated by nonlinear controllers, it increases the complexity of the electronics [50]. Micromirrors with vertical combdrive actuators reported first by Conant et al. [51] offer many advantages. They have much larger torque, which one can use to reduce the operating voltage as well as increase the resonant frequency. In addition, they are free from the pull-in effect, further increasing the stable tilt angles. It should be mentioned that lateral pull-in between comb fingers is a potential issue, but could be mitigated by MEMS design (such as V-shaped torsion beam [52] or off-centered combs [53]). Several variations of vertical combdrive mirrors have been reported, including self-aligned vertical combs [54, 55], angular vertical combs [56, 57], electrostatically assembled vertical combs [58], and thick vertical combs (100 μm) attached to mirror edges on double-sided silicon-on-insulator (SOI) wafers [36, 52].

19.3 WAVELENGTH-SELECTIVE MEMS COMPONENTS

19.3.1 Spectral Equalizers

A spectral equalizer can be realized by combining a free-space imaging spectrometer with a MEMS variable reflector array at the focal plane. An input fiber is imaged through a diffraction grating so that each spectral channel is laterally shifted to illuminate one modulator in a linear array. The reflected signal, attenuated to the desired value, is collected into a single-output fiber by a second pass through the imaging spectrometer. The first such MEMS spectral equalizer used a continuous etalon membrane [59]. This approach was later implemented in the compact package shown in Figure 19.6, which located the MEMS device array next to a single-I/O fiber. A single lens is to collimate the multiwavelength beam onto a blazed reflective grating, and refocus the spectrally separated signals with a second pass onto the MEMS array. A third and fourth pass through the lens reintegrates the signal into the I/O fiber, where it is separated by an external optical circulator. The use of such equalizers is illustrated by the before and after spectral traces at the bottom of Figure 19.6, showing the improvement in uniformity of 36 channels sent through a two-stage amplifier. The equalizer setting was generated by an iterative algorithm running on the computer controller [60].

Dynamic spectral equalization went quickly from an option to a practical requirement as the channel transmission rate increased from 2.5 to 10 and then 40 Gb/s. The simplest and least expensive dynamic gain equalizers (DGEs) use a mid-amplifier filter, which can be spectrally uniform or provide a constant spectral

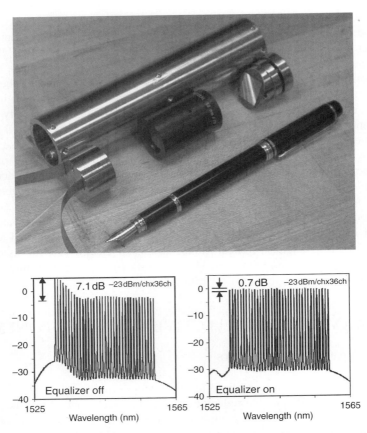

Figure 19.6 Dynamic spectral equalizer package (above) and transmission spectra showing the improvement in channel uniformity for a 36-channel DWDM transmission (this figure may be seen in color on the included CD-ROM).

slope [61]. Two distinct categories of spectral equalizers emerged. DGEs provide a smoothly varying spectral profile used to compensate for the varying gain profiles in amplifiers, while dynamic channel equalizers (DCEs) provide the discrete channel-by-channel power adjustment needed to compensate for nonuniform transmission source intensity or path-dependent loss. Channel equalizers are preferable in general, but require accurate matching of the equalizer passband to the transmission grid to avoid passband narrowing.

Channel equalizers were implemented using discrete variable optical attenuators (VOAs) attached to waveguide spectral multiplexers [62] and using an oversampled array of digital tilt mirrors [63]. However, the best performance in channel equalizers was achieved by combining the type of free-space grating demultiplexer shown in Figure 19.6 either with diffractive MEMS modulators [64] or with analog tilt mirrors [65]. The optical setup is similar to that of Figure 19.7 except without

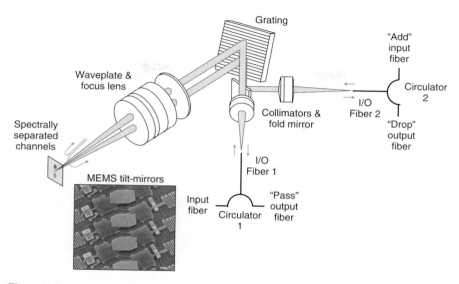

Figure 19.7 Optical schematic for a 2 × 2 MEMS wavelength add/drop switch (this figure may be seen in color on the included CD-ROM).

circulators. These components typically have 40–80 channels spaced at 100 or 50 GHz, with 6–7-dB insertion loss and 20–30-dB dynamic range. The most advantageous characteristic of MEMS equalizers is the extremely flat passband transmission profile, along with low chromatic dispersion at the edges. This performance was achieved after studying the effects of various mirror geometries [66].

After understanding the effects of mirror profile on dispersion, it became possible to use the same basic component structure as the equalizer to provide channel-by-channel dispersion compensation, although this functionality is not yet adopted in the deployed network [67].

19.3.2 Reconfigurable Optical Add–Drop Multiplexers

Wavelength switching allows network operators to use optically transparent components to pass through a network node without detecting and regenerating the data signal, the components that enable these above have been the subject of intense research and development. The most basic wavelength switch is the dynamically reconfigurable optical add–drop multiplexer (ROADM), also called wavelength add–drop multiplexer (WADM). It is essentially a 1 × 2 or 2 × 2 optical switch operating independently of each wavelength channel. ROADMs enable services convergence in next-generation optical networks. They provide a cost-effective solution for mixed services and unpredictable traffic demands.

ROADM was a natural extension of MEMS equalizers, and the first demonstration of a MEMS add/drop switch based on digital tilt mirrors occurred almost simultaneously with the equalizer [68, 69]. Add/drop requires four ports, twice as many as the equalizer, and so the basic structure is slightly more complex (Figure 19.7). The system is still based on a blazed diffraction grating, now illuminated with an upper and lower beam path. The active device is a linear array of 16 digital tilt mirrors, fabricated with surface micromachining in the multiuser multiprogramming system (MUMPS) process. Each mirror defines a DWDM channel, and in switching, directs the reflected signal back along the input direction or tilted into a new path. Optical circulators on the two I/O fibers separate the forward and reverse propagating signals. The mirrors in this switch tilted by $\pm 5°$ under a 20 V signal, switching in 20 μs. A quarter-wave plate is used to achieve 0.2 dB polarization dependence on a total insertion loss of 7.5 dB.

The dynamic channel equalizer is closely related to the ROADM, and in fact it is possible to use high-contrast equalizers as 1×1 switches in a "broadcast and select" architecture [70]. The primary disadvantage of this architecture is that it is intrinsically lossy: signals are power-split, and then unwanted signals are blocked before combining into the output fiber. This does allow multicasting, duplicating signals to multiple output fibers. Broadcast and select was actually the first to be implemented in the network, but is generally expected to be phased out in favor of multiport WSS, which in addition to switching also provide channel equalization [71] with no additional cost or complexity.

19.3.3 Wavelength-Selective Switches

A $1 \times N$ WSS enables individual wavelength channels to be routed to any of the N output channels optically. It consists of a wavelength demultiplexer, an array of $1 \times N$ space division switches followed by N VOAs, and N wavelength multiplexers. WSS can be implemented in free-space optical systems, hybrid or monolithic PLC-MEMS technologies. They are discussed in the following sections.

Free Space-Based Wavelength-Selective Switches

A schematic of the free space-based WSS is shown in Figure 19.8. The input light is dispersed by a grating onto an array of micromirrors. Unlike the digital micromirrors in WADM, analog micromirrors are used in WSS to direct the individual wavelengths to different output ports. A detailed review paper on WSS was published recently [72].

Experimentally, 1×4 WSSs with 128 channels spaced on a 50-GHz grid and 64 channels spaced on a 100-GHz grid have been demonstrated by Lucent. Typical optical insertion loss ranges from 3 to 5 dB. The channel passband is directly related to the confinement factor, defined as the ratio of the beam size to the mirror size. A confinement factor of >2.7 is needed to produce a flat-top spectral response with $>74\%$ passband width measured at -1-dB point. JDSU has reported a 1×4 WSS with 3.5-dB insertion loss [73]. Using vertical combdrive micromirrors (to be discussed below), UCLA has reported a WSS with excellent open-loop stability ($<\pm 0.0035$ dB over 3.5 h) [74].

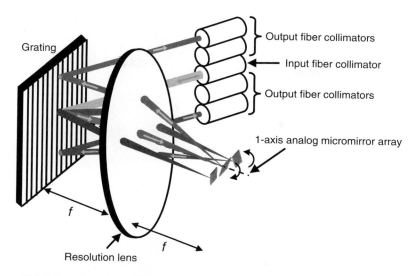

Figure 19.8 Schematic optical setup of 1×4 wavelength-selective switch (this figure may be seen in color on the included CD-ROM).

The analog micromirror array plays a key role in the performance of the WSS. High fill factor is desired to minimize the spectral gaps between channels. Large scan angle is needed to address more output ports. Oversized mirrors (several times larger than the focused optical beams) help achieve wide and flat passbands for minimal signal distortion; at the same time, the pitch of the mirror array should match that of the channel spacing. Several types of WSS micromirror arrays have been reported, including electrostatic [75, 76] and electromagnetic [77] actuations. Lucent employed a fringe-field actuated SOI micromirror array [75] and achieved a mechanical tilt angle of 9.2° at 175 V. The resonant frequency is 3.8 kHz for 80-μm-wide mirrors.

More efficient actuation has been obtained using vertical combdrive actuators. Hah et al. [76] reported a low-voltage analog micromirror array for WSS. The schematic and the SEM of the micromirror are shown in Figure 19.9. The mechanical structures are completely covered by the mirror so high fill factor is achieved along the array direction. The actuation voltage is as low as 6 V for mechanical tilt angles of ±6°. High resonant frequency (3.4 kHz) and high fill factor (98%) are also achieved [78]. The excellent stability of the mirror (±0.00085°) enables open-loop operation of the switch with insertion loss variation <±0.0035 dB over 3.5 h [74].

An alternative to MEMS micromirrors is liquid crystal-on-silicon (LCOS). Originally developed for projection displays, LCOS can control the optical phase shift of each pixel. Beamsteering is achieved through the optical phased array. Figure 19.10 shows the schematic of an LCOS-based 1 × 9 WSS [79]. In addition to wavelength switching, LCOS offers some additional features, such as

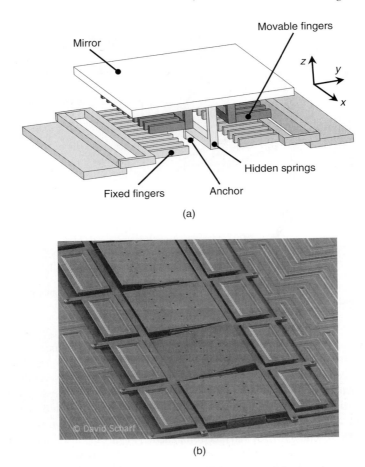

Figure 19.9 (a) Schematic and (b) SEM of the analog micromirror with hidden vertical comb drive actuators (SEM © David Scharf) (this figure may be seen in color on the included CD-ROM).

programmable bandwidth and channel spacing, or a mixture of channel spacing (e.g., 50 and 100 GHz). It also supports the "drop-and-continue" function [80].

Scaling of WSS has been analyzed in Ref. [78]. The figure of merit is the ratio of the port count and channel spacing ($N/\Delta\lambda_{ch}$). It is proportional to the product of the effective aperture of the resolution lens and the grating dispersion. A larger port count ($N \geq 8$) is desirable for mesh optical networks where it is necessary to provide two-way links to three or four adjacent neighboring nodes. N is usually limited by optical diffraction and the optical scan angle of the micromirrors. The port count can be increased from N to N^2 by arranging the output collimator in a 2D array, as shown in Figure 19.11. This is referred to as $1 \times N^2$ WSS [78, 81, 82]. It is enabled by two-axis micromirrors [83, 84]. Two types of two-axis analog micromirror arrays have been reported for WSS applications. The first is a parallel

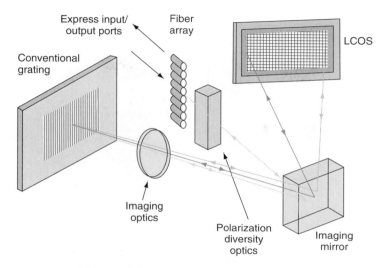

Figure 19.10 Schematic of LCOS-based 1 × 9 WSS.

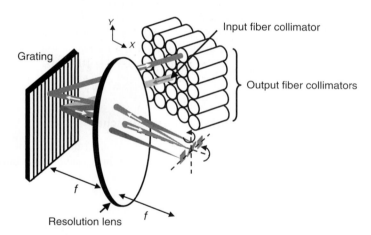

Figure 19.11 Schematic setup of 1 × N^2 using two-axis micromirror array (this figure may be seen in color on the included CD-ROM).

plate-actuated micromirror with cross-coupled torsion springs underneath the mirror [83]. Mechanical scan angles of ±4.4° and ±3.4° have been achieved at actuation voltages of ∼90 V. A 1 × 14 WSS (3 × 5 collimator array) with 50-GHz channel spacing has been constructed using this mirror array. Larger scan angles have been demonstrated using vertical motion amplifying levers [84]. The schematic of the mirror is shown in Figure 19.12. The mirror is supported by four

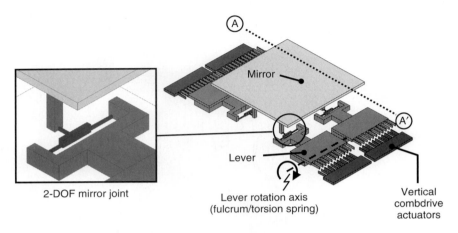

Figure 19.12 Schematic of the two-axis analog micromirror array for WSS.

levers through compliant two-axis torsion hinges. The levers amplified the vertical displacement by 3.3 times. Using four vertical combdrive actuators, scan angles of ±6.7° have been achieved for both axes at 75 V actuation voltages, with a fill factor of 98%. The resonant frequency is 5.9 kHz. By combining this micromirror array with a densely packed 2D collimator array, a WSS scalable to a port count of 1 × 32 has been demonstrated [81]. The channel spacing is 100 GHz, and the fiber-to-fiber insertion loss is 5.6 dB.

Hybrid PLC-MEMS Wavelength-Selective Switches

The discussion so far has focused on free-space optical systems. PLCs, in contrast, allow many WDM functions to be monolithically integrated on a chip. For example, 2 × 2 wavelength-selective crossconnects (WSXC) with 16 wavelength channels [85] and 1 × 9 WSS with 8 channels and 200 GHz spacing [86] have been reported using thermal-optic switches. The main drawback of the thermal-optic switch is high-power consumption and slow switching time. These are the areas where MEMS offer significant advantages. Therefore, hybrid integration of PLC and MEMS could lead to more compact, higher functional systems with low-power consumption and fast switching time.

Marom et al. [87, 88] reported a 1 × 3 hybrid WSS at 100-GHz spacing by combining the silica PLC and the MEMS tilting mirror array. The system consists of five silica PLCs arranged in a vertical stack, each containing an arrayed waveguide grating (AWG) with one star coupler terminated at the PLC edge. The bottom one is used as a demultiplexer for detection of locally dropped channels. An external spherical lens focuses the dispersed light to the micromirror array. The mirrors tilt in the vertical plane for switching of signals between the

19. Microelectromechanical Systems for Lightwave Communication

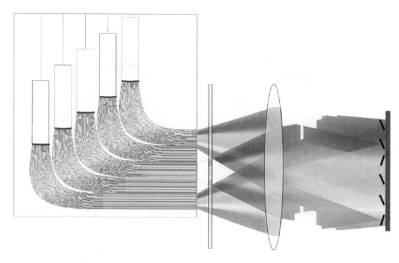

Figure 19.13 Schematic of a 1 × 9 WSS with hybrid integration of PLC and MEMS (reprinted from Ducellier et al. [89] with permission) (this figure may be seen in color on the included CD-ROM).

PLCs. An insertion loss ranging from 5 to 6.8 dB was measured using a bulk mirror. The hybrid WSS reported by Ducellier et al. [89] employs a two-axis micromirror array to steer optical beams both within a PLC (horizontally) and across stacked PLCs (vertically), as illustrated in Figure 19.13. Using two PLCs with five AWGs each, a 1 × 9 WSS has been realized. The WSS has an insertion loss of 2.8–4.3 dB for the best port and 5.6–7.8 dB for the worst port. The PDL of the device is typically 0.3 dB, and the isolation is typically greater than 35 dB over ±12.5 GHz. Wavelength-independent 1 × N optical switches with external [90] and monolithically integrated cylindrical lens [91] have also been demonstrated using a one-axis tilting mirror.

Monolithic Wavelength-Selective Switches

The hybrid integrated systems still require bulk lenses between the PLC and the MEMS micromirrors for collimation and focusing. Free-space propagation is needed to perform Fourier transformation of the optical beams [87–89]. Furthermore, optical alignment is still needed to assemble the hybrid-integrated switch. More compact system can be achieved by monolithically integrating the PLC and the MEMS micromirrors on the same substrate.

Recently, a fully integrated 1 × 4 MEMS WSS has been reported for coarse wavelength-division-multiplexing (CWDM) networks with 20-nm channel spacing [92]. The schematic of the WSS is shown in Figure 19.14. Like its free-space counterpart discussed earlier, light from waveguide is first collimated by a parabolic mirror, dispersed by a transmission micrograting, and then focused onto the integrated MEMS micromirrors. The only difference is that light is confined in

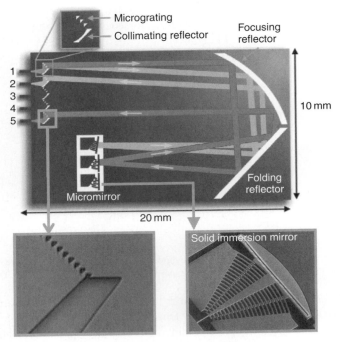

Figure 19.14 Schematic of the monolithic 1 × 4 wavelength-selective switch (WSS) realized in SOI PLC-MEMS platform (this figure may be seen in color on the included CD-ROM).

the silicon slab. The lenses in free-space systems are replaced by TIR mirrors. The etched sidewalls form the surfaces of the MEMS micromirrors. All optical and MEMS components are monolithically integrated on an SOI substrate with a 5-μm-thick device layer. The SOI platform is attractive because they are compatible with the Si PLC [93] as well as the SOI-MEMS [14] technologies. All optical paths are defined by photolithography and no optical alignment is necessary. Theoretical calculation shows that 4.1-dB insertion loss is achievable. The 1 × 4 CWDM WSS chip with eight channels has an area of 1.4×2 cm^2. Switching time of less than 1 ms has been achieved.

19.3.4 Wavelength-Selective Crossconnects

Wavelength-selective crossconnects (WSXC) are desired for mesh-based optical networks. They can reduce the cost of the networks by eliminating the optical-electrical-optical (OEO) conversions. There are several approaches to implement WSXC using MEMS technologies. One approach is to combine separate wavelength demultiplexers such as planar AWG components with wavelength-independent

$N \times N$ switches. All the channels can flow though a single large switching fabric ($>100 \times 100$ ports, such as the 3D MEMS crossconnect) [94, 95]. An alternative approach is to use a smaller switch (8×8 or 16×16 ports, such as the MEMS 2D switch) for each wavelength [19, 24]. In both cases, some of the ports of the transparent switching fabric can be connected to a conventional OEO router to enable higher level network functionality, such as packet switching, for a limited number of channels. A more power efficient approach to WSXC is to integrate wavelength multiplexing and MEMS multiport switching. This can be realized in a single monolithic component [31, 32].

The most effective approach to WSXC, however, is to build upon the WSS discussed earlier [96]. An $N \times N$ WSXC can be realized by interconnecting N modules of $1 \times N$ WSSs and N modules of $N \times 1$ WSSs [Figure 19.15(a)].

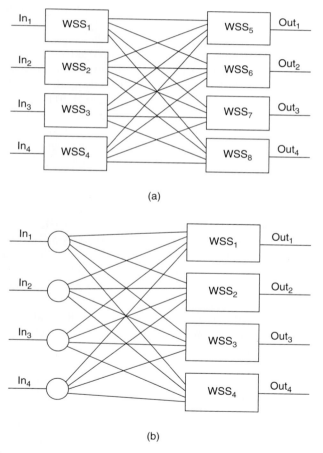

Figure 19.15 4×4 WSXC realized by (a) four 1×4 WSSs and four 4×1 WSSs, and (b) four 1×4 passive splitters and four 4×1 WSSs.

Alternatively, we can replace the WSSs in the first stage with $1 \times N$ passive optical splitters [Figure 19.15(b)]. The latter implementation has a fundamental $1/N$ splitting loss, but it allows broadcast and multicast functions. This approach was used in the 4×4 WSXC with 64 channels and on a 100-GHz grid [96]. The total insertion loss is 10.5 dB, of which 6.5 dB comes from the splitter (0.5 dB excess loss plus 6-dB splitting loss). In addition to crossconnect, their implementation also provides dynamic spectral equalization and channel-blocking capabilities. This approach to WSXC is favored by network operators because it allows flexible provisioning: a fiber node which begins as a simple spectral equalizer can be upgraded to add/drop, then to a full degree-four wavelength crossconnect, without interrupting traffic.

Using PLC-MEMS, a 2×2 WSXC with 36 wavelength channels was realized by butt-coupling four stacked PLC chips with a 36×36 array of two-axis micromirrors [95]. The MEMS array would allow 18×18 ports with 36 wavelengths. However, the optical loss is high (20 dB) because the optical axis of the steered beam is not aligned with the receiving PLC waveguide. Another drawback is the upfront investment for a large switching fabric. Another approach for PLC-MEMS WSXC employed a single arrayed waveguide lens (AWL) with three diffraction order outputs $(-1, 0, +1)$ in conjunction with an array of MEMS piston mirrors [97]. A 2×2 WSXC with 16 channels on 100-GHz grid was achieved using circulators for both I/O waveguides. The insertion loss is 10.6 dB.

Using the SOI PLC-MEMS technology described earlier, the entire WSXC can also be monolithically integrated on a chip [98]. A 4×4 WSXC is realized by integrating four 1×4 splitters, four 4×1 WSSs, and a waveguide shuffle network on a 5-μm-thick SOI [Figure 19.16(a)].

The 1×4 multimode interference (MMI) splitter is 890-μm long and 40-μm wide. The shuffle network employs 90°-waveguide bend and crossing to minimize loss and crosstalk. The 4×4 WSXC with CWDM grid has an area of 3.2×4.6 cm^2. The fiber-to-fiber insertion loss was measured to be 24 dB, which includes the 6-dB splitting loss. The excess loss can be reduced to below 3 dB by improving the fabrication process. Monolithic integration also greatly simplifies the packaging process. Since the entire switch is fully integrated, on-chip packaging can be employed to simultaneously encapsulate the MEMS actuators and protect the optical structures. A glass superstrate with an indented cavity is directly bonded on the SOI-PLC. The MEMS actuators are hermetically sealed in the cavity. Electrical interconnects can be routed through via holes in the glass. A photograph of a packaged WSXC is shown in Figure 19.16(b).

19.3.5 Spectral Intensity Filters

Wavelength control is critical to the operation of optical communication systems. WDM fiber optical systems require sources, (de)multiplexers, dispersion compensators, channel monitors, and receivers with accurate center wavelengths and bandwidths. Optical MEMS adds much needed flexibility to wavelength control

19. Microelectromechanical Systems for Lightwave Communication 731

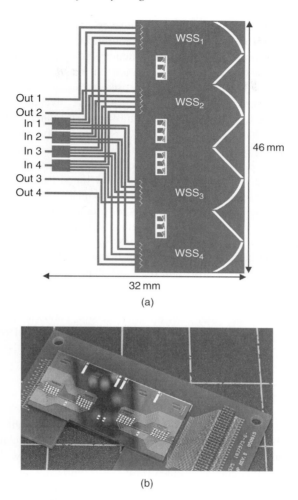

Figure 19.16 (a) Schematic of the monolithic 4 × 4 wavelength-selective crossconnect (WSXC) on SOI PLC-MEMS platform. (b) Photograph of a 4 × 4 WSXC with on-chip glass encapsulation. The MEMS actuators are connected through the via holes in glass (this figure may be seen in color on the included CD-ROM).

by providing tunable optical devices that enable better utilization of the spectrum, reduces the required number of different components to build a system, facilitates communication between different systems, and simplifies upgrades. Ultimately, the wavelength agility provided by tunable optical MEMS components, combined with the advantages of miniaturization, integration, and parallel processing, lead to communication systems with better performance and lower cost.

Optical MEMS filters and spectrometers come in a large variety of designs. Most traditional optical filters and spectrometers have MEMS counterparts, and in addition MEMS enable a number of devices that were impractical, if not impossible, to implement in traditional technologies. In this section we will describe MEMS implementations of traditional filter and spectrometer architectures, as well as several designs that rely for their operation on the characteristics of MEMS technology. The objective is not a comprehensive coverage of all Optical MEMS filters and spectrometers. Instead, the emphasis is on the advantages and challenges that are unique to the MEMS implementation of the device architectures.

A tunable Fabry–Perot (F-P) is relatively simple to implement in MEMS technology. Two single-layer dielectric or semiconductor mirrors, or a movable single-layer mirror combined with a stationary, highly reflective, multilayer mirror are sufficient to create a low-finesse F-P that can be tuned by moving the mirrors relative to each other by electrostatic or other types of MEMS actuators. This type of F-P is of limited use due to the broad reflection and transmission bands resulting from the low reflectivity of single-layer dielectric and semiconductor mirrors. In principle, we can reduce the filter bandwidth by making the cavity longer, but that is counterproductive because miniaturization is one of the major motivations for using MEMS technology. In addition, there are many applications, for example, channel filters in WDM fiber optic communication systems, where the important figure of merit is the finesse, or the ratio of free spectral range (FSR) to the transmittance bandwidth, rather than the transmission bandwidth. The finesse is determined solely by the mirror, as can be seen from the standard formula for lossless F-Ps:

$$\text{Finesse} \equiv \frac{\Delta f_{\text{FSR}}}{\Delta f_{\text{FWHM}}} = \frac{\pi \sqrt{\sqrt{R_1}\sqrt{R_2}}}{1 - \sqrt{R_1}\sqrt{R_2}} \approx \frac{\pi}{1 - R}$$

where $R_{1,2}$ are the reflectivities of the two mirrors. Most applications require higher finesse than can be obtained with single-layer dielectric mirrors to achieve acceptable specifications. Until the arrival of photonic crystals, which will be discussed in a later section of this chapter, high-finesse F-Ps could only be fabricated using multilayer dielectric mirrors. To be movable by MEMS actuators, these multilayer dielectric mirrors have to be free-standing and therefore not supported by the rigid substrates that are traditionally used. This presents challenges in MEMS fabrication, due to the thermal stresses that build up in the mirrors stacks, leading to temperature-dependent mirror curvature that is unacceptable for high-finesse applications. This fabrication challenge has been met through a variety of approaches. Early work [99] used the full thickness of silicon wafer to provide a solid substrate. These devices were fabricated by wafer bonding and were relatively bulky. Smaller devices have been created by using free-standing Si–SiO$_2$ mirror stacks, but these mirrors have some problems with curvature [100]. By careful compensation of the material stress in the dielectric stack, silicon-compatible, free-standing, dielectric mirrors with better than 99% reflectivity have been demonstrated [101].

A very elegant and powerful approach is to grow lattice-matched semiconductor mirrors, most typically using molecular beam epitaxy (MBE). Early work in AlGaAs [102, 103] has led to rapid development of this field with several important contributors [104–106], and it has also led to the creation of MEMS-tunable vertical cavity surface emitting semiconductor lasers (VCSELs) (for an in-depth description of MEMS-tunable VCSELs see Ref. [107]). This type of fabrication process results in excellent mirrors, but the process is not compatible with silicon technology.

An approach that avoids the complications of bending due to thermal stress in free-standing dielectric stacks is to tune the filters thermally, rather than by mechanical motion. In such thermally tuned devices, the dielectric mirrors are deposited directly on a silicon substrate with an intermediate film of thermooptical material. The temperature and, therefore, the effective optical thickness of the material between the dielectric mirrors is controlled by thermal dissipation in integrated resistors. This approach has been used to create tunable channel-dropping WDM filters with narrow transition bands [108].

19.3.6 Dispersion Compensators

In contrast to channel selection, dispersion compensation in WDM systems does not require high out-of-band suppression, so low-finesse F-P provides sufficient dispersion for most fiber communication systems. To avoid unwanted amplitude variations, dispersion compensation is typically carried out with Gires–Tournois (G-T) interferometers [109]. The G-T interferometer is an F-P with a highly reflective back mirror. In the ideal case of plane wave incidence and a 100% reflective back mirror (r_2), the reflectance is always unity, so the ideal G-T is an all-pass filter with a strong phase variation around resonance.

Figure 19.17 shows a G-T based on the mechanical antireflection switch (MARS) device (to be discussed in Section 19.5.1) [5]. The MARS device is a low-finesse G-T with a highly reflective dielectric stack as the back mirror and a single, free-standing, $\lambda/4$ silicon-nitride film as the front mirror. This device performs very well as a dispersion-slope compensator in spite of the relatively low finesse. A linear dispersion tunable from −100 to 100 ps/nm over 50 GHz in C-band has been demonstrated experimentally [110].

A variation of the G-T interferometer operates on oblique incidence so that the optical beam follows a zigzag pattern and the reflections from the back mirror are spatially separated as shown in Figure 19.18. The output from this device is the interference pattern of the first reflected beam and the partially transmitted beams from the front mirror. This geometry allows the phase of the reflections to be individually modulated and enables tuning of a variety of filter characteristics. Tunable (de)interleavers [111], amplitude filters [112], and dispersion compensators with linear dispersion tunable from −130 to 150 ps/nm over 40 GHz in C-band [113] have been demonstrated. This variation of the G-T interferometer is not an all-pass filter, even in the idealized case, so careful attention has to be paid to avoid parasitic amplitude modulation when the phase is tuned.

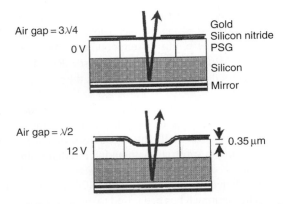

Figure 19.17 Microelectromechanical systems all-pass filter schematic showing the change in air gap with applied voltage (reprinted from Goossen et al. [5] with permission).

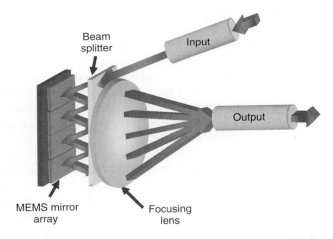

Figure 19.18 Schematic diagram of the MEMS Gires–Tournois interferometer (this figure may be seen in color on the included CD-ROM).

Alternatively, tunable dispersion compensator can also be realized by hybrid-integrated PLC-MEMS [114]. A dispersion compensator with ± 500 ps/nm tuning range and 100 GHz FSR was realized using MEMS-deformable membrane.

19.3.7 Transform Spectrometers

Transform spectrometers also lend themselves to MEMS implementations, and several different architectures have been demonstrated. Figure 19.19 illustrates a design that uses a traditional Michelson interferometer, in which the movable mirror

19. *Microelectromechanical Systems for Lightwave Communication* 735

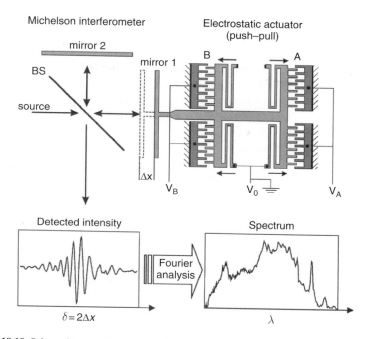

Figure 19.19 Schematic of a Fourier transform spectrometer based on a traditional Michelson interferometer with a MEMS electrostatic actuator (reprinted from Manzardo et al. [115] with permission).

is actuated by an electrostatic combdrive [115]. The light from the source is split into two parts by a beam splitter and the two parts are reflected from two different mirrors, one of which is movable to create a variable path length for the two parts of the incident light. After reflection, the two parts of the incident light recombine and interfere on the beam splitter. Each wavelength of the detected optical power or intensity, P_{detected}, shows a harmonic dependence on the path-length difference, Δx:

$$P_{\text{detected}} = P_{\text{incident}} \cos\left[\frac{4\pi \Delta x}{\lambda}\right]$$

where P_{incident} is the incident optical power and λ is the wavelength. The Fourier transform of the optical spectrum is obtained by varying the path-length difference, Δx, and the spectrum is found through an inverse Fourier transform.

Common to all transform spectrometers, the spectral resolution, $\Delta\lambda_{\text{FWHM}}$, is determined by the total range of motion, Δx_{max}, of the moving mirror [116]:

$$\frac{\Delta\lambda_{\text{FWHM}}}{\lambda} = 0.5 \frac{\lambda}{\Delta x_{\text{max}}}$$

This simple equation highlights the main challenge in MEMS implementations of transform spectrometers. Because the spectral resolution is inversely proportional to the maximum actuation distance that can be achieved, long-travel actuators are required. The micron-scale displacements that are sufficient for many MEMS applications are not useful here, and even long-range MEMS actuators, for example electrostatic combdrives with several tens of microns of motion, achieve only modest resolutions. The challenge in implementing MEMS transform spectrometers with good resolution therefore boils down to the creation of fast, accurate, and reliable long-range actuators. Transform spectrometers are also relatively complex systems with several optical components that must be well aligned. This represents both a challenge and an opportunity for MEMS. It is difficult to fabricate several very different optical devices in the same MEMS process, but if it can be done, the accuracy of MEMS technology simplifies alignment and packaging.

One approach to high-resolution transform spectroscopy with MEMS is based on "microjoinery" [117]. The microjoinery spectrometer utilizes the precision of bulk micromachining to establish a very accurate and long-range path for a slider that carries the moving reference mirror of the interferometer. The strength of this solution is that the reference mirror can be moved over long distances to create a spectrometer with very good spectral resolution. The challenge is to integrate a suitable actuator that provides the motion over the full range of the track established by the microjoined slider. Using magnetic actuation with external magnetic fields, motion of several centimeters has been demonstrated, resulting in fractional resolution on the order of 10^{-5} in the visible wavelength range.

Transform spectrometers with modest resolution can be integrated on a single chip by using vertical mirrors with integrated actuators [118]. The single-chip, integrated transform spectrometer shown in Figure 19.20 is implemented through a combination of anisotropic etching and deep reactive ion etching (DRIE). Anisotropic etching is, as the name implies, dependent on crystalline orientation, that is, it etches different crystalline planes at different etch rates, resulting in very smooth surfaces that can be used as optical interfaces and mirrors. DRIE is used to create electrostatic actuators and fiber grooves that should not be restricted by the crystalline orientation of the silicon. In the implementation of the architecture shown in Figure 19.20, the beam splitter and the movable mirror are both defined using anisotropic etching, while the fixed mirror is defined by DRIE. It is also possible to use a combination of two anisotropically etched mirrors instead of the DRIE-defined fixed mirror.

The transform spectrometers described so far are of the traditional Michelson-interferometer design. The characteristic advantages and challenges of MEMS technology have inspired nontraditional solutions of different kinds. One such MEMS architecture is shown in Figure 19.21 that depicts a reflection phase grating with a variable grating amplitude [119]. The reflected optical power from the grating has a harmonic dependence on the grating amplitude, just like the

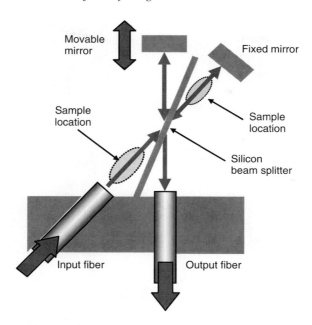

Figure 19.20 Schematic of single-chip integrated transform spectrometer based on vertical micromirrors with integrated MEMS actuators. The nonnormal incidence on the beam splitter is due to the restrictions of the surfaces that can be defined by anisotropic etching of Si (this figure may be seen in color on the included CD-ROM).

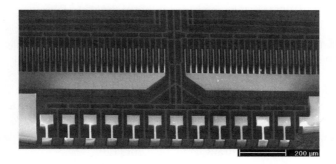

Figure 19.21 Transform spectrometer based on a diffraction phase grating with a tunable grating amplitude. The grating consists of alternating fixed (light) and movable (dark) mirror elements. The movable mirrors are displaced by an electrostatic actuator to create a variable path-length difference (reprinted from Manzardo et al. [119] with permission).

dependence of traditional Fourier transform spectrometers on the optical path-length difference. The grating transform spectrometer maps readily onto the more traditional Michelson structure. The main conceptual difference is that the grating

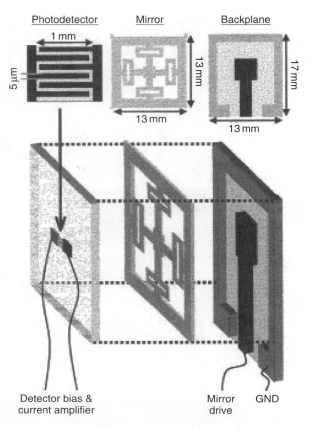

Figure 19.22 Transform spectrometer using a semitransparent detector in a standing-wave cavity (reprinted from Kung et al. [120] with permission).

acts both as a beam splitter and as a two-beam interferometer with a variable pathlength difference.

In the spectrometer shown in Figure 19.22, a standing wave is being sampled at one spatial location by a semitransparent detector [120]. The period of the standing wave pattern is varied by moving the rear mirrors of the standing-wave cavity. The response of this spectrometer is again a harmonic function of the mirror displacement, leading to the same dependence of resolution on mirror displacement as in other transform spectrometers.

The two implementations of Figures 19.21 and 19.22 show how the flexibility of MEMS technology enables nontraditional solutions. Both these implementations are very compact, thereby facilitating miniaturization, integration, and packaging. Neither of the two achieves better than modest resolution due to the limited maximum displacement of their actuators.

19.3.8 Diffractive Spectrometers and Spectral Synthesis

The nontraditional implementations of transform spectrometers described above illustrate one of the major strengths of optical MEMS. The flexibility, complexity, and accuracy afforded by lithography enables architectures that cannot practically be created by traditional manufacturing technologies. Another advantage of optical MEMS is the ability to create SLMs and other devices that require large numbers of identical components. This attribute has been exploited to expand the functionality of grating spectrometers. A traditional grating spectrometer measures spectral amplitude by dispersing the wavelengths of the incoming light over a range of angles. The spectral amplitude can be measured by using an array of detectors, or by rotating the grating and using a single detector. A variation of this traditional concept is to place an SLM in the back focal plane of the lens that captures the dispersed light from the grating as shown in Figure 19.23.

In this device, the spectral components of the incident are dispersed by the grating and modulated by the SLM. The SLM may modulate the amplitude or phase, or both, of the dispersed light. This very versatile configuration can therefore be used for Hadamard spectroscopy [121], optical pulse shaping [122, 123], spectral phase measurements [122], adjustable time delays [124–126], wavelength selective optical WDM switches [32], WDM add–drop filters [69], and a wide variety of other applications. The SLMs used in the architecture must be tailored to the specific applications. The flexibility in size, form, and function of optical MEMS has made it the technology of choice for a large number of these applications.

A variation of the grating spectrometer that uses optical MEMS not as an SLM to modulate dispersed light, but as the dispersing element is shown in Figure 19.24. The idea here is to deform the SLM, which here acts as a grating or dispersive

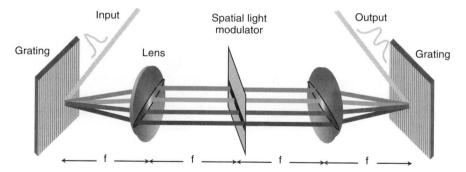

Figure 19.23 Basic optical system for spectral synthesis and measurements. The incident light is collimated onto a diffraction grating and dispersed on a spatial light modulator. The modulated spectral components from the SLM are recombined on the grating and focused on the output, which can be an optical fiber, a detector, or a detector array. The SLM is shown here as a transmission device, but it is more common to use a reflective SLM in MEMS applications (this figure may be seen in color on the included CD-ROM).

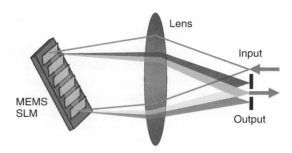

Figure 19.24 Optical MEMS SLM as a diffractive element for synthesis of spectral amplitude and phase. The SLM is deformed to create a surface that diffracts the desired spectral components of the incident light into a specific output (this figure may be seen in color on the included CD-ROM).

element, to dynamically change the characteristics of the filter or spectrometer. This architecture is neither as powerful in terms of spectral manipulation, nor as efficient in terms of optical throughput as the one shown in Figure 19.23. It does, however, require fewer components and it is more compact, which makes it preferable for many practical systems, including displays [127–129], WDM variable attenuators [130], interferometric displacement sensors for a variety of applications [131–135], spectral synthesis [136], and compact optical filters [137] and pulse shapers [138].

The diffractive MEMS device shown in Figure 19.24 is conceptually similar to an adaptive optics (AO) mirror [139]. In AO a deformable surface is employed to compensate for aberrations imposed on an optical wavefront by inhomogeneities in the transport medium between the source and detector. Most, if not all, filter applications require much more wavelength dispersion than can be provided by AO mirrors that are designed for wavefront corrections. This can be understood by considering the impulse response of the filter; the output is an impulse train corresponding to the height distribution of the individual reflectors of the diffractive surface. Neglecting weak wavelength dependencies in diffraction efficiency and output coupling, the transmission of the filter is given by the Fourier transform of the impulse response, which in turn is determined by the height distribution of the diffractive surface [140].

This simple conceptual picture of diffractive filter operation leads to three insights that are of importance to MEMS implementations. First, the filter transfer function is the Fourier transform of a nonnegative sequence, which means that in principle any transfer function can be synthesized to within a constant (see Ref. [141] for details on the restrictions on the synthesized transfer functions).

Another observation we can make from Figure 19.24 is that the total length of the impulse response is given by the maximum difference of positions of the reflectors of the diffractive MEMS along the optical axis. The spectral resolution of the filter is therefore inversely proportional to the height difference of the

MEMS SLM along the optical axis. For most applications, the resolution specifications require the height to be much larger than the height of practical MEMS structures by themselves, so grazing incidence and large diffraction angles are necessary. Early MEMS-diffractive filters that were designed for normal incidence [142] are therefore useful only for low-resolution applications. Better resolution can be obtained by adding another diffractive element [143], or by creating a diffractive structure with high diffraction angles [144].

The third characteristic of diffractive filters that is illustrated by Figure 19.24 is that the loss of the filter is proportional to its complexity. The incident light is split into N spatially separate channels and then recombined into a single-output channel, leading to a $1/N$ loss. This is true for any optical systems that separate the input into N equal channels that are incoherently recombined to create a single output. This means that only wavelengths that are reflected in phase from all N grating elements are completely transmitted by the filter. In other words, to have high-optical throughput, the filter must essentially act as a grating with all grating elements acting in-phase in the optical pass band. If good spectral resolution is also required, the diffractive element must have a high-diffraction angle as discussed before. To achieve both high throughput and good resolution, the diffractive element should behave much like a blazed grating and operated such that all the reflectors of the MEMS device are in-phase in the optical pass band. Such diffractive MEMS have been demonstrated as amplitude filters [145] and tunable WDM interleavers [146].

19.4 TUNABLE LASERS

In the filter implementations described in this section, MEMS provides a means to fabricate optical components, as well as a substrate for integration and packaging. It is clear from these filter implementations that one of the main advantages of optical MEMS is the opportunity for system-level integration. One of the successful systems applications that utilize optical filters is tunable lasers. We have mentioned above the VCSELs with tunable cavity length. Here we will describe MEMS implementations of traditional external cavity semiconductor diode lasers (ECSDLs).

A typical ECSDL has a semiconductor gain medium with a single-mode waveguide. The front facet of the gain medium is antireflection (AR) coated, and the output of the single-mode waveguide is collimated onto a diffraction grating. The incident optical mode on the grating is retroreflected back into the waveguide to form an external cavity. This setup is the traditional Littrow configuration as shown in Figure 19.25(a). An alternative design, the Littman configuration, is shown in Figure 19.25(b). Here the incident light on the grating is diffracted onto a mirror that retroreflects the light through the grating back to the waveguide to create the optical cavity. The advantage of the Littman configuration is that the light is diffracted from the grating twice per round trip of the cavity, leading to better out-of-band suppression in the grating filter.

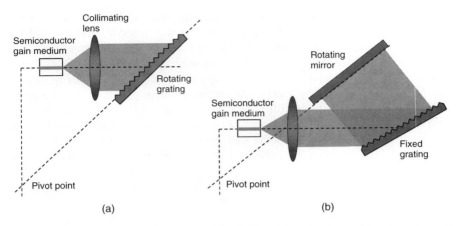

Figure 19.25 Schematic diagrams of the traditional Littrow (a) and Littman (b) configurations of tunable external cavity lasers (this figure may be seen in color on the included CD-ROM).

The laser systems of Figure 19.25 create two interacting filters: the cavity itself with an FSR that is determined by the cavity length, and the grating that only reflects one wavelength in the correct direction to establish retroreflection. To obtain lasing without an excessively high pumping threshold, these two filters must be aligned in wavelength, which means that the cavity length has to be controlled with subwavelength accuracy. Accurate alignment and cavity length control are therefore necessary and motivates the use of MEMS technology.

To tune the laser wavelength, the grating is rotated so that the center wavelength of the grating filter is changed. To achieve continuous, mode-hop-free wavelength tuning, the grating must be rotated and translated so that the cavity mode stays aligned with the grating filter. It is well known that if the grating (or mirror in the case of the Littman configuration) is rotated around a pivot point located at the intersection of the line through the rotating surface and the normal to the optical axis at a point that is a distance $n\lambda_{vac}$ from the rotating element along the optical axis, where n is the number of wavelengths in the cavity and λ_{vac} is the vacuum wavelength, then the cavity mode and grating filter stays aligned during rotation [147, 148].

In principle, an ECSLD can therefore be controlled by an actuator with one degree of freedom of motion. In practice, at least one extra degree of freedom is required to initially align the cavity mode and grating filter, and to compensate for dispersion in the optical components of the cavity. Academic research on MEMS implementations of Littrow [149, 150] and Littman [151] ECSDLs has focused on development of accurate one-degree-of-freedom actuators that can provide stable, mode-hop-free tuning after initial alignment. An interesting alternative is to use a diffractive element with separate phase and amplitude control to avoid macroscopic motion in the external cavity [152].

In contrast to academic research, commercial developments, which have mostly adopted the Littman configuration to achieve better side-mode suppression, have incorporated two or more degrees of freedom in the actuator design so that both initial alignment and compensation for cavity dispersion can be controlled by the MEMS structure [153, 154]. These types of lasers have excellent stability and optical characteristics. The complexity of the optical hardware and the control circuitry lead to costs that are significantly higher than for fixed-wavelength semiconductor lasers. Systems solutions that use single fixed-wavelength lasers are therefore still preferred, even in application that would benefit from wavelength tuning. One solution with intermediate complexity and proven market potential is to use a set of fixed-wavelength semiconductor lasers and select the output of the one with the most appropriate wavelength, using a MEMS mirror for the selection [155].

19.5 OTHER OPTICAL MEMS DEVICES

19.5.1 Data Modulators

The first practical application of MEMS devices in fiber communications was as an optical data modulator, originally intended for a low-cost fiber-to-the-home network. A modulator is essentially a 1×1 switch, operated either in transmission (two fibers) or in reflection (single fiber). The optical power is provided by a constant intensity remote source, and the modulator imprints a data signal by opening and closing in response to an applied voltage. Signaling in DWDM fiber networks usually requires an expensive wavelength-controlled laser at each remote terminal. Passive data modulators offered a potentially inexpensive solution, but waveguide modulators were too expensive and too narrow in optical spectral bandwidth to be practical. MEMS offered a new and practical solution.

The MARS modulator is a variable air-gap etalon operated in reflection. The basic structure is a quarter-wave dielectric AR coating suspended above a silicon substrate [5]. The quarter-wave layer is made of silicon nitride with $\frac{1}{4}\lambda$ optical path (index times thickness), which is roughly 0.2 µm for the 1550-nm telecom wavelength. The mechanically active silicon nitride layer is suspended over an air gap created by a $\frac{3}{4}\lambda$-thick phosphosilicate glass sacrificial layer (0.6 µm). Without deformation, the device acts as a dielectric mirror with about 70% (−1.5 dB) reflectivity. Voltage applied to electrodes on top of the membrane creates an electrostatic force and pulls the membrane closer to the substrate, while membrane tension provides a linear restoring force. When the membrane gap is reduced to $\lambda/2$, the layer becomes an AR coating with close to zero reflectivity. A switching contrast ratio of 10 dB or more was readily achieved over a wide (30 nm) spectral bandwidth.

The initial MARS device, shown in Figure 19.26, consisted of a 22-µm optical window supported by X-shaped arms, and had a resonant frequency of 1.1 MHz. Later devices used a higher yield structure with a symmetric "drum head" geometry

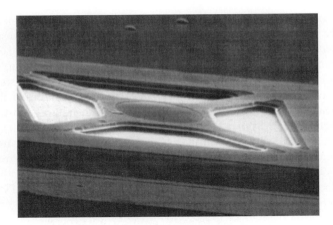

Figure 19.26 Microelectromechanical systems etalon modulator used for digital data modulation at over 1 Mb/s. The circular optical aperture is 22 mm in diameter.

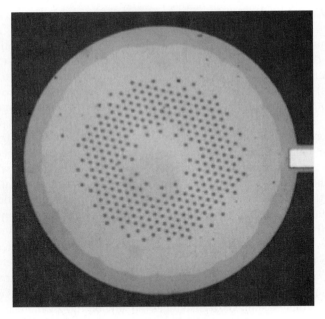

Figure 19.27 Microelectromechanical systems etalon variable attenuator using a 0.5-mm-diameter drumhead geometry. The lighter area covers an air gap between the silicon substrate, the hexagonally distributed spots are etch access holes.

[156, 157] as shown in Figure 19.27. These devices were capable of relatively high-speed operation: by optimizing the size and spacing of the etch access holes provide critical mechanical damping, digital modulation above 16 Mb/s was demonstrated

19. Microelectromechanical Systems for Lightwave Communication

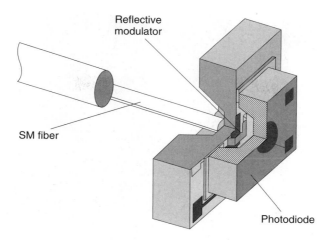

Figure 19.28 Package configuration for a MEMS data transceiver.

[158]. While such data rates are no longer relevant for telecom, even for fiber to the home, related modulators are useful for low-power dissipation telemetry from remote sensors using free-space optical communications.

These early devices provided a proving ground for the reliability and packaging of optical MEMS telecom components. Initial skepticism from conservative telecom engineers was combated by parallel testing of device array operated for months to provide trillions of operating cycles. The packaging of optical MEMS devices provided new challenges for MEMS engineers, but the simple end-coupled configuration was relatively straightforward to implement. Figure 19.28 shows the configuration for a duplex modulator incorporating a MEMS etalon, where data can be received by a photodiode and transmitted by modulating the etalon reflectivity [159].

19.5.2 Variable Attenuators

Data modulators are operated with digital signals, but the fundamental response of an etalon modulator is analog. Electrically controlled VOAs at that time were constructed with bulk optical components with electromechanical actuation, with 10–100-ms response. Erbium fiber amplifiers can use VOA to suppress transient power surges, but the timescale required was 10 µs, much slower than the data modulation rate. MEMS provided an attractive replacement for optomechanical VOAs, and this turned out to the first volume application for MEMS devices in telecom networks.

The first MEMS VOA was fabricated by scaling the optical aperture of a MARS modulator from 25 to 300 µm, so that it could be illuminated with a collimated beam. The reflected signal was focused into a separate output fiber, avoiding the need for external splitters or circulators to separate the output signal [160].

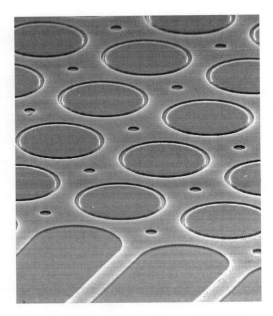

Figure 19.29 Lightconnect's diffractive MEMS VOA.

The first such VOA device is shown in Figure 19.29. The wavelength dependence of a simple etalon was reduced using a more complex three-layer dielectric stack as the mechanically active structure, where the original ¼λ silicon nitride layer is sandwiched between conductive polysilicon top (½λ thickness) and bottom (¼λ thickness) layers. This attenuator provided fast (3 μs) response with 30-dB controllable attenuation over the 40-nm operating bandwidth, with 0.06-dB polarization-dependent loss, and also supported the 100-mW power level present in amplifiers. However, the 3-dB insertion loss was excessive.

The most direct possible approach to attenuation is to use a MEMS actuator to insert an optical block between an I/O fiber. This was implemented with surface micromachining (MUMPS process) [161] and with a comb-driven SOI device [162]. Such VOAs offered excellent dynamic range (measurement limited at 90 dB), but polarization-dependent loss could be large ($\gg$1 dB) at high attenuations.

Further improvement was needed, and was made. Combining the collimated beam geometry with a first-surface torsion mirror reflector provided a low-insertion loss structure with excellent spectral and polarization performance. For example, the device demonstrated by Isamoto et al. [163] achieved 40 dB attenuation with a 600-μm mirror driven with 5 V to tilt up to 0.3°. Similar configurations were commercialized, though the specific designs have not been published.

Another commercial MEMS VOA is based on a diffractive MEMS device [4] also used with a collimated beam. This device provides excellent optical performance as well as high speed, stable operation with 30 dB contrast, and less than 40-μs response time, using an 8 V drive. A novel structure with circularly symmetric features, shown in Figure 19.27, was used to suppress polarization-dependent loss to under 0.2 dB [130]. This device was one of the first Telcordia-qualified MEMS components, with 40,000 units reportedly shipped by 2005 [164].

19.6 EMERGING MEMS TECHNOLOGIES AND APPLICATIONS

19.6.1 MEMS-Tunable Microdisk/Microring Resonators

Microdisk or microring resonators offer another order of magnitude size reduction for a wide range of WDM functions, such as add–drop multiplexers [165], dispersion compensators [166], modulators [167], and WDM lasers [168]. Semiconductor microresonators with high-index contrast can further reduce the resonator dimensions, producing wide FSRs and small footprints [169]. Integrating MEMS with microresonators will enable a host of tunable WDM functions [170–172]. Compared with other tuning mechanisms (thermal tuning [173, 174], electrical carrier injection [175], electroabsorption [176], or gain trimming [177] in III–V semiconductors), MEMS tuning is more efficient and consumes much less power. The ability to physically change the spacing between the waveguide and the microresonator enables us to control the coupling coefficient, an important tuning parameter for most signal processing functions but difficult to achieve by conventional means.

Lee and Wu [171] reported a silicon-tunable microdisk resonator with tunable optical coupling using MEMS actuators. The SEM of the device is shown in Figure 19.30. This is a vertically coupled microdisk resonator with suspended waveguides around the microdisk. The optical coupling coefficient is controlled by pulling the waveguide toward the microdisk. The quality factor of microdisk is measured to be 10^5, thanks to the sidewall smoothing process by hydrogen annealing [178]. The initial gap spacing between the waveguide and the microdisk is 1 μm. At zero bias, there is literally no coupling and the microdisk is effectively "turned off." With a voltage applied, the microresonator can switch among under-, critical, or over-coupling regimes dynamically. At critical coupling, the optical transmittance of the through waveguide is suppressed by 30 dB. In the overcoupling regime, the transmission intensity is nearly 100%, while the phases are perturbed around the resonance, similar to the all-pass optical filters discussed in the dispersion compensation section. This tunable microdisk resonator has many applications. The group delay and group velocity dispersion can be tuned by varying the gap spacing. A delay time tunable from 27 to 65 ps and dispersion from 185 to 1200 ps/nm have been experimentally demonstrated [171]. By actuating I/O waveguides, a reconfigurable optical add–drop multiplexer (ROADM) [179] has also been realized. Multiple tunable microdisks can be integrated to form WSS and WSXC. For telecom applications, high-order resonators are needed to achieve flat-top spectral response [173].

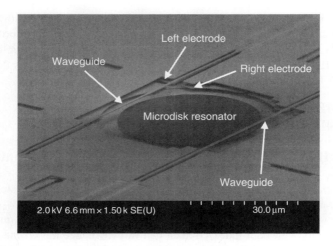

Figure 19.30 Scanning electron micrograph of MEMS microdisk resonator with variable optical couplers (reprinted from Lee and Wu [171] with permission). The suspended waveguides can be deformed by electrostatic actuation, which change the gap spacing between the waveguide and the microdisk.

Another MEMS microring ROADM was reported by Nielson et al. [172]. They used MEMS actuators to move an optically lossy film to cover the microring. When the film is in contact with the resonator, the quality factor (Q) is significantly lowered, and the resonant wavelength is no longer switched to the drop port. A 60-μs response time has been measured experimentally.

19.6.2 Photonic Crystals with MEMS Actuators

Photonic crystals and photonic bandgap materials afford unprecedented control over optical fields. These types of materials and structures are already having impact on optical MEMS, and, because this photonic crystal technology is in a very early stage, it is likely that the impact will become more significant in the future. Here we will point out some of the developments that are most exciting to designers of optical MEMS.

On the basis of the mechanism of guided resonance [180], photonic crystals can be designed to provide high reflectivity in a single-semiconductor film of sub-wavelength thickness [181–186]. These types of mirrors open up for more compact Optical MEMS devices with better temperature characteristics and more robust surfaces than for devices with the metal mirrors used in most Optical MEMS applications. High-reflectivity mirrors that do not suffer from the optical field penetration of dielectric stacks also enable compact optical cavities for optical modulators, sources, and sensors.

Photonic crystals can also be dynamically modified by MEMS actuators to create novel optical devices. A variety of different approaches has been proposed and demonstrated. Stretchable photonic crystals [187] allow the complete crystal to be dynamically altered. Photonic crystal waveguide devices have been modulated through evanescent coupling [188] and by atomic force microscopy (AFM) tips [189], and by optical carrier injection [190], and waveguide switches with electrostatic actuation have been demonstrated [191, 192]. Near-field coupling between photonic crystals has been shown to create strong modulation as a function of small relative displacements [193, 194], and the usefulness of this effect has been demonstrated in displacement sensors [195, 196] and optical filters/modulators [197, 198]. The technology is very much in an embryonic stage, and the experimental devices are proof-of-concept demonstrations that are far from ready for commercialization. The field is, however, developing very fast. New and improved application concepts are introduced at a high rate, so the opportunities for commercial development in the relatively near future seem very promising.

To take full advantage of these opportunities will require developments in MEMS technology. The very same properties of photonic crystals that make them useful for optical devices also make them extremely sensitive to pattern irregularities and surface defects. Commercial development will therefore require improved MEMS surface treatments and much better lithography than is commonly used for commercial MEMS today.

19.7 CONCLUSION

We have reviewed the recent progresses in optical MEMS for lightwave communication applications. In the past decade, we have witnessed an explosive growth and accelerated maturation of the MEMS technologies. Many innovative MEMS devices and optical designs have been introduced. Several components have been transformed from laboratory prototypes into packaged products that meet Telcordia reliability qualifications. Significant progresses have been made in optical switches and crossconnects, and various wavelength-selective devices such as filters, spectral equalizers and tunable dispersion compensators, wavelength add/drop multiplexers, WSS and WSXC, and tunable lasers. In addition to the original purposes, the technologies and expertise developed in the last decade are also available for new emerging applications.

REFERENCES

[1] K. E. Petersen, "Micromechanical light modulator array fabricated on silicon [laser display applications]," *Appl. Phys. Lett.*, 31, 521–523, 1977.
[2] K. E. Petersen, "Silicon torsional scanning mirror," *IBM J. Res. Develop.*, 24, 631–637, 1980.
[3] L. J. Hornbeck, "Projection Displays and MEMS: Timely Convergence for a Bright Future," in Proc. *SPIE-Int. Soc. Opt. Eng.*, USA, 2639, 1995, p. 2.

[4] O. Solgaard, F. S. A. Sandejas, and D. M. Bloom, "Deformable grating optical modulator," *Opt. Lett.*, 17, 688–690, 1992.
[5] K. W. Goossen, J. A. Walker, and S. C. Arney, "Silicon modulator based on mechanically-active antireflection layer with 1 Mbit/Sec capability for fiber-in-the-loop applications," *IEEE Photon. Technol. Lett.*, 6, 1119–1121, September 1994.
[6] S. S. Lee, L. Y. Lin, and M. C. Wu, "Surface-micromachined free-space fibre-optic switches," *Electron. Lett.*, 31, 1481, 1995.
[7] M. C. Wu, "Micromachining for Optical and Optoelectronic Systems," in Proc. *IEEE*, 85, 1997, pp. 1833–1856.
[8] R. S. Muller and K. Y. Lau, "Surface-Micromachined Microoptical Elements and Systems," in Proc. *IEEE*, 86, 1998, pp. 1705–1720.
[9] L. Y. Lin, E. L. Goldstein, and R. W. Tkach, "Free-space micromachined optical switches for optical networking," *IEEE J. Sel. Top. Quantum Electron.*, 5, 4–9, 1999.
[10] D. J. Bishop, C. R. Giles, and S. R. Das, "The rise of optical switching," *Sci. Am.*, 284, 88–94, January 2001.
[11] J. E. Ford, "Optical MEMS: Legacy of the Telecom Boom," *Solid State Sensor, Actuator and Microsystems Workshop*, Hilton Head SC, June 6–10, 2004, pp. 1–3.
[12] H. Toshiyoshi and H. Fujita, "Electrostatic micro torsion mirrors for an optical switch matrix," *J. Microelectromech. Syst.*, 5, 231–237, 1996.
[13] R. A. Miller, Y. C. Tai, G. Xu et al., "An Electromagnetic MEMS 2×2 Fiber Optic Bypass Switch," in Proc. *International Conference on Solid-State Sensors and Actuators*, paper 1A4, 1997.
[14] C. Marxer, C. Thio, M. A. Gretillat et al., "Vertical mirrors fabricated by deep reactive ion etching for fiber-optic switching applications," *J. Microelectromech. Syst.*, 6, 277–285, 1997.
[15] C. Marxer and N. F. de Rooij, "Micro-opto-mechanical 2×2 switch for single-mode fibers based on plasma-etched silicon mirror and electrostatic actuation," *J. Lightw. Technol.*, 17, 2–6, 1999.
[16] B. Hichwa, M. Duelli, D. Friedrich et al., "A Unique Latching 2×2 MEMS Fiber Optics Switch," in Proc. *IEEE Conference Proceedings – Optical MEMS 2000*, 2000.
[17] W. Noell, P. A. Clerc, F. Duport et al., "Novel Process-Insensitive Latchable 2×2 Optical Cross Connector for Single- and Multimode Optical MEMS Fiber Switches," in Proc. *2003 IEEE/LEOS International Conference on Optical MEMS* (Cat. No. 03EX682), Piscataway, NJ, USA, 2003, pp. 49–50.
[18] L. Y. Lin, E. L. Goldstein, and R. W. Tkach, "Free-space micromachined optical switches with submillisecond switching time for large-scale optical crossconnects," *IEEE Photon. Technol. Lett.*, 10, 525–527, 1998.
[19] L. Fan, S. Gloeckner, P. D. Dobblelaere et al., "Digital MEMS Switch for Planar Photonic Crossconnects," in Proc. *OFC. Postconference Technical Digest* (IEEE Cat. No. 02CH37339). Opt Soc. America. Part vol.1, 2002, Washington, DC, USA, 1, 2002, pp. 93–94.
[20] B. Behin, K. Y. Lau, and R. S. Muller, "Magnetically Actuated Micromirrors for Fiber-Optic Switching," in Proc. *Technical Digest. Solid-State Sensor and Actuator Workshop. Transducer Res. Found. 1998*, Cleveland, OH, USA, 1998, pp. 273–276.
[21] R. L. Wood, R. Mahadevan, and E. Hill, "MEMS 2D Matrix Switch," in Proc. *OFC. Postconference Technical Digest* (IEEE Cat. No. 02CH37339). Opt Soc. America. Part vol.1, 2002, Washington, DC, USA, 1, 2002, pp. 91–92.
[22] R. T. Chen, H. Nguyen, and M. C. Wu, "A high-speed low-voltage stress-induced micromachined 2×2 optical switch," *IEEE Photon. Technol. Lett.*, 11, 1396–1398, 1999.
[23] J.-N. Kuo, G.-B. Lee, and W.-F. Pan, "A high-speed low-voltage double-switch optical crossconnect using stress-induced bending micromirrors," *IEEE Photon. Technol. Lett.*, 16, 2042, 2004.
[24] L.-Y. Lin, E. L. Goldstein, and R. W. Tkach, "On the expandability of free-space micromachined optical cross connects," *J. Lightw. Technol.*, 18, 482–489, 2000.
[25] M. C. Wu and P. R. Patterson, "Free-space optical MEMS," in *MEMS: A Practical Guide to Design, Analysis, and Applications* (J. G. Korvink and O. Paul, eds), William Andrew Publishing, 2005, pp. 345–402.

[26] P. M. Dobbelaere, S. Gloeckner, S. K. Patra et al., "Design, Manufacture and Reliability of 2D MEMS Optical Switches," in Proc. *SPIE-Int. Soc. Opt. Eng.*, USA, 4945, 2003, pp. 39–45.
[27] T. Goh, M. Yasu, K. Hattori et al., "Low loss and high extinction ratio strictly nonblocking 16 × 16 thermooptic matrix switch on 6-in wafer using silica-based planar lightwave circuit technology," *J. Lightw. Technol.*, 19, 371–379, 2001.
[28] T. Watanabe, T. Goh, M. Okuno et al., "Silica-Based PLC 1 × 128 Thermo-Optic Switch," in Proc. *27th ECOC* (Cat. No. 01TH8551). *IEEE. Part vol.2, 2001*, Piscataway, NJ, USA, 2, 2001, pp. 134–135.
[29] J. E. Fouquet, "Compact Optical Cross-Connect Switch Based on Total Internal Reflection in a Fluid-Containing Planar Lightwave Circuit," in Proc. *OFC. Technical Digest Postconference Edition. Trends in Optics and Photonics Vol.37* (IEEE Cat. No. 00CH37079). *Opt. Soc. America. Part vol.1, 2000*, Washington, DC, USA, 1, 2000, pp. 204–206.
[30] T. Sakata, H. Togo, M. Makihara et al., "Improvement of switching time in a thermocapillarity optical switch," *J. Lightw. Technol.*, 19, 1023–1027, 2001.
[31] P. M. Hagelin, U. Krishnamoorthy, C. M. Arft et al., "Scalable Fiber Optic Switch Using Micromachined Mirrors," in Proc. *1999 International Conference on Solid-State Sensors and Actuators (Transducer '99)*, Sendai, Japan, June 7–10, 1999, pp. 782–785.
[32] P. M. Hagelin, U. Krishnamoorthy, J. P. Heritage, and O. Solgaard, "Scalable optical cross-connect switch using micromachined mirrors," *IEEE Photon. Technol. Lett.*, 12, 882–884, July 2000.
[33] R. R. A. Syms, "Scaling laws for MEMS mirror-rotation optical cross connect switches," *J. Lightw. Technol.*, 20, 1084–1094, 2002.
[34] D. T. Neilson, R. Frahm, P. Kolodner et al., "256 × 256 port optical cross-connect subsystem," *J. Lightw. Technol.*, 22, 1499–1509, 2004.
[35] T. Yamamoto, J. Yamaguchi, N. Takeuchi et al., "A three-dimensional MEMS optical switching module having 100 input and 100 output ports," *IEEE Photon. Technol. Lett.*, 15, 1360–1362, 2003.
[36] M. Yano, F. Yamagishi, and T. Tsuda, "Optical MEMS for photonic switching-compact and stable optical crossconnect switches for simple, fast, and flexible wavelength applications in recent photonic networks," *IEEE J. Sel. Top. Quantum Electron.*, 11, 383–394, 2005.
[37] X. Zheng, V. Kaman, Y. Shifu et al., "Three-dimensional MEMS photonic cross-connect switch design and performance," *IEEE J. Sel. Top. Quantum Electron.*, 9, 571–578, 2003.
[38] A. Fernandez, B. P. Staker, W. E. Owens et al., "Modular MEMS Design and Fabrication for an 80 × 80 Transparent Optical Cross-Connect Switch," in Proc. *SPIE-Int. Soc. Opt. Eng.*, USA, 5604(1), 2004, pp. 208–217.
[39] J. Kim, C. J. Nuzman, B. Kumar et al., "1100 × 1100 port MEMS-based optical crossconnect with 4-dB maximum loss," *IEEE Photon. Technol. Lett.*, 15, 1537–1539, 2003.
[40] R. Ryf, J. Kim, J. P. Hickey et al., "1296-Port MEMS Transparent Optical Crossconnect with 2.07 petabit/s Switch Capacity," in Proc. *OFC 2001. Optical Fiber Communication Conference and Exhibit. Technical Digest Postconference Edition*. Postdeadline Papers (IEEE Cat. No. 01CH37171). *Opt. Soc. America. Part vol.4, 2001*, Washington, DC, USA, 4, 2001, pp. PD28-1-3.
[41] M. Kozhevnikov, N. R. Basavanhally, J. D. Weld et al., "Compact 64 × 64 micromechanical optical cross connect," *IEEE Photon. Technol. Lett.*, 15, 993–995, 2003.
[42] V. A. Aksyuk, S. Arney, N. R. Basavanhally et al., "238 × 238 micromechanical optical cross connect," *IEEE Photon. Technol. Lett.*, 15, 587, 2003.
[43] L. Fan and M. C. Wu, "Two-Dimensional Optical Scanner with Large Angular Rotation Realized by Self-Assembled Micro-Elevator," in Proc. *IEEE/LEOS Summer Topical Meeting. Digest. Broadband Optical Networks and Technologies: An Emerging Reality. Optical MEMS. Smart Pixels. Organic Optics and Optoelectronics* (Cat. No. 98TH8369), New York, NY, USA, 1998, pp. 107–108.
[44] V. A. Aksyuk, F. Pardo, D. Carr et al., "Beam-steering micromirrors for large optical cross-connects," *J. Lightw. Technol.*, 21, 634–642, 2003.
[45] D. S. Greywall, P. A. Busch, F. Pardo et al., "Crystalline silicon tilting mirrors for optical cross-connect switches," *J. Microelectromech. Syst.*, 12, 708–712, 2003.

[46] R. Sawada, J. Yamaguchi, E. Higurashi et al., "Single Si Crystal 1024 ch MEMS Mirror Based on Terraced Electrodes and a High-Aspect Ratio Torsion Spring for 3-D Cross-Connect Switch," in Proc. *2002 IEEE/LEOS International Conference on Optical MEMS* (Cat. No. 02EX610), Piscataway, NJ, USA, 2002, pp. 11–12.
[47] N. Kouma, O. Tsuboi, Y. Mizuno et al., "A Multi-Step DRIE Process for a 128 × 128 Micromirror Array," in Proc. *2003 IEEE/LEOS International Conference on Optical MEMS* (Cat. No. 03EX682), Piscataway, NJ, USA, 2003, pp. 53–54.
[48] T. D. Kudrle, C. C. Wang, M. G. Bancu et al., "Single-crystal silicon micromirror array with polysilicon flexures," *Sensors and Actuators A-Physical,* 119, 559–566, 2005.
[49] S. D. Senturia, *Microsystem Design,* Springer, 2004, pp. 130–138.
[50] J. Dadap et. al., "Modular MEMS-based optical cross-connect with large port-count optical switch," *IEEE Photon. Technol. Lett.,* 15, 1773–1775, 2003.
[51] R. A. Conant, J. T. Nee, K. Y. Lau, and R. S. Muller, "A Flat High-Frequency Scanning Micromirror," in Proc. *Technical Digest. Solid-State Sensor and Actuator Workshop* (TRF Cat. No. 00TRF-0001). *Transducers Res. Found. 2000,* Cleveland, OH, USA, 2000, pp. 6–9.
[52] O. Tsuboi, Y. Mizuno, N. Kouma et al., "A 2-axis comb-driven micromirror array for 3D MEMS optical switch," *Trans. Inst. Elect. Eng. Jpn, Part E,* 123-E, 398–402, 2003.
[53] H. Obi, T. Yamanoi, H. Fujita, and H. Toshiyoshi, "A new design for improving stability of electrostatic vertical comb mirrors and its application to fiber optic variable attenuators," *Rev. Laser Eng.,* 33, 766–771, 2005.
[54] U. Krishnamoorthy, L. Daesung, and O. Solgaard, "Self-aligned vertical electrostatic combdrives for micromirror actuation," *J. Microelectromech. Syst.,* 12, 458–464, 2003.
[55] K.-H. Jeong and L. P. Lee, "A novel microfabrication of a self-aligned vertical comb drive on a single SOI wafer for optical MEMS applications," *J. Micromech. Microeng.,* 15, 277–281, 2005.
[56] W. Piyawattanametha, P. R. Patterson, D. Hah et al., "Surface- and bulk- micromachined two-dimensional scanner driven by angular vertical comb actuators," *J. Microelectromech. Syst.,* 14, 1329–1338, 2005.
[57] K. Jongbaeg, D. Christensen, and L. Lin, "Monolithic 2-D scanning mirror using self-aligned angular vertical comb drives," *IEEE Photon. Technol. Lett.,* 17(1), 2307–2309, 2005.
[58] M. Yoda, K. Isamoto, C. Chong et al., "A MEMS 1D Optical Scanner for Laser Projection Display Using Self-Assembled Vertical Combs and Scan-Angle Magnifying Mechanism," in Proc. *TRANSDUCERS '05. The 13th International Conference on Solid-State Sensors, Actuators and Microsystems. Digest of Technical Papers* (IEEE Cat. No. 05TH8791). *IEEE. Part Vol. 1, 2005,* Piscataway, NJ, USA, 1, 2005, pp. 968–971.
[59] J. E. Ford and J. A. Walker, "Dynamic spectral power equalization using micro-opto-mechanics," *IEEE Photon. Technol. Lett.,* 10, 1440–1442, October 1998.
[60] J. E. Ford, K. W. Goossen, J. A. Walker et al., "Interference-based micromechanical spectral equalizers," *IEEE J. Sel. Top. Quantum Electron.,* 10, 579–587, May/June 2004.
[61] K. W. Goossen, J. A. Walker, D. T. Neilson et al., "Micromechanical gain slope compensator for spectrally linear optical power equalization," *IEEE Photon. Technol. Lett.,* 12, 831–833, July 2000.
[62] C. R. Giles, V. Aksyuk, B. Barber et al., "A silicon MEMS optical switch attenuator and its use in lightwave subsystems," *IEEE J. Sel. Top. Quantum Electron.,* 5, 18–25, January/February 1999.
[63] N. A. Riza, and M. J. Mughal, "Broadband optical equalizer using fault-tolerant digital micromirrors," *Opt. Express,* 11, 1559–1565, June 30, 2003.
[64] J. I. Trisnadi, C. B. Carlisle, and R. J. Monteverde, "High-Performance Dynamic Gain Equalizer for Advanced DWDM Optical Networks," in Proc. *SPIE,* 4870, 2002, pp. 101–105.
[65] J. J. Bernstein, M. R. Dokmeci, G. Kirkos et al., "MEMS tilt-mirror spatial light modulator for a dynamic spectral equalizer," *J. Microelectromech. Syst.,* 13, 272–278, April 2004.
[66] S. H. Oh and D. M. Marom, "Attenuation mechanism effect on filter shape in channelized dynamic spectral equalizers," *Appl. Opt.,* 43, 127–131, January 1, 2004.
[67] D. T. Neilson, R. Ryf, F. Pardo et al., "MEMS-based channelized dispersion compensator with flat passbands," *J. Lightw. Technol.,* 22, 101–105, January 2004.

[68] J. E. Ford, J. A. Walker, V. Aksyuk, and D. J. Bishop, "Wavelength-Selectable Add/Drop with Tilting Micromirrors," in Proc. *IEEE Lasers and Electro-Optics Society Annual Meeting*, Postdeadline paper PD2.3, November 1997.

[69] J. E. Ford, V. A. Aksyuk, D. J. Bishop, and J. A. Walker, "Wavelength add-drop switching using tilting micromirrors," *J. Lightw. Technol.*, 17, 904–911, May 1999.

[70] J. K. Rhee, I. Tomkos, and M. J. Li, "A broadcast-and-select OADM optical network with dedicated optical-channel protection," *J. Lightw. Technol.*, 21, 25–31, January 2003.

[71] D. T. Neilson, H. Tang, D. S. Greywall et al., "Channel equalization and blocking filter utilizing microelectromechanical mirrors," *IEEE J. Sel. Top. Quantum Electron.*, 10, 563–569, May/June 2004.

[72] D. M. Marom, D. T. Neilson, D. S. Greywall et al., "Wavelength-selective $1 \times K$ switches using free-space optics and MEMS micromirrors: theory, design, and implementation," *J. Lightw. Technol.*, 23, 1620–1630, 2005.

[73] T. Ducellier, J. Bismuth, S. F. Roux et al., "The MWS 1×4: A High Performance Wavelength Switching Building Block," 2002, p. 1.

[74] J. Tsai, S. Huang, D. Hah et al., "Open-loop operation of MEMS-based $1 \times N$ wavelength-selective switch with long-term stability and repeatability," *IEEE Photon. Technol. Lett.*, 16, 1041, 2004.

[75] D. S. Greywall, P. Chien-Shing, O. Sang-Hyun et al., "Monolithic fringe-field-activated crystalline silicon tilting-mirror devices," *J. Microelectromech. Syst.*, 12, 702–707, 2003.

[76] D. Hah, H. S.-Y. Huang, J.-C. Tsai et al., "Low-voltage, large-scan angle MEMS analog micromirror arrays with hidden vertical comb-drive actuators," *J. Microelectromech. Syst.*, 13, 279–289, 2004.

[77] W. P. Taylor, J. D. Brazzle, A. B. Osenar et al., "A high fill factor linear mirror array for a wavelength selective switch," *J. Micromech. Microeng.*, 14, 147–152, 2004.

[78] J. Tsai, H. Sophia Ting-Yu, D. Hah, and M. C. Wu, "$1 \times N^2$ wavelength-selective switch with two cross-scanning one-axis analog micromirror arrays in a 4-f optical system," *J. Lightw. Technol.*, 24, 897–903, 2006.

[79] G. Baxter, S. Frisken, D. Abakoumov et al., "Highly Programmable Wavelength Selective Switch Based on Liquid Crystal on Silicon Switching Elements," in Proc. *OFCNFOEC 2006, IEEE 2006*, Piscataway, NJ, USA, 2006, p. 3.

[80] S. Frisken, Z. Hao, D. Abakoumov et al., "High Performance 'Drop and Continue' Functionality in a Wavelength Selective Switch," in Proc. *2006 OFCNFOEC, OFC 2006*, 2006, pp. 1–3.

[81] J. C. Tsai and M. C. Wu, "A high port-count wavelength-selective switch using a large scan-angle, high fill-factor, two-axis MEMS scanner array," *IEEE Photon. Technol. Lett.*, 18, 1439, 2006.

[82] J. C. Tsai, L. Fan, C. H. Chi et al., "A Large Port-Count 1×32 Wavelength-Selective Switch Using a Large Scan-Angle, High Fill-Factor, Two-Axis Analog Micromirror Array," in Proc. *30th ECOC 2004. Royal Inst. of Technol. Part vol.2, 2004*, 2, Kista, Sweden, 2004, pp. 152–153.

[83] J. Tsai and M. C. Wu, "Gimbal-less MEMS two-axis optical scanner array with high fill-factor," *J. Microelectromech. Syst.*, 14, 1323, 2005.

[84] J.-C. Tsai, L. Fan, D. Hah, and M. C. Wu, "A High Fill-Factor, Large Scan-Angle, Two-Axis Analog Micromirror Array Driven by Leverage Mechanism," in Proc. *IEEE/LEOS International Conference on Optical MEMS and Their Applications*, Takamatsu, Japan, 2004, pp. Paper C-2.

[85] K. Okamoto, M. Okuno, A. Himeno, and Y. Ohmori, "16-channel optical add/drop multiplexer consisting of arrayed-waveguide gratings and double-gate switches," *Electron. Lett.*, 32, 1471–1472, 1996.

[86] C. R. Doerr, L. W. Stulz, D. S. Levy et al., "Wavelength add-drop node using silica waveguide integration," *J. Lightw. Technol.*, 22, 2755–2762, 2004.

[87] D. M. Marom, C. R. Doerr, N. R. Basavanhally et al., "Wavelength-Selective 1×2 Switch Utilizing a Planar Lightwave Circuit Stack and a MEMS Micromirror Array," in Proc. *IEEE/LEOS Optical MEMS 2004, Paper C-1*, 2004.

[88] D. M. Marom, C. R. Doerr, M. Cappuzzo et al., "Hybrid Free-Space and Planar Lightwave Circuit Wavelength-Selective 1 × 3 Switch with Integrated Drop-Side Demultiplexer," in Proc. *31st ECOC IEEE. Part vol.4, 2005,* Stevenage, UK, 4, 2005, pp. 993–994.

[89] T. Ducellier, A. Hnatiw, M. Mala et al., "Novel High Performance Hybrid Waveguide-MEMS 1 × 9 Wavelength Selective Switch in a 32-Cascade Loop Experiment," in Proc. *ECOC 2004,* paper Th4.2.2, 2004.

[90] M. Kozhevnikov, P. Kolodner, D. T. Neilson et al., "Integrated array of 1 × N optical switches for wavelength-independent and WDM applications," *J. Lightw. Technol.,* 24, 884–890, 2006.

[91] C.-H. Chi, J. Yao, J.-C. Tsai et al., "Compact 1 × 8 MEMS Optical Switches Using Planar Lightwave Circuits," in Proc. *OFC 2004,* paper THQ4, 2, Los Angeles, California, February 23–27, 2004, p. 3.

[92] C. H. Chi, J. Tsai, M. C. Lee et al., "Integrated 1 × 4 Wavelength-Selective Switch with On-Chip MEMS Micromirrors," in Proc. *2005 CLEO, Part vol. 3,* 3, 2005, pp. 1732–1734.

[93] B. Jalali, S. Yegnanarayanan, T. Yoon et al., "Advances in silicon-on-insulator optoelectronics," *IEEE J. Sel. Top. Quantum Electron.,* 4, 938, 1998.

[94] V. Kaman, Z. Xuezhe, Y. Shifu et al., "A 32 × 10 Gb/s DWDM metropolitan network demonstration using wavelength-selective photonic cross-connects and narrow-band EDFAs," *IEEE Photon. Technol. Lett.,* 17, 1977–1979, 2005.

[95] R. Ryf, P. Bernasconi, P. Kolodner et al., "Scalable Wavelength-Selective Crossconnect Switch Based on MEMS and Planar Waveguides," in Proc. *27th ECOC, Post-Deadline Papers,* (Cat. No. 01TH8551). *IEEE. Part vol.6,* Piscataway, NJ, USA, 6, 2001, pp. 76–77.

[96] D. M. Marom, D. T. Neilson, J. Leuthold et al., "64 Channel 4 × 4 Wavelength-Selective Cross-Connect for 40 Gb/s Channel Rates with 10 Tb/s Throughput Capacity," in Proc. *ECOC'2003,* We4.P.130, Rimini, Italy, September 2003.

[97] D. T. Fuchs, C. R. Doerr, V. A. Aksyuk et al., "A hybrid MEMS-waveguide wavelength selective cross connect," *IEEE Photon. Technol. Lett.,* 16, 99–101, 2004.

[98] C.-H. Chi, J.-C. Tsai, D. Hah et al., "Silicon-Based Monolithic 4 × 4 Wavelength-Selective Cross Connect with On-Chip Micromirrors," in Proc. *OFC,* March 05–10, 2006, p. 1.

[99] J. H. Jerman and S. R. Mallinson, "A miniature Fabry-Perot Interferometer fabricated using silicon micromachining techniques," *Solid State Sensor and Actuator Workshop Technical Digest,* Hilton Head Island, SC, USA, June 6–9, 1988, pp. 16–18.

[100] A. Tran, Y. H. Lo, Z. H. Zhu et al., "Surface micromachined Fabry-Perot tunable filter," *IEEE Photon. Technol. Lett.,* 8, 393–395, March 1996.

[101] K. Cao, W. Liu, and J. J. Talghader, "Curvature compensation in micromirrors with high-reflectivity optical coatings," *J. Microelectromech. Syst.,* 10, 409–417, September 2001.

[102] E. C. Vail, M. S. Wu, G. S. Li et al., "Gaas micromachined widely tunable Fabry-Perot filters," *Electron. Lett.,* 31, 228–229, February 2, 1995.

[103] M. C. Larson, B. Pezeshki, and J. S. Harris, "Vertical coupled-cavity microinterferometer on gaas with deformable-membrane top mirror," *IEEE Photon. Technol. Lett.,* 7, 382–384, April 1995.

[104] A. Spisser, R. Ledantec, C. Seassal et al., "Highly selective and widely tunable 1.55-μm InP/air-gap micromachined Fabry-Perot filter for optical communications," *IEEE Photon. Technol. Lett.,* 10, 1259–1261, September 1998.

[105] P. Tayebati, P. Wang, M. Azimi et al., "Microelectromechanical tunable filter with stable half symmetric cavity," *Electron. Lett.,* 34, 1967–1968, October 1, 1998.

[106] M. Garrigues, J. Danglot, J. L. Leclercq, and O. Parillaud, "Tunable high-finesse InP/Air MOEMS filter," *IEEE Photon. Technol. Lett.,* 17, 1471–1473, July 2005.

[107] C. J. Chang-Hasnain, "Tunable VCSEL," *IEEE J. Sel. Top. Quantum Electron.,* 6, 978–987, November/December 2000.

[108] D. Hohlfeld and H. Zappe, "An all-dielectric tunable optical filter based on the thermo-optic effect," *J. Opt. A-Pure Appl. Opt.,* 6, 504–511, June 2004.

[109] F. Gires and P. Tournois, "Interferometer utilizable pour la compression d'impulsions lumineuses modules en frequence," *C. R. Academie Sci.,* 258, 6112–6115, 1964.

[110] C. K. Madsen, J. A. Walker, J. E. Ford et al., "A tunable dispersion compensating MEMS all-pass filter," *IEEE Photon. Technol. Lett.*, 12, 651–653, June 2000.

[111] K. Yu and A. Solgaard, "MEMS optical wavelength deinterleaver with continuously variable channel spacing and center wavelength," *IEEE Photon. Technol. Lett.*, 15, 425–427, March 2003.

[112] K. Yu and O. Solgaard, "Tunable optical transversal filters based on a Gires-Tournois interferometer with MEMS phase shifters," *IEEE J. Sel. Top. Quantum Electron.*, 10, 588–597, May/June 2004.

[113] K. Yu and O. Solgaard, "Tunable Chromatic Dispersion Compensators Using MEMS Gires-Tournois Interferometers," in Proc. *IEEE/LEOS International Conference on Optical MEMS*, Lugano, Switzerland, 2002, pp. 181–182.

[114] D. M. Marom, C. R. Doerr, M. A. Cappuzzo et al., "Compact colorless tunable dispersion compensator with 1000-ps/nm tuning range for 40-gb/s data rates," *J. Lightw. Technol.*, 24, 237–241, 2006.

[115] O. Manzardo, H. P. Herzig, C. R. Marxer, and N. F. de Rooij, "Miniaturized time-scanning Fourier transform spectrometer based on silicon technology," *Opt. Lett.*, 24, 1705–1707, December 1, 1999.

[116] D. Knipp, H. Stiebig, S. R. Bhalotra et al., "Silicon-based micro-Fourier spectrometer," *IEEE Trans. Electron Devices*, 52, 419–426, March 2005.

[117] S. D. Collins, R. L. Smith, C. Gonzalez et al., "Fourier-transform optical microsystems," *Opt. Lett.*, 24, 844–846, June 15, 1999.

[118] K. Yu, D. Lee, U. Krishnamoorthy et al., *Transducers, Seoul* June 2005.

[119] O. Manzardo, R. Michaely, F. Schadelin et al., "Minature lamellar grating interferometer based on silicon technology," *Opt. Lett.*, 29, 1437–1439, July 1, 2004.

[120] H. L. Kung, S. R. Bhalotra, J. D. Mansell et al., "Standing-wave transform spectrometer based on integrated MEMS mirror and thin-film photodetector," *IEEE J. Sel. Top. Quantum Electron.*, 8, 98–105, January/February 2002.

[121] R. A. DeVerse, R. M. Hammaker, and W. G. Fateley, "Realization of the Hadamard multiplex advantage using a programmable optical mask in a dispersive flat-field near-infrared spectrometer," *Applied Spectroscopy*, 54, 1751–1758, December 2000.

[122] J. P. Heritage, A. M. Weiner, and R. N. Thurston, "Picosecond pulse shaping by spectral phase and amplitude manipulation," *Opt. Lett.*, 10, 609–611, 1985.

[123] A. M. Weiner, J. P. Heritage, and J. A. Salehi, "Encoding and decoding of femtosecond pulses," *Opt. Lett.*, 13, 300–302, April 1988.

[124] G. J. Tearney, B. E. Bouma, and J. G. Fujimoto, "High-speed phase- and group-delay scanning with a grating-based phase control delay line," *Opt. Lett.*, 22, 1811–1813, December 1, 1997.

[125] Y. T. Pan, H. K. Xie, and G. K. Fedder, "Endoscopic optical coherence tomography based on a microelectromechanical mirror," *Opt. Lett.*, 26, 1966–1968, December 15, 2001.

[126] K. T. Cornett, P. M. Hagelin, J. P. Heritage et al., "Miniature Variable Optical Delay using Silicon Micromachined Scanning Mirrors," in Proc. *CLEO 2000*, San Francisco, CA, 2000, pp. 383–384.

[127] O. Solgaard, F. S. A. Sandejas, and D. M. Bloom, "Deformable grating optical modulator," *Opt. Lett.*, 17, 688–690, May 1, 1992.

[128] R. Apte, F. Sandejas, W. Banyai, and D. Bloom, "Grating Light Valves for High Resolution Displays," *Proceedings of the 1994 Solid-State Sensors and Actuators Workshop*, Hilton Head, South Carolina, June 13–16, 1994, pp. 1–6.

[129] S. Kubota, "The grating light valve projector," *Opt. Photon. News*, 13, 50–53, 2002.

[130] A. Godil, "Diffractive MEMS technology offers a new platform for optical networks," *Laser Focus World*, 38, 181–185, May 2002.

[131] G. G. Yaralioglu, A. Atalar, S. R. Manalis, and C. F. Quate, "Analysis and design of an interdigital cantilever as a displacement sensor," *J. Appl. Phys.*, 83, 7405–7415, June 15, 1998.

[132] N. C. Loh, M. A. Schmidt, and S. R. Manalis," Sub-10 cm (3) interferometric accelerometer with nano-g resolution," *J. Microelectromech. Syst.*, 11, 182–187, June 2002.

[133] C. A. Savran, T. P. Burg, J. Fritz, and S. R. Manalis, "Microfabricated mechanical biosensor with inherently differential readout," *Appl. Phys. Lett.*, 83, 1659–1661, August 25, 2003.

[134] N. A. Hall and F. L. Degertekin, "Self-Calibrating Micromachined Microphone with Integrated Optical Displacement Detection," in Proc. *Transducer'01*, Munich, Germany, June 10–14, 2001.

[135] T. Perazzo, M. Mao, O. Kwon et al., "Infrared vision using uncooled micro-optomechanical camera," *Appl. Phys. Lett.*, 74, 3567–3569, June 7, 1999.

[136] M. B. Sinclair, M. A. Butler, A. J. Ricco, and S. D. Senturia, "Synthetic spectra: A tool for correlation spectroscopy," *Appl. Opt.*, 36, 3342–3348, May 20, 1997.

[137] R. Belikov, X. Li, and O. Solgaard, "Programmable Optical Wavelength Filter Based on Diffraction From a 2 D MEMS Micromirror Array," in Proc. *CLEO, Technical Digest*, Baltimore, MD, June 1–6, 2003.

[138] R. Belikov, C. Antoine-Snowden, and O. Solgaard, "Femtosecond Direct Space-to-Time Pulse Shaping with MEMS Micromirror Arrays," in Proc. *2003 IEEE/LEOS International Conf. on Optical MEMS*, Waikoloa, Hawaii, August 18–21, 2003, pp. 24–25.

[139] T. G. Bifano, J. Perreault, R. K. Mali, and M. N. Horenstein, "Microelectromechanical deformable mirrors," *IEEE J. Sel. Top. Quantum Electron.*, 5, 83–89, January/February 1999.

[140] R. Belikov and O. Solgaard, "Optical wavelength filtering by diffraction from a surface relief," *Opt. Lett.*, 28, 447–449, March 15, 2003.

[141] R. Belikov, "Diffraction-Based Optical Filtering: Theory and Implementation with MEMS," *Doctoral Dissertation, Stanford University, UMI Dissertation Publishing*, 2005.

[142] G. B. Hocker, D. Younger, E. Deutsch et al., "The Polychromator: A Programmable MEMS Diffraction Grating for Synthetic Spectra," *Tech. Dig., Solid-State Sensor and Actuator Workshop*, Hilton Head, SC, June 4–8, 2000, pp. 89–92.

[143] M. A. Butler, E. R. Deutsch, S. D. Senturia et al., "A MEMS Based Programmable Diffraction Grating for Optical Holography in the Spectral Domain," *International Electron Devices Meeting, Technical Digest*, 2–5 December 2001, Washington, DC, USA, pp. 41.1.1–41.1.4.

[144] H. Sagberg, M. Lacolle, I.-R. Johansen et al., "Micromechincal gratings for visible and near-infrared spectroscopy," *IEEE J. Sel. Top. Quantum Electron.*, 10, 604–613, 2004.

[145] X. Li, C. Antoine, D. Lee et al., "Tunable blazed gratings," *J. Microelectromech. Syst.*, 13, 597–604, June 2006.

[146] C. Antoine, X. Li, J.-S. Wang, and O. Solgaard, "A Reconfigurable Optical Demultiplexer Based on a MEMS Deformable Blazed Grating," in Proc. *IEEE/LEOS International Conference on Optical MEMS and their Applications*, Big Sky, Montana, 21–24 August 2006, pp 183–184.

[147] M. G. Littman and H. J. Metcalf, "Spectrally narrow pulsed dye laser without a beam expander," *Appl. Opt.*, 17, 2224–2227, 1978.

[148] W. R. Trutna and L. F. Stokes, "Continuously tuned external-cavity semiconductor-laser," *J. Lightw. Technol.*, 11, 1279–1286, August 1993.

[149] R. R. A. Syms and A. Lohmann, "MOEMS tuning element for a Littrow external cavity laser," *J. Microelectromech. Syst.*, 12, 921–928, December 2003.

[150] X. M. Zhang, A. Q. Liu, C. Lu, and D. Y. Tang, "Continuous wavelength tuning in micromachined Littrow external-cavity lasers," *IEEE J. Quantum Electron.*, 41, 187–197, February 2005.

[151] W. Huang, R. R. A. Syms, J. Stagg, and A. Lohmann, "Precision MEMS Flexure Mount for a Littman Tunable External Cavity Laser," in Proc. *IEE Proceedings-Science Measurement And Technology*, 151, March 2004, pp. 67–75.

[152] R. Belikov, C. Antoine-Snowden, and O. Solgaard, "Tunable External Cavity Laser with a Stationary Deformable MEMS Grating," in Proc. *CLEO, Technical Digest*, paper CWL3, San Francisco, CA, May 16–21, 2004.

[153] H. Jerman and J. D. Grade, "A Mechanically-Balanced, DRIE Rotary Actuator For A High-Power Tunable Laser," *Proceedings of the 2002 Solid-State Sensors and Actuators Workshop*, Hilton Head, South Carolina, June, 2002.

[154] D. Anthon, D. King, J. D. Berger et al., "Mode-Hop Free Sweep Tuning of a MEMS Tuned External Cavity Semiconductor Laser," in Proc. *CLEO, Technical Digest,* paper CWL2, San Francisco, CA, May 16–21, 2004.

[155] B. Pezeshki, J. K. E. Vail, G. Yoffe et al., "20 mW Widely Tunable Laser Module Using DFB Array and MEMS Selection," *IEEE Photonics Technology Letters,* 14(10), 1457–1459, October 2002.

[156] J. A. Walker, K. W. Goossen, and S. C. Arney, "Fabrication of a mechanical antireflection switch for fiber-to-the-home systems," *J. Microelectromech. Syst.,* 5, 45–51, March 1996.

[157] C. Marxer, M. A. Gretillat, V. P. Jaecklin et al., "Megahertz opto-mechanical modulator," *Sensors And Actuators A-Physical,* 52, 46–50, March–April 1996.

[158] D. S. Greywall, P. A. Busch, and J. A. Walker, "Phenomenological model for gas-damping of micromechanical structures," *Sensors And Actuators A-Physical,* 72, 49–70, January 8, 1999.

[159] C. Marxer, M. A. Gretillat, N. F. de Rooij et al., "Reflective duplexer based on silicon micromechanics for fiber-optic communication," *J. Lightw. Technol.,* 17, 115–122, January 1999.

[160] J. E. Ford, J. A. Walker, D. S. Greywall, and K. W. Goossen, "Micromechanical fiber-optic attenuator with 3 microsecond response," *IEEE J. Lightw. Technol.,* 16, 1663–1670, 1998.

[161] B. Barber, C. R. Giles, V. Askyuk et al., "A fiber connectorized MEMS variable optical attenuator," *IEEE Photon. Technol. Lett.,* 10, 1262–1264, September 1998.

[162] C. Marxer, P. Griss, and N. F. de Rooij, "A variable optical attenuator based on silicon micromechanics," *IEEE Photon. Technol. Lett.,* 11, 233–235, February 1999.

[163] K. Isamoto, K. Kato, A. Morosawa et al., "A 5-V operated MEMS variable optical attenuator by SOI bulk micromachining," *IEEE J. Sel. Top. Quantum Electron.,* 10, 570–578, May/June 2004.

[164] http://www.lightconnect.com/news/news_release030205.shtml, 2005.

[165] B. E. Little, S. T. Chu, W. Pan, and Y. Kokubun, "Microring resonator arrays for VLSI photonics," *IEEE Photon. Technol. Lett.,* 12, 323–325, 2000.

[166] C. K. Madsen, G. Lent, A. T. Bruce et al., "Multistage dispersion compensator using ring resonators," *Opt. Lett.,* 24, 1555–1557, 1999.

[167] X. Qianfan, B. Schmidt, S. Pradhan, and M. Lipson, "Micrometre-scale silicon electro-optic modulator," *Nature,* 435, 325–327, 2005.

[168] C. Seung June, K. Djordjev, C. Sang Jun, and P. D. Dapkus, "Microdisk lasers vertically coupled to output waveguides," *IEEE Photon. Technol. Lett.,* 15, 1330–1332, 2003.

[169] M. Lipson, "Guiding, modulating, and emitting light on Silicon-challenges and opportunities," *J. Lightw. Technol.,* 23, 4222–4238, 2005.

[170] M. C. Lee and M. C. Wu, "MEMS-actuated microdisk resonators with variable power coupling ratios," *IEEE Photon. Technol. Lett.,* 17, 1034–1036, 2005.

[171] M.-C. M. Lee and M. C. Wu, "Tunable coupling regimes of silicon microdisk resonators using MEMS actuators," *Opt. Express,* 14, 4703–4712, 2006.

[172] G. N. Nielson, D. Seneviratne, F. Lopez-Royo et al., "Integrated wavelength-selective optical MEMS switching using ring resonator filters," *IEEE Photon. Technol. Lett.,* 17, 1190–1192, 2005.

[173] B. E. Little, S. T. Chu, P. P. Absil et al., "Very high-order microring resonator filters for WDM applications," *IEEE Photon. Technol. Lett.,* 16, 2263–2265, 2004.

[174] D. Geuzebroek, E. Klein, H. Kelderman et al., "Compact wavelength-selective switch for gigabit filtering in access networks," *IEEE Photon. Technol. Lett.,* 17, 336–338, 2005.

[175] T. A. Ibrahim, W. Cao, Y. Kim et al., "Lightwave switching in semiconductor microring devices by free carrier injection," *J. Lightw. Technol.,* 21, 2997–3003, 2003.

[176] K. Djordjev, C. Seung-June, C. Sang-Jun, and P. D. Dapkus, "Vertically coupled InP microdisk switching devices with electroabsorptive active regions," *IEEE Photon. Technol. Lett.,* 14, 1115–1117, 2002.

[177] K. Djordjev, C. Seung-June, C. Sang-Jun, and P. D. Dapkus, "Gain trimming of the resonant characteristics in vertically coupled InP microdisk switches," *Appl. Phys. Lett.,* 80, 3467–3469, 2002.

[178] M. C. Lee and M. C. Wu, "Thermal annealing in hydrogen for 3-D profile transformation on silicon-on-insulator and sidewall roughness reduction," *J. Microelectromech. Syst.,* 15, 338–343, 2006.

[179] M. C. M. Lee and M. C. Wu, "A Reconfigurable Add-Drop Filter Using MEMS-Actuated Microdisk Resonator," in Proc. *IEEE/LEOS Optical MEMs 2005* (IEEE Cat. No. 05EX1115), Piscataway, NJ, USA, 2005, pp. 67–68.
[180] S. H. Fan and J. D. Joannopoulos, "Analysis of guided resonances in photonic crystal slabs," *Phys. Rev. B*, 65, June 15, 2002.
[181] C. F. R. Mateus, M. C. Y. Huang, Y. F. Deng et al., "Ultrabroadband mirror using low-index cladded subwavelength grating," *IEEE Photon. Technol. Lett.*, 16, 518–520, February 2004.
[182] O. Kilic, S. Kim, W. Suh et al., "Photonic crystal slabs demonstrating strong broadband suppression of transmission in the presence of disorders," *Opt. Lett.*, 29, 2782–2784, December 1, 2004.
[183] L. Chen, M. C. Y. Huang, C. F. R. Mateus et al., "Fabrication and design of an integrable subwavelength ultrabroadband dielectric mirror," *Appl. Phys. Lett.*, 88, January 16, 2006.
[184] V. Lousse, W. Suh, O. Kilic et al., "Angular and polarization properties of a photonic crystal slab mirror," *Opt. Express*, 12, 1575–1582, April 19, 2004.
[185] K. B. Crozier, V. Lousse, O. Kilic et al., "Air-bridged photonic crystal slabs at visible and near-infrared wavelengths," *Phys. Rev. B*, 73, March 2006.
[186] E. Bisaillon, D. Tan, B. Faraji et al., "High reflectivity air-bridge subwavelength grating reflector and Fabry-Perot cavity in AlGaAs/GaAs," *Opt. Express*, 14, 2573–2582, April 3, 2006.
[187] W. Park and J. B. Lee, "Mechanically tunable photonic crystal structure," *Appl. Phys. Lett.*, 85, 4845–4847, November 22, 2004.
[188] X. Letartre, J. Mouette, J. L. Leclercq et al., "Switching devices with spatial and spectral resolution combining photonic crystal and MOEMS structures," *J. Lightw. Technol.*, 21, 1691–1699, July 2003.
[189] I. Marki, M. Salt, S. Gautsch et al., "Tunable Microcavities in Two Dimensional Photonic Crystal Waveguides," in Proc. *IEEE/LEOS Optical MEMs 2005*, Oulu, Finland, August 1–4, 2005, pp. 109–110.
[190] I. Marki, M. Salt, H. P. Herzig et al., "Optically tunable microcavity in a planar photonic crystal silicon waveguide buried in oxide," *Opt. Lett.*, 31, 513–515, February 15, 2006.
[191] M. C. M. Lee, D. Y. Hah, E. K. Lau et al., "MEMS-actuated photonic crystal switches," *IEEE Photon. Technol. Lett.*, 18, 358–360, January/February 2006.
[192] S. Iwamoto, M. Tokushima, A. Gomyo et al., "Optical Switching in Photonic Crystal Waveguide Controlled by Micro Electro Mechanical System," in Proc. *CLEO/Pacific Rim*, August 30–02 September, 2005, pp. 233–234.
[193] W. Suh, M. F. Yanik, O. Solgaard, and S. H. Fan, "Displacement-sensitive photonic crystal structures based on guided resonance in photonic crystal slabs," *Appl. Phys. Lett.*, 82, 1999–2001, March 31, 2003.
[194] W. Suh, O. Solgaard, and S. Fan, "Displacement sensing using evanescent tunneling between guided resonances in photonic crystal slabs," *J. Appl. Phys.*, 98, August 1, 2005.
[195] D. W. Carr, J. P. Sullivan, and T. A. Friedmann, "Laterally deformable nanomechanical zeroth-order gratings: anomalous diffraction studie by rigorous coupled-wave analysis," *Opt. Lett.*, 28, 1636–1638, September 2003.
[196] B. E. N. Keeler, D. W. Carr, J. P. Sullivan et al., "Experimental demonstration of a laterally deformable optical nanoelectromechanical system grating transducer," *Opt. Lett.*, 29, 1182–1184, June 1, 2004.
[197] J. Provine, J. Skinner, and D. A. Horsley, "Subwavelength Metal Grating Tunable Filter," in Proc. *Technical Digest of 19th IEEE International Conference on MEMS 2006*, Istanbul, Turkey, January 22–26, 2006, pp. 854–857.
[198] S. Nagasawa, T. Onuki, Y. Ohtera, and H. Kuwano, "MEMS Tunable Optical Filter Using Auto-Cloned Photonic Crystal, " in Proc. *Technical Digest of 19th IEEE International Conference on MEMS 2006*, Istanbul, Turkey, January 22–26, 2006, pp. 858–861.

20

Nonlinear optics in communications: from crippling impairment to ultrafast tools

Stojan Radic[*], David J. Moss[†], and Benjamin J. Eggleton[†]

[*] *Department of Electrical and Computer Engineering, University of California, San Diego*
[†] *ARC Centre of Excellence for Ultrahigh-bandwidth Devices for Optical Systems (CUDOS), School of Physics, University of Sydney, NSW Australia*

20.1 INTRODUCTION

It is perhaps somewhat paradoxical that optical nonlinearities in fiber, while having posed significant challenges and limitations for long haul wavelength division multiplexed (WDM) systems [1], also offer the promise of addressing the bandwidth bottleneck for signal processing for future ultrahigh-speed optical networks as they evolve beyond 40 to 160 Gb/s and ultimately to 1 Tb/s and beyond [2]. New technologies and services such as voice over Internet protocol and streaming video are already driving global bandwidth and traffic demand which in turn has driven research and development on ultrahigh bandwidth optical transmission capacities (Figure 20.1). The clusters of points in Figure 20.1 represent different generations of lightwave communication systems, from the original 0.8-μm sources and multimoded fiber to today's erbium-doped fiber amplifier (EDFA) WDM systems and onward. The resulting "optical Moore's law" corresponds to a ×10 increase in "capacity × distance" every 4 years, making it faster than the original Moore's law for integrated circuits (ICs)! This drive to higher bandwidths is being realized on many fronts—by opening up new wavelength bands (S, L, etc.), to ever higher WDM channel counts, density, and spectral efficiency [2], to higher bit rates through optical and/or electrical time division multiplexing (OTDM, ETDM). In parallel with this is the drive toward increasingly optically transparent and agile networks, toward full "photonic networks," in which ultrafast optical signals—independent of bit rate and modulation format—will

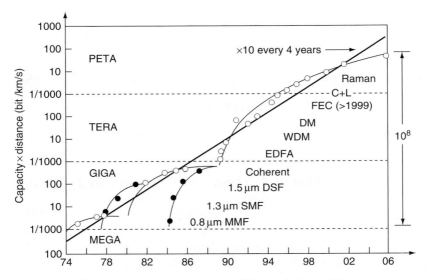

Figure 20.1 Growth in optical transmission link capacity-bandwidth product.

be transmitted and processed from end to end without costly, slow, and bulky optical–electrical–optical conversion. The result will be a critical future demand for high-performance, cost effective, ultrahigh-speed, all-optical signal processing devices.

Nonlinear optical signal processing is seen as a potential solution to this demand, offering huge benefits in cost, speed, simplicity, and footprint. In particular, all-optical devices based on the third-order $\chi^{(3)}$ optical nonlinearity offer a significant promise in this regard, not only because the intrinsic nonresonant $\chi^{(3)}$ is nearly instantaneous—relying as it does on virtual electron or hole transitions (rather than generating real carrier densities) with a response time typically < 10 fs—but also because, being a third-order nonlinearity, $\chi^{(3)}$ is responsible for a wide range of phenomena from third harmonic generation to stimulated Raman gain, four-wave mixing (FWM), optical phase conjugation, two-photon absorption (TPA), and the nonlinear refractive index (Kerr effect), which in turn gives rise to both self-phase modulation (SPM) and cross-phase modulation (XPM).

This plethora of physical processes generated by $\chi^{(3)}$ has been the basis for a wide range of activity on all-optical signal processing devices for applications such as optical performance monitoring (OPM) [3, 4], parametric amplification (PA) [5], 2R (reamplification and reshaping) and 3R (reamplification, reshaping, and retiming) optical regeneration [6–9], wavelength conversion [10–12], optical buffering and delay [13], demultiplexing [14], and the list goes on. The recent blossoming in interest in all-optical signal processing and monitoring has been in a large part fuelled by progress in nonlinear optical materials such as nonlinear glasses including chalcogenide glasses (ChG) [15] and bismuth glass [16], as well as the well-known explosion of interest in silicon as a nonlinear material for all-optical devices, arguably initiated in earnest with the observation in 2003 of stimulated Raman gain [17].

In parallel with this has been significant progress in platform technology for both fiber and integrated waveguide devices. This includes the realization of true silicon nanowires [13] that can greatly enhance nonlinear optical efficiencies through the drastic reduction in mode field area, to the development of high-quality fabrication processes (with the aim of achieving semiconductor industry standards) for some of these novel nonlinear materials that for many years have resisted this—a key example being the ChG [18]. It also includes significant progress in novel highly nonlinear fibers (HNLF) including bismuth oxide [19] and ChG [20]—the latter in particular exhibiting the largest material nonlinearity of any fiber that has been used for all-optical signal processing [7]. Like nanophotonic ICs, nonlinear fibers also have the potential to advance all-optical device performance through novel methods of mode field area reduction such as chalcogenide fiber nanotapers [21], as well as highly nonlinear glass photonic crystal fibers (PCF) [22]. All of these developments are arguably in their infancy, and the prospect for reducing peak power requirements of photonic integrated circuit (PIC)-based all-optical circuits below the current levels of a few watts [10, 23] to tens of milliwatts or even lower, will go a long way to ensure that all-optical nonlinear devices will play an integral part in future telecommunications systems.

The full scope of all-optical signal processing devices is too large to comprehensively review in a single chapter. Here we focus primarily on breakthroughs in the past few years on approaches based on highly nonlinear silica fiber as well as ChG-based fiber and waveguide devices. In Section 20.2, we briefly contrast two qualitatively different approaches to all-optical signal processing—those based on nonphase-matched processes and those on phase-matched processes. In Section 20.3, we briefly review material and platform developments in the material systems relevant here. In Section 20.4, we begin the main body of our review by considering optical parametric amplification (OPA)—the primary phase-matched process for all-optical signal processing, followed in the next section by a review of recent progress on all-optical 2R and 3R regeneration based on Kerr nonlinearities, with a focus on our work based on different nonlinear material platforms in highly nonlinear silica fiber and ChG (both in fiber and waveguides). In Section 20.6, we discuss optical phase conjugation and its role in signal processing, followed by a review of all-optical wavelength conversion in Section 20.7. Following this we successively discuss optical switching and multicasting (Section 20.8), optical performance monitoring (Section 20.9), and ending with a review of nonlinear optical methods of generating tunable optical delays and buffering.

20.2 PHASE-MATCHED VS NONPHASE-MATCHED PROCESSES

As the name implies, nonphase-matched processes do not rely on critical matching of waveguide refractive indices between in-coming and out-going signals. This includes processes such as SPM, XPM, TPA, Raman gain and others, and has been exploited to demonstrate a wide range of novel all-optical functions such as OPM

[3, 4], PA [5], 2R and 3R optical regeneration [6–9], wavelength conversion [10–12], optical buffering and delay [13], demultiplexing [14] and others. Phase-matched processes, however, are probably best represented by optical parametric signal processing techniques that have had a long and diversified history spanning more than two decades [24–31]. In the simplest configuration, efficient amplification is achieved by combining an optical pump and a signal that meet the phase-matching condition in a nonlinear waveguide [26]. A conventional parametric amplifier (PA) uses a single pump whose wavelength lies in the anomalous dispersion regime. Its operation can be physically described as degenerate four-photon mixing (FPM): two-pump photons are annihilated to generate a signal and an idler photon (see Figure 20.2 in Section 20.4). Phase-sensitive signal amplification has been the most investigated parametric process [26], partly due to its promise of noiseless signal regeneration [32]. However, referring to a parametric fiber device as an amplifier is a simplification that does not do justice to its versatility and needs to be accompanied by an understanding of its considerably more complex nature. A fiber parametric processor, besides providing phase-sensitive optical amplification, also provides wavelength conversion [25], signal conjugation [27], and optical limitation [33], each of which is an important function in all-optical networking. The standard scheme for phase-sensitive amplification (and squeezed-state generation) is degenerate PA in a $\chi^{(2)}$ medium. It is also possible in a $\chi^{(3)}$ medium: some schemes are based on degenerate backward [34, 35] and near-forward FWM a three-dimensional (3D) medium. In these schemes, the idlers are generated within the medium. Phase sensitivity (and squeezing) is obtained by the use of a beamsplitter [28, 34] or a mirror to combine the signal and idler. The amplification of vacuum fluctuations by nondegenerate FPM (MI) in a 1D medium, such as a fiber, produces squeezed states [36–38]. This chapter will not focus on phase-sensitive processing of the optical signals because this application deserves considerably more attention in a more specialized review. The relative advantages and disadvantages of phase-matched vs nonphase-matched processes are discussed in some detail at the end of the chapter. Next, we begin with a brief look at recent developments in material and platform technology.

20.3 PLATFORMS

This section reviews and contrasts the different material and embodiment (i.e., fiber vs waveguide) platforms for nonlinear all-optical devices, with an emphasis on the material system that has been the focus of much of the work reviewed in this chapter—namely ChG waveguides and fibers.

Chalcogenide glasses have attracted significant attention in recent years as a promising nonlinear material for all-optical devices [39–41]. They are amorphous materials containing at least one of the nonoxide group VI element (S, Se, and Te)—a chalcogen—combined with other elements such as Ge, As, Sb, and Ga [41, 42]. Some of the crucial properties that have made them attractive for all-optical devices include: (1) *transparency in the near-IR* as well as into the far-IR due to the relatively high atomic masses and weak bond strengths resulting in low phonon energies [41, 43]; (2) *photosensitivity to band gap light*. The weak bonding arrangements and inherent structural flexibility result in many distinct photo-induced phenomena [41]. Photo darkening and photo bleaching, which shift the absorption edge and induce changes in the refractive index [41], are particularly noteworthy and have been used to write both channel waveguides [42, 44–48] and periodic structures [49–55]; (3) *high refractive index* ($n \approx 2\text{--}3$) providing strong mode field confinement [18], tight bending radii, and allows complete band gaps to be engineered in photonic crystal structures [56]; (4) *large third-order nonlinearities* with reported values for the nonlinear refractive index (n_2) up to three orders of magnitude larger than silica, and even more for the Raman gain coefficient (g_R) [39, 41, 57]. Since these nonlinearities are based on nonresonant virtual electronic transitions, the intrinsic response time is in the 10s of fs, making devices based on these materials potentially much faster than those based on real carrier dynamics such as semiconductor optical amplifiers (SoAs) [39, 57]; and (5) *low TPA* [39], of particular importance as it has been known since 1989 [58] that TPA poses a fundamental limitation for nonlinear switching, which inspired the definition of a nonlinear figure of merit (FOM):

$$\text{FOM} = n_2/\beta\lambda \qquad (20.1)$$

where β is the TPA coefficient and n_2 is the Kerr coefficient. Since then it has come to be accepted that, for nonlinear all-optical switching at least, an FOM > 1 is required for efficient operation [59]. It is important to note that the FOM, essentially the ratio of the real to imaginary parts of $\chi^{(3)}$, is a fundamental property of a material, varying only with wavelength [60].

Probably, the most significant development in the past few years for ChG, in terms of their use for practical all-optical signal processing, has been the realization of low loss single-mode (SM) chalcogenide fibers [61], and for waveguides, the development of deposition techniques for high-quality low loss thin films, such as pulsed laser deposition (PLD) [62], as well as high-quality fabrication processes such as dry etching [18]. Together with the demonstration of sophisticated writing techniques to produce Bragg grating filters in chalcogenide waveguides of high enough quality to be useful for signal processing applications [63], the door has been opened for ChG fiber and waveguide devices to join mainstream materials such as silicon as the basis for potential solutions for integrated all-optical signal processing.

While this review does not focus explicitly on silicon, it is useful to contrast chalcogenide devices against the backdrop of intense activity that has taken place in the past few years on integrated silicon all-optical devices [64–70]. Silicon has, in many ways, become the benchmark material platform with which to compare new materials and devices in a wide range of areas, and nonlinear optics is perhaps no exception. Since the observation of Raman gain in 2003 [17], there has been intense interest worldwide in developing silicon as a material platform for all-optical devices. This has been highlighted by the achievement of ultrasmall nanowires [71] to greatly reduce operating powers [72]. In addition, together with techniques such as ion implantation [73] and the use of p–n junctions to sweep out carriers [74], nanowires have helped in solving the problem of two-photon-induced free carrier absorption that has been known since 2004 [74, 75] to pose a significant obstacle to realizing net Raman gain in silicon. A critical point, however, is that even if free-carrier effects are *completely* eliminated, one is still left with the intrinsic material TPA coefficient, which determines the nonlinear FOM, and which is nonnegligible in silicon. Table 20.1 lists the key parameters for silicon as well as typical ChG where we see that, purely from the nonlinear FOM viewpoint, silicon is rather poor with an FOM of ∼0.3–0.5 (depending on which sets of experimental results one takes [76–79]). Purely from their nonlinear material characteristics, ChG arguably represent a more promising platform, with nonlinearities ranging from being comparable to, to several times larger than, silicon, along with an FOM ranging from ∼2 to >15. Having said this though, it has recently been realized that for some other applications such as 2R regeneration [80] a modest degree of TPA can actually enhance performance, in sharp contrast with all-optical switching. For silicon, however, it is not clear if its FOM is adequate even under these circumstances, and it will be interesting to watch continuing developments following the recent groundbreaking report of 2R regeneration in silicon-on-insulator (SOI) nanowires [23] to see if this will ultimately pose an obstacle.

For novel nonlinear fibers, probably the two most promising contenders are SM selenide-based chalcogenide fiber [57] and bismuth oxide fiber [19, 81]. Table 20.2 shows some of the key material properties of both of these, along with highly nonlinear silica fiber. The key contrasts between the bismuth and chalcogenide fiber are that, whereas the chalcogenide fiber has an intrinsically higher material nonlinearity n_2 (up to ×10 that of bismuth oxide), because the core area of available SM chalcogenide fiber [61] is still quite large, the effective nonlinearity

$$\gamma = n_2 \omega / c A_{\text{eff}} \tag{20.2}$$

for both fibers is comparable at this point in time. For chalcogenide fiber, however, the hope is that further improvement in fabrication methods or the introduction of novel geometries, such as fiber tapers or photonic crystal chalcogenide fibers, will yield mode

Table 20.1

Comparison of key 3rd order nonlinear optical parameters between chalcogenide glasses and silicon.

Material	n_2 (x silica)	β (TPA)(cm/GW)	FOM
Chalcogenide glasses	100–400	<0.03–0.25	From 2 to 15 and higher
Silicon	270	0.9	0.5

Table 20.2
Parameters for some key nonlinear fibers including silica highly nonlinear fiber (HNLF), bismuth oxide (Bi_2O_3) fiber, and arsenic tri-selenide (As_2Se_3) single mode fiber.

Parameter	Units	SiO_2 DSF	Bi_2O_3 fiber	As_2Se_3 fiber
Nonlinear index (n_2)	n_2 of silica*	1	42	470
Effective core area (A_{eff})	mm^2	60	3.3	37
Nonlinearity coefficient (γ)	W/km	1.9	1360	1200
Length (L)	M	5000	1	1
Dispersion (D)	ps/nm/km	−0.69	−260	−560
Figure of merit (FOM)	–	High**	High**	2.3

* n_2 of silica $= 2.6 \times 10^{-20}$ m^2/W.
** No direct measurement found. Literature suggests an FOM greatly in excess of unity.

field areas comparable to (or even smaller) the current bismuth fiber, which would yield $\gamma > 10{,}000$ W/m, well and true in the regime of practical device requirements.

Highly nonlinear fiber is by now recognized as the most mature platform in parametric processing due to its exceptional characteristics combining low loss (<0.5 dB/km), high nonlinearity (∼30 W/km) and precise dispersion slope control (∼0.03 ps/km/nm^2). While recent advances in alternative platforms, such as PCFs and nonsilica single-mode fiber (SMF) indeed show great promise, none of these can compete at the present time, when considering the combination of the effective interaction length, nonlinearity, and phase-matching (chromatic dispersion) control.

Conventional phase-sensitive processing techniques can be classified by the nature of the photon generation mechanism. While $\chi^{(2)}$ processes have been extensively studied [82, 83] in periodically poled lithium niobate (PPLN), potassium titanyl phosphate (PPKTP) and, more recently, in gallium arsenide (GaAs) structures, the PPLN platform is distinguished both by a lack of mature fabrication and reliability issues that remain unresolved even after a two-decade long effort [82]. Intentional or parasitic visible generation in PPLN leads to optical damage, drastically reducing the device lifetime—material heating and doping has produced limited results. Recent advances in orientation-patterned growth have made quasi-phase matching (QPM) possible in GaAs with almost an order of magnitude higher nonlinear FOM [83]. Unfortunately, the short-interaction length, combined with practical (W-scale) continuous wave (CW) pump power is insufficient to generate appreciable efficiencies. Similar considerations apply to semiconductor-based wavelength converters [84].

In contrast, parametric process in high-confinement fibers (HCF) offer long interaction lengths, compatible with practical CW pump powers (W-scale). Parametric interaction in silica is by now recognized as one of the fastest and most efficient means for all-optical signal processing—used in its simplest form for optical amplification. The early history of fiber parametric processing (FPP) is characterized by pulsed experimental efforts [85]. The optical fibers used in this early period had only modest nonlinearities ($\gamma \sim 2$–5 W/km) and required high peak pump powers to observe measurable parametric gain. Recent advances in single-frequency, high-power optical pumping, high-confinement fiber, and suppression techniques for stimulated Brillouin scattering (SBS) have created the conditions necessary for CW parametric processing. Indeed, a number of record results have indicated not only the practicality of the technology, but, in certain cases, its superiority with respect to established technologies. Unlike ion-based (i.e., erbium) techniques, fiber parametric amplifier (FPA) is not confined to any spectral window, and, in principle, can provide gain bandwidths exceeding that of the combined C- and L-bands. Furthermore, PAs can be constructed to exceed the bandwidths provided by conventional Raman amplifiers.

For a more comprehensive review of the current field of nonlinear optical materials as it relates to requirements for all-optical signal processing devices, the reader is referred elsewhere [86]. To be sure, we are at a very exciting juncture that promises to yield many groundbreaking achievements in the coming years—both in terms of new materials and all-optical device demonstrations, the latter of which is the focus of the remainder of this chapter.

20.4 PARAMETRIC AMPLIFICATION

An efficient parametric exchange can be performed using a low-dispersion fiber that allows optical phase matching over a broad spectral range. However, a conventional (one-pump) PA performance is limited even in an ideally designed fiber waveguide, as illustrated in Figure 20.2(b). The single degree of freedom, provided by the pump position relative to the zero-dispersion wavelength, can be used to trade operating bandwidth for the peak signal (idler) gain. This fundamental limitation can, in principle, be addressed, by complex fiber dispersion profiles, either by specialized waveguide fabrication or by quasi-phase-matched techniques requiring many dissimilar fiber segments [29].

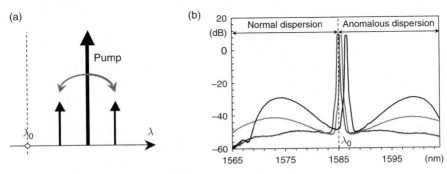

Figure 20.2 (a) Conventional parametric amplification: two-pump photons are annihilated to create signal and idler photons. λ_0 is the zero-dispersion wavelength. (b) Peak parametric gain increases by moving the pump deeper into the anomalous dispersion regime (this figure may be seen in color on the included CD-ROM).

A two-pump PA represents a four-band optical processing device that is fundamentally different from the conventional one-pump device [87, 88]. The operating principle can be described by considering two optical pumps positioned in the anomalous (unstable) and normal (stable) dispersion regimes, as shown in Figure 20.3(a). When either pump is operated in solitary mode (one-pump operation), only a weak parametric signature can be observed, or none at all, as shown in Figure 20.3(b): the anomalous pump produces weak modulational instability (MI) sidelobes (thick grey curve), whereas the normal pump exhibits no significant gain (thin curve). Combined pump operation, however, results in a dramatic effect across all four parametric bands. Two-pump FPP balances three dissimilar processes that can be precisely controlled to synthesize the desired parametric response [30, 87, 88]. First, each pump will generate near sidebands (ω_{1+}, ω_{2+}) through MI: $2\omega_{1,2} \rightarrow \omega_{1-,2+} + \omega_{1-,2+}$. Second, Bragg scattering (BS) enables the inherently stable exchanges $\omega_1 + \omega_{2+} \rightarrow \omega_{1+} + \omega_2$ and $\omega_{1-} + \omega_2 \rightarrow \omega_1 + \omega_{2-}$. Finally, phase conjugation (PC) allows for symmetric, nondegenerate FWM: $\omega_1 + \omega_2 \rightarrow \omega_{1-} + \omega_{2+}$ and $\omega_1 + \omega_2 \rightarrow \omega_{1+} + \omega_{2-}$. The true superiority of two-pump PA is in the ability to balance modulational instability coupling (MI, PC, and BR processes) by simple pump spectral tuning and to achieve the desired parametric response. Physically, two-pump PA combines degenerate and nondegenerate photon creation, while single-pump PA is limited to a single degenerate process. The new degree of freedom has a considerable implication in practical parametric device design. Higher order HNLF dispersion can be successfully matched by asymmetric pump tuning, thus overcoming one of the fundamental limitations in broadband PA construction.

The effect of pump spectral tuning [89] is illustrated in Figure 20.4(a): a two-pump amplifier was constructed using a 1-km-long HNLF segment with zero-dispersion wavelength at 1578 nm. A distant pump configuration, with pumps at

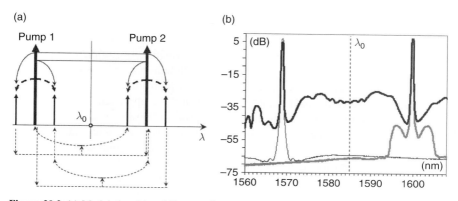

Figure 20.3 (a) Modulational instability coupling in two-pump FPP: single degenerate FPM process (MI—dashed arrows) is balanced with two nondegenerate processes (PC—solid arrows, BS—light dashed arrows), (b) Amplified spontaneous emission generated by solitary pump operating in normal (thin curve) and anomalous (thick grey curve) dispersion regime. Heavy black curve indicates combined pump operation (S. Radic et al., *IEEE Photon. Technol. Lett.*, 14, 1406 (2002)).

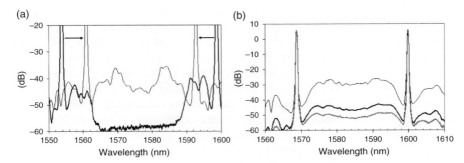

Figure 20.4 (a) Effect of pump spectral tuning in two-pump FPP; (b) Effect of pump power variation in spectrally tuned two-pump FPP. Pump powers: 189/85 mW (lower curve), 220/107 mW (middle curve), and 380/178 mW (upper curve). Higher powers correspond to a 1569-nm pump (S. Radic and C. J. McKinstrie, *Opt. Fiber Technol.*, 9, 7 (2003)).

1552 and 1599 nm (upper curve) results in an almost complete absence of parametric gain within the inner bands. A simple pump tuning (to 1561 and 1593 nm), as indicated by the lower curve, generates a considerable gain increase across the entire spectral region. Properly selected pump positions can be successfully used to generate a flat, wideband response. The pump powers represent an additional degree of freedom that can be used to control the device operation. After selecting the proper spectral positions, the pump powers can be easily adjusted to scale the magnitude of the FPP response, as shown in Figure 20.4(b).

Polarization invariance represents an important aspect of any fiber communication device. One-pump PA inherently possesses a high degree of polarization (DOP) sensitivity, as an efficient pump–signal power transfer requires copolarized signal–pump states. It is, in principle, possible to either track the polarization states in single-channel applications, or use a polarization-diversity scheme that requires polarization-maintaining fiber. The two-pump PA configuration, however, offers a simple and elegant solution to this problem. Orthogonally multiplexed parametric pumps can, in principle, provide polarization-insensitive operation [25, 27, 90]. A number of investigators have examined PA architectures with near polarization independence [91] both in fiber and semiconductor platforms. The use of polarization multiplexed (orthogonal) pumps, common in these approaches, results in the near polarization invariance of the parametric process. However, a reduced polarization dependence is necessarily accompanied by parametric gain that is significantly lower than that associated with the copolarized two-pump scheme [87].

Two-pump parametric gain has been observed in long fibers, in which PMD averages out the polarization dependence. Low-birefringence fibers are typically characterized by well-controlled, symmetric transverse geometries. This is difficult to achieve in HNLF, which has an effective area that is approximately eight times smaller than that of the standard SMF, and for which precise transverse control is correspondingly more difficult. It has been shown theoretically that a birefringent, two-pump PA generally has polarization-sensitive gain.

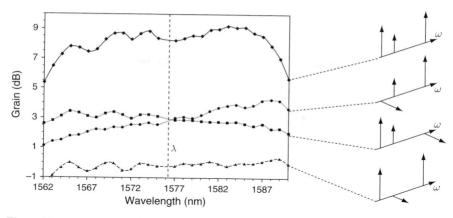

Figure 20.5 Parametric gain from a short (200 m long) HNLF. Solid rhomboids: pumps and signal copolarized and aligned with a principal axis; solid squares: C-band pump and signal aligned along a principal axis, L-band pump in the orthogonal state; solid circles: C-band pump aligned along a principal axis, signal, and L-band pump in the orthogonal state; solid triangles: both pumps aligned along a principal axis, signal in the orthogonal state. The C-band (1557 nm) pump power is 380 mW, L-band (1595.45 nm) pump power is 320 mW, and the input signal power is −25 dBm.

Figure 20.5 illustrates the limits of polarization-invariant PA behavior. Relatively large differences between the copolarized and the orthogonal pumping condition are observed along the entire scanned bandwidth.

20.5 OPTICAL REGENERATION

Noise and interference sources in optical communication systems take many forms, the most obvious of which is the amplified spontaneous emission (ASE) noise from amplifiers, which grows proportionally with the number of cascaded amplifiers in the link. If not compensated for, ASE noise is one of the most significant impairments in communication systems reaching thousands of kilometers by interfering with the signal. To limit signal degradation, all-optical signal regenerators including 2R and 3R regenerators [92] are expected to play a major role both in ultralong signal transmission and in terrestrial photonic networks. The advantages of all-optical regeneration over conventional techniques include huge potential savings in cost (by avoiding O/E/O conversion and demultiplexing), speed, and footprint.

A wide range of 2R and 3R optical regeneration techniques [90] have been studied that rely on optical nonlinear effects, to replace the widespread optoelectronic regenerators currently used primarily for long-haul and ultralong haul systems. The aim of this section is not to review all of this work but to focus on our research which has centered around all-optical 2R and 3R regeneration methods that exploit SPM and XPM processes [7, 9, 80, 93, 94].

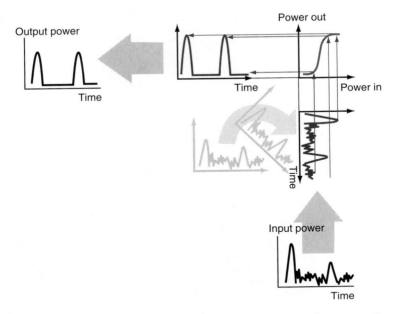

Figure 20.6 Principle of noise compression by a 2R optical regenerator that operates through an instantaneous nonlinear power transfer curve (this figure may be seen in color on the included CD-ROM).

Typical 2R regenerators achieve their functionality from a nonlinear power transfer function (PTF), governing the *instantaneous output vs input power* of the regenerator (Figure 20.6). Such transfer functions have been realized through nonlinear phenomena such as SPM [95–98], interference within an optical loop mirror [99–101], FWM [102], PA [8], saturable absorption [103], or gain dynamics in SOAs [104]. Numerous papers have experimentally and theoretically demonstrated a reduction in bit error rate (BER) degradation, or a decrease in the Q factor degradation, at the end of systems with cascaded EDFAs, by the addition of optical regenerators between the EDFAs [97, 98]. These regenerators prevent the BER degradation by subsequent noise sources (e.g., EDFAs) in an optical link, by periodically compressing the noise fluctuations on the logical 1s and 0s, which prevents a sudden noise buildup with its associated BER degradation. These 2R optical regenerators that operate on both noise and signal equally do not directly reduce the BER of a noisy signal, however. That is, the BER just before and after these regenerators is theoretically identical [93].

Contrasting this are 2R regenerators based on SPM followed by filtering, as first proposed by Mamyshev [95]. These regenerators have been the focus of significant interest, and, amongst other things, have been shown to be capable of direct BER improvement [93, 94]. We begin the next section by reviewing our recent work on optical 2R regeneration based on this approach in both fiber and integrated waveguides, followed by a section on 3R regeneration through XPM, and finally FWM in fiber.

20.5.1 2R Regeneration Through SPM in Chalcogenide Fiber

Optical 2R regeneration (re-shaping) based on SPM followed by spectral filtering [95] has a number of key advantages. These include simplicity, a bandwidth limited only by the intrinsic material nonlinear (Kerr) response, and the fact that they have been shown [94] to be capable of directly improving not only the Q-factor but the BER of input data. "Mamyshev" regenerators have been demonstrated in highly nonlinear silica-based fiber [93, 96] and have even been used to achieve 1 million kilometers of error-free transmission without electrical conversion [98]. Recently [7], we investigated 2R regeneration through the Mamyshev technique, in highly nonlinear chalcogenide fiber. Highlights of this work included achieving operation at peak powers below 10 W, and demonstrating that high material dispersion as well as moderate TPA—in contrast to previous expectations—can actually enhance the device performance by improving the transfer function. For the case of TPA (which determines the nonlinear FOM) this is rather remarkable since it has been known for 20 years [58] that for nonlinear switching, it can only degrade, not improve, device performance.

The principle of operation of a Mamyshev regenerator is shown in Figure 20.7. A noisy input signal (Return-to-Zero (RZ) in this case) is passed through a nonlinear medium (which can contain dispersion) producing SPM-induced spectral broadening. Noise on the input signal, experiences low-SPM spectral broadening, and so is filtered out by the bandpass filter (BPF) (offset from the input centre wavelength), whereas signal pulses experience significant broadening and are consequently partially transmitted through the BPF. This results in a nonlinear power transfer curve, which improves [96] the optical signal-to-noise ratio (OSNR)

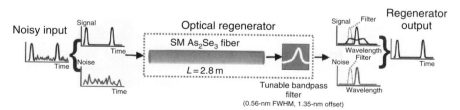

Figure 20.7 Principle of operation of 2R regenerator based on SPM and filtering. At high intensities, input signal pulses experience large SPM-induced spectral broadening and are transmitted through the (offset) bandpass filter. Noise, however, experiences little spectral broadening (even at high-average powers) and so is removed by the filter. The large n_2 in chalcogenide fiber enables operation with <3 m of nonlinear fiber (this figure may be seen in color on the included CD-ROM).

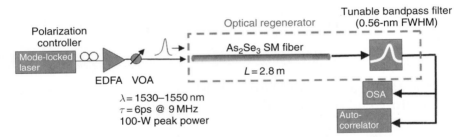

Figure 20.8 Experimental configuration for demonstrating optical regeneration. Note the EDFA is not part of the device itself but part of the laser source used to test the device (this figure may be seen in color on the included CD-ROM).

and Q-factor for modulated signals at 40 Gb/s. Further, as discussed above, it has recently been shown [93, 94] that these regenerators attenuate noise (relative to signal pulses) even more than what the nonlinear PTF would imply, the result being a direct improvement in BER rather than merely a prevention of degradation in signal to noise.

Figure 20.8 shows the experimental configuration used to demonstrate the device performance. The device was based on a 2.8-m-long As_2Se_3 SMF with an MFD of ~6.4 µm at 1550 nm, an average core/cladding refractive index of 2.7 and a numerical aperture of 0.18. Pulses from a mode-locked laser near 1550 nm with a repetition rate of 9.04 MHz and width of 5.8 ps were passed through a polarization controller, and then through a custom-built amplifier designed to keep extraneous SPM-induced spectral broadening to a minimum. The signal was then butt-coupled into the chalcogenide fiber from the SMF fiber, with typical coupling losses being 2.5–3.5 dB per facet. This consists of 0.8-dB Fresnel loss, less than 1 dB of mode overlap loss, with the rest being due to the facet cleave quality. Propagation loss in the fiber was 0.9 dB/m. The output of the fiber was passed through a tunable bandpass filter (TBF) with an FWHM bandwidth of 0.56-nm (70 GHz) offset from the input center frequency by 1.3 nm, and then directed either into an optical spectrum analyzer (OSA) or into an optical auto-correlator for temporal measurements, being careful to ensure that no detectable cladding modes were excited. The peak power levels of 10 W used in these experiments corresponded to a maximum intensity of 100 MW/cm^2, much less than the photo-darkening damage threshold reported for this material [39].

Figure 20.9 shows the output pulse spectra of the chalcogenide fiber with no filter present for different peak input powers, and clearly illustrates the principle of operation of this device—as the peak input power increases, the SPM broadens the signal spectrum so that it overlaps more with the BPF, creating a nonlinear PTF, a key benchmark of the performance of a 2R regenerator. The theoretical results (light dashed) show good agreement with experiment and were calculated using a split-step Fourier transform method with a value of n_2 of $\sim 1.1 \times 10^{-14}$ cm^2/W, or $\times \sim 400$ silica, in line with earlier measurements of n_2 in this Se-based ChG

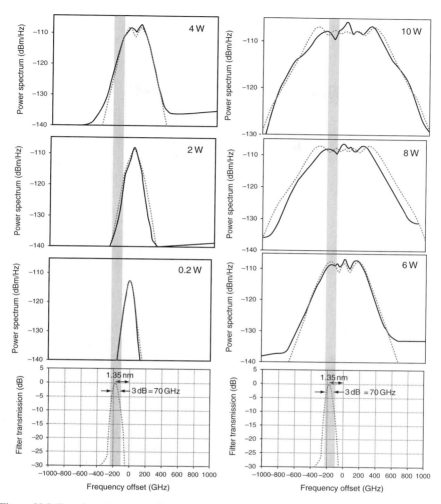

Figure 20.9 Experimental output pulse spectra for different input pulse peak powers, theory (dashed) and experiment (solid). Bottom—transmission bandpass filter transmission spectrum, offset by 1.35 nm from the input center wavelength, and with a 3-dB bandwidth of 0.56 nm (70 GHz) (this figure may be seen in color on the included CD-ROM).

fiber [39]. Also shown in Figure 20.9 is the TBF (0.56 nm FWHM) transmission spectrum, offset by 1.35 nm from the input center frequency. Note that the device performance was not exhaustively optimized in terms of filter offset.

Figure 20.10 shows the resulting experimental and theoretical nonlinear power transfer curves as a function of peak in-fiber input power of the full device with the filter in place. The curves show a clear output power limiting function at ~5 W peak input power, as well as a clear delayed "turn-on." The former is effective in

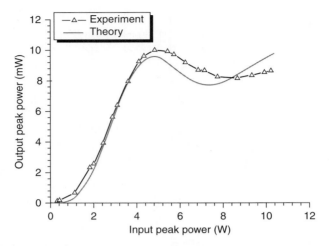

Figure 20.10 Theoretical (light) and experimental (solid with triangles) power transfer curves, for a filter offset of 1.35 nm (this figure may be seen in color on the included CD-ROM).

suppressing the noise in the logical "1s" whilst the latter contributes to suppressing noise in the "0s." There is still a slight oscillation at high-peak power levels but this is much smaller than what would result [96] without the presence of the large dispersion of -504 ps/nm/km at 1550 nm of the 2.8-m long As_2Se_3 fiber, with an average dispersion slope of $+3$ ps/nm^2/km. This is not only much larger than standard SMF fiber ($+17$ ps/nm/km) but of opposite sign—normal vs anomalous in SMF. This is ideal for SPM-based optical regenerators which play a critical role in the success of our device. SPM-induced spectral broadening, without the presence of linear dispersion, produces an oscillating spectrum at very high peak powers corresponding to nonlinear phase shifts $\gg 2\pi$, which results in an oscillating power transfer curve. The large linear normal dispersion in As_2Se_3 linearizes the induced chirp, which averages out the spectral oscillations, thus smoothing out the nonlinear transfer curve. Note that even though the total dispersion of 1–2 ps/nm present in the device is comparable with that of typical lengths of silica-based HNLF (one to several kilometers) used in regeneration experiments [96], the key is that for ChG, this level of dispersion is present in very short lengths of a few meters.

The overall success of this device was largely a result of optimizing the parameters such as length and input pulsewidth. It has been shown in Ref. [96] that the optimum device length is $L_{opt} = 2.4 L_D/N$, where $N = \sqrt{L_D/L_{NL}} \propto \sqrt{P}$ is the soliton number (although clearly with normal dispersion solitons are not present) and P is the input launch power. For this As_2Se_3 fiber, L_D is ~ 13 m and $N = \sim 13$, yielding an optimum length of $L_{opt} = 2.7$ m, very close to the length used (2.8 m).

For these proof of concept demonstrations low-duty cycle pulses were used. Ultimately, this device would have to operate on a high-duty cycle signal, such as 33% RZ at 40 Gb/s and higher—potentially up to 1 Tb/s and above, and at practical power levels. Figure 20.10 indicates that this device already operates at power levels commensurate with typical powers achievable in a 33% RZ signal at 40 Gb/s of 5 W, corresponding to ~800 mW of average power. This is within about a factor of 2 of typical peak powers used in highly nonlinear silica fiber [96]. Recently, we have succeeded in demonstrating operation at full data bit rates at 10 Gb/s in both fiber [105] and waveguides [106] for devices operating through XPM, rather than SPM.

One of the more remarkable findings of our work on these 2R regenerators was that a modest amount of nonlinear, or two-photon, absorption can actually enhance the device performance. This is in stark contrast with nonlinear switching applications, where it has been known for almost 20 years [58] that TPA can only degrade device performance and that it represents a fundamental limit which in fact inspired the definition of the nonlinear FOM ($=n_2/\beta\lambda$) in the first place. Figure 20.11 shows theoretical power transfer curves for this device obtained by varying the FOM of the material from infinite (no TPA) to 0.25 (very high TPA). Materials with an FOM in the range of 1–5 produce a flatter PTF than those with no TPA. This saturation of output power vs input power (or "step-like" function) will result in a reduction in noise on the logical "1s" and on the logical "0s"—an important requirement of 2R regenerators. For FOM <1, the transfer curves lose their sharp edge characteristics and round off. The experimental data points in Figure 20.11 agree well with the theoretical curve for an FOM = 2.5. This value is close to the experimental value of FOM = 2.8 for As_2Se_3 [40, 57, 60, 107].

To better understand the performance of this 2R regenerator in a system, simulations of a pseudo-random bit sequence (PRBS) at 40 Gb/s with an RZ signal made of 33% duty-cycle pulses, including ASE noise to a level of OSNR = 15 dB, were

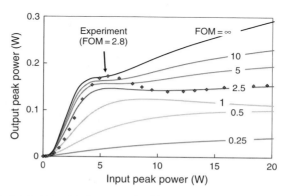

Figure 20.11 Calculated power transfer curves for 2R regenerator for different FOM with fixed n_2, including experimental data for a device with an FOM of 2.8 (black diamonds) (this figure may be seen in color on the included CD-ROM).

performed. A series of results were calculated by varying the material FOM, keeping other parameters constant. Figure 20.12 shows the input and output eye diagrams of the sequence while Figure 20.13 shows the calculated Q-factor at the output as a function of peak input power, for various FOMs. It is clear from Figures 20.12 and 20.13 that the improvement in PTF characteristics for moderate FOM of 2.5 seen in Figure 20.11, also translate into improved eye diagrams and Q-factors.

From Figure 20.13, it is evident that the performance of the regenerators depends strongly not only on the FOM but also on the absolute input peak power as well: 7 W for a device with no TPA and 14 W for an FOM of 2.5. The optimum input peak power occurs above the step-threshold power in the PTF and so the lower power pulses remain in the higher output power plateau. It is also clear from Figure 20.13 that while a device with an FOM of less than unity does not effectively increase the Q-factor, an infinite FOM is clearly not optimum either. The best device performance is obtained with an intermediate FOM and with slightly higher input power, although still well within practical power levels. Figure 20.14 summarizes these results by plotting the maximum optimized (by adjusting the input peak power) output Q-factor as a function of FOM. It shows that there is clearly an optimum range of FOM, from 2 to 4, where the

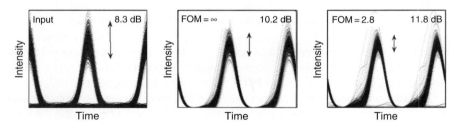

Figure 20.12 Calculated input and output eye diagrams for different FOMs.

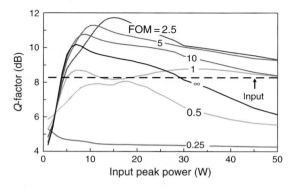

Figure 20.13 Calculated Q-factor for 40-Gb/s RZ signal with OSNR of 15 dB, for different figures of merit (FOM) as a function of input signal power. The horizontal dashed line represents the input Q-factor (this figure may be seen in color on the included CD-ROM).

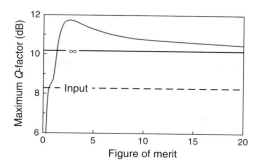

Figure 20.14 Optimized Q-factor improvement vs figure of merit, showing that modest TPA can enhance device performance (this figure may be seen in color on the included CD-ROM).

improvement in Q-factor is maximized and noticeably better than a device with no TPA (infinite FOM). It is interesting to note that As_2Se_3 ChG has an FOM of 2.8 which is near the optimum value.

A key requirement for all-optical regenerators is to keep the signal pulse distortion by the regenerator to a minimum. The spectral and temporal responses of the output pulses were measured [108] by frequency-resolved optical gating, showing explicitly that the spectral and temporal pulse distortion, in both phase and amplitude, is very small—less than a few percent.

In summary, these results show promising performance from As_2Se_3 chalcogenide fiber-based 2R all-optical regenerators, already with advantages over silica-based HNLF devices such as a reduction in fiber length from kilometers to meters, and an intrinsic *nonlinear* FOM—a fundamental quality that can intrinsically limit device performance—that is nearly ideal for 2R regeneration. In addition, there is significant room for future improvements in, e.g., reducing operating power—currently comparable with the best silica-based devices. A reduction in operating power by over an order of magnitude would be achieved by reducing the mode field diameter as well as possibly increasing the intrinsic nonlinearity of the ChG by optimizing stoichiometry. In addition, because our fiber lengths are so short, there is also scope to further reduce the required operating power further by using slightly longer lengths (up to 5–10 m). The passive propagation loss of this fiber, currently at 0.9 dB/m (resulting in a total loss for 2.8 m of fiber comparable with that for 2- to 3-km long silica HNLF devices) is also expected to decrease significantly with improvement in fabrication processes. Losses due to spectral filtering (in this case ~13 dB) together with losses from additional components (tunable filter, VOA, coupling loss at exit face, etc., in this case ~12 dB) can also be significantly reduced by optimizing filter design, improving fiber coupling (cleave quality and AR coatings), and reducing the insertion loss of all of the components. Ultimately, however, the goal is to move on beyond fiber-based devices to fully integrated all-optical photonic integrated circuits (PICS), and we turn to this topic next.

20.5.2 2R Regeneration in Integrated Waveguides

Clearly, there is compelling motivation to develop fully integrated all-optical PICs, capable of operating from 40 Gb/s up to potentially > 1 Tb/s, in the same manner as electronic ICs have revolutionized our world in the past 40 years. To date, while most all-optical signal processing devices capable of operating at speeds of >40 Gb/s and above have been based on optical fiber, all optical PICs have been reported recently. Devices based on SOAs [109–111] or electroabsorption modulators [112] have attracted significant attention, with recent results even reaching 640 Gb/s [113]. These devices exploit ever increasingly subtle ultrahigh-speed carrier dynamics to achieve the nonlinear optical response. Our work has focused rather on devices based on pure Kerr nonlinearities where the intrinsic device speed is in theory almost unlimited ($\gg 1$ Tb/s). Recently [6, 114], we reported the first fully integrated all-optical 2R regenerator based on Kerr nonlinearities, and even more recently [23] a similar approach has been reported in silicon, where the use of nanowire waveguides was employed to achieve an order of magnitude reduction in peak operating power.

Figure 20.15 shows a schematic of the device. It consists of a 5-cm-long rib waveguide in (As_2S_3) ChG integrated with a 5-mm-long dual grating that acts as a BPF. This device is a result of recent advances in low loss rib waveguides [18] and high-quality Bragg gratings in ChG (As_2S_3) [63, 115]. The waveguides were fabricated by pulsed laser deposition (PLD) [62] of a 2.4-µm-thick As_2S_3 film, with a refractive index of 2.38, followed by photolithography and reactive-ion etching to form a 5-cm-long, 4-µm-wide, rib waveguide with a rib height of 1.1 µm, and then overcoated with a polymer film transparent in the visible. The fiber-waveguide-fiber insertion loss (with high-NA (numerical aperture) fiber,

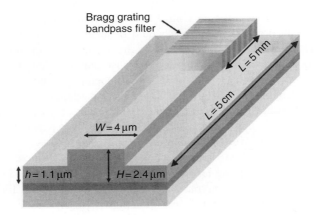

Figure 20.15 Integrated all-optical 2R regenerator in As_2S_3 chalcogenide glass. The device consists of a 5-cm-long ridge waveguide integrated with a 5-mm-long double grating bandpass filter written by a modified Sagnac interferometer (this figure may be seen in color on the included CD-ROM).

20. Nonlinear Optics in Communications

4-μm mode field diameter) was 10.5 dB, of which ~2.5 dB was propagation loss, equating to a coupling efficiency of ~4.0 dB per facet (including about 0.8 dB of Fresnel reflection loss from the material refractive index mismatch).

Waveguide gratings were written [63, 115] near the exit facet of the waveguide using a Sagnac interferometer along with a CW 532-nm doubled Nd:YAG laser having a coherence length of 4 mm, which provided a naturally compensated refractive index apodization. The sample was exposed with 10 mW for 60 s, resulting in gratings with an estimated index change $\Delta n = 0.004$–0.005. The extremely high quality of the gratings in terms of both width (> 6 nm each), depth (> 25 dB), and very low out of band sidelobes was critical for the successful performance of this device.

Figure 20.16 shows the transmission spectrum of the waveguide Bragg grating for Te (tellurium) polarization. The grating in fact consisted of two gratings offset from each other to produce an overall rejection bandwidth of 16.3 nm with a 2.8-nm wide pass band in the middle. The Tm (thulium) polarized grating was separated by ~8 nm from the Te grating due to waveguide birefringence, and so these experiments were carried out with Te-polarized light. Te-polarized pulses could be launched to better than 20-dB extinction ratio by adjusting the polarization controller before the amplifier.

Figure 20.17 shows the experimental setup to demonstrate device performance. Pulses from a mode-locked "Figure of 20.8" laser (1.5 ps, full-width, half-maximum) were passed through a polarization controller and then a custom optical amplifier (designed to minimize excess spectral broadening), resulting in 8.75-MHz repetition rate pulses at peak powers up to 1.2 kW, nearly transform limited with a spectral width of 1.9 nm, tunable from 1530 to 1560 nm. The pulse spectra were tuned to slightly offset wavelengths relative to the grating transmission pass band. The output of the amplifier was butt-coupled into the waveguide using a short (to minimize spectral broadening within the fiber) 20-cm length of standard SMF followed by 5 mm of high-NA fiber spliced on the end to improve

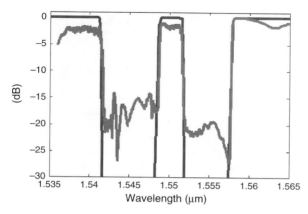

Figure 20.16 Transmission spectrum of double-integrated waveguide bandpass grating filter. Dark curve is theory and lighter curve is experiment (this figure may be seen in color on the included CD-ROM).

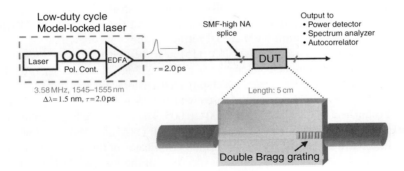

Figure 20.17 Experimental setup for testing integrated waveguide 2R regenerator (this figure may be seen in color on the included CD-ROM).

coupling efficiency to the waveguide. The output of the waveguide was then directed into either a power meter, an OSA, or an optical autocorrelator.

Figure 20.18 shows the pulse spectra for 2-ps pulses after passing through a bare waveguide with no grating present showing significant broadening with increasing input peak power, due to SPM equivalent to a nonlinear optical phase shift (ϕ_{NL}) of approximately $3\pi/2$, which was spectrally symmetric. The maximum peak power in the waveguide was 97 W, corresponding to a maximum intensity of $\sim$1.8 GW/cm^2. We verified that SPM-induced spectral broadening by the amplifier was negligible and that no photo darkening occurred.

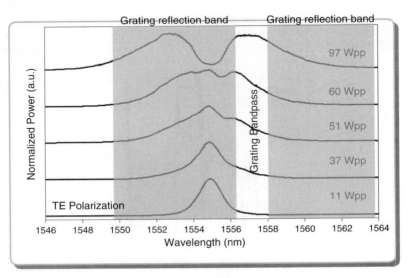

Figure 20.18 Unfiltered pulse spectra after propagation through chalcogenide waveguide along with grating passband.

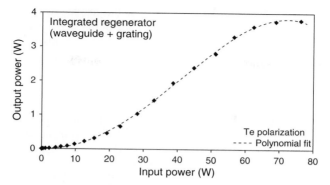

Figure 20.19 Nonlinear power transfer function from integrated waveguide/grating 2R regenerator (this figure may be seen in color on the included CD-ROM).

Figure 20.19 shows the resulting power transfer curve of the device integrated with the waveguide Bragg grating, obtained by tuning the pulse center wavelength to within the rejection band of the Bragg grating to the longer wavelength side of the pass band, for Te polarization. As seen in Figure 20.19 this resulted in low transmission at low-input power because the filter rejection bandwidth was much wider than the 3-dB spectral bandwidth of the input pulse. As the input pulse power was increased, SPM broadened the spectrum so that power was transmitted through the pass band of the grating, resulting in a clear nonlinear "S-shape," required for optical regeneration.

The temporal transfer characteristics of this device are clearly critical. The autocorrelation of the input and output pulses at high power is shown in Figure 20.20, indicating that the output pulses were broadened slightly to ∼3 ps. This broadening is not due to either waveguide dispersion or material dispersion in the passive

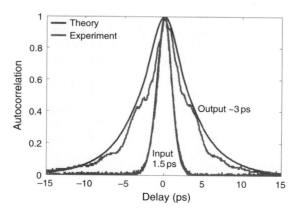

Figure 20.20 Input and output pulse autocorrelation traces, showing broadening from 1.5 to 3 ps (this figure may be seen in color on the included CD-ROM).

waveguide section (both of which are negligible on this length scale), but rather is due to a combination of grating dispersion near the edge of the stop-band and a comparatively wide pass band (3-nm square filter) relative to the input pulse spectrum (1.9 nm FWHM). Optimizing the grating pass band shape and width is expected [116] to reduce this and can provide flexibility, even tunability [117], for reconfigurable regeneration.

While this device operates at peak powers from 10s to 76 W, and was tested with low-duty cycle optical pulses, practical devices would need to operate at sub-Watt power levels and with high-duty cycle (e.g., 33% RZ) optical pulses. By increasing the device length to 50 cm through the use of serpentine or spiral structures, increasing the material nonlinearity (by using As_2Se_3 rather than As_2S_3, for example) and by decreasing the waveguide area by a factor of 10 (to 1 μm^2), a reduction in operating power by two orders of magnitude should be achievable, resulting in sub-Watt power level operation. Furthermore, the increased (normal) dispersion of longer waveguides would linearize the frequency chirp of the output pulses, reducing pulse distortion and improving transfer function characteristics [7].

In conclusion, this integrated all-optical signal regenerator consisted of a nonlinear waveguide followed by an integrated Bragg grating filter in ChG operating through the virtual material nonlinear (Kerr) response in ChG, and is therefore capable of Tb/s bit rates. We demonstrated a clear nonlinear power transfer curve using 1.5-ps optical pulses, which would result in OSNR reduction and BER improvement in optical links.

20.5.3 3R Regeneration

As useful as 2R regeneration is, it suffers from the limitation that it does not retime the signal and therefore cannot correct for timing jitter. Full regeneration (3R) requires the capability to retime signals, and it has been demonstrated experimentally that this can have a dramatic impact on system link performance, enabling up to 1 million kilometers of unrepeatered transmission [98]. Recently [9, 118, 119], an approach to 3R regeneration has been proposed and demonstrated, based on similar principles to the 2R regeneration scheme, using SPM and XPM. This approach offers many of the advantages of the 2R scheme, including the ability for direct BER reduction [93, 94, 120], while adding the capability of correcting for signal timing jitter.

The architecture and principle of a 3R regenerator (Figure 20.21), proposed first by Suzuki et al. [118], consists of an optical clock source, an EDFA, a HNLF, and a BPF. The clock, generated from a clock-recovery circuit, produces pulses shorter than the signal, either at the same repetition rate for regeneration or potentially at submultiples for demultiplexing. As the signal bits (λs) and clock pulses (λc) temporally overlap in the HNLF (Figure 21(b)), the clock spectrum experiences an XPM-induced frequency shift $\delta\omega_{XPM}$ proportional to the time derivative of the overlapping signal, $\delta\omega_{XPM} = -2\gamma L_{eff} \, dP(t)/dt$, where γ is the fiber nonlinearity, L_{eff} is the effective fiber length, and $P(t)$ is the input signal. The spectral components that have been shifted by XPM are filtered out by BPF2, which keeps the original frequency components at λc. Although this regeneration architecture inverts the

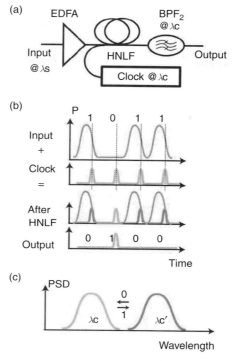

Figure 20.21 (a) Schematic of the 3R regenerator based on XPM-induced spectral shift followed by filtering. (b) The low-power clock pulses are overlapping with the edge of the high-power signal pulses, which induce a spectral shift on the clock. At the output of the regenerator, data have been imprinted at the clock wavelength, and inverted. (c) The clock spectrum is composed of two lobes after XPM (i.e., after HNLF). One lobe at λc is identical to the clock spectrum before XPM and a second lobe at $\lambda c'$ is a spectrally shifted replica of the clock spectrum. The first lobe is formed by input logical 0s whereas the second lobe is formed by input logical 1s. $\lambda s/c$, wavelength for signal/clock; P, power; PSD, power spectral density (this figure may be seen in color on the included CD-ROM).

bits, a noninverting mode can easily be achieved by offsetting the center frequency of BPF2 away from λc to transmit only XPM-generated frequencies. In both cases, the power transferred through the regenerator depends on $\delta\omega_{XPM}$ and hence on the temporal characteristics—as we have seen, a key requirement to discriminate between 0s and 1s which in turn can lead to direct BER improvement.

Figure 20.22 shows the setup used to demonstrate that the 3R regenerator can improve BER. A pattern generator sends a PRBS of $2^{23} - 1$ bits at 40 Gb/s that is transformed by cascaded modulators into 8-ps pulses at 1534.25 nm and with a bandwidth of 0.5 nm, respectively. The pulsed signal and ASE noise from an EDFA, combined to a desired OSNR using a variable attenuator (VA1) and a 3 dB coupler, are spectrally filtered with a 0.5-nm BPF filter. The 3R architecture follows the schematic of Figure 20.22 with 4-ps clock pulses at 1558.0 nm, a HNLF length of

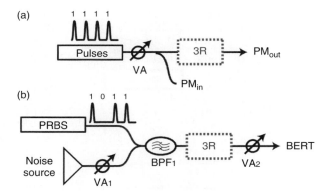

Figure 20.22 (a) Experimental setup for the power transfer function measurement of the 3R regenerator. (b) Setup to demonstrate BER improvement. VA, variable attenuator; PM_{in}/PM_{out}, power meter at the input/output of the O3R; BPF, band-pass filter; PRBS, pseudo random bit sequence; BERT, bit error ratio tester (this figure may be seen in color on the included CD-ROM).

850 m (nonlinearity $\gamma = 20$ W/1 km), BPF2 width of 0.5 nm centered at 1558.0 nm. The amplifier boosts the noisy pulses up to a peak power level of 500 mW, sufficient to induce XPM on the clock. The regenerator is followed by a variable attenuator VA2 to ensure the peak optical power is constant at a relatively high 2 mW, where shot and thermal noises are negligible as verified by the BER being $<<10^{-13}$ (no errors) with no ASE added. In contrast, with ASE noise we obtain $10^{-10} <$ BER $< 10^{-2}$, depending on the OSNR. Signal spontaneous beat noise is the most prominent source of noise when ASE is present. Figure 20.23 shows an improvement in BER from 3×10^{-6} to 2×10^{-10} at an OSNR = 15 dB, or 4 dB in OSNR at 10^{-10} BER, by adding the regenerator. Further BER improvement is possible by optimizing the regenerator parameters such as the fiber length, peak power, and the spectral shift of BPF2.

20.5.4 Optical Signal Regeneration Through Phase Sensitive Techniques

Another important approach to 2R regeneration is based on phase-matched higher order parametric processes in optical fiber using one-pump PA [33, 121]. A single, intense optical pump is combined with a signal to generate higher order FWM with a steep transfer characteristic, as illustrated in Figure 20.24. Like other all-optical schemes in fiber, efficient regeneration requires intense optical pumping and, inevitably, spectral broadening to suppress SBS. Unfortunately, pump broadening necessarily results in excessively broadened higher order FWM waves. Even for the case of a CW signal, the width of the higher order FWM tones will be a multiple of the pump spectral width. The pump-induced spectral broadening of the signal fundamentally limits the use of the proposed 2R scheme, both for inline and predetection 2R applications. The excessive phase broadening transferred from the pump to the

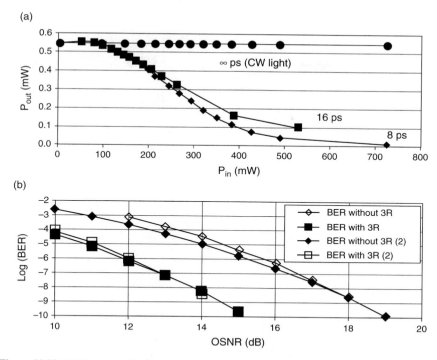

Figure 20.23 (a) Power transfer functions for 8-ps pulses, 16-ps pulses, and continuous wave (CW) light. The input and output powers, P_{in} and P_{out}, are expressed in terms of peak power. (b) BER with and without the regenerator as a function of OSNR and constant received power. The 3R regenerator improves the BER for constant OSNR or, equivalently, the 3R regenerator increases the OSNR margin for a given BER (this figure may be seen in color on the included CD-ROM).

idler is readily converted by chromatic dispersion to amplitude modulation, manifested in the form of random signal distortion. For the case in which the broadened signal is directly received, rather than retransmitted, the filter bandwidth required prior to the optical detector needs to be considerably broader than the original data rate to avoid significant spectral clipping. This, in turn, allows excessive noise to be received along with the signal, again limiting the usefulness of single-pump PA regeneration. Idler broadening is difficult to avoid in any single-pump parametric device. However, two-pump PA can be used to generate narrow idlers anywhere within the four associated bands by synchronous pump-phase modulation [122]. In a widely used method, the pump phases are dithered in opposition to each other, maintaining the average pump frequency, thus allowing narrow idler generation.

It is then natural to expect that narrow higher order idlers can also be produced using a similar scheme in a two-pump PA. In the small-signal regime, two-pump PA generates only three idlers. Figure 20.25(a) illustrates the inner-band signal and idler generated using a 1-km-long HNLF. With an increase of the input signal power,

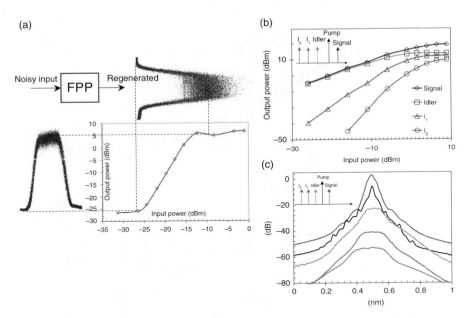

Figure 20.24 (a) Generic 2R processing scheme critically depends on steepness of the transfer characteristic; (b) Higher order 2R processing in one-pump parametric device; and (c) Excessive idler broadening (this figure may be seen in color on the included CD-ROM).

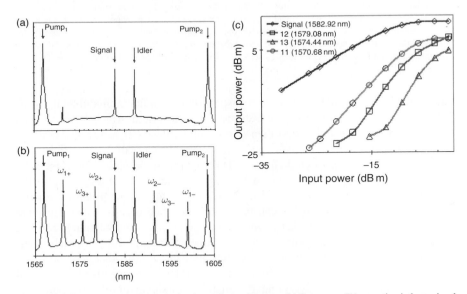

Figure 20.25 (a) Two-pump PA operating in small-signal regime; (b) Two-pump PA operating in large-signal regime; and (c) Transfer characteristics corresponding to first four significant parametric orders in two-pump FPP (S. Radic et al., *IEEE Photon. Technol. Lett.*, 2003) (this figure may be seen in color on the included CD-ROM).

20. Nonlinear Optics in Communications

several higher parametric orders appear, as shown in Figure 20.25(b). A rigorous analysis of higher order generation in two-pump device reveals a complex, cascaded FWM. Degenerate and nondegenerate FWM processes driven by the existing tones seed a newly generated tone. The growth of this new tone is driven by the aforementioned seeding processes, together with other degenerate and nondegenerate FWM processes made possible by its existence. Which subset of FWM processes dominates the generation of the new tone depends on the frequencies of the new and existing tones, the powers of the existing tones and the dispersion characteristics of the fiber. It can be shown that, when the pumps are either co or counterphased, a subset of narrow higher order tones exists. Figure 20.25(c) illustrates the transfer characteristics of the first three significant higher order terms generated in a counterphased two-pump FPP. The tone I_2, which has a steep transfer characteristic (squares curve), is generated as a narrow idler by a fifth-order cascaded FPM process.

20.6 OPTICAL-PHASE CONJUGATION

The present-day optical networks are far from the idealized concept of the optical "ether" that has been advocated for at least two decades now. The idea of all-optical layer calls for the ability of any user in the network to communicate with any other user, regardless of his distance or position, *without ever leaving the optical domain*. An important set of issues that include, but are not limited to, network transparency, wavelength provisioning, and random-path signal integrity are yet to be addressed by all-optical means. Optical processing holds the potential of unlimited, or at the very least, very high network capacities. Figure 20.26 illustrates some of the fundamental difficulties one faces in attempting to construct all-optical network. Assume that user A needs to send data to user C who is not far away, so that the signal integrity can easily be maintained along the path A–C. The node A might be able (or required) to transmit only using a specific wavelength (indicated by the line from A to Q in Figure 20.26), while the user C might require

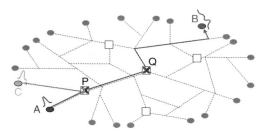

Figure 20.26 Generic all-optical networking layer: any user communicates with any other node by using randomly assigned optical path and variable or restrictive spectral constraints. P, Q indicate network locations requiring wavelength routing/conversion (P) and signal integrity maintenance (Q). Empty blocks indicate selected network locations that require either of these functions (this figure may be seen in color on the included CD-ROM).

reception/transmission using a different wavelength. It is clear that, even for a case when signal regeneration is not required, wavelength conversion must occur somewhere along the signal path. This situation is greatly complicated by introducing an arbitrary path requirement: the end user (node B in Figure 20.26) might be a long distance away, thus requiring the signal regeneration, in addition to already needed wavelength conversion. In conventional networks, described tasks are exclusively performed in electronic domain. Data paths A–B and A–C are assembled out of point-to-point photon pipelines held together and managed by electronic routers and O/E/O interfaces.

The effects of its strategic, if sparse, phase conjugation placement in the network are yet to be anticipated and fully understood. Broadband PC provided by PA will surely deserve particular attention in the coming years. It is well known that if the (complex) amplitude of an optical field satisfies the nonlinear Schrödinger (NLS) equation, the conjugate amplitude also satisfies an NLS equation in which z is replaced by $-z$, i.e., the sense of propagation is reversed (time reversal).

Consequently, any deterministic impairment that increases before a PC decreases after the PC [123]. In the first demonstration [124] of multichannel dispersion/penalty compensation, we used a single two-pump FPP to transmit five nonreturn-to-zero (NRZ) optical channels over four SMF spans, as shown in Figure 20.27(a).

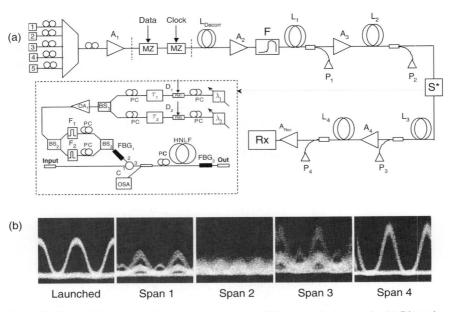

Figure 20.27 Feasibility of sparsely positioned two-pump FPP in all-optical networks: (a) Dispersion compensation and penalty reversal using a single, asymmetrically placed two-pump FPP in WDM transmission over SMF fiber; (b) Received signal represents a conjugate, partially regenerated copy of the launched waveform, exhibiting higher extinction ratio. The dispersion map and power evolution of the signal are far from optimum parameters (S. Radic et al., *OFC 2003*, PD-12).

20. Nonlinear Optics in Communications

An asymmetric dispersion map and unequalized power evolution were chosen to test the robustness of the FPP compensation scheme. Figure 20.27(b) illustrates a particularly promising result: not only that the end-of-link waveform gets reassembled to near-launch form, but it also exhibits a higher extinction factor due to partial signal regeneration.

20.7 WAVELENGTH CONVERSION

Of the many approaches to wavelength conversion, all-optical solutions based on the ultrafast Kerr nonlinearity arguably have the greatest potential for both speed and simplicity, and of these, both phase- and nonphase-matched approaches have been reported. We begin by a discussion of nonphase-matched approaches.

Nonphase-matched wavelength conversion was first demonstrated by Olsson [125] in HNLF, using XPM. This technique has subsequently been employed in bismuth fiber [126], chalcogenide fiber [105], and recently in chalcogenide waveguides [10]. The primary advantage of this technique is that, because it is nonphase-matched, it does not require dispersion-engineered fiber or waveguides. Pump–probe walk-off is certainly a concern [105] but this has not prevented demonstrations of up to 25 nm of conversion range even devices with quite high material dispersion [10, 105].

20.7.1 Chalcogenide Fiber-based Wavelength Conversion

Figure 20.28 shows the principle of operation of a wavelength converter based on XPM [125]. A CW probe experiences XPM from copropagating signal pump pulses, which generate optical sidebands. This is converted to amplitude modulation by using a BPF to select a single sideband. Importantly, the interaction length between the signal pump and the CW probe, i.e., pulse walk-off, is the limiting factor in terms of wavelength conversion range and not stringent phase matching conditions. Recently [105], this approach was used to demonstrate wavelength

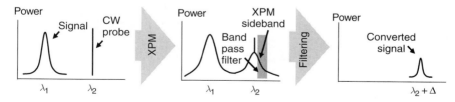

Figure 20.28 Principle of XPM wavelength conversion. Amplified pulsed pump signal (at λ_1) imposes a nonlinear frequency chirp onto a copropagating wavelength-tunable CW probe (at λ_2) through the nonlinear refractive index. Filtering one of the XPM-generated sidebands results in wavelength conversion (to $\lambda_2 + \Delta$).

conversion in As_2Se_3 ChG fiber—the fiber with the highest Kerr nonlinearity to date ($n_2 = 400 \times$ silica) for which any nonlinear signal processing has been reported. An important aspect of this work was demonstrating device operation with a full data bitstream rather than just low-duty cycle optical pulses where achieving very high peak powers (for typical system average signal powers) is much easier. Error-free conversion showing full system BER measurements was demonstrated at 10 Gb/s RZ near 1550 nm over a 10 nm wavelength range with an excess system penalty of 1.4 dB in only 1 m of fiber and 2.1 W of peak optical pump power.

The experimental setup used to demonstrate λ-conversion is shown in Figure 20.29. The 1-m length of As_2Se_3 fiber [effective core area of 37 μm² and nonlinear index of $n_2 \sim 1.1 \times 10^{-13}$ cm²/W ($\sim 400 \times$ silica)] had an effective nonlinearity coefficient $\gamma \sim 1200$ W/km [7], with a fiber dispersion (dominated by material dispersion) of $D = -560$ ps/nm/km at 1550 nm. The total insertion loss was ~ 5.9 dB, of which 1 dB was due to propagation loss, 0.6 dB due to splices to the higher NA fiber resulting in ~ 2.2 dB coupling loss per As_2Se_3 fiber facet. Data were encoded onto 7 ps pulses (equivalent to 40 Gb/s RZ) at 10 GHz using a Mach–Zehnder (MZ) modulator with an actively mode-locked fiber laser (MML). The optical data was amplified and a 1.3-nm TBF was used to remove out-of-band ASE. The pulses were combined with a CW probe from a wavelength tunable amplified laser diode and coupled into the As_2Se_3 fiber. In-line polarization controllers ensured that the polarization state of the pump and probe were aligned. The output of the As_2Se_3 fiber was then sent through a sharp (0.56 nm) tunable grating filter offset to longer wavelengths by 0.55–0.70 nm to remove the pump and select a single-XPM sideband. This amplified signal was filtered using a second 1.3-nm TBF to remove out-of-band ASE. An in-line, 200 pm wide, fiber Bragg grating (FBG) notch filter was used to further suppress the residual CW carrier prior to the signal being measured with a 30-GHz optical bandwidth receiver.

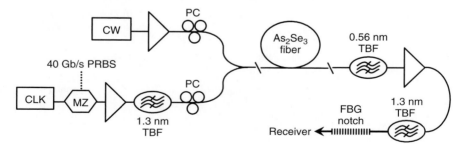

Figure 20.29 System setup for demonstrating XPM-based wavelength conversion. CLK: 10 GHz actively mode-locked, fiber laser, FBG notch: fiber Bragg grating notch filter, MZ: Mach–Zehnder modulator, PC: polarization controlled, PRBS: pseudo-random bit sequence, TBF: tunable bandpass filter.

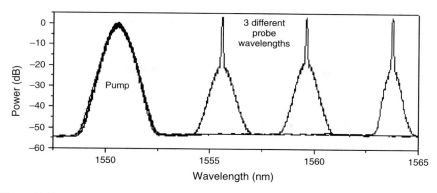

Figure 20.30 Spectra from As_2Se_3 fiber after XPM has broadened the spectra of three different CW probe wavelengths. Resolution bandwidth = 60 pm.

Figure 20.30 shows the measured spectra of the As_2Se_3 fiber output with the CW probe laser tuned to three wavelengths. The large residual CW component which is suppressed by the FBG is due to the low-duty cycle (<12 dB) of the pulses, and would be greatly reduced in a full 33% RZ signal. The relative group velocity difference between the pump and probe (i.e., walk-off) reduces the amount of the bandwidth generated through XPM as a function of wavelength offset. This is evident for the XPM sidebands for the probe at 1555.57 nm, which corresponds to a walk-off length ($L_W = T_0/D \, \Delta\lambda$, where T_0 is pulse width, D is the dispersion parameter, and $\Delta\lambda$ is the pump–probe offset) of 0.35 m. Figure 20.31 shows the BER vs received optical power for the converter as well as for the back-to-back system and shows a penalty of $\sim$1.4 dB for the three converted wavelengths at a BER = 10^{-9}. The inset receiver traces show a clear and open eye for the wavelength converted signals. Although no BER penalty was measured for increasing the wavelength offset, the effect of the walk-off reduced bandwidth lead to a longer converted pulse that could result in intersymbol interference in a full 40 Gb/s 33% RZ signal. This demonstration of λ-conversion over a 10 nm range used 12.3 dBm of probe power and 18.6 dBm of pump average power, corresponding to only 2.1 W peak power.

This approach to λ-conversion has also been reported [126] in bismuth oxide fiber over a 15-nm wavelength range at 160 Gb/s. Both fiber types offer advantages and disadvantages, and the fiber parameters are compared in Table 20.1. The significantly larger n_2 for As_2Se_3—an order of magnitude greater than Bi_2O_3—is a key advantage, although because of a much larger effective core area of 37 um^2 [and hence similar nonlinear coefficients $\gamma = n_2\omega/(cA_{\text{eff}})$] the fiber length in both demonstrations was the same (1 m)—which is still excessive for future integrated devices. Because of the higher material nonlinearity and larger mode area of the As_2Se_3 fiber, there is arguably much more potential for future improvement through the reduction in mode area by exploiting the much stronger confinement

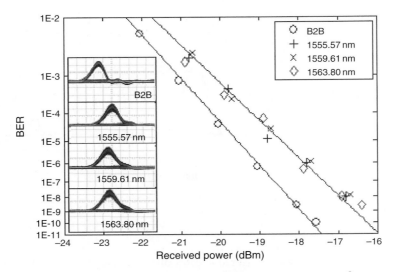

Figure 20.31 BER vs received optical power showing ∼1.4 dB penalty at BER = 10^{-9} for wavelength conversion over 10 nm compared to the back-to-back (B2B) system measurement. For clarity, the linear regression for conversion at 1555.57 and 1559.61 nm are not shown here. Inset shows eye diagrams (10 ps per division) for B2B and converted pulses.

achievable by the large linear refractive index (2.7) of As_2Se_3. This would result in a greatly reduced core area and consequently device length, which would in turn reduce both fiber losses and dispersion-related impairments.

Although dispersion is not as critical a factor as in phase-matched devices (see the next section), the conversion bandwidth is still limited by the difference in pump and probe group velocity, commonly referred to as walk-off length [105]. In fact, this is also an issue for many pump–probe devices such as signal demultiplexing and Raman modulation [127]. While the dispersion parameter of As_2Se_3 is high relative to Bi_2O_3, the potentially shorter device lengths just described make significant conversion bandwidth a possibility. For example, an effective core area of the As_2Se_3 equivalent to the Bi_2O_3 fiber would result in $\gamma = 11,100$ W/km, requiring only 12 cm of As_2Se_3 fiber which would yield > 40 nm of conversion bandwidth compared to 10 nm for 1 m of Bi_2O_3 (for 8 ps pulses). Both fibers (As_2Se_3 and Bi_2O_3) have relatively constant dispersion D over the entire communications band [7] in contrast with dispersion-shifted silica fiber.

Lastly, the nonlinear FOM ($= n_2/\lambda\beta$) of As_2Se_3, while quite good compared to many nonlinear materials (at 2.3) is not as good as silica and Bi_2O_3. As discussed in Section 20.5, however, recently [128] it has been shown that moderate TPA can have a beneficial impact on performance for some applications such as 2R regeneration.

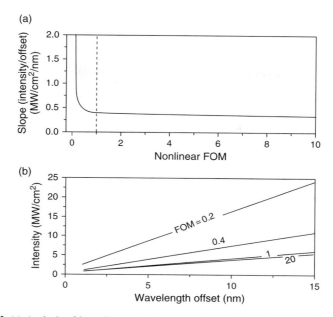

Figure 20.32 (a) Analysis of intensity required to generate a nonlinear phase shift of π by XPM vs pump–probe wavelength offset for varying FOMs. (b) The gradient of (a) vs FOM. The dotted line designates the FOM = 1 threshold required for efficient device operation.

Figure 20.32 shows the required calculated pump intensity as a function of probe–pump offset for various FOMs [105]. As the wavelength offset increases, the length over which the required phase shift must occur decreases, requiring greater pump intensities, which eventually limits the achievable conversion range. Figure 20.32(a) shows that the slope of Figure 20.32(b) increases dramatically for FOM<1, resulting in bandwidth-limited devices. A similar analysis obtained by varying both the nonlinear phase shift and fiber dispersion also indicates that an FOM >1 is needed for efficient operation. With an FOM of 2.3 for As_2Se_3 [7], TPA is not expected to limit device bandwidth.

The ultimate route to realizing the full potential of ChG (or any material) all-optical devices lies in exploiting their high-refractive index (2.4–2.8) and high n_2 by reducing the waveguide mode area. In fact, this has already begun by reports of chalcogenide planar waveguides [6], chalcogenide fiber tapering [129], or microstructured optical fiber [130].

20.7.2 Wavelength Conversion in Chalcogenide Waveguides

To date, all-optical signal processing has been mainly limited to fiber-based devices—only very recently have there been reports on integrated devices based on pure Kerr nonlinearities (2R regeneration) [6, 23]. Further, there have been even fewer reports of system BER measurements for integrated all-optical devices operating at full data rates of 10 Gb/s or higher, primarily because of peak power requirements. These include (nonphase-matched) XPM-based wavelength conversion [10] in As_2S_3 ChG waveguides and wavelength conversion through FWM in silicon nanowires [12, 131]. The lowest peak operating power for nonphase-matched integrated all-optical devices is comparable in silicon (2R regeneration) and ChG at ∼6 W. Ultimately, for practical operation devices with peak power requirements well below 1 W would be needed. Here, we focus on a recent report [106] of system penalty measurements of a waveguide all-optical wavelength converter based on XPM that achieved error-free wavelength conversion at 10 Gb/s in a 5-cm-long As_2S_3 ChG planar waveguide, over a 25-nm wavelength range with a Q-factor penalty of 2.3 dB.

The experimental configuration (Figure 20.33) consists of the input pump provided by a 10-GHz clock signal (FWHM 2.1 ps), modulated with a $2^{29} - 1$ long PRBS, and then amplified and filtered to remove out-of-band ASE noise. This was combined through a 50/50 coupler with a CW probe (also amplified and filtered for out-of-band ASE). The polarizations of both the pump and probe were aligned to the lower loss Te-mode of the waveguide to maximize XPM. The combined signals are butt-coupled to the waveguide through SMF fiber pigtails (with index matching fluid) with a short length of UHNA-4 spliced onto the end for better mode matching, resulting in a loss of ∼2.2 dB per facet. The 5-cm long, 3-μm wide As_2S_3 ChG ridge waveguide was fabricated using pulsed-laser deposition and dry etching, resulting in low-propagation loss of ∼0.3 dB/cm [18]. The pump was centered at 1541 nm with an average power of 115 mW (∼5.8 W peak powers in the waveguide). The CW probe was 25 mW coupled in and set to three different wavelengths: 1555.4, 1559.5, and 1563.6 nm.

Figure 20.34 shows the XPM spectral broadening of the probe at the output of the waveguide, which is independent of the signal–probe wavelength offset due to the short device length, that ensures the dispersive pulse walk-off between the

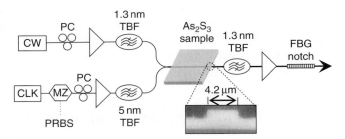

Figure 20.33 Experimental setup for demonstrating cross phase modulation-based wavelength conversion in chalcogenide ridge waveguides.

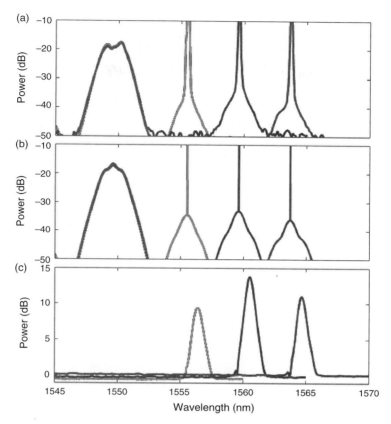

Figure 20.34 Spectra showing cross phase modulation sidebands with the probe set to three different wavelengths (this figure may be seen in color on the included CD-ROM).

signal and probe is negligible. Note that a significant portion of the CW probe remains due to the comparatively low-signal duty cycle. The BERs of the converted pulses were measured at all three probe wavelengths and the corresponding Q-factors were calculated [132] (Figure 20.35). The converted pulses were centered at 1557.0, 1560.6, and 1565.0 nm with Q-factors of 20.2, 20.7, and 20.7 dB, respectively. Comparing these Q-factors to that of the back-to-back signal at 1541 nm we obtain a Q-factor penalty of 2.3 dB, achieved over a range of 25 nm. Due to the short device length, group velocity mismatch will potentially allow a conversion range of up to ±45 nm. As mentioned, this is one of very few demonstrations of error-free wavelength conversion at telecommunication bit rates (10 Gb/s and above) based on pure Kerr (n_2) nonlinearities—the others being in HNLF [125, 126] and through FWM in silicon [12]. Next we turn to all-optical switching in fiber.

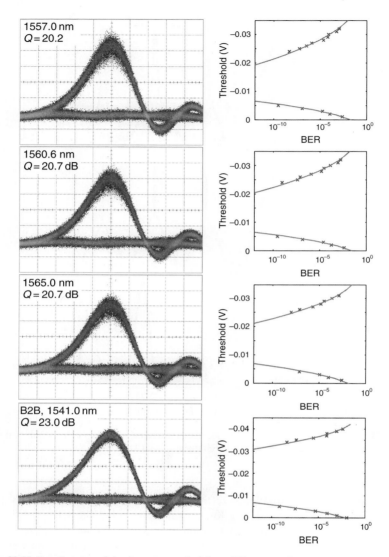

Figure 20.35 Eye diagrams of the data converted to three different wavelengths and the back-to-back pulse for comparison.

20.8 OPTICAL SWITCHING

Figure 20.36 suggests a simple method for highly efficient all-optical switching and band-level manipulation. The existence of three idler bands in a two-pump parametric device depends on the presence of *both* optical pumps: if the pump

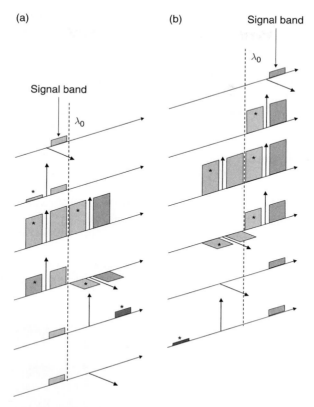

Figure 20.36 Twelve out of 24 possible pump switched configurations in two-pump FPP. Relative signal and idler magnitudes are plotted neglecting the spectral width limitations associated with one-pump interactions. HNLF birefringence is assumed to be negligible. The signal is either co or crosspolarized with the pump; any intermediate signal polarization can be considered as a superposition of the cases illustrated here. (a) Signal band positioned in vicinity of the pump operating in the normal dispersion regime. (b) Outer signal band positioned in vicinity of the pump operating in the anomalous dispersion regime. ∗ Denotes a complex-conjugate band (this figure may be seen in color on the included CD-ROM).

operating in the anomalous regime were to disappear, the signal would experience negligible change in passing through the device. A similar effect would occur in case that normal-dispersion pump is switched. The speed of the proposed scheme would depend solely on one's ability to switch the pumps on and off, because parametric process in silica can be regarded as nearly instantaneous. This method not only offers the means for ultrafast band manipulation, but also allows the user to choose whether the converted wave is a true or a conjugate replica of the original signal. The number and wavelength range of band-switching geometries can be increased further if one introduces pump–signal polarization diversity. In the orthogonally multiplexed pump configuration, band switching can be, in principle, achieved for any signal polarization state. Conversely, copolarized parametric pumps would guarantee a high-DOP selectivity that can also be used to discriminate signals with differing states. We categorize the available switching states by considering all possible pump–pump and signal–pump polarization states, shown in Figure 20.35.

In a recent demonstration, a two-pump experimental setup, with counterphased pump modulation similar to that described in Ref. [133] was constructed to study the feasibility of banded switching in a two-pump parametric device. Tunable laser sources for the parametric pumps were positioned at 1567.0 and 1596.8 nm and phase-modulated using 5 Gb/s $2^{31}-1$ PRBS to increase the Brillouin threshold to over 400 mW. A zero-chirp MZ amplitude modulator was used to modulate the signal source (1557.7 nm) with a programmed 10 Gb/s NRZ sequence. The pumps and signal were combined at the input of the 1-km-long HNLF with zero-dispersion wavelength at 1580 nm, dispersion slope of 0.03 ps/nm^2 and nonlinear coefficient $\gamma = 10$ km/W.

A small signal (Pin = −20 dBm) was RZ-modulated at 10 Gb/s and positioned within the outer parametric band ($\lambda = 1557.7$ nm), as shown in Figure 20.37.

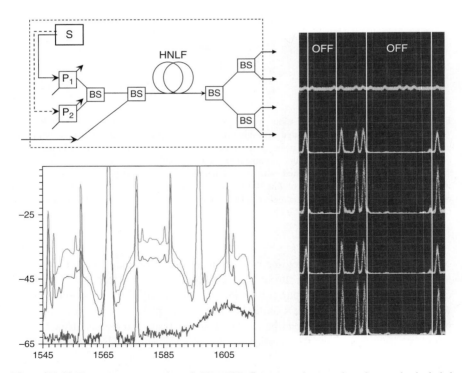

Figure 20.37 Two-pump parametric switching (this figure may be seen in color on the included CD-ROM).

The signal was simultaneously amplified with an HNLF input to output gain of 21.2 dB and converted to 1576.1, 1587.2, and 1606.2 nm with efficiencies of 23.5, 24.2, and 21.4 dB, respectively relative to the signal power input to the HNLF. The extinction ratio between ON and OFF states was 19 dB at the signal wavelength, 29 dB at 1576.1 and better than 50 dB at 1587.2 and 1606.2 nm. Inset of Figure 20.37 shows switching at bit-rate speeds in a two-pump device performing simultaneous wavelength conversion and switching. The signal was modulated using the 10 Gb/s sequence 1100110111010111001, while the L-band pump was pulsed using 10 Gb/s 1000011110000000001 sequence. A variable electrical delay line was used to synchronize the signal and pump patterns to produce a switched signal/idler sequence that is the Boolean AND of the signal and pump patterns: 1000010110000000001. The performance of the switching scheme is limited by the speed and the extinction ratio of the pump modulation. Assume, for simplicity, that signal amplification and idler generation is dominated by the nondegenerate phase-conjugated process [88], i.e., the signal-pump and idler-pump spectral separation is large. As a worst-case extinction estimate, assume that (a) partially degenerate (one pump) phase conjugation provides gain in the OFF pump state, and (b) the signal and idler experience maximal gain (an ideal phase matching condition) for both ON and OFF states. The extinction ratio is then given by

$$\frac{P_{SIG}^{ON}}{P_{SIG}^{OFF}} \sim \frac{\exp\left(2\gamma\sqrt{P_C^{ON} P_L^{ON}} L_{eff}\right)}{\exp\left(2\gamma\sqrt{P_C^{OFF} P_L^{OFF}} L_{eff}\right)} = \exp\left[2\gamma L_{eff}\left(\sqrt{P_C^{ON} P_L^{ON}} - \sqrt{P_C^{OFF} P_L^{OFF}}\right)\right]$$

where $P_{C,L}^{ON,OFF}$ are the pump powers in the ON and OFF positions and L_{eff} is the effective HNLF interaction length [27]. When only the L-band pump is switched and OFF state is long enough, a single booster with constant output power ($P_{OUT} = 2P$) will result in ON/OFF pump states $P_C^{OFF} \cong P_C^{ON} \cong 2P$ and $P_L^{OFF} \cong aP_L^{ON} \cong aP$, where a is the extinction ratio of the L-band pump. The interpump Raman interaction is neglected. The worst-case extinction for the switched signal is now estimated as $e^{2\gamma L_{eff} P(1-\sqrt{2a})} \sim e^{1.3\gamma L_{eff} P}$. For the aforementioned experimental parameters ($\gamma = 10$ km/W, $a = -12$ dB, $L_{eff} \sim 1$ km, $P_{1,2} = 250$ mW) the worst-case extinction ratio is estimated to be -14 dB. In case when C-band pump level is maintained constant, this estimate changes to -16.3 dB. The maximal gain assumption is *inherently achromatic* and should be used for lower bound estimate of the extinction ratio. The above analysis does not take into account any fast fluctuations in pump powers, which can lead to severe signal impairments and extinction degradation due to exponential nature of parametric gain. In case of dispersion-broadened pumps, internal fiber parametric device (FPD) dispersion and spectral filtering dominates pump power fluctuations. The speed of the scheme was recently demonstrated by OC-768 packet switching scheme, illustrated in Figure 20.38.

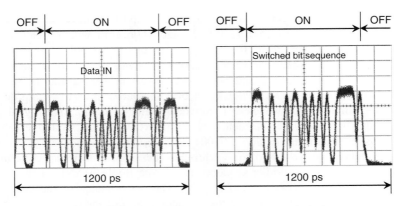

Figure 20.38 OC-768 Switching in HNLF parametric device.

20.9 OPTICAL PERFORMANCE MONITORING

As optical communication data rates increase from 40 to 160 Gb/s and beyond, signals become much more sensitive to transmission impairments such as ASE noise, chromatic dispersion, and polarization mode dispersion (PMD), loss and others. This has driven significant interest in developing all-optical techniques for OPM [134] in systems and networks. Without monitoring signal quality, a reduction in network performance is only discovered once actual data are lost. Performance monitoring of the optical signal can provide information about the signal quality and identify network fault locations and generate failure warnings. In addition, specific signal quality information such as OSNR, chromatic dispersion, PMD, and others, can be monitored and provide feedback to signal conditioning devices, such as optical regenerators and TDCs [135–137] to allow continuous real-time optimization of signal quality [138, 139]. This is critical since changes in optical links due to weather [140] and network reconfiguration, will require continuous monitoring and subsequent correction of these impairments. All-optical techniques for OPM share the same advantages of other all-optical processes over other approaches such as electronic methods—particularly as systems move from 40 to 160 Gb/s and beyond. These include significantly greater simplicity and cost effectiveness, as well as obviating the need for optical to electronic conversion, with the result that all-optical techniques are also generally much faster.

All-optical techniques were first reported for residual dispersion monitoring using SPM [138] and have since been demonstrated for in-band OSNR monitoring [3, 141, 142], PMD monitoring [143–146], chromatic dispersion monitoring [3, 4, 147], and others, using a variety of both phase- [139] and nonphase-matched techniques. We review some of our recent work here and compare their advantages and disadvantages at the end of the section.

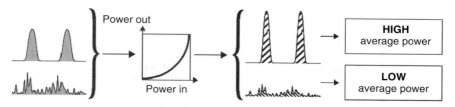

Figure 20.39 Principle of device operation for optical performance monitoring using a nonlinear power transfer function. A clean pulse train and a pure noise signal are transformed by a nonlinear power transfer function. They have the same average input power but different average output powers.

Most techniques for all-OPM ultimately exploit a nonlinear PTF in a similar manner as the 2R regeneration devices discussed earlier. In Figure 20.39 two input signals with the same average power are shown; one is a clean pulse train and the other is pure noise. They are both transformed through a nonlinear PTF, which results in the output for the clean pulse train emerging with a greater average power than the output for the pure noise because the pulse train spends more time at high powers than the noise. A simple average power measurement with a slow detector can then distinguish between pure noise and a clean pulsed input, or the noise fraction in a real signal, which affects the degree of eye-closure (Q-factor) in a data stream, and which in turn is sensitive to many underlying parameters such as residual dispersion.

The first embodiment of this principle was applied [138] to measure residual dispersion—a key requirement to provide feedback for continuous real-time dispersion control in high-bit rate systems—through SPM in a Kerr medium followed by filtering. This method resembles 2R optical regenerator schemes [95] and in fact can be incorporated in these devices. Figure 20.40 demonstrates the principle. An amplitude-modulated RZ signal will become spectrally broadened when propagating through a Kerr medium, and a measure of this *spectral* broadening can be used to determine the degree to which pulses were *temporally* broadened at the input of the fiber by residual dispersion. Figure 20.41 shows the experimental

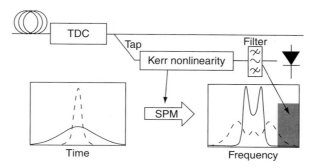

Figure 20.40 Principle of residual dispersion monitoring through SPM and filtering.

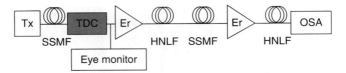

Figure 20.41 Experimental setup for dispersion monitoring through SPM and filtering. The eye is monitored at the input to the HNLF. The OSA (optical spectrum analyzer) records the SPM-broadened spectrum.

configuration—an optical bit stream is amplified, and then propagated through highly nonlinear SMF, followed by a filter and a slow detector. The detector measures the integrated power that is sensitive to the degree of spectral broadening and hence residual pulse dispersion at the fiber input. The experimental results in Figure 20.42 were achieved by recording the optical spectrum of a 40-Gb/s RZ data stream at the output of a nonlinear fiber while varying the dispersion induced pulse broadening at the input using a tunable fiber grating dispersion compensator

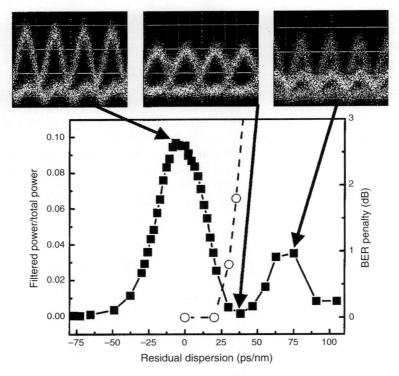

Figure 20.42 Comparison of long-pass spectral monitoring signal (solid squares) with typical 40-Gb/s RZ receiver BER penalty (open circles), both plotted vs residual dispersion. Eye diagrams are indicated for three residual dispersion values. Lines are a guide to the eye.

(TDC), and show clear correlation between spectral broadening at the output and residual dispersion at the input of the nonlinear fiber.

This clearly demonstrates that spectral broadening is a good measure of residual dispersion of an RZ data stream, and is directly correlated to BER impairments. This nonphase-matched method works well over a range of relevant BER penalties, limited only by interpulse interference at large pulse dispersion, and so has potential as an all-optical feedback element for TDCs.

Since this initial work, other methods such as phase-matched-cascaded FWM [139] and OPA [142] have been employed to demonstrate all-optical monitoring of dispersion [14] and in-band OSNR. Both OPA and FWM techniques also exploit nonlinear PTF—the shape of which determine their signal processing properties. For instance, while a linear power transfer function between the signal frequency and an FWM-generated frequency is required for frequency conversion, a quadratic (or even more nonlinear) PTF can be used to perform the more complicated function of dispersion and OSNR monitoring. Thus, by using a PA that can provide both a linear and a quadratic PTF, many functions can be achieved from a single device with only the addition of a few simple components, such as the simultaneous performance monitoring along with wavelength conversion [10], necessary in high-speed optical networks to provide all-optical routing [148], fast recovery from fiber cable damage [149], and prevent blocking [150]. Furthermore, these approaches are modulation-format independent and, as in all processes based on Kerr non-linearities, ultrafast and therefore suitable for bandwidths well beyond 40 Gb/s.

Turning next to in-band OSNR monitoring in WDM systems, we note that this can be particularly challenging since simple approaches to channel monitoring such as using an OSA to interpolate out-of-band noise (Figure 20.43) [151] to estimate the in-band noise do not work well because in these systems channels experience significant filtering that preferentially suppresses out-of-band noise [134].

Performing in-band OSNR measurements electronically using a fast detector is the most direct method, but is impractical, expensive or even impossible at very

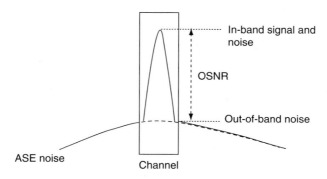

Figure 20.43 Traditional optical spectrum analyzer methods to measuring in-band OSNR, which assume the in-band noise is identical with the out-of-band noise.

high bit rates. True in-band OSNR monitoring (not interpolation-based) using all-optical techniques generally rely on ultrafast (Kerr) nonlinearities, and so have the advantage of being faster, cheaper, and simpler than fast electronics [134, 152] and a number have been reported such as polarization-nulling [153], SOAs [154], highly birefringent fiber [155], and others. OPA and FWM methods have achieved up to a 30% change in measured output power for a 20-dB change in the OSNR of the signal [3]. In addition, the OPA could simultaneously provide 19 dB of gain to the original signal and 14.1 nm of wavelength conversion. The experimental setup is shown in Figure 20.44. The ASE noise source is produced by cascading several EDFAs. The pulsed source provides a 10-GHz pulse train of 13.2 ps pulses (FWHM). Both the pulsed and ASE sources are filtered before being combined using a 50/50 coupler. Variable attenuators in the pulsed and ASE arms allow the OSNR of the combined signal to be varied. The CW OPA pump has a wavelength of 1549.8 nm, launch power of 1.01 W, and is dithered with a phase modulator to suppress SBS. The nonlinear medium is 800 m of HNLF with a nonlinearity of 9 W/km, zero-dispersion wavelength of 1548.1 nm, and dispersion slope of 0.018 ps/nm^2/km. The output of the nonlinear fiber is filtered to isolate the cascade wave and both the average power and optical spectrum are measured.

Figure 20.45 shows the optical spectrum of an OPA output showing cascaded FWM. A strong pump and a weak signal are launched into a nonlinear fiber. When phase-matching conditions are met, FWM transfers power from the pump wavelength to the signal wavelength amplifying the signal and providing efficient frequency conversion onto a generated idler wave [139]. At higher signal input powers, this process generates even more wavelengths through cascaded FWM with powers that are increasingly nonlinearly dependent on the signal power P_s [139]. Thus, by filtering out the cascaded wave a device with a nonlinear transfer function is achieved.

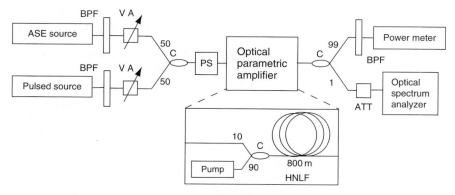

Figure 20.44 Experimental setup for measuring in-band OSNR through optical parametric amplification. BPF, bandpass filter; VA, variable attenuator; C, coupler; PS, polarization scrambler; HNLF, highly nonlinear fiber; ATT, attenuator.

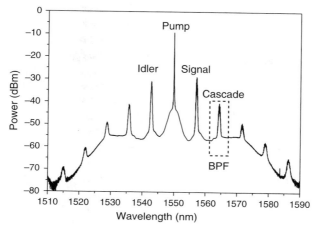

Figure 20.45 The optical spectrum at the output of the optical parametric amplifier. The dashed box indicates the filtered output.

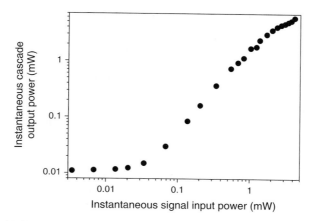

Figure 20.46 Nonlinear power transfer curve of OPA-based in-band OSNR monitor.

The instantaneous PTF of the optical PA was obtained by replacing the ASE source with a CW source at 1557 nm (pulsed source off). The relationship between the input signal power and the output cascade power is shown in Figure 20.46.

Figure 20.47 shows the measured output average cascade power as a function of changing OSNR [OSNR = $10 \log_{10}(S/N)$, where S is the average power of the pulsed signal without noise and N is the noise power in a 0.1-nm bandwidth centered at the signal wavelength]. The average combined signal launch power was kept constant at 0.27 mW, equivalent to a peak pulse power of 1.92 mW when there was no ASE. The experimental results agree well with the simulations (solid line) with more than 30% change in average cascade power over a 20-dB OSNR

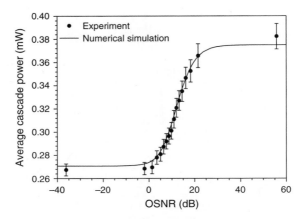

Figure 20.47 OSNR monitoring curve for OPA-based in-band OSNR monitor.

range. The sensitivity of the monitor could be improved by choosing a nonlinear transfer function with a steeper slope.

Another technique for in-band OSNR monitoring relies on a nonphase-matched process to generate the nonlinear power transfer curve, through a nonlinear loop mirror [141]. This technique also exhibits desirable features of a monitor including the ability to monitor multiple impairments simultaneously, transparency to different modulation formats and signal power, and scalability to high data rates. The device (Figure 20.48) consists of a nonlinear optical loop-mirror (NOLM) and a power meter. The key mechanism that enables OSNR monitoring with the NOLM is the profile of its PTF. The PTF of the NOLM nonlinearly maps input powers to output powers through the optical Kerr effect. This device was demonstrated at

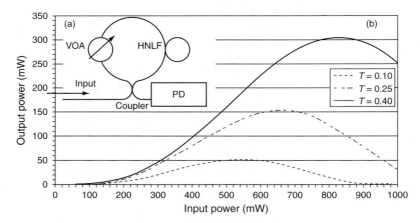

Figure 20.48 Device configuration for nonlinear optical loop mirror (NOLM)-based OSNR monitoring, along with the power transfer curves for three different settings of the VOA (T = loop transmission).

40 Gb/s by providing results with NRZ, 66% duty cycle carrier-suppressed return-to-zero (CSRZ), and 33% duty cycle RZ modulation formats. The PTF can be adjusted to maximize performance for various input signal power levels.

The experimental setup is very similar to that for the previous experiments (Figure 20.44) where the in-band OSNR monitor employs both a PRBS signal at 40 Gb/s and ASE at its input. The transform-limited pulses that represent the signal are produced by using chirp-free MZ modulators. VOAs are used to vary the OSNR of the noisy signal, while maintaining a constant average power into the OSNR monitor. An EDFA is used to boost the noisy signal power up to a level where nonlinear effects occur. The HNLF is similar to that used in the previous technique. For illustrative purposes, we chose a VOA setting of 4.6 dB inside the NOLM, thus resulting in a total NOLM attenuation of 7.8 dB.

The OSNR sensitivity results (Figure 20.49) are shown for NRZ, CSRZ, and RZ modulation formats. The EDFA was adjusted to align the peak power of the noise-free signal to the maximum of the PTF to maximize the contrast between signal alone and noise alone. Experimentally, this resulted in maximum output powers of 502 mW (NRZ), 493 mW (CSRZ), and 570 mW (RZ), in reasonable agreement with the experimental PTF local maximum of 460 mW. The maximum output contrast measured between the extreme OSNR values of 0 and 35 dB are 1.84 dB (NRZ), 3.09 dB (CSRZ), and 4.37 dB (RZ) in good agreement with simulations.

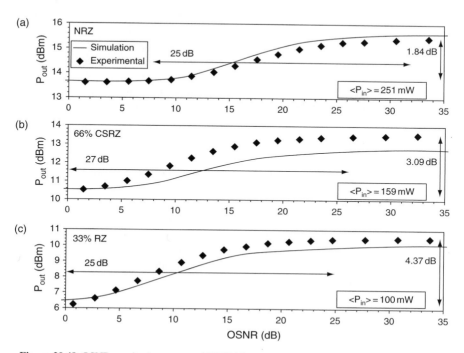

Figure 20.49 OSNR monitoring curves of NOLM-based device for different modulation formats.

This device could also perform chromatic dispersion monitoring using the same principle as OSNR monitoring [6]. As a pulse disperses due to dispersion it degrades the signal quality and so the average output power of the NOLM decreases nonlinearly. If the dispersion can then be compensated through a closed-loop TDC, the average output power will then relate to the in-band OSNR as before.

In summary, NOLM-based approach then provides an effective means to monitor the in-band OSNR. The monitor discriminates the OSNR of 40-Gb/s signals over a dynamic range of 25 dB for NRZ, 27 dB for CSRZ, and 25 dB for RZ.

Apart from residual dispersion and OSNR monitoring, PMD monitoring has attracted significant interest since this quantity in particular is extremely problematic for high bit rate systems and its sensitivity to deployed legacy fiber as well as environmental conditions and even components. While electronic RF spectrum-based methods have been demonstrated [156–158], in systems operating at 40 Gb/s this is difficult and expensive, and at 160 Gb/s and beyond, impossible. All-optical PMD monitors have been demonstrated based on frequency-resolved state-of-polarization (FRSOP) [159] and the DOP [160]. However, there are still challenges to overcome. The FRSOP method requires 10 or more state-of-polarization measurements at different frequencies, making it too slow for the real-time monitoring that requires millisecond response times, while the DOP method suffers from poor correlation with the BER when the differential group delay (DGD) becomes large compared to a bit period [161].

The recent demonstration [162, 163] of an all-optical technique to measure the power spectrum with a bandwidth (800 GHz) of ultrafast optical signals was a milestone development that enabled the first [143] fiber-based all-optical first-order PMD (i.e., DGD) monitor. This technique is especially attractive because (1) it enables the RF spectrum of signals to be determined, even if the signals are too fast (e.g., 160 Gb/s) to be measured on an electrical spectrum analyzer, and (2) it has a fast response time suitable for feedback control and gives a monitor signal strongly correlated with BER, unlike the FRSOP and the DOP methods.

The power spectrum of an optical signal at ω_s can be obtained all-optically by using it to modulate a weak CW probe (at ω_o) and measuring the optical spectrum of that probe, centered on ω_o. This is achieved by launching both the amplified signal and probe into a length of nonlinear fiber. The signal modulates the probe through XPM [163] which, for copolarized signal and probe is $\sim\exp[2i\gamma L P(t)]$, where γ is the fiber nonlinearity, L is the effective interaction length, and $P(t)$ is the signal power. If the modulation is sufficiently weak it is proportional to the instantaneous signal power, and thus the optical power spectrum around the probe frequency, as measured by an OSA [ignoring the residual probe which is (essentially) a delta function], is

$$S(\omega) \propto \left| \int_{-\infty}^{\infty} P(t) \exp[j(\omega - \omega_o)t] dt \right|^2 \qquad (20.3)$$

which is the same as the power spectrum of the original signal $P(t)$. The *key point* is that because DGD and/or dispersion are linear effects they only affect the phase and not the power spectrum of an optical signal and so cannot be probed directly by OSAs. However, because the act of intensity modulating (in time) a probe—by any method—is inherently a nonlinear process, it *does* affect its optical power spectrum. Therefore, if this probe modulation is proportional to the instantaneous signal power (determined in turn by the signal phase), then an OSA *can* be used to probe the signal phase and hence DGD or dispersion.

Figure 20.50 illustrates the experimental setup. The signal source is a 10-GHz MML with pulsewidth 10 ps at 1556 nm. The linearly polarized pulses pass through a half-wave plate and then through one of several lengths of polarization maintaining (PM) birefringent fiber to introduce DGD. Using an optical splitter, a fraction of the signal is measured using a commercial 26-GHz bandwidth photodiode (PD) connected to a 50-GHz sampling oscilloscope (OSC). The remaining signal is boosted by an EDFA, combined with a CW probe at 1535 nm, and then launched into 900 m of HNLF ($\gamma = 20$ W/km, dispersion-zero at 1551 nm, 0.54 dB/km loss). The probe polarization is initially aligned with the signal by maximizing the XPM. The average signal and probe powers launched into the HNLF are 53 and 0.1 mW, respectively.

The signal's optical power spectrum measured using OSA 1 (Figure 20.51) did not change for different PANDA fibers, or amounts of DGD, as expected. (Note the 10 GHz pulse repetition rate is reflected in the 10 GHz line spacing, although this would be present even in the absence of mode locking.)

The optical spectra recorded by OSA 2 (i.e., after modulation of the probe by the signal in the HNLF) (Figure 20.51) for different values of signal DGD [(a) 27.1; (b) 16.3; (c) 13.0; (d) 9.8; and (e) 8.4 ps] show clearly that varying the DGD of the signal *does* affect the optical spectrum of the probe—there is a distinct signal frequency minimum that is a signature of DGD. This occurs because the signal travels at different speeds on the slow and fast axis of the PANDA fiber and some frequencies become π radians out of phase. The frequency minimum is given by $f_{min} = 1/2\Delta\tau$, where $\Delta\tau$ is the DGD. There is excellent agreement with the expected frequency minima shown by the arrows. This effect has also been

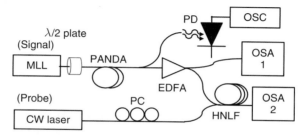

Figure 20.50 Experimental setup for DGD monitor based on all-optical spectral analysis.

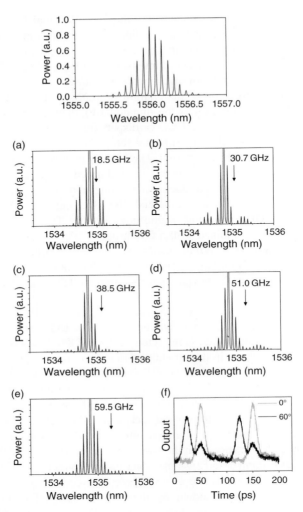

Figure 20.51 Top center: the optical power spectrum of the signal measured on OSA 1 after 27.1 ps of DGD. The shape is invariant for all lengths of PANDA fiber. The OSA traces are for (a) 27.1; (b) 16.3; (c) 13.0; (d) 9.8; and (e) 8.4 ps of DGD. The predicted signal frequency minimum, indicated by an arrow, agrees very well with the experimental result. Plot (f) is the sampling scope trace for a DGD of 27.1 ps and signal polarization of 60°.

observed with electrical spectrum analyzers [2–4]. Figure 20.51(f) shows OSC traces of the signal for two different DGD settings (0 and 27.1 ps).

In practice, maximizing any one sideband will minimize the DGD [156–158], and so for practical devices the OSA can be replaced with a cheap BPF and power meter with a millisecond response time. In terms of sensitivity, the sideband minimum must be within the power spectrum's envelop which yields [143] (for Gaussian pulses) $\Delta\tau > 2.3\Delta t$, where $\Delta\tau$ is DGD and Δt is the pulsewidth.

In summary, DGD monitoring through XPM and optical spectral analysis will enable closed-loop DGD compensation at ultrafast data rates. Future improvements include reducing device length by using the high-nonlinearity glass fiber and waveguides discussed.

In conclusion, all of these different approaches to OPM can be broadly divided into phase-matched and nonphase-matched. All have advantages and disadvantages—there is not necessarily a single best solution. They all rely on a nonlinear PTF and so have the same basic principles and performance. The NOLM-based methods offer some degree of flexibility in engineering/tailoring the nonlinear PTF. Being a nonphase matched method, the NOLM is not as sensitive to dispersion alone but the drawback is that, being an interferometer, it tends to be more sensitive to stability. This could be mitigated in the long term by integration of the device into a PIC. Solutions employing parametric processes, however, are phase-matched and so require dispersion engineered fibers as well as strong pump sources. The advantage to OPA-based methods is that they offer other functionalities such as wavelength conversion, amplification, and possibly 2R regeneration.

20.10 OPTICAL DELAYS AND BUFFERS

The ability to slow down and delay optical pulses is an intriguing physical phenomenon with significant applications, such as in telecommunications. Slow light has been observed in a wide variety of physical systems, varying from ultracold vapors [164], quantum-well structures [165], optical fibers [166–169], and systems of ring resonators [170, 171] to various photonic crystal geometries [172–175]. For all of these approaches, it is essential to provide enough bandwidth to accommodate the very short light pulses that will carry information in future telecommunications systems. However, the Kramers–Kronig (causality) relations that apply to linear resonant systems ensure that the group velocity scales with spectral width (the largest delays being achieved only for very long pulses) with the result that ultrashort pulses can be delayed only by a proportionately short time. This delay-bandwidth trade-off is inherent to linear, resonant systems [176] and although there have been investigations on how to circumvent it [177–180], these schemes remain theoretical and tend to either eliminate only the lowest orders of dispersion or require dynamical tuning mechanisms.

One way of circumventing this delay-bandwidth trade-off is to move beyond linear systems to nonlinear systems. A recent demonstration of this [169] exploited the nonlinear behavior of Bragg gratings in fibers, where the pulse can travel slowly but still remain undistorted over arbitrarily long propagation lengths, by the formation of a "gap" soliton [181]. In that work, such solitons were observed in an FBG and shown to travel at 16% of the speed of light, without broadening. In another approach [68], techniques very similar to that discussed above in the section of 2R regeneration through SPM were employed to achieve slow light. For a general review of slow light techniques the reader is referred to the chapter on slow light in this volume. This section focuses specifically on approaches to generating and controlling slow light by the use of optical nonlinearities.

The physical principle behind slow-light generation through gap-solitons is that the light is sufficiently intense for the glass in the fiber to respond nonlinearly. Intuitively one can think of the pulse as "tunneling" its way through the grating reflection band by altering the local refractive index (and hence grating wavelength) through the Kerr effect. On the basis of this physical picture, therefore, one would expect that the pulse wavelength, located within the reflection bandgap, would need to be very close to the short-wavelength edge (for most materials), so that as the local refractive index is dynamically increased (for a positive Kerr constant) the local grating edge is shifted to longer wavelength, creating a local transparency "hole" for the soliton to tunnel through. One would also expect that the pulse would probably need to have a very narrow spectral width, as this would also reduce the local dynamic self-shifting required for the pulse to shift the band edge out of its way. Both are in fact the case.

Figure 20.52 shows the grating transmission spectrum of the apodized silica FBG ($L = 10$ cm, 118-pm FWHM bandgap, >40 dB in strength), as well as the pulse wavelengths needed to observe gap solitons. Optical pulses from a microchip Nd:YAG Q-switched laser (width 680 ps, $\lambda = 1064.2$ nm, 6.6-kHz repetition rate),

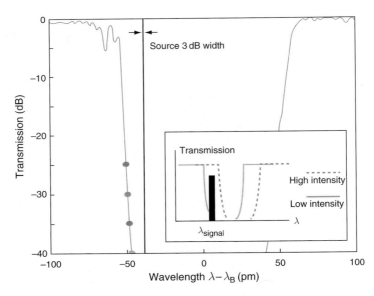

Figure 20.52 Transmission spectrum of the 10-cm apodized FBG, measured with a strain-tuning technique, with a resolution of ~ 0.2 pm. The four circles show the detuning values in the experiments. The gray line represents the spectral width of the laser source. The inset illustrates the nonlinear shift of the FBG spectrum with respect to the signal wavelength λ_{signal} (this figure may be seen in color on the included CD-ROM).

with a very narrow 3-dB spectral width of 1.9 pm (0.15 cm^{-1}, time–bandwidth product = 0.34 – close to transform limited) are launched into the FBG. The pulse detuning is controlled by applying strain to the FBG and the reflected and transmitted outputs are monitored with power meters, OSAs and OSCs (response time of 22 ps).

Figure 20.53 shows both the measured and the simulated nonlinear pulse transmission as a function of input peak power for different pulse wavelength detunings from the grating center frequency. In all cases, the results show a large increase in transmission at high powers due to the bandgap shift. The fact that the

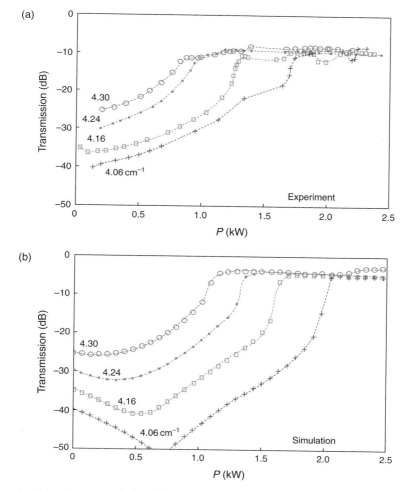

Figure 20.53 Nonlinear transmission. The transmission increases with incident peak power P due to the nonlinear bandgap shift. Four values of the detuning are shown: (a) Experimental results; (b) Simulations (this figure may be seen in color on the included CD-ROM).

transmission saturates around −10 dB (the rest being reflected) indicates that near the pulse peak the bandgap is shifted well away from the pulse spectrum, and so it cannot affect the pulse transmission. Also, as expected, the switching threshold required to reach 50% maximum transmission decreases with increasing detuning—i.e., as the pulse wavelength gets closer to the grating band edge, where less intensity is required to shift the bandgap clear of the pulse spectrum to create the tunneling window.

For slow light applications it is not the nonlinear transmission, but the pulse delay (and grating dispersion) that is most relevant. Propagation of short high-intensity light pulses through an FBG is described by the nonlinear coupled mode (NLCM) equations [181] which, although they do not have explicit dispersion terms, still contain dispersion of arbitrarily high-order through terms proportional to κ (grating coupling strength in cm^{-1}). In contrast, the dispersion of the fiber in which the grating is written is negligible over the relatively short propagation distances in Bragg gratings. In fact, both "Gap"-solitons (pulse wavelength inside the bandgap) and grating-solitons (outside bandgap) solutions to the NLCM have been found [182, 183] and both can travel at any velocity between $vg = 0$ and $vg = c/n$ without broadening or changing shape. This is because they are exact solutions to the NLCM, and so grating dispersion is countered to all orders. Intuitively, the energy in the low-intensity wings of the pulse is constantly Bragg reflected by the grating toward the central, high-intensity part, and so with dispersion eliminated the distance over which the pulse can propagate is limited only by the grating length and possibly absorption. The delay can be tuned by varying the intensity or the mechanical strain on the grating—important for applications of slow light such as optical buffers and optical delay lines [13, 167].

Figure 20.54 shows the dramatic variation in output pulse delay as a function of input power for various detunings. The minimum group velocity achieved in these experiments was $vg = 0.23c/n$. In general, while the pulse width does vary as the delay or power are arbitrarily tuned, it is possible to find the right balance of combined detuning and power (Figure 20.52, inset) to achieve a tunable delay of ∼0.5 ns with nearly constant output pulse width of 0.45 ns. Note that just outside the bandgap where the grating reflectivity is small, light can propagate slowly even at low intensities as the group velocity vg due to the FBG can be very small. However, because this is in the linear regime it is subject to the Kramers–Kronig-related bandwidth/delay trade-off, and as a result these pulses experience significant broadening. Ultimately, even if dispersion is completely eliminated, propagation loss limits the maximum achievable delay. Assuming a minimum UV-induced grating loss of $0.5\,m^{-1}$ [184] gives a practical maximum grating length of ≈ 2 m [185], then the lowest group velocity, $vg = 0.23c/n$, obtained in the experiments would yield a delay of 32 ns, or 46 pulse widths—much greater than in any known linear system.

The vision is to implement this slow-light scheme in chalcogenide-based planar waveguides, which will bring key advantages, beyond the obvious benefits of device integration, as discussed earlier. It has recently been shown [186] that

20. Nonlinear Optics in Communications

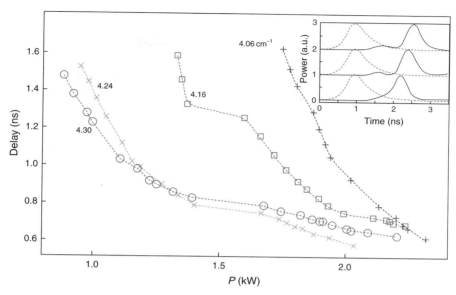

Figure 20.54 Tunable delay with power or strain. Delay of the transmitted pulses vs input peak power P for the four values of detuning considered. For example, a tunable delay of approximately 1 ns can be achieved at roughly 1.8 kW by varying the detuning by applying different levels of strain to the FBG. The inset shows that, by varying both P and δ simultaneously, it is possible to achieve tunable delay with nearly constant output pulse width of 0.45 ns. Dotted lines represent the reference pulse and solid lines represent the delayed pulses. Top: $\delta = 4.06\,\text{cm}^{-1}$ and $P = 1.75\,\text{kW}$. Middle: $\delta = 4.16\,\text{cm}^{-1}$ and $P = 1.35\,\text{kW}$. Bottom: $\delta = 4.24\,\text{cm}^{-1}$ and $P = 1.05\,\text{kW}$ (this figure may be seen in color on the included CD-ROM).

specially designed nonlinear Bragg grating arrays in integrated waveguides can open up novel ways of nondiffracting, dispersionless control of spatial and temporal properties of slow-light control. In addition, because of the very high Kerr-induced index changes available in ChG, as well as the possibility of realizing ultrasmall waveguide cross-sectional areas (due to the very large refractive index) peak powers required to achieve lower velocities and larger delays can be drastically reduced.

Another new class of devices displaying near-ideal features required by optical delay module—namely fast, continuous tuning with a wide operational range, operable in a wavelength-transparent mode—is based on the Brillouin effect in fiber. This was recently used to demonstrate a true 47-ps delay supporting a 12-GHz signal bandwidth [187]. Burzio et al. [188] suggested the use of wavelength conversion and dispersion to achieve a widely tunable delay. This technique was recently used, in combination with a one-pump parametric fiber converter (PFC), to demonstrate a single-pulse delay of 800 ps [189]. Unfortunately, a one-pump PFC cannot provide a format-independent optical delay. Indeed, CW FPC requires a phase-modulated pump to suppress Brillouin backscattering [190]. As a consequence, the pump phase modulation is transferred to the idler, and is

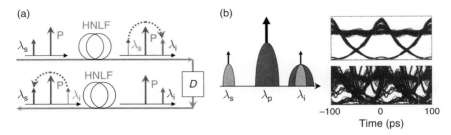

Figure 20.55 (a) One-pump parametric delay; (b) Phase dithering in one-pump device generates excessively broadened idler λ_i (left) preventing the use of highly dispersive elements. The impairment extent is illustrated by transmitting narrow (upper right) and phase-broadened (lower right) idlers through small length of dispersive fiber (this figure may be seen in color on the included CD-ROM).

converted to amplitude fluctuations in the presence of dispersion. The one-pump parametric delay scheme is illustrated in Figure 20.55(a); an input signal (λ_s) is converted to the idler wavelength (λ_i) and transmitted through the dispersive line (ΔL). The second-stage process converts the idler back to the original wavelength, resulting in an optical delay, which is the product of the wavelength shift ($\lambda_i - \lambda_s$) and the total dispersion.

The pump phase modulation required to suppress Brillouin scattering in the HNLF, results in excessive idler broadening, and is illustrated in Figure 20.55(b). In practice, an efficient FPC device requires a modulation bandwidth in excess of 3 GHz [190]. This is a nonnegligible and leads to phase fluctuations that are readily converted to amplitude fluctuations by propagation through the dispersive element D, as illustrated in inset of Figure 20.55(b).

To overcome this basic impairment, we have recently proposed and demonstrated unimpaired idler generation in counterphased and cophased dual-pump parametric devices [190]. Figure 20.56 illustrates the use of counterphased

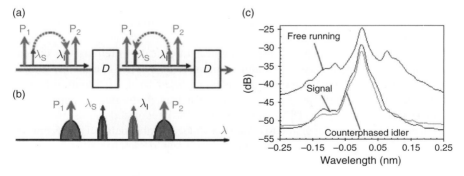

Figure 20.56 (a) Dual-pump parametric delay; (b) Counterphased pump modulation can generate unimpaired idler in any of four available bands; (c) The effect of pump synchronization on idler generation: 10 Gb/s modulated signal, idler generated by counterphased pumps and free-running pumps.

two-pump parametric device: a narrow idler is generated by synchronously modulating the pump phases and is invariant of the pump dithering speed. More importantly, the unimpaired idler can be chosen as either conjugate or nonconjugate signal copy, providing the flexibility in compensating the pulse spreading within the dispersive element. In the example shown in Figure 20.56, the pulse spreading is exactly reversed by conjugating the idler and passing through the identical dispersive block.

The experimental configuration shown in Figure 20.57 uses parametric pumps positioned in the C- and L-bands with maximal powers of 22 dBm. A signal at 1560 nm was modulated at 10 Gb/s, using the RZ format and combined with co-propagating parametric pumps. A continuously tunable idler was generated within the 1590–1605 nm band designed to provide an equalized power response while tuning the parametric pumps. The idler was filtered after the first HNLF pass and transmitted through the dispersion compensating coil (–504 ps/nm at 1550 nm). A counterpropagating parametric process within the same HNLF was used to convert back to the original wavelength (1560 nm), which eliminated the need for a separate conversion stage. The back-converted wave was passed through the nearly identical dispersion coil (–509 ps/nm at 1550 nm) and received by the BER tester. Both coils had a dispersion slope of –0.31 ps/km/nm^2.

Figure 20.58 illustrates the signal integrity as the delay is continuously tuned over the range of 12.47 ns. The BER measurement was performed using a $2^{31}-1$ long PRBS word and indicates wide performance margin at the long end of the operating range ($\tau = 12.47$ ns). The observed impairment did not scale with the increase of the tuning range, a clear indication that the system was not limited by

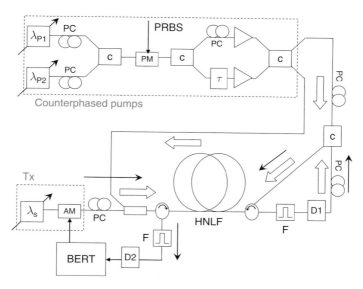

Figure 20.57 Two-pump counterphased delay line.

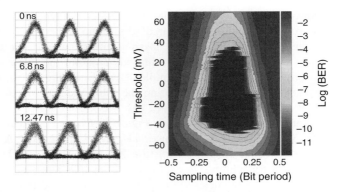

Figure 20.58 System performance of continuously tunable dual-pump parametric delay. Baseline signal (0 ns) integrity is nearly indistinguishable from signal tuned over 10 ns or more. The system margin (right), indicated by central dark area error-free floor, is maintained across the entire operating range (this figure may be seen in color on the included CD-ROM).

dispersive spreading. Instead, the observed penalty is attributed to the use of a single-phase modulator in counterphasing pump scheme and nonideal suppression of the idler spectral broadening.

20.11 FUTURE PROSPECTS

Clearly, there are still challenges to be met by all-optical signal processing devices to be fully viable in high bit rate systems. A key issue is the required operating peak power. For self-activated devices (where the signal itself drives the device) this is absolutely critical, but even for externally pumped devices (such as for FWM), it is still important. Devices based on kilometer scale lengths of HNLF are arguably already in this regime, but for both fiber devices based on novel nonlinear glasses as well as integrated devices, some progress still needs to be achieved. In particular, the lowest peak powers that have been achieved in integrated devices, for nonphase-matched processes [10, 23], are in the range of 5 W, which is still probably two orders of magnitude larger than the ideal. Phase-matched wavelength conversion has been achieved with roughly 100 mW of pump power [12], but even this ideally should be reduced. A key question is whether or not there are any fundamental limitations based on material or platform issues. Ultimate solutions will undoubtedly lie with integrated approaches, with the challenges being to greatly reduce waveguide cross-sectional area (to enhance the nonlinear parameter γ), and increase device length, while maintaining low linear and nonlinear losses. Figure 20.59 shows a layout of a future concept for a device that could potentially meet all of these requirements—based on ChG. The key features include the use of very small mode area (300 nm × 300 nm) waveguides—achievable because of the large refractive index of ChG—as well as the

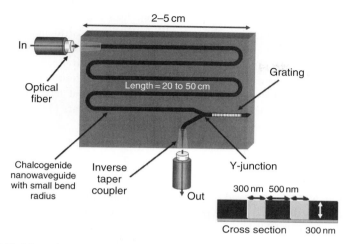

Figure 20.59 Schematic layout for future nanowire-based integrated all-optical regenerator in chalcogenide glass.

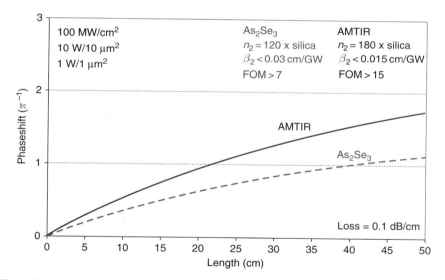

Figure 20.60 Nonlinear-induced phaseshift vs length for a pulse with peak intensity of $100\,\text{MW/cm}^2$ for both As_2Se_3 AMTIR-based nanowire waveguides (this figure may be seen in color on the included CD-ROM).

long-waveguide length (and correspondingly low propagation loss requirements). Figure 20.60 shows the nonlinear self-induced phase change for an optical pulse of peak intensity $100\,\text{MW/cm}^2$, equivalent to $1\,\text{W/}\mu\text{m}^2$ (ignoring pulse dispersion). At this intensity, then, to achieve sub-Watt peak power operation, it is clear one needs to go to sub-micrometer waveguide dimensions—otherwise referred to as nanowires. From Figure 20.60 we see that if one uses a nonlinear-induced phase-shift of $3\pi/2$ as a rough benchmark for nonlinear device performance, then for two common ChG, usable phaseshifts can be achieved at this intensity for waveguide lengths of a few tens of centimeters, with a propagation loss requirement of <0.1 dB/cm. While these are certainly challenging requirements they are not beyond the realm of feasibility given the great strides that have been made in nonlinear glass waveguide fabrication over the past few years [18].

20.12 CONCLUSIONS

In conclusion, we have reviewed some of the recent progress on all-optical nonlinear signal processing devices over the past few years that relates to both ChG (in fiber and waveguides) and HNLF, using both phase- and nonphase-matched processes.

It is clear to many of us working in the field that all-optical signal processing devices are steadily coming of age, and the prospect for reducing peak power requirements of PIC-based all-optical circuits below the current levels of a few Watts [10, 23] to tens of milliwatts or even lower, will go a long way to ensure that all-optical nonlinear devices will play an integral part in future telecommunications systems.

REFERENCES

[1] N. S. Bergano, "Wavelength division multiplexing in long-haul transoceanic transmission systems," *J. Lightw. Technol.*, 23(12), 4125–4139, 2005.

[2] H. G. Weber, R. Ludwig, S. Ferber et al., "Ultrahigh-speed OTDM-transmission technology," *J. Lightw. Technol.*, 24(12), 4616–4627, 2006.

[3] T. T. Ng, J. L. Blows, M. Rochette et al., "In-band OSNR and chromatic dispersion monitoring using a fiber optical parametric amplifier," *Opt. Express*, 13(14), 5542–5552, 2005.

[4] T. Luo, C. Yu, Z. Pan et al., "All-optical chromatic dispersion monitoring of a 40-Gb/s RZ signal by measuring the XPM-generated optical tone power in a highly nonlinear fiber," *IEEE Photon. Technol. Lett.*, 18(1–4), 430–432, 2006.

[5] S. Radic and C. J. McKinstrie, "Optical amplification and signal processing in highly nonlinear optical fiber," *IEICE Trans. Electron.*, E88C(5), 859–869, 2005.

[6] V. G. Ta'eed, M. Shokooh-Saremi, L. B. Fu et al., "Integrated all-optical pulse regenerator in chalcogenide waveguides," *Opt. Lett.*, 30(21), 2900–2902, 2005.

[7] L. B. Fu, M. Rochette, V. G. Ta'eed et al., "Investigation of self-phase modulation based optical regeneration in single mode As_2Se_3 chalcogenide glass fiber," *Opt. Express*, 13(19), 7637–7644, 2005.

[8] S. Radic, C. J. McKinstrie, R. M. Jopson et al., "All-optical regeneration in one- and two-pump parametric amplifiers using highly nonlinear optical fiber," *IEEE Photon. Technol. Lett.*, 15(7), 957–959, 2003.

[9] M. Rochette, J. L. Blows, and B. J. Eggleton, "3R optical regeneration: An all-optical solution with BER improvement," *Opt. Express*, 14(14), 6414–6427, 2006.

[10] V. G. Ta'eed, M. R. E. Lamont, D. J. Moss et al., "All optical wavelength conversion via cross phase modulation in chalcogenide glass rib waveguides," *Opt. Express*, 14(23), 11242–11247, 2006.

[11] R. Jiang, R. E. Saperstein, N. Alic et al., "Continuous-wave band translation between the near-infrared and visible spectral ranges," *J. Lightw. Technol.*, 25(1), 58–66, 2007.

[12] K. Yamada, H. Fukuda, T. Tsuchizawa et al., "All-optical efficient wavelength conversion using silicon photonic wire waveguide," *IEEE Photon. Technol. Lett.*, 18(9–12), 1046–1048, 2006.

[13] J. T. Mok and B. J. Eggleton, "Photonics: Expect more delays," *Nature*, 433(7028), 811–812, 2005.

[14] J. H. Lee and K. Kikuchi, "All fiber-based 160-Gbit/s add/drop multiplexer incorporating a 1-m-long Bismuth Oxide-based ultra-high nonlinearity fiber," *Opt. Express*, 13(18), 6864–6869, 2005.

[15] V. G. Ta'eed, N. J. Baker, L. Fu et al., "Ultrafast all-optical chalcogenide glass photonic circuits," *Opt. Express*, 15(15), 9205–9221, 2007.

[16] J. H. Lee, K. Kikuchi, T. Nagashima et al., "All-fiber 80-Gbit/s wavelength converter using 1-m-long Bismuth Oxide-based nonlinear optical fiber with a nonlinearity gamma of 1100 W(−1)km(−1)," *Opt. Express*, 13(8), 3144–3149, 2005.

[17] R. Claps, D. Dimitropoulos, V. Raghunathan et al., "Observation of stimulated Raman amplification in silicon waveguides," *Opt. Express*, 11(15), 1731–1739, 2003.

[18] Y. L. Ruan, W. T. Li, R. Jarvis et al., "Fabrication and characterization of low loss rib chalcogenide waveguides made by dry etching," *Opt. Express*, 12(21), 5140–5145, 2004.

[19] N. Sugimoto, T. Nagashima, T. Hasegawa et al., "Bismuth-based Optical Fiber with Nonlinear Coefficient of 1360 $W^{-1}.km^{-1}$," in Proc. *OFC*, Los Angeles, California, 2004.

[20] I. D. Aggarwal and J. S. Sanghera, "Development and applications of chalcogenide glass optical fibers at NRL," *J. Optoelectron. Adv. Mater.*, 4(3), 665–678, 2002.

[21] H C. Nguyen, E. C. Mägi, D.-I. Yeom et al., "Enhanced Kerr Nonlinearity in As_2Se_3 Chalcogenide Fibre Tapers with Sub-Wavelength Diameter," in Proc. *SPIE*, 2007, p. 6588.

[22] J. Le Person, F. Smektala, T. Chartier et al., "Light guidance in new chalcogenide holey fibers from GeGaSbS glass," *Mater. Res. Bull.*, 41(7), 1303–1309, 2006.

[23] M. A. F. Reza Salem, A. C. Turner, D. F. Geraghty et al., "All-optical regeneration on a silicon chip," *Opt. Express*, 15(12), 7802, 2007.

[24] J. Hansryd and P. A. Andrekson, "Broad-band continuous-wave-pumped fiber optical parametric amplifier with 49-dB gain and wavelength-conversion efficiency," *IEEE Photon. Technol. Lett.*, 13(3), 194–196, 2001.

[25] K. Inoue, "Polarization-independent wavelength conversion using fiber 4-wave-mixing with 2 orthogonal pump lights of different frequencies," *J. Lightw. Technol.*, 12(11), 1916–1920, 1994.

[26] R. H. Stolen and J. E. Bjorkholm, "Parametric amplification and frequency-conversion in optical fibers," *IEEE J. Quantum Electron.*, 18(7), 1062–1072, 1982.

[27] R. M. Jopson and R. E. Tench, "Polarization-independent phase-conjugation of lightwave signals," *Electron. Lett.*, 29(25), 2216–2217, 1993.

[28] P. Kumar and J. H. Shapiro, "Squeezed-state generation via forward degenerate 4-wave mixing," *Phys. Rev. A*, 30(3), 1568–1571, 1984.

[29] M. E. Marhic, F. S. Yang, M. C. Ho, and L. G. Kazovsky, "High-nonlinearity fiber optical parametric amplifier with periodic dispersion compensation," *J. Lightw. Technol.*, 17(2), 210–215, 1999.

[30] S. Radic, C. J. McKinstrie, A. R. Chraplyvy et al., "Continuous-wave parametric gain synthesis using nondegenerate pump four-wave mixing," *IEEE Photon. Technol. Lett.*, 14(10), 1406–1408, 2002.

[31] R. D. Li, P. Kumar, and W. L. Kath, "Dispersion compensation with phase-sensitive optical amplifiers," *J. Lightw. Technol.*, 12(3), 541–549, 1994.

[32] J. A. Levenson, I. Abram, T. Rivera, and P. Grangier, "Reduction of quantum-noise in optical parametric amplification," *J. Opt. Soc. Am. B-Opt. Phys.*, 10(11), 2233–2238, 1993.

[33] E. Ciaramella and S. Trillo, "All-optical signal reshaping via four-wave mixing in optical fibers," *IEEE Photon. Technol. Lett.*, 12(7), 849–851, 2000.

[34] H. P. Yuen and J. H. Shapiro, "Generation and detection of 2-photon coherent states in degenerate 4-wave mixing," *Opt. Lett.*, 4(10), 334–336, 1979.

[35] B. Yurke, "Use of cavities in squeezed-state generation," *Phys. Rev. A*, 29(1), 408–410, 1984.

[36] M. J. Potasek and B. Yurke, "Squeezed-light generation in a medium governed by the nonlinear Schrodinger-equation," *Phys. Rev. A*, 35(9), 3974–3977, 1987.

[37] M. D. Levenson, R. M. Shelby, A. Aspect et al., "Generation and detection of squeezed states of light by nondegenerate 4-wave mixing in an optical fiber," *Phys. Rev. A*, 32(3), 1550–1562, 1985.

[38] T. A. B. Kennedy and S. Wabnitz, "Quantum propagation—squeezing via modulational polarization instabilities in a birefringent nonlinear medium," *Phys. Rev. A*, 38(1), 563–566, 1988.

[39] R. E. Slusher, G. Lenz, J. Hodelin et al., "Large Raman gain and nonlinear phase shifts in high-purity As_2Se_3 chalcogenide fibers," *J. Opt. Soc. Am. B-Opt. Phys.*, 21(6), 1146–1155, 2004.

[40] M. Asobe, T. Kanamori, and K. Kubodera, "Applications of highly nonlinear chalcogenide glass-fibers in ultrafast all-optical switches," *IEEE J. Quantum Electron.*, 29(8), 2325–2333, 1993.

[41] A. Zakery and S. R. Elliott, "Optical properties and applications of chalcogenide glasses: a review," *J. Non-Crystal. Solids*, 330(1–3), 1–12, 2003.

[42] S. Ramachandran and S. G. Bishop, "Photoinduced integrated-optic devices in rapid thermally annealed chalcogenide glasses," *IEEE J. Sel. Top. Quantum Electron.*, 11(1), 260–270, 2005.

[43] R. G. DeCorby, N. Ponnampalam, M. M. Pai et al., "High index contrast waveguides in chalcogenide glass and polymer," *IEEE J. Sel. Top. Quantum Electron.*, 11(2), 539–546, 2005.

[44] S. Ramachandran and S. G. Bishop, "Low loss photoinduced waveguides in rapid thermally annealed films of chalcogenide glasses," *Appl. Phys. Lett.*, 74(1), 13–15, 1999.

[45] A. V. Rode, A. Zakery, M. Samoc et al., "Laser-deposited As_2S_3 chalcogenide films for waveguide applications," *Appl. Surf. Sci.*,. 197, 481–485, 2002.

[46] A. Zakery, "Low loss waveguides in pulsed laser deposited arsenic sulfide chalcogenide films," *J. Phys. D: Appl. Phys.*, 35(22), 2909–2913, 2002.

[47] A. Zoubir, M. Richardson, C. Rivero et al., "Direct femtosecond laser writing of waveguides in As_2S_3 thin films," *Opt. Lett.*, 29(7), 748–750, 2004.

[48] J. F. Viens, C. Meneghini, A. Villeneuve et al., "Fabrication and characterization of integrated optical waveguides in sulfide chalcogenide glasses," *J. Lightw. Technol.*, 17(7), 1184–1191, 1999.

[49] K. Tanaka, N. Toyosawa, and H. Hisakuni, "Photoinduced Bragg gratings in As_2S_3 optical fibers," *Opt. Lett.*, 20(19), 1976–1978, 1995.

[50] M. Asobe, T. Ohara, I. Yokohama, and T. Kaino, "Fabrication of Bragg grating in chalcogenide glass fibre using the transverse holographic method," *Electron. Lett.*, 32(17), 1611–1613, 1996.

[51] A. Saliminia, K. Le Foulgoc, A. Villeneuve et al., "Photoinduced Bragg reflectors in As-S-Se/As-S based chalcogenide glass multilayer channel waveguides," *Fiber Integr. Opt.*, 20(2), 151–158, 2001.

[52] S. Ramachandran, S. G. Bishop, J. P. Guo, and D. J. Brady, "Fabrication of holographic gratings in As_2S_3 glass by photoexpansion and photodarkening," *IEEE Photon. Technol. Lett.*, 8(8), 1041–1043, 1996.

[53] T. G. Robinson, R. G. DeCorby, J. N. McMullin et al., "Strong Bragg gratings photoinduced by 633-nm illumination in evaporated AS(2)Se(3) thin films," *Opt. Lett.*, 28(6), 459–461, 2003.

[54] N. Ponnampalam, R. G. DeCorby, H. T. Nguyen et al., "Small core rib waveguides with embedded gratings in As_2Se_3 glass," *Opt. Express*, 12(25), 6270–6277, 2004.

[55] R. Vallee, S. Frederick, K. Asatryan et al., "Real-time observation of Bragg grating formation in As_2S_3 chalcogenide ridge waveguides," *Opt. Commun.*, 230(4–6), 301–307, 2004.

[56] D. Freeman, S. Madden, and B. Luther-Davies, "Fabrication of planar photonic crystals in a chalcogenide glass using a focused ion beam," *Opt. Express*, 13(8), 3079–3086, 2005.

[57] M. Asobe, "Nonlinear optical properties of chalcogenide glass fibers and their application to all-optical switching," *Opt. Fiber Technol.*, 3(2), 142–148, 1997.

[58] V. Mizrahi, K. W. Delong, G. I. Stegeman et al., "2-photon absorption as a limitation to all-optical switching," *Opt. Lett.*, 14(20), 1140–1142, 1989.

[59] G. Lenz, J. Zimmermann, T. Katsufuji et al., "Large Kerr effect in bulk Se-based chalcogenide glasses," *Opt. Lett.*, 25(4), 254–256, 2000.

[60] H. C. Nguyen, K. Finsterbusch, D. J. Moss, and B. J. Eggleton, "Dispersion in nonlinear figure of merit of As_2Se_3 chalcogenide fiber," *Electron. Lett.*, 42(10), 571–572, 2006.

[61] J. S. Sanghera, I. D. Aggarwal, L. B. Shaw et al., "Nonlinear properties of chalcogenide glass fibers," *J. Optoelectron. Adv. Mater.*, 8(6), 2148–2155, 2006.

[62] A. Zakery, Y. Ruan, A. V. Rode et al., "Low-loss waveguides in ultrafast laser-deposited As_2S_3 chalcogenide films," *J. Opt. Soc. Am. B-Opt. Phys.*, 20(9), 1844–1852, 2003.

[63] M. Shokooh-Saremi, V. G. Ta'eed, N. J. Baker et al., "High-performance Bragg gratings in chalcogenide rib waveguides written with a modified Sagnac interferometer," *J. Opt. Soc. Am. B-Opt. Phys.*, 23(7), 1323–1331, 2006.

[64] Q. F. Xu and M. Lipson, "All-optical logic based on silicon micro-ring resonators," *Opt. Express*, 15(3), 924–929, 2007.

[65] S. R. Preble, Q. F. Xu, B. S. Schmidt, and M. Lipson, "Ultrafast all-optical modulation on a silicon chip," *Opt. Lett.*, 30(21), 2891–2893, 2005.

[66] V. R. Almeida, C. A. Barrios, R. R. Panepucci et al., "All-optical switching on a silicon chip," *Opt. Lett.*, 29(24), 2867–2869, 2004.

[67] V. R. Almeida, C. A. Barrios, R. R. Panepucci, and M. Lipson, "All-optical control of light on a silicon chip," *Nature*, 431(7012), 1081–1084, 2004.

[68] Y. Okawachi, M. A. Foster, J. E. Sharping et al., "All-optical slow-light on a photonic chip," *Opt. Express*, 14(6), 2317–2322, 2006.

[69] C. Manolatou and M. Lipson, "All-optical silicon modulators based on carrier injection by two-photon absorption," *J. Lightw. Technol.*, 24(3), 1433–1439, 2006.

[70] T. K. Liang, L. R. Nunes, M. Tsuchiya et al., "High speed logic gate using two-photon absorption in silicon waveguides," *Opt. Commun.*, 265(1), 171–174, 2006.

[71] Y. A. Vlasov and S. J. McNab, "Losses in single-mode silicon-on-insulator strip waveguides and bends," *Opt. Express*, 12(8), 1622–1631, 2004.

[72] R. L. Espinola, J. I. Dadap, R. M. Osgood et al., "Raman amplification in ultrasmall silicon-on-insulator wire waveguides," *Opt. Express*, 12(16), 3713–3718, 2004.

[73] Y. Liu and H. K. Tsang, "Nonlinear absorption and Raman gain in helium-ion-implanted silicon waveguides," *Opt. Lett.*, 31(11), 1714–1716, 2006.

[74] T. K. Liang and H. K. Tsang, "On Raman Gain in Silicon Waveguides: Limitations from Two-Photon-Absorption Generated Carriers," in Proc. *Lasers and Electro-Optics (CLEO)*. San Francisco, 2004.

[75] T. K. Liang and H. K. Tsang, "Role of free carriers from two-photon absorption in Raman amplification in silicon-on-insulator waveguides," *Appl. Phys. Lett.*, 84(15), 2745–2747, 2004.

[76] H. K. Tsang, C. S. Wong, T. K. Liang et al., "Optical dispersion, two-photon absorption and self-phase modulation in silicon waveguides at 1.5 μm wavelength," *Appl. Phys. Lett.*, 80(3), 416–418, 2002.

[77] M. Dinu, F. Quochi, and H. Garcia, "Third-order nonlinearities in silicon at telecom wavelengths," *Appl. Phys. Lett.*, 82(18), 2954–2956, 2003.

[78] A. R. Cowan, G. W. Rieger, and J. F. Young, "Nonlinear transmission of 1.5 μm pulses through single-mode silicon-on-insulator waveguide structures," *Opt. Express*, 12(8), 1611–1621, 2004.

[79] G. W. Rieger, K. S. Virk, and J. F. Young, "Nonlinear propagation of ultrafast 1.5 μm pulses in high-index-contrast silicon-on-insulator waveguides," *Appl. Phys. Lett.*, 84(6), 900–902, 2004.

[80] M. R. E. Lamont, M. Rochette, D. J. Moss, and B. J. Eggleton, "Two-photon absorption effects on self-phase-modulation-based 2R optical regeneration," *IEEE Photon. Technol. Lett.*, 18(9–12), 1185–1187, 2006.

[81] N. Sugimoto, H. Kanbara, S. Fujiwara et al., "Third-order optical nonlinearities and their ultrafast response in Bi_2O_3-B_2O_3-SiO_2 glasses," *J. Opt. Soc. Am. B-Opt. Phys.*, 16(11), 1904–1908, 1999.

[82] Titterton D. H. and Terry, J. A. C., "Study of Laser-Induced Damage in PPLN," in Proc. *SPIE*, 4679, 2002, pp. 312–320.

[83] T. Skauli, K. L. Vodopyanov, T. J. Pinguet et al., "Measurement of the nonlinear coefficient of orientation-patterned GaAs and demonstration of highly efficient second-harmonic generation," *Opt. Lett.*, 27(8), 628–630, 2002.

[84] T. Akiyama, H. Kuwatsuka, N. Hatori et al., "Symmetric highly efficient (similar to 0 dB) wavelength conversion based on four-wave mixing in quantum dot optical amplifiers," *IEEE Photon. Technol. Lett.*, 14(8), 1139–1141, 2002.

[85] R. L. Byer, "Nonlinear optics and solid-state lasers: 2000," *IEEE J. Sel. Top. Quantum Electron.*, 6(6), 911–930, 2000.

[86] V. G. Ta'eed, "All-Optical Signal Processing in Chalcogenide Glass," in *Physics* 2007, Sydney.

[87] C. J. McKinstrie and S. Radic, "Parametric amplifiers driven by two pump waves with dissimilar frequencies," *Opt. Lett.*, 27(13), 1138–1140, 2002.

[88] C. J. McKinstrie, S. Radic, and A. R. Chraplyvy, "Parametric amplifiers driven by two pump waves," *IEEE J. Sel. Top. Quantum Electron.*, 8(3), 538–547, 2002.

[89] S. Radic and C. J. McKinstrie, "Two-pump fiber parametric amplifiers," *Opt. Fiber Technol.*, 9(1), 7–23, 2003.

[90] O. Leclerc, B. Lavigne, E. Balmefrezol et al., "Optical regeneration at 40 Gb/s and beyond," *J. Lightw. Technol.*, 21(11), 2779–2790, 2003.

[91] J. P. R. Lacey, S. J. Madden, and M. A. Summerfield, "Four-channel polarization-insensitive optically transparent wavelength converter," *IEEE Photon. Technol. Lett.*, 9(10), 1355–1357, 1997.

[92] S. Boscolo, S. K. Turitsyn, and V. K. Mezentsev, "Performance comparison of 2R and 3R optical regeneration schemes at 40 Gb/s for application to all-optical networks," *J. Lightw. Technol.*, 23(1), 304–309, 2005.

[93] M. Rochette, L. B. Fu, V. Ta'eed et al., "2R optical regeneration: An all-optical solution for BER improvement," *IEEE J. Sel. Top. Quantum Electron.*, 12(4), 736–744, 2006.

[94] M. Rochette, J. N. Kutz, J. L. Blows et al., "Bit-error-ratio improvement with 2R optical regenerators," *IEEE Photon. Technol. Lett.*, 17(4), 908–910, 2005.

[95] P. V. Mamyshev, "All-Optical Data Regeneration Based on Self-Phase Modulation Effect," in Proc. *ECOC*, Madrid, Spain, 1998.

[96] T. H. Her, G. Raybon, and C. Headley, "Optimization of pulse regeneration at 40 Gb/s based on spectral filtering of self-phase modulation in fiber," *IEEE Photon. Technol. Lett.*, 16(1), 200–202, 2004.

[97] N. Yoshikane, I. Morita, T. Tsuritani et al., "Benefit of SPM-based all-optical reshaper in receiver for long-haul DWDM transmission systems," *IEEE J. Sel. Top. Quantum Electron.*, 10(2), 412–420, 2004.

[98] G. Raybon, Y. Su, J. Leuthold et al., "40 Gbit/s pseudo-Linear Transmission Over One Million Kilometers," in Proc. *Technical Digest of OFC 2002*, Anaheim, California, USA, 2002.

[99] Z. J. Huang, A. Gray, I. Khrushchev, and I. Bennion, "10-Gb/s transmission over 100 mm of standard fiber using 2R regeneration in an optical loop mirror," *IEEE Photon. Technol. Lett.*, 16(11), 2526–2528, 2004.

[100] M. Meissner, K. Spionsel, K. Cvecek et al., "3.9-dB OSNR gain by an NOLM-based 2-R regenerator," *IEEE Photon. Technol. Lett.*, 16(9), 2105–2107, 2004.

[101] R. Ludwig, A. Sizmann, U. Feiste et al., "Experimental Verification of Noise Squeezing by an Optical Intensity Filter in High-Speed Transmission," in Proc. *ECOC*, 2001.

[102] E. Ciaramella, "A New Scheme for All-Optical Signal Reshaping Based on Wavelength Conversion in Optical Fibers," in Proc. *OFC*, 2000.

[103] D. Rouvillain, F. Seguineau, L. Pierre et al., "40 Gbit/s Optical 2R Regenerator Based on Passive Saturable Absorber for WDM Long-Haul Transmission," in Proc. *OFC*, 2002.

[104] F. Ohman, S. Bischoff, B. Tromborg, and J. Mork. "Semiconductor Devices for All-Optical Regeneration," in Proc. *International Conference on Transparent Optical Networks*, Warsaw, Poland, 2003.

[105] V. G. Ta'eed, L. B. Fu, M. Pelusi et al., "Error free all optical wavelength conversion in highly nonlinear As-Se chalcogenide glass fiber," *Opt. Express*, 14(22), 10371–10376, 2006.

[106] M. R. E. Lamont, V. G. Ta'eed, M. A. F. Roelens et al., "Error-free wavelength conversion via cross phase modulation in 5 cm of As_2S_3 chalcognide glass rib waveguide," *Electron. Lett.*, 43, 945, 2007.

[107] M. Asobe, K. Suzuki, T. Kanamori et al., "Nonlinear refractive-index measurement in chalcogenide-glass fibers by self-phase modulation," *Appl. Phys. Lett.*, 60(10), 1153–1154, 1992.
[108] M. R. E. Lamont, L. B. Fu, M. Rochette et al., "2R optical regenerator in AS(2)Se(3) chalcogenide fiber characterized by a frequency-resolved optical gating analysis," *Appl. Opt.*, 45(30), 7904–7907, 2006.
[109] I. T. Monroy, F. Ohman, K. Yvind et al., "Monolithically integrated reflective SOA-EA carrier re-modulator for broadband access nodes," *Opt. Express*, 14(18), 8060–8064, 2006.
[110] M. van der Poel, J. Mork, A. Somers et al., "Ultrafast gain and index dynamics of quantum dash structures emitting at 1.55 μm," *Appl. Phys. Lett.*, 89(8), 081102, 2006.
[111] F. Ohman, S. Bischoff, B. Tromborg, and J. Mork, "Noise and regeneration in semiconductor waveguides with saturable gain and absorption," *IEEE J. Quantum Electron.*, 40(3), 245–255, 2004.
[112] H. Murai, M. Kagawa, H. Tsuji, and K. Fujii, "EA-modulator-based optical time division multiplexing/demultiplexing techniques for 160-Gb/s optical signal transmission," *IEEE J. Sel. Top. Quantum Electron.*, 13(1), 70–78, 2007.
[113] E. Tangdiongga, Y. Liu, H. de Waardt et al., "All-optical demultiplexing of 640 to 40 Gbits/s using filtered chirp of a semiconductor optical amplifier," *Opt. Lett.*, 32(7), 835–837, 2007.
[114] V. G. Ta'eed, M. Shokooh-Saremi, L. B. Fu et al., "Self-phase modulation-based integrated optical regeneration in chalcogenide waveguides," *IEEE J. Sel. Top. Quantum Electron.*, 12(3), 360–370, 2006.
[115] M. Shokooh-Saremi, V. G. Ta'eed, I. C. M. Littler et al., "Ultra-strong, well-apodised Bragg gratings in chalcogenide rib waveguides," *Electron. Lett.*, 41(13), 738–739, 2005.
[116] G. Lenz, B. J. Eggleton, C. R. Giles et al., "Dispersive properties of optical filters for WDM systems," *IEEE J. Quantum Electron.*, 34(8), 1390–1402, 1998.
[117] I. C. M. Littler, M. Rochette, and B. J. Eggleton, "Adjustable bandwidth dispersionless bandpass FBG optical filter," *Opt. Express*, 13(9), 3397–3407, 2005.
[118] J. Suzuki, T. Tanemura, K. Taira et al., "All-optical regenerator using, wavelength shift induced by cross-phase modulation in highly nonlinear dispersion-shifted fiber," *IEEE Photon. Technol. Lett.*, 17(2), 423–425, 2005.
[119] M. Daikoku, N. Yoshikane, T. Otani, and H. Tanaka, "Optical 40-Gb/s 3R regenerator with a combination of the SPM and XAM effects for all-optical networks," *J. Lightw. Technol.*, 24(3), 1142–1148, 2006.
[120] M. Rochette, J. L. Blows, and B. Eggleton, "An All-Optical Regenerator that Discriminates Noise from Signal," in Proc. *31st ECOC*, Glasgow, Scotland, 2005.
[121] K. Inoue, "Suppression of level fluctuation without extinction ratio degradation based on output saturation in higher order optical parametric interaction in fiber," *IEEE Photon. Technol. Lett.*, 13(4), 338–340, 2001.
[122] S. Radic, C. J. M. R. M. Jopson, J. C. Centanni, and A. R. Chraplyvy, "All-optical regeneration in one- and two-pump parametric amplifiers using highly nonlinear optical fiber, to appear in," *IEEE Photon. Technol. Lett.*, 15, 957–959, 2003.
[123] R. A. Fisher, *Optical Phase Conjugation*, San Diego: Academic Press, 1983.
[124] S. Radic, R. J. J. C. McKinstrie, A. Gnauck et al., "Wavelength Division Multiplexed Transmission Over Standard Single Mode Fiber Using Polarization Insensitive Signal Conjugation in Highly Nonlinear Optical Fiber," in Proc. *OFC*, Postdeadline Paper #12, Atlanta, Georgia, 2003.
[125] B. E. Olsson, P. Ohlen, L. Rau, and D. J. Blumenthal, "A simple and robust 40-Gb/s wavelength converter using fiber cross-phase modulation and optical filtering," *IEEE Photon. Technol. Lett.*, 12(7), 846–848, 2000.
[126] J. H. Lee, T. Nagashima, T. Hasegawa et al., "Wavelength conversion of 160 Gbit/s OTDM signal using bismuth oxide-based ultra-high nonlinearity fibre," *Electron. Lett.*, 41(16), 918–919, 2005.
[127] G. Burdge, S. U. Alam, A. Grudinin et al., "Ultrafast intensity modulation by Raman gain for all-optical in-fiber processing," *Opt. Lett.*, 23(8), 606–608, 1998.

[128] M. Lamont, L. B. Fu, M. Rochette et al., "Two Photon Absorption Effects on 2R Optical Regeneration," *IEEE Photon. Technol. Lett.*, 18, 1185, 2006.
[129] Y. K. Lize, E. C. Magi, V. G. Ta'eed et al., "Microstructured optical fiber photonic wires with subwavelength core diameter," *Opt. Express*, 12(14), 3209–3217, 2004.
[130] T. M. Monro, Y. D. West, D. W. Hewak et al., "Chalcogenide holey fibers," *Electron. Lett.*, 36(24), 1998–2000, 2000.
[131] H. Fukuda, K. Yamada, T. Shoji et al., "Four-wave mixing in silicon wire waveguides," *Opt. Express*, 13(12), 4629–4637, 2005.
[132] N. S. Bergano, F. W. Kerfoot, and C. R. Davidson, "Margin measurements in optical amplifier systems," *IEEE Photon. Technol. Lett.*, 5(3), 304–306, 1993.
[133] S. Radic, C. J. McKinstrie, R. M. Jopson et al., "Multiple-band bit-level switching in two-pump fiber parametric devices," *IEEE Photon. Technol. Lett.*, 16(3), 852–854, 2004.
[134] D. C. Kilper, R. Bach, D. J. Blumenthal et al., "Optical performance monitoring," *J. Lightw. Technol.*, 22(1), 294–304, 2004.
[135] L. M. Lunardi, D. J. Moss, S. Chandrasekhar et al., "Tunable dispersion compensation at 40-Gb/s using a multicavity etalon all-pass filter with NRZ, RZ, and CS-RZ modulation," *J. Lightw. Technol.*, 20(12), 2136–2144, 2002.
[136] D. J. Moss, M. Lamont, S. McLaughlin et al., "Tunable dispersion and dispersion slope compensators for 10 Gb/s using all-pass multicavity etalons," *IEEE Photon. Technol. Lett.*, 15(5), 730–732, 2003.
[137] B. J. Eggleton, B. Mikkelsen, G. Raybon et al., "Tunable dispersion compensation in a 160-Gb/s TDM system by a voltage controlled chirped fiber Bragg grating," *IEEE Photon. Technol. Lett.*, 12(8), 1022–1024, 2000.
[138] P. S. Westbrook, B. J. Eggleton, G. Raybon et al., "Measurement of residual chromatic dispersion of a 40-Gb/s RZ signal via spectral broadening," *IEEE Photon. Technol. Lett.*, 14(3), 346–348, 2002.
[139] T. T. Ng, J. L. Blows, J. T. Mok et al., "Cascaded four-wave mixing in fiber optical parametric amplifiers: Application to residual dispersion monitoring," *J. Lightw. Technol.*, 23(2), 818–826, 2005.
[140] W. H. Hatton and M. Nishimura, "Temperature-dependence of chromatic dispersion in single-mode fibers," *J. Lightw. Technol.*, 4(10), 1552–1555, 1986.
[141] R. Adams, M. Rochette, T. T. Ng, and B. J. Eggleton, "All-optical in-band OSNR monitoring at 40 Gb/s using a nonlinear optical loop mirror," *IEEE Photon. Technol. Lett.*, 18(1–4), 469–471, 2006.
[142] T. T. Ng, J. L. Blows, and B. J. Eggleton, "In-band OSNR monitoring using fiber optical parametric amplifier," *Electron. Lett.*, 41(6), 352–353, 2005.
[143] J. L. Blows, P. F. Hu, and B. J. Eggleton, "Differential group delay monitoring using an all-optical signal spectrum-analyser," *Opt. Commun.*, 260(1), 288–291, 2006.
[144] S. Nezam, Y. W. Song, C. Y. Yu et al., "First-order PMD monitoring for NRZ data using RF clock regeneration techniques," *J. Lightw. Technol.*, 22(4), 1086–1093, 2004.
[145] T. Luo, Z. Pan, S. Nezam et al., "PMD monitoring by tracking the chromatic-dispersion-insensitive RF power of the vestigial sideband," *IEEE Photon. Technol. Lett.*, 16(9), 2177–2179, 2004.
[146] Y. K. Lize, L. Christen, J. Y. Yang et al., "Independent and simultaneous monitoring of chromatic and polarization-mode dispersion in OOK and DPSK transmission," *IEEE Photon. Technol. Lett.*, 19(1), 3–5, 2007.
[147] Q. Yu, Z. Q. Pan, L. S. Yan, and A. E. Willner, "Chromatic dispersion monitoring technique using sideband optical filtering and clock phase-shift detection," *J. Lightw. Technol.*, 20(12), 2267–2271, 2002.
[148] T. Yamamoto, T. Imai, T. Komukai et al., "High-speed optical-path routing by using 4-wave-mixing and a wavelength router with fiber gratings and optical circulators," *Opt. Commun.*, 120(5–6), 245–248, 1995.
[149] G. Conte, M. Listanti, M. Settembre, and R. Sabella, "Strategy for protection and restoration of optical paths in WDM backbone networks for next-generation Internet infrastructures," *J. Lightw. Technol.*, 20(8), 1264–1276, 2002.

[150] N. Antoniades, S. J. B. Yoo, K. Bala et al., "An architecture for a wavelength-interchanging cross-connect utilizing parametric wavelength converters," *J. Lightw. Technol.*, 17(7), 1113–1125, 1999.
[151] C. K. Madsen, J. Wagener, T. A. Strasser et al., "Planar waveguide optical spectrum analyzer using a UV-induced grating," *IEEE J. Sel. Top. Quantum Electron.*, 4(6), 925–929, 1998.
[152] S. Wielandy, M. Fishteyn, and B. Y. Zhu, "Optical performance monitoring using nonlinear detection," *J. Lightw. Technol.*, 22(3), 784–793, 2004.
[153] J. H. Lee, D. K. Jung, C. H. Kim, and Y. C. Chung, "OSNR monitoring technique using polarization-nulling method," *IEEE Photon. Technol. Lett.*, 13(1), 88–90, 2001.
[154] P. Vorreau, D. C. Kilper, and J. Leuthold, "Optical noise and dispersion monitoring with SOA-based optical 2R regenerator," *IEEE Photon. Technol. Lett.*, 17(1), 244–246, 2005.
[155] E. Wong, K. L. Lee, and A. Nirmalathas, "Novel In-Band Optical Signal-to-Noise Ratio Monitor for WDM Networks, Paper 6B3-5," in Proc. *OECC*, Seoul, South Korea, 2005.
[156] R. Noe, D. Sandel, M. Yoshida-Dierolf et al., "Polarization mode dispersion compensation at 10, 20, and 40 Gb/s with various optical equalizers," *J. Lightw. Technol.*, 17(9), 1602–1616, 1999.
[157] T. Takahashi, T. Imai, and M. Aiki, "Automatic compensation technique for timewise fluctuating polarization mode dispersion in in-line amplifier systems," *Electron. Lett.*, 30(4), 348–349, 1994.
[158] H. Y. Pua, K. Peddanarappagari, B. Y. Zhu et al., "An adaptive first-order polarization-mode dispersion compensation system aided by polarization scrambling: Theory and demonstration," *J. Lightw. Technol.*, 18(6), 832–841, 2000.
[159] G. Bosco, B. E. Olsson, and D. J. Blumenthal, "Pulsewidth distortion monitoring in a 40-Gb/s optical system affected by PMD," *IEEE Photon. Technol. Lett.*, 14(3), 307–309, 2002.
[160] S. Lanne, W. Idler, J. P. Thiery, and J. P. Hamaide, "Fully automatic PMD compensation at 40 Gbit/s," *Electron. Lett.*, 38(1), 40–41, 2002.
[161] A. E. Willner, S. M. R. Motaghian, L. S. Yan et al., "Monitoring and control of polarization-related impairments in optical fiber systems," *J. Lightw. Technol.*, 22(1), 106–125, 2004.
[162] C. Dorrer and X. Liu, "Noise monitoring of optical signals using RF spectrum analysis and its application to phase = shift-keyed signals," *IEEE Photon. Technol. Lett.*, 16(7), 1781–1783, 2004.
[163] C. Dorrer and D. N. Maywar, "RF spectrum analysis of optical signals using nonlinear optics," *J. Lightw. Technol.*, 22(1), 266–274, 2004.
[164] L. V. Hau, S. E. Harris, Z. Dutton, and C. H. Behroozi, "Light speed reduction to 17 metres per second in an ultracold atomic gas," *Nature*, 397(6720), 594–598, 1999.
[165] P. Palinginis, F. Sedgwick, S. Crankshaw et al., "Room temperature slow light in a quantum-well waveguide via coherent population oscillation," *Opt. Express*, 13(24), 9909–9915, 2005.
[166] S. Longhi, D. Janner, G. Galzerano et al., "Optical buffering in phase-shifted fiber gratings," *Electron. Lett.*, 41(19), 1075–1077, 2005.
[167] Y. Okawachi, M. S. Bigelow, J. E. Sharping et al., "Tunable all-optical delays via Brillouin slow light in an optical fiber," *Phys. Rev. Lett.*, 94(15), 153902, 2005.
[168] K. Y. Song, M. G. Herraez, and L. Thevenaz, "Observation of pulse delaying and advancement in optical fibers using stimulated Brillouin scattering," *Opt. Express*, 13(1), 82–88, 2005.
[169] J. T. Mok, C. M. de Sterke, I. C. M. Littler, and B. J. Eggleton, "Dispersionless slow light using gap solitons," *Nat. Phys.*, 2(11), 775–780, 2006.
[170] J. E. Heebner, V. Wong, A. Schweinsberg et al., "Optical transmission characteristics of fiber ring resonators," *IEEE J. Quantum Electron.*, 40(6), 726–730, 2004.
[171] J. K. S. Poon, L. Zhu, G. A. DeRose, and A. Yariv, "Transmission and group delay of microring coupled-resonator optical waveguides," *Opt. Lett.*, 31(4), 456–458, 2006.
[172] H. Gersen, T. J. Karle, R. J. P. Engelen et al., "Real-space observation of ultraslow light in photonic crystal waveguides," *Phys. Rev. Lett.*, 94(7), 073903, 2005.
[173] X. Letartre, C. Seassal, C. Grillet et al., "Group velocity and propagation losses measurement in a single-line photonic-crystal waveguide on InP membranes," *Appl. Phys. Lett.*, 79(15), 2312–2314, 2001.

[174] D. Mori and T. Baba, "Dispersion-controlled optical group delay device by chirped photonic crystal waveguides," *Appl. Phys. Lett.*, 85(7), 1101–1103, 2004.

[175] Y. A. Vlasov, M. O'Boyle, H. F. Hamann, and S. J. McNab, "Active control of slow light on a chip with photonic crystal waveguides," *Nature*, 438(7064), 65–69, 2005.

[176] G. Lenz, B. J. Eggleton, C. K. Madsen, and R. E. Slusher, "Optical delay lines based on optical filters," *IEEE J. Quantum Electron.*, 37(4), 525–532, 2001.

[177] J. B. Khurgin, "Expanding the bandwidth of slow-light photonic devices based on coupled resonators." *Opt. Lett.*, 30(5), 513–515, 2005.

[178] M. L. Povinelli, S. G. Johnson, and J. D. Joannopoulos, "Slow-light, band-edge waveguides for tunable time delays," *Opt. Express*, 13(18), 7145–7159, 2005.

[179] S. Sandhu, M. L. Povinelli, M. F. Yanik, and S. H. Fan, "Dynamically tuned coupled-resonator delay lines can be nearly dispersion free," *Opt. Lett.*, 31(13), 1985–1987, 2006.

[180] M. F. Yanik and S. H. Fan, "Stopping light all optically," *Phys. Rev. Lett.*, 92(8), 083901, 2004.

[181] C. M. Desterke and J. E. Sipe, "Gap solitons," in *Progress in Optics*, (E. Wolf, ed.), 33, pp. 203–260, Elsevier, 1994.

[182] D. N. Christodoulides and R. I. Joseph, "Slow Bragg solitons in nonlinear periodic structures," *Phys. Rev. Lett.*, 62(15), 1746–1749, 1989.

[183] A. B. Aceves and S. Wabnitz, "Self-induced transparency solitons in nonlinear refractive periodic media," *Phys. Lett. A*, 141(1–2), 37–42, 1989.

[184] I. C. M. Littler, T. Grujic, and B. J. Eggleton, "Photothermal effects in fiber Bragg gratings," *Appl. Opt.*, 45(19), 4679–4685, 2006.

[185] R. S. Tucker, P. C. Ku, and C. J. Chang-Hasnain, "Slow-light optical buffers: Capabilities and fundamental limitations," *J. Lightw. Technol.*, 23(12), 4046–4066, 2005.

[186] A. A. Sukhorukov and Y. S. Kivshar, "Slow-light optical bullets in arrays of nonlinear Bragg-grating waveguides," *Phys. Rev. Lett.*, 97(23), 233901, 2006.

[187] Z. Zhu, A. M. C. Dawes, D. J. Gauthier et al., "12GHz bandwidth SBS slow light in optical fibers," in Proc. OFC, Postdeadline Paper 1, Anaheim, CA, 2006.

[188] M. Burzio et al., in Proc. *ECOC*, 1994, p. 581.

[189] J. E. Sharping, Y. Okawachi, J. van Howe et al., "All-optical, wavelength and bandwidth preserving, pulse delay based on parametric wavelength conversion and dispersion," *Opt. Express*, 13(20), 7872–7877, 2005.

[190] S. Radic, C. J. McKinstrie, R. M. Jopson et al., "Selective suppression of idler spectral broadening in two-pump parametric architectures," *IEEE Photon. Technol. Lett.*, 15(5), 673–675, 2003.

21

Fiber-optic quantum information technologies

Prem Kumar[*], Jun Chen[†], Paul L. Voss[†], Xiaoying Li[†], Kim Fook Lee[†], and Jay E. Sharping[‡]

[*]*Technological Institute, Northwestern University, Evanston, IL, USA*
[†]*Center for Photonic Communication and Computing, EECS Department, Northwestern University, Evanston, IL, USA*
[‡]*University of California, Merced, CA*

21.1 INTRODUCTION

Quantum mechanics (QM), born almost a century ago, is one of the most astonishing pieces of knowledge that human beings have ever discovered about Nature. Its rules are surprisingly simple: linear algebra and first-order partial differential equations, and yet its predictions are so unimaginably precise and unbelievably accurate when compared with experimental data. Such a successful theory, however, is not without its own imperfections (or mysteries). For example, the orthodox interpretation of QM—the Copenhagen interpretation—does not give a satisfactory explanation about how and why the wavefunction of a particle (e.g., an electron) would suddenly collapse once some measurement has been made on it (e.g., an electron has been registered by a particle detector). This so-called measurement problem has not been properly understood since the very early days of QM, until recently when the process of quantum measurement is thoroughly studied, and the concept of "decoherence" is proposed [1].

Optical Fiber Telecommunications V A: Components and Subsystems
Copyright © 2008, Elsevier Inc. All rights reserved.
ISBN: 978-0-12-374171-4

Another "mysterious" feature of QM, which we will explain in further detail, is the superposition principle and the ensuing quantum entanglement. The former allows a quantum mechanical system to be in any state that is spanned by the basis vectors of its Hilbert space. For example, if the system can be in two orthogonal states $|0\rangle$ and $|1\rangle$ (i.e., $\langle 0|1\rangle = 0$), it can also be in a linear combination of these two states—$\alpha|0\rangle + \beta|1\rangle$, where α and β are complex numbers satisfying $|\alpha|^2 + |\beta|^2 = 1$. This innocuous-looking principle is, in fact, the origin of a lot of "quantum weirdness" not observed in our classical everyday experiences. Entanglement is such a counterintuitive example. Consider the following bipartite state for particles A and B:

$$|\Psi\rangle = \frac{1}{\sqrt{2}}(|0\rangle_A|0\rangle_B + |1\rangle_A|1\rangle_B), \qquad (21.1)$$

which states that whenever particle A is detected to be in state $|0\rangle$, particle B must also be found in $|0\rangle$ (in its own subspace), and vice versa. The same rule applies if particle A is found in state $|1\rangle$. That is, we are assured to find particle B in its own $|1\rangle$ state with unit probability. This may not sound so surprising at first sight; after all, classical objects sometimes exhibit this kind of correlation too. For example, we can take a coin and split it into half. Then we put the two half-coins into two separate envelopes, which are sealed afterward. Suppose we do it in such a way that nobody, not even we, know exactly which half of the coin ends up in which envelope. It is obvious that if we open one of the envelopes and find that we get the "head" portion of the coin, we can infer with 100% certainty that the other envelope contains the "tail" portion of the coin, and vice versa. The two halves of the same coin, just like the two particles in Eqn (21.1), can be spatially separated. The deterministic correlation between the two parties in both cases remain the same, no matter how far they are from each other. An obvious question naturally arises: as the coin game is something we can play everyday, what is so shocking about the correlation that we find in the entanglement example?

The fundamental difference between quantum entanglement and classical correlation lies in the fact that particles are quantum-mechanical objects which can exist not only in states $|0\rangle$ and $|1\rangle$ but also in states described by $\alpha|0\rangle + \beta|1\rangle$ (allowed by the superposition principle!), while half-coins, being classical objects, can only live in one of the two deterministic states ("head" or "tail"), and not something in between. To put it more bluntly, in the case of coins, even though we do not know *a priori* which portion of the coin ends up in which envelope, we are confident that one of the envelopes must contain the "head" portion of the coin, the other envelope the "tail" portion of the coin. This "confident ignorance" about the results of classical correlation *cannot* be safely extended into the regime of quantum entanglement. In fact, before we decide which basis to use for measuring the states of the particles in Eqn (21.1), we do not know anything about what results we will eventually get. We cannot even say, before measurement, each particle is in some deterministic state, we just do not know which is which. It is not that we do not have the knowledge about the states of individual particles; they, in fact, do not come into being until we make the measurement. In other words, the individual particle in quantum entanglement does not have a well-defined pure state before measurement. Each of them are in a mixed state; the joint state of both particles constitutes the pure state in Eqn (21.1), which we call "entanglement."

It was Schrödinger who first realized the strangeness of entanglement, or "Verschrankung" as it was originally coined in German. He pondered the quantum-to-classical-transition problem at the same time, and extended the concept of entanglement to the ill-defined boundary between quantum and classical worlds, where the contrast is most extreme. He imagined a macroscopic object, e.g., a cat (which later becomes the notorious Schrödinger's cat), is somehow entangled with a microscopic object, e.g., an atom. The poor cat"s fate depends solely on the decaying property of the atom. If the atom decays, the cat dies; if the atom does not decay, the cat lives. As discussed before, the atom can live in a superposition state "decay–not decay," and as the two are entangled, the cat is then forced to be living in a state "dead–alive." This is very counterintuitive, since normally we do not observe a half-dead, half-alive cat in our daily lives.

This problem was brought into focus by Einstein, Podolsky, and Rosen (EPR) [2] in a famous paper in 1935, in which they pointed out the incompatibility between QM and local realism. The latter notion consists of two parts: locality is a very reasonable assumption that directly follows our everyday physical intuition, which postulates the nonexistence of "action-at-a-distance"; realism demands the existence of "elements of physical reality" in every physical system, which should take definite values prior to any conceivable measurements. In their example, EPR considered a quantum system composed of two particles such that neither one of them has well-defined position or momentum, but the sum of their positions (their center-of-mass) and the difference of their momenta (their individual momenta in the center-of-mass system) are precisely defined. It then follows that measurement of either particle's momentum (position) would immediately determine the measurement outcome for the other particle's momentum (position), without even interacting with that particle. Since the two particles can be separated by arbitrary distances, and properties like position and momentum of a particle are "elements of reality" according to EPR that must assume definite values before any measurement, EPR then suggest this "spooky action-at-a-distance" must imply that QM is at least incomplete, if not incorrect; and that there should be a deeper theory, possibly with some hidden degrees of freedom (later known as "hidden variables"), which can faithfully reproduce every result that QM has achieved, and hopefully retain our familiar deterministic classical world view—local realism. Niels Bohr [3] replied by arguing that the two particles in the EPR case are always parts of one quantum system, and thus measurement on one particle changes the possible predictions that can be made for the entire system and consequently on the other particle; QM is indeed complete and there is no need for a "more complete" theory.

While for a long time, the famous Einstein–Bohr debate has been widely regarded as merely philosophical, David Bohm [4] in 1951 introduced spin-entangled systems, as a discrete version of the original continuous EPR-entangled systems. In 1964, John Bell [5] pointed out that for such spin-entangled systems, classical hidden-variable theories would make different predictions from QM on measurements of correlated quantities. The theorem he published, later known as Bell's theorem, quantified just how strongly quantum particles were correlated than would be classically allowed. This effectively opened up the possibility of experimentally testing quantum mechanical predictions against those of classical hidden variable theory. By now a number of experiments have been performed, and the results are almost universally accepted to be fully in favor of QM [6–10]. However, from a strictly logical point of view the problem is not completely closed yet, because some loopholes in these existing experiments still make it at least logically possible to uphold a local realist world view [11, 12].

More recently, since the beginning of 1990s, the field of quantum information and quantum communication has opened up and expanded rapidly [13, 14]. Quantum entanglement, once the core concept and the sole mystery of the decade-long Einstein–Bohr debate, has begun to take on a new look. It is still an unresolved mystery, philosophically speaking, as it forces us to abandon either one of the two familiar notions that we hold dear since the beginning of modern-day science: *locality* and *realism*. But this sacrifice we have to make, much in the spirit of Niels Bohr's comment that we have no right to tell God what to do; we are only entitled to discover what God's plans are and accept them, or better yet, utilize them to our full advantage. After taking this more humble and more practical point of view, a whole new world of "quantum ideas" have been ignited and are actively being pursued. Quantum teleportation [15] and quantum cryptography [16] are two prominent examples. Quantum entanglement plays a central role in the former and can lead to many advantages in the latter.

21.2 FIBER NONLINEARITY AS A SOURCE FOR CORRELATED PHOTONS

Efficient generation and transmission of quantum-correlated photon pairs, especially in the 1550-nm fiber-optic communication band, is of paramount importance for practical realization of the quantum communication and cryptography protocols [17]. The workhorse source employed in all implementations thus far [18] has been based on the process of spontaneous parametric down-conversion (SPDC) in second-order ($\chi^{(2)}$) nonlinear crystals. Such a source, however, is not compatible with optical fibers as large coupling losses occur when the pairs are launched into the fiber. This severely degrades the correlated photon-pair rate coupled into the fiber, because the rate depends quadratically on the coupling efficiency. From a practical standpoint it would be advantageous if a photon-pair source could be developed that not only produces photons in the communication band but also can be spliced to standard telecommunication fibers with high efficiency. Over the past few years various attempts have been made to

21. Fiber-Optic Quantum Information Technologies

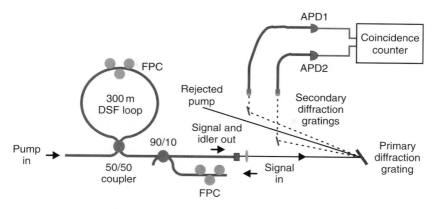

Figure 21.1 Diagram of the experimental setup; FPC, fiber polarization controller (this figure may be seen in color on the included CD-ROM).

develop more efficient photon-pair sources, but all have relied on the $\chi^{(2)}$ down-conversion process [19–25]. Of particular note is Ref. [26], in which the effective $\chi^{(2)}$ of periodically poled silica fibers was used. In this chapter, we report the first, to the best of our knowledge, photon-pair source that is based on the Kerr nonlinearity ($\chi^{(3)}$) of standard fiber. Quantum-correlated photon pairs are observed and characterized in the parametric fluorescence of four-wave mixing (FWM) in dispersion-shifted fiber (DSF).

The FWM process takes place in a nonlinear-fiber Sagnac interferometer (NFSI), shown schematically in Figure 21.1. Previously, we have used this NFSI to generate quantum-correlated twin beams in the fiber [27]. The NFSI consists of a fused-silica 50/50 fiber coupler spliced to 300 m of DSF having zero-dispersion wavelength $\lambda_0 = 1537$ nm. It can be set as a reflector with proper adjustment of the intraloop fiber polarization controller (FPC) to yield a transmission coefficient <-30 dB. When the injected pump wavelength is slightly greater than λ_0, FWM in the DSF is phase-matched [19]. Two-pump photons of frequency ω_p scatter into a signal photon and an idler photon of frequencies ω_s and ω_i, respectively, where $\omega_s + \omega_i = 2\omega_p$. Signal/Idler separations of $\simeq 20$ nm can be easily obtained with use of commercial DSF [27]. The pump is a mode-locked train of $\simeq 3$ ps long pulses that arrive at a 75.3-MHz repetition rate. The pulsed operation serves two important purposes: (i) the NFSI amplifier can be operated at low-average powers (typical values are ≤ 2 mW, corresponding to ≤ 9-W peak powers) and (ii) the production of the fluorescence photons is confined in well-defined temporal windows, allowing a gated detection scheme to be used to increase the signal-to-noise ratio. A 10% (90/10) coupler is employed to inject a weak signal, which is parametrically amplified, and the output signal and the generated idler are used for alignment purposes. For the photon-counting measurements described in this chapter, the input signal is blocked.

After passing through the 90/10 coupler, the fluorescence photons are directed toward free-space filters that separate the signal and the idler photons from each other and from the pump photons. To measure the nonclassical (i.e., quantum) correlations between the signal and the idler photons, one must effectively suppress the pump photons from reaching the detectors. Because a typical pump pulse contains $\simeq 10^8$ photons and we are interested in detecting $\simeq 0.01$ photons/pulse, a pump-to-signal (idler) rejection ratio in excess of 100 dB is required. To meet this specification, we constructed a dual-band spectral filter based on a double-grating spectrometer. A primary grating (holographic, 1200 lines/mm) is first employed to spatially separate the signal, the pump, and the idler photons. Two secondary gratings (ruled, 600 lines/mm) are then used to prevent the pump photons that are randomly scattered by the primary grating (owing to its nonideal nature) from going toward the signal and idler directions. The doubly diffracted signal and idler photons are then recoupled into fibers, which function as the output slits of the spectrometer.

Transmission spectrum of the dual-band filter, measured with a tunable source and an optical spectrum analyzer (OSA), is shown in Figure 21.2. The shape is Gaussian in the regions near the maxima of the two transmission bands, which are centered at 1546 nm (signal) and 1528 nm (idler), respectively, and the full-width at half-maximum (FWHM) is $\simeq 0.46$ nm. For pulse trains separated by 9 nm, which is the wavelength difference between the pump and the signal (or idler), this filter is able to provide an isolation ≥ 75 dB; the measurement being limited by the intrinsic noise of the OSA. The combined effect of the Sagnac loop and the double-grating filter thus provides an isolation ≥ 105 dB from the pump photons in the signal and idler channels. The maximum transmission efficiency in the

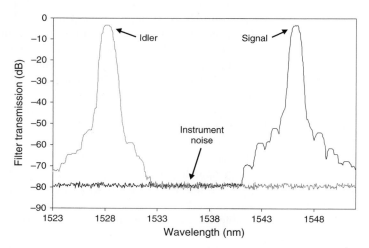

Figure 21.2 Transmission curves of the signal and idler channels in the dual-band filter (this figure may be seen in color on the included CD-ROM).

signal channel is 45% and that in the idler channel is 47%. The total collection efficiency for the signal (idler) photons is thus 33% (35%), with inclusion of the losses in the Sagnac loop (18%) and at the 90/10 coupler (10%).

The separated and filtered signal and idler photons are directed toward fiber-pigtailed InGaAS/InP avalanche photodiodes (APDs, Epitaxx EPM239BA). In recent years, the performance of InGaAs APDs as single-photon detectors for use in the fiber communication window around 1550 nm has been extensively studied by several groups [28–31]. The pulsed nature of the photon pairs allows us to use the APDs in a gated Geiger mode. In addition, the quality of our APDs permits room-temperature operation with results comparable to those obtained by other groups at cryogenic temperatures. A schematic of the electronic circuit used with the APDs is shown in the inset in Figure 21.3. A bias voltage V_B ($\simeq -60$ V), slightly below the avalanche breakdown voltage, is applied to each diode and a short-gate pulse (-8 V, 1 ns FWHM) brings the diodes into the breakdown region. The gate pulse is synchronized with the arrival of the signal and idler photons on the photodiodes. Due to limitations of our gate-pulse generator, the detectors are gated once every 128 pump pulses, giving a photon-pair detection rate of 75.3 MHz/128 = 588 kHz. We expect this rate to increase by more than an order of magnitude with use of a better pulse generator. The electrical signals produced by the APDs in response to the incoming photons (and dark events) are reshaped into 500-ns-wide transistor-transistor logic (TTL) pulses that can be individually counted or sent to a TTL AND gate for coincidence counting.

In Figure 21.3 we show a plot of the quantum efficiency vs the dark-count probability for the two APDs used in our experiments. A figure of merit for the

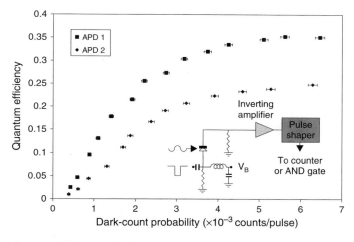

Figure 21.3 Quantum efficiency vs dark-count probability for the two APDs used in the experiments. The inset shows a schematic of the electronic circuit used with the APDs.

APDs can be introduced through the noise-equivalent power NEP $= (h\nu/\eta)(2R_D)^{1/2}$ [20], where h is the Planck constant, ν is the frequency of light, η is the detector quantum efficiency, and R_D is the dark-count rate measured during the gate time. The best values of NEP obtained by optimizing V_B are 1.0×10^{-15} W/Hz$^{-1/2}$ for APD1 and 1.6×10^{-15} W/Hz$^{-1/2}$ for APD2. These values are comparable to those reported in Refs [28–31] for cryogenically cooled APDs. Under the optimized conditions, the efficiency of APD1 (APD2) is 25% (20%) and the corresponding dark-count probability is 2.2×10^{-3}/pulse (2.7×10^{-3}/pulse).

As a first test of our photon-pair source, and of the filtering process, we measure the number of scattered photons detected in the signal (idler) channel, N_S (N_I), as a function of the number of pump photons, N_P, injected into the NFSI. The results for the idler channel are shown in the inset in Figure 21.4. We fit the experimental data with $N_S = N_D + s_1 N_P + s_2 N_P^2$, where N_D is the number of dark counts during the gate interval, and s_1 and s_2 are the linear and quadratic scattering coefficients, respectively. The fit clearly shows that the quadratic scattering owing to FWM in the fiber can dominate over the residual linear scattering of the pump due to imperfect filtering.

In Figure 21.4 we present the coincidence counting results. The diamonds represent the rate of coincidence counts as a function of the rate of the signal and idler photons generated during the same pump pulse. For convenience, we

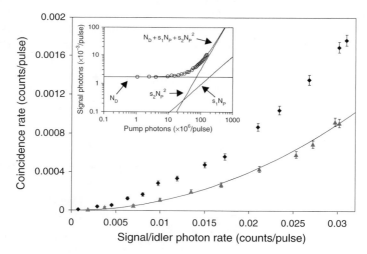

Figure 21.4 Coincidence rates as a function in the single-photon rates in two different cases: signal/idler fluorescence produced by a pump pulse (diamonds) and signal/idler fluorescence produced by two consecutive pump pulses (triangles). The line represents the calculated "accidental" counts. The inset shows a plot of the detected idler photons as a function of the injected pump photons (hollow circles). A second-order polynomial is shown to fit the experimental data. The contributions of the dark counts, linear scattering, and quadratic scattering are plotted separately as well (this figure may be seen in color on the included CD-ROM).

have plotted the coincidence rate as a function of the geometric mean of the signal and idler count rates; in fact, because the efficiency of the two detectors is different, we measure different single-photon count rates in the two channels. Dark counts have been subtracted from the plotted count rates. For the coincidence rates, both dark–dark and photon–dark coincidences have to be taken in account, but for the counting rates in our experiment the former are negligible. Thus far, we have achieved a maximum coincidence rate of 10^3 counts/s (= coincidence rate/pulse × gate-pulse rate), which is expected to go up by at least a factor of 10 with use of a higher repetition-rate gate-pulse generator.

We have performed two independent experiments to demonstrate the nonclassical nature of the coincidences. Results of the first experiment are shown by the triangles in Figure 21.4, which represent the measured coincidence rate as a function of the signal-photon count rate when the signal is delayed with respect to the idler by one pulse period. The delay was achieved by inserting a fiber patch-cord of appropriate length in the signal path from the output of the filter to APD1. For two independent photon sources, each with a count rate $R_S \ll 1$, the "accidental" coincidence rate R_C is given by $R_C = R_S^2$, regardless of the photon statistics of the sources. This quadratic relation is plotted as the solid line in Figure 21.4, which fits the delayed-coincidence data (triangles) very well. These measurements then show that while the fluorescence photons produced by the adjacent pump pulses are independent, those coming from the same pump pulse show a strong correlation, which is a signature of their nonclassical behavior.

In the second experiment, measurements were performed to demonstrate the nonclassicality test described in Ref. [24]. It can be shown that the inequality

$$R_C - R_C^{(a)} - 2\left(R_{S/2} - R_{S/2}^{(a)} + R_{I/2} - R_{I/2}^{(a)}\right) \leq 0 \quad (21.2)$$

is valid for two classical light sources, where R_C is the coincidence-count rate for the two sources, $R_C^{(a)}$ is the calculated "accidental" coincidence-count rate corresponding to the same photon-count rate for the two sources, $R_{S/2}$ and $R_{I/2}$ are the coincidence-count rates measured by passing the light from each of the two sources through a 50/50 splitter and detecting the two halves independently, and $R_{S/2}^{(a)}$ and $R_{I/2}^{(a)}$ are the calculated "accidental" coincidence-count rates in the 50/50 splitting measurements. When we substitute the experimental data, Eqn (21.2) yields $(64 \pm 9) \, 10^{-6} \leq 0$, where the error is statistical. The inequality for classical sources is thus violated by over seven standard deviations.

In conclusion, we have demonstrated, for the first time to our knowledge, a source of quantum-correlated photon pairs that is based on FWM in a fiber near 1550 nm. We have also developed and tested a room-temperature coincidence detector for the photons in that window. The photon-pair detection rate ($\simeq 10^3$ coincidence counts/s) at present is limited by the electronics employed in our setup. In addition, we believe that the spectral filter used for rejecting the pump photons can be implemented with fiber Bragg gratings, making this source integrable with the existing fiber-optic infrastructure.

21.3 QUANTUM THEORY OF FOUR-WAVE MIXING IN OPTICAL FIBER

Four-wave mixing has long been studied, especially in the context of isotropic materials, e.g., optical fibers [25, 32]. Generally speaking, it is a photon–photon scattering process, during which two photons from a relatively high-intensity beam, called pump, scatter through the third-order nonlinearity ($\chi^{(3)}$) of the material (silica glass in the case of optical fibers) to generate two daughter photons, called signal and idler photons, respectively. The frequencies of the daughter photons are symmetrically displaced from the pump frequency, satisfying the energy conservation relation $\omega_s + \omega_i = 2\omega_p$, where ω_j (j = p, s, i) denotes the pump/signal/idler frequency, respectively. They are predominantly copolarized with the pump beam, owing to the isotropic nature of the optical Kerr nonlinearity: $\chi^{(3)}_{xxxx} = \chi^{(3)}_{xxyy} + \chi^{(3)}_{xyxy} + \chi^{(3)}_{xyyx} = 3\chi^{(3)}_{xxyy}$. The daughter photons also form a time–energy-entangled state, in the sense that the two-particle wavefunction cannot be factorized into products of single-particle wavefunctions: $\Psi(\omega_s, \omega_i) \neq \zeta(\omega_s) \cdot \varphi(\omega_i)$. This four-photon scattering (FPS) process is intrinsically interesting and particularly useful when applied to the field of quantum information processing (QIP), in which generation of entangled states and test of Bell's inequalities play an important role.

A great amount of original work, both theoretical and experimental, has been done in the rapidly expanding field of QIP (see, e.g., Ref. [33] for a general review). The workhorse process for generating entangled states is the process of SPDC in second-order ($\chi^{(2)}$) nonlinear crystals, which has been studied exhaustively during the past decades. However, unlike its $\chi^{(2)}$ counterpart, the $\chi^{(3)}$ process of FWM has received relatively lesser theoretical attention in the quantum mechanical framework, despite its apparent benefits in the applications of QIP. To name a few, the ubiquitous readily available fiber plant serves as a perfect transmission channel for the FWM-generated entangled qubits, whereas it remains a technical challenge to efficiently couple $\chi^{(2)}$-generated entangled photons into optical fibers due to mode mismatch. Besides, the excellent single-mode purity of the former makes it suitable for applications that require multiple quantum interactions. Furthermore, it is also possible to wavelength multiplex several different entangled channels from the broadband parametric spectrum of FWM by utilizing the advanced multiplexing/demultimplexing devices developed in connection with the modern fiber-optic communications infrastructure. The only drawback of this scheme that has been identified is the process of spontaneous Raman scattering (SRS), which inevitably occurs in any $\chi^{(3)}$ medium and generates uncorrelated photons into the detection bands, leading to a degradation in the quality of the generated entanglement [28]. Various efforts have been made to minimize the negative effect that SRS imposes [29, 30].

In this section, we present a quantum theory that models the FWM process in an optical fiber, without inclusion of the Raman effect. The pump is treated as a classical narrow (picosecond-duration) pulse due to its experimental relevance. The signal and idler fields form a quantum mechanical two-photon (or "biphoton" [31, 34]) state at

21. Fiber-Optic Quantum Information Technologies

the output of the fiber. From the experimental point of view, what we are mostly interested in is the nonclassicality that the two-photon state exhibits. It is this unique quantum feature that makes the two-photon state a valid candidate for various quantum-entanglement-related experiments, including quantum cryptography [35], quantum teleportation [15], etc. Coincidence-photon counting, or second-order coherence measurement of the optical field [36], serves as a measurement technique that distinguishes a quantum mechanically entangled state from a classically correlated state, which will form a central part of our investigation.

A sample coincidence-counting result from Ref. [37] is shown in Figure 21.5. The top (bottom) series of data points represents the total (accidental) coincidence-count rate as a function of the single-channel count rate. SRS and dark counts from the detectors account for the major part of the accidental coincidence counts. Our to-be-developed theory, however, only takes into account the photon counts generated by the FWM process. To reconcile the theory with experiments, the contributions from SRS and dark counts from the detectors are independently measured [29, 30], and subsequently subtracted from both the single counts and the total coincidence counts. Overall quantum efficiencies of detection in both the signal and idler channels are also separately measured. The single-count rates are divided by the respective quantum efficiencies and the coincidence-count rate by the product of the efficiencies in the signal and idler channels to arrive at rates at the output of the fiber for comparison with the prediction of our theory. The dependence of the photon-counting results on various system parameters, for instance, the pump power, pump bandwidth, filter bandwidth, etc., can be studied.

Polarization entanglement has also been generated by time and polarization multiplexing two such FWM processes [38, 39]. However, the theory for that

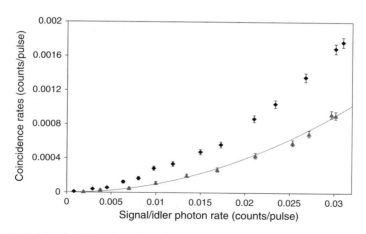

Figure 21.5 (Color online) Experimental results. Diamonds, total coincidences; triangles, accidental coincidences; curve, theoretical fit $y = x^2$ for statistically independent photon sources (this figure may be seen in color on the included CD-ROM).

particular experiment is a straightforward extension of our current theory, and therefore will not be included in the analysis to follow.

Having described the experiment in the previous section, we are ready to start building up the theoretical model for that experiment. We take the standard approach of modern quantum optics, i.e., finding out the interaction Hamiltonian and calculating the evolution of the state vector using the Schrödinger picture. We accomplish the first task by seeking connections with the well-known classical FWM theory in optical fibers [32]. The coupled classical-wave equations for the pump, signal, and idler fields are

$$\frac{\partial A_p}{\partial z} = i\gamma |A_p|^2 A_p,$$
$$\frac{\partial A_s}{\partial z} = i\gamma \left[2|A_p|^2 A_s + A_p^2 A_i^* e^{-i\Delta k z} \right], \quad (21.3)$$
$$\frac{\partial A_i}{\partial z} = i\gamma \left[2|A_p|^2 A_i + A_p^2 A_s^* e^{-i\Delta k z} \right],$$

where the usual undepleted-pump approximation has been made, and we only keep terms that are significant, i.e., to $O(A_p^2)$. Fiber loss is neglected from the above equations. The A_j (j = p, s, i) denote electric-field amplitudes for the pump, signal, and idler, respectively, and all of them have been normalized such that their unit is $\sqrt{W} \Delta k = k_s + k_i - 2k_p$ is the magnitude of the wave-vector mismatch. $\gamma = 2\pi n_2 / \lambda A_{\text{eff}}$ is the nonlinear parameter of interaction, wherein $n_2 = (3/4n^2 \epsilon_0 c) Re(\chi_{xxxx}^{(3)})$ is the nonlinear-index coefficient, ϵ_0 is the vacuum permittivity, A_{eff} is the effective mode area of the optical fiber, and $\lambda \approx \lambda_{p,s,i}$ is the wavelength involved in the FWM interaction.

Due to the highly nonresonant nature of FWM in optical fibers, we expect the quantum equations of motion, which describe the interplay between and evolution of the fields at the photon level, to fully correspond with their classical counterparts. In light of this correspondence principle, we write the quantum equations of motion by replacing the classical amplitudes in Eqn (21.3) with electric-field operators:

$$\frac{\partial E_p^{(+)}}{\partial z} = i\eta \, E_p^{(-)} E_p^{(+)} E_p^{(+)},$$
$$\frac{\partial E_s^{(+)}}{\partial z} = i\eta \, [2 E_p^{(-)} E_p^{(+)} E_s^{(+)} + E_i^{(-)} E_p^{(+)} E_p^{(+)}], \quad (21.4)$$
$$\frac{\partial E_i^{(+)}}{\partial z} = i\eta \, [2 E_p^{(-)} E_p^{(+)} E_i^{(+)} + E_s^{(-)} E_p^{(+)} E_p^{(+)}],$$

where $E_j^{(+)} = \sqrt{(\hbar \omega_j / 2\epsilon_0 V_Q)} \, a_j$ (j = p, s, i) are the positive-frequency electric-field operators, corresponding to photon annihilation operators, and V_Q is the quantization volume. Here we omit the Hermitian-conjugate equations corresponding to Eqn (21.4) for simplicity. We have assumed that the photon fields phase match, i.e., $\Delta k = 0$. In Eqn (21.4), $\eta = -(\chi^{(3)} A_{\text{eff}} n L\omega / 2 V_Q c)$ is a constant similar to γ in the classical Eqn (21.3); the exact form of this constant differs from its classical cousin to compensate for the unit discrepancy between the two sets of equations (note that the operator $E_j^{(+)}$ is of unit V/m, and the amplitude A_j is of unit $\sqrt{\text{W}}$). The correct form of the interaction Hamiltonian that we are seeking should lead to Eqn (21.4) through the Heisenberg equation of motion for the field operators, namely, $i\hbar(\partial \hat{E}/\partial t) = [\hat{E}, H_I]$, where $\hat{E}$ stands for any electric-field operator. Utilizing the mathematical facts $\partial/\partial t \equiv (c/n)(\partial/\partial z)$ and $[E_j^{(+)}(z), E_k^{(-)}(z')] = (\hbar\omega / 2\epsilon_0 V_Q)\, \delta(z - z')\, \delta_{jk}$, we arrive at the following form for our interaction Hamiltonian:

$$H_I = \beta\, \epsilon_0\, \chi^{(3)} \int_V dV \Big[E_p^{(-)} E_p^{(-)} E_p^{(+)} E_p^{(+)} + 2 E_s^{(-)} E_i^{(-)} E_p^{(+)} E_p^{(+)}$$
$$+ 2 E_p^{(-)} E_p^{(-)} E_s^{(+)} E_i^{(+)} + 4 E_p^{(-)} E_p^{(+)} E_s^{(-)} E_s^{(+)} + 4 E_p^{(-)} E_p^{(+)} E_i^{(-)} E_i^{(+)} \Big], \quad (21.5)$$

where β is an overall unknown constant related to the specific experimental details, which will be determined later when we compare our theory with the experiment; $\chi^{(3)}$ is the nonlinear electric susceptibility whose tensorial nature is ignored as all the optical fields are assumed to be linearly copolarized. The integral is taken over the entire volume of interaction, namely, the effective volume of the optical fiber. We label the first term in the integrand of Eqn (21.5) as the self-phase modulation (SPM) of the pump field, the next two terms as the FPS among the optical fields, and the last two terms as the cross-phase modulation (XPM) between pump and signal (idler) fields.

After obtaining the Hamiltonian responsible for the quantum FWM process, we are ready to tackle our next task: calculate the state vector evolution. It is worthwhile, at this point, to define the various electric field operators appearing in the Hamiltonian, in accordance with the experiment we are trying to model. The pump field is taken to be a classical narrow pulse, which is linearly polarized, propagating in the z direction (parallel with the fiber axis), with a central frequency Ω_p and an envelope of arbitrary shape $\widetilde{E}_p$. Mathematically, it can be written as

$$E_p^{(+)} = e^{-i\Omega_p t} \widetilde{E}_p(z, t)$$
$$= e^{-i\Omega_p t} \int d\nu_p\, \overline{E}_p(\nu_p)\, e^{ik_p z - i\nu_p t}, \quad (21.6)$$

wherein the bandwidth of the pump field is much smaller than Ω_p, satisfying the quasi-monochromatic approximation. The signal and idler fields are quantized electromagnetic fields, copolarized and copropagating with the pump, as given by the following multimode expansion:

$$E_{\rm s}^{(-)} = \sum_{\omega_{\rm s}} \sqrt{\frac{\hbar\omega_{\rm s}}{2\epsilon_0 V_{\rm Q}}} \frac{a_{k_{\rm s}}^\dagger}{n(\omega_{\rm s})} {\rm e}^{-{\rm i}[k_{\rm s}(\omega_{\rm s})z-\omega_{\rm s}t]}, \qquad (21.7)$$

$$E_{\rm i}^{(-)} = \sum_{\omega_{\rm i}} \sqrt{\frac{\hbar\omega_{\rm i}}{2\epsilon_0 V_{\rm Q}}} \frac{a_{k_{\rm i}}^\dagger}{n(\omega_{\rm i})} {\rm e}^{-{\rm i}[k_{\rm i}(\omega_{\rm i})z-\omega_{\rm i}t]}, \qquad (21.8)$$

where $a_{k_{\rm s}}^\dagger$ is the creation operator for the signal mode with frequency $\omega_{\rm s}$, $k_{\rm s}(\omega_{\rm s}) = n(\omega_{\rm s})\omega_{\rm s}/c$ is its wave-vector magnitude. The idler field is defined in an analogous fashion. The central frequencies of the signal and idler fields are individually denoted by $\Omega_{\rm s}$ and $\Omega_{\rm i}$, which are symmetrically distanced from the central frequency of the pump field $\Omega_{\rm p}$, satisfying the energy conservation relation $\Omega_{\rm s} + \Omega_{\rm i} = 2\Omega_{\rm p}$.

To simplify our calculation and to compare our results with the experiments, two assumptions are further made about the pump field: it has a Gaussian spectral envelope and its SPM is included in a straightforward manner, i.e.,

$$E_{\rm p}^{(+)} = {\rm e}^{-{\rm i}\Omega_{\rm p}t}{\rm e}^{-{\rm i}\gamma P_{\rm p}z} E_{\rm p0} \int {\rm d}\nu_{\rm p}\, {\rm e}^{-(\nu_{\rm p}^2/2\sigma_{\rm p}^2)}\, {\rm e}^{{\rm i}k_{\rm p}z-{\rm i}\nu_{\rm p}t}, \qquad (21.9)$$

where $P_{\rm p} \equiv 2\sqrt{\pi}A_{\rm eff}\epsilon_0\, c\, n\, \sigma_{\rm p}^2 E_{\rm p0}^2$ is the peak power of the pump pulse, which is treated as a constant under the undepleted pump approximation, and $\sigma_{\rm p}$ is the optical bandwidth of the pump. The first assumption is justified by the fact that our experimental optical filter for the pump can be well approximated by a Gaussian function in the frequency domain. The validity of the second assumption can be seen when we solve the classical equation of motion for the pump field, namely, the complex conjugate form of the first equation in Eqn (21.3), which reads

$$\frac{\partial A_{\rm p}^*}{\partial z} = -{\rm i}\gamma |A_{\rm p}|^2 A_{\rm p}^*. \qquad (21.10)$$

We choose to study the complex conjugate form of the equation because it is $A_{\rm p}^*$ that corresponds to $E_{\rm p}^{(+)}$. Straightforward calculations show that the solution to Eqn (21.10) is $A_{\rm p}^*(z) = A_{\rm p}^*(0)\,{\rm e}^{-{\rm i}\gamma P_{\rm p}z}$, where $P_{\rm p} = |A_{\rm p}|^2$ is the same undepleted peak power of the pump pulse. The SPM term of the pump, ${\rm e}^{-{\rm i}\gamma P_{\rm p}z}$, which is the nonlinear phase factor in the classical FWM theory that determines the phase-matching condition [32], now manifests itself in our quantum mechanical calculation as a "phase tag" for the pump field through its propagation along the optical fiber. Finally, the undepleted-pump approximation holds because the loss in the fiber is negligible and only a few photons are scattered ($\sim$1 out of 10^8) through the nonlinear interaction.

The two-photon state at the output of the fiber is calculated by means of first-order perturbation theory, i.e.,

$$|\Psi\rangle = |0\rangle + \frac{1}{i\hbar} \int_{-\infty}^{\infty} H_{\rm I}(t)\,dt|0\rangle. \tag{21.11}$$

Retaining of higher order terms in the perturbation series involves generation of multiphoton states, which will be ignored in our calculation owing to their smallness. We can see that only the FPS terms in the interaction Hamiltonian contribute to the formation of the signal/idler two-photon state. This is because all terms vanish when acting on the vacuum state $|0\rangle$ with the exception of $E_{\rm s}^{(-)} E_{\rm i}^{(-)} E_{\rm p}^{(+)} E_{\rm p}^{(+)}$ + h.c., which we denote as

$$H_{\rm FPS} \equiv \alpha \epsilon_0 \chi^{(3)} \int_V dV (E_{\rm s}^{(-)} E_{\rm i}^{(-)} E_{\rm p}^{(+)} E_{\rm p}^{(+)} + \text{h.c.}), \tag{21.12}$$

where $\alpha = 2\beta$, and h.c. stands for Hermitian conjugate.

The state vector is then given by

$$|\Psi\rangle = |0\rangle + \frac{1}{i\hbar} \int_{-\infty}^{\infty} H_{\rm FPS}\,dt|0\rangle, \tag{21.13}$$

which is a superposition of the vacuum and the two-photon state. Substituting Eqns (21.7–21.9) and (21.12), into Eqn (21.13), after some algebra, leads to the following form of the state vector:

$$|\Psi\rangle = |0\rangle + \sum_{k_{\rm s}, k_{\rm i}} F(k_{\rm s}, k_{\rm i})\, a_{k_{\rm s}}^{\dagger} a_{k_{\rm i}}^{\dagger} |0\rangle, \tag{21.14}$$

$$F(k_{\rm s}, k_{\rm i}) = g \int_{-L}^{0} dz \frac{1}{\sqrt{1 - ik''(\Omega_{\rm p})\sigma_{\rm p}^2 z}} \exp\left\{-\frac{ik''(\Omega_{\rm p})z}{4}(\nu_{\rm s} - \nu_{\rm i} + \Delta)^2 \right.$$

$$\left. -2i\gamma P_{\rm p} z - \frac{(\nu_{\rm s} + \nu_{\rm i})^2}{4\sigma_{\rm p}^2}\right\}, \tag{21.15}$$

$$g = \frac{\alpha \pi^2 \chi^{(3)}}{i\epsilon_0 V_Q n^3 \lambda_{\rm p} \sigma_{\rm p}} P_{\rm p}. \tag{21.16}$$

The function $F(k_{\rm s}, k_{\rm i})$ is called the two-photon spectral function [34]. Here $k''(\Omega_{\rm p}) = d^2 k/d\omega^2|_{\omega=\Omega_{\rm p}}$ is the second-order dispersion at the pump central frequency [also known as the group-velocity dispersion (GVD)], which can be obtained from $k''(\Omega_{\rm p}) = -(\lambda_{\rm p}^2/2\pi c)D_{\rm slope}(\lambda_{\rm p} - \lambda_0)$, where λ_0 is the zero-dispersion wavelength of the DSF, $D_{\rm slope} = 0.06\,\text{ps/nm}^2\,\text{km}$ is the experimental value of the dispersion slope in the vicinity of λ_0. $\Delta \equiv \Omega_{\rm s} - \Omega_{\rm i}$ is the central-frequency difference between signal and idler fields. ν_s and ν_i are related to ω_s and ω_i through the following relation: $\nu_s = \omega_s - \Omega_s$, $\nu_i = \omega_i - \Omega_i$.

In lieu of giving the detailed derivation of the two-photon state (which is lengthy), we highlight several noteworthy mathematical maneuvers along the way. The following identification of the Dirac δ-function is useful in handling the time integral:

$$\int_{-\infty}^{\infty} e^{i(\omega+\omega'-2\Omega_p-\nu_p-\nu'_p)t} dt = 2\pi\delta(\omega + \omega' - 2\Omega_p - \nu_p - \nu'_p), \quad (21.17)$$

which reinforces the energy conservation requirement in the FPS process. The volume integral $\int dV$ is reduced to a length integral $\int dz$ by using $\int\int dx\,dy \to A_{\text{eff}}$, which is a valid approximation for single spatial-mode propagation and interaction in optical fibers. Taylor expansion of the various wave-vector magnitudes k_p, k_s, k_i around the pump central frequency Ω_p has been used to simplify their relationship. In terms of the mathematical structure of the two-photon spectral function, we note that the GVD term $k''(\Omega_p)$ as well as the pump SPM term $\gamma P_p z$ play important roles in shaping the two-photon state, in contrast with the observation that the pump SPM term is virtually nonexistent in the $\chi^{(2)}$-generated two-photon states. The appearance of the pump SPM is therefore a unique signature of the $\chi^{(3)}$ two-photon state, when comparing with its $\chi^{(2)}$ counterparts.

In this and the next section, we will make use of the previously derived formulas for the two-photon state [Eqns (21.14)–(21.16)] to obtain the photon-counting formulas for the single channels as well as for the coincidences. The mathematics involved for the two cases are similar to each other, so it suffices to present a detailed version for the former. The signal-band single-photon counting rate can be calculated using the following formula [36]:

$$S_c = \int_0^\infty \langle \Psi | E_s^{(-)} E_s^{(+)} | \Psi \rangle \, dT. \quad (21.18)$$

It is obvious that an analogous approach can be applied to the idler band as well.

As S_c denotes single-photon counting probability for one pump pulse, it is by definition a dimensionless quantity. It is customary, in this case, to use the photon-number unit for the electric field operator [40]. In this unit, the electric field operator has dimensionality $1/\sqrt{\text{sec}}$, as shown below:

$$E_s^{(+)} = \sum_{k_s} \sqrt{\frac{c A_{\text{eff}}}{4 V_Q}} a_{k_s} e^{-i\omega_s t} e^{-[(\omega_s-\Omega_s)^2/2\sigma_0^2]}, \quad (21.19)$$

where the Gaussian filter in front of the detector has been included. The integrand in Eqn (21.18) can be written as

$$\langle\Psi|E_s^{(-)}E_s^{(+)}|\Psi\rangle = \frac{cA_{\text{eff}}}{4V_Q}\sum_{k_i,k_i'}\langle 0|a_{k_i}a_{k_i'}^\dagger|0\rangle \sum_{k_1,k_2,k_s,k_s'}\langle 0|a_{k_s}a_{k_1}^\dagger a_{k_2}a_{k_s'}^\dagger|0\rangle e^{i\omega_s t} \qquad (21.20)$$
$$\times e^{-[(\omega_s-\Omega_s)^2/2\sigma_0^2]}e^{-i\omega_s' t}e^{-[(\omega_s'-\Omega_s)^2/2\sigma_0^2]} F^*(k_s,k_i)F(k_s',k_i').$$

Nonvanishing results emerge only when the wave vectors observe the following restrictions:

$$k_i = k_i', \quad k_s = k_1, \quad k_s' = k_2. \qquad (21.21)$$

The integrand may be further simplified into

$$\langle\Psi|E_s^{(-)}E_s^{(+)}|\Psi\rangle = \frac{cA_{\text{eff}}}{4V_Q}\sum_{k_i}\left[\sum_{k_s}e^{i\omega_s t}e^{-[(\omega_s-\Omega_s)^2/2\sigma_0^2]}F^*(k_s,k_i)\right]$$
$$\cdot\left[\sum_{k_s'}e^{-i\omega_s' t}e^{-[(\omega_s'-\Omega_s)^2/2\sigma_0^2]}F(k_s',k_i)\right]$$
$$= \frac{cA_{\text{eff}}}{32\pi^3 u(\omega_i)u^2(\omega_s)}\int d\omega_i\left|\int d\omega_s\, e^{-i\omega_s t}e^{-[(\omega_s-\Omega_s)^2/2\sigma_0^2]}F(\omega_s,\omega_i)\right|^2, \qquad (21.22)$$

where in the last step we have invoked the following identity to transform wave-vector summations into angular frequency integrals:

$$\sum_{k_j} \to \frac{V_Q^{1/3}}{2\pi}\int dk_j = \frac{V_Q^{1/3}}{2\pi}\int \frac{d\omega_j}{u(\omega_j)}. \qquad (21.23)$$

Here $u(\omega_j) = d\omega_j/dk_j$, $j =$ s, i, is the group velocity of the jth mode, and is to be taken as a constant c/n in our simplified calculation.

Equation (21.18) can be written in the following form after all the above steps have been absorbed:

$$S_c = \frac{\pi^2\alpha^2[\chi^{(3)}]^2 A_{\text{eff}} P_p^2}{16\epsilon_0^2 V_Q^2 n^3 c^2 \lambda_p^2 \sigma_p^2}\int_{-L}^0 dz_1 \int_{-L}^0 dz_2 \frac{e^{-2i\gamma P_p(z_1-z_2)}}{\sqrt{(1-ik''\sigma_p^2 z_1)(1+ik''\sigma_p^2 z_2)}}\int d\nu_s\int d\nu_i$$
$$\times \exp\left\{-\frac{(\nu_s+\nu_i)^2}{2\sigma_p^2}-\frac{\nu_s^2}{\sigma_0^2}-i\frac{k''}{4}(z_1-z_2)(\nu_s-\nu_i+\Delta)^2\right\}. \qquad (21.24)$$

The frequency double integral can be analytically integrated through a change of variables and completion of squares, namely, let

$$\nu_+ = \frac{\nu_s + \nu_i}{2}, \quad (21.25)$$

$$\nu_- = \nu_s - \nu_i. \quad (21.26)$$

The frequency double integral can be rewritten in terms of the new variables as

$$\int d\nu_s \int d\nu_i \, e^{-[(\nu_s+\nu_i)^2/2\sigma_p^2] - \frac{\nu_s^2}{\sigma_0^2} - i\frac{k''}{4}(z_1-z_2)(\nu_s-\nu_i+\Delta)^2}$$

$$= \int d\nu_+ \exp\left\{-\frac{2\sigma_0^2 + \sigma_p^2}{\sigma_0^2 \sigma_p^2}\left[\nu_+ + \frac{\sigma_p^2 \nu_-}{2(2\sigma_0^2 + \sigma_p^2)}\right]^2\right\}$$

$$\int d\nu_- \exp\left\{-\frac{\nu_-^2}{2(2\sigma_0^2 + \sigma_p^2)} - \frac{ik''(z_1-z_2)(\nu_-+\Delta)^2}{4}\right\}. \quad (21.27)$$

The first part of the integral, concerning only Gaussian functions with real variable as arguments, is easily integrated as

$$\int d\nu_+ \exp\left\{-\frac{2\sigma_0^2 + \sigma_p^2}{\sigma_0^2 \sigma_p^2}\left[\nu_+ + \frac{\sigma_p^2 \nu_-}{2(2\sigma_0^2 + \sigma_p^2)}\right]^2\right\} = \frac{\sqrt{\pi}\sigma_p \sigma_0}{\sqrt{2\sigma_0^2 + \sigma_p^2}}. \quad (21.28)$$

The second part of the integral, having a Gaussian function with complex argument as integrand, has a closed analytical form by using the integral formula from Ref. [41], i.e.,

$$\int d\nu_- \exp\left\{-\frac{u_-^2}{2(2\sigma_0^2 + \sigma_p^2)} - \frac{ik''(z_1-z_2)(\nu_-+\Delta)^2}{4}\right\}$$

$$= \frac{\sqrt{\pi}}{\sqrt{a}\sqrt[4]{1+b^2}} \exp\left[-\frac{cb^2}{1+b^2} + \frac{i}{2}\arctan(b) + \frac{ir}{1+b^2}\right], \quad (21.29)$$

where

$$a = \frac{1}{2(2\sigma_0^2 + \sigma_p^2)}, \quad b = -\frac{k''(z_1-z_2)(2\sigma_0^2 + \sigma_p^2)}{2},$$

$$c = \frac{\Delta^2}{2(2\sigma_0^2 + \sigma_p^2)}, \quad \text{and} \quad r = -\frac{k''(z_1-z_2)\Delta^2}{4}.$$

21. Fiber-Optic Quantum Information Technologies

We therefore obtain the following final form of the single-photon counting formula:

$$S_c = A_1(\gamma P_p L)^2 \frac{\sigma_0}{\sigma_p} I_{sc}, \qquad (21.30)$$

$$A_1 = \frac{\alpha^2 \pi n A_{eff}^3}{18\sqrt{2} V_Q^2}, \qquad (21.31)$$

$$I_{sc} = \frac{1}{L^2} \int_{-L}^{0} dz_1 \int_{-L}^{0} dz_2 \frac{\exp\left[-2i\gamma P_p(z_1-z_2) - \frac{cb^2}{1+b^2} + \frac{i}{2}\arctan(b) + \frac{ir}{1+b^2}\right]}{\sqrt{(1-ik''\sigma_p^2 z_1)(1+ik''\sigma_p^2 z_2)}\sqrt[4]{1+b^2}}, \qquad (21.32)$$

where A_1 is an unknown constant with α and V_Q as fitting parameters, and I_{sc} is a double-length integral that has to be investigated numerically. Despite the seemingly complicated form of the single-counts formula, the physics behind it is quite clear. Apart from a small contribution from the double integral, the single counts scale quadratically with pump power, which coincides with the intuitive FPS picture that requires two pump photons to scatter into the signal/idler modes. It also scales linearly with the ratio of the filter bandwidth to pump bandwidth. This makes sense in that if one broadens the filter bandwidth, more photons will be collected; and conversely if the filter bandwidth is narrowed, one would expect to count less photons. The dependence on pump bandwidth is more clearly seen in the time domain. As the pulse width becomes wider (thus the pump bandwidth narrower) while maintaining the peak power to be the same, the probability of FPS increases linearly with pulse width (thus decreases linearly with pump bandwidth) simply because there is more time for the pump photons to interact; the reverse is also true. The more intricate dependence on pump power, pump bandwidth, and filter bandwidth is described by the double integral I_{sc}, which takes into account phase matching, SPM of the pump, and the Gaussian shapes of pump and filter spectrum.

Calculations of the coincidence-counting rate with Gaussian filters can be performed in a similar way to those of the single counting rate. We start with the probability of getting a coincidence count for each pulse [36]:

$$C_c = \int_0^\infty dT_1 \int_0^\infty dT_2 \langle \Psi | E_1^{(-)} E_2^{(-)} E_2^{(+)} E_1^{(+)} | \Psi \rangle. \qquad (21.33)$$

The electric fields are free fields propagating through Gaussian filters evaluated at detectors 1 and 2, defined in the photon-number unit:

$$E_1^{(+)} = \sum_{k_1} \sqrt{\frac{c A_{eff}}{4 V_Q}} a_{k_1} e^{-i\omega_s t_1} e^{-[(\omega_s - \Omega_s)^2/2\sigma_0^2]}, \qquad (21.34)$$

$$E_2^{(+)} = \sum_{k_2} \sqrt{\frac{cA_{\text{eff}}}{4V_Q}} a_{k_2} e^{-i\omega_i t_2} e^{-[(\omega_i - \Omega_i)^2/2\sigma_0^2]}, \tag{21.35}$$

where the Gaussian filters take the form $f(\omega_j - \Omega_j) = f_j e^{-[(\omega_j - \Omega_j)^2/2\sigma_j^2]}$, $f_j = 1$ and $\sigma_j = \sigma_0$ for $j = s, i$ are assumed to simplify the calculation. $t_i = T_i - l_i/c$ is the time at which the biphoton wavepacket leaves the output tip of the fiber, which in our case is almost the same for the signal and idler as there is negligible group-velocity difference between the two closely spaced (in wavelength), copolarized fields. l_i denotes the optical path length from the output tip of the fiber to the detector i, $i = 1, 2$, and can be carefully path matched to be the same.

The integrand in Eqn (21.33) can be written in the following form:

$$\langle \Psi | E_1^{(-)} E_2^{(-)} E_2^{(+)} E_1^{(+)} | \Psi \rangle = |\langle 0 | E_2^{(+)} E_1^{(+)} | \Psi \rangle|^2$$
$$= |A(t_1, t_2)|^2, \tag{21.36}$$

where $A(t_1, t_2)$ is the *biphoton amplitude* introduced in Refs [31, 34]. While the concept of a biphoton amplitude plays an important role in the study of frequency and wave-number entanglement inherent in the two-photon state, it serves merely as a calculational shorthand for our purpose in determining the coincidence counting rate. It is straightforward to show that

$$A(t_1, t_2) = \frac{cA_{\text{eff}}}{4V_Q} \sum_{k_s, k_i} F(k_s, k_i) e^{-i(\omega_s t_1 + \omega_i t_2)} e^{-[\nu_s^2 + \nu_i^2/2\sigma_0^2]}. \tag{21.37}$$

The fact that the biphoton cannot be written as a function of t_1 times a function of t_2 may be readily observed from the form that Eqn (21.37) takes. It is also nonfactorable in the wave numbers k_s and k_i, displaying its entangled nature in those degrees of freedom. However, it is not entangled in polarization, due to the fact that all the fields involved are collinear with respect to one another and the polarization states can be factored out.

When everything is taken into account, after some similar steps shown in Section 21.4, we arrive at the following form for the coincidence counting formula:

$$C_c = A_2 (\gamma P_p L)^2 \frac{\sigma_0^2}{\sigma_p \sqrt{\sigma_p^2 + \sigma_0^2}} I_{cc}, \tag{21.38}$$

$$A_2 = \frac{\alpha^2 \pi n^2 A_{\text{eff}}^4}{144 V_Q^{8/3}}, \tag{21.39}$$

$$I_{cc} = \frac{1}{L^2} \int_{-L}^{0} dz_1 \int_{-L}^{0} dz_2$$

$$\frac{\exp\left[-2i\gamma P_p(z_1-z_2) - (c'b'^2/1+b'^2) + (i/2)\arctan(b') + (ir'/1+b'^2)\right]}{\sqrt{(1-ik''\sigma_p^2 z_1)(1+ik''\sigma_p^2 z_2)}\sqrt[4]{1+b^2}}, \quad (21.40)$$

where

$$b' = -[k''(z_1-z_2)\sigma_0^2/2], \quad c' = (\Delta^2/2\sigma_0^2), \quad \text{and} \quad r' = -[k''(z_1-z_2)\Delta^2/4].$$

From Eqns (21.30) and (21.38), we can see that the single counts and the coincidence counts both scale quadratically with the pump peak power. This is a distinct feature of the $\chi^{(3)}$ interaction, in contrast to the linear dependence on pump power in $\chi^{(2)}$ SPDC. Whereas one might expect to see an exact linear relation between the single counts and the coincidence counts under ideal detection conditions (unity quantum efficiency of the detectors, no loss, no dark counts), the linearity is absent due to the broadband nature of the pump field and the presence of the filters. Some of the correlated twin photons are lost during the filtering process, and some uncorrelated photons are detected instead. The explicit dependence of C_c on the quantity $\sigma_0^2/\sigma_p\sqrt{\sigma_p^2+\sigma_0^2}$ can be understood from its limiting cases. When the pump bandwidth is wide compared with the filter bandwidth, i.e., $\sigma_p \gg \sigma_0$, every individual frequency component of the pump spectrum will generate its own energy-conserving signal/idler pairs. The filters, being narrow, are only effective at collecting a small portion of the correlated photons. Therefore, the coincidence counts should be proportional to σ_0^2/σ_p^2. However, if the pump bandwidth is sufficiently narrow, i.e., $\sigma_p \ll \sigma_0$, the photons being filtered (and subsequently collected by the detectors) are more likely to be correlated with each other, in which case the coincidence counts should scale with σ_0/σ_p. Both cases are verified when we look at the asymptotic limits:

$$\lim_{\sigma_p \gg \sigma_0} \frac{\sigma_0^2}{\sigma_p\sqrt{\sigma_p^2+\sigma_0^2}} = \frac{\sigma_0^2}{\sigma_p^2}, \quad (21.41)$$

$$\lim_{\sigma_p \ll \sigma_0} \frac{\sigma_0^2}{\sigma_p\sqrt{\sigma_p^2+\sigma_0^2}} = \frac{\sigma_0}{\sigma_p}. \quad (21.42)$$

To pin down the unknown parameters α and V_Q in Eqns (21.31) and (21.39), we fit our theory to two sets of experimental data, where the ratio of pump bandwidth to filter bandwidth is varied. The commonly used least-squares fitting technique has been employed. The results are shown in Figure 21.6, where the central solid curve corresponds to the optimum fitting parameters, which are determined to be

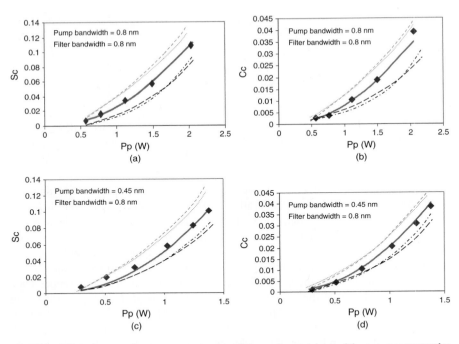

Figure 21.6 Experiment vs theory: squares correspond to experimental data and the curves correspond to theoretical predictions. (a) Single counts with $\sigma_p = 0.8$ nm and $\sigma_0 = 0.8$ nm; (b) Coincidence counts with $\sigma_p = 0.8$ nm and $\sigma_0 = 0.8$ nm; (c) Single counts with $\sigma_p = 0.45$ nm and $\sigma_0 = 0.8$ nm; (d) Coincidence counts with $\sigma_p = 0.45$ nm and $\sigma_0 = 0.8$ nm. The central solid curve represents the theoretical fit with the optimum fitting parameters ($\alpha = 0.237$, $V_Q = 1.6 \times 10^{-16}$ m^3), whereas the other curves correspond to fits with nonoptimum fitting parameters: dotted, $\alpha = 0.237$, $V_Q = 1.5 \times 10^{-16}$ m^3; dot-dashed, $\alpha = 0.237$, $V_Q = 1.7 \times 10^{-16}$ m^3; short-dashed, $\alpha = 0.250$, $V_Q = 1.6 \times 10^{-16}$ m^3; long-dashed, $\alpha = 0.220$, $V_Q = 1.6 \times 10^{-16}$ m^3 (this figure may be seen in color on the included CD-ROM).

$\alpha = 0.237$ and $V_Q = 1.6 \times 10^{-16}$ m^3. k'' has also been found to be -0.116 ps^2/km, corresponding to the wavelength difference $\lambda_p - \lambda_0 = 1.52$ nm, which agrees well with the measured experimental value. We also show the robustness of the fit by perturbing either one of the fitting parameters around its optimum value by as small as 5%. For example, the dotted curve corresponds to the case where we set $V_Q = 1.5 \times 10^{-16}$ m^3 while keeping α optimum, and the dot-dashed curve corresponds to the case where we set $V_Q = 1.7 \times 10^{-16}$ m^3 while keeping α optimum. The remaining two curves are generated when we keep V_Q optimum and set $\alpha = 0.250$ (short-dashed curve), or $\alpha = 0.220$ (long-dashed curve), respectively. The large discrepancies between the experiment and the theory induced by this operation are shown in the same figure, which boosts our confidence in the correctness of the theory.

We have provided a detailed discussion of the two-photon state originating from the third-order nonlinearity in optical fibers. This $\chi^{(3)}$ two-photon state shares some similar features with the $\chi^{(2)}$ two-photon state generated from SPDC, yet it also has some distinct characteristics. Coincidence photon-counting rate, which is a significant nonclassical figure of merit of the two-photon state, has been shown to depend heavily upon various experimental parameters. The dependence on the ratio of the pump bandwidth to filter bandwidth is of practical importance, because it serves as a guideline for optimizing the measurement of coincidence counts. Single-photon counting rate has also been studied, and both fit to the experimental data reasonably well. While in this chapter we are only concerned with parametric fluorescence from a single pump pulse, the current theory can be readily extended to include multiphoton-state generation from one pulse [42], and multiple two-photon-states generation from adjacent pulses [38, 39] to study polarization entanglement. The effect of SRS can also be included in our model by taking into account the noninstantaneous nature of the third-order nonlinearity in optical fiber.

21.4 FIBER NONLINEARITY AS A SOURCE FOR ENTANGLED PHOTONS

Quantum entanglement refers to the nonclassical interdependency of physically separable quantum subsystems. In addition to being at the heart of the most fundamental tests of QM [43–46], it is an essential resource that must be freely available for implementing many of the novel functions of QIP [47, 48]. In photonic systems, the ongoing developments in lasers, optical-fiber technology, single-photon detectors, and nonlinear optical materials have led to enormous experimental progress in both the fundamental [49–53] and applied domains [54–56]. A popular approach to generating entangled pairs of photons is based on the nonlinear process of parametric down-conversion in $\chi^{(2)}$ crystals [57–59]. Although much progress has been made using this approach, formidable engineering problems remain in coupling the entangled photons into standard optical fibers [60] for transmission, storage, and manipulation over long distances.

The coupling problem can be obviated if the entangled photons can be generated in the fiber itself and, desirably, in the fiber's low-loss propagation window near 1.5 μm, since that would minimize losses during transmission as well. Apart from the inherent compatibility with the transmission medium, a fiber-based source of entangled photons would have other advantages over its crystal counterparts [57, 58, 59, 61–63]. Particularly, the spatial mode of the photon pair would be the guided transverse mode of the fiber, which is a very pure Gaussian-like single spatial mode in modern fibers. A well-defined mode is highly desirable for realizing complex networks involving several entangling operations. In this chapter, we describe the first, to the best of our knowledge, optical fiber source of polarization-entangled photon pairs in the 1550-nm telecom band. A variety of biphoton interference experiments are presented that show the nature of the entanglement generated with this source. All four Bell states can be prepared with our setup and the CHSH form of Bell's inequality is violated by up to 10 standard deviations of measurement uncertainty.

Recently, our group has demonstrated that parametric fluorescence accompanying nondegenerate FWM in standard optical fibers is an excellent source of quantum-correlated photon pairs [64, 65]. The quantum correlation arises from FPS events, wherein two-pump photons at frequency ω_p scatter through the Kerr nonlinearity of the fiber to simultaneously create a signal photon and an idler photon at frequencies ω_s and ω_i, respectively, such that $\omega_s + \omega_i = 2\omega_p$. For a linearly polarized pump with wavelength close to the zero-dispersion wavelength of the fiber, the FWM process is phase-matched and the accompanying parametric fluorescence is predominantly copolarized with the pump. Two such parametric scattering processes can be time and polarization multiplexed to create the desired polarization entanglement. For example [see Figure 21.7(a)], when the fiber is pumped with two orthogonally polarized, relatively delayed pulses, the signal/idler photon pairs scattered from each pulse are copolarized with that pump pulse and relatively delayed by the same amount. The distinguishing time delay between the orthogonally polarized photon pairs, however, can be removed by passing the pairs through a piece of birefringent fiber of appropriate length, wherein the photon-pair traveling along the fast axis of the fiber catches up with the other pair traveling along the slow axis. When the emerging signal and idler photons are separated based on their wavelength, each stream of photons is completely unpolarized because any polarizer/detector combination is unable to determine which pump pulse a detected photon originated from. When the signal and idler photons are detected in coincidence, it is still impossible to determine which pump pulse created the detected pair. This indistinguishability gives rise to polarization entanglement in our experiment.

A schematic of the experimental setup is shown in Figure 21.7(b). Signal and idler photon pairs at wavelengths of 1547.1 and 1525.1 nm, respectively, are produced in an NFSI [65, 66]. The NFSI consists of a fused-silica 50/50 fiber coupler spliced to 300 m of DSF that has a zero-dispersion wavelength at $\lambda_0 = 1535 \pm 2$ nm. Because the Kerr nonlinearity is weak, for this length of fiber only about 0.1 photon-pair is produced with a typical 5-ps duration pump pulse containing $\sim 10^7$ photons. Thus, to reliably detect the correlated photon pairs, a pump-to-signal rejection ratio in excess of 100 dB is required. We achieve this by first exploiting the mirror-like property of the Sagnac loop, which provides a pump rejection of >30 dB, and then sending the transmitted fluorescence photons along with the leaked pump photons through a free-space double-grating spectral filter (DGSF) that provides a pump rejection ratio in excess of 75 dB [65]. The doubly diffracted signal and idler photons are then recoupled into fibers, whose numerical apertures along with the geometrical settings of the gratings determine the pass bands for the signal and idler channels. The FWHM bandwidth for both the channels is 0.6 nm.

During the experiment, for alignment and phase-control purposes, input signal and reference pulses are also needed that are temporally synchronized with the

21. Fiber-Optic Quantum Information Technologies

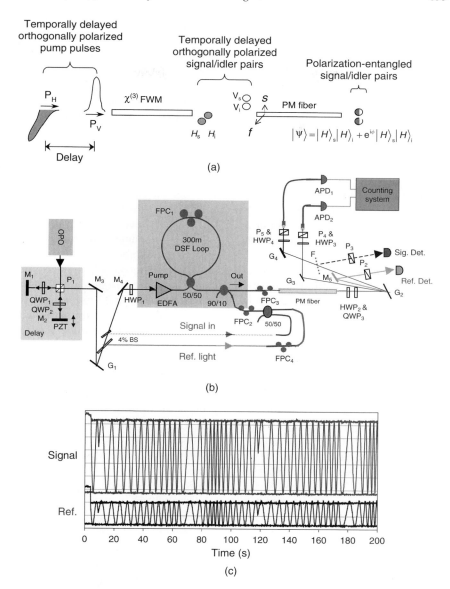

Figure 21.7 (a) Conceptual representation of the multiplexing scheme used to create polarization entanglement. (b) Schematic of the experimental setup. P_1–P_5, polarization beam splitters; G_1–G_4, diffraction gratings; M_1–M_5, mirrors; FPC_1–FPC_4, fiber polarization controllers; QWP, quarter-wave plate; HWP, half-wave plate; F, flipper mirror. (c) Sinusoidal variations (or constancy at the peaks and troughs) of the photocurrents obtained from the signal (top traces) and the reference detectors (bottom traces) upon linearly sweeping the voltage (or maintaining a fixed voltage) on the PZT, piezoelectric transducer. The clarity of the traces demonstrates minutes-long stability of the polarization interferometer formed between P_1 and P_3 (P_2) for signal (reference) light (this figure may be seen in color on the included CD-ROM).

pump pulses. The main purpose of the signal pulses is to ensure that the time distinguishability between the orthogonally polarized photon pairs is effectively removed. By spectrally carving [66] the ~150-fs pulse train from an optical parametric oscillator (OPO) [Coherent Inc., model Mira-OPO], we obtain trains of 5-ps pump pulses, 2.8-ps signal pulses, and 4-ps reference pulses at central wavelength of 1536, 1547, and 1539 nm, respectively. The pump pulses are then amplified by an erbium-doped fiber amplifier (EDFA) to achieve the required average pump power. Light at the signal and idler wavelengths from the OPO that leaks through the spectral-carving optics and the amplified spontaneous emission (ASE) from the EDFA are suppressed by passing the pump pulses through a 1-nm bandwidth tunable optical filter (Newport, TBF-1550-1.0).

A 30-ps relative delay between the two orthogonally polarized pump pulses is introduced by adding separate free-space propagation paths for the two pulses with use of a polarization beam splitter (PBS) P_1, quarter-wave plates (QWP) QWP_1 and QWP_2, and mirrors M_1 and M_2. Mirror M_2 is mounted on a piezoelectric transducer (PZT)-driven translation stage, which allows precise adjustment of the relative delay and phase difference between the orthogonally polarized pump-pulse pairs. After the NFSI, the delay is compensated by propagating the scattered photon pairs along the fast- and slow-polarization axes of a 20-m-long polarization-maintaining (PM) fiber. A careful alignment procedure is implemented to properly orient the axes of the PM fiber, taking into consideration the change of polarization state incurred by an input signal-pulse pair upon maximally amplified reflection from the NFSI [67]. Alignment is performed prior to the actual experiment by injecting weak path-matched signal-pulse pairs, having identical temporal and polarization structure as the pump pulses, into the NFSI through the 50/50 and 90/10 couplers. First the signal amplification is maximized by adjusting FPC_2, while monitoring the signal gain on a detector (ETX500) placed after P_3. Then the fringe visibility of the polarization interferometer formed between P_1 and P_3 is maximized by adjusting FPC_3, HW_2, and QWP_3, while observing the fringes in real time upon periodic scanning of M_2. Once the alignment is completed, the injected signal is blocked and further measurements are made only on the parametric fluorescence.

After compensation of the time delay, the following polarization-entangled state is generated at the output of the PM fiber: $|\Psi\rangle = |H\rangle_s|H\rangle_i + e^{2i\phi_p}|V\rangle_s|V\rangle_i$, where ϕ_p is the relative phase difference between the two delayed, orthogonally polarized pump pulses. This source can produce all four polarization-entangled Bell states. When $\phi_p = 0$, $\pi/2$, the states $|\Psi^{\pm}\rangle = |H\rangle_s|H\rangle_i \pm |V\rangle_s|V\rangle_i$ are created. The other two Bell states $|\Phi^{\pm}\rangle = |H\rangle_s|V\rangle_i \pm |V\rangle_s|H\rangle_i$ can be prepared by inserting a properly oriented HWP in the idler channel. Nonmaximally entangled pure states with an arbitrary degree of polarization entanglement can also be created with our setup by choosing the two-pump pulses to have unequal powers.

To actively monitor and control the relative phase ϕ_p during the course of data-taking, weak reference-pulse pairs of about 50 µW average power are injected into the NFSI through the 50/50 and 90/10 couplers. The reference-pulse pairs have identical temporal and polarization structure as the pump pulses, except the temporal location of the reference-pulse pairs is mismatched with respect to the pump-pulse pairs and their wavelength is slightly detuned, so that they neither interact with the pump pulses nor are seen by the single-photon detectors used in the signal and idler channels. During the course of measurements on the polarization-entangled states, the relative phase between the reference-pulse pairs, ϕ_{ref}, is monitored by measuring the photocurrent from a low-bandwidth reference detector placed after P_2 to make observations on one output port of the polarization interferometer [see Figure 21.7(b)]. The voltage created by this photocurrent is compared to a reference voltage and the difference is used to stabilize ϕ_{ref} by feeding back on the PZT through an electronic circuit. The excellent overall stability of the system is shown by the near-perfect classical interference fringes displayed in the inset in Figure 21.7(c), which were simultaneously obtained with injected signal light and with reference light while scanning ϕ_{ref} by ramping the voltage on the PZT. The relative phase between the reference-pulse pairs, ϕ_{ref}, is related to the relative phase between the pump-pulse pairs through $\phi_p = \phi_{\text{ref}} + \delta$, where δ results from dispersion in the DSF owing to slightly different wavelengths of the pulse pairs.

The photon-counting modules used for detecting the signal and idler photons consist of InGaAs/InP APDs (Epitaxx, EPM 239BA) operated in a gated Geiger mode [65]. The measured quantum efficiencies for the two detectors are 25% and 20%, respectively. The overall detection efficiencies for the signal and idler photons are about 9% and 7%, respectively, when the transmittance of the Sagnac loop (82%), 90/10 coupler, DGSF (57%), and other optical components (90%) are included. Given a parametric scattering probability of $\simeq 0.1$ pairs/pulse in the DSF, corresponding to 0.39 mW of average pump power in each direction around the Sagnac loop, and the gate rate of 588 kHz, we typically observe $\simeq 4000$ counts/s in the signal and idler channels when detecting the parametric fluorescence.

The polarization-entanglement generation scheme described here uses the fact that the FPS efficiency does not depend on the pump-polarization direction. We verify this by monitoring the parametric fluorescence while varying the polarization direction of the injected pump pulses with use of a half-wave plate (HWP_1). The individual counts for the signal and idler photons, and their coincidence counts, vs the HWP_1 angle are shown in Figure 21.8(a). The slight variation observed in the count rates is due to polarization-dependent transmission of the DGSF. Note that for the measurements shown in Figure 21.8(a) the input pump delay, the PM-fiber delay compensation, and the detection analyzers were removed.

Polarization correlations are measured by inserting adjustable analyzers in the paths of signal and idler photons, each consisting of a PBS (P_4, P_5) preceded by an adjustable HWP (HWP_3, HWP_4). For the state $|\Phi\rangle = |H\rangle_s|V\rangle_i + e^{i2\phi_p}|V\rangle_s|H\rangle_i$, when the polarization analyzers in the signal and idler channels are set to θ_1 and

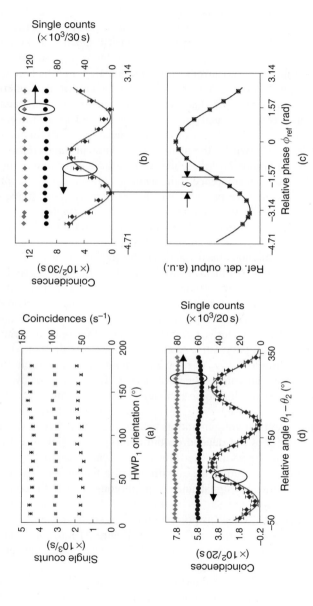

Figure 21.8 (a) Observed polarization (in)dependence of parametric fluorescence in the DSF. (b) Coincidence counts and single counts detected over 30 s when the relative phase ϕ_{ref} is varied. The solid curve is a fit to Eqn (21.43). (c) Output from the reference detector vs ϕ_{ref} showing the ordinary one-photon interference with twice the fringe spacing as in (b). (d) Measurement of polarization entanglement: Coincidence counts and single counts detected over 20 s as the analyzer angle in the idler channel is varied while keeping the signal-channel analyzer fixed at 45° relative to vertical (this figure may be seen in color on the included CD-ROM).

θ_2, respectively, the single-count probability for the signal and idler photons is $R_i = \alpha_i/2$ ($i = 1, 2$) and the coincidence-count probability R_{12} is given by

$$R_{12} = 2^{-1}\alpha_1\alpha_2[\sin^2\theta_1\cos^2\theta_2 + \cos^2\theta_1\sin^2\theta_2 \\ + 2\cos(2\phi_p)\sin\theta_1\cos\theta_1\sin\theta_2\cos\theta_2], \quad (21.43)$$

where α_i is the total detection efficiency in each channel.

We performed three sets of experiments to evaluate the degree of polarization entanglement of our source. The first measurement consisted of setting both analyzers at 45° and slowly scanning ϕ_{ref} by applying a voltage ramp on the PZT. As shown in Figure 21.8(b), the coincidence counts reveal sinusoidal variation with a fringe visibility of 93% (dark counts and accidental-coincidence counts have been subtracted), while the single counts remain unchanged. The output from the reference detector is also recorded simultaneously, which is shown in Figure 21.8(c). The relative shift of the sinusoidal variation of two-photon interference in Figure 21.8(b) from that of reference-light interference in Figure 21.8(c) is a direct measure of the phase shift δ, which is used below to properly set ϕ_p for measurements of the violation of Bell's inequality.

In the second set of measurements on polarization entanglement, we locked the generated state to $|\Phi^-\rangle = |H\rangle_s|V\rangle_i - |V\rangle_s|H\rangle_i$ by applying an appropriate feedback on the PZT, fixed the angle of the polarization analyzer in the signal channel to 45°, and varied the analyzer angle in the idler channel by rotating HWP$_4$. The result is shown in Figure 21.8(d). As expected, the coincidence-count rate displays sinusoidal interference fringes with a visibility of 92%, whereas the variation in the single-count rate is only 4% (once again, dark counts and accidental coincidences have been subtracted).

In the third set of experiments, we characterized the quality of polarization entanglement produced with our source through measurements of Bell's inequality violation. By recording coincidence counts for 16 different combinations of analyzer settings with $\theta_1 = 0°, 90°, -45°, 45°$ and $\theta_2 = -22.5°, 67.5°, 22.5°, 112.5°$, we measured the quantity S in the CHSH form of Bell's inequality [46], which satisfies $|S| \leq 2$ for any local realistic description of our experiment. The results, which are presented in Table 21.1, show that (a) the CHSH

Table 21.1

Measured values of S for the four Bell states.

Bell state	S	Violation (σ)
$\|H\rangle_s\|H\rangle_i + \|V\rangle_s\|V\rangle_i$	2.75±0.077	10
$\|H\rangle_s\|H\rangle_i - \|V\rangle_s\|V\rangle_i$	2.55±0.070	8
$\|H\rangle_s\|V\rangle_i + \|V\rangle_s\|H\rangle_i$	2.48±0.078	6
$\|H\rangle_s\|V\rangle_i - \|V\rangle_s\|H\rangle_i$	2.64±0.076	8

inequality is violated, i.e., $|S| > 2$, for all four Bell states produced with our setup and (b) the violation occurs by up to 10 standard deviations (σ) of measurement uncertainty.

To ascertain the degree of entanglement produced by the true FPS events in our setup, the accidental coincidences resulting from the uncorrelated background photons and the dark counts in the detectors were measured for each set of data acquired in the three polarization-entanglement experiments described above. The rate of accidental coincidences was as large as the rate of "true" coincidences, plotted in Figure 21.8 by subtracting the accidental coincidences, and the raw visibility of TPI was only $\simeq$30%. We believe, the majority of background photons in our setup arise from SRS as verified by our recent measurement of the noise figure of fiber-optic parametric amplifiers [68–70]. Our recent measurement with a modified DGSF has shown that the contribution of accidental coincidences can be made <10% of the total measured coincidences [64]. With these improvements, a raw TPI visibility of >85% would be obtained, i.e., without any postmeasurement corrections.

In conclusion, we have developed and characterized a fiber-based source of polarization-entangled photon pairs. The pair-production rate can be dramatically increased by using state-of-the-art pulsed lasers that have been developed for fiber-optic communications. These lasers operate at 10–40-GHz repetition rates and can have the requisite peak-pulse powers with use of medium-power EDFAs. Bulk-optic implementations of the pump delay apparatus and the detection filters were used in these proof-of-principle experiments for purposes of tunability and control. All-fiber versions of these subsystems can be readily realized with use of PM fibers, wavelength-division-multiplexing filters, and fiber polarizers. Finally, we have understood the origin of the large number of accidental coincidences in the experiment and subsequent system improvements are expected to significantly improve the degree of entanglement produced with our system. Therefore, we believe that such fiber-based entangled-photon pairs will prove to be an efficient source for developing quantum communication technologies.

21.5 HIGH-FIDELITY ENTANGLEMENT WITH COOLED FIBER

For many QIP applications it is desirable to produce entangled photon-pairs at telecom wavelengths directly in the fiber by use of the fiber's Kerr nonlinearity. Previously, our group has developed the use of an NFSI to generate quantum-correlated twin beams [66], correlated twin-photon pairs [71], and polarization entanglement [72–74]. We have pointed out that SRS, which gives rise to the majority of background photons, prevents us from observing TPI with unit visibility [74–76]. In these experiments, we have reduced the contribution of Raman photons by lowering pump power, selecting signal/idler photon-pairs at small detuning from the pump (≃5 nm), and using polarizers to remove the cross-polarized Raman photons. In this chapter, we present our measurements of the ratio between coincidence counts (coincidences arising from the same pulse) and accidental-coincidence counts (coincidences arising from adjacent pulses) as a function of pump power, and with the DSF at three different temperatures: ambient (≃300 K), dry ice (195 K), and liquid nitrogen (77 K). Then, we create polarization entangled photon-pairs with the fiber at the above three temperatures using a compact counterpropagating scheme (CPS) [77, 78]. We observe TPI with visibility >98% and Bell's inequality violation by >8 standard deviations of measurement uncertainty at 77 K.

As shown in Figure 21.9, the pump is a mode-locked pulse train with pulse duration ≃5 ps and repetition rate of 75.3 MHz, which is obtained by using a diffraction grating to spatially disperse the output of an optical parametric oscillator (Coherent Inc., model Mira-OPO). The pump central wavelength is set at 1538.7 nm. To achieve the required power, the pump pulses are further amplified by an EDFA. The ASE from the EDFA is suppressed by passing the pump through a tunable filter (Newport, model TBF-1550-1.0) with 1-nm FWHM passband. A fused-silica 90/10 fiber coupler is used to split 10% of the total pump power to a power meter for monitoring the power and stability of the pump pulses. The remaining 90% of the pump power goes through ε fiber polarization controller (FPC1) and a linear polarizer (LP), whose purpose is to adjust the input pump field

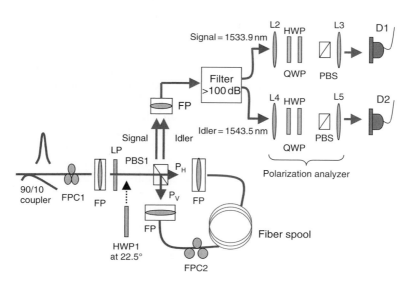

Figure 21.9 A schematic of the experimental setup. FP, fiber-port; LP, linear polarizer; L2, L3, L4, L5, fiber-to-free space collimators; PBS, polarization beam splitter; HWP, QWP, half- and quarter-wave plates; FPC, fiber polarization controller (this figure may be seen in color on the included CD-ROM).

to be horizontally polarized. A half-wave plate (HWP1) can be inserted in the pump path to split the pump pulse into horizontally and vertically polarized components for creating polarization entanglement. Initially, we use a single pump with horizontal polarization to pump a straight 300 m of DSF whose zero-dispersion wavelength is at $\lambda_0 = 1538.7$ nm. The fiber is wound on a spool of 9.32 cm diameter. A fiber polarization controller (FPC2) is used to compensate the bend-birefringence induced changes in the polarization states in the fiber. To reliably detect the scattered photon-pairs, an isolation between the pump and signal/idler photons in excess of 100 dB is required. We achieve this by using two cascaded WDM filters with FWHM of about 1 nm in the signal and idler channels, which provide total pump isolation greater than 110 dB [74]. The selected signal and idler wavelengths are 1543.5 and 1533.9 nm, respectively, corresponding to $\simeq 4.8$ nm detuning from the pump's central wavelength. Two sets of polarization analyzers, each composed of a quarter-wave plate (QWP), a half-wave plate (HWP), and a PBS are constructed and are individually inserted into the signal and idler channels. With proper settings of the QWP and HWP in each channel, the signal/idler photons with horizontal polarization can be made to arrive at the detectors with negligible loss. Raman photons that are copolarized with the pump also reach the detectors, while cross-polarized Raman photons are blocked by the PBSs. The copolarized Raman photons are inevitably detected, and hence contribute to the background photons in coincidence detection. It is known that Stokes and anti-Stokes Raman-scattering noise photons are emitted at a rate proportional to $n_{th} + 1$ and n_{th} [79, 86], respectively, where $n_{th} = 1/[\exp(h\nu/kT) - 1]$ is the Bose population factor, ν is the frequency shift of Stokes and anti-Stokes from the pump frequency, T is the temperature of the fiber, h is Planck's constant, and k is Boltzmann's constant.

The photon-counting modules consist of InGaAs/InP APDs (Epitaxx, Model EPM239BA) operated in the gated Geiger mode at room temperature [71]. The 1-ns-wide gate pulses with the FHWM window of about 300 ps arrive at a rate of 588 kHz, which is downcounted by 1/128 from the original pump pulses. The gate pulses are adjusted by an electronic delay generator to coincide with the arrival of the signal and idler photons at the APDs. The quantum efficiency of APD1 (APD2) is about 25% (20%), with a corresponding dark-count probability of 2.2×10^{-3} (2.7×10^{-3}) per pulse. The total detection efficiencies for the signal and idler photons are about 7% and 9%, respectively. Because the size of the fiber spool used in our experiment is smaller than the regular spool provided by Corning, it is convenient to cool the DSF by dry ice in a small homemade box or by liquid nitrogen in a small dewar. The loss due to fiber bending and handling, even in the low-temperature environment, is negligible. The loss due to cooling the fiber is less than 1% (4%) at 195 K (77 K). Cooling the fiber also causes contraction of the fiber length, which in turn shortens the propagation time for the photon-pairs to arrive at the detectors. The advancement of the arrival time is 1.6 ns (3.0 ns) at temperature of 195 K (77 K).

With the fiber at each temperature, we record coincidence and accidental-coincidence counts with an integration time of 60 s as the pump power. Note that accidental coincidence is the measured coincidence counts as the signal channel is delayed with respect to the idler channel by one pulse period. After subtracting the detector dark counts, we plot the ratio between the coincidence and accidental coincidence vs signal (or idler) counts per pulse. The plot at each temperature has a similar shape as those observed by other groups [81, 82]. The reason for such shape is not yet clearly understood. At high pump powers, the ratios are similar for the cooled and uncooled fibers. This may be due to multiphoton effects or leakage of the pump photons through the filters leading to increasing accidental-coincidence counts. As the pump power is decreased, we observe a ratio as high as 111:1 at 77 K and 60:1 at 195 K compared to a ratio of 28:1 at room temperature (300 K). The reason is that as the fiber is cooled, the Raman scattering is reduced as given by n_{th}, and leads to a decrease in the accidental-coincidence counts. The relative increase of the ratio is commensurate with the temperature dependence of n_{th}, which is about a factor of 4.5 (1.6) for the fiber at 77 K (195 K) compared to its room temperature value. At 300 and 195 K, the peaks in the ratio occur at an average pump power of about 50 μW (133 mW peak power). At 77 K, the peak occurs at a higher average pump power about 75 μW (200 mW peak power) because of cooling-induced loss in the fiber. At these peaks, the photon-pair production rate is ≃0.01/pulse.

To further test the fidelity of the generated photon pairs, we create polarization entanglement with the cooled fiber by using a CPS [77, 78]. The single horizontally polarized pump pulse is now split into two equally powered, orthogonally polarized components P_H and P_V by inserting and properly setting the half-wave plate (HWP1) in front of the polarization beam splitter (PBS1). For low-FPS efficiencies, where the probability for each pump pulse to scatter more than one pair is low, the clockwise (CW) and counterclockwise (CCW) pump pulses scatter signal/idler photon pairs with probability amplitudes $|H_i\rangle|H_s\rangle$ and $|V_i\rangle|V_s\rangle$, respectively. After propagating through the DSF, these two amplitudes of the photon pair are then coherently superimposed through the same PBS1. This common-path polarization interferometer has good stability for keeping zero-relative phase between horizontally and vertically polarized pumps, and hence is capable of creating polarization entanglement of the form $|H_i\rangle|H_s\rangle + e^{i2\phi_p}|V_i\rangle|V_s\rangle$ at the output of PBS1, where the relative pump phase ϕ_p is set to 0 by setting HWP1 to 22.5°. The polarization analyzers in signal and idler channels are used to set the detection polarization angles θ_1 and θ_2 of the entangled two-photon state. In our experiment, we set $\theta_1 = 45°$, vary θ_2, and record single counts for both signal and idler channels as well as coincidence counts between the two channels for each value of θ_2. We repeat these measurements with the fiber spool at 300, 195, and 77 K with corresponding integration times of 10, 50, and 60 s, respectively. At room temperature, we observe TPI with 91% visibility. As the fiber is cooled to 195 and 77 K, the TPI visibility increases to 95% and 98%,

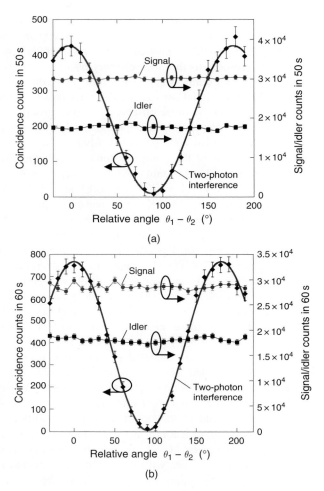

Figure 21.10 Two-photon interference with the fiber at (a) 195 K and (b) 77 K. The observed visibility is about 95% and 98.3%, respectively (this figure may be seen in color on the included CD-ROM).

respectively, as shown in Figure 21.10. All these TPI fringes are obtained when the pump power is adjusted to match the peak values of the ratio and are recorded without subtraction of accidental-coincidences. The observed higher visibility at low temperatures can be attributed to the suppression of Raman photons. The fitting function used in the above figures is $\cos^2(\theta_1 - \theta_2)$. We believe that the imperfection in spatial-mode matching of the correlated nondegenerate photon pairs at the PBS1 and the remaining copolarized Raman photons prevent this scheme from achieving unit-visibility TPI fringes.

Table 21.2
Violation of Bell's inequality for the state $|H_i\rangle|V_s\rangle - |V_i\rangle|H_s\rangle$.
σ is the standard deviation in the measurement of $|S|$.

Temperature (K)	$|S|$	Violation (σ)
300	2.22 ± 0.06	4
77	2.76 ± 0.09	>8

We further confirm the nonlocal behavior of the polarization entangled photon pairs generated from this source by making Bell's inequality measurements. For this purpose, we prepare the singlet state $|H_i\rangle|V_s\rangle - |V_i\rangle|H_s\rangle$ (where $\phi_p = \pi/2$) by inserting a QWP at 0° after HWP1 and by adding a HWP at 45° ($H \rightleftharpoons V$) in the signal channel. By recording coincidence counts for 16 different combinations of analyzer settings with $\theta_1 = 0°, 90°, -45°, 45°$ and $\theta_2 = -22.5°, 67.5°, 22.5°, 112.5°$, we measure the quantity $|S|$ in the CHSH form of Bell's inequality [46], wherein $|S| \leq 2$ holds for any local realistic system. The results are shown in Table 21.2. At 77 K we measure $|S| = 2.76 \pm 0.09$, which amounts to Bell's inequality violation by over eight standard deviations of measurement uncertainty. All these measurements are made without subtraction of the background Raman photons. Given these achievements, we believe we can reliably implement Ekert's QKD protocol with our polarization-entangled photon-pairs source.

21.6 DEGENERATE PHOTON PAIRS FOR QUANTUM LOGIC IN THE TELECOM BAND

The seminal paper by Knill et al. [83] has rekindled considerable amount of interest in the field of linear optical quantum computing (LOQC). For real applications of LOQC, single-photon as well as entangled-photon sources are indispensable [84]. Quantum interference arising from two indistinguishable photons, such as the well-known Hong-Ou-Mandel (HOM) interference [85], lies at the heart of LOQC. Therefore, it is of great significance to develop photon sources that generate *identical* photons in well-defined spatiotemporal modes. Additionally, for distributed LOQC it is desirable to produce such photons in the 1550-nm telecom band.

21.6.1 Polarization-Entangled Degenerate Photon-Pair Generation in Optical Fiber

Entangled photons in well-defined time slots are usually obtained using pulsed SPDC in $\chi^{(2)}$ crystals [86], wherein a high-frequency pump photon (ω_p) fissions into two identical daughter photons ($\omega_1 = \omega_2 = \omega_p/2$). Recently, Fan et al. [87]

have also produced identical correlated photons at the mean frequency (ω_c) of two pump frequencies ($\omega_{p1} + \omega_{p2} = 2\omega_c$) by using a reverse degenerate FWM process in a piece of $\chi^{(3)}$ microstructure fiber (MF). In that experiment, the generated photons reside in the visible/near-infrared wavelength region ($\lambda < 800$ nm). The spatial-mode profile of the identical photons is thus incompatible with standard single-mode fiber, making them unsuitable for distributed LOQC. Here, we describe a source useful for distributed LOQC by producing identical photons at a telecom-band wavelength using standard DSF. As a further endeavor, we use a novel dual-pump, CPS [78, 88] to make the otherwise identical photons entangled in polarization. We thus present the first, to the best of our knowledge, telecom-band, degenerate, correlated/entangled photon source based on a spool of standard DSF, which constitutes a promising step toward practical implementation of LOQC using fiber-based devices.

Figure 21.11 depicts our experimental setup. Out of the broadband spectrum of a femtosecond laser (repetition rate $\simeq$50 MHz), we spectrally carve out our desired pump central wavelengths ($\lambda_{p1} = 1545.95$ nm, and $\lambda_{p2} = 1555.92$ nm, pulse width $\simeq$ 5 ps) by cascading two free-space double-grating filters (DGF1 and DGF2, FWHM $\simeq$0.8 nm for each passband, see inset in Figure 21.12). An EDFA is sandwiched in between the two DGFs to provide pump power variability. The out-of-band ASE photons from the EDFA are suppressed by DGF2. Here we utilize the same nonlinear process as in Ref. [87], namely, reverse degenerate

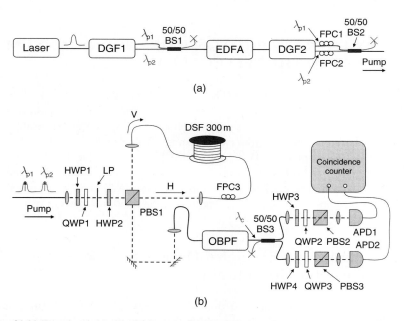

Figure 21.11 Experimental layout: (a) Pump preparation; and (b) CPS scheme. See text for details (this figure may be seen in color on the included CD-ROM).

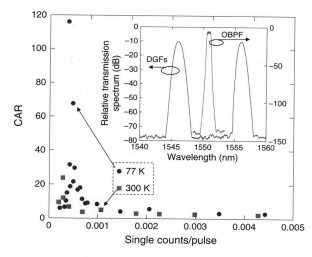

Figure 21.12 CAR plotted as a function of single counts/pulse at two different temperatures [77 K (dots) and 300 K (squares)]. Inset: spectral shapes of the cascaded DGFs and OBPF (this figure may be seen in color on the included CD-ROM).

FWM. The pump central wavelengths are selected such that their mean wavelength ($\lambda_c = 1550.92$ nm) is located near the zero-dispersion wavelength of the 300-m-long DSF. The two-pump pulses emerging from the second 50/50 beam splitter (BS2), before being launched into the CPS, have to satisfy the following criteria to maximize the FWM efficiency and the resulting entanglement in the two-photon state: (i) overlapped in time, (ii) parallel polarization, and (iii) equal power. The first criterion is met by careful path matching, and the resulting overlapped pulses from the unused port of BS2 can be monitored on a high-speed oscilloscope. The second criterion is satisfied by adjusting the fiber polarization controllers (FPC1 and FPC2) and the half-wave plate/quarter-wave plate combination (HWP1 and QWP1) such that the power exiting the LP is maximized. The last criterion is fulfilled by balancing the transmission efficiencies in the two arms of the cascaded DGFs. SRS in the DSF is suppressed by cooling the DSF to 77 K [89, 90]. Compared with the cascade-MF approach in Ref. [87], this setup is straightforward and simpler in design.

We first characterize our degenerate correlated-photon source by measuring temporal coincidences between the two identical photons. We call a "coincidence" count when the two detectors fire in the same triggered time slot, and an "accidental-coincidence" count when they fire in the adjacently triggered time slots. In the literature [87, 89, 90], a figure of merit for such a source has been established, i.e., the coincidence-to-accidental ratio (CAR). We abbreviate it as CAR hereafter. Intuitively, CAR is a measure of the purity of a correlated photon source, and a high CAR value indicates a relatively high purity of the source in coincidence basis, i.e., coincidence events due to uncorrelated noise photons are very rare. To measure the CAR, we set HWP2 so that the two-frequency pump field maintains its horizontal polarization upon entering the CPS. After propagating through the DSF, the pump and its accompanying degenerate FWM photons are made to keep the same polarization by proper adjustment of FPC3. We use an optical bandpass filter (OBPF) composed of two cascaded 100-GHz-spacing wavelength-division-multiplexing filters (FWHM $\simeq$0.3 nm, see Figure 21.12, inset) with a central wavelength of λ_c at the output of the CPS to collect the correlated FWM photon pairs, and to provide the $>$100 dB pump isolation needed to effectively detect those photons. The OBPF is followed by a 50/50 beam splitter (BS3) to probabilistically split the two daughter photons. We are only interested in the cases where the two daughter photons split up (which happen with 50% probability) to give coincidence counts; the other equally probable cases where the two photons bunch together are not recorded in coincidence detection. Nevertheless, the latter contribute to the single-count measurement results. For this part of the experiment, the polarization analyzers (HWP3/QWP2/PBS2 and HWP4/QWP3/PBS3) shown in Figure 21.11 are taken out, as the polarization properties of the detected photons are not of concern in this part of the experiment. Each photon is directed to a fiber-coupled APD1 and APD2 (Epitaxx EPM 239BA) operating in the gated Geiger mode, whose detection results are recorded and analyzed by a "coincidence counter" software. For the measurement reported in this chapter, the detection rate is 1/64 of the laser's repetition rate ($\simeq$780 kHz), which is limited by the electronics used to arm and trigger the detectors. The overall efficiencies of the two detectors, including propagation losses and detector quantum efficiencies, are 7% and 9%, respectively. We subtract detector dark-count contributions from all of our measurement results.

A sample experimental result is shown in Figure 21.12, in which we plot the CAR as a function of the single-channel photon-detection rate. A preliminary version of this result has been reported previously [91]. The CAR is measured at the ambient temperature (300 K) and when the DSF is immersed in liquid nitrogen (77 K). A CAR value as high as 116 is obtained at 77 K, whereas the peak of the CAR at 300 K is around 25. This is consistent with the expectation that SRS is more severe at a higher temperature, resulting in degraded purity of the source. The shape of the CAR function is similar to those reported previously [87, 89, 90] and can be qualitatively explained by a quantum model [92], which correctly predicts the CAR's temperature dependence. However, further work is required for a quantitative comparison between experiment and theory.

After determining the single-count rate for the optimum CAR value, we adjust HWP2 to split the dual-frequency pump field into two orthogonally polarized dual-frequency components with equal power (more specifically, $P_{1H} = P_{1V} = P_{2H} = P_{2V}$). Each pump component probabilistically produces its own degenerate signal/idler photon pair, and the two probability amplitudes superpose upon each other at the output of the CPS, as there is no distinguishability between them. As a result of reverse degenerate FWM and the CPS, polarization entanglement of the form $|H_s H_i\rangle + |V_s V_i\rangle$ is generated, wherein the signal/idler photons are of identical

21. Fiber-Optic Quantum Information Technologies

wavelength λ_c. After passing through the 50/50 beam splitter BS3, the photons are then detected in coincidence as before, except we add in the polarization analyzers shown in Figure 21.11(b), because it is the polarization entanglement properties of the photon pairs that are to be examined here. In the experiment, the polarization analyzer in one photon's path is set to a fixed linear polarization angle ϕ_0, and we stepwise rotate the HWP in the other photon's path to go from an initially parallel polarization angle ϕ_0 to a final polarization angle $\phi_0 + \pi$, which corresponds to a HWP rotation angle of $\Delta\theta \in [0, \pi/2]$. Both single counts and coincidence counts are recorded as a function of $\Delta\theta$.

A sample experimental result is shown in Figure 21.13. We demonstrate the high purity of our entanglement source by exhibiting TPI with visibility >97% together with polarization-independent single counts. These results are obtained when we pump the CPS with relatively low power [$P_{1H} = P_{1V} = P_{2H} = P_{2V} \simeq 90\,\mu\text{W}$ (peak power $\simeq 0.36\,\text{W}$)] and the DSF is cooled to 77 K. The high-TPI visibility means that we are indeed generating the maximally entangled state $|H_s\,H_i\rangle + |V_s V_i\rangle$. Nonmaximally entangled states can also be generated if we change the relative pump-power ratio between the CW and the CCW paths. The relatively large error bars associated with the TPI data points can be attributed to the low-data-collection rate in our current implementation, which can be improved by either using an OBPF with a wider passband, or using faster detection electronics [93]. As is the case with other CPS entanglement sources [78, 88], the dual-pump CPS source presented here can be easily configured to produce all four Bell states ($|H_s H_i\rangle \pm |V_s V_i\rangle, |H_s V_i\rangle \pm |V_s H_i\rangle$) and the Bell's inequality violation can be inferred from the observed >71% TPI visibility.

We further investigate the scenario when only a single-frequency pump (at either λ_{p1} or λ_{p2}) is injected into the CPS. As shown in Figure 21.14, we

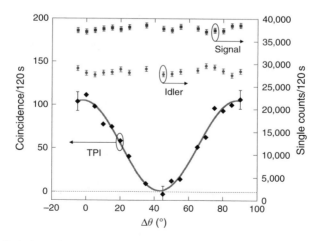

Figure 21.13 Two-photon interference (TPI) with visibility >97% is shown, while single counts in both channels exhibit no polarization dependence. Solid curve is a cosine fit to the data (this figure may be seen in color on the included CD-ROM).

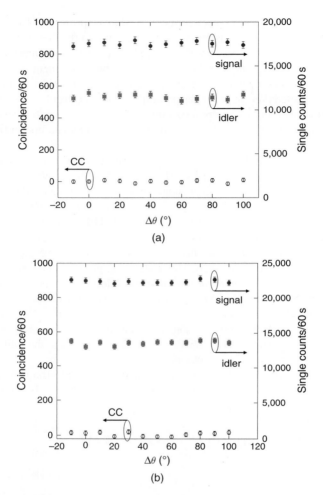

Figure 21.14 Single counts and coincidence counts (CC) show no dependence on the HWP angle $\Delta\theta$ when only a single-frequency pump at $\lambda_{p1} = 1545.95$ nm (a), or at $\lambda_{p2} = 1555.92$ nm (b), is used (this figure may be seen in color on the included CD-ROM).

observe polarization-independent behavior for both the single counts and the coincidence counts. In fact, the coincidence counts after subtraction of the dark-count contributions in both cases are zero within their error-bar ranges. This is expected because the FWM photons in this case are produced not by the reverse degenerate FWM process, but by the conventional FWM process involving a degenerate-frequency pump. As the filters in the experiment are configured to look at coincidences between photons in a single band (i.e., without its conjugate), the pairwise nature of the conventional FWM photon production is not manifested. Hence all the recorded coincidence counts are light–dark coincidences, which after subtraction of the dark counts lead to zero light–light coincidence counts. This serves as a further evidence that the degenerate-frequency polarization-entangled photon pairs are indeed produced by the dual-frequency pumps, and not by the single-frequency pumps. The polarization-independent single counts in Figure 21.14 are a result of the CPS-induced polarization indistinguishability.

To conclude this section, we have presented an efficient, high-purity source of correlated/entangled photon pairs in a single spatiotemporal mode in the telecom band. It is foreseeable that further development of such fiber-based photon sources will lead to practical applications of various quantum-communication protocols and LOQC.

21.6.2 Hong-Ou-Mandel Interference with Fiber-Generated Indistinguishable Photons

The nascent field of quantum information science, motivated by the extraordinary computing power of a full-fledged quantum computer, naturally selects photons—the fundamental energy packets of the electromagnetic field—as the carrier of quantum information from one computing node to another, a task typically associated with quantum communication. The omnipresent telecommunication fibers constitute the quantum channels for quantum cryptography [16], quantum gambling [94], and quantum games [95], which utilize remote sharing of quantum entanglement as a resource. In addition, indistinguishable photons—photons having identical wave packets—play a major role in the arena of LOQC [96]. Therefore, a telecom-band indistinguishable photon-pair source is particularly useful for the above QIP tasks. The traditional method of producing such photons—SPDC in second-order ($\chi^{(2)}$) nonlinear crystals [97]—faces formidable engineering challenges when these photons are launched into a single-mode optical fiber for long-distance transmission or mode cleansing. Large coupling losses inevitably occur due to mode mismatch [98], limiting the usefulness of such a source. Recently, we demonstrated a fiber-based source of indistinguishable photon pairs at a telecom-band wavelength near 1550 nm that utilizes the third-order ($\chi^{(3)}$) nonlinear process of FWM [99] in the fiber itself. This approach automatically takes care of the aforementioned mode-matching issue, because the photonic spatial mode of the generated photon pair is the same as that of standard optical fiber.

Four-wave mixing is a third-order process mediated by the Kerr nonlinearity of optical fiber, wherein two-pump photons annihilate to give birth to a pair of time-energy entangled daughter photons, usually denoted as signal and idler. Energy conservation as well as momentum conservation are obeyed during the FWM process: $\omega_{p1} + \omega_{p2} = \omega_s + \omega_i$, and $\vec{k}_{p1} + \vec{k}_{p2} = \vec{k}_s + \vec{k}_i$, where ω_j and $\vec{k}_j$ stand for the frequency and wavevector of the jth photon, and subscripts p1, p2, s, and i denote the two pump photons, the signal photon, and the idler photon, respectively. Our group has previously demonstrated generation [29, 30], distribution [100, 101], and storage [38, 39] of *nondegenerate* photon pairs (s ≠ i) with a single-frequency pump pulse (p1 = p2). More recently, we have also achieved success in generating *degenerate* photon pairs (s = i) with dual-frequency pump pulses (p1 ≠ p2) [99]. While it is straightforward to separate the nondegenerate photon pairs by means of their different wavelengths through a wavelength-division (de)multiplexer, it remains unclear how to *deterministically* separate the fully degenerate photon pairs, as the two photons share the same properties in all degrees of freedom: spatial, temporal (frequency), and polarization. This situation is analogous to a type-I down-conversion with collinear output photons [102]. Up to now, researchers have used a regular 50/50 beam splitter (BS) to probabilistically split the identical photons [99, 102, 103], which in actuality produces the following state [schematically shown in Figure 21.15(a)]:

$$|\Psi\rangle_{\text{in}} = |2\rangle_a |0\rangle_b,$$

$$|\Psi\rangle_{\text{out}} = \frac{1}{\sqrt{2}} \frac{|2\rangle_d |0\rangle_c - |0\rangle_d |2\rangle_c}{\sqrt{2}} + \frac{i}{\sqrt{2}} |1\rangle_c |1\rangle_d$$

$$\equiv \frac{1}{\sqrt{2}} \Psi_{2002} + \frac{i}{\sqrt{2}} \Psi_{11}. \qquad (21.44)$$

The presence of the Ψ_{2002} component limits the usefulness of such a probabilistic identical photon source, which has a maximum attainable HOM dip visibility of only 50% [102].

However, if the output wavefunction only consists of the Ψ_{11} component, the HOM dip visibility can in principle reach 100%.

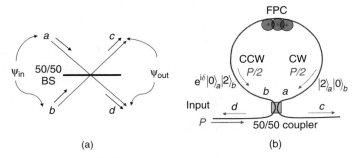

Figure 21.15 (Color online) Quantum interference at a beam splitter. (a) Schematic of input/output state transformation for a 50/50 beam splitter (BS). (b) Illustration of quantum interference in a Sagnac loop between photon pairs with different phase shift δ. CW, clockwise; CCW, counterclockwise; FPC, fiber polarization controller (this figure may be seen in color on the included CD-ROM).

One way to produce a "clean" Ψ_{11} output is to utilize a cross-polarized dual-frequency pump to excite the χ_{xyxy} component of the Kerr nonlinear susceptibility tensor [104, 105]. This is analogous to a type-II collinear down-conversion [106], where the output orthogonally polarized degenerate-frequency photon pairs are split by a PBS in a deterministic fashion. However, the intrinsically weaker nature of χ_{xyxy}, which is 1/3 of the collinear tensor component χ_{xxxx} in a Kerr nonlinear medium like fused-silica glass, makes this approach much less efficient than the copolarized FWM process. Here we introduce a new type (or topology) of copolarized identical-photon source, called the "50/50 Sagnac-loop" source, which is capable of producing a clean Ψ_{11} state. Due to its intended use to deterministically split up the identical photons, it is also given the name "quantum splitter," or QS for short.

The idea behind the QS source comes from the realization that, not only can a fiber Sagnac loop be configured as a total reflector [107] (TR) or a total transmitter (TT), it also can be set in an equally transmissive and reflective state (50/50). The difference between the above operational modes lies in the different settings of the intraloop fiber polarization controller (FPC), which results in different relative phase shifts between the CW and the CCW paths. As shown in Figure 21.15(b), the pump is injected from port d into the Sagnac loop, which is composed of a 50/50 fiber coupler, a piece of DSF of suitable length, and an FPC. The pump peak power is assumed to be P, which is split into two equally powered pulses ($P/2$) by the coupler. The two-pump pulses traverse the DSF in a counterpropagating manner, each of which probabilistically scatters copolarized FWM photon pairs. Here we neglect the case where both pumps undergo FWM scattering, as it corresponds to a higher order process of multiphoton generation, whose probability is vanishingly small when the pump power is low. The two identical-photon probability amplitudes, with a differential phase δ controlled by the setting of the FPC, are then recombined at the coupler before coming out of the Sagnac loop. The input state in this case is written as

$$|\Psi\rangle_{in} = \frac{|2\rangle_a|0\rangle_b + e^{i\delta}|0\rangle_a|2\rangle_b}{\sqrt{2}}. \tag{21.45}$$

The corresponding output state is obtained from the standard BS input/output relationship, and is given by

$$|\Psi\rangle_{out} \frac{1-e^{i\delta}}{2}\Psi_{2002} + \frac{i(1+e^{i\delta})}{2}\Psi_{11}. \tag{21.46}$$

We can readily see, from Eqn (21.46), that a pure Ψ_{11} state is obtained when we set the differential phase δ to be 0, which corresponds classically to the case of a 50/50 Sagnac loop. *This can be physically interpreted as time-reversed HOM interference.* One can also verify that when δ is set to be π (-π), the Sagnac loop is totally reflective (transmissive), and one obtains the pure state Ψ_{2002}.

We now demonstrate the identical nature of the photons produced by the QS source. The observation of a HOM dip of high visibility has been established as a figure of merit for identical-photon sources [108], and has been demonstrated using a variety of devices, including identical photon pairs from the same SPDC source [109, 110], independent indistinguishable photons from separate SPDC sources [111–113], and indistinguishable photons from a quantum-dot single-photon source [114]. Our scheme bears some resemblance to the first category, but yet is different enough in that it is fiber-based and thus easily integrable into fiber-optic networks [115]. The experimental setup is depicted in Figure 21.16 with considerable amount of detail. Figure 21.16(a) shows how the dual-frequency copolarized pump is prepared [99]. Out of the broadband spectrum of a femtosecond laser (repetition rate $\simeq 50$ MHz), we spectrally carve out our desired pump central wavelengths ($\lambda_{p1} = 1545.95$ nm and $\lambda_{p2} = 1555.92$ nm, pulse width $\simeq 5$ ps) by cascading two free-space double-grating filters [DGF1 and DGF2, FWHM $\simeq 0.8$ nm for each passband; see Figure 21.17(a)]. An EDFA is

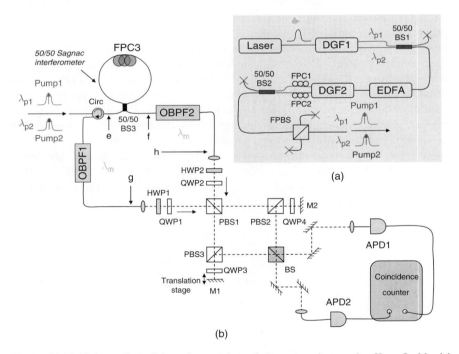

Figure 21.16 (Color online) Schematic experimental setup to observe the Hong-Ou-Mandel interference between identical photons generated from the QS source. (a) Preparation of the dual-frequency copolarized pump. BS, beam splitter; DGF, double-grating filter; FPBS, fiber polarization beam splitter; FPC, fiber polarization controller; EDFA, erbium-doped fiber amplifier. (b) Hong-Ou-Mandel experiment. PBS, polarization beam splitter; HWP, half-wave plate; QWP, quarter-wave plate; OBPF, optical band-pass filter; Circ, circulator; APD, avalanche photodiode (this figure may be seen in color on the included CD-ROM).

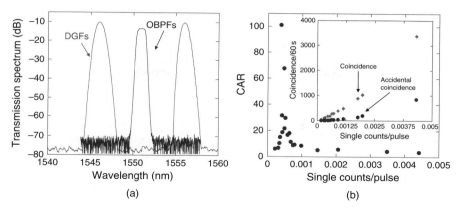

Figure 21.17 (Color online) Experimental parameters. (a) Transmission spectra of the OBPFs and DGFs. (b) Experimentally measured CAR function of the QS source with the fiber at 77 K. Inset shows the measured coincidence and accidental-coincidence counts as a function of single counts/pulse (this figure may be seen in color on the included CD-ROM).

sandwiched in between the two DGFs to provide pump power variability. The out-of-band ASE photons from the EDFA are suppressed by DGF2. The pump central wavelengths are selected such that their mean wavelength ($\lambda_m = 1550.92$ nm) is located near the zero-dispersion wavelength of the 300-m-long DSF in the Sagnac loop to maximize the FWM efficiency [116]. The two-pump pulses emerging from the second 50/50 beam splitter (BS2) are in turn passed through a fiber polarization beam splitter (FPBS) to ensure their copolarized property. The other necessary properties of the two-pump pulses, namely, temporal overlapping and equal power, are individually addressed by careful path-matching and transmission-efficiency balancing for the two pulses. Figure 21.16(b) shows the QS source and its intended use in a HOM experiment. The Sagnac loop is preceded by a circulator (Circ), which redirects the Sagnac-loop reflected photons to a separate spatial mode. The output degenerate FWM photons from the Sagnac loop are selected by two optical bandpass filters (OBPF1 and OBPF2), whose transmission spectrum is shown in Figure 21.17(a) (center wavelength = 1550.92 nm, passband $\simeq 0.8$ nm). The OBPFs also provide the necessary >100-dB isolation from the pump to effectively detect those filtered photons. The alignment of the Sagnac loop to its 50/50 state consists of using a continuous-wave laser source with its wavelength set to λ_m as the input to the Sagnac loop, and adjusting FPC3 so that the transmitted and reflected powers are equal (i.e., $P_e = P_f$). In practice, we measure the powers at the output of the OBPFs (P_g and P_h), and demand that $P_g/\eta_{eg} = P_h/\eta_{fh}$, where η_{eg} (η_{fh}) is the transmission efficiency from point e (f) to point g (h). The experimental values of η_{eg} and η_{fh} are 0.749 and 0.768, respectively. SRS in the DSF is suppressed by cooling the DSF to 77 K using liquid nitrogen [117, 118].

Before the actual HOM experiment, we first characterize the QS source by measuring its CAR [99], which has been established to be a figure of merit of such correlated-photon sources [103, 104, 117, 118]. To do that, the path-matched outputs of the QS source [points g and h in Figure 21.16(b)] are directly connected to two APD1 and APD2 (Epitaxx EPM 239BA) for coincidence detection, bypassing all the intermediate free-space optics. The APDs are operated in a gated Geiger mode, with detection rate $\simeq$780 kHz and overall detection efficiencies of 7% and 9%, respectively. Single-channel counts as well as coincidence counts are recorded by a "coincidence counter" software. Detector dark-count contributions are subtracted from all of our measurement results. We call a "coincidence" count when the two detectors fire in the same triggered time slot, and an "accidental- coincidence" count when they fire in the adjacently triggered time slots. We then vary the pump power and record the corresponding coincidence and accidental- coincidence counts. Their ratio, CAR, is plotted as a function of the single-channel count rate in Figure 21.17(b). It has a similar shape to those reported previously [99, 118]. The CAR peak value of about 100 occurs at a single-count rate of around 4×10^{-4}/pulse (photon-pair production rate $\simeq 2.2 \times 10^{-3}$/pulse), and decreases rather rapidly on both sides.

We then proceed with our HOM experiment. After the identical photons come out from g, and h, they are collimated into free space through lenses, and two sets of polarization compensators composed of half-wave plates and quarter-wave plates (HWP1/QWP1 and HWP2/QWP2) are used to restore each photon's polarization to horizontal before they enter PBS1. Each photon then traverses one arm of the Mach–Zehnder-like interferometer before being combined at the 50/50 beamsplitter (BS) cube. Quarter-wave plates QWP3 and QWP4 are each set at 45°, so that when combined with a mirror behind, they function as 45°-oriented half-wave plates, rotating the horizontal polarization of the incident light by 90°. In principle, the combinations PBS3/QWP3/M1 and PBS2/QWP4/M2 can be replaced with just two mirrors to direct the photons to the input ports of the BS. In practice, we choose to implement the more complicated version, because it is less susceptible to translation-induced misalignment of the photon wavepackets at the BS. Careful path-matching is done to ensure that the two photons reach the BS at approximately the same time (i.e., their arrival-time difference is within the tuning range of the translation stage placed under M1). The identical photons, before hitting the BS, are both of linear vertical polarization. The output photons from the BS are each coupled into single-mode fibers, which are connected to the two APDs for coincidence detection. The HOM experiment is performed by recording the coincidence counts at each setting of the translation stage, which is equivalent to recording the coincidence counts as a function of the overlap between the two identical photon wavepackets. The two scales are related by $\delta \tau = 2\Delta s/c$, where $\delta \tau$ is the temporal difference between the photon wavepackets, Δs is the difference in readings of the translation stage, and $c = 3 \times 10^8$ m/s is the speed of light in vacuum.

We first pump the QS source with relatively high pump power, and obtain a sample experimental result as shown in Figure 21.18(a). The single-count rate is $\simeq 3.4 \times 10^{-3}$/pulse, corresponding to a low-CAR value of around 2 according to Figure 21.17(b). A HOM dip of visibility $\simeq$50% is observed. This 50% dip

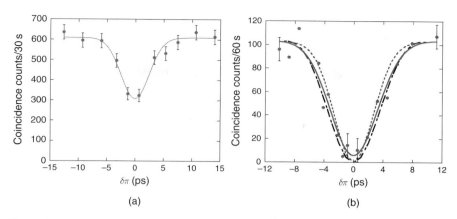

Figure 21.18 Hong-Ou-Mandel experimental results. (a) Hong-Ou-Mandel dip visibility of 50% is observed when the pump power is high. The solid curve is a least-squares Gaussian fit to the data. (b) Hong-Ou-Mandel dip visibility of 94.3% is observed when the pump power is low. The solid curve is a least-squares Gaussian fit to the data. The dotted curve is a theoretical fitting for Gaussian OBPFs, while the dot-dashed curve is that for super-Gaussian OBPFs (see Ref. [119] for details) (this figure may be seen in color on the included CD-ROM).

visibility is the upper bound for the case of two classically random photon sources as inputs to the BS, and can be explained by using a simple classical electromagnetic theory [112]. It is an expected result, because in the high-pump condition, the photon source emits mainly spontaneously scattered Raman photons, along with multiple FWM photon pairs, whose random behavior masks the true correlated nature of a single pair of identical FWM photons. We then lower down the pump power to observe a much higher HOM dip visibility of $94 \pm 1\%$, as shown in Figure 21.18(b). This result is obtained at a single-count rate of around 4×10^{-4}/pulse, which corresponds to the CAR peak in Figure 21.17(b). The near-unity visibility of the HOM dip is well beyond the classical limit of 50%, and clearly demonstrates the high indistinguishability of the QS-generated identical photons. Combined with its telecom-band operation and fiber-optic networkability, we expect to be able to achieve more complicated QIP applications with this fiber-based QS source.

Theoretical modeling of the QS source and its HOM experiment will be provided elsewhere [119]. Here we show theoretical simulation results of the HOM dip in Figure 21.18(b), along with the experimental data. The two theory curves, generated for different OBPF spectral shapes (see Figure 21.18 caption for details), appear to fit the experimental data remarkably well. Both fits agree on the ideally attainable HOM dip visibility of 100%, which corresponds to the fact that the two-photon probability distribution function is symmetric with respect to its two frequency arguments [106]. The missing $6(\pm 1)\%$ visibility can be explained by taking into account the following real-life imperfections: (i) The BS used in the experiment ($R = 0.474$, $T = 0.526$, where R and T are the BS intensity coefficients of reflection and transmission, respectively) deviates from its ideal performance, but this gives rise to a negligible correction factor [110] $2RT/(R^2 + T^2) \simeq 0.994$. (ii) The spatial-mode mismatch of the two photon wavepackets at the BS results in some distinguishing information between the two coincidence-generating probability amplitudes. A simple calculation [119] shows that a small angular mismatch of around 30 μrad can bring the HOM dip visibility down to 94%. (iii) Some remaining Ψ_{2002} component due to nonideal alignment of the 50/50 Sagnac loop may lead to a degradation of the dip visibility [102]. (iv) Existence of unsuppressed noise photons, such as Raman photons and FWM photons induced by a single-frequency pump, may also degrade the attainable HOM dip visibility.

21.7 CONCLUDING REMARKS

Because this volume deals with advances in Optical Fiber Telecommunications, our focus in this chapter has been on the generation of correlated and entangled photons in the telecom band with use the Kerr nonlinearity in DSF. Over the past decade, microstructure or holey fibers (MFs) have come on the scene, which because of their tailorable dispersion properties allow phase-matching to be obtained over a wide range of wavelengths. Generation of correlated and entangled photons in MFs has been demonstrated and rapid progress is taking place.

REFERENCES

[1] D. Heiss, *Fundamentals of Quantum Information: Quantum Computation, Communication, Decoherence, and all that*, Berlin Heidelberg: Springer-Verlag, 2002.
[2] A. Einstein, B. Podolsky, and N. Rosen, "Can quantum-mechanical description of physical reality be considered complete?" *Phys. Rev.*, 47, 777–780, 1935.
[3] N. Bohr, "Can quantum-mechanical description of physical reality be considered complete?" *Phys. Rev.*, 48, 696–702, 1935.
[4] D. Bohm, *Quantum Theory*, Prentice-Hall, 614, 1951.
[5] J. S. Bell, 1964, *Physics* (Long Island City, N.Y.) 1, reprinted in J. S. Bell, *Speakable and Unspeakable in Quantum Mechanics* (Cambridge: Cambridge University, 1987).
[6] S. J. Freedman and J. S. Clauser, *Phys. Rev. Lett.*, 28, 938–941, 1972.
[7] A. Aspect, P. Grangier, and G. Roger, *Phys. Rev. Lett.*, 47, 460, 1981.
[8] A. Aspect, P. Grangier, and G. Roger, *Phys. Rev. Lett.*, 49, 91, 1982.
[9] A. Aspect, J. Dalibard, and G. Roger, *Phys. Rev. Lett.*, 49, 1804, 1982.
[10] P. G. Kwiat, K. Mattle, H. Weinfurter et al., *Phys. Rev. Lett.*, 75, 4337, 1995.
[11] P. M. Pearle, *Phys. Rev. D*, 2, 1418, 1970.
[12] J. F. Clauser and A. Shimony, *Rep. Prog. Phys.*, 41, 1881, 1978.
[13] M. A. Nielsen and I. L. Chuang, *Quantum Computation and Quantum Information*, Cambridge: Cambridge University Press, 2000.
[14] M. Nielsen and I. Chuang, *Quantum Computation and Quantum Information*, Cambridge: Cambridge University Press, 2002.

[15] C. H. Bennett, G. Brassard, C. Crépeau et al., "Teleporting an unknown quantum state via dual classical and einstein-poldosky-rosen channels," *Phys. Rev. Lett.*, 70, 1895, 1993.
[16] N. Gisin, G. Ribordy, W. Tittel, and H. Zbinden, "Quantum cryptography," *Rev. Mod. Phys.*, 70, 145–195, 2002.
[17] C. H. Bennett and P. W. Shor, "Quantum information theory," *IEEE Trans. Inf. Theory*, 44, 2724–2742, October 1998.
[18] *J. Mod. Opt.*, 48(13), November 2001.
[19] R. H. Stolen and J. E. Bjorkholm, "Parametric amplification and frequency conversion in optical fibers," *IEEE J. Quantum Electron.*, 18, 1062–1072, July 1982.
[20] A. Lacaita, P. A. Francese, F. Zappa, and S. Cova, "Single-photon detection beyond 1 μm: performance of commercially available InGaAs/InP detectors," *Appl. Opt.*, 35, 2986–2996, June 1996.
[21] G. Ribordy, J.-D. Gautier, H. Zbinden, and N. Gisin, "Performance InGaAs/InP avalanche photodiodes as gated-mode photon counters," *Appl. Opt.*, 37, 2272–2277, April 1998.
[22] J. G. Rarity, T. E. Wall, K. D. Ridley et al., "Single-photon counting for the 1300–1600-nm range by use of Peltier-cooled and passively quenced InGaAs avalanche photodiodes," *Appl. Opt.*, 39, 6746–6753, December 2000.
[23] P. A. Hiskett, G. S. Buller, A. Y. Loudon et al., "Performance and design of InGaAs/InP photodiodes for single-photon counting at 1.55 μm," *Appl. Opt.*, 39, 6818–6829, December 2000.
[24] X. Y. Zou, L. J. Wang, and L. Mandel, "Violation of classical probability in parametric down-conversion," *Opt. Commun.*, 84, 351–354, August 1991.
[25] Y. R. Shen, *IEEE J. Quantum Electron.*, QE-22(8), 1196, 1986.
[26] G. Bonfrate, V. Pruneri, P. G. Kazansky et al., "Parametric fluorescence in periodically poled silica fibers," *Appl. Phys. Lett.*, 75, 2356–2358, October 1999.
[27] J. E. Sharping, M. Fiorentino, and P. Kumar, "Observation of twin-beam-type quantum correlation in optical fiber," *Opt. Lett.*, 26, 367–369, March 2001.
[28] P. L. Voss and P. Kumar, *Opt. Lett.*, 29, 445, 2004.
[29] X. Li, J. Chen, P. L. Voss et al., *Opt. Express*, 12, 3737, 2004.
[30] X. Li et al., *Phys. Rev. Lett.*, 94, 053601, 2005.
[31] M. H. Rubin, D. N. Klyshko, Y. H. Shih, and A. V. Sergienko, *Phys. Rev. A*, 50, 5122, 1994.
[32] G. P. Agrawal, *Nonlinear Fiber Optics*, San Diego: Academic Press, 2001, chapter 10.
[33] D. Bouwmeester, A. Ekert, and A. Zeilinger, *The Physics of Quantum Information: Quantum Cryptography, Quantum Teleportation, Quantum Computation*, 1st edn, Springer Verlag, 2000.
[34] T. E. Keller and M. H. Rubin, *Phys. Rev. A*, 56, 1534, 1997.
[35] A. K. Ekert, *Phys. Rev. Lett.*, 67, 661, 1991.
[36] R. J. Glauber, *Phys. Rev.*, 130, 2529, 1963; 131, 2766, 1963.
[37] M. Fiorentino, P. L. Voss, J. E. Sharping, and P. Kumar, *IEEE Photon. Technol. Lett.*, 14, 983, 2002.
[38] X. Li et al., *Opt. Lett.*, 30, 1201, 2005.
[39] X. Li, P. L. Voss, J. E. Sharping, and P. Kumar, *Phys. Rev. Lett.*, 94, 053601, 2005.
[40] J. H. Shapiro, *IEEE J. Quantum Electron.*, QE-21(3), 237, 1985.
[41] I. S. Gradshteyn and I. M. Ryzhik, *Table of Integrals, Series, and Products*.
[42] M. Eibl, S. Gaertner, M. Bourennane et al., *Phys. Rev. Lett.*, 90, 200403, 2003.
[43] A. Einstein, B. Podolsky, and N. Rosen, *Phys. Rev.*, 47, 777, 1935.
[44] J. S. Bell, *Physics*, 1, 195, 1964.
[45] D. M. Greenberger et al., *Am. J. Phys.*, 58, 1131, 1990.
[46] J. F. Clauser, M. A. Horne, A. Shimony, and R. A. Holt, *Phys. Rev. Lett.*, 23, 880, 1969.
[47] C. H. Bennett and S. J. Wiesner, *Phys. Rev. Lett.*, 69, 2881, 1992.
[48] C. H. Bennett et al., *Phys. Rev. Lett.*, 70, 1895, 1993.
[49] A. Aspect, P. Grangier, and G. Roger, *Phys. Rev. Lett.*, 49, 91, 1982.
[50] Z. Y. Ou and L. Mandel, *Phys. Rev. Lett.*, 61, 50, 1988.
[51] Z. Y. Ou et al., *Phys. Rev. A*, 41, 566, 1990.
[52] Y. H. Shih and C. O. Alley, *Phys. Rev. Lett.*, 61, 2921, 1988.

[53] J. W. Pan et al., *Nature*, 403, 515, 2000.
[54] K. Mattle et al., *Phys. Rev. Lett.*, 76, 4656, 1996.
[55] D. Bouwmeester et al., *Nature*, 390, 575, 1997.
[56] N. Gisin et al., *Rev. Mod. Phys.*, 74, 145, 2002.
[57] J. G. Rarity and P. R. Tapster, *Phys. Rev. Lett.*, 64, 2495, 1990.
[58] J. Brendel, E. Mohler, and W. Martiennsen, *Europhys. Lett.*, 20, 575, 1992.
[59] P. G. Kwiat, K. Mattle, H. Weinfurter et al., "New high-intensity source of polarization-entangled photon pairs," *Phys. Rev. Lett.*, 75, 4337–4341, 1995.
[60] F. A. Bovino et al., *Opt. Commun.*, 227, 343, 2003.
[61] K. Sanaka, K. Kawahara, and T. Kuga, *Phys. Rev. Lett.*, 86, 5620, 2001.
[62] J. Kim et al., *Nature*, 397, 500, 1999.
[63] C. Kurtsiefer et al., *Phys. Rev. Lett.*, 85, 290, 2000.
[64] X. Li et al., *Opt. Express*, 12, 3337, 2004.
[65] M. Fiorentino, P. L. Voss, J. E. Sharping, and P. Kumar, "All-fiber photon-pair source for quantum communications," *IEEE Photon. Technol. Lett.*, 14, 983–985, 2002.
[66] J. E. Sharping, M. Fiorentino, and P. Kumar, "Observation of twin-beam-type quantum correlation in optical fiber," *Opt. Lett.*, 26, 367–369, 2001.
[67] D. B. Mortimore, "Fiber loop reflectors," *J. Lightw. Technol.*, 6, 1217–1224, 1988.
[68] P. L. Voss, R. Tang, and P. Kumar, *Opt. Lett.*, 28, 549, 2003.
[69] P. L. Voss and P. Kumar, *Opt. Lett.*, 29, 445, 2004.
[70] P. L. Voss and P. Kumar, *J. Opt. B: Quantum Semiclass. Opt.*, 6, S762, 2004.
[71] M. Fiorentino, P. L. Voss, J. E. Sharping, and P. Kumar, *IEEE Photon. Technol. Lett.*, 14, 983, 2002.
[72] X. Li, P. L. Voss, J. E. Sharping, and P. Kumar, *Phys. Rev. Lett.*, 94, 053601, 2005.
[73] X. Li, P. L. Voss, J. Chen et al., *Opt. Lett.*, 30, 1201, 2005.
[74] X. Li, C. Liang, K. F. Lee et al., *Phys. Rev. A*, 2006; to appear. See arXiv:quant-ph/0601087
[75] X. Li, J. Chen, P. L. Voss et al., *Opt. Express*, 12, 3737, 2004.
[76] X. Li, P. L. Voss, J. Chen et al., *Opt. Express*, 13, 2236, 2005.
[77] P. Kumar, M. Fiorentino, P. L. Voss, and J. E. Sharping, U.S. Patent No. 6897434, 2005.
[78] H. Takesue and K. Inoue, *Phys. Rev. A*, 70, 031802, 2004.
[79] S. A. E. Lewis, S. V. Chernikov, and J. R. Taylor, *Opt. Lett.*, 24, 1823, 1999.
[80] R. H. Stolen and G. D. Bjorkholm, "Parametric amplification and frequency conversion in optical fibers," *IEEE J. Quantum Electron.*, 18, 1062–1072, 1982.
[81] H. Takesue and K. Inoue, *Opt. Express*, 13, 7832, 2005.
[82] J. Fan, A. Dogariu, and L. J. Wang, *Opt. Lett.*, 30, 1530, 2005.
[83] E. Knill, R. Laflamme, and G. J. Milburn, *Nature*, 409, 46, 2001.
[84] P. Kok, W. J. Munro, K. Nemoto et al., *arXiv*: quant-ph/0512071
[85] C. K. Hong, Z. Y. Ou, and L. Mandel, *Phys. Rev. Lett.*, 59, 2044, 1987.
[86] D. Branning, W. Grice, R. Erdmann, and I. A. Wamsley, *Phys. Rev. A*, 62, 013814, 2000.
[87] J. Fan, A. Dogariu, and L. J. Wang, *Opt. Lett.*, 30, 1530, 2005.
[88] P. Kumar, M. Fiorentino, P. L. Voss, and J. E. Sharping, U.S. Patent No. 6897434.
[89] K. F. Lee, J. Chen, C. Liang et al., to appear in *Opt. Lett.* (2006). See also post deadline paper presented at the *Frontiers in Optics* 2005, paper PDP-A4.
[90] H. Takesue and K. Inoue, *Opt. Express*, 13, 7832, 2005.
[91] J. Chen, K. F. Lee, C. Liang, and P. Kumar, in Proc. *IEEE/LEOS Summer Topical Meetings*, paper TuB2.3, 2006.
[92] J. Chen, P. L. Voss, M. Medic, and P. Kumar, "On the coincidence-to-accidental-coincidence ratio of fiber-generated photon pairs," to be submitted.
[93] C. Liang, K. F. Lee, P. L. Voss et al., in Proc. *SPIE*, 5893, 2005, p. 282.
[94] L. Goldenberg, L. Vaidman, and S. Wiesner, *Phys. Rev. Lett.*, 82, 3356, 1999.
[95] J. Eisert, M. Wilkens, and M. Lewenstein, *Phys. Rev. Lett.*, 83, 3077, 1999.
[96] E. Knill, R. Laflamme, and G. J. Milburn, *Nature*, 409, 46–52, 2001.
[97] Articles in *J. Mod. Opt.*, 48(13), November 2001.

[98] S. Castelletto et al., *New J. Phys.*, 6, 87, 2004.
[99] J. Chen et al., *Opt. Lett.*, 31, 2798, 2006.
[100] C. Liang et al., presented at *OFC/NFOEC 2005*, post-deadline paper, paper PDP35, 2005.
[101] C. Liang et al., *Opt. Express*, 15, 1322, 2007.
[102] M. Halder et al., *Phys. Rev. A*, 71, 042335, 2005.
[103] J. Fan, A. Dogariu, and L. J. Wang, *Opt. Lett.*, 30, 1530, 2005.
[104] J. Fan and A. Migdall, *Opt. Express*, 13, 5777, 2005.
[105] J. Chen et al., presented at Frontiers in Optics 2006/Laser Science XXII, paper FTuR4.
[106] W. P. Grice and I. A. Walmsley, *Phys. Rev. A*, 56, 1627–1634, 1997.
[107] D. B. Mortimore, *J. Lightw. Technol.*, 6, 1217, 1988.
[108] P. Kok et al., ArXiv: quant-ph/05 12071, 2006.
[109] W. P. Grice et al., *Phys. Rev. A*, 57, R2289, 1998.
[110] C. K. Hong, Z. Y. Ou, and L. Mandel, *Phys. Rev. Lett.*, 59, 2044–2046, 1987.
[111] H. de Riedmatten et al., *Phys. Rev. A*, 67, 022301, 2003.
[112] J. G. Rarity, P. R. Tapster, and R. Loudon, ArXiv: quant-ph/9702032, 1997.
[113] R. Kaltenbaek et al., *Phys. Rev. Lett.*, 96, 240502, 2006.
[114] C. Santori et al., *Nature*, 419, 594, 2002.
[115] X. Li et al., *Phys. Rev. A*, 73, 052301, 2006.
[116] J. Fan, A. Dogariu, and L. J. Wang, *Opt. Lett.*, 30, 1530, 2005.
[117] H. Takesue, and K. Inoue, *Opt. Express*, 13, 7832, 2005.
[118] K. F. Lee et al., *Opt. Lett.*, 31, 1905, 2006.
[119] J. Chen, K. F. Lee, and P. Kumar, ArXiv:quant-ph/0702176, 2007.
[120] P. G. Kwiat, E. Waks, A. G. White et al., "Ultrabright source of polarization-entangled photons," *Phys. Rev. A*, 60, R773–R776, August 1999.
[121] S. Tanzilli, H. De Riedmatten, W. Tittel et al., "Highly efficient photon-pair source using periodically poled lithium niobate waveguide," *Electron. Lett.*, 37, January 2001.
[122] C. E. Kuklewicz, E. Keskiner, F. N. C. Wong, and J. H. Shapiro, "A high-flux entanglement source based on a doubly-resonant optical parametric amplifier," *J. Opt. B: Quantum and Semiclass. Opt.*, 4, June 2002.
[123] J. Chen, X. Li, and P. Kumar, *Phys. Rev. A*, 72, 033801, 2005.
[124] J. L. O'Brien et al., *Nature*, 426, 264–267, 2003.
[125] N. K. Langford et al., *Phys. Rev. Lett.*, 95, 210504, 2005.
[126] N. Kiesel et al., *Phys. Rev. Lett.*, 95, 210505, 2005.
[127] R. Okamato et al., *Phys. Rev. Lett.*, 95, 210506, 2005.
[128] P. Walther and A. Zeilinger, *Phys. Rev. A*, 72, 010302, 2005.
[129] Z. Zhao et al., *Phys. Rev. Lett.*, 94, 030501, 2005.
[130] J. Chen, K. F. Lee, and P. Kumar, submitted to *Phys. Rev. Lett.*
[131] J. Chen et al., *Opt. Lett.*, 31, 2798, 2006.
[132] M. A. Jaspan et al., *Appl. Phys. Lett.*, 89, 031112, 2006.
[133] R. H. Hadfield et al., *Opt. Express*, 13, 10846, 2005.
[134] G. N. Gol'tsman et al., *Appl. Phys. Lett.*, 79, 705, 2001.
[135] Y. J. Lu, R. L. Campbell, and Z. Y. Ou, *Phys. Rev. Lett.*, 91, 163602, 2003.
[136] D. F. V. James et al., *Phys. Rev. A*, 64, 052312, 2001.
[137] J. B. Altepeter, E. R. Jeffrey, and P. G. Kwiat, *Advances in Atomic, Molecular, and Optical Physics*, Elsevier, 52, 2005, chapter 3.
[138] W. J. Munro, K. Nemoto, and A. G. White, *J. Mod. Opt.*, 48, 1239, 2001.
[139] N. Patel, *Nature*, 445, 144–146, 2007.
[140] D. Bouwmeester, A. Ekert, and A. Zeilinger, *The Physics of Quantum Information*, Berlin: Springer-Verlag, 2000.
[141] D. C. Burnham and D. L. Weinberg, "Observation of simultaneity in parametric production of optical photon pairs," *Phys. Rev. Lett.*, 25, 84–87, 1970.

[142] X. Li, P. Voss, J. E. Sharping, and P. Kumar, "Violation of Bell's Inequality Near 1550 nm Using an All-Fiber Source of Polarization-Entangled Photon Pairs," in Proc. *QELS'2003*, Optical Society of America, QTuB4, Washington, DC, 2003.

[143] X. Li, P. Voss, J. E. Sharping, and P. Kumar, "Optical-fiber source of polarization-entangled photon pairs in the 1550 nm telecom band," Quantum Physics Archive, 0402191 (2004), http://xxx.lanl.gov/abs/quant-ph/0402191

[144] P. Russell, "Photonic crystal fibers," *Science*, 299, 358–362, 2003.

[145] M. Fiorentino, J. E. Sharping, P. Kumar et al., "Soliton squeezing in microstructure fiber," *Opt. Lett.*, 27, 649–651, 2002.

[146] A. Dogariu, J. Fan, L. J. Wang, and J. A. West, "Photon-Pairs Generation in Micro-Structured Fiber," in Proc. *QELS'2003*, Optical Society of America, QTuB6, Washington, DC, 2003.

[147] J. E. Sharping, M. Fiorentino, A. Coker et al., "Four-wave mixing in microstructure fiber," *Opt. Lett.*, 26, 1048–1050, 2001.

[148] J. E. Sharping, M. Fiorentino, P. Kumar, and R. S. Windeler, "Optical-parametric oscillator based on four-wave mixing in microstructure fiber," *Opt. Lett.*, 19, 1675–1677, 2002.

[149] S. Lloyd, M. S. Shahriar, J. H. Shapiro, and P. R. Hemmer, "Long distance unconditional teleportation of atomic states via complete Bell state measurements," *Phys. Rev. Lett.*, 87, 167903, 2001.

[150] G. P. Agrawal, *Nonlinear Fiber Optics*, 3rd edn, San Diego, CA: Academic Press, 2000.

[151] D. Ouzounov, D. Homoelle, W. Zipfel et al., "Dispersion measurements of microstructured fibers using femtosecond laser pulses," *Opt. Commun.*, 192, 219–223, 2001.

[152] P. L. Voss and P. Kumar, "Raman-noise-induced noise-figure limit for chi (3) parametric amplifiers," *Opt. Lett.*, 29, 445–447, 2004.

[153] J. D. Harvey, R. Leonhardt, S. Coen et al., "Scalar modulation instability in the normal dispersion regime by use of a photonic crystal fiber," *Opt. Lett.*, 28, 2225–2227, 2003.

Index to Volumes VA and VB

1 × 1 WSS, A:317
1 × 2 MMI coupler, A:271, 273
1 × 2 thermo-optic switch, A:393
1 × 4 wavelength-selective switch (WSS), A:723, 727
1 × 9 WSS, A:317
   waveguide layout, A:317
1 × K WSS, A:317
1 × N^2 using two-axis micromirror array, A:720, 721
100-Gb/s transmitter and receiver PIC, A:352
10GBASE-LRM, A:699
1480-nm pumps/14xx-nm high-power lasers, A:122–124
2 × 2 MEMS wavelength add/drop switch, A:721
2 × 2 MMI coupler, A:273
2 × 2 star coupler, A:273, 297
   mutual coupling between waveguides, A:300
20 Gb/s dual XFP transceiver, A:416–417
2D MEMS switches, A:714
2R (reamplification and reshaping), A:760
2R regenerators, B:722
3D MEMS switches, A:716–719
   Fujitsu's 80 × 80 OXC, A:718
   Lucent's optical system layout for OXC, A:718
   structure of, A:715
3D photonic crystals, B:8
3R (reamplification, reshaping, and retiming), A:760
3R regeneration, A:782–784
3rd harmonic generation, B:11
40-Gb/s WDM transceiver, A:417
   integrated driver and MZ modulator, A:418
   receiver performance, A:419

8-channel CWDM (de)multiplexer, A:304
8/9 factor, A:656
9xx-nm MM pump lasers, classification of
   heat transport classification, A:125–126
   optical classification, A:126–127

A

access-layer networks, B:514
acknowledgment pulse, B:780, 784
acoustooptic tunable filters, B:253
active optical device modeling, B:825
adaptive equalization techniques, B:122, 157
adaptive filter, A:679
Adaptive unconstrained routing, B:649
add/drop
   filters, B:505
   multiplexers, B:306, 487, 524, 635
Address resolution protocol, B:393
adiabatic directional coupler, A:274
advanced modulation formats, B:12, 23, 28, 52, 64, 70, 74, 75, 837
"Advanced Research and Development Activity" (ARDA) agency, A:32
aerial cable terminal, B:424
aerospace applications, limitations, A:375
alignment accuracy, B:135
all-optical signal processing devices, A:11, 760, 761, 766, 778, 818
   future prospects
      nanowire-based integrated all-optical regenerator in chG, A:819
      nonlinear-induced phaseshift *vs* length for a pulse, A:819
alternate chirp return-to-zero, B:58
alternate mark inversion, B:41
alternate phase, B:41

alternate polarization
  modulation, B:41
  signal, B:207, 602
American National Standards
  Institute, B:415
amplification scheme, B:71
amplified spontaneous emission (ASE),
  A:6, 57, 58, 63, 114, 127, 132, 154,
  158, 159, 231, 393, 505, 662, 769, 854
amplified spontaneous emission, B:6, 25,
  30, 96, 137, 235, 310, 714, 753, 796,
  831, 841
amplifier
  cascades, B:498
  chip, B:188, 189
  control, B:243, 325, 326, 331
  design, B:18, 568
  gain, B:236, 242, 253, 326, 330, 331, 753
  noise, B:247
  spacing, B:73, 274, 496
amplitude
  distortion, B:651
  impairments, B:426, 428
  implementations, B:153
  measurement techniques, B:245
  modulated pilot tone, B:256
  phase modulation (chirp), 44
  power spectrum, B:245
  shift keying, B:34, 123, 132
  signal processing, B:147
analog micromirror, A:723, 724
  two-axis for WSS, A:724
analog transport system *vs* digital
  transport system, A:344
analog-to-digital conversion, B:106, 131
anisotropic etching, A:736
antenna remoting, B:758
APD, *see* avalanche photodiodes (APDs)
Apollo Cablevision, B:459
application-specific integrated circuits,
  B:165
ARDA, *see* "Advanced Research and
  Development Activity" (ARDA)
  agency
arrayed waveguide grating (AWG), A:273,
  279–311, B:6, 311, 661, 697, 807
  aberrations, A:299–301

athermal (temperature-insensitive),
  A:305–307
  channels' peak transmissions shift,
    A:306
  silicone-filled, A:306
  specifications and athermal
    condition, A:307
band de/multiplexer using, A:303
chirped, A:309–311
component in WDM systems, A:394
conventional *vs* silicone-filled
  athermal, A:307
creating flat spectral-response, A:288
crosstalk, A:284–286
demultiplexing properties of
  32-ch 10-GHz, A:285
  32-ch 50-GHz, A:284
  32-ch 100-GHz, A:289
  32-ch 100-GHz, parabolic horns,
    A:289
  400-ch 25-GHz, A:285
  sinusoidal phase fluctuations, A:286
dense InP, A:349
diffraction order, A:281
dispersion, A:286–287
  characteristics of three kinds of,
    A:287
flat spectral-response, A:287–290
flattening passbands, techniques, A:288
  see also arrayed waveguide grating
    (AWG), periodic stationary imaging
gaussian spectral response, A:283–284
grating efficiency, improving, A:300, 301
grating order of zero, *see* arrayed
  waveguide lens (AWL)
high spectral sampling, A:297–299
  sampling theorem, A:298–299
  use, A:299
interference condition, A:281
parabola-type flat, A:289
  dispersion characteristics of 32-ch
    100-GHz, A:290, 291
  phase retardations, A:290
passband, A:283
  and achieving multiple zero-loss
    maxima, A:290–291
  "flat" passbands, A:287

Index

periodic stationary imaging, A:290
  achieving multiple zero-loss maxima, A:290
  possible ways of laying out the MZI–AWG, A:295
  second star coupler transmissivity, A:300
  spatial range of, A:282
  tandem configuration, A:308–309
  three-arm interferometer input, A:296–297
  two-arm-interferometer input, A:293–296
    creating two zero-loss maxima per FSR, A:293–294
  waveguides, A:279
    in first slab, A:279
    path-length difference, A:282
    in second slab, A:280
    wavelength for fixed light output position, A:282
    wavelength spacing, A:282
arrayed waveguide grating (AWG) demultiplexer, A:297
  16-Channel, consisting of three-arm interferometer, A:298
arrayed waveguide lens (AWL), A:301
Artificial defects, A:470–471
ASE noise, signal impairment, A:769
ASE, *see* amplified spontaneous emission (ASE)
asymmetric directional coupler, A:273, 274
asymmetric Fabry–Perot modulator, B:751
asymmetric-bandwidth interleavers, B:168
asynchronous amplitude histogram, B:280
asynchronous digital subscriber line, B:457
asynchronous histogram monitoring techniques, B:280
asynchronous optical packets, switching of, B:631
asynchronous transfer mode, B:416, 479, 516
  adaptation layer, B:523
AT&T, B:16, 293, 410, 415, 443, 448, 451, 454, 461, 465, 466, 468, 511, 526, 528, 530, 549
  long lines business unit, B:530
  Project Lightspeed, B:443

athermal (temperature-insensitive) AWGs, A:305–307
Atlantic and the Pacific Oceans, B:564
Attenuation mechanisms, PCF, A:495–498
  absorption/scattering, A:494–495
  bend loss, A:495–496
  confinement loss, A:498
Autocorrelation and first-order PMD, A:633
automatic dispersion compensation, B:213
automatic protection switch, B:487
automatic switched optical network, B:615, 642
avalanche photodiodes (APDs), A:835, B:5, 6, 46
  low-excess noise APDs, *see* multiplication regions, low-excess noise APDs
  performance of InGaAs APDs as single-photon detectors, A:835
  separate absorption/charge/multiplication APDs, A:245–247
  InP/InGaAsP/InGaAs SACM APDs, A:246
  "long-wavelength" photodetectors, A:245
  SACM structure, "C" (charge layer), A:246
  single-photon avalanche detectors, A:256–258
  Geiger-mode operation, A:256
  quantum cryptography/quantum key distribution, A:256
  single-photon-counting avalanche photodetector (SPAD), A:256
average effective refractive index, B:713
AWG multiplexer
  distributed feedback (DFB) lasers spectrum with, A:353
AWG, *see* arrayed waveguide grating (AWG)
AWG-based switch fabrics, B:727
AWL, *see* arrayed waveguide lens (AWL)

B

back-to-back
  bit error rate, B:113
  configuration, B:225, 582, 603, 605
  measurements, B:126, 224

back-to-back (*continued*)
  performance, B:67, 215
  sensitivity, B:38, 53, 54, 154
backbone routers, B:531, 546
Baker–Hausdorff formula, B:832
balanced detection, B:47, 61, 72, 73, 100, 138, 141, 144, 147, 168, 170, 218, 225, 633, 759
balanced driving *see* push-pull; operation
band edge engineering, A:456
  broad area coherent laser action, A:473–476
  lattice point control and polarization mode control, A:476
  resonance action at band edge, A:473
band engineering, A:456, 476–477
  group velocity of light, control of, A:477–478
  light propagation, angular control of, A:478–479
band gap/defect engineering, A:456–457
  three-dimensional photonic crystals, A:467
    fabrication technology, A:467–468
    light propagation, control of, A:472–473
    spontaneous emission, control of, A:468–472
  two-dimensional photonic crystal slabs, A:456–457
    heterostructures, effect of introducing, A:459–463
    high-Q nanocavities, A:463–466
    line/point-defect systems, control of light by combined, A:457–459
    nonlinear and/or active functionality, introduction of, A:466–467
band multiplexer, A:303
  5-band 8-skip-0 multiplexer, A:304
bandwidth, B:26, 200, 454, 753, 769, 851
  limitations, B:193, 705, 720
  -on-demand, B:302, 304, 549, 556
  -time trading, B:781
  -times-distance, B:445, 446, 458, 461, 468
banyan networks, B:773, 774, 778
BASE, *see* broad-area single-emitter (BASE)

Baud, B:29, 165, 166
beam propagation method (BPM), A:282
  central wavelength, A:282
  finite-difference (FD-BPM), A:299, 300
  Fourier transform (FT-BPM), A:299–300
  intensity distribution, double-peaked, A:288
  mutual coupling between waveguides, A:300
  shorter wavelength component, A:282
beam propagation method, B:805
beam-steering mirrors, B:316
Bell company, B:451, 462
Bell system, B:522, 530
Bellcore, B:522, 530
bending losses, A:387
  guide configurations, A:387
BER measurement of QD laser module, A:37
Bernoulli traffic, B:785, 786
BERTS, *see* bit-error rate test set (BERTS)
bidirectional line-switched rings, B:523
binary differential phase shift keying, B:58
binary electronic drive signals, B:62
binary optical intensity eye diagram, B:54
binary phase shift keyed, B:757
binary tree-like configuration, B:794
bipolar feedback signal, B:211
birefringence, A:275, 494, 607–609
birefringence's derivative, A:612
bit error rate, B:26, 113, 133, 180, 238, 385, 494, 533, 567, 571, 582, 596, 600–602, 604, 804, 830
  measurements, B:65, 215, 219, 224, 676
  tester, B:219
bit-error rate test set (BERTS), A:204
BitTorrent, B:457
blind
  equalization, B:122
  estimator, B:160
block length, B:164
blocker, *see* 1 × 1 WSS
blocking/nonblocking, B:653
Boltzmann's constant, B:742
boundary-value problem, B:822

BPM, see beam propagation method (BPM)
Bragg diffraction, A:473
Bragg gratings, A:327, 512–513,
  B:253, 586
Bragg reflection, A:464
Bragg wavelength, B:632
branch stations, B:569
Bridge Protocol Data Units, B:362
bridging, B:346, 358, 362, 568, 604
Brillouin gain
  coefficient, B:746
  spectra, B:427, 428
Brillouin scattering, A:507–509
  backward, A:507
  forward, A:508–509, B:831
broad-area single-emitter (BASE), A:127
broadband
  access, B:417
  asynchronous histogram, B:276
  deployment, B:409, 615
  digital cross-connect system, B:525
  over power line, B:466
  storm, B:362, 365
broadbanding techniques
  principles of, A:682–683
  RC broadbanding, A:683–687
  series shunt-peaked broadbanding,
    A:694–697
  series-peaked broadbanding,
    A:692–693
  shunt-peaked broadbanding,
    A:687–692
Broglie wavelength, B:3
bubble-and-bust, B:2, 3
buffer
  delay, B:709
  memory, B:656
  parameters, B:710
  technologies, comparison of, B:719
buffering, B:226, 536, 629, 630, 637, 643,
  650, 651, 665, 667–679, 671, 677, 683,
  695, 696, 698, 699, 701, 707, 710, 715,
  729, 733, 734, 771, 772, 777, 778, 781,
  792, 793
  electronic messages, B:792
  mechanisms, B:630
  space, B:781

bufferless photonic switching nodes, B:778
built-in optical buffer, B:649
buried oxide (BOX) layer, A:384
burst assembly, B:648
burst header cell, B:645
burst mode
  operation, B:385
  receivers, B:170, 359
burst scheduling, B:651
butt-joint regrowth, A:345–346, 348
butterfly network see banyan networks

C

cabinet installation, B:425
cabinet size reduction, B:425
cable management technology, B:422
cable modem, B:406, 414, 440, 460,
  461, 466
cable television, B:17, 305, 378, 403, 739,
  748, 758
cables with PF polymer-based POFs,
  A:599
canonical format indicator, B:366
capability index (Cpk), A:356
capacity upgrade, B:13, 131, 171, 596
capital expenditure, B:482, 484, 506, 507
capital investment, B:418
card failure, B:544
carrier craft personnel, B:423
carrier density, B:749, 825, 829
carrier phase-estimation method, B:102
carrier suppressed return-to-zero, B:40,
  56, 57, 597
carrier-grade telecommunication
  equipment, B:65
  see also Ethernet
carrier-to-noise ratio, B:426, 744
Cartesian grid, B:34
cascaded filtering, B:320
cascaded optical injection locking
  (COIL), A:153, 158
cascaded-ring-optical-waveguide filters,
  A:397, 398
catastrophic optic mirror damage
  (COMD), A:119
catastrophic problems, B:244

CD *see* chromatic dispersion
cellular-mobile telephony, B:449, 454
chalcogenide fiber-based wavelength
    conversion, A:789–793
  BER *vs* received optical power, A:791
  demonstrating XPM-based wavelength
    conversion, A:790
  principle of XPM wavelength
    conversion, A:789
  required calculated pump
    intensity, A:793
  spectra of As_2Se_3 fiber output, A:791
chalcogenide fibers
  high quality fabrication processes, A:763
  nonlinear material characteristics,
    A:764
  PLD development, A:763
  and waveguide devices, A:763
chalcogenide waveguides, A:793–796
  BERs of converted pulses, A:795
  cross phase modulation-based, A:794
channel conditioning, B:308
channel demultiplexing, B:123, 776
channel filtering effects, B:500
channel frequencies, B:125
channel identification, B:243
channel power
  attenuation, B:330
  monitoring scheme, B:252
channel spacing, B:52, 53, 71, 123, 124,
    126, 127, 195, 218, 234, 238, 247, 322,
    390, 391, 479, 484, 492, 493, 495, 499,
    500, 596, 604, 605, 675, 835
channel spectra, B:169, 564, 596, 604, 605
channel telemetry, B:309
channel-degrading effects, B:233, 236, 285
channel-to-channel cross-polarization,
    B:605
chip cost *vs* assembly plus testing, A:345
chirp, B:4, 5, 42–45, 51, 54, 56, 58, 59,
    135, 205, 235, 245, 247, 264, 265, 266,
    277, 319, 493, 651, 653
  effects, B:264
  fluctuation, B:265
  managed laser, B:42
  multiprocessors, B:765
  parameter, B:264

Chirped AWG, A:309
  interleave Chirping, A:310–311
  WSS using two and AWL, A:311
  linear chirping, A:310
chirped fiber Bragg gratings, B:181, 195
chirped return-to-zero, B:58, 596
chromatic dispersion, B:2, 15, 24, 31, 37,
    42, 54, 63, 68–70, 72, 77, 110, 118,
    120, 121, 123, 131, 133, 134, 137, 150,
    155, 165, 166, 168–170, 180–182, 193,
    202, 203, 208, 212–214, 223, 225, 234,
    235, 243, 245, 247, 256, 257, 260, 261,
    263–271, 273, 275, 281–284, 348, 359,
    365, 366, 379, 380, 384, 391, 414, 565,
    566, 571–573, 576–581, 593, 596,
    601–603, 616, 630, 682, 775, 816, 818,
    828, 836, 844, 846, 847
  compensation of, B:212
chromatic dispersion, transmission
    impairments, A:800, 808
circuit-switched
  layer, B:522, 530, 536, 551
  telephone network, B:19, 542
cladding-pumped fiber, A:556
class of service, B:545, 648
cleaving/splicing, PCF, A:500
client layer protection requirements, B:852
client optoelectronic, B:485
client service interfaces, B:491
clock distribution circuits, B:187
clock modulation, B:211, 224
clock recovery, B:108, 138, 192, 193, 203,
    204, 208, 210, 211, 216, 217, 219,
    224–226, 267, 280, 282, 348, 356, 660,
    673, 681, 781, 788
  circuit, B:108, 216
  device, B:203, 204, 211, 216, 225, 226
clock-phase margin, B:192
CMOS integration and integrated silicon
    photonics
  examples, A:416–420
  microwave OE oscillators, A:420–421
  microwave spectrum analyzer, A:421
  SOI transistor, A:414–416
CMOS technology, state of art in, A:700
CMOS, *see* complementary metal oxide
    semiconductor (CMOS)

Index

CMOS-like gate dielectric, charge accumulation, A:404
coarse WDM, A:304, B:478, 516
coaxial cable, B:349, 410, 413, 419, 551, 617
code division multiplexing, B:656
coherent crosstalk, B:131, 170, 241
coherent demodulation, B:12, 28, 46, 69, 116
   see also delay demodulation
coherent detection, A:705–706, B:13, 23, 24, 26, 33, 49, 52, 63, 64, 76, 78, 79, 97, 98, 118, 121, 131, 137, 145, 148, 149, 154, 195, 196, 268, 272, 606, 743, 745
   performance of, B:145
   principle of, B:98
coherent optical communications, B:95, 96
coherent receiver, B:46, 47, 48, 64, 96–99, 102, 121, 132, 147, 150, 152, 162, 166, 170, 171, 319, 745
   see also digital coherent receiver
coherently connected interferometers (FIR-based TODCs), A:329
   two-arm interferometers, A:329–330
   multiple controls, A:329, 330
   single controls, A:329–330
COIL, *see* cascaded optical injection locking (COIL)
coincidence-counting rate
   biphoton amplitude, A:844
coincidence-to-accidental ratio (CAR), A:865
collective blindness, new dawn, A:25–26
   coherent three-dimensional clusters, A:26
   "Frank–van der Merwe" growth mode, A:26
collimators, B:111
collision avoidance, B:695
colorless drop, B:313
combiner, *see* optical coupler
COMD, *see* catastrophic optic mirror damage (COMD)
commercial packet switches, B:20
committed information rate, B:378
common gate cascode shunt-peaked amplifier, A:690

communication engineering techniques, B:23, 28
communication symbols, B:28
community antenna television, B:440
Compact resonant ring modulator, A:409
competitive local exchange carrier, B:448, 524
complementary metal oxide semiconductor (CMOS), A:381, B:166
composite triple beat, B:426
consent decree, B:519
constant bit rate, B:549
constant modulus algorithm, B:161
   see also Godard algorithm
constant-amplitude format, B:150
constellation diagram, B:46, 121, 133, 148, 159
constrained coding, B:39–41
continuous-time DFE using analog feedback, A:698
control devices
   inter-signal, A:312
      large channel count ROADMs, A:316–317
      reconfigurable optical add/drop multiplexers, A:313–316
      wavelength-selective switches, A:317–318
   intra-signal, A:318
      optical chromatic dispersion compensators, A:325–335
      optical equalizers, A:322–325
      temporal pulse waveform shapers, A:319–322
coplanar waveguide (CPW), A:243
core segment, B:555
core-layer networks, B:529
correlative coding, B:41, 53, 57
cost effective-related technologies, B:502
cost reduction, B:13, 131, 171, 186, 401, 421, 422, 431
cost-effective technologies, B:502
cost-effectiveness of silicon, B:20
counter-pumped amplifiers, B:333
coupled mode theory, A:458
coupled nonlinear Schrodinger equations, A:652–655

coupled resonator optical waveguides (CROW), A:444
coupled-resonator waveguides, B:713
coupling, B:8, 9, 36, 44, 191, 246, 316, 324, 326, 328, 330, 332, 338, 416, 421, 422, 428, 504, 505, 671, 683, 752, 757, 807, 814, 819, 824, 825, 837
   waveguides, B:422
CPW, see coplanar waveguide (CPW)
Crank–Nicholson scheme, B:807
cross phase modulation, B:52, 60, 64, 72, 79, 123, 124, 126, 127, 137, 162, 166, 167, 170, 171, 209, 211, 241, 273, 274, 566, 567, 572, 578, 831, 833
cross-phase modulation (XPM), A:537, 657, 761
crossbar, B:778, 782, 790
crosstalk, B:33, 253, 321, 353, 651, 653, 728
crystal and waveguide lasers, B:748
current injection, B:825
current-controlled oscillator, B:749
customer premises equipment, B:521
cybersphere, B:457
cyclic redundancy check value, B:359

D

data burst transmission, B:627
data communication channel, B:645
data communication network, B:645
data modulators (MEMS devices), A:743–745
Data Over Cable Interface Specification, B:414
data signal processing, B:13
data stream intensity autocorrelation, B:268
data traffic and evolution of the optical network, B:19
data transmission
   upgrading to 10Gb/s, A:672
Data Vortex see topology
data warehousing, B:481
data-aided MSPE scheme, B:164
datacom applications with POFs, A:602
DC properties of long-wavelength, QW lasers

DC measurements of optical gain/refractive index change/LEF, A:56–58
   E-TEK Dynamics LOFI, A:57
   Hakki–Paoli method, A:57
DC-measured experimental data/theory, comparison, A:59–63
   Auger recombination process, A:61
   LEFs of InAlGaAs and InGaAsP systems, comparison of ASE spectra, A:64
optical gain/refractive index change/LEF, theory of
   non-Markovian gain model, A:55
DDF, see dispersion-decreasing fiber (DDF)
decision feedback equalizer (DFE), A:675–676, B:214
   better performance than LE, A:675
   designing of, A:676
   low SNR, A:675
   structure of, A:675
decision flip-flops, B:191
decisive break-throughs, A:26–27
defect budgeting, A:361
defect density and functionality per chip, A:361–363
   position of PIC technology, A:363
deflection, B:651, 725, 774, 775, 777, 792–796
   nodes, B:783
   routing, B:725, 777, 792, 793
   signal, B:774, 775, 793, 795, 796
degenerate mode group, B:838
degenerate photon pairs for quantum logic in telecom band
   photon-pair generation in optical fiber, A:863–869
      CAR plotted as function of single counts/pulse, A:865
      pump preparation and CPS scheme, A:864
      single counts and coincidence counts, A:868
      two-photon interference, A:867
delay demodulation, B:12, 46–49, 64
delay interferometer, B:46, 133, 144, 193, 194

delay line
 buffers, B:710
 multiplexer, B:223
demodulation, B:28, 46, 48, 49, 58, 62, 69, 96, 97, 100, 106, 107, 115–117, 123, 137, 147, 601, 603
demodulator phase error, B:152, 153
demultiplexer, B:123, 160, 165, 183, 191–193, 203, 208, 210, 215–218, 224, 226, 314, 353, 415, 618, 671, 676, 699, 718, 724, 790, 851
demultiplexing, A:782, 792, B:26, 53, 123, 125, 137, 180, 183–185, 192, 208, 209, 210, 215–218, 222, 226, 256, 296, 314, 320, 415, 526, 527, 561, 604, 656, 676
 applications, B:216
 components, B:226
Dense InP AWG (25GHz)-based 64-channel O-CDMA encoder/decoder chip, A:349
dense wavelength division multiplexed, B:4, 119, 351
dense wavelength-division multiplexing (DWDM), A:4, 109, 116, 172
design of experiments (DOE), A:357
destination address, B:358, 360, 362, 375, 653, 682, 774, 792, 795, 796
detection techniques, B:18, 28, 745
DGD *see* differential group delay
DGD changes *vs* outside temperature, A:618
dial-up modem, B:440
dielectric thin-film filters, B:253
differential binary phase-shift keying, B:134
differential delays, B:381
differential detection, B:58, 61, 116, 164, 598, 602
differential group delay (DGD), A:607, B:70, 160, 162, 168, 214, 219, 222, 223, 262–266, 268–272, 570, 571, 582–583, 603
differential modal dispersion (DMD), A:699
differential mode delay, B:837
differential phase-shift keying, B:33, 49, 50, 58, 62, 131, 203, 256, 269, 598, 625

differential quadrature phase-shift keying, B:184, 390
differential quadrature PSK (DQPSK) demodulator, A:278
differential-detection penalty, B:141, 145, 148, 153, 155
differential-phase distance, B:148
differentially coherent demodulation, B:46
diffraction order, A:281
diffractive spectrometers and spectral synthesis
 characteristics of, A:739–741
 optical MEMS, A:740
 optical system, A:739
digital coherent receiver, B:97, 98, 105–107, 111, 114, 123, 125, 127, 128, 132, 157, 158, 162, 164, 165
 advantages of, B:123
digital communication receiver, B:30
digital cross-connect, B:623
digital equalization, B:157
digital homodyne receiver, B:123, 124
digital line module (DLM), A:366
digital micromirror devices, successful MEMS product, A:713
digital network, constructing, A:344–345
digital optical network (DON), A:344, 376
 architecture, A:367–368
 monolithically integrated TX and RX PICs, A:351
digital radio-frequency, B:12
digital ROADMs, A:368
 data ingress/egress in, A:369–370
digital self-coherent receivers, B:131, 147, 150
digital signal processing, B:23, 48, 63, 64, 69, 70, 76, 78, 79, 97, 107, 131, 163, 196
digital subscriber line, B:410, 438, 515
 multiplexer, B:462
digital TV, B:413
 evolution of, B:449
digital video recorder, B:554
diode-pumped crystal, B:748
diodepumped solid-state lasers (DPSSL), A:108
direct-contact wafer bonding, A:346
direct-detection schemes, B:149

directional coupler, A:272
  adiabatic, A:274
  asymmetric, A:273, 274
  curved, A:274, 275
  multi-section, A:274
directional separability, B:306
directly coupled resonator configuration, A:432, 433, 437, 439
directly modulated lasers, B:42, 722, 757
Discovery Gate, B:381
discovery window, B:381
discrete event simulations, B:852
discrete multitone, B:415
dispersion compensation fibers, A:526–539
  design examples, A:532–534
  experimental results, A:534–536
  fiber profile designs, A:531–532
  future developments, A:538–539
  mode field diameter/effective area, A:536–537
  nonlinear properties, A:537
  principle, A:527–531
    multiple path interference (MPI), A:531
    parameter κ, A:529
    PMD, see polarization mode dispersion (PMD)
    Rayleigh scattering, A:530
  splicing loss, A:538
  technologies, A:526–527
dispersion compensation, B:9, 10, 12, 13, 58, 71, 79, 97, 121, 165, 166, 181, 194, 195, 208, 213, 214, 219, 234, 236, 255, 261, 263, 296, 317, 318, 319, 322, 353, 390, 495, 505, 571, 706, 713, 716, 757
  devices, B:571
  fiber, B:181–183, 495, 581, 741
  inaccuracies in, B:234
  technique, B:213, 214
  unit, B:836
dispersion compensators, B:8, 195, 213, 261, 356, 573, 625, 634, 714
  all-pass filter, A:733
  MEMS Gires–Tournois interferometer, A:734
dispersion equalization capability, B:150

dispersion management, B:319, 836
dispersion map, B:52, 54, 58, 60, 63, 64, 74–76, 79, 170, 244, 274, 317–320, 334–336, 339, 390, 577, 835
  effect of, B:75
dispersion slope, B:58, 213, 214, 222, 223, 335, 576, 577, 579, 581, 582, 836
dispersion tolerance, B:55, 69, 70, 182, 319, 352, 388, 493, 495, 496
dispersion-decreasing fiber (DDF), A:548, 553–554
dispersion-limited propagation distance, B:51
dispersion-managed fiber cable systems, A:538
dispersion-managed fiber, B:213
dispersion-managed transmission, B:167
dispersion-related impairment, B:281
dispersion-shifted fiber, B:12, 213, 576
dispersive impairments monitoring, B:260
distortion, B:247
distributed feedback (DFB) lasers, A:347
  L–I–V characteristics, A:363
  typical light vs current/voltage (L–I–V), A:352
distributed feedback, B:62, 110, 388, 663, 749, 826
  lasers, B:826
distribution network, B:784
DLM, see digital line module (DLM)
DMD, see differential modal dispersion (DMD)
donut-shaped far-field pattern, A:476
doped fiber amplifier, B:4, 6, 96, 235, 413, 561, 593, 748, 752, 836
double sideband, B:264
double-clad fibers, fiber lasers/amplifiers by OVD, A:555–572
  concept, A:555–556
  double-clad fiber 101, A:556–562
  requirements, A:558
double-clad fibers, high-performance Yb-Doped
  Al/Ge counter-graded composition design/fabrication, A:568–570
  reduced-overlap, A:569
  introduction, A:565

Index 891

SBS effect in optical fiber, A:565–566
SBS-managed fiber composition design, A:566–568
SBS-Managed LMA, characteristics of, A:570–572
double-gate switch
 "off" state, A:313, 314
 "on" state, A:313–314
double-gate thermooptic switches (TOSWs), A:313
double-stage polarization, B:181
DPSSL, *see* diodepumped solid-state lasers (DPSSL)
DRAM (Dynamic Random Access Memory)/flash memory cells, A:23
driver amplifier, B:45, 183, 185–191, 193, 195
drop spectra, A:462
drop-and-continue function, B:305
Drude model, A:399
dual-band filter, transmission spectrum of measured with tunable source and OSA, A:834
dual-frequency pumps
 degenerate frequency polarization-entangled photon pairs produced by, A:869
duobinary, B:41, 45, 53, 54, 133, 256, 258, 357, 390, 495, 834, 836
DWDM, *see* dense wavelength-division multiplexing (DWDM)
dynamic bandwidth services, B:549
dynamic channel equalizers (DCEs), A:720
dynamic gain equalization filter, A:301–303
 arrangement with lower insertion loss, A:302
 consisting of two AWGs connected by AWL, A:302
dynamic gain equalizers (DGEs), A:719
dynamic optical circuit switching, B:626
dynamic photonic crystals, A:443
 creating, A:445
 future prospects, A:450–451
 structure used to stop light, A:447
dynamic photonic crystals, stopping light in, A:443
 conditions for stopping light, A:444–445

 numerical demonstration in photonic crystal, A:448–450
 from tunable bandwidth filter to light-stopping system, A:447–448
 tunable Fano resonance, A:445–447
 tuning spectrum of light, A:443–444
dynamic random-access memory, B:710
dynamic spectral equalizer, A:720

E
e-commerce, B:481
e-Field Domain Signal Processing, A:705–706
e-Health, B:614
E-Line, B:378
e-science applications, B:555
EAM, *see* electroabsorption modulator (EAM)
EDC, *see* electronic dispersion compensation (EDC)
EDFA, *see* erbium-doped fiber amplifier (EDFA)
edge emitting lasers, B:4, 826
edge routers, B:649
EEQ, *see* electrical equalization (EEQ)
effective-index microstructured optical fibers (EI-MOF), A:575–579
 air-filling fraction, A:576
 attenuation, A:579
 basic concepts, A:575–576
 dispersion, A:578–579
 endlessly single-mode (ESM), A:577
 small modal area/nonlinearity, A:577–578
 V-number, A:576–577
EI-MOF, *see* effective-index microstructured optical fibers (EI-MOF)
EIT, *see* Electromagnetically Induced Transparency (EIT)
electrical amplifiers, B:758, 830
electrical equalization (EEQ), A:322
electrical multiplexing, B:14, 201
electrical phase shifters, B:192
electrical sampling systems, B:205
electrical signal processing, B:214
electrical spectrum analyzer, B:276

electrical switching fabric, B:502
electrical synchronous demodulation, B:100
electrical time-division multiplexing, B:179
electro-absorption
   devices, B:5
   modulator, B:5, 42, 43, 186, 204, 388, 636, 663
electro-optic
   bandwidth, B:191
   coefficient, B:188
   effect, B:807
   modulators, B:5
   phase comparator, B:210
electroabsorption (EA), A:192
electroabsorption modulator (EAM), A:185, 189, 198, 208, 210, 348
   small-signal frequency response, A:353
electroabsorption-modulated lasers (EML), A:347
electromagnetically induced transparency (EIT), A:445
electron charge, B:99, 825
electronic and optical properties, A:28–31
   biexciton-binding energy, A:31
   excitonic/biexcitonic recombination, A:31
   GaAs-based QD technology, A:29
   piezoelectric potentials, A:29
   shape/composition, QDs, A:28
   tuneability of the emission of (InGa)As/GaAs QDs, A:29, 30
electronic bottleneck, B:705, 706
electronic buffers, B:710, 718, 733
electronic compensation for 10-GB/S applications
   LAN: transmission over MMF, A:699
   electronic equalization for 10GBASE-LRM, A:700–701
   metro and Long-Haul transmission over SMF
      E-field domain signal processing, A:705–706
      electronic compensation of PMD, A:704–705
      electronic equalization to compensate chromatic dispersion, A:703–704
      modeling of transmission channel and impairments, A:701–703

electronic demodulator error compensation, B:150
electronic demultiplexer circuits, B:192
electronic dispersion compensation (EDC), A:539, B:12, 158, 353, 356, 495
electronic equalization, B:107
   and adaptation techniques, A:678–681
   BER, adapting coefficients, A:679
   implementation considerations, A:673
   LMS adaptive transversal filter, A:680
   low bit-error rate (BER) application, A:673
   minimize BER, A:679
   nonclassical channel, A:673
   performance metrics for adaptive filter, A:679
   synchronous histogram monitor, A:679
   to compensate chromatic dispersion, A:703–704
   compensation of PMD, A:704–705
   for 10GBASE-LRM, A:700–701
electronic logic gates, B:787
electronic multiplexer, B:26, 32, 184, 185, 187–189, 191, 192, 415
electronic packet switch, B:642, 650, 656, 696–699, 701, 705, 720, 721, 725, 731, 732, 735
electronic polarization demultiplexing, B:157, 159
electronic predistortion, B:76, 78
electronic signal processing in optical networks
   adaptation techniques, A:678–681
   data center applications, A:707
   electronic compensation for 10-GB/S applications, A:699–707
   electronic equalization
      decision feedback equalizer, A:675–676
      linear equalizer, A:673–675
      maximum A posteriori equalization, A:677–678
      maximum likelihood sequence estimation, A:676–677
      variants on theme, A:678

Index

high-speed electronic implementation
architectural considerations,
A:697–699
broadbanding techniques,
A:682–697
prospects and trends for next-
generation systems, A:707–708
role of, A:671–672
storage area networks, A:706–707
electronic signal processing, B:7, 23, 24,
37, 180, 184, 185, 195
electronic signal-processing, adaptive,
A:681, 701, 705, 708
high-speed, A:672, 681
receiver-based electronic dispersion
compensation (EDC), A:672, 701
signal impairments, mitigate, A:671–672
upgrading data transmission to 10Gb/s
over installed fiber, A:672
electronic switch fabrics, B:729
electronic time division multiplexing,
B:13, 14, 32, 179, 180–183, 185–195,
197, 199, 201, 202, 210, 218, 224–226,
705, 706
electronic transmission mitigation, B:391
electronically predistorted optical
waveforms, B:44
electrooptical effects in silicon, A:398
electrostatic actuation, A:715
embedded DRAM (eDRAM), B:718
EML, *see* electroabsorption-modulated
lasers (EML)
emulation, A:641–652
emulators, A:645–647
first-order, limitations, A:642–645
sources, A:647–650
with variable DGD sections,
A:650–652
encircled flux, B:837
encode information, B:28
end-to-end
connection, B:708
networked information, B:23, 26
optical data path, B:791
endlessly single mode (ESM), A:490, 577
entertainment video service, B:411
equalization filter coefficients, B:161

equalizers, A:322
erbium-doped fiber amplifier (EDFA),
A:107, 116, 124, 135, 161, 170, 343,
B:1, 2, 4, 13, 96, 113, 117, 125, 169,
180, 195, 200, 215, 216, 219, 221–223,
225, 235, 241, 248, 253, 257, 263, 265,
294, 325, 326, 330–334, 338, 388, 389,
413, 496, 497, 561, 564, 565, 572,
584–593, 790, 830, 836, 851
gain spectrum, B:338
systems, B:331, 591
technologies, B:96
error correction, B:564, 582, 583, 706
coding, B:28, 38, 39
error correlation information, B:277
error distribution, B:182
error-free
operation, B:217
performance, B:672, 679
ESM, *see* endlessly single mode (ESM)
Ethernet, B:14–16, 19, 132, 169, 179, 196,
200, 225, 345, 346–353, 355, 357–363,
365–383, 385–387, 389–397, 399, 406,
417, 464, 477, 480, 481, 483–486, 490,
491, 493, 502, 506, 507, 513, 515–518,
520, 526, 528, 529, 532, 543, 550, 553,
620, 623, 624, 638, 642, 682, 703, 837
carrier, B:16, 169, 369, 638
frame format, B:358
layer spanning tree protocol, B:543
layering, B:347, 379
passive optical network, B:379,
387, 703
physical layer, B:351
service models, B:378
switches, B:486, 517, 528, 553, 623
transport, B:169, 179, 370, 528, 529
virtual circuits, B:378
Euclidean distance, B:148, 149
European Commission, B:452, 466, 615
evanescent coupler, *see* directional coupler
evanescent coupling, B:824
excess burst rate, B:378
excess information rate, B:378
explicit telemetry, B:309
extinction ratio, B:43, 56, 59, 67, 493,
651, 663

extra link capacity, B:545
eye diagram, B:51, 54, 55, 57, 60, 63, 152, 191, 193, 194, 219, 243, 245, 276, 277, 280–283, 581, 582, 594, 663, 676, 839
eye monitoring techniques, B:279

F

fabric failure, B:544
fabrication technology, A:467–468
Fabry–Perot, B:253, 352, 440, 681, 751, 826, 827
  filters, B:253
  lasers, B:352, 440, 827
factory automation, A:357
Fano-interference, A:445
  configuration, A:432–433, 438
Faraday rotation, principle, A:350
Fast Ethernet lines, B:518
  see also Ethernet
fast mixing assumption, A:615
fast mixing model, A:616
fault management, B:243, 244, 278, 856
FAX see dial-up modem
FBG, see Fiber Bragg Grating (FBG)
FDTD, see finite-difference time-domain (FDTD)
federal communications commission, B:17, 402, 403, 407, 408, 413, 453, 458, 468, 519, 528
feed-forward
  arrangement, B:710
  pump, B:333
  scheme, B:164
feedback loop, B:77, 78, 214, 497
fiber attenuation, B:246, 388, 591
Fiber Bragg Grating (FBG), A:112, 114, 116, 117, 121, 129, B:9, 181, 253, 661, 586, 632
Fiber Channel, A:706–707, B:485, 486
fiber coupling, B:316, 428
fiber lasers, B:747
fiber lasers/amplifiers, A:503
fiber nonlinearities, B:24, 26, 28, 38, 40–42, 52, 57, 58, 63–65, 69, 70, 73–76, 79, 123, 133, 166, 202, 208, 213, 245, 245, 248, 260, 268, 273–276, 325, 330, 338, 760

monitoring techniques, B:275
as source for correlated photons
  coincidence counting results, A:836
  coincidence rates as function in single-photon rates, A:835
  correlations between signal and idler photons, measuring, A:834–835
  dual-band spectral filter based on double-grating spectrometer, A:834
  fiber polarization controller, A:833
  photon-counting measurements, A:833
  photon-pair source, efficiency of, A:833
  quantum efficiency vs dark-count probability, A:835–836
  spectral filter with fiber Bragg gratings, implementation of, A:837
  transmission curves, A:834
  transmission efficiency in, A:834
  two-pump photons, A:833
as source for entangled photons
  coincidence counts and single counts, A:856
  degree of polarization entanglement, A:584, 857
  efficient source for developing quantum communication technologies, A:858
  measured values of S for four Bell states, A:857
  measurement of polarization entanglement, A:855–858
  nature of entanglement generated, A:851
  observed polarization (in) dependence of parametric fluorescence, A:856
  parametric fluorescence, source of quantum-correlated photon pairs, A:852
  photon-counting modules, A:855
  polarization correlations, A:855–856
  polarization- entangled states, measurements, A:855
  polarization entanglement, experimental setup and sinusoidal variations, A:853

Index

polarization-entangled photon pairs
  in 1550-nm telecom band, A:851
quality of polarization
  entanglement, A:857
quantum efficiencies, detection
  of, A:855
signal and idler photon pairs, A:852
fiber optics applications
  communication, B:47, 50, 71
  current status of, B:442
  history of, B:445
fiber performance technology, B:426
fiber polarization controller, A:833
fiber refractive index, B:566, 567
fiber to the node, B:410
fiber transmission
  impairments, B:24
  line, B:212
  medium, advantages of, B:739
fiber-based
  access, B:403, 405, 408, 418, 446
  broadband, B:16, 401, 403, 404, 410, 431, 438, 453, 469
  optical signal processing, B:215, 217
  solutions, B:403
fiber-drawing techniques, A:620
fiber-fed base stations, B:760
fiber-in-the-loop deployments, B:468
fiber-optic quantum information
  technologies
    degenerate photon pairs for quantum
      logic in telecom band
      Hong-Ou-Mandel interference, A:869–876
      photon-pair generation in optical fiber, A:863–869
    fiber nonlinearity as source for
      correlated photons, A:832–837, 851–858
    high-fidelity entanglement with cooled
      fiber, A:585–863
    quantum theory of four-wave mixing in
      optical fiber, A:838–851
fiber-to-the-building, B:438, 464
fiber-to-the-curb, B:16, 404, 408, 410, 413, 415, 417, 419–421, 429–431, 438, 442, 443, 458, 462

fiber-to-the-home, B:5, 16, 17, 251, 401, 404, 410, 413, 416, 417, 419, 421, 422, 426, 428, 430, 440–442
fiber-to-the-premises, B:410, 416, 418, 437–444, 446, 448, 452, 453, 461, 466–469, 613–615, 624
fiber's birefringence, A:607
field programmable gate array, B:165, 493
field tests and PMD, A:616–621
figure of merit, A:334–335
  for dispersion compensators, A:335
  for TODC, A:334–335
filter
  characteristics, B:53, 67, 68, 320
  detuning, B:277
  offset, B:283
finite element method, B:812
finite impulse response, B:159, 834
finite linewidth, B:143
  see also frequency jitter
finite measurement bandwidths, A:617
finite-difference time-domain (FDTD), A:441, 442, B:809
  switching dynamics, A:442
first-in-first-out buffer, B:711
first-order PMD, A:633
  autocorrelation, A:634
  penalty, A:629, 638
fixed filter based nodes, B:499
fixed frequency grid, B:183
fixed wavelength converters, B:725
"flat" passbands, A:287
flat-top filter construction theorem, A:293
flip-chip bonding, A:346
flip-chip, B:189, 422
forgetting factor, B:154
forward error correction, B:10, 12, 23, 28, 39, 60, 63, 65, 67, 125, 145, 167, 180, 205, 247, 306, 372, 385, 493, 564, 571, 582, 583, 596, 604
forward propagating slab waveguide, A:391
four-dimensional signal space, B:50
four-point constellation, B:96
four-stage distribution network, B:786, 787

four-wave mixing (FWM), A:760, 762, 833
  four-photon scattering (FPS) process, A:838
  photon–photon scattering process, A:838
  signal and idler photons, A:838
four-wave mixing, B:40, 52, 166, 167, 209, 211, 215–217, 241, 253, 260, 273, 274, 567, 573, 578, 668, 673, 675, 676, 831
Fourier space, B:818
Fourier transforms, B:818
frame check sequence, B:359
frame relay, B:479, 481
France Telecom, B:458
Franz–Keldysh modulators, A:399
Franz-Keldysh effect (FK), A:195, 196, 213, B:5
free spectral range, B:194, 257
free-space optics, B:139
frequency drift, B:62, 113, 139, 247
frequency estimation, B:158, 162, 165
  see also phase estimation
frequency jitter, B:143
frequency offset compensation, B:163
frequency-domain
  explanation, B:54
  split-step, B:833
frequency-encoding code division multiplexing (FE-CDM), A:320
frequency-shift keying, B:33, 133, 663
Fresnel reflections, B:246
Fujitsu's 80 × 80 OXC, A:717
Full Service Access Network Group, B:415
full width at half maximum (FWHM), A:39, 114
full-duplex Ethernets, B:348
fusion splices, B:246
FWHM, *see* full width at half maximum (FWHM)
FWM, *see* four-wave mixing (FWM)

G
GaAs-based QD technology, A:29
gain error, B:330, 332, 333
gain peaking, B:189

gain ripple, B:308, 330, 338
  impact of, B:330
gain-flattening filter, B:326, 584, 585
Galerkin method, B:812
gallium arsenide, B:44
Gas-Based nonlinear optics
  high-harmonic generation, A:510–511
  induced transparency, electromagnetically, A:511
  stimulated Raman scattering, A:510
Gate message, B:381
gate oxides, carrier accumulation on, A:403–404
gate switching energy, B:707
Gaussian approximation, B:842, 845
Gaussian AWG, A:283–284
  wide passband, A:288
Gaussian distribution, B:115
Gaussian noise, B:30, 115, 259
Gaussian random variables, B:849
Gemini network interface, B:792
generalized multi-protocol label switching (GMPLS), A:367, B:347, 615, 642
generic framing procedure, B:373
germanium photodetectors, A:409–414
  benefits, A:411–412
  and photoreceivers for integrated silicon photonics, A:409–414
  p-type Ge contacts, A:413
  single heterojunction, A:413
  waveguide (WPDs), A:412
GI-POF *see* graded-index plastic optical fiber (GI-POF)
Gigabit Ethernet, B:10, 200, 346, 348, 349, 351, 353, 355, 357, 370, 372, 373, 385, 392, 394, 485, 521, 703
  link, B:346, 392
  physical layer, B:349
  private line, B:521
  see also Ethernet
Gigabit interface convertor, B:346
Gires–Tournois etalons, A:326–327
Global Lambda Integrated Facility, B:617
global system for mobile communications (GSM), A:179
Godard algorithm, B:122
Google, B:389

Index 897

Gordon-Mollenauer phase noise, B:162, 167
graded-index plastic optical fiber (GI-POF), A:597, 707
gradient-type optimization algorithm, B:161
grating coupler, A:389–390
  backward scattering, A:391–392
  fiber-coupling efficiency of, A:392
grating mirrors, reflectivity spectrum, A:348
Gray coding, B:35
grid computing, B:549, 555, 611, 613, 617, 620, 629
grid spacing, B:253
grooming, B:502, 518, 526, 528
Group Velocity Dispersion (GVD), A:494–495
  hollow core, A:495
  solid core, A:494–495
group-delay ripple, B:320
group-velocity dispersion, B:120, 495, 831
Groupe speciale mobile, B:756
grouped routing, B:851
GSM, *see* global system for mobile communications (GSM)
guard time, B:656, 660, 670, 681, 775, 776, 781, 784, 786, 788
GVD, *see* group velocity dispersion (GVD)

H

half mirror, B:112
half-rate decision, B:192
harmonic mixer, B:192
head-end switch, B:393
header error control, B:373
Hello message, B:376
Helmholtz equation, B:805, 806, 817, 838, 839
Hermite–Gaussian modes, B:838
Hermitian matrix, B:848
heterodyne efficiency, B:47
heterodyne receiver, B:13, 96, 100, 102, 105
  see also homodyne receiver
heterodyne signal, B:104, 746
heterojunction bipolar transistor, B:179
heterostructure photo-transistor, B:755

high speed optical modulators, recent developments
  EAMs, *see* traveling-wave EAMs
  external modulators, key performance data, A:186
  high-speed modulation, *see* high-speed modulation, optical modulators
  intensity modulators based on absorption changes, A:195–197
    Electroabsorption (EA), A:195
    FK effect, A:195
    IS Stark-shift modulator, A:197
    Kramers–Krönig relation, A:185
  modulators based on phase changes and interference, A:193–195
    LiNbO$_3$ technology, A:195
    linear Pockels effect, A:193
    Mach–Zehnder configuration, A:194–195
  novel types of modulators, *see* modulators, novel types
  on-off keying (OOK), A:184
  principles and mechanisms of external optical modulation, A:185–186
  quantum-confined Stark effects (QCSE), A:185
high-Q nanocavities, A:463–466
high-bit-rate digital subscriber line, B:443
high-capacity packet switches, B:696, 697
high-definition television, B:406, 413, 457, 461–463, 551, 552
high-fidelity entanglement with cooled fiber
  measurements of ratio between coincidence counts and accidental-coincidence counts, A:861
  photon-counting modules, A:860
  two-photon interference, A:862
  violation of Bell's inequality, A:863
high-index-contrast waveguide, A:386
  types and performance on SOI, A:384–388
high-performance optical channels, B:244
high-power photodetectors
  charge-compensated uni-traveling carrier photodiodes, A:239
  saturation currents/RF output powers, A:239

high-power photodetectors (*continued*)
  high-power PDA, A:240
    Partially Depleted Absorber (PDA) photodiodes, A:240
  high-power waveguide photodiodes, A:242–244
    coplanar waveguide (CPW), A:243
    modified uni-traveling carrier photodiodes, A:240–242
    modified UTC (MUTC), A:240
  saturation, impact of physical mechanisms, A:234
  thermal considerations
    Au bonding, advantage, A:244
  transimpedance amplifier (TIA), A:233
  uni-traveling carrier photodiodes, A:235–238
    advantages, compared with PIN structure, A:235
    optical preamplifiers, advantage, A:236
    pseudo-random bit sequence (PRBS) signal, A:237
    space-charge effect, A:238
high-quality video, B:553, 554
high-speed direct modulation of strained QW lasers
  InAlGaAs and InGaAsP systems/ comparison with theory, experimental results, A:66–69
  microwave modulation experiment, A:66
  small-signal modulation response, theory, A:65–66
  K factor, A:65
high-speed edge emitters, tunnel injection resonances, A:43–44
high-speed electronic implementation
  architectural considerations, A:697–699
  broadbanding techniques
    common source amplifier, A:684
    compensated degenerative amplifier, A:686
    implementation of adaptive, A:681
    principles of, A:682–683
    RC broadbanding, A:683–687
    series shunt-peaked broadbanding, A:694–697
    series-peaked, A:692–693
    shunt-peaked, A:687–692
    small signal, high-frequency equivalent circuit of amplifier, A:684, 695
    small signal, high-frequency model of amplifier, A:684, 692
    use of CMOS technology, A:682
  electronic compensation for 10-GB/S applications, A:699–707
    data center applications, A:707
    electronic equalization for 10GBASE-LRM, A:700–701
    LAN: transmission over MMF, A:699–701
    metro and long-haul transmission over SMF, A:701–706
    storage area networks, A:706–707
  prospects and trends for next-generation systems, A:707–708
high-speed low-chirp semiconductor lasers, A:53
  gain and differential gain, A:54
  quantum dot lasers, *see* quantum dots (QD), lasers
  QW lasers, DC properties of long-wavelength, *see* DC properties of long-wavelength, QW lasers
  strained QW lasers, high-speed direct modulation, *see* high-speed direct modulation of strained QW lasers
high-speed modulation, optical modulators
  chirp issues, modulators
    sinusoidal modulation, A:191
    issues/principles/limitations, A:187–190
    factors, A:187
    flip-chip mounting, A:188
    optoelectronic integrated circuits (OEICs), A:187
    "traveling-wave," A:190
high-speed nanophotonics
  device structure, A:37
  directly modulated QD lasers, A:36–37

eye pattern/bit-error rate measurements, QD laser module, A:37–38
  nonreturn-to-zero (NRZ), A:37
  saturation of radio frequency (RF) amplifier, A:38
  temperature-dependent BER measurements, A:38
high-Speed
  Ethernet, B:391, 411
  photodiode, B:185, 191, 192, 270
  transmission experiments, B:213
Higher Speed Study Group, B:179, 390
higher-layer protocol identifiers, B:389
highly nonlinear fibers (HNLF), A:761
hinge model, A:615–616, 620, 631–632
histogram techniques, B:277
hitless Operation, B:306
hollow-core PBGF, A:579–585
  definition, A:581
  dispersion, A:578–579
    Bessel function, zeroth-order, A:582
  five transverse mode intensity profiles, A:584
  modal properties, A:584–585
  origin, A:580–581
homodyne detection, B:13, 48, 95, 97, 100, 133
homodyne phase-diversity receiver, B:105, 112, 113, 115, 117
homodyne receiver, B:96–98, 100, 102, 106, 111, 123, 124, 127
  optical circuit, B:111
Hong-Ou-Mandel interference with fiber-generated indistinguishable photons
  copolarized identical-photon source, 50/50 Sagnac-loop, A:871
  dual-frequency copolarized pump, preparation of, A:872
  experimental parameters, A:873
  experimental results, A:874
  experimental setup, A:872
  fiber-based source of photon pairs at telecom-band wavelength near 1550 nm, A:869
  four-wave mixing, A:869
  photons produced by QS source, A:871
  QS source and use in HOM experiment, A:873
  QS source by measuring CAR, features of, A:872–875
  quantum interference at beam splitter, A:870
  real-life imperfections, A:875
  theoretical simulation results of HOM dip, A:875
hubbed rings, B:302
hybrid communication network, B:226
hybrid integration platform, B:787
Hybrid Optical Packet Infrastructure, B:617
hybrid transmission, B:169
hybrid-fiber coaxial, B:16, 404, 460

I

IEEE Higher Speed Study Group, B:179
III–V photonic integrated circuits, A:343–345, 376–377
  future of OEO networks enabled by III–V VLSI, A:372–373
  on chip amplifiers offer additional bandwidth, A:373–374
  mobile applications for optical communications, A:375–376
  power consumption and thermal bottleneck challenge, A:374–375
  manufacturing advances for III-V fabrication implying scalability
    defect density and functionality per chip, A:361–363
    design for manufacturability, A:356–358
    in-line testing, A:358–359
    yield management methodologies, A:359–361
  manufacturing advances for III–V fabrication implying scalability, A:355–356
  network architecture impact of LSI PICs
    component consolidation advantages, A:366–367
    data ingress/egress in digital ROADM, A:369–370

III–V photonic integrated circuits (*continued*)
  DON architecture, A:367–368
  network management advantages, A:370–372
  Q-improvement cost, A:369
  photonic material integration methods, A:345–346
  small-scale integration
    large-scale III–V photonic integration, A:351–355
    multi-channel interference-based active devices, A:348–351
    single-channel multi-component chips, A:347–348
III–V semiconductor arena
  Franz–Keldysh modulators, A:399
  Quantum-Confined Stark Effect (QCSE) modulators, A:399
impairment mitigation, B:18, 75, 196
impairment of the transmitted signal, B:308
impairment-aware first-fit algorithm, B:240
impairment-aware routing, B:237
impairment-unaware algorithms, B:240
implementation of MLSE, challenges, A:700
in-band OSNR monitoring, A:800
In-Fiber devices, A:501–502
in-line testing, A:358–359
in-phase and quadrature, B:96
incumbent local exchange carriers, B:419, 448
indium phosphide, B:44, 186
  modulators, B:45
  semiconductors, B:6
individual bulk devices, B:253
information spectral density, B:604
InGaAs/GaAs semiconductor, A:474
ingression paths, B:794
injection-locked lasers, B:5
inner ring, B:538
InP PLCs, A:275
  geometry as source of birefringence, A:275
InP-based transmitter devices
  data capacity for, A:365
  number of functions per chip for, A:364

InP/InGaAsP quantum well structure, A:468
InP/InGaAsP/InGaAs SACM APDs, A:246
input–output coupling, A:388–389
  gratings, A:389–392
  tapers, A:389, 390
insertion loss, B:654
"instantaneous frequency" of electric field, A:444
instruction-level parallelism, B:768
integrated silicon photonics chips, examples, A:416
  20 Gb/s dual XFP transceiver, A:416–417
  40-Gb/s WDM transceiver, A:417
  integrated driver and MZ modulator, A:418
  receiver performance, A:419
Integrated yield management (IYM) triangle, A:359
Intel, B:769
intelligent optical switch, B:528
intensity eye diagrams, B:50, 53, 57, 60
  see also eye diagram
intensity modulation, B:30, 60, 95, 133, 137, 625, 663, 739, 745, 749, 752, 834, 837
  direct detection, B:95, 739
intensity-dependent refractive index coefficient, B:273
inter-nodal management, B:243
inter-symbol interference, B:31, 135, 349, 565, 582, 843
interchannel effects, B:166
interconnects, B:720, 721, 732
interframe gaps, B:373
Interior Gateway Protocols, B:532
intermediate frequency, B:48, 95, 743
intermodulation distortion (IMD), A:166, 168, B:426
international telecommunication union, B:347, 569
Internet Engineering Task Force, B:532
Internet Group Management Protocol, B:553
Internet lingo peer-to-peer, B:555

Internet Protocol, B:345, 346, 448, 480, 516, 695, 701
   television, B:15, 19, 295, 345, 389, 391, 441, 448, 449, 457, 461, 462, 464, 529, 535, 553, 555
Internet service provider, B:389, 515, 516, 518, 523, 526, 528, 551–555
interoperability, B:620
interstage, B:585, 787
intersymbol interference (ISI), A:322
   effects, A:672
intra-fiber device (cutting/joining), A:499–502
   cleaving/splicing, A:500
   in-fiber devices, A:501–502
   mode transformers, A:500–501
intrachannel dispersion slope, B:581–583
intrachannel effects, B:167, 579, 580, 597
intrachannel four-wave mixing, B:40, 597, 600, 601
   see also four-wave mixing
intrachannel nonlinearities, B:41, 71, 73, 75
intradyne detection, B:63, 64
intradyne receiver, B:102, 164–166
inverse dispersion fiber, B:125
ISO standards organization, B:513
ITRS Semiconductor Roadmap, B:706

J

J × K WSS, A:317
Japanese Ministry of International Trade, B:458
JET signaling, B:646–648, 679
jitter, B:143, 204, 206, 218, 237, 283, 533, 554, 651, 653
Jones matrix, B:159, 160
Jones matrix formalism, A:541
Jones vector, A:654, 658
Joule heating, B:755
just-enough-time, B:645
just-in-time, B:645, 646

K

Karhunen–LoŠve Technique, B:847
"KEISOKU", A:82
Kerr effect, B:12, 24, 25, 70, 566, 567, 572

Kerr nonlinear material, A:438–439
   transmitted *vs* input power, A:440
Kerr nonlinearities, A:499, 778, 789, 790, 794
Kerr-related nonlinear effects, PCF, A:504–507
   correlated photon pairs, A:505–506
   parametric amplifiers/oscillators, A:505
   soliton self-frequency shift cancellation, A:507
   supercontinuum generation, A:504–505
killer application, B:461, 551

L

label switched paths, B:365, 541
labor-intensive changes, B:430
Laguerre–Gaussian modes, B:838
   see also Hermite–Gaussian modes
LAN, *see* local area networks (LAN)
LAN/MAN Standard Committee, B:346
lanthanum-doped lead zirconium titanate, B:655
Large Channel Count ROADMs, A:316–317
large effective area core (LMA), A:543, 558, 562, 570–572
Large Hadron Collider, B:614
large-scale photonic integrated circuits (LS-PICs), A:351–352
   100 Gb/s DWDM transmitter module, A:366
   challenges to production, A:356
   design, A:357
   future, A:373
   manufacturing, A:359–360, 362
   network architecture impact of component consolidation advantages, A:366–367
   data ingress/egress in digital ROADM, A:369–370
   DON architecture, A:367–368
   network management advantages, A:370–372
   Q-improvement cost, A:369
   output power distributions on 40-channel, A:364

large-scale photonic integrated circuits (LS-PICs) (*continued*)
   process capability (Cp)/capability index (Cpk), A:356
   response *vs* frequency curves, A:353
   upper control limit (UCL)/lower control limit (LCL), design, A:357
   wafer processing, A:358
   wafers yields *vs* chip size for time periods of production, A:362
   and yield improvement, A:359–361
   see also Integrated yield management (IYM) triangle
laser chips, B:422
laser frequency offset, B:142, 143, 146
laser optical fiber interface (LOFI), A:57
laser rangefinders, B:423
latching/nonlatching, B:653
latency, B:21, 310, 374, 504, 533, 549, 550, 554, 646, 648–650, 664, 766–769, 772, 773, 775, 780–785, 787–789, 791, 792, 795, 796
lattice constants, A:461
Laudable characteristics, B:244
Layer, 2 functions, B:16, 358
Layer, 3 intelligence, B:507
LCOS, *see* liquid crystal-on-silicon (LCOS)
LCOS-based 1 × 9 WSS, A:725
LE *see* linear equalizer
legacy transport networks, B:371
LFF, *see* low-fill factor (LFF)
light emitter, A:468
   substance, A:469
light transmission, input port to output port, A:314
light-emitting diode, B:811
Lightguide Cross Connect, B:520
lightpath labeling, B:309
lightwave systems, historic evolution of, B:26
line coding, B:23, 24, 28, 38, 39, 53, 72, 348, 349, 355, 370, 373
line defect, A:457, 458
line-coded modulation, B:53
line/point-defect systems, A:457, 458
linear equalization, B:582

linear equalizer, A:673–675
   designing transversal filter adaptation algorithm for tap weights, A:675, 676–678
   analog *vs* digital implementation, A:675
   number of taps, A:674–675
   removal of ISI, linear filter, A:673, 676
   structure of, A:674
linear optical quantum computing (LOQC), A:863
linear refractive index of silica, B:273
linearly polarized electromagnetic wave, A:443
linewidth enhancement factor (LEF), A:54
link aggregation group, B:374
link capacity adjustment scheme, B:347
link state advertisements, B:540
link utilization, B:782
liquid crystal-on-silicon (LCOS), A:723
   features of, A:723–724
lithium niobate, B:5, 44, 187, 206, 726, 748, 751
LMA, *see* large effective area core (LMA)
loading coils, B:10
Local Access and Transport Area, B:519
local area networks (LAN), A:668, B:236, 345
local fault, B:357
"local field" model, A:247
local loop unbundling, B:453
local oscillator power, B:95
localization of fiber failure, B:250
LOFI, *see* laser optical fiber interface (LOFI)
logically inverted data pattern, B:61
long-distance
   monopolies, B:452
   optical communications, B:565
long-haul applications, B:51, 166
long-wavelength VCSELS, A:83–87
loop timing, B:385
Lorentzian distribution, B:143
loss of signal, B:357, 533
low-density parity check, B:353
low-dispersion, B:2

Index

low-fill factor (LFF), A:132
low-pass filtering effects, B:190
lower sideband, B:265
Lucent's optical system layout for OXC, A:718

M

M × M star coupler, ideal, A:297
Mach–Zehnder delay
  interferometer (MZDI), A:276, 277, 382, 401, 402
  demodulators, A:278
  interleavers, A:279
  plot of optical power from, A:295
  possible ways of laying out the MZI–AWG, A:295
  switches, usage of SAG, A:348
Mach-Zehnder
  characteristic, B:194
  delay line interferometer, B:265
  demodulator, B:599
  electro-optic amplitude, B:598
  electro-optic modulator, B:594
  intensity, B:663
  interferometer, B:193, 206, 209, 256, 598, 602, 634, 750
  modulator, A:405, B:42, 43, 126, 134, 184, 206, 594, 758, 807
    based on CMOS gate configuration, A:406
    lumped-element, A:408
    traveling-wave, A:408
  push–pull modulators, B:96
Management Information Base, B:386
Manakov equation, A:656
Manakov-PMD equation, A:655–656
MAP, see Maximum A Posteriori Equalization (MAP)
Marcum Q-function, B:140, 142
master-oscillator power-amplifier (MOPA), A:112
master-planned communities, B:441
matched filter, B:30, 31
matrix coefficients, B:162
matrix inversion, B:160
Maxichip, A:133

maximal-ratio combining, B:105, 109
Maximum A Posteriori Equalization (MAP), A:677–678
  with BCJR algorithm, A:677–678
  to improve system performance with FEC, A:678
maximum likelihood estimation, B:97
  sequence estimation, B:23, 150, 182
maximum likelihood sequence estimation (MLSE), A:676–677
  whitened matched filter (WMF), transversal filter, A:676–677
Maximum Refractive Index, A:488–489
maximum transport unit, B:682
Maxwell distribution, A:613, 614–615
Maxwell's equations, B:804, 805, 809, 816, 820, 822
McIntyre's local-field avalanche theory, A:247
MCVD, see modified chemical vapor deposition (MCVD)
MDSMT, see modified dead-space multiplication theory (MDSMT)
Mean time between failure, B:534
Mean time to repair, B:535
measured BER *vs* average optical power, A:700
media-independent interface, B:347, 348
medium access control, B:346
memory elements, B:20, 766, 767, 772
memory-less modulation, B:37
MEMS technologies and applications, emerging
  MEMS-tunable microdisk/microring resonators, A:747–748
  lightconnect's diffractive MEMS VOA, A:746
  photonic crystals with MEMS actuators, A:748–749
MEMS, see Microelectromechanical system (MEMS)
metalorganic chemical vapor deposition (MOCVD), A:27, 29, 35, 82, 83, 87, 88
metalorganic vapor phase epitaxy (MOVPE), A:346

metro and long-haul transmission
over SMF, A:701–707
e-field domain signal processing,
A:705–706
electronic compensation of PMD,
A:704–705
electronic equalization to compensate
chromatic dispersion, A:703–704
modeling of transmission channel and
impairments, A:701–703
Metro Ethernet
Forum, B:347
services, B:481
metro layer networks, B:518
metro networks, B:17, 196, 307, 484, 487,
495, 496, 498, 502, 505, 514, 524, 548
architectures, evolution of, B:482
geography of, B:505
modulation formats for, B:493
metro nodes, B:496
metro optical transport convergence, B:484
metro segment, B:552
metropolitan area networks, B:346, 483, 511
micro-ring and micro-disk resonators, A:349
fabricated eight-channel active, A:350
vertically coupled to I/O bus lines, A:350
microelectro mechanical systems, B:10,
249, 312
microelectromechanical system (MEMS),
A:83, 89, 99
emerging MEMS technologies and
applications
MEMS-tunable microdisk/microring
resonators, A:747–748
photonic crystals with MEMS
actuators, A:748–749
optical switches and crossconnects
3D MEMS switches, A:716–719
two-dimensional MEMS switches,
A:714–716
other optical MEMS devices
data modulators, A:743–745
variable attenuators, A:745–747
tunable lasers, A:741–743
wavelength-selective mems components
diffractive spectrometers and
spectral synthesis, A:739–741

dispersion compensators, A:733–734
reconfigurable optical add–drop
multiplexers, A:721–722
spectral equalizers, A:719–721
spectral intensity filters, A:730–733
transform spectrometers, A:734–738
wavelength-selective crossconnects,
A:728–730
wavelength-selective switches,
A:722–728
microoptoelectromechanical systems
(MOEMS), A:713
microstructured optical fibers (MOF)
effective-index microstructured optical
fibers (EI-MOF), A:575–579
attenuation, A:579
basic concepts, A:575–576
dispersion, A:578–579
endless single mode/large effective
area/coupling, A:576–577
small modal area/nonlinearity,
A:577–578
hollow-core PBGF, A:579–585
waveguiding, A:573–575
microwave monolithic integrated circuit,
B:745
microwave OE oscillators, A:420–421
spectrum analyzer, A:421
Miller multiplication of transistor, A:685
misalignment, B:837
mitigation, B:70
MLSE, *see* multiple likelihood sequence
estimation (MLSE)
MMF, *see* multimode fiber (MMF)
MMI, *see* multimode interference (MMI)
mobile applications for optical
communications, A:375–376
MOCVD, *see* metalorganic chemical vapor
deposition (MOCVD)
modal transmission line theory, B:820
mode division multiplexing, B:36
mode transformers, A:500–501
mode-locked fiber lasers, B:204
mode-locked laser diodes, B:204
mode-locked pulse train, frequency
spectrum of, A:321
mode-locked QD lasers

Index

device structure, A:39
hybrid mode-locking, A:41–42
passive mode-locking, A:39–40
full width at half maximum
 (FWHM), A:39
modes *see* sub-pulses
modified chemical vapor deposition
 (MCVD), A:491, 539, 562–564
modified dead-space multiplication
 theory (MDSMT), A:254
modified uni-traveling carrier
 photodiodes (MUTC), A:240
modular deployment, B:306
modulation format, B:12, 13, 18, 24, 27,
 28–30, 34, 36, 37, 41, 42, 46, 49, 50,
 52, 55–58, 61, 62, 64, 65, 67–76, 97,
 100, 108, 132–135, 161, 169, 171, 181,
 183, 184, 194, 195, 202, 203, 215–217,
 222, 223, 226, 237–239, 256, 257, 259,
 260, 264, 272, 309, 321, 390, 413, 415,
 493–495, 569, 572, 583, 593, 596–600,
 604–606, 660, 663, 834–837
 auxiliary polarization, B:40
 binary, B:28, 37, 63, 70, 97, 185
 coded, B:28, 39, 41, 53
 duobinary, B:41, 45, 53, 495
 response, B:748, 750
 tone techniques, B:253
 waveforms, B:28–30, 33, 36, 46, 59
 with memory, B:37, 38
modulator devices, A:404–409
 reducing modulators size, A:407
 variable attenuator waveguide, A:404
modulator, B:134, 183, 206, 258, 662
 bias, B:44, 56, 190
 extinction ratios, B:67
modulators, novel types
 Intersubband Electroabsorption
  Modulators, A:208–211
  effects, A:209
  negative chirp/high-optical
   transmission, A:209
 Silicon-Based Modulators, A:211–213
  complementary metal–oxide–
   semiconductor (CMOS), A:212
  G–L /Ge intervalley scattering
   time, A:213

MOEMS, *see* microoptoelectromechanical
 systems (MOEMS)
molecular beam epitaxy (MBE), A:37, 44,
 82, 87, 110, 252
moment generating function, B:849
monitoring physical-layer impairments,
 B:245
monolithic integration, B:684
Monte Carlo methods, B:841, 842
Moore's law, B:735, 765
MOPA, *see* master-oscillator power-
 amplifier (MOPA)
Mosaic browser, B:404
MPEG-2 and, B:4 compression, B:462
multi-Gbaud rates, B:64
multi-mode fiber model, B:757, 839
multi-mode interference devices, B:807
multi-mode Systems, B:837
multi-processor chips, B:20
multi-protocol label switching, B:365,
 513, 643
multi-section directional coupler, A:274
multi-service platform, B:528
multichannel entertainment digital
 video, B:535
multicore architectures, emergence of, B:797
multifamily dwelling units, B:464
multifiber wavelength-selective switch,
 B:296
multilayer network stack, B:537
multilevel coding, B:40, 123
multilevel decision circuitry, B:149
multilevel modulation, B:13, 24, 37, 38,
 40, 78, 96, 97, 226
multimode fiber (MMF), A:672
multimode fiber-coupled 9xx nm pump lasers
 active-cooled pump diode packages
  beam symmetrization coupling
   optics, A:132–133
  Maxichip, A:133
  "wallplug efficiency," A:133
 BA pump diode laser
  broad-area single-emitter
   (BASE), A:127
  frequency stabilization, A:129
  slow-axis BPP,causes for NA
   degradation, A:127

multimode fiber-coupled 9xx nm
    pump lasers (*continued*)
  classification of 9xx-nm mm pump lasers
    heat transport classification,
      A:125–126
    optical classification, A:126–127
  combined power, A:133
  passive-cooled 9xx-nm MM pumps
    beam symmetrization coupling
      optics, A:131–132
    direct coupling, A:129–131
    low-fill factor (LFF), A:132
    simple lens coupling, A:131
    Single Emitter Array Laser (SEAL),
      A:132, 133
  pulse operation, A:133
multimode interference (MMI), A:112,
  232, 271, 393, 730
multiparty gaming, B:480
multipath interference, B:322
multiple access interference, B:633
multiple input multiple-output, B:36, 161
multiple likelihood sequence estimation
  (MLSE), A:626
multiple service operators, B:378, 413
multiple-quantum-well, B:749
multiple-wavelength transmission
  schemes, B:767
multiplexing, B:20, 26, 28, 29, 31–33, 36,
  47, 49, 53, 64, 78, 119, 131, 132, 180,
  183, 185, 186, 193, 195, 201, 203, 204,
  206, 207, 208, 215, 217, 218, 222, 223,
  226, 238, 242, 247, 251, 253, 294, 296,
  297, 304, 311–313, 320, 322, 346, 347,
  353, 368, 372, 415, 417, 444, 478, 484,
  485, 490, 491, 498–500, 514, 517, 522,
  526, 527, 604, 617, 626, 629, 635, 642,
  644, 656, 661, 665, 669, 695, 803
  see also demultiplexing
multiplication regions, low-excess noise
  APDs
  AlInAs, A:250–252
    active region, definition, A:251
    heterojunction, A:252–256
      electrons gain energy, A:254
      "initial-energy effect," A:254
      MDSMT, A:254

thin, A:248–250
  "dead space," A:248
  noise reduction in thin APDs, A:250
  "quasi-ballistic," A:249
multipoint-to-multipoint, B:359, 378,
  380, 481
multiport switch, B:500, 501
multisource agreement, B:346
multistage topologies, B:773, 782, 788
multisymbol phase estimation, B:49, 132
MUX *see* multiplexing

N
nanocavity with various lattice point
  shifts, A:465
Napster, B:468
*narrowband digital cross-connect
  system*, B:522
National LambdaRail, B:617
national research and educational
  networks, B:611
National Television System Committee, B:411
Negative Core-Cladding Index Difference
  all-solid structures, A:493
  higher refractive index glass,
    A:492–493
  hollow-core silica/air, A:492
  low leakage guidance, A:493
  surface states, core-cladding
    boundary, A:493
network
  architectures, B:297, 410, 482
  automation, B:503
  component failures, B:522, 537
  diameter, B:553, 767, 769, 789
  dimensioning, B:850
  element, B:293
  evolution, B:511, 549
  fault management, B:244
  management, B:2, 234, 239, 305, 310,
    324, 336, 339, 366, 440, 445, 483,
    556, 766
  modeling, B:850
  network interface, B:620
  -on-chip, B:797
  optimization problem, B:547

optimization process, B:547
premise equipment, B:528
restoration, B:537
scaling, B:308
segments, B:513
survivability, B:487
switching, B:623
NOBEL project, B:615
node object, B:853
node-sequenced channel, B:339
node-to-node demand connection, B:489
noise spectral density, B:137
non-adiabatic parabolic horn, A:288
non-Gaussian beat noise, B:61
non-return-to-zero, B:35, 50–56, 58–63, 69–72, 76, 108, 123, 135, 168–170, 184–187, 189, 190, 194, 195, 200, 203, 256, 258, 261, 265, 267, 269–272, 282, 322, 493, 495, 594, 596, 599, 834–837, 839
  on-off keying, B:256, 493
  phase-shift keying, B:108
non-zero dispersion-shifted fiber, B:195, 213, 255
nonlinear distortion, B:24, 60, 76, 77, 121
nonlinear effect, B:2, 10, 64, 73, 137, 159, 162, 166, 167, 181, 235, 241, 243, 260, 268, 272–274, 276, 317, 495, 566, 567, 572, 573, 577, 580, 586, 593, 594, 596, 597, 600–602, 605, 625, 743, 745, 755, 759, 832
nonlinear fibers, A:546–555, 765
  designs, A:548–555
    core refractive index change /core radius, A:549
    W-profile, A:551–555
  requirements, A:547–548
    effective interaction length, A:547
    figure of merit, A:547–548
    nonlinear efficiency, A:547–548
nonlinear impairments, B:24, 170, 578, 589, 590, 597
nonlinear optical loop mirror, B:209, 222
nonlinear optical materials, A:760, 766
  bismuth glass, A:760, 765, 789
  chalcogenide glasses (ChG), A:762–764
  silicon, A:763–764, 794

nonlinear optical methods
  of generating tunable optical delays and buffering, A:811–818
Nonlinear optical signal processing, A:760
Nonlinear optics in communications
  future prospects, A:818–820
  optical delays and buffers, A:811–818
  optical performance monitoring, A:800–811
  optical regeneration, A:769–771
    optical signal regeneration through phase sensitive techniques, A:784–787
    3R regeneration, A:782–784
    2R regeneration in integrated waveguides, A:778–782
    2R regeneration through SPM in chalcogenide fiber, A:771–777
  optical switching, A:796–800
  optical-phase conjugation, A:787–789
  parametric amplification, A:766–769
  phase-matched vs nonphase-matched processes, A:761–762
  platforms, A:762–766
  wavelength conversion
    chalcogenide fiber-based wavelength conversion, A:789–793
    wavelength conversion in chalcogenide waveguides, A:793–796
nonlinear phase
  fluctuation, B:127, 276
  noise compensation, B:157, 162, 274
  noise, B:60, 72, 73, 131, 151, 154, 157, 162, 164, 167, 274, 276, 601
  rotation, B:274
nonlinear polarization scattering, B:137, 259
nonlinear refractive index ((Kerr effect), A:760
nonlinear signal-noise interactions, B:72
nonlinear taper, B:808
nonlinear transmission, B:41, 52, 54–58, 60, 63, 64, 71, 79, 135, 154, 155, 274
nonlinearity compensation, B:79
nonperiodic problems, B:819
nonreturn-to-zero (NRZ), A:37, 42, 45, 95, 198, 204, 237, 324

nonstatic nature of optical networks, B:234
nonzero dispersion, B:2, 41, 72, 255, 576, 836
nonzero dispersionshifted fibers (NZDSF), A:525–526, 533–534
normalization constant, B:149
Nyquist limit, B:97, 104, 114
NZDSF, *see* nonzero dispersionshifted fibers (NZDSF)

O

OEICs, *see* optoelectronic integrated circuits (OEICs)
OMM's 16 × 16 switch, A:716
on-off keying (OOK), A:184 B:13, 30, 123, 131, 203, 211, 214–218, 256, 493, 594, 599, 601, 663
OOK, *see* on-off keying (OOK)
open shortest path first, B:532
open systems interconnection, B:513
operation, administration, and maintenance, B:347
operational costs, B:482–485, 506, 507
operational savings, B:420
OPM, *see* optical performance monitoring (OPM)
optical add and drop multiplexers, B:15, 253, 295, 487, 498, 850
optical amplifier, B:3, 25, 26, 30, 30, 65, 73, 74, 96, 138, 170, 182, 183, 209, 211, 217, 241, 242, 247, 248, 260, 274, 294, 310, 314, 479, 484, 495, 496, 505, 533, 565–269, 580, 625, 654, 739, 742, 743, 745, 752, 753, 757, 760
optical bandpass filter, B:54, 219, 223, 257
optical buffers, B:665
optical burst switching, B:170, 556, 615, 627, 649, 650, 655, 679, 701
optical channel
  configuration, B:101
  monitor, B:248, 249
  switching, B:642
optical chromatic dispersion compensators, A:325–335
  tunable ODC, A:325
optical clock recovery, B:216, 219

optical code division multiple access, B:36, 617, 625, 632, 633
optical communication systems, fibers speciality in
  dispersion compensation, A:526–538
  double-clad fibers for fiber lasers, A:555–572
  nonlinear fibers, A:546–555
  single polarization fibers, A:539–546
optical communication technologies, B:26
optical coupler, A:271
optical crossconnect (OXC), A:716, B:615, 642, 645, 647, 850
optical data modulator, B:185
optical delay interferometer, B:133
optical delays and buffers, A:811–818
  counterphased pump modulation, impairment, A:816
  dual-pump parametric delay, A:816
  effect of pump synchronization, A:816
  nonlinear transmission, A:813
  one-pump parametric delay, A:816
  phase dithering in one-pump device, A:816
  system performance of tunable dual-pump parametric delay, A:818
  transmission spectrum of FBG, A:812–813
  tunable delay with power or strain, A:815
  two-pump counterphased delay line, A:817
Optical demodulators, A:278
optical demultiplexer, B:123, 203
optical dispersion compensation, B:71
optical duobinary, B:53, 54, 62, 139
optical equalizers (OEQ), A:322–325, B:181, 182, 190, 193–195
  ineffectiveness, A:324
  symmetric and asymmetric ISI impulse responses, A:324
  two-tap, A:323–324
    band-limited photodiode and its equalization, A:324
    107-Gb/s NRZ bandwidth before and after equalization, A:324
    using serial arrangement of MZIs, A:328
    using finite-impulse response (FIR) filters, A:325

Index 909

optical fibers
  polarization effects in, A:539–542
  polarization-maintaining, A:542–544
  single-polarization, A:544–546
optical field vectors, B:29
optical field, change in, A:295
optical filters, A:275, B:30, 31, 46, 52, 54,
    62, 68, 69, 97, 119, 123, 131, 133, 139,
    143, 166, 195, 211, 223, 235, 245, 256,
    265, 267, 268, 284, 295, 320, 479, 496,
    499, 506, 605, 661, 776, 830
  designs example, A:294
  Fourier filter-based interleaver,
      A:294
  nonlinear phase response, A:276
  represented
    in frequency domain by Fourier
        transform, A:292
    in time domain by impulse
        response, A:292
  zero-loss maxima per FSR
    multiple, A:293, 294
    with one, A:293, 294
    see also flat-top filter construction
        theorem
optical frequency, B:268, 272, 712, 713
optical gates, B:209
optical header, B:630, 644, 657, 658,
    660–665, 678, 684
  processing technologies, B:658
  swapping, B:678
  technique, B:663
optical hybrid, B:101, 103, 144
optical impairments, B:239, 244
optical index, B:570, 805
Optical injection locking (OIL), A:145
optical injection phase-locked loop
    (OIPLL), A:179, B:757
optical intensity, B:30, 34, 46, 50, 51,
    54, 59, 60, 135, 273, 274, 664
  profile, A:288
optical interconnection networks,
    B:766, 771, 777
Optical Internet Forum, B:347
Optical Internetworking Forum, B:620
Optical isolators, A:350
  TE mode waveguide, A:351

optical label
  encoding, B:31
  switching, B:19, 643, 678, 682
optical line terminal, B:251, 379
optical local oscillator, B:133
optical memory, B:637
optical MEMS devices, other
  data modulators, A:743–745
  etalon modulator, A:744
  etalon variable attenuator, A:744
  microdisk resonator, A:747
optical MEMS, developments and
    fabrication technologies of,
    A:730–732, 748–749
Optical Mesh Service, B:549
optical metropolitan networks, evolution
    of, B:477
optical microelectromechanical systems
    (MEMS), A:713
  applications of, A:713
  developments of, A:714
  see also microoptoelectromechanical
      systems (MOEMS)
optical modulation amplitude,
    photodetector capacitance,
    A:410–411
optical modulation formats, B:12, 23, 26,
    27, 39, 42, 45, 49, 132, 134
  overview of, B:133
optical modulator, B:217, 309, 757
optical monitoring, B:623, 636
optical multiplexing, B:14, 36, 201
optical network unit (ONU), A:172, 174,
    176, B:250, 379, 460, 464, 633
optical networking, B:617
optical noise, B:31, 137, 138, 148, 164,
    166, 245, 253, 254, 309, 330, 593
optical nonlinearity, A:760
optical packet switching, B:170, 615, 617,
    623, 624, 627–633, 635–638, 642,
    644, 646, 649–651, 654–658, 660,
    662–665, 667, 669–671, 675–677,
    679, 684, 696, 697, 699, 701, 712,
    715, 720, 725, 733, 775, 776
optical packet synchronizer, B:665
optical parametric amplification
    (OPA), A:761, B:260

optical passbands, A:287
optical performance monitoring (OPM), A:760, 800–811
  advantages and disadvantages of phase-matched and nonphase-matched processes, A:800, 804, 811
  configuration of, A:800
  DGD monitor, A:808
  measuring in-band OSNR, A:804
  nonlinear optical loop mirror (NOLM), A:806
  nonlinear power transfer curve of OPA-based in-band OSNR monitor, A:805
  optical power spectrum of signal measured on OSA, A:810
  optical spectrum analyzer methods, A:803
  optical spectrum at output of OPA, A:805
  OSNR monitoring curves
    of NOLM-based device, A:807
    for OPA-based in-band OSNR monitor, A:805
  principle of device operation, A:801
  principle of residual dispersion monitoring, A:800, 801
  provides signal quality and identify network fault, A:800
  spectral monitoring signal vs 40-Gb/s RZ receiver BER penalty, A:802
optical performance monitoring, B:15, 233
optical phase conjugation, role in signal processing, A:787
optical phase, B:11, 26, 43–46, 57, 59, 72, 97, 100, 111, 150, 151, 154, 170, 207, 208, 210, 211, 316, 746, 825
  conjugation, B:11, 26
  diversity homodyne receiver, B:111
  lock loop, B:97, 100, 150, 746
optical power loss, B:246
optical preamplification, B:105, 113, 114, 753
optical protection mechanisms, B:568
optical pulses, B:50, 56, 181, 204, 205, 208, 214, 572
  compressor, B:215

optical regeneration, A:769–771
  ASE noise, signal impairment, A:769
  experimental configuration, A:772
  limiting signal degradation, using 2R and 3R regenerators, A:769–770
  optical signal regeneration through phase sensitive techniques
    generic 2R, higher order 2R processing, excessive idler broadening, A:786
    power transfer functions, A:785
    two-pump PA operating in small-signal, large-signal and transfer characteristics, A:786
  optical-phase conjugation, A:787–789
    all-optical networking layer, A:787
    feasibility of two-pump FPP in all-optical networks, A:788
  principle of noise compression, A:770
  3R regeneration, A:782–784
    BER improvement, optimizing parameters, A:784
    capability to retime signal timing jitter, A:782
    operating principle of, A:782–783
    power transfer function, A:784
    structural design and principle of, A:782–783
  2R regeneration in integrated waveguides, A:778–782
    autocorrelation of input and output pulses, A:781
    capable of operating T/s bit rates, A:782
    power transfer function, resulting in OSNR reduction and BER improvement, A:782
    structure of, A:776
    testing integrated waveguide 2R regenerator, A:780
    transmission spectrum, A:779
    unfiltered pulse spectra, A:780
  2R regeneration through SPM in chalcogenide fiber, A:771–777
    better performance, A:779
    calculated Q-factor, A:776
    input and output eye diagrams, A:776
    optimized Q-factor improvement vs FOM, A:777

output pulse spectra of the chG, A:773
power transfer curves, A:771, 773–774
principle of operation, A:771, 772
wavelength conversion
advantage of, A:789
chalcogenide fiber-based wavelength conversion, A:789–793
chalcogenide waveguides, A:793–796
optical regenerators, B:204, 520, 721, 723
optical resonators, B:717
optical return loss, B:246
optical ring resonator, B:717
optical routers, B:683
optical sampling technique, B:205
optical sensors, A:514
optical signal processing based on VCSEL technologies, A:94–97
optical signal to noise ratio, B:42, 65, 131, 181, 234, 319, 566, 572, 587, 594, 600, 636, 653, 796, 831
optical signal, representation of, B:837
  processing, B:4, 11, 201, 202, 204, 208, 214, 215, 217, 225, 697
optical signal-to-noise ratio (OSNR) penalty, A:623, 703–705
optical spectral filters, B:498
optical spectrum analyzer (OSA), A:150
optical splitter splits, A:271
  1 × 2 couplers, A:271–272
  2 × 2 couplers, A:272–275
optical subcarrier monitoring, B:255
optical supervisory channel, B:242
optical switch fabrics, B:724
optical switches and crossconnects
  2D MEMS switch, A:714–716
optical switching, A:796–800, B:19, 209, 239, 322, 506, 507, 620, 636, 637, 643–645, 651, 654, 655, 665, 669, 679, 684, 725, 731, 772, 773, 778, 789
  devices, B:772
  OC-768 Switching in HNLF parametric device, A:800
  signal band of pump operating in normal and anomalous dispersion, A:797
  two-pump parametric switching, A:798
optical terminal multiplexers, B:850

optical time division multiplexing, B:14, 32, 56, 119, 120, 180, 191, 193–195, 201–209, 211, 213, 215–219, 221–223, 225–227, 229, 231, 249, 617, 664, 705
  transmitter, B:14, 195, 201–204, 222, 225
optical time-domain reflectometry (OTDR), conventional, A:176
optical transmission link capacity-bandwidth product, growth in, A:760
optical transmitter, B:421
optical transponder, B:132, 520, 533, 536
optical transport network, B:483, 487
  evolution of, B:132
optical wavelength domain reflectometry, B:250
optical-electrical-optical conversion, B:202, 615
Optical-electrical-optical (OEO) consumption vs 2.5G (Sonet/SDH), A:371
optical-optical-optical, B:615
optical-to-electrical conversion, B:245, 695, 731
optically transparent mesh transmission, B:334, 336
optimal receiver sensitivity, B:139, 145
optimal signal-to-noise ratio, B:42, 52, 55, 56, 59, 63–70, 72–76, 125, 126, 131, 134, 136–138, 142, 143, 146, 148, 154, 155, 166, 168, 169, 181, 190, 235, 239, 243, 245, 248, 249, 253–260, 263, 266, 274, 281–284, 319, 321–324, 335, 336, 390, 494, 495, 566, 582, 587, 600, 603, 653, 831, 836
  monitoring, B:253, 255, 257–260, 266
  penalty, B:68–70, 76, 142, 143, 146, 166, 181, 582, 603
  performance, B:63, 169
optimization algorithms, B:505
optoelectronic bandwidth, B:188
optoelectronic integrated circuits (OEICs), A:187
optoelectronic oscillator, A:420
optoelectronics, B:426, 440, 461, 491, 501
  base rate receiver, B:211
  buffer, B:678
  components, B:26, 37, 188, 734
  conversion stages, B:569

912 Index

optoelectronics (*continued*)
  demodulation, B:46
  integrated circuits, B:754
  integration, B:64
  switch fabrics, B:697, 733
  switching, B:653
Optoplex's high-isolation, B:249
orthogonal codes, B:36
orthogonal frequency division
    multiplexing, B:32
*Orthogonal modulation header
    technique*, B:663
orthogonal polarization, B:30–33, 46, 47,
    157, 159, 161, 183, 254, 259, 262, 263
orthogonal transverse modes, B:838
oscillator laser, B:164, 745
  voltage-controlled, B:192, 193, 210
oscilloscope
  digital sampling, B:196
  digital storage, B:170
  electrical sampling, B:191, 205
Osmosis, B:789, 790, 791
outage map, A:628
outer ring, B:538, 539
Outside plant cabinet technology, B:424
outside vapor deposition (OVD), A:539,
    552, 562–565
OVD, *see* outside vapor deposition (OVD)
overlay switching, B:454
OXC, *see* optical crossconnect (OXC)

P
P–N Junction
  in depletion, A:399–402
  in forward bias, A:402–403
p-type doping, B:4
P2P (E-Line), B:378
packaged photodiodes, B:421
packaged switch, A:716
packet buffering, B:643
packet dropping, B:664
packet loss, B:378, 379, 533, 536, 554,
    667, 669
packet switch architectures, B:697
packet switching nodes, B:21, 774
packet synchronization processing, B:657

packet transmission, protection scheme
    in, B:853
parabolic horn, width, A:288–289
parabolic refractive index profile, B:838
Parabolic waveguide horns, A:288
  straight multimode waveguide, A:291
paradigm changes in semiconductor
    physics and technology, A:27–28
parallel loss elements, B:253
parallelism and synchronization, B:853
parametric amplification (PA), A:760,
    766–769
  amplified spontaneous emission, A:767
  effect of pump spectral tuning,
    A:767–768
  modulational instability, A:767
  operating principle, A:767
  parametric gain, A:768
parametric fluorescence of FWM, A:833
parametric gain, A:766, 768
parametric processor, functions of,
    A:760, 762
parasitic effects, B:830
partial response, B:53
Partially Depleted Absorber (PDA)
    photodiodes, A:240
partially depleted-absorber photodiode,
    B:755
passband ripple, B:320
passive device modeling methods, B:805
passive optical network, B:246, 379,
    624, 703
passive optics, B:787
passive waveguide devices/resonators
  filters, A:393–395
  highly compact splitters, A:393
  photonic crystal resonators, A:395
  ring resonators, A:395–397
  splitters and couplers, A:393
passive/active components for POFs,
    A:599–601
path adjustments, B:784
  iterations, B:785, 786, 788
path-protected ring, B:852
path-routing hand offs, B:236
pattern-dependent performance, B:188
PBG, *see* photonic band gap (PBG)

PBGF, *see* photonic band-gap fibers (PBGFs)
PCF, *see* Photonic crystal fibers (PCF)
PDCD, *see* polarization-dependent chromatic dispersion (PDCD)
peer-to-peer, B:345, 479, 480, 555
per-channel dispersion
  compensation, B:318
  management, B:317, 319
per-channel equalization, B:311, 312
per-channel symbol rate, B:23, 24, 71
perfectly matched layer, B:813
performance
  analysis and metrics, B:781
  estimation, B:841
  evaluation, B:785
  functions, B:535
  monitoring techniques, B:243
periodic dispersion maps, B:317
permanent virtual circuit, B:523
Petri Nets, B:852, 853
PF polymer POFS, connectors for, A:599–600
PHASARs (phased arrays), *see* arrayed waveguide grating (AWG)
phase demodulator, B:224
phase difference, B:44, 48, 112, 116, 133, 139, 163, 164, 192, 206, 210–212, 264, 272, 599
phase distortions, B:59
phase error, B:142, 143, 146, 153, 163
phase estimation, B:49, 102, 110, 113–117, 132, 158, 162–165
  algorithms, B:162
phase modulation, B:34, 38, 42, 44, 45, 47, 49, 58, 59, 60, 100, 109, 110, 118, 134–136, 164, 184, 204, 205, 214, 223, 276, 572, 596, 597, 744–746, 759, 831
phase noise, B:76, 100, 102, 107, 110, 111, 115, 118, 127, 162, 164, 167, 206, 274, 748, 841
  tolerance, B:115
phase portrait, B:282, 283
phase shaped binary transmission, B:41, 53, 190
phase wrapping, B:163
phase-diversity homodyne receiver, B:97, 101, 102

phase-encoded data, B:163
phase-locked loop, B:97, 192, 210
phase-matched *vs* nonphase-matched processes, A:757–759
phase-modulated data signal, B:211
phase-modulated formats, B:60, 64
  characteristic of, B:60
phase-sensitive detection, B:267
phase-shift keyed (PSK), A:278
phase-shift keying, B:34, 96, 108, 123, 133, 161, 269, 593, 601
  formats, B:599
phased amplitude-shift signaling, B:53
phasor diagrams, B:167, 274
"phonon bottleneck," A:25
photo sharing, B:480
photocurrents, B:104, 108, 112, 113
photodetector capacitance, A:410
  optical modulation amplitude as function of, A:411
photodetector, B:5, 7, 46, 47, 54, 206, 211, 212, 219, 253, 264, 268, 349, 636, 745, 753–755, 758, 760, 787, 826, 830
photodetectors, advances in
  avalanche, *see* avalanche photodiodes
  high-power, *see* high-power photodetectors
  waveguide, *see* waveguide photodiodes
photodiode, B:5, 6, 46, 95, 97, 99, 100, 105, 112, 113, 138, 180, 184, 191–193, 205, 210, 259, 270, 325, 351, 421, 422, 561, 565, 566, 582, 598, 741–746, 754–756, 848
  responsivity, B:742, 746, 848
photon
  density, B:825
  electron resonance, B:748
  lifetime, B:4, 825
photonic band gap (PBG), A:485, 490
photonic band-gap fibers (PBGFs), A:573, 575, 579–581
Photonic crystal add–drop filter, A:395
photonic crystal cladding
  characteristics, A:487–490
    maximum refractive index, A:488–489
    photonic band gaps, A:490
    transverse effective wavelength, A:489

photonic crystal fibers (PCF), A:525–526, 761
  applications, A:502–515
    Bragg gratings, A:512–513
    Brillouin scattering, A:507–509
    fiber lasers/amplifiers, A:503
    gas-based nonlinear optics, A:509–511
    high power/energy transmission, A:502–503
    Kerr-related nonlinear effects, A:504–507
    laser tweezers in hollow-core PCF, A:513–514
    optical sensors, A:514
    telecommunications, A:511–512
  design approach, A:487
  fabrication techniques, A:485–487
  guidance, characteristics, A:490–498
    attenuation mechanisms, A:495–498
    birefringence, A:494
    classification, A:490
    Group Velocity Dispersion, A:494–495
    Kerr Nonlinearities, A:499
    Negative Core-Cladding Index Difference, A:492–493
    Positive Core-Cladding Index Difference, A:490–492
  intra-fiber devices (cutting/joining), A:499–502
    cleaving/splicing, A:500
    in-fiber devices, A:501–502
    mode transformers, A:500–501
  photonic crystal cladding, characteristics, A:487–490
    maximum refractive index, A:488–489
    photonic band gaps, A:490
    transverse effective wavelength, A:489
photonic crystal resonators, A:395, 440, 441
photonic crystal theory: temporal coupled-mode formalism, A:433–434, 451
  stopping light in dynamic photonic crystals, A:445
    conditions for stopping light, A:447–448
    future prospects, A:450–451
    numerical demonstration in photonic crystal, A:448–450
    from tunable bandwidth filter to light-stopping system, A:447–448
    tunable Fano resonance, A:445–447
    tuning spectrum of light, A:443–444
  temporal coupled-mode theory
    for optical resonators, A:432–438
    to predict optical switching, A:438–442
photonic crystal, A:456–457, 461, 478
  light-stopping process in, A:449
  numerical demonstration in, A:448–450
  point and line defect states in, A:432
  switches, A:438
Photonic double heterostructure, A:466
photonic heterostructures, A:456
photonic integrated circuit (PIC) transmitter (TX), A:347
photonic integrated circuit (PIC), A:347, 354, 365
  future, A:373
  100-Gb/s transmitter and receiver, A:352
  Q measurement vs wavelength, A:355
  ultralong-haul transmission, A:354–355
  using hybrid Raman/EDFA amplifier, A:355
  see also planar lightwave circuits (PLCs)
photonic material integration methods
  butt-joint regrowth, A:345–346, 348
  direct-contact wafer bonding, A:346
  quantum well intermixing (QWI), A:346, 348
  selective area growth (SAG), A:346, 348
photonic, B:6–9, 20, 21, 202, 226, 296, 310, 339, 483, 507, 532, 615, 617, 632, 634, 636, 637, 668, 706, 713, 755, 766, 772, 774, 776–778, 781, 782, 797, 798, 807, 811, 814, 816, 817, 820
  band, B:8, 811, 816, 817, 820
  bottleneck, B:705, 706
  cross-connect, B:296, 304, 311, 312, 532

Index

crystal, B:7–9, 636, 637, 668, 713, 811, 814, 817
integrated circuit, B:6, 7, 778, 807
per bit, B:137
photoreceiver, B:46, 191, 192, 839
physical coding sublayer, B:348
physical layer, B:346, 513
  development, B:348
  fault management, B:243
  measurement parameters, B:245
  optical encryption, B:170
physical medium
  attachment, B:348
  dependent, B:348
PIC-based all-optical circuits, A:761, 777
pilot-tone-based monitoring technique, B:256
pin semiconductor structures, B:43
planar holographic bragg reflectors, A:311–312
  4-Channel CWDM multiplexer using, A:312
Planar lightwave circuits (PLCs), A:269, B:6, 139, 181, 206, 207, 217, 223, 253, 296, 655
  and birefringence, A:275
    PDW shift, A:275
  parameter "delta," A:270
  polarization conversion, A:276
  surface and oscillating electric field, A:275
  wavelength shift and refractive index, A:305
planar lightwave circuits in fiber-optic communications, A:269–270, 336
  basic waveguide theory and materials index contrast, A:270
    optical couplers/splitters, A:271–275
    polarization effects, A:275–277
  inter-signal control devices, A:312
    large channel count ROADMs, A:316–317
    reconfigurable optical add/drop multiplexers, A:313–316
    wavelength-selective switches, A:317–318
  intra-signal control devices, A:318
    optical chromatic dispersion compensators, A:325–335

optical equalizers, A:322–325
temporal pulse waveform shapers, A:319–322
passive optical filtering, demodulating, and demultiplexing devices, A:277
arrayed waveguide gratings, A:279–311
Mach–Zehnder interferometers, A:277–279
planar holographic bragg reflectors, A:311–312
plane wave expansion, B:816
plasmonic VCSELS, A:91–93
plastic optical fiber (POF), A:672, 707
  datacom applications, A:602
  development, A:595–598
    in attenuation of, A:595–596
    in high-speed transmission, A:597–598
  enabling 10-Gb/s transmission, A:672
  passive/active components, A:599–601
    connectors/termination for PF Polymer, A:601
    PMMA, connectors/termination methods, A:599–600
    transceivers for, A:601
  varieties, A:598–599
    product-specifications of, A:598
    typical construction of PMMA, A:598–599
plastic optical fiber, B:9
platform developments, A:762–766
  advances in, A:765–766
  chalcogenide glasses, nonlinear material for all-optical devices, A:762
  parameters for silicon and ChG with nonlinear FOM, A:760
  properties of chalcogenide, A:762–763
  properties of SM selenide-based chG and bismuth oxide, A:764
PLC, *see* silicon planar lightwave circuits (PLCs)
pluggable transmitters, next-generation, B:493
PMD emulator (PMDE), A:642, 644–644
PMD source (PMDS), A:642
  cautions about, A:650

PMD Taylor expansion, A:632–633
  validity of, A:633–634
PMD, *see* polarization mode dispersion (PMD)
PMF, *see* polarization maintaining fibers (PMF)
PMMA, *see* polymethylmethacrylate (PMMA)
PMMA-Based POFs, connectors for, A:599–600
POF, *see* Plastic optical fiber (POF)
point defect, A:457, 458
  quality factors of, A:458–459
Point of Interface, B:526
point-of-presence, B:519
point-to-multipoint, B:348, 379, 512
point-to-point
  connections, B:512, 553, 641
  emulation, B:382, 384
  Ethernet development, B:16, 347
  links, B:325, 569, 767, 771, 804
  transmission, B:204, 225, 316
  WDM systems, B:524, 532
Poisson, B:534, 805, 826
polarization alignment, B:108, 114, 152
polarization beam splitter, B:46, 103, 112, 157, 183, 223, 269
polarization control, B:9, 47, 125, 182–184, 582, 746
polarization controller, B:125, 182–184, 582
polarization conversion, A:276
polarization crosstalk, A:276
polarization demultiplexer, B:165, 183, 184, 224
polarization demultiplexing, B:137, 157, 165, 166, 222
polarization dependence, B:102, 111, 389, 651, 757
polarization diversity, B:33, 47, 102, 104, 105, 111, 112, 114, 151, 152, 155, 162, 746
polarization effects, optical fibers, A:539–542
  birefringence B, A:540
  Jones matrix formalism, A:541
  in telecommunication, A:541

polarization entangled photon-pairs
  with compact counterpropagating scheme (CPS), creating, A:859
  Ekert's QKD protocol, implementing, A:863
polarization filtering, B:139–142, 145, 146
polarization fluctuation, B:114
polarization hole burning, B:263
polarization independent gate, B:210
polarization interleaving, B:32
polarization maintaining fibers (PMF), A:523, 525, 542–544
polarization management, B:33
polarization matching, B:746
polarization mode dispersion (PMD), A:531, 537–538, 605–606, 609–610
  birefringence, A:607–608
  emulation, A:641–642
    emulators, A:645–647, 651
    emulators with variable DGD sections, A:650–652
    first-order emulation and its limitations, A:642–645
    sources, A:647–650
  environmental fluctuations, A:617
  fiber models, A:612
  fiber plant, elementary model of installed, A:614–616
  field tests, survey of, A:616–621
  high-order effects
    experimental and numerical studies, A:639–641
    PMD Taylor expansion, A:632–637
    practical consideration, A:637–639
  notation convention, A:606
  and optical nonlinearities, A:652
    coupled nonlinear Schrodinger equations, A:652–655
    effect of nonlinear polarization rotation on PMDCs, A:659–661
    interactions between third-order nonlinearities and PMD, A:662
    Manakov-PMD equation, A:655–656
    nonlinear polarization rotation, A:656–659
  polarization evolution and, A:609

Index   917

propagation, A:606–607
properties of, in fibers, A:611–614
transmission impairments caused by
   the first-order, A:621
   first-order PMD penalties from real
      receivers, A:624–627
   PMD-induced outages, implication of
      the hinge model on, A:631–632
   PMD-induced outages, probability
      of, A:627–631
   Poole's approximation, A:622–625
polarization mode dispersion, B:9, 10, 15,
   24, 37, 63–65, 68, 70, 71, 77, 79, 123,
   131, 133, 134, 137, 150, 152, 155,
   159–162, 166–170, 180–182, 184,
   196, 202, 203, 212, 214, 215, 219, 222,
   224–226, 234, 235, 239, 243, 245,
   255–257, 259, 262–266, 268–272,
   279, 281–284, 309, 321, 346, 348–351,
   356, 357, 383, 385, 390, 570–573, 582,
   583, 593, 603, 654, 831, 841
   compensation of, B:214
   effects of, B:160
   mitigation, B:167, 168, 219, 268
   monitoring technique, B:270
polarization mode dispersion (PMD),
   transmission impairments, A:800, 808
polarization modes, A:544–546
polarization modulator, B:602, 603
polarization multiplexing, B:24, 33, 47,
   63, 64, 78, 79, 123, 137, 183, 184,
   195, 196, 226
polarization nulling method, B:254, 255
polarization orientation, B:165
polarization parameters, B:109
polarization related impairment, B:263
polarization scramblers, B:168, 182, 583
polarization sensitivity, B:114
polarization shift keying, B:30, 133
polarization-dependent chromatic
   dispersion (PDCD), A:641, 643
polarization-dependent loss (PDL), A:354
polarization-dependent wavelength
   (PDW), A:275
   shift resulting in PMD, A:276
polarization-dependent wavelength shift
   (PDWS), A:354

polarization-dependent
   frequency, B:62, 154
   gain, B:263
   loss, B:160, 184, 255, 308, 321, 654
polarization-division multiplexing, B:131
Polarization-entangled degenerate photon-
   pair generation in optical fiber
CAR
   as function of single-channel
      photon-detection rate, A:866
   measuring, A:866
   plotted as function of single counts/
      pulse, A:865
   correlated/entangled photon pairs in
      single spatiotemporal mode, A:869
   degenerate correlated-photon, A:865
   determining single-count rate for
      optimum CAR value, A:866
   experimental layout, A:864
   polarization-independent behavior for
      single counts and coincidence
      counts, A:864
   properties of photon pairs, A:867
   two-photon interference (TPI) with
      visibility, A:867
Polarizationindependent (PI) SOAs,
   A:373
polymer modulators, B:751
polymethylmethacrylate (PMMA), A:595
   POF Cords, A:598–599
Poole's approximation, A:622–624
Poole's formalism, A:639
Positive Core-Cladding Index Difference,
   A:490–492
   multiple cores, fibers, A:492
   ultralarge area single mode, A:491
potential dispersion monitoring, B:283
power amplifiers, B:329, 330
power consumption and thermal
   bottleneck challenge, A:374–375
power coupling, B:326, 328, 330, 336–339
power detection, B:268
power dissipation, B:798
power fading effect, B:264
power penalties, B:118, 121, 234, 672, 740
power splitters, B:305, 314, 316, 393
power splitting ratio, A:622

power transients, B:247
predicted end of life, B:323
predistortion algorithms, B:77
prehistoric era, A:24
   single QD, δ-function, A:24
      "phonon bottleneck," A:25
      threshold current density of laser, A:24
      zero dimensional structures, A:24
"principal state transmission," B:214, 222
Principal States of Polarization (PSP), A:607–608
Principle of dispersion compensating fibers, A:527–531
   dispersion compensation, A:528–531
   fiber dispersion, A:527–528
printed circuit boards, B:355
Private Branch Exchange, B:518
probability density functions, B:847
Process capability (Cp), A:356
Process monitor (PM) wafers, A:358
product sensitivity analysis (PSA), A:361
propagation delay, B:757, 780, 857, 858
propagation solution technique, B:809
protection switching, B:374
protocol data units, B:359
Protocol independent multicast, B:553
provider backbone bridge, B:369
pseudo-level modulation, B:40
pseudo-linear transmission regime, B:213
pseudo-multilevel format, B:40, 55, 57
pseudo-multilevel modulation, B:40, 41
pseudo-random binary sequence, B:59, 113, 116, 167, 207
pseudo-wire setup, B:550
Public-Switched Telephone Network, B:526
pull-tapes, B:423
pulse amplitude modulation, B:133, 353
pulse carver, B:56–59, 62, 180, 264, 599
pulse carving schemes, B:599
Pulse code modulation, B:522
pulse compression unit, B:223
pulse generation, B:135, 216, 274
pulse pattern generator, B:113
pulse position modulation, B:32, 663
pulse sources, B:204
pulse-to-pulse overlap, B:579, 580, 601
Pump diode lasers

high-radiance diode laser technologies, A:133–134
multimode fiber-coupled 9xx nm, *see* multimode fiber-coupled 9xx nm pump lasers
1480-nm pumps and 14xx-nm high-power, A:122–124
optical power supply, A:107–108
power photonics, A:108
   "all fiber" telecom technology, A:108
   DPSSLs, A:108
   VCSELs, A:108
single-mode fiber 980-nm pumps, *see* single-mode fiber 980-nm pumps
status/trends/opportunities
   "loss budget," A:135–136
   status, Raman amplification, A:135
   trends and opportunities, A:136
   telecom optical amplifiers, A:108
   VCSEL pump and high-power diode lasers, A:134–135
pump-diode breaks, B:585
pump-to-pump interactions, B:591
push-pull
   configurations, B:635
   modulator, B:96, 113
   operation, B:44

Q

Q-factor, B:243, 245, 248, 250, 263, 278, 281, 600–605, 653, 841–847
Q-tag, B:366, 369, 370
Q/Bit error rate monitoring techniques, B:278
Q factor, A:458–459
   of ideal LE, A:673–674
   of point-defect cavity, A:459, 464
Q-improvement cost, A:369
   for optical and electronic performance, A:370
QCSE, *see* quantum-Confined Stark Effect (QCSE)
quadratic impact, B:260
quadrature amplitude modulation, B:34, 110, 150, 392, 416
quadrature phase shift keying, B:38, 62, 96, 203, 583

Index

Quadrature point, B:51
quadrature signal space, B:50, 78
quality of service (QoS), 18, 237, 238, 248, 253, 275, 346, 417, 482, 484, 485, 523, 524, 531, 533, 535–537, 545–547, 549, 615, 627, 628, 638, 645, 648, 657, 679, 852
quantum dots (QD) amplifier modules
    device structure, A:42
    eye pattern and BER measurements, A:42–43
        error-free modulation, A:42–43
        external cavity laser (ECL), A:42–43
quantum dots (QD), A:23, 47, B:3, 4, 637
    lasers
        optical gain and LEF p-doped and tunneling injection QDs, A:70–74
        p-type doping and tunneling injection, A:74–76
            limitations, A:75
    utilization of, A:35–36
quantum efficiency, B:99, 105, 726, 754, 755
quantum entanglement, A:830–831, 851
    incompatibility with realism, A:831
    measurements of correlated quantities, A:832
    measuring states of particles, A:830
quantum information processing (QIP), A:838
    generation of entangled states and Bell's inequalities, A:838
quantum measurement, process of, A:829
quantum mechanics (QM), A:829
quantum theory of four-wave mixing in optical fiber, A:838–851
    applications of QIP, A:838
    coincidence counting formula, A:848–849
    coincidence-counting rate with Gaussian filters, calculation of, A:847–848
    coincidence-counting result, A:839
    coincidence-photon counting, A:839
    coupled classical-wave equations for pump, A:840
    dependence of photon-counting results, A:847, 848–851
    dependence on ratio of pump bandwidth to filter bandwidth, A:851
    experiment *vs* theory, A:850
    Hamiltonian in quantum FWM process, A:841
    interaction Hamiltonian, A:840–841
    measurement techniques, coincidence-photon counting, A:839
    negative effect of SRS, A:838
    polarization entanglement, A:839
    process of FWM, A:838
    quantum efficiencies of detection in signal and idler channels, A:839
    single-photon counting rate, formula, A:844, 846–848
    state vector evolution, calculating, A:841–844
    theoretical model, A:839–840
    two-photon state, calculated by first-order perturbation theory, A:843
    two-photon states, A:844–845, 850–851
Quantum well intermixing (QWI), A:346, 348
quantum wells (QWs), A:110
Quantum-Confined Stark Effect (QCSE), A:185, 196, 208, 210, 212
    modulators, A:399, 400
quantum-confined Stark effect, B:749
quantum-mechanical, B:11
quasi-linear transmission, B:71

R

radio frequency communication, B:23, 79
radio frequency tone, measurement of, B:263, 268
Raman gain coefficient, A:761, 763, 764, 792
Raman
    amplifiers, B:4, 325, 332, 333, 496, 580, 586–593
    couplers, B:222
    effects, B:592, 832
    gain, B:11, 587, 588, 591, 592
    interactions, B:333, 592, 593

Raman (*continued*)
  preamplifier, B:589
  pump modules, B:221
  scattering, B:273, 325, 330, 389, 586–588, 831
  tilt errors, B:333
random access memories, B:772
random nonlinear phase shift, B:274
rapid spanning tree protocol, B:364
rate-equation approach, B:825, 826
Rayleigh
  backscattering, B:589
  scattering, B:246
real-time optical Fourier transformation, B:272
Real-Time Protocol, B:516
Real-valued baseband, B:52
Rearrangeably nonblocking topologies, B:782
received signal r(t), A:669
receiver (RX) PICs, A:351, 366
  determining PDL and PDWS, A:354
  with integrated PI-SOA, A:374
  key DC performance characteristics, A:354
receiver impairments, B:142, 146
receiver sensitivity
  enhancement, B:153
  penalty, B:38, 148–150
  sensitivity, B:113, 139, 145, 153
Receiver side equalization, B:495
receiver technology advances, B:421
Receiver-based electronic dispersion compensation (EDC), A:701
recirculating loop buffer, B:
reconfigurable optical add–drop multiplexer (ROADM), A:313–316, 721–722
  athermal 16-ch ROADM, A:315
  40-channel reflective, A:316
  64-channel reflective, A:317
  designs, A:316–317
  modular 80-channel, A:318
  transmission spectra of athermal, A:315
  waveguide of 16-channel optical, A:313

reconfigurable optical add/drop multiplexer, B:6, 11, 15, 24, 65, 68, 132, 168–170, 196, 293–297, 299, 300–317, 319–325, 327, 329, 331, 333–339, 341, 343, 499, 500, 502, 520, 521, 524, 526, 528, 529, 532, 533, 545, 548, 622, 623, 636, 642, 642
  cascade penalties, B:319
  layer, B:521, 522, 526, 545, 548
  network layers, B:521
  technologies, B:293, 543
  transmission system design, B:316
Reed–Solomon codes, B:180
refractive index, A:460–461, B:9, 11, 24, 273, 274, 633, 712, 713, 715, 749, 806, 807, 818, 819, 838
  of defect, A:466
regenerators, B:720
Regional Bell Operating Companies, B:468
relative group delay, measurement of, B:266
relative sensitivity penalty, B:145
remote antenna units, B:756
remote fault, B:357, 394
remote mammography, B:614
remote node, B:379
residual dispersion, B:125, 213, 222, 317, 571, 580, 581
resilient packet ring, B:484
resonance spectra, tuning of, A:466
resonant grating filter, B:824
resonant mode, amplitude of, A:435
resonant ring modulator, A:408
resonator buffers, B:717
restoration, B:537, 538, 542, 852, 856
Restore Management, B:856
return-to-zero, B:40, 55, 56, 58, 108, 135, 180, 203, 208, 493, 593, 594, 596, 635
  dispersion maps, B:319
  format, B:572
  modulation, B:50, 55
  phase-shift keying, B:108
  quadrature phase shift keying, B:606
ridge waveguide, A:385
ring hopping, B:525
ring resonators, A:326, 395–397

Index

configured add–drop filter, A:396
  ring add–drop filters, A:396–397
ring-based network, B:489
ring-to-ring interconnections, B:489
Ritz method, B:812, 813
ROADM, *see* reconfigurable optical add–drop multiplexer (ROADM)
round trip
  latency, B:787, 788
  propagation time, B:779
  time, B:77, 381
router common cards, B:536
routers, B:649, 683, 701, 707
routing
  latency, B:788
  power, B:300–305
  properties, B:300
  technologies, B:19
row decoder, B:718
rule of thumb, B:71, 181, 582, 602, 669

S

S-tag, B:369
Sagnac interferometer, B:209
sampled symbol values, B:35
sampling of the signal, B:108
sampling theorem, A:298–299
  coefficient, A:299
SAN, *see* storage area networks (SAN)
satellite carriers, B:553
satellite-based technologies, B:466
SBS, *see* stimulated Brillouin Scattering (SBS)
scalability, B:369, 653, 675
scalability: chip size *vs* cost, A:363–365
Scalable Photonic Integrated Network, B:778
scalar electric field, B:806
scattering losses, A:388
scattering, B:567, 586–588, 783, 813
SEAL, *see* Single Emitter Array Laser (SEAL)
second-order PMD (SOPMD), A:633, 640, 646
selective area growth (SAG), A:346, 348
self-coherent

modulation formats, B:132, 134, 169, 171
  optical transport systems, B:171
  receivers, B:157, 166
  signals, B:131, 133, 137
Self-homodyne detection, B:133
self-managed network, B:235
self-phase modulation (SPM), A:657, 756
  B:123, 136, 273, 317, 566, 567, 572, 637, 831
semianalytical techniques, B:847
semiconductor optical amplifier (SOA), A:348
  and isolator, A:351
semiconductor quantum dots, future applications devices
  collective blindness, A:25–26
  electronic/optical properties, A:28–31
  nanophotonics, *see* high-speed nanophotonics
  prehistoric era, *see* prehistoric era
  QD amplifier modules, *see* quantum dots (QD) amplifier modules
  QD Lasers, *see* Mode-Locked QD lasers
semiconductor
  devices, B:5, 44, 215
  lasers, B:748–750, 752
  optical amplifier, B:43, 170, 248, 388, 496, 635, 654, 715, 774, 826
sensitivity penalties, B:148
separate absorption and multiplication (SAM) APD structures, A:245
serial filters, B:253
serial-to-parallel demultiplexer, B:718
series-peaked amplifier, A:692
server consolidation, B:481
service level agreement, B:18, 347, 378, 536
Session Initiation Protocol, B:518
severe electrical low-pass filtering, B:53
SG-DBR laser, wavelength-tunable, A:348
  integration with EAM and SOA, A:348
shared restoration, B:300, 542
shared-risk-link group, B:521, 537
SHEDS (super high-efficiency diode source) program, A:108, 114, 124
Shortest Path First Fit, B:649
shunt peaking, A:688

shunt-peaked amplifier utilizing common base cascode, A:689
shunt-peaked common source amplifier, A:688
Si-CMOS fabrication processes, B:188
side-coupled resonator configuration, A:432, 433, 437–438
 nonlinear, A:440
sideband instability, B:831
signal amplitude, B:102, 134, 148, 150, 152, 159, 167
signal bandwidth, B:24, 37, 52, 58, 63, 247, 320, 322, 746, 753, 758
signal bit rate, B:96, 282
signal conditioner, B:356
signal distortions, B:26, 31, 41, 42, 54, 70, 75, 76, 79
signal launch power, B:26, 60, 64, 73, 76
signal photocurrent, B:98, 105
signal processing applications
 Bragg grating filters in chG, importance of, A:763
signal quality assessment, B:243
signal space, B:28–31, 34, 35, 37, 39, 50
signal-dependent noise, B:38
Signal-noise
 interaction, B:72
 nonlinearities, B:79
signal-signal interactions, B:72
signal-spontaneous beat noise, B:753
signal-to-noise ratio, B:12, 15, 26, 26, 65, 67, 96, 104, 105, 113, 115, 116, 131, 137, 143, 236, 248, 254, 255, 278, 308, 309, 313, 392, 415, 498, 572, 580, 587, 593, 651, 653, 719, 728, 731, 743–746, 749, 753, 754, 758, 759, 774, 804
silica planar lightwave circuits, B:253
silica PLCs, A:275
 strain as source of birefringence, A:275
silicon electronics, B:188
silicon germanium bipolar, B:166, 182, 186, 187
silicon microelectronics industry, R&D investments, A:382
silicon nanowires
 development of high quality fabrication processes, A:763

enhance nonlinear optical efficiencies, A:761
 as material platform, A:763–764
 mode field area reduction, A:761
silicon photonic waveguides, nonlinear effects, A:421–423
 Raman amplification, A:423
 optically pumped CW silicon laser, A:422
 two-photon absorption, A:421
silicon photonics, A:381–382, B:20, 188, 636
 achieving electrically pumped laser, approaches, A:423
 epitaxial approach, A:424–425
 extrinsic gain, A:424
 hybrid wafer-bonded III–V epitaxial structures, A:423–424
 active modulation silicon photonics
 carrier accumulation on gate oxides, A:403–404
 modulation effects, A:397–399
 modulator devices, A:404–409
 P–N junction in depletion, A:399–402
 P–N junction in forward bias, A:402–403
 CMOS integration and examples of integrated silicon photonics chips, A:416–420
 microwave OE oscillators, A:420–421
 microwave spectrum analyzer, A:421
 SOI transistor description, A:414–416
 germanium photodetectors and photoreceivers, A:409–414
 high-index-contrast waveguide types and performance on SOI, A:384–388
 implemented on SOI, A:381–382
 input–output coupling, A:388–389
 gratings, A:389–392
 tapers, A:389
 integration, B:636
 interleaver and demuxing performance, A:419

Index

manufacture in CMOS process, A:417
nonlinear effects, A:421–423
passive waveguide devices and
   resonators
   filters, A:393–394
   photonic crystal resonators, A:395
   ring resonators, A:395–397
   splitters and couplers, A:393
SOI wafer technology, A:383–384
toward silicon laser, A:423–425
trends and applications, A:425–426
waveguide, A:393
silicon planar lightwave circuits
   (PLCs), A:715
silicon processing technology, B:7
silicon strip, A:385
silicon–InP lasers and detectors, laser
   structure, A:349, 350
silicon-grating coupler, A:390
Silicon-On-Insulator (SOI), A:381
   advantages, A:384
   created by SMARTCUT technique,
      A:383
   integration of photonics, A:414–415
   loss mechanisms, A:387
   requiring BOX, A:383–384
   waveguide types, A:384
simplex/duplex PMMA POF cords,
   A:598–599
Single Emitter Array Laser (SEAL),
   A:132
single multiplexer circuit, B:187
single polarization fibers (SPF), A:544–546
single QDS, A:32–35
   ARDA agency, A:32
   "International Technology Road Map
      for Semiconductors," A:33
   nanoflash, obstacles, A:33
   nanoflash memory cell based on QDs,
      future, A:34
   purcell effect, A:32
   "Quantum Information Science and
      Technology Road Map," A:32
   single-photon emitter (SPE), A:32
single sideband, B:52, 206, 265
   measurement of, B:206
   modulation, B:45, 52

single-channel
   fiber optic, B:13
   optical systems, B:576
   transmission, B:52, 73, 76, 125
single-ended detection, B:61, 72, 73
single-fiber capacity, B:295
single-mode fiber (SMF), A:37, 42, 82, 86,
   161, 488, 490, 494–496, 498, 500–501,
   506, 509, 672
single-mode fiber 980-nm pumps
   DWDM, A:109
   failure rate, A:120–122
      critical process parameters, A:120–122
   materials for 980-nm pump diodes,
      A:109–110
      InGaAs QW, A:109
      low-loss large optical cavity (LOC),
         A:110, 114
   optical beam, A:110–112
      alpha-distributed feedback (DFB)
         lasers, A:112
      broad-area (BA) diode lasers, A:112
      mode filters, A:112
      MOPA, A:112
      narrow stripe technology, A:111–112
      Fiber Bragg Grating (FBG),
         A:111–112
   output power scaling
      amplified spontaneous emission
         (ASE), A:114
      constant mirror reflectivity
         scaling, A:114
      "constant power ratio scaling"/
         "constant photon lifetime
         scaling," A:113
      length scaling of laser diode, A:112–114
      SHEDS (super high-efficiency diode
         source) program, A:114
      vertical epitaxial structure, A:114–115
   packaging, A:117
   reliability
      catastrophic junction meltdown,
         A:118–119
      catastrophic optic mirror damage
         (COMD), A:119
      electron beam-induced current (EBIC)
         charge thermography, A:120

single-mode fiber 980-nm
  pumps (*continued*)
    failure modes, CMD/NRR/TAD, A:118
      nonabsorbing mirror (NAM), A:119
      NRR heating, A:119
      TAD at facets, A:119–120
    spectral stability, A:115–117
      Fabry–Perot laser, A:116
single-mode fiber, B:49, 203, 255, 756
single-photon emitter (SPE), A:32
single-photon-counting avalanche photodetector (SPAD), A:256
single-polarization, B:9, 63, 64, 152, 157, 203
single-rack router, B:704
single-waveguide phase modulator, B:134, 136, 781
sinusoidal transmission, B:61
size, weight, and power (SWaP), A:375
slot efficiency, B:781, 785, 786, 788
slot waveguide, A:386
slow-light waveguide, B:712, 714
small form factor pluggable, B:346
small-scale integration, III-V photonic integrated circuit
  large-scale III–V photonic integration, A:351–355
    multi-channel interference-based active devices, A:348
      multi-channel ring resonator devices, A:349
      novel devices and integration methods, A:349–351
    single-channel multi-component chips, A:347–348
SMARTCUT technique (for SOI), A:383
  processing sequence, A:384
SMF, *see* single-mode fiber (SMF)
SOI CMOS process, A:409
SOI transistor, A:414–416
  fabricated at Freescale Semiconductor, A:414
space division multiplexing, B:36, 789
space domain, B:650
space switching, B:670
SPAD, *see* single-photon-counting avalanche photodetector (SPAD)

spanning tree protocol, B:358, 362, 364, 374, 529, 543
spatial dispersion
  of focal position, A:281
  of input-side position, A:282
spatial effects, B:837
specialty fibers
  active fibers, A:525
  coupling fibers or bridge fibers, A:525
  dispersion compensation fibers, A:525
  high numerical aperture (NA) fiber, A:525
  highly nonlinear optical fiber, A:525
  photonic crystal fibers (PCFs), A:525
  photo-sensitive fibers, A:525
  polarization control fibers, A:525
spectral compactness, B:63
spectral density, B:65, 67, 137, 604
spectral efficiency, B:12, 13, 24, 97, 123, 124, 127, 131, 133, 136, 137, 146, 150, 155, 169, 183, 184, 195, 208, 217, 218, 226, 391
spectral intensity filters
  4×4 WSXC, A:729
  advantages and challenges, A:732
  tunable Fabry–Perot (F–P), A:732
spectrally narrow formats, B:69
spectrum monitoring, B:268, 272, 593
speed digital signal processing, B:28, 97
SPF, *see* single polarization fibers (SPF)
splicing attributes, B:428
splicing loss, A:538
split-step method, B:832–834
splitter size reduction, B:425
splitters and couplers, A:393
splitting of packets, B:699
splitting ratios, B:388
SPM, *see* self-phase modulation (SPM)
spontaneous emission factor, B:138, 753
spontaneous-spontaneous beat noise, B:753
spoofing, B:486
spun fiber, A:613
spur-free dynamic range (SFDR), A:146, 152, 166, 168, 169, 178, 179
spurious free dynamic ranges, B:758
square-law detection, B:58, 95, 96, 150
square-shaped optical pulse, A:320–321
  waveforms and frequency spectra, A:320

squelch tables, B:540
SRS, *see* stimulated Raman scattering (SRS)
standard estimation theory, B:843
standard single-mode fiber, B:26, 169, 565
standardized terrestrial optical communications, B:28
star coupler, B:144, 808
star coupler, second transmissivity, A:300
star-shaped hub-and-spoke architecture, B:348
Stark effect, B:5, 749
start frame delimiter, B:359
state diagrams, B:39
state of polarization (SOP), A:607
steady-state channel power control, B:324
stimulated Brillouin Scattering (SBS), A:530, 548, 555, 565–568, B:273, 389, 426, 745, 831
stimulated Raman scattering (SRS), A:499, 510
Stochastic Differential Equation techniques, A:613
storage area networks (SAN), A:672, 706–707, B:196
straight-line phase modulator, B:59, 60
strick-sense nonblocking, B:651
strictly nonblocking topologies, B:782
sub-pulses, B:570
subcarrier multiplexing header technique, B:661
submarine
 applications, B:585
 repeaters, B:568
Subscriber Loop Carrier, B:410
subtending rings, B:526
subwavelength granularity, B:622, 624, 642, 651, 684
superprism dispersion, B:820
surface emitting lasers, recent advances
 long-wavelength VCSELS, A:83–87
 optical signal processing based on VCSEL technologies, A:94–97
 optical Kerr effect, A:95
 plasmonic vcsels, A:91–93
 VCSEL-based slow light devices, A:97–99
surface-emitting laser (VCSEL), A:597

swept acousto-optic tunable filters, B:253
switch controller, B:630, 657
switch fabrics, B:724
 technologies, B:730
switching
 nodes, B:21, 773–776, 778–780, 782, 783, 787, 792, 793, 796
 speed, B:310
 technologies, B:625, 651
 voltage, B:44, 45, 206, 727
synchronization, B:58, 184, 187, 359, 385, 416, 643, 657, 660, 665–668, 678, 697, 757, 781, 788, 853
 inaccuracies, B:781
synchronized digital hierarchy, B:32, 350, 850
Synchronizer architectures, B:667
synchronous optical network (SONET), B:32, 296, 306, 308, 310, 350–352, 355–357, 365, 370, 372, 374, 377, 378, 390, 392, 478, 482–484, 487–491, 493, 498, 502, 507, 513–518, 521–528, 530–532, 536, 538–540, 542–546, 548, 618, 620, 622, 623, 627, 637, 643, 646, 682, 850
 cross-connect layers, B:532
 interfaces, B:392, 524
 rings, B:516, 518, 522–524, 526, 528, 531, 532, 542, 543, 545, 548
system penalties, B:68
systematic errors, B:335

T

Taguchi method, A:357
tail-end switch, B:542
tandem AWG, A:308–309
 10 GHz-spaced 1010-channel, A:308
 demultiplexing properties of all the channels, A:309
tandem switch, B:526
tap
 coefficients, B:121, 122
 delay, B:282
tapped delay line, A:322–323, 327
 parallel arrangement, A:328–329
 serial arrangement, A:327–328

Taylor series, A:632–637
TCP/IP, B:483, 617, 701, 707
Telco metro network, B:519, 548
telecom packaging technology, A:130
tell-and-wait protocols, B:646
temperature variations, B:70, 181
temporal coupled-mode theory for optical resonators, A:432–438
    applied to photonic crystal structure, A:433
    directly coupled resonator configuration, A:432, 437
    Fano-interference configuration, A:433, 438
    light from one port to other
        direct pathway, A:433
        resonant pathway, A:433
    perfect crystal with photonic band gap, A:431
    physical implementation, A:440
    side-coupled resonator configuration, A:432, 433, 437–438
        nonlinear, A:440
    using to predict optical switching, A:438–442
temporal pulse waveform shapers, A:319–322
    experimental original spectra and auto-correlated pulse waveforms, A:321
    square-shaped optical pulse, A:320–321
    synthesized spectra and corresponding cross-correlated pulse waveforms, A:321
TeraFlops, B:769
ternary symbol alphabet, B:39
terrestrial cables, B:568, 569
terrestrial networks, B:295, 511, 564, 567–569, 581
terrestrial systems *see* undersea systems
Thermo-optic device, dependence on temperature, A:398
thin-film filter technology, B:249, 421
    components, B:421
third-order nonlinearities and PMD, A:662

three-dimensional photonic crystals, A:467
    fabrication technology, A:467–468
    light propagation, control of, A:472–473
    spontaneous emission, control of, A:468–472
three-dimensional
    photonic crystals, B:8
    signal space, B:30
three-modulator transmitter architecture, B:58
threshold crossing alert, B:533
TIA, *see* transimpedance amplifier (TIA)
tilt error, B:333, 339
time delay units, B:654
time division multiplexing, B:17, 32, 179, 180, 201, 217, 347, 415, 426, 481, 484, 514, 656, 705
time domain header technique, B:660
time slot interchanger, B:667
time switch, B:665
Time-domain split-step, B:833
timing extraction devices, B:210
timing jitter, B:206, 245
TIR, *see* total internal reflection (TIR)
titanium indiffused lithium niobate, B:748
tone-fading techniques, B:264
topology, B:21, 362–366, 376, 486, 487, 514, 521, 540, 541–543, 768, 769, 773, 777, 778, 782–785, 788, 790, 792–794, 796, 850
    banyan, B:778, 782
    Data Vortex, B:789, 792–796
    deflection routing, B:792
    hub-and-spoke, B:528
    modifications, B:782
    omega, B:778, 782–785
    tree-like, B:543
total internal reflection (TIR), A:573
total internal reflection mirror, B:671
TPA, *see* two-photon absorption (TPA)
traditional best-path algorithm, B:240
traditional first-fit algorithm, B:240
traffic congestion, B:537
traffic generating node separation, B:505
traffic generation modules, B:782

Index

traffic matrix, B:524, 547, 555
training sequence, B:160
transceivers, POFs, A:601
  IEC standards for, A:601
transform spectrometers
  advantages and challenges of MEMS, A:736
  Fourier transform, A:735
  grating, A:737
  semitransparent detector, A:738
  single-chip integrated, A:736
transient control, B:326, 331
transimpedance amplifier (TIA), A:233
transimpedance amplifier configuration, A:410
Transmission Control Protocol, B:669, 701
transmission distances, B:42, 43, 77, 213, 222, 234, 317, 325, 351–353, 388, 596
transmission experiments, B:123, 194, 215
transmission fiber, B:33, 102, 103, 213, 226, 274, 332, 333, 389, 495, 565, 570, 581, 582, 587–589, 592
transmission impairments, B:12, 13, 24, 38, 41, 63, 65, 67, 78, 98, 122, 123, 131–133, 152, 157, 166, 171, 221, 240, 325, 348, 565
  compensation of, B:77, 78, 151, 157
  impact of, B:571
  origin of, B:122
transmission line methods, B:822
transmission of optical data signals, B:180
transmission penalty, B:127, 133, 320
transmission spectra
  TO switches are "off," A:314
  TO switches are "on," A:315
transmission speed, B:625
transmission system model, B:742, 743
transmission technologies, B:507, 513, 771
transmitted signal, B:97, 99, 108, 122, 237, 308, 319, 542
transmitter (TX) PIC, A:351, 355
  10-channel working at elevated temperature, A:375
  channels operating at 40 Gb/s, A:363
  key performance metrics, A:352

transmitter chirp, B:268, 277, 493
transmitter side modulation formats, B:495
transoceanic
  distances, B:564, 565, 568, 572, 573, 601
  transmissions, B:572
transpacific distance, B:577, 601, 604
transparency, B:237, 621
transparent bridging, B:362, 367
transponder, B:37, 38, 64, 132, 299, 300, 302, 303, 308, 314, 315, 353, 355, 377, 378, 495, 564, 605
transport network, B:32, 131, 132, 167, 168, 182, 239, 370–372, 374, 377, 480, 483, 487, 511, 529, 553, 554, 615, 618, 622, 850
transversal filters, B:122, 214
transverse effective wavelength, A:489
transverse electric (TE) polarizations, birefringence, A:275
transverse magnetic (TM) polarizations, birefringence, A:275
transverse mode, A:473
traveling-wave EAMs
  future development, A:207–208
  high-efficiency modulators for 100 Gb/s and beyond, A:205–207
  ridge-waveguide eam structure, A:198–200
    beam-propagation method (BPM), A:199
  TWEAM structures for increased impedance, A:200–205
    benzocyclobutene (BCB) layer, A:202
    bit-error rate test set (BERTS), A:204
    frequency limitations, A:201
    LiNbO3-based intensity Mach–Zehnder interferometer (MZI) modulators, A:200
travelling-wave amplifier, B:191
tributary adapter module (TAM), A:367
tunable bandwidth filter, A:447
  band diagrams calculation, A:448
  photonic bands for structure, A:447
  two-cavity structure, A:446–447
Tunable laser technologies, B:492

tunable lasers
  commercial uses, A:743
  external cavity semiconductor diode lasers, A:741, 742
  Littrow and Littman configuration, A:741
tunable ODC (TODC), A:325
  AWG-based
    polymer thermooptic lens, A:334
    polymer thermooptic lens, measured response, A:335
    silica thermooptic lens, A:333
  chirped bragg grating, A:327
  coherently coupled MZIs, A:330
  figure of merit, A:334–335
  finite impulse response (FIR), A:325, 327–328
    tapped delay lines, *see* tapped delay line
  Gire–Tournois etalons, A:327
  gratings, A:331–334
    arrayed waveguide grating, A:331–334
    diffraction grating, A:331
    virtually imaged phased array (VIPA), A:331, 332
  infinite impulse response (IIR), A:325, 326–327
    Bragg Gratings, A:327
    Gires–Tournois etalons, all-pass filters, A:326
    ring resonators, all-pass filters, A:326
  ring resonators, A:326
  3-Stage, A:330
  4-Stage, A:330, 331
  4-Stage version realized in silica PLC, A:330, 331
  transversal filter
    in InP waveguides, A:329
    in silica waveguides, A:328
  types of, A:325
tuneable polarization controller, B:183
tuneable wavelength converter, B:630, 665, 670, 672, 701
Tuning and switching functions, B:316
tunneling injection, B:4

TWEAM (traveling-wave EAMs), A:198, 200–206, 209, 210
twisted pair, B:10, 345, 349, 410, 413, 415, 460
two to four fiber pairs, B:564
two-dimensional photonic crystal slabs, A:456–457
  heterostructures, effect of introducing, A:459–463
  high-Q nanocavities, A:463–466
  line/point-defect systems, control of light by combined, A:457–459
  nonlinear and/or active functionality, introduction of, A:466–467
two-photon absorption (TPA), A:760, B:11, 268
TX optical sub-assemblies (TOSAs), A:347
typical fiber performance, A:555

U

Ultra-compact 4-channel arrayed-waveguide grating filter, A:394
ultra-high-performance systems, B:2
Ultra-long-haul
  distance, B:601
  transmission, B:181, 221ultra-low latency propagation, B:780
ultra-low-polarization-dependent fibers, B:2
ultrafast nonlinear interferometer, B:676
ultrahigh-speed laser modulation by injection locking
  applications
    analog optical communications, A:177–179
    global system for mobile communications (GSM), A:179
    metropolitan area networks, A:177
    optical time-domain reflectometry (OTDR), conventional, A:176
    passive optical networks, A:171–176
    Rayleigh backscattering effects, A:175
    wireless local area networks (WLANs), A:179

Index

ultrahigh-speed laser modulation by injection locking (*continued*)
  Fabry–Perot (FP) laser, A:145
  OIL, basic principle, A:146–151
    optical spectrum analyzer (OSA), A:150
  OIL VCSELs, modulation properties, A:151–160
    advantest dynamic chirp measurement system, A:163
    cascaded optical injection locking (COIL) structure, A:153
    cavity resonance shift in wavelength, A:155
    COIL, promising approach, A:160
    erbium-doped fiber amplifier (EDFA), A:161
    large signal modulation, A:160–163
    small-signal modulation response, A:151–160
  OIL VCSELs, nonlinearity/dynamic range
    experiments, A:166–168
    intermodulation distortion (IMD), A:166
    simulations, A:168–169
  Optical injection locking (OIL), A:145
  relative intensity noise of OIL VCSELs, A:169–171
  RF link gain enhancement of OIL VCSELs, A:163–165
  spur-free dynamic range (SFDR), A:146
  vertical-cavity surface- emitting lasers (VCSELs), A:146
Ultrawave, B:220, 221
undersea cable, B:561, 564, 571, 581, 601, 606
undersea systems, B:565, 569
undirectional ethernet broadcast, B:392
uni-travelling carrier, B:191, 755
unidirectional link routing protocol, B:393
unidirectional path-switched rings, B:523
universal ethernet interface, B:346
Universalmobile telecommunication system, B:756
unshielded twisted pair, B:349
US Congress, B:451
US Telecommunications policy, B:520

User datagram protocol, B:725
user network interface, B:347, 620

V

vapor-doping OVD process, A:562–563
  advantage, A:563–565
variable attenuators
  MEMS data transceiver, A:745
  waveguide, A:404
variable optical attenuator (VOA), A:276, 352, B:308
  power transfer function of, A:353
variants on theme
  equalization architectures for better performance
    DFE with precursor canceller, A:678
    fixed delay tree search, A:678
VCSEL-based slow light devices, A:97–99
VCSELs diodes, A:44–47, 108
vector modulation/demodulation architecture, B:106
Verizon, B:403, 408, 410, 416, 438, 442, 443, 448, 459, 461, 465–468
vertical-cavity surface-emitting lasers (VCSELs), A:146, B:4, 5, 668, 814, 827, 828, 839
very high-speed digital subscriber line, B:415
Very-long-baseline interferometry, B:614
vestigial sideband, B:52, 266
Video hub office, B:553
video-on-demand, B:413, 552–555
virtual channel identifier, B:523
virtual communities, B:480
virtual concatenation, B:347, 526
virtual private network, B:546, 613
Viterbi algorithm, A:676, B:164, 182
Voice over internet protocol, B:345, 455, 479, 516, 518, 551, 637
voice-telephony connections, B:454
VSB signals, B:52, 267

W

wafer bonding (integration technique), A:349
wafer-bonded hybrid laser, A:423
  CW light-current curve, A:424

"wallplug efficiency," A:132
waveform distortion, B:25, 244, 274, 493, 566
waveguide
　single side cavity, transmission, A:445–446
　transmission spectra of one- and two-cavity structures, A:446
　two cavities
　　side-coupled, A:445
　　transmission spectrum, A:446–447
waveguide-grating couplers, A:391
waveguide modulators, B:751
waveguide parameters, A:280
　in first and second slab, A:281
　path-length difference, A:282
waveguide photodiodes
　edge-coupled waveguide photodiodes, A:222–224
　"double core," A:223
　evanescently coupled waveguide photodiodes, A:224–230
　　electro-optic sampling, A:228
　　enhancements/modifications, A:225
　　MMI, A:232
　　negative aspects, A:224
waveguides, AWG, A:279
　in first slab, A:279
　improve coupling efficiency to highly sampled output, A:300
　layout and results from demultiplexer consisting of MZI, A:296
　path-length difference, A:282
　in second slab, A:280
wavelength add/drop, B:295
wavelength blocker, B:296, 500
wavelength coded tag, B:250
wavelength conversion, A:789–796, B:11, 204, 300, 499, 625, 632, 634–636, 638, 651, 654, 655, 663, 669, 671–673, 675, 676, 677, 682, 682, 683, 696, 696, 697, 701, 720–723, 725–727, 731, 734
wavelength division multiplexed, B:1, 2, 6, 7, 13, 14, 15, 17–19, 24, 28, 32, 33, 42, 53, 58, 63, 68, 71, 76, 95–97, 118, 125, 126, 132, 137, 162, 166–170, 181–184, 194–197, 200–202, 204, 208, 215–218, 226, 239, 241–243, 247, 248, 250–252, 256, 263, 268, 273, 274, 284, 294, 295–297, 300, 304, 309, 312, 314, 316, 321, 322, 324–325, 329, 336, 337, 346, 351, 370, 379, 388–393, 430, 477, 481, 484–487, 491–493, 502, 503, 506, 507, 514, 516, 521, 522, 524, 531–533, 535, 536, 543–549, 551, 556, 561, 564–567, 572, 573, 576, 578, 579, 581, 594, 596, 597, 604, 615, 618, 622, 625, 633, 642, 643, 656, 662, 675, 676, 678, 699, 703, 706, 720, 749, 771, 772, 778, 789, 803, 833–836, 850–853, 857
wavelength division multiplexing (WDM), A:43, 82, 172, 270, 393, 432, 511, 524
　CWDM, A:304
　multiplexing channels, A:287
wavelength domain, B:650
wavelength filters, B:422, 506
wavelength independent modulation, B:45
wavelength independent power couplers, B:312
wavelength integration and control, surface emitting lasers, A:87–91
wavelength locked distributed feedback, B:62
wavelength management, B:132, 183, 196
wavelength monitoring techniques, B:248
wavelength multiplexing header technique, B:662
Wavelength, polarization, and fabrication (WPF) changes, A:272, 273
wavelength routing, B:335, 336
　switching fabric, B:672
wavelength selective
　couplers, B:193
　crossconnect, B:242
Wavelength-selective crossconnects (WSXC), A:728
　4 × 4 WSXC, A:729
wavelength-selective MEMS components spectral equalizers, A:719–721
wavelength-selective switches (WSS), A:317–318

Index 931

free space-based wavelength-selective switches
   1×4 wavelength-selective switch, A:723
   $1 \times N^2$ using two-axis micromirror array, A:725
   analog micromirror, A:722
   free space-based wavelength-selective switches, A:722–726
   LCOS-based 1×9 WSS, A:723
   two-axis analog micromirror array for WSS, A:726
   hybrid PLC-MEMS wavelength-selective switches, A:726–727
   1×9 WSS with hybrid integration of PLC and MEMS, A:727
   advantages of, A:726
   monolithic wavelength-selective switches, A:727–728
   1×4 WSS, A:722
wavelength services, B:521
wavelength spacing, A:282
wavelength speedup, B:780
wavelength striped message format, B:775
wavelength stripes, B:780
WDM *see* wavelength division multiplexed; wavelength division multiplexing (WDM)
WDM technology, A:82
Wentzel-Kramers-Brillouin, B:838
whitened matched filter (WMF), A:676
Wi-Fi, B:236, 418, 466, 614
Wide area network (WAN), A:371
   router-to-router bandwidth compared to, A:372
Wide Area Network, B:528
   interface sublayer, B:352
"wide-band" (de)multiplexers, A:287
Wideband digital crossconnect system, B:522
WiMax, B:614, 754
wire waveguide, *see* silicon strip
wireless-fidelity, B:418
wireless local area networks (WLANs), A:179, B:739, 756, 757

wireless-over-fiber systems, B:756
WLANs, *see* wireless local area networks (WLANs)
WMF *see* whitened matched filter (WMF)
World Wide Web, B:468, 617
world-wide web and Internet, A:372
WPF changes, *see* Wavelength, polarization, and fabrication (WPF) changes
W-profile design, A:551–555
WSS, *see* wavelength-selective switches (WSS)
WSXC *see* Wavelength-selective crossconnects (WSXC)

X
XAUI interface, B:353, 356
xDSL, B:421, 462, 516, 551
XPM, *see* cross-phase modulation (XPM)
XPolM, A:652, 657, 658, 659, 660, 661, 662

Y
Yahoo, B:389
Y-branch coupler, A:271, 272
yield management methodologies, A:359–361
   cluster analysis and negative binomial model, A:360
   in-line analysis, A:361
   limited yield analysis, A:360
   refined yield analysis, A:360
Y-junction, B:776, 777
YouTube, B:457, 458, 468, 552

Z
zero phase difference, B:211
zero transmission, B:59, 206
zone analysis, A:361
Zoomy communications, B:441